WATCH THE NEW
MIS VIDEOS

Want to see how the information systems concepts you're learning apply to the real world? Use the password card in this textbook to check out relevant MIS video clips online at **www.course.com/mis/stair**. Click on the video clip that corresponds to the chapter you're studying and discover more today.

If your book does not contain a password card, please ask your professor to contact their Thomson Course Technology representative.

Clip	Chapter
Go Inside Krispy Kreme	Chapter 1: An Introduction to Information Systems
Ben Casnocha	Chapter 2: Information Systems in Organizations
Getting Started	Chapter 3: Hardware: Input, Processing, and Output Devices
P2P's Cloudy Future	Chapter 4: Software: Systems and Application Software
Predicting Huge Surf	Chapter 5: Organizing Data and Information
Free Wi-Fi Blankets	Chapter 6: Telecommunications and Networks
Online Storage	Chapter 7: The Internet, Intranets, and Extranets
Find the Best Deals Online	Chapter 8: Electronic Commerce
Get News Delivered to Your Desktop	Chapter 9: Transaction Processing and Enterprise Resource Planning
Army's Virtual World	Chapter 10: Information and Decision Support Systems
Predicting the Future of AI by Marvin Minsky	Chapter 11: Specialized Business Information Systems: Artificial Intelligence, Expert Systems, Virtual Reality, and Other Specialized Systems
Design Your Own Video Game	Chapter 12: Systems Investigation and Analysis
Making Flowcharts Using Office XP	Chapter 13: Systems Design, Implementation, Maintenance, and Review
Controversial Digital Bouncers	Chapter 14: Security, Privacy, and Ethical Issues in Information Systems and the Internet

LOOK FOR THESE OTHER POPULAR THOMSON COURSE TECHNOLOGY
MIS TITLES

Database Systems: Design, Implementation, and Management, Sixth Edition

by Peter Rob and Carlos Coronel
ISBN: 0-619-21323-X

Systems Analysis and Design in a Changing World, Third Edition

by John W. Satzinger, Robert B. Jackson, and Stephen D. Burd
ISBN: 0-619-21325-6

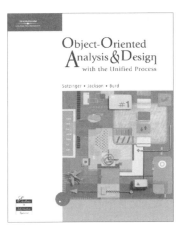

Object-Oriented Analysis and Design with the Unified Process

by John W. Satzinger, Robert B. Jackson, and Stephen D. Burd
ISBN: 0-619-21643-3

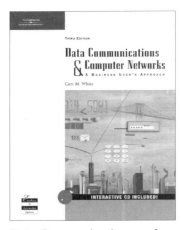

Data Communications and Computer Networks: A Business User's Approach, Third Edition

by Curtis M. White
ISBN: 0-619-16035-7

Information Technology Project Management, Fourth Edition

by Kathy Schwalbe
ISBN: 0-619-21526-7

Principles of Information Security, Second Edition

by Michael Whitman and Herbert Mattord
ISBN: 0-619-21625-5

VIEW OUR ENTIRE COLLECTION OF PRODUCTS ONLINE AT WWW.COURSE.COM/MIS.

Principles of Information Systems

A Managerial Approach

Seventh Edition

Ralph M. Stair
Florida State University

George W. Reynolds
The University of Cincinnati

THOMSON

COURSE TECHNOLOGY

Australia · Canada · Mexico · Singapore · Spain · United Kingdom · United States

THOMSON
★
COURSE TECHNOLOGY

Principles of Information Systems, A Managerial Approach, Seventh Edition
by Ralph M. Stair and George W. Reynolds

Senior Vice President, Publisher:
Kristen Duerr
Executive Editor:
Mac Mendelsohn
Senior Acquisitions Editor:
Maureen Martin
Senior Product Manager:
Eunice Yeates-Fogle
Development Editor:
Karen Hill, Elm Street Publishing Services, Inc.
Associate Product Manager:
Mirella Misiaszek

Editorial Assistant:
Jennifer Smith
Production Editor:
Kelly Robinson
Senior Marketing Manager:
Karen Seitz
Text Designer:
Jennifer McCall
Cover Designer:
Lisa Rickenbach
Composition House:
Digital Publishing Solutions

Photo Researcher:
Abby Reip
Copyeditor:
Jeri Freedman, Foxxe Editorial Services
Proofreader:
John Bosco, Green Pen Quality Assurance
Indexer:
Liz Cunningham

Copyright © 2006 Thomson Course Technology, a division of Thomson Learning, Inc.
Thomson Learning™ is a trademark used herein under license.

Printed in the United States
2 3 4 5 6 7 8 9 QWV 09 08 07 06 05

Library of Congress Cataloging-in-Publication Data
Stair, Ralph M.
    Principles of information systems: a managerial approach / Ralph M. Stair, George W. Reynolds.
    p. cm.
    Includes bibliographical references and index.
    ISBN 0-619-21561-5 (alk. paper)
    1. Management information systems. I. Reynolds, George Walter, 1944– II. Title.

    T58.6 .S72 2001
    658.4'038—dc21
    00-047561

For more information contact Thomson Course Technology, 25 Thomson Place, Boston, Massachusetts, 02210.

Or find us on the World Wide Web at: www.course.com

Disclaimer
Thomson Course Technology reserves the right to revise this publication and make changes from time to time in its content without notice.

ISBN: 0-619-21561-5. International Student Edition ISBN: 0-619-21525-9. Instructor Edition ISBN: 0-619-21373-6.

For Lila and Leslie
—Ralph M. Stair

To my grandchildren: Michael, Jacob, Jared, Fievel, and Aubrey Danielle
—George W. Reynolds

BRIEF CONTENTS

PART 1   **An Overview**   1

    Chapter 1   An Introduction to Information Systems   2

    Chapter 2   Information Systems in Organizations   46

PART 2   **Information Technology Concepts**   87

    Chapter 3   Hardware: Input, Processing, and Output Devices   88

    Chapter 4   Software: Systems and Application Software   140

    Chapter 5   Organizing Data and Information   192

    Chapter 6   Telecommunications and Networks   240

    Chapter 7   The Internet, Intranets, and Extranets   294

    World Views Case: Australian Film Studio Benefits from Broadband Networks—and
Location   342

    World Views Case: Virtual Learning Environment Provides Instruction
Flexibility   344

PART 3   **Business Information Systems**   347

    Chapter 8   Electronic Commerce   348

    Chapter 9   Transaction Processing and Enterprise Resource Planning Systems   398

    Chapter 10   Information and Decision Support Systems   452

    Chapter 11   Specialized Business Information Systems: Artificial Intelligence, Expert
Systems, Virtual Reality, and Other Specialized Systems   506

    World Views Case: Kulula.com: The Trials and Tribulations of a South African Online
Airline   549

PART 4   **Systems Development**   553

    Chapter 12   Systems Investigation and Analysis   554

    Chapter 13   Systems Design, Implementation, Maintenance, and Review   610

    World Views Case: Brandon Trust Develops MIS Capability for Improved Operations
and Services   664

    World Views Case: Strategic Enterprise Management at International Manufacturing
Corporation (IMC)   666

PART 5   **Information Systems in Business and Society**   669

    Chapter 14   Security, Privacy, and Ethical Issues in Information Sytems and the
Internet   670

    World Views Case: Efforts to Build E-Government and an Information Society in
Hungary   722

    Glossary   725

    Subject Index   743

    Name and Company Index   753

CONTENTS

Preface xxi

PART 1   An Overview   1

Chapter 1   An Introduction to Information Systems   2
Boehringer Ingelheim, GMBH, Germany   3
Information Concepts   5
Data Versus Information   5
The Characteristics of Valuable Information   7
The Value of Information   7
System and Modeling Concepts   8
System Components and Concepts   9
System Performance and Standards   11
System Variables and Parameters   11
Information Systems @ Work: Delta Sings a New Song with a Focus on Information Systems   12
Modeling a System   13
What Is an Information System?   15
Input, Processing, Output, Feedback   15
Manual and Computerized Information Systems   16
Computer-Based Information Systems   17
Business Information Systems   21
Electronic and Mobile Commerce   21
Ethical and Societal Issues: Phishing for Visa Card Customers   23
Transaction Processing Systems and Enterprise Resource Planning   25
Information and Decision Support Systems   26
Specialized Business Information Systems: Artificial Intelligence, Expert Systems, and Virtual Reality   29
Systems Development   32
Systems Investigation and Analysis   33
Systems Design, Implementation, Maintenance, and Review   33
Information Systems in Society, Business, and Industry   33
Security, Privacy, and Ethical Issues in Information Systems and the Internet   33
Computer and Information Systems Literacy   35
Information Systems in the Functional Areas of Business   36
Information Systems in Industry   36
Case One: Tyndall Federal Credit Union Explores New ATM Services   42
Case Two: The Queen Mary 2 and partner   43
Case Three: MyFamily Comforts Its Members   44

Chapter 2   Information Systems in Organizations   46
Whirlpool Corporation Worldwide, United States   47
Organizations and Information Systems   48

Organizational Structure 51

Organizational Culture and Change 54

Reengineering 57

Continuous Improvement 58

Technology Diffusion, Infusion, and Acceptance 59

Total Quality Management 59

Outsourcing, On-Demand Computing, and Downsizing 60

Ethical and Societal Issues: The U.S. Offshore Outsourcing Dilemma 61

Organizations in a Global Society 62

Competitive Advantage 63

Factors That Lead Firms to Seek Competitive Advantage 63

Strategic Planning for Competitive Advantage 64

Performance-Based Information Systems 67

Productivity 68

Return on Investment and the Value of Information Systems 68

Information Systems @ Work: OneBeacon Focuses on Information Systems 69

Risk 70

Careers in Information Systems 71

Roles, Functions, and Careers in the IS Department 72

Typical IS Titles and Functions 75

Other IS Careers 77

Case One: Ingersoll-Rand and the 21st Supplier 83

Case Two: The Bank of Montreal Learns from Past Mistakes 83

Case Three: Shell Oil Saves with Contingent Workforce Management System 84

PART 2 Information Technology Concepts 87

Chapter 3 Hardware: Input, Processing, and Output Devices 88
Porsche AG, Germany 89

Computer Systems: Integrating the Power of Technology 90

Hardware Components 92

Hardware Components in Action 92

Processing and Memory Devices: Power, Speed, and Capacity 93

Processing Characteristics and Functions 94

Memory Characteristics and Functions 98

Multiprocessing 100

Secondary Storage 102

Access Methods 103

Secondary Storage Devices 103

Enterprise Storage Options 107

Input and Output Devices: the Gateway to Computer Systems 109

Characteristics and Functionality 110

Input Devices 111

Information Systems @ Work: Banks Weigh Move to Improved Check-Clearing Process 115

Ethical and Societal Issues: Medication Management Systems 117

Output Devices 118

Special-Purpose Input and Output Devices   121

Computer System Types, Selecting, and Upgrading   122

Computer System Types   122

Selecting and Upgrading Computer Systems   126

Case One: Hilton Hotels Implements "OnQ"   134

Case Two: U. S. Post Office Information System Initiative   135

Case Three: PeopleSoft Offers Hosting Services   136

Chapter 4   Software: Systems and Application Software   140
Jones Lang Lasalle, United States   141

An Overview of Software   143

Systems Software   143

Application Software   143

Supporting Individual, Group, and Organizational Goals   144

Systems Software   145

Operating Systems   145

Current Operating Systems   149

Workgroup Operating Systems   152

Enterprise Operating Systems   154

Operating Systems for Small Computers and Special-Purpose Devices   154

Utility Programs   156

Middleware   158

Application Software   158

Types and Functions of Application Software   158

Personal Application Software   160

Information Systems @ Work: Telstra Benefits from E-Training for Managers   161

Workgroup Application Software   168

Enterprise Application Software   169

Application Software for Information, Decision Support, and Specialized
Purposes   171

Programming Languages   171

The Evolution of Programming Languages   171

Selecting a Programming Language   175

Software Issues and Trends   176

Software Bugs   176

Copyrights and Licenses   176

Open-Source Software   177

Shareware, Freeware, and Public Domain Software   178

Multiorganizational Software Development   178

Ethical and Societal Issues: Weighing the Benefits of Open Source   179

Software Upgrades   180

Global Software Support   180

Case One: PACS at North Bronx Healthcare Network   186

Case Two: Alaska Airlines Drops Sabre for Speed   187

Case Three: UPS Provides Shipping Management Software to Customers   188

Chapter 5   Organizing Data and Information   192
DDB Worldwide, United States   193

Data Management   196

The Hierarchy of Data   196

Data Entities, Attributes, and Keys   197

The Traditional Approach Versus the Database Approach   198

Data Modeling and the Relational Database Model   201

Data Modeling   201

The Relational Database Model   202

Database Management Systems (DBMS)   206

Overview of Database Types   206

Providing a User View   207

Creating and Modifying the Database   208

Storing and Retrieving Data   210

Manipulating Data and Generating Reports   211

Database Administration   213

Popular Database Management Systems   214

Special-Purpose Database Systems   215

Selecting a Database Management System   215

Using Databases with Other Software   217

Database Applications   217

Linking the Company Database to the Internet   217

Information Systems @ Work: Web-Based DBMS Empowers Cruise Line Personnel   218

Data Warehouses, Data Marts, and Data Mining   219

Ethical and Societal Issues: The Growing Cost of Data-Related Regulations   223

Business Intelligence   224

Distributed Databases   225

Online Analytical Processing (OLAP)   227

Open Database Connectivity (ODBC)   228

Object-Oriented and Object-Relational Database Management Systems   229

Visual, Audio, and Other Database Systems   229

Case One: Brazilian Grocer Gets Personal with Customers   236

Case Two: DBMS Upgrade Faces Employee Opposition   237

Case Three: Hyundai Shoots for #1 with Executive DBMS   238

Chapter 6   Telecommunications and Networks   240
The Barilla Group, Italy   241

An Overview of Communications Systems   243

Communications   243

Telecommunications   244

Communications Channels   245

Basic Communications Channel Characteristics   245

Channel Bandwidth and Information-Carrying Capacity   246

Types of Media   246

Modems 250

Special-Purpose Modems 250

Carriers and Services 252

Local Exchange Carriers 252

Competitive Local Exchange Carriers 253

Long-Distance Carriers 253

Switched and Dedicated Lines 254

Voice and Data Convergence 255

Phone and Dialing Services 255

WATS 256

ISDN 256

T-Carrier System 256

Digital Subscriber Line (DSL) 257

Wireless Mobile 257

Networks 259

Network Types 259

Network Topology 262

Terminal-to-Host, File Server, and Client/Server Systems 263

Interconnecting Networks 265

Communications Protocols 265

Wireless Communications Protocols 267

Ethical and Societal Issues: Manufacturers and Retailers Grapple with RFID and Privacy 270

Network Switching Devices 271

Network Basics 273

Basic Processing Strategies 273

Communications Software 274

Telecommunications Applications 275

Linking Personal Computers to Mainframes and Networks 275

Voice Mail 276

Electronic Software Distribution 276

Electronic Document Distribution 276

Call Centers 276

Telecommuting 277

Videoconferencing 278

Information Systems @ Work: Live, Work, and Play in Luxury at the Heliopolis Complex 279

Electronic Data Interchange 280

Public Network Services 281

Electronic Funds Transfer 281

Distance Learning 283

Specialized Systems and Services 283

Case One: Stop & Shop Customers Make a New Buddy 289

Case Two: Application Performance Monitoring at JCPenney 290

Case Three: Hospitals Turn Burden into Benefit 291

Chapter 7   **The Internet, Intranets, and Extranets   294**
Colliers International, USA   **295**
Use and Functioning of the Internet   297
How the Internet Works   299
Ethical and Societal Issues: Entrepreneurs Work to Lessen the Global Divide   301
Internet Service Providers   302
The World Wide Web   304
Web Browsers   307
Search Engines   307
Web Programming Languages   309
Business Uses of the Web   310
Developing Web Content   311
Web Services   312
Internet and Web Applications   313
E-Mail and Instant Messaging   313
Internet Cell Phones and Handheld Computers   316
Career Information and Job Searching   316
Telnet and FTP   317
Web Log (Blog)   317
Information Systems @ Work: Inscene Embassy Changes Marketing Focus with
Blogs   318
Usenet and Newsgroups   319
Chat Rooms   320
Internet Phone and Videoconferencing Services   320
Content Streaming   322
Shopping on the Web   322
Web Auctions   323
Music, Radio, and Video on the Internet   323
Office on the Web   324
Internet Sites in Three Dimensions   324
Free Software and Services   324
Other Internet Services and Applications   325
Intranets and Extranets   326
Net Issues   328
Management Issues   328
Service and Speed Issues   328
Privacy, Fraud, Security, and Unauthorized Internet Sites   329
Case One: Eastman Chemical Revamps with Web Services   338
Case Two: Sentry Insurance Provides Secure VPN Access   339
Case Three: Best Western: First to Provide Free Internet   340
Part 2 World Views Case: Australian Film Studio Benefits from Broadband
Networks—and Location   342
Part 2 World Views Case: Virtual Learning Environment Provides Instruction
Flexibility   344

PART 3   Business Information Systems   347

Chapter 8   Electronic Commerce   348
Sprint, United States; And Sony, Japan   349
An Introduction to Electronic Commerce   350
Multistage Model for E-Commerce   352
E-Commerce Challenges   355
The E-Commerce Supply Chain   356
Business-to-Business (B2B) E-Commerce   357
Ethical and Societal Issues: Canadian Prescription Drug Web Sites   358
Business-to-Consumer (B2C) E-Commerce   359
Consumer-to-Consumer (C2C) E-Commerce   359
Global E-Commerce   359
Mobile Commerce   362
Mobile Commerce in Perspective   362
Technology Needed for Mobile Commerce   362
Anywhere, Anytime Applications of Mobile Commerce   363
E-Commerce Applications   363
Retail and Wholesale   363
Manufacturing   364
Marketing   366
Investment and Finance   367
Auctions   369
Technology Infrastructure   370
Hardware   370
Web Server Software   371
E-Commerce Software   372
E-Commerce Transaction Processing   373
Electronic Payment Systems   374
Secure Sockets Layer   374
Information Systems @ Work: Marketing Meets Search Engine   375
Electronic Cash   376
Electronic Wallets   377
Smart, Credit, Charge, and Debit Cards   377
Threats to E-Commerce   378
E- and M-Commerce Incidents   378
Theft of Intellectual Property   379
Fraud   380
Invasion of Consumer Privacy   382
Strategies for Successful E-Commerce   383
Developing an Effective Web Presence   384
Putting Up a Web Site   384
Building Traffic to Your Web Site   385
Maintaining and Improving Your Web Site   385
Case One: Izumiya of Japan: E-Commerce Poster Child   392
Case Two: Dylan's Candy Bar Sweetens Its Sales with E-Commerce   393

Case Three: RealEstate.com: Buying and Selling Homes Online   394

Chapter 9   Transaction Processing and Enterprise Resource Planning
                Systems   398
            McDonald's, United States   399
            An Overview of Transaction Processing Systems   401
                    Traditional Transaction Processing Methods and Objectives   402
            Transaction Processing Activities   407
                    Data Collection   408
                    Data Editing   409
                    Data Correction   409
                    Data Manipulation   409
                    Data Storage   409
                    Document Production and Reports   410
            Control and Management Issues   410
                    Business Continuity Planning   410
                    Disaster Recovery   411
                    Transaction Processing System Audit   411
            Traditional Transaction Processing Applications   412
                    Order Processing Systems   413
                    Order Entry   413
            Ethical and Societal Issues: TPSs for Everyone   415
                    Sales Configuration   416
                    Shipment Planning   417
                    Shipment Execution   418
                    Inventory Control   419
                    Invoicing   420
                    Customer Relationship Management (CRM)   421
                    Routing and Scheduling   423
            Purchasing Systems   424
                    Inventory Control   424
                    Purchase Order Processing   424
                    Receiving   426
                    Accounts Payable   426
            Accounting Systems   428
                    Budget   428
                    Accounts Receivable   429
                    Payroll   430
                    Asset Management   432
                    General Ledger   432
            International Issues   434
                    Different Languages and Cultures   434
                    Disparities in Information System Infrastructure   434
                    Varying Laws and Customs Rules   434
                    Multiple Currencies   435
            Enterprise Resource Planning   436

An Overview of Enterprise Resource Planning   436

Advantages and Disadvantages of ERP   438

Information Systems @ Work: ERP Consolidation Opens Doors for College Lender   439

Case One: Blue Cross Blue Shield CRM Wins BIG Customers   447

Case Two: Luxury Jet Service Develops TPS to Speed Customers on Their Way   448

Case Three: Point and Pay in Japan with KDDI   449

Chapter 10   Information and Decision Support Systems   452
Kiku-Masamune Sake Brewing Co., Ltd, Japan   453

Decision Making and Problem Solving   455

Decision Making as a Component of Problem Solving   455

Programmed Versus Nonprogrammed Decisions   456

Optimization, Satisficing, and Heuristic Approaches   457

An Overview of Management Information Systems   458

Management Information Systems in Perspective   458

Inputs to a Management Information System   459

Outputs of a Management Information System   460

Characteristics of a Management Information System   463

Functional Aspects of the MIS   464

Financial Management Information Systems   465

Manufacturing Management Information Systems   467

Marketing Management Information Systems   472

Information Systems @ Work: A "TaylorMade" Information and Decision Support System   473

Human Resource Management Information Systems   476

Other Management Information Systems   479

An Overview of Decision Support Systems   480

Characteristics of a Decision Support System   481

Capabilities of a Decision Support System   482

A Comparison of DSS and MIS   484

Components of a Decision Support System   484

The Database   485

The Model Base   486

The Dialogue Manager   486

Group Support Systems   486

Characteristics of a GSS That Enhance Decision Making   487

GSS Software   489

GSS Alternatives   490

Executive Support Systems   491

Ethical and Societal Issues: CEOs "Called to Action" Regarding Information Security   492

Executive Support Systems in Perspective   493

Capabilities of Executive Support Systems   494

Case One: Sensory Systems Provide Better-Tasting Products for Kraft   501

Case Two: The Ups and Downs of the Ladder Industry   502

Case Three: Sanoma Magazines Follows Key Performance Indicators to Success
503

Chapter 11   Specialized Business Information Systems: Artificial Intelligence,
              Expert Systems, Virtual Reality, and Other Specialized
              Systems   506
              Amazon.Com, United States   507
              An Overview of Artificial Intelligence   509
                     Artificial Intelligence in Perspective   509
                     The Nature of Intelligence   509
                     The Difference Between Natural and Artificial Intelligence   511
                     The Major Branches of Artificial Intelligence   512
                     Expert Systems   513
                     Robotics   513
                     Vision Systems   514
                     Natural Language Processing   514
                     Learning Systems   516
                     Neural Networks   516
              **Ethical and Societal Issues: Biologically Inspired Algorithms Fight Terrorists and
              Guide Businesses   517**
                     Other Artificial Intelligence Applications   518
              An Overview of Expert Systems   519
                     Characteristics and Limitations of an Expert System   520
                     When to Use Expert Systems   521
                     Components of Expert Systems   521
                     The Inference Engine   524
                     The Explanation Facility   525
                     The Knowledge Acquisition Facility   525
                     The User Interface   525
                     Expert Systems Development   526
                     The Development Process   526
                     Participants in Developing and Using Expert Systems   527
                     Expert Systems Development Tools and Techniques   527
                     Expert Systems Development Alternatives   529
                     Applications of Expert Systems and Artificial Intelligence   530
              Information Systems @ Work: Moving Data—and Freight—Efficiently   531
                     Integrating Expert Systems   533
              Virtual Reality   533
                     Interface Devices   534
                     Forms of Virtual Reality   535
                     Virtual Reality Applications   536
              Other Specialized Systems   538
              Case One: French Burgundy Wines: The Sweet Smell of Success   545
              Case Two: Expert System Provides Safety in Nuclear Power Plants   546
              Case Three: BankFinancial Corp. Gets a Lesson in Predictive Analytics   547

Part 3 World Views Case: Kulula.com: The Trials and Tribulations of a South African
   Online Airline   549

PART 4   Systems Development   553

Chapter 12   Systems Investigation and Analysis   554
BMW, Germany   555

An Overview of Systems Development   557
   Participants in Systems Development   557
   Initiating Systems Development   559
   Information Systems Planning and Aligning Corporate and IS Goals   561
**Information Systems @ Work: Fakta Designs Financial Management System to
   Support Corporate Goals   562**
   Establishing Objectives for Systems Development   564
   Web-Based Systems Development: The Internet, Intranets, Extranets, and
      E-Commerce   566
   Systems Development and Enterprise Resource Planning   567
Systems Development Life Cycles   568
   The Traditional Systems Development Life Cycle   569
   Prototyping   571
   Rapid Application Development, Agile Development, Joint Application
      Development, and Other Systems Development Approaches   573
   The End-User Systems Development Life Cycle   574
   Outsourcing and On Demand Computing   575
Factors Affecting Systems Development Success   577
   Degree of Change   577
Ethical and Societal Issues: Complexity, Interoperability, and Control   579
   Quality and Standards   580
   The Capability Maturity Model (CMM)   581
   Use of Project Management Tools   582
   Use of Computer-Aided Software Engineering (CASE) Tools   584
   Object-Oriented Systems Development   585
Systems Investigation   586
   Initiating Systems Investigation   586
   Participants in Systems Investigation   586
   Feasibility Analysis   587
   Object-Oriented Systems Investigation   588
   The Systems Investigation Report   588
Systems Analysis   589
   General Considerations   589
   Participants in Systems Analysis   590
   Data Collection   590
   Data Analysis   592
   Requirements Analysis   595
   Object-Oriented Systems Analysis   598
   The Systems Analysis Report   598

Case One: Hackensack University Medical Center Consolidates Systems for Fast Information Access 606

Case Two: PepsiCo Implements New Procurement System to Minimize Costs 607

Case Three: Segway Stays Light and Nimble with Outsourced Systems 608

Chapter 13   **Systems Design, Implementation, Maintenance, and Review** 610

The Home Depot, United States 611

Systems Design 613

     Logical and Physical Design 613

Information Systems @ Work: AM General Designs RFID System for Just-in-Time Deliveries 614

     Object-Oriented Design 618

     Interface Design and Controls 619

     Design of System Security and Control 621

     Generating Systems Design Alternatives 626

     Evaluating and Selecting a System Design 628

     Evaluation Techniques 629

     Freezing Design Specifications 630

     The Contract 631

     The Design Report 631

Systems Implementation 632

     Acquiring Hardware from an IS Vendor 632

     Acquiring Software: Make or Buy? 634

     Acquiring Database and Telecommunications Systems 640

     User Preparation 641

     IS Personnel: Hiring and Training 642

     Site Preparation 642

     Data Preparation 642

     Installation 643

     Testing 643

     Start-Up 644

     User Acceptance 645

Systems Operation and Maintenance 645

     Reasons for Maintenance 646

Ethical and Societal Issues: Finding Trust in Computer Systems 647

     Types of Maintenance 648

     The Request for Maintenance Form 648

     Performing Maintenance 648

     The Financial Implications of Maintenance 649

     The Relationship between Maintenance and Design 650

Systems Review 650

     Types of Review Procedures 651

     Factors to Consider during Systems Review 651

     System Performance Measurement 652

Case One: MetLife Selects the Best Technologies Around the Globe 659

Case Two: The Hudson River Park Trust Turns to Construction ASP 660

Case Three: Haworth Upgrades Supply Chain Systems 661

Part 4 World Views Case: Brandon Trust Develops MIS Capability for Improved
    Operations and Services 664

Part 4 World Views Case: Strategic Enterprise Management at International
    Manufacturing Corporation (IMC) 666

PART 5   **Information Systems in Business and Society 669**

Chapter 14   Security, Privacy, and Ethical Issues in Information Sytems and the
    Internet 670
    Computer Assisted Passenger Prescreening System, United
    States 671

Computer Waste and Mistakes 673

    Computer Waste 673

    Computer-Related Mistakes 673

Ethical and Societal Issues: CAN-SPAM: Deterrent or Accelerant? 674

Preventing Computer-Related Waste and Mistakes 675

    Establishing Policies and Procedures 675

    Implementing Policies and Procedures 676

    Monitoring Policies and Procedures 677

    Reviewing Policies and Procedures 677

Computer Crime 678

The Computer as a Tool to Commit Crime 680

    Cyberterrorism 680

    Identity Theft 681

The Computer as the Object of Crime 682

    Illegal Access and Use 682

    Data Alteration and Destruction 683

    Using Antivirus Programs 686

    Information and Equipment Theft 688

    Software and Internet Software Piracy 688

    Computer-Related Scams 689

    International Computer Crime 690

Preventing Computer-Related Crime 691

    Crime Prevention by State and Federal Agencies 691

    Crime Prevention by Corporations 691

    Using Intrusion Detection Software 694

    Using Managed Security Service Providers (MSSPs) 694

    Internet Laws for Libel and Protection of Decency 695

    Preventing Crime on the Internet 697

Privacy Issues 698

    Privacy and the Federal Government 698

Information Systems @ Work: UK BioBank Raises Privacy Issues 699

    Privacy at Work 700

    E-Mail Privacy 700

    Privacy and the Internet 701

    Fairness in Information Use 702

Federal Privacy Laws and Regulations   702

State Privacy Laws and Regulations   704

Corporate Privacy Policies   705

Individual Efforts to Protect Privacy   705

The Work Environment   706

Health Concerns   706

Avoiding Health and Environmental Problems   707

Ethical Issues in Information Systems   709

The AITP Code of Ethics   710

The ACM Code of Professional Conduct   711

Case One: Working to Reduce the Number of Software Vulnerabilities   717

Case Two: Beware Spyware!   718

Case Three: Cyberstalking   719

**Part 5 World Views Case: Efforts to Build E-Government and an Information Society
   in Hungary   722**

Glossary   725

Subject Index   743

Name and Company Index   753

People in nearly every line of work and in every type of organization make use of information systems. Chances are, regardless of your future role, you will need to understand what information systems can and cannot do and be able to use them to help you accomplish your work. You will be expected to discover opportunities to use information systems and to participate in the design of solutions to business problems employing information systems. You will be challenged to identify and evaluate IS options. To be successful, you must be able to view information systems from the perspective of business and organizational needs. For your solutions to be accepted, you must recognize and address their impact on fellow workers, customers, suppliers, and other key business partners. For these reasons, a course in information systems is essential for students in today's high-tech world.

Principles of Information Systems: A Managerial Approach, Seventh Edition, continues the tradition and approach of the previous editions. Our primary objective is to provide the best IS text and accompanying materials for the first information technology course required of all business students. Through surveys, questionnaires, focus groups, and feedback that we have received from current and past adopters, as well as others who teach in the field, we have been able to develop the highest-quality set of teaching materials available.

Principles of Information Systems: A Managerial Approach, Seventh Edition, stands proudly at the beginning of the IS curriculum and remains unchallenged in its position as the only IS principles text offering the basic IS concepts that every business student must learn to be successful. In the past, instructors of the introductory course faced a dilemma. On one hand, experience in business organizations allows students to grasp the complexities underlying important IS concepts. For this reason, many schools delayed presenting these concepts until students completed a large portion of the core business requirements. On the other hand, delaying the presentation of IS concepts until students have matured within the business curriculum often forces the one or two required introductory IS courses to focus only on personal computing software tools and, at best, merely to introduce computer concepts.

This text has been written specifically for the introductory course in the IS curriculum. *Principles of Information Systems: A Managerial Approach, Seventh Edition*, treats the appropriate computer and IS concepts together with a strong managerial emphasis on meeting business and organizational needs.

APPROACH OF THE TEXT

Principles of Information Systems: A Managerial Approach, Seventh Edition, offers the traditional coverage of computer concepts, but it places the material within the context of addressing business and organizational needs. Placing IS concepts in this context and taking a general management perspective has always set the text apart from general computer books; thus, it appeals not only to MIS majors but also to students from other fields of study. The text isn't overly technical. Instead, it deals with the role that information systems play in an organization and the key principles a manager needs to grasp to be successful. These principles of information systems are distilled and presented in a way that is both understandable and relevant. In addition, this book offers an overview of the entire IS discipline, while giving students a solid foundation for further study in advanced IS courses such as programming, systems analysis and design, project management, database management, data communications, Web site and systems development, electronic commerce applications, and decision support. As such, it serves the needs of both general business students and those who will become IS professionals.

The overall vision, framework, and pedagogy that made the previous editions so popular have been retained in the seventh edition, offering a number of benefits to students. We continue to present IS concepts with a managerial emphasis. While the fundamental vision

of this market-leading text remains unchanged, the seventh edition more clearly highlights established principles and draws out new ones that have emerged as a result of business, organizational, and technological change.

IS Principles First, Where They Belong

Exposing students to fundamental IS principles is an advantage for students who do not later return to the discipline for advanced courses. Since most functional areas in business rely on information systems, an understanding of IS principles helps students in other course work. In addition, introducing students to the principles of information systems helps future business function managers avoid mishaps that often result in unfortunate consequences. Furthermore, presenting IS concepts at the introductory level creates interest among general business students who may later choose information systems as a field of concentration.

Author Team

Ralph Stair and George Reynolds have teamed up again for the seventh edition. Together, they have more than 60 years of academic and industrial experience. Ralph Stair brings years of writing, teaching, and academic experience to this text. He has written more than 22 books and a large number of articles while at Florida State University. George Reynolds brings a wealth of computer and industrial experience to the project, with more than 30 years of experience working in government, institutional, and commercial IS organizations. He has also authored 14 texts and is an assistant professor at the University of Cincinnati, where he teaches the introductory IS course. The Stair and Reynolds team brings a solid conceptual foundation and practical IS experience to students.

GOALS OF THIS TEXT

Because *Principles of Information Systems: A Managerial Approach, Seventh Edition* is written for all business majors, we believe it is important not only to present a realistic perspective on information systems in business but also to provide students with the skills they can use to be effective business leaders in their organization. To that end, this book has four main goals:

1. To provide a core of IS principles with which every business student should be familiar
2. To offer a survey of the IS discipline that will enable all business students to understand the relationship of IS courses to their curriculum as a whole
3. To present the changing role of the IS professional
4. To show the value of the discipline as an attractive field of specialization

By achieving these goals, this text will enable students, regardless of their major, to understand and use fundamental IS principles so that they can function more efficiently and effectively as workers, managers, decision makers, and organizational leaders.

IS Principles

Principles of Information Systems: A Managerial Approach, Seventh Edition, although comprehensive, cannot cover every aspect of the rapidly changing IS discipline. The authors recognize this and provide students an essential core of guiding IS principles to use as they face the career challenges ahead. Think of principles as basic truths, rules, or assumptions that remain constant regardless of the situation. As such, they provide strong guidance in the face of tough decisions. A set of IS principles is highlighted at the beginning of each chapter, as are the chapter's learning objectives. Then these principles are applied to solve real-world problems from the opening vignettes to the end-of-chapter material. The ultimate goal of *Principles of Information Systems* is to develop effective, thinking employees by instilling them with principles to help guide their decision making and actions.

Survey of the IS Discipline

This text not only offers the traditional coverage of computer concepts but also provides students with a solid grounding in the business uses of technology. In addition to serving general business students, this book offers an overview of the entire IS discipline and solidly prepares future IS professionals for advanced IS courses and their careers in the rapidly changing IS discipline.

Changing Role of the IS Professional

As business and the IS discipline have changed, so too has the role of the IS professional. Once considered a technical specialist, today the IS professional operates as an internal consultant to all functional areas of the organization, being knowledgeable about their needs and competent in bringing the power of information systems to bear throughout the organization. The IS professional views issues through a global perspective that encompasses the entire organization and the broader industry and business environment in which it operates.

The scope of responsibilities of an IS professional today is not confined to just his/her employer but encompasses the entire interconnected network of suppliers, customers, competitors, regulatory agencies, and other entities—no matter where they are located. This broad mission creates a new challenge: how to help an organization survive in a highly interconnected, highly competitive global environment. In accepting that challenge, the IS professional plays a pivotal role in shaping the business itself and ensuring its success. To survive, businesses must now strive for the highest level of customer satisfaction and loyalty through competitive prices and ever-improving product and service quality. The IS professional assumes the critical responsibility of determining the organization's approach to both overall cost and quality performance and therefore plays an important role in the continued survival of the organization. This new duality in the role of the IS employee—a professional who exercises a specialist's skills with a generalist's perspective—is reflected throughout the book.

IS as a Field for Further Study

Despite the downturn in the economy at the start of the 21st century, especially in technology-related sectors, the outlook for computer and IS managers is bright. In fact, employment of computer and IS managers is expected to grow much faster than the average occupation through the year 2012. Technological advancements are boosting the employment of computer-related workers; in turn, this will create demand for managers to direct these workers. In addition, job openings will result from the need to replace managers who retire or move into other occupations.

A career in information systems can be exciting, challenging, and rewarding! This text shows the value of the discipline as an appealing field of study and the IS graduate as an integral part of today's organizations. Perhaps more than ever before, the IS professional must be able to align IS and organizational goals and to ensure that IS investments are justified from a business perspective. So, bright and interested students are needed in the IS discipline. Upon graduation, IS graduates at many schools are among the highest paid of all business graduates. Throughout this text, we prepare students for their careers by highlighting the many challenges and opportunities available to IS professionals.

CHANGES IN THE SEVENTH EDITION

We have implemented a number of exciting changes to the text in response to user feedback on how the text can be aligned even more closely with IS principles and concepts and the ways the course is now being taught. A summary of these changes follows:

- *Unifying Theme.* In this edition, we stress the global aspects of information systems as a major theme. As organizations increasingly find themselves competing in a global

marketplace, they must recognize the resulting implications on their information systems. Globalization is profoundly changing businesses, markets, and society. With its years of service to the IS discipline, this text retains the traditions and strengths of past successes while helping future managers and decision makers face tomorrow's global challenges.

- *All New World Views Cases.* While the text has always stressed the global factors affecting information systems, these factors are emphasized even more in this edition through the World Views Cases. These cases, written by instructors outside the United States and about real organizations outside the United States, provide the reader with solid insight into the IS issues facing foreign-based or multinational companies.

- *All New Vignettes Emphasize International Aspects.* In addition to the World Views Cases, all of the chapter opening vignettes raise actual issues from foreign-based or multinational companies.

- *Why Learn About Features.* Each chapter has a new "Why Learn About" section at the beginning of the chapter to pique student interest. The section sets the stage for students by briefly describing the importance of the chapter's material to business students—whatever their chosen field.

- *Information Systems @ Work Special Interest Boxes.* Highlighting current topics and trends in today's headlines, these boxes show how information systems are used in a variety of business career areas.

- *Career Exercises.* New end-of-chapter Career Exercises ask students to research how a topic discussed in the chapter relates to a business area of their choice. Students are encouraged to use the Internet, the college library, or interviews to collect information about business careers.

- *All New Videos and Video Questions.* New video segments are provided for each chapter of the seventh edition. These segments demonstrate key chapter concepts. Students can actually see IS principles at work in a variety of settings and then answer questions to help them apply what they have learned.

- *Thoroughly Revised End-of-Chapter Material.* The material at the end of each chapter has been thoroughly updated. Summaries linked to the principles, key terms, self-assessment questions, review questions, discussion questions, problem-solving exercises, team activities, and Web exercises have been replaced and revised to reflect the theme of the seventh edition and to give students the opportunity to explore the latest technology in a business setting.

- *All New Cases.* Three new end-of-chapter cases provide a wealth of practical information for students and instructors. Each case explores a chapter concept or problem that a real-world company or organization has faced. The cases can be assigned as individual homework exercises or serve as a basis for class discussion.

CHAPTER CHANGES

Each chapter has been completely updated with the latest topics and examples. Here is a summary of some of the changes.

Chapter 1, An Introduction to Information Systems

This chapter is full of new boxes, photos, figures, tables, examples, and more than 40 current references. The new opening vignette focuses on Boehringer Ingelheim, which is among the world's 20 largest pharmaceutical companies, with $7.6 billion in revenue and 32,000

employees in 60 nations. The new "Why Learn About" section motivates students by showing them the importance of information systems to achieve their career goals. The new "Information Systems @ Work" box stresses the use of information systems by Delta's low-fare, all-digital spin-off, Song airline, which uses technology, state-of-the-art information systems, and good old-fashioned customer service to win customers. Figure 1.6 contains new examples of models, and Figure 1.8 contains new images. The section on "Computer-Based Information Systems" includes many new examples and photos. Table 1.4 is new to this edition, revealing the many powerful uses of the Internet. The section on "Business Information Systems" includes a new figure that shows how these systems evolved from early transaction processing systems of the 1950s to the advanced decision support and special-purpose systems of today. This section also includes new examples and a new photo of the use of B2B (business-to-business) applications. Mobile commerce (m-commerce) is introduced in this chapter. The new "Ethical and Societal Issues" box includes information on the dangers of *phishing*—the use of e-mail and Web sites as bait to lure consumers into revealing private information. There is a new photo on the use of ERP. The section on "Virtual Reality" contains new examples and new photos of a head-mounted display and a data glove. Figure 1.18 is new to this edition, showing the most common attacks to information systems today, and Figure 1.19 reveals the huge costs of these attacks. The all-new end-of-chapter cases include information on Tyndall Federal Credit Union, which provides banking services to military personnel at Tyndall Air Force Base in Panama City, Florida; the Queen Mary 2 (QM2), the largest and most expensive cruiseship ever built; and My-Family.com, Inc., a leading online subscription business for researching family history.

Chapter 2, Information Systems in Organizations

The material on organizational change has been enhanced to include sustaining and disruptive change. Christiansen, who wrote the business best-sellers *Inventor's Dilemma* and the more recent *Inventor's Solution*, discusses these types of change. The six-sigma quality program has been introduced and demonstrated with several examples. The material on competitive advantage has been strengthened with material from Jim Collin's best-selling business book *Good to Great*. Specifically, we introduce the notion of technology acceleration with new examples and a new table. The productivity paradox has been introduced, as recommended by a reviewer. The careers in information systems section has been updated with new positions, such as the chief technology officer (CTO), and several examples and direct quotes from CTOs. The Clinger-Cohen Act, which requires CIOs for certain federal agencies, has also been included, as requested by a reviewer.

Chapter 3, Hardware: Input, Processing, and Output Devices

The new "Why Learn About Hardware?" and "Information Systems @ Work" features help demonstrate why a business major needs to understand this chapter's concepts. In addition, the chapter has been broadened to a managerial focus on what a non-IS decision maker needs to know about hardware. The latest information on processors, main memory, secondary storage devices, and input/output devices is covered, including such topics as smart phones, tablet PCs, and MP3 players. There is added coverage of grid computing and utility computing. More than 60 new references and examples appear throughout the chapter. The material on computer system types has been modified to conform to the Gartner Group's industry standard definitions for the various types of computer systems. The section on selecting/upgrading computer systems has been updated to include a subsection on printers and DVD burners. Also, the need for proper disposal of computer hardware is discussed.

Chapter 4, Software: Systems and Application Software

New to this edition is a section that divides operating systems into four general categories: single computer with a single user, single computer with multiple users, multiple computers, and special-purpose computers. The section on common hardware functions has been streamlined, and the section on file management now describes file types and conventions.

The section on personal computer operating systems has been revamped to include current operating systems and developments—including Windows XP, Apple, and Linux—and to reduce material on older operating systems. The section on Windows Server has been updated to include a discussion of Windows Server 2003, and the discussion of enterprise operating systems has also been updated and streamlined. Reflecting industry trends, a new section on operating systems for small computers and special-purpose devices has been included, with coverage of Palm OS, Windows Embedded, and Windows Mobile developments. Examples of utility programs have been thoroughly revised. A new section on middleware has been included. The material on proprietary and off-the-shelf software has been revised to address reviewer comments. The material on personal application software has been updated. For example, the material on Excel now includes Solver. The material on software suites now includes more information about open-source suites. There is a new section on other personal application software, which focuses on project management, financial management and tax preparation, educational and reference, desktop publishing, computer-aided design, and statistical software. There is a new section on application software for information, decision support, and specialized purposes. The section on fifth generation languages has been changed to emphasize visual and object-oriented languages and is now forward thinking in its title, "Languages beyond the Fourth Generation." Finally, the section on open-source software near the end of the chapter has been expanded and includes a new table listing different open-source software products.

Chapter 5, Organizing Data and Information

The material on the traditional approach to database management has been reduced slightly. The material in the section on the database approach has been updated, including Table 5.2 on the disadvantages of the database approach. Because hierarchical and network models are no longer widely used, that material has been cut drastically. The section on data cleanup has been updated with an example. Instructors who want to cover normalization will be able to use the example as a starting point. The section on database management systems has been updated to include a discussion of flat files, single-user, and multiuser databases. The section on manipulating data and generating reports includes new material on query-by-example (QBE) and new examples of SQL. Based on a reviewer suggestion, SQL commands have been introduced in a new table to show students how SQL can be used for a variety of database purposes, such as selecting, projecting, joining, and security. The section on popular database management systems has been updated, and a new section on special-purpose database systems has been added, which contains a number of new examples of real organizations using specialized-database systems. A new section on using databases with other software has also been added. Finally, the section on database applications has been updated with new material and new examples, such as Oracle's Warehouse Management software, which can incorporate data from radio-frequency identification (RFID) technology. There is also a new section on visual, audio, and other databases near the end of the chapter. As with other chapters, the end-of-chapter material has been updated.

Chapter 6, Telecommunications and Networks

An expanded section on voice and data convergence discusses the "big picture" of how we are changing from a telecommunications infrastructure based on POTS (plain old telephone service) using twisted pair copper wires intended for analog voice communication signals, circuit switching, and "dumb" voice telephones to a new infrastructure for transmitting digital data signals based on packet switching through fiber-optic links with intelligent user devices that provide addressing information. The focus is on converting all signals to a digital form and using a single digital network to carry all communications (including voice and data). The coverage of cellular transmission has been expanded and updated to include a discussion of the features and characteristics of 1G, 2G, 2.5G, 3G, and Multichannel Multipoint Distribution System cellular communications. Personal area networks and metropolitan area networks have been added under the discussion of network types. The

discussion of communications protocols has been simplified and streamlined with additional protocols including frame relay, FireWire, IEEE 802.11 g, IEEE 802.16 (WiMax), and IEEE 802.20 (Mobile Broadband Wireless Access). A discussion on the role of a hub has been added under the section on "Network Switching Devices." The use of call centers and electronic funds transfer was added to the discussion of telecommunications applications.

Chapter 7, The Internet, Intranets, and Extranets

Chapter 7 was significantly revamped and reorganized to reflect the dynamic nature of the Internet. There are more than 70 new references for current examples throughout the chapter. The major section on Internet services was moved later in the chapter, expanded to include Web applications, and retitled "Internet and Web Applications." This important change allowed us to cover the technology first and the applications of the technology next in one unified section. A new table on the use of broadband to connect to the Internet has been included, comparing the transmission times for a full-length movie: two weeks over a standard modem, two days over cable or satellite, and about 20 seconds over a T-1 line. In the section on the Web, we have introduced the notion of Web portals and provided examples. The section on Web browsers has been updated and enhanced, and the section on search engines now includes more sophisticated search parameters and approaches. Additional languages and applications have been included in the section on Java and Internet programming languages. The discussion of developing Web content has been updated to include new Web authoring software and the importance of content management systems (CMS). An important addition to the seventh edition is a new section on Web services, which discusses the approach and standards that are used. The section on "Internet and Web Applications" has been totally updated. We have new material on Internet cell phones and handheld computers. There are also new sections on career information and online job searches and Web logs, or blogs. The new material on voice over Internet Protocol is introduced by a powerful quote from the chairman of the U.S. Federal Communications Commission that stresses the increasing importance of this technology and its implications to the traditional long-distance phone companies. There is a new table of sales on Web sites. The section on music, radio, and video on the Internet has been updated. The sections on intranets and extranets have been updated, along with the section on net issues. For example, we now have a new table on the source of Internet attacks expressed in percentages. Of course, the end-of-chapter material has been totally updated to reflect these changes.

Chapter 8, Electronic Commerce

A brief summary of the status of e-commerce around the world has been added. M-commerce is introduced as a new business model. In addition to examples of m-commerce, the technology required is discussed, and the anywhere/anytime capability of m-commerce is addressed. Web services are discussed, including a definition, use of standards, examples, and a Gartner Group forecast of the growth of Web services. Phishing is discussed as a serious threat to e-commerce. The role of the Federal Trade Commission in regulating Internet activities is briefly addressed. The use of Web site customer experience technology to analyze the usability of a site is discussed.

Chapter 9, Transaction Processing and Enterprise Resource Planning Systems

New material has been added concerning the various laws to control the operation and use of these systems: Sarbanes-Oxley, Graham-Leach-Bliley, and the Health Insurance Portability and Accountability Act. Current examples of companies being affected by these acts have been added. The material on business resumption planning and disaster recovery has been modified so that the definition of these terms is consistent with current industry thinking. Radio-frequency identification (RFID) technology is mentioned as a new technology that is revolutionizing TPSs. The list of vendors providing customer relationship management (CRM) systems has been updated for currency. An interesting example of the difficulties GM had implementing a CRM system in China is presented. The list of vendors providing ERP

systems has been updated to reflect recent industry acquisitions. There is brief discussion of the competition in the ERP arena between Oracle and PeopleSoft, as well as PeopleSoft's recent acquisition of JD Edwards. New examples of companies trying to deal with international systems are added. The section on ERP has been modified to focus on just what is an ERP and what business processes are involved and affected.

Chapter 10, Information and Decision Support Systems

The section on inputs to a management information system emphasizes supply chain management to a great extent in this new edition. Data mining is included in the section on outputs of a management information system. The section on the uses and management of funds discusses return on investment (ROI). The manufacturing MIS section includes a discussion of smart labels. Design and engineering aspects of the manufacturing MIS discuss the involvement of customers and also include the importance of scheduling software and the notion of 3D computer-aided design (CAD) tools. In the section on quality of and testing for the manufacturing MIS, we discuss total quality management and continuous improvement. The marketing MIS section describes the importance of group sales meetings and contains increased coverage and examples of CRM software and systems. The section on promotion and advertising includes an example of Choice Stream, which sends people e-mail about entertainment choices according to their preferences. This section also includes a discussion of richness and reach, two important marketing concepts. Rich media advertising and the increasing use of Internet advertising are also included. A new section on outplacement has been included in the material on the human resource MIS. There are a number of new examples on the use of a geographic information system (GIS). The section on group support systems (GSSs) and groupware has been enhanced to include the use of newer wireless systems, including Blackberry and other mobile communications devices and systems.

Chapter 11, Specialized Information Systems

The title of the chapter has been changed from "Specialized Business Information Systems" to "Specialized Information Systems" to reflect the unique applications in nonprofit and military organizations. The chapter begins with a new example of chess master Garry Kasparov competing against an artificial intelligence (AI) software package that runs on a PC called Deep Junior. The section on robotics has many new examples, from NASA to entertainment. The section on vision systems also has new examples, including the use of vision systems to inspect wine bottles in California and attach windscreens on Jaguar S-Type cars. The chapter also discusses the use of natural language processing by a hardware company to develop a Web site that allows customers to find what they need. The section on neural networks has examples on improving motor coordination in robots, reading bar codes, and preventing fraud and terrorism. The chapter also has a new section on other AI applications, which highlights the use of genetic algorithms and intelligent agents and provides numerous examples of these applications. There are also many new examples of expert systems. The section on virtual reality has been updated with new examples such as the following: the use of virtual reality by an automotive company to help design cars and factories; the use of a virtual reality interface device to help people play chess with computers; SimCity, a virtual reality game; and a virtual reality Web site by Disney. The last section on other specialized systems has been completely revised with new examples that range from small microchips planted in the brain to the use of game theory and the development of small networks, called smart dust, by the University of California at Berkeley. There is also new material on informatics, including bioinformatics and medical informatics.

Chapter 12, Systems Investigation and Analysis

The material on project management has been strengthened with new examples and quotes. The importance of good project leadership has been emphasized. The effect of the Sarbanes-Oxley Act on systems development is also been emphasized with several examples. The section on developing a competitive advantage also has new examples, including the efforts of Fuji

and Kodak to develop Web sites for storing and sharing photos. The importance of scalability has been emphasized in the section on performance objectives. In the section on Web-based systems development, the discussion of HTML, XML, and other Web tools is new. We have increased the scope of the former "Systems Development and E-Commerce" section to "Systems Development, the Internet, Intranets, Extranets, and E-Commerce." New tools and techniques have been introduced, and the section has been updated to reflect the changes to Chapter 7. The material on prototyping has also been updated. Rapid application development (RAD) tools by IBM and others have been highlighted in the section on RAD. The section on outsourcing has been expanded to include on-demand computing, along with new material and examples. The section on factors affecting systems development success has new material and examples. The material on CASE tools has been streamlined and updated to include new tools, such as VRCASE. As requested by a reviewer, we have deleted the discussion of upper and lower CASE tools. New material on scalability has been added to the objectives of a systems development project.

Chapter 13, Systems Design, Implementation, Maintenance, and Review

In the section on software design, we have added information about software developed to comply with the Sarbanes-Oxley Act. In the section on personnel design, the use of outsourcing jobs to India by a London, England, travel agency is discussed. The elements of good interactive dialogue are now summarized in a table to streamline the discussion. The section on disaster planning has been updated and integrated into a new section titled "Disaster Planning and Recovery." There is a new table on systems controls that highlights input, processing, output, database, telecommunications, and personnel controls. There is also new material on "on-demand" or "utility" computing, with several new examples. How individuals and organizations can dispose of older systems is also discussed. Reusable software is also emphasized, with new examples. The section on cross-platform development includes new tools, including Web services and .NET by Microsoft®. The material on integrated development environments (IDEs) has been enhanced to include newer programming tools and approaches, including Microsoft's Visual Studio .NET. The software acquisition section now emphasizes that software is increasingly being viewed as a utility or a service, not a product to be purchased. A new section on systems operation has been added that includes information about support, training, and help desks. The section on systems maintenance includes new information on legacy systems.

Chapter 14, Security, Privacy, and Ethical Issues in Information Systems and the Internet

Many new topics have been added including the following: the Computer Assisted Passenger Prescreening System for airline safety, the Controlling the Assault of Non-Solicited Pornography and Marketing (CAN-SPAM) Act, the Sarbanes-Oxley Act, the Health Insurance Portability and Accountability Act, the 2003 Computer Crime and Security Survey, cyberterrorism, identity theft, virus variants, biometrics, the Children's Internet Protection Act, the Child Online Protection Act, the Gramm-Leach-Bliley Act, software vulnerabilities, spyware, and cyberstalking.

WHAT WE HAVE RETAINED FROM THE SIXTH EDITION

The seventh edition builds on the strengths of past editions; it retains the focus on IS principles and strives to be the most current text on the market.

- *Overarching Principle.* This book continues to stress a single-all-encompassing theme: The right information, if it is delivered to the right person, in the right fashion, and at the right time, can improve and ensure organizational effectiveness and efficiency.

- *Information Systems Principles.* Information System Principles summarize key concepts that every student should know. This important feature is a convenient summary of key ideas presented at the start of each chapter.

- *Learning Objectives Linked to Principles.* Carefully-crafted learning objectives are included with every chapter. The learning objectives are linked to the Information Systems Principles and reflect what a student should be able to accomplish after completing a chapter.

- *Summary Linked to Principles.* Each chapter includes a detailed summary, and each section of the summary is tied to an Information System Principle.

- *Ethical and Societal Issues Special Interest Boxes.* Each chapter includes an "Ethical and Societal Issues" box that presents a timely look at the ethical challenges and the societal impact of information systems. Ethics remains a compelling issue for today's business and IS students, and they gain exposure to ethical and societal issues by grappling with the in-depth questions related to the company scenarios. All boxes relate to the issues discussed in the chapters.

- *Current Examples, Boxes, Cases, and References.* As we have in each edition, we take great pride in presenting the most recent examples, boxes, cases, and references throughout the text. Some of these were developed at the last possible moment, literally weeks before the book went into publication. Information on new hardware and software, the latest operating systems, application service providers, the Internet, electronic commerce, ethical and societal issues, and many other current developments can be found throughout the text. Our adopters have come to expect the best and most recent material. We have done everything we can to meet or exceed these expectations.

- *Self-Assessment Tests.* This popular feature helps students review and test their understanding of key chapter concepts.

STUDENT RESOURCES

MIS Companion CD

We are pleased to include in every textbook a free copy of Thomson Course Technology's MIS Companion CD, which is composed of training lessons in Excel, Access, and MIS concepts. The Companion CD's content is integrated throughout the book. Wherever you see the CD icon in the chapter margins, you know that you can find additional related material on the CD.

Student Online Companion Web Site

We have created an exciting online companion, password protected for students to utilize as they work through the seventh edition of *Principles of Information Systems*. In the front of this text you will find a key code that provides full access to a robust Web site, located at *www.course.com/mis/stair*. This Web resource includes the following features:

- **Videos**
  Links to 14 topical video clips, one relating to every chapter in the book, can be found on this Web site. Questions corresponding to the respective video clips are

featured at the end of each chapter in the book. These exercises reinforce the concepts taught and provide the students with more critical thinking opportunities.

- **PowerPoint Slides**
  Direct access is offered to the book's PowerPoint presentations, which cover the key points from each chapter. These presentations are a useful study tool.

- **Classic Cases**
  A frequent request from adopters is that they wish to have a broader selection of cases to choose from. To meet this need, a set of over 85 cases from the fourth, fifth, and sixth editions of the text are included here. These are the authors' choices of the "best cases" from these editions and span a broad range of companies and industries.

- **Links to Useful Web Sites**
  Chapters in *Principles of Information Systems, Seventh Edition*, reference many interesting Web sites. This resource takes you to links you can follow directly to the home pages of those sites so that you can explore them. There are additional links to Web sites that the authors, Ralph Stair and George Reynolds, think you would be interested in checking out.

- **Hands-On Activities**
  Use the Hands-On Activities to test your comprehension of IS topics and enhance your skills using Microsoft Office applications and the Internet. Using these links, you can access three critical-thinking exercises per chapter; each activity asks you to work with an Office tool or do some research on the Internet.

- **Test Yourself on IS**
  This tool allows you to access 20 multiple-choice questions for each chapter, test yourself, and then submit your answers. You will immediately find out which questions you got right and which you got wrong. For each question that you answer incorrectly, you are given the correct answer and the page in your text where that information is covered. Special testing software randomly compiles 20 questions from a database of 50 questions, so you can quiz yourself multiple times on a given chapter and get some new questions each time.

- **Glossary of Key Terms**
  The glossary of key terms from the text is available to search.

- **Online Readings**
  This feature provides you access to a computer database that contains articles relating to hot topics in information systems.

INSTRUCTOR RESOURCES

The teaching tools that accompany this text offer many options for enhancing a course. And, as always, we are committed to providing one of the best teaching resource packages available in this market.

Instructor's Manual

An all-new *Instructor's Manual* provides valuable chapter overviews; highlights key principles and critical concepts; offers sample syllabi, learning objectives, and discussion topics; and features possible essay topics, further readings and cases, and solutions to all of the end-of-chapter questions and problems, as well as suggestions for conducting the team activities. Additional end-of-chapter questions are also included.

Sample Syllabus

A sample syllabus with sample course outlines is provided to make planning your course that much easier.

Solutions

Solutions to all end-of-chapter material are provided in a separate document for your convenience.

Test Bank and Test Generator

ExamView® is a powerful objective-based test generator that enables instructors to create paper-, LAN- or Web-based tests from test banks designed specifically for their Thomson Course Technology text. Instructors can utilize the ultra-efficient Quick Test Wizard to create tests in less than five minutes by taking advantage of Thomson Course Technology's question banks or customizing their own exams from scratch. Page references for all questions are provided so you can cross-reference test results with the book.

PowerPoint Presentations

A set of impressive Microsoft PowerPoint slides is available for each chapter. These slides are included to serve as a teaching aid for classroom presentation, to make available to students on the network for chapter review, or to be printed for classroom distribution. Our presentations help students focus on the main topics of each chapter, take better notes, and prepare for examinations. Instructors can also add their own slides for additional topics they introduce to the class.

Figure Files

Figure Files allow instructors to create their own presentations using figures taken directly from the text.

DISTANCE LEARNING

Thomson Course Technology, the premiere innovator in management information systems publishing, is proud to present online courses in WebCT and Blackboard.

- *Blackboard and WebCT Level 1 Online Content.* If you use Blackboard or WebCT, the test bank for this textbook is available at no cost in a simple, ready-to-use format. Go to *www.course.com* and search for this textbook to download the test bank.

- *Blackboard and WebCT Level 2 Online Content.* Blackboard 5.0 and 6.0 as well as Level 2 and WebCT Level 2 courses are also available for *Principles of Information Systems, Seventh Edition.* Level 2 offers course management and access to a Web site that is fully populated with content for this book.

For more information on how to bring distance learning to your course, instructors should contact their Thomson Course Technology sales representative.

ACKNOWLEDGMENTS

A book of this size and undertaking requires a strong team effort. We would like to thank all of our fellow teammates at Thomson Course Technology and Elm Street Publishing

Services for their dedication and hard work. Special thanks to Eunice Yeates-Fogle, our Senior Product Manager. Our appreciation goes out to all the many people who worked behind the scenes to bring this effort to fruition, including Mirella Misiaszek, our Associate Product Manager. Kelly Robinson, our Production Editor, shepherded the book through the production process, and Beth Paquin drove the Student Online Companion. We would also like to acknowledge and specially thank Elm Street Publishing Services. Karen Hill, our development editor, deserves special recognition for her tireless effort and help in all stages of this project.

We are grateful to the sales force at Thomson Course Technology, whose efforts make this all possible. You helped to get valuable feedback from current and future adopters. As Thomson Course Technology product users, we know how important you are.

We would especially like to thank Ken Baldauf for his excellent help in writing most of the boxes and cases for this edition. Ken also provided invaluable feedback for many topics discussed in the book.

Ralph Stair would like to thank the Department of Management Information Systems, College of Business Administration, at Florida State University for their support and encouragement. He would also like to thank his family, Lila and Leslie, for their support.

George Reynolds would like to thank the Department of Information Systems, College of Business, at the University of Cincinnati for their support and encouragement. He would also like to thank his family, Ginnie, Tammy, Kim, Kelly, and Kristy, for their patience and support in this major project. He would also like to thank Kristen Duerr and Ralph Stair for asking him to join the writing team back in 1997.

OUR PREVIOUS ADOPTERS AND POTENTIAL NEW USERS

We sincerely appreciate our loyal adopters of the previous editions and welcome new users of *Principles of Information Systems: A Managerial Approach, Seventh Edition*. As in the past, we truly value your needs and feedback. We can only hope the seventh edition continues to meet your high expectations.

We would especially like to thank reviewers of the seventh edition, focus group members, and reviewers of previous editions

In addition, Ralph Stair would like to thank Mike Jordan, Senior Lecturer, Division of Business Information Management Glasgow Caledonia University for his hospitality and help at the UKAIS conference.

Reviewers for the Seventh Edition

We are indebted to the following individuals for their perceptive feedback on early drafts of this text:

Chang E. Koh, *University of North Texas*
Howard Sundwall, *West Chester University*

Reviewers for the First, Second, Third, Fourth, Fifth, and Sixth Editions

The following people shaped the book you hold in your hands by contributing to previous editions:

Jill Adams, *Navarro College*
Robert Aden, *Middle Tennessee State University*
A. K. Aggarwal, *University of Baltimore*
Sarah Alexander, *Western Illinois University*
Beverly Amer, *University of Florida*
Noushin Asharfi, *University of Massachusetts*
Yair Babad, *University of Illinois—Chicago*

Cynthia C. Barnes, *Lamar University*
Charles Bilbrey, *James Madison University*
Thomas Blaskovics, *West Virginia University*
John Bloom, *Miami University of Ohio*
Warren Boe, *University of Iowa*
Glen Boyer, *Brigham Young University*
Mary Brabston, *University of Tennessee*

Jerry Braun, *Xavier University*

Thomas A. Browdy, *Washington University*

Lisa Campbell, *Gulf Coast Community College*

Andy Chen, *Northeastern Illinois University*

David Cheslow, *University of Michigan—Flint*

Robert Chi, *California State University—Long Beach*

Carol Chrisman, *Illinois State University*

Miro Costa, *California State University—Chico*

Caroline Curtis, *Lorain County Community College*

Roy Dejoie, *USWeb Corporation*

Sasa Dekleva, *DePaul University*

Pi-Sheng Deng, *California State University—Stanislaus*

Roger Deveau, *University of Massachusetts—Dartmouth*

John Eatman, *University of North Carolina*

Juan Esteva, *Eastern Michigan University*

Gordon Everest, *University of Minnesota*

Badie Farah, *Eastern Michigan University*

Karen Forcht, *James Madison University*

Carroll Frenzel, *University of Colorado—Boulder*

John Gessford, *California State University—Long Beach*

Terry Beth Gordon, *University of Toledo*

Kevin Gorman, *University of North Carolina—Charlotte*

Costanza Hagmann, *Kansas State University*

Bill C. Hardgrave, *University of Arkansas*

Al Harris, *Appalachian State University*

William L. Harrison, *Oregon State University*

Dwight Haworth, *University of Nebraska—Omaha*

Jeff Hedrington, *University of Wisconsin—Eau Claire*

Donna Hilgenbrink, *Illinois State University*

Jack Hogue, *University of North Carolina*

Joan Hoopes, *Marist College*

Donald Huffman, *Lorain County Community College*

Patrick Jaska, *University of Texas at Arlington*

G. Vaughn Johnson, *University of Nebraska—Omaha*

Grover S. Kearns, *Morehead State University*

Robert Keim, *Arizona State University*

Karen Ketler, *Eastern Illinois University*

Mo Khan, *California State University—Long Beach*

Michael Lahey, *Kent State University*

Jan de Lassen, *Brigham Young University*

Robert E. Lee, *New Mexico State University—Carlstadt*

Joyce Little, *Towson State University*

Herbert Ludwig, *North Dakota State University*

Jane Mackay, *Texas Christian University*

Al Maimon, *University of Washington*

James R. Marsden, *University of Connecticut*

Roger W. McHaney, *Kansas State University*

Lynn J. McKell, *Brigham Young University*

John Melrose, *University of Wisconsin—Eau Claire*

Michael Michaelson, *Palomar College*

Ellen Monk, *University of Delaware*

Bertrad P. Mouqin, *University of Mary Hardin-Baylor*

Bijayananda Naik, *University of South Dakota*

Pamela Neely, *Marist College*

Leah R. Pietron, *University of Nebraska—Omaha*

John Powell, *University of South Dakota*

Maryann Pringle, *University of Houston*

John Quigley, *East Tennessee State University*

Mahesh S. Raisinghani, *University of Dallas*

Mary Rasley, *Lehigh-Carbon Community College*

Earl Robinson, *St. Joseph's University*

Scott Rupple, *Marquette University*

Dave Scanlon, *California State University—Sacramento*

Werner Schenk, *University of Rochester*

Larry Scheuermann, *University of Southwest Louisiana*

James Scott, *Central Michigan University*

Vikram Sethi, *Southwest Missouri State University*

Laurette Simmons, *Loyola College*

Janice Sipior, *Villanova University*

Anne Marie Smith, *LaSalle University*

Harold Smith, *Brigham Young University*

Patricia A. Smith, *Temple College*

Herb Snyder, *Fort Lewis College*

Alan Spira, *University of Arizona*

Tony Stylianou, *University of North Carolina*

Bruce Sun, *California State University—Long Beach*

Hung-Lian Tang, *Bowling Green State University*

William Tastle, *Ithaca College*

Gerald Tillman, *Appalachian State University*

Duane Truex, *Georgia State University*

Jean Upson, *Lorain County Community College*

Misty Vermaat, *Purdue University—Calumet*

David Wallace, *Illinois State University*

Michael E. Whitman, *University of Nevada—Las Vegas*

David C. Whitney, *San Francisco State University*

Goodwin Wong, *University of California—Berkeley*

Amy Woszczynski, *Kennesaw State University*

Judy Wynekoop, *Florida Gulf Coast University*

Myung Yoon, *Northeastern Illinois University*

Focus Group Contributors for the Third Edition

Mary Brabston, *University of Tennessee*

Russell Ching, *California State University—Sacramento*

Virginia Gibson, *University of Maine*

Bill C. Hardgrave, *University of Arkansas*

Al Harris, *Appalachian State University*

Stephen Lunce, *Texas A & M International*

Merle Martin, *California State University—Sacramento*

Mark Serva, *Baylor University*

Paul van Vliet, *University of Nebraska—Omaha*

OUR COMMITMENT

We are committed to listening to our adopters and readers and to developing creative solutions to meet their needs. The field of IS continually evolves, and we strongly encourage your participation in helping us provide the freshest, most relevant information possible.

We welcome your input and feedback. If you have any questions or comments regarding *Principles of Information Systems: A Managerial Approach, Seventh Edition*, please contact us through Thomson Course Technology or your local representative, via e-mail at ct.mis@thomson.com, via the Internet at *www.course.com*, or address your comments, criticisms, suggestions, and ideas to:

Ralph Stair

George Reynolds

Thomson Course Technology

25 Thomson Place

Boston, MA 02210

PART
· 1 ·

An Overview

Chapter 1 An Introduction to Information Systems
Chapter 2 Information Systems in Organizations

CHAPTER
· 1 ·

An Introduction to Information Systems

PRINCIPLES	LEARNING OBJECTIVES
■ The value of information is directly linked to how it helps decision makers achieve the organization's goals.	■ Discuss why it is important to study and understand information systems. ■ Distinguish data from information and describe the characteristics used to evaluate the quality of data.
■ Models, computers, and information systems are constantly making it possible for organizations to improve the way they conduct business.	■ Name the components of an information system and describe several system characteristics. ■ Identify the basic types of models and explain how they are used.
■ Knowing the potential impact of information systems and having the ability to put this knowledge to work can result in a successful personal career, organizations that reach their goals, and a society with a higher quality of life.	■ List the components of a computer-based information system. ■ Identify the basic types of business information systems and discuss who uses them, how they are used, and what kinds of benefits they deliver.
■ System users, business managers, and information systems professionals must work together to build a successful information system.	■ Identify the major steps of the systems development process and state the goal of each.
■ Information systems must be applied thoughtfully and carefully so that society, business, and industry can reap their enormous benefits.	■ Describe some of the threats to security and privacy that information systems and the Internet can pose. ■ Discuss the expanding role and benefits of information systems in business and industry.

INFORMATION SYSTEMS IN THE GLOBAL ECONOMY

BOEHRINGER INGELHEIM, GMBH, GERMANY

Lean and Mean with Information Systems

If a business owner could foresee the future, decisions would be easy to make and involve little risk. Unfortunately, crystal balls, tarot cards, and tea leaves have proven useless to business decision makers. Since they can't see into the future, their goal is to find out what is currently happening in their business. For a small-business owner, one who is involved in every transaction, staying abreast of the state of the business is fairly straightforward. But as businesses grow and diversify, it becomes increasingly difficult to access current information on which to base decisions. Information systems provide a solution—opening lines of communication between all business units and providing a continuous stream of up-to-the-second information to support fast strategic decision making.

Boehringer Ingelheim is among the world's 20 largest pharmaceutical companies. A giant company with $7.6 billion in revenue and 32,000 employees in 60 nations, Boehringer has diversified into segments that include manufacturing and marketing pharmaceuticals (such as prescription medicines and consumer healthcare products), products for industrial customers (such as chemicals and biopharmaceuticals), and animal health products.

The sheer size of the company was slowing the flow of information to decision makers in the organization. "I want to be told where I stand and where we are heading," says Holger Huels, chief financial officer, "I like to [be able to] see negative trends and counter them as fast as possible." With each of the company's segments using diverse information systems, it took a significant amount of time to collect and combine all of the financial records. Each month the accounting department would spend three days collecting and analyzing printed reports to create the company's monthly report.

Top managers decided to totally revamp the company's systems with state-of-the-art information systems from SAP, the world's largest enterprise software company. It took 14 months to roll out the new system, and many employees needed intensive training. In the end, the results were well worth the investment in time and money. The software provided a standard system used across all of Boehringer's business segments and offered convenient Web access to current information. Boehringer is now able to complete monthly reports just two hours after the close of business at the end of each month. The new system has made the accounting department much more productive, allowing staff to run up-to-date reports whenever needed.

Boehringer is committed to providing employees at all levels of the company with access to the applications and information they need to meet their objectives. About one third of Boehringer's employees do their work outside the office. To provide its mobile workforce with up-to-the-minute data, the company deployed software from BackWeb Technologies, which allows access to current sales information through a Web portal and a custom Web interface, wherever employees travel. With the new system, Boehringer's employees can access and change information presented in the portal when they are offline, with updates later when they log on.

By the time Boehringer was finished with its technology makeover, the company had implemented over seven new interconnected information systems and invested millions in hardware, software, databases, telecommunications, and training. But the investment has paid off. Employees can now access up-to-date organization-wide information, wherever they may be, with the click of a mouse. And decision makers can react as nimbly and quickly to changes as many of Boehringer's smaller competitors.

As you read this chapter, consider the following:

- In designing its new information systems, what do you think were Boehringer's most critical goals and considerations?
- How are hardware, software, databases, telecommunications, people, and procedures used in Boehringer's information system to provide valuable data?

Why Learn About Information Systems?

Information systems are used in almost every imaginable career area. Sales representatives use information systems to advertise products, communicate with customers, and analyze sales trends. Managers use them to make multimillion-dollar decisions, such as deciding to build a new manufacturing plant or research a new cancer drug. Corporate lawyers use information systems to develop contracts and other legal documents for their firm. From a small music store to huge multinational companies, businesses of all sizes could not survive without information systems to perform accounting and finance operations. Regardless of your college major or chosen career, you will find that information systems are indispensable tools to help you achieve your career aspirations. Learning about information systems can help you get your first job, obtain promotions, and advance your career. Why learn about information systems? What is in it for you? Learning about information systems will help you achieve your goals! In addition, most sections in this chapter are covered in complete chapters later in this book. For example, the sections in this chapter on hardware, software, databases, telecommunications, e-commerce and m-commerce, transaction processing and enterprise resource planning, information and decision support, special purpose systems, systems development, and ethical and societal issues become complete chapters in the rest of the book. Let's get started by exploring the basics of information systems.

information system (IS)

A set of interrelated components that collect, manipulate, store, and disseminate data and information and provide a feedback mechanism to meet an objective.

An **information system (IS)** is a set of interrelated components that collect, manipulate, store, and disseminate data and information and provide a feedback mechanism to meet an objective. The feedback mechanism helps organizations achieve their goals, such as increasing profits or improving customer service.

We all interact daily with information systems, both personally and professionally. We use automatic teller machines at banks, checkout clerks scan our purchases using bar codes and scanners, we access information over the Internet, and we get information from kiosks with touchscreens. Major *Fortune* 500 companies are spending in excess of $1 billion per year on information technology. In the future, we will depend on information systems even more. Knowing the potential of information systems and having the ability to put this knowledge to work can result in a successful personal career, organizations that reach their goals, and a society with a higher quality of life.

Computers and information systems are constantly changing the way organizations conduct business. They are becoming fully integrated into our lives, businesses, and society. They can help organizations carry on daily operations (operational systems). For example, WalMart uses operational systems to pull supplies from distribution centers and ultimately suppliers, stock shelves, and push out products and services through customer purchases. Computer and information systems also act as command and control systems that monitor processes and help supervisors control them. For example, air traffic control centers use computers and information systems as command and control centers to monitor and direct planes in their air space.

Today we live in an information economy. Information itself has value, and commerce often involves the exchange of information, rather than tangible goods. Systems based on computers are increasingly being used to create, store, and transfer information. Investors are using information systems to make multimillion-dollar decisions, financial institutions are employing them to transfer billions of dollars around the world electronically, and manufacturers are using them to order supplies and distribute goods faster than ever before. Computers and information systems will continue to change our society, our businesses, and our lives. In this chapter, we present a framework for understanding computers and information systems and discuss why it is important to study information systems. This understanding will help you unlock the potential of properly applied IS concepts.

INFORMATION CONCEPTS

Information is a central concept throughout this book. The term is used in the title of the book, in this section, and in almost every chapter. To be an effective manager in any area of business, you need to understand that information is one of an organization's most valuable and important resources. This term, however, is often confused with the term *data*.

Data Versus Information

Data consists of raw facts, such as an employee's name and number of hours worked in a week, inventory part numbers, or sales orders. As shown in Table 1.1, several types of data can be used to represent these facts. When these facts are organized or arranged in a meaningful manner, they become information. **Information** is a collection of facts organized in such a way that they have additional value beyond the value of the facts themselves. For example, a particular manager might find the knowledge of total monthly sales to be more suited to his or her purpose (i.e., more valuable) than the number of sales for individual sales representatives. Providing information to customers can also help companies increase revenues and profits.

data
Raw facts, such as an employee's name and number of hours worked in a week, inventory part numbers, or sales orders.

information
A collection of facts organized in such a way that they have additional value beyond the value of the facts themselves.

Table 1.1

Types of Data

Data	Represented By
Alphanumeric data	Numbers, letters, and other characters
Image data	Graphic images and pictures
Audio data	Sound, noise, or tones
Video data	Moving images or pictures

Data represents real-world things. As we have stated, data—simply raw facts—has little value beyond its existence. For example, consider data as pieces of railroad track in a model railroad kit. In this state, each piece of track has little value beyond its inherent value as a single object. However, if some relationship is defined among the pieces of the track, they will gain value. By arranging the pieces of track in a certain way, a railroad layout begins to emerge (see Figure 1.1a). Information is much the same. Rules and relationships can be set up to organize data into useful, valuable information.

Figure 1.1

Defining and Organizing Relationships Among Data Creates Information

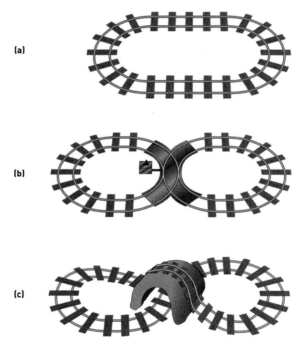

The type of information created depends on the relationships defined among existing data. For example, the pieces of track could be rearranged to form different layouts (see Figure 1.1b). Adding new or different data means relationships can be redefined and new information can be created. For instance, adding new pieces to the track can greatly increase the value—in this case, variety and fun—of the final product. We can now create a more elaborate railroad layout (see Figure 1.1c). Likewise, our manager could add specific product data to his sales data to create monthly sales information broken down by product line. This information could be used by the manager to determine which product lines are the most popular and profitable.

Turning data into information is a **process**, or a set of logically related tasks performed to achieve a defined outcome. The process of defining relationships among data to create useful information requires knowledge. **Knowledge** is an awareness and understanding of a set of information and the ways that information can be made useful to support a specific

process
A set of logically related tasks performed to achieve a defined outcome.

task or reach a decision. Part of the knowledge needed for building a railroad layout, for instance, is understanding how large an area is available for the layout, how many trains will run on the track, and how fast they will travel. The act of selecting or rejecting facts according to their relevancy to particular tasks is also based on a type of knowledge used in the process of converting data into information. Therefore, information can be considered data made more useful through the application of knowledge. In some cases, data is organized or processed mentally or manually. In other cases, a computer is used. In the earlier example, the manager could have manually calculated the sum of the sales of each representative, or a computer could calculate this sum. What is important is not so much where the data comes from or how it is processed but whether the results are useful and valuable. This transformation process is shown in Figure 1.2.

knowledge
An awareness and understanding of a set of information and ways that information can be made useful to support a specific task or reach a decision.

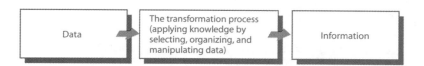

Figure 1.2

The Process of Transforming Data into Information

The Characteristics of Valuable Information

To be valuable to managers and decision makers, information should have the characteristics described in Table 1.2. These characteristics also make the information more valuable to an organization. Many organizations and shipping companies, for example, are able to determine the exact location of inventory items and packages in their systems. Recently, the U.S. Army Materiel Command tagged all its cargo and food shipments with radio-frequency identification chips for shipment to the Middle East. Because of the easy electronic retrieval of information from the tags, the time needed to take inventory of the cargo when it arrived was reduced from the usual 2 to 3 days to just 22 minutes.[1] In addition, if an organization's information is not accurate or complete, people can make poor decisions, costing organizations and individuals thousands, or even millions, of dollars. Many believe, for example, that the collapse of energy-trading firm Enron in the early 2000s was a result of inaccurate accounting and reporting information, which led investors and employees alike to misjudge the actual state of the company's finances and suffer huge personal losses. Some believe that Enron's inaccurate accounting was intentional and designed to deceive employees and investors and not a simple accounting mistake. As another example, if an inaccurate forecast of future demand indicates that sales will be very high when the opposite is true, an organization can invest millions of dollars in a new plant that is not needed. Furthermore, if information is not pertinent to the situation, not delivered to decision makers in a timely fashion, or too complex to understand, it may be of little value to the organization.

Useful information can vary widely in the value of each of these quality attributes. For example, with market-intelligence data, some inaccuracy and incompleteness is acceptable, but timeliness is essential. Market intelligence may alert us that our competitors are about to make a major price cut. The exact details and timing of the price cut may not be as important as being warned far enough in advance to plan how to react. On the other hand, accuracy, verifiability, and completeness are critical for data used in accounting for the use of company assets such as cash, inventory, and equipment.

The Value of Information

The value of information is directly linked to how it helps decision makers achieve their organization's goals. For example, the value of information might be measured in the time required to make a decision or in increased profits to the company. Consider a market forecast that predicts a high demand for a new product. If market forecast information is used to develop the new product and the company is able to make an additional profit of $10,000, the value of this information to the company is $10,000 minus the cost of the information. Valuable information can also help managers decide whether to invest in additional information systems and technology. A new computerized ordering system may cost $30,000,

Characteristics	Definitions
Accurate	Accurate information is error free. In some cases, inaccurate information is generated because in accurate data is fed into the transformation process (this is commonly called garbage in, garbage out [GIGO]).
Complete	Complete information contains all the important facts. For example, an investment report that does not include all important costs is not complete.
Economical	Information should also be relatively economical to produce. Decision makers must always balance the value of information with the cost of producing it.
Flexible	Flexible information can be used for a variety of purposes. For example, information on how much inventory is on hand for a particular part can be used by a sales representative in closing a sale, by a production manager to determine whether more inventory is needed, and by a financial executive to determine the total value the company has invested in inventory.
Reliable	Reliable information can be depended on. In many cases, the reliability of the information depends on the reliability of the data collection method. In other instances, reliability depends on the source of the information. A rumor from an unknown source that oil prices might go up may not be reliable.
Relevant	Relevant information is important to the decision maker. Information that lumber prices might drop may not be relevant to a computer chip manufacturer.
Simple	Information should also be simple, not overly complex. Sophisticated and detailed information may not be needed. In fact, too much information can cause information overload, whereby a decision maker has too much information and is unable to determine what is really important.
Timely	Timely information is delivered when it is needed. Knowing last week's weather conditions will not help when trying to decide what coat to wear today.
Verifiable	Information should be verifiable. This means that you can check it to make sure it is correct, perhaps by checking many sources for the same information.
Accessible	Information should be easily accessible by authorized users to be obtained in the right format and at the right time to meet their needs.
Secure	Information should be secure from access by unauthorized users.

Table 1.2

Characteristics of Valuable Data

but it may generate an additional $50,000 in sales. The *value added* by the new system is the additional revenue from the increased sales of $20,000. Most corporations have cost reduction as a primary goal. Using information systems, some manufacturing companies have been able to slash inventory costs by millions of dollars.

SYSTEM AND MODELING CONCEPTS

system
A set of elements or components that interact to accomplish goals.

Like information, another central concept of this book is that of a system. A **system** is a set of elements or components that interact to accomplish goals. The elements themselves and the relationships among them determine how the system works. Systems have inputs, processing mechanisms, outputs, and feedback (see Figure 1.3). For example, consider an automatic car wash. Obviously, tangible *inputs* for the process are a dirty car, water, and the various cleaning ingredients used. Time, energy, skill, and knowledge are also needed as inputs to the system. Time and energy are needed to operate the system. Skill is the ability to successfully operate the liquid sprayer, foaming brush, and air dryer devices. Knowledge is used to define the steps in the car wash operation and the order in which those steps are executed.

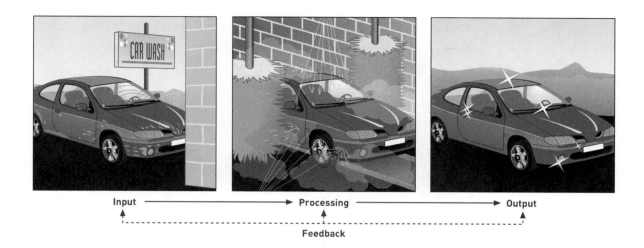

Input ——————————→ Processing ——————————→ Output

- -
Feedback

The *processing mechanisms* consist of first selecting which of the cleaning options you want (wash only, wash with wax, wash with wax and hand dry, etc.) and communicating that to the operator of the car wash. Note that there is a *feedback mechanism* (your assessment of how clean the car is). Liquid sprayers shoot clear water, liquid soap, or car wax depending on where your car is in the process and which options you selected. The *output* is a clean car. It is important to note that independent elements or components of a system (the liquid sprayer, foaming brush, and air dryer) interact to create a clean car.

System Components and Concepts

Figure 1.3 shows a typical system diagram—a simple automatic car wash. The primary purpose of the car wash is to clean your automobile. The **system boundary** defines the system and distinguishes it from everything else (the environment).

The way system elements are organized or arranged is called the *configuration*. Much like data, the relationships among elements in a system are defined through knowledge. In most cases, knowing the purpose or desired outcome of a system is the first step in defining the way system elements are configured. For example, the desired outcome of our system is a clean car. Based on past experience, we know that it would be illogical to have the liquid sprayer element precede the foaming brush element. The car would be rinsed and then soap would be applied, leaving your car a mess. As you can see from this example, knowledge is needed both to define relationships among the inputs to a system (your dirty car and instructions to the operator) and to organize the system elements used to process the inputs (the foaming brush must precede the liquid sprayer). Figure 1.4 shows a few systems with their elements and goals.

System Types
Systems can be classified along numerous dimensions. They can be simple or complex, open or closed, stable or dynamic, adaptive or nonadaptive, and permanent or temporary. Table 1.3 defines these characteristics.

Classifying Organizations by System Type
Most companies can be described using the classification scheme in Table 1.3. For example, a janitorial company that cleans offices after business hours most likely represents a simple, stable system because there is a constant and fairly steady need for its services. A successful computer manufacturing company, however, is typically complex and dynamic because it operates in a changing environment. If a company is nonadaptive, it may not survive very long. Many of the early computer companies, including Osborne Computer, which manufactured one of the first portable computers, and VisiCorp, which developed the first spreadsheet program, did not change rapidly enough with the changing market for computers and software. As a result, these companies did not survive. On the other hand, IBM was able

Figure 1.3

Components of a System

A system's four components consist of input, processing, output, and feedback.

system boundary

The limits of the system; it distinguishes it from everything else (the environment).

to reinvent itself from a manufacturer of large, mainframe computers to a manufacturer of all classes of computers and a software and services provider.

Figure 1.4

Examples of Systems and Their Goals and Elements

(Source: © Steve Smith/Getty Images; © Stanley Rowin / Index Stock Imagery; © PhotoDisc/Getty Images.)

System	Elements			
	Inputs	**Processing mechanisms**	**Outputs**	**Goal**
Coffee Shop 	Coffee beans, tea bags, water, sugar, cream, spices, pastries, other ingredients, labor, management	Brewing equipment	Coffee, tea, pastries, other beverages and food items	Quickly prepared delicious coffees, teas, and various food items
College 	Students, professors, administrators, textbooks, equipment	Teaching, research, service	Educated students; meaningful research; service to community, state, and nation	Acquisition of knowledge
Movie 	Actors, director, staff, sets, equipment	Filming, editing, special effects, film distribution	Finished film delivered to movie theaters	Entertaining movie, film awards, profits

Table 1.3

Systems Classifications and Their Primary Characteristics

Simple ⬌	Complex
Has few components, and the relationship or interaction between elements is uncomplicated and straightforward	Has many elements that are highly related and interconnected
Op ⬌	**Closed**
Interacts with its environment	Has no interaction with the environment
Stable ⬌	**Dynamic**
Undergoes very little change over time	Undergoes rapid and constant change over time
Adaptive ⬌	**Nonadaptive**
Is able to change in response to changes in the environment	Is not able to change in response to changes in the environment
Permanent ⬌	**Temporary**
Exists for a relatively long period of time	Exists for only a relatively short period of time

System Performance and Standards

System performance can be measured in various ways. **Efficiency** is a measure of what is produced divided by what is consumed. It can range from 0 to 100 percent. For example, the efficiency of a motor is the energy produced (in terms of work done) divided by the energy consumed (in terms of electricity or fuel). Some motors have an efficiency of 50 percent or less because of the energy lost to friction and heat generation.

Efficiency is a relative term used to compare systems. For example, a gasoline engine is more efficient than a steam engine because, for the equivalent amount of energy input (gas or coal), the gasoline engine produces more energy output.

Effectiveness is a measure of the extent to which a system achieves its goals. It can be computed by dividing the goals actually achieved by the total of the stated goals. For example, a company may have a goal to reduce damaged parts by 100 units. A new control system may be installed to help achieve this goal. Actual reduction in damaged parts, however, is only 85 units. The effectiveness of the control system is 85 percent (85/100 = 85%). Effectiveness, like efficiency, is a relative term used to compare systems. See the "Information Systems @ Work" box, which describes how Delta's low-fare, all-digital spin-off airline called Song is attempting to make Delta more efficient and effective with enhanced services for passengers. Evaluating system performance also calls for the use of performance standards. A **system performance standard** is a specific objective of the system. For example, a system performance standard for a particular marketing campaign might be to have each sales representative sell $100,000 of a certain type of product each year (see Figure 1.5a). A system performance standard for a certain manufacturing process might be to have no more than 1 percent defective parts (see Figure 1.5b). Once standards are established, system performance is measured and compared with the standard. Variances from the standard are determinants of system performance.

System Variables and Parameters

Parts of a system are under direct management control, while others are not. A **system variable** is a quantity or item that can be controlled by the decision maker. The price a company charges for its product is a system variable because it can be controlled. A **system parameter** is a value or quantity that cannot be controlled, such as the cost of a raw material. The number of pounds of a chemical that must be added to produce a certain type of plastic is another example of a quantity or value that is not controlled by management; it is controlled by the laws of chemistry.

efficiency
A measure of what is produced divided by what is consumed.

effectiveness
A measure of the extent to which a system achieves its goals; it can be computed by dividing the goals actually achieved by the total of the stated goals.

system performance standard
A specific objective of the system.

system variable
A quantity or item that can be controlled by the decision maker.

system parameter
A value or quantity that cannot be controlled, such as the cost of a raw material.

Delta Sings a New Song with a Focus on Information Systems

Times are tough for U.S. airlines. With the economy in a slump, heightened security in airports, and a cautious public less willing to fly, many airlines are struggling at the brink of bankruptcy. The old airline giants, which maintain large fleets, are particularly at a disadvantage as smaller and more streamlined airlines enter the market and win over customers with lower fares. Airlines are experimenting with new business models and practices hoping to find one that works in this highly competitive environment. Delta's low-fare, all-digital spin-off airline called Song is one such experiment. Song uses technology, state-of-the-art information systems, and good old-fashioned customer service to win customers.

Song caters to its passengers through a number of in-flight amenities. Each of its thirty-six 757s is furnished with roomy leather seats throughout the aircraft. Passengers are offered an extensive menu of beverages and snacks from well-known vendors such as Pizzeria Uno, Cinnabon, Lender's, and Yoplait—all of which they can pay for with the swipe of a credit card. But the most innovative and appealing of Song's services to passengers is located in the seatback above the fold-down tray: a personal video monitor with "touchscreen" technology and credit card "swipe" capability. Passengers can choose from 24 free, all-digital live DISH Network TV channels or pay-per-view programming. They can use the display to design a personal playlist of favorite songs delivered as streaming MP3 music. They can even join in an interactive multiplayer video game with other passengers on the flight. Other features of the LCD display include an interactive map program to view landmarks below, connecting gate information at the upcoming stop, or shopping from an online version of the *Sky-mall* catalog.

To make such luxurious accommodations available to passengers at a low fare, Song has streamlined its information systems to reduce overhead and turn flights around more quickly. Song prides itself on spending 23 percent more time in the air than aircraft in Delta's main line. Song's planes require only 50 minutes between landing and taking off. More time in the air and less on land allows for more flights and more income for the company. Song's information screens at airport gates double as movie screens, where mock horror films such as *The Thing That Wouldn't Get Out of the Aisle* are featured. These minifilms entertain passengers as they wait for their flight and educate them on how to make the boarding process run more smoothly and quickly.

Song provides an innovative online reservation system for booking flights. Unlike most online flight reservation systems, Song customers can easily compare prices across a variety of departure and return days and times. Reservations made through Song's Web site save Song $4 per flight over those made through traditional reservation systems. An e-ticketing kiosk allows customers to check themselves in when they arrive at the terminal and automatically dispenses a boarding pass, saving the company $5 per customer

per flight. Automated kiosks shorten lines in the terminal and reduce the need for customer service agents.

Technology is deeply infused into the Song business model to provide both customer benefits and reduced costs for the airline. Delta considers Song to be a testing ground for new innovations. Dozens of streamlined information systems and new technologies are being tested. The effects of each new system are analyzed in terms of time saved, money saved, and customer satisfaction. Every second or penny saved is crucial. Ideas that prove to be successful with Song will be integrated into Delta's primary fleet. Song is far from unique in its use of technology and information systems to gain an advantage in a challenging marketplace. Businesses in all markets rely on information systems and technology to assist them in accomplishing more for less.

The implementation of new and improved information systems are, in many cases, dramatically changing the way employees do their jobs. The goal is to allow business processes to proceed as smoothly as possible, with as little inconvenience to employees, management, and customers as possible. By allowing people to do what they need to do without delay, they are able to accomplish more in a given period. Frustration is minimized, and interaction between management, employees, and customers becomes less stressed and more enjoyable.

Discussion Questions

1. How have the information systems used at Song affected the duties of check-in agents and flight attendants?
2. What do you think are the biggest challenges for check-in agents and flight attendants in dealing with customers in the traditional airline business model? How do the information systems Song uses allow these workers to be more efficient and effective?

Critical Thinking Questions

3. Consider the hardships of air travel: the search for cheap rates, long check-in lines at the terminal, backups at security checkpoints, delayed and canceled flights, crowded seating, and hours of waiting time. How has Song used technology and information systems to soften these hardships? What more could the company do?
4. Consider other transportation modes: train, ship, taxi, and bus. Do these other modes share any of the same challenges that Song has addressed with information systems? Could other transportation areas benefit from Song's approach?

SOURCES: Kathleen Melymuka, "Delta's Test Pilot: IT at Song, Delta's New Low-Cost Airline Unit," *Computerworld*, August 18, 2003, *www.computerworld.com*; "Song Unveils the World's Most Sophisticated Single Aisle In-Flight Entertainment System," *PR Newswire*, November 17, 2003; Song Web site, *www.flysong.com*, accessed January 16, 2004.

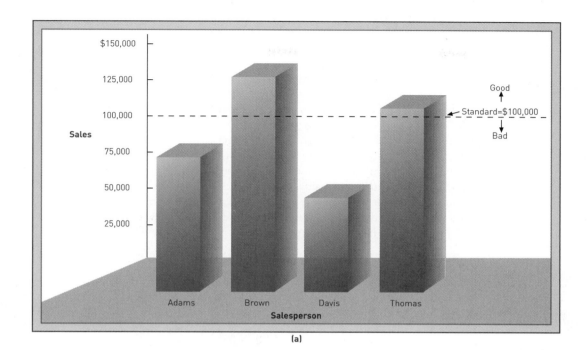

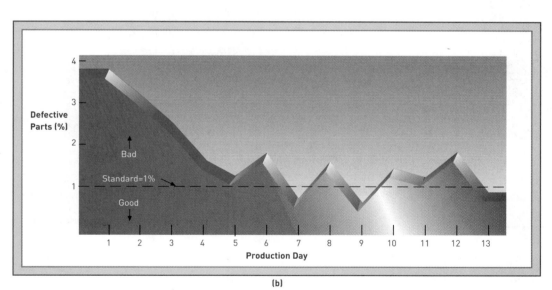

(b)

Modeling a System

The real world is complex and dynamic. So when we want to test different relationships and their effects, we use models of systems, which are simplified, instead of real systems. A **model** is an abstraction or an approximation that is used to represent reality. Models enable us to explore and gain an improved understanding of real-world situations.

Since the beginning of recorded history, people have used models. A written description of a battle, a physical mock-up of an ancient building, and the use of symbols to represent money, numbers, and mathematical relationships are all examples of models. Today, managers and decision makers use models to help them understand what is happening in their organizations and make better decisions.

There are various types of models. The major ones are narrative, physical, schematic, and mathematical, as shown in Figure 1.6. A *narrative model*, as the name implies, is based on words; thus, it is a logical and not a physical model. Both verbal and written descriptions of

Figure 1.5

System Performance Standards

model
An abstraction or an approximation that is used to represent reality.

reality are considered narrative models. In an organization, reports, documents, and conversations concerning a system are all important narratives. A *physical model* is a tangible representation of reality. Many physical models are computer designed or constructed. An engineer may develop a physical model of a chemical reactor to gain important information about how a large-scale reactor might perform, or a builder may develop a scale model of a new shopping center to give a potential investor information about the overall appearance and approach of the development. A *schematic model* is a graphic representation of reality. Graphs, charts, figures, diagrams, illustrations, and pictures are all types of schematic models. Schematic models are used extensively in developing computer programs and systems. A blueprint for a new building, a graph that shows budget and financial projections, electrical wiring diagrams, and graphs that show when certain tasks or activities must be completed to stay on schedule are examples of schematic models used in business. A *mathematical model* is an arithmetic representation of reality. Computers excel at solving mathematical models. Retail chains, for example, have developed mathematical models to identify all the activities, effort, and time associated with planning, building, and opening a new store so that they can forecast how long it will take to complete a store.

Figure 1.6

Four Types of Models

Narrative (words, spoken or written), physical (tangible), schematic (graphic), and mathematical (arithmetic) models.

(Source: © Bob Daemmrich/ PhotoEdit; © PhotoDisc/Getty Images)

Narrative

Physical

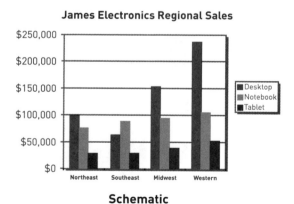

Schematic

Mathematical

In developing any model, accuracy is critical. An inaccurate model will usually lead to an inaccurate solution to a problem. Most models contain many assumptions, and it is important that they be as realistic as possible. Potential users of the model must be aware of the assumptions under which the model was developed.

WHAT IS AN INFORMATION SYSTEM?

As mentioned previously, an information system (IS) is a set of interrelated elements or components that collect (input), manipulate (process) and store, and disseminate (output) data and information and provide a feedback mechanism to meet an objective (see Figure 1.7). The feedback mechanism helps organizations achieve their goals, such as increasing profits or improving customer service.

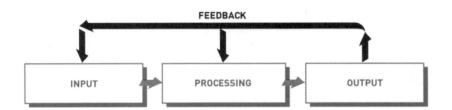

Figure 1.7

The Components of an Information System

Feedback is critical to the successful operation of a system.

Input, Processing, Output, Feedback

Input

In information systems, **input** is the activity of gathering and capturing raw data. In producing paychecks, for example, the number of hours every employee works must be collected before paychecks can be calculated or printed. In a university grading system, individual instructors must submit student grades before a summary of grades for the semester or quarter can be compiled and sent to the students.

Input can take many forms. In an information system designed to produce paychecks, for example, employee time cards might be the initial input. In a 911 emergency telephone system, an incoming call would be considered an input. Input to a marketing system might include customer survey responses. Car manufacturers are experimenting with a fingerprint identification input device in their car security systems. You may soon be able to gain entry to a car and start it with the touch of a finger. This unique input device will also adjust mirrors, the steering-wheel position, the temperature, and the radio for an individual's size and preferences. Regardless of the system involved, the type of input is determined by the desired output of the system.

Input can be a manual or automated process. A scanner at a grocery store that reads bar codes and enters the grocery item and price into a computerized cash register is a type of automated input process. Regardless of the input method, accurate input is critical to achieve the desired output.

input
The activity of gathering and capturing raw data.

Processing

In information systems, **processing** involves converting or transforming data into useful outputs. Processing can involve making calculations, making comparisons and taking alternative actions, and storing data for future use. Processing data into useful information is critical in business settings.

Processing can also be done manually or with computer assistance. In the payroll application, each employee's number of hours worked must be converted into net, or take-home, pay. Other inputs often include employee ID number and department. The required processing can first involve multiplying the number of hours worked by the employee's hourly pay rate to get gross pay. If weekly hours worked exceed 40 hours, overtime pay may also be included. Then deductions—for example, federal and state taxes, contributions to health and life insurance or savings plans—are subtracted from gross pay to get net pay.

Once these calculations and comparisons are performed, the results are typically stored. *Storage* involves keeping data and information available for future use, including output, discussed next.

processing
Converting or transforming data into useful outputs.

output
Production of useful information, usually in the form of documents and reports.

Output

In information systems, **output** involves producing useful information, usually in the form of documents and reports. Outputs can include paychecks for employees, reports for managers, and information supplied to stockholders, banks, government agencies, and other groups. In some cases, output from one system can become input for another. For example, output from a system that processes sales orders can be used as input to a customer billing system. Often, output from one system can be used as input to control other systems or devices. For instance, the design and manufacture of office furniture is complicated with many variables. The salesperson, customer, and furniture designer can go through several design iterations to meet the customer's needs. Special computer programs and equipment create the original design and allow the designer to rapidly revise it. Once the last design mock-up is approved, the computer creates a bill of materials that goes to manufacturing to produce the order.

Output can be produced in a variety of ways. For a computer, printers and display screens are common output devices. Output can also be a manual process involving handwritten reports and documents.

Feedback

feedback
Output that is used to make changes to input or processing activities.

In information systems, **feedback** is output that is used to make changes to input or processing activities. For example, errors or problems might make it necessary to correct input data or change a process. Consider a payroll example. Perhaps the number of hours an employee worked was entered into a computer as 400 instead of 40 hours. Fortunately, most information systems check to make sure that data falls within certain ranges. For number of hours worked, the range might be from 0 to 100 hours because it is unlikely that an employee would work more than 100 hours for any given week. So, the information system would determine that 400 hours is out of range and provide feedback, such as an error report. The feedback is used to check and correct the input on the number of hours worked to 40. If undetected, this error would result in a very high net pay on the printed paycheck! Some blame the August 14, 2003, power blackout in the U.S.'s Northeast on a faulty computer system that wasn't able to provide second-by-second feedback.[2]

Feedback is also important for managers and decision makers. For example, a bedding maker used a computerized feedback system to link its suppliers and plants. The output from an information system might indicate that inventory levels for a few items are getting low—a potential problem. A manager could use this feedback to decide to order more inventory from a supplier. The new inventory orders then become input to the system. In addition to this reactive approach, a computer system can also be proactive—predicting future events to avoid problems. This concept, often called **forecasting**, can be used to estimate future sales and order more inventory before a shortage occurs.

forecasting
Predicting future events to avoid problems.

Manual and Computerized Information Systems

As discussed earlier, an information system can be manual or computerized. For example, some investment analysts manually draw charts and trend lines to assist them in making investment decisions. Tracking data on stock prices (input) over the last few months or years, these analysts develop patterns on graph paper (processing) that help them determine what stock prices are likely to do in the next few days or weeks (output). Some investors have made millions of dollars using manual stock analysis information systems. Of course, today many excellent computerized information systems have been developed to follow stock indexes and markets and to suggest when large blocks of stocks should be purchased or sold (called *program trading*) to take advantage of market discrepancies.

Many information systems begin as manual systems and become computerized. For example, consider the way the U.S. Postal Service sorts mail. At one time, most letters were visually scanned by postal employees to determine the ZIP code and were then manually placed in an appropriate bin. Today the bar-coded addresses on letters passing through the postal system are read electronically and automatically routed to the appropriate bin via conveyors. The computerized sorting system results in speedier processing time and provides management with information to help plan transportation needs. It is important to stress, however, that simply computerizing a manual information system does not guarantee

improved system performance. If the underlying information system is flawed, the act of computerizing it might only magnify the impact of these flaws.

Computer-Based Information Systems

A **computer-based information system (CBIS)** is a single set of hardware, software, databases, telecommunications, people, and procedures that are configured to collect, manipulate, store, and process data into information. For example, a company's payroll systems, order entry system, or inventory control systems are examples of a CBIS. The components of a CBIS are illustrated in Figure 1.8. (*Information technology, IT*, is a related term. For our purposes, IT refers to the technology components of hardware, software, databases, and telecommunications.) A business's **technology infrastructure** includes all the hardware, software, databases, telecommunications, people, and procedures that are configured to collect, manipulate, store, and process data into information. The technology infrastructure is a set of shared IS resources that form the foundation of each individual computer-based information system.

computer-based information system (CBIS)
A single set of hardware, software, databases, telecommunications, people, and procedures that are configured to collect, manipulate, store, and process data into information.

technology infrastructure
All the hardware, software, databases, telecommunications, people, and procedures that are configured to collect, manipulate, store, and process data into information.

Figure 1.8

The Components of a Computer-Based Information System

Hardware

Hardware consists of computer equipment used to perform input, processing, and output activities.[3] Input devices include keyboards, automatic scanning devices, equipment that can read magnetic ink characters, and many other devices. Investment firms often use voice response to allow customers to get their balances and other information using ordinary spoken

hardware
Computer equipment used to perform input, processing, and output activities.

sentences. The Scripps Institution of Oceanography developed a special underwater computer optical input device to allow a diver as deep as 100 feet to control an underwater camera, which was formerly controlled by a computer system and mouse on the surface.[4] Processing devices include the central processing unit and main memory.[5] Processor speed is important in creating video images.[6] Lifelike movie characters such as Gollum in the *Lord of the Rings* show what is possible with today's fast processors. Mental Images of Germany and Pixar of the United States have used such award-winning image-rendering techniques. The technology is also used to help design cars, such as the sleek shapes of Mercedes Benz vehicles. Specialized, inexpensive hardware has also been used in schools to help students learn a variety of subjects.[7]

There are many output devices, including secondary storage devices, printers, and computer screens. One company, for example, uses computer hardware in its stores to allow customers to order items that are not on store shelves. The hardware helps the company "save the sale" and increase revenues. Michael Dell, founder of Dell Inc., believes that hardware will increasingly include very small devices that are connected to other hardware devices. He said, "Nanotechnology and communications will be in everything. All kinds of other devices will attach and link together, centered, I think, with the PC."[8] Nanotechnology can involve molecule-sized hardware devices.[9] There are also many special-purpose hardware devices. Computerized event data recorders (EDRs) are now being placed into vehicles. Like an airplane's black box, EDRs record a vehicle's speed, possible engine problems, a driver's performance, and more. The technology is being used to monitor vehicle operation, determine the cause of accidents, and investigate whether truck drivers are taking required breaks. In Florida, an EDR was used to help convict a driver of vehicular homicide.[10] A surgeon watching a 3-D computer screen and using a joystick can view up to 2,000 slices or cross-sectional areas of a patient's body before precisely removing a cancerous tumor. Sophisticated input and output devices are making surgery more precise, which can save lives.[11]

Hardware is a component of a Computer-Based Information System and includes input, processing, and output.

(Source: Courtesy of Acer America Inc.)

Software

software
The computer programs that govern the operation of the computer.

Software consists of the computer programs that govern the operation of the computer. These programs allow a computer to process payroll, send bills to customers, and provide managers with information to increase profits, reduce costs, and provide better customer service.[12] With software, people can work anytime at any place. On a trip back to the United States from Australia and New Zealand, Steve Ballmer, CEO of Microsoft commented, "I could carry my slides, I could carry my e-mail. I could carry anything I needed to read. I could carry my life with me. It was very powerful."[13] There are two basic types of software: system software, such as Windows XP, which controls basic computer operations such as start-up and printing, and applications software, such as Office 2003, which allows specific tasks to be accomplished, such as word processing or tabulating numbers.

Databases

A **database** is an organized collection of facts and information, typically consisting of two or more related data files. An organization's database can contain facts and information on customers, employees, inventory, competitors' sales information, online purchases, and much more. Most managers and executives believe a database is one of the most valuable and important parts of a computer-based information system. Increasingly, organizations are placing important databases on the Internet, discussed next.[14]

Telecommunications, Networks, and the Internet

Telecommunications is the electronic transmission of signals for communications, which enable organizations to carry out their processes and tasks through effective computer networks.[15] Large restaurant chains, for example, can use telecommunications systems and satellites to link hundreds of restaurants to plants and corporate headquarters to speed credit card authorization and report sales and payroll data. **Networks** are used to connect computers and computer equipment in a building, around the country, or around the world to enable electronic communication. Investment firms can use wireless networks to connect thousands of people with their corporate offices. Hotel Commonwealth in Boston uses wireless telecommunications to allow guests to connect to the Internet, get voice messages, and perform other functions without plugging their computers or mobile devices into a wall outlet.[16] Wireless transmission is also allowing drones, like Boeing's Scan Eagle, to monitor power lines, buildings, and other commercial establishments.[17] The drones are smaller and less-expensive versions of the Predator and Global Hawk drones that were used successfully in the Afghanistan and Iraq conflicts by the U.S. military. One company uses a private network to connect offices in the United States, Germany, China, Korea, and other companies. It doesn't use a public network available to everyone, such as the Internet, discussed next.

The **Internet** is the world's largest computer network, actually consisting of thousands of interconnected networks, all freely exchanging information. Research firms, colleges, universities, high schools, and businesses are just a few examples of organizations using the Internet.

The Internet is used by most businesses and industries. ChemConnect, for example, is an Internet site that allows companies to buy and sell chemicals and chemical products online.[18] The Internet site is the largest online chemical trading site with more than $8 billion in annual sales. In addition to being able to cheaply download music, audio software and the Internet allow people to change a song's tempo, create mixes of their favorite tunes, and modify soundtracks to suit their personal taste. It is even possible to play two or more songs simultaneously, called *mashing*.[19] Mortgage companies use the Internet to help make loans to customers. Lending through the Internet has grown dramatically from about $260 billion in 2001 to more than $800 billion in 2003, representing almost 30 percent of the total loan market.[20] Medco Health Online, a subsidiary of Merck & Company, received more than a billion dollars in sales through its Internet site.[21] The Internet has also been used in public elections.[22] Aniers, a suburb of Geneva, was the first city in Switzerland to use the Internet in a public, binding vote. Businesses are increasingly using *instant messaging (IM)* to communicate.[23] IM allows managers and employees to instantly communicate over the Internet using text messages, video, and even sound. Today, bedrooms, studies, kitchens, and living rooms are being connected to the Inte rnet.[24] In addition, travelers who can gain access to the Internet can communicate with anyone else on the Internet, including those who are in flight. Some airline companies are starting Internet service on their flights to allow people to send and receive e-mail, check investments, and browse the Internet. According to Vinton Cerf, one of the pioneers of the Internet, "There will be a very large number of devices on the Net—appliances, things you wear and carry around, and things that are embedded in passive things like wine corks and your socks."[25] Table 1.4 lists companies that have used the Internet to their advantage.

database
An organized collection of facts and information.

telecommunications
The electronic transmission of signals for communications; enables organizations to carry out their processes and tasks through effective computer networks.

networks
Computers and computer equipment in a building, around the country, or around the world to enable electronic communications.

Internet
The world's largest computer network, actually consisting of thousands of interconnected networks, all freely exchanging information.

Organization	Objective	Description of Internet Usage
Godiva Chocolatier	Increase sales and profits	The company developed a very profitable Internet site that allows customers to buy and ship chocolates. According to Kim Land, director of Godiva Direct, "This was set up from the beginning to make money." In two years, online sales have soared by more than 70% each year.
Environmental Defense	Alert the public to environmental concerns	The organization, formerly the Environmental Defense Fund, successfully used the Internet to alert people to the practice of catching sharks, removing their fins for soup, and returning them to the ocean to die. The Internet site also helped people fax almost 10,000 letters to members of Congress about the practice. According to Fred Krupp, the executive director of the Environmental Fund, "The Internet is the ultimate expression of 'think global, act local.'"
Buckman Laboratories	Better employee training	The company used the Internet to train employees to sell specialty chemicals to paper companies, instead of bringing them to Memphis for training. According to one executive, "Our retention rate is much higher, and we removed a week [of training] in Memphis, which meant big savings." Using the Internet lowered the hourly cost of training an employee from $1,000 to only $40.
Siemens	Reduce costs	Using the Internet, the company, which builds and services power plants, was able to reduce the cost of entering orders and serving customers. The Internet solution cost about $60,000 compared with a traditional solution that would have cost of $600,000.
Goldman Industrial Group	Save time	The company makes machine tools and was able to slash the time it takes to fill an order from 3 or 4 months to about a week using the Internet to help coordinate parts and manufacturing with its suppliers and at its plants.
Partnership America	Make better decisions	The company developed an Internet site for wholesalers of computer equipment and supplies. The wholesalers use the site to make better decisions about the features and prices of various pieces of computer equipment. The system allows wholesalers to connect to Partnership America's site using cell phones. "When many of our customers need information, they're not at their desks," says one company representative.
Altra Energy Technologies	Get energy to companies that need it	The company developed an Internet site to help companies buy oil, gas, and wholesale power over the Internet.

Table 1.4

Uses of the Internet

intranet

An internal network based on Web technologies that allows people within an organization to exchange information and work on projects.

extranet

A network based on Web technologies that allows selected outsiders, such as business partners and customers, to access authorized resources of the intranet of a company.

The *World Wide Web (WWW)* or the Web is a network of links on the Internet to documents containing text, graphics, video, and sound. Information about the documents and access to them are controlled and provided by tens of thousands of special computers called Web servers. The Web is one of many services available over the Internet and provides access to literally millions of documents.

The technology used to create the Internet is now also being applied within companies and organizations to create an **intranet**, which allows people within an organization to exchange information and work on projects. The Virgin Group, for example, uses an intranet to connect its 200 global operating companies and 20,000 employees.[26] According to Ashley Stockwell of the Virgin Group, "One of our key challenges at Virgin is to provide high-quality service to our family of companies. One key tool to help us provide this was the development of an intranet and extranet." An **extranet** is a network based on Web technologies that allows selected outsiders, such as business partners and customers, to access authorized resources of the intranet of a company. Companies can move all or most of their business activities to an extranet site for corporate customers. Many people use extranets every day without realizing it—to track shipped goods, order products from their suppliers, or access customer assistance from other companies. Log on to the FedEx site to check the status of a package, for example, and you are using an extranet.

People

People are the most important element in most computer-based information systems. Information systems personnel include all the people who manage, run, program, and maintain the system. Large banks can hire hundreds of IS personnel to speed up the development of computer-related projects. Users are people who use information systems to get results. Users include financial executives, marketing representatives, manufacturing operators, and many others. Certain computer users are also IS personnel.

Procedures

Procedures include the strategies, policies, methods, and rules for using the CBIS. For example, some procedures describe when each program is to be run or executed. Others describe who can have access to facts in the database. Still other procedures describe what is to be done in case a disaster, such as a fire, an earthquake, or a hurricane that renders the CBIS unusable.

Now that we have looked at computer-based information systems in general, we will briefly examine the most common types used in business today. These IS types are covered in more detail in Part 3.

procedures
The strategies, policies, methods, and rules for using a CBIS.

BUSINESS INFORMATION SYSTEMS

The most common types of information systems used in business organizations are electronic and mobile commerce systems, transaction processing systems, management information systems, and decision support systems. In addition, some organizations employ special-purpose systems such as artificial intelligence systems, expert systems, and virtual reality systems. Together, these systems help employees in organizations accomplish both routine and special tasks—from recording sales, to processing payrolls, to supporting decisions in various departments, to providing alternatives for large-scale projects and opportunities. Figure 1.9 gives a simple overview of the development of important business information systems discussed in this section.

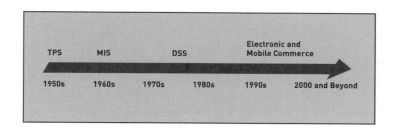

Figure 1.9

The Development of Important Business Information Systems

Electronic and Mobile Commerce

E-commerce involves any business transaction executed electronically between parties such as companies (business-to-business, B2B), companies and consumers (business-to-consumer, B2C), consumers and other consumers (consumer-to-consumer, C2C), business and the public sector, and consumers and the public sector. People may assume that e-commerce is reserved mainly for consumers visiting Web sites for online shopping. But Web shopping is only a small part of the e-commerce picture; the major volume of e-commerce—and its fastest-growing segment—is business-to-business (B2B) transactions that make purchasing easier for corporations. This growth is being stimulated by increased Internet access, growing user confidence, better payment systems, and rapidly improving Internet and Web security. Corporate Express, an office-supply company located in Broomfield, Colorado, uses a sophisticated B2B system to coordinate billions of dollars of office supplies that flow from its suppliers, through its offices, to its customers.[27] Today, more than half of its 75,000 daily orders arrive electronically through B2B on the Internet. E-commerce offers opportunities for small businesses, too, by enabling them to market and sell at a low cost worldwide, thus

e-commerce
Any business transaction executed electronically between parties such as companies (business-to-business), companies and consumers (business-to-consumer), consumers and other consumers (consumer-to-consumer), business and the public sector, and consumers and the public sector.

IBM PartnerWorld® is an example of B2B (business-to-business) e-commerce that provides member companies with resources for product marketing, technical support and training.

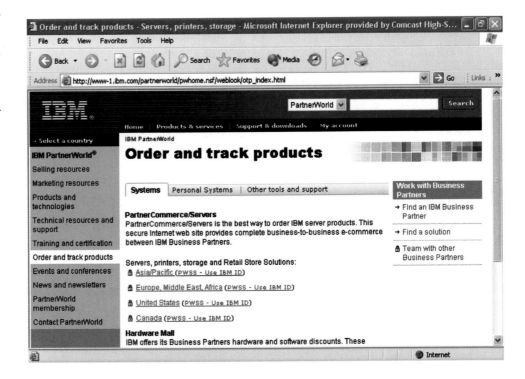

mobile commerce (m-commerce)
Transactions conducted anywhere, anytime.

allowing them to enter the global market right from start-up. **Mobile commerce (m-commerce)** are transactions conducted anywhere, anytime. M-commerce relies on the use of wireless communications to allow managers and corporations to place orders and conduct business using handheld computers, portable phones, laptop computers connected to a network, and other mobile devices.

Consumers who have tried online shopping appreciate the ease of e-commerce. They can avoid fighting crowds in the malls, shop online at any time from the comfort of their home, and have goods delivered to them directly. As a result, advertisers plan to increase spending by 6.3 percent online versus 4.7 percent in conventional media.[28] In addition, current laws governing online purchases exempt purchasers from paying state sales taxes. However, e-commerce is not without its downside. Consumers continue to have concerns about sending credit card information over the Internet to sites with varying security measures, where high-tech criminals could obtain it. In addition, denial-of-service attacks that overwhelm the capacity of some of the Web's most established and popular sites have raised new concerns for continued growth of e-commerce. Privacy is an additional concern. Individuals want to know what data is gathered when a consumer visits a Web site and what companies do with the collected data; some have sold data to multiple sources, leading marketing companies to know more than we would like. See the "Ethical and Societal Issues" box, which discusses other potential problems of e-commerce.

Phishing for Visa Card Customers

A new type of Internet fraud is becoming increasingly prevalent—and costing consumers their money and identity. This latest scam is called *phishing* because it uses e-mail and Web sites as bait to lure consumers into revealing private information.

E-commerce systems rely on the trust of the participants. If they do not trust the technology to provide safe and secure transactions, e-commerce would have no future. While network research has produced more secure connections between two parties over the Internet, no foolproof systems exist to guarantee that the participants are who they claim to be. Phishing scams exploit this system vulnerability.

A phishing scam was recently launched against Visa card customers and serves as a textbook example of the technique. A mass e-mail was sent to Internet users with an official-looking Visa return address, claiming to have come from Visa International Services. Sending e-mail with a forged return address is a common practice in Internet fraud and is formally referred to as *spoofing*. The e-mail stated that Visa had implemented a new "security system to help you to avoid possible fraud actions" and asked users to click a link to "reactivate your account." The link was printed as *www.visa.com*, but when users clicked the link, it took them to a Web site that resembled the Visa Web site—with an official Visa logo, artwork, and design—but was not owned by Visa. The site asked customers to enter personal information, including their Visa credit card number. The scam artists then had both a customer's account number and his or her name.

The 2003 holiday season saw a 400 percent increase in phishing scams, with 60 unique attacks launched and more than 60 million fraudulent e-mails sent out. It is estimated that 5 to 20 percent of recipients respond to phishing scams. In the Visa scam, the owners of the fraudulent site shut down and disappeared prior to discovery, taking with them an unknown quantity of customer records. The information they stole could be sold in the underground credit card market and used by crooks and thieves to assume the identity of the victims and make illegal purchases.

Phishing scams are increasingly difficult to detect. The fraudulent e-mails and Web sites look identical to original corporate correspondence and Web sites. Web addresses appear legitimate and may even employ secure connections (identified by the closed-lock icon at the bottom of the browser window). Such scams make it difficult for legitimate businesses to communicate electronically with their customers and to conduct business online. "At stake is our very trust that the Internet can be relied upon for safe and secure commerce and communications," says Dave Jevans, chairman of the Anti-Phishing Working Group (*www.antiphishing.org*).

Software tools designed to detect phishing scams typically identify only 50 to 70 percent of all phony systems. The only defense consumers have against such scams is education—and caution. Be leery of any e-mail from a company that asks you to visit a Web page to provide private information. Check with the company at its official Web site to confirm that such requests are legitimate before complying.

Critical Thinking Questions

1. How can people protect themselves from becoming a victim of a phishing scam?
2. What action can people take if they discover that their private information has been stolen?

What Would You Do?

You've received an e-mail from your college's Financial Aid Department that congratulates you on being the recipient of funds from a newly launched grant program. To receive your $2,000 for this semester, you are required to visit the Financial Aid Web site (*www.financial-aid.yourschool.com*) and submit a brief online application form. After filling out the form, which collects information such as your name, address, phone, date of birth, school ID number, Social Security number, and bank-account number (for automatic deposit), you click the *Submit* button and head out to celebrate your good fortune. After a week, the money has yet to be deposited, and you are getting concerned.

3. What in this scenario suggests that this might be a phishing scam?
4. If you were responsible for information security at your school, what system might you design to assure students that official school correspondence really comes from the school and not from an imposter?

SOURCES: Paul Roberts, "Latest 'Phishing' Scam Targets Visa Customers," *Computerworld*, December 26, 2003; "Growth in Internet Fraud to Be Key Concern In 2004," *Electronic Commerce News*, January 5, 2004; the Anti-Phishing group Web site, *www.anti-phishing.org*, accessed January 17, 2004.

Yet, in spite of the concerns, e-commerce offers many advantages for streamlining work activities. Figure 1.10 provides a brief example of how e-commerce can simplify the process for purchasing new office furniture from an office-supply company. Under the manual system, a corporate office worker must get approval for a purchase that costs more than a certain amount. That request goes to the purchasing department, which generates a formal purchase order to procure the goods from the approved vendor. Business-to-business e-commerce automates the entire process. Employees go directly to the supplier's Web site, find the item in its catalog, and order what they need at a price set by the employee's company. If approval is required, the approver is notified automatically. As the use of e-commerce systems grows, companies are phasing out their traditional systems. The resulting growth of e-commerce is creating many new business opportunities.

E-commerce can have a positive impact on stock prices and the market value of firms. Today, several e-commerce firms have teamed up with more traditional brick-and-mortar firms to draw from each other's strengths. Some e-commerce customers can order products through an Internet site and pick them up at a local store close by.

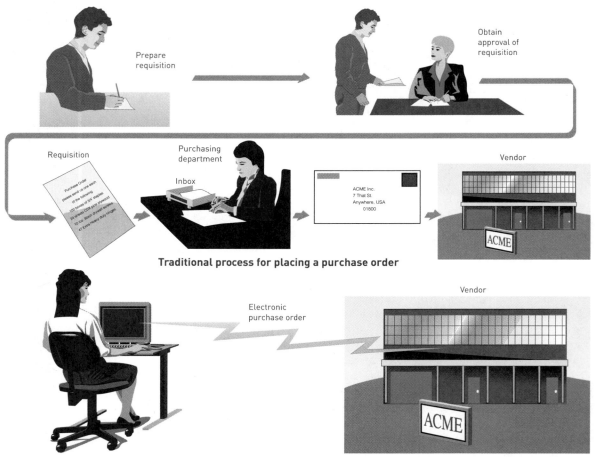

Traditional process for placing a purchase order

E-commerce process for placing a purchase order

Figure 1.10

E-Commerce Greatly Simplifies Purchasing

In addition to e-commerce, business information systems include the use of telecommunications and the Internet to perform many related tasks. *Electronic procurement (e-procurement)*, for example, involves using information systems and the Internet to acquire parts and supplies using information systems and the Internet. *Electronic business (e-business)* goes beyond e-commerce to include the use of information systems and the Internet to perform all business-related tasks and functions, such as accounting, finance, marketing, manufacturing, and human resources activities. *Electronic management (e-management)* involves the use of

information systems and the Internet to manage profit and nonprofit organizations, including governmental agencies, the military, and religious and charitable organizations. E-management includes all aspects of staffing and hiring, directing, controlling, and other management tasks.

Transaction Processing Systems and Enterprise Resource Planning

Transaction Processing Systems

Since the 1950s computers have been used to perform common business applications. The objective of many of these early systems was to reduce costs by automating many routine, labor-intensive business systems. A **transaction** is any business-related exchange such as payments to employees, sales to customers, or payments to suppliers. Thus, processing business transactions was the first application of computers for most organizations. A **transaction processing system** (TPS) is an organized collection of people, procedures, software, databases, and devices used to record completed business transactions. To understand a transaction processing system is to understand basic business operations and functions.

One of the first business systems to be computerized was the payroll system (see Figure 1.11). The primary inputs for a payroll TPS are the numbers of employee hours worked during the week and pay rate. The primary output consists of paychecks. Early payroll systems were able to produce employee paychecks, along with important employee-related reports required by state and federal agencies, such as the Internal Revenue Service. Other routine applications include sales ordering, customer billing and customer relationship management, inventory control, and many other applications. Some automobile companies, for example, use their TPS to buy billions of dollars of needed parts each year through Internet sites. Because these systems handle and process daily business exchanges, or transactions, they are all classified as TPSs.

transaction
Any business-related exchange, such as payments to employees, sales to customers, and payments to suppliers.

Transaction processing system (TPS)
An organized collection of people, procedures, software, databases, and devices used to record completed business transactions.

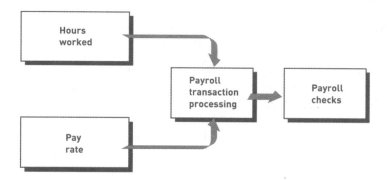

Figure 1.11

A Payroll Transaction Processing System

The inputs (numbers of employee hours worked and pay rates) go through a transformation process to produce outputs (paychecks).

In improved forms, these TPSs are still vital to most modern organizations: Consider what would happen if an organization had to function without its TPS for even one day. How many employees would be paid and paid the correct amount? How many sales would be recorded and processed? Transaction processing systems represent the application of information concepts and technology to routine, repetitive, and usually ordinary business transactions that are critical to the daily functions of that business.

Enterprise Resource Planning

An **enterprise resource planning (ERP) system** is a set of integrated programs that is capable of managing a company's vital business operations for an entire multisite, global organization. ERP systems can replace many applications with one unified set of programs. Sutter Health, a large network of 33 hospitals with over 4 million patients in northern California, uses an ERP system to process medical transactions and to exchange information between hospitals, physicians, and employees.[29] Although the scope of an ERP system may vary from company to company, most ERP systems provide integrated software to support the manufacturing and finance business functions of an organization. In such an environment, a forecast is prepared that estimates customer demand for several weeks. The ERP system checks what is

enterprise resource planning (ERP) system
A set of integrated programs capable of managing a company's vital business operations for an entire multisite, global organization.

SAP AG, a German software company, is one of the leading suppliers of ERP software. The company employs nearly 30,000 people in more than 50 countries.

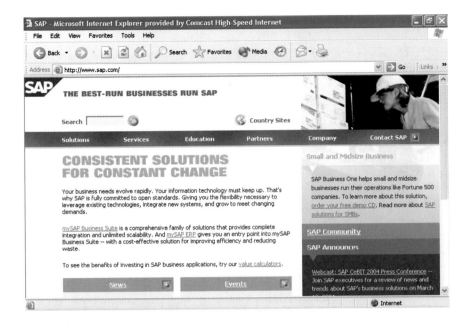

already available in finished product inventory to meet the projected demand. Manufacturing must then produce inventory to eliminate any shortcomings. In developing the production schedule, the ERP system checks the raw-material and packing-material inventories and determines what needs to be ordered to meet the schedule. Most ERP systems also have a purchasing subsystem that orders the needed items. In addition to these core business processes, some ERP systems can support additional business functions, such as human resources, sales, and distribution. Customer relationship management (CRM) features, for example, help organizations manage all aspects of customer interactions, including inquiries, sales, delivery of products and services, and support after the sale. Today, ERP companies have Internet-based systems to manage customer relationships. The primary benefits of implementing an ERP system include easing adoption of improved work processes and improving access to timely data for decision making. An ERP system can take a large number of separate systems developed over a number of years by an organization and replace them with one unified set of programs, making the system easier to use and more effective.

Information and Decision Support Systems

The benefits provided by an effective TPS are tangible and justifies their associated costs in computing equipment, computer programs, and specialized personnel and supplies. They speed business activities and reduce clerical costs. Although early accounting and financial TPSs were already valuable, companies soon realized that the data stored in these systems could be used to help managers make better decisions in their respective business areas, whether human resource management, marketing, or administration. Satisfying the needs of managers and decision makers continues to be a major factor in developing information systems.

Management Information Systems

management information system (MIS)
An organized collection of people, procedures, software, databases, and devices used to provide routine information to managers and decision makers.

A **management information system (MIS)** is an organized collection of people, procedures, software, databases, and devices used to provide routine information to managers and decision makers. The focus of an MIS is primarily on operational efficiency. Marketing, production, finance, and other functional areas are supported by MISs and linked through a common database. Management information systems typically provide standard reports generated with data and information from the TPS (see Figure 1.12).

Management information systems were first developed in the 1960s and are characterized by the use of information systems to produce managerial reports. In most cases, these early reports were produced periodically—daily, weekly, monthly, or yearly. Gambling casinos generate daily reports that tell their staff what specific customers like. Some casinos know

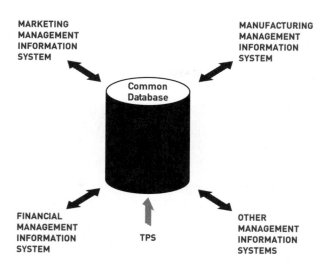

Figure 1.12

Functional management information systems draw data from the organization's transaction processing system.

whether a certain customer likes flowers in her room or a beverage in his hand and can accommodate individual needs or desires. Periodic reports are printed regularly, so they are called *scheduled reports*. Scheduled reports help managers perform their duties. For example, a summary report of total payroll costs might help an accounting manager control future payroll costs. Because of their value to managers, MISs have proliferated throughout the management ranks. For instance, the total payroll summary report produced initially for an accounting manager might also be useful to a production manager to help monitor and control labor and job costs. Other scheduled reports are used to help managers from a variety of departments control customer credit, payments to suppliers, the performance of sales representatives, inventory levels, and more.

Other types of reports were also developed during the early stages of MISs. *Demand reports* were developed to give decision makers certain information on request. For example, prior to closing a sale, a sales representative might seek a demand report on how much inventory exists for a particular item. This report would tell the representative whether enough inventory is on hand to fill the customer's order. *Exception reports* describe unusual or critical situations, such as low inventory levels. An exception report is produced only if a certain condition exists—in this case, inventory falling below a specified level. For example, in a bicycle manufacturing company, an exception report might be produced by the MIS if the number of bicycle seats is too low and more should be ordered.

Decision Support Systems

By the 1980s, dramatic improvements in technology resulted in information systems that were less expensive but more powerful than earlier systems. People at all levels of organizations began using personal computers to do a variety of tasks; they were no longer solely dependent on the IS department for all their information needs. So, people quickly recognized that computer systems could support additional decision-making activities. A **decision support system (DSS)** is an organized collection of people, procedures, software, databases, and devices used to support problem-specific decision making. The focus of a DSS is on decision-making effectiveness. Whereas an MIS helps an organization "do things right," a DSS helps a manager "do the right thing." Oxford Bookstore, located in Calcutta, uses a DSS and the Internet to allow book lovers in India to purchase their favorite books at Oxford's traditional retail stores or through its Internet site. The Internet site provides a wealth of information to help people make better book-purchasing decisions.[30] Blue Cross of Pennsylvania uses a DSS from InterQual to help it support level-of-care decisions.[31]

A DSS supports and assists all aspects of problem-specific decision making. A DSS can also support customers by rapidly responding to their phone and e-mail inquiries. A DSS goes beyond a traditional MIS by providing immediate assistance in solving complex problems. Many of these problems are unique and not straightforward, and information is often difficult to obtain. For instance, an auto manufacturer might try to determine the best location to build a new manufacturing facility, or an oil company might want to discover

decision support system (DSS)
An organized collection of people, procedures, software, databases, and devices used to support problem-specific decision making.

the best place to drill for oil. Some big oil companies use a DSS to track and manage projects and employees in countries around the world. Traditional MISs are seldom used to solve these types of problems; a DSS can help by suggesting alternatives and assisting in final decision making.

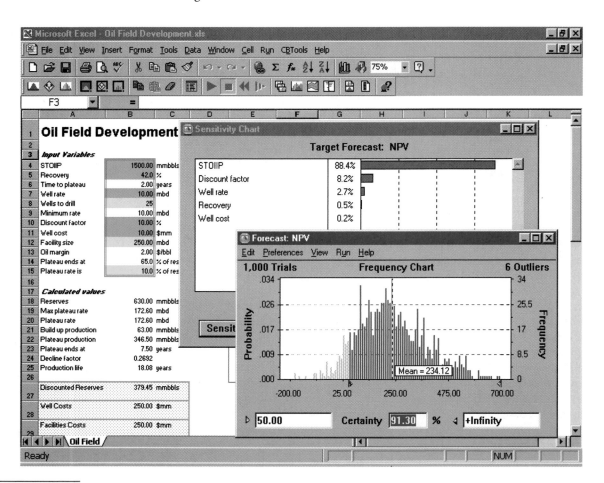

Decisioneering provides decision support software called Crystal Ball, which helps business people of all types assess risks and make forecasts. Shown here is the Standard Edition being used for oil field development.

(Source: Crystal Ball screenshot courtesy of Decisioneering, Inc.)

Decision support systems are used when the problem is complex and the information needed to make the best decision is difficult to obtain and use. So, a DSS also involves managerial judgment and perspective. Managers often play an active role in the development and implementation of the DSS. A DSS recognizes that different managerial styles and decision types require different systems. For example, two production managers in the same position trying to solve the same problem might require different information and support. The overall emphasis is to support, rather than replace, managerial decision making.

The essential elements of a DSS include a collection of models used to support a decision maker or user (model base), a collection of facts and information to assist in decision making (database), and systems and procedures (dialogue manager) that help decision makers and other users interact with the DSS (see Figure 1.13). Software is often used to manage the database (the database management system, DBMS) and the model base (the model management system, MMS).

In addition to DSSs that support individual decision making, there are group decision support systems and executive support systems that use the same overall approach of a DSS. A group decision support system, also called a *group support system*, includes the DSS elements just described and software, called *groupware*, to help groups make effective decisions. An executive support system, also called an *executive information system*, helps top-level managers, including a firm's president, vice presidents, and members of the board of directors, make better decisions. An executive support system can be used to assist with strategic planning, top-level organizing and staffing, strategic control, and crisis management.

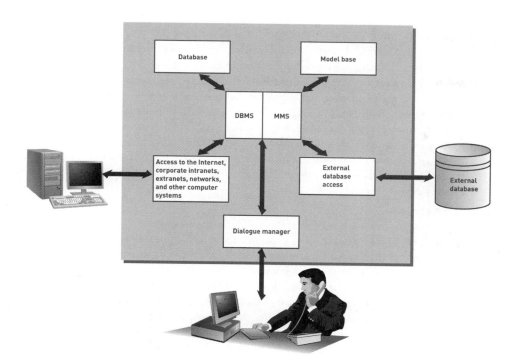

Figure 1.13

Essential DSS Elements

Specialized Business Information Systems: Artificial Intelligence, Expert Systems, and Virtual Reality

In addition to TPSs, MISs, and DSSs, organizations often use specialized systems. One of these systems is based on the notion of **artificial intelligence (AI)**, where the computer system takes on the characteristics of human intelligence. The field of artificial intelligence includes several subfields (see Figure 1.14).

artificial intelligence (AI)
A field in which the computer system takes on the characteristics of human intelligence.

Figure 1.14

The Major Elements of Artificial Intelligence

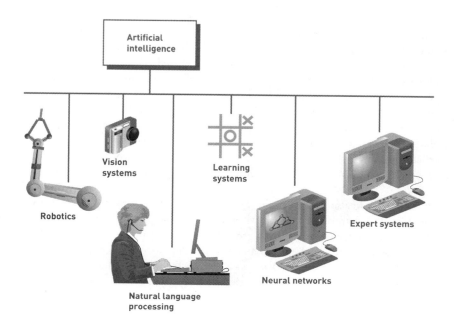

Artificial Intelligence

Robotics is an area of artificial intelligence in which machines take over complex, dangerous, routine, or boring tasks, such as welding car frames or assembling computer systems and components. Vision systems allow robots and other devices to see, store, and process visual images. Natural language processing involves the ability of computers to understand and act on verbal or written commands in English, Spanish, or other human languages. Learning systems give computers the ability to learn from past mistakes or experiences, such as playing games or making business decisions, and neural networks is a branch of artificial intelligence that allows computers to recognize and act on patterns or trends. Some successful stock, options, and futures traders use neural networks to spot trends and make them more profitable with their investments.

Expert Systems

expert system
A system that gives a computer the ability to make suggestions and act like an expert in a particular field.

Expert systems give the computer the ability to make suggestions and act like an expert in a particular field. The unique value of expert systems is that they allow organizations to capture and use the wisdom of experts and specialists. Therefore, years of experience and specific skills are not completely lost when a human expert dies, retires, or leaves for another job. Expert systems can be applied to almost any field or discipline. Expert systems have been used to monitor complex systems such as nuclear reactors, perform medical diagnoses, locate possible repair problems, design and configure IS components, perform credit evaluations, and develop marketing plans for a new product or new investment strategies. The collection of data, rules, procedures, and relationships that must be followed to achieve value or the proper outcome is contained in the expert system's **knowledge base**.

knowledge base
The collection of data, rules, procedures, and relationships that must be followed to achieve value or the proper outcome.

The end of the twentieth century brought advances in both artificial intelligence and expert systems. More and more organizations are using these systems to solve complex problems and support difficult decisions. However, many issues remain to be resolved, and more work is needed to refine their meaningful uses.

Virtual Reality

virtual reality
The simulation of a real or imagined environment that can be experienced visually in three dimensions.

Virtual reality is the simulation of a real or imagined environment that can be experienced visually in three dimensions. Originally, virtual reality referred to immersive virtual reality, which means the user becomes fully immersed in an artificial, 3-D world that is completely generated by a computer. The virtual world is presented in full scale and relates properly to the human size. It may represent any 3-D setting, real or abstract, such as a building, an archaeological excavation site, the human anatomy, a sculpture, or a crime scene reconstruction. Virtual worlds can be animated, interactive, and shared. Through immersion, the user can gain a deeper understanding of the virtual world's behavior and functionality.

A variety of input devices such as head-mounted displays (see Figure 1.15), data gloves (see Figure 1.16), joysticks, and handheld wands allow the user to navigate through a virtual environment and to interact with virtual objects. Directional sound, tactile and force feedback devices, voice recognition, and other technologies are used to enrich the immersive experience. Several people can share and interact in the same environment. Because of this ability, virtual reality can be a powerful medium for communication, entertainment, and learning.

Virtual reality can also refer to applications that are not fully immersive, such as mouse-controlled navigation through a 3-D environment on a graphics monitor, stereo viewing from the monitor via stereo glasses, stereo projection systems, and others. Some virtual reality applications allow views of real environments with superimposed virtual objects. Motion trackers monitor the movements of dancers or athletes for subsequent studies in immersive virtual reality. Telepresence systems (e.g., telemedicine, telerobotics) immerse a viewer in a real world that is captured by video cameras at a distant location and allow for the remote manipulation of real objects via robot arms and manipulators. Many believe that virtual reality is reshaping the interface between people and information technology by offering new ways to communicate information, visualize processes, and express ideas creatively.

Useful applications of virtual reality include training in a variety of areas (military, medical, equipment operation, etc.), education, design evaluation (virtual prototyping), architectural walk-throughs, human factors and ergonomic studies, simulation of assembly sequences and maintenance tasks, assistance for the handicapped, study and treatment of

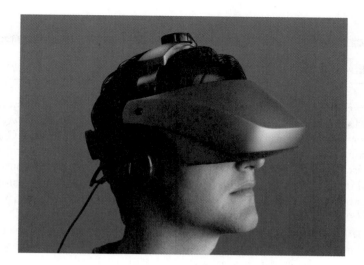

Figure 1.15

A Head-Mounted Display

The head-mounted display (HMD) was the first device of its kind providing the wearer with an immersive experience. A typical HMD houses two miniature display screens and an optical system that channels the images from the screens to the eyes, thereby presenting a stereo view of a virtual world. A motion tracker continuously measures the position and orientation of the user's head and allows the image-generating computer to adjust the scene representation to the current view. As a result, the viewer can look around and walk through the surrounding virtual environment.

(Source: Image Courtesy of 5DT, Inc., *www.5DT.com*.)

phobias (fear of flying), entertainment, and, of course, virtual reality games. Students taking Television and the Modern Presidency at the University of Denver, are able to chat with former presidents, White House administrators, and Washington insiders.[32]

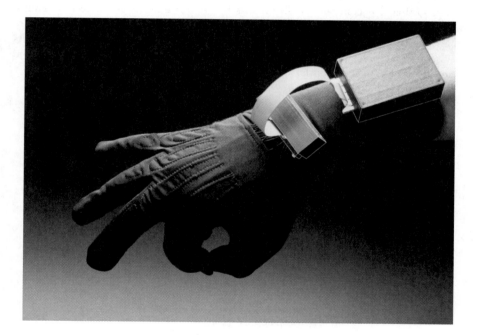

Figure 1.16

A Data Glove

Realistic interactions with virtual objects via such devices as a data glove that senses hand position allow for manipulation, operation, and control of virtual worlds.

(Source: Image Courtesy of 5DT, Inc., *www.5DT.com*.)

It is difficult to predict where information systems and technology will be in 10 to 20 years. It seems, however, that we are just beginning to discover the full range of their usefulness. Technology has been improving and expanding at an increasing rate; dramatic growth and change are expected for years to come. Without question, a knowledge of the effective use of information systems will be critical for managers both now and in the long term. But how are these information systems created?

SYSTEMS DEVELOPMENT

systems development
The activity of creating or modifying existing business systems.

Systems development is the activity of creating or modifying existing business systems. People inside a company can develop systems, or companies can use *outsourcing*, hiring an outside company to perform some or all of a systems development project. Outsourcing allows a company to focus on what it does best and delegate other functions to companies with expertise in systems development. Cox Insurance Holdings, for example, outsourced its commercial underwriting operations to another company.[33] Outsourcing enabled Cox Insurance to streamline its operations and reduce costs. Outsourcing, however, is not the best alternative for all companies. Toyota recently stopped outsourcing its financial services and started to perform the financial services function internally.[34] According to the director of Toyota Financial Services, " You depend on that service provider. you worry about whether or not it will be in business next year, and whether or not it will be able to service you consistently throughout the terms of the agreement and beyond." Other companies have used outsourcing for software development, database development, and other aspects of systems development.

Developing information systems to meet business needs is highly complex and difficult, so much so that it is common for IS projects to overrun budgets and exceed scheduled completion dates. Business managers would like the development process to be more manageable, especially with predictable costs and timing. One strategy for improving the results of a systems development project is to divide it into several steps, each with a well-defined goal and set of tasks to accomplish (see Figure 1.17). These steps are summarized next.

Figure 1.17

An Overview of Systems Development

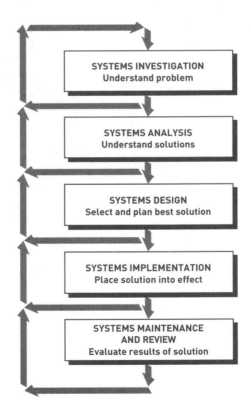

Systems Investigation and Analysis

The first two steps of systems development are systems investigation and analysis. The goal of the *systems investigation* is to gain a clear understanding of the problem to be solved or opportunity to be addressed. A cruise line company, for example, may launch a systems investigation to determine whether a development project is feasible to automate purchasing at ports around the world. Once an organization understands the problem, the next question to be answered is, "Is the problem worth solving?" Given that organizations have limited resources—people and money—this question deserves careful consideration. If the decision is to continue with the solution, the next step, *systems analysis*, defines the problems and opportunities of the existing system.

Systems Design, Implementation, Maintenance, and Review

Systems design determines how the new system will work to meet the business needs defined during systems analysis. *Systems implementation* involves creating or acquiring the various system components (hardware, software, databases, etc.) defined in the design step, assembling them, and putting the new system into operation. The purpose of *systems maintenance and review* is to check and modify the system so that it continues to meet changing business needs.

INFORMATION SYSTEMS IN SOCIETY, BUSINESS, AND INDUSTRY

Information systems have been developed to meet the needs of all types of organizations and people, and their use is spreading throughout the world to improve the lives and business activities of many citizens. But to provide their enormous benefits, information systems must be implemented thoughtfully and carefully. The speed and widespread use of information systems opens organizations and individuals to a variety of threats from unethical people.

Security, Privacy, and Ethical Issues in Information Systems and the Internet

Although information systems can provide enormous benefits, there are a number of potential negative aspects to their use.[35] Figure 1.18, for example, shows the percentage of businesses and other organizations attacked by various means in a one-year period. Figure 1.19 reveals the cost of losses during a one-year period of about 250 organizations that responded to a survey.

In addition to attacks on information systems, computer resources can be wasted, and computer-related mistakes and misuse have cost organizations millions of dollars. In an act of revenge, one fired employee in Australia used a computer to hack into a sewerage system and released millions gallons of raw waste into rivers and parks.[36] He released raw waste more than 40 times before he was caught. Computer crime and the invasion of privacy are also potential problems.[37]

Increasingly, the ethical use of systems has also been highlighted in the news. Ethical issues concern what is generally considered right or wrong. Some IS professionals believe that computers may create new opportunities for unethical behavior. For example, a faculty member of a medical school falsified computerized research results to get a promotion—and a higher salary. In another case, a company was charged with using a human resource information system to time employee layoffs and firings to avoid paying pensions. More and more, the Internet is also associated with unethical behavior. Unethical investors have placed false rumors or wrong information about a company on the Internet and tried to influence its

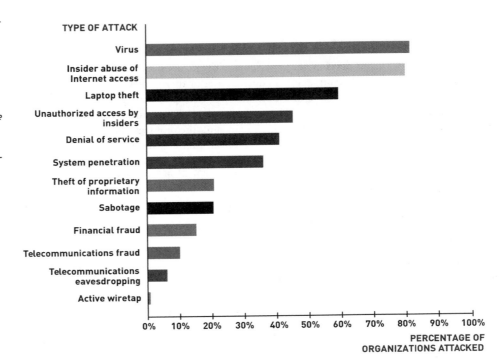

Figure 1.18

Attacks on Businesses and Other Organizations in One Year

(Source: Data from Riva Richmond, "How to Find Your Weak Spots," *The Wall Street Journal*, September 29, 2003, p. R3.)

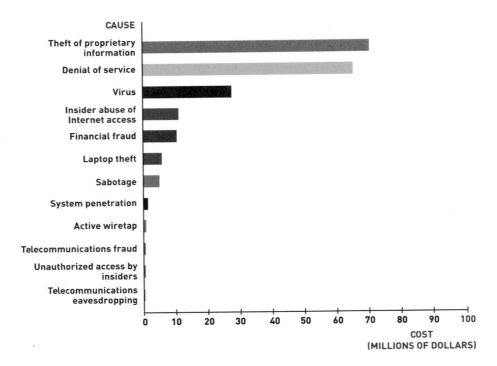

Figure 1.19

The Cost and Cause of Computer Attacks

(Source: Data from Riva Richmond, "How to Find Your Weak Spots," *The Wall Street Journal*, September 29, 2003, p. R3.)

stock price to make money. Stolen property is another issue. A few weeks before the $100-million film, *Hulk*, first hit the big screen, pirated copies were available on the Internet, after a man in New Jersey got an illegal copy of the film before its formal release.[38]

Many organizations have codes of conduct to foster ethical behavior in the use of information systems. Some other security and control measures are controversial, however. The USA Patriot Act, enacted as a result of the September 11, 2001, terrorist attacks, requires companies to respond to a subpoena in five days or fewer. The USA Patriot Act can require organizations to supply financial and personal information, such as books checked out at a library. Like many companies, Sumitomo Mitsui Bank uses a new computer system to collect

any necessary data and make it available to government agencies. Without the new information system, it could take weeks just to collect the needed data.[39] Some believe that provisions in this act may violate an individual's privacy.

Individuals and organizations can install security and control measures to protect themselves against some of the potential negative aspects of computerization. For example, a number of software products have been developed to detect and remove viruses from computer systems. Software can also protect you from spam, unwanted e-mail.[40] Information systems can help reduce crime as well.[41] A free computer center in Wellington, New Zealand, has cut vandalism by keeping young people off the street and giving people in the community a sense of pride. When a pair of headphones disappeared from the center, the community rallied to make sure that they were promptly returned.

Some individuals and companies install *firewalls* (software and hardware that protect a computer system or network from outside attacks) to avoid viruses and prevent unauthorized people from gaining access to the computer system. Identification numbers and passwords can also be used. One individual has proposed that Web cameras be put in critical areas and that "citizen spotters" be hired to monitor the Webcams.[42] In response to possible abuses, a number of laws have been passed to protect people from invasion of their privacy, including The Privacy Act, enacted in the 1970s.

Use of information systems also raises a number of work concerns, including job loss through increased efficiency and some potential health problems from making repetitive motions and other workplace issues. *Ergonomics*, the study of designing and positioning computer systems, can help people and companies avoid health-related problems of using computer systems.

Computer and Information Systems Literacy

In the twenty-first century, business survival and prosperity continue to become more difficult. For example, increased mergers among former competitors to create global conglomerates, continued downsizing of corporations to focus on their core businesses and to improve efficiencies, efforts to reduce trade barriers, and the globalization of capital all point to the increased internationalization of business organizations and markets. In addition, business issues and decisions are becoming more complex and must be made faster. Whatever career path you take, an understanding of information systems will help you cope, adapt, and prosper in this challenging environment.

The Internet is one means for advancing your knowledge of information systems and other professional study. Increasingly, courses or entire degrees are being offered over the Internet. One Harvard University professor, for example, stays up late at night to teach a class on clinical design to doctors in Asia.[43] In addition to traditional colleges and universities, a number of online degree programs are offered over the Internet.[44] Online degree programs are especially attractive to people with full-time jobs because they can learn during nontraditional hours from home.[45] While some people question the value of these online programs, others believe they foster motivation and determination in working professionals.

A knowledge of information systems will help you make a significant contribution on the job. It will also help you advance in your chosen career or field. Managers are expected to identify opportunities to implement information systems to improve their business. They are also expected to lead IS projects in their areas of expertise. To meet these personal and organizational goals, you must acquire both computer literacy and information systems literacy. **Computer literacy** is a knowledge of computer systems and equipment and the ways they function. It stresses equipment and devices (hardware), programs and instructions (software), databases, and telecommunications.

Information systems literacy goes beyond a knowledge of the fundamentals of computer systems and equipment. **Information systems literacy** is a knowledge of how data and information are used by individuals, groups, and organizations. It includes knowledge of not only computer technology but also aspects of the broader range of information technology.

computer literacy
Knowledge of computer systems and equipment and the ways they function; it stresses equipment and devices (hardware), programs and instructions (software), databases, and telecommunications.

information systems literacy
Knowledge of how data and information are used by individuals, groups, and organizations.

Most important, however, it encompasses how and why this technology is applied in business. Knowing about various types of hardware and software is an example of computer literacy. Knowing how to use hardware and software to increase profits, cut costs, improve productivity, and increase customer satisfaction is an example of information systems literacy. Information systems literacy can involve a knowledge of how and why people (managers, employees, stockholders, and other individuals) use information technology; a knowledge of organizations, decision-making approaches, management levels, and information needs; and a knowledge of how organizations can use computers and information systems to achieve their goals. Knowing how to deploy transaction processing, management information, decision support, and special-purpose systems to help an organization achieve its goals is a key aspect of information systems literacy.

Information Systems in the Functional Areas of Business

Studies have shown that the involvement of managers and decision makers in all aspects of information systems is a major factor for organizational success, including higher profits and lower costs. Information systems are used in all functional areas and operating divisions of business. In *finance* and *accounting*, information systems are used to forecast revenues and business activity, determine the best sources and uses of funds, manage cash and other financial resources, analyze investments, and perform audits to make sure that the organization is financially sound and that all financial reports and documents are accurate. In *sales* and *marketing*, information systems are used to develop new goods and services (product analysis), determine the best location for production and distribution facilities (place or site analysis), determine the best advertising and sales approaches (promotion analysis), and set product prices to get the highest total revenues (price analysis).

In *manufacturing*, information systems are used to process customer orders, develop production schedules, control inventory levels, and monitor product quality. In addition, information systems are used to design products (*computer-assisted design*, or *CAD*), manufacture items (*computer-assisted manufacturing*, or *CAM*), and integrate multiple machines or pieces of equipment (*computer-integrated manufacturing*, or *CIM*). Information systems are also used in *human resource management* to screen applicants, administer performance tests to employees, monitor employee productivity, and more. *Legal information systems* are used to analyze product liability and warranties and to develop important legal documents and reports.

Information Systems in Industry

Information systems are used in almost every industry or field. The *airline industry* employs Internet auction sites to offer discount fares and increase revenue. *Investment firms* use information systems to analyze stocks, bonds, options, the futures market, and other financial instruments, as well as to provide improved services to their customers. *Banks* use information systems to help make sound loans and good investments, as well as to provide online check payment for account holders. The *transportation industry* uses information systems to schedule trucks and trains to deliver goods and services at the lowest cost. *Publishing companies* use information systems to analyze markets and to develop and publish newspapers, magazines, and books. *Healthcare organizations* use information systems to diagnose illnesses, plan medical treatment, track patient records, and bill patients. HMOs use Web technology to access patients' insurance eligibility and other information held in databases to cut patient costs. *Retail companies* are using the Web to take customer orders and provide customer service support. Retail companies also use information systems to help market products and services, manage inventory levels, control the supply chain, and forecast demand. *Power management* and *utility companies* use information systems to monitor and control power generation and usage. *Professional services* firms employ information systems to improve the speed and quality of services they provide to customers. Management consulting firms use intranets and extranets to provide information on products, services, skill levels, and past engagements to its consultants. These industries will be discussed in more detail as we continue through the book.

SUMMARY

Principle

The value of information is directly linked to how it helps decision makers achieve the organization's goals.

Information systems are used in almost every imaginable career area. Regardless of your college major or chosen career, you will find that information systems are indispensable tools to help you achieve your career aspirations. Learning about information systems can help you get your first job, obtain promotions, and advance your career.

Data consists of raw facts; information is data transformed into a meaningful form. The process of defining relationships between data requires knowledge. Knowledge is an awareness and understanding of a set of information and the way that information can be made useful to support a specific task. To be valuable, information must have several characteristics: It should be accurate, complete, economical to produce, flexible, reliable, relevant, simple to understand, timely, verifiable, accessible, and secure. The value of information is directly linked to how it helps people achieve their organization's goals.

Principle

Models, computers, and information systems are constantly making it possible fororganizations to improve the way they conduct business.

A system is a set of elements that interact to accomplish a goal or set of objectives. The components of a system include inputs, processing mechanisms, and outputs. Systems also contain boundaries that separate them from the environment and each other. Feedback is used by the system to monitor and control its operation to make sure that it continues to meet its goals and objectives. Systems may be classified in many ways. They may be considered simple or complex. A stable, nonadaptive system does not change over time, while a dynamic, adaptive system does. Open systems interact with their environments; closed systems do not. Some systems exist temporarily; others are considered permanent.

System performance is measured by its efficiency and effectiveness. Efficiency is a measure of what is produced divided by what is consumed; effectiveness is a measure of the extent to which a system achieves its goals. A systems performance standard is a specific objective. A system variable is a quantity or item that can be controlled by the decision maker, such as how much of a product to produce, while a system parameter is a value or quantity that cannot be controlled, such as the cost of raw material.

There are four basic types of models: narrative, physical, schematic, and mathematical. These models serve as an abstraction or an approximation that is used to represent reality. Models enable us to explore and gain an improved understanding of real-world situations. The narrative model provides a verbal description of reality. A physical model is a tangible representation of reality, often computer designed or constructed. A schematic model is a graphic representation of reality such as a graph, chart, figure, diagram, illustration, or picture. A mathematical model is an arithmetic representation of reality.

Principle

Knowing the potential impact of information systems and having the ability to put this knowledge to work can result in a successful personal career, organizations that reach their goals, and a society with a higher quality of life.

Information systems are sets of interrelated elements that collect (input), manipulate and store (process), and disseminate (output) data and information. Input is the activity of capturing and gathering new data; processing involves converting or transforming data into useful outputs; and output involves producing useful information. Feedback is the output that is used to make adjustments or changes to input or processing activities.

The components of a computer-based information system include hardware, software, databases, telecommunications and the Internet, people, and procedures. CBISs play an important role in today's businesses and society. The key to understanding the existing variety of systems begins with learning their fundamentals. The types of business information systems used within organizations can be classified into four basic groups: (1) e-commerce and m-commerce, (2) TPS and ERP systems, (3) MIS and DSS, and (4) specialized business information systems.

E-commerce involves any business transaction executed electronically between parties such as companies (business-to-business), companies and consumers (business-to-consumer), business and the public sector, and consumers and the public sector. The major volume of e-commerce and its fastest-growing segment is business-to-business transactions that make purchasing easier for big corporations. E-commerce also offers opportunities for small businesses by enabling them to market and sell at a low cost worldwide, thus allowing them to enter the global market right from start-up. M-commerce involves anytime, anywhere computing that relies on wireless networks and systems.

The most fundamental system is the transaction processing system (TPS). A transaction is any business-related exchange. The TPS handles the large volume of business transactions that occur daily within an organization. An enterprise resource planning (ERP) system is a set of integrated programs that is capable of managing a company's vital business operations for an entire multisite, global organization.

The management information system (MIS) uses the information from a TPS to generate information useful for management decision making. Management information systems produce a variety of reports. Scheduled reports contain prespecified information and are generated regularly. Demand reports are generated only at the request of the user. Exception reports contain listings of items that do not meet a predetermined set of conditions.

A decision support system (DSS) is an organized collection of people, procedures, databases, and devices used to support problem-specific decision making. A DSS differs from an MIS in the support given to users, the decision emphasis, the development and approach, and system components, speed, and output.

The specialized business information systems include artificial intelligence systems, expert systems, and virtual reality systems. Artificial intelligence (AI) includes a wide range of systems, in which the computer system takes on the characteristics of human intelligence. Robotics is an area of artificial intelligence in which machines take over complex, dangerous, routine, or boring tasks, such as welding car frames or assembling computer systems and components. Vision systems allow robots and other devices to have "sight" and to store and process visual images. Natural language processing involves the ability of computers to understand and act on verbal or written commands in English, Spanish, or other human languages. Learning systems give computers the ability to learn from past mistakes or experiences, such as playing games or making business decisions, while neural networks is a branch of artificial intelligence that allows computers to recognize and act on patterns or trends. An expert system (ES) is designed to act as an expert consultant to a user who is seeking advice about a specific situation. Originally, the term *virtual reality* referred to immersive virtual reality, in which the user becomes fully immersed in an artificial, 3-D world that is completely generated by a computer. Virtual reality can also refer to applications that are not fully immersive, such as mouse-controlled navigation through a three-dimensional environment on a graphics monitor, stereo viewing from the monitor via stereo glasses, stereo projection systems, and others.

Principle

System users, business managers, and information systems professionals must work together to build a successful information system.

Systems development involves creating or modifying existing business systems. The major steps of this process and their goals include systems investigation (gain a clear understanding of what the problem is), systems analysis (define what the system must do to solve the problem), systems design (determine exactly how the system will work to meet the business needs), systems implementation (create or acquire the various system components defined in the design step), and systems maintenance and review (maintain and then modify the system so that it continues to meet changing business needs).

Principle

Information systems must be applied thoughtfully and carefully so that society, business, and industry can reap their enormous benefits.

Information systems play a fundamental and ever-expanding role in society, business, and industry. But their use can also raise serious security, privacy, and ethical issues. Effective information systems can have a major impact on corporate strategy and organizational success. Businesses around the globe are enjoying better safety and service, greater efficiency and effectiveness, reduced expenses, and improved decision making and control because of information systems. Individuals who can help their businesses realize these benefits will be in demand well into the future.

Computer and information systems literacy are prerequisites for numerous job opportunities, not just in the IS field. Computer literacy (a knowledge of computer systems and equipment) and information systems literacy (a knowledge of how data and information are used by individuals, groups, and organizations) is needed to get the most from any information system. Today, information systems are used in all the functional areas of business, including accounting, finance, sales, marketing, manufacturing, human resource management, and legal information systems. Information systems are also used in every industry, such as airlines, investment firms, banks, transportation companies, publishing companies, healthcare, retail, power management, professional services, and more.

CHAPTER 1: SELF-ASSESSMENT TEST

The value of information is directly linked to how it helps decision makers achieve the organization's goals.

1. A (an) _____ is a set of interrelated components that collect, manipulate, and disseminate data and information and provide a feedback mechanism to meet an objective.

2. Numbers, letters, and other characters are represented by _____.

   a. image data
   b. numeric data
   c. alphanumeric data
   d. symmetric data

3. Knowledge is an awareness and understanding of a set of information. True or false?

Models, computers, and information systems are constantly making it possible for organizations to improve the way they conduct business.

4. A (an) _____ is a set of elements or components that interact to accomplish a goal.

5. Which of the following is a way to classify systems?

   a. permanent—temporary
   b. simple—dynamic
   c. input—output
   d. open—adaptive

6. Graphs, charts, and figures are examples of physical models. True or false?

Knowing the potential impact of information systems and having the ability to put this knowledge to work can result in a successful personal career, organizations that reach their goals, and a society with a higher quality of life.

7. A (an) _____ consists of hardware, software, databases, telecommunications, people, and procedures.

8. Computer programs that govern the operation of a computer system are called _____.

   a. feedback
   b. feedforward
   c. software
   d. transaction processing system

9. Payroll and order processing are examples of a computerized management information system. True or false?

10. What type of system is used when the problem is complex and the information needed to make the best decision is difficult to obtain?

    a. TPS
    b. MIS
    c. DSS
    d. AI

11. _____ involves anytime, anywhere commerce that uses wireless communications.

System users, business managers, and information systems professionals must work together to build a successful information system.

12. What determines how a new system will work to meet the business needs defined during systems investigation?

    a. systems implementation
    b. systems review
    c. systems development
    d. systems design

Information systems must be applied thoughtfully and carefully so that society, business, and industry can reap their enormous benefits.

13. _____ literacy is a knowledge of how data and information are used by individuals, groups, and organizations.

CHAPTER 1: SELF-ASSESSMENT TEST ANSWERS

(1) information system (2) c (3) True (4) system (5) a (6) False (7) computer-based information system (CBIS) (8) c (9) False (10) c (11) Mobile commerce (m-commerce) (12) d (13) Information systems

KEY TERMS

artificial intelligence (AI) 29
computer-based information system (CBIS) 17
computer literacy 35
data 05

database 19
decision support system (DSS) 27
e-commerce 21
effectiveness 11
efficiency 11

enterprise resource planning (ERP) system 25
expert system 30
extranet 20
feedback 16

forecasting 16
hardware 17
information 05
information system (IS) 04
information systems literacy 36
input 15
Internet 19
intranet 20
knowledge 07
knowledge base 30
management information system (MIS) 26

mobile commerce (m-commerce) 22
model 13
networks 19
output 16
procedures 21
process 06
processing 15
software 18
system 08
system boundary 09

system parameter 11
system performance standard 11
system variable 11
systems development 32
technology infrastructure 17
telecommunications 19
transaction 25
transaction processing system (TPS) 25
virtual reality 30

REVIEW QUESTIONS

1. What is an information system? What are some of the ways information systems are changing our lives?
2. How would you distinguish data and information? Information and knowledge?
3. Identify at least six characteristics of valuable information.
4. Define the term *system*. What is the difference between a stable system and a dynamic system?
5. What are the components of any information system?
6. What is feedback? What are possible consequences of inadequate feedback?
7. How is system performance measured?
8. What is a model? What is the purpose of using a model?
9. What is a computer-based information system? What are its components?
10. Identify three functions of a transaction processing system.
11. What is the difference between an intranet and an extranet?
12. What is m-commerce? Describe how it can be used.
13. What are the most common types of computer-based information systems used in business organizations today? Give an example of each.
14. Identify three elements of artificial intelligence.
15. What are computer literacy and information systems literacy? Why are they important?
16. What are some of the benefits organizations seek to achieve through using information systems?
17. Identify the steps in the systems development process and state the goal of each.

DISCUSSION QUESTIONS

1. Why is the study of information systems important to you? What do you hope to learn from this course to make it worthwhile?
2. What is a database? Why is it an important part of a computer-based information system?
3. What is the difference between e-commerce and m-commerce?
4. What is the difference between an MIS and a DSS?
5. Suppose that you are a teacher assigned the task of describing the learning processes of preschool children. Why would you want to build a model of their learning processes? What kinds of models would you create? Why might you create more than one type of model?
6. Describe the "ideal" automated auto license plate renewal system for the drivers in your state. Describe the input, processing, output, and feedback associated with this system.
7. How is it that useful information can vary widely from the quality attributes of valuable information?
8. Discuss the potential use of virtual reality to enhance the learning experience for new automobile drivers. How might such a system operate? What are the benefits and potential disadvantages of such a system?
9. Discuss how information systems are linked to the business objectives of an organization.
10. What are your career goals and how can a computer-based information system be used to achieve them?

PROBLEM-SOLVING EXERCISES

1. Prepare a data disk and a backup disk for the problem-solving exercises and other computer-based assignments you will complete in this class. Create one directory for each chapter in the textbook (you should have 14 directories). As you work through the problem-solving exercises and complete other work using the computer, save your assignments for each chapter in the appropriate directory. On the label of each disk, be sure to include your name, course, and section. On one disk write "Working Copy"; on the other write "Backup."

2. Search through several business magazines (*Business Week, Computerworld, PC Week*, etc.) for a recent article that discusses the use of information technology to deliver significant business benefits to an organization. Now use other resources to find additional information about the same organization (*Reader's Guide to Periodical Literature*, online search capabilities available at your school's library, the company's public relations department, Web pages on the Internet, etc.). Use word processing software to prepare a one-page summary of the different resources you tried and their ease of use and effectiveness.

3. Create a table that lists all the courses you are taking in the first column. The other columns of the table should be the weeks of the semester or quarter, such as Week 1, Week 2, and so on. The body of the table should contain the actual assignments, quizzes, exams, the final exam, and so forth for each course. Place the table into a database and print the results. Create a table in the database for the first three weeks of class and print the results. Create another table in the database for your two hardest classes for all weeks and print the results.

4. Do some research to obtain estimates of the rate of growth of e-commerce and m-commerce. Use the plotting capabilities of your spreadsheet or graphics software to produce a bar chart of that growth over a number of years. Share your findings with the class.

TEAM ACTIVITIES

1. Before you can do a team activity, you need a team! The class members may self-select their teams, or the instructor may assign members to groups. Once your group has been formed, meet and introduce yourselves to each other. You will need to find out the first name, hometown, major, and e-mail address and phone number of each member. Find out one interesting fact about each member of your team, as well. Come up with a name for your team. Put the information on each team member into a database and print enough copies for each team member and your instructor.

2. With the other members of your group, use word processing software to write a one-page summary of what your team hopes to gain from this course and what you are willing to do to accomplish these goals. Send the report to your instructor via e-mail.

WEB EXERCISES

1. Throughout this book, you will see how the Internet provides a vast amount of information to individuals and organizations. We will stress the World Wide Web, or simply the Web, which is an important part of the Internet. Most large universities and organizations have an address on the Internet, called a Web site or home page. The address of the Web site for this publisher is *www.course.com*. You can gain access to the Internet through a browser, such as Internet Explorer or Netscape. Using an Internet browser, go to the Web site for this publisher. What did you find? Try to obtain information on this book. You may be asked to develop a report or send an e-mail message to your instructor about what you found.

2. Go to an Internet search engine, such as *www.yahoo.com*, and search for information about a company, including its Web site. Write a report that summarizes the size of the company, number of employees, its products, the location of its headquarters, and its profits (or losses) for last year. Would you want to work for this company?

3. Using the Internet, search for information on the use of information systems in a company or organization that interests you. How does the organization use technology to help it accomplish its goals?

CAREER EXERCISES

1. In the Career Exercises found at the end of every chapter, you will explore how material in the chapter can help you excel in your college major or chosen career. Write a brief report on the career that appeals to you the most. Do the same for two other careers that interest you.

2. Research the three career areas you selected and describe the job opportunities, job duties, and the possible starting salaries for each in a report.

VIDEO QUESTIONS

Watch the video clip **Go Inside Krispy Kreme** and answer these questions:

1. Provide a description of how Krispy Kreme is using each of the elements of an information system: hardware, software, databases, telecommunications, people, and procedures to provide services for its employees.

2. How have information systems assisted the many Krispy Kreme franchises in providing consistent products and services for their customers?

CASE STUDIES

Case One

Tyndall Federal Credit Union explores new ATM services

Tyndall Federal Credit Union has provided banking services to military personnel at Tyndall Air Force Base in Panama City, Florida, since 1956. Recently the credit union has struggled to keep up with the demand for services from its 80,000 members. Waiting lines in its six branch locations increased in size, and customer aggravation was beginning to show. Also, the credit union's highly mobile members were unable to carry out transactions when they were out of the country. Tyndall Federal's information systems needed to expand to keep up with its growing membership.

Tyndall Federal hired IS specialists from IBM to assist in developing a solution. IBM partnered with Wincor Nixdorf, designer of bank ATM machines, to develop what they called the compact-BANK—an ATM machine that offers all of the services a teller provides in a branch location. The compact-BANKs go beyond the standard ATM services by dispensing both cash and coins; scanning and cashing personal checks; processing passbook transactions; printing statements,

cashier's checks, and other documents; and processing stop payments. These "super ATMs" also provide cardless transactions, such as creating new bank accounts and delivering targeted marketing messages. They even have a microphone and speakers to provide personal assistance. "Our members love the compact-BANKs. They provide fast service during peak hours and are beginning to fulfill Tyndall's vision of 24x7 service," says Janet Turner, vice president of Interactive Services at Tyndall FCU.

By installing compact-BANK stations around the Panama City area, Tyndall Federal was able to reduce traffic at its branch locations and improve local member satisfaction. But what about those members overseas? To address their needs, Tyndall Federal hired FundsXpress Financial Network, a leading provider of online financial services, to design a Web-based banking service for its members. Now members stationed overseas can access bank services such as real-time account balances, account transfers, extended online account history, e-mail payments, electronic statements, and check imaging.

While Tyndall Federal has solved problems for its members, has it created management problems for itself? No.

Tyndall Federal's system development project was managed wisely and connects all of its systems—its branch office, ATM, and Web-based services—to a centralized server for easy management. From the main branch, a manager can use remote reporting tools to view and monitor transactions at any location. Also, customers have access to consistent interfaces and services from any of the systems.

Discussion Questions

1. If you were deciding on a credit union or bank with which to do business, would the services provided by Tyndall Federal influence your decision? How?
2. If you were employed as a bank teller at Tyndall Federal, how would you react to news that your employer was deploying automated compact-BANKs around town?

Case Two
The Queen Mary 2 and partner

The Queen Mary 2 (QM2) is the largest and most expensive cruise ship ever built. It includes five swimming pools, a planetarium, a two-story theater that seats 1,000, a casino, a gym, luxurious kennels, a nursery staffed with British nannies, and the largest ballroom, library, and wine collection at sea. Of all its amenities, the one considered most valuable to the crew and management and key to the functioning of the vessel is the integrated network and information system accessible in every cabin.

The $800 million QM2, constructed in the shipyard at Chantiers de l'Atlantique, France, and owned by Miami-based Cunard Line Ltd., made her maiden voyage in early 2004. Passengers in each of her 1,310 cabins had access to digital entertainment such as on-demand movies and interactive television. Each cabin is also wired with Internet access and network services. For example, passengers use the network to make shore excursion reservations and dinner plans.

Upon checking in, passengers are presented with a plastic bar-coded card. The card is used while on board to make purchases, which are then billed to the customer's account. It is also swiped as guests leave and return to the ship to track passenger location. The ship's massive data network brings order where there once was chaos. Ship managers can run reports showing which passengers are on board, how many will be attending the morning exercise class, and which entrée was most popular at last night's dinner. The network and database are backed up by redundant systems that automatically take over if the primary system fails.

The information system, called The Ship Partner, is used to track security, billing, telephone service, onboard television, and other operations. It was designed by Discovery

Critical Thinking Questions

3. If you were a bank officer in charge of security at Tyndall Federal, what security concerns do you think are the most important to address and solve in designing its compact-BANK?
4. Why might credit-union members prefer banking in a branch office over banking with an ATM?

SOURCES: "Tyndall Federal Credit Union: Banking on Next-Generation ATMs from IBM and Wincor Nixdorf," *IBM Success Stories*, www.306.ibm.com/software/success/, accessed January 17, 2004; "FundsXpress to Provide Online Financial Services to Tyndall Federal Credit Union," FundsXpress press release,www.fundsxpress.com/press/2003/11-17-2003.html, accessed January 17, 2004; Tyndall Federal Credit Union Web site, www.tyndallfcu.org/, accessed January 17, 2004.

Travel Systems LP (DTS). John Broughan, president of DTS, says that the IT needs of cruise ship operators differ from those of typical hotel property management companies, so specialized systems had to be created to better serve cruise companies.

The Queen Mary 2 provides yet another example of how information systems assist with management functions, providing valuable information and offering services to customers.

Discussion Questions

1. What conveniences does The Ship Partner information system provide to passengers of the Queen Mary 2? What entertainment services could be made available to passengers through this digital network?
2. How does The Ship Partner information system assist ship managers with their duties and responsibilities?

Critical Thinking Questions

3. How does The Ship Partner information system assist Cunard in competing in the travel industry? What other travel and leisure industries would benefit from a system like The Ship Partner?
4. Why is it important for The Ship Partner to have a backup system? How would a systemwide failure affect the functioning of the ship?

SOURCES: Todd R. Weiss, "New Queen Mary 2 Offers High Tech on the High Seas," *Computerworld*, January 12, 2004, *www.computerworld.com*; Eric Thomas, "Queen Mary 2, World's Biggest Liner, Awaits Its Champagne Moment," *Agence France Presse*, January 8, 2004; The Chantiers de l'Atlantique Web site, *www.chantiersatlantique.com/UK/index_UK.htm*, accessed January 18, 2004.

Case Three

MyFamily Comforts Its Members

MyFamily.com, Inc., is a leading online subscription business for researching family history, and the site also allows families to set up their own Web site and share photos with other family members.

MyFamily.com was one of the rare companies that survived the hardships of the dot-com bust. Through smart business management and providing a highly valued service to its customers, MyFamily.com actually grew its business when many others lost theirs. From 1999 through 2003 MyFamily.com doubled its subscribers each year, finishing 2003 with 1.6 million customers.

The rapid growth of the company presented MyFamily.com with challenges in customer relationship management (CRM). The company was hiring many customer service representatives just to respond to customer e-mail. Much of the e-mail involved simple questions that employees answered repeatedly day in and day out. What MyFamily.com needed was a system to help organize its customer support function and allow the company to make better use of its employees. The solution lay in a self-service CRM application from RightNow Technologies called eService Center.

The eService Center provides Web-based customer support for routine customer inquiries, freeing up customer service representatives to handle more difficult problems. It uses artificial intelligence contained in a single self-learning knowledge base that can be accessed from the Web, e-mail, chat room, or telephone. The system makes it easy for customers to find answers to questions by presenting the most successful solutions first and refining the solution based on customer responses. The eService Center also includes analytics and the ability to measure customer satisfaction through surveys.

Within 30 days after MyFamily.com implemented Right-Now's system, the number of e-mails that employees had to answer fell 30 percent, according to Mary Kay Evans, spokeswoman for MyFamily.com. Calculating the savings in employee time, MyFamily.com has received over two and a half times as much as it invested in the system over nine months—a 260 percent return on investment (ROI). The new system earned MyFamily.com two awards in 2003: SearchCRM honored MyFamily.com with the Customer Touch award, and *CRM Magazine* awarded MyFamily its 2003 CRM Elite award.

Discussion Questions

1. What type of information system is RightNow's eService Center, a TPS, MIS, DSS, or some other specialized system? Present the rationale for your answer.
2. Besides cost savings, what other benefits does the eService Center provide for the upper-level managers of MyFamily.com?

Critical Thinking Questions

3. Have you had any experience with automated customer service systems? Do you think that these services benefit the company or the customer more? Why?
4. The types of questions that this automated system assists customers with are described as typical customer inquiries. Do you think handling frequently asked questions (FAQs) is a job better suited for humans or machines? Why?

SOURCES: Linda Rosencrance, "CRM with a Family Touch," *Computerworld*, April 7, 2003, *www.computerworld.com*; "RightNow Customer MyFamily.com Wins SearchCRM.com's Customer Touch Award for Effective Service & Support," *PR Newswire*, September 11, 2003; MyFamily.com Web site, *www.myfamily.com*, accessed January 17, 2004.

NOTES

Sources for the opening vignette: Marc Songini, "Case Study: Boehringer Cures Slow Reporting," *Computerworld*, July 21, 2003, *www.computerworld.com*; "Boehringer Ingelheim Deploys BackWeb's Offline Solution for the Plumtree Corporate Portal," *PR Newswire*, December 15, 2003; the Boehringer Ingelheim Web site, *www.boehringer-ingelheim.com/corporate/home/home.asp*, accessed January 22, 2004.

1. Booth-Thomas, Cathy, "The See-It-All Chip," *Time* magazine special technology section, October 2003, p. A12.
2. Nussbaum, Bruce, "Technology: Just Make It Simpler," *Business Week*, September 8, 2003, p. 38.
3. Wildstrom, Stehen, "Tablet PCs," *Business Week*, August 4, 2003, p. 22.

4. Heun, Christopher, "Marine Mouse Takes IT to New Depths," *InformationWeek*, November 5, 2002, p. 20.
5. Clark, Don, "A 64-Bit Bet on Its Future," *The Wall Street Journal*, April 21, 2003, p. B1.
6. Goldsmigh, Charles, "German Visual Image Firm Is Honored for Film Graphics," *The Wall Street Journal*, February 26, 2003, p. B1.
7. Wildstrom, Stephen, "A Dana for Every Schoolkid," *Business Week*, April 21, 2003, p. 26.
8. Brandel, S. "35 Years of Leadership," *Computerworld*, September 30, 2003, p. 55
9. Beauprez, Jennifer, "State Urged to Think Small," *The Denver Post*, July 13, 2003, p. K1.
10. Smith, Jeff, "Snitch or Savior?" *Rocky Mountain News*, June 28, 2003, p. 1C.

11. Demaitre, Eugene, "Doctors Bring 3-D into the Operating Room", *Computerworld*, June 2, 2003, p. 25.
12. Cowley, Stacy, "Software Market Hit by Purchasing Delays ," *Computerworld*, July 14, 2003, p. 12.
13. Brandel, S. "35 Years of Leadership," *Computerworld*, September 30, 2003, p. 55.
14. King, Julia, "Open for Inspection," *Computerworld*, July 21, 2003, p. 39.
15. Hamblen, Matt, "Compression Relives Congestion," *Computerworld*, March 10, 2003, p. 30.
16. Hamblen, Matt, "Hotel Goes Wireless," *Computerworld*, July 14, 2003, p. 16.
17. Sitch, Stephane, "Invasion of the Drones," *Forbes*, March 17, 2003, p. 52.
18. Angwin, Julia, "Top Online Chemical Exchange Is Likely Success Story," *The Wall Street Journal*, January 8, 2004, p. A15.
19. Sternstein, Aliya, "Mashups," *Forbes*, July 21, 2003, p. 145.
20. Barta, Patrick, "What Happened to the Paper Mortgage," *The Wall Street Journal*, p. R4.
21. Cohen, Alan, "Online Prescriptions," *PC Magazine*, August 19, 2003, p. 68.
22. "Swiss Town Leads Way with Internet Voting," *CNN Online*, January 20, 2003.
23. Mossberg, Walter, "Instant Messages That Come with Sights, Sounds," *The Wall Street Journal*, August 13, 2003, p. D10.
24. Brady, Diane, "Net Hookups Are Spreading from the Study to the Living Room, Bedroom, and Kitchen," *Business Week*, July 21, 2003, p. 58.
25. Brandel, S. "35 Years of Leadership," *Computerworld*, September 30, 2003, p. 55.
26. Kelly, Lisa, "Virgin Sets Up Global Intranet," *Computing*, November 6, 2003, p. 15.
27. Anthes, Gary, "Corporate Express Goes Direct," *Computerworld*, September 1, 2003, p. 17.
28. Chabrow, Eric, "Online Ad Sales Rebounding," *Information Week*, January 6, 2003, p.16.
29. Cuneo, Wileen Colkin, "Uptick in Care," *Information Week*, November 3, 2003, p. H18.
30. Vijayan, Jaikumar, "Bookseller Expands Its Reach with Integrated Internet Platform, *Computerworld*, June 2, 2003, p. 29.
31. Staff, "IBC Supports Decision Support," *Health Management Technology*, September, 2003, p.10.
32. Jones, Rebecca, "Distance Learning Brings D.C. to Denver," *Rocky Mountain News*, February 10, 2003, pp. 12A.
33. Staff, "The Rise of Outsourcing," *Insurance Day*, November 5, 2003.
34. Nash, Emma, "Toyota Puts Brakes on Outsourcing," *Computing*, November 6, 2003, p. 4.
35. Regalado, Antonio, "Greenpeace Warns of Pollutants from Nanotechnology, " *The Wall Street Journal*, July 25, 2003, p. B1.
36. Schwartz, Mathew, "Wanted: Security Tag Team," *Computerworld*, June 30, 2003, p. 38.
37. Staff, "The Net Detectives," *Business and Finance*, January 30, 2003, p.85.
38. Grover, Ronald and Green, Heather, "Hollywood Heist," *Business Week*, July 14, 2003, p. 73.
39. Thibodeaqu, Patrick, "Bank Users Online Workflow to Comply with USA Patriot Act," *Computerworld*, June 2, 2003, p. 26.
40. Mangalindan, Mylene, "Didn't Get E-Mail? That Could Be Spam's Fault, Too," *The Wall Street Journal*, August 4, 2003, p. B1.
41. Strecker, Tom, "Computers Cut Vandalism," *New Zealand Infotech Weekly*, May 12, 2003, p. 6.
42. Gomes, Lee, "Is Antiterror Plan by Priceline Founder Genius or Just Goofy?" *The Wall Street Journal*, June 30, 2003, p. B1.
43. Sandberg, Jared, "Elite Colleges Finally Embrace Online Degree Courses," *The Wall Street Journal*, January 15, 2003, p. B1.
44. Dunham, Kemba, "Online-Degree Programs Surge," *The Wall Street Journal*, January 28, 2003, p. B8.
45. Fillion, Roger, "ECollege Records Its First Profit," *The Rocky Mountain News*, July 23, 2003, p. 2b.

CHAPTER
· 2 ·

Information Systems in Organizations

INFORMATION SYSTEMS IN THE GLOBAL ECONOMY ➤
WHIRLPOOL CORPORATION WORLDWIDE, UNITED STATES

Bringing Innovation and Quality to Every Home . . . Everywhere

Whirlpool's motto, used for the title of this article, reflects the company's strong global ambitions. Whirlpool is the world's leading manufacturer and marketer of major home appliances. With nearly 50 manufacturing and research centers around the world—and 68,000 employees—Whirlpool rakes in more than $11 billion in sales each year. The company markets dozens of brand names, including Whirlpool, KitchenAid, Brastemp, Bauknecht, and Consul, to consumers in more than 170 countries.

Whirlpool attributes its success in the global marketplace to product development and information systems that "help our operations reduce costs, improve efficiencies, and introduce a continuous stream of relevant innovation to consumers." The person responsible for Whirlpool's streamlined global information systems is Chief Information Officer (CIO) David Butler, who took on Whirlpool systems as a final challenge prior to retirement.

In 1998, when Butler joined Whirlpool, the company supported a variety of diverse and complex systems in several countries. His goal was to lower operating costs by simplifying the systems and creating a more integrated global technology platform. The key to meeting his goal, Butler decided, was to standardize all Whirlpool information systems throughout the organization on those developed by the biggest and best vendors: SAP, the largest interenterprise software company, and IBM. His theory was that consolidating to one IS platform would lower operating costs and let business units share information more easily. Butler understood the important lessons taught in this chapter: information systems play an important role within an organization—that of increasing efficiency and productivity for greater market share and competitive advantage. He selected the SAP-IBM alliance to develop and maintain Whirlpool's information systems, knowing that the cutting-edge innovations of these companies would keep Whirlpool ahead of the competition over the long haul.

Butler had little hope or expectation that Whirlpool would be willing to invest in a new information system because "no one funds simplification for simplification's sake." He planned to fund the effort by cutting the cost of parts of the business. The savings from an implementation in one phase would fund the next phase of the project. He set a goal of a 5 percent annual productivity increase to fund the effort. The result? The project paid for itself!

Volker Loehr, IBM's vice president of the IBM-SAP alliance, says that consolidation is a key strategy for many global organizations today. Companies are increasingly moving from a country-by-country approach to a single operating platform.

David Butler is now enjoying his retirement while new Whirlpool CIO, Esat Sezer, oversees the final stages of the four-year execution of Butler's plan. Sezer is capitalizing on the new standardized platform by globalizing systems development projects to encourage efficiency. Sezer's strategy is to develop a packaged application in one country and deploy it in many. The trick is to

develop systems that address the sometimes diverse needs of different global regions.

Whirlpool's unified global information system has more than paid for itself. It has provided the company with the ability to manage global operations from a single location. Even more important, it has broken down cultural barriers and allowed Whirlpool to take advantage of talent within the organization—wherever that talent may be.

As you read this chapter, consider the following:

- Why is it important for an organization to continually evaluate and improve its information systems?
- What role do information systems play in providing an organization with a competitive advantage?

Why Learn About Information Systems in Organizations?

Organizations of all types use information systems to record and track their daily operations. After graduating, a management major might be hired to work with computerized employee files and records for a shipping company. A marketing major might work for a large retail store analyzing customer needs with a computer. An accounting major might work for an accounting or consulting firm using a computer to audit other companies' financial records. A real estate major might use the Internet and work within a loose organizational structure with clients, builders, and a legal team located around the world. A biochemist could work for a large drug company and use a computer to analyze the potential of a new cancer drug.

While your career might be different from your classmates', you will almost certainly be working with computers and information systems to help your company or organization become more efficient, effective, productive, and competitive in its industry. In this chapter, you will see how information systems can help organizations produce higher-quality products and increase their return on investment. We begin by investigating organizations and information systems.

Technology's impact on business is growing steadily. Once used to automate manual processes, technology has now transformed the nature of work and the shape of organizations themselves. During the late 1960s and early 1970s, many computerized information systems were developed to provide reports for business decision makers. The information in these reports helped managers monitor and control business processes and operations. For example, reports that listed the quantity of each inventory item in stock could be used to monitor inventory levels. Unfortunately, many of these early computer systems did not take the overall goals of the organization and managerial problem-solving styles into consideration. In this chapter, we will explore the use of information systems in today's organizations.

ORGANIZATIONS AND INFORMATION SYSTEMS

organization
A formal collection of people and other resources established to accomplish a set of goals.

An **organization** is a formal collection of people and other resources established to accomplish a set of goals. The primary goal of a for-profit organization is to maximize shareholder value, often measured by the price of the company stock. Nonprofit organizations include social groups, religious groups, universities, and other organizations that do not have profit as the primary goal.

An organization is a system. Money, people, materials, machines and equipment, data, information, and decisions are constantly in use in any organization. As shown in Figure 2.1, resources such as materials, people, and money are input to the organizational system from the environment, go through a transformation mechanism, and are output to the environment. The outputs from the transformation mechanism are usually goods or services.

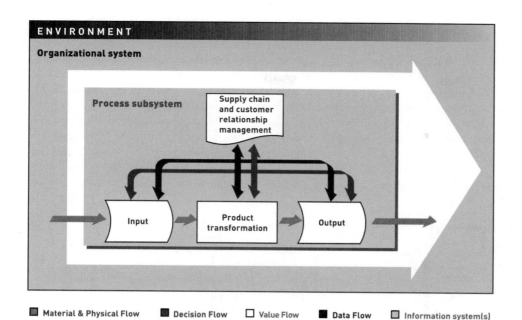

Figure 2.1

A General Model of an Organization

Information systems support and work within all parts of an organizational process. Although not shown in this simple model, input to the process subsystem can come from internal and external sources. Just prior to entering the subsystem, data is external. Once it enters the subsystem, it becomes internal. Likewise, goods and services can be output to either internal or external systems.

The goods or services produced by the organization are of higher relative value than the inputs alone. Through adding value or worth, organizations attempt to achieve their goals.

How does this increase in value occur? Within the transformation mechanism, various subsystems contain processes that help turn specific inputs into goods or services of increasing value. These processes increase the relative worth of the combined inputs on their way to becoming final outputs of the organization. Let us reconsider our simple car wash example from Chapter 1 (see Figure 1.3). The first process might be identified as washing the car. The output of this system—a clean, but wet, car is worth more than the mere collection of ingredients (soap and water), as evidenced by the popularity of automatic car washes. Consumers are willing to pay for the skill, knowledge, time, and energy required to wash their car. The second process is drying—transforming the wet car into a dry one with no water spotting. Again, consumers are willing to pay for the additional skill, knowledge, time, and energy required to accomplish this transformation. Another example of how inputs can be transformed into increased value for an organization is software that looks for new business opportunities. Software that analyzes social networks collects data from an employee's computerized phone list, address book, e-mails, instant-message communications, and calendars. The data is then used to build relationships to identify potential new customers and new business opportunities for increased sales.[1] So, organizations establish these processes to achieve their goals—to exploit opportunities and solve problems.

All business organizations contain a number of processes. Providing value to a stakeholder—customer, supplier, manager, or employee—is the primary goal of any organization. The value chain, first described by Michael Porter in a 1985 *Harvard Business Review* article, is a concept that reveals how organizations can add value to their products and services. The **value chain** is a series (chain) of activities that includes inbound logistics, warehouse and storage, production, finished product storage, outbound logistics, marketing and sales, and customer service (see Figure 2.2). Each of these activities is investigated to determine what can be done to increase the value perceived by a customer. Depending on the customer, value may mean lower price, better service, higher quality, or uniqueness of product. The value comes from the skill, knowledge, time, and energy invested by the company. By adding a significant amount of value to their products and services, companies will ensure further organizational success. Cessna Aircraft, for example, has used supply chain management to improve the quality of supplies by about 86 percent and increase material availability by 28 percent.[2]

Supply chain and customer relationship management are two key aspects of managing the value chain just described and shown in Figure 2.2. *Supply chain management (SCM)*

value chain
A series (chain) of activities that includes inbound logistics, warehouse and storage, production, finished product storage, outbound logistics, marketing and sales, and customer service.

helps determine what supplies are required, what quantities are needed to meet customer demand, how the supplies are to be processed (manufactured) into finished goods and services, and how the shipment of supplies and products to customers is to be scheduled, monitored, and controlled. For an automotive company, for example, SCM is responsible for identifying key supplies and parts, negotiating with supply and parts companies for the best prices and support, making sure that all supplies and parts are available when they are needed to manufacture cars and trucks, and sending finished products to dealerships around the country when and where they are needed. Hewlett-Packard, for example, saved about $130 million by using SCM to streamline several businesses, including its inkjet supplies operations.[3] Using the PowerChain software program, which minimizes total inventory costs, the company was able to eliminate regional delivery centers and ship directly to retailers.

Figure 2.2

The Value Chain of a Manufacturing Company

The management of raw materials, inbound logistics, and warehouse and storage facilities is called *upstream management*, and the management of finished product storage, outbound logistics, marketing and sales, and customer service is called *downstream management*.

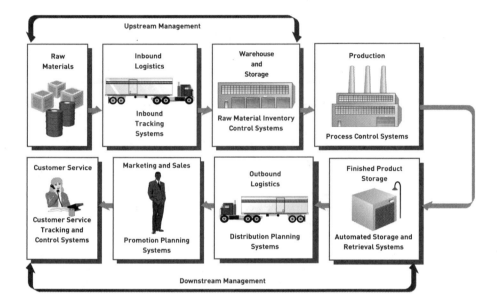

Wal-Mart's use of information systems is an integral part of its operation. It gives suppliers access to its inventory system, so the suppliers are able to monitor the database and automatically send another shipment when stocks are low, eliminating the need for purchase orders, which speeds delivery time, lowers Wal-Mart's inventory carrying costs, and reduces stockout costs.

Customer relationship management (CRM) programs help a company manage all aspects of customer encounters, including marketing and advertising, sales, customer service after the sale, and programs to help keep and retain loyal customers.[4] CRM can help a company collect customer data, contact customers, educate customers on new products, and actively sell products to existing and new customers. Often, CRM software uses a variety of information sources, including sales from different retail stores, surveys, e-mails, and Internet browsing habits, to compile comprehensive customer profiles. CRM can also be used to get customer feedback to help design new products and services.

What role does an information system play in these processes? A traditional view of information systems holds that they are used by organizations to control and monitor processes to ensure effectiveness and efficiency. An information system can turn feedback from the subsystems into more meaningful information for employees' use within an organization. This information might summarize the performance of the systems and be used to change the way that the system operates. Such changes could involve using different raw materials (inputs), designing new assembly-line procedures (product transformation), or developing new products and services (outputs). In this view, the information system is external to the process and serves to monitor or control it.

A more contemporary view, however, holds that information systems are often so intimately intertwined that they are best considered *part of* the process itself. From this perspective, the information system is internal to and plays an integral role in the process, whether providing input, aiding product transformation, or producing output. Consider a phone directory business that creates phone books for international corporations. A corporate customer requests a phone directory listing all steel suppliers in Western Europe. Using its information system, the directory business can sort files to find the suppliers' names and phone numbers and organize them into an alphabetical list. The information system itself is an integral part of this process. It does not just monitor the process externally but works as part of the process to transform raw data into a product. In this example, the information system turns raw data input (names and phone numbers) into a salable output (a phone directory). The same system might also provide the input (data files) and output (printed pages for the directory).

The latter view brings with it a new perspective on how and why information systems can be used in business. Rather than searching to understand information systems independently of the organization, we consider the potential role of information systems within the process itself, often leading to the discovery of new and better ways to accomplish the process.

Organizational Structure

Organizational structure refers to organizational subunits and the way they relate to the overall organization. Depending on the goals of the organization and its approach to management, a number of structures can be used. An organization's structure can affect how information systems are viewed and used. Although there are many possibilities, organizational structure typically falls into one of these categories: traditional, project, team, multidimensional, or virtual.

Traditional Organizational Structure

In the type of structure known as **traditional organizational structure**, also called a *hierarchical structure*, a managerial pyramid shows the hierarchy of decision making and authority from the strategic management to operational management and nonmanagement employees. The strategic level, including the president of the company and vice presidents, has a higher degree of decision authority, more impact on corporate goals, and more unique and one-of-a-kind problems to solve (see Figure 2.3). In most cases, major department heads report to a president or top-level manager. The major departments are usually divided according to function and can include marketing, production, information systems, finance and accounting, research and development, and so on (see Figure 2.4). The positions or departments that are directly associated with making, packing, or shipping goods are called *line positions*. A

organizational structure
Organizational subunits and the way they relate to the overall organization.

traditional organizational structure
Organizational structure in which major department heads report to a president or top-level manager.

production supervisor who reports to a vice president of production is an example of a line position. Other positions may not be directly involved with the formal chain of command but may assist a department or area. These are *staff positions*, such as a legal counsel reporting to the president.

Figure 2.3

A simplified model of the organization, showing the managerial pyramid from top-level managers to nonmanagement employees.

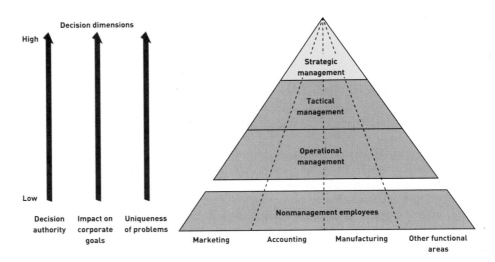

Figure 2.4

A Traditional Organizational Structure

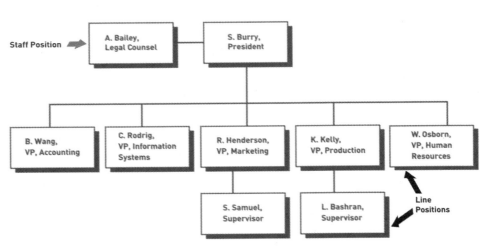

Today, the trend is to reduce the number of management levels, or layers, in the traditional organizational structure. A structure with a reduced number of management layers, often called a **flat organizational structure**, empowers employees at lower levels to make decisions and solve problems without needing permission from midlevel managers. **Empowerment** gives employees and their managers more responsibility and authority to make decisions, take certain actions, and in general have more control over their jobs. For example, an empowered salesclerk would be able to respond to certain customer requests or problems without needing permission from a supervisor. On a factory floor, empowerment can mean that an assembly-line worker has the ability to stop the production line to correct a problem or defect before the product is passed to the next station.

Information systems can be a key element in empowering employees. Often, information systems make empowerment possible by providing information directly to employees at lower levels of the hierarchy. The employees may also be empowered to develop or use their own personal information systems, such as a simple forecasting model or spreadsheet. Office Depot, for example, developed a new computer system to empower 60,000 employees in about 1,000 stores. The system used an existing database and a powerful reporting system to give employees better information and increased decision-making capabilities.[5]

flat organizational structure
Organizational structure with a reduced number of management layers.

empowerment
Giving employees and their managers more responsibility and authority to make decisions, take certain actions, and have more control over their jobs.

Project Organizational Structure

A **project organizational structure** is centered around major products or services. For example, in a manufacturing firm that produces baby food and other baby products, each line is produced by a separate unit. Traditional functions such as marketing, finance, and production are positioned within these major units (see Figure 2.5). Many project teams are temporary—when the project is complete, the members go on to new teams formed for another project.

project organizational structure
Structure centered on major products or services.

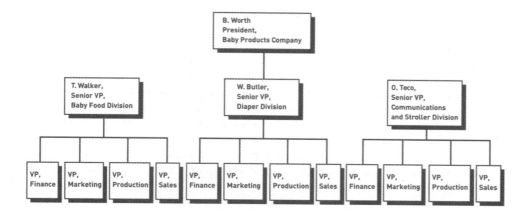

Figure 2.5

A Project Organizational Structure

Team Organizational Structure

The **team organizational structure** is centered on work teams or groups. In some cases, these teams are small; in others, they are very large. Typically, each team has a leader who reports to an upper-level manager in the organization. Depending on the tasks being performed, the team can be either temporary or permanent. AmeriHealth, for example, used small teams to obtain big returns.[6] The company, which specializes in providing healthcare to the poor and disadvantaged, used a team structure to organize its administrators, physicians, and others associated with the healthcare provider. According to the company president, "No one, and I mean no one, can do it on his own. The key is to find people whose goals and vision are similar to yours, but whose backgrounds are in different areas. It is the only way to grow and learn how to do your own job better."

team organizational structure
Structure centered on work teams or groups.

Multidimensional Organizational Structure

A **multidimensional organizational structure**, also called a matrix organizational structure, may incorporate several structures at the same time. For example, an organization might have both traditional functional areas and major project units. When diagrammed, this structure forms a matrix, or grid (see Figure 2.6).

A potential disadvantage is multiple lines of authority. Employees have two bosses or supervisors: one functional boss and one project boss. As a result, conflicts may occur when one boss wants one thing and the other boss wants something else. For example, the

multidimensional organizational structure
Structure that may incorporate several structures at the same time.

Figure 2.6

A Multidimensional Organizational Structure

Employees in each group may have two bosses—a project boss and a functional boss.

functional boss might want the employee to work on a new product in the next two days, while the project boss might want the employee to fly to a two-day meeting. Obviously, the employee cannot do both. One way to resolve this problem is to give one boss priority if there are problems or conflicts.

Virtual Organizational Structure and Collaborative Work

virtual organizational structure

Structure that employs individuals, groups, or complete business units in geographically dispersed areas.

A **virtual organizational structure** employs individuals, groups, or complete business units in geographically dispersed areas. These people may be in different countries, operating in different time zones. They may never meet face to face in the same room, which explains the use of the word virtual. Despite this separation, they can collaborate on any aspect of a project, such as supplying raw materials, producing goods and services, and delivering goods and services to the marketplace. In some cases, a virtual organization is temporary, lasting only a few weeks or months. In others, it can last for years or decades.

A virtual organizational structure can be used within a firm. A company that makes lottery equipment, for example, can use a virtual organizational structure with dispersed workers who have distinct skills and abilities to save millions of dollars. Often, however, virtual structures are formed with individuals or groups outside a company. One company, for example, formed an organization called the Virtual Corporate Management System to assemble several small manufacturing businesses to compete with larger businesses. One small business that manufactured military communications equipment saw a dramatic increase in revenues when it joined the Virtual Corporate Management System.

In addition to reducing costs or increasing revenues, a virtual organizational structure can provide an extra level of security. Many companies are now dispersing employees and using a virtual structure in case of a terrorist attack or a disaster. If a disaster strikes at the primary location, the company still has sufficient employees at other locations to keep the company running. Procter & Gamble, a pioneer in virtual work structure, now defines the workplace as "anywhere someone is trying to be productive, whether it's a P&G location or not."[7] Today's workers are getting company work done at home, at a customer's location, in coffee shops, on pleasure boats, and at convenient work centers in suburbia. It also allows people to work at any time. Using the Internet and e-mail, workers can put the finishing touches on a new business proposal in Europe or Asia, while coworkers in North America are sleeping.

There are a number of keys to successful virtual organization structures. One strategy is to have in-house employees concentrate on the firm's core businesses and use virtual employees, groups, or businesses to do everything else. Using information systems to coordinate the activities of a virtual structure is essential. Even with sophisticated IS tools, face-to-face meetings are usually needed, especially at the beginning of new projects. One virtual team had a problem when a Russian member of the team thought her coworkers said her work was "awful."[8] Her team members actually said her work was "awesome," but the word was not translated correctly. The incident resulted in negative behavior, in which coworkers insulted each other by sending nasty e-mails, called *flaming*.

A virtual organizational structure allows *collaborative work*, where managers and employees can effectively work in groups around the world. A management team, for example, can include executives from Australia and England. A programming team can consist of people in the United States and India. Collaborative work can also include all aspects of the supply chain and customer relationship management. An automotive design team, for example, can include critical parts suppliers, engineers from the company, and important customers.

Organizational Culture and Change

culture

Set of major understandings and assumptions shared by a group.

organizational culture

The major understandings and assumptions for a business, a corporation, or an organization.

Culture is a set of major understandings and assumptions shared by a group, for example, within an ethnic group or a country. **Organizational culture** consists of the major understandings and assumptions for a business, a corporation, or an organization. The understandings, which can include common beliefs, values, and approaches to decision making, are often not stated or documented as goals or formal policies. Employees, for example, might be expected to be clean-cut, wear conservative outfits, and be courteous in dealing with all customers. Sometimes organizational culture is formed over years. In other cases, it can be

formed rapidly by top-level managers—for example, implementation of a "casual Friday" dress policy.

Like organizational structure, organizational culture can significantly affect the development and operation of information systems within an organization. A procedure associated with a newly designed information system, for example, might conflict with an informal procedural rule that is part of organizational culture. Organizational culture might also influence a decision maker's perception of the factors and priorities that must be considered in setting objectives. For example, there might be an unwritten understanding that all inventory reports must be prepared before ten o'clock Friday morning. Because of this understood time deadline, the decision maker may reject a cost-reduction option that requires compiling the inventory report over the weekend.

Organizational change deals with how for-profit and nonprofit organizations plan for, implement, and handle change. Change can be caused by internal or external factors. Internal factors include activities initiated by employees at all levels. External factors include activities wrought by competitors, stockholders, federal and state laws, community regulations, natural occurrences (such as hurricanes), and general economic conditions. Many European countries, for example, adopted the euro, a single European currency, which changed how financial companies do business and how they use their information systems.

organizational change
The responses that are necessary for for-profit and nonprofit organizations to plan for, implement, and handle change.

Cingular's planned acquisition of AT&T Wireless will combine the strengths of the two companies and is expected to create customer benefits and growth prospects neither company could have achieved on its own. Together, the companies can provide better coverage, improve reliability, enhance call quality, and offer a wide array of new and innovative services for consumers.

(Source: AP/Wide World Photos.)

Change can be sustaining or disruptive.[9] *Sustaining change* can help an organization improve raw materials supply, the production process, and the products and services offered by the organization. New manufacturing equipment to make disk drives is an example of a sustaining change. The new equipment might reduce the costs of producing the disk drives and improve overall performance. *Disruptive change*, on the other hand, often harms an organization's performance or even puts it out of business. The 3.5-inch hard-disk drive was

a disruptive technology for companies that produced the 5.25-inch hard-disk drive. When it was first introduced, the 3.5-inch drive was slower and had lower capacity and lower demand than the existing 5.25-inch disk drives. Over time, however, the 3.5-inch drive improved and replaced the 5.25-inch drive in performance and demand. Some companies that produced the 5.25-inch drives that didn't change didn't survive. Today, many of the 5.25-inch drive companies are out of business. In general, disruptive technologies may not originally have good performance, low cost, or even strong demand. Over time, however, they often replace existing technologies. They can cause good, stable companies to fail when they don't change or adopt the new technology.

Overcoming resistance to change, especially disruptive change, can be the hardest part of bringing information systems into a business. Occasionally, employees even attempt to sabotage a new information system because they do not want to learn the new procedures and commands. In most of these instances, the employees were not involved in the decision to implement the change, nor were they fully informed about the reasons the change was occurring and the benefits that would accrue to the organization.

The dynamics of change can be viewed in terms of a change model. A **change model** is a representation of change theories that identifies the phases of change and the best way to implement them. Kurt Lewin and Edgar Schein propose a three-stage approach for change (see Figure 2.7). *Unfreezing* is the process of ceasing old habits and creating a climate receptive to change. *Moving* is the process of learning new work methods, behaviors, and systems. *Refreezing* involves reinforcing changes to make the new process second nature, accepted, and part of the job.[10] When a company introduces a new information system, a few members of the organization must become agents of change—champions of the new system and its benefits. Understanding the dynamics of change can help them confront and overcome resistance so that the new system can be used to maximum efficiency and effectiveness.

Organizational learning is closely related to organizational change. According to the concept of **organizational learning**, organizations adapt to new conditions or alter their practices over time. So, assembly-line workers, secretaries, clerks, managers, and executives learn better ways of doing business and incorporate them into their day-to-day activities. Collectively, these adjustments based on experience and ideas are called *organizational learning*. In some cases, the adjustments can be a radical redesign of business processes, often called *reengineering*. In other cases, these adjustments can be more incremental, a concept called *continuous improvement*.

change model
Representation of change theories that identifies the phases of change and the best way to implement them.

organizational learning
Adaptations to new conditions or alterations of organizational practices over time.

Figure 2.7

A Change Model

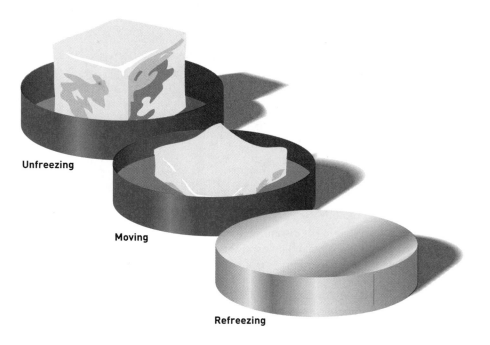

Unfreezing

Moving

Refreezing

Reengineering

To stay competitive, organizations must occasionally make fundamental changes in the way they do business. In other words, they must change the activities, tasks, or processes that they use to achieve their goals. **Reengineering**, also called **process redesign**, involves the radical redesign of business processes, organizational structures, information systems, and values of the organization to achieve a breakthrough in business results (see Figure 2.8). Reengineering can reduce delivery time, increase product and service quality, enhance customer satisfaction, and increase revenues and profitability. As a result of increased electronic trading, the Securities Industry Association is radically reengineering the communications systems used for stock and securities trading. The new communications systems will be fully implemented by 2005. The reengineering also calls for a new settlement system, called Straight-Through Processing.

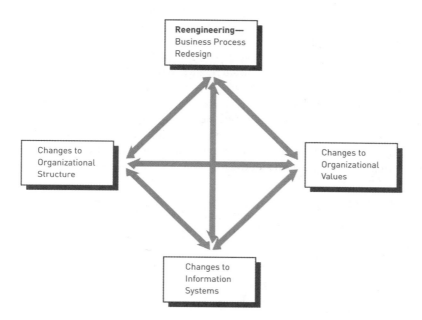

reengineering (process redesign)
The radical redesign of business processes, organizational structures, information systems, and values of the organization to achieve a breakthrough in business results.

Figure 2.8

Reengineering

Reengineering involves the radical redesign of business processes, organizational structure, information systems, and values of the organization to achieve a breakthrough in business results.

With the increased volume of securities trading in the global marketplace, the Securities Industry Association has released a new model to reengineer the process for settling and clearing transactions that reduces costs and saves time.

(Source: AP/Wide World Photos.)

A business process includes all activities, both internal (such as thinking) and external (such as taking action), that are performed to transform inputs into outputs. It defines the way work gets done. A few companies still process a customer order manually using several different people. The order moves from one step to the next, allowing people to make numerous errors and create misunderstandings. Today, most companies have computerized

customer ordering, saving money and reducing possible errors. This simple example illustrates the fundamental changes reengineering creates, often across multiple departments. But asking people to work differently often meets with stiff resistance, and change is difficult to maintain—the values of the organization and its employees must be changed also. In the previous example of order processing, the original work process may have evaluated employees on how many orders were entered each day. Under the reengineered process, they may be evaluated on different factors associated with customer service—percentage of orders delivered on time or accuracy of customer bills. Helping employees understand the benefits of the new system is a major hurdle.

In contrast to simply automating the existing work process, reengineering challenges the fundamental assumptions governing their design. It requires finding and vigorously challenging old rules blocking major business process changes. These rules are like anchors weighing a firm down and keeping it from competing effectively. Examples of such rules are given in Table 2.1. The Tennessee Valley Authority (TVA), for example, embarked on its largest and most-aggressive IS reengineering project in history to improve efficiency and reduce power costs. The efficiency from the new system allowed TVA to purchase in volume and save the company $23.5 million. According to the senior vice president of information systems, "From the sheer magnitude, it was the largest undertaking of the agency's history."[11] Other companies and organizations are also using reengineering to improve processes and reduce errors. Tiny radio-frequency microchips are being placed onto products to track their location and reduce errors.[12] The Johns Hopkins hospital in Baltimore, for example, will be one of the first hospitals to use radio-frequency chips on its drugs. In addition to cutting costs and reducing errors, the chips should also curb drug counterfeiting. Today, many companies use reengineering to increase their competitive position in the marketplace. Reengineering, however, can be disruptive, expensive, and time-consuming to implement.[13]

Table 2.1

Selected Business Rules That Affect Business Processes

Rule	Original Rationale	Potential Problem
Small orders must be held until full-truckload shipments can be assembled.	Reduce delivery costs.	Customer delivery is slowed—lost sales.
No order can be accepted until customer credit is approved.	Reduce potential for bad debt.	Customer service is poor—lost sales.
All merchandising decisions are made at headquarters.	Reduce number of items carried in inventory.	Customers perceive organization has limited product selection—lost sales.

Continuous Improvement

continuous improvement
Constantly seeking ways to improve the business processes to add value to products and services.

The idea of **continuous improvement** is to constantly seek ways to improve the business processes to add value to products and services. This continual change, in turn, will increase customer satisfaction and loyalty and ensure long-term profitability. Manufacturing companies make continual product changes and improvements. Service organizations regularly find ways to provide faster and more effective assistance to customers. By doing so, these companies increase customer loyalty, minimize the chance of customer dissatisfaction, and diminish the opportunity for competitive inroads. ArvinMeritor, Inc., for example, has a vice president of continuous improvement and quality to constantly improve the quality of its products and services.[14] The company is a large, $7-billion global supplier of automotive parts and supplies.

Organizational commitment to goals such as continuous improvement can be supported by the strategic use of information systems. Continuous improvement involves constantly improving and modifying products and services to remain competitive and to keep a strong customer base. Table 2.2 compares reengineering and continuous improvement.

Business Process Reengineering	Continuous Improvement
Strong action taken to solve serious problem	Routine action taken to make minor improvements
Top-down driven by senior executives	Worker driven
Broad in scope; cuts across departments	Narrow in scope; focus is on tasks in a given area
Goal is to achieve a major breakthrough	Goal is continuous, gradual improvements
Often led by outsiders	Usually led by workers close to the business
Information system integral to the solution	Information systems provide data to guide improvement team

Technology Diffusion, Infusion, and Acceptance

To be effective, reengineering and continuous improvement efforts must be accepted and used throughout an organization. Even if a company buys or develops new computerized systems, managers and employees may never use them. Or new systems may not be used to their potential. Millions of dollars can be wasted as a result. The extent to which new computerized systems are used throughout an organization can be measured by the amount of technology diffusion, infusion, and acceptance. **Technology diffusion** is a measure of how widely technology is spread throughout an organization. An organization in which computers and information systems are located in most departments and areas has a high level of technology diffusion.[15] Some online merchants, such as Amazon.com, have a high level of diffusion and use computer systems to perform most of their business functions, including marketing, purchasing, and billing. **Technology infusion**, on the other hand, is the extent to which technology permeates an area or department. In other words, it is a measure of how deeply imbedded technology is in an area of the organization. Some architectural firms, for example, use computers in all aspects of designing a building or structure. This design area thus has a high level of infusion. Of course, it is possible for a firm to have a high level of infusion in one aspect of its operations and a low level of diffusion overall. The architectural firm may use computers in all aspects of design (high infusion in the design area) but may not use computers to perform other business functions, including billing, purchasing, and marketing (low diffusion).

Although an organization may have a high level of diffusion and infusion, with computers throughout the organization, this does not necessarily mean that information systems are being used to their full potential. In fact, the assimilation and use of expensive computer technology throughout organizations varies greatly.[16] One reason is a low degree of acceptance and use of the technology among some managers and employees. Research has attempted to explain the important factors that enhance or hinder the acceptance and use of information systems.[17] A number of possible explanations of technology acceptance and usage have been studied. The **technology acceptance model (TAM)** specifies the factors that can lead to higher acceptance and usage of technology in an organization, including the perceived usefulness of the technology, the ease of its use, the quality of the information system, and the degree to which the organization supports the use of the information system.[18] Companies hope that a high level of diffusion, infusion, and acceptance will lead to greater performance and profitability.[19]

Total Quality Management

The definition of the term *quality* has evolved over the years. In the early years of quality control, firms were concerned with meeting design specifications—that is, conformance to standards. If a product performed as designed, it was considered a high-quality product. A product can perform its intended function, however, and still not satisfy customer needs. Today, **quality** means the ability of a product (including services) to meet or exceed customer expectations. For example, a computer that not only performs well but is easy to maintain and repair would be considered a high-quality product. Increasingly, customers expect good support after the sale. As a result, high-quality help desks are becoming an important strategic

Table 2.2

Comparing Business Process Reengineering and Continuous Improvement

technology diffusion
A measure of how widely technology is spread throughout the organization.

technology infusion
The extent to which technology is deeply integrated into an area or department.

technology acceptance model (TAM)
A model that describes the factors that lead to higher levels of acceptance and usage of technology.

quality
The ability of a product (including services) to meet or exceed customer expectations.

component for many companies. According to the help-desk solutions manager for the BT Group, a London-based company, "It's not so much the money but how you spend it, how you maximize your support while not overstretching your resources."[20] This view of quality is completely customer oriented. A high-quality product will satisfy customers by functioning correctly and reliably, meeting needs and expectations, and being delivered on time with courtesy and respect.

Increasingly, quality is often achieved through reduced errors. Cardinal Health initiated a $100-million redesign of its computer system to reduce errors. The existing system was 99 percent accurate, but that still meant 10,000 errors per million shipments. The new system dramatically reduced this already low error rate.[21]

To help them deliver high-quality goods and services, some companies have adopted continuous improvement strategies that require each major business process to follow a set of total quality management and six-sigma guidelines. **Total quality management** (TQM) consists of a collection of approaches, tools, and techniques that fosters a commitment to quality throughout the organization. TQM involves developing a keen awareness of customer needs, adopting a strategic vision for quality, empowering employees, and rewarding employees and managers for producing high-quality products. As a result, processes may be redefined and restructured. The U.S. Postal Service uses the Mail Preparation Total Quality Management (MPTQM) program to certify leading mail sorting and servicing companies.[22]

Another quality tool is *six sigma*, a statistical term that means products and services will meet quality standards 99.9997 percent of the time.[23] In a normal distribution curve used in statistics, six standard deviations (six sigma) is 99.9997 percent of the area under the curve. According to a report by the National Institute of Medicine, drug errors occur about 7 percent of the time. Today, many hospitals and healthcare facilities strive to achieve six-sigma quality to reduce errors and improve patient care. Other industries also use six sigma. Checkpoint Systems, a manufacturing company with headquarters in Puerto Rico, used six sigma to improve product quality by 50 percent and reduce scrap by 30 percent.[24]

Information systems are fully integrated into business processes in organizations that adhere to TQM and six-sigma strategies. Capturing and analyzing customer feedback and expectations and designing, manufacturing, and delivering high-quality products and services to customers around the world are only a few ways computers and information systems are helping companies pursue their goals of quality.

Outsourcing, On-Demand Computing, and Downsizing

A significant portion of an organization's expenses are used to hire, train, and compensate talented staff. So organizations today are trying to control costs by determining the number of employees they need on the payroll to maintain high-quality goods and services. With fierce competition in the marketplace, it is critical for organizations to use their resources wisely. Strategies to contain costs are outsourcing, on-demand computing, and downsizing (sometimes called *rightsizing*).

Outsourcing involves contracting with outside professional services to meet specific business needs. Often, a specific business process is outsourced, such as employee recruiting and hiring, development of advertising materials, product sales promotion, or global telecommunications network support. One reason that organizations outsource a process is to enable them to focus more closely on their core business—and target limited resources to meet strategic goals.

Other reasons for outsourcing are to obtain cost savings or to benefit from the expertise of a service provider. A computer company, for example, can outsource the manufacturing of its personal computers and save hundreds of millions of dollars while its in-house staff concentrates on developing new designs and features. Procter & Gamble saved about $1 billion by transferring some work from the United States to lower-wage countries such as Costa Rica and the Philippines.[25] Companies that are considering outsourcing to cut the cost of their IS operations need to review this decision carefully, however. A growing number of organizations are finding that outsourcing does not necessarily lead to reduced costs. One of the primary reasons for cost increases is poorly written contracts that tack on charges from the outsourcing vendor for each additional task. The "Ethical and Societal Issues" feature discusses the trend toward offshore outsourcing in more depth.

total quality management (TQM)
A collection of approaches, tools, and techniques that fosters a commitment to quality throughout the organization.

outsourcing
Contracting with outside professional services to meet specific business needs.

The U.S. Offshore Outsourcing Dilemma

The U.S. government has worked to create a democratic world and to empower Third-World countries through industrialization and capitalism. The result has been an increasingly global economy in which companies from around the world compete in one huge marketplace. Since developing nations typically have lower standards of living, and thus lower wages, they are able to offer considerably lower prices for their goods and services. Consider China, which prides itself on being a leader in manufacturing; it is able to produce large quantities of products more quickly and inexpensively than any other nation.

This trend has now expanded beyond manufacturing to service industries. Countries such as India, Bulgaria, Russia, and Romania are able to provide many high-level services—call center operations (telephone customer support), software development, engineering, software design, and research and development—at significantly lower rates than U.S. companies and those in other more-developed countries can. One study found that U.S. companies are charging $80 to $120 per hour for programming work, while the same work can be done in India for about $40.

With the highly competitive state of the U.S. technology market and the continuous pressure to reduce costs, U.S. companies are turning to foreign providers to compete and survive. Companies outside the United States are often referred to as *offshore companies*, and immediate U.S. neighbors are called *near offshore*. Contracting offshore companies for a portion of operations is called offshore outsourcing. Companies such as Connect Offshore are emerging to assist businesses in locating companies that provide offshore staffing, help establish offshore offices, and find offshore project management. For a flat rate of $1,750 per month, Connect Offshore promises to save companies as much as 60 percent in staffing costs.

Recently, IBM announced that it will move as many as 4,730 U.S. programming jobs to India, China, and elsewhere. It is not alone. Corio, Inc., a leading enterprise application service provider, maintains 20 percent of its workforce in Bangalore, India. Industry watcher Forrester Research has described this trend as an exodus and predicts that 3.3 million U.S. service industry jobs over the next 15 years will move offshore. One U.S. Senator estimates that in the United States alone, the value of information technology (IT) services provided by offshore labor will double to $16 billion in 2005 and triple to $46 billion by 2007. The Gartner Research Group predicts that by the end of 2005, 1 out of every 20 corporate IT jobs in the United States will be moved offshore, along with 10 percent of the positions at U.S.-based IT vendors and technology service firms. Also, estimates project that only 40 percent of the IT workers whose jobs are lost will be shifted to other responsibilities; the rest will be out of work.

With mounting evidence of the negative impact of offshore outsourcing on U.S. jobs and such dire predictions, outsourcing has turned into a hot political issue—one that balances free trade against job protection laws and taxes. Several bills have been proposed at both the state and federal levels to protect U.S. technology jobs. Senators Craig Thomas of Wyoming and George Voinovich of Ohio proposed a bill that would prohibit government contractors from performing work outside the United States. In Indiana, a bill was proposed to require government contracts to be awarded only to U.S. citizens or people authorized to work in the United States. Proposed legislation in North Carolina would require a call center operator to disclose his or her location upon request.

Some worried IT managers, concerned about their company's increased relations with offshore companies and their own job security, have worked to thwart the outsourced projects. Some have even sabotaged their offshore partners' work. With so much at stake, it's easy to sympathize with these individuals.

Critical Thinking Questions

1. What effect do you think an organization's offshore outsourcing has on the morale of its employees?
2. Do you agree with some politicians that we need laws to limit offshore outsourcing?

What Would You Do?

You're the second in command of the IS group for a midsized company, Tempo Industries. Your boss has just informed you that the company is considering outsourcing 50 percent of its systems development work to a company in Romania that can provide high-quality services at half the cost of your current workforce. You must prepare a report that lists which systems development projects in your organization should be kept at home and which should be outsourced. You are also asked to include a list of employees who will no longer be needed and the total savings to the company, if it adopts this plan. In addition, you are asked to provide any alternative plans for reducing development costs by at least 25 percent.

3. What should you consider when deciding which systems to outsource and which people to lay off?
4. What arguments and alternatives could you present that might dissuade the company from the move to outsource? Is offshore outsourcing an inevitable necessity?

SOURCES: Declan McCullagh, "Offshoring and the 2004 Elections," *C/Net News*, January 12, 2004, *www.cnet.com*; Patrick Thibodeau, "Offshore Outsourcing Is Relentless," *Computerworld*, June 27, 2003, *www.computerworld.com*; Patrick Thibodeau, "State, U.S. Lawmakers Pushing to Hinder Offshore Outsourcing," *Computerworld*, December 15, 2003, *www.computerworld.com*; Connect Offshore Web site, *www.connectoffshore.com*, accessed January 24, 2004; Patrick Thibodeau, "Internal Resistance Can Doom Offshore Projects," *Computerworld*, January 23, 2004, *www.computerworld.com*.

on-demand computing
(on-demand business, utility computing) Contracting for computer resources to rapidly respond to an organization's varying workflow.

On-demand computing is an extension of the outsourcing approach, and many companies offer on-demand computing to business clients and customers. **On-demand computing,** also called *on-demand business* and *utility computing*, involves rapidly responding to the organization's flow of work as the need for computer resources varies. It is often called utility computing because the organization pays for computing resources from a computer or consulting company just as it pays for electricity from a utility company. This approach treats the computer-based information system—including hardware, software, databases, telecommunications, personnel, and other components—more as a service than as separate products. In other words, instead of purchasing hardware, software, and database systems, the organization only pays for the systems it needs at peak times, normally by paying a fee to another company. The approach can be less expensive because the organization does not pay for systems that it doesn't routinely need. It also allows the organization's IS staff to concentrate on other, more-strategic issues.

A number of companies have benefited from on-demand computing. Goodyear, for example, hired a computer company to develop and run an on-demand computing system to allow its dealers to view its catalogue and place product orders on the Internet.[26] Goodyear, a $14.5-billion tire and rubber company, has about 5,000 dealers around the world. The new Web-based system, called XPLOR, provided service to its dealers while cutting costs. Saks, the large retail company, also used on-demand computing to pay for a procurement system to save on supplies, such as tissue paper, janitorial supplies, and shopping bags, only when the procurement system was needed. By using on-demand computing, Saks paid for the procurement of only the services it actually used as demand for supplies fluctuated. Saks estimates that it saved about $500,000 on light bulbs alone using on-demand computing.

downsizing
Reducing the number of employees to cut costs.

Downsizing involves reducing the number of employees to cut costs. The term *rightsizing* is also used. Rather than pick a specific business process to be downsized, companies usually look to downsize across the entire company. Downsizing clearly drives down total payroll costs. Employee morale, however, can suffer. One study reported that a 3 percent reduction of a company's workforce can result in a 9 percent increase in its stock price.[27] A 10 percent reduction of a company's workforce, however, can reduce a company's stock price. A workforce cut of more than 10 percent can result in the company's stock price falling up to 40 percent.

Employers need to be open to alternatives for reducing the number of employees but use layoffs as the last resort. It's much simpler to encourage people to leave voluntarily through early retirement or other incentives. Voluntary downsizing programs include a "buyout package" offered to certain classes of employees (e.g., those over 50 years old). The buyout package offers employees certain benefits and cash incentives if they voluntarily retire from the company. Other options are job sharing and transfers.

Organizations in a Global Society

Increasingly, organizations operate in a global society. As companies rely on virtual structures and outsourcing to a greater extent, businesses can operate around the world. An American company, for example, can get inputs for products and services from Europe, assemble them in Asia, and ship them to customers in Australia and New Zealand. After-the-sale support can be given by a call center in India. The Internet and telecommunications make this trend possible.

However, there are many challenges to operating in a global society. Some countries have seen high-paying jobs transferred to other countries, where labor and production costs are lower. In addition, every country has a set of customs, cultures, standards, politics, and laws that can make it difficult for businesses operating around the world. Language can also be a potential problem. Some companies that outsourced their call centers to foreign countries are now moving them back because of customer complaints. It can also be more difficult to manage and control operations in different countries. Many of today's organizations operate globally to give them a competitive advantage, discussed next.

COMPETITIVE ADVANTAGE

A **competitive advantage** is a significant and (ideally) long-term benefit to a company over its competition. Establishing and maintaining a competitive advantage is complex, but a company's survival and prosperity depend on its success in doing so. In his book *Good to Great*, Jim Collins outlined how technology can be used to accelerate companies from good to great.[28] Table 2.3 shows how a few companies accomplished this move. Ultimately, it is not how much a company spends on information systems but how investments in technology are made and managed. Companies can spend less and get more value.

competitive advantage
A significant and (ideally) long-term benefit to a company over its competition.

Company	Business	Competitive Use of Information Systems
Circuit City	Consumer electronics	Developed sophisticated sales and inventory-control systems to deliver a consistent experience to customers
Gillette	Shaving products	Developed advanced computerized manufacturing systems to produce high-quality products at low cost
Walgreens	Drug and convenience stores	Developed satellite communications systems to link local stores to centralized computer systems
Wells Fargo	Financial services	Developed 24-hour banking, ATMs, investments, and increased customer service using information systems

Table 2.3

How Some Companies Used Technologies to Move from Good to Great

(Source: Data from Jim Collins, *Good to Great*, Harper Collins Books, 2001, p. 300.)

Factors That Lead Firms to Seek Competitive Advantage

A number of factors can lead to the attainment of competitive advantage. Michael Porter, a prominent management theorist, suggested a now widely accepted competitive forces model, also called the **five-forces model**. The five forces include (1) rivalry among existing competitors, (2) the threat of new entrants, (3) the threat of substitute products and services, (4) the bargaining power of buyers, and (5) the bargaining power of suppliers. The more these forces combine in any instance, the more likely firms will seek competitive advantage and the more dramatic the results of such an advantage will be.

Rivalry Among Existing Competitors

The rivalry among existing competitors is an important factor that leads firms to seek competitive advantage. Typically, highly competitive industries are characterized by high fixed costs of entering or leaving the industry, low degrees of product differentiation, and many competitors. Although all firms are rivals with their competitors, industries with stronger rivalries tend to have more firms seeking competitive advantage. To compete with existing competitors, companies are constantly analyzing how their resources and assets are used. The *resource-based view* is an approach to acquiring and controlling assets or resources that can help the company achieve a competitive advantage.[29] Using the resource-based view, for example, a transportation company might decide to invest in radio-frequency technology to tag and trace products as they move from one location to another.

five-forces model
A widely accepted model that identifies five key factors that can lead to attainment of competitive advantage including (1) rivalry among existing competitors, (2) the threat of new entrants, (3) the threat of substitute products and services, (4) the bargaining power of buyers, and (5) the bargaining power of suppliers.

Threat of New Entrants

The threat of new entrants is another important force leading an organization to seek competitive advantage. A threat exists when entry and exit costs to the industry are low and the technology needed to start and maintain the business is commonly available. For example, consider a small restaurant. The owner does not require millions of dollars to start the business, food costs do not go down substantially for large volumes, and food processing and preparation equipment is commonly available. When the threat of new market entrants is high, the desire to seek and maintain competitive advantage to dissuade new market entrants is usually high.

In the restaurant industry, competition is fierce because entry costs are low. Therefore, a small restaurant that enters the market can be a threat to existing restaurants.

(Source: © Owen Franken/CORBIS.)

Threat of Substitute Products and Services

The more consumers are able to obtain similar products and services that satisfy their needs, the more likely firms are to try to establish competitive advantage. For example, consider the photographic industry. When digital cameras started to become more popular, traditional film companies had to respond to stay competitive and profitable. Traditional film companies, such as Kodak and others, started to offer additional products and enhanced services, including digital cameras, the ability to produce digital images from traditional film cameras, and Web sites that could be used to store and view pictures.

Bargaining Power of Customers and Suppliers

Large buyers tend to exert significant influence on a firm. This influence can be diminished if the buyers are unable to use the threat of going elsewhere. Suppliers can help an organization obtain a competitive advantage. In some cases, suppliers have entered into strategic alliances with firms. When they do so, suppliers act as a part of the company. Suppliers and companies can use telecommunications to link their computers and personnel to obtain fast reaction times and get the parts or supplies when they are needed to satisfy customers. Government agencies are also using strategic alliances.[30] The investigative units of the U.S. Customs and Immigration and Naturalization Service entered into a strategic alliance to streamline operations and to place all investigative operations into a single department.

Strategic Planning for Competitive Advantage

To be competitive, a company must be fast, nimble, flexible, innovative, productive, economical, and customer oriented.[31] It must also align its IS strategy with general business strategies and objectives.[32] Given the five market forces just mentioned, Porter proposed three general strategies to attain competitive advantage: altering the industry structure, creating new products and services, and improving existing product lines and services. Subsequent research into the use of information systems to help an organization achieve a competitive advantage has confirmed and extended Porter's original work to include additional strategies, such as forming alliances with other companies, developing a niche market, maintaining competitive cost, and creating product differentiation.[33]

Altering the Industry Structure

Altering the industry structure is the process of changing the industry to become more favorable to the company or organization. The introduction of low-fare airline carriers, such as Southwest Airlines, has forever changed the airline industry, making it difficult for traditional airline companies to make high profit margins. To fight back, airline companies like Delta are launching their own low-fare flights.[34] Delta claims that its Song airline will be one of the first "all-digital" airlines. The approach will include a host of services on the flights, including entertainment, satellite TV, and many similar services.

Obtaining a competitive advantage can also be accomplished by gaining more power over suppliers and customers. Some automobile manufacturers, for example, insist that their suppliers be located close to major plants and manufacturing facilities and that all business transactions be accomplished using electronic data interchange (EDI, which is direct computer-to-computer communications with minimal human effort). This system helps the automobile company control the cost, quality, and supply of parts and materials.

A company can also attempt to create barriers to new companies entering the industry. An established organization that acquires expensive new technology to provide better products and services can discourage new companies from getting into the marketplace. Creating strategic alliances may also have this effect. A **strategic alliance**, also called a **strategic partnership**, is an agreement between two or more companies that involves the joint production and distribution of goods and services. Samsung Electronics and Echelon Corporation, for example, signed a strategic alliance agreement to develop and market electronic devices that can be connected to each other and the Internet.[35] The alliance is oriented toward the home networking market.

strategic alliance (strategic partnership)

An agreement between two or more companies that involves the joint production and distribution of goods and services.

Creating New Products and Services

Creating new products and services is always an approach that can help a firm gain a competitive advantage, and it is especially true of the computer industry and other high-tech businesses. If an organization does not introduce new products and services every few months, the company can quickly stagnate, lose market share, and decline. Companies that stay on top are constantly developing new products and services. A large U.S. credit-reporting agency, for example, can use its information system to help it explore new products and services in different markets. Delta Airlines created a new service by installing hundreds of self-service kiosks to reduce customer check-in times. The new kiosks allow Delta customers to check in, get boarding passes, change seats, and sign up for standby flights or upgrades. On average, the kiosks save flyers from 5 to 15 minutes for each check-in at an airport terminal.[36]

Our fast-moving society is highly competitive. To maintain a competitive advantage, companies continually innovate and create new products. Think Outside was the first to market the Stowaway keyboard, a full-size keyboard that folds to pocket size for use with handheld computers.

(Source: Courtesy of Think Outside, Inc.)

Improving Existing Product Lines and Services

Improving existing product lines and services is another approach to staying competitive. The improvements can be either real or perceived. Manufacturers of household products are always advertising new and improved products. In some cases, the improvements are more perceived than real refinements; usually, only minor changes are made to the existing product. Many food and beverage companies are introducing "Healthy" and "Light" product lines. Some companies are now starting to put radio-frequency ID (RFID) tags on their products that identify products and track their location as the products are shipped from one location to another.[37] Customers and managers can instantly locate products as they are shipped from suppliers, to the company, to warehouses, and finally to customers. In another case, Metro, the third largest retail store in Europe, used portable computers to show shoppers where products are located in the store and to display discounted prices and any specials.[38]

Using Information Systems for Strategic Purposes

The first IS applications attempted to reduce costs and to provide more efficient processing for accounting and financial applications, such as payroll and general ledger. These systems were seen almost as a necessary evil—something to be tolerated to reduce the time and effort required to complete previously manual tasks. As organizations matured in their use of information systems, enlightened managers began to see how they could be used to improve organizational effectiveness and support the fundamental business strategy of the enterprise. Combining the improved understanding of the potential of information systems with the growth of new technology and applications has led organizations to use information systems to gain a competitive advantage. In simplest terms, competitive advantage is usually embodied either in a product or service that has the most added value for consumers and that is unavailable from the competition or in an internal system that delivers benefits to a firm not enjoyed by its competition.

Although it can be difficult to develop information systems to provide a competitive advantage, some organizations have done so with success. A classic example is SABRE, a sophisticated computerized reservation system installed by American Airlines and one of the first information systems recognized for providing competitive advantage. Travel agents used this system for rapid access to flight information, offering travelers reservations, seat assignments, and ticketing. The travel agents also achieved an efficiency benefit from the SABRE system. Because SABRE displayed American Airline flights whenever possible, it also gave the airline a long-term, significant competitive advantage. Today, SABRE is aggressively

seeking a competitive advantage by investing heavily in e-commerce technology and developing Internet travel sites. It invested more than $200 million in technology recently. Much of the investment was in the company's Travelocity.com site (the second largest online travel agency) and GetThere.com.[39] Increasingly, companies are using e-commerce as part of a strategy to achieve a competitive advantage.

Quite often, the competitive advantage a firm gains with a new information system is only temporary—competitors are quick to copy a good idea. So, although the SABRE system was the first online reservation system, other carriers soon developed similar systems. However, SABRE has maintained a leadership position in the past because it was the first system available, has been aggressively marketed, and has had continual upgrades and improvements over time. Maintaining a competitive advantage takes effort and is not guaranteed. SABRE's competitive advantage, for example, is being challenged with the many online travel sites available to today's travelers.

The extent to which companies are using computers and information technology for competitive advantage continues to grow. Many companies have even instituted a new position—chief knowledge officer—to help them maintain a competitive advantage. Forward-thinking companies must constantly update or acquire new systems to remain competitive in today's dynamic marketplace. In addition to using information systems to help a company achieve a competitive advantage internally, companies are increasingly investing in information systems to support their suppliers and customers. Investments in information systems that result in happy customers and efficient suppliers can do as much to achieve a competitive advantage as internal systems, such as payroll and billing. Table 2.4 lists several examples of how companies have attempted to gain a competitive advantage.

Table 2.4

Competitive Advantage Factors and Strategies

Factors That Lead to Attainment of a Competitive Advantage	Alter Industry Structure	Create New Products and Services	Improve Existing Product Lines and Services
Rivalry among existing competitors	Netflix changes the industry structure with its use of online ordering for DVDs.	Apple, Dell, and other PC makers develop computers that excel at downloading Internet music and playing the music on high-quality speakers.	Food and beverage companies offer "healthy" and "light" product lines.
Threat of new entrants	HP and Compaq merge to form a large Internet and media company.	Apple Computer introduces an easy-to-use iMac computer that can be used to create and edit home movies.	Starbucks offers new coffee flavors at premium prices.
Threat of substitute products and services	Ameritrade and other discount stockbrokers offer low fees and research on the Internet.	Wal-Mart uses technology to monitor inventory and product sales to determine the best mix of products and services to offer at various stores.	Cosmetic companies add sunscreen to their product lines.
Bargaining power of buyers	Ford, GM, and others require that suppliers locate near their manufacturing facilities.	Investors and traders of the Chicago Board of Trade (CBOT) put pressure on the institution to implement electronic trading.	Retail clothing stores require manufacturing companies to reduce order lead times and improve materials used in the clothing.
Bargaining power of suppliers	American Airlines develops SABRE, a comprehensive travel program used to book airline, car rental, and other reservations.	Broadcom develops a chip for wireless computing used in notebook PCs from Apple, Dell, Hewlett-Packard, and Gateway.	Hayworth, a supplier of office furniture, has a computerized-design tool that helps it design new office systems and products.

PERFORMANCE-BASED INFORMATION SYSTEMS

There have been at least three major stages in the business use of information systems. The first stage started in the 1960s and was oriented toward cost reduction and productivity. This stage generally ignored the revenue side, not looking for opportunities to increase sales via the use of information systems. The second stage started in the 1980s and was defined by Porter and others. It was oriented toward gaining a competitive advantage. In many cases, companies spent large amounts on information systems and ignored the costs. Today, we are seeing a shift from strategic management to performance-based management in many IS organizations.[40] This third stage carefully considers both strategic advantage and costs. This stage uses productivity, return on investment (ROI), net present value, and other measures of performance. Figure 2.9 illustrates these stages.

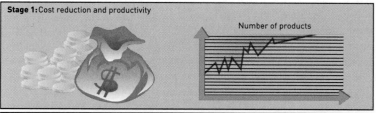
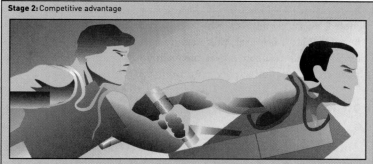
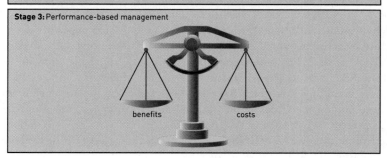

Stage 1: Cost reduction and productivity

Number of products

Stage 2: Competitive advantage

Stage 3: Performance-based management

benefits   costs

Figure 2.9

Three Stages in the Business Use of Information Systems

This balanced approach attempts to reduce costs and increase revenues. Aviall, an aviation parts company, for example, invested over $3 million in a new Web site—Aviall.com—that slashed inventory ordering costs from $9 to $.39 per order.[41] Harrah's in Las Vegas uses ROI to project and monitor the costs of new information systems.[42] According to Chuck Atwood, the chief financial officer of the large casino company, "By setting projects with specific return criteria expected, then monitoring achievement to those objectives, our IT team has built credibility within the organization."

Productivity

productivity
A measure of the output achieved divided by the input required.

Developing information systems that measure and control productivity is a key element for most organizations. **Productivity** is a measure of the output achieved divided by the input required. A higher level of output for a given level of input means greater productivity; a lower level of output for a given level of input means lower productivity. Consider a tax preparation firm, where productivity can be measured by the hours spent on preparing tax returns divided by the total hours the employee worked. For example, in a 40-hour week, an employee may have spent 30 hours preparing tax returns. The productivity is thus equal to 30/40, or 75 percent. With administrative and other duties, a productivity level of 75 may be excellent. The numbers assigned to productivity levels are not always based on labor hours—productivity may be based on factors like the amount of raw materials used, resulting quality, or time to produce the goods or service. In any case, what is important is not the value of the productivity number but how it compares with other time periods, settings, and organizations.

$$Productivity = (Output/Input) \times 100\%$$

Once a basic level of productivity is measured, an information system can monitor and compare it over time to see whether productivity is increasing. Then, corrective action can be taken if productivity drops below certain levels. In addition to measuring productivity, an information system can also be used within a process to significantly increase productivity. Thus, improved productivity can result in faster customer response, lower costs, and increased customer satisfaction. See the "Information Systems @ Work" feature for an example of how one company used information systems to streamline its business processes.

In the late 1980s and early 1990s, overall productivity did not seem to increase with increases in investments in information systems. Often called the *productivity paradox*, this situation troubled many economists who were expecting to see dramatic productivity gains.[43] In the early 2000s, however, productivity again seemed on the rise. According to IDC, a marketing research company, investments in information systems will contribute an estimated 80 percent of the productivity gains from 2002 through 2010.[44]

Return on Investment and the Value of Information Systems

return on investment (ROI)
One measure of IS value that investigates the additional profits or benefits that are generated as a percentage of the investment in IS technology.

One measure of IS value is **return on investment (ROI)**. This measure investigates the additional profits or benefits that are generated as a percentage of the investment in IS technology. A small business that generates an additional profit of $20,000 for the year as a result of an investment of $100,000 for additional computer equipment and software would have a return on investment of 20 percent ($20,000/$100,000). One study investigated the ROI for computer-related training and certification.[45] According to the study, the Microsoft Certified Solution Developer for Microsoft.NET and the Microsoft Certified Database Administrator received ROI values of 170 percent and 122 percent for large organizations. For smaller organizations, the Check Point Certified Security Administrator received an ROI value of 98 percent.

Because of the importance of ROI, many computer companies provide ROI calculators to potential customers. ROI calculators are typically found on a vendor's Web site and can be used to estimate returns.

OneBeacon Focuses on Information Systems

"It's not uncommon for insurers to have multiple policy systems that are 15 or 20 years old and don't talk to one another," says Larry Goldberg, senior vice president of Sapiens Americas, a leading insurance IS provider. OneBeacon Insurance Group was in such a predicament. Until recently, the process for policy writing took OneBeacon agents a long, arduous week. Field agents filled out application forms and faxed them to the main office. Data-entry clerks would then input the information into a system. The system would provide a quotation, which was then relayed to the agent via telephone. If the customer decided to accept the quote, he or she would have to wait one week before the coverage would be issued.

This wasteful system was not only time-consuming but also expensive. OneBeacon was losing $50 million a month due to inefficiencies. The expense of the inefficiencies was passed on to customers in the form of higher premiums. The cost of creating new policies was so high that OneBeacon all but gave up its efforts to insure small businesses; the cost of the quotes didn't justify the return.

Under the management of a new chief information officer (CIO), Mike Natan, OneBeacon reengineered its business. The restructuring was based on a new $15 million Web-based policy administration system purchased from Sapiens International. The system resides on servers at the insurance company's home office and is accessed from Web-based applications downloaded to an agent's desktop or notebook computers. Agents use standard Web forms to send customer information to company headquarters, and the system scores the request for risk and approves or disapproves the policy. Any questionable applications are flagged and forwarded for human inspection. Quotes are provided in a matter of seconds, and a policy is issued within 15 minutes. What used to take a week can now be accomplished in the time it takes to sip a cup of coffee. What's more, the cost of providing a quote was more than halved—from $15 per policy to $7.

The new system has allowed OneBeacon once again to offer insurance coverage to small commercial businesses. Between the money saved by the new system and the additional revenue coming in from small businesses, OneBeacon expects to realize a return on its investment in less than a year.

The change in information systems has led to changes in organizational structure. The company has reduced the number of data-entry clerks—those who collected faxed quotation requests and entered them into the old system. But computers aren't replacing human employees at OneBeacon. The company has been hiring hundreds of agents to keep up with the increased demand for its insurance policies. The end result is that OneBeacon has moved its workforce from its back office to the field, where they can generate more revenue for the company rather than burn it up through inefficiencies.

OneBeacon's project follows an industry trend of automating the underwriting process, says analyst Janie Bisker. "You're going to see more of these automated systems," he says. "The general idea is to reduce cost but also increase accuracy and the convenience for the broker and consumer." Where once brokers waited uncomfortably for quotations while customers lost patience, requests are now filled almost instantaneously. The uninterrupted flow of the sale allows brokers to hold a customer's attention and build momentum for increased sales.

Discussion Questions

1. How does OneBeacon's new system improve the productivity of its field agents?
2. How does the new system improve OneBeacon's standing in terms of return on investment, earnings growth, market share, and customer satisfaction?

Critical Thinking Questions

3. If you were a field agent for OneBeacon, what type of information would you need delivered to your portable computer in order to write a policy?
4. It took a considerable amount of courage for OneBeacon to agree to invest $15 million in a new system and reengineer its business. If you were CIO Mike Natan, how would you have sold the company on your idea for the new system in light of what you have learned about performance-based information systems?

SOURCES: Lucas Mearian, "Sticking to Policy at OneBeacon," *Computerworld*, July 7, 2003, *www.computerworld.com*; Matt Glynn, "OneBeacon Adds 130 Jobs, Doubles Space in Amherst," *The Buffalo News*, August 5, 2003, Business Section, p. B6; OneBeacon Web site, *www.onebeacon.com*, accessed January 23, 2004.

Ace Hardware's new Web-based inventory management system benefits customers by speeding checkout time. The new system enables real-time viewing and tracking of sales data, replacing a week-long lag time under the old system.

(Source: AP/Wide World Photos.)

total cost of ownership (TCO)
Measurement of the total cost of owning computer equipment, including desktop computers, networks, and large computers.

Earnings Growth

Another measure of IS value is the increase in profit, or earnings growth, it brings. For instance, suppose a mail-order company, after installing an order-processing system, had a total earnings growth of 15 percent compared with the previous year. Sales growth before the new ordering system was only about 8 percent annually. Assuming that nothing else affected sales, the earnings growth brought by the system, then, was 7 percent. Aviall, an aviation parts company, invested over $30 million in a new computer system to improve inventory control and earnings growth. According to the vice president of information services, "Our competitors thought we were insane. Some investors asked for my resignation." The investment in the new information system was very successful, however. Net earnings rose almost 75 percent.[46]

Market Share

Market share is the percentage of sales that one company's products or services have in relation to the total market. If installing a new online Internet catalog increases sales, it might help a company increase its market share by 20 percent.

Customer Awareness and Satisfaction

Although customer satisfaction can be difficult to quantify, about half of today's best global companies measure the performance of their information systems based on feedback from internal and external users. Some companies use surveys and questionnaires to determine whether the IS investment has increased customer awareness and satisfaction.

Total Cost of Ownership

In addition to such measures as return on investment, earnings growth, market share, and customer satisfaction, some companies are also tracking total costs. One measure, developed by the Gartner Group, is the **total cost of ownership (TCO)**. This approach breaks total costs into such areas as the cost to acquire the technology, technical support, administrative costs, and end-user operations. Other costs in TCO include retooling and training costs. TCO can be used to get a more accurate estimate of the total costs for systems that range from small PCs to large mainframe systems. Market research groups often use TCO to compare different products and services. For example, a survey of large global enterprises ranked messaging and collaboration software products using the TCO model.[47] The survey analyzed acquisition, maintenance, administration, upgrading, downtime, and training costs. In this survey, Oracle had the lowest TCO of about $65 per user for messaging and collaboration.

The preceding are only a few measures that companies have used to plan for and maximize the value of their investments in IS technology. In many cases, it is difficult to be accurate with ROI measures. For example, an increase in profits may be caused by an improved information system or other factors, such as a new marketing campaign or a competitor that was late in delivering a new product to the market. Regardless of the difficulties, organizations must attempt to evaluate the contributions that information systems make to be able to assess their progress and plan for the future. Information technology and personnel are too important to leave to chance.

Risk

In addition to the return-on-investment measures of a new or modified system discussed in Chapter 1 and this chapter, managers must also consider the risks of designing, developing, and implementing these systems. In addition to successes, there are also costly failures. Some companies, for example, have attempted to implement ERP systems and failed, costing some companies many millions of dollars. In other cases, e-commerce applications have been implemented with little success. The costs of development and implementation can be greater than the returns from the new system. The risks of designing, developing, and implementing new or modified systems will be covered in more detail in Chapters 12 and 13, which discuss systems development.

CAREERS IN INFORMATION SYSTEMS

Realizing the benefits of any information system requires competent and motivated IS personnel, and many companies offer excellent job opportunities. Numerous schools have degree programs with such titles as information systems, computer information systems, and management information systems. These programs are typically in business schools and within computer science departments. Degrees in information systems have provided high starting salaries for many students after graduation from college. In addition, students are increasingly looking at business degrees with a global or international orientation.[48]

The job market in the early twenty-first century has been tight, however. Many jobs have been lost in U.S. companies as firms merged, outsourced certain jobs overseas, or went bankrupt. This tight market triggered some extreme job-hunting strategies.[49] One person drove his car 250 miles to buy Krispy Kreme doughnuts and delivered them with his resume to a potential supervisor, who was looking for someone with drive and initiative. The supervisor liked the doughnuts and the candidate and hired him. Job experts warn, however, that extreme job-hunting tactics work only if the tactics are related to the work situation.

Many companies, such as FedEx, are joining with colleges and universities to help prepare students for careers.[50] FedEx has opened its $23 million four-story FedEx Technology Institute in Memphis, Tennessee. Jim Phillips, chairman of the institute, hopes the new facility will show "mind-blowing" technologies and applications to college students and instructors. The institute is part of the University of Memphis. Programs include the Center for Managing Emerging Technology, Center for Supply Chain Management, Center for Multimedia Arts, Center for Digital Economic and Regional Development, Center for Spatial Analysis, Center for Artificial Intelligence, Center for Life Sciences, and the Advanced Learning Center.

Today, companies are rebounding and looking for IS talent. Online job listings for IS positions, for example, increased in 2003.[51] Demand for IS professionals has grown also in nonprofit organizations and in government. In addition to salary, IS workers seek paid vacation, health insurance, stock options, and flexible hours as important job factors. A study done by *Computerworld* listed the top places to work in information systems (see Table 2.5).

Table 2.5

The Top Places to Work in Information Systems

(Source: Data from Steve Ulfelder, "100 Best Places to Work in IT," *Computerworld*, June 9, 2003, p. 23.)

Company Name	Business
Hershey Foods Corp.	Maker of candies
Harley-Davidson, Inc.	Motorcycle manufacturer
University of Miami	Florida university
Network Appliances, Inc.	Hardware storage company
Vision Service Plan	Eye-care provider
Harrah's Entertainment	Casino operator
Saint Luke's Health System	Nonprofit health organization, including eight hospitals
Rich Products, Inc.	Family-owned food company
Discover Financial Services	Credit card company
Software Performance Systems, Inc.	Provider of financial management software

On the job, computer systems are also making IS professionals' work easier. Called *autonomics* by some, the use of advanced computer systems can help IS professionals spend less time maintaining existing systems and more time solving problems or looking for new opportunities. Colgate-Palmolive Co., for example, uses autonomics to keep its computer systems running better in more than 50 countries.

Opportunities in information systems are not confined to single countries. Some companies seek skilled IS employees from foreign countries, including Russia and India. The U.S. H-1B and L-1 visa programs seek to allow skilled employees from foreign lands into the United States. But not everyone is happy with these programs. Some companies may be firing U.S. workers and hiring less-expensive workers under the H-1B program. Because of a difficult economy, some companies may be abusing the H1-B visa program.[52] The L-1 visa program is often used for intracompany transfers for multinational companies. Some people fear, however, that the L-1 visa program could also be used to bring cheap IS personnel into the United States to replace more expensive American workers.[53] The Internet also makes it easier to export IS jobs to other countries.[54] Procter & Gamble estimates that it has reduced costs by about $1 billion by exporting IS jobs to Costa Rica, the Philippines, and Great Britain.

Roles, Functions, and Careers in the IS Department

Information systems personnel typically work in an IS department that employs Web developers, computer programmers, systems analysts, computer operators, and a number of other personnel. They may also work in other functional departments or areas in a support capacity. In addition to technical skills, IS personnel also need skills in written and verbal communication, an understanding of organizations and the way they operate, and the ability to work with people. According to George Voutes, enterprise technology programs manager for Deutsche Asset Management Technology, "We have to get away from strict programming and systems development. Those are skills to get into the field, but we have to train our technology people more like business people and arm them with strong communications skills."[55] IS personnel also need group-oriented skills.[56] Today, many good business and computer science schools require business and communications skills of their graduates. In general, IS personnel are charged with maintaining the broadest perspective on organizational goals. For most medium-to large-sized organizations, information resources are typically managed through an IS department. In smaller businesses, one or more people may manage information resources, with support from outside services—outsourcing. Outsourcing is also popular with larger organizations. According to a study by Gartner, Inc., a technology consulting company, "By 2004, 80 percent of U.S. executive boardrooms will have discussed offshore outsourcing, and more than 40 percent will have completed some type of pilot."[57] As shown in Figure 2.10, the IS organization has three primary responsibilities: operations, systems development, and support.

Operations

The operations component of a typical IS department focuses on the use of information systems in corporate or business unit computer facilities. It tends to focus more on the *efficiency* of IS functions rather than their effectiveness.

The primary function of a system operator is to run and maintain IS equipment. System operators are responsible for starting, stopping, and correctly operating mainframe systems, networks, tape drives, disk devices, printers, and so on. System operators are typically trained at technical schools or through on-the-job experience. Other operations include scheduling, hardware maintenance, and preparation of input and output. Data-entry operators convert data into a form the computer system can use. They may use terminals or other devices to enter business transactions, such as sales orders and payroll data. Increasingly, data entry is being automated—captured at the source of the transaction rather than being entered later. In addition, companies may have local area network and Web or Internet operators who are responsible for running the local network and any Internet sites the company may have.

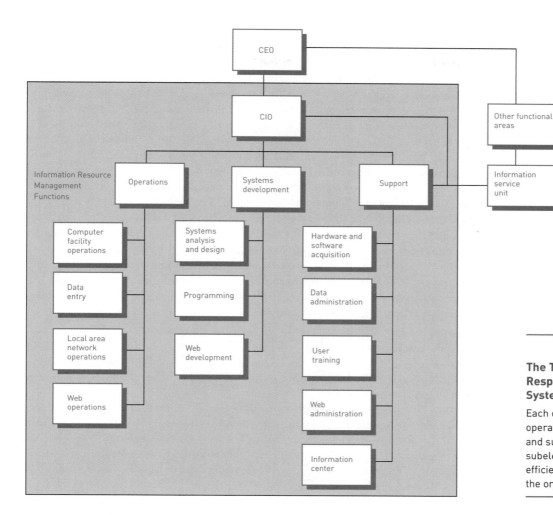

Figure 2.10

The Three Primary Responsibilities of Information Systems

Each of these elements—operations, systems development, and support—contains subelements that are critical to the efficient and effective operation of the organization.

System operators focus on the efficiency of IS functions, rather than their effectiveness. Their primary responsibility is to run and maintain IS equipment.

(Source: © Royalty-Free/CORBIS.)

Systems Development

The systems development component of a typical IS department focuses on specific development projects and ongoing maintenance and review. Systems analysts and programmers, for example, focus on these concerns. The role of a systems analyst is multifaceted. Systems analysts help users determine what outputs they need from the system and construct the plans needed to develop the necessary programs that produce these outputs. Systems analysts then work with one or more programmers to make sure that the appropriate programs are purchased, modified from existing programs, or developed. The major responsibility of a computer programmer is to use the plans developed by the systems analyst to develop or adapt one or more computer programs that produce the desired outputs. The main focus of systems analysts and programmers is to achieve and maintain IS effectiveness. To help companies select the best analysts and programmers, companies such as TopCoder offer tests to evaluate the proficiency and competence of existing IS employees or job candidates. TopCoder Collegiate Challenge allows programming students to compete with other programmers around the world.[58] The competition gives $25,000 to the best collegiate programmer. Some companies, however, are skeptical of the usefulness of these types of tests.[59]

With the dramatic increase in the use of the Internet, intranets, and extranets, many companies have Web or Internet developers who are responsible for developing effective and attractive Internet sites for customers, internal personnel, suppliers, stockholders, and others with a business relationship with the company.

Support

The support component of a typical IS department focuses on providing user assistance in the areas of hardware and software acquisition and use, data administration, user training and assistance, and Web administration. In many cases, the support function is delivered through an information center.

Because IS hardware and software are costly, especially if purchase mistakes are made, the acquisition of computer hardware and software is often managed by a specialized support group. This group sets guidelines and standards for the rest of the organization to follow in making purchases. Gaining and maintaining an understanding of available technology is an important part of the acquisition of information systems. Also, developing good relationships with vendors is important.

A database administrator focuses on planning, policies, and procedures regarding the use of corporate data and information. For example, database administrators develop and disseminate information about the corporate databases for developers of IS applications. In addition, the database administrator is charged with monitoring and controlling database use.

User training is a key to get the most from any information system. The support area insures that appropriate training is available to users. Training can be provided by internal staff or from external sources. For example, internal support staff may train managers and employees in the best way to enter sales orders, to receive computerized inventory reports, and to submit expense reports electronically. Companies also hire outside firms to help train users in other areas, including the use of word processing, spreadsheets, and database programs.

Web administration is another key area of the support function. With the increased use of the Internet and corporate Web sites, Web administrators are sometimes asked to regulate and monitor Internet use by employees and managers to make sure that it is authorized and appropriate. Web administrators also are responsible for maintaining the corporate Web site. Keeping corporate Web sites accurate and current can require substantial resources.

The support component typically operates the information center. An **information center** provides users with assistance, training, application development, documentation, equipment selection and setup, standards, technical assistance, and troubleshooting. Although many firms have attempted to phase out information centers, others have changed the focus of this function from technical training to helping users find ways to maximize the benefits of the information resource.

information center
A support function that provides users with assistance, training, application development, documentation, equipment selection and setup, standards, technical assistance, and troubleshooting.

Information Service Units

An **information service unit** is basically a miniature IS department attached and directly reporting to a functional area. Notice the information service unit shown in Figure 2.10. Even though this unit is usually staffed by IS professionals, the project assignments and the resources necessary to accomplish these projects are provided by the functional area to which it reports. Depending on the policies of the organization, the salaries of IS professionals staffing the information service unit may be budgeted to either the IS department or the functional area.

information service unit
A miniature IS department.

Typical IS Titles and Functions

The organizational chart shown in Figure 2.10 is a simplified model of an IS department in a typical medium-sized or large organization. Many organizations have even larger departments, with increasingly specialized positions such as librarian, quality assurance manager, and the like. Smaller firms often combine the roles depicted in Figure 2.10 into fewer formal positions.

The Chief Information Officer

The overall role of the chief information officer (CIO) is to employ an IS department's equipment and personnel in a manner that will help the organization attain its goals. The CIO is usually a manager at the vice presidential level concerned with the overall needs of the organization. He or she is responsible for the corporate-wide policy making, planning, management, and acquisition of information systems. Some of the CIO's top concerns include integrating IS operations with corporate strategies, keeping up with the rapid pace of technology, and defining and assessing the value of systems development projects. The high level of the CIO position is consistent with the idea that information is one of the organization's most important resources. This individual works with other high-level officers of an organization, including the chief financial officer (CFO) and the chief executive officer (CEO), in managing and controlling total corporate resources. CIOs must work closely with advisory committees, stressing effectiveness and teamwork and viewing information systems as an integral part of the organization's business processes—not an adjunct to the organization. Thus, CIOs need both technical and business skills. For federal agencies, the Clinger-Cohen Act of 1996 required the establishment of a CIO to coordinate the purchase and management of information systems.[60]

Depending on the size of the IS department, there may be several people at senior IS managerial levels. Some of the job titles associated with IS management are the CIO, vice president of information systems, manager of information systems, and chief technology officer (CTO). A central role of all these individuals is to communicate with other areas of the organization to determine changing needs. Often these individuals are part of an advisory or steering committee that helps the CIO and other IS managers with their decisions about the use of information systems. Together they can best decide what information systems will support corporate goals. The chief technology officer (CTO), for example, typically works under a CIO and specializes in hardware and related equipment and technology.[61] According to John Voeller, a CTO for the Black & Veatch engineering and construction company, "I don't just look at the technology of my enterprise. I look far beyond information technology at nanotechnologies, biotech, and other domains." The CTO position also exists in federal agencies. According to Debra Stouffer, CTO of the Environmental Protection Agency (EPA), "I can carve out what the CTO position will be. I was very attracted to the mission—protecting human health and safeguarding the environment."[62]

LAN Administrators

Local area network (LAN) administrators set up and manage the network hardware, software, and security processes. They manage the addition of new users, software, and devices to the network. They isolate and fix operations problems. LAN administrators are in high demand and often solve both technical and nontechnical problems.

Internet Careers

The recent bankruptcy of some Internet start-up companies, called the *dot-gone era* by some, has resulted in layoffs for some firms. Some executives of these bankrupt start-up Internet companies lost hundreds of millions of dollars in a few months. Yet, the growth in the use of the Internet to conduct business continues and has caused a steady need for skilled personnel to develop and coordinate Internet usage. As seen in Figure 2.10, these careers are in the areas of Web operations, Web development, and Web administration. As with other areas in IS, there are a number of top-level administrative jobs related to the Internet. These career opportunities are found in both traditional companies and companies that specialize in the Internet.

Internet jobs within a traditional company include Internet strategists and administrators, Internet systems developers, Internet programmers, and Internet or Web site operators. The Internet has become so important to some companies that some have suggested a new position, chief Internet officer, with responsibilities and a salary similar to the CIO's.

In addition to traditional companies, there are many exciting career opportunities in companies that offer products and services over the Internet. These companies include Amazon.com, Yahoo!, eBay, and many others. Systest, for example, specializes in finding and eliminating digital bugs that could halt the operation of a computer system.

A number of Internet sites, such as Monster.com, post job opportunities for Internet careers and more traditional careers. Most large companies have job opportunities listed on their Internet sites. These sites allow prospective job hunters to browse job opportunities, job locations, salaries, benefits, and other factors. In addition, some of these sites allow job hunters to post their resumes.

Internet job sites such as Monster.com allow job hunters to browse job opportunities and post their resumes.

Quite often the people filling IS roles have completed some form of certification.[63] **Certification** is a process for testing skills and knowledge resulting in an endorsement by the certifying authority that an individual is capable of performing a particular job. Certification frequently involves specific, vendor-provided or vendor-endorsed coursework. There are a number of popular certification programs, including the following:

- Certified Project Manager
- Microsoft Certified Systems Engineer—Certification on Microsoft Windows that requires passing six exams[64]

certification
Process for testing skills and knowledge that results in a statement by the certifying authority that says an individual is capable of performing a particular kind of job.

- Microsoft Certified Professional Systems Engineer
- Certified Information Systems Security Professional (CISSP)—The federal government is helping military personnel get IS certification. Some GI bill beneficiaries, for example, can now be reimbursed for technology certification through the Computing Technology Industry Association.
- Certified Information Systems Auditor
- Cisco Certified Internetwork Expert
- Novel Certified Professional
- Novell Certified Network Engineer
- Oracle Certified Professional

One of the greatest fears of every IS manager is spending several thousand dollars to help an employee get certified and then losing that person to a higher-paying position with a new firm. As a consequence, some organizations request a written commitment from individuals to stay a certain time after obtaining their certification. Needless to say, this requirement can create some ill will with the employee. To provide newly certified employees with incentives to remain, other organizations provide salary increases based on their additional credentials.

Other IS Careers

There are many other exciting IS careers. With the increase in attacks on computers, there are new and exciting careers in security and fraud detection and prevention. The University of Denver, for example, offers a masters program in cyber security.[65] Like many other universities, the University of Denver is hoping to get accreditation from the National Security Agency for the new degree program that specializes in security.

Insurance fraud and vehicle theft are no longer perpetrated primarily by small-time crooks but by organized crime rings using computers to falsify claim receipts, ship stolen vehicles throughout the world, and commit identity theft and fraud. The National Insurance Crime Bureau, a nonprofit organization supported by roughly 1,000 property and casualty insurance companies, uses computers to join forces with special investigation units and law enforcement agencies, as well as to conduct online fraud-fighting training to investigate and prevent these types of crimes.

In addition to working for an IS department in an organization, IS personnel can work for one of the large consulting firms, such as Accenture, IBM, EDS, and others. These jobs often entail a large amount of travel, because consultants are assigned to work on various projects wherever the client is. Such roles require excellent people and project management skills in addition to IS technical skills.

Other IS career opportunities also exist, including being employed by a hardware or software vendor developing or selling products. Such a role enables an individual to work on the cutting edge of technology, which can be extremely challenging and exciting! As some computer companies cut their services to customers, new companies are being formed to fill the need. With names such as Speak With a Geek and the Geek Squad, these companies are helping people and organizations with their computer-related problems that computer vendors are no longer solving.

SUMMARY

Principle

The use of information systems to add value to the organization is strongly influenced by organizational structure, culture, and change.

Organizations use information systems to support organizational goals. Because information systems typically are designed to improve productivity, methods for measuring the system's impact on productivity should be devised. An organization is a formal collection of people and various other resources established to accomplish a set of goals. The primary goal of a for-profit organization is to maximize shareholder value. Nonprofit organizations include social groups, religious groups, universities, and other organizations that do not have profit as the primary goal. Organizations are systems with inputs, transformation mechanisms, and outputs. Value-added processes increase the relative worth of the combined inputs on their way to becoming final outputs of the organization. The value chain is a series (chain) of activities that includes (1) inbound logistics, (2) warehouse and storage, (3) production, (4) finished product storage, (5) outbound logistics, (6) marketing and sales, and (7) customer service.

Organizational structure refers to organizational subunits and the way they relate to the overall organization. Several basic organization structures exist: traditional, project, team, multidimensional (also called *matrix structure*), and virtual organizational structure. A virtual organizational structure employs individuals, groups, or complete business units in geographically dispersed areas. The individuals, groups, or complete business units can involve people in different countries operating in different time zones with different cultures.

Organizational culture consists of the major understandings and assumptions for a business, corporation, or organization. Organizational change deals with how profit and nonprofit organizations plan for, implement, and handle change. Change can be caused by internal or external factors. The change model consists of these stages: unfreezing, moving, and refreezing. According to the concept of organizational learning, organizations adapt to new conditions or alter practices over time.

Principle

Because information systems are so important, businesses need to be sure that improvements or completely new systems help lower costs, increase profits, improve service, or achieve a competitive advantage.

Business process reengineering involves the radical redesign of business processes, organizational structures, information systems, and values of the organization to achieve a breakthrough in results. Continuous improvement constantly seeks ways to improve business processes to add value to products and services. The extent to which technology is used throughout an organization can be a function of technology diffusion, infusion, and acceptance. Technology diffusion is a measure of how widely technology is in place through an organization. Technology infusion is the extent to which technology permeates an area or department. The technology acceptance model (TAM) investigates factors, such as perceived usefulness of the technology, the ease of use of the technology, the quality of the information system, and the degree to which the organization supports the use of the information system, to predict IS usage and performance. Total quality management consists of a collection of approaches, tools, and techniques that fosters a commitment to quality throughout the organization. Six sigma is often used in quality control. It is a statistical term that means products and services will meet quality standards 99.9997 percent of the time. Outsourcing involves contracting with outside professional services to meet specific business needs. This approach allows the company to focus more closely on its core business and to target its limited resources to meet strategic goals. Downsizing involves reducing the number of employees to reduce payroll costs; however, it can lead to unwanted side effects.

Competitive advantage is usually embodied in either a product or service that has the most added value to consumers and that is unavailable from the competition or in an internal system that delivers benefits to a firm not enjoyed by its competition. A five-forces model covers factors that lead firms to seek competitive advantage: rivalry among existing competitors, the threat of new market entrants, the threat of substitute products and services, the bargaining power of buyers, and the bargaining power of suppliers. Three strategies to address these factors and to attain competitive advantage include altering the industry structure, creating new products and services, and improving existing product lines and services.

Developing information systems that measure and control productivity is a key element for most organizations. A useful measure of the value of an IS project is return on investment (ROI). This measure investigates the additional profits or benefits that are generated as a percentage of the investment in IS technology. Total cost of ownership (TCO) can also be a useful measure.

Principle

Information systems personnel are the key to unlocking the potential of any new or modified system.

Information systems personnel typically work in an IS department that employs a chief information officer, chief

technology officer, systems analysts, computer programmers, computer operators, and a number of other personnel. The overall role of the chief information officer (CIO) is to employ an IS department's equipment and personnel in a manner that will help the organization attain its goals. The chief technology officer (CTO) typically works under a CIO and specializes in hardware and related equipment and technology. Systems analysts help users determine what outputs they need from the system and construct the plans needed to develop the necessary programs that produce these outputs. Systems analysts then work with one or more programmers to make sure that the appropriate programs are purchased, modified from existing programs, or developed. The major responsibility of a computer programmer is to use the plans developed by the systems analyst to build or adapt one or more computer programs that produce the desired outputs.

Computer operators are responsible for starting, stopping, and correctly operating mainframe systems, networks, tape drives, disk devices, printers, and so on. LAN administrators set up and manage the network hardware, software, and security processes. There is also an increasing need for trained personnel to set up and manage a company's Internet site, including Internet strategists, Internet systems developers, Internet programmers, and Web site operators. Information systems personnel may also work in other functional departments or areas in a support capacity. In addition to technical skills, IS personnel also need skills in written and verbal communication, an understanding of organizations and the way they operate, and the ability to work with people (users). In general, IS personnel are charged with maintaining the broadest enterprise-wide perspective.

In addition to working for an IS department in an organization, IS personnel can work for one of the large consulting firms, such as Accenture, IBM, EDS, and others. Another IS career opportunity is to be employed by a hardware or software vendor developing or selling products.

CHAPTER 2: SELF-ASSESSMENT TEST

The use of information systems to add value to the organization is strongly influenced by organizational structure, culture, and change.

1. The value chain is a series of activities that includes inbound logistics, warehouse and storage, production, finished product storage, outbound logistics, marketing and sales, and customer service. True or False?
2. A(n) _____ is a formal collection of people and other resources established to accomplish a set of goals.
3. A virtual organizational structure is centered on major products or services. True or False?
4. The concept in which organizations adapt to new conditions or alter their practices over time is called

    a. organizational learning
    b. organizational change
    c. continuous improvement
    d. reengineering

Because information systems are so important, businesses need to be sure that improvements or completely new systems help lower costs, increase profits, improve service, or achieve a competitive advantage.

5. _____ involves contracting with outside professional services to meet specific business needs.
6. Today, quality means

    a. achieving production standards
    b. meeting or exceeding customer expectations
    c. maximizing total profits
    d. meeting or achieving design specifications

7. Technology diffusion is a measure of how widely technology is spread throughout an organization. True or False?
8. Reengineering is also called _____ .
9. What is a measure of the output achieved divided by the input required?

    a. efficiency
    b. effectiveness
    c. productivity
    d. return on investment

10. _____ is a measure of the additional profits or benefits generated as a percentage of the investment in IS technology.

Information systems personnel are the key to unlocking the potential of any new or modified system.

11. Who is involved in helping users determine what outputs they need and constructing the plans needed to produce these outputs?

    a. the CIO
    b. the applications programmer
    c. the systems programmer
    d. the systems analyst

12. The systems development component of a typical IS department focuses on specific development projects and ongoing maintenance and review. True or False?

13. The _____ is typically in charge of the IS department or area in a company.

KEY TERMS

certification 76
change model 56
competitive advantage 63
continuous improvement 58
culture 54
downsizing 62
empowerment 52
five-forces model 63
flat organizational structure 52
information center 74
information service unit 75
multidimensional organizational structure 53

on-demand computing (on-demand business, utility computing) 62
organization 48
organizational change 55
organizational culture 54
organizational learning 56
organizational structure 51
outsourcing 60
productivity 68
project organizational structure 53
quality 59
reengineering (process redesign) 57
return on investment (ROI) 68

strategic alliance (strategic partnership) 64
team organizational structure 53
technology acceptance model (TAM) 59
technology diffusion 59
technology infusion 59
total cost of ownership (TCO) 70
total quality management (TQM) 60
traditional organizational structure 51
value chain 49
virtual organizational structure 54

REVIEW QUESTIONS

1. What is the value chain?
2. What is the difference between a virtual organizational structure and a traditional organizational structure?
3. What role does an information system play in today's organizations?
4. What is reengineering? What are the potential benefits of performing a process redesign?
5. What is the difference between reengineering and continuous improvement?
6. What is the technology acceptance model (TAM)?
7. What is quality? What is total quality management (TQM)?
8. What are organizational change and organizational learning?
9. List and define the basic organizational structures.

10. Sketch and briefly describe the three-stage organizational change model.
11. What is downsizing? How is it different from outsourcing?
12. What are some general strategies employed by organizations to achieve competitive advantage?
13. What are the five common justifications for implementation of an information system?
14. Define the term *productivity*. Why is it difficult to measure the impact that investments in information systems have on productivity?
15. Briefly define *technology diffusion* and *infusion*.
16. What is the total cost of ownership?
17. What is the role of a systems analyst? What is the role of a programmer?

DISCUSSION QUESTIONS

1. You have been hired to work in the IS area of a manufacturing company that is starting to use the Internet to order parts from its suppliers and offer sales and support to its customers. What types of Internet positions would you expect to see at the company?

2. You have decided to open an Internet site to buy and sell used music CDs to other students. Describe the value chain for your new business.

3. What sort of IS career would be most appealing to you—working as a member of an IS organization, consulting, or working for an IT hardware or software vendor? Why?

4. What are the advantages of using a virtual organizational structure? What are the disadvantages?

5. As part of a TQM project initiated three months ago, you decided that your company needed a new information system. The computer systems were brought in over the weekend. The first notice your employees received about the new information system was the computer located on each desk. How might the new system affect the culture of your organization? What types of behaviors might employees exhibit in response? As a manager, how should you have prepared the employees for the new system?

6. You have been asked to participate in the preparation of your company's strategic plan. Specifically, your task is to analyze the competitive marketplace using Porter's five-forces model. Prepare your analysis, using your knowledge of a business you have worked for or have an interest in working for.

7. Based on the analysis you performed in Discussion Question 6, what possible strategies could your organization adopt to address these challenges? What role could information systems play in these strategies? Use Porter's strategies as a guide.

8. There are many ways to evaluate the effectiveness of an information system. Discuss each method and describe when one method would be preferred over another method.

9. You have been hired as a sales representative for a sporting goods store. You would like the IS department to develop new software to give you reports on which types of customers are spending the most at your store. Describe your role in getting the new software developed. Describe the roles of the systems analysts and the computer programmers. Discuss how the change model can be applied to breaking a bad habit—say, smoking or eating fatty foods. Some people have also related the stages in the change model to the changes one must go through to deal with a major life crisis—like divorce or the loss of a loved one. Explain.

PROBLEM-SOLVING EXERCISES

1. For an industry of your choice, find the number of employees, total sales, total profits, and earnings growth rate of 15 firms. Using a database program, enter this information for the last year. Use the database to generate a report of the three companies with the highest earnings growth rate. Use your word processor to create a document that describes the 15 firms. What other measures would you use to determine which is the best company in terms of future profit potential?

2. A new IS project has been proposed that will produce not only cost savings but also an increase in revenue. The initial costs to establish the system are estimated to be $500,000.

The rest of the cash flow data is presented in the following table.

	Year 1	Year 2	Year 3	Year 4	Year 5
Increased Revenue	$0	$100	$150	$200	$250
Cost Savings	$0	$ 50	$ 50	$ 50	$ 50
Depreciation	$0	$ 75	$ 75	$ 75	$ 75
Initial Expense	$500				

Note: All amounts in 000s.

a. Using your spreadsheet program, calculate the return on investment (ROI) for this project. Assume that the cost of capital is 7 percent.

b. How would the rate of return change if the project were able to deliver $50,000 in additional revenue and generate cost savings of $25,000 in the first year?

TEAM ACTIVITIES

1. With your team, interview one or more people who were either outsourced or downsized from a position in the last few years. Find out how the process was handled and what justification was given for taking this action. Also try to get information from an objective source (financial reports, investment brokers, industry consultants) on how this action has affected the organization.

2. With your team, research a firm that has achieved a competitive advantage. Write a brief report that describes how the company was able to achieve its competitive advantage.

WEB EXERCISES

1. This book emphasizes the importance of information. You can get information from the Internet by going to a specific address, such as *www.ibm.com*, *www.whitehouse.gov*, or *www.fsu.edu*. This will give you access to the home page of the IBM corporation, the White House, or Florida State University. Note that "com" is used for businesses or commercial operations, "gov" is used for governmental offices, and "edu" is used for educational institutions. Another approach is to use a search engine. Yahoo!, developed by two Tulane University students, was one of the first search engines on the Internet. A search engine is a Web site that allows you to enter key words or phrases to find information. You can also locate information through lists or menus. The search engine will return other Web sites (hits) that correspond to a search request. Using Yahoo! at www.yahoo.com, search for information about a company or topic discussed in Chapter 1 or 2. You may be asked to develop a report or send an e-mail message to your instructor about what you found.

2. Use the Internet to search for information about a company that you think will become a great company in the next 5 to 10 years. You can use a search engine, such as Yahoo!, or a database at your college or university. Write a brief report describing the company and why it will succeed.

CAREER EXERCISES

1. Organizations can use traditional, project, virtual, and other organizational structures. For the career of your choice, describe which organizational structure or structures are likely to be used and how computers and information systems can help you communicate and work with others in your organization.

2. Pick the five best companies for your career. Describe how each company uses information systems to help achieve a competitive advantage.

VIDEO QUESTIONS

Watch the video clip **Ben Casnocha** and answer these questions:

1. How does Ben Casnocha's company, Comcate, epitomize today's information economy—the trend towards profiting from non-tangible, information-based services.

2. Comcate provides an e-mail and tracking system for individuals and organizations to use for communicating with government agencies. Why has this business prospered? What demand does it fulfill?

CASE STUDIES

Case One

Ingersoll-Rand and the 21st Supplier

Successful manufacturing companies spend a lot of time and energy finding suppliers to provide the best raw materials at the best price. If a supplier overcharges, or otherwise wastes the time and resources of the manufacturer, the product costs rise, resulting in lost customers. The money that manufacturers spend on raw materials for their products is called *procurement expenditures*.

It is an age-old rule of thumb that a manufacturer's 20 largest suppliers account for 80 percent of its procurement expenditures. Based on this rule, manufacturers have traditionally concentrated on their 20 largest suppliers in an effort to control procurement expenditures.

Ingersoll-Rand Co. is a large manufacturer of industrial equipment, with $8 billion in sales. Like most manufacturers, Ingersoll-Rand focused on its top 20 suppliers—to the point where they were bending over backward to reduce prices. Having done all they could with their top 20, Ingersoll-Rand began focusing on its smaller suppliers—what the company referred to as its "21st supplier." In investigating these smaller supply companies, Ingersoll-Rand found that many were increasing prices as much as 10 percent annually—during years when inflation was practically nonexistent. Smaller suppliers were also more likely to deliver their materials late, costing Ingersoll-Rand in express-delivery charges. In general, smaller suppliers were much less organized, reliable, and cost efficient than the larger suppliers that typically make up the top 20.

In communicating with other manufacturers, Ingersoll-Rand found that the problem with small supply companies was pervasive across a variety of manufacturing areas. So, Ingersoll-Rand saw an opportunity to create a service company to address the issue. Partnering with SupplyWorks, Inc., a vendor of supplier-relationship management software, and Worldwide Logistics Solutions (part of Roberson Transportation), Ingersoll-Rand created a new company called The 21st Supplier.

The slogan for the new company is, "Focus on your 'Top 20' key suppliers; let us manage the rest." With Ingersoll's new system, manufacturers turn over the relationships of their many small suppliers to The 21st Supplier, which then works to optimize the transactions. The 21st Supplier works to "continuously improve the process of ordering and replenishing materials, while identifying new ways to bring supplier innovation to customers." The 21st Supplier utilizes software and networking solutions and information systems created by SupplyWorks to automate the ordering process to improve inventory management, to reduce transaction time, and to streamline administrative activities.

As with most businesses in a tight economy, Ingersoll-Rand is investigating every opportunity to save a penny. In its effort to move forward in tough times, the company has created additional revenue through its new business, which generates its income using information systems and business practices inspired by the tough economy. Now that's smart business!

Discussion Questions

1. How did Ingersoll-Rand turn a problem into an advantage?
2. How did the establishment of The 21st Supplier provide Ingersoll-Rand with a competitive advantage over other manufacturers?

Critical Thinking Questions

3. When Ingersoll-Rand discovered the significance of "the 21st supplier," it could have simply applied that insight to its own business practices in an attempt to gain a competitive advantage. Why do you think the company decided to start a new business and provide this service and knowledge to other manufacturers—including its competitors?
4. It's clear that Ingersoll-Rand's service benefits manufacturers. How do you think supplier-relationship management software benefits the suppliers?

SOURCES: Steve Ulfelder, "Managing the Other 20 Percent," *Computerworld*, March 24, 2003, *www.computerworld.com*; Peter Strozniak, "New Tools Help Companies Turn Inventory from a Problem Child to a Competitive Advantage," *Frontline Solutions*, February 1, 2003, *www.frontlinetoday.com/*; The 21st Supplier Web site, *www.the21stsupplier.com*, accessed January 18, 2004.

Case Two

The Bank of Montreal Learns from Past Mistakes

The Bank of Montreal was founded in 1817 and later grew into the BMO Financial Group. This financial service provider offers a wide range of services across Canada and the United States. BMO is a large institution, with average assets of $268 billion as of July 31, 2003, and more than 34,000 employees. The company is represented in the United States by its subsidiary, Chicago-based Harris Bank.

As with most businesses in the mid to late 1990s, BMO was swept up in the enthusiasm for and the opportunities presented by the Internet. BMO developed North America's first standalone, full-service e-banking option, which it called

"Mbanx." Targeting the service to tech-savvy young customers eager to bank in cyberspace, BMO launched a huge marketing campaign for the service, centered around folk legend Bob Dylan's hippie anthem "The Times They Are A-changin'." What the company discovered, though, was that the times weren't a-changin' as fast as it thought.

Assuming that its young clientele would leap at the opportunity to bank online, BMO failed to support its new banking system in its brick and mortar branches. In no time, Mbanx was scrubbed as a failure. Customers were simply not willing to give up traditional banking practices for online banking.

Following the Mbanx disaster, BMO broadened its focus, providing the convenience of all its online systems over the Web to regular customers. BMO continues to regard technological innovation as a source of competitive advantage, but its view of technological leadership has changed. It now aspires to stay on the cutting edge of innovation without pushing to the bleeding edge of technology deployment. BMO employs a five-step model to improve the productivity of its information technology and increase its value to the bank.

Step one is to centralize the IT function. Information systems should be centralized and communicate across the entire organization. BMO follows the operating principle of "one organization; one set of principles"—what it refers to as the Power of One. CIO Lloyd Darlington states that "by centralizing, we are able to leverage scale, enforce standards, and aggressively manage our vendors; this has allowed us to absorb 30 percent growth in computing resources annually while keeping our costs flat."

Step two is to effectively manage supply and demand. Today, businesses are much less willing to invest in new information systems, so IS managers must be prudent in acquiring new components.

Step three is standardize, standardize, standardize. BMO standardizes the technology it uses, its IT practices, and project management and IT applications. "We are only doing what manufacturing companies have understood for sixty years. You can increase productivity enormously and mini-

mize the frequency of errors by developing and implementing standard ways of doing work that represent best practices," says Darlington.

Step four is benchmark, measure, and reward. New systems and their delivery are rigorously benchmarked (tested in action) and evaluated with detailed score cards. Score cards are designed to clearly communicate the company's business plan objectives for the year and its progress in meeting those objectives. Score cards are linked with compensation, a powerful motivator that drives employees to excel.

Step five, the final step, is to change the culture. Darlington has worked to reorganize its IS staff into a business group sharply focused on the bottom line. "We introduced corporate values to the organization," said Darlington. "We call it 'SPIRIT,' an acronym for *s*ervice and solutions; *p*ersonal excellence; *i*ntegrity; *r*espect and trust; *i*nnovation; and *t*eamwork." These values represent the fabric of our Technology and Solutions culture."

Discussion Questions

1. What was the lesson learned from the Mbanx disaster? State your answer in a manner that can be applied to any IS development initiative.
2. What type of organization or industry might benefit from staying on the "bleeding edge" of technology deployment? Why did BMO decide against it?

Critical Thinking Questions

3. Are the five steps that Darlington and BMO apply appropriate for companies of all sizes, in all industries?
4. What are the disadvantages of locking an entire international organization into a standardized system?

SOURCES: David Carey, "BMO's Five Steps to Delivering IT Value," *CIO Canada*, January 1, 2004, *www.itworldcanada.com*; "BMO Bank of Montreal Redesigns Its Online Banking Web Site," *Canada NewsWire*, November 6, 2003; BMO Financial Group Web site, *www.bmo.com*, accessed January 25, 2004.

Case Three

Shell Oil Saves with Contingent Workforce Management System

As the U.S. economy improves, many companies are beginning to hire again. However, many are opting for contract laborers rather than full-time employees. Usually, contract laborers are hired through employment agencies and are called contingent workers. For large companies such as Shell Oil Co., which spends nearly $100 million annually on contract laborers to fill short-term jobs ranging from accounting to IT consulting, the task of finding workers can be tedious, time-consuming, and wasteful.

Recently, Shell decided on a more cost-effective approach to manage its contingent workforce. A new Web-based system from Denver-based IQNavigator, Inc., now assists Shell with its contract hiring. The system automates many of the processes, including qualification of suppliers, requests for proposals, time-and-expense entries, and invoicing.

Prior to acquiring the new system, Shell employed a hodgepodge of manual processes that were costly and time-consuming. So, it created a task force to reorganize its process and reduce costs by 8 percent annually. Switching to the IQNavigator system allowed the task force to surpass its goal—in a little more than one month. Shell cut the number of contract labor agencies it worked with from more than 20

to just 4. Through consolidation, automation, and process improvements, the company has cut its payments to new contractors by an average of 28 percent.

The beauty of the IQNavigator system is that Shell doesn't have to pay for it; the contract labor agencies pick up the bill. Having suppliers pay access fees for contingent workforce management software "has become the norm in the industry," says Pamela O'Rourke, president of Icon Information Consultants, one of Shell's preferred contract labor agencies. The agencies pay "system access fees," which range from 3 to 5 percent of an invoice. The four preferred agencies don't mind picking up the tab, considering the number of invoices Shell provides. Contingent workforce management systems such as IQNavigator's allow companies to hire locally for low rates, without much hassle, and with guarantees that the worker will be qualified. Some companies are even looking to increase the use of such systems as an alternative to offshore outsourcing.

Discussion Questions

1. How did the contingent workforce management system save Shell money?

2. How do you think Shell recognized the need for a new system?

Critical Thinking Questions

3. What are some of the advantages of hiring locally rather than outsourcing jobs to inexpensive overseas companies?

4. What are the disadvantages, besides saving some money, of hiring locally rather than outsourcing jobs overseas?

SOURCES: Thomas Hoffman, "Contingent Workforce: Managing the Temporary Players," *Computerworld*, June 30 2003, *www.computerworld.com*; Renee Boucher Ferguson and John S. McCright, "Getting a Grip on Spending," *eWeek*, November 17, 2003, News &Analysis, p. 38; IQNavigator Inc Web site, *www.iqnavigator.com/news.html*, accessed January 25, 2004.

NOTES

Sources for opening vignette: Chris Murphy, "A Simple, Global Ambition," *InformationWeek*, November 24, 2003, Leadership Section, p. 61; Meridith Levinson, "On the Move," *CIO Magazine*, Columns Section, *www.cio.com*; Whirlpool Web site, *www.whirlpool.com*, accessed January 25, 2004.

1. Bulkeley, Wailin, "Six Degrees of Exploitation," *The Wall Street Journal*, August 4, 2003, p. B1.
2. Avery, Susan, "Cessna Soars," *Purchasing*, September 4, 2003, p. 25.
3. Billington, Corey, et al., "Accelerating The Profitability of Hewlett-Packard's Supply Chains," *Interfaces*, January-February 2004, p. 59.
4. Gentle, Michael, "CRM: Ready or Not," *Computerworld*, August 18, 2003, p. 40.
5. Anthes, Gary, "Best In Class: Data Warehouse Boosts Profits by Empowering Sales Force," *Computerworld*, February 24, 2003, p. 46.
6. McCue, Michael, "Small Teams, Big Returns," *Managed Health Care Executive*, January, 2003, p. 20.
7. King, Julia, "On-The Fly," *Computerworld*, September 8, 2003, p. 35.
8. Prashad, Sharda, "Building Trust Tricky for Virtual Teams," *The Toronto Star*, October 23, 2003, p. K06.
9. Christensen, Clayton, *The Innovator's Dilemma*, Harvard Business School Press, 1997, p. 225; and *The Inventor's Solution*, Harvard Business School Press, 2003.
10. Schein, E.H., *Process, Consultation: Its Role in Organizational Development*, (Reading, MA: Addison-Wesley, 1969). See also Peter G.W. Keen, "Information Systems and Organizational Change," *Communications of the ACM*, vol. 24, no. 1, January 1981, pp. 24–33.
11. Songini, Marc, "Best In Class: Re-Engineering Drives Down Cost of Power," *Computerworld*, February 24, 2003, p. 46.
12. Schoenberger, Chana, "Drugs Tagged with Radio Chips Could Cut Costs and Curb Counterfeiting and Medical Errors," *Forbes*, September 15, 2003, p. 126.
13. Davenport, Thomas, et al., "What Went Wrong with the Business-Process Reengineering Fad? And Will It Come Back?" *Computerworld*, June 23, 2003, p. 48.
14. Troy, Mitch, "ArvinMeritor Appoints Vice Presidents of Continuous Improvement and Quality," *PR Newswire*, September 19, 2003.
15. Loch, Christoph and Huberman, Berndao, "A Punctuated-Equilibrium Model of Technology Diffusion," *Management Science*, February, 1999, p. 160.
16. Armstrong, Curtis and Sambamurthy, V. "Information Technology Assimilation in Firms," *Information Systems Research*, December, 1999, p. 304.
17. Agarwal, Ritu and Prasad, Jayesh, "Are Individual Differences Germane to the Acceptance of New Information Technology," *Decision Sciences*, Spring 1999, p. 361.
18. Kwon, et al., "A Test of the Technology Acceptance Model," *Proceedings of the Hawaii International Conference on System Sciences*, January 4–7, 2000.
19. Watts, Stephanie, et al., "Informational Influence in Organizations," *Information Systems Research*, March, 2003, p. 47.
20. Gilhooly, Kym, "First Help Is Best," *Computerworld*, September 8, 2003, p. 23.
21. Thibodeau, Patrick, "Best In Class: Automated Warehouse Reduces Errors," *Computerworld*, February 24, 2003, *www.computerworld.com*.
22. Staff, "Anacora Awarded Quality Mail Partner by U.S. Postal Service," *PR Newswire*, October 14, 2003.
23. Watson, Mark, "It Works for Cars; It Can Work for Health Care," *The Commercial Appeal*, October 25, 2003, p. C1.
24. Boyle, Edward, "Six Sigma Success," *Paper, Film, and Foil Converter*, January 1, 2004, p. 1.
25. Kirkpatrick, David, "The Net Makes It All Easier," *Fortune*, May 26, 2003, p. 146.
26. Staff, "On Demand Vision," *www.IBM.com*, Accessed on January 25, 2004.
27. Cornel, Andrew, "Cutting from the Top," *Australian Financial Review*, May 17, 2003, p. 23.
28. Collins, Jim, *Good to Great*, Harper Collins Books, 2001, p. 300.
29. Slotegraaf, Rebecca, et al., "The Role of Firm Resources," *Journal of Marketing Research*, August 2003, p. 295.
30. Saccomano, Ann, "Bureau of Immigration and Customs Reorganizes," *The Journal of Commerce Online Edition*," May 21, 2003, p. 1.

31. Kurtz, et al., "The New Dynamics of Strategy," *IBM Systems Journal*, vol. 42, No. 3, 2003, p. 462.

32. Kearns, et al., "A Resource-Based View of Strategic IT Alignment," *Decision Sciences*, Winter 2003, p. 1.

33. M. Porter and V. Millar, "How Information Systems Give You Competitive Advantage," *Journal of Business Strategy*, Winter 1985. See Also M. Porter, *Competitive Advantage* (New York: Free Press, 1985).

34. Melymuka, Kathleen, "Delta's New All-Digital Song," *Computerworld*, August 8, 2003, p. 37.

35. Staff, "Echelon and Samsung Sign Strategic Alliance," *Business Wire*, June 23, 2003.

36. Rosencrane, Linda, "Best In Class: Self-Service Check-In Kiosks Give Travelers More Control," *Computerworld*, February 24, 2003, p. 48.

37. Khermouch, Gerry, "Bar Codes Better Watch Their Backs," *Business Week*, July 14, 2003, p. 42.

38. Moes, Annick, "Germans Put Stock in Store of the Future," *The Wall Street Journal*, June 19, 2003, p. A10.

39. Trottman, Melanie, "Sabre Looks to Web Ventures," *The Wall Street Journal*, June 24, 2003, p. B6.

40. Kohli, Rajiv, et al., "Measuring Information Technology Performance," *Information Systems Research*, June 2003, p. 127.

41. Alexander, Steve, "Best In Class: Web Site Adds Inventory Control and Forecasting," *Computerworld*, February 24, 2003, p.45.

42. Melymuka, Kathleen, "Betting on IT Value," *Computerworld*, May 3, 2004, p. 33.

43. Staff, "Paradox Lost," *The Economist*, September 13, 2003, *www.economist.com*.

44. Park, Andrew, "Computers Get Their Groove Back," *Business Week*, January 12, 2004, p. 96.

45. Staff, "Certification–Return On Investment," *Computer Reseller News*, August 25, 2003, p. 54.

46. Alexander, Steve, "Best In Class: Web Site Adds Inventory Control and Forecasting," *Computerworld*, February 24, 2003, p.45.

47. Staff, "Detailed Comparison of a Range of Factors Affecting Total Cost of Ownership (TCO) for Messaging and Collaboration," *M2 Presswire*, September 5, 2003.

48. Staff, "An International Internet Research Assignment," *Journal of Education for Business*, January 2003, p. 158.

49. Lublin, Joann, "X-Treme Job Hunting," *The Wall Street Journal*, July 29, 2003, p. B1.

50. Brandel, Mary, "Home-Schooling–IT Talent," *Computerworld*, January 27, 2003, p. 36.

51. Hoffman, Thomas, "Online Tech Job Positions Increase, But IT Execs Don't Expect a Jump in Hiring," *Computerworld*, September 8, 2003, p. 11.

52. Grow, Brian, "Skilled Workers or Indentured Servants," *Business Week*, June 16, 2003, p. 54.

53. Grow, Brian, "A Loophole as Big as a Mainframe," *Business Week*, March 10, 2003, p. 82.

54. Kirkpatrick, David, "The Net Makes It All Easier," *Fortune*, May 26, 2003, p. 146.

55. Hoffman, Thomas, "Preparing Generation Z," *Computerworld*, August 25, 2003, p. 41.

56. Tesch, et al., "The Impact of Information System Personnel Skill Discrepancies on Stakeholder Satisfaction," *Decision Science*, Winter 2003, p. 107.

57. Gongloff, Mark, "U.S. Jobs Jumping Ship," *CNN Money Online*, May 2, 2003.

58. Staff, "And The Champion Coders Are," *Business Week*, May 10, 2004, p. 16.

59. Hoffman, Thomas, "Programmer Testing Services Get Wary Reception from IT," *Computerworld*, April 14, 2003, p. 6.

60. Staff, "The Clinger-Cohen Act of 1996," *The Governments Accounts Journal*, Winter 1997, p. 8.

61. Melymuks, K, "So You Want to Be a CTO," *Computerworld*, March 24, 2003, p. 42.

62. Prencipe, Loretta, " Chief Technology Officers," *InfoWorld*, January 6, 2003, p. 44.

63. Staff, "Highest-Paying IT Certifications, Q1 2004," *Computerworld*, May 24, 2004, p. 44.

64. Sliwa, Carol, "Microsoft to Introduce Security Certifications," *Computerworld*, June 2, 2003, p. 6.

65. Fillion, Roger, "DU Program Seeks NSA's Nod," *The Rocky Mountain News*, June 3rd, 2004, p. 3B.

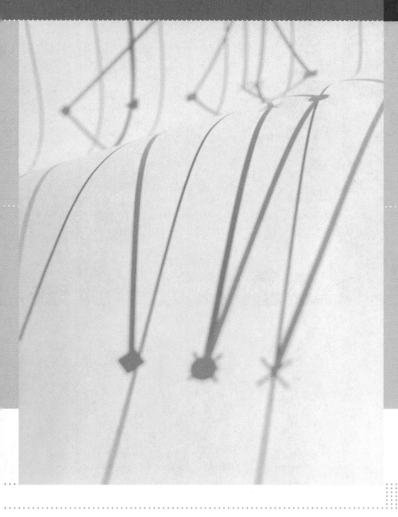

PART • 2 •

Information Technology Concepts

Chapter 3 Hardware: Input, Processing, and Output Devices

Chapter 4 Software: Systems and Application Software

Chapter 5 Organzing Data and Information

Chapter 6 Telecommunications and Networks

Chapter 7 The Internet, Intranets, and Extranets

CHAPTER
· 3 ·

Hardware:
Input, Processing, and Output Devices

PRINCIPLES

LEARNING OBJECTIVES

PRINCIPLES	LEARNING OBJECTIVES
▪ Assembling an effective, efficient computer system requires an understanding of its relationship to the information system and the organization. The computer system objectives are subordinate to, but supportive of, the information system and the needs of the organization.	▪ Describe how to select and organize computer system components to support information system (IS) objectives and business organization needs.
▪ When selecting computer devices, you also must consider the current and future needs of the information system and the organization. Your choice of a particular computer system device should always allow for later improvements.	▪ Describe the power, speed, and capacity of central processing and memory devices. ▪ Describe the access methods, capacity, and portability of secondary storage devices. ▪ Discuss the speed, functionality, and importance of input and output devices. ▪ Identify popular classes of computer systems and discuss the role of each.

INFORMATION SYSTEMS IN THE GLOBAL ECONOMY ▶
PORSCHE AG, GERMANY

Auto Manufacturer Upgrades Its Computer Hardware

Porsche is known for manufacturing world-class sports cars, among them its famous 911 and Boxster automobiles. The company built on its reputation when it recently introduced the four-wheel-drive Cayenne, the world's fastest SUV, and the Carerra GT, a super sports car with a powerful V10 engine. Porsche's headquarters are in Stuttgart, Germany; it maintains production facilities in Stuttgart and Leipzig, and it has dealerships and sales offices worldwide. Annual sales are at record levels—approaching a jaw-dropping $7 billion—and its employees number just under 10,000.

To maintain its market leadership, Porsche must employ information systems as cutting edge as its cars. The company was one of the first auto makers to introduce enterprise resource planning (ERP) software to support its accounting, finance, purchasing, and material-management business processes. It implemented its ERP system in the early 1990s and chose then state-of-the-art Hewlett-Packard V-class servers as the underlying computer hardware to handle some 1.5 million daily transactions quickly and reliably. Over the years, however, the volume of transactions its system must handle has more than doubled. This increase comes from four factors: (1) Porsche plans to use new ERP software modules to support additional business processes, (2) the company introduced new automobile models, (3) it built a new production plant for the Cayenne and Carerra GT, and (4) the number of users of the ERP system exceeded 4,000.

Porsche Information Kommunikation Services (PIKS) GmbH is a wholly owned subsidiary of Porsche AG, with headquarters in Stuttgart. Its 86 employees are responsible for planning and operating the IS infrastructure for the entire Porsche group, including networks, servers, security systems, and storage systems. For over a year, PIKS evaluated several different computer manufacturers and hardware options to meet the new processing requirements. It was critical that the new hardware not increase the company's hardware budget. Obviously, the new hardware must work well with existing components of the infrastructure (software, network, and other computer hardware). Importantly, the new hardware must be extremely reliable and available.

To meet the company's computing needs, PIKS decided to replace its HP V-class servers with two HP Superdome servers, each with 24 processors and 28 gigabytes of RAM. This hardware upgrade also provides for increased processing power to meet future needs; the PA-8700 processors in the current version of the Superdome server can be upgraded to processors from the Intel Itanium processor family to double the processing capability.

As you read this chapter, consider the following:

- How are companies using computer hardware to compete and meet their business objectives?
- How do organizations go about selecting computer hardware and what must you know to assist in this process?

Why Learn About Hardware?

Organizations invest in computer hardware to improve worker productivity, increase revenue, reduce costs, and provide better customer service. Those that don't may be stuck with outdated hardware that often fails and cannot take advantage of the latest software advances. As a result, obsolete hardware can place an organization at a competitive disadvantage. Managers, no matter what their career field and educational background, are expected to know enough about hardware to ask tough questions to invest wisely for their area of the business. Managers in marketing, sales, and human resources often help IS specialists assess opportunities to apply computer hardware and evaluate the options and features specified for the hardware. Managers in finance and accounting especially must also keep an eye on the bottom line, guarding against overspending, yet be willing to invest in computer hardware when and where business conditions warrant it.

Today's use of technology is practical—intended to yield real business benefits, as seen with Porsche. Employing information technology and providing additional processing capabilities can increase employee productivity, expand business opportunities, and allow for more flexibility. As we already discussed, a computer-based information system (CBIS) is a combination of hardware, software, database(s), telecommunications, people, and procedures—all organized to input, process, and output data and information. In this chapter, we concentrate on the hardware component of a CBIS. **Hardware** consists of any machinery (most of which use digital circuits) that assists in the input, processing, storage, and output activities of an information system. The overriding consideration in making hardware decisions in a business should be how hardware can be used to support the objectives of the information system and the goals of the organization.

hardware

Any machinery (most of which use digital circuits) that assists in the input, processing, storage, and output activities of an information system.

COMPUTER SYSTEMS: INTEGRATING THE POWER OF TECHNOLOGY

A computer system is a special subsystem of an organization's overall information system. It is an integrated assembly of devices—centered on at least one processing mechanism utilizing digital electronics—that are used to input, process, store, and output data and information.

Putting together a complete computer system, however, is more involved than just connecting computer devices. In an effective and efficient system, components are selected and organized with an understanding of the inherent trade-offs between overall system performance and cost, control, and complexity. For instance, in building a car, manufacturers try to match the intended use of the vehicle to its components. Racing cars, for example, require special types of engines, transmissions, and tires. The selection of a transmission for a racing car, then, requires not only consideration of how much of the engine's power can be delivered to the wheels (efficiency and effectiveness) but also how expensive the transmission is (cost), how reliable it is (control), and how many gears it has (complexity). Similarly, organizations assemble computer systems so that they are effective, efficient, and well suited to the tasks that need to be performed.

As we saw in the opening vignette, people involved in selecting their organization's computer hardware must have a clear understanding of the business requirements so they can make good acquisition decisions. Here are several examples of applying business understanding to reach critical hardware decisions.

ARZ Allgemeines Rechenzentrum GmbH (ARZ) is one of Austria's leading providers of information services, processing more than 8 million transactions per day for financial and medical institutions. ARZ is keenly interested in reducing processing costs, maintaining

sufficient capacity to handle an increasing workload, and providing highly reliable processing. Peter Gschirr, information technology director at ARZ, selected two large, extremely powerful IBM zSeries mainframe computers to handle the workload.[1]

As a result of declining ticket sales, Continental Airlines management mandated that all system development projects must pay back their costs in less than one year. In addition, the Continental Technology Group needed a safe, reliable, and user-friendly development environment to write new application software. Jack Wang, managing director of the Technology Group, acquired a Hewlett-Packard HP NonStop S8600 server to run customer-service applications and a NonStop S74000 server for software development and backup in the event the primary server fails. These high-end, fault-tolerant servers are frequently used for critical financial transactions—in the financial services and electronic commerce industries.[2]

Air Products and Chemicals, Inc., is an international supplier of industrial gases and related equipment, as well as specialty chemicals. The company employs unique software that requires powerful, high-performance computers to conduct simulations and computations. Eric Werley of Air Products' technical computing group recognized several shortcomings in using

Porsche needed cost-effective, reliable, powerful computers on which to run its ERP software.

Source: Getty Images.)

servers from three different manufacturers: The acquisition and ongoing maintenance of these server systems were costly, and the servers were difficult to manage due to lack of standardization. Air Products migrated to Dell computer hardware, including powerful Dell Dual Precision 410 and Dell GxPro workstations and Dell PowerEdge servers to eliminate these problems.[3]

Specialized Bicycles is a pioneer in designing and manufacturing high-performance bicycles, helmets, and other cycling accessories. When Specialized Bicycles wanted to offer its products over the Web, it selected computer hardware that could rapidly increase computing capacity, provide high reliability so that the Web site was always available, and easily be integrated with the rest of the organization's hardware. To meet these requirements, Ron Pollard, chief information officer, went with Sun Microsystems Enterprise 450 Server.[4]

Sullivan Street Bakery, started in 1994 by an art student and anthropologist, is today one of the most popular Italian bakeries in New York. The owners chose Apple's Mac computers because they are efficient and easy to use. The Macs make managing production almost effortless, and they support billing and inventory control. Workers don't have to worry about how to use the computers and can concentrate instead on maintaining the quality of their hand-crafted traditional Italian-style breads.[5]

As each of these examples demonstrates, assembling the right computer hardware requires an understanding of its relationship to the information system and the needs of the organization. Remember that the computer hardware objectives are subordinate to, but supportive of, the information system and the needs of the organization.

The components of all information systems—such as hardware devices, people, and procedures—are interdependent. Because the performance of one system affects the others, all of these systems should be measured according to the same standards of effectiveness and efficiency, given the constraints of cost, control, and complexity.

When selecting computer hardware, you also must consider the current and future uses to which these systems will be put. Your choice of a particular computer system should always allow for later improvements in the overall information system. Reasoned forethought—a

trait required for dealing with computer, information, and organizational systems of all sizes—is the hallmark of a true business professional.

Hardware Components

Computer system hardware components include devices that perform the functions of input, processing, data storage, and output (see Figure 3.1). To understand how these hardware devices work together, consider an analogy from a paper-based office environment. Imagine a one-room office occupied by a single individual. The human being (the processor) is capable of organizing and manipulating data. The person's mind (register storage) and the desk occupied by the human being (primary storage) are places to temporarily store data. Filing cabinets fill the need for a more permanent form of storage (secondary storage). In this analogy, the incoming and outgoing mail trays can be understood as sources of new data (input) or as places to put the processed paperwork (output).

central processing unit (CPU)
The part of the computer that consists of three associated elements: the arithmetic/logic unit, the control unit, and the register areas.

arithmetic/logic unit (ALU)
Portion of the CPU that performs mathematical calculations and makes logical comparisons.

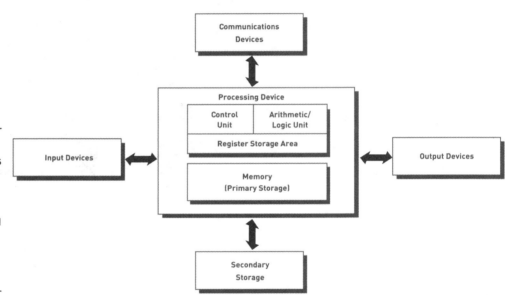

Figure 3.1

Computer System Components

These components include input devices, output devices, communications devices, primary and secondary storage devices, and the **central processing unit (CPU)**. The control unit, the arithmetic/logic unit (ALU), and the register storage areas constitute the CPU.

control unit
Part of the CPU that sequentially accesses program instructions, decodes them, and coordinates the flow of data in and out of the ALU, the registers, primary storage, and even secondary storage and various output devices.

register
High-speed storage area in the CPU used to temporarily hold small units of program instructions and data immediately before, during, and after execution by the CPU.

primary storage (main memory; memory)
Part of the computer that holds program instructions and data.

The ability to process (organize and manipulate) data is a critical aspect of a computer system, in which processing is accomplished by an interplay between one or more of the central processing units and primary storage. Each **central processing unit** (CPU) consists of three associated elements: the arithmetic/logic unit, the control unit, and the register areas. The **arithmetic/logic unit** (ALU) performs mathematical calculations and makes logical comparisons. The **control unit** sequentially accesses program instructions, decodes them, and coordinates the flow of data in and out of the ALU, the registers, primary storage, and even secondary storage and various output devices. **Registers** are high-speed storage areas used to temporarily hold small units of program instructions and data immediately before, during, and after execution by the CPU.

Primary storage, also called **main memory** or just **memory**, is closely associated with the CPU. Memory holds program instructions and data immediately before or immediately after the registers. To understand the function of processing and the interplay between the CPU and memory, let's examine the way a typical computer executes a program instruction.

Hardware Components in Action

The execution of any machine-level instruction involves two phases: the instruction phase and the execution phase. During the instruction phase, the following takes place:

- *Step 1: Fetch instruction.* The fetch stage reads a program's instructions and any necessary data into the processor.
- *Step 2: Decode instruction.* The instruction is decoded and is passed to the appropriate processor execution unit. There are several execution units: The arithmetic/logic unit

performs all arithmetic operations, the floating-point unit deals with noninteger operations, the load/store unit manages the instructions that read or write to memory, the branch processing unit predicts the outcome of a branch instruction in an attempt to reduce disruptions in the flow of instructions and data into the processor, the memory-management unit translates an application's addresses into physical memory addresses, and the vector processing unit handles vector-based instructions that accelerate graphics operations.

Steps 1 and 2 are called the instruction phase, and the time it takes to perform this phase is called the **instruction time (I-time)**.

The second phase is the execution phase. During the execution phase, the following steps are performed:

- *Step 3: Execute the instruction.* The execution stage is where the hardware element, now freshly fed with an instruction and data, carries out the instruction. This could involve making an arithmetic computation, logical comparison, bit shift, or vector operation.
- *Step 4: Store results.* During this step, the results are stored in registers or memory.

Steps 3 and 4 are called the execution phase. The time it takes to complete the execution phase is called the **execution time (E-time)**.

After both phases have been completed for one instruction, they are again performed for the second instruction, and so on. The instruction phase followed by the execution phase is called a **machine cycle** (see Figure 3.2). Some processing units can speed up processing by using **pipelining**, whereby the processing unit gets one instruction, decodes another, and executes a third at the same time. The Pentium 4 processor, for example, uses two execution unit pipelines. This gives the processing unit the ability to execute two instructions in a single machine cycle.

instruction time (I-time)
The time it takes to perform the fetch-instruction and decode-instruction steps of the instruction phase.

execution time (E-time)
The time it takes to execute an instruction and store the results.

machine cycle
The instruction phase followed by the execution phase.

pipelining
A form of CPU operation in which there are multiple execution phases in a single machine cycle.

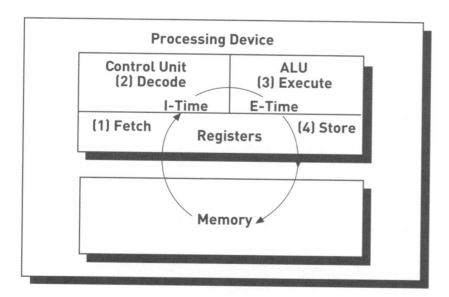

Figure 3.2

Execution of an Instruction

In the instruction phase, a program's instructions and any necessary data are read into the processor (1). Then the instruction is decoded so the central processor can understand what is to be done (2). In the execution phase, the ALU does what it is instructed to do, making either an arithmetic computation or a logical comparison (3). Then the results are stored in the registers or in memory (4). The instruction and execution phases together make up one machine cycle.

PROCESSING AND MEMORY DEVICES: POWER, SPEED, AND CAPACITY

The components responsible for processing—the CPU and memory—are housed together in the same box or cabinet, called the *system unit*. All other computer system devices, such as the monitor and keyboard, are linked either directly or indirectly into the system unit housing. As discussed previously, achieving IS objectives and organizational goals should be the primary consideration in selecting processing and memory devices. In this section, we investigate the characteristics of these important devices.

Processing Characteristics and Functions

Because having efficient processing and timely output is important, organizations use a variety of measures to gauge processing speed. These measures include the time it takes to complete a machine cycle and clock speed.

Machine Cycle Time

As we've seen, the execution of an instruction takes place during a machine cycle. The time in which a machine cycle occurs is measured in fractions of a second. Machine cycle times are measured in *microseconds* (one-millionth of one second) for slower computers to *nanoseconds* (one-billionth of one second) and *picoseconds* (one-trillionth of one second) for faster ones. Machine cycle time also can be measured in terms of how many instructions are executed in a second. This measure, called **MIPS**, stands for millions of instructions per second. MIPS is another measure of speed for computer systems of all sizes.

MIPS
Millions of instructions per second.

Clock Speed

Each CPU produces a series of electronic pulses at a predetermined rate, called the **clock speed**, which affects machine cycle time. The control unit portion of the CPU controls the various stages of the machine cycle by following predetermined internal instructions, known as **microcode**. You can think of microcode as predefined, elementary circuits and logical operations that the processor performs when it executes an instruction. The control unit executes the microcode in accordance with the electronic cycle, or pulses of the CPU "clock." Each microcode instruction takes at least the same amount of time as the interval between pulses. The shorter the interval between pulses, the faster each microcode instruction can be executed (see Figure 3.3).

clock speed
A series of electronic pulses produced at a predetermined rate that affects machine cycle time.

microcode
Predefined, elementary circuits and logical operations that the processor performs when it executes an instruction.

Figure 3.3

Clock Speed and the Execution of Microcode Instructions

A faster clock speed means that more microcode instructions can be executed in a given time period.

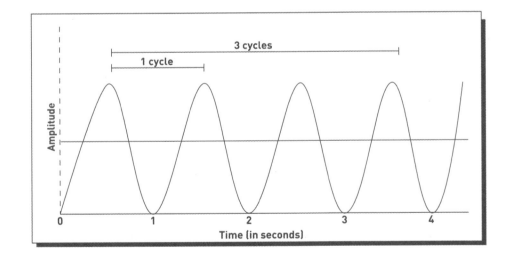

Clock speed is often measured in megahertz. As seen in Figure 3.3, a **hertz** is one cycle or pulse per second. **Megahertz (MHz)** is the measurement of cycles in millions of cycles per second, and **gigahertz (GHz)** stands for billions of cycles per second. The clock speed for personal computers can range from 200 MHz for computers bought in the mid-1990s to well over 3.2 GHz for the most advanced systems.[6]

hertz
One cycle or pulse per second.

megahertz (MHz)
Millions of cycles per second.

gigahertz (GHz)
Billions of cycles per second.

Because the number of microcode instructions needed to execute a single program instruction—such as performing a calculation or printing results—can vary, there is no direct relationship between clock speed measured in megahertz and processing speed measures such as MIPS and milliseconds. Although widely touted, clock speed is only meaningful when making speed comparisons between computer chips in the same family from the same manufacturer; comparing one Intel Pentium 4 chip with another, for example.

Wordlength and Bus Line Width

Data is moved within a computer system in units called bits. A **bit** is a binary digit—0 or 1. Another factor affecting overall system performance is the number of bits the CPU can

bit
Binary digit—0 or 1.

process at one time, or the **wordlength** of the CPU. Early computers were built with CPUs that had a wordlength of 4 bits, meaning that the CPU was capable of processing 4 bits at one time. The 4 bits could be used to represent actual data, an instruction to be processed, or the address of data to be accessed. The 4-bit limitation was quite confining and greatly constrained the power of the computer. Over time, CPUs have evolved to 8-, 16-, 32-, and 64-bit machines with dramatic increases in power and capability. Computers with larger wordlengths can transfer more data between devices in the same machine cycle. They can also use the larger number of bits to address more memory locations and hence are a requirement for systems with certain large memory requirements. A 64-bit machine allows the CPU to directly address 18 quintillion (a billion billion) unique address locations compared with 4.3 billion for a 32-bit processor. The ability to directly access a larger address space is critical for multimedia, imaging, and database applications; however, the computer's operating system and related application software must also support 64-bit technology to achieve the full benefit of the 64-bit architecture.

Data is transferred from the CPU to other system components via **bus lines**, the physical wiring that connects the computer system components. The number of bits a bus line can transfer at any one time is known as bus line width. Bus line width should be matched with CPU wordlength for optimal system performance. It would be of little value, for example, to install a new 64-bit bus line if the system's CPU had a wordlength of only 16. Assuming compatible wordlengths and bus widths, the larger the wordlength, the more powerful the computer.

Because all these factors—machine cycle time, clock speed, wordlength, and bus line width—affect the processing speed of the CPU, comparing the speed of two different processors even from the same manufacturer can be confusing. Although the megahertz rating has important consequences for the design of a computer system and is therefore important to the computer design engineer, it is not necessarily a good measure of processor performance, especially when comparing one family of processors with the next or when making comparisons between manufacturers. Chip makers such as Intel, Advanced Micro Devices, and Sun Microsystems have developed a number of benchmarks for speed. To ensure objective comparisons, many people prefer to use general computer system benchmarks such as SYSmark, distributed by a consortium called the Business Application Performance Corporation, whose members include hardware and software manufacturers such as AMD, Dell, Hewlett-Packard Co., IBM, Intel, and Microsoft, and industry publications such as Computer Shopper and ZDNet. For less-technical measures of performance, popular computer journals (such as *PC Magazine* and *PC World*) often rate personal computers on price, performance, reliability, service, and other factors.

Physical Characteristics of the CPU

CPU speed is also limited by physical constraints. Most CPUs are collections of digital circuits imprinted on silicon wafers, or chips, each no bigger than the tip of a pencil eraser. To turn a digital circuit within the CPU on or off, electrical current must flow through a medium (usually silicon) from point A to point B. The speed at which it travels between points can be increased by either reducing the distance between the points or reducing the resistance of the medium to the electrical current.

Reducing the distance between points has resulted in ever-smaller chips, with the circuits packed closer together. In the 1960s, shortly after patenting the integrated circuit, Gordon Moore, former chairman of the board of Intel (the largest microprocessor chip maker), formulated what is now known as **Moore's Law**. This hypothesis states that transistor (the microscopic on/off switches, or the microprocessor's brain cells) densities on a single chip will double every 18 months. Moore's Law has held up amazingly for nearly four decades. In 2003, Moore himself forecast that the steady growth in the density and performance of microprocessors may go on only for the next 8 to 12 years. A key problem will be the need to narrow the minimum width of basic circuit features of a chip, which today are 90 nanometers (one billionth of a meter, 10^{-9} meter) wide in the most advanced chip manufacturing process.[7] (The 90-nanometer designation refers to the width of the smallest circuit lines on the chip. The actual features on 90-nanometer chips can be quite small, down to around 45 nanometers for some of the smallest structures on the chip.)[8] Chip makers then will have to

find a new technology to replace the semiconductor if they are to keep up this rate of improvement. As silicon transistors grow smaller—there will be a billion on a single chip by 2008—power dissipation becomes nearly impossible to control, and even cosmic rays can cause random processing errors.[9] According to Nathan Brockwood, principal analyst at research company Insight 64, "Every generation [of new chips] requires greater investment in [research and development] and manufacturing to make it work, because the low-hanging fruit in terms of semiconductor production was harvested years ago."[10] In addition to increased processing speeds, Moore's Law has had an impact on costs and overall system performance. As seen in Figure 3.4, the number of transistors on a chip continues to climb.

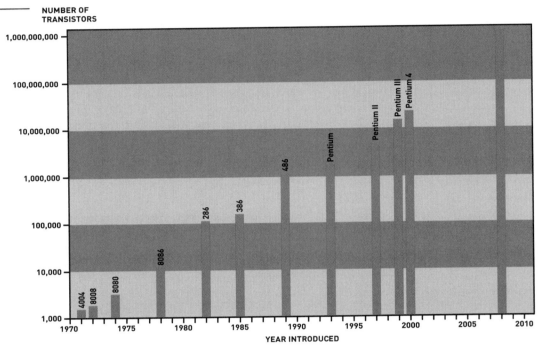

superconductivity
A property of certain metals that allows current to flow with minimal electrical resistance.

optical processors
Computer chips that use light waves instead of electrical current to represent bits.

Researchers are taking many approaches to continue to improve the performance of computers. One approach is to substitute superconductive material for the silicon in computer chips. **Superconductivity** is a property of certain metals that allows current to flow with minimal electrical resistance. Traditional silicon chips create some electrical resistance that slows processing. Chips built from less-resistant superconductive metals offer increases in processing speed. Materials other than silicon, including carbon and gallium arsenide (GaAs), are used in the development of special-purpose chips.

Several companies are experimenting with chips called **optical processors**, which use light waves instead of electrical current to represent bits. The primary advantage of optical processors is their speed. Lenslet, an Israeli-based start-up firm, has developed an optical computer that is capable of performing 8 trillion operations per second (teraflops). The company's prototype is fairly large and bulky, but the goal is to shrink it to a single chip by 2008. The powerful processor will be put to use in such applications as high-resolution radar, electronic warfare, luggage screening at airports, video compression, weather forecasting, and cellular phone base stations.[11]

IBM and other companies are turning to strained silicon, a technique that boosts performance and lowers power consumption by stretching silicon molecules farther apart, thus allowing electrons to experience less resistance and flow up to 70 percent faster, which can lead to chips that are up to 35 percent faster—without having to shrink the size of transistors.[12]

Another advancement is the development of the carbon nanotube, so called because it is a pure carbon tube made of hexagonal structures 1 to 3 nanometers in diameter. The

PMC-Sierra's RM9000 family of MIPS-RISC processors provides high-performance and low-power solutions for embedded applications such as networking, printing, workstation, and consumer devices.

(Source: Courtesy of PMC-Sierra, Inc.)

nanotubes can be used to form the tiny circuits for computer components. The practical use of nanotubes may not occur until sometime in the 2010 decade because of manufacturing difficulties and other complications.

Scientists at the Los Alamos National Laboratory have taken miniaturization to the extreme. They are experimenting with using radio waves to manipulate individual atoms into executing a simple computer program. Their goal is to be able to manipulate thousands of atoms and build a computer many times smaller yet more powerful than any computer currently in existence.

Complex and Reduced Instruction Set Computing

Processors for many personal computers are designed based on **complex instruction set computing (CISC)**, which places as many microcode instructions into the central processor as possible. In the mid-1970s John Cocke of IBM recognized that most of the operations of a CPU involved only about 20 percent of the available microcode instructions. This led to an approach to chip design called **reduced instruction set computing (RISC)**, which involves reducing the number of microcode instructions built into a chip to this essential set of common microcode instructions. RISC chips are faster than CISC chips for processing activities that predominantly use this core set of instructions because each operation requires fewer microcode steps prior to execution. Most RISC chips use pipelining, which, as mentioned earlier, allows the processor to execute multiple instructions in a single machine cycle. With less sophisticated microcode instruction sets, RISC chips are also less expensive to produce and are quite reliable.

Gates Corporation, which makes belts, hoses, and automotive products, moved 42 business applications off its aging mainframe computers onto two RISC-based HP Superdome servers running an Oracle ERP program and HP systems-management software. Gates has 64 processors installed in each Superdome, which can be expanded to 256 processors. The move was made to boost processing capacity and system reliability.[13]

In June 2003, Apple introduced the PowerPC G5, the world's first 64-bit RISC processor for desktop computers. By almost any benchmark, RISC processors run faster than Intel's Pentium processor. And because RISC chips have a simpler design and require less silicon, they are cheaper to produce. The PowerPC chip is designed to provide portable and desktop personal computers the processing power normally associated with much more expensive computers. For example, the Macintosh PowerPC G5 has the ability to make functions such as voice recognition, dictation, pen input, and touchscreens practical. The PowerPC G5 can come with two IBM PowerPC 970 microprocessors with a 64-bit architecture. Sun Microsystems' Sparc chip is another example of a RISC processor.

When selecting a CPU, organizations must balance the benefits of speed with cost. CPUs with faster clock speeds and machine cycle times are usually more expensive than slower ones. This expense, however, is a necessary part of the overall computer system cost, for the CPU

complex instruction set computing (CISC)
A computer chip design that places as many microcode instructions into the central processor as possible.

reduced instruction set computing (RISC)
A computer chip design based on reducing the number of microcode instructions built into a chip to an essential set of common microcode instructions.

is typically the single largest determinant of the price of many computer systems. CPU speed can also be related to complexity. Having a less complex code, as in the case of RISC chips, not only can increase speed and reliability but can also reduce chip manufacturing costs.

Memory Characteristics and Functions

Main memory is located physically close to the CPU, but not on the CPU chip itself. It provides the CPU with a working storage area for program instructions and data. The chief feature of memory is that it rapidly provides the data and instructions to the CPU.

Storage Capacity

byte (B)
Eight bits that together represent a single character of data.

Like the CPU, memory devices contain thousands of circuits imprinted on a silicon chip. Each circuit is either conducting electrical current (on) or not (off). Data is stored in memory as a combination of on or off circuit states. Usually 8 bits are used to represent a character, such as the letter *A*. Eight bits together form a **byte** (**B**). Following is a list of storage capacity measurements. In most cases, storage capacity is measured in bytes, with one byte usually equal to one character. The contents of the Library of Congress, with over 126 million items and 530 miles of bookshelves, would require about 20 petabytes of digital storage.

Name	Abbreviation	Number of Bytes
Byte	B	1
Kilobyte	KB	2^{10} or approximately 1,024 bytes
Megabyte	MB	2^{20} or 1,024 kilobytes (about 1 million)
Gigabyte	GB	2^{30} or 1,024 megabytes (about 1 billion)
Terabyte	TB	2^{40} or 1,024 gigabytes (about 1 trillion)
Petabyte	PB	2^{50} or 1,024 terabytes (about 1 quadrillion)
Exabyte	EB	2^{60} or 1,024 petabytes (about 1 billion billion, or 1 quintillion)

Types of Memory

random access memory (RAM)
A form of memory in which instructions or data can be temporarily stored.

There are several forms of memory, as shown in Figure 3.5. Instructions or data can be temporarily stored in **random access memory** (**RAM**). RAM is temporary and volatile—RAM chips lose their contents if the current is turned off or disrupted (as in a power surge, brownout, or electrical noise generated by lightning or nearby machines). RAM chips are mounted directly on the computer's main circuit board or in other chips mounted on peripheral cards that plug into the computer's main circuit board. These RAM chips consist of millions of switches that are sensitive to changes in electric current.

Figure 3.5

Basic Types of Memory Chips

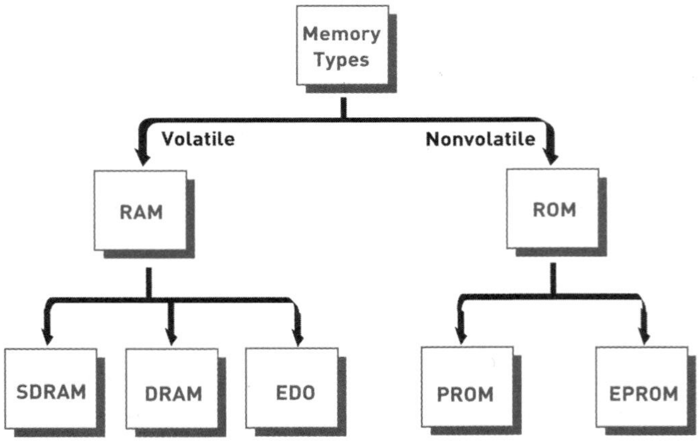

RAM comes in many different varieties. One version is extended data out, or EDO RAM, which is faster than older types of RAM memory. Another kind of RAM memory is called dynamic RAM (DRAM) and is based on single-transistor memory cells. SDRAM, or synchronous DRAM employs a minimum of four transistors per memory cell and needs high or low voltages at regular intervals—every two milliseconds (two one-thousands of a second)—to retain its information. Compared with EDO RAM, SDRAM provides a faster transfer speed between the microprocessor and the memory.

Another type of memory, **ROM**, an acronym for **read-only memory**, is usually non-volatile. In ROM, the combination of circuit states is fixed, and therefore its contents are not lost if the power is removed. ROM provides permanent storage for data and instructions that do not change, such as programs and data from the computer manufacturer, including the instructions that tell the computer how to start up when power is turned on.

There are other types of nonvolatile memory as well. Programmable read-only memory (PROM) is a type in which the desired data and instructions—and hence the desired circuit state combination—must first be programmed into the memory chip. Thereafter, PROM behaves like ROM. PROM chips are used in situations in which the CPU's data and instructions do not change, but the application is so specialized or unique that custom manufacturing of a true ROM chip would be cost prohibitive. A common use of PROM chips is for storing the instructions to popular video games, such as those for GameBoy and Xbox. Game instructions are programmed onto the PROM chips by the game manufacturer. Instructions and data can be programmed onto a PROM chip only once.

Erasable programmable read-only memory (EPROM) is similar to PROM except, as the name implies, the memory chip can be erased and reprogrammed. EPROMs are used when the CPU's data and instructions change, but only infrequently. An automobile manufacturer, for example, might use an industrial robot to perform repetitive operations on a certain car model. When the robot is performing its operations, the nonvolatility and rapid accessibility to program instructions offered by EPROM is an advantage. Once the model year is over, however, the EPROM controlling the robot's operation will need to be erased and reprogrammed to accommodate a different car model.

Over the past decade, microprocessor speed has doubled every 18 months, but memory performance has not kept pace. In effect, memory has become the principal bottleneck to system performance. Thus, microprocessor manufacturers are working with memory vendors to create memory that can keep up with the performance of faster processors and bus architectures. One approach that has been taken is to employ **cache memory**, a type of high-speed memory that a processor can access more rapidly than main memory (see Figure 3.6). Frequently used data is stored in easily accessible cache memory instead of slower memory such as RAM. Because there is less data in cache memory, the CPU can access the desired data and instructions more quickly than if it were selecting from the larger set in main memory. The CPU can thus execute instructions faster, and the overall performance of the computer system is improved. There are three types of cache memory. The Level 1 (L1) cache is on the CPU chip. The Level 2 (L2) cache memory can be accessed by the CPU over a high-speed dedicated bus interface. The latest processors go a step further and place the L2 cache directly on the CPU chip itself and provide high-speed support for a tertiary Level 3 (L3) external cache. Deerfield, often called the Itanium 2 chip, is the low-power version of Intel's 64-bit servers and was introduced with 1.5 MB of Level 3 cache.[14]

read-only memory (ROM)
A nonvolatile form of memory.

cache memory
A type of high-speed memory that a processor can access more rapidly than main memory.

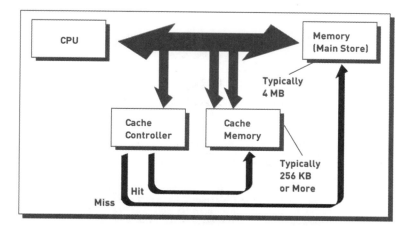

Figure 3.6

Cache Memory

Processors can access this type of high-speed memory faster than main memory. Located on or near the CPU chip, cache memory works in conjunction with main memory. A cache controller determines how often the data is used and transfers frequently used data to cache memory, then deletes the data when it goes out of use.

When the processor needs to execute an instruction, it looks first in its own data registers. If the needed data is not there, it looks to the L1 cache, then to the L2 cache, then to the L3

cache. If the data is not in any cache, the CPU requests the data from main memory. It might not even be there, in which case the system has to retrieve the data from secondary storage. It can take from one to three clock cycles to fetch information from the L1 cache, while the CPU waits and does nothing. It takes 6 to 12 cycles to get data from an L2 cache on the processor chip. It can take dozens of cycles to fetch data from an L3 cache and hundreds of cycles to fetch data from secondary storage. This hierarchical arrangement of memory helps bridge a widening gap between processor speeds, which are increasing at roughly 50 percent per year, and DRAM access rates, which are climbing at only 5 percent per year.

Costs for memory capacity continue to decline. When considered on a megabyte-to-megabyte basis, memory is still considerably more expensive than most forms of secondary storage. Memory capacity can be important in the effective operation of a CBIS. The specific applications of a CBIS determine the amount of memory required for a computer system. For example, complex processing problems, such as computer-assisted product design, re-quire more memory than simpler tasks such as word processing. Also, because computer systems have different types of memory, other programs may be needed to control how memory is accessed and used. In other cases, the computer system can be configured to maximize memory usage. Before additional memory is purchased, all these considerations should be addressed.

Multiprocessing

multiprocessing
The simultaneous execution of two or more instructions at the same time.

A number of forms of **multiprocessing** involve the simultaneous execution of two or more instructions at the same time. One form of multiprocessing involves coprocessors. A **copro-cessor** speeds processing by executing specific types of instructions while the CPU works on another processing activity. Coprocessors can be internal or external to the CPU and may have different clock speeds than the CPU. Each type of coprocessor performs a specific func-tion. For example, a math coprocessor chip can be used to speed mathematical calculations, and a graphics coprocessor chip decreases the time it takes to manipulate graphics.

coprocessor
Part of the computer that speeds processing by executing specific types of instructions while the CPU works on another processing activity.

Massively Parallel Processing

Another form of multiprocessing, called **massively parallel processing**, speeds processing by linking hundreds and even thousands of processors to operate at the same time, or in parallel. Each processor includes its own bus, memory, disks, copy of the operating system, and ap-plication software. With parallel processing, a business problem (such as designing a new product or piece of equipment) is divided into several parts. Each part is "solved" by a separate processor. The results from each processor are then assembled to get the final output (see Figure 3.7).

massively parallel processing
A form of multiprocessing that speeds processing by linking hun-dreds or thousands of processors to operate at the same time, or in par-allel, with each processor having its own bus, memory, disks, copy of the operating system, and applications.

Massively parallel processing systems can coordinate large amounts of data and access them with greater speed than was previously possible. The most frequent business uses for massive parallel processing include modeling, simulation, and the analysis of large amounts

Figure 3.7

Massively Parallel Processing

Massively parallel processing involves breaking a problem into various subproblems or parts, then processing each of these parts independently. The most difficult aspect of massively parallel processing is not the simultaneous processing of the subproblems but the logical structuring of the prob-lem into independent parts.

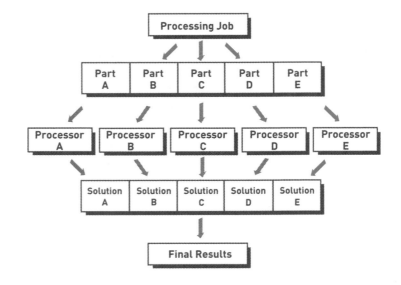

of data. In today's challenging marketplace, consumers are demanding increased product features and a whole array of new services. These consumer demands have forced companies to find more effective and insightful ways of gathering and analyzing information, not just about existing customers but also potential customers. Collecting and organizing this enormous amount of data is difficult. Massively parallel processing can access and analyze the data to create the information necessary to build an effective marketing program that can give the company a competitive advantage. Ford Motor Company is interested in the use of massively parallel processing to predict driver and passenger injuries in accident scenarios. The computer could compute and predict the damage done to human organs. Today's analyses with test dummies are very crude. They can determine at a basic level whether a certain type of crash is survivable. But occupant injury analysis takes much more computing power than is available now.[15]

Symmetrical multiprocessing (SMP) is another form of parallel processing in which multiple processors run a single copy of the operating system and share the memory and other resources of one computer. Sharing resources creates more overhead than a single-processor system or the massively parallel processing system. As a result, the processing capability of SMP systems isn't proportionally greater than that of single processor systems (i.e., the capability of an SMP processor with two processors is less than twice the speed of a single processor). SMP has been implemented in the Sun Microsystems' UltraSparc and Sparcserver, IBM Alpha, Macintosh PowerPC, and Intel chips.

Cendant Corporation operates Internet-based travel sites, including CheapTickets.com and Galileo, a corporate reservations site. It recently moved its airline fare system to IBM e-Server SMP systems to deliver fares to customers faster. In addition, the project saved the company tens of millions of dollars in hardware maintenance and programming costs. The Galileo 360 eFares system now runs on more than 100 clustered IBM eServer x440 and x445 systems. These computers are high-performance four-way and eight-way SMP servers that use Intel Xeon processors, which Cendant links together in powerful clusters. With the new SMP systems, Cendant reduced the hours of preprocessing work necessary to post the new fares issued by airlines six times a day. End users can access airline updates immediately and get the new fares faster.[16]

Grid Computing

Grid computing is the use of a collection of computers, often owned by multiple individuals or organizations, to work in a coordinated manner to solve a common problem. Grid computing is one low-cost approach to massively parallel processing. The grid can include dozens, hundreds, or even thousands of computers that run collectively to solve extremely large parallel processing problems. Key to the success of grid computing is a central server that acts as the grid leader and traffic monitor. This controlling server divides the computing task into subtasks and assigns the work to computers on the grid that have (at least temporarily) surplus processing power. The central server also monitors the processing, and if a member of the grid fails to complete a subtask, it will restart or reassign the task. When all the subtasks are completed, the controlling server combines the results and advances to the next task until the whole job is completed.

The types of computing problems most compatible with grid computing are those that can be divided into problem subsets that can run in parallel. Such problems are frequently encountered in scientific and engineering computing. Mission-critical or time-sensitive applications should not be considered for grid computing because security, reliability, and performance cannot be guaranteed currently. Business issues also arise. For example, if an application is to run on 25 computers owned by 15 different organizations, how do you define who pays the costs and how those costs are allocated among the organizations?

In a grid experiment nicknamed "the Big Mac," Virginia Tech students and staff linked 1,100 Macintosh G5 Power Mac computers to form the world's third-fastest supercomputer, capable of performing 10.3 trillion operations per second. The processors are linked by a high-speed network called *Infiniband* that allows them to break up major calculations and analyze each part at the same time. The entire system cost about $7 million.[17]

symmetrical multiprocessing (SMP)
Another form of parallel processing in which multiple processors run a single copy of the operating system and share the memory and other resources of one computer.

grid computing
The use of a collection of computers, often owned by multiple individuals or organizations, to work in a coordinated manner to solve a common problem.

By installing the SETI@home screen-saver program on their personal computers, millions of people worldwide contribute their idle CPU time to analyzing radio data from space for signs of intelligent life. With SETI, a central server doles out the work by giving each computer on the grid data from a tiny slice of the sky to analyze.

(Source: Copyright ©2003 SETI@home.)

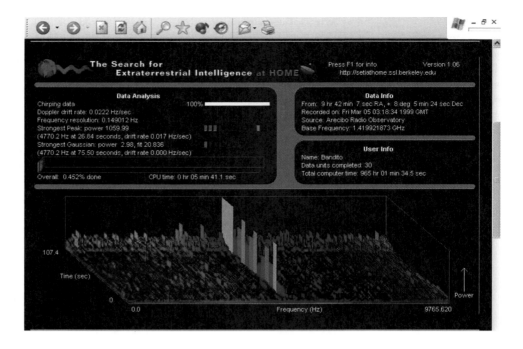

SECONDARY STORAGE

Driven by such factors as the need to retain more data longer to meet government regulatory concerns, the need to store new forms of digital data such as audio and video data, and the need to keep systems running under the onslaught of ever-increasing volumes of e-mail, the amount of data that companies store digitally is increasing at a rate of 85 to 95 percent annually![18] Organizations need a way to store large amounts of data and instructions more permanently than main memory allows. **Secondary storage**, also called *permanent storage*, serves this purpose.

Compared with memory, secondary storage offers the advantages of nonvolatility, greater capacity, and greater economy. As previously noted, on a cost-per-megabyte basis, most forms of secondary storage are considerably less expensive than memory (see Table 3.1). Because of the electromechanical processes involved in using secondary storage, however, it is considerably slower than memory. The selection of secondary storage media and devices requires an understanding of their primary characteristics—access method, capacity, and portability.

Table 3.1

Cost Comparison for Various Forms of Data Storage

All forms of secondary storage cost considerably less per megabyte of capacity than SDRAM, although they have slower access times. A data tape cartridge costs about $.16 per gigabyte, while SDRAM can cost around $900 per gigabyte.

Data Storage Media	Capacity	Cost/GB
Data tape cartridge	500 GB	$.16
DVD-RW	4.7 GB	$.29
CD-ROM	740 MB	$.34
Floppy diskette	1.44 MB	$347
Compact flash memory	128 MB	$367
SDRAM	128 MB	$874

As with other computer system components, the access methods, storage capacities, and portability required of secondary storage media are determined by the information system's objectives. An objective of a credit card company's information system, for example, might be to rapidly retrieve stored customer data to approve customer purchases. In this case, a fast access method is critical. In other cases, such as equipping a sales force with laptop computers, portability and storage capacity might be major considerations in selecting and using secondary storage media and devices.

Storage media that provide faster access methods are generally more expensive than slower media. The cost of additional storage capacity and portability vary widely, but they are also factors to consider. In addition to cost and portability, organizations must address security issues to allow only authorized people access to sensitive data and critical programs. Because the data and programs kept in secondary storage devices are so critical to most organizations, all of these issues merit careful consideration.

Access Methods

Data and information access can be either sequential or direct. **Sequential access** means that data must be accessed in the order in which it is stored. For example, inventory data stored sequentially may be stored by part number, such as 100, 101, 102, and so on. If you want to retrieve information on part number 125, you must read and discard all the data relating to parts 001 through 124.

Direct access means that data can be retrieved directly, without the need to pass by other data in sequence. With direct access, it is possible to go directly to and access the needed data—say, part number 125—without having to read through parts 001 through 124. For this reason, direct access is usually faster than sequential access. The devices used to sequentially access secondary storage data are simply called **sequential access storage devices (SASDs)**; those used for direct access are called **direct access storage devices (DASDs)**.

Secondary Storage Devices

The most common forms of secondary storage include magnetic tapes, magnetic disks, virtual tapes, and optical discs. Some of these media (magnetic tapes) allow only sequential access, while others (magnetic and optical discs) provide direct and sequential access. Figure 3.8 shows some different secondary storage media.

Magnetic Tape

One common secondary storage medium is **magnetic tape**. Similar to the tape found in audio- and videocassettes, magnetic tape is a Mylar film coated with iron oxide. Portions of the tape are magnetized to represent bits. Magnetic tape is an example of a sequential access storage medium. If the computer needs to read data from the middle of a reel of tape, all the tape before the desired piece of data must be passed over sequentially—one disadvantage of magnetic tape. When information is needed, it can take time for a tape operator to load the magnetic tape on a tape device and get the relevant data into the computer. Despite the falling prices of hard-disk drives, tape storage is still a popular choice for low-cost data backup for off-site storage in the event of a disaster.

---

secondary storage (permanent storage)
Devices that store larger amounts of data, instructions, and information more permanently than allowed with main memory.

sequential access
Retrieval method in which data must be accessed in the order in which it is stored.

direct access
Retrieval method in which data can be retrieved without the need to read and discard other data.

sequential access storage device (SASD)
Device used to sequentially access secondary storage data.

direct access storage device (DASD)
Device used for direct access of secondary storage data.

Figure 3.8

Types of Secondary Storage

Secondary storage devices such as magnetic tapes and disks, optical discs, CD-ROMs, and DVDs are used to store data for easy retrieval at a later date.

(Source: Courtesy of Imation Corp.)

magnetic tape
Common secondary storage medium; Mylar film coated with iron oxide with portions of the tape magnetized to represent bits.

Table 3.2

Comparison of Tape Formats

(Source: Data from Russell Kay, "Tape Types," *Computerworld*, August 25, 2003, accessed at *www.computerworld.com*.)

Technology is improving to provide tape storage devices with greater capacities and faster transfer speeds. In addition, the large, bulky tape drives used to read and write on large diameter reels of tapes in the early days of computing have been replaced with much smaller tape cartridge devices measuring a few millimeters in diameter that take up much less floor space and allow hundreds of tape cartridges to be stored in a small area. Several competing cartridge tape formats are summarized in Table 3.2.

Tape Format	Manufacturers	Specific Model	Storage Capacity (GB)	Data Transfer Rate (GB/Second)	Approximate Cost of Cartridge per GB of Storage (Winter 2004)
S-AIT Advanced Intelligent Tape	Sony Electronics	SAIT-1	500	30	$.16
LTO-2 Linear Tape Open	Hewlett-Packard IBM Certance	Ultrium-2	200	35	$.60
SDLT Super Digital Linear Tape	Quantum	SDLT 320	160	16	$.81
VXA	Exabyte	VXA-2	80	6	$1.00
Mammoth	Exabyte	Mammoth-2	60	12	$1.00

Magnetic Disks

magnetic disk

Common secondary storage medium, with bits represented by magnetized areas.

Magnetic disks are also coated with iron oxide; they can be thin steel platters (hard disks; see Figure 3.9) or Mylar film (diskettes). As with magnetic tape, magnetic disks represent bits by small magnetized areas. When reading from or writing data onto a disk, the disk's read/ write head can go directly to the desired piece of data. Thus, the disk is a direct access storage medium. Although disk devices can be operated in a sequential mode, most disk devices use direct access. Because direct access allows fast data retrieval, this type of storage is ideal for companies that need to respond quickly to customer requests, such as airlines and credit card firms. For example, if a manager needs information on the credit history of a customer or the seat availability on a particular flight, the information can be obtained in a matter of seconds if the data is stored on a direct access storage device.

Figure 3.9

Hard Disk

Hard disks give direct access to stored data. The read/write head can move directly to the location of a desired piece of data, dramatically reducing access times, as compared with magnetic tape.

(Source: Courtesy of Seagate Technology.)

Magnetic disk storage varies widely in capacity and portability. Standard diskettes are portable but have a slower access time and lower storage capacity (1.44 MB for some computers) than fixed hard disks. Hard-disk storage, while more costly and less portable, has a greater storage capacity and quicker access time.

Making magnetic media hold more data has traditionally been a matter of making the grains in the recording medium, and the spots that hold the recorded bit, ever smaller and closer together. During the 1990s, the storage density of magnetic disks increased 75 percent per year, outpacing even Moore's Law. Between the fall of 1999 and the fall of 2003, the hard-disk drive industry increased capacity by more than 20 times and dropped the cost per unit of memory by half.[19] IBM's Millipede is a 1 terabit-per-square-inch recording device that holds the equivalent of 25 DVDs on a surface the size of a postage stamp.[20] Toshiba and others are developing coin-sized hard drives that can hold up to 3 GB.[21]

RAID

Putting an organization's data online involves a serious business risk—the loss of critical business data can put a corporation out of business. The concern is that the most critical mechanical components inside a disk storage device—the disk drives, the fans, and other input/output devices—can break (like most things that move).

Organizations now require that their data-storage devices be fault tolerant—have the ability to continue with little or no loss of performance in the event of a failure of one or more key components. A **redundant array of independent/inexpensive disks (RAID)** is a method of storing data that generates extra bits of data from existing data, allowing the system to create a "reconstruction map" so that if a hard drive fails, it can rebuild lost data. With this approach, data is split and stored on different physical disk drives using a technique called *stripping* to evenly distribute the data. Since being developed at the University of Berkeley in 1987, RAID technology has been applied to storage systems to improve system performance and reliability.

RAID can be implemented in several ways. In the simplest form, RAID subsystems duplicate data on drives. This process, called **disk mirroring**, provides an exact copy that protects users fully in the event of data loss. However, if full copies are always to be kept current, organizations need to double the amount of storage capacity that is kept online. Thus, disk mirroring is expensive. Other RAID methods are less expensive because they only partly duplicate the data, allowing storage managers to minimize the amount of extra disk space (or overhead) they must purchase to protect data.

The effective use of RAID storage hardware and software can lead to savings from improved efficiency. Quicken Loans, Inc., an online mortgage-lending company, moved to a centralized RAID storage system with automated management software from EMC Corp. and reduced the number of IT staffers devoted to managing storage from 15 to 3. The easier access to information also helps cut the time to create detailed financial reports from one to three days to two to eight hours. [22]

Virtual Tape

Virtual tape is a special storage technology that manages less frequently needed data so that it appears to be stored entirely on tape cartridges, although some parts of it may actually be located in faster, hard-disk storage. The software associated with a virtual tape system is sometimes called a *virtual tape server*. Virtual tape can be used with a sophisticated storage-management system in which data is moved as it falls through various usage thresholds to slower but less costly forms of storage media. Use of this technology can reduce data access time, lower the total cost of ownership, and reduce the amount of floor space consumed by tape operations. IBM and Storage Technology are well-established vendors of virtual tape systems.

Telecom company Qualcomm, Inc., needed to rapidly back up and recover data stored in an Oracle database. The company turned to virtual tape methods and special software to reduce its backup window from 2 hours to 15 minutes, and it has tripled the number of complete backups it does each day. [23]

Optical Discs

Another type of secondary storage medium is the **optical disc**. Similar in concept to a ROM chip, an optical disc is simply a rigid disk of plastic onto which data is recorded by special lasers that physically burn pits in the disk. Data is directly accessed from the disk by an optical disc device, which operates much like a stereo's compact disc player. This optical disc device

redundant array of independent/inexpensive disks (RAID)
Method of storing data that generates extra bits of data from existing data, allowing the system to create a "reconstruction map" so that if a hard drive fails, the system can rebuild lost data.

disk mirroring
A process of storing data that provides an exact copy that protects users fully in the event of data loss.

virtual tape
Storage device that manages less frequently needed data so that it appears to be stored entirely on tape cartridges although some parts of it may actually be located in faster, hard-disk storage.

optical disc
A rigid disk of plastic onto which data is recorded by special lasers that physically burn pits in the disk.

compact disc read-only memory (CD-ROM)
A common form of optical disc on which data, once it has been recorded, cannot be modified.

CD-recordable (CD-R) disk
An optical disc that can be written on only once.

CD-rewritable (CD-RW) disk
An optical disc that allows personal computer users to replace their diskettes with high-capacity CDs that can be written on and edited.

Figure 3.10

Digital Versatile Disk and Player

DVDs look like CDs but have a much greater storage capacity and can transfer data at a much faster rate.

(Source: Courtesy of Sony Electronics.)

---

digital versatile disk (DVD)
Storage medium used to store digital video or computer data.

magneto-optical (MO) disk
A hybrid between a magnetic disk and an optical disc.

uses a low-power laser that measures the difference in reflected light caused by a pit (or lack thereof) on the disk.

A common form of optical disc is called **compact disc read-only memory (CD-ROM)**. Once data has been recorded on a CD-ROM, it cannot be modified—the disk is "read-only." A CD burner is the informal name for a CD recorder, a device that can record data to a compact disc. **CD-recordable (CD-R)** and **CD-rewritable (CD-RW)** are the two most common types of drives that can write CDs, either once (in the case of CD-R) or repeatedly (in the case of CD-RW). CD-RW technology allows personal computer users to replace their diskettes with high-capacity CDs that can be written on and edited. The CD-RW disk can hold 740 MB of data—roughly 500 times the capacity of a 1.4-MB diskette.

Digital Versatile Disk

A **digital versatile disk (DVD)** is a five-inch diameter CD-ROM look-alike with the ability to store about 135 minutes of digital video or several gigabytes of data (see Figure 3.10). Software programs, video games, and movies are common uses for this storage medium. At a data transfer rate of 1.25 MB/second, the access speed of a DVD drive is faster than that of the typical CD-ROM drive.

DVDs are replacing recordable and rewritable CD discs (CD-R and CD-RW) as the preferred format for sharing movies and photos. Where a CD can hold about 740 MB of data, a single-sided DVD can hold 4.7 GB, with double-sided DVDs having a capacity of 9.4 GB. Unfortunately, DVD manufacturers haven't agreed on a standard, so there are several types of recorders and disks. Recordings can be made on record-once disks (DVD-R and DVD+R) or on rewritable disks (DVD-RW, DVD+RW, and DVD-RAM). Rewritable disks are less widely compatible than others. Dell and Hewlett-Packard use DVD+RW; Apple, Gateway, and IBM offer DVD-R.

Magneto-Optical Disk

A **magneto-optical (MO) disk** is a type of disk drive that combines magnetic disk technologies with CD-ROM technologies. Like magnetic disks, MO disks can be read from and written to. And like diskettes, they are removable. However, their storage capacity can exceed 5 GB, much greater than CDs. This type of disk uses a laser beam to change the molecular configuration of a magnetic substrate on the disk, which in turn creates visual spots. In conjunction with a photodetector, another laser beam reflects light off the disk and measures the size of the spots; the presence or absence of a spot indicates a bit.

A new magneto-optical format based on blue laser light (traditional MO, CD, and DVD formats all use red lasers) has been developed. Because the wavelength of blue light is shorter than that of red light, the beam from a blue laser makes a much smaller spot on the recording layer of a disk. A smaller spot means less space is needed to record one bit of data, so more data can be stored on a disk.[24] Sony Corp.'s new blue laser optical disc format allows for storage of up to 23.3 GB per disk with a 9 MB/second transfer rate. By 2005, Sony expects to double the capacity and data transfer rate to 50 GB and 18 MB/second. Plasmon PLC, Toshiba Corp., and NEC Corp. are also developing blue-laser MO products.[25]

Memory Cards

A group of computer manufacturers formed the Personal Computer Memory Card International Association (PCMCIA) to create standards for a peripheral device known as a *PC memory card*. These PC memory cards are credit-card-sized devices that can be installed in an adapter or slot in many personal computers. To the rest of the system, the PC memory card functions as though it were a fixed hard-disk drive. Although the cost per megabyte of

storage is greater than for traditional hard-disk storage, these cards are less prone to fail than hard disks, are portable, and are relatively easy to use. Software manufacturers often store the instructions for their program on a memory card for use with laptop computers.

Flash Memory

Flash memory is a silicon computer chip that, unlike RAM, is nonvolatile and keeps its memory when the power is shut off. It gets its name from the fact that the microchip is organized so that a section of memory cells (called a *block*) is erased or reprogrammed in a single action or "flash." Flash memory chips are small and can be easily modified and reprogrammed, which makes them popular in computers, cell phones, and other products. Flash memory is also used in some handheld computers to store data and programs, in MP3 players to hold music, in digital cameras to store photos, and in airplanes to store flight information in the cockpit. Compared with other types of secondary storage, flash memory can be accessed more quickly, consumes less power, and is smaller in size. The primary disadvantage is cost. Flash memory chips can cost almost three times more per megabyte than a traditional hard disk. Nonetheless, the market for flash memory has exploded in recent years.

A 256-MB memory device called Migo is available from Forward Solutions. It is a keychain device that comes with software that captures files and settings from your e-mail, word processing, Web browser, and presentation programs. These preferences are stored in flash memory, and when you plug the Migo device into another computer, it transfers them to this machine so that you can work as if you were at your own PC. When you are done working and unplug the Migo, the software wipes all traces of the files and settings off the PC. Then, when you plug the Migo back into your own PC at home, it automatically synchronizes all data and moves all changes to your desktop or laptop. The device eliminates the need for people to lug their laptop computers with them everywhere they go.[26]

Expandable Storage

Expandable storage devices use removable disk cartridges (see Figure 3.11). When your storage needs to increase, you can use more removable disk cartridges. The storage capacity can range from less than 100 MB to several gigabytes per cartridge. In recent years, the access speed of expandable storage devices has increased. Some devices are about as fast as an internal disk drive.

Expandable storage devices can be internal or external. A few personal computers are now including internal expandable storage devices as standard equipment. Zip by Iomega is an example. Of course, a CD-RW drive by Hewlett-Packard, Iomega, and others can also be used for expandable storage. Although more expensive than fixed hard disks, removable disk cartridges combine hard-disk storage capacity and diskette portability. Some organizations prefer removable hard-disk storage for the portability and control it provides. For example, a large amount of data can be taken to any location, or it can be secured so that access is controlled.

Enterprise Storage Options

Businesses increasingly need to store large amounts of data created throughout the organization. Such large secondary storage is called *enterprise storage*. There are three forms of enterprise data storage: attached storage methods, network-attached storage (NAS), and storage area networks (SAN).

flash memory
A silicon computer chip that, unlike RAM, is nonvolatile and keeps its memory when the power is shut off.

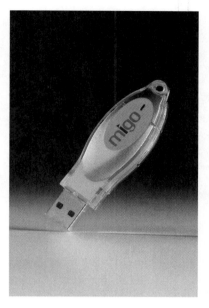

The Migo key-chain device comes with software that captures files and settings from your e-mail, word processing, Web browser, and presentation programs.

(Source: Courtesy of PowerHouse Technologies Group, Inc.)

expandable storage devices
Storage that uses removable disk cartridges to provide additional storage capacity.

Figure 3.11

Expandable Storage

Expandable storage drives allow users to add more storage capacity simply by plugging in a removable disk or cartridge. The disks can be used to back up hard-disk data or to transfer large files to colleagues.

(Source: Courtesy of Iomega.)

Attached Storage

Attached storage methods include the tape, hard disks, and optical devices discussed previously, which are connected directly to a single computer. Attached storage methods, while simple and cost-effective for individuals and small groups of users, do not allow systems to share storage, and they make it difficult to back up data.

Because of the limitations of attached storage, firms are turning to network-attached storage (NAS) and storage area networks (SAN). These alternative forms of enterprise data storage enable an organization to share data-storage resources among a much larger number of computers and users, resulting in improved storage efficiency and greater cost-effectiveness. In addition, they simplify data backup and reduce the risk of downtime. Industry experts estimate that up to 30 percent of system downtime is a direct result of data-storage failures, so eliminating storage problems as a cause of downtime is a major advantage.[27]

Network-Attached Storage

network-attached storage (NAS)
Storage devices that attach to a network instead of to a single computer.

Network-attached storage (NAS) employs storage devices that attach to a network instead of to a single computer. NAS includes software to manage storage access and file management and relieve the users' computers of those responsibilities. The result is that both application software and files can be served faster because they are not competing for the same processor resources. Computer users can share and access the same information, even if they are using different types of computers. Common applications for NAS include consolidated storage, Internet and e-commerce applications, and digital media. NAS vendors include EMC Corp., Hewlett-Packard Co., IBM, Procom Technology, and others.

Los Alamos National Laboratory installed a 1,400-node cluster of Pentium-based servers that uses a new object-based file system technology for data storage. The lab used network-attached storage (NAS) technology to spread file management capabilities across the servers to achieve 4 GB/sec. throughput and to store up to 600 TB on the cluster. The hardware is used to run simulations of nuclear weapons tests.[28]

SAN

storage area network (SAN)
Technology that provides high-speed connections between data-storage devices and computers over a network.

A **storage area network (SAN)** is a special-purpose high-speed network that provides direct connections between data-storage devices and computers across the enterprise (see Figure 3.12) and enables different types of storage subsystems, such as multiple RAID storage devices and magnetic tape backup systems, to be integrated into a single storage system. Use of a SAN offloads the network traffic associated with storage onto a separate network. This allows data to be easily copied to a remote location for disaster recovery, making it easier for companies to create backups and implement disaster recovery. Although SAN has been around for several years, its use is mainly limited to organizations with large IS budgets due to its cost and complexity.[29]

Implementing a SAN enables an organization to centralize the people, policies, procedures, and practices for managing storage, thus enabling a data-storage manager to apply the data policies consistently across an entire enterprise. This centralization eliminates inconsistent local treatment of data by different system administrators and end users, providing efficient and cost-effective data-storage practices.

A fundamental difference between NAS and SAN is that NAS uses file input/output, which defines data in terms of complete containers of information, while SAN deals with block input/output, which is based on subsets of data smaller than a file. SAN manufacturers include EMC, Hitachi Data Systems Corp., and IBM. Hitachi Data Systems Corp.'s Thunder 9580V is the newest member of the company's product line, which is targeted at small and midsize enterprises looking to consolidate storage and servers in the data center. It has a capacity of 64 TB.[30]

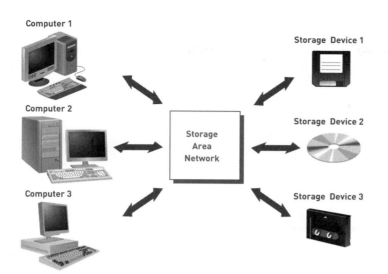

Figure 3.12

Storage Area Network

A SAN provides high-speed connections between data-storage devices and computers over a network.

Wildman, Harrold, Allen & Dixon LLP employs 550 people, and the amount of litigation data it must store has been growing rapidly—from 90 GB to 600 GB in just one year. As a result, the firm decided to change its storage infrastructure, moving from a group of high-capacity disk drives to an automated storage area network. The firm spent less than $80,000 on implementing the SAN.[31]

Total System Services, Inc., a credit card transaction processing company, uses two Cisco network switches to connect two IBM SAN systems in data centers 17 miles apart using Internet communication protocols. Total Systems Services mirrors up to 2 TB of data between the arrays.[32]

As large-scale SANs are deployed throughout organizations, the resulting increase in computers and network connections becomes difficult to manage. In response, software tools designed to automate storage using previously defined policies are finding a place in the enterprise. Known as **policy-based storage management**, the software products from industry leaders such as Veritas Software Corp., Legato Systems, Inc. EMC Corp., and IBM automatically allocate storage space to end users, balance the loads on servers and disks, and reroute networks when systems go down—all based on policies set up by system administrators.

The overall trend in secondary storage is toward direct-access methods, higher capacity, increased portability, and automated storage management. Organizations should select a specific type of storage based on their needs and resources. In general, the ability to store large amounts of data and information and access it quickly increases organizational effectiveness and efficiency by allowing users to access desired information in a timely fashion. There is also increasing interest in pay-per-use services, in which organizations rent storage on massive storage devices housed either at a service provider (e.g., Hewlett-Packard or IBM) or on the customer's own premises, paying only for the amount of storage they use. This approach is sensible for organizations with wildly fluctuating storage needs, such as those involved in the testing of new drugs or developing software.

policy-based storage management
Automation of storage using previously defined policies.

INPUT AND OUTPUT DEVICES: THE GATEWAY TO COMPUTER SYSTEMS

A user's first experience with computers is usually through input and output devices. Through these devices—the gateways to the computer system—people provide data and instructions

to the computer and receive results from it. Input and output devices are part of the overall user interface, which includes other hardware devices and software that allow human beings to interact with a computer system.

As with other computer system components, the selection of input and output devices depends on organizational goals and IS objectives. For example, many restaurant chains use handheld input devices or computerized terminals that let waiters enter orders to ensure timely and accurate data input. These systems have cut costs by making inventory tracking more efficient and marketing to customers more effective.

Characteristics and Functionality

Rapidly getting data into a computer system and producing timely output is very important for today's organizations. The form of the output desired, the nature of the data required to generate this output, and the required speed and accuracy of the output and the input determine the appropriate output and input devices. Some organizations have very specific needs for output and input, requiring devices that perform specific functions. The more specialized the application, the more specialized the associated system input and output devices.

The speed and functions performed by the input and output devices selected and used by the organization should be balanced with their cost, control, and complexity. More specialized devices might make it easier to enter data or output information, but they are generally more costly, less flexible, and more susceptible to malfunction.

The Nature of Data

Getting data into the computer—input—often requires transferring human-readable data, such as a sales order, into the computer system. Human-readable data is data that can be directly read and understood by human beings. A sheet of paper containing inventory adjustments is an example of human-readable data. By contrast, machine-readable data can be understood and read by computer devices (e.g., the universal bar code that grocery scanners read) and is typically stored as bits or bytes. Data on inventory changes stored on a diskette is an example of machine-readable data.

Data can be both human readable and machine readable. For example, magnetic ink on bank checks can be read by human beings and computer system input devices. Most input devices require some human interaction, because people most often begin the input process by organizing human-readable data and transforming it into machine-readable data. Every keystroke on a keyboard, for example, turns a letter symbol of a human language into a digital code that the machine can understand.

Data Entry and Input

Getting data into the computer system is a two-stage process. First, the human-readable data is converted into a machine-readable form through a process called **data entry**. The second stage involves transferring the machine-readable data into the system. This is **data input**.

Today, many companies are using online data entry and input—the immediate communication and transference of data to computer devices directly connected to the computer system. Online data entry and input places data into the computer system in a matter of seconds. Organizations in many industries require the instantaneous updating offered by this approach. For example, an airline clerk may need to enter a last-minute reservation. Online data entry and input is used to record the reservation as soon as it is made. Reservation agents at other terminals can then access this data to make a seating check before they make another reservation.

Source Data Automation

Regardless of how data gets into the computer, it should be captured and edited at its source. **Source data automation** involves capturing and editing data where the data is originally created and in a form that can be directly input to a computer, thus ensuring accuracy and timeliness. For example, using source data automation, salespeople enter sales orders into the computer at the time and place they take the order. Any errors can be detected and corrected immediately. If any item is temporarily out of stock, the salesperson can discuss options with the customer. Prior to source data automation, orders were written on a piece of paper and

data entry
Process by which human-readable data is converted into a machine-readable form.

data input
Process that involves transferring machine-readable data into the system.

source data automation
Capturing and editing data where the data is initially created and in a form that can be directly input to a computer, thus ensuring accuracy and timeliness.

entered into the computer later (often by someone other than the person who took the order). Often the handwritten information wasn't legible or, worse yet, got lost. If problems occurred during data entry, it was necessary to contact the salesperson or the customer to "recapture" the data needed for order entry, leading to further delays and customer dissatisfaction.

Input Devices

Literally hundreds of devices can be used for data entry and input. They range from special-purpose devices used to capture specific types of data to more general-purpose input devices. Some of the special-purpose data entry and input devices will be discussed later in this chapter. First, we will focus on devices used to enter and input more general types of data, including text, audio, images, and video for personal computers.

Personal Computer Input Devices

A keyboard and a computer mouse are the most common devices used for entry and input of data such as characters, text, and basic commands. Some companies are developing newer keyboards that are more comfortable, more easily adjusted, and faster to use. These keyboards, such as the split keyboard by Microsoft and others, are designed to avoid wrist and hand injuries caused by hours of keyboarding. Using the same keyboard, you can enter sketches on the touchpad and text using the keys. Another innovation is the development of wireless mouses and keyboards.

A computer mouse is used to "point to" and "click on" symbols, icons, menus, and commands on the screen. This causes the computer to take a number of actions, such as placing data into the computer system.

A keyboard and mouse are two of the most common devices for computer input. Wireless mice and keyboards are now readily available.

(Source: Courtesy of Gateway, Inc.)

Voice-Recognition Devices

Another type of input device can recognize human speech. Called **voice-recognition devices**, these tools use microphones and special software to record and convert the sound of the human voice into digital signals. Speech recognition can be used on the factory floor to allow equipment operators to give basic commands to machines while they are using their hands to perform other operations. Voice recognition is also used by security systems to allow only authorized personnel into restricted areas. Voice recognition has been used in many products, including automobiles. Voice-recognition systems now available on many makes of autos and trucks allow a driver to activate radio programs and CDs. The systems can even tell you the time. Asking "What time is it?" will get a response such as "Eleven thirty-four." In another application, Voice Command software from Microsoft allows users to issue voice commands to their personal digital assistants or cell phones. The software understands English and is integrated with other Microsoft software so that users can successfully tell devices to "Call the office" or even ask them "What's my next appointment?"[33]

Voice-recognition devices analyze and classify speech patterns and convert them into digital codes. Some systems require "training" the computer to recognize a limited vocabulary

voice-recognition device
An input device that recognizes human speech.

of standard words for each user. Operators train the system to recognize their voices by repeating each word to be added to the vocabulary several times. Other systems allow a computer to understand a voice it has never heard. In this case, the computer must be able to recognize more than one pronunciation of the same word—for example, recognizing the use of the phrase "Please?" spoken by someone from Cincinnati as meaning the same as "Huh?" spoken by someone from the Bronx, or "I beg your pardon?" spoken by a British person.

Burlington Northern and Santa Fe Railway Co. (BNSF) used to track its trains the old-fashioned way, through two-way voice radios located in every locomotive cab. Train crews dropped off cars and then radioed that information back to a dispatcher at BNSF's network operations center. The dispatcher then had to type the information into databases running on IBM mainframe computers. Recently, BNSF launched a project to use its voice radios as the interface to an interactive voice response (IVR) system to provide input direct to the company's databases. It chose ScanSoft, Inc., to provide it with speech-recognition software. The project proved quite a challenge because two-way radio systems have lower fidelity than the phone lines traditionally used with IVR. Plus, the noisy environment of the locomotive cab compounded the fidelity problem. ScanSoft built the BNSF IVR application on its SpeechWorks software and added noise filters. ScanSoft also sampled engineer radio calls to teach the software to recognize speech generated in the noisy environment. In the end, BNSF got a system that automatically integrates radio calls with its back-end systems, providing it with current and accurate information on its trains and individual cars. BNSF is now able to provide customers with more frequent information on car movement and better estimates of expected time of arrival.[34]

Digital Computer Cameras

digital computer camera
Input device used with a PC to record and store images and video in digital form.

Digital computer cameras look very similar to regular cameras but record and store images or video in digital form (see Figure 3.13). When you take pictures, the images are electronically stored in the camera. A cable is then connected from the camera to a port on the computer, and the images can be downloaded. Or you can simply remove a flash memory card or diskette from the camera and insert it into your personal computer. Once on the computer's hard disk, the images can be edited, sent to another location, pasted into another application, or printed. For example, a photo of the project team captured by a digital computer camera can be downloaded and then pasted into a project status report. Digital cameras now rival cameras used by professional photographers for photo quality and such features as zoom, flash, exposure controls, special effects, and even video-capture capabilities. With the right software, you can even add sound and handwriting to the photo.

Figure 3.13

A Digital Camera

Digital cameras save time and money by eliminating the need to buy and process film.

(Source: Courtesy of Casio, Inc.)

More than two dozen camera manufacturers, including Canon, Kodak, Fuji, Hewlett-Packard, Minolta, Olympus, Pentax, Sony, and Toshiba, offer at least one digital camera

model for under $280 with sufficient resolution to produce high-quality 5" × 7" photos.[35] Some manufacturers offer a digital camera that records full-motion video.

The number one advantage of digital cameras is saving time and money by eliminating the need to process film. In fact, digital cameras that can easily transfer images to CDs increasingly endanger the consumer film business of Kodak and Fujitsu. As a result, Kodak is now allowing photographers to have it both ways. When Kodak print film is developed, Kodak offers the option of placing pictures on a CD in addition to the traditional prints. Once stored on the CD, the photos can be edited, placed on an Internet site, or sent electronically to business associates or friends around the world.

FedEx is testing high-speed, camera-based scanners that read all six sides of a package to capture digital images of each bar code. An array of cameras looks at all six sides of a package as it goes down the St. Louis facility's sorting conveyors. The digital camera images detect the bar code that is on one of the box's six sides and decodes the bar code the way a laser scanner decodes bar codes in other FedEx facilities. The package is then sorted according to the information on the bar code and placed on the appropriate truck by a worker. While laser scanners read 95 percent of boxes the first time, the camera scanners can read approximately 98 percent of the packages. Given the hundreds of thousands of packages that must be sorted, the 3 percent improvement is significant, reducing the amount of human intervention and shortening the amount of time FedEx spends sorting boxes from its inbound arrivals to its outbound departures. This system enables FedEx to cut costs and improve service to the customer.[36]

Terminals

Inexpensive and easy to use, terminals are input devices that perform data entry and data input at the same time. A terminal is connected to a complete computer system, including a processor, memory, and secondary storage. General commands, text, and other data are entered via a keyboard or mouse, converted into machine-readable form, and transferred to the processing portion of the computer system. Terminals, normally connected directly to the computer system by telephone lines or cables, can be placed in offices, in warehouses, and on the factory floor.

Scanning Devices

Image and character data can be input using a scanning device. A page scanner is like a copy machine. The page to be scanned is typically inserted into the scanner or placed face down on the glass plate of the scanner, covered, and scanned. With a handheld scanner, the scanning device is moved or rolled manually over the image to be scanned. Both page and handheld scanners can convert monochrome or color pictures, forms, text, and other images into machine-readable digits. It has been estimated that U.S. enterprises generate over 1 billion pieces of paper daily. To cut down on the high cost of using and processing paper, many companies are looking to scanning devices to help them manage their documents.

Optical Data Readers

A special scanning device called an *optical data reader* can also be used to scan documents. The two categories of optical data readers are for optical mark recognition (OMR) and optical character recognition (OCR). OMR readers are used for test scoring and other purposes when test takers use pencils to fill in boxes on OMR paper, which is also called a "mark sense form." OMR is used in standardized tests, including the SAT and GMAT tests. In comparison, most OCR readers use reflected light to recognize various characters. With special software, OCR readers can convert handwritten or typed documents into digital data. Once entered, this data can be shared, modified, and distributed over computer networks to hundreds or thousands of individuals.

Magnetic Ink Character Recognition (MICR) Devices

In the 1950s, the banking industry became swamped with paper checks, loan applications, bank statements, and so on. The result was the development of magnetic ink character recognition (MICR), a system for reading this data quickly. With MICR, data is placed on the bottom of a check or other form using a special magnetic ink. Data printed with this ink using a character set is readable by both people and computers (see Figure 3.14). Read the "Information Systems @ Work" special feature to learn more about how banks are combining

the capabilities of old technology such as MICR with new technologies to improve their operations.

Figure 3.14

MICR Device

Magnetic ink character recognition is a process in which data is coded on the bottom of a check or other form using special magnetic ink, which is readable by both computers and human beings. For an example, look at the bottom of a bank check or most utility bills.

(Source: Courtesy of NCR Corporation.)

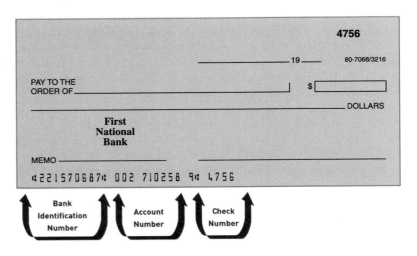

point-of-sale (POS) device

Terminal used in retail operations to enter sales information into the computer system.

Point-of-Sale (POS) Devices

Point-of-sale (POS) devices are terminals used in retail operations to enter sales information into the computer system. The POS device then computes the total charges, including tax. Many POS devices also use other types of input and output devices, such as keyboards, bar-code readers, printers, and screens. A large portion of the money that businesses spend on computer technology involves POS devices.

Automated Teller Machine (ATM) Devices

Another type of special-purpose input/output device, the automated teller machine (ATM), is a terminal used by most bank customers to perform withdrawals and other transactions with their bank accounts. The ATM, however, is no longer used only for cash and bank receipts. Companies use various ATM devices, sometimes called *kiosks,* to support their business processes. Some can dispense tickets for airlines, concerts, and soccer games. Some colleges use them to output transcripts. For this reason, the input and output capabilities of ATMs are quite varied. Like POS devices, ATMs may combine other types of input and output devices. Unisys, for example, has developed an ATM kiosk that allows bank customers to make cash withdrawals, pay bills, and also receive advice on investments and retirement planning.

Banks Weigh Move to Improved Check-Clearing Process

The effective use of information systems is critical to the success of the banking industry. Basic transaction processing, debit card processing, automated teller machines, online bill payment, automated bill payment, and check imaging and processing all rely on information systems for efficiency and high reliability.

Until recently, the check-clearing and settlement process had changed little since the 1950s, when banks introduced magnetic-ink encoding. The clearance and settlement of checks took from one to four days and required paper checks to be shipped by plane and truck to the banks that issued them. This process also involved error-prone manual data entry.

Recent technological advances such as image-capture hardware and high-speed networks created an opportunity to store and exchange check images in lieu of actual checks. However, banks could not be forced to accept electronic check images. So, until recently, banks that didn't invest in imaging systems could force other banks to continue the costly practice of hiring air couriers to transport bundles of checks across the country.

The Check Clearing for the 21st Century Act, or Check 21, advocates the use of a device called an image replacement document (IRD), a paper facsimile that is the legal equivalent of an original check. Here's how it works. Suppose that Bank of America wishes to transmit a check image to a bank in North Dakota, but the North Dakota bank isn't equipped to receive digital images. Bank of America would transmit the image to a processing site near the bank in North Dakota, which would print out the image as an IRD and ship it to the North Dakota bank. From then on, the process would proceed exactly as in the past. This process enables the Bank of America to reduce its check transportation costs, but the North Dakota bank is not forced to invest in imaging technology.

Use of this improved process will speed the collection of checks, improve the availability of funds, reduce fraud losses, and lower collection costs. According to Small Value Payment Company, a bank-owned provider of electronic-payment services, banks can achieve a net savings of $2.1 billion a year through the use of substitute checks. In addition, customers receive the convenience of viewing cleared checks via a PC connected to the Internet or from a link on an institution's Web page. The use of IRDs allows banks to use more efficient delivery channels for statements via CDs, DVDs, and e-mail.

Most large banks already have much of the basic computer hardware required for check imaging and processing. Many smaller banks, however, will instead depend on their larger counterparts to perform the check image processing. Another approach that some may take is to outsource check clearing and settlement, relying on a third-party provider to perform these functions.

Imagine that you are the manager in charge of the check-clearing and settlement operation in a midsized bank that is considering investing in information systems and hardware to take advantage of Check 21. Senior bank managers have asked for your recommendation on whether the bank should make the necessary investments to convert to this new process.

Discussion Questions

1. What are the potential benefits for your bank to convert to the new process?
2. What are some of the issues and factors that may complicate this move?

Critical Thinking Questions

3. How would you decide whether your bank should convert to the new Check 21 process?
4. What factors should you consider in deciding whether you should outsource the check-clearing and settlement functions rather than continue to perform them in house?

SOURCES: Steven Marlin, "Check Clearing to Get Electronic Overhaul," *InformationWeek*, June 10, 2003, *www.informationweek.com*; Stessa B. Cohen, "New Sterling Spinoff Focuses on Check 21 Support," Gartner Group, September 17, 2003, at *www4.gartner.com*; Lucas Mearian, "Check 21 Becomes Law, Allows Speedier Electronic Settlements," *Computerworld*, November 23, 2003, *www.computerworld.com*; Carey Richardson, "Check 21: Check Clearing for the 21st Century," de novo banks.com, January 22, 2004, *www.denovobanks.com*.

Pen Input Devices

By touching the screen with a pen input device, it is possible to activate a command or cause the computer to perform a task, enter handwritten notes, and draw objects and figures. Pen input requires special software and hardware. Handwriting recognition software can convert handwriting on the screen into text. The Tablet PC from Microsoft and its various hardware partners can transform handwriting into typed text and store the "digital ink" just the way a person writes it. Users can use a pen to write and send e-mail, add comments to Word documents, mark up PowerPoint presentations, and even hand draw charts in a document. That data can then be moved, highlighted, searched, and converted into computer readable data. If perfected, this interface is likely to become widely used. Providing such a simple input method is highly attractive to users who are uncomfortable using a keyboard. The keys to the success of this means of input are the accuracy at which handwriting can be read and translated into digital form and cost.

Light Pens

A light pen uses a light cell in the tip of a pen. The cell recognizes light from the screen and determines the location of the pen on the screen. Like pen input devices, light pens can be used to activate commands and place drawings on the screen.

Touch-Sensitive Screens

Advances in screen technology allow display screens to function as input as well as output devices. By touching certain parts of a touch-sensitive screen, you can execute a program or cause the computer to take an action. Touch-sensitive screens are popular input devices for some small computers because they preclude the necessity of keyboard input devices that consume space in storage or in use. They are frequently used at gas stations for customers to select grades of gas and request a receipt, on photocopy machines to enable users to select various options, at fast-food restaurants for order clerks to enter customer choices, at information centers in hotels to allow guests to request facts about local eating and drinking establishments, and at amusement parks to provide directions to patrons. They also are used in kiosks at airports and department stores.

Bar-Code Scanners

A bar-code scanner employs a laser scanner to read a bar-coded label. This form of input is used widely in grocery store checkouts and in warehouse inventory control. Often, bar-code technology is combined with other forms of technology to create innovative ways for capturing data. For example, in-store promotions are essential for pumping up the sales of popular DVDs. London-based Twentieth Century Fox Home Entertainment provided 75 of its sales reps with Hewlett-Packard iPaq handheld computers and digital cameras to document that individual retail stores fulfilled their agreements to install banners and displays to promote specific films. The system also includes a bar-code scanner for accurate capture of item counts. The sales rep visits a store, captures photos of the in-store promotions, determines product inventory, and sends all this data back to Fox over a wireless communications network.[37]

The Uniform Code Council (UCC), which assigns the 12-digit universal product codes (UPCs), determined that the numbers would eventually run out if more digits weren't added. In 1997, the council notified manufacturers, retailers, and distributors that as of January 1, 2005, it would introduce 13-digit UPCs and that companies would have to be able to process them. This seemingly simple task has required a myriad of changes to database fields, computer hardware devices, and software applications, resulting in a major effort for IS organizations.[38]

When organizations consider implementing the newest computer hardware to streamline input operations and reduce errors, they often run into budget limitations and resistance to change, as illustrated in the "Ethical and Societal Issues" feature.

Medication Management Systems

The U.S. Department of Health and Human Services (HHS) has mandated that pharmaceutical companies put bar codes on all drugs dispensed in hospitals to reduce medication errors. While estimates vary widely, some believe that literally tens of thousands of people die each year from medical errors from dispensing wrong medicines, wrong dosages, or incomplete patient data. Although these regulations apply only to drug manufacturers and not to hospitals, it is expected that hospital pharmacists and vendors will use the bar codes and that networks will be needed to support them.

In addition, the Food and Drug Administration (FDA) has mandated the use of the National Drug Code to identify the type of medication and dose. The goal is to ensure that the right patient receives the right drug in the right dose at the right time. These bar-code regulations will go into effect three years after the FDA publishes its final rules sometime in late 2004. By setting the regulations, the FDA broke an impasse among drug manufacturers, resellers, and hospitals over the use of bar codes. The manufacturers didn't want to use bar codes because the hospitals didn't have readers, and the hospitals didn't want to install the technology because so few drugs had the bar codes.

A medication-management system based on bar codes provides multiple checks to ensure that a patient receives the correct drug. When nurses dispense medications, they first scan a bar code on their badge, then the code on the patient's bracelet, and finally the code on the drug. All this information is sent to a database that contains patient and prescription information. The information is checked against the patient's electronic medical records to ensure that the correct medicine and dose is being given at the correct time and to double-check possible allergies or adverse drug interactions. The system can also be used to check blood types before transfusions and to track lab specimens. If something fails to match up, an audible alert sounds.

It will be expensive for hospitals to deploy the bar-code technology. The FDA has estimated the cost to exceed $7 billion, with an additional $1 billion needed to install wireless LAN technology to connect nurses dispensing drugs at patients' bedsides. The FDA estimates a one time cost of $250,000 to equip a 125-bed hospital with bar-code technology and systems. A hospital of that size with no wireless LAN will need to spend an additional $50,000 for a network. A very large hospital could spend a couple million dollars.

Despite the advances that innovative healthcare companies have made using business technology, the industry still lags in its adoption of patient safety technologies, according to a survey by the Healthcare Information and Management Systems Society. Nearly three-fourths of the respondents said a lack of funding is the biggest reason for the slow adoption of patient safety technology. The second-biggest factor, cited by 45 percent, is physician resistance to new systems.

Critical Thinking Questions

1. Under what conditions should adoption of new patient safety information systems be mandatory for all participants in the healthcare industry?
2. Why would doctors resist adoption of new patient safety information systems? What could be done to overcome this resistance?

What Would You Do?

You are the manager of quality and patient services of the only hospital in a 50-mile radius. The 250-bed facility is known for the high quality of its service and its strong staff. However, the hospital is typically slow to adopt new technology. You have heard rumors that the hospital's board of directors wishes to move slowly on implementation of a new medication management system. They want to be sure that the new technology is reliable before considering any changes. Furthermore, budget considerations would make it advantageous to extend the implementation over two or three years.

3. Do you agree with the board's position on the need to implement a medication management system? Why or why not?
4. Once this project receives board approval to proceed, what actions can you take to help accelerate the implementation of this system and avoid resistance on the part of any of those affected by the new system?

SOURCES: Bob Brewin, "HHS Mandates Bar Codes on All Hospital Drugs," *Computerworld*, March 13, 2003, *www.computerworld.com*; Bob Brewin, "Billions Needed to Meet Drug Bar-Code Mandate," *Computerworld*, March 17, 2003, *www.computerwold.com*; Marianne Kolbasuk McGee, "Drug Tracking: FDA Could Make Bar Codes Practical," *InformationWeek*, May 19, 2003, *www.informationweek.com*; Stephanie Stahl, "Editor's Note: IT: Good for What Ails You," *InformationWeek*, May 19, 2003, *http://informationweek.com*; Marianne Kolbasuk McGee, "Eye on Patient Safety," *InformationWeek*, November 3, 2003, *www.informationweek.com*.

radio-frequency identification (RFID)

A technology that employs a microchip with an antenna that broadcasts its unique identifier and location to receivers.

Radio-Frequency Identification

Radio-frequency identification (RFID) technology employs a microchip, called a *smart tag*, with an antenna that broadcasts its unique 96-bit identifier and location to corresponding receivers. The receiver relays the data to a computer, which decodes the information and processes it. One application of RFID is to place a microchip on retail items and install in-store readers that constantly count the inventory on the shelves to restock them automatically.

Smart tags can be also embedded in items that are difficult to bar code, such as a bunch of grapes. They can even be embedded in raw materials, extending their benefits to manufacturers and suppliers. The difference between bar-code scanners and RFID scanners is that the latter records data from each item at a distance of 4 to 5 feet. It does not require a human to manually pass the item over a scanner, so transactions can be completed faster. As a result, RFID can produce more accurate inventory counts than bar codes.

The sale of counterfeit drugs is a $30 billion problem for the drug industry and is causing it to explore item-level use of radio-frequency identification technology. In a test of the technology, five pharmaceutical manufacturers including Abbott Laboratories, Johnson & Johnson, Pfizer, and Procter & Gamble began shipping bottles of two types of pills with RFID labels. The bottles are being tracked as they move from manufacturers' plants to their distribution centers, then to distributors' facilities, retailers' distribution centers, and finally to CVS and Rite-Aid retail pharmacies. McKesson Corp. and Cardinal Health are the participating distributors. If the pills move to destinations outside the normal flow, this will be detected. Long-term, improved inventory management will be a benefit of item-level RFID tagging, pharmaceutical executives say. It will help avoid product stock-outs and make it easier to trace drugs that have been recalled.[39]

Output Devices

Computer systems provide output to decision makers at all levels of an organization to solve a business problem or capitalize on a competitive opportunity. In addition, output from one computer system can be used as input into another computer system. The desired form of this output might be visual, audio, or even digital. Whatever the output's content or form, output devices function to provide the right information to the right person in the right format at the right time.

Display Monitors

The display monitor is a TV-screen-like device on which output from the computer is displayed. Because some monitors use a cathode ray tube to display images, they are sometimes called *CRTs*. Such a monitor works much the same way a traditional TV screen does—one or more electron beams are generated from cathode ray tubes. As the beams strike a phosphorescent compound (phosphor) coated on the inside of the screen, a dot on the screen called a pixel lights up. A **pixel** is a dot of color on a photo image or a point of light on a display screen. It can be in one of two modes: on or off. The electron beam sweeps back and forth across the screen so that as the phosphor starts to fade, it is struck again and lights up again.

pixel

A dot of color on a photo image or a point of light on a display screen.

With today's wide selection of monitors, price and overall quality can vary tremendously. The quality of a screen is often measured by the number of horizontal and vertical pixels used to create it. A larger number of pixels per square inch means a higher resolution, or clarity and sharpness of the image. For example, a screen with a 1,024 × 768 resolution (786,432 pixels) has a higher sharpness than one with a resolution of 640 × 350 (224,000 pixels). The distance between one pixel on the screen and the next nearest pixel is known as *dot pitch*. The common range of dot pitch is from .25 mm to .31 mm. The smaller the dot pitch, the better the picture. A dot pitch of .28 mm or smaller is considered good. Greater pixel densities and smaller dot pitches yield sharper images of higher resolution.

A monitor's ability to display color is a function of the quality of the monitor, the amount of RAM in the computer system, and the monitor's graphics adapter card. The color graphics adapter (CGA) was one of the first technologies to display color images on the screen. Today, super video graphics array (SVGA) displays are standard, providing vivid colors and superior resolution.

Liquid Crystal Displays (LCDs)

Because CRT monitors use an electron gun, there must be a distance of at least one foot between the gun and screen, causing them to be large and bulky. Thus, a different technology, flat-panel display, is used for some personal computers and all notebooks. One common technology used for flat-screen displays is the same liquid crystal display (LCD) technology used for pocket calculators and digital watches. LCD monitors are flat displays that use liquid crystals—organic, oil-like material placed between two polarizers—to form characters and graphic images on a backlit screen. These displays are much easier on your eyes because they are flicker-free, far lighter, less bulky, and don't emit the type of radiation that makes some CRT users worry.

The primary choices in LCD screens are passive-matrix and active-matrix LCD displays. In a passive-matrix display, the CPU sends its signals to transistors around the borders of the screen, which control all the pixels in a given row or column. In an active-matrix display, each pixel is controlled by its own transistor attached in a thin film to the glass behind the pixel. Passive-matrix displays are typically dimmer, slower, but less expensive than active-matrix ones. Active-matrix displays are bright, clear, and have wider viewing angles than passive-matrix displays. Active-matrix displays, however, are more expensive and can increase the weight of the screen.

CRT monitors are large and bulky in comparison with LCD monitors (flat displays).

(Source: Courtesy of ViewSonic Corporation.)

LCD technology is also being used to create thin and extremely high-resolution monitors for desktop computers, so-called *flat-screen monitors*. Although the screen may measure just 13 inches from corner to corner, the display's extremely high resolution—1,280 × 1,280 pixels—lets it show as much information as a conventional 20-inch monitor. And while cramming more into a smaller area causes text and images to shrink, you can comfortably sit much closer to an LCD screen than the conventional CRT monitor. The IBM ThinkVision L150 flat-screen monitor has a base that swivels a full 360 degrees. The Samsung SynchMaster 153T has a screen that does not swivel but can rotate from the horizontal to the vertical position. At the high end are dazzling plasma screens from Fujitsu, NEC, and Samsung exceeding 50 inches but with a price of nearly $10,000.[40] The iMac computer can be ordered with a 20-inch flat screen.

Organic Light-Emitting Diodes

Organic light-emitting diode (OLED) technology is based on research done by Eastman Kodak Co. and is just reaching the market in small electronic devices. OLEDs use the same base technology as LCDs, with one key difference: whereas LCD screens contain a fluorescent backlight and the LCD acts as a shutter to selectively block that light, OLEDs directly emit light. OLEDs can provide sharper and brighter colors than LCDs and CRTs, and since they don't require a backlight, the displays can be half the thickness of OCDs and can be used in

flexible displays. Another big advantage is that OLEDs don't break when dropped. OLEDs are currently limited to use in cell phones, car radios, and digital cameras but may be used in computer displays—if the average display lifetime can be extended beyond the current 8,000 hours.[41]

Printers and Plotters

One of the most useful and popular forms of output is called *hard copy*, which is simply paper output from a printer. Printers with different speeds, features, and capabilities are available. Some can be set up to accommodate different paper forms, such as blank check forms and invoice forms. Newer printers allow businesses to create customized printed output for each customer from standard paper and data input using full color.

The speed of the printer is typically measured by the number of pages printed per minute (ppm). Like a display screen, the quality, or resolution, of a printer's output depends on the number of dots printed per inch. A 600-dpi (dots-per-inch) printer prints more clearly than a 300-dpi printer. A recurring cost of using a printer is the inkjet or laser cartridge that must be replaced every few thousand pages of output. Figure 3.15 shows a laser printer.

Figure 3.15

Laser Printer

Laser printers, available in a wide variety of speeds and price ranges, have many features, including color capabilities. They are the most common solution for outputting hard copies of information.

(Source: Courtesy of Lexmark International.)

Researchers are working on specialized commercial inkjet printers that can be used to create 3-D parts. These printers spray out layers of inklike polymers, to "print" a new circuit board including all the electronic connections and transistors laid out in the proper arrays. Plastic Logic Ltd. creates inkjet-printed plastic transistors for flat-panel displays. It is also developing flexible, plastic "smart labels" that could be used on consumer products. Such labels would contain electronic circuitry that could trigger warnings to notify consumers when food in a package is no longer fresh.[42]

plotter
A type of hard-copy output device used for general design work.

Plotters are a type of hard-copy output device used for general design work. Businesses typically use these devices to generate paper or acetate blueprints, schematics, and drawings of buildings or new products onto paper or transparencies. Standard plot widths are 24 inches and 36 inches, and the length can be whatever meets the need—from a few inches to feet.

Computer Output Microfilm (COM) Devices

Companies that produce and store significant numbers of paper documents often use computer output microfilm (COM) devices to place data from the computer directly onto microfilm for future use. The traditional photographic phase of conversion to microfilm is eliminated. Once this is done, a standard microfilm reader can access the data. Newspapers and journals typically place their past publications on microfilm using COM, giving readers the ability to view past articles and news items.

Music Devices

MP3
A standard format for compressing a sound sequence into a small file.

MP3 (MPEG-1 Audio Layer-3) is a standard format for compressing a sound sequence into a very small file while preserving the original level of sound quality when it is played. By compressing the sound file, it requires less time to download the file and less storage space

on a hard drive. Although MP3 is currently the leading digital audio format, it is facing competition from Windows Media from Microsoft, RealNetwork's Real Audio, and an upgraded version of MP3 called mp3PRO, which cuts the storage space required in half.

Many different **music devices** about the size of a cigarette pack can be used to download music from the Internet and other sources. These devices have no moving parts and can store hours of music. A number of computer manufacturers—including Dell, Hewlett-Packard, NEC, and others—offer computers that make downloading and playing music from the Internet in the MP3 format easier with better sound quality. These specialized computers offer easy downloading and superior speakers for playing music. Apple expanded into the digital music market not only with an MP3 player (the iPod) but also with its iTunes Music Store that allows users to find music online, preview it, and get it in a way that is safe, legal, and affordable. It's possible that Apple will expand this market in other ways, perhaps eventually partnering with companies to develop a way to sell video content in a similar medium.[43]

music device
A device that can be used to download music from the Internet and play the music.

Apple's iPod, a digital music player for Mac and Windows, weighs just 5.6 ounces and can store up to 10,000 songs.

(Source: Courtesy of Apple Computer, Inc.)

Special-Purpose Input and Output Devices

Many additional input and output devices are used for specialized or unique applications. A **multifunction device** can combine a printer, fax machine, scanner, and copy machine into one device. Multifunction devices are less expensive than buying these devices separately, and they take less space on a desktop compared with separate devices.

Special-purpose hearing devices can be used to detect manufacturing or equipment problems. The Georgia Institute of Technology has developed a hardware device that can "listen to" equipment to detect worn or damaged parts. Voice-output devices, also called *voice-response devices*, allow the computer to send voice output in the form of synthesized speech over phone lines and other media. Some banks and financial institutions use voice recognition and response to give customers account information over the phone.

On-board computer-based navigation systems are an option on many luxury cars, including the Lexus, Acura, Infiniti, and Cadillac. The system is able to pinpoint the position of the auto from location data received from a global positioning satellite (GPS) received through an on-board antenna. It uses vehicle speed information to determine how far you are traveling and a gyro sensor to tell when you turn. Then the device compares all this information with the system's CD-ROM road map database to show your position within a few feet. The Garmin Street Pilot III Deluxe is a portable positioning system for your car that costs under $700. It provides turn-by-turn voice prompts that guide you to your destination. Just enter a street address or business location to access the shortest and fastest route directly to the door. Downloadable maps focus in on your region of choice.

multifunction device
A device that can combine a printer, fax machine, scanner, and copy machine into one device.

COMPUTER SYSTEM TYPES, SELECTING, AND UPGRADING

special-purpose computers
Computers used for limited applications by military and scientific research groups.

general-purpose computers
Computers used for a wide variety of applications.

Table 3.3

Types of Computer Systems

In general, computers can be classified as either special purpose or general purpose. **Special-purpose computers** are used for limited applications by military and scientific research groups such as the CIA and NASA. Other applications include specialized processors found in appliances, cars, and other products. Special-purpose computers are increasingly being used by businesses. For example, automobile repair shops connect special-purpose computers to your car's engine to identify specific performance problems.

General-purpose computers are used for a variety of applications and to perform the business applications discussed in this text. They combine processors, memory, secondary storage, input and output devices, a basic set of software programs, and other components. General-purpose computer systems can be broken into two major groups: systems used by a single user at a time and systems used by multiple concurrent users. Table 3.3 shows the general ranges of capabilities for various types of computer systems.

Table 3.3

Types of Computer Systems

Factor	Single-User Systems					Multiuser Systems (Servers)		
	Handheld	Portable	Thin Client	Desktop	Workstation	Server	Mainframe	Supercomputer
Cost Range	$200 to $1,500	$1,000 to $3,500	$250 to $1,000	$600 to $3,500	$4,000 to $40,000	$500 to $50,000	> $100,000	> $250,000
Weight	< 24 oz.	< 7 lbs.	< 15 lbs.	< 25 lbs.	< 25 lbs.	> 25 lbs.	> 200 lbs.	> 200 lbs.
Typical Size	Palm size	Size of a 3-ring notebook	Fits on desktop	Fits on desktop	Fits on desktop	3-drawer filing cabinet	Refrigerator	Refrigerator and larger
CPU Speed	> 200 MHz	> 2 GHz	> 200 MHz	> 3 GHz	> 3 GHz	> 2 GHz	> 300 MIPS	> 2 teraflops
Typical Use	Personal organizer	Improvement of worker productivity	Data entry and Internet access	Improvement of worker productivity	Engineering, CAD, software development	Support for network and Internet applications	Computing for large organization; provides massive data storage	Scientific applications; intensive number crunching
Example	Handspring Treo 600 smart phone	Motion Computing M1300 Mainstream Tablet PC	Max-speed Max-Term 8400	iMac Power PC G4	Sun Microsystems Sun Blade 2500 Workstation	Hewlett-Packard HP ProLiant BL	Unisys ES5000	IBMs RS/6000 SP

Computer System Types

Computer systems can range from small handheld computers to massive supercomputers that require entire rooms. We start first with the smallest computers.

Handheld Computers

handheld computer
A single-user computer that provides ease of portability because of its small size.

Handheld computers are single-user computers that provide ease of portability because of their small size—some are as small as a credit card. These systems often include a wide variety of software and communications capabilities. Most are compatible with and able to communicate with desktop computers over wireless networks. Some even add a built-in global positioning system receiver with software that can integrate the location data into the application. For example, if you click on an entry in the address book, the device displays a map and directions from your current location. Such a computer can also be mounted in your car and serve as a navigation system. One of the shortcomings of handheld computers is that they require lots of power relative to their size.

PalmOne (formerly Palm, Inc.) is the company that invented the Palm Pilot organizer in 1996. The Palm Personal Digital Assistant (PDA) enables its user to track appointments, addresses, and tasks. PalmOne has now signed licensing agreements with Handspring, IBM, Sony, and many other manufacturers, permitting them to make what amounts to Palm clones. As a result of the popularity of the Palm Personal Digital Assistant, handheld computers are often referred to as PDAs.

Smart phones combine the functions of a telephone, PDA, game console, and a wireless-data device—and sometimes digital cameras. They enable a user to browse the Internet while on the move, store and play music, jot down brief messages, and place and receive phone calls. With their greater functionality, smart phones are expected to outsell PDAs very soon.[44] The Handspring Treo 600 has a bright color display, a fast processor, and the latest operating software from PalmSource. It is about 4.5 inches in length and 1 inch thick. It comes with a tiny keyboard. Five directional buttons allow the user to select an item and then click the center button to select it. The Treo can receive wireless data, and Sprint offers unlimited data use for a $10 monthly surcharge on a voice plan. Web pages load quickly, although the small display limits the usefulness of browsing. The user can download games, applications, and different ring tones wirelessly from Sprint's PCS Vision service. Sprint also offers a $5 per month Business Connection service, which can forward e-mail from corporate accounts to wireless devices and fetch mail from standard Internet accounts.

The Treo 600 is a smart phone that enables users to browse the Internet, place and receive phone calls, send e-mail and text messages, and even take a picture with a built-in camera.

(Source: Courtesy of PalmOne, Inc.)

Portable Computers

A variety of **portable computers**, those that can be carried easily, are now available—from laptops, to notebooks, to subnotebooks, to tablet computers. A *laptop computer* is a small, lightweight PC about the size of a three-ring notebook. The even smaller and lighter *notebook* and *subnotebook* computers offer similar computing power. Some notebook and subnotebook computers fit into docking stations of desktop computers to provide additional storage and processing capabilities.

Tablet PCs are portable, lightweight computers that allow users to roam the office, home, or factory floor carrying the device like a clipboard. They come in two varieties, slate and convertible. The slate devices have no keyboard, and users enter data with a writing stylus directly on the display screen. The convertible tablet PC comes with a swivel screen and can be used as both a traditional notebook or as a pen-based portable tablet. Acer, Fujitsu, Toshiba, ViewSonic, and others offer tablet PCs that weigh under 4 pounds and cost under $2,000.

While the use of tablet PCs has been slow to catch on, hardware manufacturers and computer industry analysts think that the device has a bright future. Indeed, Bill Gates estimated that sales of the devices would soon exceed that of desktop and laptop computers.[45] They are quite popular in the health services field. HealthSouth, a company that provides outpatient surgery and other healthcare services, ordered 5,000 tablet PCs equipped with wireless LAN connections from Motion Computing, Inc. The tablet devices are used by physical therapists at the company's 1,400 rehabilitation centers to provide access to patient records and enable them to document clinical progress.[46]

portable computer
Computer small enough to be carried easily.

The ViewSonic V1250 has a 12.1-inch screen and comes with a small navigation pad that protrudes from the right of the screen, with buttons for tasks such as scrolling, toggling between applications, and launching Internet Explorer. It also comes with a docking station that doubles as a battery charger.

(Source: Courtesy of ViewSonic Corporation.)

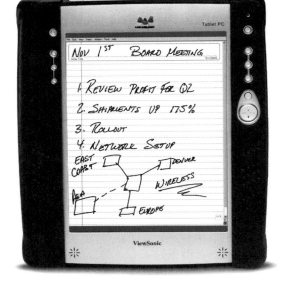

thin client

A low-cost, centrally managed computer with essential but limited capabilities that is devoid of a DVD player, diskette drive, and expansion slots.

desktop computer

A relatively small, inexpensive single-user computer that is highly versatile.

workstation

A more powerful personal computer that is used for technical computing, such as engineering, but still fits on a desktop.

server

A computer designed for a specific task, such as network or Internet applications.

Thin Client

A **thin client** is a low-cost, centrally managed computer that is devoid of a DVD player, diskette drive, and expansion slots. These computers have limited capabilities and perform only essential applications, so they remain "thin" in terms of the client applications they include. These stripped-down versions of desktop computers do not have the storage capacity or computing power of typical desktop computers, nor do they need it for the role they play. With no hard disk, they never pick up viruses or experience a hard-disk crash. Unlike personal computers, thin clients download software from a network when needed, making support, distribution, and updating of software applications much easier and less expensive. Their primary market is small businesses and educational institutions.

Desktop Computers

Desktop computers are relatively small, inexpensive single-user computer systems that are highly versatile. Named for their size—they are small enough to fit on an office desk—*desktop computers* can provide sufficient memory and storage for most business computing tasks. Desktop computers have become standard business tools; more than 30 million are in use in large corporations.

In addition to traditional PCs that use Intel processors and Microsoft software, there are other options. One of the most popular is the iMac by Apple Computer.

Workstations

Workstations are more powerful than personal computers but still small enough to fit on a desktop. They are used to support engineering and technical users who perform heavy mathematical computing, computer-aided design (CAD), and other applications requiring a high-end processor. Such users need very powerful CPUs, large amounts of main memory, and extremely high-resolution graphic displays to meet their needs.

Servers

A computer **server** is a computer used by many users to perform a specific task, such as running network or Internet applications. Servers typically have large memory and storage capacities, along with fast and efficient communications abilities. A Web server is used to handle Internet traffic and communications. An Internet caching server stores Web sites that are frequently used by a company. An enterprise server stores and provides access to programs that meet the needs of an entire organization. A file server stores and coordinates program and data files. A transaction server is used to process business transactions. Server systems consist of multiuser computers including supercomputers, mainframes, and servers.

Servers offer great **scalability**, the ability to increase the processing capability of a computer system so that it can handle more users, more data, or more transactions in a given period. Scalability is increased by adding more, or more powerful, processors. *Scaling up* adds more powerful processors, and *scaling out* adds many more equal (or even less powerful) processors to increase the total data processing capacity.

Mainframe Computers

A **mainframe computer** is a large, powerful computer shared by dozens or even hundreds of concurrent users connected to the machine over a network. The mainframe computer must reside in a data center with special heating, venting, and air-conditioning (HVAC) equipment to control the temperature, humidity, and dust levels around the computer. In addition, most mainframes are kept in a secure data center with limited access to the room through some kind of security system. The construction and maintenance of such a controlled-access room with HVAC can add hundreds of thousands of dollars to the cost of owning and operating a mainframe computer. Mainframe computers also require specially trained individuals (called *system engineers* and *system programmers*) to care for them.

scalability
The ability to increase the capability of a computer system to process more transactions in a given period by adding more, or more powerful, processors.

mainframe computer
Large, powerful computer often shared by hundreds of concurrent users connected to the machine via terminals.

Mainframe computers have been the workhorses of corporate computing for more than 50 years. They can support hundreds of users simultaneously and handle all of the core functions of a corporation.

(Source: Courtesy of IBM Corporation.)

The role of the mainframe is undergoing some remarkable changes as lower-cost, single-user computers become increasingly powerful. Many computer jobs that used to run on mainframe computers have migrated onto these smaller, less expensive computers. This information processing migration is called *computer downsizing*. The new role of the mainframe is as a large information processing and data-storage utility for a corporation—running jobs too large for other computers, storing files and databases too large to be stored elsewhere, and storing backups of files and databases created elsewhere. The mainframe is capable of handling the millions of daily transactions associated with airline, automobile, and hotel/motel reservation systems. It can process the tens of thousands of daily queries necessary to provide data to decision support systems. Its massive storage and input/output capabilities enable it to play the role of a video computer, providing full-motion video to multiple, concurrent users.

CartaSi S.p.A is one of Europe's largest credit card companies, with 7.5 million credit cards outstanding. Its 350 call-center service agents must process some 50,000 customer calls per day for 16 banks in Italy. The agents retrieve customer data located in databases in the company's mainframe computers to answer questions related to a customer's account.[47] The processing speed and large data-storage capacity of these machines make this an efficient and cost-effective solution.

supercomputers

The most powerful computer systems, with the fastest processing speeds.

Supercomputers

Supercomputers are the most powerful computer systems, with the fastest processing speeds. They are designed for applications that require extensive and rapid computational capabilities. With an entry-level cost of $250,000, not all organizations can afford such a computer.

Originally, supercomputers were used primarily by government agencies to perform the high-speed number crunching needed in weather forecasting and military applications. With recent improvements in the cost and performance (lower cost and faster speeds) of these machines, they are being used more broadly for commercial purposes today. For example, France's Compagnie Générale de Géophysique SA is a global oil-services company that has linked more than 3,000 Dell PowerEdge servers to act as a supercomputing cluster to analyze seismic data. The company uses various server clusters to process seismic data to identify new oil and gas reservoirs, as well as to model existing reservoirs to optimize production.[48]

This 512-processor supercomputer at NASA's Ames Research Center is used for performing scalability studies, benchmarking, and solving large-scale problems for the Information Power Grid. The system has 192 GB of main memory, contains 2 TB of disk storage, and can reach a peak processing speed of 307 GFLOPS.

(Source: Tom Trower, NASA Ames Research Center.)

Selecting and Upgrading Computer Systems

computer system architecture

The structure, or configuration, of the hardware components of a computer system.

The structure, or configuration, of the hardware components of a computer system is called the **computer system architecture**. This architecture can include a mixture of components, including processing, memory, storage, and input and output devices. A computer system that was once effective may need to be enhanced or upgraded to support new business activities and a changing environment. The ability to upgrade a system can be an important factor in selecting the best computer hardware. Computer systems can be upgraded by installing additional memory, additional processors (such as a math coprocessor), more hard-disk storage, a memory card, or various input and output devices.

Given typical failure rates, most organizations can safely plan to keep a desktop computer four or more years. However, if the system fails after the standard three-year warranty period, it is probably more cost-effective to replace it with an upgraded model rather than spending time to fix it. For notebooks, which face rougher handling by users, most companies plan on replacing them every three or four years. As workstations and notebooks are replaced, consider rolling over the most capable personal computers to less-demanding users. Companies are sometimes forced to upgrade otherwise usable machines because of the memory and CPU requirements of newer versions of resource intensive software.

Many large corporations set their own internal architectures by selecting specific computer configurations from a small set of manufacturers. The goal is to reduce hardware support costs and increase the organization's flexibility. Business units within an organization that adopt different hardware complicate future corporate IS projects. For example, the

installation of software is made much easier if similar equipment from the same manufacturer is involved, rather than new and different systems at each installation site.

Tim Link, CIO at Ohio State University Newark and Central Ohio Technical College admits, "We had the 'Noah's ark syndrome'—we had two of everything." Link carefully planned how to minimize the number of configurations he must support. He decided to buy personal computers with a 3-GHz processor and a full gigabyte of memory to avoid the need for memory or hard drive upgrades. Over a three-year period, he replaced an unwieldy mix of 1,300 personal computers with one of three standard configurations from Dell. As a result of standardization on both hardware and software, the number of personal computer support calls dropped over 22 percent. In the future, he plans to replace one-third of his systems each year. He's reluctant to go to a four-year life cycle for PCs because although that would reduce PC purchase costs, it would mean more individual models, which would complicate support issues.[49]

An important factor to consider in upgrading to new computer hardware, whether it be for a single individual or for a large multinational corporation, is the proper disposal of the old equipment. The National Recycling Coalition estimates nearly 500 million personal computers will have become obsolete in the period from 1997 to 2007.[50] As a result, the EPA has tagged e-waste as the fastest-growing stream of waste in the United States.[51] A typical CRT monitor contains 3 to 9 pounds of lead. Printed circuit boards contain beryllium, cadmium, flame retardants, and other compounds that can contaminate the air and ground-water, exposing human beings to carcinogens and other toxins. Unfortunately, few individuals or organizations realize the potential for harm from improper disposal of their computer hardware, nor have they budgeted for the costs associated with proper disposal. Three major computer hardware vendors, Dell, IBM, and Hewlett-Packard, offer disposal programs for any computer hardware, regardless of brand. They follow environment-friendly recycling practices and do not export the waste. There are also a few reputable vendors who offer e-waste disposal and recycling services. In some cases the computer hardware may still have a useful life and can be donated to schools or charitable organizations.

Hard Drive Considerations

The optimal hard drive for a computer is a function of several overlapping features. Since its main role is to serve as a long-term data store, capacity, speed, and media capabilities are key features. Today's business software applications and large video, audio, and graphics files require lots of storage, so a hard drive storage capacity of at least 80 GB is recommended. Other considerations are access speed (look for 10 milliseconds or less), amount of RAM, and hard drive cache size. Many single-user computers employ fans to help cool the internal components. These fans can cause the case to resonate and make the system seem loud. It is wise to select a system with fans, hard drive, and power supply that are classified as "quiet."

Main Memory Considerations

Main memory stores software code, while the processor reads and executes the code. Having more RAM main memory means you can run more software programs at the same time. Systems with 512 MB are well suited to take advantage of today's advanced personal productivity software (word processing, spreadsheet, graphics, and database) and multimedia programs.

As discussed earlier, your system's processor, main memory, and cache memory depend heavily on each other to achieve optimal system functionality. The original manufacturer of your computer considers this interdependency when designing and choosing the parts for the system. If you plan to upgrade your system's main memory above 512 MB, you should consult your supplier to understand your system's main memory limits on size of cache and the implication of exceeding those limits.

Printer Considerations

Laser printers and inkjet printers are the two primary choices for printers, and the differences between the two are becoming smaller and smaller. While most inkjet printers are color and laser printers monochrome, there are also color laser printers. All produce sharp images, with resolutions of 600 × 600 dots per inch (dpi) to 1,800 × 1,800 dpi now common. The major differences are in price, color, and speed.

There are two cost factors to consider when purchasing a printer. The first is the purchase price of the printer. Prices for laser printers range from $200 to over $2,000, while prices for inkjet printers range from $50 to $300. The second cost to consider is operating cost. Laser printers provide greater printing duty cycle (volume of pages printed per month) and longer life for the ink/toner products, giving them a much lower operating cost than inkjet printers. Laser printers typically have operating costs of from $.01 to $.04 per page. Inkjet printers can have operating costs of from $0.03 to $0.08 per page for black-and-white pages and $0.10 to $0.20 for color pages. The cost for printing photos on special paper can exceed $.50 per page for either type of printer because of the high cost of the paper.

Laser printers are generally faster than inkjet printers and capable of handling higher-volume printing than the inkjet printers. Laser printers print 15 to 30 pages per minute (ppm), and inkjet printers print 10 to 20 ppm for black and white and 2 to 10 ppm for color.

For color printing, inkjet printers print vivid hues and cost much less than color laser printers. Inkjet printers can produce high-quality banners, graphics, greeting cards, letters, text, and prints of photos. For most people who require color printing capability, inkjet printers are the most cost-effective solution.

DVD Burners

CD burners have been supplanted by DVD burners. Today's drives can burn both DVDs and CDs. A DVD burner that mounts inside a computer costs under $200. Making sure that your DVD player will handle the discs you create is the most difficult part of choosing a DVD writer. Most players can handle DVD-R and DVD+R, but not all work with the various rewritable formats. Check the DVD player manufacturer's Web site to be sure. When comparing DVD writers, you'll see numbers like 2.4×/2.4×/8×. These are the numbers for one-time recordable (-R or +R), rewritable (-RW or +RW), and read (ROM) formats. DVD-R drives run from 1× (about 1.3 MB per second) to 4×. DVD+R runs up to 2.4 MB per second.

As mentioned throughout this chapter, a computer system's components and architecture should be chosen to support fundamental objectives, current business processes, and future needs of the organization and information system. Each computer system component—processing, memory, storage, input, and output devices—has a critical role in the successful operation of the computer system, the information system, and the organization. A thorough understanding of the broader system goals and the characteristics of the hardware as they relate to these goals will be an important guide for the future IS professional or business user.

SUMMARY

Principle

Assembling an effective, efficient computer system requires an understanding of its relationship to the information system and the organization. The computer system objectives are subordinate to, but supportive of, the information system and the needs of the organization.

Hardware includes any machinery (often using digital circuitry) that assists with the input, processing, and output activities of a computer-based information system (CBIS). A computer system is an integrated assembly of physical devices with at least one central processing mechanism; it inputs, processes, stores, and outputs data and information.

Computer system hardware should be selected and organized to effectively and efficiently attain computer system objectives. These objectives should in turn support IS objectives and organizational goals. Balancing specific computer system objectives in terms of cost, control, and complexity will guide selection.

Processing is performed by cooperation between the central processing unit (CPU) and memory. The CPU has three main components: the arithmetic/logic unit (ALU), the control unit, and register areas. The ALU performs calculations and logical comparisons. The control unit accesses and decodes instructions and coordinates data flow. Registers are temporary holding areas for instructions to be executed by the CPU.

Principle

When selecting computer devices, you must also consider the current and future needs of the information system and the organization. Your choice of a particular computer system device should always allow for later improvements.

Instructions are executed in a two-phase process. In the instruction phase, instructions are brought into the central processor and decoded. In the execution phase, the computer executes the instruction and stores the result. The completion of this two-phase process is a machine cycle. Processing speed is often measured by the time it takes to complete one machine cycle, which is measured in fractions of seconds.

Computer system processing speed is also affected by clock speed, which is measured in megahertz (MHz). Speed is further determined by a CPU's wordlength, the number of bits it can process at one time. (A bit is a binary digit, either 0 or 1.) A 64-bit CPU has a wordlength of 64 bits and will process 64 bits of data in one machine cycle.

Moore's Law is a hypothesis that states that the number of transistors on a single chip will double every 18 months. This hypothesis has held up amazingly well.

Processing speed is also limited by physical constraints, such as the distance between circuitry points and circuitry materials. Advances in superconductive metals, optical processors, strained silicon, and carbon nanotubes will result in faster CPUs. Many processors are complex instruction set computing (CISC) chips, which have many microcode instructions placed in them. With reduced instruction set computing (RISC) chips, only essential instructions are included, so processing is faster.

Primary storage, or memory, provides working storage for program instructions and data to be processed and provides them to the CPU. Storage capacity is measured in bytes. A common form of memory is random access memory (RAM). RAM is volatile—loss of power to the computer will erase its contents—and comes in many different varieties. The mainstream type of RAM is extended data out, or EDO RAM, which is faster than older types of RAM memory. Two other variations of RAM memory include dynamic RAM (DRAM) and synchronous DRAM. SDRAM also has the advantage of a faster transfer speed between the microprocessor and the memory. DRAM chips need high or low voltages applied at regular intervals—every two milliseconds (two one-thousandths of a second) or so—if they are not to lose their information.

Read-only memory (ROM) is nonvolatile and contains permanent program instructions for execution by the CPU. Other nonvolatile memory types include programmable read-only memory (PROM) and erasable programmable read-only memory (EPROM). Cache memory is a type of high-speed memory that CPUs can access more rapidly than RAM.

Processing done using several processing units is called *multiprocessing*. One form of multiprocessing uses coprocessors; coprocessors execute one type of instruction while the CPU works on others. Massively parallel processing involves linking several processors to work together to solve complex problems. Symmetrical multiprocessing involves using multiple processors which share a single copy of the operating system, memory, and other resources of one computer. Grid computing is the use of a collection of computers, often owned by multiple individuals or organizations, to work in a coordinated manner to solve a common problem.

Computer systems can store larger amounts of data and instructions in secondary storage, which is less volatile and has greater capacity than memory. The primary characteristics of secondary storage media and devices include access method, capacity, and portability. Storage media can implement either sequential access or direct access. Sequential access requires data to be read or written in sequence. Direct

access means that data can be located and retrieved directly from any location on the media.

Common forms of secondary storage include magnetic tape, magnetic disk, virtual disk, optical disc, compact disc, digital versatile disk, magneto-optical disc, memory cards, flash memory, and expandable storage devices. Redundant array of independent/inexpensive disks (RAID) is a method of storing data that generates extra bits of data from existing data, allowing the system to more easily recover data in the event of a hardware failure. Network-attached storage (NAS) and storage area networks (SAN) are alternative forms of data storage that enable an organization to share data resources among a much larger number of computers and users for improved storage efficiency and greater cost-effectiveness. Software tools known as policy-based storage management automate storage using previously defined policies. The overall trend in secondary storage is toward direct access methods, higher capacity, increased portability, and automated storage management. There is also increasing interest in renting space on massive storage devices.

Input and output devices allow users to provide data and instructions to the computer for processing and allow subsequent storage and output. These devices are part of a user interface through which human beings interact with computer systems. Input and output devices vary widely, but they share common characteristics of speed and functionality.

Data is placed in a computer system in a two-stage process: data entry converts human-readable data into machine-readable form; data input then transfers it to the computer. Online data entry and input immediately converts and transfers data from devices to the computer system. Source data automation involves automating data entry and input so that data is captured close to its source and in a form that can be input directly to the computer.

There is a wide range of computer input devices, including a keyboard, a mouse, voice recognition, digital cameras, and terminals. Scanners are input devices that convert images and text into binary digits. Specialized scanners include optical mark recognition (OMR) devices, optical character recognition (OCR) devices, and magnetic ink character recognition (MICR) devices. Some input and output devices combine several functions. Point-of-sale (POS) devices are terminals with bar-code scanners that read and enter codes into computer systems. Automated teller machines (ATMs) are terminals with keyboards used for transactions. Pen input devices allow users to activate a command, enter handwritten notes, and draw objects and figures. Radio-frequency identification technology employs a microchip with an antenna that broadcasts its unique identifier and location to strategically placed receivers.

Output devices provide information in different forms, from hard copy to sound to digital format. Display monitors are standard output devices; monitor quality is determined by size, color, and resolution. Liquid crystal display and organic light-emitting diode technology is enabling improvements in the resolution and size of computer monitors. Other output devices include printers, plotters, and computer output microfilm. Printers are popular hard-copy output devices whose quality is measured by speed and resolution. Plotters output hard copy for general design work. Computer output microfilm (COM) devices place data from the computer directly onto microfilm. MP3 is a standard format for compressing the amount of data it takes to accurately represent sound. Numerous music devices based on the compressed music are available.

Computers may be classified as special purpose or general purpose. General-purpose computers are used for numerous applications and can be broken into two major groups—systems used by a single user and systems used by multiple concurrent users.

Single-user systems include handheld, portable, thin client, desktop, and workstation computers. Handheld computers provide ease of portability because of their small size. They include palmtop computers, personal digital assistants, and smart phones. A variety of portable computers are available, including the laptop, notebook, subnotebook, and tablet computer. The thin client is a diskless, inexpensive computer used for accessing server-based applications and the Internet. Desktop computers are relatively small, inexpensive computer systems that are highly versatile. Workstations are advanced PCs with greater memory, processing, and graphics abilities.

Multiuser systems include servers, mainframes, and supercomputers. A computer server is a computer designed for a specific task, such as network or Internet applications. Servers typically have large memory and storage capacities, along with fast and efficient communications abilities. Mainframe computers have greater processing capabilities, while supercomputers are extremely fast computers used to solve the most intensive computing problems.

The configuration of computer system hardware components is the computer system architecture. Computer systems can be upgraded by changing hard drives, memory, printers, DVD burners, and other devices. Care must be taken in the disposal of obsolete systems because computer hardware contains many potentially harmful metals and chemicals.

CHAPTER 3: SELF-ASSESSMENT TEST

Assembling an effective, efficient computer system requires an understanding of its relationship to the information system and the organization. The computer system objectives are subordinate to, but supportive of, the information system and the needs of the organization.

1. Non-IS managers have little need to understand computer hardware. True or False?
2. The information system solutions chosen to be implemented at most nonprofit organizations are virtually identical. True or False?
3. _____ is any machinery (most of which use digital circuits) that assists in the input, processing, storage, and output activities of an information system.
4. Which represents a larger amount of data—a terabyte or gigabyte?
5. Which of the following performs mathematical calculations and makes logical comparisons?
   a. Control unit
   b. Register
   c. ALU
   d. Main memory
6. What term describes the number of bits the CPU can process at one time?
   a. register
   b. bus line
   c. microcode
   d. wordlength
7. _____ involves capturing and editing data when the data is originally created and in a form that can be directly input to a computer, thus ensuring accuracy and timeliness.

When selecting computer devices, you also must consider the current and future needs of the information system and the organization. Your choice of a particular computer system device should always allow for later improvements.

8. Having a collection of computers, often owned by multiple individuals or organizations, that work in a coordinated manner to solve a common problem is _____.
   a. massively parallel processing
   b. symmetrical multiprocessing
   c. grid computing
   d. all of the above
9. Three fundamental strategies for providing data storage are _____.
   a. expandable, nonexpandable, and static
   b. attached storage, network-attached storage, and storage area networks
   c. sequential, direct, indirect
   d. hard drive, CD-ROM, DVD
10. The relative clock speed of two CPUs from different manufacturers is a good indicator of their relative processing speed. True or False?
11. A technology that employs a microchip to broadcast its unique identifier and location to receivers is _____.
   a. radio-frequency ID
   b. bar-code scanning
   c. biometrics
   d. cryptography

CHAPTER 3: SELF-ASSESSMENT TEST ANSWERS

(1) False (2) False (3) Hardware (4) terabyte (5) c (6) d (7) Source data automation (8) d (9) b (10) False (11) a

KEY TERMS

arithmetic/logic unit (ALU) 92
bit 94
bus line 95
byte (B) 98
cache memory 99
CD-recordable (CD-R) disk 106
CD-rewritable (CD-RW) disk 106
central processing unit (CPU) 92
clock speed 94
compact disc read-only memory (CD-ROM) 106

complex instruction set computing (CISC) 97
computer system architecture 126
control unit 92
coprocessor 100
data entry 110
data input 110
desktop computer 124
digital computer camera 112
digital versatile disk (DVD) 106
direct access 103

direct access storage device (DASD) 103
disk mirroring 105
execution time (E-time) 93
expandable storage devices 107
flash memory 107
general-purpose computers 122
gigahertz (GHz) 94
grid computing 101
handheld computers 122
hardware 90

hertz 94
instruction time (I-time) 93
machine cycle 93
magnetic disk 104
magnetic tape 103
magneto-optical (MO) disk 106
mainframe computer 125
massively parallel processing 100
megahertz (MHz) 94
microcode 94
MIPS 94
Moore's Law 95
MP3 120
multifunction device 121
multiprocessing 100
music device 121
network-attached storage (NAS) 108
optical disc 105

optical processors 96
pipelining 93
pixel 118
plotters 120
point-of-sale (POS) device 114
policy-based storage management 109
portable computer 123
primary storage (main memory; memory) 92
radio-frequency identification (RFID) 118
random access memory (RAM) 98
read-only memory (ROM) 99
reduced instruction set computing (RISC) 97
redundant array of independent/ inexpensive disks (RAID) 105
register 92

scalability 125
secondary storage (permanent storage) 103
sequential access 103
sequential access storage device (SASD) 103
server 124
source data automation 110
special-purpose computers 122
storage area network (SAN) 108
supercomputers 126
superconductivity 96
symmetrical multiprocessing 101
thin client 124
virtual tape 105
voice-recognition device 111
wordlength 95
workstation 124

REVIEW QUESTIONS

1. What role does the mainframe computer play in today's large organization?
2. How would you distinguish between a desktop computer and a workstation?
3. What is MP3?
4. What is RFID technology? Identify three practical uses for this technology.
5. What does the resolution of a digital camera convey? How does this help you select the appropriate camera for your use?
6. Give three practical examples of the use of voice-recognition devices.
7. What is the difference between data entry and data input?
8. What is policy-based storage management?
9. When it comes to recording data on a storage medium, what is the advantage of using a blue laser over a red laser?
10. Explain the two-phase process for executing instructions.
11. Why is it said that the components of all information systems are interdependent?
12. Identify the three components of the CPU and explain the role of each.
13. What is the difference between sequential and direct access of data?
14. Identify several types of secondary storage media in terms of access method, capacity, and portability.
15. Identify and briefly describe the various classes of personal computers.
16. What is the difference between cache memory and main memory?
17. What is source data automation?
18. What is the overall trend in secondary storage devices?

DISCUSSION QUESTIONS

1. Identify and briefly describe three forms of multiprocessing.
2. Briefly outline some of the issues and some of the solutions associated with the discarding of old computer hardware.
3. Imagine that you are the business manager for your university. What type of computer would you recommend for broad deployment in the university's computer labs—a standard desktop personal computer or a thin client? Why?
4. Which would you rather have—a PDA or smart phone? Why?
5. Which form of monitor would you prefer and why—CRT, LCD, or OLED?
6. Why is there a need for standards groups such as the Uniform Code Council, which is behind the universal product code? What role do they play?
7. Identify and briefly describe the three fundamental approaches to data storage.
8. Describe how you would select the best DVD burner to meet your needs.

9. Discuss fully: What advantages does a 64-bit processor have over a 32-bit processor? Are there any disadvantages?
10. What is Moore's Law? What are some of its implications?

11. If cost were not an issue, describe the characteristics of your ideal computer. What would you use it for? Would you choose a handheld, portable, desktop, or workstation computer? Why?

PROBLEM-SOLVING EXERCISES

1. Some believe that the information technology industry has driven the economy and in large measure determines stock market prices—not just technology stocks but other stocks as well. Do some research to find an index that measures the stock performance of the largest technology companies. Plot that index versus the index for the S&P 500 for the past five years. What is your conclusion?

2. Over the upcoming year, your department is expected to add six people to its staff. You will need to acquire six computer systems and two additional printers for the new employees to share. Standard office computers have a Pentium (3-GHz) processor with 256 MB of RAM, 15-inch SVGA color monitor, and a 40-GB hard-disk drive. At least two of the new people will use their computers more than three hours per day. You would like to provide high-resolution monitors and special wireless mice and keyboards for these people—if it fits within your budget. You are not sure whether you want to upgrade the machines to 512 MB of RAM and 120 GB hard drives.

   Your department budget will allow a maximum of $6,000 for computer hardware purchases this year, and you want to select only one vendor for all of the hardware. A price list from three vendors appears in the following table, with prices for a single unit of each component. Use a spreadsheet to find the department's best solution; write a short memo explaining your rationale. Specify which vendor to choose and which items to be ordered as well as the total cost.

Component	Expert Solutions Ltd.	Business Processing Enterprises	Super Systems Inc.
3 GHz Pentium with 256 MB RAM with 40 GB hard drive	$525	$505	$550
Upgrade to 512 MB RAM	60	65	50
Upgrade to 120 GB hard drive	90	125	105
Standard 15-inch .28 dpi SVGA monitor	200	210	215
High-resolution 17-inch flat screen monitor	400	420	450
Wireless mouse and keyboard	65	60	55
12 ppm color inkjet printer	75	55	65
Surge protector/power strip	35	32	35
Three-year warranty (parts and labor)	340	300	320

3. Do research on three different digital cameras. Identify the factors (cost, size, weight, resolution, features, etc.) that distinguish each one. Use a spreadsheet to enter these factors in one column. Next enter the value of this factor for each of the three digital cameras in a different column. Now write a brief memo recommending which one you would buy and why. Cut and paste the spreadsheet into the memo.

TEAM ACTIVITIES

1. With two or three of your classmates, interview someone from the computer operations or support group of your college or university. Document the role that the mainframe and the servers play in terms of the types of data they hold and the software applications they run.

2. With one or two of your classmates, visit several local car dealers to identify what makes and models of automobiles come equipped with on-board computer-based navigation systems. Ask to see a demo of the capabilities of these systems. Write a brief reporting summarizing your findings, including costs and features.

WEB EXERCISES

1. Do research on the Web to identify the current status of the use of radio-frequency ID chips in the consumer goods industry. Write a brief report summarizing your findings.

2. Do research on the Web to find the current status of implementing the new 13-digit bar code. Are companies struggling to do this or has it already been successfully accomplished? Write a brief report summarizing your findings.

CAREER EXERCISES

1. Imagine that you are going to buy some sort of handheld computer device to support you in your school activities. What tasks could it help you perform? What sort of features would you look for in this device? Visit a computer store or a consumer electronics store and see whether you can purchase such a device for under $400.

2. Your company's finance department is planning to acquire 50 new computers and monitors, plus several new printers.

Managers have asked you to be a member of the hardware selection team to help define user computer hardware needs. How would you go about documenting users' needs? How would you determine the number of printers needed? What advice would you offer about the disposal of old computer assets and planning for the future disposal of the new equipment?

VIDEO QUESTIONS

Watch the video clip **Getting Started** and answer these questions:

1. What reason is given for buying a good high-quality keyboard and monitor?

2. After seeing what it takes to build your own computer, do you think that it is more economical to build or buy a pre-built computer? What do you think are the advantages and disadvantages of both?

CASE STUDIES

Case One

Hilton Hotels Implements "OnQ"

Hilton Hotels Corporation owns, manages, and develops hotels, resorts, and timeshare properties totaling 337,000 rooms at 2,084 properties worldwide. Hilton hotel brands include Hilton, Hilton Garden Inn, Doubletree, Embassy Suites, Hampton, Homewood Suites by Hilton, and Conrad. Hilton also develops and operates timeshare resorts through Hilton Grand Vacations Company and its related entities.

In a strategic move, Hilton acquired the Promus Hotel Corporation, with some 1,600 hotels and 45,000 employees. Although the acquisition made perfect business sense, it created many problems for the company's IS organization. Even

before the purchase, the two companies had a hodgepodge of computer hardware, including laptop and desktop computers from various vendors, plus HP-9000 and IBM AS/400 servers. Following the acquisition, Hilton was left with a hopeless jumble of IS hardware, software, and databases. Not only was the resulting tower of Babel expensive to operate and maintain, but it was not as reliable as needed to support the business.

Hilton decided to implement a common IS platform to link all its brands and all its hotels. Such a common platform would also improve the reliability and efficiency of computer operations by simplifying and standardizing the architecture. Hilton replaced the various computers and servers with Dell computers connected to powerful Dell servers holding common software and data about employees and customers. The

total cost of the conversion was close to $4 million, but Hilton expects to save about $5 million of its $150 million IS budget.

Now with "OnQ," as the system is known, when a frequent guest walks into a Hilton hotel, reservation specialists can talk to that guest about his or her last stay at a Hampton or Embassy Suites hotel, report that the guest's special room requirements and amenities have been taken care of, and see that his or her reservation next week at the Homewood Suites, Hilton, or Doubletree down the road or in another city is all set. The ability to cross-sell among the entire Hilton family of brands helps ensure that Hilton keeps business within its system when a customer's first choice is not available. In fact, cross-selling accounted for more than $300 million per year in incremental systemwide booked revenue. None of Hilton's competitors has this capability, and with technology advancements taking on increased importance in the years ahead, it gives Hilton a real competitive advantage. In addition, this common platform has enhanced the efficiency and productivity of back-office operations, such as payroll, purchasing, and financial forecasting to maximize revenues.

Hilton has been able to create a distinct advantage through the implementation of new information systems to operate the business efficiently and maximize customer service and loyalty. In short, information systems are helping Hilton bring customer service to a new and even better level.

Discussion Questions

1. What would you identify as the biggest benefit that Hilton gained by implementing its "OnQ" information system?
2. What do you suppose were some of the factors that were considered in selecting the vendor for the computer hardware?

Critical Thinking Questions

3. Do you think that Hilton will be able to maintain a permanent competitive advantage with its "OnQ" system? Why or why not?
4. To what degree should senior executives consider the compatibility of two firms' information systems when considering a merger or acquisition?

SOURCES: Gary H. Anthes, "Hilton Checks into New Suite," *Computerworld*, June 30, 2003, *www.computerworld.com*; "PeopleSoft: Hilton Customer Success Story," accessed at PeopleSoft Web site, *www.peoplesoft.com*, January 20, 2004; "Room to Grow," Dell Web site, February 2003, *www1.us.dell.com*; "Hilton Hotels 2002 Annual Report—Dear Fellow Shareholders," accessed at Hilton Web site, *www.hilton.com*, January 20, 2004.

Case Two

U. S. Post Office Information System Initiative

Flat mail—newspapers, catalogs, magazines, and other periodicals—represent close to 25 percent of total mail volume and generate over $16 billion per year in revenue for the U.S. Postal Service (USPS). But flat mail varies greatly in size and shape and must either be sorted on various machines or presorted by mailers to carrier routes, requiring letter carriers to check through as many as five separate flat bundles at the customer's mailbox before delivering the mail.

The Corporate Flat Strategy calls for bringing flat mail up to the same level of sophistication achieved for letter mail processing—to automate sorting down to the order that letter carriers deliver their routes. This reduction of effort is projected to generate $4 billion per year in labor savings. Achieving this goal will require that flat mail be received at the post office already bar-coded by the shipper. In addition, the USPS must implement technologies to read bar codes, track packages, and automate the sorting process. The USPS prefers the use of 11-digit ZIP codes and gives significant rate discounts for their use.

The Flat Sequencing System (FSS), the first initiative in this strategy, will automate flat sorting to the "walk" sequence of all carriers. The old equipment deployed for the sorting of letters came with three bar-code readers and keying stations. The USPS will work with vendors to modify this equipment to replace one keying station with an automated high-speed feeder, replace the three bar-code readers with one optical character reader (OCR), and provide a flat feeder to sequence flat mail into the correct delivery order.

The FSS must be able to handle all types of flat mail with a minimum machine throughput of 40,000 pieces per hour for a single pass sorting operation or a cumulative 16,350 pieces per hour for a multipass operation. Additional performance criteria include achieving a minimum of 95 percent sort rate to the correct delivery point with an error rate not to exceed 1 percent of the volume sorted.

Eventually, officials hope to sort letters and flat mail through one stream that is then packaged together in one batch of mixed letters and flats into the correct delivery sequence for the carrier. This larger goal, known as "delivery point" packaging, could be achieved by 2007.

Discussion Questions

1. How long will it take to sort 240,000 pieces of flat mail through the new FSS assuming a single-pass operation? How many flats will not be sorted to the delivery point? How many flats will be sorted to the incorrect delivery point?
2. Which is the more serious error—not sorting to the delivery point or sorting to an incorrect delivery point? What will be required to correct each type of error?

Critical Thinking Questions

3. Visit the FedEx and United Parcel Service Web sites and identify characteristics that distinguish their customer requirements and operations from that of the USPS.

4. Visit the USPS Web site and read about the current status of the Corporate Flat Strategy. Write a brief report summarizing your findings.

SOURCES: Linda Rosencrance, "Postal Service Wants Feedback on Automating 'Flats' Mail," *Computerworld*, July 17, 2003, *www.computerworld.com*; "United States Postal Service Corporate Flats Strategy," prepared by the Strategic Operations Planning Group, May 2003, accessed at *www.ribbs.usps.gov*, February 5, 2004; Sara Michael, "USPS Seeks Customer Input," FCW.com, August 8, 2003, accessed at *www.fcw.com*.

Case Three

PeopleSoft Offers Hosting Services

PeopleSoft is the world's second-largest provider of enterprise software, with nearly $3 billion in revenue and 12,000 employees. The company has 11,000 customers in 150 countries worldwide. PeopleSoft acquired J.D. Edwards, a competitor, in August 2003 and is currently fighting a hostile takeover by Oracle Corporation.

In early 2004, PeopleSoft announced new hosting and application management services for customers with annual revenues under $1 billion. With the new hosting service, customers who license their software from PeopleSoft can receive access to third-party data storage and mainframe computers and servers. Three levels of service are available: (1) the value level is a basic level of service, which guarantees 99.5 percent availability of all hardware and services, plus limited access to PeopleSoft support services; (2) the enhanced level of service guarantees 99.5 percent availability and gives expanded access to the support services; and (3) the ultimate level of service guarantees 99.9 percent availability and around-the-clock support services.

Surebridge, Inc., in Lexington, Massachusetts, provides PeopleSoft's hosting services. Surebridge provides the computing environment and the service level customers need and takes care of all staffing and resource requirements, according to a customer's agreement. Its data center facilities include a 24×7×365 Network Operations Command Center that monitors all of its computer hardware. The data center itself is equipped with numerous redundant systems to ensure reliability, including power systems, HVAC, fire-suppression systems, and cabling systems. In addition, multiple security measures are in place for the data center, including a closed-circuit television system used to monitor all access.

Surebridge provides customers with the flexibility of deploying PeopleSoft software in many ways, installed on computers located at the customer's own facilities, remotely hosted at the customer's location, or dedicated hosting at Surebridge's own state-of-the-art computer facilities.

Hosting at the customer's facilities allows companies to maintain full internal control of their IS investment and infrastructure while outsourcing the day-to-day management and support of the PeopleSoft software. Surebridge manages the PeopleSoft applications whether they run on the client company's computers or at a third party's facilities. The client organization maintains complete control over the physical infrastructure, hardware, communications, data, and system access. Surebridge proactively manages and supports the PeopleSoft application through its specialized application management systems and customer service experts to ensure that the systems are running optimally. This keeps costs down.

The option of remotely hosting at a customer's location allows companies to maintain full internal control of their computer hardware and software while outsourcing the day-to-day management and support of their application environment. Surebridge manages the PeopleSoft application, whether it resides at the customer's premise or at a third-party facility, remotely through a secure connection that Surebridge assists in designing and implementing. This outsourcing allows the customer's IS organization to focus on serving its customers. The customer maintains complete control over the physical infrastructure, hardware, communications, data, and system access. Surebridge proactively manages and supports the customer's application through its specialized management systems and customer service experts to ensure the systems are running optimally. This arrangement also keeps costs down.

With the dedicated hosting service, Surebridge runs companies' applications in its data center, managing the complete set of computer hardware technology required to support their business application environment. This approach allows firms to get up and running faster and with the lowest initial hardware cost. It provides a high level of security and performance so that customers can focus on their business without having to worry about training, personnel turnover, upgrades, and systems management. Surebridge provides a single contact person who knows his or her customers' business and responds to their questions and issues. In addition, customers have 24×7 access to Surebridge's Web-based customer portal, so they can monitor the capacity, utilization, availability, and reliability of their systems.

Assume that your manufacturing firm with $500 million in annual sales is considering investing $3 million in the purchase of PeopleSoft software to support several critical business functions, including human resources, finance and accounting, and supply chain management. You are the chairperson of a task force put together to decide which of the three Surebridge hosting service options would best meets your company's needs.

Discussion Questions

1. What are the key advantages and disadvantages of each of the three hosting service options?
2. What other information do you need to know about the service options to make a choice?

Critical Thinking Questions

3. What additional information do you need to know about your company's business needs in order to make a good choice?

4. If you were to visit the Surebridge facilities, what would you want to see? What questions would you ask?

SOURCES: Gillian Law, "PeopleSoft Aims Hosting Services at Midmarket," *Computerworld*, January 27, 2004, *www.computerworld.com*; PeopleSoft Web page, "Hosting Services," accessed at *www.peoplesoft.com/en/hosting* February 5, 2004; "PeopleSoft Introduces Mid Market Hosting Solutions," PeopleSoft press release, January 26, 2004, accessed at *www.peoplesoft.com/en/about/press*; Surebridge Web page, "Services," *www.surebridge.com*, accessed on February 9, 2004.

NOTES

Sources for the opening vignette. "Porsche's CEO Talks Shop, *Business-Week Online*, December 22, 2003, accessed at *www.businessweekonline.com*; Daren Fonda, "A New Porsche for Purists," *Time Magazine*, July 28, 2003, accessed at *www.time.com*; "About Porsche," "Finances," "Corporate Divisions," and "Cayenne" accessed at the Porsche Web site at *www3.us.porsche.com/english/usa/home.htm*, January 19, 2004; "Porsche AG, SAP Customer Success Story," accessed at *www.sap.com*, January 19, 2004; "Porsche, HP Success Story," March 10, 2003, accessed at *www.hp.com*, January 19, 2004.

1. "ARZ Allgemeines Reschenzentrum Enjoys Reduced Costs through Workload License Charges for IBM eServer," the IBM Web site—Success Stories, *www-306.ibm.com/software/success/cssdb.nsf/cs/DNSD-5UZKY7?OpenDocument&Site=eserverzseries*, accessed January 19, 2004.
2. "Wind Beneath Its Wings—NonStop Technology Gives Continental Airlines a Competitive Edge," December 12, 2003, the Hewlett Packard Web site, *www.hp.com/cgi-bin/pf-new.cgi?IN=http://h71033.www7.hp.com/object/CONTAIRSS.html*, accessed January 19, 2004.
3. "Air Products and Chemicals, Inc." May 2003, the Dell Web site, *www1.us.dell.com/content/topics/global.aspx/casestudies/en/2003_air?c=us&cs=555&l=en&s=biz*, accessed January 19, 2004.
4. "Specialized Bicycles Manufacturing," the Sun Web site, *www.sun.com/desktop/success/specialized.html*, accessed January 19, 2004.
5. Levy, David, "A Bakery on the Rise," the Apple Web site, *www.apple.com/business/profiles/sullivanstreet/index2.html*, accessed January 19, 2004.
6. Krazil, Tom, "Intel to Ring in 2004 with Delayed Prescott Launch," *Computerworld*, December 15, 2003, accessed at *www.computerworld.com*.
7. Lawson, Stephan, "Moore: Innovation Key to Keeping His Law Alive," *Computerworld*, February 11, 2003, accessed at *www.computerworld.com*.
8. Krazit, Tom, "Chip Vendors Prepare for 90 Nanometer Era," *InfoWorld*, September 24, 2003, accessed at *www.infoworld.com*.
9. Anthes, Gary H., "Microprocessors March On," *Computerworld*, March 10, 2003, accessed at *www.computerworld.com*.
10. Lawson, Stephan, "Moore: Innovation Key to Keeping His Law Alive," *Computerworld*, February 11, 2003, accessed at *www.computerworld.com*.
11. Reuters, "Processing at the Speed of Light," *Wired News*, October 29, 2003, accessed at *www.wired.com*.
12. Anthes, Gary H., "Microprocessors March On," *Computerworld*, March 10, 2003, accessed at *www.computerworld.com*.
13. Dunn, Darrell, "Fast Times Demand Flexible Systems," *InformationWeek*, June 21, 2004, accessed at *www.informationweek.com*.
14. Krazit, Tom, "Fall IDF: Deerfield to Appear Before Intel Developer Forum Gets Underway," *Computerworld*, September 3, 2003, accessed at *www.computerworld.com*.
15. Anthes, Gary H., "U.S. Losing Lead in Supercomputing, User Says," *Computerworld*, September 29, 2003, accessed at *www.computerworld.com*.
16. Rosencrance, Linda, "IBM Taps IBM eServer Systems, Linux for Expanded Airfare System," *Computerworld*, May 6, 2004, accessed at *www.computerworld.com*.
17. Kahn, Chris, Associated Press, "Supercomputer Strings 1,100 Macs Right Off the Shelf," *The Cincinnati Enquirer*, November 9, 2003, p. A18.
18. Kharif, Olga, "Everybody Has More to Store," *Business Week Online*, October 28, 2003, accessed at *www.businessweek.com*.
19. Salkever, Alex, "Consumers: Thanks for the Memory," *Business Week Online*, October 28, 2003, accessed at *www.businessweek.com*.
20. Vinas, Tonya, "IBM's Millipede on the March to Smaller, Cheaper Data Storage," *Small Times Magazine*, November 26, 2003, accessed at *www.smalltimes.com*.
21. Williams, Martyn, "Toshiba Prepares Coin-Size Hard-Disk Drive," *Computerworld*, December 16, 2003, accessed at *www.computerworld.com*.
22. Dunn, Darrell, "Automating the IT Factory," *InformationWeek*, July 12, 2004, accessed at *www.informationweek.com*.
23. Garvey, Martin J., "Avamar Speeds Data Recovery," *InformationWeek*, April 26, 2004, accessed at *www.informationweek.com*.
24. Apicella, Mario, "Betting on a Blue Laser," *Computerworld*, November 25, 2003 accessed at *www.computerworld.com*.
25. Williams, Martyn, "Sony to Launch 23GB Optical Data Disk in November," *Computerworld*, October 22, 2003, accessed at *www.computerworld.com*.
26. "About Migo," *www.4migo.com*, accessed December 26, 2003.
27. McKendrick, Joe, "SANS Provide A Data 'Dial Tone' Across the Business," *Insurance Networking News*, January 2, 2004, accessed at *www.insurancenetworking.com*.
28. Mearian, Lucas, "Los Alamos Tries Object-Based Storage," *ComputerWorld*, October 20, 2003, accessed at *www.computerworld.com*.
29. McKendrick, Joe, "SANS Provide A Data 'Dial Tone' Across the Business," *Insurance Networking News*, January 2, 2004, accessed at *www.insurancenetworking.com*.
30. Shafer, Scott Tyler, "Hitachi Rolls Out Thunder Array," *Computerworld*, October 7, 2003, accessed at *www.computerworld.com*.

31. Garvey, Martin J., "Legal Storage Upgrade," *InformationWeek*, June 28, 2004, accessed at *www.informationweek.com*.

32. Mearian, Lucas, "Hitachi, Cisco Plan Key Rollouts at Storage Networking Show," *Computerworld*, April 13, 2003, accessed at *www.computerworld.com*.

33. Ewait, David M., "Don't Talk, Just Listen," *InformationWeek*, November 10, 2003, accessed at *www.informationweek.com*.

34. Brewin, Bob, "A Railroad Finds Its Voice," *Computerworld*, January 26, 2004, accessed at *www.computerworld.com*.

35. Information from Deal Time Web site, *www.dealtime.com*, accessed January 28, 2004.

36. Rosencrance, Linda, "New FedEx Ground Facility Features Scanning Technology," *ComputerWorld*, April 23, 2003, accessed at *www.computerworld.com*.

37. Brewin, Bob, "Fox UK Using Handhelds to Manage DVD Sales, Inventory," *Computerworld*, February 26, 2003, accessed at *www.computerworld.com*.

38. Melymuka, Kathleen, "Another Digit, Another Deadline: Retailers Need Longer Bar Codes by 2005," *Computerworld*, January 28, 2004, accessed at *www.computerworld.com*.

39. Whiting, Rick, "Drugmakers 'Jumpstart' RFID Tagging of Bottles," *InformationWeek*, July 26, 2004, accessed at *www.informationweek.com*.

40. "Monitors with an Edge," *BusinessWeek Online*, November 10, 2003, accessed at *www.businessweekonline.com*.

41. Robb, Drew, "Displays Go for Sharper Image," *Computerworld*, January 12, 2004, accessed at *www.computerworld.com*.

42. Weiss, Todd R., "Printer Magic," *Computerworld*, January 26, 2004, accessed at *www.computerworld.com*.

43. Faas, Ryan, "A Look Ahead: Apple in 2004," *Computerworld*, December 22, 2003, accessed at *www.computerworld.com*.

44. Kessler, Michelle, "Sales of Smart Phones Leave PDAs in the Dust," *USA Today*, January 29, 2004, p. B1.

45. Brewin, Bob, "After a Year, Tablet Still Niche," *Computerworld*, November 24, 2003, accessed at *computerworld.com*.

46. Krazit, Tom, "Motion Computing Adds Celeron to Low-Cost Tablet PC," *Computerworld*, December 4, 2003, accessed at *www.computerworld.com*.

47. Babcock, Charles, "Italian Company Finds Web Service," *InformationWeek*, January 7, 2004, accessed at *www.informationweek.com*.

48. Greenemeier, Larry, "Dell Adds French Accent to Supercomputing Clusters," *InformationWeek*, September 10, 2003, accessed at *www.informationweek.com*.

49. Scheier, Robert L., "Refreshing the Desktop," *Computerworld*, March 15, 2004, accessed at *www.computerworld.com*.

50. By the Associated Press, "Groups Promote Recycling of Tech Gadgets," *InformationWeek*, December 29, 3003, accessed at *www.informationweek.com*.

51. Johnson, Mary Fran, "Cleaning IT's Basement," *Computerworld*, February 2, 2004, accessed at *http://computerworld.com*.

CHAPTER

• 4 •

Software:

Systems and Application Software

PRINCIPLES	LEARNING OBJECTIVES
▪ When selecting an operating system, you must consider the current and future requirements for application software to meet the needs of the organization. In addition, your choice of a particular operating system must be consistent with your choice of hardware.	▪ Identify and briefly describe the functions of the two basic kinds of software. ▪ Outline the role of the operating system and identify the features of several popular operating systems.
▪ Do not develop proprietary application software unless doing so will meet a compelling business need that can provide a competitive advantage.	▪ Discuss how application software can support personal, workgroup, and enterprise business objectives. ▪ Identify three basic approaches to developing application software and discuss the pros and cons of each.
▪ Choose a programming language whose functional characteristics are appropriate for the task at hand, taking into consideration the skills and experience of the programming staff.	▪ Outline the overall evolution of programming languages and clearly differentiate among the five generations of programming languages.
▪ The software industry continues to undergo constant change; users need to be aware of recent trends and issues to be effective in their business and personal life.	▪ Identify several key issues and trends that have an impact on organizations and individuals.

INFORMATION SYSTEMS IN THE GLOBAL ECONOMY ⟩⟩
JONES LANG LASALLE, UNITED STATES

Software Provides Best Practices Framework for International Business

Jones Lang LaSalle is the world's leading real estate services and investment management firm, serving clients in more than 100 markets on 5 continents. LaSalle's 17,600 employees provide property-management services, implementation services—matching properties with buyers—and investment-management services on a local, regional, and global level to property owners, tenants, and investors.

The company has built a significant presence in the Asia Pacific region over the past 45 years. But the company's rapid growth in the region resulted in a lack of standardized practices and technology—each of the offices conducted business in its own fashion. International director and head of management services for LaSalle Asia, James Wong, blames the patchwork of disparate systems and software for problems in managing the region. Disjointed systems naturally lead to disjointed business practices that impede the flow of information throughout an organization. Wong set out to standardize the way each business unit operated.

He found a solution in a business unit of Intuit, Inc., called MRI Real Estate Solutions. (Intuit produces well-known financial software Quicken, QuickBooks, and TurboTax.) The MRI software is designed to handle the critical business and financial operations of real estate management—streamlining business operations, increasing operational efficiency, and enhancing customer service.

When choosing software, a business must find a product that matches its management philosophies, approaches, and strategies. James Wong puts it this way: "Jones Lang LaSalle was looking for an enterprise technology provider that understood our business environment. MRI stood out from other solutions providers because its offering combines industry-specific features and services with a global presence to support our business."

Most software packages marketed to real estate and property-management companies are generic packages that can be used for any one of a variety of industries. MRI's software is designed specifically for real estate and property management and includes "best practices" features important to day-to-day operations. The term *best practices* refers to information management practices that have been confirmed to achieve the best results within an industry. Packaging best practices into software implies that the software does not just automate existing routines but instead helps businesses operate in the most efficient and effective manner possible.

Jones Lang LaSalle capitalized on the best practices built into MRI Real Estate Solutions to reengineer its Asia Pacific operations. The software contains features to support tenancy management, accounts receivable, accounts payable, general ledger, and executive management reporting—nearly every vital area of LaSalle's organization. Instead of dealing with a patchwork of differing software systems, LaSalle employees now interact with a single, integrated system that provides real-time access to their business data and is accessible from almost anywhere at any time. LaSalle's managers use the

software's advanced reporting and analytical tools to access up-to-the-second information, vastly improving their forecasting and budgeting, as well as enhancing customer service. The software also is flexible enough to meet the unique needs of the Hong Kong market—currency, regulations, and culture—while supporting the organization's international standards.

Jones Lang LaSalle continues to deploy MRI Real Estate Solutions throughout its Asia Pacific operations. So far, the company has installed the software in its offices in Hong Kong, Singapore, Tokyo, and Korea. Can software affect the success of a business? The world's largest real estate services and investment management firm is banking on it. While in years past software may have been designed to support specific but limited aspects of a company's business practices, today's software is at the core of those processes, defining best practices and leading companies to success.

As you read this chapter, consider the following:

- How do companies such as Jones Lang LaSalle acquire software? What options are available, and what considerations affect software purchasing decisions?
- Jones Lang LaSalle decided to purchase software from an outside company. Why didn't it develop the software itself? What benefits and drawbacks are involved in purchasing software as opposed to developing your own?

Why Learn About Software?

Software is indispensable for any computer system and the people using it. In this chapter, you will learn about systems and application software. Without systems software, computers would not be able to input data from a keyboard, make calculations, or print results. Application software is the key to helping you achieve your career goals. Sales representatives use software to enter sales orders and help their customers get what they want. Stock and bond traders use software to make split-second decisions involving millions of dollars. Scientists use software to analyze the threat of global warming. Regardless of your job, you most likely will use software to help you advance in your career and earn higher wages. Today, most organizations could not function without accounting software to print payroll checks, enter sales orders, and send out bills. You can also use software to help you prepare your personal income taxes, keep a budget, and play amazing games. Software can truly advance your career and enrich your life. We begin by giving you an overview of software and programming languages that are used to develop software.

Software has had a profound impact on individuals and companies. It can make the difference between profits and losses, survival and bankruptcy. SmartSignal, for example, makes software that can predict when machines are likely to fail.[1] The software can be used to predict when aircraft engines might fail or power plants might malfunction, preventing potential disasters.

In the 1950s, when computer hardware was relatively rare and expensive, software costs were a comparatively small percentage of total information systems (IS) costs. Today, the situation has dramatically changed. Software can represent 75 percent or more of the total cost of a particular information system for three major reasons: Advances in hardware technology have dramatically reduced hardware costs, increasingly complex software requires more time to develop and so is more costly, and salaries for software developers have increased. In the future, as suggested in Figure 4.1, software is expected to make up an even greater portion of the cost of the overall information system. The critical functions software serves, however, make it a worthwhile investment.

Figure 4.1

The Importance of Software in Business

Since the 1950s, businesses have greatly increased their expenditures on software compared with hardware.

AN OVERVIEW OF SOFTWARE

One of software's most critical functions is to direct the workings of the computer hardware. As we saw in Chapter 1, software consists of computer programs that control the workings of computer hardware. **Computer programs** are sequences of instructions for the computer. **Documentation** describes the program functions to help the user operate the computer system. The program displays some documentation on screen, while other forms appear in external resources, such as printed manuals. There are two basic types of software: systems software and application software.

Systems Software

Systems software is the set of programs designed to coordinate the activities and functions of the hardware and various programs throughout the computer system. A particular systems software package is designed for a specific CPU design and class of hardware. The combination of a particular hardware configuration and systems software package is known as a **computer system platform.**

Application Software

Application software consists of programs that help users solve particular computing problems. Both systems and application software can be used to meet the needs of an individual, a group, or an enterprise. Before an individual, a group, or an enterprise decides on the best approach for acquiring application software, goals and needs should be analyzed carefully.

computer programs
Sequences of instructions for the computer.

documentation
Text that describes the program functions to help the user operate the computer system.

systems software
The set of programs designed to coordinate the activities and functions of the hardware and various programs throughout the computer system.

computer system platform
The combination of a particular hardware configuration and systems software package.

Application software has the greatest potential to affect processes that add value to a business because it is designed for specific organizational activities and functions.

(Source: © Charlie Westerman/ Getty Images.)

application software
Programs that help users solve particular computing problems.

Lotus Notes is an application that enables a workgroup to schedule meetings and coordinate activities.

(Source: Screen Captures or other materials © 2004 IBM Corporation. Used with permission of IBM Corporation. IBM, Lotus, and Notes are trademarks of IBM Corporation in the United States, other countries, or both.)

Supporting Individual, Group, and Organizational Goals

Every organization relies on the contributions of individuals, groups, and the entire enterprise to achieve its business objectives. And conversely, the organization also supports individuals, groups, and the entire enterprise with specific application software and information systems. As the power and reach of information systems expand, they promise to reshape every aspect of our lives: how we work and play, how we are educated, how we interact with others, how our businesses and governments conduct their work, and how scientists perform research. Apple Computer, for example, was one of the first PC companies to develop sophisticated photo-processing software.[2] The software from Apple, called iPhoto, allows digital camera users to find, organize, and share digital images. After the success of iPhoto, several companies are now developing or marketing similar products, such as Picasa and Photoshop Album, for Windows PCs with Intel chips. Information systems are changing virtually every method for capturing, storing, transmitting, and analyzing knowledge, including books, newspapers, magazines, movies, television, phone calls, musical recordings, and architectural drawings. One useful way of classifying the many potential uses of information systems is to identify the scope of the problems and opportunities addressed by a particular organization. This is called the **sphere of influence**. For most companies, the spheres of influence are personal, workgroup, and enterprise, as shown in Table 4.1.

sphere of influence
The scope of problems and opportunities addressed by a particular organization.

Table 4.1

Classifying Software by Type and Sphere of Influence

Software	Personal	Workgroup	Enterprise
Systems software	Personal computer and workstation operating systems	Network operating systems	Midrange computer and mainframe operating systems
Application software	Word processing, spreadsheet, database, graphics	Electronic mail, group scheduling, shared work	General ledger, order entry, payroll, human resources

Information systems that operate within the **personal sphere of influence** serve the needs of an individual user. These information systems enable their users to improve their personal effectiveness, increasing the amount of work they can do and its quality. Such software is often referred to as **personal productivity software**. There are many examples of such applications within the personal sphere of influence—a word processing application to enter, check spelling, edit, copy, print, distribute, and file text material; a spreadsheet application to manipulate numeric data in rows and columns for analysis and decision making; a graphics application to perform data analysis; and a database application to organize data for personal use.

A **workgroup** is two or more people who work together to achieve a common goal. A workgroup may be a large, formal, permanent organizational entity such as a section or department or a temporary group formed to complete a specific project. The human resource department of a large firm is an example of a formal workgroup. It consists of several people, is a formal and permanent organizational entity, and appears on a firm's organization chart. An information system that operates in the **workgroup sphere of influence** supports a workgroup in the attainment of a common goal. Users of such applications are operating in an environment where communication, interaction, and collaboration are critical to the success of the group. Applications include systems that support information sharing, group scheduling, group decision making, and conferencing. These applications enable members of the group to communicate, interact, and collaborate.

Information systems that operate within the **enterprise sphere of influence** support the firm in its interaction with its environment. The surrounding environment includes customers, suppliers, shareholders, competitors, special interest groups, the financial community, and government agencies. Every enterprise has many applications that operate within the enterprise sphere of influence. The input to these systems is data about or generated by basic business transactions with someone outside the business enterprise. These transactions include customer orders, inventory receipts and withdrawals, purchase orders, freight bills, invoices, and checks. One of the results of processing transaction data is that the records of the company are instantly updated. For example, processing the number of hours employees work updates individuals' payroll records used to generate their checks. The order entry, finished product inventory, and billing information systems are examples of applications that operate in the enterprise sphere of influence. These applications support interactions with customers and suppliers.

personal sphere of influence
Sphere of influence that serves the needs of an individual user.

personal productivity software
Software that enables users to improve their personal effectiveness, increasing the amount of work they can do and its quality.

workgroup
Two or more people who work together to achieve a common goal.

workgroup sphere of influence
Sphere of influence that serves the needs of a workgroup.

enterprise sphere of influence
Sphere of influence that serves the needs of the firm in its interaction with its environment.

SYSTEMS SOFTWARE

Controlling the operations of computer hardware is one of the most critical functions of systems software. Systems software also supports the application programs' problem-solving capabilities. Different types of systems software include operating systems, utility programs, and middleware.

Operating Systems

An **operating system (OS)** is a set of computer programs that controls the computer hardware and acts as an interface with application programs (see Figure 4.2). Operating systems can control one computer or multiple computers, or they can allow multiple users to interact with one computer. The various combinations of OSs, computers, and users include:

- *A Single Computer with a Single User.* This system is commonly used in a personal computer or a handheld computer that allows one user at a time.
- *A Single Computer with Multiple Users.* This system is typical of larger, mainframe computers that can accommodate hundreds or thousands of people, all using the computer at the same time.

operating system (OS)
A set of computer programs that controls the computer hardware and acts as an interface with application programs.

- *Multiple Computers.* This system is typical of a network of computers, such as a home network that has several computers attached or a large computer network with hundreds of computers attached around the world.
- *Special-Purpose Computers.* This system is typical of a number of special-purpose systems that control sophisticated military aircraft, the space shuttle, some home appliances, and a variety of other special-purpose computers.

Figure 4.2

The role of the operating system and other systems software is as an interface or buffer between application software and hardware.

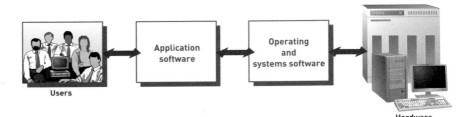

The OS, which plays a central role in the functioning of the complete computer system, is usually stored on disk. After a computer system is started, or "booted up," portions of the OS are transferred to memory as they are needed. Many computers allow a *dual-boot* feature, which allows you to select the OS you want to use before the boot process is completed. Some OSs for small computers use an "Instant On" feature that significantly reduces the time needed to boot a computer. A computer can also be rebooted by using the keyboard. David Bradley, an IBM employee, developed the procedure of using the Ctrl+Alt+Delete keys to reboot a personal computer decades ago.[3] "I may have invented it, but Bill [Gates] made it famous," says Bradley. The collection of programs that make up the OS performs a variety of activities, including:

- Performing common computer hardware functions
- Providing a user interface and input/output management
- Providing a degree of hardware independence
- Managing system memory
- Managing processing tasks
- Providing networking capability
- Controlling access to system resources
- Managing files

kernel
The heart of the operating system, which controls the most critical processes.

The **kernel**, as its name suggests, is the heart of the OS and controls the most critical processes. The kernel ties all of the components of the OS together and regulates other programs.

Common Hardware Functions

All application programs must perform certain tasks. For example:

- Get input from the keyboard or some other input device
- Retrieve data from disks
- Store data on disks
- Display information on a monitor or printer

Each of these basic functions requires a detailed set of instructions. The OS converts a simple, basic instruction into the set of detailed instructions required by the hardware. In effect, the OS acts as intermediary between the application program and the hardware. The typical OS performs hundreds of such functions, each of which is translated into one or more instructions for the hardware. The OS notifies the user if input/output devices need attention, if an error has occurred, and if anything abnormal happens in the system.

User Interface and Input/Output Management

user interface
Element of the operating system that allows individuals to access and command the computer system.

One of the most important functions of any OS is providing a **user interface**. A user interface allows individuals to access and command the computer system. The first user interfaces for mainframe and personal computer systems were command based. A **command-based user**

interface requires that text commands be given to the computer to perform basic activities. For example, the command ERASE 00TAXRTN would cause the computer to erase or delete a file called 00TAXRTN. RENAME and COPY are other examples of commands used to rename files and copy files from one location to another. Many mainframe computers use a command-based user interface. In some cases, a specific *job control language (JCL)* is used to control how jobs or tasks are to be run on the computer system. The use of command-based user interfaces has declined over time.

A **graphical user interface (GUI)** uses pictures (called **icons**) and menus displayed on screen to send commands to the computer system. Many people find that GUIs are easier to use because users intuitively grasp the functions. Today, the most widely used graphical user interface is Windows by Microsoft. Alan Kay and others at Xerox PARC (Palo Alto Research Center, located in California) were pioneers in investigating the use of overlapping windows and **icons** as an interface. As the name suggests, Windows is based on the use of a window, or a portion of the display screen dedicated to a specific application. The screen can display several windows at once. The use of GUIs has contributed greatly to the increased use of computers because users no longer need to know command-line syntax to accomplish tasks.

In addition to providing a user interface, today's OSs manage all aspects of computer input and output. Input management includes controlling a keyboard, mouse, touchscreen, and many other input devices. Output management includes controlling the display screen, printers, plotters, and other output devices. Many OSs, for example, send output from the CPU to a disk before it is printed. This process, called *spooling*, frees the CPU to perform other tasks, which typically results in getting more done in less time.

Hardware Independence

The applications make use of the OS by making requests for services through a defined **application program interface (API)**, as shown in Figure 4.3. Programmers can use APIs to create application software without having to understand the inner workings of the OS.

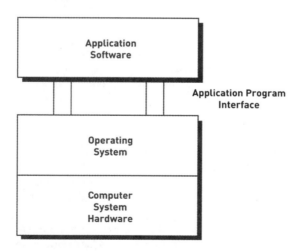

command-based user interface
A user interface that requires that text commands be given to the computer to perform basic activities.

graphical user interface (GUI)
An interface that uses icons and menus displayed on screen to send commands to the computer system.

icon
Picture.

application program interface (API)
Interface that allows applications to make use of the operating system.

Figure 4.3

Application Program Interface Links Application Software to the Operating System

Suppose that a computer manufacturer designs new hardware that can operate much faster than before. If the same OS for which an application was developed can run on the new hardware, minimal (or no) changes are needed to the application to enable it to run on the new hardware. If APIs did not exist, the application software developers might have to completely rewrite the application program to take advantage of the new, faster hardware.

Memory Management

The purpose of memory management is to control how memory is accessed and to maximize available memory and storage. Newer OSs typically manage memory better than older OSs. The memory-management feature of many OSs allows the computer to execute program instructions effectively and speed processing. One way to increase the performance of an old computer is to upgrade to a newer OS and increase the amount of memory.[4]

Controlling how memory is accessed allows the computer system to efficiently and effectively store and retrieve data and instructions and to supply them to the CPU. Memory-management programs convert a user's request for data or instructions (called a *logical view* of the data) to the physical location where the data or instructions are stored. A computer understands only the physical view of data—that is, the specific location of the data in storage or memory and the techniques needed to access it. This concept is described as logical versus physical access. For example, the current price of an item, say, a Texas Instruments BA-35 calculator with an item code of TIBA35, might always be found in the logical location "TIBA35$." If the CPU needed to fetch the price of TIBA35 as part of a program instruction, the memory-management feature of the OS would translate the logical location "TIBA35$" into an actual physical location in memory or secondary storage (see Figure 4.4).

Figure 4.4

An Example of the Operating System Controlling Physical Access to Data

The user prompts the application software for specific data. The OS translates this prompt into instructions for the hardware, which finds the data the user requested. Having successfully completed this task, the OS then relays the data back to the user via the application software.

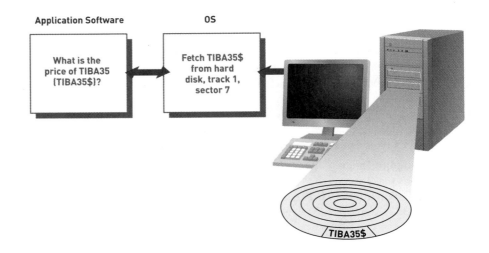

virtual memory
Memory that allocates space on the hard disk to supplement the immediate, functional memory capacity of RAM.

paging
Process of swapping programs or parts of programs between memory and one or more disk devices.

Most OSs support **virtual memory**, which allocates space on the hard disk to supplement the immediate, functional memory capacity of RAM. Virtual memory works by swapping programs or parts of programs between memory and one or more disk devices—a concept called **paging**. This reduces CPU idle time and increases the number of jobs that can run in a given time span.

Processing Tasks

Managing all processing activities is accomplished by the task-management features of today's OSs. Task management allocates computer resources to make the best use of each system's assets. Task-management software can permit one user to run several programs or tasks at the same time (multitasking) and allow several users to use the same computer at the same time (time-sharing).

multitasking
Capability that allows a user to run more than one application at the same time.

An OS with **multitasking** capabilities allows a user to run more than one application at the same time. Without having to exit a program, you can work in one application, easily pop into another, and then jump back to the first program, picking up where you left off. Better still, while you're working in the *foreground* in one program, one or more other applications can be churning away, unseen, in the *background*, sorting a database, printing a document, or performing other lengthy operations that otherwise would monopolize your computer and leave you staring at the screen unable to get other work done. Multitasking can save users a considerable amount of time and effort.

time-sharing
Capability that allows more than one person to use a computer system at the same time.

Time-sharing allows more than one person to use a computer system at the same time. For example, 15 customer service representatives may be entering sales data into a computer system for a mail-order company at the same time. In another case, thousands of people may be simultaneously using an online computer service to get stock quotes and valuable business news.

The ability of the computer to handle an increasing number of concurrent users smoothly is called **scalability**. This feature is critical for systems expected to handle a large number of users such as a mainframe computer or a Web server. Because personal computer OSs usually are oriented toward single users, the management of multiple-user tasks often is not needed.

Networking Capability

The OS can provide features and capabilities that aid users in connecting to a computer network. For example, Apple computer users have built-in network access through the AppleShare feature, and the Microsoft Windows OSs come with the capability to link users to other devices and the Internet.

Access to System Resources

Computers often handle sensitive data that can be accessed over networks. The OS needs to provide a high level of security against unauthorized access to the users' data and programs. Typically, the OS establishes a logon procedure that requires users to enter an identification code and a matching password. If the identification code is invalid or if the password does not go with the identification code, the user cannot gain access to the computer. The OS also requires that user passwords be changed frequently—say, every 20 to 40 days. If the user is successful in logging on to the system, the OS records who is using the system and for how long. In some organizations, such records are also used to bill users for time spent using the system. The OS also reports any attempted breaches of security.

File Management

The OS performs a file-management function to ensure that files in secondary storage are available when needed and that they are protected from access by unauthorized users. Many computers support multiple users who store files on centrally located disks or tape drives. The OS keeps track of where each file is stored and who may access it. The OS must be able to resolve what to do if more than one user requests access to the same file at the same time. Even on stand-alone personal computers with only one user, file management is needed to track where files are located, what size they are, when they were created, and who created them.

Operating systems have file-management conventions that specify how files can be named and organized (see Table 4.2). In Windows, for example, filenames can be 256 characters long, both numbers and spaces can be included in the filename, and upper and lowercase letters can be used. Windows and Linux don't permit certain characters, such as \, *, >, <, and some other characters, to be used in filenames. Operating systems also specify how files can be organized in folders or subdirectories. Organizing files into multiple folders or subdirectories makes it much easier to locate files, instead of having all files in one large folder or directory.

scalability
The ability of the computer to handle an increasing number of concurrent users smoothly.

Convention (Rule)	Windows	Macintosh	Linux
Length in characters	256	31	255
Case sensitive?	No	Yes	Yes
Can numbers be used?	Yes	Yes	Yes
Can spaces be used?	Yes	Yes	No

Table 4.2

Conventions, or Rules, for Filenames

Current Operating Systems

Early OSs were very basic. Today, however, more advanced OSs have been developed, incorporating some features previously available only with mainframe OSs. Table 4.3 classifies a few current OSs by sphere of influence.

Table 4.3

Popular Operating Systems cross All Three Spheres of Influence

Personal	Workgroup	Enterprise
Windows XP, Windows Mobile, and Windows Embedded	Windows NT Server	Windows NT Server
Mac OS	Windows 2003 Server	Windows 2003 Server
Mac OS X	Mac OS Server	Windows Advanced Server, Limited Edition
UNIX	UNIX	UNIX
Solaris	Solaris	Solaris
Linux	Linux	Linux
RedHat Linux	RedHat Linux	RedHat Linux
Palm OS	Netware	
	IBM OS/390	IBM OS/390
	IBM z/OS	IBM z/OS
	HP MPE/iX	HP MPE/iX

Microsoft PC Operating Systems

Ever since a then-small company called Microsoft developed PC-DOS and MS-DOS to support the IBM personal computer introduced in the 1970s, there has been a continuous and steady evolution of personal computer OSs. *PC-DOS* and *MS-DOS* had command-driven interfaces that were difficult to learn and use. Each new version of OS has improved the ease of use, processing capability, reliability, and ability to support new computer hardware devices.

Windows 1.0 was introduced in 1985. *Windows 95* evolved from this early OS in the summer of 1995. Windows 98 was developed and introduced in 1998. Windows 98 Second Edition followed. The *Windows New Technology (NT) Workstation* OS was designed to take advantage of 32-bit processors, and it featured multitasking and advanced networking capabilities. Microsoft later renamed the Windows NT line of OSs *Windows 2000. Windows Millennium Edition (ME)* was designed for home use and enables even novice computer users to organize photos, make home movies and records, and play music, as well as perform the usual computer tasks such as accessing the Internet, playing games, and word processing.

In a blow to Microsoft, in February 2004, it was reported that up to 15 percent of Microsoft's source code for Windows 2000 and Windows NT was leaked on the Internet.[5] The leaked software code could make it easier for criminal hackers to develop viruses and worms to attack computers with these OSs.[6]

Windows XP (XP reportedly stands for the wonderful *experience* that you will have with your personal computer) was released in fall 2001. Previous consumer versions of Windows were notably unstable and crashed frequently, requiring frustrating and time-consuming reboots. With XP, Microsoft hopes to bring reliability to the consumer. The OS requires more than 2 GB (gigabytes) of hard drive space and more than an hour to install. It only works well on personal computers with at least 128 MB of RAM and a 400 MHz or faster processor. Its redesigned icons, taskbar, and window borders make for more pleasant viewing. The Start menu is two columns wide with recently used programs in the left column and everything else (e.g., My Documents, My Computer, and Control Panel) in the right column. It comes with Internet Explorer 6 browser software, which boasts improved security and reliability features, including a one-way firewall that blocks hacker invasions coming in from the Internet.

Newer OSs also come with media capabilities. Media Player by Microsoft, for example, can record TV programs, play music, and organize photos. The software also acts as a TV

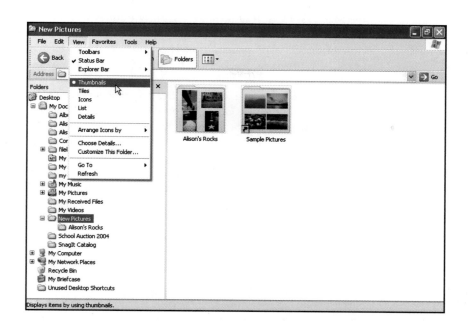

tuner. Microsoft Windows XP Media Center Edition incorporates additional multimedia features.[7] Windows XP can be run in the *classic style*, which is similar to previous versions of Windows, or the *Web style*, which is similar to how Internet sites run, including the ability to run programs or open folders with one click of the mouse. Windows XP can show small images of photos and documents, called *thumbnails*, in folders by clicking the Thumbnails view in the View Menu (see Figure 4.5). Windows XP also has Windows Error Reporting (WER), which sends error reports to Microsoft when errors occur or the OS crashes.[8] WER can also inform users of fixes or patches to avoid future problems. Radio Shack, for example, is using Windows XP in more than 5,000 stores to help run point-of-sale terminals and other devices.[9] The next major revision of the Windows OS (code named Longhorn) is expected in 2005.[10] Today, Microsoft has about 93 percent of the PC OS market.[11] Apple has 3 percent of the market, Linux has 3 percent, and other companies account for about 1 percent of the PC OS market.

Apple Computer Operating Systems

While IBM system platforms traditionally use one of the Windows OSs and Intel microprocessors (often called *WINTEL* for this reason), Apple computers typically use non-Intel microprocessors designed by Apple, IBM, and Motorola and a proprietary Apple OS—the Mac OS. Although IBM and IBM-compatible computers hold the largest share of the business PC market, Apple computers are also quite popular, especially in the fields of publishing, education, graphic arts, music, movies, and media. GarageBand, for example, is Macintosh software that allows people to create their own music the way a professional does, and it can sound like a small orchestra.[12] Pro Tools is another software program used to edit digital music.[13] While some people believe that this type of software produces "fake" music, others believe software that manipulates digital music will dominate music production and editing in the future.

The Apple OSs have also evolved over a number of years and often provide features not available from Microsoft. The Mac OS 9 had a Multiple Users feature, which allowed you to safely share your Macintosh computer with other people. Starting in July 2001, the Mac OS X was installed on all new Macs. It includes an entirely new user interface, which provides a new visual appearance for users—including luminous and semitransparent elements, such as buttons, scroll bars, windows, and fluid animation to enhance the user's experience. Since then, OS X has been upgraded with additional releases, nicknamed Jaguar (OS X.2) and Panther (OS X.3). One goal of the OS is to provide an even more stable computing

environment than the Mac OS 9 OS. It also comes with new features such as automatic networking and an instant wake-from-sleep capability for portable computers. The new, more modular structure of the OS X programming code permits easier changes and faster improvements. Unfortunately, programs developed to run in earlier Mac OS environments will not run with OS X. They need to run in what Apple calls the Classic environment, so when you open one of the old programs, you wait while this environment is established and essentially run the older OS within the new OS X platform. Cline, Davis & Mann, Inc., a New York advertising agency, specializes in healthcare industry campaigns with major pharmaceutical clients such as Pfizer (maker of Viagra), GlaxoSmithKline (maker of Serevent, an asthma inhaler), and Janssen Pharmaceutica (maker of Risperdal for schizophrenia and Reminyl for Alzheimer's disease). About 90 percent of the computers for the ad agency's 360 employees are Macs, which are used mainly for graphics and creative work. The firm is anxious to move to Mac OS X to take advantage of the improved reliability and new features. Newer versions of Mac OS X include 3-D effects and better management of windows on a display screen.[14] Fewer virus attacks, compared with Microsoft's Windows OS, are another advantage of using Mac OS X.[15]

Linux

Linux is an OS developed by Linus Torvalds in 1991 as a student in Finland. The OS is under the GNU General Public License, and its source code is freely available to everyone. This doesn't mean, however, that Linux and its assorted distributions are free—companies and developers may charge money for it as long as the source code remains available. Linux is actually only the *kernel* of an OS, the part that controls hardware, manages files, separates processes, and so forth. Several combinations of Linux are available, with various sets of capabilities and applications to form a complete OS. Each of these combinations is called a *distribution* of Linux.

Linux is available over the Internet and from other sources, including RedHat Linux and Caldera OpenLinux. Many individuals and organizations are starting to use Linux. Galileo, the large travel and airline ticketing company, for example, uses Linux to run its Internet site.[16] "We are looking to deploy Linux wherever we can," says the chief technology officer for the company. Panasonic is using Linux to run one of its high-speed television tuners.[17] A survey revealed that many CIOs are considering switching to Linux and open-source software because of security concerns with Microsoft software.[18] Linux, however, is not without controversy.[19] One company, for example, is claiming that leading Linux distributors may be infringing on copyright and trade secret rights.[20] A number of companies, including Intel and IBM, support Linux and open software and are contributing funds to defend Linux against copyright-infringement lawsuits.[21]

Workgroup Operating Systems

To keep pace with today's high-tech society, the technology of the future must support a world in which network usage, data-storage requirements, and data-processing speeds increase at a dramatic rate. This rapid increase in communications and data-processing capabilities pushes the boundaries of computer science and physics. Powerful and sophisticated OSs are needed to run the servers that meet these business needs for workgroups. Small businesses, for example, often use workgroup OSs to run networks and perform critical business tasks.[22]

Windows Server

Microsoft designed *Windows Server 2003* to do a host of new tasks that are vital for Web sites and corporate Web applications on the Internet.[23] For example, Microsoft Windows Server 2003 can be used to coordinate large data centers. The OS also works with other Microsoft products. It can be used to prevent unauthorized disclosure of information by blocking text and e-mails from being copied, printed, or forwarded to other people.[24] Besides being more reliable than Windows NT, this OS is capable of handling extremely demanding computer tasks, such as order processing. It can be tuned to run on machines with up to 32 microprocessors—satisfying the needs of all but the most demanding of Web operators. Four machines can be clustered to prevent service interruptions, which are disastrous for Web sites.

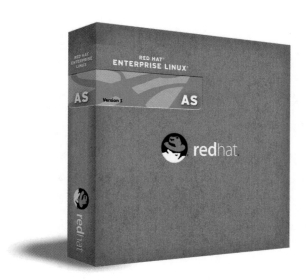

Microsoft *Windows Advanced Server, Limited Edition,* was the first 64-bit version of the Windows Server family. Introduced in August 2001, it was designed to run on the 64-bit Itanium processor from Intel (also known as the IA64). This OS enables Microsoft to begin competing with rival Linux vendors (RedHat, Caldera, SuSE, and TurboLinux), which already have 64-bit Itanium versions of their Linux distributions. In addition, Sun Microsystems and IBM have had 64-bit UNIX OSs for years.

UNIX

UNIX is a powerful OS originally developed by AT&T for minicomputers. UNIX can be used on many computer system types and platforms, from personal computers to mainframe systems. UNIX also makes it much easier to move programs and data among computers or to connect mainframes and personal computers to share resources. There are many variants of UNIX—including HP/UX from Hewlett-Packard, AIX from IBM, UNIX SystemV from UNIX Systems Lab, Solaris from Sun Microsystems, and SCO from Santa Cruz Operations.

NetWare

NetWare is a network OS sold by Novell that can support end users on Windows, Macintosh, and UNIX platforms. NetWare provides directory software to track computers, programs, and people on a network, making it easier for large companies to manage complex networks. NetWare users can log in from any computer on the network and still get their own familiar desktop with all their applications, data, and preferences.

RedHat Linux

RedHat Software offers a Linux network OS that taps into the talents of tens of thousands of volunteer programmers who generate a steady stream of improvements for the Linux OS. The *RedHat Linux* network OS is very efficient at serving up Web pages and can manage a cluster of up to eight servers. The film *Lord of the Rings* used Linux and hundreds of servers to deliver many of the special effects seen in the finished film.[25] Burlington Coat Factory needed a new OS for more than 1,250 personal computers to support back-office functions such as shipping, receiving, and order processing at 250 stores. RedHat Linux was chosen because the OS is inexpensive and runs on standard industry hardware, so risks were minimized. If RedHat Linux didn't work as expected, Burlington could have kept the hardware and bought another OS. In addition, Linux environments typically have fewer virus and security problems than other OSs. RedHat Linux has proven to be a very stable and efficient OS.

Mac OS X Server

The *Mac OS X Server* is the first modern server OS from Apple Computer. It provides UNIX-style process management. Protected memory puts each service in its own well-guarded chunk of dynamically allocated memory, preventing a single process from going awry and bringing

down the system or other services. Under preemptive multitasking, a computer OS uses some criteria to decide how long to allocate to any one task before giving another task a turn to use the OS. Preempting is the act of taking control of the OS from one task and giving it to another. A common criterion for preempting is simply elapsed time. In more sophisticated OSs, certain applications can be given higher priority than other applications, giving the higher-priority programs longer processing times. Preemptive multitasking ensures that each process gets the right amount of CPU time and the system resources it needs for optimal efficiency and responsiveness.

Enterprise Operating Systems

New mainframe computers provide the computing and storage capacity to meet massive data-processing requirements and serve a large number of users with high performance, excellent system availability, strong security, and scalability. In addition, a wide range of application software has been developed to run in the mainframe environment, making it possible to purchase software to address almost any business problem. As a result, mainframe computers remain the computing platform of choice for mission-critical business applications for many companies. z/OS from IBM, MPE/iX from Hewlett-Packard, and Linux are three examples of mainframe OSs.

z/OS

The *z/OS* is IBM's first 64-bit enterprise OS. It supports IBM's z900 and z800 lines of mainframes that can come with up to sixteen 64-bit processors. (The z stands for zero downtime.) It provides several new capabilities to make it easier and less expensive for users to run large mainframe computers. The OS has improved workload management and advanced e-commerce security. The IBM zSeries mainframe, like previous generations of IBM mainframes, lets users subdivide a single computer into multiple smaller servers, each of which is capable of running a different application. In recognition of the widespread popularity of a competing OS, z/OS allows partitions to run a version of the Linux OS. An apparel maker, for example, can upgrade to a mainframe that runs the Linux OS.

MPE/iX and HP-UX

Multiprogramming Executive with integrated POSIX (MPE/iX) is the Internet-enabled OS for the Hewlett-Packard e3000 family of computers using RISC processing. MPE/iX is a robust OS designed to handle a variety of business tasks, including online transaction processing and Web applications. It runs on a broad range of HP e3000 servers—from entry-level to workgroup and enterprise servers within the data centers of large organizations. *HP-UX* is a mainframe OS from Hewlett-Packard. The OS is designed to support Internet, database, and a variety of business applications. It can work with Java programs and Linux applications. The OS comes in four different versions: foundation, enterprise, mission critical, and technical. HP-UX supports Hewlett-Packard's computers and those designed to run Intel's Itanium processors.

Linux

RedHat Software announced the availability of *RedHat Linux* for *IBM* mainframe computers in December 2001. This version of RedHat Linux means that the company has Linux versions for everything from handheld devices to the largest enterprise mainframes.

Operating Systems for Small Computers and Special-Purpose Devices

New OSs and other software are changing the way we interact with personal digital assistants (PDAs), cell phones, digital cameras, TVs, and other appliances. These OSs are also called *embedded operating systems* because they are typically embedded in a computer chip. Embedded software is a $21 billion industry.[26] Some of these OSs allow handheld devices to be synchronized with PCs using cradles, cables, and wireless connections. Cell phones also use embedded OSs (see Figure 4.6). In addition, there are OSs for special-purpose devices, such as TV set-top boxes, computers on the space shuttle, computers in military weapons, and computers in some home appliances. Here are some of the more popular OSs for such devices.

Figure 4.6

Many cell phones such as this one from Nokia also have an integrated imaging device. Point, use the color display as a viewfinder, and snap a picture. Images can be stored on the device and sent to a friend.

(Source: Courtesy of Nokia.)

Palm OS

The strategy Palm has taken with its *Palm OS* operating system is to extend the capabilities of these devices to more generalized purposes for its own computers and for those made by other companies. Today, the company has two major businesses. The *PalmOne* line of products produces a number of innovative handheld computers and smart phones, including the Treo, Palm, Zire, and Tungsten. These products run the Palm operating system and a variety of applications. PalmSource makes the Palm operating system that is used on over 30 million handheld computers and smart phones by PalmOne and other companies. The company also develops and supports applications, including business, multimedia, games, productivity, reference and education, hobbies and entertainment, travel, sports, utilities, and a variety of wireless applications.

Palm has added features to allow better integration with desktop PCs and enabled users to add applications to the device. Many Apple Mac users, for example, prefer Palm operating systems for PDAs because they are easy to integrate with Mac programs and applications.[27] Such flexibility has enabled the Palm to remain relatively easy to use while adding some expandability and capability as a general-purpose computing platform. An office supplies store, for example, can use wireless devices containing the Palm OS to scan shipments, create an electronic manifest as deliveries are loaded on a truck, and capture customer signatures electronically. Customers can track the status of their orders via its Web site. Today, Palm has about 51 percent of the market for PDA or handheld OSs.[28] Microsoft has about 32 percent of the market, Linux has 3 percent, and other companies account for about 14 percent (see Figure 4.7).

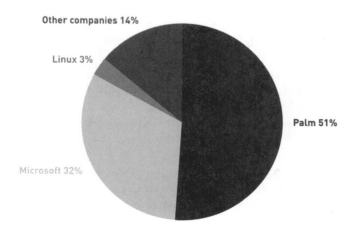

Other companies 14%

Linux 3%

Microsoft 32%

Palm 51%

Figure 4.7

Market Share for Personal Digital Assistants

Windows Embedded

Windows Embedded is a family of Microsoft OSs included with or embedded into small computer devices. Windows Embedded includes Windows CE.Net and Windows XP Embedded. *Windows CE.Net* is a key step in taking Microsoft closer to its vision of anywhere, anytime access to Web-based content and services. It is an embedded OS for use in mobile devices, such as smart phones and PDAs, and it can also be used in a variety of other devices, such as digital cameras, thin clients, TV set-top boxes, and automotive computers. PDAs with Windows CE try to bring as much of the functionality of a desktop PC as possible to a handheld device. Such a PDA is a programmable computer that performs most of the functions of a dedicated device.

The *Windows XP Embedded* OS is used in devices such as handheld computers, TV set-top boxes, and automated industrial machines. It is based on Microsoft's Windows XP Professional desktop OS and includes more than 10,000 software components, including such features as a built-in chat feature that is Microsoft's answer to America Online Inc.'s popular Instant Messaging. In addition, it is designed to run applications in their own memory spaces so that they don't interfere with or corrupt one another. As with Windows CE.Net, hardware developers can choose the pieces of the OS they need for certain devices.

Windows Mobile

Windows Mobile is a family of Microsoft OSs for mobile or portable devices. Windows Mobile includes Pocket PC, Pocket PC Phone Edition, and SmartPhone.[29] These OSs have many features, including handwriting recognition, the ability to beam information to devices running either Pocket PC or Palm Inc.'s competing OS, Microsoft's instant messaging technology, and support for more secure Internet connections. They also have advanced telecommunications capabilities, discussed in more detail in Chapter 6. Motorola, Samsung, Dell, Hewlett-Packard, and Toshiba have products that run Windows Mobile. Wireless services from Verizon, Cingular, and others are often available with devices that use Windows Mobile.

Utility Programs

utility programs
Programs used to merge and sort sets of data, keep track of computer jobs being run, compress data files before they are stored or transmitted over a network, and perform other important tasks.

Utility programs are used to merge and sort sets of data, keep track of computer jobs being run, compress files of data before they are stored or transmitted over a network (thus saving space and time), and perform other important tasks.[30] One utility program is even used to make computer systems run better and longer without problems.[31] Another type of utility program allows people and organizations to tap into unused computer power over a network.[32] Often called *grid computing*, the approach can be very efficient and less expensive than purchasing additional hardware or computer equipment. Novartis, a large drug company, uses grid computing to get more done for less. According to a company spokesperson, "The grid has opened up a number of opportunities for us which were just not there before." The company saved about $2 million using grid computing by not having to buy additional hardware and equipment. In the future, grid computing could become a common feature of OSs and provide inexpensive, on-demand access to computer power and resources.

Utility programs often come installed on computer systems, but a number of utility programs can also be purchased. Let's look at some common types of utilities.

Hardware Utilities

A number of hardware utilities are available from companies such as Symantec, which produces Norton Utilities. Hardware utilities can check the status of all parts of the PC including hard disks, memory, modems, speakers and printers.[33] Disk utilities check the hard disk's boot sector, file allocation tables, and directories and analyze them to ensure that the hard disk hasn't been tampered with. Disk utilities can also optimize the placement of files on a crowded disk.

Virus-Detection and -Recovery Utilities

Computer viruses from the Internet and other sources can be a nuisance—and sometimes can completely disable a computer. Virus-detection and virus-recovery software can be installed to constantly monitor and protect the computer. If a virus is found, the software can eliminate the virus, or "clean" the virus from the computer system. To keep the virus-detection and recovery software current and make sure that the software checks for the latest

viruses, it can be easily updated through the Internet. Symantec and McAfee are two companies that make virus-detection and -recovery software.

File-Compression Utilities

File-compression programs can reduce the amount of disk space required to store a file or reduce the time it takes to transfer a file over the Internet. A popular program on Windows PCs is WinZip (*www.winzip.com*), which generates zip files. A zip file has a .zip extension, and the file can be easily unzipped to the original file. *MP3* (*Motion Pictures Experts Group-Layer 3*) is a popular file-compression format used to store, transfer, and play music. It can compress files 10 times smaller than original with near-CD-quality sound. Software, such as iTunes from Apple, can be used to store, organize, and play MP3 music files.

Spam and Pop-Up Blocker Utilities

Getting unwanted e-mail (spam) and having annoying and unwanted ads pop up on your display screen while you are on the Internet can be frustrating and a big waste of time. A number of utility programs can be installed to help block unwanted e-mail spam and pop-up ads, including Cloudmark SpamNet, IhateSpam, Spamnix, McAfee SpamKiller, and Ad-aware.[34]

Network and Internet Utilities

A broad range of network and systems-management utility software is available to monitor hardware and network performance and trigger an alert when a Web server is crashing or a network problem occurs. Although these general management features are helpful, what is needed is a way to pinpoint the cause of the problem. Topaz from Mercury Interactive is an example of software called an *advanced Web-performance monitoring utility*. It is designed not only to sound an alarm when there are problems but also to let network administrators isolate the most likely causes of the problems. Its Auto RCA (root-cause analysis) module uses statistical analysis with built-in rules to measure system and Web performance. Actual performance data is compared with the rules, and the results can help pinpoint where trouble originated—in the application software, database, server, network, or the security features.

Server and Mainframe Utilities

Some utilities enhance the performance of servers and mainframe computers. IBM has created systems-management software that allows a support person to monitor the growing number of desktop computers in a business attached to a server or mainframe computer. With this software, the support people can sit at their personal computers and check or diagnose problems, such as a hard-disk failure on a computer on a network. The support people can even repair individual systems anywhere on the organization's network, often without having to leave their desks. The direct benefit is to the system manager, but the business gains from having a smoothly functioning information system. Utility programs can meet the needs of an individual, a workgroup, or an enterprise, as listed in Table 4.4. They perform useful tasks—from tracking jobs to monitoring system integrity.

Table 4.4

Examples of Utility Programs

Personal	Workgroup	Enterprise
Software to compress data so that it takes less hard disk space	Software to provide detailed reports of workgroup computer activity and status of user accounts	Software to archive contents of a database by copying data from disk to tape
Screen saver	Software that manages an uninterruptible power source to do a controlled shutdown of the workgroup computer in the event of a loss of power	Software that compares the content of one file with another and identifies any differences
Virus detection software	Software that reports unsuccessful user log on attempts	Software that reports the status of a particular computer job

Other Utilities

Utility programs exist for almost every conceivable task or function. Microsoft Windows Rights Management Services, for example, can be used with Microsoft's Office programs to

manage and protect important corporate documents.[35] ValueIT is a utility that can help a company verify the value of investments in information systems and technology.[36] According to the chief financial officer for Ticonderoga, a pencil manufacturer, "Looking at how your financial performance and IT spending compares with other companies in your industry lets you see whether you have a competitive advantage." Widgit Software has developed an important software utility that helps people with visual disabilities use the Internet.[37] The software converts icons and symbols into plain text that can be easily seen. There is even a software utility that allows a manager to see every keystroke a worker makes on a computer system.[38] Monitoring software can catalogue the Internet sites that employees visit and the time that employees are working at their computer.

Middleware

middleware
Software that allows different systems to communicate and transfer data back and forth.

Middleware is software that allows different systems to communicate and transfer data back and forth. It is often used as an interface between client and server computers on a network. Middleware can also be used as an interface between the Internet and older, legacy systems. For example, middleware can be used to transfer a request for information from a corporate customer on the corporate Web site to a traditional database on a mainframe computer and return the results to the individual on the Internet.

APPLICATION SOFTWARE

As discussed earlier in this chapter, the primary function of application software is to apply the power of the computer to give individuals, workgroups, and the entire enterprise the ability to solve problems and perform specific tasks. When you need the computer to do something, you use one or more application programs. The application programs interact with systems software, and the systems software then directs the computer hardware to perform the necessary tasks.

The functions performed by application software are diverse. Suppose a manager is concerned that too many employees are getting overtime pay by working more than 40 hours each week, even though many others are working less than 40 hours per week. She would like to have those working below the 40-hour threshold replace those over the threshold, and hence avoid the time-and-a-half overtime pay rate. The manager can enlist a computer to print the names of all employees working significantly more or significantly less than 40 hours per week on average over the last three months. Also, voice stress software is being developed to help detect fraud in the insurance industry.[39] Pilot programs have been very successful in detecting people who try to make false claims or in detecting scam insurance companies. Philips has developed application software that can send video clips over mobile phones.[40] The software is being used by the British Broadcasting Corporation for breaking news stories or as a backup to broadcast-quality video. Primerica Life Insurance uses an application software package to capture and submit insurance policy applications using small handheld computers.[41] According to Tom Swift, senior vice president of Field Technology, "With 30,000 to 35,000 life insurance applications being processed each month, the potential cost savings are significant." Other application programs can complete sales orders, control inventory, pay bills, write paychecks to employees, and provide financial and marketing information to managers and executives. Most of the computerized business jobs and activities discussed in this book involve the use of application software. We begin by investigating the types and functions of application software.

Types and Functions of Application Software

Application software can be broken into two major categories: proprietary software and off-the-shelf software (see Figure 4.8). A company can develop a one-of-a-kind program for a

specific application (called **proprietary software**). Proprietary software is usually developed and owned by the company or organization that will use the software. When an outside company builds or develops the software, it is often called *contract software*. A company can also purchase or acquire an existing software program (sometimes called **off-the-shelf software** because it can literally be purchased or acquired "off the shelf" in a store.) The relative advantages and disadvantages of proprietary software and off-the-shelf software are summarized in Table 4.5.

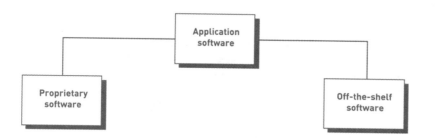

Many companies use off-the-shelf software to support business processes. Key questions for selecting off-the-shelf software include: (1) Will the software run on the OS and hardware you have selected? (2) Does the software meet the essential business requirements that have been defined? (3) Is the software manufacturer financially solvent and reliable? and (4) Does the total cost of purchasing, installing, and maintaining the software compare favorably to the expected business benefits?

proprietary software
A one-of-a-kind program for a specific application, usually developed and owned by a single company.

off-the-shelf software
Existing software program that is purchased.

Figure 4.8

Types of Application Software
Some off-the-shelf software may be modified to allow some customization.

Table 4.5

A Comparison of Proprietary and Off-the-Shelf Software

Proprietary Software		Off-the-Shelf Software	
Advantages	**Disadvantages**	**Advantages**	**Disadvantages**
You can get exactly what you need in terms of features, reports, and so on.	It can take a long time and significant resources to develop required features.	The initial cost is lower since the software firm is able to spread the development costs over a large number of customers.	An organization might have to pay for features that are not required and never used.
Being involved in the development offers a further level of control over the results.	In-house system development staff may become hard pressed to provide the required level of on going support and maintenance because of pressure to get on to other new projects.	There is a lower risk that the software will fail to meet the basic business needs—you can analyze existing features and the performance of the package.	The software may lack important features, thus requiring future modification or customization. This can be very expensive because users must adopt future releases of the software as well.
There is more flexibility in making modifications that may be required to counteract a new initiative by one of your competitors or to meet new supplier and/or customer requirements. A merger with another firm or an acquisition also will necessitate software changes to meet new business needs.	There is more risk concerning the features and performance of the software that has yet to be developed.	Package is likely to be of high quality since many customer firms have tested the software and helped identify many of its bugs.	Software may not match current work processes and data standards.

It is possible to modify some off-the-shelf programs, in effect blending the off-the-shelf and customized approaches. For example, a software developer may write a collection of programs to be used in an auto body shop that includes features to generate estimates, order parts, and process insurance. Body shops of all types have these needs. Designed properly— and with provisions for minor tailoring for each user—the same software package can be sold

to many users. However, since each body shop has slightly different requirements, some modifications to the software may be needed. As a result, software vendors often provide a wide range of services, including installation of their standard software, modifications to the software required by the customer, training of the end users, and other consulting services.

Another approach to obtaining a customized software package is the use of an application service provider. An **application service provider** (ASP) is a company that can provide software, end-user support, and the computer hardware on which to run the software from the user's facilities. An ASP can also take a complex corporate software package and simplify it so that it is easier for the users to set up and manage. ASPs provide contract customization of off-the-shelf software, and they speed deployment of new applications while helping IS managers avoid implementation headaches, reducing the need for many skilled IS staff members and reducing project start-up expenses. Such an approach allows companies to devote more time and resources to more important tasks. The use of an ASP makes the most sense for relatively small, fast-growing companies with limited IS resources. It is also a good strategy for companies looking to deploy a single, functionally focused application quickly, such as setting up an e-commerce Web site or supporting expense reporting. The "Information Systems @ Work" feature discusses an online training system IBM offers and supports. Contracting with an ASP may make less sense, however, for larger companies with major systems that have their technical infrastructure already in place.

Using an ASP is not without risks—sensitive information could be compromised in a number of ways, including unauthorized access by employees or computer hackers; the ASP might not be able to keep its computers and network up and running as consistently as is needed; or a disaster could disable the ASP's data center, temporarily putting an organization out of business. These are legitimate concerns that an ASP must address.

Personal Application Software

There are literally hundreds of computer applications that can help individuals at school, home, and work. The features of personal application software are summarized in Table 4.6. In addition to these general-purpose programs, there are literally thousands of other personal computer applications to perform specialized tasks: to help you do your taxes, get in shape, lose weight, get medical advice, write wills and other legal documents, make repairs to your computer, fix your car, write music, and edit your pictures and videos (see Figures 4.9 and 4.10). This type of software, often called *user software* or *personal productivity software*, includes the general-purpose tools and programs that support individual needs.

---

application service provider (ASP)

A company that provides software, end-user support, and the computer hardware on which to run the software from the user's facilities.

Telstra Benefits from E-Training for Managers

Companies often look to outsiders for help in areas beyond their expertise. Australian telecommunications giant Telstra turned to IBM Learning Solutions for management skills training. In a program Telstra named "Frontline Management Foundation Program (FLM)," managers from across the organization receive state-of-the-art management training, primarily through online resources.

Managers progress through the course at their own pace, investing only a few hours each week. The course contains a variety of engaging online learning modules covering a range of managerial topics such as leadership, management fundamentals, staffing, teamwork, and coaching. One module titled "The Team Building Process" presents a tutorial on identifying the symptoms of ineffective teams. Managers read about symptoms of an ineffective team and see examples of each symptom. They answer questions as the tutorial progresses from screen to screen to check their understanding. The tutorial concludes with a video of an actual team meeting, and managers must evaluate the team's effectiveness in an online questionnaire. Some questionnaires are graded automatically; others are submitted to a live e-facilitator to assist the student. Each module contains several lessons, and each concludes with an online examination. Students may retake any module at any time until they pass.

Since managers can do most of their class work in the office, at home, or anywhere they have access to the Internet, the training is part of their normal daily routine. Previous classroom training programs required Telstra managers to take four weeks off work for group training sessions. By delivering 75 percent of the course online, the new e-training system requires only four days of classroom work, allowing managers to remain on the job while in training.

After the online training and the classroom workshops, the program progresses to a third phase, in which managers apply their knowledge to their own jobs. Even after completing the course, managers can return to the training software at any time to brush up their skills.

E-training was the ideal software solution for Telstra's management training challenge. Alan Bedford, general manager of Telstra's Leadership and Management Development, said the diversity of Telstra's work environments—ranging from isolated teams doing field work to call center and office managers—presented unique training challenges for the company. "Our overall objective was to provide a learning program flexible enough to meet the needs of highly diverse work environments and demands, while providing a consistency of structure and content for these fundamental skills."

The training program was originally developed by IBM for its own managers. Since its creation, IBM has saved $80 million in training costs over four years, while producing better-trained managers. IBM packages its successful learning solutions software and customizes it for the unique needs of its clients. IBM provides the software, the back-end servers that store and run the software, and training specialists that administer and teach the training program, as well as providing online support for students. "We believe that this type of program is certainly the answer to the complex training needs that many Australian companies are facing," says Richard Matthewman of IBM Learning Solutions, Australia and New Zealand.

Telstra plans to train 3,300 managers in four years with the system. To date, the program has received highly satisfactory feedback from individual trainees, as well as evidence of a change in management approaches among managers who have taken the course.

Discussion Questions

1. How has IBM's custom software solution assisted in making Telstra's training more effective?
2. As a manager in training, would you prefer to attend a formal, extended training seminar or work at your own pace through an e-training system? Why?

Critical Thinking Questions

3. As a trainer, would you prefer to teach students through face-to-face meetings, in multiweek seminars, or would you prefer to act as an e-facilitator guiding students through the e-training system and assisting them with trouble areas when needed. Why?
4. Why would IBM be interested in packaging its in-house course to other companies?

SOURCES: "IBM Success Stories: Telstra," IBM Web site, *www.ibm.com*, accessed February 21, 2004; "IBM Helps Telstra to Teach Staff," News. com.au, *www.news.com.au*, accessed February 22, 2004; Telstra Web site, *www.telstra.com*, accessed February 21, 2004; IBM Australia Web site, *www.ibm.com/au/*, accessed February 21, 2004; IBM Learning Solutions, *www.306.ibm.com/services/learning*, accessed February 21, 2004.

Table 4.6

Examples of Personal Productivity Software

Type of Software	Explanation	Example	Vendor
Word processing	Create, edit, and print text documents	Word WordPerfect	Microsoft Corel
Spreadsheet	Provide a wide range of built-in functions for statistical, financial, logical, database, graphics, and date and time calculations	Excel Lotus 1-2-3	Microsoft Lotus/IBM
Database	Store, manipulate, and retrieve data	Access Approach dBASE	Microsoft Lotus/IBM Borland
Online information services	Obtain a broad range of information from commercial services	America Online CompuServe MSN	America Online CompuServe Microsoft
Graphics	Develop graphs, illustrations, and drawings	Illustrator FreeHand	Adobe Macromedia
Project management	Plan, schedule, allocate, and control people and resources (money, time, and technology) needed to complete a project according to schedule	Project for Windows On Target Project Schedule Time Line	Microsoft Symantec Scitor Symantec
Financial management	Provide income and expense tracking and reporting to monitor and plan budgets (some programs have investment portfolio management features)	Managing Your Money Quicken	Meca Software Intuit
Desktop publishing (DTP)	Works with personal computers and high-resolution printers to create high-quality printed output, including text and graphics; various styles of pages can be laid out; art and text files from other programs can also be integrated into "published" pages	QuarkXPress Publisher PageMaker Ventura Publisher	Quark Microsoft Adobe Corel
Creativity	Helps generate innovative and creative ideas and problem solutions. The software does not propose solutions, but provides a framework conducive to creative thought. The software takes users through a routine, first naming a problem, then organizing ideas and "wishes," and offering new information to suggest different ideas or solutions	Organizer Notes	Macromedia Lotus

Figure 4.9

TurboTax

Tax-preparation programs can save hours of work and are typically more accurate than doing a tax return by hand. Programs can check for potential problems and give you help and advice about what you may have forgotten to deduct.

(Source: Turbo Tax Deluxe 2003 screenshot courtesy of Intuit.)

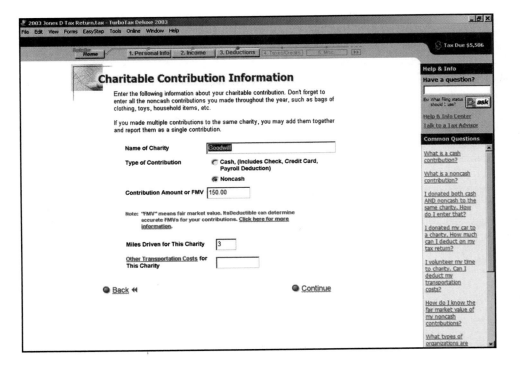

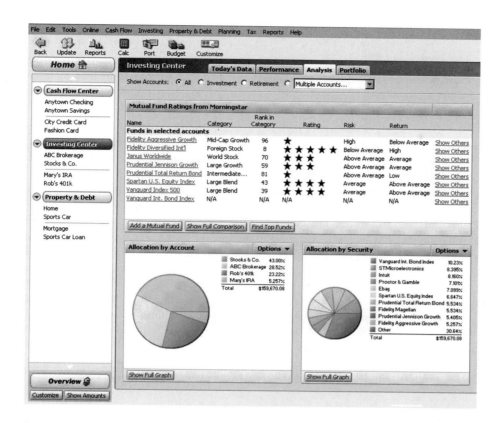

Figure 4.10

Quicken

Off-the-shelf financial-management programs are useful for paying bills and tracking expenses.

(Source: Courtesy of Intuit.)

Word Processing

If you write reports, letters, or term papers, word processing applications can be indispensable. The majority of personal computers in use today have word processing applications installed. Such applications can be used to create, edit, and print documents. Most come with a vast array of features, including those for checking spelling, creating tables, inserting formulas, creating graphics, and much more (see Figure 4.11). This book (and most like it) was entered into a word processing application using a personal computer.

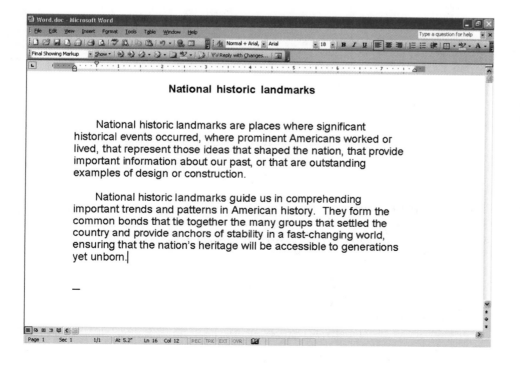

Figure 4.11

Word Processing Program

Word processing applications can be used to write letters, holiday greeting cards, work reports, and term papers.

Word processing programs can be used with a team or group of people collaborating on a project. The authors and editors who developed this book, for example, used the "track changes" and "reviewing" features of Microsoft Word to track and make changes to chapter files. You can insert comments in or make revisions to a document that a coworker can review and either accept or reject.

Spreadsheet Analysis

People use spreadsheets to prepare budgets, forecast profits, analyze insurance programs, summarize income tax data, and analyze investments. Whenever numbers and calculations are involved, spreadsheets should be considered. Woodward Aircraft Engine Systems, for example, uses spreadsheets to compute the inventory levels it needs to manufacture engine parts.[42] The calculations made in the spreadsheet have helped the company reduce inventory levels to one third of their original values for some parts and components, reducing inventory costs. Features of spreadsheets include graphics, limited database capabilities, statistical analysis, built-in business functions, and much more (see Figure 4.12). The business functions include calculation of depreciation, present value, internal rate of return, and the monthly payment on a loan, to name a few. Optimization is another powerful feature of many spreadsheet programs. *Optimization* allows the spreadsheet to maximize or minimize a quantity subject to certain constraints. For example, a small furniture manufacturer that produces chairs and tables might want to maximize its profits. The constraints could be a limited supply of lumber, a limited number of workers that can assemble the chairs and tables, or a limited amount of various hardware fasteners that may be required. Using an optimization feature, such as Solver in Microsoft Excel, the spreadsheet can determine what number of chairs and tables to produce with labor and material constraints in order to maximize profits.

Figure 4.12

Spreadsheet Program

Spreadsheet programs should be considered when calculations are required.

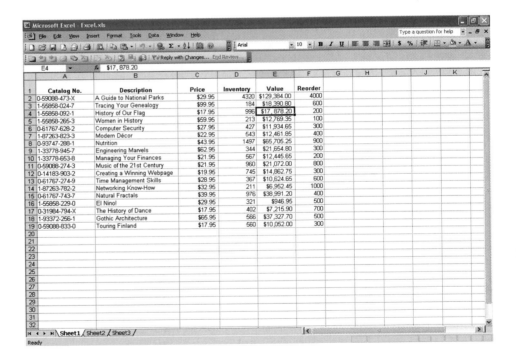

Database Applications

Database applications are ideal for storing, manipulating, and retrieving data. These applications are particularly useful when you need to manipulate a large amount of data and produce reports and documents. Database manipulations include merging, editing, and sorting data. The uses of a database application are varied. You can keep track of a CD collection, the items in your apartment, tax records, and expenses. A student club can use a database to store names, addresses, phone numbers, and dues paid. In business, a database application can help process sales orders, control inventory, order new supplies, send letters to customers, and pay employees. Database management systems can be used to track and

analyze stock and bond prices, analyze weather data to make forecasts for the next several days, and summarize medical research results. A database can also be a front end to another application. For example, a database application can be used to enter and store income tax information; the stored results can then be exported to other applications, such as a spreadsheet or tax-preparation application (see Figure 4.13).

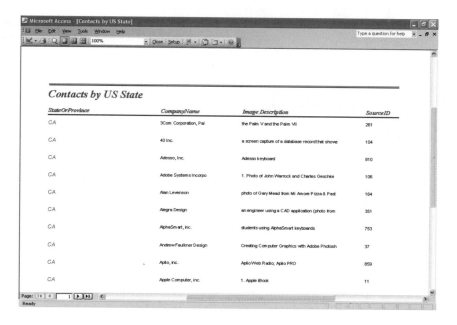

Figure 4.13

Database Program

Once entered into a database application, information can be manipulated and used to produce reports and documents.

Graphics Program

It is often said that a picture is worth a thousand words. With today's graphics programs, it is easy to develop attractive graphs, illustrations, and drawings. Graphics programs can be used to develop advertising brochures, announcements, and full-color presentations. If you are asked to make a presentation at school or work, you can use a graphics program to develop and display slides while you are making your talk. A graphics program can be used to help you make a presentation, a drawing, or an illustration (see Figure 4.14). Because of their popularity, many colleges and departments require students to become proficient at using presentation graphics programs.

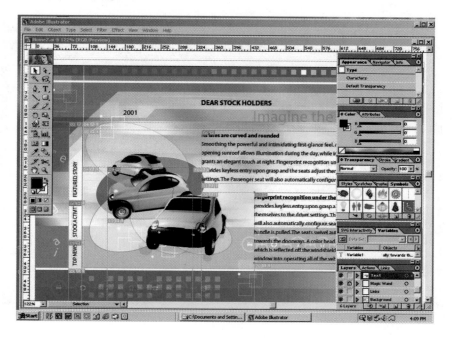

Figure 4.14

Graphics Program

Graphics programs can help you make a presentation at school or work. They can also be used to develop attractive brochures, illustrations, drawings, and maps.

(Source: Courtesy of Adobe Systems Incorporated.)

Many graphics programs, such as PowerPoint by Microsoft, consist of a series of slides. Each slide can be displayed on a computer screen, printed as a handout, or (more commonly) projected onto a large viewing screen for audiences. Powerful built-in features allow you to develop attractive slides and complete presentations. You can select a template for a type of presentation, such as recommending a strategy for managers, communicating bad news to a sales force, giving a training presentation, or facilitating a brainstorming session. The presentation graphics program takes you through the presentation step by step, including applying color and attractive formatting. Of course, you can also custom-design your own presentation. All types of charts, drawings, and formatting are available. Most presentation graphics programs come with many pieces of *clip art*, such as drawings and photos of people meeting, medical equipment, telecommunications equipment, entertainment, and much more.

Personal Information Managers

Personal information managers (PIMs) help individuals, groups, and organizations store useful information, such as a list of tasks to complete or a list of names and addresses. They usually provide an appointment calendar and a place to take notes. In addition, information in a PIM can be linked. For example, you can link an appointment with a sales manager that appears in the calendar with information on the sales manager in the address book. When you click the appointment in the calendar, information on the sales manager from the address book is automatically opened and displayed on the computer screen. Microsoft Outlook is an example of a PIM software package.[43]

Some personal information managers allow people to schedule and coordinate group meetings. If a computer or handheld device is connected to a network, the PIM data can be uploaded and coordinated with the calendar and schedule of others using the same PIM software on the network. Some PIMs can also be used to coordinate e-mails sent and received over the Internet.

Online Information Services

Online information services allow you to connect a personal computer to the outside world through phone lines, cable, satellite, or power lines in some cases. Using an online service, you can get investment information, make travel plans, and check news from around the world. You can also get prices and features for most consumer items, learn about companies, send e-mail to friends and family, learn about degree programs offered by colleges and universities around the world, and search for job openings in your area (see Figure 4.15).

Figure 4.15

Online Information Services

Online services provide instant access to information. Prices of cars and trucks, travel discounts, information about companies, stock market data, and much more is available with a few keystrokes using an online service.

(Source: The AOL.com triangle logo, the Running Man icon, and AOL are registered trademarks of America Online, Inc. The AOL.com screenshot is © 2004 by America Online, Inc. The America Online content, name, icons and trademarks are used with permission.)

Software Suites and Integrated Software Packages

A **software suite** is a collection of single application programs packaged in a bundle.[44] Software suites can include word processors, spreadsheets, database management systems, graphics programs, communications tools, organizers, and more. Some suites support the development of Web pages, note taking, and speech recognition, where applications in the suite can accept voice commands and record dictation.[45] There are a number of advantages to using a software suite. The software programs have been designed to work similarly, so once you learn the basics for one application, the other applications are easy to learn and use. Buying software in a bundled suite is cost-effective; the programs usually sell for a fraction of what they would cost individually.

Microsoft Office, Corel's WordPerfect Office, Lotus SmartSuite, and Sun Microsystems's StarOffice are examples of popular general-purpose software suites for personal computer users (see Figure 4.16).[46] The Free Software Foundation offers software similar to Sun Microsystems's StarOffice that includes word processing, spreadsheet, database, presentation graphics, and e-mail applications for the Linux OS.[47] OpenOffice is another Office suite for Linux.[48] Microsoft Office via Wine can be used to run Microsoft Office applications on Linux, although some features may not work as well as with a Microsoft OS. Each of these software suites includes a spreadsheet program, word processor, database program, and graphics package with the ability to move documents, data, and diagrams among them (see Table 4.7). Thus, a user can create a spreadsheet and then cut and paste that spreadsheet into a document created using the word processing application.

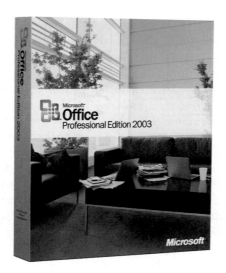

software suite
A collection of single application programs packaged in a bundle.

Figure 4.16

Software Suite

A software suite, such as Microsoft Office 2003, offers a collection of powerful programs, including word processing, spreadsheet, database, graphics, and other programs. The programs in a software suite are designed to be used together. In addition, the commands, icons, and procedures are the same for all programs in the suite.

(Source: Courtesy of Microsoft Corporation.)

Personal Productivity Function	Microsoft Office	Lotus SmartSuite Millennium Edition	Corel WordPerfect Office	Sun Microsystems
Word Processing	Word	WordPro	WordPerfect	Writer
Spreadsheet	Excel	Lotus 1-2-3	Quattro Pro	Calc
Presentation Graphics	PowerPoint	Freelance Graphics	Presentations	Impress
Database	Access	Lotus Approach	Paradox	

Table 4.7

Major Components of Leading Software Suites

There are more than a hundred million users worldwide of the Microsoft Office software suite, with Office 2003 representing the latest version of the productivity software.[49] The updated suite offers many enhancements over earlier versions of Office, including new

collaboration features. A SharePoint feature, for example, lets people post files and messages on a central site, making it easier for a group of people to work on common Office documents. People can work together, viewing documents and making changes in real time. As one person is making changes, other people are locked out from changing the same document—their mouse pointer disappears from their display screen. In addition, information from Office can be integrated with corporate systems and applications. With InfoPath 2003, for example, information can be saved to corporate databases. Office X is Microsoft's version of Office for the Mac OS X OS.

Microsoft Office goes beyond its role as a mainstream package of ready-to-run applications with the extensive custom development facilities of Visual Basic for Applications (VBA)—a built-in facility that is part of every Office application. VBA provides a means for users to enhance off-the-shelf applications to tailor the programs for special tasks.

Since one or more applications in a suite may not be as desirable as the others, some people still prefer to buy separate packages. Another issue with the use of software suites is the large amount of main memory required to run them effectively. For example, many users find that they must spend hundreds of dollars for additional internal memory to upgrade their personal computer to be able to run a software suite. Continual debates rate one vendor's spreadsheet superior to another vendor's, and yet a third vendor may have the best word processing package. Thus, some users prefer using individual software packages from different vendors rather than a software suite from a single vendor.

In addition to suites, some companies produce *integrated application packages* that contain several programs. For example, Works 2003 is one program that contains basic word processing, spreadsheet, database, address book, calendar, and other applications. Although not as powerful as standalone software included in software suites, integrated software packages offer a range of capabilities for less money. Some integrated packages cost about $100.

Other Personal Application Software

In addition to the software already discussed, there are a number of other interesting and powerful application software tools for individuals. In some cases, the features and capabilities of these application software tools can more than justify the cost of an entire computer system. TurboTax, for example, is a popular tax-preparation program. A number of exciting software packages have been developed for training and distance learning. University professors often believe that colleges and universities must invest in distance learning for their students.[50] Some universities offer complete degree programs using this type of software over the Internet. High-school teachers can use software, such as Plato, to determine which students need to perform better on state education examinations.[51] Engineers, architects, and designers often use computer-aided design (CAD) software to design and develop buildings, electrical systems, plumbing systems, and more. Autosketch, CorelCAD, and AutoCad are examples of CAD software. There are programs that perform a wide array of statistical tests. Colleges and universities often have a number of courses in statistics that use this type of application software. Two popular applications in the social sciences are SPSS and SAS.

Workgroup Application Software

workgroup application software

Software that supports teamwork, whether in one location or around the world.

Workgroup application software is designed to support teamwork, whether people are in the same location or dispersed around the world. This support can be accomplished with software known as *groupware* that helps groups of people work together more efficiently and effectively. Microsoft Exchange Server 2003, for example, has groupware and e-mail features.[52] Monsanto, a chemical manufacturing company, used groupware to let members of its agricultural chemical group work together to develop new products.[53] According to one company official, "We determined that this application has the potential to significantly increase the productivity of our engineering group by making programs more accessible to our young engineers." Also called collaborative software, the approach allows a team of managers to work on the same production problem, letting them share their ideas and work via connected computer systems. The "Three Cs" rule for successful implementation of groupware is summarized in Table 4.8.

Convenient	If it's too hard to use, it doesn't get used; it should be as easy to use as the telephone.
Content	It must provide a constant stream of rich, relevant, and personalized content.
Coverage	If it isn't close to everything you need, it may never get used.

Table 4.8

Ernst & Young's "Three Cs" Rule for Groupware

Examples of workgroup software include group scheduling software, electronic mail, and other software that enables people to share ideas. Lotus Notes from IBM, for example, gives companies the capability of using one software package, and one user interface, to integrate many business processes. It can allow a global team to work together from a common set of documents, have electronic discussions using threads of discussion, and schedule team meetings. As Lotus Notes matured, Lotus added services to it and renamed it Domino (Lotus Notes is now the name of the e-mail package), and now an entire third-party market has emerged to build collaborative software based on Domino. These products, which include Changepoint's Involv, remove the burden of Notes administration and broaden the application's scope to better support the Internet. For example, Domino.Doc is a Domino-based document management application with built-in workflow and archiving capabilities. Its "life cycle" feature tracks a document through the review, approval, publishing, and archiving processes. Similarly, the workflow integration adds support for multiple roles, log tracking, and distributed approval.

Workgroup application software is found in most industries. IBM, for example, is developing software to help Sesame Street Productions convert thousands of episodes of the popular *Sesame Street* TV show from the old analog film format to a digital format.[54] The project "will save both time and money while it opens up new avenues for improving our abilities to create and generate revenues," says Sherra Pierre of Sesame Workshop. Landmark Theaters and Microsoft have teamed up to convert about 200 screens to a digital cinema system that makes film distribution and playback easier.[55] When completely implemented, Landmark will be the largest digital cinema theater chain in the United States. Application software is becoming so popular that there are contests for people who use the application software to create and display art.[56] In Britain, new workgroup application software has been developed for Camelot, the national lottery, to allow the sale of lottery tickets on the Internet with cell phones.[57]

Enterprise Application Software

Software that benefits an entire organization can also be developed or purchased. A fast-food chain, for example, might develop a materials ordering and distribution program to make sure that each fast-food franchise gets the necessary raw materials and supplies during the week. This materials ordering and distribution program can be developed internally using staff and resources in the IS department or purchased from an external software company. One of the first enterprise applications was a payroll program for Lyons Bakeries in England, developed in 1954 on the Leo 1 computer.[58] According to David Caminer, who helped develop the payroll application, "There was nothing to go on, no well-understood good practice, so we had to work out what to do from scratch." Table 4.9 lists a number of applications that can be addressed with enterprise software. Many organizations are moving to integrated enterprise software that supports supply chain management (movement of raw materials from suppliers through shipment of finished goods to customers), as shown in Figure 4.17.

Organizations can no longer respond to market changes using nonintegrated information systems based on overnight processing of yesterday's business transactions, conflicting data models, and obsolete technology. As a result, many corporations are turning to **enterprise resource planning (ERP)** software, a set of integrated programs that manage a company's

enterprise resource planning (ERP)

A set of integrated programs that manage a company's vital business operations for an entire multisite, global organization.

Table 4.9

Examples of Enterprise Application Software

Accounts receivable	Sales ordering
Accounts payable	Order entry
Airline industry operations	Payroll
Automatic teller systems	Human resource management
Cash-flow analysis	Check processing
Credit and charge card administration	Tax planning and preparation
Manufacturing control	Receiving
Distribution control	Restaurant management
General ledger	Retail operations
Stock and bond management	Invoicing
Savings and time deposits	Shipping
Inventory control	Fixed asset accounting

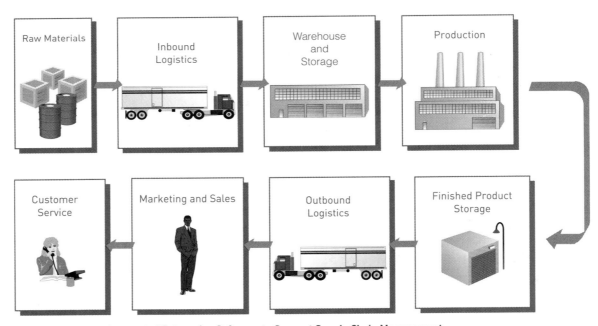

Integrated Enterprise Software to Support Supply Chain Management

Figure 4.17

Use of Integrated Supply Chain Management Software

vital business operations for an entire multisite, global organization. Thus, an ERP system must be able to support multiple legal entities, multiple languages, and multiple currencies. Although the scope of an ERP system may vary from vendor to vendor, most ERP systems provide integrated software to support manufacturing and finance. In addition to these core business processes, some ERP systems may be capable of supporting additional business functions such as human resources, sales, and distribution. The primary benefits of implementing ERP include eliminating inefficient systems, easing adoption of improved work processes, improving access to data for operational decision making, standardizing technology vendors and equipment, and enabling the implementation of supply chain management.

Application Software for Information, Decision Support, and Specialized Purposes

Specialized application software for information, decision support, and other purposes is available in every industry. Genetic researchers, for example, are using GenVision to visualize and analyze the human genome.[59] Music executives use decision support software to help pick the next hit.[60] One music producer in Barcelona says that the Hit Song Science software product picked Norah Jones as a success before she won a number of Grammy awards. Sophisticated decision support software is now being used to increase the cure rate for cancer by analyzing about 100 different scans of the cancer tumor to create a 3-D view of the tumor. Software can then consider thousands of angles and doses of radiation to determine the best radiation program. The software analysis takes only minutes, but the results can save years or decades of life for the patient. As you will see in future chapters, information, decision support, and specialized systems are used in businesses of all sizes and types to increase profits or reduce costs. But how are all these systems actually developed or built? The answer is through the use of programming languages, discussed next.

PROGRAMMING LANGUAGES

Both OSs and application software are written in coding schemes called *programming languages*. The primary function of a programming language is to provide instructions to the computer system so that it can perform a processing activity. IS professionals work with **programming languages**, which are sets of keywords, symbols, and a system of rules for constructing statements by which humans can communicate instructions to be executed by a computer. Programming involves translating what a user wants to accomplish into a code that the computer can understand and execute. *Program code* is the set of instructions that signal the CPU to perform circuit-switching operations. In the simplest coding schemes, a line of code typically contains a single instruction such as, "Retrieve the data in memory address X." As discussed in Chapter 3, the instruction is then decoded during the instruction phase of the machine cycle. Like writing a report or a paper in English, writing a computer program in a programming language requires the programmer to follow a set of rules. Each programming language uses a set of symbols that have special meaning. Each language also has its own set of rules, called the **syntax** of the language. The language syntax dictates how the symbols should be combined into statements capable of conveying meaningful instructions to the CPU. A rule that "variable names must start with a letter" is an example. A variable is a quantity that can take on different values. Program variable names such as SALES, PAYRATE, and TOTAL follow the rule because they start with a letter, whereas variables such as %INTEREST, $TOTAL, and #POUNDS do not.

Programming languages were developed to help solve particular problems. Since they were each designed for different problems, they contain different attributes. Each of the attributes in Table 4.10 represents two extremes, with most languages falling somewhere between these extremes.

programming languages
Sets of keywords, symbols, and a system of rules for constructing statements by which humans can communicate instructions to be executed by a computer.

syntax
A set of rules associated with a programming language.

The Evolution of Programming Languages

The desire to use the power of information processing efficiently in problem solving has pushed the development of newer programming languages. The evolution of programming languages is typically discussed in terms of generations of languages (see Table 4.11).

Table 4.10

Programming Language Attributes

Extreme 1	Extreme 2
Supports programming of batch processing systems with data collected into a set and processed at one time.	Supports programming of real-time systems with each data transaction processed when it occurs.
Requires programmer to write procedure-oriented code, describing step by step each action the computer must take.	Enables a programmer to write non-procedure-oriented code, describing the end result desired without having to specify how to accomplish it.
Supports business applications that require the ability to store, retrieve, and manipulate alphanumeric data and process large files.	Supports sophisticated scientific computations.
Programmers write code with a relatively high level of errors.	Programmers write code with a relatively low level of errors.
Programmers are less productive and able to create only a small amount of code per unit of time.	Programmers are more productive and are able to create a large amount of code per unit time.

Generation	Language	Approximate Development Date	Sample Statement or Action
First	Machine language	1940s	00010101
Second	Assembly language	1950s	MVC
Third	High-level language	1960s	READ SALES
Fourth	Query and database languages	1970s	PRINT EMPLOYEE NUMBER IF GROSS PAY>1000
Beyond Fourth	Natural and intelligent languages	1980s	IF gross pay is greater than 40, THEN pay the employee overtime pay.

Table 4.11

The Evolution of Programming Languages

First Generation

The first generation of programming languages is *machine language*, which required the use of binary symbols (0s and 1s). Because this is the language of the CPU, text files that are translated into binary sets can be read by almost every computer system platform.

Second Generation

Developers of programming languages attempted to overcome some of the difficulties inherent in machine language by replacing the binary digits with symbols that programmers could more easily understand. These second-generation languages use codes like A for add, MVC for move, and so on. Another term for these languages is *assembly language*, which comes from the programs (called *assemblers*) used to translate it into machine code. Systems software programs such as OSs and utility programs are often written in an assembly language.

Third Generation

Third-generation languages continued the trend toward greater use of symbolic code and away from specifically instructing the computer how to complete an operation. BASIC, COBOL, C, and FORTRAN are examples of third-generation languages that use English-like statements and commands. This type of language is easier to learn and use than machine and assembly languages because it more closely resembles everyday human communication and understanding.

With third-generation and higher-level programming languages, each statement in the language translates into several instructions in machine language. A special software program called a **compiler** converts the programmer's source code into the machine-language instructions consisting of binary digits, as shown in Figure 4.18. A compiler creates a two-stage process for program execution. First, it translates the program into a machine language; second, the CPU executes that program. Another approach is to use an interpreter, which is a language translator that converts each statement in a programming language into machine language and executes the statement, one at a time. An interpreter does not produce a complete machine-language program. After the statement executes, the machine-language statement is discarded, the process continues for the next statement, and so on.

compiler

A special software program that converts the programmer's source code into the machine-language instructions consisting of binary digits.

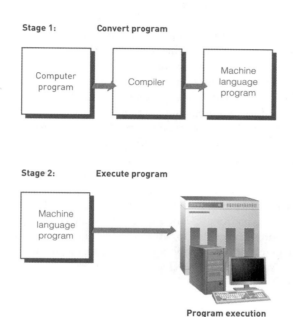

Stage 1:   Convert program

Computer program → Compiler → Machine language program

Stage 2:   Execute program

Machine language program →

Program execution

Figure 4.18

How a Compiler Works

A compiler translates a complete program into a complete set of binary data (Stage 1). Once this is done, the CPU can execute the converted program in its entirety (Stage 2).

Fourth Generation

Fourth-generation programming languages emphasize what output results are desired rather than how programming statements are to be written. As a result, many managers and executives with little or no training in computers and programming are using fourth-generation languages (4GLs). Languages for accessing information in a database are often fourth-generation languages. Prime examples include PowerBuilder, Delphi, Essbase, Forte, Focus, Powerhouse, SAS, and many others. Natural is a 4GL that can be used with Windows, UNIX, or Linux.[61] Another popular fourth-generation language is called Structured Query Language (SQL), which is often used to perform database queries and manipulations.

Languages Beyond the Fourth Generation

After the fourth generation, it becomes more difficult to classify programming languages. Languages beyond the fourth generation include artificial intelligence, visual, and object-oriented languages. In general, these languages are easier for nonprogrammers to use compared with older generation languages.

Programming languages used to create artificial intelligence or expert systems applications are often called *fifth-generation languages (5GLs)*. Fifth-generation languages are sometimes called natural languages because they use even more English-like syntax than 4GLs. They allow programmers to communicate with the computer by using normal sentences. For example, computers programmed in fifth-generation languages can understand queries such as, "How many athletic shoes did our company sell last month?" Fifth-generation languages, for example, can be used to determine where to explore for oil and natural gas.

Visual languages use a graphical or visual interface for program development. Unlike earlier languages that depended on writing detailed programming statements, visual languages allow programmers to "drag and drop" programming objects onto the computer

screen. Many of these languages are used to develop applications on the Internet. *Visual Basic* was one of the first visual programming languages. Other languages with visual development interfaces include Visual Basic .NET and Visual C++ .NET. *Visual Basic* and *Visual Basic .NET* can be used to develop applications that run under the Windows OS. C++ is a powerful and flexible programming language used mostly by computer systems professionals to develop software packages. *Java* is a programming language developed by Sun Microsystems that can run on any OS and on the Internet. Java can be used to develop complete applications or smaller applications, called *Java Applets*. Visual Basic .NET, C++, and Java are also examples of objected-oriented languages, which are discussed next.

The preceding programming languages separate data elements from the procedures or actions that will be performed on them, but another type of programming language ties them together into units called *objects*. An object consists of data and the actions that can be performed on the data. For example, an object could be data about an employee and all the operations (such as payroll calculations) that might be performed on the data. Programming languages that are based on objects are called *object-oriented programming languages*.

Building programs and applications using object-oriented programming languages is like constructing a building using prefabricated modules or parts. The object containing the data, instructions, and procedures is a programming building block. The same objects (modules or parts) can be used repeatedly. One of the primary advantages of an object is that it contains reusable code. In other words, the instruction code within that object can be reused in different programs for a variety of applications, just as the same basic prefabricated door can be used in two different houses. An object can relate to data on a product, an input routine, or an order-processing routine. An object can even direct a computer to execute other programs or to retrieve and manipulate data. So, a sorting routine developed for a payroll application could be used in both a billing program and an inventory control program. By reusing program code, programmers are able to write programs for specific application problems more quickly (see Figure 4.19). By combining existing program objects with new ones, programmers can easily and efficiently develop new object-oriented programs to accomplish organizational goals.

Figure 4.19

Reusable Code in Object-Oriented Programming

By combining existing program objects with new ones, programmers can easily and efficiently develop new object-oriented programs to accomplish organizational goals. Note that these objects can be either commercially available or designed internally.

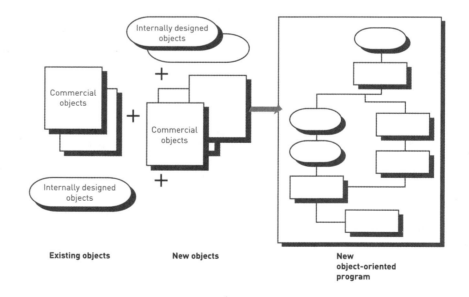

There are several object-oriented programming languages; some of the most popular include Smalltalk, Visual Basic .NET, C++, and Java. Some old languages, such as COBOL, have been modified to support the object-oriented approach.[62] As mentioned earlier, Java is an Internet programming language from Sun Microsystems that can run on a variety of computers and OSs, including UNIX, Windows, and Macintosh OSs.

Object-oriented programs often use *methods*, which are instructions to perform a specific task in the program.[63] The following instructions in C++ can be used to compute the area of a rectangle, given the width and length.

```
// Method to Compute the Area of a Rectangle Given the Width and Length
double Rectangle: :ComputeArea ( )
{
    return width * length;
}
    // End of the ComputeArea Method
```

Once developed as part of a C++ program, the instructions or method can be used in other programs to compute the area of a picture frame, a living room, a front lawn, or any other application that requires the area of a rectangle. Below are a few instructions in another C++ program that show how the above ComputeArea method can be used to compute the area of a picture frame.

```
// Assign Data and Compute Area
frame Object -> SetDimensions (frameWidth, frameLength);
frameArea = frameObject -> ComputeArea( );
```

Selecting a Programming Language

Selecting the best programming language to use for a particular program involves balancing the functional characteristics of the language.

Machine and assembly languages provide the most direct control over computer hardware. For this reason, many vendors of popular application software programs take the time and effort to code portions of their leading programs in assembly language to maximize their speed. When a programmer requires a high degree of control over how various hardware components are used, these languages should be used. In selecting any programming language, the amount of direct control that is needed over the operation of the hardware can be an important factor to consider.

More recent programming languages are typically more complex than earlier programming languages. Although these newer languages appear to be simpler because they are more English-like, each command can drive complex routines and functions that operate behind the scenes. It takes less time to develop computer programs using higher-level languages than with lower-level languages. This means that the cost to develop computer programs can be substantially less with these more recent programming languages. Although training programmers to use these higher-level programming languages may produce high up-front costs, using higher-level languages can reduce the total costs to develop computer programs in the long run.

C++ and Java both have advantages and disadvantages, but Java may be the future of programming. Java is far easier to learn, and as a result, people become productive much sooner. Programmers who learn C++ must spend a lot of time debugging rather than learning software engineering techniques. An increasing number of colleges in the United States are using Java as their first programming language. Java is also more portable—with the ability to run on more OSs and hardware. However, C++ will not disappear anytime soon. A large base of C++ programs is installed, and there are a large number of users because Microsoft uses it for programming. The American National Standards Institute (ANSI) and International Organization for Standardization (ISO) standards committees have also been working on C and C++ since 1990, and it is apparent that people will continue to use C++ within or outside a Microsoft environment.

SOFTWARE ISSUES AND TRENDS

Because software is such an important part of today's computer systems, issues such as software bugs, licensing, upgrades, and global software support have received increased attention. We highlight several major software issues and trends in this section: software bugs, copyright, software licensing, open-source software, shareware and other public domain software, multiorganizational software development, software upgrades, and global software support.

Software Bugs

software bug
A defect in a computer program that keeps it from performing in the manner intended.

A **software bug** is a defect in a computer program that keeps it from performing in the manner intended. Some software bugs are obvious and cause the program to terminate unexpectedly. Other bugs are subtler and allow errors to creep into your work. Computer and software vendors tell us that since humans design and program hardware and software, bugs are inevitable. In fact, according to the Pentagon and the Software Engineering Institute at Carnegie Mellon University, there are typically 5 to 15 bugs in every 1,000 lines of code—the software instructions that make sense only to computers and programmers.

Most software bugs arise because manufacturers release new software as early as possible instead of waiting until all bugs are identified and removed. They are under intense pressure from customers to deliver the software they have announced and from shareholders to begin selling the new product to increase sales. Meanwhile, the software manufacturer's quality-assurance people fight a losing battle for more testing time to identify and remove bugs. Although the decision of when to release new software is based on a fine line, the industry clearly favors releasing software early and with defects. After all, software companies make money on upgrades, so there is little incentive to achieve a perfect first release. Table 4.12 summarizes tips for reducing the impact of software bugs.

Copyrights and Licenses

Most software products are protected by law using copyright or licensing provisions. Those provisions can vary, however. In some cases, you are given unlimited use of software on one or two computers. This is typical with many applications developed for personal computers. In other cases, you pay for your usage—if you use the software more, you pay more. This approach is becoming popular with software placed on networks or larger computers. Most of these protections prevent you from copying software and giving it to others without restrictions. Some software now requires that you *register* or *activate* it before it can be fully used. Registration and activation sometimes put software on your hard disk that monitors activities and changes to your computer system.

Table 4.12

How to Deal with Software Bugs

Register all software so that you receive bug alerts, fixes, and patches.
Check the manual or read-me files for workarounds.
Access the support area of the manufacturer's Web site for patches.
Install the latest software updates.
Before reporting a bug, make sure that you can re-create the circumstances under which it occurs.
Once you can re-create the bug, call the manufacturer's tech support line.
Avoid buying the latest release of software for several months or a year until the software bugs have been discovered and removed.

In general, software manufacturers want to license their software to lock in a steady, predictable stream of revenue from customers. Software manufacturers also want to encourage customers to move to the latest releases of their software products to minimize the effort required to support out-of-date products. There are numerous types of software licenses to help accomplish these objectives, including the following:

- *Usage-based License.* In this arrangement, software fees are based on the actual usage of the manufacturer's products. Licensees are charged in much the same way that customers of utility firms are charged—by increased fees for increased use of power or water.
- *Capacity-based License.* With capacity-based licenses, the fees paid to the software manufacturer are based on the processing power of the computer on which the software is installed. Users who run their software on a more powerful processor pay more for the software. The fees charged do not relate to the actual use of the products.
- *Software as a Network Service.* When software is accessed as a network service, the software manufacturer makes its products available through the Internet. The advantages of this form of usage is that the software manufacturer automatically offers users bug fixes, enhancements, and other updates over the Web and charges a subscription fee for the software and associated services.
- *Subscription Licensing.* With subscriptions, user companies sign a multiyear deal with a manufacturer for individual products or a collection of products and then pay annual subscription fees based on the number of PCs they have. As discussed next, some software doesn't have restrictive copyright or licensing agreements.

Open-Source Software

Open-source software is software that is freely available to anyone in a form that can be easily modified. The Weather.com Web site, for example, uses open-source software to cut costs by 50 percent or more compared with traditional software.[64] According to Dan Agronow, CIO for Weather.com, "Where it makes sense, we will always look at open-source alternatives." The Open Source Initiative is a nonprofit corporation dedicated to the development and promotion of open-source software (see the OSI's Web site at *www.opensource. org* for more information on the group's efforts). Users can download the source code and build the software themselves, or the software's developers can make executable versions available along with the source. Open-source software development is a collaborative process—developers around the world use the Internet to keep in close contact via e-mail and to download and submit new software. Major software changes can occur in days rather than weeks or months. Open-source software is at the heart of many of the Internet's most popular services, including e-mail and the Web. A number of open-source software packages are widely used, including the Linux system; Free BSD, another OS; Apache, a popular Web server; Sendmail, a program that delivers e-mail for most systems on the Internet; and Perl, a programming language used to develop Internet application software. With heightened concerns about the security of computer systems, Microsoft decided to show government agencies the source code used to develop its Windows OS to convince government agencies that the software is secure and safe.[65] See Table 4.13 for some examples of open-source software.

open-source software
Software that is freely available to anyone in a form that can be easily modified.

Software Type	Example
Operating system	Linux
Application software	Open Office
Database software	MySQL
Internet browser	Mozilla
Internet messaging	Jabber

Table 4.13

Examples of Open-Source Software

Why would an organization run its business using software that's free? How can something that's given away over the Internet be stable or reliable or sufficiently supported to place at the core of a company's day-to-day operations? The answer is surprising—open-source software is often *more* reliable than commercial software. How can this be? First, by making a program's source code readily available, users can fix any problems they discover. A fix is often available within hours of the problem's discovery. Second, with the source code for a program accessible to thousands of people, the chances of a bug being discovered and fixed before it does any damage are much greater than with traditional software packages. Read the "Ethical and Societal Issues" feature to learn about the trend toward government use of open-source software.

The question of software support is the biggest stumbling block to the acceptance of open-source software at the corporate level. Getting support for traditional software packages is easy—you call a company's toll-free support number or access its Web site. But how do you get help if an open-source package doesn't work as expected? Since the open-source community lives on the Internet, you look there for help. Through use of Internet discussion areas, you can communicate with others who use the same software, and you may even reach someone who helped develop it. Users of popular open-source packages can get correct answers to their technical questions within a few hours of asking for help on the appropriate Internet forum. Another approach is to contact one of the many companies emerging to support and service such software—for example, RedHat for Linux, C2Net for Apache, and Sendmail, Inc., for Sendmail. These companies offer high-quality, for-pay technical assistance.

Shareware, Freeware, and Public Domain Software

shareware and freeware
Software that is very inexpensive or free, but whose source code cannot be modified.

Many software users are doing what they can to minimize software costs. Some are turning to **shareware and freeware**—software that is very inexpensive or free, usually for use in personal computers, but whose source code cannot be modified. Shareware may not be as powerful as commercial software, but some people get what they need at a good price. In some cases, you are given the opportunity to try the software before sending a nominal fee to the software developer. Some shareware and freeware is in the public domain, often called *public domain software*. This software is not protected by copyright laws and can be freely copied and used. Although shareware and freeware can be free or inexpensive to acquire, it can be more expensive to use and maintain over time compared with software that is purchased.[66] If the software is hard to use and doesn't perform all the required functions, the cost of wasted time and lost productivity can be far greater than the cost of purchasing better software. Shareware, freeware, and public domain software is often not open source—that is, the source code is not available and cannot be modified.

Multiorganizational Software Development

The use of the Internet to spur development of open-source software has led to extending software development beyond a single organization by finding others who share the same problem and involving them in a common development effort. Such an approach spreads development costs across multiple organizations and increases the opportunity for people with highly specialized expertise to contribute. It also increases the number of developers and users who can identify and eliminate bugs in software and fosters the growth of a community of people who can support the software. In addition, colleges and universities sometimes develop software that is made available to companies. The University of Iowa, for example, developed a production-planning software package for agricultural products, such as seed-corn hybrids.[67] One company was able to use the software to increase profits by approximately $5 million.

Weighing the Benefits of Open Source

Recently, the city of Munich decided to switch its 14,000 computers used by local-government employees from Microsoft Windows to Linux, a free open-source OS. Open-source software is distributed along with its source code so that users can edit and change it as desired. When Microsoft's CEO, Steve Ballmer, heard about Munich's plans, he flew there to lobby the mayor. In the negotiations, Ballmer even went as far as matching Linux's price—essentially giving the city Microsoft Windows for free—and still the mayor turned him down. The rationale? City officials said the decision was a matter of principle: The municipality wanted to control its technological destiny. It did not wish to place the functioning of government in the hands of a commercial vendor with proprietary standards that is accountable to shareholders rather than to citizens.

This sentiment is growing among governments around the world. Brazil also recommended that all its government agencies and state enterprises buy open-source software. China is working to adopt Linux to become self-sufficient and secure. India, Japan, and South Korea are also aggressively pursuing open-source alternatives to Microsoft's software. In the United States, state, county, and city governments are equal in their interest in open-source software. In Florida, Miami-Dade County is considering swapping out its 15,000 copies of Microsoft Windows for Linux.

Some in private industry are following government's lead. Such big players in the technology industry as IBM, Novell, and Sun Microsystems are switching or have switched their internal operations to Linux. Many other nontech companies are implementing Linux in portions of their operations. Ford Motor Company is switching to Linux for much of its server computing. With large corporations blazing the trail for Linux, industry watchers anticipate that many other businesses of all sizes will follow.

Why all this fuss over software? As we saw in Munich, it isn't just a matter of price. The reason governments and businesses like open-source software is its accessibility. Open-source software can be tweaked, edited, and dissected whenever needed; it also has no secrets. Traditional, proprietary, licensed software is just the opposite—it is one big secret. Using software packages such as Microsoft Windows is similar to owning a car with no hood to access to the engine. You are provided with a slick user interface—a dashboard, steering wheel, and pedals—but if you are curious about how it works or need to repair it, you are out of luck.

Supporters of open-source software development have a simple philosophy: When programmers can read, redistribute, and modify the source code, the software evolves for everyone's benefit. People improve it, people adapt it, and people fix bugs. Development can happen at a speed that, compared with the slow pace of conventional software development, seems astonishing.

Opponents of the open-source movement, such as Microsoft, have sought to discredit the software. They have suggested that the very openness of the software makes it vulnerable to hackers and terrorists. Microsoft has also funded studies that have found that Windows has a lower cost of ownership than Linux.

While working to discredit Linux, Microsoft has made moves to open up some of its code under an initiative called "shared source." The shared source effort allows certain approved governments and large corporate clients to access most of the Windows software source code. This limited access helps Microsoft assure purchasers that Windows doesn't contain secret security backdoors. Microsoft has also made available portions of the source code for Windows CE, the Microsoft OS for handheld PCs and mobile phones, so others can more easily develop applications for it.

The one major drawback of open-source software—and the major challenge for the Linux OS—is the lack of application programs for personal computing. Since Microsoft holds a monopoly in the desktop PC OS market, the vast majority of software is written for Microsoft Windows. The software developed for Linux is much smaller in quantity, variety, and some may argue in quality. While an inexpensive alternative to Microsoft Office called Star Office was written to run on Linux, and a Windows-like interface, called Lindows for Linux is available, most users are more comfortable with the familiar Microsoft software. Since a solid business model for open-source development is still in the blueprint stages, developers have little incentive to create software that will be given away for free. Companies such as IBM, however, are finding success in distributing Linux to their customers, and profiting from the design, training, and support aspects of Linux-based information systems.

Critical Thinking Questions

1. Why do you think Ford is using Linux on its back-end servers but not on its employees' desktop PCs?
2. How can software companies like Microsoft compete against the trend toward open-source software?

What Would You Do?

As head of the technology group at Seabreeze Security Systems Corporation, Tom Gaskins had experienced a lot of pressure to cut costs over the past two years. He was exiting a meeting with the CIO when he was given his biggest challenge yet: to reduce the technology budget by one-third over the next year. The odds of being able to meet that challenge seemed particularly small, given that this was the year the organization was scheduled to upgrade to the latest version of Microsoft Windows and Office. That software investment would put him 20 percent over last year's budget. Tom could put off the upgrade for another year or two, but it didn't look as though the budget would be any better in the foreseeable future. How could he reduce costs? Should he lay off staff? Should he investigate alternatives to Microsoft? Changing platforms would mean quite a bit of work and significant staff training.

3. If you were Tom, how would you calculate the cost of changing to an alternative platform such as Linux against the cost of the typical Windows/Office upgrade?

4. How would the costs of changing platforms differ over the next five years and multiple upgrades of Windows and Office?

SOURCES: "Microsoft at the Power Point," *Economist.com*, September 11, 2003, *www.economist.com*; John Lettice, "Motor Giant Ford to Move to Linux," *The Register*, September 9, 2003, *www.theregister.com*; Linux Online, *www.linux.org/*, accessed February 8, 2004; Open Source Initiative Web site, *www.opensource.org/*, accessed February 8, 2004; Sun Microsystems Star Office Web site, *wwws.sun.com/software/star/staroffice/6.0/*, accessed February 8, 2004.

Software Upgrades

Software companies revise their programs and sell new versions periodically. In some cases, the revised software offers new and valuable enhancements. In other cases, the software uses complex program code that offers little in terms of additional capabilities. In addition, revised software can contain bugs or errors. Deciding whether to purchase the newest software can be a problem for corporations and individuals with a large investment in software. Should the newest version be purchased when it is released? Some organizations and individuals do not always get the most current software upgrades or versions, unless there are significant improvements or capabilities. Instead, they may upgrade to newer software only when there are vital new features. Software upgrades usually cost much less than the original purchase price.

Global Software Support

Large, global companies have little trouble persuading vendors to sell them software licenses for even the most far-flung outposts of their company. But can those same vendors provide adequate support for their software customers in all locations? Supporting local operations is one of the biggest challenges IS teams face when putting together standardized, company-wide systems. In slower technology growth markets, such as Eastern Europe and Latin America, there may be no official vendor presence at all. Instead, large vendors such as Sybase, IBM, and Hewlett-Packard typically contract out support for their software to local providers.

One approach that has been gaining acceptance in North America is to outsource global support to one or more third-party distributors. The software-user company may still negotiate its license with the software vendor directly, but it then hands over the global support contract to a third-party supplier. The supplier acts as a middleman between software vendor and user, often providing distribution, support, and invoicing. American Home Products Corporation handles global support for both Novell NetWare and Microsoft Office applications this way—throughout the 145 countries in which it operates. American Home Products, a pharmaceutical and agricultural products company, negotiated the agreements directly with the vendors for both purchasing and maintenance, but fulfillment of the agreement is handled exclusively by Philadelphia-based Softsmart, an international supplier of software and services.

In today's computer systems, software is an increasingly critical component. Whatever approach individuals and organizations take to acquire software, it is important for everyone to be aware of the current trends in the industry. Informed users are wiser consumers, and they can make better decisions.

SUMMARY

Principle

When selecting an operating system, you must consider the current and future requirements for application software to meet the needs of the organization. In addition, your choice of a particular operating system must be consistent with your choice of hardware.

Software consists of programs that control the workings of the computer hardware. There are two main categories of software: systems software and application software. Systems software is a collection of programs that interacts between hardware and application software. Systems software includes operating systems, utility programs, and middleware. Application software enables people to solve problems and perform specific tasks. Application software may be proprietary or off the shelf.

An operating system (OS) is a set of computer programs that controls the computer hardware to support users' computing needs. OS hardware functions by converting an instruction from an application into a set of instructions needed by the hardware. The OS also serves as an intermediary between application programs and hardware, allowing hardware independence. Memory management involves controlling storage access and use by converting logical requests into physical locations and by placing data in the best storage space, perhaps expanded or virtual memory.

Task management allocates computer resources through multitasking and time-sharing. With multitasking, users can run more than one application at a time. Time-sharing allows more than one person to use a computer system at the same time.

The ability of a computer to handle an increasing number of concurrent users smoothly is called *scalability*, a feature critical for systems expected to handle a large number of users.

An OS also provides a user interface, which allows users to access and command the computer. A command-based user interface requires text commands to send instructions; a graphical user interface (GUI), such as Windows, uses icons and menus.

Software applications make use of the OS by making requests for services through a defined application program interface (API). Programmers can use APIs to create application software without having to understand the inner workings of the OS. APIs also provide a degree of hardware independence so that the underlying hardware can change without necessarily requiring a rewrite of the software applications.

Over the years, several popular OSs have been developed. These include several proprietary OSs used primarily on mainframes. MS-DOS is an early OS for IBM-compatibles. Older Windows OSs are GUIs used with DOS. Newer versions, like Windows XP, are fully functional OSs that do not need DOS. Apple computers use proprietary OSs like the Mac OS and Mac OS X. UNIX is a powerful OS that can be used on many computer system types and platforms, from personal computers to mainframe systems. Use of the UNIX OS makes it easy to move programs and data among computers or to connect mainframes and personal computers to share resources. Linux is the kernel of an OS whose source code is freely available to everyone. Several variations of Linux are available, with sets of capabilities and applications to form a complete OS, for example, RedHat Linux. z/OS and MPE/iX are OSs for mainframe computers. A number of OSs have been developed to support consumer appliances such as Palm OS, Windows CE.Net, Windows XP Embedded, Pocket PC, and variations of Linux.

Utility programs are used to perform many useful tasks and often come installed on computers along with the OS. This software is used to merge and sort sets of data, keep track of computer jobs being run, compress files of data, protect against harmful computer viruses, and monitor hardware and network performance. Middleware is software that allows different systems to communicate and transfer data back and forth.

Principle

Do not develop proprietary application software unless doing so will meet a compelling business need that can provide a competitive advantage.

Application software applies the power of the computer to solve problems and perform specific tasks. One useful way of classifying the many potential uses of information systems is to identify the scope of problems and opportunities addressed by a particular organization or its sphere of influence. For most companies, the spheres of influence are personal, workgroup, and enterprise.

User software, or personal productivity software, includes general-purpose programs that enable users to improve their personal effectiveness, increasing the amount of work that can be done and its quality. Software that helps groups work together is often referred to as workgroup application software. Examples of such software include group scheduling software, electronic mail, and other software that enables people to share ideas. Enterprise software that benefits the entire organization can also be developed or purchased. Many organizations are turning to enterprise resource planning software, a set of integrated programs that manage a company's vital business operations for an entire multisite, global organization.

Three approaches to developing application software are as follows: build proprietary application software, buy existing

programs off the shelf, or use a combination of customized and off-the-shelf application software. Building proprietary software (in-house or contracted out) has the following advantages: the organization will get software that more closely matches its needs; by being involved with the development, the organization has further control over the results; and the organization has more flexibility in making changes. The disadvantages include the following: it is likely to take longer and cost more to develop, the in-house staff will be hard pressed to provide ongoing support and maintenance, and there is a greater risk that the software features will not work as expected or that other performance problems will occur.

Purchasing off-the-shelf software has many advantages. The initial cost is lower, there is a lower risk that the software will fail to work as expected, and the software is likely to be of higher quality than proprietary software. Some of the disadvantages are that the organization may pay for features it does not need, the software may lack important features requiring expensive customization, and the system may work in such a way that work process reengineering is required.

Some organizations have taken a third approach—customizing software packages. This approach usually involves a mixture of the preceding advantages and disadvantages and must be carefully managed.

An application service provider (ASP) is a company that can provide software, end-user support, and the computer hardware on which to run the software from the user's facilities. ASPs provide contract customization of off-the-shelf software, and they speed deployment of new applications while helping IS managers avoid implementation headaches. Use of ASPs reduces the need for many skilled IS staff members and also lowers a project's start-up expenses.

Although there are literally hundreds of computer applications that can help individuals at school, home, and work, the primary applications are word processing, spreadsheet analysis, database, graphics, and online services. A software suite, such as SmartSuite, WordPerfect, StarOffice, or Office, offers a collection of powerful programs.

Principle

Choose a programming language whose functional characteristics are appropriate for the task at hand, taking into consideration the skills and experience of the programming staff.

All software programs are written in coding schemes called *programming languages*, which provide instructions to a computer to perform some processing activity. There are several classes of programming languages, including machine, assembly, high-level, query and database, object-oriented, and visual programming languages.

Programming languages have gone through changes since their initial development in the early 1950s. In the first generation, computers were programmed in machine language, or binary code, a series of statements written in 0s and

1s. The second generation of languages was termed *assembly languages*; these languages support the use of symbols and words rather than 0s and 1s. The third generation consists of many high-level programming languages that use English-like statements and commands. They also must be converted to machine language by special software called a compiler but are easier to write than assembly or machine-language code. These languages include BASIC, COBOL, FORTRAN, and others. A fourth-generation language is less procedural and more English-like than third-generation languages. The fourth-generation languages include database and query languages such as SQL.

Beyond fourth-generation languages, it becomes more difficult to classify programming languages. Fifth-generation programming languages combine rules-based code generation, component management, visual programming techniques, reuse management, and other advances. These languages offer the greatest ease of use yet. Visual and object-oriented programming languages—such as Smalltalk, C++, and Java—use groups of related data, instructions, and procedures called *objects*, which serve as reusable modules in various programs. These languages can reduce program development and testing time. Java can be used to develop applications on the Internet.

Selecting the best programming language to use for a particular program involves balancing the functional characteristics of the language with cost, control, and complexity issues.

Principle

The software industry continues to undergo constant change; users need to be aware of recent trends and issues to be effective in their business and personal life.

Software bugs, software licensing and copyrighting, open-source software, shareware and freeware, multiorganizational software development, software upgrades, and global software support are all important software issues and trends.

A software bug is a defect in a computer program that keeps it from performing in the manner intended. Software bugs are common, even in key pieces of business software.

Software manufacturers are developing new approaches to licensing their software to lock in a steady, predictable stream of revenue from their customers. Some of these new approaches include usage-based licenses, capacity-based licenses, software as a network service, and subscription licensing.

Open-source software is software that is freely available to anyone in a form that can be easily modified. Open-source software development and maintenance is a collaborative process with developers around the world using the Internet to keep in close contact via e-mail and to download and submit new software. Shareware and freeware can reduce the cost of software, but sometimes they may not be as powerful as

commercial software. Also, their source code usually cannot be modified.

Multiorganizational software development is the process of extending software development beyond a single organization by finding others who share the same business problem and involving them in a common development effort.

Software upgrades are an important source of increased revenue for software manufacturers and can provide useful new functionality and improved quality for software users.

Global software support is an important consideration for large, global companies putting together standardized, company-wide systems. A common solution is outsourcing global support to one or more third-party software distributors.

CHAPTER 4: SELF-ASSESSMENT TEST

When selecting an operating system, you must consider the current and future requirements for application software to meet the needs of the organization. In addition, your choice of a particular operating system must be consistent with your choice of hardware.

1. What is the heart of the OS that controls the most critical processes?

   a. platform
   b. instruction set
   c. kernel
   d. CPU

2. A command-based user interface requires that text commands be given to the computer to perform basic activities. True or False?

3. _____ is the process of swapping programs or parts of programs between memory and disk.

4. The file manager component of the OS controls how memory is accessed and maximizes available memory and storage. True or False?

Do not develop proprietary application software unless doing so will meet a compelling business need that can provide a competitive advantage.

5. The primary function of application software is to apply the power of the computer to give individuals, workgroups, and the entire enterprise the ability to solve problems and perform specific tasks. True or False?

6. Software that enables users to improve their personal effectiveness, increasing the amount of work they can do and its quality is called

   a. personal productivity software
   b. operating system software
   c. utility software
   d. graphics software

7. Optimization can be found in which type of application software?

   a. spreadsheets
   b. word processing programs
   c. database programs
   d. presentation graphics programs

8. Software used to solve a unique or specific problem that is usually built in-house, but can also be purchased from an outside company is called _____.

9. What type of software has the greatest potential to affect the processes that add value to a business because it is designed for specific organizational activities and functions?

   a. personal productivity software
   b. operating system software
   c. utility software
   d. applications software

Choose a programming language whose functional characteristics are appropriate for the task at hand, taking into consideration the skills and experience of the programming staff.

10. A built-in scripting facility that is part of every Microsoft Office application and provides a means of enhancing off-the-shelf applications to allow users to tailor the programs is called _____.

    a. Visual Basic
    b. SmallTalk
    c. Norton Utilities
    d. Java

11. A class of applications software that helps groups work together and collaborate is called _____.

12. Each programming language has its own set of rules, called the _____ of the language.

13. A special software program called a *compiler* performs the conversion from the programmer's source code into the machine-language instructions consisting of binary digits. True or False?

CHAPTER 4: SELF-ASSESSMENT TEST ANSWERS

(1) c (2) True (3) Paging (4) False (5) True (6) a (7) a (8) proprietary software (9) d (10) a (11) workgroup application software (12) syntax (13) True

KEY TERMS

application program interface (API) 147
application service provider (ASP) 160
application software 143
command-based user interface 147
compiler 173
computer programs 143
computer system platform 143
documentation 143
enterprise resource planning (ERP) 169
enterprise sphere of influence 145
graphical user interface (GUI) 147

icon 147
kernel 146
middleware 158
multitasking 148
off-the-shelf software 159
open-source software 177
operating system (OS) 145
paging 148
personal productivity software 145
personal sphere of influence 145
programming languages 171
proprietary software 159
scalability 149

shareware and freeware 178
software bug 176
software suite 167
sphere of influence 144
syntax 171
systems software 143
time-sharing 148
user interface 146
utility programs 156
virtual memory 148
workgroup 145
workgroup application software 168
workgroup sphere of influence 145

REVIEW QUESTIONS

1. What is the difference between systems and application software? Give four examples of personal productivity software.
2. How do software bugs arise?
3. Identify and briefly discuss two types of user interfaces provided by an operating system.
4. What are the two basic types of software? Briefly describe the role of each.
5. Name four operating systems that support the personal, workgroup, and enterprise spheres of influence.
6. What is middleware?
7. What is multitasking?
8. Define the term *utility* software and give two examples.

9. Identify the two primary sources for acquiring application software.
10. What is an application service provider? What issues arise in considering the use of one?
11. What is open-source software? What is the biggest stumbling block with the use of open-source software?
12. What does the acronym API stand for? What is the role of an API?
13. Briefly discuss the advantages and disadvantages of frequent software upgrades.
14. Describe the term *enterprise resource planning (ERP) system*. What functions does such a system perform?
15. Identify and briefly discuss four different types of software licenses.

DISCUSSION QUESTIONS

1. Assume that you must take a computer programming course next semester. What language do you think would be best for you to study? Why? Do you think that a professional programmer needs to know more than one programming language? Why or why not?
2. Identify the three spheres of influence and briefly discuss the software needs of each.
3. Identify the three fundamental types of applications software. Discuss the advantages and disadvantages of each type.

4. Describe how the OS can manage the computer's memory.
5. You are using a new release of an application software package. You think that you have discovered a bug. Outline the approach that you would take to confirm that it is indeed a bug. What actions would you take if it truly were a bug?
6. How can application software improve the effectiveness of a large enterprise? What are some of the benefits associated with implementation of an enterprise resource planning system? What are some of the issues that could keep the use

of enterprise resource planning software from being successful?

7. Define the term *application service provider*. What are some of the advantages and disadvantages of employing an ASP? What precautions might you take to minimize the risk of using one?

8. Briefly outline the evolution of programming languages. Use your imagination and creativity to develop a brief description of the sixth generation of programming languages. How would they work? What sort of features might be included?

9. Contrast and compare two popular OSs for personal computers.

10. If you were the IT manager for a large manufacturing company, what issues might you have with the use of open-source software? What advantages might there be for use of such software?

11. Identify four types of software licenses frequently used. Which approach does the best job of ensuring a steady, predictable stream of revenue from customers? Which approach is most fair for the small company that makes infrequent use of the software?

PROBLEM-SOLVING EXERCISES

1. Choose an application software package that might be useful for a career that interests you and develop a six-slide presentation of its history, current level of usage, typical applications, ease of use, etc.

2. Use a spreadsheet package to prepare a simple monthly budget and forecast your cash flow—both income and expenses—for the next six months (make up numbers rather than using actual ones). Now use a graphics package to plot the total monthly income and monthly expenses for six months. Cut and paste both the spreadsheet and the graph into a word-processing document that summarizes your financial condition.

TEAM ACTIVITIES

1. Form a group of three or four classmates. Find articles from business periodicals, search the Internet, or interview people on the topic of software bugs. How frequently do they occur, and how serious are they? What can software users do to encourage defect-free software? Compile your results for an in-class presentation or a written report.

2. Form a group of three or four classmates. Identify and contact an individual with a local firm. Interview the individual and describe the application software the company uses and the importance of the software to the organization. Write a brief report summarizing your findings.

WEB EXERCISES

1. Microsoft, IBM/Lotus, and Corel are the important providers of personal productivity software suites. Do research to assess the relative success of these three products in terms of sales of their software suites. Do you think it is possible that Microsoft will become the only provider of such software? Would this be good or bad? Why? Write a brief report summarizing your findings and conclusions.

2. Do research on the Web and develop a two-page report summarizing the latest consumer appliance OSs. Which one seems to be gaining the most widespread usage? Why do you think this is the case?

3. Do research on the Web about application software that is used in an industry that is of interest to you. Write a brief report describing how the application software can be used to increase profits or reduce costs.

CAREER EXERCISES

1. What personal computer OS would help you the most in the first job you would like to have after you graduate? Why? What features are the most important to you?

2. Think of your ideal job. Describe five software packages that could help you advance in your career. If the software package doesn't exist, describe the kinds of software packages that could help you in your career.

VIDEO QUESTIONS

Watch the video clip **P2P's Cloudy Future** and answer these questions:

1. Software, such as Morpheus and Grokster, provides users with convenient methods for carrying out legal and illegal (or unethical) activities and violations of copyright laws. Should such software and the manufacturers of the software bear any of the responsibility for how it is used? Why or why not?

2. If the courts were to make it illegal to produce and distribute P2P software such as Morpheus, Grokster, and Kazaa, or place restrictions on such software, how might it affect innovation within the software industry?

CASE STUDIES

Case One

PACS at North Bronx Healthcare Network

"I have never done an implementation of any system that has so dramatically impacted the way we do business," says Dan Morreale, CIO at North Bronx Healthcare Network, one of six regional networks established by New York City Health and Hospitals Corp. Morreale is referring to the North Bronx's new $6 million picture archiving and communications software (PACS). The software captures, stores, and displays patient x-rays and other images in digital form.

You might wonder how a medical facility justifies a $6 million technology investment. Morreale could clearly show how the system would pay for itself in the matter of a few years. In his calculations, he showed that the new software would save North Bronx Healthcare Network almost $2 million per year by eliminating costs associated with film- and paper-based reports, such as:

- $1 million from reduced film-processing costs
- $400,000 from eliminating labor costs for film processing and storage
- $130,000 from lower real estate costs, because the organization no longer needs 5,000 square feet of floor space to store film
- $400,000 from eliminating manually produced reports

Not only could the software pay for itself in three years, but North Bronx would enjoy additional rewards that are more difficult to calculate.

The PACS implementation was the final step in digitizing all of North Bronx Healthcare Network's record keeping. Being able to advertise itself as fully digital helps a hospital to attract the best medical personnel. This advantage is particularly important in tight labor markets experiencing a shortage of good radiologists. The best professionals want to work with PACS because it allows them to be more effective in their work.

Digitizing records and images has many other advantages. Digital images can be sent to multiple physicians simultaneously. Traditional x-rays travel by courier from doctor to doctor, often keeping doctors and patients waiting for days or weeks for results. Physicians can manipulate and zoom in on the images to focus on medical problems. Digital x-rays are easy to copy and less likely to be lost. Also, through computer networks, doctors can share x-rays with experts around the world and benefit from consultations.

Although PACS have been used since the mid-1990s, about two-thirds of U.S. hospitals haven't purchased the system yet, estimates Jocelyn Young, a healthcare industry analyst at market research firm IDC. But the advantages of such software for cost savings, improved workflow, better patient

diagnoses, and a competitive advantage over other hospitals are sure to make these high priced, advanced software systems increasingly popular in years to come.

Discussion Questions

1. How can medical facilities save money by moving to digital x-rays?
2. How does switching from a traditional x-ray system to PACS improve the effectiveness of a medical facility?

Critical Thinking Questions

3. What other photography and imaging industries could benefit from the digitization of traditional paper-based processing?
4. How can PACS affect the medical services patients receive from multiple hospitals, doctors, and specialists?

SOURCES: John Webster, "Hospital Imaging Systems: A Tough Sell," *Computerworld*, February 2, 2004, www.computerworld.com; "North Bronx Healthcare Network to Reduce Digital Medical Image Management Costs with EMC Centera," *techinfocenter.com*, June 9, 2003, *www.techinfocenter.com*; AGFA Radiology Solutions Web site, *www.afga.com*, accessed February 8, 2004.

Case Two

Alaska Airlines Drops Sabre for Speed

Alaska Airlines, the ninth-largest airline in the United States, has traditionally used travel planning and pricing services from Fort Worth, Texas—based Sabre Holdings Corp. Sabre developed the first computerized airline reservation systems in the 1960s and has evolved over time to include three business units:

1. Travelocity, the most popular online travel service
2. Sabre Travel Network, which includes the world's largest global distribution system (GDS), connecting travel agents and travel suppliers with travelers
3. Sabre Airline Solutions, the leading provider of decision support tools, reservations systems, and consulting services for airlines

Sabre's evolution has not been swift enough, however, for Alaska Airlines and others who have opted to drop Sabre for software from ITA Software, Inc. ITA claims that its QPX software is the world's first modern airfare pricing, airfare shopping, and seat availability management system. Can this upstart company dethrone Sabre, the industry giant? ITA has already signed Orbitz, Air Canada, Alaska Airlines, America West Airlines, Continental Airlines, Galileo, Accovia, and others. What does QPX software provide that Sabre is unable to compete with?

Speed! QPX can do in seconds what Sabre does in minutes. ITA Software's CEO and founder, Jeremy Wertheimer, says, "It processes and confirms availability for [trip] pricing in less than one-tenth of a second, single airline airfare shopping queries in less than two seconds, and comprehensive airfare shopping queries across all airlines in less than 15 seconds." How does QPX software operate at these record-breaking transaction speeds? It makes use of two increasingly popular technologies and methodologies: (1) Linux-based, distributed servers and (2) object-oriented programming.

The IBM mainframe computers that Sabre uses are very slow by today's standards. Steve Jarvis, Alaska Airlines' vice president of e-commerce and distribution, says that when the airline was using the mainframe-based Sabre system, it often had to make more than 40 different data requests to produce one screen of itinerary options—"now we do it all with one trip to the data source." ITA's PC-server-based system runs algorithms that allow it to analyze airfares and routing options more efficiently.

The QPX software uses an object-oriented design to provide reliability to the software. Each component is programmed as an independent, self-contained object. The objects of the software interact, with the output of one acting as input for the next. Each object is tweaked for maximum efficiency, speed, and reliability. Rather than data requests competing with each other in one large software matrix, the object-oriented approach creates copies of the objects to handle each data request for maximum speed. Problems that may arise are more easily identified in the small module in which they occur. Debugging in this type of program architecture requires significantly less effort than large mainframe systems.

QPX also reduces costs. Jarvis says that a Web site, airline, travel agency, or reservation system using QPX can book and ticket any itinerary for a fraction of a dollar per ticket, compared with the $15 or more for bookings made through travel agents. Alaska Airlines earned $600 million in passenger revenue through its Web site in 2003, 30 percent of its total revenue. Using QPX software, it hopes to increase this amount to $1 billion in 2005.

Sabre has recognized the benefits of the technology implemented by ITA. In late 2001 it announced a similar effort to transition from IBM transaction processing mainframes. It estimated that its new system would take at least three years to complete. Asked why Alaska Airlines couldn't wait for Sabre's new system, Steve Jarvis replied, "We couldn't wait on Sabre. ITA's algorithms are widely regarded as the best in the industry, and we needed to move."

Discussion Questions

1. How can ITA's new travel-planning software assist Alaska Airlines in satisfying its passengers?
2. How can ITA's new travel-planning software assist Alaska Airlines in increasing its revenue from online sales from $600 million per year to $1 billion per year?

Critical Thinking Questions

3. Why do you think Steve Jarvis of Alaska Airlines believed his company couldn't wait for Sabre's new system and "needed to move"?
4. After Sabre upgrades its systems, what advantage will it have over ITA? What advantage does ITA have over Sabre even after Sabre upgrades its systems?

SOURCES: Dan Verton, "Alaska Airlines Opts for Linux-based Booking Engine," *Computerworld*, January 27, 2004, *www.computerworld.com*; Sabre Holdings Web site, *www.sabreholdings.com*, accessed February 10, 2004; ITA Software's Web site, *www.itasoftware.com*, accessed February 10, 2004.

Case Three

UPS Provides Shipping Management Software to Customers

Because of the popularity of the Web, and thanks to recent innovations in software development tools, many of today's software programs are written to run on Web servers using Web browsers as the user interface. Businesses are particularly fond of Web-delivered services, since they provide universal accessibility and are easy to deploy and maintain. UPS provides many Web-based software applications for its customers, the most recent of which is called Quantum View Manage.

Quantum View Manage is a component of the larger Quantum View system, which includes three Web-based applications that help businesses control their supply chains. Quantum View Manage is an application that allows UPS small package shippers to track package movements within their own supply chains. It is designed for customers who ship, on average, 30 or more packages per week and wish to track packages coming in and going out.

Quantum View Manage offers more convenience than traditional tracking software by putting shipping status information at users' fingertips. UPS spokeswoman Laurie Mallis says that one of the primary benefits of the software is that customers don't have to dig to find or enter tracking information. The software maintains and displays all shipping information for an organization in a number of useful report formats. Customers can configure the software to view shipment information for multiple accounts, so they can see when a package is processed, where it is in transit, when it arrives, or whether and why it is delayed, according to Mallis.

The software provides many features for ultimate customer flexibility. It shows both an inbound view of packages being shipped by your vendors and suppliers to designated locations and an outbound view about packages billed to your UPS account. "Customers can customize the information that they're receiving, because different departments within different companies have different needs," Mallis explained. For example, a customer service department might want to know whether a package was shipped or, if there's a delay, what the cause is. Shippers can be notified via e-mail, so they can see problems before their customers do, enabling better service. In addition, a shipper's finance department could use the confirmed delivery notice to automatically trigger an invoice and then view all outstanding C.O.D. orders. Some customers use Quantum View Manage to help manage inventory.

UPS customers seem to love this new free service. All 622 customers who tested the product have adopted it. In addition, another 300 customers are asking for it.

Quantum View Manage is an excellent example of how a company can leverage software to win a competitive advantage. UPS has invested heavily in software over the years and maintains its own international computer network. UPS employs over 4,450 technology employees and was voted as one of the best places to work in IT by *Computerworld* magazine.

Using automated and handheld bar-code scanners, UPS is able to track the 13 million-plus packages it handles each day. With this sophisticated tracking system and network in place, it was a simple matter for UPS to organize and deliver pertinent information to customers and assist them in managing the flow of goods through their supply chains.

Discussion Questions

1. What advantage does Web-based software provide to businesses such as UPS?
2. What risks are involved for companies that rely on Web-based software such as Quantum View Manage in managing their own businesses?

Critical Thinking Questions

3. UPS advertises its overall goal as "enabling commerce around the globe." How does Quantum View Manage assist UPS in achieving that goal?
4. Consider the components of a supply chain for a typical manufactured product. How could Quantum View Manage improve the management of shipping between supply chain units?

SOURCES: Linda Rosencrance, "UPS Launches Quantum View Manage," *Computerworld*, February 4, 2004, *www.computerworld.com*; "UPS Opens Supply Chain 'Window' with Quantum View Manage," *Business Wire*, February 4, 2004, *www.lexis-nexis.com*.

NOTES

Sources for the opening vignette: "Jones Lang LaSalle Selects Intuit MRI Real Estate Solutions for Its Operations in Hong Kong," *Business Wire*, February 2, 2004, *www.lexisnexis.com*; Intuit Web site, *www.intuit.com*, accessed February 7, 2004; Jones Lang LaSalle's Web site, *www.jones-langlasalle.com*, accessed February 7, 2004.

1. Prince, Marcelo, "SmartSignals' Software Searches for Trouble," *The Wall Street Journal*, September 3, 2003, p. B4A.
2. Mossberg, Walter, "Two Windows Programs Make Flawed Attempts to Catch up to iPhoto," *The Wall Street Journal*, January 9, 2003, p. B1.
3. Staff, "Meet The Inventor of CtrlAltDelete," *CNN Online*, January 29, 2004, *www.cnn.com*.
4. Husted, Bill, "Operating System Good Way to Coax New Life from Old PC," *Rocky Mountain News*, January 20. 2003, p. 6B.
5. Staff, "Microsoft Grapples with Leak of Source Code Online," *CNN Online*, February 13, 2004.
6. Legon, Jeordan, "Profanity, Partner's Name Hidden in Leaked Microsoft Code," *CNN Online*, February 13, 2004, *www.cnn.com*.
7. Howard, Bill, "Second-Generation Media Center Edition," *PC Magazine*, October 28, 2003, p. 43.
8. Canter, Sheryl, "Windows XP Error Reports," *PC Magazine*, February 3, 2004, p. 64.
9. Bacheldor, Beth, "Retail Innovation Starts at Store," *InformationWeek*, January 19, 2004, p. 45.
10. Miller, Michael, "Where Is Windows Going," *PC Magazine*, November 11, 2003, p. 97.
11. Spanbauer, Scott, "After Antitrust," *PC World*, May 2004, p. 30.
12. Mossberg, Walter, "How to Become a Rock Star," *The Wall Street Journal*, February 4, 2004, p. D1.
13. Brown, Mark, "Stars For An Hour," *Rocky Mountain News*, March 6, 2004, p. 1D.
14. Wildstrom, Stephen, "Apple Gets the Little Things Right," *BusinessWeek*, November 17, 2003, p. 30.
15. Mossberg, Walter, "If You're Getting Tired of Fighting Viruses, Consider a New Mac," *The Wall Street Journal*, October 23, 2003, p. B1.
16. Staff, "Travel Web Services Take an Open Source Route," *Computer Weekly*, April 8, 2003, p. 4.
17. Staff, "Panasonic to Supply Linux Development Platform and Operating System for Its Latest Broadband TV Tuner," *Electronic News*, April 21, 2003.
18. Staff, "CIOs Eyeing Open-Source Software," *Information Security*, January 2004, p. 22.
19. Bank, Rick, et al., "RedHat Seeks U.S. Ruling on Linux," *The Wall Street Journal*, August 5, 2003, p. B5.
20. Port, Otis, "Will This Feud Choke the Life Out of Linux?" *Business Week*, July 7, 2003, p. 81.
21. Bank, David, "Intel, IBM to Back New Linux Fund," *The Wall Street Journal*, January 12, 2004, p. B3.
22. Janowske, Davis, et al., "Taking Care of Small Business," *PC Magazine*, February 3, 2004, p. 12.
23. Dragan, Richard, "Windows Server 2003 Delivers Improvements All Around," *PC Magazine*, May 27, 2003, p. 28.
24. Clark, Don, "Microsoft Offers New Lock for Files," *The Wall Street Journal*, February 24, 2003.
25. Grimes, Brad, "Linux Goes to the Movies," *PC Magazine*, May 27, 2003, p. 70.
26. Krishnadas, K. C., "India Pursues Global Role in Embedded Software," *Electronic Engineering Times*, April 14, 2003, p. 23.
27. Drier, Troy, "Mac On The Go," *PC Magazine*, February 17, 2004, p. 35.
28. Spanbauer, Scott, "After Antitrust," *PC World*, May 2004, p. 30.
29. Brown, Bruce and Magre, "Redone Pocket PC OS Gets a New Name," *PC Magazine*, August 5, 2003, p. 32.
30. Muchmore, Michael, "System Utilities," *PC Magazine*, June 17, 2003, p. 122.
31. Staff, "Computers Predict Their Own Future," *ORMS Today*, April 2003, p. 20.
32. Anthes, Gary, "Grids Extend Reach," *Computerworld*, October 13, 2003, p. 29.
33. Spector, Lincoln, "The Trouble-Free PC", *PC World*, February 2004, p. 74.
34. Tynan, Daniel, "Natural-Born Killers," *PC World*, May 2003, p. 113.
35. Dragan, Richard, "Maintain Control of Business Content," *PC Magazine*, February 17, 2004, p. 42.
36. Staff, "Value Assessment," *MSI*, January 1, 2004, p. 51.
37. Staff, "Web Access Software," *Telecomworldwire*, January 12, 2004.
38. Richmond, Riva, "Do You Know Where Your Workers Are?" *The Wall Street Journal*, January 12, 2004, p. R1.
39. Staff, "Fraud Software Set to Take Off," *Post Magazine*, January 22, 2004, p. 2.
40. Staff, "Philips Software for Mobiles in Use by the BBC," *New Media Age*, January 29, 2004, p. 10.
41. Rodgers, Johnnah, "Field-Force Mobility," *Insurance & Technology*, February 1, 2004, p. 17.
42. Srinivasan, M., et al., "Woodward Aircraft Engine Systems Sets Work-In-Process Levels," *Interfaces*, July—August, 2003, p. 61.
43. Dragan, Richard, "Microsoft Outlook 2003," *PC Magazine*, February 17, 2004, p. 64.
44. Mendelson, Edward, "Microsoft Office 2003," *PC Magazine*, April 22, 2003, p. 34.
45. Sliwa, Carol, "Microsoft Expands Its Office Family by Two," *Computerworld*, March 10, 2003, p. 6.
46. O'Reilly, "The Other Office Suite," *PC World*, July 2003, p. 60.
47. Mendelson, Edward, "StarOffice 7 Makes A Run At Office," *PC Magazine*, February 3, 2004, p. 40.
48. Albro, Edward, "The Linux Experiment," *PC World*, February 2004, p. 105.
49. Graven, Matthew, "Microsoft Office 2003: A New Strategy," *PC Magazine*, October 28, 2003, p. 86.
50. Staff, "E Learning: The Challenges," *Computing*, May 1, 2003. pp. 29.
51. Forelle, Charles, "The Never-Ending Exam," *The Wall Street Journal*, May 1, 2003, p. B1.
52. Lipschutz, Robert, "Exchange 2003: More Approachable, More Affordable," *PC Magazine*, August 19, 2003, p. 37.
53. Staff, "Monsanto Puts a Friendly Face on Hard Software," *KMWorld*, February 2003, p. 28.
54. Crupi, Anthony, "Turning Muppets into Digital Assets," *Cable World*, April 14, 2003, p. 15.
55. Staff, "Landmark Theaters and Microsoft Create the Largest Digital Cinema in the United States," *Millimeter*, April 4, 2003.
56. Staff, "Digimations Art Contest," *Millimeter*, April 17, 2003.
57. Nash, Emma, "Camelot Installs New IBM System," *Computing*, May 1, 2003, pp. 1.
58. Aeberhard, John, "Fifty Years of Business Software," *Computing*, January 29, 2004, p. 8.
59. Staff, "Genomic Data Visualization Software," *Drug Discovery and Development*, February 1, 2003, p. 72.
60. Hecht, Jeff, "Talent-Spotting Software Predicts Cart Toppers," *New Scientist*, March 15, 2003, p. 17
61. Staff, "Software's Natural," *VARBusiness*, January 26, 2004, p. 46.
62. Babcock, Charles, "COBOL Enters the 21st Century," *Information Week*, May 5, 2003, p. 47.

63. Zak, Diane, "An Introduction to Programming with C++," *Course Technology*, 2003, p. 629.

64. King, Julia, "A Sunny Forecast for Open Source," *Computerworld*, April 26. 2004, p. 19.

65. Bank, David, "Microsoft to let Governments See Code," *The Wall Street Journal*, January 15, 2003, p. B5.

66. MacCormack, Alan, "The True Costs of Software," *Computerworld*, August 18, 2003, p. 44.

67. Jones, Philip, et al., "Managing the Seed-Corn Supply Chain," *Interfaces*, January—February 2003, p. 80. 4-52

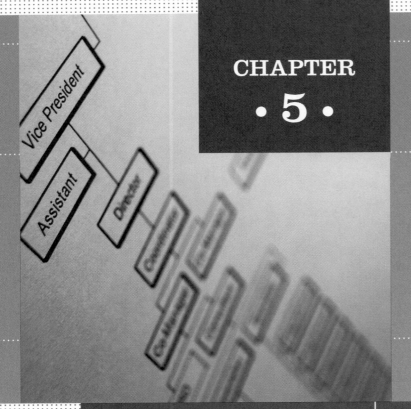

CHAPTER
• 5 •

Organizing Data
and Information

PRINCIPLES

- The database approach to data management provides significant advantages over the traditional file-based approach.

- A well-designed and well-managed database is an extremely valuable tool in supporting decision making.

- The number and types of database applications will continue to evolve and yield real business benefits.

LEARNING OBJECTIVES

- Define general data management concepts and terms, highlighting the advantages of the database approach to data management.

- Describe the relational database model and outline its basic features.

- Identify the common functions performed by all database management systems and identify popular end-user database management systems.

- Identify and briefly discuss current database applications.

INFORMATION SYSTEMS IN THE GLOBAL ECONOMY
DDB WORLDWIDE, UNITED STATES

Using Databases to Understand Gut Feelings

DDB is a global advertising agency based in New York, with 206 offices in 96 countries. As an advertising agency, DDB's primary objective is to generate ideas to sell its clients' products. To be successful, an advertiser must thoroughly understand prospective customers—what interests and motivates buyers. One of the founders of DDB, Bill Burnboch (the B in DDB), put it this way: "You can say the right thing about a product and nobody will listen. You've got to say it in such a way that people will feel it in their gut. Because if they don't feel it, nothing will happen."

To understand what customers feel "in their gut," every two years DDB conducts marketing research on 21,000 consumers in 23 countries. The participants respond to 600 questions relating to brand interests and purchases, lifestyle attitudes, interests, and opinions—even seemingly irrelevant information such as how well they think they'd do in a fist fight.

"Traditionally, a lot of advertising strategy was prepared based on things that were easy to manage, such as man-on-the-street interviews and focus groups," explains Janice Riggs, a project director at DDB. "While those are interesting and useful, they're not necessarily statistically valid." Riggs is in charge of DDB's Brand Capital project. Brand Capital is DDB's worldwide brand perception and consumer lifestyle project designed to help clients position and market their brands globally. Brand Capital generates mountains of data, which can only be analyzed and managed with a powerful database management system. DDB uses a database management system from leading database company SAS to "mine" the 12.5 million consumer responses for nuggets of valuable information.

By using sophisticated data-mining tools, DDB is able to turn terabytes of consumer data into useful business intelligence. An example of the types of valuable insights produced by DDB's Brand Capital is the discovery of a correlation between sports and greeting cards. DDB discovered that people who attend at least a dozen professional sporting events a year are also very likely to send greeting cards—useful information for a card company. DDB used this information as a springboard to design different greeting card advertising strategies to build on customers' loyalty with team sporting events, the team's city, and even local historical events to generate excitement about a product—something that's going to be talked about tomorrow.

The advertising industry has historically shied away from such intense and thorough marketing research because agencies don't have the time, talent, or money to turn raw data into intelligence and to present results in an easy-to-read report. Systems such as those provided by SAS are making reliable and thorough research convenient and affordable. The SAS system makes the centralized database available to DDB's 300 strategists, analysts, and client-services directors worldwide through an easy-to-use Web interface that requires no knowledge of code or complicated queries. "We've taken out the pain-in-the-neck part of it, and we free people up to apply creativity and imagination to the information they're given," Riggs says.

Since all DDB employees are working from the same information, they speak to clients with a unified voice. The easily accessible huge data store has significantly altered the way DDB employees work. When they sign a new client, rather than rushing off to create consumer focus groups to collect their opinions, DDB employees sit down at their computers and run queries on the database containing the responses that consumers have already provided in the target market and location.

Before implementing the SAS database system, a single market study request could take two days and cost a few hundred thousand dollars. "But now we turn those requests around instantaneously and at no extra cost," says Denny Merritt, the DDB software developer who helps develop software tools for interacting with brand and lifestyle data.

Being able to collect, store, and evaluate large quantities of consumer opinions has given DDB rare insight into human behavior and an edge over competing agencies, which are simply unable to provide the level of intelligence that DDB offers. "There is never a time when we do not find cool stuff in all of this," Riggs says. "People really are interesting."

As you read this chapter, consider the following:

- How do databases and database management systems allow businesses to do things that they couldn't previously do?
- What considerations and precautions are necessary when developing databases to ensure that they fully support an organization's requirements and lead, rather than mislead, the decision making process?

Why Learn About Database Systems?

A huge amount of data is entered into computer systems every day. Where does all this data go and how is it used? How can it help you on the job? In this chapter, you will learn about database management systems and how they can help you. If you become a marketing manager, you can have access to a vast store of data on existing and potential customers from surveys, their Web habits, and their past purchases from different stores. This information can help you sell products and services. If you become a corporate lawyer, you will have access to past cases and legal opinions from sophisticated legal databases. This information can help you win cases and protect your organization legally. If you become a human resource (HR) manager, you will be able to use databases to analyze the impact of raises, employee insurance benefits, and retirement contributions on long-term costs to your company. Regardless of your major in school, you likely will find that using database management systems will be a critical part of your job. You will see in this chapter how you can use data mining to get valuable information to help you succeed. We start this chapter by introducing basic concepts of database management systems.

Like other components of a computer-based information system, the overall objective of a database is to help an organization achieve its goals. A database can contribute to organizational success in a number of ways, including the ability to provide managers and decision makers with timely, accurate, and relevant information based on data. Databases also help companies generate information to reduce costs, increase profits, track past business activities, and open new market opportunities. For example, to achieve better customer satisfaction, Hilton Hotels is using its vast database system to customize service.[1] The new database system provides detailed information about Hilton customers. A receptionist at a Hilton Hotel in New York, for example, might apologize to a customer for not having her room cleaned up as she wanted during a recent stay at a Hilton in Orlando. The new system, called OnQ, allows Hilton to store and retrieve a tremendous amount of detailed customer satisfaction information. OnQ will also help Hilton make better decisions to meet customer needs. As a result of the data contained in OnQ, the receptionist at the Hilton in New York could

decide to offer a customer a special rate or provide additional service based on information from the customer's stay in Orlando. Hilton hopes that this information support will improve customers' stays and increase profits in the long run. The Panzano restaurant in downtown Denver uses a database to store customer information, such as birthday, anniversary, and preferences for certain types of food and beverages.[2]

The ability of an organization to gather data, interpret it, and act on it quickly can distinguish winners from losers in a highly competitive marketplace. In his book *Spinning Straw into Gold: The Magic of Turning Data into Gold*, John Miglautsch describes how database marketing can result in higher profits.[3] Miglautsch stresses the importance of marketing only to customers who have a high profit potential. A database can also help scientific organizations obtain their goals.[4] According to Joanna Batstone, director of IBM's Life Sciences Solutions Department, "Data can reside on different continents and departments. Not only do scientists generate more of their own data, they have to look at other people's as well." Britain, which has the largest national store of forensic DNA samples, will now have the largest DNA medical database. More than 20 universities will be involved in collecting DNA samples in England.[5] Databases can even help religious organizations. For example, with over a million entries, the American Theological Library Association database is one of the largest databases containing articles and references on religion and theology.[6] UPS, a large shipping company, uses a database to process almost 60 million database transactions every day.[7]

A database is critical for many aspects of an organization's information system. It provides an essential foundation for an organization's information and decision support system. Without a good database, executives and managers will not get the information they need to make good decisions. A database is also the foundation of most systems development projects. If the database is not designed properly, the systems development effort can be like a house of cards, collapsing under the weight of inaccurate and inadequate data. It is also critical to the success of an organization that database capabilities be aligned with the company's goals. Because data is so critical to an organization's success, many firms develop databases to help them access data more efficiently and use it more effectively. This typically requires a good database management system and a good database administrator.

A **database management system (DBMS)** consists of a group of programs that manipulate the database and provide an interface between the database and its users and other application programs. A DBMS is normally purchased from a database company. A DBMS provides a single point of management and control over data resources, which can be critical to maintaining the integrity and security of the data. A database, a DBMS, and the application programs that utilize the data in the database make up a database environment. A **database administrator (DBA)** is a skilled and trained IS professional who directs all activities related to an organization's database, including providing security from intruders.[8]

Understanding basic database system concepts can enhance your ability to use the power of a computerized database system to support IS and organizational goals. Direct-mail-order companies, for example, can use a DBMS to send out new catalogues to customers. These companies face the difficult question of how many times to send a new catalogue to current and potential customers.[9] Printing and mailing costs can be 25 percent or more of total expenses, so if a company has too many mailings, it costs more for the revenue generated. Yet, if it schedules too few mailings, sales could suffer. Deciding how many times to mail catalogues has a direct impact on profitability. German company Rhenania faced this important decision. The company, which sells CDs, books, and related products, mails catalogues to customers 18 times a year. The company used a database system to obtain important customer preference information. The information from the database was fed into a statistical analysis program, which revealed that profits could be maximized if the company sent out catalogues between 20 and 25 times each year. In contrast, Port City Metals Services in Tulsa, Oklahoma, uses a database management system to identify its best customers and to identify potential customers with the same characteristics.[10] The company fabricates steel parts for manufacturing companies.

database management system (DBMS)
A group of programs that manipulate the database and provide an interface between the database and the user of the database and other application programs.

database administrator (DBA)
A skilled IS professional who directs all activities related to an organization's database.

DATA MANAGEMENT

Without data and the ability to process it, an organization would not be able to successfully complete most business activities. It would not be able to pay employees, send out bills, order new inventory, or produce information to assist managers in decision making. As you recall, data consists of raw facts, such as employee numbers and sales figures. For data to be transformed into useful information, it must first be organized in a meaningful way.

The Hierarchy of Data

Data is generally organized in a hierarchy that begins with the smallest piece of data used by computers (a bit) and progresses through the hierarchy to a database. As discussed in Chapter 3, a bit (a binary digit) represents a circuit that is either on or off. Bits can be organized into units called *bytes*. A byte is typically 8 bits. Each byte represents a **character**, which is the basic building block of information. A character may consist of uppercase letters (A, B, C, . . . , Z), lowercase letters (a, b, c, . . . , z), numeric digits (0, 1, 2, . . . , 9), or special symbols (., !, [+], [-], /, . . .).

Characters are put together to form a field. A **field** is typically a name, number, or combination of characters that describes an aspect of a business object (e.g., an employee, a location, a truck) or activity (e.g., a sale). In addition to being entered into a database, fields can be computed from other fields. *Computed fields* include the total, average, maximum, and minimum value. A collection of related data fields is a **record**. By combining descriptions of various aspects of an object or activity, a more complete description of the object or activity is obtained. For instance, an employee record is a collection of fields about one employee. One field would be the employee's name, another her address, and still others her phone number, pay rate, earnings made to date, and so forth. A collection of related records is a **file**—for example, an employee file is a collection of all company employee records. Likewise, an inventory file is a collection of all inventory records for a particular company or organization. Some database software refers to files as tables.

At the highest level of this hierarchy is a *database*, a collection of integrated and related files. Together, bits, characters, fields, records, files, and databases form the **hierarchy of data** (see Figure 5.1). Characters are combined to make a field, fields are combined to make a record, records are combined to make a file, and files are combined to make a database. A database houses not only all these levels of data but the relationships among them.

character
Basic building block of information, consisting of uppercase letters, lowercase letters, numeric digits, or special symbols.

field
Typically a name, number, or combination of characters that describes an aspect of a business object or activity.

record
A collection of related data fields.

file
A collection of related records.

hierarchy of data
Bits, characters, fields, records, files, and databases.

Figure 5.1

The Hierarchy of Data

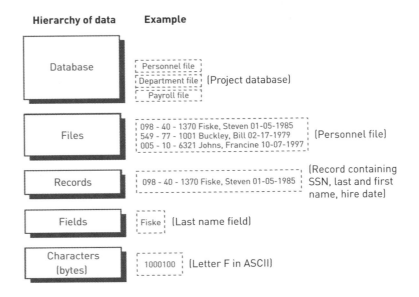

Data Entities, Attributes, and Keys

Entities, attributes, and keys are important database concepts. An **entity** is a generalized class of people, places, or things (objects) for which data is collected, stored, and maintained. Examples of entities include employees, inventory, and customers. Most organizations organize and store data as entities.

An **attribute** is a characteristic of an entity. For example, employee number, last name, first name, hire date, and department number are attributes for an employee (see Figure 5.2). Inventory number, description, number of units on hand, and the location of the inventory item in the warehouse are examples of attributes for items in inventory. Customer number, name, address, phone number, credit rating, and contact person are examples of attributes for customers. Attributes are usually selected to capture the relevant characteristics of entities such as employees or customers. The specific value of an attribute, called a **data item**, can be found in the fields of the record describing an entity.

KEY FIELD				
Employee #	Last name	First name	Hire date	Dept. number
005-10-6321	Johns	Francine	10-07-1997	257
549-77-1001	Buckley	Bill	02-17-1979	632
098-40-1370	Fiske	Steven	01-05-1985	598

ENTITIES (records)

ATTRIBUTES (fields)

Attributes and data items are used by most organizations. Many governments use attributes and data items to help identify and locate possible terrorists. As a part of its Homeland Security effort, the U.S. State Department is sharing its database of 50 million visa applications with the FBI to help identify possible terrorists.[11] Partially as a response to the threat of terrorism, some governments are increasingly using databases to track and prevent unwanted people from entering their country. The U.S. government, for example, is using fingerprinting databases to track tens of thousands of suspected terrorists or visitors of "national security concern" as they enter the country.[12] The database contains visual images of fingerprints. The National Security Entry-Exit Registration System attempts to close the borders to suspected terrorists by comparing the fingerprints of entering visitors against a comprehensive database. During the first year, about 200,000 visitors are expected to be fingerprinted and screened by the system. Sharing attributes and data items can also be a critical factor in coordinating responses across diverse functional areas of an organization.

As discussed, a collection of fields about a specific object is a record. A **key** is a field or set of fields in a record that is used to identify the record. A **primary key** is a field or set of fields that uniquely identifies the record. No other record can have the same primary key. The primary key is used to distinguish records so that they can be accessed, organized, and manipulated. For an employee record, such as the one shown in Figure 5.2, the employee number is an example of a primary key.

Locating a particular record that meets a specific set of criteria may require the use of a combination of secondary keys. For example, a customer might call a mail-order company to place an order for clothes. If the customer does not know his primary key (such as a customer number), a secondary key (such as last name) can be used. In this case, the order clerk enters the last name, such as Adams. If there are several customers with a last name of Adams, the clerk can check other fields, such as address, first name, and so on, to find the correct customer record. Once the correct customer record is obtained, the order can be completed and the clothing items shipped to the customer.

entity
Generalized class of people, places, or things for which data is collected, stored, and maintained.

attribute
A characteristic of an entity.

data item
The specific value of an attribute.

Figure 5.2

Keys and Attributes

The key field is the employee number. The attributes include last name, first name, hire date, and department number.

key
A field or set of fields in a record that is used to identify the record.

primary key
A field or set of fields that uniquely identifies the record.

The Traditional Approach Versus the Database Approach

Since the first use of computers to perform routine business functions in the 1950s, companies have used the traditional approach to process their transactions. This approach is based on using separate files for each application, such as payroll and billing. While some companies still use the traditional approach, most organizations today use the database approach, which utilizes a unified and integrated database for most or all of a company's transactions. In this section, we will explore both the traditional and database approach.

The Traditional Approach

One of the most basic ways to manage data is via files. Because a file is a collection of related records, all records associated with a particular application (and therefore related by the application) can be collected and managed together in an application-specific file. At one time, most organizations had numerous application-specific data files; for example, customer records often were maintained in separate files, with each file relating to a specific process completed by the company, such as shipping or billing. This approach to data management, whereby separate data files are created and stored for each application program, is called the **traditional approach to data management**. For each particular application, one or more data files are created (see Figure 5.3).

traditional approach to data management
An approach whereby separate data files are created and stored for each application program.

Figure 5.3

The Traditional Approach to Data Management

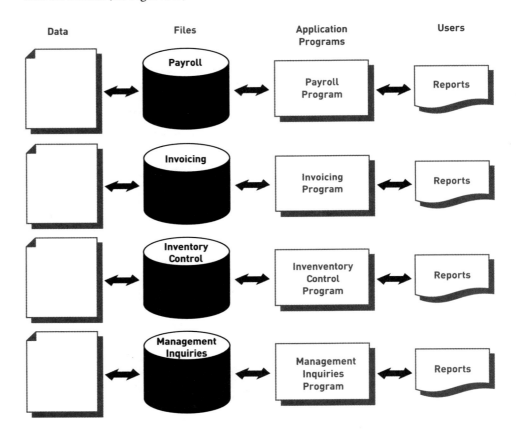

With the traditional approach, one or more data files are created and used for every application. For example, the inventory control program can have one or more files containing inventory data, such as the inventory item, number on hand, and item description. Likewise, the invoicing program can have files on customers, inventory items being shipped, and so on. With the traditional approach to data management, it is possible to have the same data, such as inventory items, in several different files used by different applications.

One of the flaws in this traditional file-oriented approach to data management is that much of the data—for example, customer name and address—is duplicated in two or more files. This duplication of data in separate files is known as **data redundancy**. The problem with data redundancy is that changes to the data (e.g., a new customer address) might be

data redundancy
Duplication of data in separate files.

made in one file and not another. The order-processing department might have updated its file to the new address, but the billing department might still be sending bills to the old address. Data redundancy, therefore, conflicts with **data integrity**—the degree to which the data in any one file is accurate. Data integrity follows from the control or elimination of data redundancy. Keeping a customer's address in only one file decreases the possibility that the customer will have two different addresses stored in different locations. The efficient operation of a business requires a high degree of data integrity, which is one of the advantages of the database approach.

The Database Approach

Because of the problems associated with the traditional approach to data management, many managers wanted a more efficient and effective means of organizing data. The result was the **database approach to data management**. In a database approach, a pool of related data is shared by multiple application programs. Rather than having separate data files, each application uses a collection of data that is either joined or related in the database.

The database approach offers significant advantages over the traditional file-based approach. One is that by controlling data redundancy, the database approach can use storage space more efficiently and increase data integrity. The database approach can also provide an organization with increased flexibility in the use of data. Because data once kept in two files is now located in the same database, it is easier to locate and request data for many types of processing. A database also offers the ability to share data and information resources. This can be a critical factor in coordinating organization-wide responses across diverse functional areas of a corporation.

To use the database approach to data management, additional software—a database management system (DBMS)—is required. As previously discussed, a DBMS consists of a group of programs that can be used as an interface between a database and the user or the database and application programs. Typically, this software acts as a buffer between the application programs and the database itself. Figure 5.4 illustrates the database approach.

data integrity
The degree to which the data in any one file is accurate.

database approach to data management
An approach whereby a pool of related data is shared by multiple application programs.

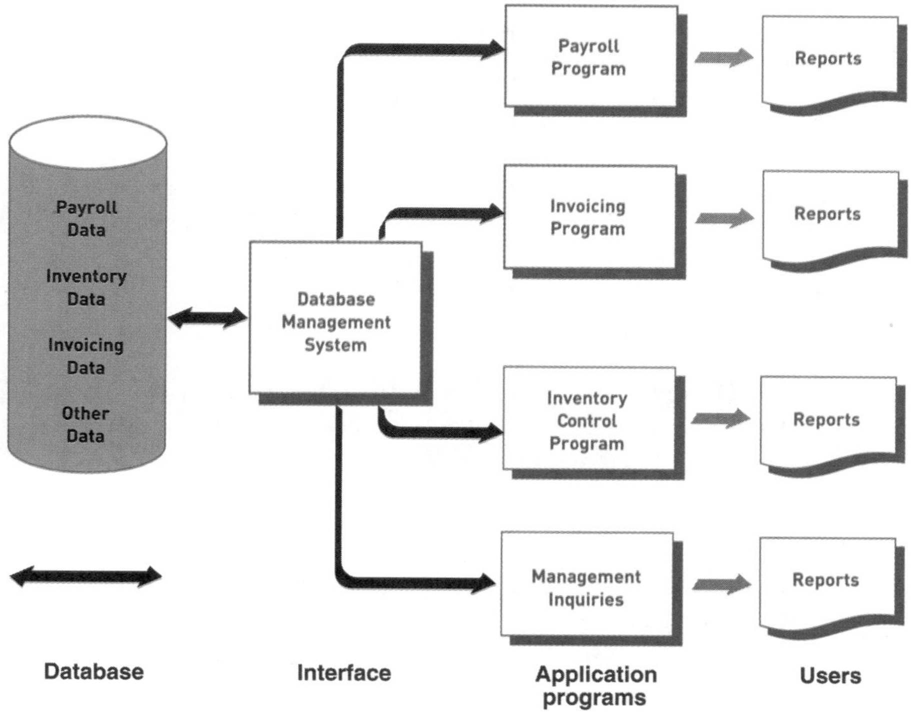

Figure 5.4

The Database Approach to Data Management

Table 5.1

Advantages of the Database Approach

Table 5.1 lists some of the primary advantages of the database approach. There are, however, disadvantages (see Table 5.2).

Advantages	Explanation
Improved strategic use of corporate data	Accurate, complete, up-to-date data can be made available to decision makers where, when, and in the form they need it.
Reduced data redundancy	The database approach can reduce or eliminate data redundancy. Data is organized by the DBMS and stored in only one location. This results in more efficient utilization of system storage space.
Improved data integrity	With the traditional approach, some changes to data were not reflected in all copies of the data kept in separate files. This is prevented with the database approach because there are no separate files that contain copies of the same piece of data.
Easier modification and updating	With the database approach, the DBMS coordinates updates and data modifications. Programmers and users do not have to know where the data is physically stored. Data is stored and modified once. Modification and updating is also easier because the data is stored at only one location in most cases.
Data and program independence	The DBMS organizes the data independently of the application program. With the database approach, the application program is not affected by the location or type of data. Introduction of new data types not relevant to a particular application does not require the rewriting of that application to maintain compatibility with the data file.
Better access to data and information	Most DBMSs have software that makes it easy to access and retrieve data from a database. In most cases, simple commands can be given to get important information. Relationships between records can be more easily investigated and exploited, and applications can be more easily combined.
Standardization of data access	A primary feature of the database approach is a standardized, uniform approach to database access. This means that the same overall procedures are used by all application programs to retrieve data and information.
A framework for program development	Standardized database access procedures can mean more standardization of program development. Because programs go through the DBMS to gain access to data in the database, standardized database access can provide a consistent framework for program development. In addition, each application program need address only the DBMS, not the actual data files, reducing application development time.
Better overall protection of the data	The use of and access to centrally located data are easier to monitor and control. Security codes and passwords can ensure that only authorized people have access to particular data and information in the database, thus ensuring privacy.
Shared data and information resources	The cost of hardware, software, and personnel can be spread over a large number of applications and users. This is a primary feature of a DBMS.

Table 5.2

Disadvantages of the Database Approach

Disadvantages	Explanation
More complexity	Database management systems can be difficult to set up and operate. Many decisions must be made correctly for the database management system to work effectively. In addition, users have to learn new procedures to take full advantage of a database management system.
More difficult to recover from a failure	With the traditional approach to file management, a failure of a file only affects a single program. With a database management system, a failure can shut down the entire database.
More expensive	Database management systems can be more expensive to purchase and operate. The expense includes the cost of the database and specialized personnel, such as a database administrator, who is needed to design and operate the database.

Many modern databases are enterprise-wide, encompassing much of the data of the entire organization. Often, distinct yet related databases are linked to provide enterprise-wide databases. Much planning and organization go into the development of such databases. For example, Best Buy is a specialty retailer of consumer electronics, personal computers, entertainment software, and appliances. It operates nearly 2,000 retail stores and commercial Web sites under the names Best Buy, Magnolia Hi-Fi, Media Play, On Cue, Sam Goody, and Suncoast. Best Buy uses information about the business and its customers to tailor the product mix to its customer base, minimize the time items are held in inventory to reduce costs, and respond quickly to customer needs. At the center of this strategic information is a database, which consolidates information from about 350 different sources across the enterprise.

TA MODELING AND THE RELATIONAL DATABASE MODEL

Because there are so many elements in today's businesses, it is critical to keep data organized so that it can be used effectively. A database should be designed to store all data relevant to the business and provide quick access and easy modification. Moreover, it must reflect the business processes of the organization. When building a database, an organization must carefully consider these questions:

- *Content:* What data should be collected and at what cost?
- *Access:* What data should be provided to which users and when?
- *Logical structure:* How should data be arranged so that it makes sense to a given user?
- *Physical organization:* Where should data be physically located?

Data Modeling

Key considerations in organizing data in a database include determining what data is to be collected in the database, who will have access to it, and how they might wish to use the data. Once these determinations are made, a database can then be created. Building a database requires two different types of designs: a logical design and a physical design. The *logical design* of a database shows an abstract model of how the data should be structured and arranged to meet an organization's information needs. The logical design of a database involves identifying relationships among the different data items and grouping them in an orderly fashion. Because databases provide both input and output for information systems throughout a business, users from all functional areas should assist in creating the logical design to ensure that their needs are identified and addressed. *Physical design* starts from the logical database design and fine-tunes it for performance and cost considerations (e.g., improved response time, reduced storage space, lower operating cost). The person who fine-tunes the physical design must have an in-depth knowledge of the DBMS to implement the database. For example, the logical database design may need to be altered so that certain data entities are combined, summary totals are carried in the data records rather than calculated from elemental data, and some data attributes are repeated in more than one data entity. These are examples of **planned data redundancy**, which is done to improve the system performance so that user reports or queries can be created more quickly.

One of the tools database designers use to show the logical relationships among data is a data model. A **data model** is a diagram of entities and their relationships. Data modeling usually involves understanding a specific business problem and analyzing the data and information needed to deliver a solution. When done at the level of the entire organization, this is called *enterprise data modeling*. **Enterprise data modeling** is an approach that starts by investigating the general data and information needs of the organization at the strategic level and then examining more specific data and information needs for the various functional areas and departments within the organization. Various models have been developed to help managers and database designers analyze data and information needs. An entity-relationship diagram is an example of such a data model.

planned data redundancy
A way of organizing data in which the logical database design is altered so that certain data entities are combined, summary totals are carried in the data records rather than calculated from elemental data, and some data attributes are repeated in more than one data entity to improve database performance.

data model
A diagram of data entities and their relationships.

enterprise data modeling
Data modeling done at the level of the entire enterprise.

entity-relationship (ER) diagrams
Data models that use basic graphical symbols to show the organization of and relationships between data.

Entity-relationship (ER) diagrams use basic graphical symbols to show the organization of and relationships between data. In most cases, boxes are used in ER diagrams to indicate data items or entities contained in data tables, and diamonds show relationships between data items and entities. In other words, ER diagrams are used to show data items in tables (entities) and the ways they are related.

ER diagrams help ensure that the relationships among the data entities in a database are correctly structured so that any application programs developed are consistent with business operations and user needs. In addition, ER diagrams can serve as reference documents once a database is in use. If changes are made to the database, ER diagrams help design them. Figure 5.5 shows an ER diagram for an order database. In this database design, one salesperson serves many customers. This is an example of a one-to-many relationship, as shown by the one-to-many symbol ("crow's-foot") shown in Figure 5.5. The ER diagram also shows that each customer can place one-to-many orders, each order includes one-to-many line items, and many line items can specify the same product (a many-to-one relationship). There can also be one-to-one relationships. For example, one order generates one invoice.

Figure 5.5

An Entity-Relationship (ER) Diagram for a Customer Order Database

Development of ER diagrams helps ensure that the logical structure of application programs is consistent with the data relationships in the database.

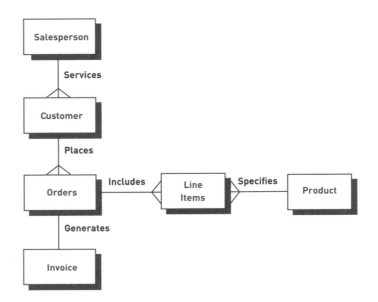

The Relational Database Model

relational model
A database model that describes data in which all data elements are placed in two-dimensional tables, called *relations*, that are the logical equivalent of files.

Although there are a number of different database models, including flat files, hierarchical, and network models, the **relational model** has become the most popular database model, and use of this model will increase in the future. The relational model describes data using a standard tabular format. In a database structured according to the relational model, all data elements are placed in two-dimensional tables, called *relations*, that are the logical equivalent of files. The tables in relational databases organize data in rows and columns, simplifying data access and manipulation. It is normally easier for managers to understand the relational model (see Figure 5.6) than other database models. Databases based on the relational model include DB2 from IBM, Oracle, Sybase, SQL Server and Access from Microsoft, and MySQL.

In the relational model, each row of a table represents a data entity, with the columns of the table representing attributes. Each attribute can take on only certain values. The allowable values for these attributes are called the **domain**. The domain for a particular attribute indicates what values can be placed in each of the columns of the relational table. For instance, the domain for an attribute such as gender would be limited to male or female. A domain for pay rate would not include negative numbers. Defining a domain can increase data accuracy. For example, a pay rate of −$5.00 could not be entered into the database because it is a negative number and not in the domain for pay rate.

domain
The allowable values for data attributes.

Data Table 1: Project Table

Project	Description	Dept. Number
155	Payroll	257
498	Widgets	632
226	Sales Manual	598

Data Table 2: Department Table

Dept.	Dept. Name	Manager SSN
257	Accounting	005-10-6321
632	Manufacturing	549-77-1001
598	Marketing	098-40-1370

Data Table 3: Manager Table

SSN	Last Name	First Name	Hire Date	Dept. Number
005-10-6321	Johns	Francine	10-07-1997	257
549-77-1001	Buckley	Bill	02-17-1979	632
098-40-1370	Fiske	Steven	01-05-1985	598

Figure 5.6

A Relational Database Model

In the relational model, all data elements are placed in two-dimensional tables, or relations. As long as they share at least one common element, these relations can be linked to output useful information.

Manipulating Data

Once data has been placed into a relational database, users can make inquiries and analyze data. Basic data manipulations include selecting, projecting, and joining. **Selecting** involves eliminating rows according to certain criteria. Suppose a project table contains the project number, description, and department number for all projects being performed by a company. The president of the company might want to find the department number for Project 226, a sales manual project. Using selection, the president can eliminate all rows but the one for Project 226 and see that the department number for the department completing the sales manual project is 598.

Projecting involves eliminating columns in a table. For example, we might have a department table that contains the department number, department name, and Social Security number (SSN) of the manager in charge of the project. The sales manager might want to create a new table with only the department number and the Social Security number of the manager in charge of the sales manual project. Projection can be used to eliminate the department name column and create a new table containing only department number and SSN.

Joining involves combining two or more tables. For example, we can combine the project table and the department table to get a new table with the project number, project description, department number, department name, and Social Security number for the manager in charge of the project.

As long as the tables share at least one common data attribute, the tables in a relational database can be **linked** to provide useful information and reports. Being able to link tables to each other through common data attributes is one of the keys to the flexibility and power of relational databases. Suppose the president of a company wants to find out the name of the manager of the sales manual project and the length of time the manager has been with

selecting
Data manipulation that eliminates rows according to certain criteria.

projecting
Data manipulation that eliminates columns in a table.

joining
Data manipulation that combines two or more tables.

linking
Data manipulation that combines two or more tables using common data attributes to form a new table with only the unique data attributes.

the company. Assume that the company has the manager, department, and project tables shown in Figure 5.6. A simplified ER diagram showing the relationship between these tables is shown in Figure 5.7. Note the crow's-foot by the project table. This indicates that a department can have many projects. The president would make the inquiry to the database, perhaps via a personal computer. The DBMS would start with the project description and search the project table to find out the project's department number. It would then use the department number to search the department table for the manager's Social Security number. The department number is also in the department table and is the common element that allows the project table and the department table to be linked. The DBMS then uses the manager's Social Security number to search the manager table for the manager's hire date. The manager's Social Security number is the common element between the department table and the manager table. The final result: the manager's name and hire date are presented to the president as a response to the inquiry (see Figure 5.8).

A Simplified ER Diagram Showing the Relationship between the Manager, Department, and Project Tables

Linking Data Tables to Answer an Inquiry

In finding the name and hire date of the manager working on the sales manual project, the president needs three tables: project, department, and manager. The project description (Sales Manual) leads to the department number (598) in the project table, which leads to the manager's SSN (098-40-1370) in the department table, which leads to the manager's name (Fiske) and hire date (01-05-1985) in the manager table.

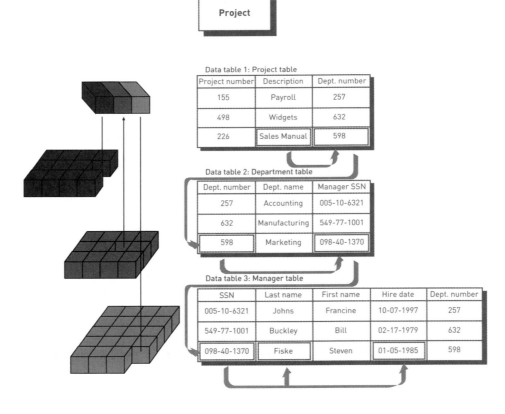

One of the primary advantages of a relational database is that it allows tables to be linked, as shown in Figure 5.8. This linkage is especially useful when information is needed from multiple tables, as in our example. The manager's Social Security number, for example, is maintained in the manager table. If the Social Security number is needed, it can be obtained by linking to the manager table.

The relational database model is by far the most widely used. It is easier to control, more flexible, and more intuitive than other approaches because it organizes data in tables. As seen in Figure 5.9, a relational database management system, such as Access, provides a number of tips and tools for building and using database tables. This figure shows the database displaying information about data types and indicating that additional help is available. The ability to link relational tables also allows users to relate data in new ways without having to redefine complex relationships. Because of the advantages of the relational model, many companies use it for large corporate databases, such as those for marketing and accounting. The relational model can be used with personal computers and mainframe systems. A travel reservation company, for example, can develop a fare-pricing system by using relational database technology that can handle millions of daily queries from online travel companies, such as Expedia, Travelocity, and Orbitz.

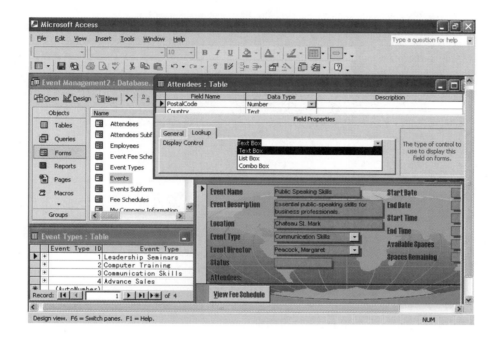

Figure 5.9

Building and Modifying a Relational Database

Relational databases provide many tools, tips, and tricks to simplify the process of creating and modifying a database.

Data Cleanup

As discussed in Chapter 1, the characteristics of valuable data include that the data is accurate, complete, economical, flexible, reliable, relevant, simple, timely, verifiable, accessible, and secure. The purpose of **data cleanup** is to develop data with these characteristics. A database can contain errors. For example, a survey of a thousand electric-utility companies found that customer databases were only 45.6 percent accurate.[13] The errors were caused by inaccurate data entry. When a database is created with data from multiple sources, those disparate sources may store different values for the same customer due to spelling errors, multiple account numbers, and address variations.

Consider a database for a fitness center designed to track member dues. The table contains the attributes name, phone number, gender, dues paid, and date paid (see Table 5.3). As the records in Table 5.3 show, Anita Brown and Sim Thomas have paid their dues in September. Sim has paid his dues in two installments. Note that no primary key uniquely identifies each record. As we will see next, this problem must be corrected.

Because Sim Thomas has paid dues twice in September, the data in the database is now redundant. The name, phone number, and gender for Thomas are repeated in two records. Notice that the data in the database is also inconsistent: Thomas has changed his phone number, but only one of the records reflects this change. Further reducing this database's reliability is the fact that no primary key exists to uniquely identify Sim Thomas's record. The first Thomas could be Sim Thomas, but the second might be Steve Thomas. These problems and irregularities in data are called *anomalies*. Data anomalies often result

data cleanup

The process of looking for and fixing inconsistencies to ensure that data is accurate and complete.

in incorrect information, causing database users to be misinformed about actual conditions. Anomalies must be corrected.

Table 5.3

Fitness Center Dues

Name	Phone	Gender	Dues Paid	Date Paid
Brown, A	468-3342	Female	$30	September 15th
Thomas, S.	468-8788	Male	$15	September 15th
Thomas, S.	468-5238	Male	$15	September 25th

To solve these problems in the fitness center's database, we can add a primary key, called *member number*, and put the data into two tables: a Fitness Center Members table with gender, phone number, and related information, and a Dues Paid table with dues paid and date paid (see Tables 5.4 and 5.5). As you can see, both tables include the member number attribute so that they can be linked.

Table 5.4

Fitness Center Members

Member No.	Name	Phone	Gender
SN123	Brown, A	468-3342	Female
SN656	Thomas, S.	468-5238	Male

Table 5.5

Dues Paid

Member No.	Dues Paid	Date Paid
SN123	$30	September 15th
SN656	$15	September 15th
SN656	$15	September 25th

With the relations in Table 5.4 and Table 5.5, we have reduced the redundancy and eliminated the potential problem of having two different phone numbers for the same member. Also note that the member number gives each record in the Fitness Center Members table a primary key. Because we have two payment entries ($15 each) listed with the same member number (SN656), we know that one person made the payments, not two different people. Formalized approaches, such as *database normalization*, are often used to clean up problems with data.

DATABASE MANAGEMENT SYSTEMS (DBMS)

Creating and implementing the right database system ensures that the database will support both business activities and goals. But how do we actually create, implement, use, and update a database? The answer is found in the database management system. As discussed earlier, a database management system (DBMS) is a group of programs used as an interface between a database and application programs or a database and the user. The capabilities and types of database systems, however, vary considerably.

Overview of Database Types

Database management systems can range from small, inexpensive software packages to sophisticated systems costing hundreds of thousands of dollars. Following are a few popular alternatives.

Flat File

A flat file is a simple database program that has no relationship between its records and is often used to store and manipulate a single table or file. Flat files do not use any of the database models discussed previously, such as the relational model. Many spreadsheet and word processing programs have flat file capabilities. These software packages can sort tables and make simple calculations and comparisons. OneNote, developed by Microsoft in 2003, was designed to let people put ideas, thoughts, and notes into a computer file.[14] Each can be placed anywhere on a page in a OneNote file. Ideas, thoughts, and notes can also be placed in a box on a page, called a *container*. Pages are organized into sections and subsections that appear as colored tabs. Once entered, the ideas, thoughts, and items in a OneNote file can be retrieved, copied, and pasted into other applications, such as word processing and spreadsheet programs.

Single User

Databases for personal computers are most often meant for a single user. Only one person can use the database at a time. Access and Quicken are examples of popular single-user DBMSs, through which users store and manipulate financial data. Microsoft's InfoPath is another example of a single-user database.[15] The database is part of Microsoft's Office suite, and it helps people collect and organize information from a variety of sources. InfoPath has built-in forms that can be used to enter expense information, time-sheet data, and a variety of other information.

Multiple Users

Large mainframe computer systems need multiuser DBMSs. These more powerful, expensive systems allow dozens or hundreds of people to access the same database system at the same time. Popular vendors for multiuser database systems include Oracle, Sybase, and IBM.

All DBMSs share some common functions, such as providing a user view, physically storing and retrieving data in a database, allowing for database modification, manipulating data, and generating reports. These DBMSs are capable of handling the most complex of data-processing tasks. A medical clinic, for example, can use a database like IBM's DB2 to develop an all-inclusive database containing patient records, physician notes, demographic data, genetic data, and proteomic (protein-related) data for millions of patients.

Providing a User View

Because the DBMS is responsible for access to a database, one of the first steps in installing and using a database involves telling the DBMS the logical and physical structure of the data and relationships among the data in the database. This description is called a **schema** (as in schematic diagram). A schema can be part of the database or a separate schema file. The DBMS can reference a schema to find where to access the requested data in relation to another piece of data.

schema
A description of the entire database.

A DBMS also acts as a user interface by providing a view of the database. A user view is the portion of the database a user can access. To create different user views, subschemas are developed. A **subschema** is a file that contains a description of a subset of the database and identifies which users can view and modify the data items in that subset. While a schema is a description of the entire database, a subschema shows only some of the records and their relationships in the database. Normally, programmers and managers need to view or access only a subset of the database. For example, a sales representative might need only data describing customers in his region, not the sales data for the entire nation. A subschema could be used to limit his view to data from his region. With subschemas, the underlying structure of the database can change, but the view the user sees might not change. For example, even if all the data on the southern region changed, the northeast region sales representative's view would not change if he accessed data on his region.

subschema
A file that contains a description of a subset of the database and identifies which users can view and modify the data items in the subset.

A number of subschemas can be developed for different users and the various application programs. Typically, the database user or application will access the subschema, which then accesses the schema (see Figure 5.10). Subschemas can also provide additional security because programmers, managers, and other users are typically allowed to view only certain parts of the database.

The Use of Schemas and Subschemas

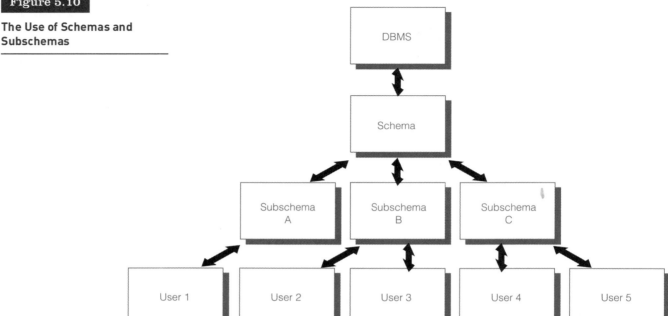

Creating and Modifying the Database

data definition language (DDL)
A collection of instructions and commands used to define and describe data and data relationships in a specific database.

Schemas and subschemas are entered into the DBMS (usually by database personnel) via a data definition language. A **data definition language (DDL)** is a collection of instructions and commands used to define and describe data and data relationships in a specific database. A DDL allows the database's creator to describe the data and the data relationships that are to be contained in the schema and the many subschemas. In general, a DDL describes logical access paths and logical records in the database. Figure 5.11 shows a simplified example of a DDL used to develop a general schema. The Xs in Figure 5.11 reveal where specific

Using a Data Definition Language to Define a Schema

```
SCHEMA DESCRIPTION
SCHEMA NAME IS XXXX
AUTHOR        XXXX
DATE          XXXX
FILE DESCRIPTION
      FILE NAME IS XXXX
        ASSIGN XXXX
      FILE NAME IS XXXX
        ASSIGN XXXX
AREA DESCRIPTION
    AREA NAME IS XXXX
RECORD DESCRIPTION
    RECORD NAME IS XXXX
    RECORD ID IS XXXX
    LOCATION MODE IS XXXX
    WITHIN XXXX AREA FROM XXXX THRU XXXX
SET DESCRIPTION
    SET NAME IS XXXX
    ORDER IS XXXX
    MODE IS XXXX
    MEMBER IS XXXX
    .
    .
    .
```

information concerning the database is to be entered. File description, area description, record description, and set description are terms the DDL defines and uses in this example. Other terms and commands can be used, depending on the particular DBMS employed.

Another important step in creating a database is to establish a **data dictionary**, a detailed description of all data used in the database. The data dictionary contains the name of the data item, aliases or other names that may be used to describe the item, the range of values that can be used, the type of data (such as alphanumeric or numeric), the amount of storage needed for the item, a notation of the person responsible for updating it and the various users who can access it, and a list of reports that use the data item. A data dictionary can also include a description of data flows, the way records are organized, and data-processing requirements. Figure 5.12 shows a typical data dictionary entry.

data dictionary
A detailed description of all the data used in the database.

```
            NORTHWESTERN MANUFACTURING

PREPARED BY:          D. BORDWELL
DATE:                 04 AUGUST 2005
APPROVED BY:          J. EDWARDS
DATE:                 13 OCTOBER 2005
VERSION:              3.1
PAGE:                 1 OF 1

DATA ELEMENT NAME:    PARTNO
DESCRIPTION:          INVENTORY PART NUMBER
OTHER NAMES:          PTNO
VALUE RANGE:          100 TO 5000
DATA TYPE:            NUMERIC
POSITIONS:            4 POSITIONS OR COLUMNS
```

Figure 5.12

A Typical Data Dictionary Entry

For example, the information in a data dictionary for the part number of an inventory item can include the name of the person who made the data dictionary entry (D. Bordwell), the date the entry was made (August 4, 2005), the name of the person who approved the entry (J. Edwards), the approval date (October 13, 2005), the version number (3.1), the number of pages used for the entry (1), the part name (PARTNO), other part names that may be used (PTNO), the range of values (part numbers can range from 100 to 5000), the type of data (numeric), and the storage required (four positions are required for the part number). The following are some of the typical uses of a data dictionary.

- *Provide a standard definition of terms and data elements.* This standardization can help in programming by providing consistent terms and variables to be used for all programs. Programmers know what data elements are already "captured" in the database and how they relate to other data elements.
- *Assist programmers in designing and writing programs.* Programmers do not need to know which storage devices are used to store needed data. Using the data dictionary, programmers specify the required data elements. The DBMS locates the necessary data. More important, programmers can use the data dictionary to see which programs already use a piece of data and, if appropriate, can copy the relevant section of the program code into their new program, thus eliminating duplicate programming efforts.
- *Simplify database modification.* If for any reason a data element needs to be changed or deleted, the data dictionary would point to specific programs that utilize the data element that may need modification.

A data dictionary helps achieve the advantages of the database approach in these ways:

- *Reduced data redundancy.* By providing standard definitions of all data, it is less likely that the same data item will be stored in different places under different names. For example, a data dictionary would reduce the likelihood that the same part number would be stored as two different items, such as PTNO and PARTNO.

- *Increased data reliability.* A data dictionary and the database approach reduce the chance that data will be destroyed or lost. In addition, it is more difficult for unauthorized people to gain access to sensitive data and information.
- *Faster program development.* With a data dictionary, programmers can develop programs faster. They don't have to develop names for data items because the data dictionary does that for them.
- *Easier modification of data and information.* The data dictionary and the database approach make modifications to data easier because users do not need to know where the data is stored. The person making the change indicates the new value of the variable or item, such as part number, that is to be changed. The database system locates the data and makes the necessary change.

Storing and Retrieving Data

As just described, one function of a DBMS is to be an interface between an application program and the database. When an application program needs data, it requests that data through the DBMS. Suppose that to calculate the total price of a new car, an auto dealer pricing program needs price data on the engine option—six cylinders instead of the standard four cylinders. The application program thus requests this data from the DBMS. In doing so, the application program follows a logical access path. Next, the DBMS, working in conjunction with various system software programs, accesses a storage device, such as disk or tape, where the data is stored. When the DBMS goes to this storage device to retrieve the data, it follows a path to the physical location (physical access path) where the price of this option is stored. In the pricing example, the DBMS might go to a disk drive to retrieve the price data for six-cylinder engines. This relationship is shown in Figure 5.13.

Figure 5.13

Logical and Physical Access Paths

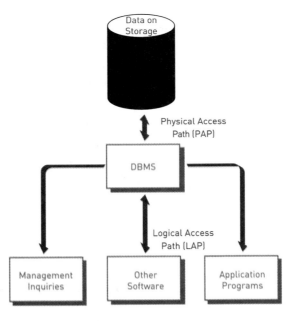

This same process is used if a user wants to get information from the database. First, the user requests the data from the DBMS. For example, a user might give a command, such as LIST ALL OPTIONS FOR WHICH PRICE IS GREATER THAN 200 DOLLARS. This is the logical access path (LAP). Then the DBMS might go to the options price sector of a disk to get the information for the user. This is the physical access path (PAP).

When two or more people or programs attempt to access the same record in the same database at the same time, there can be a problem. For example, an inventory control program might attempt to reduce the inventory level for a product by ten units because ten units were just shipped to a customer. At the same time, a purchasing program might attempt to increase the inventory level for the same product by 200 units because more inventory was just

received. Without proper database control, one of the inventory updates may not be correctly made, resulting in an inaccurate inventory level for the product. **Concurrency control** can be used to avoid this potential problem. One approach is to lock out all other application programs from access to a record if the record is being updated or used by another program.

Manipulating Data and Generating Reports

Once a DBMS has been installed, employees and managers can use it to generate reports and obtain important information. Some databases use *Query-by-Example (QBE)*, which is a visual approach to developing database queries or requests. Like Windows and other GUI operating systems, you can perform queries and other database tasks by opening windows and clicking on the data or features you want (see Figure 5.14).

concurrency control
A method of dealing with a situation in which two or more people need to access the same record in a database at the same time.

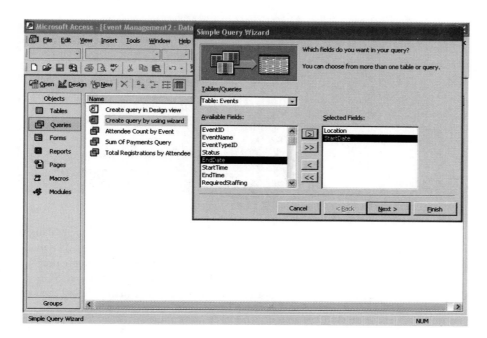

Figure 5.14

Some databases use Query-by-Example (QBE) to generate reports and information.

In other cases, database commands can be used in a programming language. For example, COBOL commands can be used in simple programs that will access or manipulate certain pieces of data in the database. Here's another example of a DBMS query: SELECT * FROM EMPLOYEE WHERE JOB_CLASSIFICATION = "C2". The * tells the program to include all columns from the EMPLOYEE table. In general, the commands that are used to manipulate the database are part of the **data manipulation language (DML)**. This specific language, provided with the DBMS, allows managers and other database users to access, modify, and make queries about data contained in the database to generate reports. Again, the application programs go through subschemas, schemas, and the DBMS before actually getting to the physically stored data on a device such as a disk.

In the 1970s, D. D. Chamberlain and others at the IBM Research Laboratory in San Jose, California, developed a standardized data manipulation language called *Structured Query Language (SQL)*, pronounced like the word *sequel* or simply spelled out as *SQL*. The EMPLOYEE query shown earlier is written in SQL. In 1986, the American National Standards Institute (ANSI) adopted SQL as the standard query language for relational databases. Since ANSI's acceptance of SQL, interest in making SQL an integral part of relational databases on both mainframe and personal computers has increased. SQL has many built-in functions, such as average (AVG), the largest value (MAX), the smallest value (MIN), and others. Table 5.6 contains examples of SQL commands.

data manipulation language (DML)
The commands that are used to manipulate the data in a database.

Table 5.6

Examples of SQL Commands

SQL Command	Description
SELECT ClientName, Debt FROM Client WHERE Debt>1000	This query displays all clients (ClientName) and the amount they owe the company (Debt) from a database table called Client for clients that owe the company more than $1,000 (WHERE Debt>1000).
SELECT ClientName, ClientNum, OrderNum FROM Client, Order WHERE Client.ClientNum=Order.ClientNum	This command is an example of a join command that combines data from two tables: the client table and the order table (FROM Client, Order). The command creates a new table with the client name, client number, and order number (SELECT ClientName, ClientNum, OrderNum). Both tables include the client number, which allows them to be joined. This is indicated in the WHERE clause that states that the client number in the client table is the same (equal to) the client number in the order table (WHERE Client.ClientNum=Order.ClientNum).
GRANT INSERT ON Client to Guthrie	This command is an example of a security command. It allows Bob Guthrie to insert new values or rows into the Client table.

SQL lets programmers learn one powerful query language and use it on systems ranging from PCs to the largest mainframe computers (see Figure 5.15). Programmers and database users also find SQL valuable because SQL statements can be embedded into many programming languages, such as the widely used C++ and COBOL languages. Because SQL uses standardized and simplified procedures for retrieving, storing, and manipulating data in a database system, the popular database query language can be easy to understand and use.

Figure 5.15

Structured Query Language

SQL has become an integral part of most relational database packages, as shown by this screen from Microsoft Access.

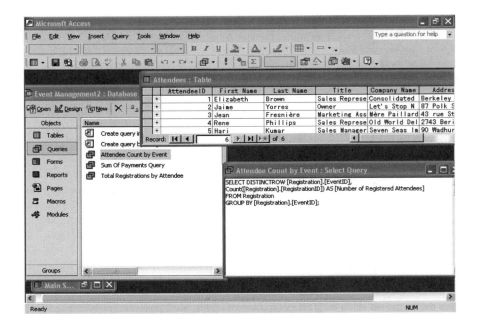

Once a database has been set up and loaded with data, it can produce desired reports, documents, and other outputs (see Figure 5.16). These outputs usually appear in screen

displays or hard-copy printouts. The output-control features of a database program allow you to select the records and fields to appear in reports. You can also make calculations specifically for the report by manipulating database fields. Formatting controls and organization options (like report headings) help you to customize reports and create flexible, convenient, and powerful information-handling tools.

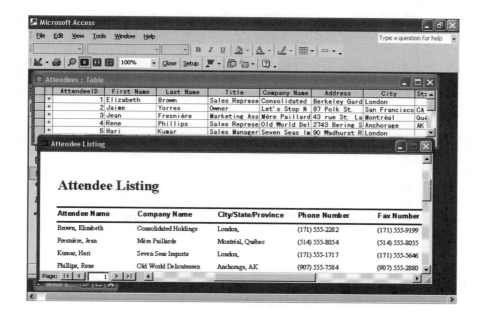

Figure 5.16

Database Output

A database application offers sophisticated formatting and organization options to produce the right information in the right format.

A database program can produce a wide variety of documents, reports, and other outputs that can help organizations achieve their goals. The most common reports select and organize data to present summary information about some aspect of company operations. For example, accounting reports often summarize financial data such as current and past-due accounts. Many companies base their routine operating decisions on regular status reports that show the progress of specific orders toward completion and delivery. Increasingly, companies are using databases to provide improved customer service.

Exception, scheduled, and demand reports, first discussed in Chapter 1, highlight events that require urgent management attention. Database programs can produce literally hundreds of documents and reports. A few examples include these:

- Form letters with address labels
- Payroll checks and reports
- Invoices
- Orders for materials and supplies
- A variety of financial performance reports

Database Administration

As mentioned earlier, a database administrator (DBA) is a highly skilled systems professional who directs or performs all activities necessary to maintain a successful database environment. The DBA's responsibilities include designing, implementing, and maintaining the database system and the DBMS; establishing policies and procedures pertaining to the management, security, maintenance, and use of the database management system; and training employees in database management and use.

A DBA is expected to have a clear understanding of the fundamental business of the organization, be proficient in the use of selected database management systems, and stay abreast of emerging technologies and new design approaches. Typically, a DBA has a degree in either computer science or management information systems and some on-the-job training with a particular database product or more extensive experience with a range of database products.

The DBA works with users to decide the content of the database—to determine exactly what entities are of interest and what attributes are to be recorded about those entities. Thus, it is important for non-IS personnel to have some idea of what the DBA does and why this function is important. The DBA can play a crucial role in the development of effective information systems to benefit the organization, its employees, and its managers.

The DBA also works with programmers as they build applications to ensure that their programs comply with database management system standards and conventions. Once the database is built and operating, the DBA monitors operations logs for security violations. Database performance is also monitored to ensure that the system's response time meets users' needs and that it operates efficiently. If there is a problem, the DBA attempts to correct it before it becomes serious.

Some organizations have also created a position called the *data administrator*, a nontechnical, but important role that ensures that data is managed as an important organizational resource. The **data administrator** is responsible for defining and implementing consistent principles for a variety of data issues, including setting data standards and data definitions that apply across the many databases that an organization may have. For example, the data administrator would ensure that a term such as "customer" is defined and treated consistently in all corporate databases. This person also works with business managers to identify who should have read and/or update access to certain databases and to selected attributes within those databases. This information is then communicated to the database administrator for implementation. The data administrator can be a high-level position reporting to top-level managers.

data administrator
A nontechnical position responsible for defining and implementing consistent principles for a variety of data issues.

Popular Database Management Systems

The latest generation of database management systems makes it possible for end users to build their own database applications. End users are using these tools to address everyday problems, such as how to manage a mounting pile of information on employees, customers, inventory, or sales. These database management systems are an important personal productivity tool, along with word processing, spreadsheet, and graphics software.

A key to making DBMSs more usable for some databases is the incorporation of "wizards" that walk you through how to build customized databases, modify ready-to-run applications, use existing record templates, and quickly locate the data you want. These applications also include powerful new features such as help systems and Web-publishing capabilities. For example, users can create a complete inventory system and then instantly post it to the Web, where it does double duty as an electronic catalogue. Some of the more popular DBMSs for end users include Microsoft's Access and Corel's Paradox. The complete database management software market encompasses software used by professional programmers and that runs on midrange, mainframe, and supercomputers. The entire market generates $10 billion per year in revenue, including IBM, Oracle, and Microsoft. Although Microsoft rules in the desktop PC software market, its share of database software on bigger computers is small.

Like other software products, there are a number of open-source database systems, including PostgreSQL and MySQL.[16] MySQL is the most popular open-source database management system used by travel agencies, manufacturing companies, and other companies.[17] The use of MySQL has increased by more than 30 percent in some years, compared with a 6 percent growth for other popular database packages.[18] According to Chares Gary of the MetaGroup, "I get calls now from customers, not [asking,] 'Should I use an open-source database?' but instead it's, 'Should I use PostgreSQL or MySQL?'"[19] In addition, many traditional database programs are now available on open-source operating systems. The popular DB2 relational database from IBM, for example, is available on the Linux operating system.[20] The Sybase IQ database is also available on the Linux operating system.[21] S&H Co., which offers S&H Greenpoints to reward loyal retail customers, is considering the use of Sybase IQ running on the Linux operating system.[22] "We find Linux to be very cost effective and powerful for what we want to do. It's the best use of financial resources to go with an open-source platform," says Frank Lundy, database director for S&H Co.

Special-Purpose Database Systems

In addition to the popular database management systems just discussed, there are a number of specialized database packages used for specific purposes or in specific industries. Summation and Concordance, for example, is a special-purpose database system used in law firms to organize legal documents.[23] CaseMap organizes information about a case, and LiveNote is used to display and analyze transcripts. These databases help law firms develop and execute good litigation strategies. The Scottish Intelligence Database (SID) is used by the Scottish Drug Enforcement Agency to share crime reports and up-to-date information about crime and criminals.[24] According to Detective Superintendent Ian McCandish, "SID will allow officers access to data from the Shetlands to Gretna Green. Criminals may travel the length and breadth of Scotland, but if we hold the intelligence on them it will be available to front-line officers irrespective of where they are." GlobalSpec is a specialized database for engineers and product designers that contains more than 45 million parts from about 9,500 parts catalogues.[25] Architects Wimberly, Allison, Gong & Goo use a special-purpose database to store and manipulate three-dimensional drawings.[26] J.P. Morgan Chase & Co. uses an "in-memory" database to speed trade orders from pension funds, hedge funds, and various institutional investors.[27] In-memory databases use a computer's memory instead of a hard disk to store and manipulate important data. The database can process more than 100,000 queries per second.

Selecting a Database Management System

The database administrator often selects the best database management system for an organization. The process begins by analyzing database needs and characteristics. The information needs of the organization affect the type of data that is collected and the type of database management system that is used. Important characteristics of databases include the size of the database, number of concurrent users, performance, the ability of the DBMS to be integrated with other systems, the features of the DBMS, vendor considerations, and the cost of the system.

Database Size

Database size depends on the number of records or files in the database. The size determines the overall storage requirement for the database. Most database management systems can handle relatively small databases of less than 100 million bytes; fewer can manage terabyte-sized databases.

Today, companies are trimming the size of their databases to maintain good performance and reduce costs.[28] According to a project manager at Kennametal, "Our overweight database was months away from crashing due to exceeding our production disk-space capacity." The company was able to save about $700,000 in additional hardware and storage costs by trimming its database.

Number of Concurrent Users

The number of simultaneous users that can access the contents of the database is also an important factor. Clearly, a database that is used by a large workgroup must be able to support a number of concurrent users; if it cannot, then the efficiency of the members will be lowered. The term *scalability* is sometimes used to describe how well a database performs as the size of the database or the number of concurrent users is increased. A highly scalable database management system is desirable to provide flexibility. Unfortunately, many companies make a poor DBMS choice in this regard and then later are forced to convert to a new DBMS when the original does not meet expectations.

Performance

How fast the database is able to update records can be the most important performance criterion for some organizations. Credit card and airline companies, for example, must have database systems that can update customer records and check credit or make a plane reservation in seconds, not minutes. Other applications, such as payroll, can be done once a week or less frequently and do not require immediate processing. If an application demands immediacy, it also demands rapid recovery facilities in the event that the computer system shuts

down temporarily. Other performance considerations include the number of concurrent users that can be supported and the amount of memory that is required to execute the database management program. One database was able to process 1,184,893 transactions per minute on average, setting a new world record for speed.[29] Organizations often undergo *performance tuning* to increase database speed and storage efficiency. Performance tuning involves making adjustments to the database to enhance overall performance.

A database management system used by online stores must be able to support a large number of concurrent users by quickly checking a customer's credit card and processing his or her order for merchandise.

Integration
A key aspect of any database management system is its ability to be integrated with other applications and databases. A key determinant here is what operating systems it can run under—such as Linux, UNIX, or Windows. Some companies use several databases for different applications at different locations. A manufacturing company with four plants in three different states might have a separate database at each location. The ability of a database program to import data from and export data to other databases and applications can be a critical consideration.

Features
The features of the database management system can also make a big difference. Most database programs come with security procedures, privacy protection, and a variety of tools. Other features can include the ease of use of the database package and the availability of manuals and documentation to help the organization get the most from the database package. Additional features such as wizards and ready-to-use templates help improve the ease of use. Because of the pressure to reduce IS budgets and the scarcity of experienced database administrators, organizations are demanding database software that comes with features that simplify database management tasks.

The Vendor
The size, reputation, and financial stability of the vendor should also be considered in making any database purchase. Some vendors are well respected in the IS industry and have a large support staff to give assistance, if necessary. A well-established and financially secure database company is more likely than others to remain in business.

Cost
Database packages for personal computers can cost a few hundred dollars, but large database systems for mainframe computers can cost hundreds of thousands of dollars. In addition to the initial cost of the database package, monthly operating costs should be considered. Some

database companies rent or lease their database software. Monthly rental or lease costs, maintenance costs, additional hardware and software costs, and personnel costs can be substantial.

Using Databases with Other Software

Database management systems are often used in conjunction with other software packages or the Internet. A database management system can act as a front-end application or a back-end application. A *front-end application* is one that directly interacts with people or users. Marketing researchers often use a database as a front-end to a statistical analysis program. The researchers enter the results of market questionnaires or surveys into a database. The data is then transferred to a statistical analysis program to determine the potential for a new product or the effectiveness of an advertising campaign. A *back-end application* interacts with other programs or applications; it only indirectly interacts with people or users. When people request information from a Web site on the Internet, the Web site can interact with a database (the back end) that supplies the desired information. For example, you can connect to a university Web site to find out whether the university's library has a book you want to read. The Web site then interacts with a database that contains a catalogue of library books and articles to determine whether the book you want is available.

DATABASE APPLICATIONS

Today, there is a shift from simply accessing the data contained in a database to managing and manipulating the content of a database to produce useful information. Common manipulations are searching, filtering, synthesizing, and assimilating the data contained in a database using a number of database applications. These applications allow users to link the company databases to the Internet, set up data warehouses and marts, use databases for strategic business intelligence, place data at different locations, use online processing and open connectivity standards for increased productivity, develop databases with the object-oriented approach, and search for and use unstructured data such as graphics, audio, and video.

Linking the Company Database to the Internet

Customers, suppliers, and company employees must be able to access corporate databases through the Internet, intranets, and extranets to meet various business needs. For example, Internet customers need to access the corporate product database to obtain product information, including size, color, type, and price details. Suppliers use the Internet and corporate extranets to view inventory databases to check the levels of raw materials and the current production schedule to determine when and how much of their products must be delivered to support just-in-time inventory management. Company employees need to be able to access corporate databases to support decision making even when they are located remotely. In such cases, they may use laptop computers and access the data via the Internet and company intranet. Read the "Information Systems @ Work" box to see how one cruise line has helped its employees access and use its data.

Web-Based DBMS Empowers Cruise Line Personnel

In these days of value-driven, performance-based information systems, many businesses are turning to new database technologies to streamline operations and save money. Recently, Holland America Cruise Lines traded its old, complex mainframe database system—one that only a computer programmer could understand—for a new system with which ordinary employees could interact directly.

The goal of the upgrade was to increase revenues by $1 million annually by speeding information to employees for more efficient sales, marketing, and revenue management. Prior to the upgrade, database programmers and tech staff would prepare and distribute weekly reports to address other employees' ad hoc inquiries and would generate scheduled reports to assess the company's revenue and inventory, according to Paul Grigsby, senior revenue manager at Holland America. The new reporting system, called WebFocus, connects to the same mainframe database as the old DBMS but provides more powerful analysis, querying, and reporting tools that are accessed through an intuitive Web-based user interface.

Fulfilling demand report requests previously took up to two days, but with the new system the IS staff can fine-tune requests and give reports to end users almost immediately. With a bit of training, some end users are interacting with the system directly to create their own reports. Those who work with the data most, such as revenue management personnel, began using the system first. The new system had its skeptics at the beginning, says Grigsby. "We are yield managers, not computer programmers, and I was frankly suspicious about putting the reporting function in our hands," he explains. "However, after training and a goodly amount of trial and error, we began to see the rewards of empowering the end users." After running a query and producing a report, employees can alter the view of the data by resorting it or introducing new fields into the inquiry—without formally requesting a new report from the IS department. Grigsby says, "It makes both my time and the information systems department's time more efficient."

Teresa Tennant is the manager of online communications for the shore excursion department at Holland America. She plans to become less dependent on the IS staff by training on the new system. With the new system, she can log on to a Web site to access reports in whatever format she wants—Excel spreadsheet, Word document, or portable document file (PDF). "It's very handy. I can drill down into complete details of the booking by travel agency and individual," she says.

The company is limited, however, because of the size of the existing mainframe DBMS. It plans to load the information into a streamlined data warehouse, where it can be accessed more quickly. Speed and accessibility of information access, after all, are the attributes that will increase productivity and revenues. "The real value of the new system at Holland America," Bill Hostmann, an analyst at Gartner Inc., says, "is how it makes it easier to develop and distribute business management information to more users in a timely fashion around the world than ever before and thereby more fully leverage existing IT investments."

Discussion Questions

1. How does the new reporting system at Holland America Line empower revenue management personnel?
2. How does the new system allow the IS staff to work more efficiently?

Critical Thinking Questions

3. The change in information access at Holland America Line is indicative of a general trend in many industries: Non-IS employees are assuming traditional IS staff responsibilities. Do you think that this trend evolved purely out of efforts to save money by reducing IS staff, or are there substantial benefits to bringing IS power to the people? What might those benefits be?
4. Is it realistic to expect nontechnical employees to acquire higher-level technical skills? Will the nontechnical staff be willing and able to assume the task?

SOURCES: Mark L Sangini, "Cruise Line Changes BI Tack," *Computerworld,* October 6, 2003, *www.computerworld.com;* Information Builder's Web site, *www.informationbuilders.com,* accessed March 5, 2004; Holland America Line Web site, *www.hollandamerica.com,* accessed March 5, 2004.

Developing a seamless integration of traditional databases with the Internet is often called a *semantic Web*.[30] According to Tim Berners-Lee, creator of the World Wide Web (WWW), "The Semantic Web is about taking the relational database and webbing it." A semantic Web allows people to access and manipulate a number of traditional databases at the same time through the Internet. Many software vendors—including IBM, Oracle, Microsoft, Macromedia, Inline Internet Systems, and Netscape Communications—are incorporating the capability of the Internet into their products. Such databases allow companies to create an Internet-accessible catalogue, which is nothing more than a database of items, descriptions, and prices. Simplest-Shop, for example, uses the Web to sell compact discs on the Internet.[31] The company offers a wealth of information on each compact disc, including product reviews. The Web site makes it appear as though the company has a large number of employees, but the Internet has allowed one person, Calin Uioreanu from Romania, to set up the site as a side business. He is a full-time software engineer. AutoTradeCenter, Inc., has linked its Web site to an Oracle database.[32] The database supports about 24,000 franchise dealerships and 80,000 independent dealerships. The database helps AutoTradeCenter customers, including Honda, DaimlerChrysler, Volkswagen, Subaru, among others.

In addition to the Internet, organizations are gaining access to databases through networks to get good prices and reliable service. Catholic Health, for example, has developed a network to allow physicians to access patient information from remote locations.[33] Perry Manufacturing in North Carolina uses a network to allow its managers to have access to e-mail and other information stored on its database. Connecting databases to corporate Web sites and networks can lead to potential problems, however. One database expert believes that up to 40 percent of Web sites that connect to corporate databases are susceptible to hackers' taking complete control of the database.[34] By typing certain characters in a form on some Web sites, a hacker is able to issue SQL commands to control the corporate database.

Data Warehouses, Data Marts, and Data Mining

The raw data necessary to make sound business decisions is stored in a variety of locations and formats. This data is initially captured, stored, and managed by transaction processing systems that are designed to support the day-to-day operations of the organization. For decades, organizations have collected operational, sales, and financial data with their online transaction processing (OLTP) systems. The data can be used to support decision making using data warehouses, data marts, and data mining.

Data Warehouses

A **data warehouse** is a database that holds business information from many sources in the enterprise, covering all aspects of the company's processes, products, and customers. The data warehouse provides business users with a multidimensional view of the data they need to analyze business conditions. Data warehouses allow managers to *drill down* to get more detail or *roll up* to take detailed data and generate aggregate or summary reports. A data warehouse is designed specifically to support management decision making, not to meet the needs of transaction processing systems. Ace Hardware Corporation, for example, uses a data warehouse to analyze pricing trends.[35] According to the company's data warehouse designer, "We had one store that only sold one wheelbarrow a year, but when he lowered the price, he sold four in one month." The price reduction was suggested by the data warehouse. Office Depot developed a powerful data warehouse to give employees better information on employee and store performance. According to a company spokesperson, "Sometimes data warehousing can be ignored by the stores unless it provides direct metrics on how the store is performing."[36] A data warehouse stores historical data that has been extracted from operational systems and external data sources (see Figure 5.17). This operational and external data is "cleaned up" to remove inconsistencies and integrated to create a new information database that is more suitable for business analysis.

Data warehouses typically start out as very large databases, containing millions and even hundreds of millions of data records. As this data is collected from the various production systems, a historical database is built that business analysts can use. To keep it fresh and accurate, the data warehouse receives regular updates. Old data that is no longer needed is

data warehouse
A database that collects business information from many sources in the enterprise, covering all aspects of the company's processes, products, and customers.

Figure 5.17

Elements of a Data Warehouse

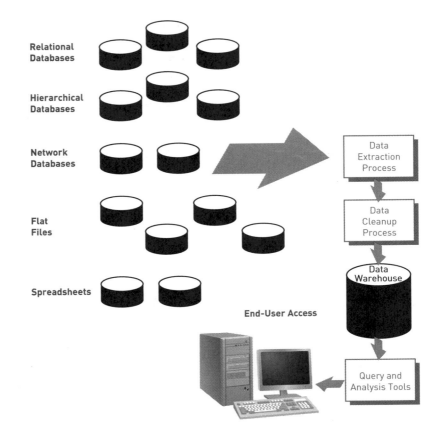

Relational Databases

Hierarchical Databases

Network Databases

Flat Files

Spreadsheets

Data Extraction Process

Data Cleanup Process

Data Warehouse

Query and Analysis Tools

End-User Access

purged from the data warehouse. Updating the data warehouse must be fast, efficient, and automated, or the ultimate value of the data warehouse is sacrificed. It is common for a data warehouse to contain from 3 to10 years of current and historical data. Data-cleaning tools can merge data from many sources into one database, automate data collection and verification, delete unwanted data, and maintain data in a database management system. Data warehouses can also get data from unique sources. Oracle's Warehouse Management software, for example, can accept information from radio-frequency identification (RFID) technology, which is being used to tag products as they are shipped or moved from one location to another.[37]

7-Eleven, a convenience retail chain with more than 25,000 stores worldwide, used Oracle's E-Business Suite to consolidate hundreds of individual business systems into an integrated system that enables the business to track purchasing, assets, costs, and payroll.

(Source: © Tom Wagner/CORBIS SABA.)

The primary advantage of data warehousing is the ability to relate data in new, innovative ways. However, a data warehouse can be extremely difficult to establish, with the typical cost exceeding $2 million. Table 5.7 compares OLTP processing and data warehousing.

Characteristic	OLTP Database	Data Warehousing
Purpose	Support transaction processing	Support decision support
Source of data	Business transactions	Multiple files, databases—data internal and external to the firm
Data access allowed users	Read and write	Read only
Primary data access mode	Simple database update and query	Simple and complex database queries with increasing use of data mining to recognize patterns in the data
Primary database model employed	Relational	Relational
Level of detail	Detailed transactions	Often summarized data
Availability of historical data	Very limited—typically a few weeks or months	Multiple years
Update process	Online, ongoing process as transactions are captured	Periodic process, once per week or once per month
Ease of process	Routine and easy	Complex, must combine data from many sources; data must go through a data cleanup process
Data integrity issues	Each individual transaction must be closely edited	Major effort to "clean" and integrate data from multiple sources

Table 5.7

Comparison of OLTP and Data Warehousing

Data Marts

A **data mart** is a subset of a data warehouse. Data marts bring the data warehouse concept—online analysis of sales, inventory, and other vital business data that has been gathered from transaction processing systems—to small and medium-sized businesses and to departments within larger companies. Rather than store all enterprise data in one monolithic database, data marts contain a subset of the data for a single aspect of a company's business—for example, finance, inventory, or personnel. In fact, a specific area in the data mart may contain more detailed data than the data warehouse would provide.

Data marts are most useful for smaller groups who want to access detailed data. A warehouse contains summary data that can be used by an entire company. Because data marts typically contain tens of gigabytes of data, as opposed to the hundreds of gigabytes in data warehouses, they can be deployed on less-powerful hardware with smaller secondary storage devices, delivering significant savings to an organization. Although any database software can be used to set up a data mart, some vendors deliver specialized software designed and priced specifically for data marts. Already, companies such as Sybase, Software AG, Microsoft, and others have announced products and services that make it easier and cheaper to deploy these scaled-down data warehouses. The selling point: Data marts put targeted business information into the hands of more decision makers.

Data Mining

Data mining is an information-analysis tool that involves the automated discovery of patterns and relationships in a data warehouse. The FBI, for example, is using the ClearForest database package to support data mining of the Terrorism Intelligence Database.[38] According to FBI director Robert Mueller, "We are now focused on implementing a data warehousing capability that can bring together our information into databases that can be accessed by agents

data mart
A subset of a data warehouse.

data mining
An information-analysis tool that involves the automated discovery of patterns and relationships in a data warehouse.

throughout the world as well as our analysts as soon as a piece of information is developed." Data mining has also been used in the airline passenger–profiling system used to block suspected terrorists from flying and the Total Information Awareness Program, which attempts to detect patterns of terrorist activity.[39] Organizations are also investing in systems for data mining to meet new government regulations, as the "Ethical and Societal Issues" feature discusses.

Data mining's objective is to extract patterns, trends, and rules from data warehouses to evaluate (i.e., predict or score) proposed business strategies, which in turn will improve competitiveness, improve profits, and transform business processes. It is used extensively in marketing to improve customer retention; cross-selling opportunities; campaign management; market, channel, and pricing analysis; and customer segmentation analysis (especially one-to-one marketing). In short, data-mining tools help end users find answers to questions they never even thought to ask.

E-commerce presents another major opportunity for effective use of data mining. Attracting customers to online Web sites is tough; keeping them can be next to impossible. For example, when online retail Web sites launch deep-discount sales, they cannot easily figure out how many first-time customers are likely to come back and buy again. Nor do they have a way of understanding which customers acquired during the sale are price sensitive and more likely to jump on future sales. As a result, companies are gathering data on user traffic through their Web sites into databases. This data is then analyzed using data-mining techniques to personalize the Web site and develop sales promotions targeted at specific customers.

Predictive analysis is a form of data mining that combines historical data with assumptions about future conditions to predict outcomes of events such as future product sales or the probability that a customer will default on a loan. Retailers use predictive analysis to upgrade occasional customers into frequent purchasers by predicting what products they will buy if offered an appropriate incentive. Genalytics, Magnify, NCR Teradata, SAS Institute, Sightward, SPSS, and Quadstone have developed predictive analysis tools. Predictive analysis software can be used to analyze a company's customer list and a year's worth of sales data to find new market segments that could be profitable.

predictive analysis
A form of data mining that combines historical data with assumptions about future conditions to predict outcomes of events such as future product sales or the probability that a customer will default on a loan.

Halfords, a U.K. retailer of car parts, bikes, and accessories, used SPSS's Clementine, a predictive analysis data-mining tool, to identify the most profitable sites to build new stores as part of its expansion program.

ETHICAL AND SOCIETAL ISSUES

The Growing Cost of Data-Related Regulations

Increasing government regulations are forcing many businesses to scrutinize their data, database systems, and storage technologies. Government agencies such as the Internal Revenue Service have been concerned about the accuracy and security of records storage since the invention of the computer. Recently, the mounting concerns of varying government agencies have been transformed into laws—laws that are costing businesses big bucks to implement.

In the wake of corporate scandals and in an effort to make investment brokers and dealers accountable for their practices, the Securities and Exchange Commission (SEC) supported passage of the Sarbanes-Oxley Act of 2002. Among the act's many provisions is a requirement for brokers and dealers to log and record all electronic communications, such as e-mail, using "write-once, read many," or WORM, technology. WORM technology ensures that stored data cannot be tampered with later.

The SEC regulatory requirements compelled Jay Cohen, corporate compliance officer at The Mony Group, to install an EMC Centera storage server system and database application called AXS-One Email and Instant Messaging Management Solution software suite to track e-mails. Cohen explains. "All the external e-mail from the sales force, either incoming or outgoing, goes into the AXS-One e-mail archival system." The system permanently stores each communication as a record with a unique identifier in a database. Administrators can then run queries on the e-mail data for content surveillance, security, and audit trails.

Another law, the USA Patriot Act, has placed a significant burden on financial service firms. They are now responsible for monitoring new customers to ensure that they are not laundering money for terrorists. Under scrutiny by regulators, such firms are employing data-mining applications that analyze risks and use complex algorithms to identify unusual customer trends within transactions.

Financial services institutions have also reacted to the threat of terrorist attacks by investing heavily in indestructible data warehouses, computer networks, and hardware redundancy, and they have transferred all paper records to electronic archives. The TowerGroup estimates that by 2007 the global financial services industry will have spent $523 billion on operational resiliency—technology upgrades for disaster recovery, business continuity, and security. Between 2003 and 2007 TowerGroup predicts that U.S. retail banks alone will spend $1.1 billion, or 4.4 percent of their technology budgets, in response to 9/11. Expensive artificial intelligence systems will drive the applications, looking for suspicious deviations from the norm.

While financial industries buckle down, medical industries have an equal share of new government regulations to meet. In 1996 the U.S. Department of Health and Human Services established national standards to protect the privacy of personal health information with HIPAA—the Health Insurance Portability and Accountability Act.

HIPAA caused health and medical organizations to scramble to secure medical records in private database systems.

More recently, the Food and Drug Administration issued a ruling that requires pharmaceutical companies to apply bar codes to thousands of prescription and over-the-counter drugs dispensed in hospitals. The FDA believes the move will save lives by reducing medical errors but estimated that it would hit the nation's 6,000-plus hospitals with a $7 billion technology bill for bar-code readers, databases, and management tools.

Information system companies are designing systems to help businesses and organizations comply with the many new laws and regulations. IBM recently unveiled an integrated system to help users preserve electronic documents for regulations such as the Sarbanes-Oxley Act and HIPAA. The system, called IBM TotalStorage Data Retention 450, combines server, storage, and software components in a secure cabinet. At a cost starting at $141,600 for a 3.5-TB configuration, systems such as this are costly but necessary to comply with government regulations.

With budgets a constant worry, financial and medical organizations are complaining that regulation should be backed with financial assistance to help them comply with the regulations. In the meantime, the costs for e-mail archiving, prescription bar-code readers, and complex and expensive systems that support secure and private record keeping will most likely be passed along to customers in the form of higher prices.

Critical Thinking Questions

1. Why has the government stepped up efforts to regulate record keeping in financial and medical industries?
2. With many medical organizations already strapped for cash, especially hospitals, is it fair for the government to force them to comply with expensive new regulations? Who should bear the financial burden of secure and private record keeping?

What Would You Do?

As CIO of the Prime Market Finance Corp, you are responsible for complying with new laws that require you to permanently store all electronic communications. Prime Market employees currently send and receive 200,000 e-mail messages per day. One third of the e-mail received is junk mail—spam. One third of the e-mail sent is not related to the business. You are considering the IBM TotalStorage Data Retention system, but first you need to determine how much data storage will be required. Estimating that 100 e-mail messages equal 1 MB of storage space, you calculate that 200,000 e-mail messages per day amount to 2,000 MB or 2 GB. That amounts to 730 GB per year, and more than 7 TB over 10 years. It appears that IBM's low-end system, which supports 3.5 TB, will not meet your storage needs over the long haul.

3. What policies and technologies might you implement to reduce the overall amount of e-mail sent and received?

4. Who should bear the financial burden of archiving data that may or may not be needed as evidence in court?

SOURCES: Lucas Mearian, "SEC Holds Fast on WORM Standard for Securities Firms," *Computerworld*, March 9, 2003, *www.computerworld.com*; Mitch Betts, "Editor's Note: The New Rules of Storage," *Computerworld*, November 17, 2003, *www.computerworld.com*; Lucas Mearian, "Sidebar: Regulations, Volume and Capacity Add Archiving Pressure," *Computerworld*, February 16, 2004, *www.computerworld.com*; Lucas Mearian, "Compliance Laws Vex IT: The USA Patriot Act Is Keeping Financial Firms Busy," *Computerworld*, September 8, 2003, *www.computerworld.com*; Thomas Hoffman, "IBM Tailors Bundle for Preserving Corporate Data: Integrated System to Aid in Regulatory Compliance Efforts," *Computerworld*, February 23, 2004, *www.computerworld.com*; Bob Brewin, "FDA Mandates Bar Codes on Drugs Used in Hospitals," *Computerworld*, February 26, 2004, *www.computerworld.com*.

Traditional DBMS vendors are well aware of the great potential of data mining. Thus, companies such as Oracle, Sybase, Tandem, and Red Brick Systems are all incorporating data-mining functionality into their products. Table 5.8 summarizes a few of the most frequent applications for data mining.

Table 5.8

Common Data-Mining Applications

Application	Description
Branding and positioning of products and services	Enable the strategist to visualize the different positions of competitors in a given market using performance (or importance) data on dozens of key features of the product in question and then to condense all that data into a perceptual map of just two or three dimensions
Customer churn	Predict current customers who are likely to go to a competitor
Direct marketing	Identify prospects most likely to respond to a direct marketing campaign (such as a direct mailing)
Fraud detection	Highlight transactions most likely to be deceptive or illegal
Market basket analysis	Identify products and services that are most commonly purchased at the same time (e.g., nail polish and lipstick)
Market segmentation	Group customers based on who they are or on what they prefer
Trend analysis	Analyze how key variables (e.g., sales, spending, promotions) vary over time

Business Intelligence

business intelligence
The process of gathering enough of the right information in a timely manner and usable form and analyzing it to have a positive impact on business strategy, tactics, or operations.

Closely linked to the concept of data mining is use of databases for business-intelligence purposes. **Business intelligence (BI)** is the process of gathering enough of the right information in a timely manner and usable form and analyzing it so that it can have a positive impact on business strategy, tactics, or operations. Business intelligence turns data into useful information that is then distributed throughout an enterprise. Companies use this information to make improved strategic decisions about which markets to enter, how to select and manage key customer relationships, and how to select and effectively promote products to increase profitability and market share.

Today, a number of companies use the business-intelligence approach. BankFinancial Corporation of Chicago uses it to help target promotions to bank customers.[40] Owens & Minor, a large medical supply company, uses business-intelligence software from Business

Objects to analyze sales data.[41] According to Scott Wiener, chief technology officer at Certive Corporation, "Today, the most innovative business-intelligence technology is able to recommend the optimal course of action based on business rules, representing the first step in automated decision making."[42] Wiener predicts that business-intelligence software may eliminate the need for a large number of mid-level managers. Companies such as Ben and Jerry's ice cream process and store huge amounts of data.[43] The company collects data on all 190,000 pints it produces in its factories each day, with all the data being shipped to the company's headquarters in Burlington, Vermont, which is a few miles from its first store that was opened over 25 years ago. In the marketing department, the massive amount of data is analyzed. Using business-intelligence software, the company is able to cut costs and improve customer satisfaction. The software allows Ben and Jerry's to match the over 200 calls and e-mails received each week with ice cream products and supplies. Today, the company can quickly determine whether a bad batch of milk or eggs was used in production or whether sales of its Chocolate Chip Cookie Dough is gaining on the No. 1 selling Cherry Garcia. An insurance and financial services company can use business intelligence to gain an in-depth understanding of its customers—from the profits they generate, their rate of retention, and the opportunities they offer to cross-sell the company's products. Employees can obtain the data needed to zero in on problem areas, obtain a detailed picture of the profitability of any customer, and see which products are selling and which are not. If sales decline, they can track those declines to specific offices or even to individual sales reps to pinpoint problems and take immediate steps to remedy them.

Competitive intelligence is one aspect of business intelligence and is limited to information about competitors and the ways that knowledge affects strategy, tactics, and operations. Competitive intelligence is a critical part of a company's ability to see and respond quickly and appropriately to the changing marketplace. Competitive intelligence is not espionage—the use of illegal means to gather information. In fact, almost all the information a competitive-intelligence professional needs can be collected by examining published information sources, conducting interviews, and using other legal, ethical methods. Using a variety of analytical tools, a skilled competitive-intelligence professional can by deduction fill the gaps in information already gathered.

The term **counterintelligence** describes the steps an organization takes to protect information sought by "hostile" intelligence gatherers. One of the most effective counterintelligence measures is to define "trade secret" information relevant to the company and control its dissemination.

Knowledge management is the process of capturing a company's collective expertise wherever it resides—in computers, on paper, or in people's heads—and distributing it wherever it can help produce the biggest payoff. The goal of knowledge management is to get people to record knowledge (as opposed to data) and then share it. Although a variety of technologies can support it, knowledge management is really about changing people's behavior to make their experience and expertise available to others. Knowledge management had its start in large consulting firms and has expanded to nearly every industry. Pharmaceutical companies, for example, must have access to various databases from different biotechnology companies to ensure they make informed decisions.

Distributed Databases

Distributed processing involves placing processing units at different locations and linking them via telecommunications equipment. A **distributed database**—a database in which the data may be spread across several smaller databases connected via telecommunications devices—works on much the same principle. A user in the Milwaukee branch of a clothing manufacturer, for example, might make a request for data that is physically located at corporate headquarters in Milan, Italy. The user does not have to know where the data is physically stored. The user makes a request for data, and the DBMS determines where the data is physically located and retrieves it (see Figure 5.18).

competitive intelligence
One aspect of business intelligence limited to information about competitors and the ways that knowledge affects strategy, tactics, and operations.

counterintelligence
The steps an organization takes to protect information sought by "hostile" intelligence gatherers.

knowledge management
The process of capturing a company's collective expertise wherever it resides—in computers, on paper, in people's heads—and distributing it wherever it can help produce the biggest payoff.

distributed database
A database in which the data may be spread across several smaller databases connected via telecommunications devices.

Figure 5.18

The Use of a Distributed Database

For a clothing manufacturer, computers may be located at corporate headquarters, in the research and development center, in the warehouse, and in a company-owned retail store. Telecommunications systems link the computers so that users at all locations can access the same distributed database no matter where the data is actually stored.

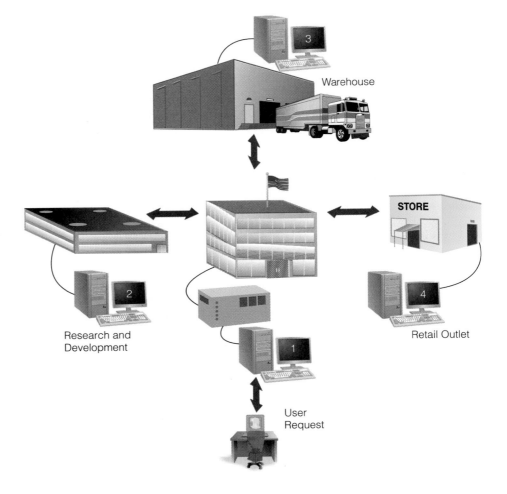

Distributed databases give corporations more flexibility in how databases are organized and used. Local offices can create, manage, and use their own databases, and people at other offices can access and share the data in the local databases. Giving local sites more direct access to frequently used data can improve organizational effectiveness and efficiency significantly.

Despite its advantages, distributed processing creates additional challenges in maintaining data security, accuracy, timeliness, and conformance to standards. Distributed databases allow more users direct access at different sites; thus, controlling who accesses and changes data is sometimes difficult. Also, because distributed databases rely on telecommunications lines to transport data, access to data can be slower.

replicated database
A database that holds a duplicate set of frequently used data.

To reduce telecommunications costs, some organizations build a replicated database. A **replicated database** holds a duplicate set of frequently used data. At the beginning of the day, the company sends a copy of important data to each distributed processing location. At the end of the day, the different sites send the changed data back to update the main database. This process, often called *data synchronization*, is used to make sure that replicated databases are accurate, up to date, and consistent with each other. A railroad, for example, can use a replicated database to increase punctuality, safety, and reliability. The primary database can hold data on fares, routings, and other essential information. The data can be continually replicated and downloaded on a read-only basis from the master database to hundreds of remote servers across the country. The remote locations can send back the latest figures on ticket sales and reservations to the main database.

Online Analytical Processing (OLAP)

For nearly two decades, multidimensional databases and their analytical information display systems have provided flashy sales presentations and trade show demonstrations. All you have to do is ask where a certain product is selling well, for example, and a colorful table showing sales performance by region, product type, and time frame automatically pops up on the screen. Called **online analytical processing (OLAP)**, these programs are now being used to store and deliver data warehouse information efficiently.[44] OLAP allows users to explore corporate data from a number of different perspectives.

OLAP servers and desktop tools support high-speed analysis of data involving complex relationships, such as combinations of a company's products, regions, channels of distribution, reporting units, and time periods. Speed is essential as businesses grow and accumulate more and more data in their operational systems and data warehouses. Long popular with financial planners, OLAP is now being put in the hands of other professionals. The leading OLAP software vendors include Cognos, Comshare, Hyperion Solutions, Oracle, Mine-Share, WhiteLight, and Microsoft.

Consumer goods companies use OLAP to analyze the millions of consumer purchase records captured by scanners at the checkout stand. This data is used to spot trends in purchases and to relate sales volume to promotions and store conditions, such as displays, and even the weather. OLAP tools let managers analyze business data using multiple dimensions, such as product, geography, time, and salesperson. The data in these dimensions, called *measures*, is generally aggregated—for example, total or average sales in dollars or units, or budget dollars or sales forecast numbers. Rarely is the data studied in its raw, unaggregated form. Each dimension also can contain some hierarchy. For example, in the time dimension, users may examine data by year, by quarter, by month, by week, and even by day. A geographic dimension may compile data from city, state, region, country, and even hemisphere.

The value of data ultimately lies in the decisions it enables. Powerful information-analysis tools in areas such as OLAP and data mining, when incorporated into a data warehousing architecture, bring market conditions into sharper focus and help organizations deliver greater competitive value. OLAP provides top-down, query-driven data analysis; data mining provides bottom-up, discovery-driven analysis. OLAP requires repetitive testing of user-originated theories; data mining requires no assumptions and instead identifies facts and conclusions based on patterns discovered. OLAP, or multidimensional analysis, requires a great deal of human ingenuity and interaction with the database to find information in the database. A user of a data-mining tool does not need to figure out what questions to ask; instead, the approach is, "Here's the data, tell me what interesting patterns emerge." For example, a data-mining tool in a credit card company's customer database can construct a profile of fraudulent activity from historical information. Then, this profile can be applied to all incoming transaction data to identify and stop fraudulent behavior, which may otherwise go undetected. Table 5.9 compares the OLAP and data-mining approaches to data analysis.

online analytical processing (OLAP)

Software that allows users to explore data from a number of different perspectives.

Table 5.9

Comparison of OLAP and Data Mining

Characteristic	OLAP	Data mining
Purpose	Supports data analysis and decision making	Supports data analysis and decision making
Type of analysis supported	Top-down, query-driven data analysis	Bottom-up, discovery-driven data analysis
Skills required of user	Must be very knowledgeable of the data and its business context	Must trust in data mining tools to uncover valid and worthwhile hypothesis

Open Database Connectivity (ODBC)

open database connectivity (ODBC)

Standards that ensure that software can be used with any ODBC-compliant database.

To help with database integration, many companies rely on **open database connectivity (ODBC)** standards. ODBC standards help ensure that software can be used with any ODBC-compliant database, making it easier to transfer and access data among different databases. For example, a manager might want to take several tables from one database and incorporate them into another database that uses a different database management system. In another case, a manager might want to transfer one or more database tables into a spreadsheet program. If all this software meets ODBC standards, the data can be imported or exported to other applications (see Figure 5.19). For example, a table in an Access database can be exported to a Paradox database. Tables and data can also be imported using ODBC. For example, a table in a FileMaker Pro database can be imported into an Access database. Linking allows an application to use data or an object stored in another application without actually importing the data or object into the application. The Access database, for example, can link to a table in the Lotus 1-2-3 spreadsheet. Applications that follow the ODBC standard can use these powerful ODBC features to share data between different applications stored in different formats.

Figure 5.19

Advantages of ODBC

ODBC can be used to export, import, or link tables between different applications.

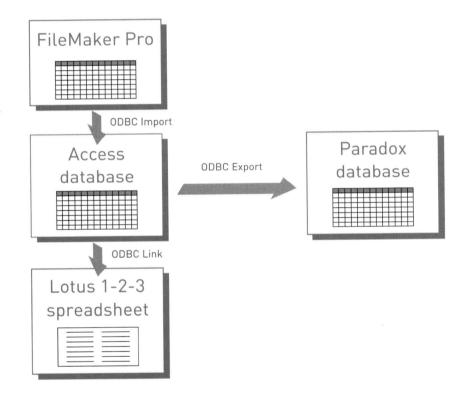

ODBC-compliant products do suffer from their all-purpose nature, however. Their overall performance is usually less efficient than that of products designed for use with a specific database. Yet, more and more vendors are building ODBC-compliant products as businesses increasingly use distributed databases. Many organizations are using such tools to allow their workers and managers easier access to a variety of databases and data sources. ODBC standards also make it easier for growing companies to integrate existing databases, to connect more users into the same database, and to move application programs from PC-oriented databases to larger, workstation-based databases, and vice versa.

Object-Oriented and Object-Relational Database Management Systems

An **object-oriented database** uses the same overall approach of objected-oriented programming that was first discussed in Chapter 4. With this approach, both the data and the processing instructions are stored in the database. For example, an object-oriented database could store both monthly expenses and the instructions needed to compute a monthly budget from those expenses. A traditional DBMS might only store the monthly expenses. In an object-oriented database, a *method* is a procedure or action. A sales tax method, for example, could be the procedure to compute the appropriate sales tax for an order or sale—for example, multiplying the total amount of an order by 5 percent, if that is the local sales tax. A *message* is a request to execute or run a method. For example, a sales clerk could issue a message to the object-oriented database to compute sales tax for a new order. Many object-oriented databases have their own query language, called *object query language (OQL)*, which is similar to SQL, discussed previously.

An object-oriented database uses an **object-oriented database management system (OODBMS)** to provide a user interface and connections to other programs. A number of computer vendors sell or lease OODBMSs, including eXcelon, Versant, Poet, and Objectivity. Object-oriented databases are used by a number of organizations. Versant's OODBMS, for example, is being used by companies in the telecommunications, financial services, transportation, and defense industries.[45] J. D. Edwards is using an object-oriented database to help its customers make fast and efficient forecasts of future sales and to determine whether they have enough materials and supplies to meet future demand for products and services.[46] With an object-oriented database, customers can quickly get a variety of reports on inventory and supplies. The *Object Data Standard* is a design standard by the *Object Database Management Group (www.odmg.org)* for developing object-oriented database systems.

An **object-relational database management system (ORDBMS)** provides a complete set of relational database capabilities plus the ability for third parties to add new data types and operations to the database. These new data types can be audio, images, unstructured text, spatial, or time series data that require new indexing, optimization, and retrieval features. Each of the vendors offering ORDBMS facilities provides a set of application programming interfaces to allow users to attach external data definitions and methods associated with those definitions to the database system. They are essentially offering a standard socket into which users can plug special instructions. DataBlades, Cartridges, and Extenders are the names applied by Oracle and IBM to describe the plug-ins to their respective products. Other plug-ins serve as interfaces to Web servers.

Visual, Audio, and Other Database Systems

In addition to raw data, organizations are increasingly finding a need to store large amounts of visual and audio signals in an organized fashion. Credit card companies, for example, input pictures of charge slips into an image database using a scanner. The images can be stored in the database and later sorted by customer, printed, and sent to customers along with their monthly statements. Image databases are also used by physicians to store x-rays and transmit them to clinics away from the main hospital. Financial services, insurance companies, and government branches are also using image databases to store vital records and replace paper documents. Music companies need the ability to store and manipulate sound from recording studios.

Visual databases can be stored in some object-relational databases or special-purpose database systems. Purdue University has developed an audio database and processing software to give singers a voice makeover.[47] The database software can correct pitch errors and modify voice patterns to introduce vibrato and other voice characteristics. According to the project director, "We look at the results from good singers and those of bad singers, and try to understand those differences." Drug companies often need the ability to analyze a large number of visual images from laboratories.[48] The PetroView database and analysis tool allows

object-oriented database
Database that stores both data and its processing instructions.

object-oriented database management system (OODBMS)
A group of programs that manipulate an object-oriented database and provide a user interface and connections to other application programs.

object-relational database management system (ORDBMS)
A DBMS capable of manipulating audio, video, and graphical data.

petroleum engineers to analyze geographic information to help them determine where to drill for oil and gas.[49] A visual-fingerprint database was used to solve a 40-year-old murder case in California.[50] The fingerprint database was a $640 million project started in 1995.

Combining and analyzing data from separate and totally different databases is an increasingly important database challenge. Global businesses, for example, sometimes need to analyze sales and accounting data stored around the world in different database systems. Companies such as IBM are developing *virtual database systems* to allow different databases to work together as a unified database system.[51] Using an IBM virtual database system, a Canadian bioresearch firm was able to integrate data from different databases that used different file formats and types. DiscoveryLink, one of IBM's projects, is allowing biomedical data from different sources to be integrated. According to Raimond Winslow, professor of biomedical engineering and computer science and director of the Center for Cardiovascular Bioinformatics and Modeling at Johns Hopkins University, "Information tools [such as DiscoveryLink] provide ways in which you can mine these huge data sets." The Centers for Disease Control (CDC) also has the problem of integrating different databases.[52] The CDC has more than 100 databases on various diseases. Searching these databases for data and information on diseases such as SARS (severe acute respiratory syndrome) can be difficult.

In addition to visual, audio, and virtual databases, there are a number of other special-purpose database systems.[53] *Spatial data technology* involves the use of a database to store and access data according to the locations it describes and to permit spatial queries and analysis. MapExtreme is spatial technology software from MapInfo that extends a user's database so that it can store, manage, and manipulate location-based data. Police departments, for example, can use this type of software to bring together crime data and map the data visually so that patterns are easier to analyze. Police officers can select and work with spatial data at a specified location, within a rectangle, a given radius, or a polygon such as a precinct. For example, a police officer can request a list of all liquor stores within a 2-mile radius of the precinct. Builders and insurance companies use spatial data to make decisions related to natural hazards. Spatial data can even be used to improve financial risk management with information stored by investment type, currency type, interest rates, and time.

Spatial data technology is used by NASA to store data from satellites and Earth stations. Location-specific information can be accessed and compared.

(Source: Courtesy of NASA.)

SUMMARY

Principle

The database approach to data management provides significant advantages over the traditional file-based approach.

Data is one of the most valuable resources that a firm possesses. It is organized into a hierarchy that builds from the smallest element to the largest. The smallest element is the bit, a binary digit. A byte (a character such as a letter or numeric digit) is made up of 8 bits. A group of characters, such as a name or number, is called a *field* (an object). A collection of related fields is a *record*; a collection of related records is called a *file*. The database, at the top of the hierarchy, is an integrated collection of records and files.

An entity is a generalized class of objects for which data is collected, stored, and maintained. An attribute is a characteristic of an entity. Specific values of attributes—called *data items*—can be found in the fields of the record describing an entity. A data key is a field within a record that is used to identify the record. A primary key uniquely identifies a record, while a secondary key is a field in a record that does not uniquely identify the record.

The traditional approach to data management has been from a file perspective. Separate files are created for each application. This approach can create problems over time: As more files are created for new applications, data that is common to the individual files becomes redundant. Also, if data is changed in one file, those changes might not be made to other files, reducing data integrity.

Traditional file-oriented applications are often characterized by program-data dependence, meaning that they have data organized in a manner that cannot be read by other programs. To address problems of traditional file-based data management, the database approach was developed. Benefits of this approach include reduced data redundancy, improved data consistency and integrity, easier modification and updating, data and program independence, standardization of data access, and more-efficient program development.

One of the tools that database designers use to show the relationships among data is a data model. A data model is a map or diagram of entities and their relationships. Enterprise data modeling involves analyzing the data and information needs of an entire organization. Entity-relationship (ER) diagrams can be employed to show the relationships between entities in the organization.

The newest, most flexible structure is the relational model, in which data is set up in two-dimensional tables. Tables can be linked by common data elements, which are used to access data when the database is queried. Each row represents a record. Columns of the tables are called attributes, and allowable values for these attributes are called *the domain*. Basic data manipulations include selecting, projecting, and joining. The relational model is easier to control, more flexible, and more intuitive than the other models because it organizes data in tables.

Principle

A well-designed and well-managed database is an extremely valuable tool in supporting decision making.

A DBMS is a group of programs used as an interface between a database and its users and other application programs. When an application program requests data from the database, it follows a logical access path. The actual retrieval of the data follows a physical access path. Records can be considered in the same way: A logical record is what the record contains; a physical record is where the record is stored on storage devices. Schemas are used to describe the entire database, its record types, and their relationships to the DBMS.

A database management system provides four basic functions: providing user views, creating and modifying the database, storing and retrieving data, and manipulating data and generating reports. Schemas and subschemas are entered into the computer via a data definition language, which describes the data and relationships in a specific database. Subschemas are used to define a user view, the portion of the database that a user can access and manipulate. Another tool used in database management is the data dictionary, which contains detailed descriptions of all data in the database.

Once a DBMS has been installed, the database may be accessed, modified, and queried via a data manipulation language. A more specialized data manipulation language is the query language, the most common being Structured Query Language (SQL). SQL is used in several popular database packages today and can be installed on PCs and mainframes.

Popular end-user DBMSs include Corel's Paradox and Microsoft's Access. These DBMSs provide "wizards" to help end users create a new database and load data and to perform many other functions. IBM, Oracle, and Microsoft are the leading DBMS vendors.

Selecting a database management system begins by analyzing the information needs of the organization. Important characteristics of databases include the size of the database, number of concurrent users, performance, the ability of the DBMS to be integrated with other systems, the features of the DBMS, vendor considerations, and the cost of the database management system.

Principle

The number and types of database applications will continue to evolve and yield real business benefits.

Traditional online transaction processing (OLTP) systems put data into databases very quickly, reliably, and efficiently, but they do not support the types of data analysis needed today. So, organizations are building data warehouses, which are relational database management systems specifically designed to support management decision making. Data marts are subdivisions of data warehouses, which are commonly devoted to specific purposes or functional business areas.

Data mining, which is the automated discovery of patterns and relationships in a data warehouse, is emerging as a practical approach to generating hypotheses about the patterns and anomalies in the data that can be used to predict future behavior.

Predictive analysis is a form of data mining that combines historical data with assumptions about future conditions to forecast outcomes of events such as future product sales or the probability that a customer will default on a loan.

Business intelligence is the process of getting enough of the right information in a timely manner and usable form and analyzing it so that it can have a positive impact on business strategy, tactics, or operations. Competitive intelligence is one aspect of business intelligence limited to information about competitors and the ways that information affects strategy, tactics, and operations. Competitive intelligence is not espionage—the use of illegal means to gather information. Counterintelligence describes the steps an organization takes to protect information sought by "hostile" intelligence gatherers. Knowledge management is the process of capturing a company's collective expertise wherever it resides—in computers, on paper, or in people's heads—and distributing it wherever it can help produce the biggest payoff. The goal of knowledge management is to get people to record knowledge (as opposed to data) and then share it.

With the increased use of telecommunications and networks, distributed databases, which allow multiple users and different sites access to data that may be stored in different physical locations, are gaining in popularity. To reduce telecommunications costs, some organizations build replicated databases, which hold a duplicate set of frequently used data.

Multidimensional databases and online analytical processing (OLAP) programs are being used to store data and allow users to explore the data from a number of different perspectives. Open database connectivity (ODBC) standards allow different database applications to share information.

An object-oriented database uses the same overall approach of objected-oriented programming, first discussed in Chapter 4. With this approach, both the data and the processing instructions are stored in the database. An object-relational database management system (ORDBMS) provides a complete set of relational database capabilities, plus the ability for third parties to add new data types and operations to the database. These new data types can be audio, video, and graphical data that require new indexing, optimization, and retrieval features.

In addition to raw data, organizations are increasingly finding a need to store large amounts of visual and audio signals in an organized fashion. There are also a number of special-purpose database systems. Spatial data technology involves the use of an object-relational database to store and access data according to the locations it describes and to permit spatial queries and analysis.

CHAPTER 5: SELF-ASSESSMENT TEST

The database approach to data management provides significant advantages over the traditional file-based approach.

1. A group of programs that manipulate the database and provide an interface between the database and the user of the database and other application programs.

   a. GUI
   b. operating system
   c. DBMS
   d. productivity software

2. A(n) _____ has no relationship between its records and is often used to store and manipulate a single table or file.

3. A primary key is a field or set of fields that uniquely identifies the record. True or False?

4. The duplication of data in separate files is known as

   a. data redundancy
   b. data integrity
   c. data relationships
   d. data entities

5. _____ is a data-modeling approach that starts by investigating the general data and information needs of the organization at the strategic level and then examining more specific data and information needs for the various functional areas and departments within the organization.

6. The most popular database model is

   a. relational
   b. network

c. normalized

d. hierarchical

A well-designed and well-managed database is an extremely valuable tool in supporting decision making.

7. A(n) _____ is a highly skilled and trained systems professional who directs or performs all activities related to maintaining a successful database environment.

8. Once data has been placed into a relational database, users can make inquiries and analyze data. Basic data manipulations include selecting, projecting, and optimization. True or False?

9. Because the DBMS is responsible for providing access to a database, one of the first steps in installing and using a database involves telling the DBMS the logical and physical structure of the data and relationships among the data in the database. This description is called a(n) _____.

10. The commands that are used to access and report information from the database are part of the

a. data definition language

b. data manipulation language

c. data normalization language

d. subschema

11. Access is a popular DBMS for

a. personal computers

b. graphics workstations

c. mainframe computers

d. supercomputers

12. The ability of a vendor to provide global support for large, multinational companies or companies outside the United States is becoming increasingly important. True or False?

The number and types of database applications will continue to evolve and yield real business benefits.

13. A(n) _____ holds business information from many sources in the enterprise, covering all aspects of the company's processes, products, and customers.

14. An information-analysis tool that involves the automated discovery of patterns and relationships in a data warehouse is called

a. a data mart

b. data mining

c. predictive analysis

d. business intelligence

15. _____ is a continuous process involving the legal and ethical collection of information (about other companies or organizations), analysis that doesn't avoid unwelcome conclusions, and controlled dissemination of that information to decision makers.

CHAPTER 5: SELF-ASSESSMENT TEST ANSWERS

(1) c (2) flat file (3) True (4) a (5) Enterprise data modeling (6) a (7) database administrator (8) False (9) schema (10) b (11) a (12) True (13) data warehouse (14) b (15) Competitive intelligence

---

KEY TERMS

attribute 197
business intelligence 224
character 196
competitive intelligence 225
concurrency control 211
counterintelligence 225
data administrator 214
data cleanup 205
data definition language (DDL) 208
data dictionary 209
data integrity 199
data item 197
data manipulation language (DML) 211
data mart 221
data mining 221
data model 201
data redundancy 198
data warehouse 219
database administrator (DBA) 195

database approach to data management 199
database management system (DBMS) 195
distributed database 225
domain 202
enterprise data modeling 201
entity 197
entity-relationship (ER) diagrams 202
field 196
file 196
hierarchy of data 196
joining 203
key 197
knowledge management 225
linking 203
object-oriented database 229
object-oriented database management system (OODBMS) 229

object-relational database management system (ORDBMS) 229
online analytical processing (OLAP) 227
open database connectivity (ODBC) 228
planned data redundancy 201
primary key 197
predictive analysis 222
projecting 203
record 196
relational model 202
replicated database 226
schema 207
selecting 203
subschema 207
traditional approach to data management 198

REVIEW QUESTIONS

1. What is an attribute? How is it related to an entity?
2. Define the term *database*. How is it different from a database management system?
3. What is a flat file?
4. What is the purpose of data cleanup?
5. How would you describe the traditional approach to data management? How does it differ from the database approach?
6. What is data modeling? What is its purpose? Briefly describe three commonly used data models.
7. What is a database schema, and what is its purpose?
8. Identify important characteristics in selecting a database management system.
9. What is the difference between a data definition language (DDL) and a data manipulation language (DML)?
10. What is a distributed database system?
11. What advantages does the open database connectivity (ODBC) standard offer?
12. What is a data warehouse, and how is it different from a traditional database used to support OLTP?
13. What is data mining? What is OLAP? How are they different?
14. What is an ORDBMS? What kind of data can it handle?
15. What is business intelligence? How is it used?
16. Give an example of a visual database.

DISCUSSION QUESTIONS

1. You have been selected to represent the student body on a project to develop a new student database for your school. What actions might you take to fulfill this responsibility to ensure that the project meets the needs of students and is successful?
2. Your company would like to increase revenues from its existing customers. How can data mining be used to accomplish this objective?
3. You are going to design a database for your cooking club to track its recipes. Identify the database characteristics most important to you in choosing a DBMS. Which of the database management systems described in this chapter would you choose? Why? Is it important for you to know what sort of computer the database will run on? Why or why not?
4. Make a list of the databases in which data about you exists. How is the data in each database captured? Who updates each database and how often? Is it possible for you to request a printout of the contents of your data record from each database? What data privacy concerns do you have?
5. You are the vice president of information technology for a large, multinational, consumer packaged goods company (e.g., Procter & Gamble, Unilever, or Gillette). You must make a presentation to persuade the board of directors to invest $5 million to establish a competitive-intelligence organization—including people, data-gathering services, and software tools. What key points do you need to make in favor of this investment? What arguments can you anticipate that others might make?
6. Briefly describe how visual and audio databases can be used by companies today.
7. Briefly discuss what impact data privacy legislation may have on the building and use of customer and employee data warehouses.

PROBLEM-SOLVING EXERCISES

1. Develop a simple data model for the members of a student club, where each row is a student. For each row, what attributes should you capture? What will be the unique key for the records in your database? Describe how you might use the database.

2. A video movie rental store is using a relational database to store information on movie rentals to answer customer questions. Each entry in the database contains the following items: Movie ID No. (primary key), Movie Title, Year Made, Movie Type, MPAA Rating, Number of Copies on Hand, and Quantity Owned. Movie types are comedy, family, drama, horror, science fiction, and western. MPAA ratings are G, PG, PG-13, R, X, and NR (not rated). Use an end-user database management system to build a data entry screen to enter this data. Build a small database with at least 10 entries.

3. To improve service to their customers, the salespeople at the video rental store have proposed a list of changes being considered for the database in the previous exercise. From this list, choose two database modifications and modify the data entry screen to capture and store this new information.

    Proposed changes:

    a. Add the date that the movie was first available to help locate the newest releases.
    b. Add the director's name.
    c. Add the names of three primary actors in the movie.
    d. Add a rating of one, two, three, or four stars.
    e. Add the number of Academy Award nominations.

TEAM ACTIVITIES

1. In a group of three or four classmates, interview a database administrator (DBA) for a company in your area. Describe this person's duties and responsibilities. What are the career opportunities of a DBA?

2. As a team of three or four classmates, interview business managers from three different businesses that use databases to help them in their work. What data entities and data attributes are contained in each database? How do they access the database to perform analysis? Have they received training in any query or reporting tools? What do they like about their database and what could be improved? Do any of them use data-mining or OLAP techniques? Weighing the information obtained, select one of these databases as being most strategic for the firm and briefly present your selection and the rationale for the selection to the class.

3. Imagine that you and your classmates are a research team developing an improved process for evaluating auto loan applicants. The goal of the research is to predict which applicants will become delinquent or forfeit their loan.

Those who score well on the application will be accepted; those who score exceptionally well will be considered for lower-rate loans. Prepare a brief report for your instructor addressing these questions:

a. What data do you need for each loan applicant?
b. What data might you need that is not typically requested on a loan application form?
c. Where might you get this data?
d. Take a first cut at designing a database for this application. Using the chapter material on designing a database, show the logical structure of the relational tables for this proposed database. In your design, include the data attributes you believe are necessary for this database, and show the primary keys in your tables. Keep the size of the fields and tables as small as possible to minimize required disk drive storage space. Fill in the database tables with the sample data for demonstration purposes (10 records). Once your design is complete, implement it using a relational DBMS.

WEB EXERCISES

1. Use a Web search engine to find information on one of the following topics: business intelligence, knowledge management, predictive analysis. Find a definition of the term, an example of a company using the technology, and three companies that provide such software. Cut graphics and text material from the Web pages and paste them into a word processing document to create a two-page report on your selected topic. At the home page of each software company, request further information from the company about its products.

2. Use a Web search engine to find three companies in an industry that interests you that use a database management system. Describe how databases are used in each company. Could the companies survive without the use of a database management system? Why?

CAREER EXERCISES

1. For a career area of interest to you, describe three databases that could help you on the job.

2. How could you use data mining to help you make better decisions at work? Give five specific examples.

VIDEO QUESTIONS

Watch the video clip **Predicting Huge Surf** and answer these questions:

1. The Maverick's Surf Contest has been hailed by Sports Illustrated as "the Super Bowl of big wave surfing." What role do databases play in Jeff Clark's ability to predict the best day and time for the contest, when the surf will be at its peak?

2. The reporter claims that the surf contest is some-times called with only hours of notice. How, would you guess, does Jeff get the word out?

CASE STUDIES

Case One

Brazilian Grocer Gets Personal with Customers

Grupo Pão de Açúcar is Brazil's largest retail company, and its brand Pão de Açúcar (Portuguese for "Sugarloaf," Rio de Janeiro's famous mountain) is Brazil's most popular supermarket. Pão de Açúcar attributes its popularity to the personalized service it offers its millions of customers from its 433 stores. You may wonder how such a large company can possibly offer personalized service to so many customers. Pão de Açúcar found the solution in databases, data mining, and database management systems.

Through a study on customer loyalty, Pão de Açúcar found that customers today are going to fewer grocery stores than they used to and becoming more loyal to a specific store. Based on this information, Miriam Salomão, market-intelligence manager for Pão de Açúcar, set out to analyze customer purchasing data so that the chain could deliver more personalized services. However, the customer data being captured and stored in the company database was not customer specific. "Although we had a lot of data from our checkouts—detailed information by product, day, and store—this data didn't bring us a true understanding of our customers," says Salomão. "We needed to create a view of the data that consolidated each one of the tickets [customer receipts] and associated those tickets with people."

The store launched a customer relationship program that centered around a "loyalty card." Customers interested in receiving personalized discounts fill out a short registration form to receive a loyalty card, which is swiped at the checkout to apply additional discounts. Within no time, Pão de Açúcar had 1.73 million households signed up for the program. The loyalty card provided customer identification, a database key to group purchases by customer, and a foundation on which to build powerful information systems and services.

Pão de Açúcar called in database specialists to assist it with the project. The specialists implemented a data-mining solution to analyze customer activity data. They began by developing a data mart that organized the data so that individual transactions of each customer could be viewed. The data mart is used to create statistical models related to customer segmentation, consumption profiles, buying propensity, and other meaningful reports. This information provides valuable customer information and insight into their interests and needs.

The new program, formally named Mais, Portuguese for "More," allowed Pão de Açúcar to relate with customers individually. Customers receive a personalized mailing indicating sales on items that they have previously purchased or offers for new products related to those the customer uses. Mais members also receive special treatment and perks while shopping. A survey indicated that customers appreciate the personal recognition even more than the financial benefits. Miriam Salomão believes that getting to know customers as individuals has provided the store with a competitive advantage over its competition. "We know our customers, and we know they prefer a more personal, customized relationship. I'm sure they recognize us as a different store. This perception, although not easily measurable, is very important—maybe the most important perception of all."

Discussion Questions

1. What types of information can retailers such as Pão de Açúcar gather from checkout data that is not associated with a customer?
2. What additional information can be acquired when customer identity is associated with checkout data?

Critical Thinking Questions

3. What privacy issues come into play when customer identity is associated with checkout data?

4. What additional benefits and issues arise when customer data is combined from several different types of retail stores, such as the information associated with a typical credit card?

SOURCES: "Pão de Açúcar Making Millions of Customers Feel Like Family with SAS," *www.sas.com/success/paodeacucar.html*, accessed March 6, 2004; Grupo Pão de Açúcar Web site, *www.cbd-ri.com.br/eng/home/index.asp*, accessed March 6, 2004; SAS Data and Text Mining, *www.sas.com/technologies/analytics/datamining*, accessed March 6, 2004.

Case Two

DBMS Upgrade Faces Employee Opposition

Often upgrading a database management system (DBMS) leads to changes in business procedures, and change is sometimes met with resistance by the workforce. This was the result when Huntington Bancshares Inc., a $28 billion regional bank holding company based in Columbus, Ohio, moved to a new DBMS.

Prior to updating its system, Huntington Bancshares delivered a paper report, called the balance sheet income report, to its hundreds of offices every month. The report included 200,000 pages—the equivalent of 40 cartons of paper—of detailed financial information. The company decided it was time to move to a paperless, Web-based system to save on costs related to processing, printing, and delivering 2.4 million pages per year. Employees should celebrate such a decision, shouldn't they? Unfortunately, Huntington Bancshares employees had a different reaction. Here is a sample of their comments:

- "My manager says I have to have these [paper] reports for my file every month."
- "I'm not going to be able to do my job anymore."
- "I can't possibly ask my people to learn this. I'll have to do it for them every month."
- "You may have saved paper, but you have just doubled my workload."
- "Who made this decision?"

In response to the employee feedback, Raymond Heizer, IS project leader for corporate profitability systems at Huntington, says, "As far as our users were concerned, the sun came up every morning, and they got their balance sheet and income statement delivered to their desk every month end."

Despite employee opposition, Huntington Bancshares went ahead with the upgrade. It implemented an Oracle Corp. database running on a UNIX server and purchased a reporting system from Crystal Decisions Inc. The rollout took four months and cost a little more than $1 million. The bank is now saving $30,000 per year in paper costs alone, says Al Werner, vice president of corporate systems. But the return on investment has been realized in areas other than paper savings.

Cost center managers can view up-to-date balance sheet income reports immediately online, rather than waiting for the first of the month. They can also more easily see and resolve exceptions—items in an account balance that don't match the credits. "We've always had a lot of data in the bank, but it was always seen in rows and columns. Now we can have bar charts of mismatches," says Al Werner, vice president of corporate systems. "Huntington now has a standard set of metrics on a single report that "each branch can use to determine the 'health' of his/her branch."

Huntington Bancshares' success in reducing its piles of paper has not been a universal experience, though. With the invention of the computer, many dreamed of a paperless society. In many cases the opposite has been true. Computers generate amazing amounts of useful information, and printed reports are the norm in most corporations. Paper consumption by U.S. companies is growing 6 to 8 percent annually, according to document technology user group Xplor International in Torrance, California.

Still, new government programs and regulations are encouraging banks to go paperless. In June 2003, Congress passed the Check Clearing for the 21st Century Act, also known as Check 21, which allows banks to voluntarily exchange electronic images over networks instead of using paper checks. Huntington is riding the crest of a wave of banks trying to go paperless with online banking statements, to save money, and to comply with new regulations, says Avivah Litan, an analyst at Gartner Inc. "The trend started two years ago, but in the last nine months, it's really been moving ahead," Litan says.

As for the disgruntled Huntington Bancshares employees, they have stopped mourning the loss of their beloved paper reports now that they see the benefits of the database and reporting tools, Werner says. Now they're saying, "Wow, these are pretty nice features."

Discussion Questions

1. What are the benefits of viewing database reports on paper?
2. What are the benefits of being able to interact with the data online?

Critical Thinking Questions

3. Why do you think Huntington employees reacted so strongly to news of the new paperless system?

4. What will it take to turn businesses away from their dependency on paper?

Case Three
Hyundai Shoots for #1 with Executive DBMS

Hyundai Motor Company has set a lofty goal of becoming the top auto producer for the twenty-first century. It is counting on its data-driven information systems to take it there. The 30-year-old company has rapidly become one of the world's largest auto producers. Its business strategy revolves around using advanced technology to produce top-quality, reliable vehicles that satisfy customers. With over 47,000 employees and capital exceeding $350 million, Hyundai Motor Company is the largest independent manufacturing plant in Korea.

Hyundai employs an enterprise-wide DBMS, called an Enterprise Information and Management System (EIMS), which supplies valuable information to its manufacturing plants in Korea, Tokyo, Peking, Detroit, and Frankfurt, and subsidiaries such as Hyundai Motor America, Hyundai Auto Technical, Hyundai Motor Finance Company, and Hyundai Motor India. Hyundai counts on the decision support information stored in its data warehouse to direct decisions made at all levels of management.

The system employs several different reporting tools to keep top managers informed of the state of the company. They receive periodic reports on the company's progress toward long-term goals. Exception reports provide a warning system, allowing executives to view all levels of sales and production volumes and to locate performance problems by comparing figures with an established warning point. Many other reports support decision making and provide timely access to management-level information.

To assist decision makers, the EIMS data warehouse stores information from all organizational divisions, including:

- Organizational charts, personnel records, and staff counts from the human resource department
- Daily and monthly sales, market share, and market analysis from domestic sales
- Foreign exports, daily exports, local sales, inventory, and competitive analysis from foreign sales
- production per factory and per model, target achievement, and factory operation from production

Hyundai has found that the best way to protect capital investments is to invest in systems to manage valuable data and information. Rather than reacting to market movements, Hyundai can be proactive and innovative—in other words, lead rather than follow.

Discussion Questions

1. How can Hyundai's EIMS assist managers in setting car sticker prices?

2. What features of a DBMS make it especially useful to high-level executives?

Critical Thinking Questions

3. In a distributed organization such as Hyundai, how can database reports be distributed uniformly and consistently across the organization?

4. No doubt Hyundai's competitors also employ sophisticated database management tools. What factors allow one organization to better utilize its data to achieve a competitive advantage?

SOURCES: "Technology Drives Decisions at Hyundai," SAS Institute Web site, www.sas.com/success/hyundai.html, accessed March 6, 2004; Hyundai Web site, www.hyundai.com, accessed March 6, 2004; "Hyundai Helps Chinese Auto Industry Mature," The Korea Herald, October 28, 2003, www.lexis-nexis.com.

SOURCES: Lucas Mearian, "Bank Tries to Break the Paper Habit," Computerworld, August 4, 2003, www.computerworld.com; Huntington Bancshares Inc Web site, www.huntington.com, accessed March 6, 2004; Lucas Mearian, "Huntington Bancshares Moves to AIX for Scalability," Computerworld, July 7, 2003, www.computerworld.com.

NOTES

Sources for the opening vignette, "The Secret of Their Success: DDB Uses SAS® to Explore Why, How Consumers Respond to Certain Brands," SAS Web site, www.sas.com/success/ddb.html, accessed March 5, 2004; DDB press kit, www.ddb.com/5_media/downloads/presskit.pdf, accessed March 5, 2004; "Bill Burnbach Said... ," www.ddb.com/5_media/downloads/bb_quotes.pdf, accessed March 5, 2004.

1. Binkley, Christina, "Soon, the Desk Clerk Will Know All About You," The Wall Street Journal, May 8, 2003, p. D4.

2. Alsever, Jennifer, "Restaurants Keep Tabs on Diners," Rocky Mountain News, February 22, 2004, p. K1.

3. Schell, Ernie, "Spinning Straw Into Gold," Catalog Age, May 1, 2003, p. 6.

4. Gynne, Peter, "Managing the Life Science Data Deluge," Geonomics and Proteomics, May 01, 2003, p. 39; Wilson, Clare, "A Healthy Investment," New Scientist, May 10, 2003, p. 25.

5. Hamilton, David, "Biological Databases Are Becoming Freely Available," The Wall Street Journal, May 19, 2003, p. R12.

6. Wunderlich, Clifford, "ALTA Religion Database," *Library Journal Reviews*, May 15, 2003, p. 137.

7. Benjamin, Matthew, "UPS Eyes a Future Going Far Beyond Package Delivery," *U.S. News & World Report*, January 26, 2004, p. ee2.

8. Fonseca, Brian, "DBA Boundaries Blurring," *eWeek*, January 26, 2004, p. 995.

9. Elsner, Ralf, et al., "Optimizing Rhenania's Mail Order Business," *Interfaces*, January 2003, p. 50.

10. Krol, Carol, "Looking to Expand," *BtoB*, May 5, 2003, p. 32.

11. Datz, Todd, "State to Share Data with FBI," *CIO Magazine*, May 15, 2003.

12. Fisher, Dennis, "New DHS Border Plan Scrutinized," *eWeek*, January 12, 2004, p. 1.

13. Staff, "Most Utilities Fall Short on Customer Data Accuracy," *Utility Weekly*, May 2003, p. 3.

14. Mossberg, Walter, "Microsoft's OneNote Turns Scribbled Ideas into Computer Files," *The Wall Street Journal*, October 30, 2003, p. B1.

15. McAmis, David, "Introducing InfoPath," *Intelligent Enterprise*, February 7, 2004, p. 36.

16. Perez, Jeanette, "Open-Source DBs Go Big Time," *Intelligent Enterprise*, February 7, 2004, p. 8.

17. Hall, Mark, "MySQL Breaks into the Data Center," *Computerworld*, October 13, 2003, p. 32.

18. Whting, Rick, "Open-Source Database Gaining," *InformationWeek*, January 12, 2004, p. 10.

19. Fonseca, Brian, "Database Opening Up," *eWeek*, January 12, 2004, p. 12.

20. Langley, Nick, "DB2 Vies for Top Database Position," *Computer Weekly*, January 27, 2004, p. 40.

21. Whiting, Rick, "Sybase and IBM Ready Databases for Linux," *InformationWeek*, January 26, 2004, p. 49.

22. Whiting, Rick, "Sybase and IBM Ready Databases for Linux," *InformationWeek*, January 26, 2004, p. 49.

23. Voorhees, Mark, "Ready to Rumble," *The American Lawyer*, February 2004.

24. Arnott, Sarah, "Scottish Police Forces to Share Data," *Computing*, January 8, 2004, p. 8.

25. Staff, "Partners in Supply," *Design News*, January 12, 2004, p. 50.

26. Foley, John, "Blue Print for Change," *InformationWeek*, January 26, 2004, p. 22.

27. Whiting, Rick, "Stock Trades Get a Boost," *InformationWeek*, January 12, 2004, p. 47.

28. Robb, Drew, "The Database Diet," *Computerworld*, March 8, 2004, p. 32.

29. Staff, "Oracle and HP Set World Record," *VARBusiness*, January 26, 2004, p. 64.

30. Thibodeau, Patrick, "The Web's Next Leap," *Computerworld*, April 21, 2003, p. 34.

31. Loftus, Peter, "Smooth Talk," *The Wall Street Journal*, March 31, 2003, p. R9.

32. Fonseca, Brian, "ATC Database Upgrade Supports Growth, Improves Reliability," *eWeek*, January 26, 2004, p. 33.

33. Radding, Alan, "SSL Virtual Private Networks are Simpler to Set Up," *Computerworld*, April 28, 2003, p. 28.

34. Saran, Cliff, "Code Issue Affects 40% of Websites," *Computer Weekly*, January 13, 2004, p. 5.

35. Betts, Mitch, "Unexpected Insights," *Computerworld*, April 14, 2003, p. 34.

36. Anthes, Gary, "Best In Class: Data Warehouse Boosts Profits by Empowering Sales Force," *Computerworld*, February 24, 2003, p. 46.

37. Sullivan, Laurie, "Oracle Embraces RFID," *Information Week*, February 2, 2004, p. 8.

38. Verton, Dan, "FBI Begins Knowledge Management FaceLift," *Computerworld*, April 21, 2003, p. 10.

39. Davis, Ann, "Data Collection Is Up Sharply Following 9/11," *The Wall Street Journal*, May 22, 2003, p. B1.

40. Anthes, Gary, "The Forrest is Clear," *Computerworld*, April 14, 2003, p. 31.

41. Leon, Mark, "Keys to the Kingdom," *Computerworld*, April 14, 2003, p. 42.

42. Siener, Scott, "Total Automation," *Computerworld*, April 14, 2003, p. 52.

43. Schlosser, Julie, "Looking for Intelligence in Ice Cream," *Fortune*, March 17, 2003, p. 114.

44. Staff, "The State of Business Intelligence," *Computer Weekly*, February 3, 2004, p. 24.

45. Vaas, Lisa, "Tools Give Insights into Databases," *eWeek*, January 27, 2003, p. 9.

46. Bacheldor, Beth, "Object-Oriented Database Speeds Queries," *InformationWeek*, March 10, 2003, p. 49.

47. Johnson, Colin, "Tech Gives Tone-Deaf a Voice Makeover," *Electronic Engineering Times*, May 5, 2003, p. 51.

48. Derra, Skip, "Image Analysis Software Shows Its Flexibility," *Drug Discovery and Development*, March 01, 2003, p. 61.

49. Staff, "Software System Allows Geographic Display, Analysis of Upstream Activity," *Offshore*, February 2003, p. 58.

50. Worthen, Ben, "Database Cracks Murder Case," *CIO Magazine*, May 1, 2003.

51. Vaas, Lisa, "Virtual Databases Make Sense Out of Varied Data," *eWeek*, March 31, 2003, p. 12.

52. Dignan, Larry, "Diagnosis Disconnected," *Baseline*, May 5, 2003.

53. Thomas, Daniel, "Online Age Verification," *Computer Weekly*, May 6, 2003, p. 14.

CHAPTER
· 6 ·

Telecommunications and Networks

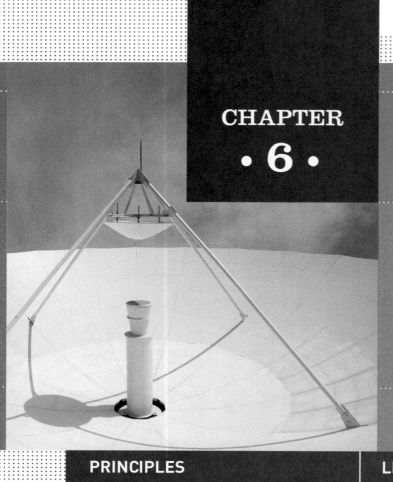

PRINCIPLES

- Effective communications are essential to organizational success.

- An unmistakable trend of communications technology is that more people are able to send and receive all forms of information over greater distances at a faster rate.

LEARNING OBJECTIVES

- Define the terms *communications* and *tele-communications* and describe the components of a telecommunications system.

- Identify broad categories of communications media and discuss the basic characteristics of specific media types.

- Describe how a modem works.

- Explain the types of telecommunications carriers today and the services they provide.

- Identify the benefits associated with a telecommunications network.

- Define the term *network topology* and identify five alternatives.

- Discuss the different communications protocols and devices used for telecommunications.

- Name three distributed processing alternatives and discuss their basic features.

- List some telecommunications applications that organizations are benefiting from today.

INFORMATION SYSTEMS IN THE GLOBAL ECONOMY ▶

THE BARILLA GROUP, ITALY

Pasta Giant Uses Wireless Networking to Liberate Employees

In 1877 Pietro Barilla opened a bread and pasta bakery in Parma, Italy. Today, 128 years later, Pietro's three great-grandsons, Guido, Luca, and Paolo, control what has become Italy's largest food-processing company: Barilla. With 29 production plants distributed around the world, providing 1,265,000 tons of food products to more than 100 countries, Barilla is undisputedly the world's leading name in pasta. Barilla is a great family-run multinational, squarely set on the shoulders of four generations of the Barilla family.

Barilla's business philosophy incorporates the metaphor of an iceberg: the visible part of a product is what is eaten, directly perceived by the consumer. But like the tip of the iceberg, it is only a part of the entire product. The hidden part is enormous, made up of research, intuition, design, control, and study—the activities that are part of the Barilla Quality System. Barilla considers its human resources its greatest asset—the key to its success—and fosters employee talent and promotes their leadership with a managerial style based on integrity in decision making and conduct.

Recently, the American branch of Barilla constructed a new facility in Bannockburn, Illinois. Barilla designed a state-of-the-art communications and computer network system to improve information flow throughout the large company campus. The new network connects the two sprawling wings of the main facility and also a research kitchen located in a separate building a mile away.

With Barilla's emphasis on people, teamwork is an important component of its organizational culture, and the new facility's design supports teamwork with several "huddle" rooms, as well as a central area where people can hold discussions or planning sessions in a congenial, open atmosphere. The new computer network adds yet another benefit, freeing employees to travel wherever they are needed within the company campus through mobile technology.

Barilla America worked with Cisco Systems to implement a wireless computer network that spanned the entire campus. Devices called "access points," which send and receive communication signals through the air, were distributed about the organization and connected to the company's servers. Special wireless adapter cards were installed in employees' notebook and handheld computers so that wherever they roamed on the corporate campus, they would be connected to the network. Cisco also installed powerful long-range wireless connections to link the headquarters and the test kitchen. Today's technology provides top-notch speed in wireless networks for a fraction of the installation expense of wired networks.

Perhaps the most interesting of the new technologies implemented at Barilla are its Internet phones. They offer standard phone services, as well as additional services that interact with computer systems. Internet phones use the same technology as the regular Internet—the Internet Protocol (IP)—to communicate over the local data network. IP phones enable voice, video, and data transmissions to converge in one device. An additional bonus of IP phones is that they are less expensive than traditional leased lines and provide a company with full control over the telephone network.

The combination of wireless computing and IP phones provides Barilla employees the ability to access any information, including database reports from the corporate database server, e-mail, and the Web, and communicate with colleagues one at a time or in groups through voice or video from any location on the campus. Such flexibility frees Barilla employees to work and interact in any number of environments and settings. "People can sit in a huddle room or anywhere else, enter data, or send and receive e-mail," says Vince Danca, infrastructure manager. "I see people in our corporate 'living room' with laptops checking sales data, discussing trends, doing research. Everybody loves using wireless." Executives use wireless-equipped PDAs to move throughout the office while staying connected to the network. They can use the PDA to check e-mail or conduct calls. "I can take calls anywhere at headquarters without running to the phone," says Danca. "It makes me and anybody else using the system immediately more responsive to customers and other callers."

Wireless networking has liberated Barilla employees from their desks and offices and allows them to work more naturally together. Productivity and employee morale is at an all-time high at Barilla's U.S. offices. It's easy to understand why so many businesses are moving to untether their employees through wireless networking.

As you read this chapter, consider the following:

- Having instant access to valuable information is the primary goal of an information system. How do computer networks support this goal?
- How are wireless networks affecting the way people do their jobs?

Why Learn About Telecommunications and Networks?

Today's decision makers need the ability to access data wherever it resides. They must be able to establish fast, reliable connections to exchange messages; download data, software, and updates; route transactions to processors; and send output to printers. Regardless of your chosen major or future career field, you will need the communications capabilities provided by telecommunications and networks. Among all business functions, perhaps supply chain management emphasizes the use of telecommunications and networks the most because it encompasses inbound logistics, warehouse and storage, production, finished product storage, outbound logistics, marketing and sales, and customer service. All members of the supply chain must work together efficiently and effectively to increase the value perceived by the customer, so partners must communicate well. Other employees in managerial, human resources, finance, research and development, marketing, and sales positions must also use communications technology to communicate with people both inside and outside the organization. As a member of any organization, you must be able to take advantage of the capabilities that these technologies offer you in order to be successful. We begin this chapter by discussing the importance of effective communications.

In today's high-speed business world, effective communication is critical to organizational success, as it is with The Barilla Group. Often, what separates good management from poor management is the ability to identify problems and solve them with available resources. Efficient communication is one of the most valuable of these resources, because it enables a company to keep in touch with its operating divisions, customers, suppliers, and stockholders. For example, Wells Fargo & Company, the global financial services firm, is investing in a major upgrade of its online application called Commercial Electronic Office (CEO) to make the services easier to use. CEO is accessed over telecommunications networks by some 140,000 end users within Wells Fargo's corporate clients. They use the system to view account information and bank statements and to access money-transfer services and an automated clearinghouse that processes nearly $6 trillion in electronic payments annually. "We are an

information-rich company, but there's so much information that it's sometimes hard to find what you need," said Steve Ellis, executive vice president of Wells Fargo's wholesale services group. "So we're grabbing it and pushing it up front to the person who needs it, so they can do their job faster."[1] Forward-thinking companies such as Wells Fargo hope to save billions of dollars, reduce time to market, and enable collaboration with their business partners through the use of telecommunications systems.

AN OVERVIEW OF COMMUNICATIONS SYSTEMS

Memos, notices on bulletin boards, and presentations are all obvious examples of continual communications within a business organization. Other not-so-obvious examples include policy and procedure manuals and even salaries (they communicate the company's perception of the value of the contribution of the person being paid). Communication also exists in other forms—for instance, warning lights on a computer system that monitors manufacturing processes and signals from a building management system that monitors temperature, humidity, lighting, and security of a building. Communication is any process that permits information to pass from a sender to one or more receivers. Communication of all types forms a major part of any business system. Therefore, managers must gain an appreciation of communication concepts, media, and devices—as well as an understanding of how these factors may best be employed to develop effective and efficient business systems.

Communications

Communications is the transmission of a signal by way of a medium from a sender to a receiver (see Figure 6.1). The signal contains a message composed of data and information. The signal goes through some communications medium, which is anything that carries a signal between a sender and receiver. In human speech, the sender transmits a signal through the transmission medium of the air. In telecommunications, the sender transmits a signal through a transmission medium such as a cable.

communications
The transmission of a signal by way of a medium from a sender to a receiver.

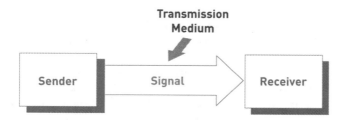

Transmission Medium

Sender — Signal → Receiver

Figure 6.1

Overview of Communications

The message (data and information) is communicated via the signal. The transmission medium "carries" the signal.

The components of communications can easily be recognized if you consider human communication (see Figure 6.2). When we talk to one another face to face, we send messages to each other. A person may be the sender at one moment and the receiver a few seconds later. The same entity, a person in this case, can be a sender, a receiver, or both. This process is typical of two-way communication. The signals we use to convey these messages are our spoken words—our language. For communication to be effective, both sender and receiver must understand the signals and agree on the way they are to be interpreted. For example, if the sender in Figure 6.2 is speaking in a language the receiver does not understand, or if the sender believes that a particular word has one meaning and the receiver believes that the word has some other meaning, effective communication will not occur.

In addition to the flow of communications, shown in Figure 6.2, communications can be synchronous or asynchronous. With **synchronous communications**, the receiver gets the message instantaneously, when it is sent. Voice and phone communications are examples of

synchronous communications
Communications in which the receiver gets the message instantaneously.

Figure 6.2

Communications and Telecommunications

In human speech, the sender transmits a signal through the transmission medium of the air. In telecommunication, the sender transmits a signal through a cable or other telecommunications medium.

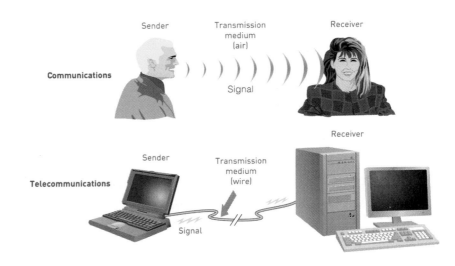

asynchronous communications
Communications in which the receiver gets the message minutes, hours, or days after it is sent.

synchronous communications. With **asynchronous communications**, the receiver gets the message later—sometimes hours or days after the message is sent. Sending a letter through the post office or e-mail over the Internet are examples of asynchronous communications. Academic researchers are actively investigating the impact of both synchronous and asynchronous communications on effectiveness, performance, and other organizational measures. Both types of communications are important in business, regardless of whether the communication is done manually or electronically.

Telecommunications

Telecommunications refers to the electronic transmission of signals for communications, by such means as telephone, radio, and television. Telecommunications has the potential to create profound changes in business because it lessens the barriers of time and distance. Telecommunications not only is changing the way businesses operate but also is altering the nature of commerce itself. As networks are connected with one another and information is transmitted more freely, a competitive marketplace is making excellent quality and service imperative for success. **Data communications**, a specialized subset of telecommunications, refers to the electronic collection, processing, and distribution of data—typically between computer system hardware devices. Data communications is accomplished through the use of telecommunications technology.

data communications
A specialized subset of telecommunications that refers to the electronic collection, processing, and distribution of data—typically between computer system hardware devices.

Figure 6.3 shows a general model of telecommunications. The model starts with a sending unit (1), such as a person, a computer system, a terminal, or another device, that originates the message. The sending unit transmits a signal (2) to a telecommunications device (3). The telecommunications device performs a number of functions, which can include converting the signal into a different form or from one type to another. A telecommunications device is a hardware component that allows electronic communications to occur, or to occur more efficiently. The telecommunications device then sends the signal through a medium (4). A **telecommunications medium** is anything that carries an electronic signal and interfaces between a sending device and a receiving device. The signal is received by another telecommunications device (5) that is connected to the receiving computer (6). The process can then be reversed, and another message can go back from the receiving unit (6) to the original sending unit (1). In this chapter, we will explore the components of the telecommunications model shown in Figure 6.3. An important characteristic of telecommunications is the speed at which information is transmitted, which is measured in bits per second (bps). Common speeds are in the range of thousands of bits per second (Kbps) to millions of bits per second (Mbps) and even billions of bits per second (Gbps).

telecommunications medium
Anything that carries an electronic signal and interfaces between a sending device and a receiving device.

Advances in telecommunications technology allow us to communicate rapidly with clients and coworkers almost anywhere in the world. Telecommunications also reduces the amount of time needed to transmit information that can drive and conclude business actions.

In addition to external communications, telecommunications technology also helps businesses coordinate activities and integrate various departments to increase operational efficiency and support effective decision making. The far-reaching developments of telecommunications are having and will continue to have a profound effect on business information systems and on society in general.

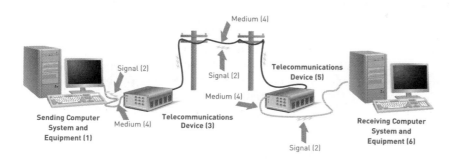

Figure 6.3

Elements of a Telecommunications System

Telecommunications devices relay signals between computer systems and transmission media.

Telecommunications technology enables business people to communicate with coworkers and clients from remote locations.

(Source: © Corbis.)

COMMUNICATIONS CHANNELS

In today's global, fast-paced business environment, the use of telecommunications can help businesses solve problems and capitalize on opportunities. But using telecommunications effectively requires careful analysis of telecommunications media and devices.

Basic Communications Channel Characteristics

The transmission medium carries messages from the source of the message to its receivers. A given transmission medium may be divided into one or more communications channels, each capable of carrying a message. Communication channels can be classified as simplex, half-duplex, or full-duplex.

A **simplex channel** can transmit data in only one direction. It is seldom used for business telecommunications. Doorbells and radio and TV broadcasting operate using a simplex channel. A **half-duplex channel** can transmit data in either direction, but not simultaneously. For example, A can begin transmitting to B over a half-duplex line, but B must wait until A is finished to transmit to A. Personal computers are usually connected to a remote computer

simplex channel
A communications channel that can transmit data in only one direction.

half-duplex channel
A communications channel that can transmit data in either direction, but not simultaneously.

full-duplex channel

A communications channel that permits data transmission in both directions at the same time, thus the full-duplex channel is like two simplex lines.

bandwidth

The range of frequencies that an electronic signal occupies on a given transmission medium.

Shannon's fundamental law of information theory

The law of telecommunications that states that the information-carrying capacity of a channel is directly proportional to its bandwidth—the broader the bandwidth, the more information that can be carried.

broadband

Telecommunications in which a wide band of frequencies is available to transmit information allowing more information to be transmitted in a given amount of time.

over a half-duplex channel. A **full-duplex channel** permits data transmission in both directions at the same time, so a full-duplex channel is like two simplex lines. Private leased lines or two standard phone lines are required for full-duplex transmission.

Channel Bandwidth and Information-Carrying Capacity

In addition to the directions that telecommunications can travel, businesses must consider the speed at which signals can be transmitted. Speed depends on the channel **bandwidth,** that is, the range of frequencies that an electronic signal occupies on a given transmission medium. **Shannon's fundamental law of information theory** states that the information-carrying capacity of a channel is directly proportional to its bandwidth—the broader the bandwidth, the more information per unit time (bits per second, or bps) can be carried. In general, **broadband** refers to telecommunications in which a wide band of frequencies is available to transmit information, allowing more information to be transmitted in a given amount of time.

Telecommunications professionals consider the capacity of the channel when they recommend transmission media for particular business needs. In general, today's organizations need more bandwidth for increased transmission speed to carry out their daily functions. Let's take a look at the different types of telecommunications media that are available.

Types of Media

Various types of communications media are available. Each type exhibits its own characteristics, including cost, capacity, and speed. In developing a telecommunications system, the selection of transmission media depends on the purpose of the overall information and organizational systems, the purpose of the telecommunications subsystems, and the characteristics of the media. The transmission media should be chosen to support the goals of the information and organizational systems at the least cost and to allow for possible modification of system goals over time. Transmission media can be divided into two broad categories: guided transmission media, in which communications signals are guided along a solid medium, and wireless media, in which the communications signal is sent over airwaves, as summarized in Table 6.1.

Table 6.1

Transmission Media Types

Guided Media Types			
Media Type	**Description**	**Advantages**	**Disadvantages**
Twisted-pair wire cable	Twisted pairs of copper wire, shielded or unshielded	Used for telephone service; widely available	Transmission speed and distance limitations
Coaxial cable	Inner conductor wire surrounded by insulation	Cleaner and faster data transmission than twisted-pair wire	More expensive than twisted-pair wire
Fiber-optic cable	Many extremely thin strands of glass bound together in a sheathing; uses light beams to transmit signals	Diameter of cable much smaller than coaxial; less distortion of signal; capable of high transmission rates	Expensive to purchase and install
Wireless Media Types			
Media Type	**Description**	**Advantages**	**Disadvantages**
Microwave	High-frequency radio signal sent through atmosphere and space (often involves use of communications satellites)	Avoids cost and effort to lay cable or wires; capable of high-speed transmission	Must have unobstructed line of sight between sender and receiver; signal highly susceptible to interception
Cellular	Divides coverage area into cells; each cell has mobile telephone subscriber office	Supports mobile users; costs are dropping	Signal highly susceptible to interception
Infrared	Signals sent through air as light waves	Devices can be moved, removed, and installed without expensive wiring	Must have unobstructed line of sight between sender and receiver; transmission effective only for short distances

Twisted-Pair Wire Cable

Twisted-pair wire cable contains two or more twisted pairs of wire, usually copper (see Figure 6.4a). Proper twisting of the wire keeps the signal from "bleeding" into the next pair and creating electrical interference. Because the twisted-pair wires are insulated, they can be placed close together and packaged in one group. Hundreds of wire pairs can be grouped into one large wire cable.

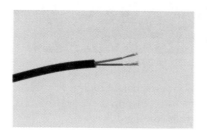

Figure 6.4

Types of Cable

(a) Twisted-pair wire cable

Source: Fred Bodin.

Twisted-pair cables come in different categories (Category 1, 2, 3, 4, 5, 5E, and 6). The lower categories are used primarily in homes. Higher categories are used as a cheaper alternative to coaxial cable for smaller networks. Category 1 is traditional telephone cable. Category 3 is the most common type of cable found in corporate settings, and it normally contains four pairs of wire. Category 5 is frequently installed in new buildings and is capable of carrying data at speeds faster than 1 gigabit/second.

Coaxial Cable

Figure 6.4b shows a typical coaxial cable, similar to that used in cable television installations. A coaxial cable consists of an inner conductor wire surrounded by insulation, called the *dielectric*. The dielectric is surrounded by a conductive shield (usually a layer of foil or metal braiding), which is in turn covered by a layer of nonconductive insulation, called the *jacket*. When used for data transmission, coaxial cable falls in the middle of the cabling spectrum in terms of cost and performance. The cable itself is more expensive than twisted-pair wire cable but less so than fiber-optic cable (discussed next). However, the cost of installation and other necessary communications equipment makes it difficult to compare the total costs of using each media. Coaxial cable offers cleaner and crisper data transmission (less noise) than twisted-pair wire cable. It also offers a higher data transmission rate.

(b) Coaxial cable

Source: Fred Bodin.

Fiber-Optic Cable

Fiber-optic cable, consisting of many extremely thin strands of glass or plastic bound together in a sheathing (a jacket), transmits signals with light beams (see Figure 6.4c). These high-intensity light beams are generated by lasers and are conducted along the transparent fibers. These fibers have a thin coating, called *cladding*, which effectively works like a mirror, preventing the light from leaking out of the fiber.

The much smaller diameter of fiber-optic cable makes it ideal in situations where there is not room for bulky copper wires—for example, in crowded conduits, which can be pipes or spaces carrying both electrical and communications wires. In such tight spaces, the smaller fiber-optic telecommunications cable is very effective. Because fiber-optic cables are immune to electrical interference, signals can be transmitted over longer distances with fewer expensive

(c) Fiber-optic cable

(Source: © Greg Pease/Getty Images.)

repeaters to amplify or rebroadcast the data. Fiber-optic cable and associated telecommunications devices are more expensive to purchase and install than their twisted-pair wire counterparts, although the cost is coming down.

Microwave Transmission

Microwave transmissions are sent through the atmosphere and space. Although these transmission media do not entail the expense of laying cable, the transmission devices needed to utilize this medium are quite expensive. Microwave is a high-frequency radio signal that is sent through the air (see Figure 6.5). Microwave transmission is line of sight, which means that the straight line between the transmitter and receiver must be unobstructed. Typically, microwave stations are placed in a series—one station will receive a signal, amplify it, and retransmit it to the next microwave transmission tower. Such stations can be from 30 to 70 miles apart (depending on the height of the towers) before the curvature of the earth makes it impossible for the towers to "see one another." Microwave signals can carry thousands of channels at the same time.

Figure 6.5

Microwave Communications

Because they are line-of-sight transmission devices, microwave dishes must be placed in relatively high locations such as atop mountains, towers, and tall buildings.

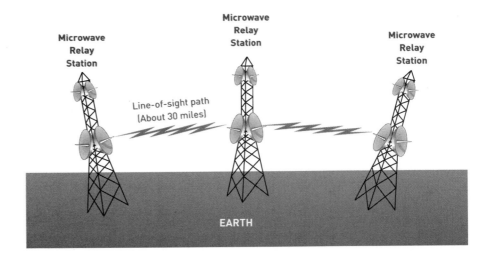

A communications satellite is basically a microwave station placed in outer space (see Figure 6.6). The satellite receives the signal from the earth station, amplifies the relatively weak signal, and then rebroadcasts it at a different frequency. The advantage of satellite communications is the ability to receive and broadcast over large geographic regions. Such problems as the curvature of the earth, mountains, and other structures that block the line-of-sight microwave transmission make satellites an attractive alternative. Geostationary, low-earth-orbit, and small mobile satellite stations are the most common forms of satellite communications.

A *geostationary satellite* orbits the earth directly over the equator, approximately 22,000 miles above the earth. At this altitude, one complete trip around the earth takes 24 hours, so the satellite remains over the same spot on the earth's surface at all times. The satellite thus stays fixed in the sky relative to any point on the earth from which it can be seen. Three such

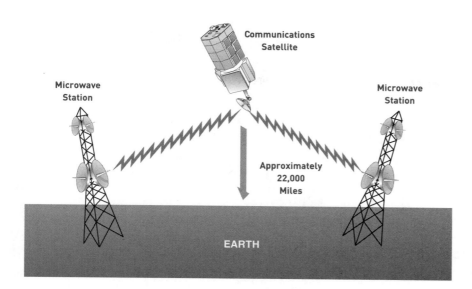

Figure 6.6

Satellite Transmission

Communications satellites are relay stations that receive signals from one earth station and rebroadcast them to another.

satellites, spaced at equal intervals (120 angular degrees apart), can provide coverage of the entire world. A geostationary satellite can be accessed using a dish antenna aimed at the spot in the sky where the satellite hovers.

A *low-earth-orbit (LEO) satellite* system employs a large number of satellites, each in a circular orbit at a constant altitude of a few hundred miles. Each orbit takes the satellites over the geographic poles, with one orbit taking roughly 90 minutes. The satellites are spaced so that, from any point on the earth at any time, at least one satellite is on a line of sight. The entire system operates similarly to the way a cellular telephone system functions, except that the wireless receivers/transmitters are moving rather than fixed and are in space rather than on the earth. LEO service subscribers can access the satellites using an antenna a little more sophisticated than old-fashioned television "rabbit ears."

Small mobile satellite stations allow people and businesses to communicate. These portable systems have a dish that is a few feet in diameter and can operate on battery power anywhere in the world. This capability is important for news organizations that need to transmit news stories from remote locations. Many people are also investing in direct satellite dish technology to receive TV and send and receive computer communications.

Cellular Transmission

With cellular transmission, a local area, such as a city, is divided into cells. As a car or vehicle with a cellular device, such as a mobile phone, moves from one cell to another, the cellular system passes the phone connection from one cell to another (see Figure 6.7). The signals from the cells are transmitted to a receiver and integrated into the regular phone system. Cellular phone users can thus connect to anyone that has access to regular phone service, such as a child at home or a business associate in London. They can also contact other cellular phone users. Because cellular transmission uses radio waves, it is possible for people with special receivers to listen to cellular phone conversations, so they are not secure.

Infrared Transmission

Another mode of transmission, called *infrared transmission*, sends signals through the air via light waves. Infrared transmission requires line-of-sight transmission and short distances—under a few hundred yards. Infrared transmission can be used to connect various small devices and computers. For example, infrared transmission has been used to allow handheld computers to transmit data and information to larger computers within the same room. Infrared transmission can also be used to connect a display screen, a printer, and a mouse to a computer. Some special-purpose phones can also use infrared transmission. This means of transmission can be used to establish a wireless network, with the advantage that devices can be moved, removed, and installed without expensive wiring and network connections.

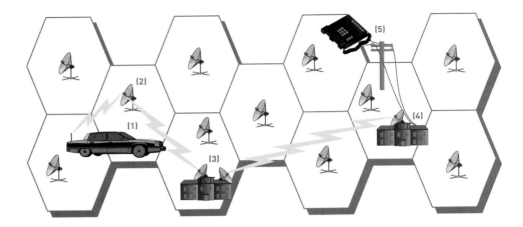

A Typical Cellular Transmission Scenario

Using a cellular car phone, the caller dials the number (1). The signal is sent from the car's antenna to the low-powered cellular antenna located in that cell (2). The signal is sent to the regional cellular phone switching office, also called the *mobile telephone subscriber office (MTSO)* (3). The signal is switched to the local telephone company switching station located nearest the call destination (4). Now integrated into the regular phone system, the call is automatically switched to the number originally dialed (5), all without the need for operator assistance.

analog signal
A continuous, curving signal.

digital signal
A signal represented by bits.

modem
A device that translates data from digital to analog and analog to digital.

How a Modem Works

Digital signals are modulated into analog signals, which can be carried over existing phone lines. The analog signals are then demodulated back into digital signals by the receiving modem.

Modems

In data telecommunications, it is not uncommon to use transmission media of differing types and capacities at various stages of the communications process. If a typical telephone line is used to transfer data, it can only accommodate an **analog signal** (a continuous, curving signal). Because a computer generates a **digital signal** representing bits, a special device is required to convert the digital signal to an analog signal, and vice versa (see Figure 6.8). Translating data from digital to analog is called *modulation*, and translating data from analog to digital is called *demodulation*. Thus, these devices are modulation/demodulation devices, or **modems**. Penril/Bay Networks, Hayes, Microcom, Motorola, and U.S. Robotics are examples of modem manufacturers.

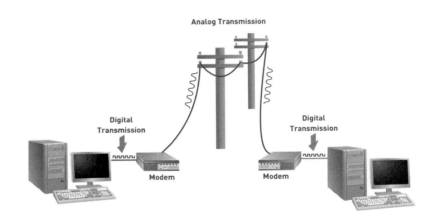

Modems can automatically dial telephone numbers, originate message sending, and answer incoming calls and messages. Modems can also perform tests and checks on how they are operating. Some modems are able to vary their transmission rates, commonly measured in bits per second. Cellular modems are placed in laptop personal computers to allow people on the go to communicate with other computer systems and devices.

Special-Purpose Modems

With a cellular modem, you can connect to other computers while in your car, on a boat, or in any area that has cellular transmission service. Expansion slots used for PC memory cards can also be used for standardized credit card–sized PC modem cards, which work like standard modems. PC modems are becoming increasingly popular with notebook and portable computer users.

Cable companies are promoting the cable modem, which has a low initial cost and transmission speeds up to 10 Mbps. A cable modem can deliver network and Internet access up to 50 times faster than a standard modem and phone line. In addition, a cable modem is

always on, so you can be on the Internet 24 hours a day, 7 days a week. Fees from $30 to $50 per month usually include unlimited service, local news, and an e-mail account. Cable service may cost only an additional $10 per month if you have existing Internet access through an Internet provider and want to upgrade from phone to cable service. @Home and Time Warner's AOL are leading companies using cable TV to bring the Internet to homes and businesses.

A cable modem can deliver network and Internet access up to 50 times faster than a standard modem and phone line.

(Source: Courtesy of D-Link Systems, Inc.)

Multiplexers

Because media and channels are expensive, devices that allow several signals to be sent over one channel have been developed. A multiplexer is one of these devices. A **multiplexer** allows several telecommunications signals to be transmitted over a single communications medium at the same time (see Figure 6.9).

multiplexer
A device that allows several telecommunications signals to be transmitted over a single communications medium at the same time.

Figure 6.9

Use of a Multiplexer to Consolidate Data Communications onto a Single Communications Link

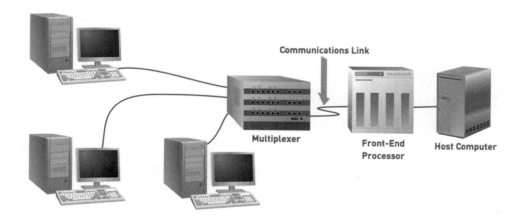

Communications Link

Multiplexer

Front-End Processor

Host Computer

Front-End Processors

Front-end processors are special-purpose computers that manage communications to and from a computer system. Like a receptionist handling visitors at an office complex, communications processors direct the flow of incoming and outgoing messages. They connect a midrange or mainframe computer to hundreds or thousands of communications lines. They poll terminals and other devices to see whether they have any messages to send. They provide automatic answering and calling, as well as perform circuit checking and error detection. Front-end processors also develop logs or reports of all communications traffic, edit data

front-end processor
A special-purpose computer that manages communications to and from a computer system.

before it enters the main processor, determine message priority, automatically choose alternative and efficient communications paths over multiple data communications lines, and provide general data security for the main system CPU. Because front-end processors perform all these tasks, the midrange or mainframe computer is able to process more work (see Figure 6.10).

Front-End Processor

A front-end processor takes the burden of communications management away from the main system processor.

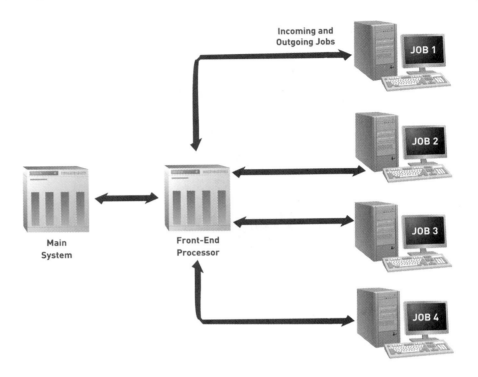

CARRIERS AND SERVICES

Telecommunications carriers organize communications channels, networks, hardware, software, people, and business procedures to provide individuals and businesses with valuable communications services. The types of carriers can be divided into three broad categories: local exchange carriers, competitive local exchange carriers, and long-distance carriers.

Local Exchange Carriers

local exchange carrier (LEC)
A public telephone company in the United States that provides service to homes and businesses within its defined geographical area, called its *local access and transport area* (*LATA*).

A **local exchange carrier** (**LEC**) is a public telephone company in the United States that provides service to homes and businesses within its defined geographical area, called its *local access and transport area* (*LATA*). LEC companies are also sometimes referred to as *telcos*. Homes and businesses from within the LEC's LATA are connected to the local exchange via what is called a *local loop*—typically, a twisted-pair copper wire (see Figure 6.11). This local loop is also called the "last mile." Once the subscriber reaches the LEC, calls can be routed literally anyplace in the world using the telco's switching equipment.

Each LEC has many local exchanges. Each local exchange (outside the United States the term *public exchange* is used) has equipment that can switch calls locally to subscribers connected to the same local exchange, connect to other local exchanges within the same LATA, or connect to interexchange carriers (IXCs) that carry traffic between LECs, such as long-distance carriers AT&T, MCI, and Sprint. The current rules for permitting a company to provide intra-LATA or inter-LATA service (or both) are based on the Telecommunications Act of 1996.

Individuals and organizations that rely on pairs of copper wires for their connection to the LEC are subject to significant limitations. First, the speed at which data can be transmitted

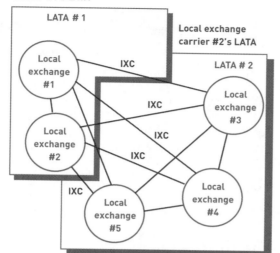

Figure 6.11

Local Exchange Carriers

over copper wires is much slower than other transmission media options. Second, the wires constitute a single point of failure that if cut (by a careless backhoe operator or even an automobile collision with a telephone pole) can result in loss of communications. Thus, many organizations are working with competitive local exchange carriers, discussed next. The telephone system was originally designed for voice transmission only using analog transmission technology. Today, your computer's modem converts signals from analog to digital. With integrated services digital network (ISDN) or digital subscriber line (DSL) service, discussed later in the chapter, the local loop can carry digital signals directly and at a much higher bandwidth than they do for voice only.

Competitive Local Exchange Carriers

Competitive local exchange carriers (CLECs) are companies that compete with the LECs. They include wireless service providers, satellite TV service providers, cable TV companies, and even power companies—the same power lines that bring electricity to our homes and businesses may become the next pathway for high-speed Internet access. This development could turn every power plug into a broadband connection. This technology is not yet commercially feasible. However, utility companies PPL Corporation in Allentown, Pennsylvania, and Ameren Corporation in St. Louis are conducting trial programs with consumers.[2] Cable companies have offered some residential phone service since 1998. At the end of 2003, there were nearly 2.4 million cable telephone subscribers, and the number of subscribers was expected to quickly increase to more than 10 million by the end of 2005.[3] Time Warner Cable expects to have digital telephone service available across the United States.[4] The competitive local exchange carriers provide valuable backup capability over the "last mile." In many cases, they also offer faster service and cheaper rates than the traditional local exchange carriers.

competitive local exchange carrier (CLEC)
A company that is allowed to compete with the LECs, such as a wireless, satellite, or cable service provider.

Long-Distance Carriers

In the past few years, the three established **long-distance carriers**—AT&T, Sprint, and MCI—have lost market share under the provisions of the Telecommunications Act of 1996 (see Table 6.2). The act allowed BellSouth, Qwest Communications, SBC Communications, and Verizon Communications to offer long-distance service within the regions they serve on a state-by-state basis after receiving approval from each state's regulators and the Federal Communications Commission (FCC). These formerly local companies now bundle their long-distance service with traditional local service in a comprehensive package that they advertise aggressively.

long-distance carrier
A traditional long-distance phone provider, such as AT&T, Sprint, or MCI.

Table 6.2

Long-Distance Market Share

(Source: Data from Todd Rosenbluth, "The Price Wars Wounding Telecom," *BusinessWeek Online*, February 13, 2004, accessed at *www.businessweek.com*.)

Long-Distance Carrier	2003 Long-Distance Market Share (Percentage of Households)
AT&T	27%
Verizon	14%
SBC	14%
MCI	12%
Sprint	7%
BellSouth	4%
Qwest	3%

Switched and Dedicated Lines

switched line

A communications line that uses switching equipment to allow one transmission device to be connected to other transmission devices.

dedicated line

A communications line that provides a constant connection between two points; no switching or dialing is needed, and the two devices are always connected.

Communications carriers typically provide the use of standard telephone lines, called **switched lines**. These lines use switching equipment to allow one transmission device (e.g., your telephone) to be connected to other transmission devices (e.g., the telephones of your friends and relatives). A switch is a circuit that directs messages along specific paths in a telecommunications system. When you make a phone call, the local telephone service provider's switching equipment connects your phone to the phone of the person you're calling. Fees for a switched business line (versus a residential line) can range from $25 to $100 or more per month. A **dedicated line**, also called a *leased line*, provides a constant connection between two points. No switching or dialing is needed; the two devices are always connected. Many firms with a high volume of data traffic between two points—say, a headquarters in Chicago and a sales office in San Diego—utilize dedicated lines. The high initial cost of purchasing or leasing such a line is offset by eliminating long-distance charges incurred with a switched line. Monthly fees for a dedicated line can range from $50 to $500 or more, but there is no additional charge for use.

Telecommunications networks require state-of-the-art computer software technology to continuously monitor the flow of voice, data, and image transmission over billions of circuit miles worldwide.

(Source: © Roger Tully/Getty Images.)

Voice and Data Convergence

The Internet protocol (IP) is the communications protocol by which data is sent from one computer to another on the Internet. Each computer on the Internet has an IP address that uniquely identifies it from all other computers on the Internet. When data is sent over the Internet, it is divided into small packets, each containing both the sender's Internet address and the receiver's address. **Voice over Internet protocol (VoIP)** is the basic *transport* of voice in the form of a data packet using the Internet protocol. *IP telephony* is the technology for transmitting voice communications over a network using an open standards-based Internet protocol. IP telephony uses VoIP but is a *software application* suite offering feature-rich applications. These often-modular applications lend themselves to cost-effective integration with other applications that share the IP network. This will be discussed in more detail in Chapter 7.

Voice and data convergence is the integration of voice and data applications in a common environment. Instead of using fixed circuits, as has been the case since the phone network was invented over a century ago, voice conversations and even video can be digitized, broken up into data packets, and transmitted the way data is routed over the Internet. The move to Internet telephony provides tremendous cost savings for both service providers and customers. New telecommunications equipment can support the routing of voice, data, and video and yet is 10 percent of the cost of older equipment. In addition, the new equipment is only half as expensive to operate. However, to capitalize on the benefits of this technology, carriers must replace their old equipment and reengineer their trunk lines that carry traffic over long distances. This transition to digitized voice, data, and video is the dominant technological issue facing the world's phone companies over the next decade. Every carrier has the same goal; the only difference is their timetable for achieving it. Telecom Italia is the carrier furthest along in this transition; it has completely rebuilt the core of its nationwide network around Internet equipment.[5]

IP telephony is just beginning to be used in some organizations. Research firm McGee-Smith Analytics estimates that 1 percent of the 80,000 call centers in the United States today are using applications built to run over converged IP-based voice/data networks. By 2008 industry experts expect that 20 percent of organizations using call centers will be running IP-based call center systems, with features such as automated call distribution, contact management, and computer/telephony integration.[6] One key advantage, of course, is cost. It costs less to run one converged IP network than it does to run separate lines for voice and data. Another advantage is that IP call centers allow remote connectivity—call center agents can move to any location with a network to receive calls, whether at home or abroad. Since call routing can take place over the corporate network, phone charges are reduced, which can vastly lower costs.

Phone and Dialing Services

Common carriers are beginning to provide more and more phone and dialing services to home and business users. Automatic number identification (ANI), or caller ID, equipment can be installed on a phone system to identify and display the number of an incoming call. In a business setting, ANI can be used to identify the caller and link that caller with information stored in a computer. For example, when a customer calls Federal Express, the customer service rep uses the ANI to identify the name and address of the customer, thus saving time when handling a request for a pickup. The ANI can be very useful in helping people screen calls before they are answered. Unwanted phone calls from other people and businesses can be identified before the phone is ever answered. ANI, however, doesn't always work well with different carriers. Common carriers offer even more services to extend the capabilities of the typical phone system. Even with all the advances in computers and telecommunications, common carrier services remain important.

voice over Internet protocol (VoIP)
The basic transport of voice in the form of a data packet using the Internet protocol.

voice and data convergence
The integration of voice and data applications in a common environment.

Some of these services are listed here:

- The ability to integrate personal computers so that the telephone number of the caller is automatically captured and used to look up information in a database about a customer
- Use of access codes to screen out junk calls, wrong numbers, and other unwanted phone calls
- Call screening priorities (e.g., only certain calls are transmitted during certain times of the day, such as from 10:00 P.M. to 7:00 A.M.)
- The ability to use one number for a business phone, home phone, personal computer, fax, and so forth
- Intelligent dialing (when a busy signal is received, the phone redials the number when your line and the line of the party you are trying to reach are both free)

WATS

Wide-area telephone service (WATS) is a fixed-rate long-distance telecommunications service for heavy users of voice services such as customer support organizations, telemarketing firms, large businesses, and government agencies. The IN-WATS service is used strictly for incoming calls and makes use of certain reserved area codes such as 800, 877, or 888. Individuals calling an IN-WATS number are not charged for the call. Instead, the recipient of the call pays a fixed monthly rate, provided that the maximum hours of usage is not exceeded. The OUT-WATS service is used by the subscriber to place outgoing calls at a fixed monthly rate. Some calling restrictions apply. For example, the caller may not be able to make a WATS call within the same state.

ISDN

integrated services digital network (ISDN)
A set of standards for integrating voice and data communications onto a single line via digital transmission over copper wire or other media.

Integrated services digital network (ISDN) is a set of standards for integrating both voice and data communications onto a single line via digital transmission over ordinary telephone copper wire; other media can be used as well. ISDN requires special adapters at both ends of the transmission line, so the access provider must also have an ISDN adapter (see Figure 6.12). The ISDN Basic Rate Interface is frequently chosen for homes and small businesses. It provides two communications channels, so there can be simultaneous transmission of voice, data, or fax. The ISDN Primary Rate Interface provides 30 channels that can each transport data at 64 Kbps, or an aggregate rate of 1,920 Kbps. ISDN is available in most urban areas of the United States; however, it is being replaced by DSL service.

T-Carrier System

The T-carrier system was introduced in the 1960s to support digitized voice transmission. The system uses four wires and provides duplex capability (two wires receiving and two sending at the same time). The four wires can be twisted-pair copper wires, coaxial cable, optical fiber, digital microwave, or other transmission media. The T-1 carrier is capable of carrying 1.544 Mbps over copper wire and is a commonly used in the United States, Japan, and Canada. Internet access providers (e.g., America Online) are frequently connected to the Internet by a T-1 line owned by one of the major communications carriers. The T-3 line is capable of transmitting data at a rate of 44.736 Mbps and is also commonly used by Internet service providers.

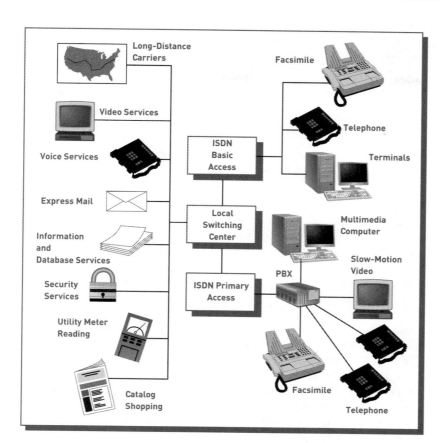

Figure 6.12

ISDN Network Switching

Digital Subscriber Line (DSL)

A **digital subscriber line (DSL)** is a telecommunications technology that delivers high-bandwidth information to homes and small businesses over ordinary copper telephone wires. If your home or small business is close enough to a telephone company's central office that offers DSL service, you may be able to receive data at rates up to 6.1 Mbps, enabling continuous transmission of video, audio, and even 3-D effects. More typically, a DSL connection provides a transmission rate of 512 Kbps to 1.544 Mbps from the central office to the subscriber and about 128 Kbps from the subscriber to the central office. A DSL line can carry both data and voice signals, and the data part of the line is continuously connected. DSL installations began in 1998 and are proceeding rapidly in a number of U.S. communities and elsewhere. DSL is expected to replace ISDN in many areas and to compete with the cable modem in bringing high-speed data communications to homes and small businesses. The number of DSL lines worldwide doubled in 2003 to 36 million and is soon expected to reach 160 million lines—or about 13 percent of the world's phone lines. Outside the United States, DSL outpaces cable modem installations by nearly 2 to 1.[7]

digital subscriber line (DSL)
A telecommunications technology that delivers high-bandwidth information to homes and small businesses over ordinary copper telephone wires.

Wireless Mobile

All the major cellular carriers offer some sort of wireless mobile data services. A few of the options are summarized in Table 6.3. With wireless mobile service, you can access the Web anyplace a cell phone network exists. Verizon Wireless runs several mobile networks using CDMA (Code-Division Multiple Access) 1xRTT, a communications technology that supports mobile data communications at speeds ranging from 144 Kbps to 2 Mbps. It is a form

	Verizon Wireless	Sprint PCS	T-Mobile	Cingular
Basic Technology Employed	CDMA 1xRTT	CDMA 1xRTT	EDGE	EDGE
Transmission Speed	70 Kbps – 2 Mbps	70 Kbps – 2 Mbps	120–200 Kbps	120–200 Kbps

Table 6.3

Some Wireless Data Communications Options

of multiplexing that allows numerous communications devices to share a single communications channel. NationalAccess, the company's primary offering, is available in thousands of cities around the country with a monthly charge from $35 to $80. AT&T, T-Mobile, and Cingular employ GSM (Global System for Mobile communication) technology, which is the de facto wireless telephone standard in Europe, with more than 120 million users worldwide in 120 countries. It is a digital mobile telephone system that digitizes and compresses data and then sends it down a communications channel with two other streams of data, each in its own time slot. Since many GSM network operators have roaming agreements with foreign operators, users can often continue to use their mobile phones when they travel to other countries. In 2003, many network operators started rolling out their next generation of wireless services, which boast an average speed around 120 Kbps and a top speed of 200 Kbps. The service is available in more than 6,500 cities and towns with an unlimited-access plan for $80 a month.

Genex Services is one of the nation's largest healthcare management service providers, with mobile workers deployed across the country. The company supports hundreds of case-management workers who visit homebound patients to coordinate their care and relay information to employers, doctors, and insurance agencies. The caseworkers must be able to update medical files easily and have uninterrupted access to the most accurate, up-to-date information on the patient. Before the caseworkers had an advanced telecommunications system, they would take handwritten notes during a patient visit or conversation with an insurance company or care provider. Then when they got home, perhaps several hours later, they had to write the information up formally, make a conventional modem call, and upload their information to update the patient information. Genex decided to try wireless Internet technology that employs PC cards that use the same communication protocol as cell phones and operate similarly to cellular modems. Caseworkers can now get online anywhere they can receive a cellular signal—and access patient information literally anywhere and anytime. The result is that data is more accurate and up to date, and the process is streamlined. Wireless Internet has worked well for Genex.[8]

Adoption of cellular data services is still in its early stages; many business users are taking a trial approach. Wireless data communications will be broadly adopted once providers can offer users enough bandwidth and connectivity to enable business users to use cellular as their sole connection.

With all of these telecommunications options, the problem for individuals and businesses is how to choose the best one. Each option has its own cost, speed, and reliability to consider. Table 6.4 shows some of the costs, advantages, and disadvantages of different lines and services offered by communications carriers.

Line/Service	Speed	Cost per Month	Advantages	Disadvantages
Standard phone service	56 Kbps	$10–$40	Low cost and broadly available	Too slow for video and downloads of large files
ISDN	64 Kbps–128 Kbps	$50–$150	Fast for video and other applications	Higher costs and not available everywhere
DSL	500 Kbps–1.544 Mbps	$20–$120 in addition to standard phone service	Fast, and the service comes over standard phone lines	Slightly higher costs and not available everywhere
Cable modem	500 Kbps–1.544 Mbps	$20–$120	Fast and uses existing cable that comes into the home	Slightly higher costs and not available everywhere
T1	1.544 Mbps	$600–$1,200	Very fast broadband service, typically used by corporations and universities	Very expensive, high installation fee, and users pay a monthly fee based on distance
Wireless data communications	70 Kbps–2Mbps	$35–$80	Provides network access for mobile worker	Areas of country may not have service yet

Table 6.4

Costs, Advantages, and Disadvantages of Several Line and Service Types

NETWORKS

A **computer network** consists of communications media, devices, and software needed to connect two or more computer systems or devices. The computers and devices on the networks are also called *network nodes*. Once connected, the nodes can share data, information, and processing jobs. More and more businesses are linking computers in networks to streamline work processes and allow employees to collaborate on projects. The effective use of networks can turn a company into an agile, powerful, and creative organization, giving it a long-term competitive advantage. Networks can be used to share hardware, programs, and databases across the organization. They can transmit and receive information to improve organizational effectiveness and efficiency. They enable geographically separated workgroups to share documents and opinions, which fosters teamwork, innovative ideas, and new business strategies.

computer network
The communications media, devices, and software needed to connect two or more computer systems and/or devices.

Network Types

Depending on the physical distance between nodes on a network and the communications and services provided by the network, networks can be classified as personal area, local area, metropolitan area, wide area, or international network.

Personal Area Networks

A **personal area network (PAN)** is a wireless network that supports the interconnection of information technology devices within a range of 33 feet or so. One device is selected to assume the role of the controller during wireless PAN initialization, and this controller device mediates communication within the PAN. The controller broadcasts a beacon that synchronizes all devices and allocates time slots for the devices.

With a PAN, a person with a laptop, digital camera, and portable printer could connect them without having to hardwire anything. Digital image data could be downloaded from the camera to the laptop and then printed on a high-quality printer—all wirelessly.

personal area network (PAN)
A network that supports the interconnection of information technology within a range of 33 feet or so.

local area network (LAN)
A network that connects computer systems and devices within the same geographic area.

Local Area Networks

A network that connects computer systems and devices within the same geographic area is a **local area network (LAN)**. Typically, local area networks are wired into office buildings and factories (see Figure 6.13). While unshielded twisted-pair wire cable is a commonly used medium with LANs, other media—including fiber-optic cable—are also popular. They can be built to connect personal computers, laptop computers, or powerful mainframe computers. When a personal computer is connected to a local area network, a network interface card (NIC) is usually required. A network interface card is a card or board that is placed in a computer's expansion slot, usually provided by the PC manufacturer when you obtain your PC, to allow it to communicate with the network. A wire or connector from the network is plugged directly into the network interface card. For example, a salesperson whose notebook computer has an interface card can establish a link to the corporate LAN. He or she can access the network while at the office and download data needed for the next sales call.

Figure 6.13

A Typical LAN

All network users within an office building can connect to each other's devices for rapid communication. For instance, a user in research and development could send a document from her computer to be printed at a printer located in the desktop publishing center.

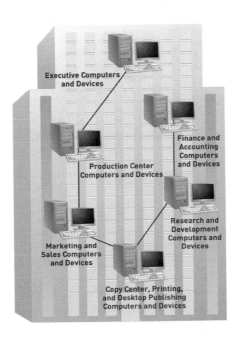

Another basic LAN is a simple peer-to-peer network that might be used for a very small business to allow the sharing of files and hardware devices such as printers. In a peer-to-peer network, each computer is set up as an independent computer, except that other computers can access specific files on its hard drive or share its printer. These types of networks have no server. Instead, each computer uses a network interface card and cabling to connect it to the next machine. Examples of peer-to-peer networks include Windows for Workgroups, Windows NT, Windows 2000, and AppleShare. Performance of the computers on a peer-to-peer network is usually slower because one computer is actually sharing the resources of another computer. These networks, however, are a good beginning network from which small businesses can grow. The software cost is minimal, and network cards can be used if the company decides to enlarge the system. In addition, peer-to-peer networks are becoming cheaper, faster, and easier to use for home-based businesses.

SchlumbergerSema, now part of France-based Atos Origin, was the IS and security integrator chosen for the 2004 Summer Olympic Games in Athens, Greece. The company spent three years and millions of dollars to implement a LAN to support 10,500 PCs and 900 servers, even though the LAN would only be in place for three weeks. One of the LAN applications was a security system that uses bar-code badge readers to keep track of the 200,000 athletes, coaches, sports media, and volunteers given admittance to authorized venues and buses. Reporters covering the Olympics used PCs and kiosks on the LAN to get event results and background information on the competitors that were stored on UNIX, Solaris, and Windows NT servers.[9]

With more people working at home, connecting home computing devices and equipment into a unified network is on the rise. Small businesses are also connecting their systems and equipment. With a home or small business network, computers, printers, scanners, and other devices can be connected. A person working on one computer, for example, can use data and programs stored on another computer's hard disk. In addition, a single printer can be shared by several computers on the network. To make home and small business networking a reality, a number of companies are offering standards, devices, and procedures.

Metropolitan Area Networks

A **metropolitan area network (MAN)** is a telecommunications network that connects users and their computers in a geographical area larger than that covered by a LAN but smaller than the area covered by a WAN. Most MANs have a range of roughly 30 miles. An example of a MAN would be redefining the many networks within a city into a single larger network or connecting several campus LANs into a single university-wide MAN.

Clark County, Nevada (which encompasses Las Vegas), is the nation's sixth-largest school district, with 289 schools. The county spent $15 million to build an IP-based metropolitan area network. An additional $16 million will be spent to outfit the district offices and every classroom with about 27,000 phone sets that can operate in both digital and IP modes. The MAN and the dual-mode phone system are designed to support the school system's explosive growth. The district, which serves 268,000 students and has 30,000 workers, is adding new schools at the rate of one per month.[10]

metropolitan area network (MAN)
A telecommunications network that connects users and their computers within a geographical area larger than that covered by a LAN but smaller than the area covered by a WAN, such as a city or college campus.

Wide Area Networks

A **wide area network (WAN)** is a telecommunications network that ties together large geographic regions. A WAN may be privately owned or rented and includes the use of public (shared users) networks. When you make a long-distance phone call or access the Internet, you are using a wide area network. WANs usually consist of computer equipment owned by the user, together with data communications equipment and telecommunications links provided by various carriers and service providers (see Figure 6.14).

wide area network (WAN)
A network that ties together large geographic regions.

North America

Figure 6.14

A Wide Area Network

Wide area networks are the basic long-distance networks used by organizations and individuals around the world. The actual connections between sites, or nodes (shown by dashed lines), may be any combination of satellites, microwave, or cabling. When you make a long-distance telephone call or access the Internet, you are using a WAN.

International Networks

Networks that link systems between countries are called **international networks**. However, international telecommunications comes with special problems. In addition to requiring

international network
A network that links systems between countries.

sophisticated equipment and software, global area networks must meet specific national and international laws regulating the electronic flow of data across international boundaries, often called *transborder data flow*. Some countries have strict laws limiting the use of telecommunications and databases, making normal business transactions such as payroll costly, slow, or even impossible. Other countries have few laws concerning telecommunications and database use. These countries, sometimes called *data havens*, allow other governments and companies to avoid their country's laws by processing data within their boundaries. International networks in developing countries can have inadequate equipment and infrastructure that can cause problems and limit the usefulness of the network.

Boehringer Ingelheim GmbH is a huge pharmaceutical manufacturer, with $7.6 billion in revenue and 32,000 employees in 60 nations. The Ingelheim, Germany–based company is using an international network to consolidate data from its widespread operations and financial applications and present key financial information on a daily, weekly, or monthly basis. "I want to be told where I stand and where we are heading," says Boehringer's chief financial officer, Holger Huels. "I like to [be able to] see negative trends and counter them as fast as possible." In addition, Boehringer is now able to close its books for most of its divisions just two hours after the close of business at the end of each month.[11]

Network Topology

Networks that link computers and computer devices provide for flexible processing. Building a network involves two types of design: logical and physical. A logical model shows how the network will be organized and arranged. A physical model describes how the hardware and software in the network will be physically and electronically linked.

The number of possible ways to logically arrange the nodes, or computer systems and devices on a network, may seem limitless. Actually, there are only five major types of **network topologies**—logical models that describe how networks are structured or configured. These types are ring, bus, hierarchical, star, and hybrid (see Figure 6.15).

network topology
Logical model that describes how networks are structured or configured.

Figure 6.15

The Basic Network Topologies

The four main topologies are (a) ring, (b) bus, (c) hierarchical, and (d) star. In addition, a hybrid configuration (e) can be formed from elements of any of these four topologies.

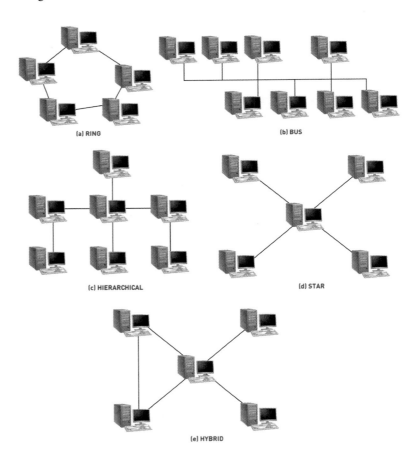

(a) RING

(b) BUS

(c) HIERARCHICAL

(d) STAR

(e) HYBRID

The **ring network** contains computers and computer devices placed in a ring or circle. With a ring network, there is no central coordinating computer. Messages are routed around the ring from one device or computer to another in one direction. A **bus network** consists of computers and computer devices on a single line. Each device is connected directly to the bus and can communicate directly with all other devices on the network. The bus network is one of the most popular types of personal computer networks. The **hierarchical network** uses a treelike structure. Messages are passed along the branches of the hierarchy until they reach their destination. Like a ring network, a hierarchical network does not require a centralized computer to control communications. Hierarchical networks are easier to repair than other topologies because you can isolate and repair one branch without affecting the others. A **star network** has a central hub or computer system. Other computers or computer devices are located at the end of communications lines that originate from the central hub or computer system. The central computer of a star network controls and directs messages. If the central computer breaks down, it results in a breakdown of the entire network. Many organizations use a **hybrid network**, which is simply a combination of two or more of the four topologies just discussed. The exact configuration of the network depends on the needs, goals, and organizational structure of the company involved.

Terminal-to-Host, File Server, and Client/Server Systems

Aside from the logical configuration of topologies, networks can also be classified according to how the computers on the network connect and interoperate. There are many ways to connect computers, including terminal-to-host, file server, and client/server architectures.

Terminal-to-Host

With **terminal-to-host** architecture, the application and database reside on one host computer, and the user interacts with the application and data using a "dumb" terminal. (Even if you use a PC to access the application, you run terminal emulation software on the PC to make it act as if it were a dumb terminal with no processing capacity.) Since a dumb terminal has no data-processing capability, all computations, data accessing and formatting, and data display are done by an application that runs on the host computer (see Figure 6.16).

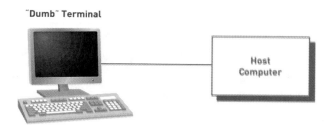

"Dumb" Terminal

Host Computer

File Server

In **file server** architecture, the application and database reside on one host computer, called the *file server*. The database management system runs on the end user's personal computer or workstation. If the user needs even a small subset of the data that resides on the file server, the file server sends the user the entire file that contains the data requested, including a lot of data the user does not want or need. The downloaded data can then be analyzed, manipulated, formatted, and displayed by a program that runs on the user's personal computer (see Figure 6.17).

Client/Server

In **client/server** architecture, multiple computer platforms are dedicated to special functions such as database management, printing, communications, and program execution. These platforms are called *servers*. Each server is accessible by all computers on the network. Servers can be computers of all sizes; they store both application programs and data files and are equipped with operating system software to manage the activities of the network. The server distributes programs and data files to the other computers (clients) on the network as they

ring network
A type of topology that contains computers and computer devices placed in a ring or circle; there is no central coordinating computer; messages are routed around the ring from one device or computer to another.

bus network
A type of topology that contains computers and computer devices on a single line; each device is connected directly to the bus and can communicate directly with all other devices on the network; one of the most popular types of personal computer networks.

hierarchical network
A type of topology that uses a treelike structure with messages passed along the branches of the hierarchy until they reach their destination.

star network
A type of topology that has a central hub or computer system, and other computers or computer devices are located at the end of communications lines that originate from the central hub or computer.

hybrid network
A network topology that is a combination of other network types.

terminal-to-host
An architecture in which the application and database reside on one host computer, and the user interacts with the application and data using a "dumb" terminal.

Figure 6.16

Terminal-to-Host Connection

file server
An architecture in which the application and database reside on one host computer, called the file server.

client/server
An architecture in which multiple computer platforms are dedicated to special functions such as database management, printing, communications, and program execution.

request them. An application server holds the programs and data files for a particular application, such as an inventory database. Processing can be done at the client or server.

Figure 6.17

File Server Connection

The file server sends the user the entire file that contains the data requested. The downloaded data can then be analyzed, manipulated, formatted, and displayed by a program that runs on the user's personal computer.

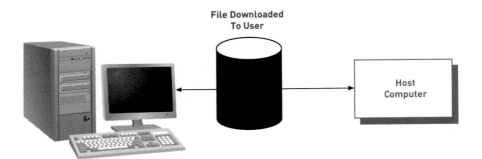

A client is any computer (often an end user's personal computer) that sends messages requesting services from the servers on the network. A client can converse with many servers concurrently. A user at a personal computer initiates a request to extract data that resides in a database somewhere on the network. A data request server intercepts the request and determines on which data server the data resides. The server then formats the user's request into a message that the database server will understand. Upon receipt of the message, the database server extracts and formats the requested data and sends the results to the client. Only the data needed to satisfy a specific query is sent—not the entire file (see Figure 6.18). As with the file server approach, once the downloaded data is on the user's machine, it can then be analyzed, manipulated, formatted, and displayed by a program that runs on the user's personal computer.

Figure 6.18

Client/Server Connection

Multiple computer platforms, called *servers*, are dedicated to special functions such as database management, data storage, printing, communications, network security, and program execution. Each server is accessible by all computers on the network. A server distributes programs and data files to the other computers (clients) on the network as they request them. The client requests services from the servers, provides a user interface, and presents results to the user. Once data is moved from a server to the client, the data may be processed on the client.

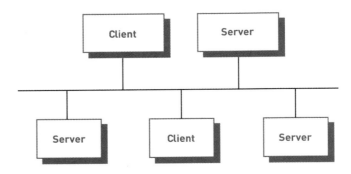

Morgan Stanley, the global financial services firm, took more than four years to overhaul its IT infrastructure with the goal of creating a new information system infrastructure based on a thin-client server model. When completed, end users will have relatively low-cost, centrally managed computers with limited capabilities. All data, applications, and even operating systems will be hosted on network servers and downloaded to the thin clients when needed. With thousands of internally developed applications and 36,500 supported PCs spread throughout 20 countries, Morgan Stanley moved to the client/server architecture to quickly deliver anything end users need, including data, applications, software patches, and system configuration changes. The payoff for Morgan Stanley has been increased flexibility, increased performance and reliability, and decreased hardware and information systems support costs.[12] Table 6.5 lists the advantages and disadvantages of client/server architecture.

Advantages	Disadvantages
Moving applications from mainframe computers and terminal-to-host architecture to client/server architecture can yield significant savings in hardware and software support costs.	Moving to client/server architecture is a major two- to five-year conversion process.
Minimizes traffic on the network because only the data needed to satisfy a user query is moved from the database to the client device.	Controlling the client/server environment to prevent unauthorized use, invasion of privacy, and viruses is difficult.
Security mechanisms can be implemented directly on the database server through the use of stored procedures.	Client/server architecture leads to a multivendor environment with problems that are difficult to identify and isolate to the appropriate vendor.

Table 6.5

Advantages and Disadvantages of Client/Server Architecture

INTERCONNECTING NETWORKS

As we have already discussed, networks are designed to meet varying business needs. So, one organization may have many different networks, and different organizations may have yet more varieties. Even with all the differences in computer networks, in today's interconnected world people still need to communicate with one another—no matter where they happen to be. To allow the free flow of information, a diverse set of data communications technologies are involved, from a common communications protocol to various message switching devices.

Communications Protocols

A **communications protocol** is a standard set of rules that control a telecommunications connection. Communications are not possible unless senders and receivers recognize and observe a common protocol. Communications protocols are critical when connecting devices from different manufacturers or separate networks.

There are myriad communications protocols. Protocols are often described at the international, national, or industry level. The International Standards Organization is a worldwide federation of standards bodies from some 100 countries. Among the standards it fosters is Open Systems Interconnection (OSI), a universal reference model for communication protocols. Many countries have national standards organizations, such as the American National Standards Institute (ANSI), that participate in and contribute to ISO standards. The Institute of Electrical and Electronics Engineers (IEEE) also furthers the development of national and international standards. Many standards are developed by groups of telecommunications hardware, software, or service providers who form a consortium. We describe a few of the more common protocols and standards here.

communications protocol
A standard set of rules that control a telecommunications connection.

OSI

The Open Systems Interconnection (OSI) model serves as a standard model for network architectures and is endorsed by the International Standards Committee. The OSI model divides data communications functions into seven distinct layers (see Table 6.6) to promote the development of modular networks that simplify the development, operation, and maintenance of complex telecommunications networks.

TCP/IP

Transmission Control Protocol/Internet Protocol (TCP/IP) is the primary communications protocol of the Internet. Its use is growing rapidly as Internet connections are becoming widespread. The TCP protocol is responsible for verifying correct delivery of data from sender to receiver. The IP protocol is responsible for moving packets of data from node to node in the network. We discuss TCP/IP in more detail in Chapter 7.

SNA

SNA is a proprietary IBM architecture and set of products for network computing within an enterprise. IBM is finding ways to combine its own SNA protocol within the enterprise with TCP/IP for applications in the global network arena.

Table 6.6

The Seven-Layer OSI Model

Layer	Function
Application	Controls user input from the terminal and executes the user's application program in the host computer.
Presentation	Formats the data so that it can be presented to the user or the host. For example, data to be displayed on the user's screen is formatted into the correct number of lines per screen and characters per line.
Session	Initiates, maintains, and ends each session. A session consists of all the frames that make up a particular activity, plus signals that identify the beginning and end.
Transport	Enables the user and host nodes to communicate with each other. It also synchronizes fast- and slow-speed devices and avoids overburdening a device with too much output.
Network	Causes the physical layer to transfer the frames from node to node.
Data link	Formats the data into a record called a *frame* and performs error detection.
Physical	Transmits the data from one node to another.

IEEE 802.3 (Ethernet)

Ethernet is the most widely installed local area network (LAN) technology. An Ethernet LAN typically uses coaxial cable or special grades of twisted-pair wires. The most commonly installed Ethernet systems are called 10BASE-T, and they provide transmission speeds up to 10 Mbps. Devices are connected to the cable and compete for access using a Carrier Sense Multiple Access with Collision Detection (CSMA/CD) protocol. Following this protocol, devices on the LAN "listen" before and during transmitting to detect whether another device on the LAN also transmitted, thus disrupting their original message. If such a "collision" is detected, the device retransmits the original message after waiting a random time interval. Fast Ethernet or 100Base-T provides transmission speeds up to 100 Mbps and is typically used for LAN systems, supporting workstations with 10BASE-T cards. Gigabit Ethernet provides an even higher level of support at 1,000 Mbps—equal to 1 gigabit (Gb), or 1 billion bits per second.

Frame Relay

Frame relay is a packet switching protocol for cost-efficient data transmission of intermittent traffic between local area networks (LANs) and between end points in a wide area network (WAN). This protocol puts data in a variable-sized unit called a *frame,* and the service provider figures out the route each frame travels to its destination and can charge based on usage. When an error is detected in a frame, it is simply thrown away or "dropped." The sending and receiving devices at the initial and final destination are responsible for detecting and retransmitting dropped frames. Frame relay networks in the United States support data transfer rates at T-1 (1.544 Mbps) and T-3 (45 Mbps) speeds. In Europe, frame relay speeds vary from 64 Kbps to 2 Mbps. In the United States frame relay is quite popular because it is relatively inexpensive. However, it is being replaced in some areas by faster technologies, such as ATM.

ATM

ATM (asynchronous transfer mode) is a switching technology that relies on dedicated connections and organizes data into 53-byte cell units. Data cells from various communications devices are then multiplexed, meaning each data cell is transmitted in different time slices over a physical transmission medium. ATM is designed to be implemented in hardware (rather than software), so faster processing and switch speeds are possible. Speeds on ATM networks can reach 10 Gbps.

FireWire

Firewire is Apple Computer's standard for connecting peripheral devices (printers, scanners, cameras, etc.) to the personal computer. FireWire provides a single plug-and-socket connection on which up to 63 devices can be attached, with data transfer speeds up to 400 Mbps. The standard describes a pathway between one or more peripheral devices and the computer's microprocessor. Many peripheral devices now come equipped to meet the FireWire specification and the similar IEEE 1394 communications standard.

Wireless Communications Protocols

With the spread of wireless network technology to support such devices as personal digital assistants, mobile computers, and cell phones, the telecommunications industry needed new protocols to connect these devices. Wireless communications protocols are still evolving as the industry matures. The IEEE 802 series of network standards sprang from the IS industry starting in 1980. In addition, several generations of communication protocols (1G, 2G, 2.5G, and 3G) were developed for the digital voice communication (cell phone) industry.

Bluetooth

Bluetooth is a telecommunications specification that describes how cellular phones, computers, faxes, personal digital assistants, printers, and other electronic devices can be interconnected using a short-range (10 to 30 feet) wireless connection. Bluetooth enables users of multifunction devices to synchronize with information in a desktop computer, send or receive faxes, print, and, in general, coordinate all mobile and fixed computer devices. The Bluetooth technology is named after the tenth century Danish King Harald Blatand, or Harold Bluetooth in English. He had been instrumental in uniting warring factions in parts of what is now Norway, Sweden, and Denmark—just as the technology is designed to allow collaboration between differing industries such as the computing, mobile phone, and automotive markets.[13]

IEEE 802.11b (Wi-Fi)

Wi-Fi (short for wireless fidelity) is the popular term for a high-frequency wireless local area network. Wi-Fi technology is rapidly gaining acceptance in many companies as an alternative to a wired LAN. It can also be installed for a home network. Wi-Fi is specified in the 802.11b specification from the Institute of Electrical and Electronics Engineers.

Intel and Microsoft are counting on Wi-Fi to spur sales of their products. Intel has introduced the Wi-Fi-ready Centrino chip in its laptop computers. Microsoft is pushing its Windows XP operating system, which has been designed to support Wi-Fi. The consumer electronics industry is counting on Wi-Fi to link home appliances, such as sending MP3 songs and videos from a computer to a TV, to sound systems, or to an MP3 player.

There are thousands of public Wi-Fi "hot spots"—places to access a wireless network. Behind every hot spot is a broadband connection such as DSL, cable, or T1. Unlike cellular technology, which can route your call from one tower to the next as you drive down the highway, Wi-Fi was not designed to hand a user off from one access point to another. And current Wi-Fi access points have a maximum range of about 300 feet.

In offering public Wi-Fi access at their restaurants, McDonald's, Schlotsky's Deli, and Starbucks hope that people will not only work, check e-mail, or chat online but also stay longer and spend more money or select them over a competitor. Pricing models vary. McDonald's offers a free hour of Wi-Fi access with each Extra Value Meal purchased in selected restaurants. After the time is up, the service costs $3 an hour. Starbucks charges $6 a day or $30 a month for unlimited use. Pittsburgh, Pennsylvania, International Airport recently completed deployment of a free wireless LAN in its food court and expanded the access to all gates. The feedback from travelers has been so positive that executives at other airports are considering following this approach. The hospitality industry is also installing Wi-Fi for their guests. Starwood, Holiday Inn, and Marriott Hotels are installing Wi-Fi, with prices varying from roughly $10 to $25 per day for customers to use both wired and wireless access services. Choice Hotels International installed free wireless Internet access in public areas and guest rooms at all of its 370 Comfort Suites and 140 Clarion properties. Best Western also plans to roll out free wireless access.[14]

Wired Equivalent Privacy (WEP) is the standard security shipped with all Wi-Fi hardware. Unfortunately, WEP provides only minimal protection and is considered next to worthless by the security community. Wi-Fi access points broadcast their existence to the world, making them easy to locate. With a range of 100 to 500 feet, access points often can be used by people in a company's parking lot or street out front. General Motors has deployed Wi-Fi in its manufacturing plants but is holding off implementing Wi-Fi in its headquarters. It is concerned that until tighter security measures are in place, guests at a Marriott Hotel across the street could log on to GM's network and steal confidential memos and data.[15]

A group of vendors, including Cisco Systems, Enterasys, Microsoft, Proxim/Agere, and Symbol Technologies, defined an improved security standard called Wi-Fi Protected Access (WPA). The new security standard was designed so that it can be implemented by simply upgrading the firmware—programming inserted into programmable read only memory (PROM), thus becoming a permanent part of the device—in existing PC cards and access points. So, the customer's current Wi-Fi equipment will not have to be discarded to upgrade to the improved security level of WPA.[16]

IEEE 802.11 g

This protocol is a faster version of the Wi-Fi technology, which enables data transmission at up to 54 Mbps. Both Wi-Fi and 802.11g operate in the 2.4 GHz range, so it possible to pick up interference from microwaves, cordless phones, and baby monitors.

IEEE 802.16 (WiMax)

Worldwide Interoperability for Microwave Access (WiMax, or 802.16) is designed to support wireless metropolitan area networks and be compatible with European standards. WiMax metropolitan area networks are supported by a backbone of base stations connected to a public network. Each base station supports hundreds of fixed subscriber stations, which can be both public Wi-Fi hot spots and enterprise networks.

With a wireless access range of up to 31 miles and a 70 Mbps data transfer rate, WiMax can extend broadband wireless access to new locations and over longer distances, as well as significantly reduce the cost of bringing broadband to new areas. WiMax may also offer a solution to what is sometimes called the "last mile" problem, referring to the expense and time needed to connect individual homes and offices to trunk lines for communications.

IEEE 802.20 (MBWA)

Mobile Broadband Wireless Access (MBWA, or 802.20) operates from existing cellular towers and promises the same coverage area as a mobile phone system with the speed of a Wi-Fi connection. Its successful development would pose a threat to the broad use of other advanced wireless technology, so it has generated much controversy among vendors who have made substantial investments in alternative technologies. No products supporting 802.20 are expected before 2006.

Table 6.7 summarizes current IEEE 802 series of wireless protocols, their ranges, and transmission speeds.

Table 6.7

Wireless Networks Based on IEEE 802.*xx* Standards

	Personal Area Network	Local Area Network		Metropolitan Area Network	Wide Area Network
Protocol	IEEE 802.15, Bluetooth	IEEE 802.11b, Wi-Fi	IEEE 802.11g	IEEE 802.16, WiMax	IEEE 802.20
Range	30 feet	300 feet	100 feet	31 miles	National or global
Transmission Speed	1 Mbps–400 Mbps	11 Mbps	54 Mbps	70 Mbps	11 Mbps

1G

The first generation of wireless networks for voice communication was based on analog transmission of voice and data and is called 1G. Many of these systems were individually tailored, country-specific solutions, including technologies such as advanced mobile phone service (AMPS) and total access communications system (TACS).

2G

With the spread of cellular phone communication in the 1990s, the telecommunications industry needed to develop faster and more reliable protocols for wireless voice transmission. The 2G wireless protocol is the result, and much of this technology is still in use today. A break from its analog predecessor, the 2G cell phone protocol features digital voice encoding.

2.5G

Since its inception, users of 2G technology wanted improved service and more functions—increased bandwidth, faster transmission rates, and support for packet routing to enable the introduction of new data services, including multimedia. The present state of mobile wireless communications is often called 2.5G—a midway point between 2G and 3G technologies. The increased data rates rise to a theoretical maximum of 384 Kbps.

3G

The 3G protocol of wireless technology will soon be available and will transmit wireless data at 144 Kbps to mobile users and 2 Mbps to fixed locations. In addition to providing enhanced multimedia transmission (voice, data, video, and remote control) services, it will also support cellular phones, e-mail, paging, fax, videoconferencing, and Web browsing. The scope of 3G is international; its proponents promise that it will keep people connected at all times and in all places. The International Telecommunication Union (ITU) coordinates 3G standards through its International Mobile Telecommunications-2000 project. An essential goal of 3G is to provide roaming capability throughout Europe, Japan, and North America. Like Wi-Fi, 3G promises to provide users with wireless Internet access. The protocol has been adopted in Europe and Asia and spread to North America. As a phone system, 3G provides much broader coverage than Wi-Fi's collection of hot spots. But Wi-Fi's hot spots are strategically placed in hotels, airports, and gathering places where mobile Net surfers are most likely to be. Thus, there is a definite overlap and competition between Wi-Fi and 3G.

MMDS

Multichannel Multipoint Distribution System (MMDS) allows phone companies to deliver T1 speeds to the homes of individual users over rooftop antennas mounted anywhere within a 35-mile radius of a powerful transmission tower. Spike Broadband Systems is using this technology to build a wireless broadband-access system for Denmark's second largest mobile-phone company, Sonofon. Spike claims it can build the infrastructure at a fraction of the cost required to build cable-modem or DSL networks. In addition, the system can also provide local telephone service.[17]

The variety of wireless communications protocols is capable of supporting a wide range of business applications. One of the applications with the greatest potential is the use of radio-frequency identification (RFID) tags. Read the "Ethical and Societal Issues" special interest feature to learn about some of the unique issues raised by this new technology.

Network Switching Devices

In addition to communications protocols, different hardware devices enable the high-speed switching of messages from one network to another. We discuss several of the most common next.

Manufacturers and Retailers Grapple with RFID and Privacy

Many manufacturers and retailers are experimenting with radio-frequency identification (RFID) tags for tracking merchandise throughout the supply chain. Tiny RFID chips, expected to replace the universal product code, wirelessly send their stored data to RFID reader devices without the need for visual scanning. By pointing a handheld RFID reader at an aisle of merchandise, a store clerk could take inventory of hundreds of items in a matter of seconds. Not only could the products "call out" their presence, but they could also inform the reader of how long they've been on the shelf, where they were manufactured, and what the URL of the Web site is to order more inventory.

It's easy to see the appeal of such money-saving technology. Many manufacturers have incorporated RFID to track merchandise through the stages of the value chain. Wal-Mart is requiring all of its suppliers to use RFID tags on the pallets of merchandise that they deliver. Such tagging automates merchandise tracking. RFID readers positioned at the sending and receiving docks of manufacturers, warehouse facilities, and transportation companies automatically gather and store in a central database the location information of merchandise as it travels through the supply chain. For example, Wal-Mart would be able to tell that four pallets of Bounty paper towels left the Procter & Gamble manufacturing plant March 5 at 10:49 a.m. and are currently aboard a truck in Lansing, Michigan.

Consumer privacy issues arise, however, when RFID tags move from the pallet level to placement on individual items. Wal-Mart had to pull the plug on its item-level RFID tag experiment over concerns raised from privacy advocacy groups. The groups wanted assurances that the tags would be removed from items at the checkout station. They were concerned that if the tags remained on items, anyone with an RFID reader could scan a person, a person's bag, car, or home to "take inventory" of what that person possessed. The Benetton Corporation was similarly criticized for its plan to sew RFID tags into its clothing product lines.

The German retail giant MetroGroup invented an interesting use for RFID chips. Metro gave out about 10,000 customer loyalty cards with embedded RFID chips as part of an effort to bring wireless technology into its stores and warehouses. Customers who carried the card with them when visiting Metro's experimental "future store" were amazed when video presentations started automatically as they approached. Automated systems were alerted to a customer's presence through wireless communication from the chip embedded in the card in their pocket or wallet. The cards sparked protests by privacy advocates, who worry that the cards could allow stores to secretly track consumers as they shop—and perhaps in other environments. Metro played down those suspicions, saying the RFID-equipped cards were never used to store or process customer behavior. "We never saw a privacy problem," Metro spokesman Albrecht von Truchsess said.

Even with all its problems, RFID still has vast potential to improve tracking and inventory systems of all kinds. The U.S. government is looking to RFID to track the origin and movement of livestock in efforts to control diseases such as mad cow disease and bird flu. RFID is also being proposed for tracking medications and ensuring that hospitals dispense the correct medication, in the prescribed dosage, to patients. But RFID also could be abused and used to infringe on people's privacy. RFID tags are so small—smaller than a grain of rice—that they can be placed on a person without the person's knowledge and used to track the person's movements and behaviors. Because of the potential for both good and evil, RFID may be the most hotly debated wireless networking technology of our day.

Critical Thinking Questions

1. We've discussed how manufacturers and retailers are using RFID. How could businesses and organizations in the fields of transportation, hospitality, education, finance, entertainment, and government utilize RFID?
2. What, if anything, can be done to ease the concerns of privacy advocates over RFID?

What Would You Do?

You have just taken a new position as the inventory manager at Prescott Auto Rentals. The last manager was fired because of his inability to track vehicles during peak hours when sometimes more than two dozen customers attempt to enter and exit the lot at the same time. You are asked to design a new, more efficient system for checking rental cars out and in. In your last job at Quality-Plus Manufacturing, you were involved with developing an RFID system for tracking pallets of merchandise. Could an RFID system be implemented for rental cars? RFID tags could be attached to the sides of the vehicles and readers installed at the gates of the lot. Perhaps the RFID chips could be programmed to store and automatically send the mileage of the vehicle to speed up the check-in/checkout process.

3. What privacy concerns might such a system raise? What is the limit of the data that you might dare to store on the tag if you were able to store any customer information that you possessed?
4. How do privacy issues associated with tags attached to rental cars compare with those of tags sewn into your blue jeans?

SOURCES: "German Chain Kills RFID Plan," *The Associated Press*, March 1, 2004, *www.lexis-nexis.com*; "Bennetton: No Microchips Present in Garments On Sale—No Decision Yet Taken on Industrial Use," Bennetton press release, April 4, 2003, *www.benetton.com/press/*; Emily Kaiser, "Wal-Mart Takes Lead in Retail Tracking Technology," *Reuters*, January 27, 2004, *www.reuters.com*; Carol Sliwa, "Retailers See RFID Ahead," *Computerworld*, January 26, 2004, *www.computerworld.com*.

Private Branch Exchange

A **private branch exchange** (PBX) is an on-premise switching system owned or leased by a private enterprise that interconnects its telephones and provides access to the public telephone systems. Users gain access to the outside world by sharing access to a limited number of trunk lines between the PBX and the local exchange carrier's central office. It reduces costs by not requiring a separate line for each user to the local telephone company's central office. Thus, there is a trade-off between the expense of the PBX equipment and the savings in the reduced number of incoming trunk lines. The PBX provides many useful services such as call transfer, call forwarding, audioconferencing, and voice mail. Centrex is a local telephone company that is an alternative to the PBX. With Centrex, each user must have a line to the central office, and the switching function is performed by equipment located there, as shown in Figure 6.19.

private branch exchange (PBX)
An on-premise switching system owned or leased by a private enterprise that interconnects its telephones and provides access to the public telephone system.

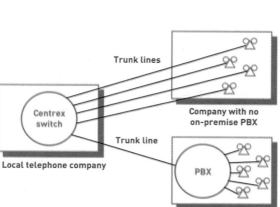

Figure 6.19

Comparison of Private Branch Exchange and Centrex

Many PBXs can switch communications among digital devices as well as telephones—in a digital format without requiring a modem. Often such PBXs are called IP-PBXs or digital PBXs. A personal computer can be wired to a digital PBX and then connect to any printer, computer, or other device that is also wired to the PBX. Of course, this represents a considerable savings in equipment costs.

Brown Brothers Harriman & Company (BBH), a New York–based banking and investment management firm with 3,000 brokers and financial services workers, recently installed four IP-PBX voice switches at its offices in Boston, New York, and Jersey City. They chose this technology because it was less expensive and could handle voice communications in either the traditional circuit-switched or the newer IP-based mode. The cost of the PBX upgrade project was $2.5 million, but the new equipment has already saved $200,000 in annual costs by eliminating voice-only communications lines. BBH hesitates to expand the use of IP telephony capabilities to its offices worldwide because of a lack of open standards and the difficulty of getting products that the company has bought from different vendors to operate together.[18]

Bridge

A **bridge** is used to connect two or more networks that use the same communications protocol. An example of a bridge is a device that decides whether a message from you to someone else is going to someone on the LAN on your floor or to another LAN that covers the floor above yours. The bridge examines each message on the LAN, passing each message known to be destined for someone on the same LAN and forwarding those known to be destined from someone on another LAN.

bridge
A device used to connect two or more networks that use the same communications protocol.

switch

A telecommunications device that routes incoming data from any one of many ports to a specific output port that will take the data toward its intended destination.

Switch

A **switch** is a telecommunications hardware device that routes incoming data from any one of many input ports (a specific place to physically connect another device or communications line) to a specific output port that will take the data toward its intended destination. There are three basic types of switches:

1. In the traditional circuit-switched telephone network, one or more *telephone switches* are used to set up a temporary connection or circuit dedicated to support the conversation between two or more parties.
2. On a LAN, a *LAN switch* determines from the physical device address in each incoming message frame which output port the message should be forwarded to.
3. In a wide area packet-switched network such as the Internet, a *network switch* determines from the IP address in each packet which output port to use for the next part of its trip to the intended destination.

In large networks, the trip from one switch to another is called a *hop*. The time a switch takes to figure out where to forward a data unit is called its *latency* and introduces some delay in the delivery of data.

router

A device or software in a computer that determines the next network point to which a data packet should be forwarded toward its destination.

Router

A **router** is a device or software in a computer that determines the next network point to which a data packet should be forwarded toward its destination. To do this, the router operates as an intelligent traffic cop. It maintains a table of the available routes and their conditions and uses this information, along with distance data and cost algorithms, to determine the best route for a given packet. Typically, a packet will travel through a number of network routers before arriving at its destination. The functionality of a router is often included in a network switch.

hub

A place of convergence where data arrives from one or more directions and is forwarded out in one or more other directions.

Hub

In data communications, a **hub** is a place of convergence where data arrives from one or more directions and is forwarded out in one or more other directions. Hubs commonly contain multiple ports used to connect segments of a network. A passive hub serves simply as a conduit for the data, enabling it to go from one device (or segment) to another. An intelligent hub includes additional features that enable a network administrator to monitor the traffic passing through the hub and to configure each port in the hub. A switching hub reads the destination address of each packet and then forwards the packet to the correct port.

gateway

A network point that acts as an entrance to another network.

Gateway

A **gateway** is a network point that acts as an entrance to another network. For example, the computers that control traffic within your company's network or at your local Internet service provider (ISP) are gateways. A gateway is often associated with both a router, which knows where to direct a given packet of data that arrives at the gateway, and a switch, which furnishes the actual path in and out of the gateway for a given packet. The gateway may link two networks that employ different telecommunications protocols. In this situation, the data received by a gateway must be restructured into a form that is usable on the receiving network.

Corrugated Supply is a small industrial supplier that converts paper into corrugated sheets that its customers then finish and assemble into boxes. To enable it to deliver fast turnaround times and superior service, the company implemented an enterprise resource planning (ERP) system integrated with the company's Web site to provide customers real-time access to their orders. When the company expanded to a new facility near its headquarters, it needed a cost-effective way to network voice and data traffic between the two buildings. A wireless connection between the facilities avoided the high cost of laying cable across public easements. In addition, the company wanted to ensure that the network could accommodate new facilities as the company grew. Corrugated Supply chose a combination of redundant Cisco routers and Cisco switches to provide data routing and switching, wireless networking, and IP communications. This solution delivers the performance and reliability the company's customers demand; it also provides flexibility for future growth.[19]

NETWORK BASICS

Businesses link their personnel and equipment to enable people to work quicker and more efficiently. Computer networks allow organizations flexibility—to accomplish work wherever and whenever it is most beneficial. To take full advantage of networks and distributed processing, you should understand basic processing strategies, communications software, and communications protocols.

Basic Processing Strategies

When an organization needs to use two or more computer systems, one of three basic processing strategies may be followed: centralized, decentralized, or distributed. With **centralized processing**, all processing occurs in a single location or facility. This approach offers the highest degree of control, since all data processing is done on a single centrally managed computer. 7-Eleven is a $9 billion convenience store chain that has implemented a centralized processing strategy to manage its 5,800 stores. It uses a proprietary Retail Information System (RIS) that runs on a single mainframe computer operated by Electronic Data Systems (EDS) and a centralized processing network to enable it to keep all its stores operating efficiently and to share information among all its suppliers. The RIS provides store managers with daily, weekly, and monthly sales tallies, which they use to create their orders. Store managers enter orders into workstations or handheld computers by 10 A.M. daily. By 11 A.M., orders have been transmitted to a central database, consolidated, and dispatched to 7-Eleven's suppliers. The consolidation takes place four times a day, once for each time zone in which 7-Eleven operates. The centralized processing network also connects the stores to McLane Company, 7-Eleven's primary wholesale distributor, and to the commissaries and bakeries that provide fresh food products so that all can view the same sales and shipment information. "There's quite a bit of information sharing," says Ruel Athey, vice president of customer service at McLane Information Systems. "We work closely with 7-Eleven and the suppliers to come up with the most efficient distribution process we can."[20]

With **decentralized processing**, processing devices are placed at various remote locations. The individual computer systems are isolated and do not communicate with each other. Decentralized systems are suitable for companies that have independent operating divisions. Some drugstore chains, for example, operate each location as a completely separate entity; each store has its own computer system that works independently of the computers at other stores.

With **distributed processing**, computers are placed at remote locations but connected to each other via a network. One benefit of distributed processing is that processing activity can be allocated to the location(s) where it can most efficiently occur. For example, the New York headquarters may have the largest computer system, but the Atlanta office might have hundreds of employees to input the data. The system's output may be most needed in Chicago, the location of the warehouse. With distributed processing, each of these offices can organize and manipulate the data to meet its specific needs, as well as share its work product with the rest of the organization. The distribution of the processing across the organizational system ensures that the right information is delivered to the right individuals, maximizing the capabilities of the overall information system by balancing the effectiveness and efficiency of each individual computer system.

Cooper Tire and Rubber Co.'s tire division plans to move more of its manufacturing offshore, with a goal of having its own manufacturing facility just outside Shanghai by 2007. Meanwhile, Cooper's tire division is implementing a two-year business plan that calls for establishing a global distributed processing network to connect the computers and decision makers at all its plants. At the same time, it is transitioning from a rudimentary paper and spreadsheet demand-forecasting process to automated on-demand forecasting. Cooper is implementing several applications from i2 Technologies, Inc., that will address the complex

centralized processing
Processing alternative in which all processing occurs in a single location or facility.

decentralized processing
Processing alternative in which processing devices are placed at various remote locations.

distributed processing
Processing alternative in which computers are placed at remote locations but connected to each other via a network.

demand, inventory, and resource planning issues of a global manufacturer. These investments will help Cooper manage its expanding tire imports and make demand-sourcing decisions within its distribution centers.[21]

The September 11, 2001, terrorist attacks sparked many companies to distribute their workers, operations, and systems much more widely, a reversal of the recent trend toward centralization. The goal is to minimize the consequences of a catastrophic event at one location while ensuring uninterrupted systems availability.

Communications Software

communications software
Software that provides a number of important functions in a network, such as error checking and data security.

Communications software provides a number of important functions in a network. Most communications software packages provide error checking and message formatting. In some cases, when there is a problem, the software can indicate what is wrong and suggest possible solutions. Communications software can also maintain a log listing all jobs and communications that have taken place over a specified period of time. In addition, data security and privacy techniques are built into most packages.

In Chapter 4, you learned that all computers have operating systems that control many functions. When an application program requires data from a disk drive, it goes through the operating system. Now consider a situation in which a computer is attached to a network that connects large disk drives, printers, and other equipment and devices. How does an application program request data from a disk drive on the network? The answer is through the network operating system.

network operating system (NOS)
Systems software that controls the computer systems and devices on a network and allows them to communicate with each other.

A **network operating system (NOS)** is systems software that controls the computer systems and devices on a network and allows them to communicate with each other. A NOS performs the same types of functions for the network as operating system software does for a computer, such as memory and task management and coordination of hardware. When network equipment (such as printers, plotters, and disk drives) is required, the NOS makes sure that these resources are used correctly. In most cases, companies that produce and sell networks provide the NOS. For example, NetWare is the NOS from Novell, a popular network environment for personal computer systems and equipment. Windows NT and Windows 2000 are other commonly used network operating systems.

network-management software
Software that enables a manager on a networked desktop to monitor the use of individual computers and shared hardware (such as printers), scan for viruses, and ensure compliance with software licenses.

Software tools and utilities are available for managing networks. With **network-management software**, a manager on a networked personal computer can monitor the use of individual computers and shared hardware (such as printers), scan for viruses, and ensure compliance with software licenses. Network-management software also simplifies the process of updating files and programs on computers on the network—changes can be made through a communications server instead of being made on individual computers. In addition, network-management software protects software from being copied, modified, or downloaded illegally and performs error control to locate telecommunications errors and potential network problems. Some of the many benefits of network-management software include fewer hours spent on routine tasks (such as installing new software), faster response to problems, and greater overall network control.

Network management is one of the most important tasks of IS managers. In fact, poor management of the network can cause a whole company to suffer. With networks now being used to communicate with customers and business partners, network outages or slow performance can even mean a loss of business. Network management includes a wide range of technologies and processes used to automate infrastructure monitoring and help IS staffs identify and address problems before they affect customers, business partners, or employees.

Fault detection and *performance management* are the two types of network-management products. Both employ the *Simple Network Management Protocol* (SNMP) to obtain key information from individual network components. SNMP is the standard management protocol used on TCP/IP networks. It allows virtually anything on the network, including switches, routers, firewalls, and even operating systems and server products and utilities, to communicate with management software about its current operations and state of health. SNMP can also be used to control these devices and products, telling them to redirect traffic, change traffic priorities, or even to shut down.

Fault management software alerts IS staff in real-time when a device is failing. Equipment vendors place traps (code in a software program for handling unexpected or unallowable conditions) on their hardware to identify the occurrence of problems. In addition, the IS staff can place agents—automated pieces of software—on networks to monitor different functions. When a device exceeds a given performance threshold, the agent sends an alarm to the company's IS fault management program. For example, if a CPU registers that it is more than 80 percent busy, an alarm may be generated.

Performance management software sends messages to the various devices (i.e., polls them) to sample their performance and to determine whether they are operating within acceptable levels. The devices reply to the management system with performance data that the system stores in a database. This real-time data is automatically correlated to historical trends and displayed graphically so that the IS staff can identify any unusual variations.

Today, most IS organizations use a combination of fault management and performance management to ensure that their network remains up and running and that every network component and application is performing acceptably. With the two technologies, the IS staff can identify and resolve fault and performance issues before they affect customers and service. The latest network-management technology even incorporates automatic fixes—the network-management system identifies a problem, notifies the IS manager, and automatically corrects the problem before anyone outside the IS department notices it.

Con Edison Communications is testing management software from Hewlett-Packard Co. on the fiber-optic network it operates in the New York metropolitan area. By installing HP OpenView Network Node Manager software, Con Edison Communications network administrators will be able to monitor all elements on its metropolitan area network from a single management console, enabling them to react more quickly when problems occur. Con Edison has already spent more than $1 million to roll out HP's OpenView TeMIP Operations Support system, which offers a set of telecommunications-infrastructure-management capabilities, including fault-management and service-activation tools. Network Node Manager will feed network alarms and fault information into the TeMIP software.[22]

TELECOMMUNICATIONS APPLICATIONS

Telecommunications and networks are a vital part of today's information systems. In fact, it is hard to imagine how organizations could function without them. For example, when a business needs to develop an accurate monthly production forecast, a manager simply downloads sales forecast data gathered directly from customer databases. Telecommunications provides the network link allowing the manager to access the data quickly and generate the production report, which in turn supports the company's objective of better financial planning.

The consumer goods giant Procter & Gamble uses LANs in all its plants to link office and plant workers to common software and shared databases and to provide e-mail services. The result is faster, more cost-effective, higher-quality product manufacturing. Other organizations transfer millions of important and strategic messages from one location to another every day. Telecommunications is a critical component of information systems. In some industries it is almost a requirement for doing business; most companies could not survive without it. This section looks at some significant business applications of networks.

Linking Personal Computers to Mainframes and Networks

One of the most basic ways that telecommunications connect an individual to information systems is by connecting personal computers to mainframe computers so that data can be downloaded or uploaded. For example, a data file or document file from a database can be downloaded to a personal computer for an individual to use. Some communications software programs instruct the computer to connect to another computer on the network, download or send information, and then disconnect from the telecommunications line. They are called

unattended systems because they perform the functions automatically, without user intervention.

Voice Mail

With **voice mail**, users can leave, receive, and store verbal messages for and from other people around the world. In some voice mail systems, a code can be assigned to a group of people instead of an individual. Suppose the code 100 stands for all 250 sales representatives in a company. If anyone calls the voice mail system, enters the number 100, and leaves a message, all 250 sales representatives will receive the same message.

Electronic Software Distribution

Electronic software distribution involves installing software on a file server for users to share by signing on to the network and requesting the software be downloaded to their computers. Electronic software distribution is quicker and more convenient than traditional ways of acquiring software and is significantly less costly and more efficient than having a network administrator constantly install upgrades.

The Marathon Oil Corporation uses Microsoft's Systems Management Server (SMS) for software distribution to reduce the time and effort required to provide users with the latest software releases and patches. Marathon uses SMS to scan workstations and servers to identify those needing an upgrade or security patch and to push the patches to the servers.[23] The software makes the entire process simpler and much more reliable so that the IS staff is free to provide other support services. With the software, IS workers can remotely install and update the operating system and applications on the PCs. Previously, Marathon had to update and repair PCs by sending technicians into the field or shipping disk drives back and forth.

One problem with electronic software distribution is the size of software programs—downloading large programs requires high-capacity telecommunications media and a lot of time. Software piracy is another issue that must be addressed.

Electronic Document Distribution

Networks also allow organizations to transmit documents without using paper. It is not known how many millions of dollars companies spend on printing, distributing, and storing documents of all types, but the amount is staggering. **Electronic document distribution** involves transmitting documents—such as sales reports, policy manuals, and advertising brochures—over communications lines and networks. Electronic document distribution software allows word processing and graphics documents to be converted into binary code and sent over networks.

Microsoft's Windows Rights Management Services (RMS) enables employees to apply rights, privileges, and protections to Office 2003 documents distributed within a business network. The service allows the author of a Word 2003 document to specify which group of users can open, modify, print, and forward an Office 2003 document and under what circumstances it can be used. For example, the author can designate a sensitive document as read-only or set an expiration date for a time-sensitive or highly confidential document. The software works with Outlook 2003, Word 2003, PowerPoint 2003, and Excel 2003. Microsoft can provide an RMS-enabled add-on for Internet Explorer for protected Web-page viewing.[24]

Call Centers

A call center is a physical location where customer and other telephone calls are handled by an organization, usually with some amount of computer automation. Call centers are used by customer service organizations, telemarketing companies, computer product help desks, charitable and political campaign organizations, and any organization that uses the telephone to sell or support products and services. An automatic call distributor (ACD) is a telephone facility that manages incoming calls, handling them based on the number called and an associated database of instructions. Call centers frequently employ an ACD to validate callers,

place outgoing calls, forward calls to the right party, allow callers to record messages, gather usage statistics, balance the workload of support personnel, and provide other services.

Complicating the role of the call center is the National Do Not Call Registry set up in 2003 by the U.S. Federal Trade Commission. Telemarketers who call numbers on the list face penalties of up to $11,000 per call, as well as possible consumer lawsuits. Some 60 million consumers have signed up for the list by logging on to the site, *www.donotcall.gov*. While the registry has greatly reduced the number of unwanted calls to consumers, it has created several compliance-related issues for direct marketing companies. For instance, apart from the national registry, 43 states maintain their own do-not-call lists, which must be combined with the National Registry.[25]

Offshore call centers that provide technical support services are a fact of life for many technology vendors and their customers. But both vendors and users agree that support operations have to balance their desire to reduce labor costs with customer satisfaction considerations. For example, in late 2003 Dell said that it was returning phone-based technical support for its corporate PCs to the United States because of complaints from some users about the quality of service they received from a call center in India. However, other major vendors, including IBM, Hewlett-Packard Co., Oracle Corp., and Computer Associates International, Inc., plan to continue using global support operations.

In early 2004, Hewlett-Packard Co. set up a 600-person contact center in Bangalore, India, to provide postsales support to U.S. customers of all of its consumer products. The center also serves as a research and development facility for new support processes and technologies. The support center employs people with varying levels of expertise so that any problems can be handled within the center itself. The center initially offered voice-based support in English, but it also plans to offer e-mail or chat-based support.[26]

Offshore call centers provide technical support services for many technology vendors and their customers.

(Source: Sondeep Shankar/ Bloomberg News/Landov.)

Telecommuting

More and more work is being done away from the traditional office setting. Many enterprises have adopted policies for **telecommuting** that enable employees to work away from the office using computing devices and networks. Companies can save money because less office and parking space and office equipment are required for their workers.

There are several reasons why telecommuting is popular among workers. Single parents find that it helps balance family and work responsibilities because it eliminates the daily commute. It also enables qualified workers who may be unable to participate in the normal workforce (e.g., those who are physically challenged or who live in rural areas too far from the city office to commute regularly) to become productive workers. Extensive use of telecommuting can lead to decreased need for office space, potentially saving a large company

telecommuting
A work arrangement whereby employees work away from the office using personal computers and networks to communicate via e-mail with other workers and to pick up and deliver results.

millions of dollars. Corporations are also being encouraged by public policy to try telecommuting as a means of reducing traffic congestion and air pollution.

Some types of jobs are better suited for telecommuting than others, including jobs held by salespeople, secretaries, real estate agents, computer programmers, and legal assistants, to name a few. Telecommuting also requires a special personality type to be effective. Telecommuters need to be strongly self-motivated, organized, able to stay on track with minimal supervision, and have a low need for social interaction. Jobs not good for telecommuting include those that require frequent face-to-face interaction, need much supervision, and have lots of short-term deadlines. Employees who choose to work at home must be able to work independently, manage their time well, and balance work and home life. Read the "Information Systems @ Work" special feature to gain further insight into state-of-the-art telecommuting facilities.

U.S. federal agencies are implementing telecommuting for government employees at a rate well behind the private sector. Only 5 percent of the government's 1.8 million civilian workers had the option to telecommute in 2002 compared with 20 percent of workers in private industry. Some of the key barriers to adoption are concerns about data security, IS technical issues, and management resistance. For example, the Patent Office decided not to offer a telecommuting program to its 800 senior patent examiners because of a concern over the potential loss of the confidential patent information. In addition, since patent examiners must comb through millions of patent records, as well as scientific and technical databases worldwide, they would require expensive broadband access at their away-from-the-office workplace. As a result, patent officials have rejected telecommuting for now. Interestingly, a federal official at the General Services Administration, which has a role in crafting government-wide telecommuting policy, said that he believes management resistance is really the key barrier. Speaking on condition of anonymity, the official said "When you are trying to make ingrained behavior changes, you have to have some very strong force."[27]

Videoconferencing

videoconferencing
A telecommunication system that combines video and phone call capabilities with data or document conferencing.

Videoconferencing enables people to hold a conference by combining voice, video, and audio transmission. Not only are travel expenses and time reduced, but managerial effectiveness is also increased through faster response to problems, access to more people, and less duplication of effort by geographically dispersed sites. Almost all **videoconferencing** systems combine video and phone call capabilities with data or document conferencing (see Figure 6.20). You can see the other person's face, view the same documents, and swap notes and drawings. With some of the systems, callers can make changes to live documents in real time. Many businesses find that the document- and application-sharing feature of the videoconference enhances group productivity and efficiency. Meeting over phone lines also fosters teamwork and can save corporate travel time and expense.

Live, Work, and Play in Luxury at the Heliopolis Complex

One of the largest construction projects of the century is taking place in Cairo, Egypt. Ten thousand workers have been laboring around the clock to build the extravagant Heliopolis Complex. The $750 million multiuse development, brainchild of the Citystars Group, will be a self-sufficient city-within-a-city. It includes:

- *Star Living:* 450 premium residential units (for full-time residents) in a wide variety of sizes and layouts.
- *Star Capital:* 70,000 square meters of prestigious office space in a dynamic environment that meets the design and operational requirements for major corporations and multinational occupiers.
- *Three 4- and 5-star hotels:* the Inter-Continental Heliopolis Hotel, Le Meridien Star, and Holiday Inn, together offering more than 1,300 rooms, suites, apartments, and chalets for guests. Also, luxurious swimming pools, health clubs, restaurants, banquet halls, and conference rooms.
- *Entertainment:* A two-level entertainment center, including a cinema multiplex with 16 screens, a family theme park, bowling, and billiards. Also, two levels of world-class exhibition space.
- *Star Centre:* A three-level shopping mall with more than 450 stores, including gigantic grocery, furniture, toy, and department stores, as well as a traditional Egyptian *souq* (market) selling jewelry, antiques, and crafts.
- *Star Care:* A medical facility to support the residents and visitors with services for routine needs and emergencies alike.

All of this is situated in large buildings, including three huge pyramids, on spacious property that includes abundant green space and convenient shuttle services.

Citystars Heliopolis is designed to provide residents and visitors with a top-of-the-line lifestyle incorporating state-of-the-art technology. "One of our visions for Citystars was to create an integrated residential, office, and shopping complex whereby all our guests, customers, and tenants would be able to communicate using the latest state-of-the-art IT networks," says Abdulrahman Sharbatly, chairman and founder of Citystars. Citystars selected Cisco Systems equipment for what is being touted as the largest network installation in Europe, the Middle East, and Africa.

The state-of-the-art LAN will be based on the popular Internet Protocol (IP) to support voice, video, and data communications. The network will provide the complex's customers with the latest information and communications technology services, including IP telephony—a telephone system that is integrated into the data network.

With 45,000 Ethernet ports distributed around the city, residents and visitors will always be within reach of a high-speed network connection. Approximately 7,000 IP phones have been installed, which communicate over the data network and connect to computers for added functionality and services. Citystars is planning to significantly expand its network by adding wireless access points to allow its residents and visitors to connect to the Heliopolis network wirelessly from any location.

Integrated communities, which link home and work environments, take telecommuting to the ultimate level. While telecommuting uses telecommunications networks to allow employees to work from home, integrated communities allow employees to completely merge work life and home life—ignoring the traditional boundaries of the workplace. Work can occur wherever and whenever you wish it to. Residents of Heliopolis and their guests can connect to private corporate networks from anywhere in the city. A businessperson can meet with out-of-town clients in a hotel suite, around the pool, or in a local café and have full access to corporate data and customer records as if he or she were at the office.

Integrated communities such as Heliopolis are growing increasingly more common. Others, less grandiose than Heliopolis but equally convenient, provide residents with living, working, shopping, and playing space in a compact geography that leaves little need for a car. Computer and communications networks in such communities are an essential selling point; they provide residents with instant access to information, services, and other residents within the community. While high-speed wireless network connectivity is only available in pockets in traditional city structures, integrated communities can provide universal coverage to citizens. Such communities might cause you to wonder if we might all someday live in employer-centered cities where the line between life and work are blurred.

Discussion Questions

1. What are the advantages of living and working in an integrated community such as Heliopolis?
2. What role does the LAN play in providing services to the residents of Heliopolis?

Critical Thinking Questions

3. Would a work-life community such as Heliopolis be beneficial to you personally in your chosen career area? Why or why not?
4. What are the dangers, if any, of blurring the distinction between work time and home/fun time?

SOURCES: "Citystars Heliopolis Cairo" Web site, *www.citystars.com.eg/index. htm*, accessed March 10, 2004; "Citystars Installs Cisco-Based Network Infrastructure across $600m Heliopolis Complex," *News@Cisco, http://newsroom.cisco.com*, accessed March 10, 2004; "Egypt's Citystars Streamlines Business with Oracle E-business Suite," *Financial Times*, December 29, 2003, *www.lexis-nexis.com*; "Cairo's Wealthy Won't Need to Leave City to Shop," *Shopping Centers Today, www.icsc.org*, accessed March 10, 2004.

Videoconferencing

Videoconferencing allows participants to conduct long-distance meetings "face to face" while eliminating the need for costly travel.

(Source: Steve Chenn/CORBIS.)

Group videoconferencing is used daily in a variety of businesses as an easy way to connect work teams. Members of a team go to a specially prepared videoconference room equipped with sound-sensitive cameras that automatically focus on the person speaking, large TV-like monitors for viewing the participants at the remote location, and high-quality speakers and microphones. Videoconferencing costs have come down steadily, while video quality and synchronization of audio to data—once weak points for the technology—have improved. Videoconferencing systems designed to support large meetings cost an average of about $20,000 each. Rooms to handle meetings for 10 or fewer people can be equipped for $7,000 or less. There are additional expenses associated with use of the telecommunications network to relay voice, video, data, and images.

Rao Vellanki, operations manager at Best Buy, moved from India to the United States in 1977. In 2003, he returned to India to manage Best Buy's offshore development in Chennai, where the firm has several hundred software developers working on more than 500 applications. While Vellanki relies on a variety of methods to communicate with the home office, he sees the use of videoconferencing as critical. "There is no question, in my view, that seeing the faces adds to the effectiveness of communications. The personal touch is there, the integrity is there—even though it's only an intangible, we find it unavoidable," he says.[28]

Dow Chemical recently completed a three-year, multimillion-dollar project that involved upgrading all of its 560 videoconferencing systems in 43 countries worldwide. Dubbed iRoom, the project's goal is to improve information sharing and speed decision making across all levels of the company. Dow also hopes to cut back on business travel and conduct more meetings.[29]

Electronic Data Interchange

electronic data interchange (EDI)

An intercompany, application-to-application communication of data in standard format, permitting the recipient to perform the functions of a standard business transaction.

Electronic data interchange (EDI) is an intercompany, application-to-application communication of data in standard format, permitting the recipient to perform a standard business transaction, such as processing purchase orders. Connecting corporate computers among organizations is the idea behind EDI, which uses network systems and follows standards and procedures that allow output from one system to be processed directly as input to other systems, without human intervention. With EDI, the computers of customers, manufacturers, and suppliers can be linked (see Figure 6.21). This technology eliminates the need for paper documents and substantially cuts down on costly errors. Customer orders and inquiries are transmitted from the customer's computer to the manufacturer's computer. The

manufacturer's computer can then determine when new supplies are needed and can automatically place orders by connecting with the supplier's computer.

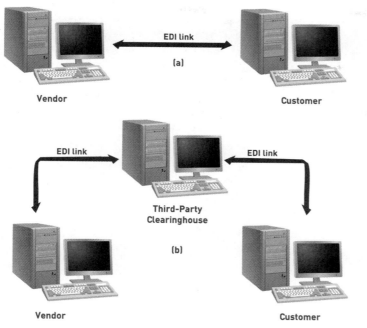

Vendor   **Customer**

Third-Party Clearinghouse

Vendor   **Customer**

Figure 6.21

Two Approaches to Electronic Data Interchange

Many organizations now insist that their suppliers operate using EDI systems. Often the EDI connection is made directly between vendor and customer (a); alternatively, the link may be provided by a third-party clearinghouse, which provides data conversion and other services for the participants (b).

Wal-Mart is the only U.S. retailer that sells Mary-kateandashley One, a new fragrance branded by twin teenage entertainment moguls Mary-Kate and Ashley Olsen. Fragrance and cosmetics manufacturer Coty uses EDI to process orders electronically and keep Wal-Mart's shelves adequately stocked with the fragrance. Coty's Sanford, North Carolina, manufacturing and distribution facility annually processes more than 1.5 million orders, 98.3 percent of which are in EDI format.[30]

As more industries demand that businesses have EDI capability to stay competitive, work activities will continue to evolve. Processes as simple as billing and ordering will change, and new industries will emerge to help build the software and networks needed to support EDI.

Public Network Services

Public network services give personal computer users access to vast databases, the Internet, and other services, usually for an initial fee plus usage fees. Public network services allow customers to book airline reservations, check weather forecasts, get information on TV programs, analyze stock prices and investment information, communicate with others on the network, play games, and receive articles and government publications. Fees, based on the services used, can range from under $15 to over $500 per month (see Figure 6.22). Providers of public network services include Microsoft, America Online, and Prodigy. These companies provide a vast array of services, including news, electronic mail, and investment information. AOL is the number one provider of public network services in terms of size.

Electronic Funds Transfer

Electronic funds transfer (EFT) is a system of transferring money from one bank account directly to another without any paper money changing hands. It is used for both credit transfers, such as payroll payments, and for debit transfers, such as mortgage payments. The benefits of EFT include reduced administrative costs, increased efficiency, simplified bookkeeping, and greater security. One of the most widely used EFT programs is Direct Deposit, in which an employee's payroll check is deposited directly into his or her bank account. The two primary components of EFT are wire transfer and automated clearing house.

public network services
Systems that give personal computer users access to vast databases and other services, usually for an initial fee plus usage fees.

electronic funds transfer (EFT)
A system of transferring money from one bank account directly to another without any paper money changing hands.

Figure 6.22

Public network services provide users with the latest information required to remain competitive. MSN, for example, enables registered users to personalize their home page.

wire transfer

An extremely fast, reliable means to move funds from one account to another.

Wire transfer is an extremely fast, reliable means to move funds from one account to another. There are three major wire-transfer networks. Fedwire is provided by the U.S. Federal Reserve Bank. It supports bank-to-bank and third-party transfers that involve banks and their corporate and retail customers. Clearing House Inter-Bank Payments System (CHIPS) is a bank-owned, privately operated, real-time, final settlement payments system for business-to-business transactions. It processes large-value payment transfers for both domestic and international financial communities. The Society for Worldwide Interbank Financial Telecommunication (SWIFT) is a system that combines messaging and funds transfer.

automated clearing house (ACH)

A secure, private network that connects all U.S. financial institutions to one another by way of the Federal Reserve Board or other ACH operators.

The **automated clearing house (ACH)** network is a secure, private network that connects all U.S. financial institutions to one another by way of the Federal Reserve Board or other ACH operators. It enables electronic payments to be handled and processed throughout the 50 states and U.S. territories and Canada. For payments, funds are transferred electronically from one bank account to the billing company's bank, usually less than a day after the scheduled payment date. The National Automated Clearing House Association (NACHA) in Herndon, Virginia, sets the rules and standards for ACH transactions. A number of banks and organizations have established payment systems based on ACH. PayPal, Inc., developed a system that lets individuals send money to one another via e-mail. Several banks also offer consumer-oriented payment services: Bank One Corporation in Chicago (prior to its merger with JP Morgan in 2004) offered eMoneyMail, Citigroup, Inc., has c2it, and Wells Fargo Bank NA in San Francisco touts BillPoint. Table 6.8 compares the services offered by ACH and wire transfer.

Table 6.8

Comparison of ACH Payments and Wire Transfers

	ACH Payments	**Wire Transfers**
Time to clear	Overnight	Immediately
Can payment be cancelled?	Yes	No
Is there a guarantee of sufficient funds?	No	Yes
Approximate cost per transaction	$.25	$10–$40

Distance Learning

Miami University, Florida, is exploring a partnership with Spanish telecommunications company Telefonica de España SA to provide cyberclasses and physician consulting services to doctors and clinics in Latin America.[31] Often called **distance learning** or cyberclasses, such electronic classes are likely to be the wave of the future. With distance learning software and systems, instructors can easily create course home pages on the Internet. Students can access the course syllabus and instructor notes on the Web page. Student e-mail mailing lists allow students and the instructor to e-mail one another for homework assignments, questions, or comments about material presented in the course. It is also possible to form chat groups so that students can work together as a "virtual team," which meets electronically to complete a group project.

distance learning
The use of telecommunications to extend the classroom.

Specialized Systems and Services

In addition to the applications just discussed, a number of other specialized telecommunications systems and services exist. For example, with millions of personal computers in businesses across the country, interest in specialized and regional information services is increasing. Specialized services, which can be expensive, include professional legal, patent, and technical information. For example, investment companies can use systems such as Quotron and Shark to get up-to-the-minute information on stocks, bonds, and other investments.

Regional services, also called *metropolitan services*, include local electronic bulletin boards and electronic mail facilities that offer information regarding local club, school, and government activities. An electronic bulletin board is a message center that displays messages in electronic form, much as a bulletin board displays paper messages in schools and offices. An electronic bulletin board can be accessed by subscribers with personal computers, network equipment, and software. In addition to regional bulletin boards, national and international bulletin boards are available for people and groups with special interests or needs. These types of bulletin boards exist for many users, such as users of certain software packages and users with certain hobbies. Many public network services, including Prodigy, America Online, and CompuServe, provide access to hundreds of different bulletin boards on a variety of topics and interest areas.

Global positioning systems (GPSs) are used to provide specialized telecommunications services. They have long been used by the military to find locations of troops, equipment, and the enemy—within yards in some cases. Today, GPSs are being used by companies to survey land and buildings and by individuals to locate their positions while camping or exploring. Some auto companies have placed GPSs in their cars to assist travelers in need. Toyota provides its G-Book in-vehicle network information service to Mazda owners in Japan. G-Book provides vehicle owners a wide range of services, including vehicle location and navigation information, maintenance notifications, music and games, emergency roadside assistance, and e-commerce opportunities. The onboard implementation of the G-Book service features a data communications module, a touchscreen navigation terminal, and a GPS antenna. Toyota and Mazda believe it is important for auto makers and the information industry to work together to promote an industry standard so they can offer customers lower-cost and higher-quality service.[32]

With all these telecommunications systems and services, it is no wonder that managers and workers are able to conduct business from remote locations. Often called *virtual workers*, these employees can conduct business at any time and at any place. Today, new telecommunications systems and services are being introduced every month. In the future, we can expect even more innovations to dramatically alter how businesses and individuals stay connected and in touch.

SUMMARY

Principle

Effective communications are essential to organizational success.

Communications is any process that permits information to pass from a sender to one or more receivers. Communications of all types form a major part of any business system. Telecommunications refers to the electronic transmission of signals for communications, including telephone, radio, and television. Telecommunications is creating profound changes in business because it lessens the barriers of time and distance.

The elements of a telecommunications system start with a sending unit, such as a person, a computer system, a terminal, or another device, that originates the message. The sending unit transmits a signal to a telecommunications device. The telecommunications device performs a number of functions, which can include converting the signal into a different form or from one type to another. A telecommunications device is a hardware component that allows electronic communication to occur or to occur more efficiently. The telecommunications device then sends the signal through a medium. A telecommunications medium is anything that carries an electronic signal and interfaces between a sending device and a receiving device. The signal is received by another telecommunications device that is connected to the receiving computer. The process can then be reversed, and another message can go back from the receiving unit to the original sending unit. With synchronous communications, the receiver gets the message instantaneously, when it is sent. Voice and phone communications are examples. With asynchronous communications, the receiver gets the message hours or days after the message is sent.

A communications channel is the transmission medium that carries a message from the source to its receivers. Communication channels can be classified as simplex, half-duplex, or full-duplex. Shannon's Law states that the information-carrying capacity of a channel is directly proportional to its bandwidth—the broader the bandwidth, the more information can be carried.

Principle

An unmistakable trend of communications technology is that more people are able to send and receive all forms of information over greater distances at a faster rate.

The telecommunications media that physically connect data communications devices can be divided into two broad categories: guided transmission media, in which communications signals are guided along a solid medium, and wireless media, in which the communications signal is sent over airwaves. Guided transmission media include twisted-pair wire cable, coaxial cable, and fiber-optic cable. Wireless media types include microwave, cellular, and infrared.

Modems convert signals from digital to analog form for transmission, then back to digital. Four types of modems are internal modems, external modems, cellular modems, and cable modems.

There are many types of telecommunications carriers. A local exchange carrier is a public telephone company in the United States that provides service to homes and businesses within its defined geographical area, called its *local access and transport area*. Homes and businesses are connected to the local exchange via what is called a local *loop*—typically, a pair of copper wires, also called the "last mile." The local exchange office has switching equipment that can switch calls locally or to long-distance carrier phone offices.

Competitive local exchange carriers compete with the local exchange carriers. These competitors include wireless service providers, satellite TV service providers, cable TV companies, and even power companies.

The three established long-distance providers include AT&T, Sprint, and MCI. Local exchange carriers have been able to compete with them to provide long-distance services due to the provisions of the Telecommunications Act of 1996.

Switched lines use switching equipment to allow one transmission device to be connected to other transmission devices. A dedicated line, also called a *leased line*, provides a constant connection between two points. No switching or dialing is needed; the two devices are always connected.

Voice over Internet Protocol (VoIP) is the basic *transport* of voice in the form of a data packet using the Internet Protocol. IP telephony is the technology for transmitting voice communications over a network using an open standards-based Internet Protocol. IP telephony uses VoIP but is a software application suite offering rich feature applications. These often-modular applications lend themselves to cost-effective integration with other applications that share the IP network.

Voice and data convergence is the integration of voice and data applications in a common environment. It offers great opportunity for cost reduction to both service providers and customers. Indeed, the transition to digitized voice, data, and video is the dominant technological issue facing the world's phone companies over the next decade.

Telecommunications carriers offer a wide array of phone and dialing services including automatic number identification, WATS, ISDN, T-Carrier system, digital subscriber line, and wireless mobile.

The effective use of networks can turn a company into an agile, powerful, and creative organization, giving it a long-term competitive advantage. Networks can be used to share

hardware, programs, and databases across the organization. They can transmit and receive information to improve organizational effectiveness and efficiency. They enable geographically separated workgroups to share documents and opinions, which fosters teamwork, innovative ideas, and new business strategies.

The physical distance between nodes on the network and the communications and services provided by the network determines whether it is called a *personal area network (PAN)*, *local area network (LAN)*, *metropolitan area network (MAN)*, or a *wide area network (WAN)*. A PAN supports the interconnection of information technology devices within a range of about 33 feet. The major components in a LAN are a network interface card, a file server, and a bridge or gateway. A MAN connects users and their computers in a geographical area larger than a LAN but smaller than a WAN. WANs tie together large geographic regions including communications between countries, linking systems from around the world. The electronic flow of data across international and global boundaries is often called *transborder data flow*.

Network topology refers to the manner in which devices on the network are physically arranged. Communications networks can be configured in numerous ways, but five designs are most prevalent: bus, hierarchical, star, ring, and hybrid (hybrid networks combine the basic designs of the four other topologies to suit the specific communication needs of an organization).

Three commonly used approaches for connecting computers include terminal-to-host, file server, and client/server. With terminal-to-host connections, the application and database reside on the same host computer, and the user interacts with the application and data using a "dumb" terminal. Since a dumb terminal has no data-processing capability, all computations, data access and formatting, and data display are done by an application that runs on the host computer.

In the file server approach, the application and database reside on the same host computer, called the *file server*. The database management system runs on the end user's personal computer or workstation. If the user needs even a small subset of the data that resides on the file server, the file server sends the user the entire file that contains the data requested, including a lot of data the user does not want or need. The downloaded data can then be analyzed, manipulated, formatted, and displayed by a program that runs on the user's personal computer.

A client/server system is a network that connects a user's computer (a client) to one or more host computers (servers). A client is often a PC that requests services from the server, shares processing tasks with the server, and displays the results. Many companies have reduced their use of mainframe computers in favor of client/server systems using midrange or personal computers to achieve cost savings, provide more control over the desktop, increase flexibility, and become more responsive to business changes. The start-up costs of these systems can be high, and the systems are more complex than a centralized mainframe computer.

When people on one network wish to communicate with people or devices in a different organization on another network, they need a common communications protocol and various network devices to do so. A communications protocol is a standard set of rules that control a telecommunications connection. There are myriad communications protocols including international, national, and industry standards.

TCP/IP is the primary communications protocol of the Internet. Ethernet is the most widely installed LAN technology. Specified in a standard, IEEE 802.3, an Ethernet LAN typically uses coaxial cable or special grades of twisted-pair wires. The most commonly installed Ethernet systems are called 10Base-T, which provide transmission speeds up to 10 Mbps. Devices are connected to the cable and compete for access using a Carrier Sense Multiple Access with Collision Detection (CSMA/CD) protocol.

ATM (asynchronous transfer mode) is a switching technology that relies on dedicated connections and organizes digital data into 53-byte cell units. These cells are then transmitted over a physical medium using digital signal technology.

Bluetooth is a communications standard whose goal is to simplify communications among cell phones, handheld computers, and other wireless devices by streaming information back and forth using short-range radio waves.

The IEEE 802.*xx* is a family of technical specifications for wireless local area networks developed by a working group of the Institute of Electrical and Electronics Engineers (IEEE). All use the Ethernet protocol and CSMA/CA (Carrier Sense Multiple Access with Collision Avoidance) for path sharing. 1G, 2G, 2.5G, and 3G refer to the various evolutions of wireless networks used to support users and their various devices and communications needs.

Multichannel Multipoint Distribution System (MMDS) allows phone companies to deliver T1 speeds to the homes of individual users over rooftop antennas mounted anywhere within a 35-mile radius of a powerful transmission tower.

In addition to communications protocols, various devices are used in telecommunications. A PBX is an on-premise switching system owned or leased by a private enterprise that connects its telephones and provides access to the public telephone system.

A bridge is used to connect two or more networks that use the same communications protocol. A switch is a telecommunications hardware device that routes incoming data from any one of many input ports to a specific output port that will take the data toward its intended destination. A router is a device or software in a computer that determines the next network point to which a data packet should be forwarded toward its destination. A hub is a place of convergence where data arrives from one or more directions and is forwarded out in one or more other directions. Hubs commonly contain multiple ports used to connect segments of a network. A gateway is a network point that acts as an entrance to another network.

When an organization needs to use two or more computer systems, one of three basic data processing strategies may

be followed: centralized, decentralized, or distributed. With centralized processing, all processing occurs in a single location or facility. This approach offers the highest degree of control. With decentralized processing, processing devices are placed at various remote locations. The individual computer systems are isolated and do not communicate with each other. With distributed processing, computers are placed at remote locations but connected to each other via telecommunications devices. The September 11 terrorist attacks sparked many companies to distribute their workers, operations, and systems much more widely, a reversal of the trend toward centralization. The goal is to minimize the consequences of a catastrophic event at one location while ensuring uninterrupted systems availability.

Communications software performs important functions, such as error checking and message formatting. A network operating system controls the computer systems and devices on a network, allowing them to communicate with one another. Network-management software enables a manager to monitor the use of individual computers and shared hardware, scan for viruses, and ensure compliance with software licenses.

Many applications of telecommunications exist today, including the following: linking personal computer to mainframes, voice mail, electronic software distribution, electronic document distribution, call centers, telecommuting, videoconferencing, electronic data interchange, public network services, electronic funds transfer, distance learning,

and specialized systems and services. Personal-computer-to-mainframe links enable people to upload and download data. Voice mail users can leave messages for and receive and store messages from other people around the world. Electronic software distribution involves installing software on a computer by sending programs over a network so they can be downloaded into individual computers. Electronic document distribution allows organizations to transmit documents without the use of paper, thus cutting costs and saving time. Call centers allow for the efficient handling of a large number of inbound and outbound telephone calls. Telecommuting uses information technology to enable employees to work away from the office. Videoconferencing brings groups together in voice, video, and audio calls. Electronic data interchange (EDI), another rapidly growing area, enables customers, suppliers, and manufacturers to exchange data electronically. EDI reduces the need for manual paper systems, while speeding up the rate at which business can be transacted. Electronic funds transfer is a system of transferring money from one bank account directly to another without any paper money changing hands. Public network services give users access to vast databases and services, usually for an initial fee plus usage fees. Distance learning is a way to support education of students who are unable to meet frequently with their instructor. Specialized services, which are more expensive, include legal, patent, and technical information. Regional services include local electronic bulletin boards that offer e-mail facilities and information regarding local activities.

CHAPTER 6: SELF-ASSESSMENT TEST

Effective communications are essential to organizational success.

1. Which of these items carries an electronic signal between sender and receiver?

    a. modem
    b. communications protocol
    c. transmission media
    d. telecommunications

2. A(an) _____ provides service to homes and businesses within its defined geographical area.

3. A simplex channel can transmit data in only one direction. True or False?

4. Two broad categories of transmission media are:

    a. guided and wireless
    b. shielded and unshielded
    c. twisted and untwisted
    d. infrared and microwave

An unmistakable trend of communications technology is that more people are able to send and receive all forms of information over greater distances at a faster rate.

5. Which law states that the information-carrying capacity of a channel is directly proportional to its bandwidth?

    a. Moore's Law
    b. Murphy's Law
    c. Shannon's law of information theory
    d. Law of Diminishing Returns

6. _____ is the basic transport of voice in the form of a data packet using the Internet Protocol.

7. A telecommunications technology that delivers high-bandwidth information to homes and businesses over ordinary copper telephone wires.

    a. WATS
    b. ISDN
    c. T-carrier system
    d. digital subscriber line

8. Switching equipment that routes phone calls and messages within a building and connects internal phone lines to a few phone company lines is called a PBX. True or False?

9. A(an) _____ is a network that supports the interconnection of information technology devices within a range of 33 feet or so.

10. Electronic software distribution involves installing software on a file server for users to share by signing on to the network and requesting the software be downloaded onto their computers. True or False?

11. Which of the following statements about call centers is true?

    a. They are used to handle incoming calls only, not outgoing calls.

    b. They frequently employ an ACD to forward calls to the right party and balance the workload of support personnel.

    c. It is illegal for charitable organizations to employ them.

    d. Do Not Call legislation has had a minimal impact on their operation.

12. Telecommuting enables employees to save money because less office space, parking space, and office equipment is required. True or False?

13. _____ is a system of transferring money from one bank account directly to another without any paper money changing hands

CHAPTER 6: SELF-ASSESSMENT TEST ANSWERS

(1) c (2) local exchange carrier (3) True (4) a (5) c (6) VoIP (7) d (8) True (9) personal area network (10) True (11) b (12) True (13) Electronic funds transfer

KEY TERMS

analog signal 250
asynchronous communications 244
automated clearing house (ACH) 282
bandwidth 246
bridge 271
broadband 246
bus network topology 263
centralized processing 273
client/server 263
communications 243
communications protocol 265
communications software 274
competitive local exchange carrier (CLEC) 253
computer network 259
data communications 244
decentralized processing 273
dedicated line 254
digital signal 250
digital subscriber line (DSL) 257
distance learning 283
distributed processing 273
electronic data interchange (EDI) 280

electronic document distribution 276
electronic funds transfer (EFT) 281
electronic software distribution 276
file server 263
front-end processor 251
full-duplex channel 246
gateway 272
half-duplex channel 245
hierarchical network topology 263
hub 272
hybrid network topology 263
integrated services digital network (ISDN) 256
international network 261
local area network (LAN) 260
local exchange carrier (LEC) 252
long-distance carrier 253
metropolitan area network (MAN) 261
modem 250
multiplexer 251
network-management software 274
network operating system (NOS) 274
network topology 262

personal area network (PAN) 259
private branch exchange (PBX) 271
public network services 281
ring network topology 263
router 272
Shannon's fundamental law of information theory 246
simplex channel 245
star network topology 263
switch 272
switched line 254
synchronous communications 243
telecommunications medium 244
telecommuting 277
terminal-to-host 263
videoconferencing 278
voice and data convergence 255
voice mail 276
voice over Internet protocol (VoIP) 255
wide area network (WAN) 261
wire transfer 282

REVIEW QUESTIONS

1. What is the difference between synchronous and asynchronous communications?

2. Describe the elements and steps involved in the telecommunications process.

3. What characteristic of a channel determines its information carrying capacity?

4. Identify three types of guided telecommunications media.

5. Define the term *computer network*.

6. What advantages and disadvantages are associated with the use of client/server computing?
7. What is a DSL line? What capabilities does it provide?
8. Identify four different competitive local exchange carriers.
9. What is the difference between a switched and a dedicated line?
10. What is VoIP?
11. What role do the bridge, router, gateway, and switch play in a network?
12. Describe a local area network and its various components.
13. What is a metropolitan area network?
14. What is EDI? Why are companies using it?
15. What is videoconferencing? Why is it being used by many companies?

DISCUSSION QUESTIONS

1. How might you use a personal area network in your home? What devices might eventually connect to such a network?
2. Why is an organization that employs centralized processing likely to have a different management decision-making philosophy than an organization that employs distributed processing?
3. Why do you think there are so many emerging wireless network standards?
4. Briefly discuss the pros and cons of e-mail versus voice mail. Under what circumstances would you would use one and not the other?
5. What issues would you expect to encounter in establishing an international network for a large, multinational company?
6. If it were available and you could afford it, which would you rather have in your home—T1, ISDN, or DSL? Why?
7. You need to get cash to a friend across county—quickly! What are your options? Which one would you choose and why? Describe the fundamental network that supports this service.
8. Do you think that this course is a good candidate for a distance learning course? Why or why not?
9. What is meant by voice and data convergence? What impact is this having?
10. Identify at least seven communications protocols. Why do you think that there are so many protocols? Will the number of protocols increase or shrink over time?

PROBLEM-SOLVING EXERCISES

1. You have been hired as a telecommunications consultant to help an organization assess the benefits and potential cost savings associated with installing videoconferencing capabilities at each of the firm's six work locations. You have determined that the cost to establish each video conference facility is $60,000 and that each facility will have an annual operating and support cost of $20,000. In the first year, you estimate that videoconferencing will save approximately $400,000 in travel expense. The savings are expected to increase at a rate of 5 percent per year. Develop a spreadsheet to analyze the costs and savings over a three-year period. Write a recommendation to management based on your findings and any other factors that might support the installation a videoconferencing system.

2. You are the leader of a group trying to interest management in providing telecommuting opportunities to its employees. Use PowerPoint or similar software to make a convincing presentation to management for adopting such a program. Your presentation must address such questions as what the benefits and disadvantages are for the company, the employees, and the local community.

TEAM ACTIVITIES

1. With a group of your classmates, develop a proposal to install videoconferencing equipment in one of your school's classrooms to enable lectures to be viewed by students at distant videoconferencing facilities. What sort of equipment is required, what vendors can provide this equipment, and what does it cost to install and operate?

2. Form a team to identify the public locations (airport, public library, Starbucks, etc.) in your area where wireless LAN connections are available. Visit at least two locations and write a brief paragraph discussing your experience at each location trying to connect to the Internet.

WEB EXERCISES

1. Do research on the Web on 3G wireless networks. What services are considered to be part of 3G? What is its current status, and what are the current major issues? Write a short report on what you found.

2. Go online to investigate how the banking industry is currently using wire transfer and ACH services. Find two different firms and compare their systems.

CAREER EXERCISES

1. Consider an industry with which you are familiar through work experience, coursework, or a study of industry performance. How could electronic data interchange be used in this industry? What limitations would EDI have in this industry?

2. There are a number of online job-search companies, including Monster.com. Investigate one or more of these companies and research the positions available in the telecommunications industry, including the Internet. You may be asked to summarize your findings for your class in a written or verbal report.

VIDEO QUESTIONS

Watch the video clip **Free Wi-Fi Blankets** and answer these questions:

1. The "last mile" problem consists of the high cost of providing high-speed Internet connections to every home. It is not financially feasible to provide fiber optic cable, the last mile, to each residence. How does Tim Cozar's system solve the last mile problem for San Bruno, California?

2. How might Tim Cozar's "Wi-Fi Blanket" save businesses and organizations significant money over installing traditional wired networks?

CASE STUDIES

Case One

Stop & Shop Customers Make a New Buddy

Dutch retail giant Royal Ahold has begun testing some radically new technology in its Stop & Shop supermarkets. In a pilot project, the company is testing a new self-service system that utilizes wireless technology to help customers locate goods, scan purchased items, and much more. The system, called "Shopping Buddy," was designed by software vendor Cuesol, Inc.

Shoppers who wish to participate in the experiment must sign up for a Stop & Shop loyalty card. The card will be used to bill the customer for items purchased in the Shopping Buddy system. Upon entering the Stop & Shop, Shopping Buddy customers are handed a portable computer, which they insert into a cart-mounted holder. The first step for shoppers is to swipe their loyalty card to sign on to the system.

The Shopping Buddy user interface consists of a flat-panel display, about the size of a magazine page, with color images of specific products. Using a keypad, shoppers can type the name of an item, and the system tells them where it's located

and how to get there. The Shopping Buddy includes a scanning wand that customers use to scan items as they place them in the cart. As the customer progresses through the store, location-sensitive advertisements inform the customer of special deals on items located nearby—deals available only to those using Shopping Buddy.

When the customer is ready to check out, the purchase data is automatically and wirelessly transferred from the Shopping Buddy device to the checkout system, and funds are automatically transferred from the customer's bank account to Stop & Shop. David O'Neill, a senior citizen who is a regular at the Stop & Shop supermarket in Quincy, Massachusetts, hasn't needed to talk to a cashier in months. He particularly appreciates the convenient checkout process. "The biggest pain in the neck is the lines at the registers," O'Neill says. Shopping Buddy has done away with long lines and delays at checkout.

The Shopping Buddy system employs several wireless technologies to support the wide range of services it offers. A Wi-Fi network is employed to assist customers in finding products on the shelves. Wi-Fi also allows customers to place deli orders anywhere in the store and pick them up as they pass the deli counter. Infrared technology is used to scan purchase items. Bluetooth transmissions provide location-specific advertisements and wireless data transfer at checkout. Since the system maintains information about a shopper's previous purchases, the Shopping Buddy system can target advertisements and promotions to customers according to individual customer tastes, for example, "You haven't purchased coffee in two weeks. Are you running low? There's a special on Folgers located on your right!"

First-time Shopping Buddy users earn $5 off their groceries for trying out the system. Most shoppers continue using the system after they've tried it. The ease of use and additional features save customers time and frustration. Nina Tobin, another recent shopper in the Quincy store, liked the fact that she could review what she'd placed in her cart to verify that she hadn't forgotten anything. The only problem she experienced was in forgetting the brand of breakfast cereal her husband had requested. "The computer can't help me with that, though." It could be that by next year, grocers will have an information system to remedy that problem as well.

Discussion Questions

1. In what ways does Shopping Buddy provide benefits and conveniences for Stop & Shop customers?
2. How does the Shopping Buddy system assist the Royal Ahold Corporation in understanding its clientele and the Stop & Shop store owners in running a smooth and effective operation?

Critical Thinking Questions

3. What issues of security and privacy are introduced through the wireless systems of Shopping Buddy?
4. In what types of businesses would you expect systems such as Shopping Buddy to begin cropping up? How might devices such as Shopping Buddy assist customers making large purchases such as automobiles or even homes?

SOURCES: Fred O'Connor, "Magic Wand Makes Checkout Lines Vanish," *Infoworld*, February 3, 2004, *www.infoworld.com*; Laurie Hibberd, "Shopping Carts with Brains," *CBS Early Show*, August 12, 2003, *www.cbsnews.com*; Cuesol Web site, *www.cuesol.com*, accessed March 9, 2004.

Case Two
Application Performance Monitoring at JCPenney

Have you ever had to wait for your computer to respond after clicking an icon, link, or command? Who hasn't? Computer delays on PCs could be the result of insufficient processing speed, memory capacity, or other hardware inadequacies. They might also be the result of flaws or bugs in the operating system, software, or the interaction between the two. When working on complex business networks, delays can be numerous, long-lasting, and enraging. The complexity of the interaction between workstations, servers, routers, and other networking hardware and software make it nearly impossible to track down the source of delays without specially designed software called *application performance monitoring software*.

JCPenney maintains a WAN that provides services to more than 1,000 department stores. For large companies such as this, getting the network to run at its peak efficiency is crucial. When it comes to troubleshooting sluggish network response times, Barry Hicks, head of performance management at JCPenney explains, "We can only approximate. Having a verifiable user response time is something we consider

to be mission-critical to our operations as we move more and more to client/server and Web-based applications."

To assist in tracking down trouble spots in its network, JCPenney has turned to software from Austin-based NetQoS called SuperAgent. The company chose this software from several available because it provides network managers with "actionable information," meaning that if performance on an application slows down, an IT manager can identify the precise port on a switch or server that is causing the problem.

At a starting price of $34,500, the software is not cheap. The price reflects the level of sophistication of the software. SuperAgent employs artificial intelligence to travel the network and test network components. It is a guardian of the network and as such measures and analyzes application response time for all user transactions, compares the response time against baselines, then automatically investigates the cause of problems as they occur. It tracks problems to specific network components, servers, or bugs in the software or database. To a company that depends on the free flow of information for survival, software like SuperAgent is a must.

Ginger Elliott, manager of network technologies at JCPenney, says that SuperAgent and a related product called ReporterAnalyzer have helped the firm track the cause of performance problems and determine whether they are the result of network or application glitches. With such information, the retailer is better able to plan for growth, especially as more demands are placed on its Web sites.

Products to measure network performance, both for internal business units and network service providers, are becoming essential. Users are beginning to look at computer networks as they do utilities such as water and electricity. Rather than just being content that the network is up, they now complain when their time is wasted from network delays.

Discussion Questions

1. What types of financial losses can occur for companies as large as JCPenney from network delays or outages?

2. What types of applications do you think run on JCPenney's WAN?

Critical Thinking Questions

3. How might network delays and problems contribute to a company's decision to outsource its network to professionals such as Sprint?

4. Some IS and telecommunications networks are becoming so complex that they require automated artificial intelligence software to maintain them. What are the inherent dangers in depending on machines to service machines?

SOURCES: Matt Hamblen, "J.C. Penney Seeks Better Grip on SLAs," *Computerworld*, February 16, 2004, *www.computerworld.com*; NetQS Solutions Web site, *www.netqos.com/solutions*, accessed March 10, 2004; JCPenney Web site, *www.jcpenney.com*, accessed March 10, 2004.

Case Three

Hospitals Turn Burden into Benefit

The recent ruling by the U.S. Food and Drug Administration requiring the use of bar codes on drugs in hospitals is causing hospitals across the country to scramble to install network infrastructures to support the new law. Many hospitals are installing wireless LANs (WLANs) to interact with the handheld bar-code readers that are used to match data on patient wristbands with the bar codes on packaged doses of drugs. The installation of wireless networks in hospitals across the country is having unexpected side effects.

Hospital administrators, burdened with the expense of the new bar-code systems, are discovering additional ways to use the wireless networks to increase their return on investment. Besides being supplied with portable scanning devices, nurses and other medical personnel will have new communications tools—tools that share the same wireless network used by the bar-code scanners.

IP wireless phones are making a tremendous impact in the healthcare market. John Hummel, CIO at Sutter Health in Sacramento, is equipping all 26 of Sutter's hospitals with extensive wireless LANs. Hummel says that Sutter has started testing wireless IP phones for use in all of its facilities.

Sutter Health is also testing a new IP-based, hands-free wireless voice communicator from Vocera Communications, Inc. The Vocera communication badge clips onto a shirt pocket or hangs on a lanyard around a person's neck. It uses speech recognition to allow for hands-free communications. A nurse can ask, "Vocera, where is Dr. Richards?" and the Vocera system, working through the WLAN with the assistance of server software, would locate Dr Richards's Vocera badge and state the location: "Dr. Richards is in room 237." The nurse might follow with the request "Vocera, connect me with Dr. Richards." Dr. Richards's badge would beep, and he could say, "Vocera, connect," and follow with a conversation with the nurse.

Indianapolis-based Community Health Network, which operates five major hospitals and numerous other medical facilities, has deployed about 100 IP phones, says Chris Cerny, the healthcare company's manager of enterprise networking. Community Health's Indiana Heart Hospital is finding the phones particularly useful. The hospital was designed to be all digital, without paper files and without nursing stations. Fully digital and wireless, the hospital staff receive all the information they need through handheld computers. IP wireless phones allow patients to call their nurse simply by pushing the call button on the headboard of the bed. A connection is established with the nurse's phone, and the patient can speak with the nurse, hands free, over the built-in speaker phone.

Using brand-new technology sometimes has its drawbacks. Community Health in Indiana had some technical difficulties when first trying out its IP phones. Technicians installed wireless access points around the facility densely to ensure that there were no dead spots. They later discovered that they had installed too many access points for voice communications. When the SpectraLink phones roamed from one access point to another, they often took too long to authenticate themselves and disconnected before the process was completed. Reducing the number of access points solved the problem.

While many hospitals are burdened by the cost of the new government regulation, it may serve as the impetus to overhaul outdated information systems. As hospitals move to digital systems for tracking drugs and patients, they are discovering the many other benefits of wireless networks and digital information systems. The new federal law may have provided a valuable nudge in moving the U.S. healthcare industry closer to an all-digital recordkeeping system.

Discussion Questions

1. How does the use of IP phones and communication devices such as the Vocera badge alter the way nurses do their job?
2. Besides drug distribution control and voice communications, how else could the medical staff benefit from a WLAN?

Critical Thinking Questions

3. What additional services might be provided for patients over a wireless hospital network?

4. What security, safety, and privacy issues do hospitals have to consider when setting up wireless networks?

SOURCES: Bob Brewin, "Hospitals Eye Wider Use Of Wireless IP Phones," *Computerworld*, March 01, 2004, *www.computerworld.com*; Vikki Lipset, "Vocera Secures Funding from Intel, Cisco," *Internet.com*, *www.wifiplanet.com/news/article.php/3073271*; Vocera Communications Web site, *www.vocera.com/products/products.shtm*, accessed March 10, 2004; Cisco IP Phones Web site, *www.cisco.com/en/US/products/hw/phones/ps379/index.html*, accessed March 10, 2004.

NOTES

Sources for the opening vignette: Stacy Williams, "Leading Pasta Maker Increases Productivity, Reduces Costs with Cisco WLAN." *Cisco Newsroom*, February 27, 2004, *http://newsroom.cisco.com/dlls/2004/ts_022704.html*; Barilla Group Web site, *www.barillagroup.com/index.htm*, accessed March 9, 2004; "IP Telephony/VOIP" Cisco Web site, *www.cisco.com/en/US/tech/tk652/tk701/tech_protocol_family_home.html*, accessed March 9, 2004.

1. Mearian, Lucas, "Wells Fargo Upgrades Online Apps," *Computerworld*, February 16, 2004, accessed at *www.computerworld.com*.
2. Ho, David, "High Speed Internet Access Over Power Lines Looks Promising, Government Says," *The San Diego Union-Tribune*, January 15, 2003, accessed at *www.signonsandiego.printthis.clickability.com*.
3. Backover, Andrew and McCarthy, Michael, "Cable Firms Wired About Offering Net Phone Calls," *USA Today*, December 10, 2003, p. 7B.
4. Boyer, Mike, "Time Warner Calling Ahead," *Cincinnati Enquirer*, December 18, 2003, p. D-1.
5. Reinhardt, Andy; Ewing, Jack; Kunii, M. Irene, "The Wireless Challenge," *BusinessWeek Online*, October 20, 2003, accessed at *www.businessweek.com*.
6. Hamblen, Matt, "IP Telephony Is Bringing New Flexibility to the Call Center," Computerworld, July 7, 2003, accessed at *www.computerworld.com*.
7. Reinhardt, Andy; Ewing Jack, and Kunii, M. Irene, "The Wireless Challenge," *BusinessWeek Online*, October 20, 2003, accessed at *www.businessweek.com*.
8. Ewalt M., David, "Cellular Data Services Promise," *InformationWeek*, January 12, 2004, accessed at *www.informationweek.com*.
9. Messmer, Ellen, "Olympic Network Gets Gold Security Protection," *Network World*, December 3, 2003, accessed at *www.computerworld.com*.
10. Hamblen, Matt, "Las Vegas Schools Mix IP, Digital Communications," *Computerworld*, December 8, 2003, accessed at www.computerworld.com.
11. Songini L., Marc, "Case Study: Boehringer Cures Slow Reporting," *Computerworld*, July 23, 2003, accessed at *www.computerworld.com*.
12. Cowley, Stacy, "New IT Model Gives Morgan Stanley Flexibility," *IDG News Service*, January 22, 2003, accessed at *www.nwfusion.com*.
13. "About The Bluetooth Name," The Official Bluetooth Membership Site at *www.bluetooth.org*, accessed on August 18, 2004.
14. Kontzer, Tony and Ewalt M., David, "Free for All Access to Wireless LANs," *InformationWeek*, February 9, 2004, accessed at *www.informationweek.com*.
15. Green, Heather and Rosenbush, Steve; Crockett O, Roger; and Holmes Stanley, "Wi-Fi Means Business," *BusinessWeek Online*, April 28, 2003, accessed at *www.businessweek.com*.
16. Levitt, Jason, "Tech Guide: Wi-Fi Security for the Masses," *InformationWeek*, June 30, 2003, accessed at *www.informationweek.com*.
17. "Broadband's Next Wave: Wireless?," *The New York Times on the Web*, May 17, 2001, accessed at *www.nytimes.com*.
18. Hamblen, Matt, "Bank Takes Hybrid Route on New Telephony System," *Computerworld*, September 8, 2003, accessed at *www.computerworld.com*.
19. "Cisco Helps Corrugated Supplies Corp. Increase Visibility, Reduce Costs, and Deliver Superior Service to Customers," accessed at Cisco Web site at *http://business.cisco.com/* on February 18, 2004.
20. Marlin, Steven, "The 24-Hour Supply Chain," *InformationWeek*, January 26, 2004, accessed at *www.informationweek.com*.
21. Sullivan, Laurie, "Tire Maker Looks for Low-Cost Efficiencies," *InformationWeek*, February 23, 2004, accessed at *www.informationweek.com*.
22. Hamblen, Matt, "Con Edison Unifies Network-Management," *Computerworld*, November 3, 2003, accessed at *www.computerworld.com*.
23. Sliwa, Carol, "Users Turn to Microsoft's SMS for Patch Management," *Computerworld*, January 5, 2004, accessed at *www.computerworld.com*.
24. Rooney, Paula, "Microsoft Ships Digital-Rights-Management Software," *Information Week*, November 4, 2003, accessed at *www.informationweek.com*.
25. Rosencrance, Linda, "Do-Not-Call Registration Web Site Bombarded with Sign-Ups," *Computerworld*, June 30, 2003, accessed at *www.computerworld.com*.
26. Ribeiro, John, "HP Sets Up India Call Center," *Computerworld*, February 11, 2004, accessed at *www.computerworld.com*.
27. Thibodeau, Patrick, "Feds Lag in Adopting Telecommuting Programs," *Computerworld*, February 25, 2003, accessed at *www.computerworld.com*.

28. Thibodeau, Patrick, "Tools for Managing Offshore IT," *Computerworld*, September 1, 2003, accessed at *www.computerworld.com*.

29. Duffy Marsan, Carolyn, "AV, IT System Marriage Proves Powerful," *Network World*, January 5, 2004, accessed at *www.nwfusion.com*.

30. Bednarz, Ann, "Internet EDI: Blending Old and New," *Network World*, February 23, 2004, accessed at *www.nwfusion.com*.

31. Hoffman, Thomas, "Best Places Profile, #3: University of Miami," *Computerworld*, June 9, 2003, accessed at *www.computerworld.com*.

32. Rosencrance, Linda, "Toyota to Provide Mazda Owners in Japan In-Vehicle Info Service," *Computerworld*, February 10, 2004, accessed at *www.computerworld.com*.

CHAPTER
· 7 ·

The Internet, Intranets, and Extranets

- **The Internet is like many other technologies—it provides a wide range of services, some of which are effective and practical for use today, others are still evolving, and still others will fade away from lack of use.**

  - Briefly describe how the Internet works, including alternatives for connecting to it and the role of Internet service providers.

- **Originally developed as a document-management system, the World Wide Web is a menu-based system that is easy to use for personal and business applications.**

  - Describe the World Wide Web and the way it works.
  - Explain the use of Web browsers, search engines, and other Web tools.
  - Identify and briefly describe the applications associated with the Internet and the Web.

- **Because the Internet and the World Wide Web are becoming more universally used and accepted for business use, management, service and speed, privacy, and security issues must continually be addressed and resolved.**

  - Identify who is using the Web to conduct business and discuss some of the pros and cons of Web shopping.
  - Outline a process for creating Web content.
  - Describe Java and discuss its potential impact on the software world.
  - Define the terms *intranet* and *extranet* and discuss how organizations are using them.
  - Identify several issues associated with the use of networks.

INFORMATION SYSTEMS IN THE GLOBAL ECONOMY
COLLIERS INTERNATIONAL, USA

Connecting 56 Companies in 51 Countries

Colliers International is an $880-million consortium of 56 independently owned real estate companies in 51 countries. Dr. Joshua Fost was hired by Colliers in 2000 as chief technology officer (CTO) and took on the tremendous challenge of building a globally coordinated Web site for the consortium. The ambitious goal of the Web site was to combine the information for all properties being sold by Colliers affiliates into one central database, accessed through a Web site and translated into 14 languages.

When Fost joined Colliers, property data was stored in separate databases owned by each affiliate and accessed through proprietary information systems. With the Internet boom of the 1990s, many Colliers affiliates developed Web sites of their own. Colliers prides itself on a business approach that encompasses both depth of knowledge in local markets and global breadth through its international reach. To reflect the company's global breadth in its Web presence, Fost recognized that the hundreds of independent Web sites needed to be centralized and standardized. If successful, the Colliers International Web site would provide all affiliates and their customers with access to information on properties around the world and would more firmly establish the Colliers brand.

To accomplish this feat, Fost turned to two outside companies: Interwoven, Inc., a provider of enterprise content management software, and a leading Internet services firm, Molecular. The content management system developed by Interwoven provided a central database for all Colliers properties. Under the guidance of Fost, Interwoven and Molecular developed Web-based interfaces that served both for input into the database and for searching the site by employees and the public.

Kevin Hayden, vice president of corporate marketing for Interwoven, says that organizations such as Colliers view their Web sites as the doorstep to their businesses. "With the right content management system in place, companies like Colliers can improve their ability to deliver appropriate, uniformly branded content to thousands of constituents—and thereby provide better customer service and bottom-line results."

Colliers's Web-based system took more than two years to develop prior to its launch at the end of 2003. Joshua Fost's efforts as CTO of Colliers so impressed the InfoWorld Media Group that they named him one of the Top 25 Chief Technology Officers for 2004. The award honors information technology executives who have shown leadership both inside and outside their companies. Margaret Wigglesworth, president and CEO of Colliers, was equally impressed by Fost's efforts, saying, "As [a member of] a partnership of independently owned firms, Joshua faced the challenge of integrating the tools and technology solutions of more than 100 offices nationwide. He focused his attention on creating solutions that would help Colliers offices communicate more efficiently and allow resources and best practices to be shared more freely." Colliers anticipates a 125 percent return on investment for the Web project within three years. The Colliers project is just one example of how companies are using the Internet and Web to connect their organizations both internally and externally with their customers.

As you read this chapter, consider the following:

- What Internet technologies provide networking benefits to companies for private internal communications and public external communications?
- What special challenges do international organizations face when creating a Web presence?

Why Learn About the Internet?

To say that the Internet has had a big impact on organizations of all types and sizes would be a huge understatement. Since the early 1990s when the Internet was first used for commercial purposes, it has had a profound impact on all aspects of business. Businesses use the Internet to sell and advertise their products and services, reaching out to new and existing customers. If you are undecided about a career, you can use the Internet to help you investigate career opportunities and salaries using sites such as *www.monster.com* and HotJobs at *www.yahoo.com*. Most companies have Internet sites that list job opportunities, descriptions, qualifications, salaries, and benefits. If you have a job, you probably use the Internet daily to communicate with coworkers and your boss. Purchasing agents use the Internet to save millions of dollars in supplies every year. Travel and events-management agents use the Internet to find the best deals on travel and accommodations. Automotive engineers use the Internet to work with other engineers around the world developing designs and specifications for new automobiles and trucks. Property managers use the Internet to find the best prices and opportunities for commercial and residential real estate. Whatever your career, you will probably use the Internet daily. This chapter starts by exploring how the Internet works and then investigates the many exciting opportunities for using the Internet to help you achieve your goals.

Internet

A collection of interconnected networks, all freely exchanging information.

The Internet is the world's largest computer network. Actually, the **Internet** is a collection of interconnected networks, all freely exchanging information (see Figure 7.1). Research firms, colleges, and universities have long been part of the Internet, and now businesses, high schools, elementary schools, and other organizations are joining up as well. Nobody knows exactly how big the Internet is because it is a collection of separately run, smaller computer networks. There is no single place where all the connections are registered. Figure 7.2 shows the staggering growth of the Internet, as measured by the number of Internet host sites or domain names. Domain names are discussed later in the chapter.

Figure 7.1

Routing Messages over the Internet

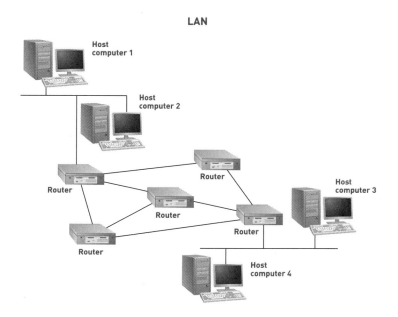

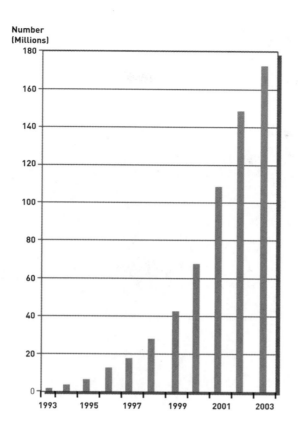

Internet Growth: Number of Internet Domain Names

(Source: Data from "The Internet Domain Survey," *www.isc.org*.)

USE AND FUNCTIONING OF THE INTERNET

The Internet is truly international in scope, with users on every continent—including Antarctica. However, the United States has the most usage by far. More than 100 million people in the United States (roughly one-third the population) have used the Internet. Although the United States still claims more Web activity than other countries, the Internet is expanding around the globe but at differing rates for different countries. In Russia, using the Internet's e-mail capabilities provides a timely mail service; it may take weeks for a Russian airmail letter to reach the United States. Most Internet usage in South Korea is through high-speed broadband connections. International use of the Internet is expected to continue its growth. International Internet usage could surge to over a billion users in a few years.[1]

The ancestor of the Internet was the **ARPANET**, a project started by the U.S. Department of Defense (DoD) in 1969. The ARPANET was both an experiment in reliable networking and a means to link DoD and military research contractors, including a large number of universities doing military-funded research. (*ARPA* stands for the Advanced Research Projects Agency, the branch of the Defense Department in charge of awarding grant money. The agency is now known as DARPA—the added *D* is for *Defense*.) The ARPANET was highly successful, and every university in the country wanted to sign up. This wildfire growth made it difficult to manage the ARPANET, particularly the large and rapidly growing number of university sites on it. So, the ARPANET was broken into two networks: MILNET, which included all military sites, and a new, smaller ARPANET, which included all the nonmilitary sites. The two networks remained connected, however, through use of the **Internet protocol** (**IP**), which enabled traffic to be routed from one network to another as needed. All the networks connected to the Internet speak IP, so they all can exchange messages.

ARPANET

Project started by the U.S. Department of Defense (DoD) in 1969 as both an experiment in reliable networking and a means to link DoD and military research contractors, including a large number of universities doing military-funded research.

Internet Protocol (IP)

Communication standard that enables traffic to be routed from one network to another as needed.

There are more than 600 million Internet users around the world, and the number of users could surge to more than 1 billion in a few years. Global firms such as IKEA use the Internet to reach consumers worldwide.

Today, several people, universities, and companies are attempting to make the Internet faster and easier to use. Robert Kahn, who managed the early development of the ARPANET, is one individual who wants to take the Internet to the next level. He is president of the nonprofit organization National Research Initiatives, which provides guidance and funding for the development of a national information infrastructure. The organization is looking into using "digital objects," which allow programs and data to be used and shared on all types of computer systems. To speed Internet access, a group of corporations and universities, the University Corporation for Advanced Internet Development (UCAID), is working on a faster, new Internet. Called Internet2 (I2), Next Generation Internet (NGI), and Abilene, depending on the universities or corporations involved, the new Internet offers the potential of faster Internet speeds, up to 2 Gbits per second or more. Another effort, the "100 by 100" initiative, received $7.5 million in funding from the National Science Foundation (NSF) to develop a system to deliver 100-MB speed to 100 million homes in the United States.[2] About 200 colleges and universities now have I2 service.[3] Some I2 connections can transmit data at 100 Mbits per second, which is about 10 times faster than many cable modems and 200 times faster than dial-up connections. The speed allows students to transfer the contents of a DVD disk in less than a minute.

The Internet is increasingly going wireless. In addition to land-based systems, the Internet is also becoming available at sea and in the air. Going online while aboard a cruise ship is becoming a reality. Some cruise lines are installing Internet appliances that allow crew members to access the Internet to get news and send and receive e-mails. Some airline companies are offering Internet service on their flights. Terry Feldman, a truck driver, uses wireless Internet access from the cab of his truck to check weather and shipments and to send and receive e-mail to and from family and friends.[4] Physicians at the Parker Adventist Hospital use wireless Internet access to speed ordering prescriptions and recording patient treatments.[5] With this more accurate record keeping, the approach also has the potential to reduce errors and harmful interactions of different drugs prescribed to patients. In another interesting wireless application, Zipcar allows people to share cars in urban areas where parking can be both scarce and expensive.[6] The company has more than 10,000 members and 200

cars in Boston, New York, and Washington, DC. Wireless Internet hot spot connections in the United States have increased from about 1,000 locations in 2001 to more than 50,000 locations in 2004, increasing the use of wireless Internet access.[7]

How the Internet Works

The Internet transmits data from one computer (called a *host*) to another (see Figure 7.1). If the receiving computer is on a network to which the first computer is directly connected, it can send the message directly. If the receiving computer is not on a network to which the sending computer is connected, the sending computer relays the message to another computer that can forward it. The message may be sent through a router (see Chapter 6) to reach the forwarding computer. The forwarding host, which presumably is attached to at least one other network, in turn delivers the message directly if it can or passes it to yet another forwarding host. It is quite common for a message to pass through a dozen or more forwarders on its way from one part of the Internet to another.

The various networks that are linked to form the Internet work pretty much the same way—they pass data around in chunks called *packets*, each of which carries the addresses of its sender and its receiver. The set of conventions used to pass packets from one host to another is known as the Internet Protocol (IP), which operates at the network layer of the seven-layer OSI model discussed in Chapter 6. Many other protocols are used in connection with IP. The best known is the **Transmission Control Protocol (TCP)**, which operates at the transport layer. TCP is so widely used as the transport layer protocol that many people refer to TCP/IP, the combination of TCP and IP used by most Internet applications. Adhering to the same technical standards allows the more than 100,000 individual computer networks owned by governments, universities, nonprofit groups, and companies to constitute the Internet. Once a network following these standards links to a **backbone**—one of the Internet's high-speed, long-distance communications links—it becomes part of the worldwide Internet community.

Each computer on the Internet has an assigned address called its **uniform resource locator, or URL**, to identify it to other hosts. The URL gives those who provide information over the Internet a standard way to designate where Internet elements such as servers, documents, and newsgroups can be found. Let's look at the URL for Thomson Course Technology, *http://www.course.com*.

The "http" specifies the access method and tells your software to access this particular file using the Hypertext Transport Protocol. This is the primary method for interacting with the Internet.

The "www" part of the address signifies that the address is associated with the World Wide Web service discussed later. The "course.com" part of the address is the domain name that identifies the Internet host site. Domain names must adhere to strict rules. They always have at least two parts separated by dots (periods). For some Internet addresses, the rightmost part of the domain name is the country code (au for Australia, ca for Canada, dk for Denmark, fr for France, jp for Japan, etc.). Many Internet addresses have a code denoting affiliation categories (Table 7.1 contains a few popular categories). The leftmost part of the domain name identifies the host network or host provider, which might be the name of a university or business.

Transmission Control Protocol (TCP)
Widely used transport layer protocol that is used in combination with IP by most Internet applications.

backbone
One of the Internet's high-speed, long-distance communications links.

Uniform Resource Locator (URL)
An assigned address on the Internet for each computer.

Affiliation ID	Affiliation
com	business organizations
edu	educational sites
gov	government sites
net	networking organizations
org	organizations

Table 7.1

U.S. Top-Level Domain Affiliations

Herndon, Virginia–based Network Solutions, Inc. (NSI) was the sole company in the world with the direct power to register addresses using .com, .net, or .org domain names. But this government contract ended in October 1998, as part of the U.S. government's move to turn management of the Web's address system over to the private sector. Today, other companies, called *registrars*, can register domain names, and additional companies are seeking accreditation to register domain names from the Internet Corporation for Assigned Names and Numbers (ICANN). Some registrars are concentrating on large corporations, where the profit margins may be higher, compared with small businesses or individuals.

There are hundreds of thousands of registered domain names. Some people, called *cyber-squatters*, have registered domain names in the hope of selling the names to corporations or people at a later date. The domain name Business.com, for example, sold for $7.5 million. In one case, a federal judge ordered the former owner of one Web site to pay the person who originally registered the domain name $40 million in compensatory damages and an additional $25 million in punitive damages. But some companies are fighting back, suing people who register domain names in hopes of trying to sell them to companies. Today, the Internet Corporation for Assigned Names and Numbers has the authority to resolve domain name disputes. Under new rules, if an address is found to be "confusingly similar" to a registered trademark, the owner of the domain name has no legitimate interest in the name. The rule was designed in part to prevent cyber-squatters.

Accessing the Internet

There are numerous ways to connect to the Internet (see Figure 7.3), but Internet access is not distributed evenly throughout the world. See the "Ethical and Societal Issues" feature for a discussion of the challenges of global Internet access. Which access method is chosen is determined by the size and capability of the organization or individual.

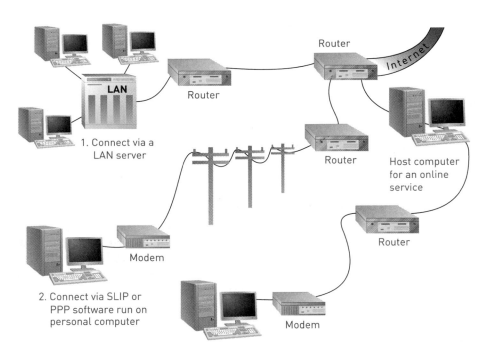

Figure 7.3

Several Ways to Access the Internet

There are several ways to access the Internet, including using a LAN server, dialing into the Internet using SLIP or PPP, or using an online service with Internet access.

1. Connect via a LAN server

2. Connect via SLIP or PPP software run on personal computer

3. Connect via an online service

ETHICAL AND SOCIETAL ISSUES

Entrepreneurs Work to Lessen the Global Divide

Despite the global spread of information and communication technologies, large parts of the world remain technologically disconnected. The United States has more computers than the rest of the world combined, and when assessed by region, Internet use is dominated by North Americans. To develop a truly global economy, those who enjoy the benefits of technology need to help those less fortunate get connected—and they are!

Many successful entrepreneurs and business leaders are investing heavily to bring technology to remote and sometimes undeveloped or underdeveloped corners of the world. Martin Varsavsky, an Argentinean telecommunications entrepreneur based in Spain, believes that entrepreneurs and philanthropists have a lot in common. "The motivation of an entrepreneur and [a] philanthropist are essentially the same. They both pinpoint problems, economic or social, and set out to solve them in a new way." In 2000, Varsavsky established the Varsavsky Foundation, an organization committed to bringing tools of learning, from basic bricks-and-mortar classrooms to online cyberlibraries, to children of all ages throughout the world. The foundation further aims to fully integrate technology and the Internet into learning.

Perhaps the most famous entrepreneur turned philanthropist is Microsoft's Bill Gates. The Bill and Melinda Gates Foundation has provided funds to bring technologies to remote areas of the world, such as the snow-swept terrain of Canada's Northwest Territories. There they have set up a system of cyberlibraries in six remote communities to connect rural residents to the world.

The World Economic Forum is an independent international organization committed to improving the state of the world. The forum allows the world's leaders to collaborate to address global issues and promote global citizenship among its corporate members. In 2000, the World Economic Forum launched the Global Digital Divide Initiative (GDDI) to develop public and private partnerships to bridge the gap between those who have information and communication technology (ICT) access, skills, and resources and those who do not.

The Jordan Education Initiative, a 2003 GDDI program, brought together leaders from the information technology and telecom industries such as Dell, HP, IBM, Intel, Microsoft, Siemens, Skillsoft, and Sun Microsystems. They worked with Jordanian authorities to improve education in the kingdom. Ninety-six "Discovery Schools" were selected to pilot the program in Jordan. They are serving as a test bed for how ICT can benefit schools and their pupils. Though focused on the advancement of learning in Jordan, the plan also stimulates the growth of the local information technology industry through infrastructure improvements and e-content development.

The digital divide occurs on many levels. While the World Economic Forum is working at the global level, the Aspira Association works on the national level in the United States. Aspira is a confederation of independent organizations throughout the

United States and Puerto Rico. Aspira plans to use a $1.7 million grant to develop a national model for training Hispanic parents and students in information technology at more than 40 community technology centers across the country.

Several Latin American governments are realizing that technology may be a life line for their people in hard economic times. In Venezuela, the government has set up 243 "infocentres," offering free access to the World Wide Web in libraries, museums, city halls, and the offices of nongovernmental organizations in the country's 23 states, and it hopes to open 100 more. In 2002, the Mexican government launched the E-Mexico program, with the aim of installing 3,200 "digital community centers" in schools, community centers, city halls, libraries, and health clinics in rural villages and towns.

Although information and communications technology may be assisting many cultures and peoples in reaching their potential, many still need assistance. Some cultures prefer to do without modern technological conveniences, which contradict religious beliefs and cultural values. To develop a global economy, global business efforts and IS design should accommodate the needs of all and act as a bridge between differing cultures.

Critical Thinking Questions

1. How do businesses such as Microsoft benefit from their philanthropic investments?
2. What advantages do Internet-connected cultures have over those without information and communications technology?

What Would You Do?

Your global business has finally reached what many would consider a comfortable level of success. After reinvesting in the company and fulfilling all other financial obligations, your chief financial officer is looking for suggestions on how to invest $200,000. Some of your competitors have made big splashes in the media with what you've previously considered to be extravagant donations to charities. You wonder whether this might be an option for some or all of your remaining profits.

3. Out of all the needs in the world, what philanthropic venture would most interest you?
4. Would the type of industry and locations in which you work have an impact on how you invest?

SOURCES: The Digital Divide Network, *www.digitaldividenetwork.org*, accessed March 12, 2004; Mugo Macharia, "ASPIRA Brings Digital Opportunity to the Latino Community," *Digital Divide Network*, *www.digitaldividenetwork. org*, accessed March 12, 2004; Humberto Márquez, "Telecentres to Narrow Digital Divide," *Terra Viva Online*, December 12, 2003, *www.ipsnews.net/ focus/tv_society/viewstory.asp?idn=74*; "Jordan Education Initiative to Roll Out e-Learning across the Kingdom and Beyond," *World Economic Forum*

press release, June 21, 2003, *www.weforum.com*; "Wired in the Wilderness: Providing Opportunities to NorthWest Territories," *Bill and Melinda Gates Foundation Story Gallery*, www.gatesfoundation.org/Story-Gallery/, accessed March 12, 2004; The Varsavsky Foundation Web site, www.varsavsky *foundation.org/*, accessed March 12, 2004.

Connect via LAN Server

This approach requires the user to install on his or her PC a network adapter card and Open Datalink Interface (ODI) or Network Driver Interface Specification (NDIS) packet drivers. These drivers allow multiple transport protocols to run on one network card simultaneously. LAN servers are typically connected to the Internet at 56 Kbps or faster. In addition, the higher cost of this service can be shared among several dozen LAN users to allow a reasonable cost per user. Additional costs associated with a LAN connection to the Internet include the cost of the software mentioned at the beginning of this section.

Connect via SLIP/PPP

Serial Line Internet Protocol (SLIP)
A communications protocol that transmits packets over telephone lines.

Point-to-Point Protocol (PPP)
A communications protocol that transmits packets over telephone lines.

This approach requires a modem and the TCP/IP protocol software plus **Serial Line Internet Protocol (SLIP)** or **Point-to-Point Protocol (PPP)** software. SLIP and PPP are two communications protocols that transmit packets over telephone lines, allowing dial-up access to the Internet. If you are running Windows, you will also need Winsock. Users must also have an Internet service provider that lets them dial into a SLIP/PPP server. SLIP/PPP accounts can be purchased for $30 a month or less from regional providers. With all this in place, a modem is used to call into the SLIP/PPP server. Once the connection is made, you are on the Internet and can access any of its resources. The costs include the cost of the modem and software, plus the service provider's charges for access to the SLIP/PPP server. The speed of this Internet connection is limited to the slower of your computer's modem and the speed of the modem of the SLIP/PPP server to which you connect.

Connect via an Online Service

This approach requires nothing more than what is required to connect to any of the online information services, such as a modem, standard communications software, and an online information service account. Increasingly, online services are offering DSL, satellite, and cable connection to the Internet, which provide faster speeds. These technologies were discussed in Chapter 6. There is normally a fixed monthly cost for basic services, including e-mail. Additional fees usually apply for DSL, satellite, or cable access, although these costs are falling. The online information services provide a wide range of services, including e-mail and the World Wide Web. America Online and Microsoft Network are examples of such services.

Other Ways to Connect

In addition to computers, many other devices can be connected to the Internet, including cell phones, PDAs, and home appliances. These devices also require specific protocols and approaches to connect. For example, *wireless application protocol (WAP)* is used to connect cell phones and other devices to the Internet.

Internet Service Providers

Internet service provider (ISP)
Any company that provides individuals or organizations with access to the Internet.

An **Internet service provider** (ISP) is any company that provides individuals and organizations with access to the Internet. ISPs do not offer the extended informational services offered by commercial online services such as America Online or MSN. There are literally thousands of Internet service providers, ranging from universities making unused communications line capacity available to students and faculty to major communications giants such as AT&T and MCI. To use this type of connection, you must have an account with the service provider and software that allows a direct link via TCP/IP.

In choosing an Internet service provider, the important criteria are cost, reliability, security, availability of enhanced features, and the service provider's general reputation. Reliability is critical because if your connection to the ISP fails, it interrupts your communication with customers and suppliers. Among the value-added services ISPs provide are electronic commerce, networks to connect employees, networks to connect with business partners, host computers to establish your own Web site, Web transaction processing,

network security and administration, and integration services. Many corporate IS managers welcome the chance to turn to ISPs for this wide range of services because they do not have the in-house expertise and cannot afford the time to develop such services from scratch. In addition, when organizations go with an ISP-hosted network, they can also tap the ISP's national infrastructure at minimum cost. That is important when a company has offices spread across the country.

To use an ISP such as MSN, you must have an account with the service provider and software that allows a direct link via TCP/IP.

In most cases, ISPs charge a monthly fee that can range from $15 to $30 for unlimited Internet connection through a standard modem. The fee normally includes e-mail. Some ISPs, however, are experimenting with low-fee or no-fee Internet access. But there are strings attached to the no-fee offers in most cases. Some free ISPs require that customers provide detailed demographic and personal information. In other cases, customers must put up with extra advertising. For example, a *pop-up ad* is a window that is displayed when someone visits a Web site. It pops up and advertises a product or service. Some e-commerce retailers have posted ads that resemble computer-warning messages and have been sued for deceptive advertising.[8] A *banner ad* appears as a banner or advertising window that you can ignore or click to go to the advertiser's Web site. Table 7.2 identifies several corporate Internet service providers.

Internet Service Provider	Web Address
AT&T's WorldNet Service	www.att.net
BellSouth	www.bellsouth.com
EarthLink	www.earthlink.net
Sprint	www.sprint.com

Table 7.2

A Representative List of Internet Service Providers

Many ISPs and online services offer broadband Internet access through digital subscriber lines (DSLs), cable, or satellite transmission. Most broadband users pay $50 or less per month for unlimited service. Broadband use has spread globally; more than 70 percent of households in South Korea have broadband access, while about 35 percent of Canadian households have

broadband connections.[9] A study by the Pew & American Life Project reports that more than 30 percent of U.S. Web households have broadband. This percentage jumped to more than 40 percent in 2004.[10] Most other countries offer broadband, including Italy (Telecom Italia), Britain (BT Group), Switzerland (Swisscom AG), France (France Telecom), and many others.[11] Some businesses and universities use the very fast T-1 lines to connect to the Internet. Table 7.3 compares the speed of modem, DSL, cable, and T-1 Internet connections to perform basic tasks. These technologies were discussed in Chapter 6.

Table 7.3

Approximate Times to Perform Basic Tasks with a Modem, DSL or Cable, or T-1 Internet Connection

Task	Modem	DSL or Cable	T-1
Send 20-page term paper	30 seconds	3 seconds	Almost instantaneous
Send a 4-minute song as an MP3 file	30 minutes	2 minutes	Almost instantaneous
Send a full-length motion picture as a compressed file	About 2 weeks	About 2 days	20 seconds

THE WORLD WIDE WEB

World Wide Web (WWW or W3)
A collection of tens of thousands of independently owned computers that work together as one in an Internet service.

The World Wide Web was developed by Tim Berners-Lee at CERN, the European Organization for Nuclear Research in Geneva. He originally conceived of it as an internal document-management system. From this modest beginning, the **World Wide Web** (the Web, WWW or W3) has grown to a collection of tens of thousands of independently owned computers that work together as one in an Internet service. These computers, called *Web servers*, are scattered all over the world and contain every imaginable type of data. Thanks to the high-speed Internet circuits connecting them and some clever cross-indexing software, users are able to jump from one Web computer to another effortlessly—creating the illusion of using one big computer. Because of its ability to handle multimedia objects, including linking multimedia objects distributed on Web servers around the world, the Web has become the most popular means of information access on the Internet today.

The Web is a menu-based system that uses the client/server model. It organizes Internet resources throughout the world into a series of menu pages, or screens, that appear on your computer. Each Web server maintains pointers, or links, to data on the Internet and can retrieve that data. However, you need the right hardware and telecommunications connections, or the Web can be painfully slow. Traditionally, graphics and photos have taken a long time to materialize on the screen, and an ordinary phone line connection may not always provide sufficient speed to use the Web effectively. Serious Web users need to connect via the LAN server, SLIP/PPP, DSL, cable, or other approaches discussed earlier. Web *plug-ins* can help provide additional features to standard Web sites. Macromedia's Flash and Real Player are examples of Web plug-ins.

home page
A cover page for a Web site that has graphics, titles, and text.

hypermedia
Tools that connect the data on Web pages, allowing users to access topics in whatever order they wish.

Data can exist on the Web as ASCII characters, word processing files, audio files, graphic and video images, or any other sort of data that can be stored in a computer file. A Web site is like a magazine, with a cover page called a **home page** that has color graphics, titles, and text. All the highlighted type (sometimes underlined) is hypertext, which links the on-screen page to other documents or Web sites. **Hypermedia** connects the data on pages, allowing users to access topics in whatever order they wish. As opposed to a regular document, which you read linearly, hypermedia documents are more flexible, letting you explore related documents at your own pace and navigate in any direction. For example, if a document mentions the Egyptian pharaohs, you can choose to see a picture of the pyramids, jump into a

description of the building of the pyramids, and then jump back to the original document. Hypertext links are maintained using URLs. Table 7.4 lists some interesting Web sites. Many PC and business magazines also publish interesting and useful Web sites, and Web sites are often evaluated and reviewed in print media and online.

Site	Description	URL
ZoomShare	A site that provides free personal Web sites that can be used to store up to 50 personal pictures and text. The site allows links to other Internet sites, has a free e-mail facility, and lets users publish messages on a Web journal.	www.zoomshare.com
Monster	A job-hunting site where you can search for a job by type or company, list your resume, and perform basic company research. One feature, Talent Market, allows people to put their skills up for bid.	www.monster.com
Centers for Disease Control (CDC)	A government site that provides a wealth of information on a wide variety of health topics.	www.cdc.gov
ICQ	A chat facility that offers free chat services for two or more people.	www.icq.com
NASA Human SpaceFlight	A site from NASA that gives information about past and present missions into space.	www.spaceflight.nasa.gov
MSN MoneyCentral	A Microsoft site that offers a large range of financial and investment information.	www.moneycentral.msn.com
Britannica	A site that provides the popular encyclopedia online.	www.britannica.com
Yahoo Maps	A service that offers street addresses and driving directions.	www.yahoo.com
eBay	A popular auction site on the Internet.	www.ebay.com
Amazon.com	A popular site that sells books, videos, music, furniture, and much more.	www.amazon.com
Travelocity	A large site that offers travel information and bargains.	www.travelocity.com
WebMD	A site that provides medical information and advice.	www.webmd.com

Table 7.4

Several Interesting Web Sites

A *Web portal* is an entry point or doorway to the Internet. Web portals include AOL, MSN, Yahoo!, and others. For example, some people use Yahoo.com as their Web portal, which means they have set Yahoo! as their starting point. When they enter the Internet, the Yahoo! Web site appears. You can use Yahoo! to search the Internet, send e-mail, get directions for a trip, buy products and services, find the address and phone number of friends or relatives, and more.

Hypertext Markup Language (HTML) is the standard page description language for Web pages. One way to think about HTML is as a set of highlighter pens in different colors that you use to mark up plain text to make it a Web page—red for the headings, yellow for bold, and so on. The **HTML tags** let the browser know how to format the text: as a heading, as a list, or as body text. HTML also tells whether images, sound, and other elements should be inserted. Users mark up a page by placing HTML tags before and after a word or words. For example, to turn a sentence into a heading, you place the <H1> tag at the start of the sentence. At the end of the sentence, you place the closing tag </H1>. When you view this page in your browser, the sentence will be displayed as a heading. So, a Web page is

Hypertext Markup Language (HTML)
The standard page description language for Web pages.

HTML tags
Codes that let the Web browser know how to format text—as a heading, as a list, or as body text—and whether images, sound, and other elements should be inserted.

made up of two things: text and tags. The text is your message, and the tags are codes that mark the way words will be displayed. All HTML tags are encased in a set of less than (<) and greater than (>) arrows, such as <H2>. The closing tag has a forward slash in it, such as </B> for closing bold. Consider the following text and tags:

<h1 align="center">Principles of Information Systems</h1>

This HTML code centers Principles of Information Systems as a major, or level 1, heading. The "h1" in the HTML code indicates a first-level heading. On some browsers, the heading might be 14-point type size with a Times Roman font. On other browsers, it might be a larger 18-point size in a different font. Figure 7.4 shows a simple document and its corresponding HTML tags.

Figure 7.4

Sample Hypertext Markup Language

Shown at the left on the screen is a document, and at the right are the corresponding HTML tags.

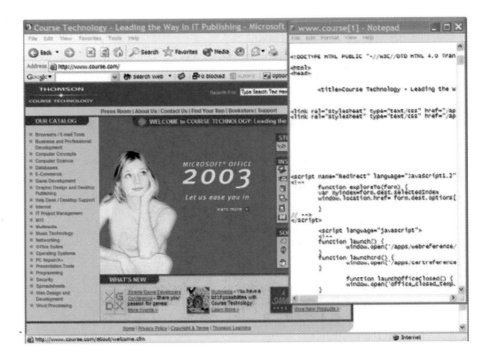

A number of newer Web standards are gaining in popularity, including Extensible Markup Language (XML), Extensible Hypertext Markup Language (XHTML), Cascading Style Sheets (CSS), Dynamic HTML (DHTML), and Wireless Markup Language (WML), which can display Web pages on small screens, such as smart phones and PDAs. XHTML is a combination of XML and HTML that has been approved by the World Wide Web Consortium (W3C).

Extensible Markup Language (XML)

Markup language for Web documents containing structured information, including words, pictures, and other elements.

Extensible Markup Language (XML) is a markup language for Web documents containing structured information, including words and pictures. With XML, there is no predefined tag set. With HTML, for example, the <H1> tag always means a first-level heading. The content and formatting are contained in the same HTML document. XML Web documents contain the content of a Web page. The formatting of the content is contained in a style sheet. A few typical instructions in XML follow:

<CHAPTER>Hardware
<TOPIC>Input Devices
<TOPIC>Processing and Storage Devices
<TOPIC>Output Devices

How the preceding content is formatted and displayed on a Web page is contained in a style sheet, such as a cascading style sheet (CSS) that follows.

Chapter: (font-size: 18pt; color: blue; font-weight: bold; display: block; font-family: Arial; margin-top: 10pt; margin-left: 5pt)

Topic: (font-size: 12pt; color: red; font-style: italic; display: block; font-family: Arial; margin-left: 12pt)

Note that the chapter title Hardware is displayed on the Web page in a large font (18 points). Hardware will appear in the color blue and bold. The Input Devices title will appear in a smaller font (12 points). Input devices will appear in the color red and italics.

XML includes the capabilities to define and share document information over the Web. A company can use XML to exchange ordering and invoicing information with its customers.[12] CSS improves Web page presentation, and DHTML provides dynamic presentation of Web content. These standards move more of the processing for animation and dynamic content to the Web browser, discussed next, and provide quicker access and displays.

Web Browsers

A **Web browser** creates a unique, hypermedia-based menu on your computer screen that provides a graphical interface to the Web. The menu consists of graphics, titles, and text with hypertext links. The hypermedia menu links you to Internet resources, including text documents, graphics, sound files, and newsgroup servers. As you choose an item or resource, or move from one document to another, you may be jumping between computers on the Internet without knowing it, while the Web handles all the connections. The beauty of Web browsers and the Web is that they make surfing the Internet fun. Just clicking with a mouse on a highlighted word or graphic whisks you effortlessly to computers halfway around the world. Most browsers offer basic features such as support for backgrounds and tables, the ability to view a Web page's HTML source code, and a way to create hot lists of your favorite sites. Web browsers enable net surfers to view more complex graphics and 3-D models, as well as audio and video material, and to run small programs embedded in Web pages called **applets**. We discuss applets in more detail in a later section. Internet Explorer from Microsoft and Netscape are examples of Web browsers.

Web browser
Software that creates a unique, hypermedia-based menu on a computer screen, providing a graphical interface to the Web.

applet
Small program embedded in Web pages.

Search Engines

Looking for information on the Web is a little like browsing in a library—without the card catalog, it is extremely difficult to find information. Web search tools—called **search engines**—take the place of the card catalog. Most search engines, such as Yahoo.com and Google.com, are free. They make money by charging advertisers to put ad banners on their search engines. Companies often pay a search engine for a *sponsored link*, which is usually displayed at the top of the list of links for an Internet search.

The Web is a huge place, and it gets bigger with each passing day, so even the largest search engines do not index all Internet pages. Even if you do find a search site that suits you, your query might still miss the mark. So when searching the Web, you may wish to try more than one search engine to expand the total number of potential Web sites of interest. In addition, searches can use words, such as AND and OR to refine the search. Searches can also use filters, such as displaying Web sites in English only. Filters limit searches to a language, certain file formats, a range of dates, and more. Some search engines also have a subject directory that allows people to get information on various industries and organizations.

Once you find a document that comes close to your goal, you can usually find related material by following the highlighted entries that take you to other Web pages when you click on them. And if you come across something you think you'll want to return to, you can add it to the "hot list" or "favorites" list on your Web browser to save time in the future.

search engine
A Web search tool.

AltaVista is a popular search engine on the Web.

Search engines that use keyword indexes produce an index of all the text on the sites they examine. Typically, the engine reads at least the first few hundred words on a page, including the title, the HTML "alt text" coded into Web-page images, and any keywords or descriptions that the author has built into the page structure. The engine throws out words such as "and," "the," "by," and "for." The engine assumes whatever words are left are valid page content; it then alphabetizes these words (with their associated sites) and places them in an index where they can be searched and retrieved. Some companies use a *meta tag* that is placed in the HTML header for search engine robots from sites such as Yahoo! and Google to find and use. Meta tags are not seen on the Web page when it is displayed; they just help a Web site be discovered and displayed by a search engine. There are a number of Web search tools to choose from, as summarized in Table 7.5.

Table 7.5

Popular Search Engines

Search Engine	Web Address
AltaVista	www.altavista.com
Ask Jeeves	www.ask.com
Google	www.google.com
HotBot	www.hotbot.lycos.com
Infoseek	http://infoseek.go.com
Northern Light	www.northernlight.com
Yahoo!	www.yahoo.com

meta-search engine

A tool that submits keywords to several individual search engines and returns the results from all search engines queried.

Another option is to use a meta-search engine. A **meta-search engine** submits keywords to several individual search engines and returns the results from all search engines queried. Ixquick (*www.ixquick.com*), ProFusion (*www.profusion.com*), and Dogpile (*www.dogpile.com*) are examples of meta-search engines. Dogpile, for example, searches Askjeeves, Google, Fast, and other search engines.[13] Meta-search engines do not query all search engines.

Web Programming Languages

Java is an object-oriented programming language from Sun Microsystems based on the C++ programming language, which allows small programs—the applets mentioned earlier—to be embedded within an HTML document. When the user clicks on the appropriate part of the HTML page to retrieve it from a Web server, the applet is downloaded onto the client workstation, where it begins executing. Today, Java is being used on computers, cell phones, and a variety of other devices.[14] Cell phones with Java, for example, can give you a weather forecast, allow you to play games, get maps, find the nearest bathroom in some cities, and find which subway to take in London, England.

Java lets software writers create compact "just-in-time" applet programs that can be dispatched across a network such as the Internet. On arrival, the applet automatically loads itself on a personal computer and runs—reducing the need for computer owners to install huge programs anytime they need a new function. And unlike other programs, Java software can run on any type of computer. Java is used by programmers to make Web pages come alive, adding splashy graphics, animation, and real-time updates. Java-enabled Web pages are more interesting than plain Web pages. A financial services company, for example, can use Java to develop a Web-based financial system.

The relationship among Java applets, a Java-enabled browser, and the Web is shown in Figure 7.5. A *Java applet* is a small program developed using the Java programming language. To develop a Java applet, the author writes the code for the client side and installs that on the Web server. The user accesses the Web page and pulls it down to his personal computer, which serves as a client. The Web page contains an additional HTML tag called APP, which refers to the Java applet. A rectangle on the page is occupied by the Java application. If the user clicks on the rectangle to execute the Java application, the client computer checks to see whether a copy of the applet is already stored locally on the computer's hard drive. If it is not, the computer accesses the Web server and requests that the applet be downloaded. The applet can be located anywhere on the Web. If the user's Web browser is Java enabled (e.g., Sun's HotJava browser or Netscape's Navigator product), then the applet is pulled down into the user's computer and is executed within the browser environment.

Java

An object-oriented programming language from Sun Microsystems based on C++ that allows small programs (applets) to be embedded within an HTML document.

Figure 7.5

Downloading an Applet from a Web Server

The user accesses the Web page from a Web server. If the user clicks on the APP rectangle to execute the Java application, the client's computer checks for a copy on its local hard drive. If the applet is not present, the client requests that the applet be downloaded.

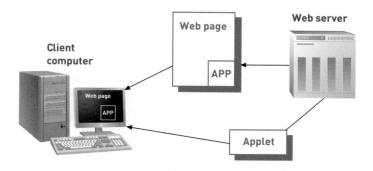

The Web server that delivers the Java applet to the Web client is not capable of determining what kind of hardware or software environment the client is running on, and the developer who creates the Java applet does not want to worry about whether it will work correctly on Windows, UNIX, and MacOS. Java is thus often described as a "cross-platform" programming language.

The development of Java has had a major impact on the software industry. Sun Microsystems's strategy is to open up Java to any and all. Any software vendors and individual developers—from development tool vendors, language compiler developers, database management system vendors, and client/server application vendors to small businesses—can then use Java to create Internet-capable, run-anywhere applications and services (see Figure 7.6). As a result, the Java community is becoming broader every day, encompassing some of the world's biggest independent software vendors, as well as users ranging from corporate CIOs, programmers, multimedia designers, and marketing professionals to educators, managers, film and video producers, and hobbyists. GE Power Systems used Java, XML, and various

Web tools to integrate a number of different systems to allow its managers and employees to share data and information. According to the CIO of GE Power, "The objective was to provide a seamless method for the selection of parts and services, with information being able to come through multiple sources, such as Web browsers, EDI, XML exchanges, or an ERP system."[15]

Figure 7.6

Web Page with a Java Applet

Free Java applets can be downloaded from this site for use on your own Web site.

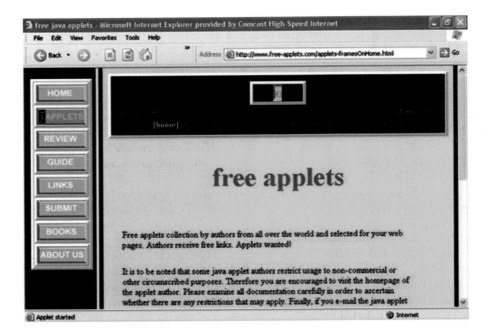

In addition to Java, companies use a variety of other programming languages and tools to develop Web sites. JavaScript, VBScript, and ActiveX (used with Internet Explorer) are Internet languages used to develop Web pages and perform important functions, such as accepting user input. *Hypertext Preprocessor*, or *PHP*, is an open-source programming language. Code or instructions from PHP can be embedded directly into HTML code. Unlike some other Internet languages, PHP can run on a Web server, with the results being transferred to a client computer. PHP can be used on a variety of operating systems, including Microsoft Windows, Macintosh OS X, HP-UX, and others. It can also be used with a variety of database management systems, such as IBM's DB2, Oracle, Informix, MySQL, and many others. PHP's ability to run on different operating systems and database management systems, along with being an open-source language, has made it popular with many Web developers.

Business Uses of the Web

In 1991, the Commercial Internet Exchange (CIX) Association was established to allow businesses to connect to the Internet. Since then, firms have been using the Internet for a number of applications. Electronic mail is a major application for most companies. Many companies also display and sell products over the Internet. Customers can place orders by keying in payment information and shipping addresses.

Businesses everywhere are making large investments to present themselves and their products on the Web, to serve existing customers, and to find new ones. The Web can also cut costs. Aviall, an aviation parts company, invested more than $3 million in a new Web site, Aviall.com. The new Internet site cut ordering costs from $9 per order to $.39 per order.[16] Increasingly, small businesses are using the Internet to boost revenues and decrease costs.[17] As more and more people gain access to the World Wide Web, its functions are changing drastically. We discuss a couple of these applications, corporate intranets and extranets, later in the chapter. Chapter 8 covers e-commerce in more detail.

Push technology is used to send information automatically over the Internet rather than make users search for it with their browsers. Frequently, the information, or "content," is customized to match an individual's needs or profile. The use of push technology is also frequently referred to as *Webcasting*. Most push systems rely on HTTP (Hypertext Transport Protocol) or Java technology to collect content from Web sites and deliver it to employees' or users' desktops. A business, for example, could use push technology to deliver important marketing information to its sales force.

Before they can be "pushed," employees or users must download and install software that acts like a TV antenna, capturing transmitted content. As with any technology, the people paying for push have yet to venture beyond rudimentary applications. Most are focusing on improving communications with employees, customers, and business partners.

A number of companies are using push technology to deliver critical information over the Internet. SAP, for example, uses push technology to deliver critical enterprise resource planning software over the Internet. Enterprise resource planning software is used by many large corporations to streamline operations. A training and performance company can use push technology to deliver critical business information. Other companies also use push technology to make priority deliveries of information over the Internet.

There are drawbacks to the use of push technology. One issue, of course, is information overload. Another is that the volume of data being broadcast is so great that push technology can clog up the Internet communications links with traffic.

push technology
Automatic transmission of information over the Internet rather than making users search for it with their browsers.

Developing Web Content

The art of Web design involves getting around the technical limitations of the Web and using a set of tools to make appealing designs. For example, an HTML converter translates a text document into an HTML document, and an HTML editor inserts HTML codes for you as you create an HTML document. The following are tips to create a Web page.

1. Your computer must be linked to a Web server, which can deliver Web pages to other browsers.
2. You will need a Web browser program to look at HTML pages you create.
3. The actual design can take one of the following approaches: (a) Write your copy with a word processor, then use an HTML converter to convert the page into HTML format complete with tags so the browser knows how it should format the page. (b) Use an HTML editor to write text and add HTML tags at the same time. (c) Edit an existing

HTML template (with all the tags ready to use) to meet your needs. (d) Use an ordinary text editor and type in the start and end tags for each item.

4. Open the page with the browser and see the result. You can correct mistakes by correcting the tags.

5. Add links to your home page to allow your readers to click on a word and be taken to a related home page. The new page may be either a part of your Web site or a home page on a different Web site.

6. To add pictures, you must first store them as a file on your hard drive. This can be done in one of several ways: draw them yourself using a graphics software package, copy pictures from other Web pages, buy a disk of clip art, scan photos, or use a digital camera.

7. You can add sound by using a microphone connected to your computer to record a sound file; adding links to the page will enable those who access your Web page to hear it.

8. Upload the HTML file to your Web site using e-mail or FTP.

9. Review the Web page to make sure that all links are correctly established to other Web sites.

10. Advertise your Web page to others and encourage them to stop, take a look, and send feedback by e-mail.

After Web content development, the next step is to place the content on a Web site or home page. Popular options include ISPs, free sites, and Web hosting. Web hosting services provide space on their Web site for individuals and businesses that don't have the financial resources to have their own Web site or don't have the time or skills to develop their own Web site. Some Internet service providers include limited Web space, typically 1 to 6 MB, as part of their monthly fee. If more disk space is needed, there are additional charges. Free sites offer limited space for an Internet site. In return, free sites often require the user to view advertising or agree to other terms and conditions. A Web host is another option. A Web host can charge $15 or more per month, depending on services. Some Web hosting sites include domain name registration, Web authoring software, and activity reporting and monitoring of the Web site.

A number of products make developing and maintaining Web content easier. In short, these products can greatly simplify the creation of a Web page. Microsoft, for example, has introduced a Web development platform called .NET. The .NET platform allows different programming languages to be used and executed. It also includes a rich library of programming code to help build XML Web applications. Once a Web site has been constructed, a *content management system (CMS)* can keep the Web site running smoothly. CMS consists of both software and support. Companies that provide CMS can charge from $15,000 to more than $500,000 annually, depending on the complexity of the Web site being maintained and the services being performed. Leading CMS vendors include BroadVision, Documentum, EBT, FileNet, Open Market, and Vignette. Many of these products are popular with a newer approach to developing and maintaining Web content called *Web services*, discussed next.

Web Services

Web services
Standards and tools that streamline and simplify communication among Web sites for business and personal purposes.

Web services consist of standards and tools that streamline and simplify communication among Web sites, promising to revolutionize the way we develop and use the Web for business and personal purposes. "Web Services is a tsunami of technology evolution," says Andre Mendes, chief technology officer at Public Broadcasting Service (PBS).[18] Internet companies, including Amazon, eBay, and Google, are now using Web services.[19] Amazon, for example, has developed Amazon Web Services (AWS) to make the contents of its huge online catalog available by other Web sites or software applications.[20] Mitsubishi Motors of North America uses Web services to link about 700 automotive dealers on the Internet. The Wells Fargo bank uses Web services to process electronic payments for its 100 largest corporate customers.[21] According to Steve Ellis, executive vice president of Wells Fargo, "Using Web services reduces [account setup] times by 30 to 50 percent for each new customer we add." Web services can also be used to develop new systems to send and receive secure messages between healthcare facilities, doctors, and patients while maintaining patient privacy.

The key to Web services is XML.[22] Just as HTML was developed as a standard for formatting Web content into Web pages, XML is used within a Web page to describe and transfer data between Web service applications. XML is easy to read and has wide industry support. Besides XML, three other components are used in Web service applications:

1. SOAP (Simple Object Access Protocol) is a specification that defines the XML format for messages. SOAP allows businesses, their suppliers, and their customers to communicate with each other. It provides a set of rules that makes it easier to move information and data over the Internet. Jonathan Pettus, a manager of NASA's Marshall Space Flight Center, envisions using SOAP and online job sites to help get resumes from potential job candidates over the Internet into NASA's computer system for evaluation, although a full implementation of Web services may not occur in the near future.[23]
2. WSDL (Web Services Description Language) provides a way for a Web service application to describe its interfaces in enough detail to allow a user to build a client application to talk to it. In other words, it allows one software component to connect to and work with another software component on the Internet.
3. UDDI (Universal Discovery Description and Integration) is used to register Web service applications with an Internet directory so that potential users can easily find them and carry out transactions over the Web.

There is strong indication that XML Web services technology is more than just a passing technical fad. Three major forces in technology, Microsoft, IBM, and Sun are all investing heavily in Web services development. Microsoft and IBM have joined to develop Web service standards to allow companies to conduct business over the Internet.[24] According to one industry expert, "It's a milestone. These two companies represent half the software industry." The Web Services Interoperability Organization (WS-I) has also been established to promote Web services interoperability across platforms, operating systems, and programming languages.[25] It includes dozens of big technology companies including HP and Oracle.

The acceptance and implementation of Web services hinge on developments in Internet security and privacy. As security and privacy issues are resolved, we will no doubt be relying on Web services to handle many of our online errands. In fact, Web services could develop into an electronic extension of ourselves, anticipating our needs and delivering information and services as we need them, freeing up our time for other endeavors.

INTERNET AND WEB APPLICATIONS

The types of Internet and Web applications available are vast and ever expanding. Some of these services are discussed next and summarized in Table 7.6.

E-Mail and Instant Messaging

E-mail is no longer limited to simple text messages. Depending on your hardware and software and the hardware and software of the recipient, you can embed sound and images in your message and attach files that contain text documents, spreadsheets, graphs, or executable programs. The authors of this book, for example, attached chapter files to e-mail messages that were sent to editors and reviewers for feedback. E-mail travels through the systems and networks that make up the Internet. Gateways can receive e-mail messages from the Internet and deliver them to users on other networks. Thus, you can send e-mail messages to literally anyone in the world if you know that person's e-mail address and if you have access to the Internet or another system that can send e-mail.

E-mail has changed the way people communicate. It improves the efficiency of communications by reducing interruptions from the telephone and unscheduled personal contacts. Also, messages can be distributed to multiple recipients easily and quickly without the inconvenience and delay of scheduling meetings. Because past messages can be saved, they can be reviewed, if necessary. And because messages are received at a time convenient to the

recipient, the recipient has time to respond more clearly and to the point. Many people have two or more e-mail addresses, including free e-mail services.[26] In an experiment, a Columbia University sociologist sent e-mail to a random group of people, trying to contact a specific person, to test the social links that e-mail provides.[27] In most cases, it only took five to seven steps to reach the desired person. The researchers were given 18 people to locate in 13 countries.

Table 7.6

Summary of Internet and Web Applications

Service	Description
E-mail	Enables you to send text, binary files, sound, and images to others.
Career information and job searches	Enables you to get up-to-date information on careers and actual jobs.
Telnet	Enables you to log on to another computer and access its public files. Users can log on to a work computer from an off-site location.
FTP	Enables you to copy a file from another computer to your computer.
Web logs (blogs)	Allows people to create and use a Web site to write about their observations, experiences, and feelings on a wide range of topics.
Usenet and newsgroups	Allows online discussion groups that focus on a particular topic.
Chat rooms	Enables two or more people to have online text conversations in real time.
Internet phone	Enables you to communicate with others around the world by linking Internet and traditional phone service.
Internet video conferencing	Supports simultaneous voice and visual communications.
Content streaming	Enables you to transfer multimedia files over the Internet so that the data stream of voice and pictures plays more or less continuously.
Instant messaging	Allows two or more people to communicate instantly on the Internet.
Shopping on the Web	Allows people to purchase products and services on the Internet.
Web auctions	Lets people bid on products and services.
Music, radio, and video on the Internet	Lets users play or download music, radio, and video.
Office on the Web	Allows people to have access to important files and information through a Web site.
Internet sites in 3-D	Allows people to view products and images at different angles in what appears to be three dimensions.
Free software and services	Allows people to obtain a wealth of free software, advice, and information on the Internet; unwanted advertising and false information are potential drawbacks.
Additional Internet services	Provides a variety of other services to individuals and companies.

For large organizations whose operations span a country or the world, e-mail allows people to work around the time zone changes. Some users of e-mail estimate that they eliminate two hours of verbal communications for every hour of e-mail use. But the person at the other end still must check the mailbox to receive messages. Table 7.7 lists some abbreviations commonly used in personal e-mail messages. These abbreviations are normally not appropriate for business correspondence.

Expressions	Abbreviations
;-] Smile with a wink	AAMOF—As a matter of fact
;-(Frown with a wink	AFAIK—As far as I know
:-# My lips are sealed	BTW—By the way
:-D Laughing	CUL8R—See you later
:-0 Shocked	F2F—Face to face
:-] Blockhead	LOL—Laughing out loud
:-@ Screaming	OIC—Oh, I see
:-& Tongue-tied	TIA—Thanks in advance
%-] Brain-dead	TTFN—Ta-Ta for now

Table 7.7

Some Common Abbreviations Used in Personal E-Mail

Some companies use bulk e-mail to send legitimate and important information to sales representatives, customers, and suppliers around the world. With its popularity and ease of use, however, some people feel they are drowning in too much e-mail. Over a trillion e-mail messages are sent from businesses in North America each year. This staggering number is up from 40 billion e-mail messages in 1995. Many e-mails are copies sent to a large list of corporate users.[28] Companies and individuals are taking a number of steps to cope with and reduce the mountains of e-mail. Some companies have banned the use of copying others on e-mails unless it is critical. Some e-mail services scan for possible junk or bulk mail, called *spam*, and delete it or place it in a separate file.[29] As much as half of all e-mail can be considered spam.[30] Some business executives receive 300 or more spam e-mails in their corporate mailboxes every morning. The Sobig.F virus of 2003 was sent in e-mail messages to thousands of computers.[31] One computer received more than 700 e-mail messages in about five hours.

Software products can help companies, individuals, and even some U.S. senators sort and answer large amounts of e-mail. This software has the ability to recognize key words and phrases and respond to them. The federal government and some states are now proposing legislation to block unwanted e-mail.[32] California, for example, has proposed legislation that would prevent companies from sending unsolicited commercial e-mail to California residents.[33] The U.S. Congress passed a federal law, called Controlling the Assault of Non Solicited Pornography and Marketing Act (CAN SPAM), to reduce spam sent by companies in the United States.[34] Unfortunately, legitimate e-mail can get lost. An informal survey of about 10,000 individuals by a columnist for *Information Week* revealed that up to 40 percent of legitimate e-mails are not getting to their proper destinations.[35]

Instant messaging is online, real-time communication between two or more people who are connected to the Internet. With instant messaging, two or more screens open up. Each screen displays what one person is typing. Because the typing is displayed on the screen in real-time, it is like talking to someone using the keyboard.

A number of companies offer instant messaging, including America Online, Yahoo!, and Microsoft. America Online is one of the leaders in instant messaging, with about 40 million users of its Instant Messenger and about 50 million people using its client program ICQ. In

instant messaging
A method that allows two or more individuals to communicate online using the Internet.

addition to being able to type messages on a keyboard and have the information instantly displayed on the other person's screen, some instant messaging programs are allowing voice communication or connection to cell phones. A wireless service provider announced that it has developed a technology that can detect when a person's cell phone is turned on. With this technology, someone on the Internet can use instant messaging to communicate with someone on a cell phone anywhere in the world. Apple Computer is experimenting with adding audio and video to its instant messaging service, called iChat.[36] The iSight camera can be used to transfer visual images through instant messaging.

Instant messaging services often use a *buddy list* that alerts people when their friends are also online. This feature makes instant messaging even more useful. Instant messaging is so popular that it helps Internet service providers and online services draw new customers and keep old ones. Buddy lists, however, can lead to inadvertent and unwanted sales pitches.[37] In one case, an instant message advertised a humorous Osama bin Laden game. When downloaded, the program sent similar advertisements to everyone on the person's buddy list.

Internet Cell Phones and Handheld Computers

Increasingly, cell phones, handheld computers, and other devices are being connected to the Internet. Some cell phones, for example, can be connected to the Internet to allow people to search for information, buy products, and chat with business associates and friends. A sales manager for a computer company can use her cell phone to check her company's Internet site to see whether there are enough desktop computers in inventory to fill a large order for an important customer. Using Short Message Service, people can send brief text messages of up to 160 characters between two or more cell phone users. The service is often called *texting*. Some cell phones also come equipped with digital cameras, FM radios, video games, and small color screens to watch TV. Using multimedia messaging service (MMS), people can send pictures, video, and audio over cell phones to other cell phones or Internet sites. An insurance investigator can use MMS to send photos of a car accident to a central office to process an insurance claim. Of course, cell phones can also be used to send e-mail messages to others. Legislation passed in fall 2003 allows cell phone users to keep their phone numbers when switching to another cell phone company.

In addition to cell phones, handheld computers and other devices can also be connected to the Internet using phone lines or wireless connections, such as WiFi, discussed in Chapter 6. Once connected, these devices have full access to the Internet and all its applications discussed in this chapter and throughout the book. Managers use handheld computers, such as the BlackBerry or Treo handheld computer, and the Internet to check business e-mail when they are out of the office; sales representatives use them to demonstrate products to customers, check product availability and pricing, and upload customer orders.

Career Information and Job Searching

The Internet is an excellent source of job-related information. People looking for their first job or seeking information about new job opportunities can find a wealth of information. Search engines, such as *www.google.com* and *www.yahoo.com*, can be a good starting point for searching for specific companies or industries. You can use a directory on Yahoo's home page, for example, to explore various industries and careers. Most medium or large companies have Internet sites that list open positions, salaries, benefits, and people to contact for further information. IBM's Web site, *www.ibm.com*, has a link to "Jobs at IBM." When you click on this link, you can get information on jobs with IBM around the world. Some Internet sites specialize in certain careers or industries. The site *www.directmarketingcareers.com* lists direct marketing jobs and careers, and the site *www.einsurancejobs.com* lists insurance jobs. Some sites can help people develop a good resume and find a good job. They can also help you develop an effective cover letter for a resume, prepare for a job interview, negotiate a good employment contract, and more. In addition, several Internet sites specialize in helping people get job information and even apply for jobs online, including *www.monster.com* and *www.careerbuilder.com*. You must be careful when applying for jobs online, however. Some bogus companies or Web sites will steal your identity by asking for personal information.

People eager to get a job often give their Social Security number, birth date, and other personal information. The result can be no job, large bills on your credit card, and ruined credit.

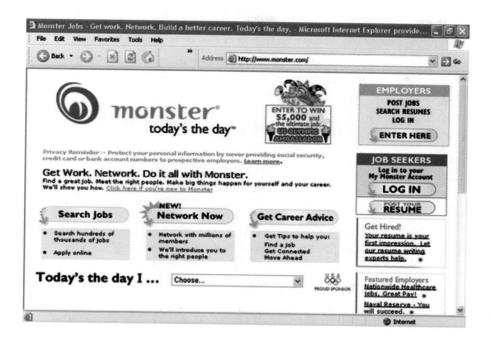

Several Internet sites specialize in helping people get job information and even apply for jobs online.

Telnet and FTP

Telnet is a terminal emulation protocol that enables you to log on to other computers on the Internet to gain access to their publicly available files. Telnet is particularly useful for perusing library card files and large databases. It is also called *remote logon.*

File Transfer Protocol (FTP) is a protocol that describes a file transfer process between a host and a remote computer. Using FTP, companies and individuals can copy files from one computer to another. Companies, for example, use it to transfer vast amounts of business transactional data to the computers of its customers and suppliers. FTP can also be used to gain access to a wealth of free software on the Internet. The authors and editors of this book used an FTP site provided by the publisher, Course Technology, to share and transfer important files around the world during the publication process.

Web Log (Blog)

A **Web log**, also called a **blog**, is a Web site that people can create and use to write about their observations, experiences, and feelings on a wide range of topics.[38] A *blogger* is a person who creates a blog, while *blogging* refers to the process of placing entries on a blog site.[39] A blog is like a journal. When people post information to a blog, it is placed at the top of the blog. Previous entries on the blog are pushed down. Blogs can be used by anyone or any organization to publish and share information. Jacob Crossman, a software engineer, uses blogs to share information about the status of engineering projects with others.[40] "One of the disadvantages of a paper-based engineering notebook is that it's hard to find things unless you want to go through it manually. So I decided to use the blog feature of the Social-text's software to keep track of my ideas," says Crossman. See the "Information Systems @ Work" feature to learn about additional uses of blogs.

Telnet
A terminal emulation protocol that enables users to log on to other computers on the Internet to gain access to public files.

File Transfer Protocol (FTP)
A protocol that describes a file transfer process between a host and a remote computer and allows users to copy files from one computer to another.

Web log (blog)
A Web site that people can create and use to write about their observations, experiences, and feelings on a wide range of topics.

Inscene Embassy Changes Marketing Focus with Blogs

The Web log, or blog, has become a popular method of Internet communication—so popular, in fact, that many believe it has the potential to change the nature of the Web. Blogs empower regular Webizens (citizens of the Web) to publish their thoughts without any formal knowledge of HTML. Some forms of blogs are similar to usenet newsgroups designed for the Web. Most blog Web sites are centered around a particular topic and may be either one person's ongoing writing or assorted postings from a group of writers or the general public.

Blogs began as underground commentaries published by and for "Internet insiders"—those who spend a substantial portion of their lives online. These original blogs took the form of Web sites centered on the often radical viewpoints of intellectuals and malcontents. Within a few years, though, blogs were discovered by the general public through postings on Web sites that catered to special-interest groups. Blogs eventually made their way into mainstream culture when they were mentioned on network TV, in magazines, and on the radio.

The power of the blog caught the attention of the business community, especially marketers. Businesses are seriously examining the potential of blogs to gather and harness public opinion regarding their products. One of the boldest experiments in product-based blogs has been launched by Inscene Embassy. Inscene is a German fashion label that appeals to the teen market, with fashions that are casual, radical, and cross-cultural. Inscene has drafted young people in several cities around the world—including Tokyo, New York, London, and Berlin—to maintain blogs as cultural ambassadors for the brand.

The blogs create a sense of community among Inscene customers, who use the blog to share their thoughts on fashion and the latest Inscene merchandise. The blog provides Inscene with invaluable information on the way its target market thinks. The paid young people, who act as moderators for the blogs, guide the conversation and filter what is posted to keep the blogs clean of profanity and abuse.

Meg Hourihan, coauthor of the book *We Blog: Publishing Online with Weblogs*, sees a powerful commercial future for blogging. "Commercial Web sites aren't inherently better than personal ones, but they have business models and budgets. They have target audiences that can benefit from the type of focused content produced by bloggers. When a blogger is being paid to maintain a Weblog, he is able to do so full-time, with all his attention focused on the topic of choice."

Blogs are transforming the Web from a medium for only those with HTML knowledge and Web server access to a form of communication and expression for anyone with an Internet connection. Blogs have a unique power and appeal for those who become involved with them. Enthusiasm is contagious, and the ideas and energy generated by discussion on blogs, if harnessed, could be very useful to commercial enterprises.

Discussion Questions

1. How has the use of blogs changed the way marketers collect marketing data at Inscene Embassy?
2. Besides being a collection point for customer opinion, how do the blogs assist Inscene Embassy in selling clothes?

Critical Thinking Questions

3. If you were head of marketing at Inscene, what instructions would you give your paid bloggers to assist in building your product's reputation and maximizing the amount of valuable information you can collect from the blogs?
4. What dangers does Inscene Embassy expose itself to in supporting a public forum centered around its products?

SOURCES: James Lewin, "Learning from Blogs," *ITWorld*, December 17, 2003, *www.itworld.com*; Inscene Embassy Web site, *www.inscene-embassy.de/*, accessed March 20, 2004; Jonathan V. Last, "World Wide Dean," *The Daily Standard*, January 9, 2004, *www.lexis-nexis.com*.

Blogs exist in a wide variety of topics and areas. For example, the Western States Information Network (WSIN) developed a blog to allow local fire departments, water departments, and similar organizations to share information online.[41] E-mails between departments are posted on the blog. The blog provides a central location where all e-mails can be posted and read. Blogs are used by journalists, people in disaster areas, soldiers in the field, and people who just want to express themselves.[42] Venture capitalists can use *www.ventureblog.com* to investigate Internet or dot.com companies.[43] Tom Daschle, the U.S. Senate minority leader, has a blog that he uses to write an online diary of his trips and experiences.[44] According to Daschle, "This new blog concept appealed to me." Howard Dean and his staff used BlogforAmerica.com to post entries related to his presidential campaign in 2004 before he dropped out of the race.

Blog sites, such as *www.blogger.com* and *www.globeofblogs.com* can include information and tools to help people create and use Web logs. Blogs are easy to set up. You can go to a blog service provider, such *www.livejournal.com*, create a user name and password, select a theme, choose a URL, follow any other instructions, and start making your first entry.

Usenet and Newsgroups

Usenet is a system closely allied with the Internet that uses e-mail to provide a centralized news service. It is actually a protocol that describes how groups of messages can be stored on and sent between computers. Following the usenet protocol, e-mail messages are sent to a host computer that acts as a usenet server. This server gathers information about a single topic into a central place for messages. A user sends e-mail to the server, which stores the messages. The user can then log on to the server to read these messages or have software on the computer log on and automatically download the latest messages to be read at leisure. Thus, usenet forms a virtual forum for the electronic community, and this forum is divided into newsgroups.

Newsgroups make up usenet, a worldwide discussion system classified by subject. Articles or messages are posted to newsgroups using newsreader software and are then broadcast to other interconnected computer systems via a wide variety of networks. A newsgroup is essentially an online discussion group that focuses on a particular topic. Newsgroups are organized into various hierarchies by general topic, and within each topic there can be many subtopics. On the Internet, there are tens of thousands of newsgroups, covering topics from astrology to zoology (see Table 7.8). Discussions take place via e-mail, which is sent to the newsgroup's address. A newsgroup may be moderated or unmoderated. If a newsgroup is moderated, e-mail is automatically routed to the moderator, a person who screens all incoming e-mail to make sure that it is appropriate before posting it to the newsgroup. Some people who are frequent newsgroup users invest in a newsgroup reader that makes reading and posting messages easier. People posting messages on newsgroups should be careful. Some Internet sites put millions of newsgroup messages that go back to 1980s online for anyone to see.

usenet

A system closely allied with the Internet that uses e-mail to provide a centralized news service; a protocol that describes how groups of messages can be stored on and sent between computers.

newsgroups

Online discussion groups that focus on specific topics.

alt.airline	biz.ecommerce
alt.aol	alt.current-events.net-abuse.spam
alt.books	alt.politics
alt.fan	alt.hackers
alt.sports.baseball	alt.music
alt.sports.basketball	news.software

Table 7.8

Selected Usenet Newsgroups

Newsgroup servers around the world host newsgroups that share information and commentary on predefined topics. Each group takes the form of a large bulletin board where

members post and reply to messages, creating what is called a *message thread*. The open nature of newsgroups encourages participation, but the discussions often become rambling and unfocused. As a result of so much active participation, newsgroups can evolve into tight communities where certain members tend to dominate the discussions.

Here are some tips to consider when accessing newsgroups:

1. When you join a newsgroup, first check its list of Frequently Asked Questions, or FAQs (pronounced "facks"), before submitting any questions to the newsgroup. The FAQ list will have answers to common questions the group receives. It is considered impolite to waste the group's time by asking common questions when FAQs are available.
2. Read messages without responding at first to get an understanding of the newsgroup. Many newsgroups include members from around the world, and in the interest of courtesy, you should pick up some sense of the audience and its culture before jumping in with questions and opinions.
3. It is impolite to jump into the middle of a conversation.
4. Avoid raising points and issues long since discussed and abandoned.
5. Be concerned about what you say and the feelings of others. Remember, a person is receiving your messages. Do not use extreme words or repeat rumors (you could risk libel or defamation lawsuits).
6. Do not post copyrighted material, and be careful how you use copyrighted material downloaded to your computer.
7. Protect yourself by not offering personal information such as your home address, employer, or phone number.
8. Remember that this global online community has fragmented into thousands of different groups for a reason—to maintain the focus of each conference. Respect the specific subject matter of the group.

Chat Rooms

chat room
A facility that enables two or more people to engage in interactive "conversations" over the Internet.

A **chat room** is a facility that enables two or more people to engage in interactive "conversations" over the Internet. When you participate in a chat room, there may be dozens of participants from around the world. Multiperson chats are usually organized around specific topics, and participants often adopt nicknames to maintain anonymity. One form of chat room, Internet Relay Chat (IRC), requires participants to type their conversation rather than speak. Voice chat is also an option, but you must have a microphone, sound card and speakers, a fast modem or broadband, and voice-chat software compatible with that used by the other participants.

Internet Phone and Videoconferencing Services

Internet phone service enables you to communicate with others around the world. This service is relatively inexpensive and can make sense for international calls. With some services, it is possible to make a call from someone using the Internet to someone using a standard phone. You can also keep your phone number when you move to another location.[45] According to one Internet phone user who moved from Madison, Wisconsin, to California, "I was so happy about that. Nothing changed for my customers. For all they knew I was still in Madison." Cost is often a big factor in use of Internet phones—a call can be as low as 1 cent a minute for calls within the United States. Low rates are also available for calling outside the United States. In addition, voice mail and fax capabilities are available. Some cable TV companies, for example, are offering cable TV, phone service, and caller ID for under $40 a month.[46]

Using *voice over IP (VoIP)* technology, as described in Chapter 6, network managers can route phone calls and fax transmissions over the same network they use for data—which means no more separate phone bills.[47] Gateways installed at both ends of the communications link convert voice to IP packets and back. With the advent of widespread, low-cost Internet telephony services, traditional long-distance providers are being pushed to either respond in kind or trim their own long-distance rates. The school system in Appleton, Wisconsin, for example, installed a phone system based on the Internet and saved about 30 percent a year

in telecommunications costs.[48] A start-up organization called Skype, founded by some of the same people who developed music-sharing site Kazaa, has developed software to allow people to make free, clear phone calls over PCs anywhere in the world.[49] According to Michael Powell, chairman of the U.S. Federal Communications Commission (FCC), Internet technology has revolutionized communications: "I knew it was over when I downloaded Skype. When the inventors of Kazaa are distributing for free a little program that you can use to talk to anybody else, and the quality is fantastic, and it is free—it's over." VoIP (pronounced *voyp*) is growing from 30 to 50 percent annually.[50] The number of VoIP lines is expected to exceed 12 million by 2006. Companies that offer VoIP include Packet8, Callserve, Dialpad, Net2Phone, and WebPhone.[51]

Figure 7.7 shows how VoIP works. Voice travels over the Internet, rather than the circuit-switched public network. Most corporate IP telephony applications use gateways that sit between the public branch exchange (PBX) and a router to convert calls into IP packets and shunt them onto the network. Using packets allows multiple parties to share digital lines, so data transmission is much more efficient than traditional phone conversations, each of which requires a line. When the packets hit the destination gateway, the message is depacketized, converted back into voice, and sent out via local phone lines. With newer multiVoIP, the PBX, seen in Figure 7.7, is not needed. Phones are directly connected to a multiVoIP box. This arrangement can be cheaper than standard VoIP, making the technology more attractive to small businesses.

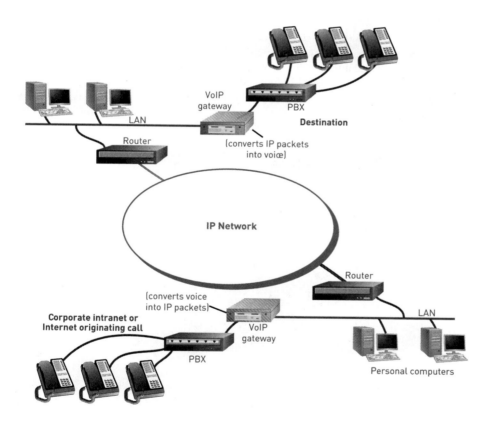

Figure 7.7

How Voice Over IP Works

A codec (*co*mpression-*dec*ompression) device prepares the voice transmission by squeezing the recorded sound data and slicing it into packets for transfer over the Internet. On the receiving end, a codec reassembles and decompresses the data for playback. Different codecs are optimized for different uses and conditions, and the characteristics of a specific codec can affect voice quality.

What is especially interesting about VoIP is the ways voice is being merged with video and data communications over the Web or a company's data network. In the long run, it's not the cost savings that will boost the market, it's the multimedia capabilities it gives us and

the smart call-management capabilities. Travel agents could use voice and video over the Internet to discuss travel plans; Web merchants could use it to show merchandise and take orders; customers could show suppliers problems with their products.

Internet videoconferencing, which supports both voice and visual communications, is another important Internet application. Microsoft's NetMeeting, a utility within Windows, is an inexpensive and easy way for people to meet and communicate on the Web. The Internet can also be used to broadcast sales seminars using presentation software and videoconferencing equipment. These Internet presentations are often called Webcasts or Webinars. Hardware and software are needed to support videoconferencing. The key here is a video codec to convert visual images into a stream of digital bits and translate back again. The ideal video product will support multipoint conferencing, in which multiple users appear simultaneously on the multiple screens.

Content Streaming

content streaming

A method for transferring multimedia files over the Internet so that the data stream of voice and pictures plays more or less continuously without a break, or very few of them; enables users to browse large files in real time.

Content streaming is a method for transferring multimedia files over the Internet so that the data stream of voice and pictures plays more or less continuously, without a break, or with very few of them. It also enables users to browse large files in real time. For example, rather than wait the half-hour it might take for an entire 5-MB video clip to download before they can play it, users can begin viewing a streamed video as it is being received.

Shopping on the Web

Shopping on the Web for books, clothes, cars, medications, and even medical advice can be convenient and easy. Some Internet shoppers are loyal to a few familiar Internet sites. This is good news for well-established and popular Internet sites. To add to their other conveniences, many Web sites offer free shipping and free pickup for returned items that don't fit or don't meet a customer's needs.[52] Table 7.9 shows the number of unique visitors to several Internet shopping sites.[53]

Table 7.9

Unique Visitors to Shopping Sites

(Source: Data from Mylene Mangalinsan, "Yahoo Hopes New Service Turns Searchers into Shoppers," *The Wall Street Journal*, September 23, 2003, p. B1.)

Company	Millions of Unique Visitors
eBay	42.4
Amazon	26.1
Yahoo Shopping	15.1
DealTime	11.9
Wal-Mart Stores	9.2
Target	7.6
AOL Shopping	7.5
Bizrate.com	7.4
Sears	5.3
MSN Shopping	4.9

bot

A software tool that searches the Web for information, products, prices, and so forth.

Increasingly, people are using bots to help them search for information or shop on the Internet. A **bot** is a software tool that searches the Web for information, products, or prices. A bot, short for *robot*, can find the best prices or features from multiple Web sites. Hotbot.com is an example of an Internet bot.

Web Auctions

A **Web auction** is a way to match people and companies that want to sell products and services with people and companies who want to buy products and services. Web auction sites are a place where businesses are growing their markets or reaching customers in a very low cost-per-transaction basis. Web auctions are transforming the customer-supplier relationship.

In addition to typical products and services, Internet auction sites excel at offering unique and hard-to-find items. Finding these items without an Internet auction site is often difficult, time-consuming, and expensive. The business-to-business application of auction sites is expected to continue. Almost anything you may want to buy or sell can be found on auction sites. One of the most popular auction sites is eBay, which often has millions of auctions occurring at the same time. Companies, such as AuctionDrop, will sell your items on eBay for a fee.[54] The eBay site is easy to use, and a large number of products and services can be found. eBay is also allowing shoppers to use some of their loyalty points from Hilton Hotels and Sprint to buy one of its 16 million items listed on its Internet site.[55] One point is worth one penny. In addition to eBay, there are a number of other auction sites on the Web. Traditional companies are even starting their own auction sites.

Although auction Web sites are excellent for matching buyers and sellers, there can be problems with their use. Auction sites on the Web are not always able to determine whether products and services listed by people and companies are legitimate. In one case, a person posted items he didn't own on an auction site and stole $120,000 from auction users.[56] The person who stole the money, who is now in jail, said, "I never knew how easy it was to manipulate people. It was like taking candy from a baby." In addition, some Web sites have had illegal or questionable items offered. Many Web sites have an aggressive fraud investigation system to prevent and help prosecute fraudulent use of their sites. Even with these potential problems, the use of Web auction sites is expected to continue to grow rapidly.

Music, Radio, and Video on the Internet

Music, radio, and video are hot growth areas on the Internet. Audio and video programs can be played on the Internet, or files can be downloaded for later use. Using music players and music formats such as MP3, discussed in Chapter 3, it is possible to download music from the Internet and listen to it anywhere using small, portable music players.

Web auction

An Internet site that matches people who want to sell products and services with people who want to purchase these products and services.

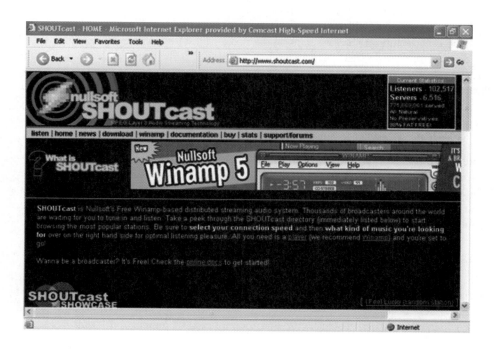

The Internet makes it possible to listen to radio broadcasts. Users can select a music genre and hear streaming audio.

The Recording Industry Association of America (RIAA) won a legal battle against Napster, preventing the company from allowing free copies of music to be shared over the Internet. More recently, the RIAA has been filing lawsuits against individuals who download music illegally.[57] But a number of companies now offer music downloads for minimal fees, usually under a dollar per song. The Internet is also being used to form music collaborations that would be difficult or impossible otherwise, allowing musicians to record music from long distance.

Radio broadcasts are now available on the Internet. Entire audio books can also be downloaded for later listening, using devices such as the Audible Mobile Player. This technology is similar to the popular books-on-tape media, except you don't need a cassette tape or a tape player.

Some corporations have also started to use Internet video to broadcast corporate messages or to advertise on the Web. Doctors can also use Internet video to monitor and even control surgical operations that take place thousands of miles away from them. As mentioned in Chapter 6, Internet video is also being used successfully for teleconferencing, which can connect employees, managers, and corporate executives around the world in private conversations. Using Internet video, it is also possible to receive TV programs, such as sports broadcasts, from an Internet site.

Office on the Web

Aside from the use of video and audio conferences to connect employees who are widely spread, employers are also offering employees the option of telecommuting to their jobs. From home offices, coworkers can connect to the workplace via a virtual office on the Web. For example, you can receive a phone call from your boss, who wants you to send a financial document to a coworker immediately. You may also want to set up conference calls or track upcoming appointments.

To help solve these problems, you can set up an Internet office. An Internet office is a Web site that contains files, phone numbers, e-mail addresses, an appointment calendar, and more. Using a standard Web browser, you can access important files and information. An Internet office allows your desktop computer, phone books, appointment schedulers, and other important information to be with you wherever you are. For those who still travel for their work, special Internet sites allow them to travel light and never leave important information behind. An Internet site can allow you to connect to an office or home PC while traveling using the Internet. It is also possible to print documents while traveling. For individuals and employees who travel, these Internet sites are invaluable.

You can also hire an office assistant on the Web.[58] The virtual office assistant can be used to do accounting tasks, help market products, or buy gifts for family and friends. Online office assistants can be located across the country. They stay connected using the Internet, fax, or phone calls.

Internet Sites in Three Dimensions

Some Web sites offer three-dimensional views of places and products. For example, a 3-D Internet auto showroom allows people to get different views of a car, simulating the experience of walking around in a real auto showroom. When looking at a 3-D real estate site on the Web, people can tour the property, go into different rooms, look at the kitchen appliances, and even take a virtual walk in the garden. Without a doubt, 3-D Internet sites will become common in the future.

Free Software and Services

The Internet has always been known as a source for free software, advice, and services.[59] The software for many of the services just discussed can be downloaded from the Internet free of charge. ICQ, the instant messaging system discussed earlier, for example, can be downloaded at no charge. Some e-mail services, such as those offered by Yahoo! and Hotmail, are also free. In addition, there is a wealth of information and advice on the Internet. Using a search engine, it is possible to obtain free information on almost any topic, ranging from investments

to dating. Table 7.10 contains some popular and useful free software and services. Note that free services can change or even stop their offerings with little or no warning.

Site	Description
www.accuweather.com	Gives detailed weather forecasts and news.
www.travelaxe.com	Searches a number of travel and hotel Web sites for the best deals.
www.davespda.com	Gives information on handheld computers and PDA devices.
zone.msn.com and www.sega.com	Offer free games on the Internet

Table 7.10

Free Internet Services

The disadvantages of free services can be many, however. Some sites bombard the user with annoying advertising to cover their costs or to make a profit. In addition, the information and advice may not always be truthful or helpful. For example, some people have posted false information on investment chat rooms, hoping to manipulate the price of a stock and make a profit. In some cases, groups or sites on the Internet have a bias. They can appear to be helpful but are actually trying to advance their own agenda by posting false or misleading information. Thus, great care must be exercised when obtaining free software or services from the Internet.

Other Internet Services and Applications

Other Internet services are constantly emerging.[60] A vast amount of information is available over the Internet from libraries.[61] Many articles that served as the basis of the boxes, cases, and examples used throughout this book were obtained from university libraries online. Movies can be ordered and even delivered over the Internet.[62] The Internet can provide critical information during times of disaster or terrorism. During a medical emergency, critical medical information can be transmitted over the Internet. People wanting to consolidate their credit card debt or to obtain lower payments on their existing home mortgages have turned to sites, such as Quicken Loan, E-Loan, and LendingTree for help.

The Internet also facilitates distance learning, which has dramatically increased in the last several years. Many colleges and universities now allow students to take courses without ever visiting campus. In fact, you may be taking this course online. Businesses are also taking advantage of distance learning through the Internet. Video cameras can be attached to computers and connected to the Internet.[63] Internet cameras can be used to conduct job interviews, hold group meetings with people around the world, monitor young children at daycare centers, check rental properties and second homes from a distance, and more. People can use the Internet to connect with friends or others with similar interests.[64] Internet sites, such as ZeroDegrees.com, Tribe.net, and Spoke.com, are examples of social networking Internet sites. Ex-Investco employees use Internet sites, such as Xinvesco.com, to connect to other ex-employees.[65] Today, manufacturers of sound systems are putting Internet and network capabilities into their devices.[66] For example, high-end Marantz video projectors and McIntosh stereo amplifiers are able to download music from the Internet. Other devices for home use allow people to view photos and see movies throughout a house using wireless networks and the Internet.[67] Gateway and Apex Digital, for example, produce devices that use wireless networks and the Internet to transfer audio and video files from a PC to a TV and back.

INTRANETS AND EXTRANETS

intranet
An internal corporate network built using Internet and World Wide Web standards and products; used by employees to gain access to corporate information.

An **intranet** is an internal corporate network built using Internet and World Wide Web standards and products. Employees of an organization use it to gain access to corporate information. After getting their feet wet with public Web sites that promote company products and services, corporations are seizing the Web as a swift way to streamline—even transform—their organizations. These private networks use the infrastructure and standards of the Internet and the World Wide Web. A big advantage of using an intranet is that many people are already familiar with Internet technology, so they need little training to make effective use of their corporate intranet.

An intranet is an inexpensive yet powerful alternative to other forms of internal communication, including conventional computer setups. One of an intranet's most obvious virtues is its ability to slash the need for paper. Because Web browsers run on any type of computer, the same electronic information can be viewed by any employee. That means that all sorts of documents (such as internal phone books, procedure manuals, training manuals, and requisition forms) can be inexpensively converted to electronic form on the Web and be constantly updated. An intranet provides employees with an easy and intuitive approach to accessing information that was previously difficult to obtain. For example, it is an ideal solution to providing information to a mobile sales force that needs access to rapidly changing information. Intranets can also do something far more important. The Bank of New Zealand allows its 16 business units to develop and manage content for its intranet.[68] According to a company spokesperson, "They've never had the ability to have that sense of ownership of the content that we can now give to the business units." The Scottish Water Utility used the same approach with its intranet, which replaced three legacy systems.[69] "Our intranet ownership was held by the internal communications team, and we wanted to get it out into the business to provide a tool for people to share communications across the organization," says Linda Fay, the digital media team leader for Scottish Water.

An intranet is an internal corporate network used by employees to gain access to company information.

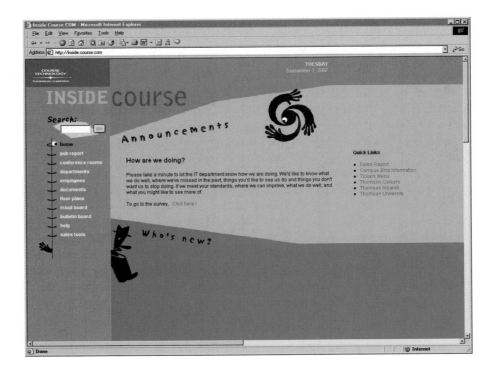

A rapidly growing number of companies offer limited network access to selected customers and suppliers. Such networks are referred to as *extranets*, which connect people who are external to the company. An **extranet** is a network that links selected resources of the intranet of a company with its customers, suppliers, or other business partners. Again, an extranet is built around Web technologies. James Stevenson, an executive for a London-based Web Agency, believes extranets are being used to a greater extent today. "Are they being used now? Absolutely. Customers are learning from the extranets of the past and engaging with re-creating the technology in a more business-like way," says Stevenson.[70] WHSmith News uses an extranet to help identify sales trends.[71] The magazine distributor uses its extranet to communicate daily sales data to retail outlets. According to Richard Webb, business systems manager for the company, "When the *Christmas Radio Times* and *TV Times* went on the shelves, we could get feedback on how they are selling or if there are hotspots for sales. There's so much data relating to this category—it's about distilling it down to meaningful information."

Security and performance concerns are different for an extranet than for a Web site or network-based intranet. User authentication and privacy are critical on an extranet so that information is protected. Obviously, performance must be good to provide quick response to customers and suppliers. Table 7.11 summarizes the differences between users of the Internet, intranets, and extranets.

extranet
A network based on Web technologies that links selected resources of a company's intranet with its customers, suppliers, or other business partners.

Type	Users	Need for User ID and Password
Internet	Anyone	No
Intranet	Employees and managers	Yes
Extranet	Business partners	Yes

Table 7.11

Summary of Internet, Intranet, and Extranet Users

Secure intranet and extranet access applications usually require the use of a virtual private network (VPN). A **virtual private network (VPN)** is a secure connection between two points across the Internet. VPNs transfer information by encapsulating traffic in IP packets and sending the packets over the Internet, a practice called **tunneling**. Most VPNs are built and run by Internet service providers. Companies that use a VPN from an Internet service provider have essentially outsourced their networks to save money on wide area network equipment and personnel. In using a VPN, a user sends data from his or her personal computer to the company's firewall, discussed later in the chapter, which also converts the data into a coded form that cannot be easily read by an interceptor. The coded data is then sent via an access line to the company's Internet service provider. From there, the data is transmitted through tunnels across the Internet to the recipient's Internet service provider and then over an access line to the receiving company's firewall, where it is decoded and sent to the receiver's personal computer (see Figure 7.8).

virtual private network (VPN)
A secure connection between two points across the Internet.

tunneling
The process by which VPNs transfer information by encapsulating traffic in IP packets over the Internet.

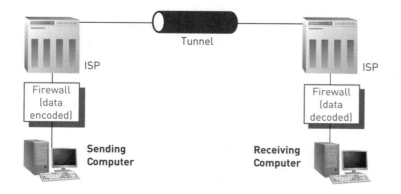

Figure 7.8

Virtual Private Network

NET ISSUES

The topics raised in this chapter apply not only to the Internet and intranets but also to LANs, private WANs, and every type of network. Control, access, hardware, and security issues affect all networks, so it is important to mention some of these management issues.

Management Issues

Although the Internet is a huge, global network, it is managed at the local level; no centralized governing body controls the Internet. Preventing attacks is always an important management issue. The number of attacks has grown dramatically over the last few years. Every month, about 100 PCs become infected with a virus for every 1,000 PCs on corporate networks. Figure 7.9 shows the typical sources of Internet attacks.[72] The percentages total more than 100 percent because those being surveyed gave multiple responses.

Figure 7.9

Typical Sources of Internet Attacks

(Source: Data from Riva Richmond, "How to Find Your Weak Spots," *The Wall Street Journal*, September 29, 2003, p. R3.)

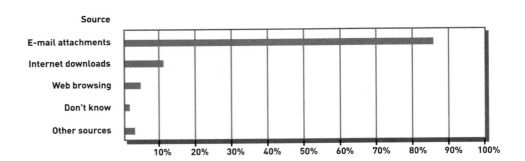

Although the U.S. federal government provided much of the early direction and funding for the Internet, the government does not own or manage it. The Internet Society and the Internet Activities Board (IAB) are the closest the Internet has to a centralized governing body. These societies were formed to foster the continued growth of the Internet. The IAB oversees a number of task forces and committees that deal with Internet issues. One of the main functions of the IAB is to manage the network protocols used by the Internet, including TCP/IP. Some universities and government agencies are investigating how the Internet can be controlled to prevent sensitive information and pornographic material from being placed on it.

Service and Speed Issues

Service and speed issues on the Internet are a function of the increasing volume of traffic and more sophisticated Web sites. The growth in Internet traffic continues to be phenomenal. Traffic volume on company intranets is growing even faster than the Internet. Companies setting up an Internet or intranet Web site often underestimate the amount of computing power and communications capacity they need to serve all the "hits" (requests for pages) they get from Web surfers. Web server computers can be overwhelmed with thousands of hits per hour. In addition, Web sites are becoming more sophisticated with video and audio clips and other features that require faster Internet speeds.

Routers, the specialized computers that send packets down the right network pathways, can also become bottlenecks. For each packet, every router along the way must scan a massive address book of about 40,000 area destinations (akin to Internet ZIP codes) to pick the right one. These routers can get overloaded and lose packets. The TCP/IP protocol compensates for this by detecting a missing packet and requesting the sending device to resend the packet. However, this leads to a vicious circle, as the network devices continually try to resend lost

packets, further taxing the already overworked routers. This leads to long response times or loss of the connection to the network.

Several actions are opening up the bottleneck. Various backbone providers have been upgrading their backbone links, installing bigger, faster "pipes" and converting to newer transmission technology, such as asynchronous transfer mode (ATM), which can send a message down the right path more quickly than standard packet-switching technology. Each ATM transmission is preaddressed with its own route, so routing addresses do not have to be looked up, and the packet can zip right through an ATM switch. Also, router manufacturers are developing improved models with increased hardware capacity and more efficient software to provide quick access to addresses. Yet a third solution is to prioritize traffic. Today, all network traffic travels through the same big backbone pipes. There is no way to make sure that your urgent message is not stalled behind someone downloading a magazine page. With prioritized service, customers could pay more for guaranteed delivery speed, much like an overnight package costs more than second-day delivery. If implemented, this solution could also affect the cost of network services that generate a lot of traffic, such as Internet phone and videoconference services. Finally, the increased availability of faster DSL, satellite, and cable connections, discussed in Chapter 6, are also speeding Internet access. The cost of these faster services is also falling.

Privacy, Fraud, Security, and Unauthorized Internet Sites

As the use of the Internet grows, privacy, fraud, and security issues become even more important. People and companies are reluctant to embrace the Internet unless these issues are successfully addressed.

Privacy

From a consumer perspective, the protection of individual privacy is essential. Yet, many people use the Internet without realizing that their privacy may be in jeopardy. A number of companies, including Jupiter Media Metrix and Nielsen/NetRatings, help other companies monitor visits to their Internet sites. These companies often call a random sample of Internet users to gain insights into the habits and desires of Internet users. Some companies hire people to visit chat rooms on the Internet to get important marketing information.

Spyware can hijack your browser, generate pop-up ads, and report your activities to someone else over the Internet. *Spyware* consists of hidden files and information trackers that install themselves secretly when you visit some Internet sites. Many Internet sites use cookies to gather information about people who visit their sites. A **cookie** is a text file that an Internet company can place on the hard disk of a computer system. These text files keep track of visits to the site and the actions people take. To help prevent this potential problem, some companies are developing software to prevent these files from being placed on computer systems. CookieCop, for example, allows Internet users to accept or reject cookies by Internet site. Microsoft's Internet Explorer 6 browser also has the ability to screen Web sites according to their privacy policy. Using the Platform for Privacy Preferences (P3P), Internet Explorer 6 can summarize the privacy policy for Web sites and prevent information from being transmitted from your computer to a Web site that doesn't meet certain criteria. In addition, preferences in Internet browsers can be set to restrict the use of cookies. Many browsers also allow users to easily delete cookies.

cookie
A text file that an Internet company can place on the hard disk of a computer system to track user movements.

Fraud

Internet fraud is another important issue. Some people have received false messages that seem to be from their Internet service providers asking them to update their personal information, including Social Security numbers and credit card information. But instead of going to the Internet service provider, the information is captured and used by online thieves. This type of Internet activity, often called *phishing*, is becoming more prevalent and dangerous. The possibility of Internet fraud has prevented many people from using the Internet. Local, state, and federal agencies are actively pursuing Internet fraud. As law enforcement agencies crack down on fraud, public confidence in using the Internet should increase.

Security with Encryption and Firewalls

When it comes to security on the Internet, it is essential to remember two things. First, there is no such thing as absolute security. Second, plenty of clever people consider it great sport to try to breach any security measures—the better your security, the greater the challenge to them.

From a corporate strategy perspective, security of data is essential. Such approaches as cryptography can help. **Cryptography** is the process of converting a message into a secret code and changing the encoded message back to regular text. The original conversion is called **encryption**. The unencoded message is called *plaintext*. The encoded message is called *ciphertext*. Decryption converts ciphertext back into plaintext (see Figure 7.10). For much of the Cold War era, cryptography was the province of military and intelligence agencies; uncrackable codes were reserved for people with security clearance only.

cryptography
The process of converting a message into a secret code and changing the encoded message back to regular text.

encryption
The conversion of a message into a secret code.

Figure 7.10

Cryptography is the process of converting a message into a secret code and changing the encoded message back into regular text.

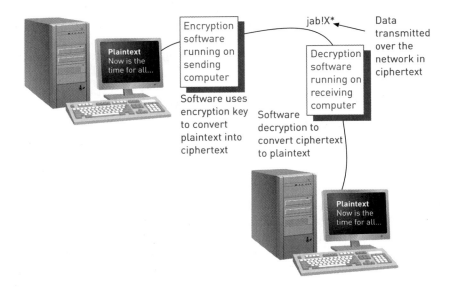

The Data Encryption Standard (DES), adopted as a federal standard in 1977 to protect unclassified communications and data, was designed by IBM and modified by the National Security Agency. It uses 56-bit keys, meaning a user must employ precisely the right combination of fifty-six 1s and 0s to decode information correctly. Other technologies offer a range of key lengths—in the case of RC5, up to 2,048 bits. The RSA protocol has no limit on key length, but it can slow transmission, since it uses separate keys for encryption and decryption. Many products mix technologies: They use a fast algorithm such as DES for the actual encryption but send the DES key through a more secure method such as RSA.

U.S. banks and brokerage houses use the federal government's DES algorithm to protect the integrity and confidentiality of fund transfers totaling trillions of dollars a day worldwide. Organizations encrypt the words and videos of their teleconferencing sessions. Individuals encode their electronic mail. And researchers use encryption to hide information about new discoveries from prying eyes.

Encryption is not just for keeping secrets. It can also be used to verify who sent a message and to tell whether the message was tampered with en route. A **digital signature** is a technique used to meet these critical needs for processing online financial transactions. Digital signatures involve a complicated technique that combines the public-key encryption method with a "hashing" algorithm that prevents reconstructing the original message. The hashing algorithm provides further encoding by using rules to convert one set of characters to another set (e.g., the letter *s* is converted to a *v*, *2* is converted to *7*, etc.). Thus, encryption also can prevent electronic fraud by authenticating senders' identities with digital signatures. A digital ID, for example, can be purchased by an individual. When the individual requests sensitive information from a Web site, the digital ID is sent to the site to confirm that individual's identity. A server certificate authenticates its site to users so they can be confident the Web

digital signature
Encryption technique used to verify the identity of a message sender for processing online financial transactions.

site is safe. Many Web sites display a padlock icon at the bottom of an Internet screen to indicate that the site is encrypted.

The most popular method of preventing unauthorized access to corporate computer data is to construct what is known as a firewall between company computers and the Internet. An Internet **firewall** is a device that sits between your internal network and the Internet. Its purpose is to limit access into and out of your network based on your organization's access policy. A firewall can be anything from a set of filtering rules set up on the router to an elaborate application gateway consisting of one or more specially configured computers that control access.

Firewalls permit desired services on the outside, such as e-mail, to pass. In addition, most firewalls allow access to the Web from inside protected networks. But firewalls deny other, unwanted access. For example, you may be able to use the Telnet utility to log in to systems on the Internet, but users on remote systems cannot log in to your local system because the firewall prevents it.

The U.S. Computer Emergency Readiness Team (*www.us-cert.gov*) responds to virus attacks, network intrusions, and other threats. It also provides security training and consulting services to individual agencies. Vulnerability and threat information is shared with the public. The *www.cert.org* Web site by Carnegie Mellon also contains information about Internet security.

firewall
A device that sits between an internal network and the Internet, limiting access into and out of a network based on access policies.

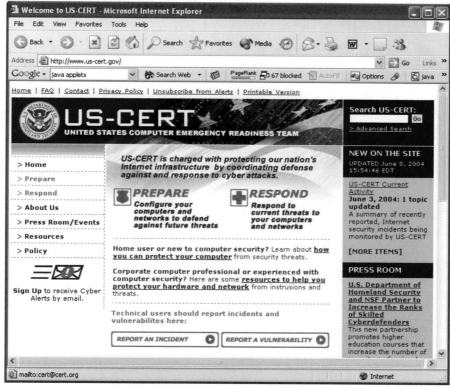

The U.S. Computer Emergency Readiness Team assists civilian government agencies with computer security incidents.

Unauthorized Sites

Unauthorized and unwanted Internet sites are also problems some companies face. A competitor or an unhappy employee can create an Internet site with an address similar to a company's. When someone searches for information about the company, he or she may find an unauthorized site instead. In some cases, the site may appear to be the legitimate, official corporate site. In others, it is obvious that the site is not sponsored or authorized by the company.

Some unauthorized or unwanted sites contain damaging information about a company. Sometimes the information is true, and sometimes it is false or misleading. A fired employee, for example, could post stories about his boss or former employer that may not be entirely

true. A competitor or an environmental watch group could also post information about why a customer should not do business with a company. As people travel the Internet, they should be aware that not all sites they see are endorsed by companies or organizations, so a wise Net surfer uses caution and some skepticism in accepting posted information as the whole truth. Unauthorized and unwanted Internet sites can be very troublesome, and it is not unusual for companies to sue those who post these sites. Network management issues will take an increasing amount of time for IS personnel, but any user needs to be aware of the basics to function effectively in business. Communications, service, and daily work are all at stake.

SUMMARY

Principle

The Internet is like many other technologies—it provides a wide range of services, some of which are effective and practical for use today, others are still evolving, and still others will fade away from lack of use.

The Internet started with ARPANET, a project started by the U.S. Department of Defense (DOD). Today, the Internet is the world's largest computer network. Actually, it is a collection of interconnected networks, all freely exchanging information. The Internet transmits data from one computer (called a *host*) to another. The set of conventions used to pass packets from one host to another is known as the Internet Protocol (IP). Many other protocols are used in connection with IP. The best known is the Transmission Control Protocol (TCP). TCP is so widely used that many people refer to TCP/IP, the combination of TCP and IP used by most Internet applications. Each computer on the Internet has an assigned address to identify it from other hosts, called its *uniform resource locator* (*URL*). There are several ways to connect to the Internet: via a LAN whose server is an Internet host, via SLIP or PPP, and via an online service that provides Internet access.

An Internet service provider is any company that provides individuals or organizations with access to the Internet. To use this type of connection, you must have an account with the service provider and software that allows a direct link via TCP/IP. Among the value-added services ISPs provide are electronic commerce, intranets, and extranets, Web-site hosting, Web transaction processing, network security and administration, and integration services.

Principle

Originally developed as a document-management system, the World Wide Web is a menu-based system that is easy to use for personal and business applications.

The Web is a collection of independently owned computers that work together as one. High-speed Internet circuits connect these computers, and cross-indexing software is employed to enable users to jump from one Web computer to another effortlessly. Because of its ability to handle multimedia objects and hypertext links between distributed objects, the Web is emerging as the most popular means of information access on the Internet today.

A Web site is like a magazine, with a cover page called a *home page* that has graphics, titles, and black and highlighted text. Web pages are loosely analogous to chapters in a book. Hypertext links are maintained using URLs, a standard way of coding the locations of the HTML (Hypertext Markup Language) documents. In addition to HTML, a number of newer Web standards are gaining in popularity, including Extensible Markup Language (XML), Extensible Hypertext Markup Language (XHTML), Cascading Style Sheets (CSS), and Dynamic HTML (DHTML).

A Web browser reads HTML and creates a unique, hypermedia-based menu on your computer screen that provides a graphical interface to the Web. The browser uses data about links to accomplish this; the data is stored on the Web server. The hypermedia menu links you to other Internet resources, not just text documents, graphics, and sound files. Internet Explorer and Netscape are examples of Web browsers. A search engine helps find information on the Internet. Popular search engines include Yahoo! and Google. A meta-search engine submits keywords to several individual search engines and returns the results. Push technology is used to send information automatically over the Internet rather than making users search for it with their browsers.

Internet and Web applications include e-mail; career information and job searching; Telnet; FTP; Web logs (blogs); usenet and newsgroups; chat rooms; Internet phone; Internet video; content streaming; instant messaging; shopping on the Web; Web auctions; music, radio, and video; office on the Web; 3-D Internet sites; free software; and other applications. E-mail is used to send messages. The Internet offers a vast amount of career and job search information. Telnet enables you to log on to remote computers. FTP is used to transfer a file from another computer to your computer. Web logs (blogs) are Internet sites that people and organizations can create and use to write about their observations, experiences, and feelings on a wide range of topics. Usenet supports newsgroups, which are online discussion groups focused on a particular topic. Chat rooms let you talk to dozens of people at one time, who can be located all over the world. Internet phone service enables you to communicate with others around the world. Internet video enables people to conduct virtual meetings. Content streaming is a method of transferring multimedia files over the Internet so that the data stream of voice and pictures plays continuously. Instant messaging allows people to communicate in real time using the Internet. Shopping on the Web is popular for a host of items and services. Web auctions are a way to match people looking for products and services with people selling these products and services. The Web can also be used to download and play music, listen to radio, and view video programs. With office on the Web, it is possible to store important files and information on the Internet. When telecommuting or traveling, these files and this information can be downloaded or sent to other people. Some Internet sites are three-dimensional, allowing people to manipulate the site to see different views of products and images on the Internet. A wealth of free software and services are available through the Internet. Some of the free

information, however, may be misleading or even false. Some of the other Internet services include information about space exploration, fast information transfer, obtaining a home loan, and distance learning.

Principle

Because the Internet and the World Wide Web are becoming more universally used and accepted for business use, management, service and speed, privacy, and security issues must continually be addressed and resolved.

A rapidly growing number of companies are doing business on the Web and enabling shoppers to search for and buy products online. For many people, it is easier to shop on the Web than search through catalogs or trek to the shopping mall.

The steps to creating a Web page include getting space on a Web server; getting a Web browser program; writing your copy with a word processor, using an HTML editor, editing an existing HTML document, or using an ordinary text editor to create your page; opening the page using a browser, viewing the result, and correcting any tags; adding links to your home page to take viewers to another home page; adding pictures and sound; uploading the HTML file to your Web site; reviewing the Web page to make sure that all links are working correctly; and advertising your Web page. A number of products make developing and maintaining Web content easier, such as Microsoft's .NET Framework. Once a Web site has been constructed, a content management system (CMS) can be used to keep the Web site running smoothly. Web services are also used to develop Web content. Web services consist of a collection of standards and tools that streamline and simplify communication among Web sites, which could revolutionize the way people develop and use the Web for business and personal purposes. In addition to XML, Web services use the Simple Object Access Protocol (SOAP), Web Services Description Language (WSDL), and Universal Discovery Description and Integration (UDDI).

Java is an object-oriented programming language from Sun Microsystems based on C++ that allows small programs—applets—to be embedded within an HTML document. When the user clicks on the appropriate part of the HTML page to retrieve it from a Web server, the applet is downloaded onto the client workstation, where it begins executing. The development of Java has had a major impact on the software industry. In addition to Java, there are a number of other programming languages and tools used to develop Web sites, including Hypertext Preprocessor (PHP).

An intranet is an internal corporate network built using Internet and World Wide Web standards and products. It is used by the employees of an organization to gain access to corporate information. Computers using Web server software store and manage documents built on the Web's HTML format. With a Web browser on your PC, you can call up any Web document—no matter what kind of computer it is on. Because Web browsers run on any type of computer, the same electronic information can be viewed by any employee. That means that all sorts of documents can be converted to electronic form on the Web and constantly be updated.

An extranet is a network that links selected resources of the intranet of a company with its customers, suppliers, or other business partners. It is also built around Web technologies. Security and performance concerns are different for an extranet than for a Web site or network-based intranet. User authentication and privacy are critical on an extranet. Obviously, performance must be good to provide quick response to customers and suppliers.

Management issues and service and speed affect all networks. No centralized governing body controls the Internet. Also, because the amount of Internet traffic is so large, service bottlenecks often occur. Privacy, fraud, and security issues must continually be addressed and resolved. Cryptography techniques and firewalls are required to combat information thieves and provide as much security as possible. Some unauthorized and unwanted Internet sites contain damaging information about companies. These sites can be placed on the Internet by unhappy employees, competitors, or other individuals and groups.

CHAPTER 7: SELF-ASSESSMENT TEST

The Internet is like many other new technologies—it provides a wide range of services, some of which are effective and practical for use today, others are still evolving, and still others will fade away from lack of use.

1. The _____ was the ancestor of the Internet. It was developed by the U.S. Department of Defense.

2. On the Internet, what enables traffic to flow from one network to another?

   a. Internet Protocol
   b. ARPANET
   c. Uniform Resource Locator
   d. LAN Server

3. Each computer on the Internet has an address called the *Transmission Control Protocol.* True or False?
4. Which of the following is NOT a way to gain access to the Internet?

    a. LAN Server
    b. usenet
    c. Point-to-Point Protocol, or PPP
    d. online service

5. _____ is a protocol that describes a file transfer process between a host and remote computer.

Originally developed as a document-management system, the World Wide Web is a menu-based system that is easy to use for personal and business applications.

6. A Web log, also called a blog, is an online Web site that people can create and use to write about their observations, experiences, and feelings on a wide range of topics. True or False?
7. What allows two or more people to engage in online, interactive "conversation" over the Internet?

    a. content streaming
    b. chat rooms
    c. newsgroups
    d. usenet

8. _____ can be used to route phone calls over networks and the Internet.

9. What is the standard page description language for Web pages?

    a. Home Page Language
    b. Hypermedia Language
    c. Java
    d. Hypertext Markup Language (HTML)

Because the Internet and the World Wide Web are becoming more universally used and accepted for business use, management, service and speed, privacy, and security issues must continually be addressed and resolved.

10. A(an) _____ is a network based on Web technology that links customers, suppliers, and others to the company.
11. A cookie is a text file that an Internet company can place on the hard drive of a computer system. True or False?
12. What sits between an internal network or computer and the Internet to prevent unauthorized access to a computer system?

    a. digital signature
    b. firewall
    c. extranet
    d. internet

13. _____ is used to send information automatically over the Internet.

CHAPTER 7: SELF-ASSESSMENT TEST ANSWERS

(1) ARPANET (2) a (3) False (4) b (5) File Transfer Protocol (FTP) (6) True (7) b (8) Voice over IP (VoIP) (9) d (10) extranet (11) True (12) b (13) Push technology

KEY TERMS

applet 307
ARPANET 297
backbone 299
bot 322
chat room 320
content streaming 322
cookie 329
cryptography 330
digital signature 330
encryption 330
Extensible Markup Language (XML) 306
extranet 327
File Transfer Protocol (FTP) 317
firewall 331
home page 304

HTML tags 305
hypermedia 304
Hypertext Markup Language (HTML) 305
instant messaging 315
Internet 296
Internet service provider (ISP) 302
Internet Protocol (IP) 297
intranet 326
Java 309
meta-search engine 308
newsgroups 319
Point-to-Point Protocol (PPP) 302
push technology 311
search engine 307

Serial Line Internet Protocol (SLIP) 302
Telnet 317
Transmission Control Protocol (TCP) 299
tunneling 327
Uniform Resource Locator (URL) 299
usenet 319
virtual private network (VPN) 327
Web auction 323
Web browser 307
Web log (blog) 317
Web services 312
World Wide Web (WWW or W3) 304

REVIEW QUESTIONS

1. What is the Internet? Who uses it and why?
2. What is ARPANET?
3. What is TCP/IP? How does it work?
4. Explain the naming conventions used to identify Internet host computers.
5. What is a domain name?
6. Briefly describe three different ways to connect to the Internet. What are the advantages and disadvantages of each approach?
7. What is an Internet service provider? What services does one provide?
8. What are the advantages and disadvantages of e-mail?
9. What is a blog?
10. What are Telnet and FTP used for?
11. What is an Internet chat room?
12. What is content streaming?
13. What is instant messaging?
14. Briefly describe a Web auction.
15. What is the Web? Is it another network like the Internet or a service that runs on the Internet?
16. What is a URL and how is it used?
17. What is HTML and how is it used?
18. What is a Web browser? How is it different from a Web search engine?
19. What is push technology?
20. What is an intranet? Provide three examples of the use of an intranet.
21. What is an extranet? How is it different from an intranet?
22. What is cryptography?
23. What are firewalls? How are they used?

DISCUSSION QUESTIONS

1. Instant messaging is being widely used today. Describe how this technology could be used in a business setting. Are there any drawbacks or limitations to using instant messaging in a business setting?
2. Your company is about to develop a new Web site. Describe how you could use Web services for your site.
3. How can you protect yourself from unwanted e-mail?
4. Briefly describe how the Internet phone service operates. Discuss the potential impact that this service could have on traditional telephone services and carriers.
5. The U.S. federal government is against the export of strong cryptography software. Discuss why this may be so. What are some of the pros and cons of this policy?
6. Identify three companies with which you are familiar that are using the Web to conduct business. Describe their use of the Web.
7. What is voice over IP (VoIP), and how could it be used in a business setting?
8. Outline a process to create a Web page. What computer hardware and software do you need if you wish to create a Web home page containing both sound and pictures?
9. Describe how push technology can be used in a business setting.
10. One of the key issues associated with the development of a Web site is getting people to visit it. If you were developing a Web site, how would you inform others about it and make it interesting enough that they would return and also tell others about it?
11. Getting music, radio, and video programs from the Internet is easier than in the past, but some companies are still worried that people will illegally obtain copies of this programming without paying the artists and producers royalties. If you were an artist or producer, what would you do?
12. How could you use the Internet if you were a traveling salesperson?
13. Briefly summarize the differences in how the Internet, a company intranet, and an extranet are accessed and used.

PROBLEM-SOLVING EXERCISES

1. Do research on the Web to find several popular Web auction sites. After researching these sites, use a word processor to write a report on the advantages and potential problems of using a Web auction site to purchase a product or service. Also discuss the advantages and potential problems of selling a product or service on a Web auction site. How could you prevent scams on an auction Web site?

2. Develop a brief proposal for creating a business Web site. How could you use Web services to make creating and maintaining the Web site easier and less expensive? Develop a simple spreadsheet to analyze the income you need to cover your Web site and other business expenses.

3. You are a manager of a small company with a new Web site. How would you avoid privacy, fraud, and security problems? Develop a brief report describing what you would do. Using a graphics program, diagram how you would protect your Internet site from outside hackers.

TEAM ACTIVITIES

1. With your teammates, identify a company that is making effective use of a company extranet. Find out all you can about its extranet. Try to speak with one or more of the customers or suppliers who use the extranet and ask what benefits it provides from their perspective.

2. Have each team member use a different search engine to find information about blogs. Meet as a team and decide which search engine was the best for this task. Write a brief report to your instructor summarizing your findings.

WEB EXERCISES

1. This chapter covers a number of powerful Internet tools, including Internet phones, search engines, browsers, e-mail, newsgroups, Java, intranets, and much more. Pick one of these topics and get more information from the Internet. You may be asked to develop a report or send an e-mail message to your instructor about what you found.

2. The Internet can be a powerful source of information about various industries and organizations. Locate several industry or organization Web sites. Which one is the best? Why?

CAREER EXERCISES

1. Use the Internet to explore the starting salaries, benefits, and job descriptions for three career areas.

2. Describe how the Internet can be used on the job for two careers that interest you.

VIDEO QUESTIONS

Watch the video clip **Online Storage** and answer these questions:

1. What are the advantages of storing files using services such as *briefcase.yahoo.com*?

2. Do you think that Internet storage may become our primary form of storage in the future, over storing our files on local hard drives? What will it take, in terms of network performance, storage capacity, and price to make this a reality?

CASE STUDIES

Case One

Eastman Chemical Revamps with Web Services

Eastman Chemical Company is the world's largest producer of polyester plastics for packaging; a leading supplier of raw materials for paints, inks, and graphic arts; and a marketer of more than 1,200 chemicals, fibers, and plastics products. The company is, by all measures, successful, with recent annual sales revenue of $5.8 billion. Eastman has approximately 15,000 employees in more than 30 countries and manufacturing sites strategically located in 17 different countries.

Eastman maintains a set of application servers that provide information services to employees worldwide. The servers contain a variety of operating systems (OSs): a large IBM server runs a proprietary OS, and others run Microsoft Windows 2000 and Windows NT OSs. Recently, Eastman Chemical decided to revamp its information systems. It had been using packaged applications and wanted to develop its own custom applications to meet all the needs of its staff. Eastman IS specialists turned to Web service technologies as the ideal solution for connecting differing computer platforms over the Internet.

Web services provide standardized interfaces between diverse computer systems so that they can automatically make and fulfill data requests over the Internet. Before Eastman's IS crew could begin the development process, they first needed to dissect the ways their current systems were being used. To do that, they studied the company's application servers, assessing what the applications did, stripping off the user interfaces and exposing the application functions as services, says Carroll Pleasant, an associate analyst in Eastman's emerging digital technologies group.

Once the group defined the "services" that the current system provided, they interviewed the users to learn how the system could be improved. "The users will be the ones deciding what the business processes will be, rather than having the applications determine the business process for them," explained Pleasant.

In developing the new Web services–based systems, the team ran into many challenges. Although Web services technology has been around for a few years, the technology is still in its infancy and is just beginning to affect businesses. While many businesses are using isolated Web service applications for special tasks, not many have been brave enough to base an entire corporate information system on the technology. Much of the work Pleasant and his team are conducting is considered pioneering work. "It's going to take a long, long time for everything to switch over to Web services and a service-oriented architecture," Pleasant says. "We see the movement going on with almost all of our vendors. We're confident they're going this route. But it takes time to get there."

In the meantime, Eastman employees from around the world are enjoying the first of the Web services rollouts. The new applications include rich graphical user interfaces that "deploy like a Web application," Pleasant says. "The user just clicks on a shortcut and points to the application running on the server." From any Eastman location in the world, the Web service communicates with the server and quickly delivers the requested information.

Web services are destined to change the landscape of the Internet. Services that support a wide variety of needs are already passing their data back and forth over the Internet in high volumes. Applications as sophisticated as storage management and customer relationship management and as simple as stock quotations are employing Web services. Web services may use either a client/server or peer-to-peer architecture. As Web services proliferate, some people are becoming concerned that the services' demands on bandwidth may overcome the ability of the Internet to support them.

Discussion Questions

1. What advantages do Web services promise Eastman Chemical Company over its previous information systems?

2. What risks are Pleasant and his crew taking by being the first to deploy the new technology? What benefits might they enjoy?

Critical Thinking Questions

3. How do Web services meet the unique challenges, such as those that Eastman faces, of running applications on the Web?

4. How does the "services" approach of Web services make sense for achieving high-quality information systems?

Case Two

Sentry Insurance Provides Secure VPN Access

In the early days of computing, businesses and organizations maintained their private corporate data and information systems within the boundaries of their physical office space. Over time, privately owned computer networks allowed businesses to extend the reach of their corporate information systems to select off-site locations such as branch and partners' offices. The global growth of networks assisted companies in establishing offices around the world and accessing data from any location.

Although traditional private networks provided a means by which data could flow throughout the world, they were still limited in their scope. For example, employees were only able to access corporate data from their workplace and not, for example, from their home or the neighborhood coffee shop. In many ways, the limitations of these private networks were considered necessary, and perhaps even a benefit. After all, if private corporate information was available from any location, how could it remain private?

When the Internet became available for commercial and public use, the benefits and dangers were all too clear to most businesses. Careful consideration was given to ways in which private networks could be connected to the Internet without sacrificing security and privacy. As Internet access made its way into private homes, so did the electronic pathways to private corporate networks. The benefits were obvious to both employers and employees. Network technicians set out to develop a method by which private data could be passed over a public network—the Internet—and still remain secure. The solution was the virtual private network (VPN). VPN technology became even more attractive as the Internet went wireless. Now employees could log in to the corporate network from any location using a wireless portable device.

Like most global companies, Sentry Insurance wanted to provide its employees with off-site access to corporate records. Until recently the Stevens Point, Wisconsin–based company only provided such remote access to its IS staff and a few select "power users." But the staff received repeated requests from managers and employees for VPN access, so they embarked on a project that would provide such access to their tens of thousands of users.

The initial impulse to provide traditional VPN software to all employees was quickly nixed by Eliot Irons, information security manager at Sentry. "The insurance industry is highly regulated, and there's a very high awareness of keeping our customers' data safe. I'm trying to give everybody peace of mind," explained Irons. A VPN requires installation on each remote workstation, and once in place it provides that person with broad network access. The traditional VPN was simply too costly, complex, and risky in terms of security. The company found an attractive alternative in a Secure Sockets Layer (SSL) VPN. This less-powerful version of a VPN allows employees to access e-mail and applications over the Web through a standard Web browser. "Instead of giving users a thick VPN client, we just point them at a URL," explained Irons.

As an ex-security officer for the U.S. military, Irons was well aware of techniques for keeping the company's data secure. Besides being encrypted through SSL, packets were checked at multiple firewalls to determine whether they had legitimate or deviant intent. Sentry's private network remains virtually unhackable, even though it is accessible over the Internet, due to sophisticated hardware used at the junction where the Internet connects to the company's LAN. The hardware is called AirGap, and it separates the internal and external networks by passing packets through momentary connections.

Irons has delivered satisfaction at two levels: convenience and security. Today, 1,500 Sentry employees and about 20,000 customers use the new VPN system to access Sentry's claims databases, company intranet, Outlook e-mail, and 401(k) program portal. Sentry's agents are able to work from home and other remote locations, accessing extremely private valuable information through a secure, password-protected interface. With most North American employees on board, Irons is now focusing on bringing Sentry's overseas offices online.

Discussion Questions

1. What advantages and disadvantages does VPN technology offer the employees of Sentry Insurance?

2. What advantages and disadvantages has VPN technology provided Sentry Insurance as a company?

Critical Thinking Questions

3. What additional security measures should an employee take when accessing a private corporate network over a VPN as opposed to accessing it from a private corporate office?

4. What traditional work habits might be altered because of VPN technology? What new working styles might emerge?

SOURCES: Carol Sliwa, "Enterprises Take Early Lead in Web Services Integration Projects," *Computerworld*, June 16, 2003, *www.computerworld.com*; Eastman Chemical Company profile, *www.eastman.com/About_Eastman/The_Company/Company_Profile.asp*, accessed March 19, 2004; W3C Web Services Activity Web site, *www.w3.org/2002/ws/*, accessed March 19, 2004.

SOURCES: Toni Kistner, "Ramping Up Remote Access for All," *Network World*, October 1, 2003, *www.networkworld.com*; Sentry Insurance Web site, *www.sentry.com*, accessed March 19, 2004; Whale Communications Web site, *www.communications.com*, accessed March 19, 2004.

Case Three

Best Western: First to Provide Free Internet

Internet access has become a necessity for today's business traveler. Those in the hospitality industry have picked up on this fact and are working to profit from it. Many well-known hotel chains are providing Internet hookups in rooms and profiting from their hourly usage rates. But Best Western has taken a unique and different approach.

Best Western is using the value of high-speed Internet access to gain a competitive advantage by offering it to customers as a standard, free service. Best Western realizes that Internet access is more popular with customers than even local phone service—particularly since most business travelers carry their own cell phones that accommodate both local and long-distance calls. High-speed Internet access provides business travelers with valued access to e-mail, the Web, and corporate VPNs.

Best Western believes that travelers are tired of paying extra to access their e-mail and will appreciate this free, no-hassle service. Research seems to support this belief. A recent survey by Yesawich, Pepperdine, Brown & Russell showed that 65 percent of business travelers agree that free access to the Internet from their room is extremely or very desirable. Many hotel chains charge $10 per day for the service, and customers suffer not only from the expense but also from the inconvenience of an "activation procedure" that connects them to the Internet and automatically updates the customer's billing record in the database.

Best Western plans to offer free access in all of its 2,300 hotels in the United States, Canada, and the Caribbean. Best Western is the first major chain to offer this service free to its customers. According to Tom Higgins, CEO of Best Western, the company is responding to popular demand. "It's the number one amenity requested by virtually everyone, especially businesspeople." Best Western's initial goal is to wire at least 15 percent of the rooms in each hotel. It also plans to create wireless access in each hotel's public areas.

The rest of the hotel industry is watching. Best Western's move will no doubt influence the other large chains to follow suit. In the near future, business travelers will be assured of Internet access wherever they may roam. Free high-speed Internet access, combined with increased use of VPN technologies, will affect the way businesspeople work. Employers are more likely to send employees on the road, and employees will be less tethered to their offices since the office will virtually travel with each employee.

The hotel industry is just one of several industries working to keep people online. The transportation industry—trains, planes, and cruise lines—is offering high-speed Internet access, and restaurants, coffee shops, airports, and most places where people congregate are providing wireless high-speed access. While most are charging for the service, any might decide to follow Best Western's lead and offer the service for free—to get a leg up on the competition.

Discussion Questions

1. Do you think Best Western's free high-speed Internet service will win over customers? Why?
2. If Best Western's plan fails and does not improve its customer base, what will Best Western have lost in the effort? How might it recoup its losses?

Critical Thinking Questions

3. What expenses are incurred when a business provides high-speed Internet access to its customers?
4. Considering the expenses of providing the service, and assuming that half the customers use the service, how much profit is made from the service at $10 per day per customer? How might Best Western make up its loss of the $10-per-day service charge?

SOURCES: James Lewin, "Will this Western Cause a Stampede?" *ITWorld*, January 28, 2004, *www.itworld.com*; "Best Western Plans Industry's Largest High Speed Internet Rollout," *PR Newswire*, November 3, 2003, *www.lexis-nexis.com*; Best Western Web site, *www.bestwestern.com*, accessed March 20, 2004.

NOTES

Sources for the opening vignette: Howard Baldwin, "Joshua Fost Remains Undaunted by Vast Global Infrastructure," *InfoWorld*, January 23, 2004, *www.infoworld.com*; "Colliers International Goes Live with Interwoven," *PR Newswire*, November 10, 2003, *www.lexis-nexis.com*; Interwoven Web site, *www.interwoven.com*, accessed March 13, 2004; Molecular Web site, *www.molecular.com*, accessed March 13, 2004.

1. Green, Heather, "The Underground Internet," *Business Week*, September 15, 2003, p. 80.
2. Gomes, Lee, "Growth of the Internet," *The Wall Street Journal*, December 22, 2003, p. B1.
3. Jones, Adam, "Colleges Put Web On Warp Speed," *Rocky Mountain News*, February 16, 2004, p. 1B.
4. Fillion, Roger, "Wireless Internet Access is Popping Up in Unexpected Places," *Rocky Mountain News*, January 19, 2004, p. 1B.
5. Brand, Rachel, "Wireless to Wellness," *Rocky Mountain News*, February 2, 2004, p. 1B.
6. Grimes, Brad, "Leave the Driving to Zipcar," *PC Magazine*, February 17, 2004, p. 60.
7. Fillion, Roger, "Wireless Internet Access is Popping Up in Unexpected Places," *Rocky Mountain News*, January 19, 2004, p. 1B.
8. Morrissey, Brian, "DoubleClick Hit With Deceptive Ad Suit," *InternetNews.com*, July 22, 2003.

9. Rosenbush, Steve, "How The U.S. Lags," *Business Week*, March 1, 2004, p. 39.

10. Mullaney, Timothy, "At Last, The Web Hits 100 MPH," *Business Week*, June 23, 2003, p. 80.

11. Pringle, David, "Europe's Broadband Battle Spans Borders," *The Wall Street Journal*, January 15, 2004, p. B4.

12. Howes, Tim, "Managing Data Centers Through XML," *Computerworld*, January 26, 2004, p. 30.

13. Taylor, Josh, "Web Stars," *PC World*, February 2004, p. 97.

14. Jesdanum, Anick, "Phones Juiced for Java," *Rocky Mountain News*, July 14, 2003, p. 6B.

15. Vijayan, Jaikumar, "Best In Class: Application Framework Allows Easy Portal Access," *Computerworld*, February 24, 2003, p. 51.

16. Alexander, Steve, "Best In Class: Web Site Adds Inventory Control and Forecasting," *Computerworld*, February 24, 2003, p.45.

17. Pflughoeft, Kurt, et al., "Multiple Conceptualization of Small Business Web Use and Benefit," *Decision Sciences*, Summer 2003, p. 467.

18. Violino, Bob, "Waves of Change," *Computerworld*, May 19, 2003, p. 28.

19. Miller, Michael, "Web Services Build Momentum," *PC Magazine*, September 2, 2003, p. 8.

20. Akin, Jim, "Amazon Everywhere," *PC Magazine*, September 16, 2003, p. 70.

21. Gralla, Preston, "Web Services in Action," *Computerworld*, May 19, 2003, p. 36.

22. Coyle, Frank, "Web Services, Simply Put," *Computerworld*, May 19, 2003, p. 38.

23. Sliwa, Carol, "Users Proceed Cautiously On Web Services Track," *Computerworld*, December 31, 2003, p. 10.

24. Bulkeley, William, "Microsoft, IBM Set Standards Pact," *The Wall Street Journal*, September 18, 2003, p. B10.

25. Sliwa, Carol, "Group Releases Let of Guidelines for Web Services," *Computerworld*, August 18, 2003, p. 14.

26. Mossberg, Walter, "Checking Out E-Mail for Cheapskates," *The Wall Street Journal*, August 27, 2003, p. D1.

27. Lauerman, John, "E-Mail Experiment Supports 6 Degrees of Separation Theory," *Bloomberg News*, reprinted in *The Rocky Mountain News*, August 8, 2003, p. 27A.

28. Zaslow, Jeffery, "The Politics of the CC Line," *The Wall Street Journal*, May 28, 2003, p. D1.

29. Blackman, Andrew, "Spam's Easy Target," *The Wall Street Journal*, August 19, 2003, p. B1.

30. Metz, Cade, "Can E-Mail Survive," *PC Magazine*, February 17, 2004, p. 65.

31. Guth, Robert, "Sobig.F Virus Clogs PCs with E-Mail," *The Wall Street Journal*, August 20, 2003, p. B5.

32. Tynan, Daniel, "Uncle Sam Vs. Spam," *PC World*, August 2003, p. 123.

33. Mangalindan, Mylene, "California Gets Serious About Spam," *The Wall Street Journal*, September 24, 2003, p. A3.

34. Metz, Cade, "Can E-Mail Survive," *PC Magazine*, February 17, 2004, p. 65.

35. Langa, Fred, "E-Mail—Hideously Unreliable," *InformationWeek*, January 12, 2004.

36. Mossberg, Walter, "Apple Out to Raise Bar for Audio, Video," *Rocky Mountain News*, August 18, 2003, p. 5B.

37. Staff, "AOL Service Hit with Aggressive Pitch Tied to Game Link," *The Wall Street Journal*, February 12, 2004, p. D5.

38. Goldsborough, Reid, "Blogs," *Arrivals*, August-September 2003, p. 42.

39. Staff, "Wacky Questions," *Rocky Mountain News*, August 5, 2003, p. 2D.

40. Rosencrance, Linda, "Blogs Bubble Into Business," *Computerworld*, January 26, 2004, p. 23.

41. Verton, Dan, "Blogs Play a Role in Homeland Security," *Computerworld*, May 12, 2003, p. 12.

42. Ernst, Warren, "Building Blogs," *PC Magazine*, June 30, 2003, p. 58.

43. Dreier, Troy, "Blog On," *PC Magazine*, September 2, 2003, p. 154.

44. Taranto, James, "They Can Run, But Can They Blog?" *The Wall Street Journal*, August 6, 2003, p. A12.

45. Becker, Tom, "Internet Protocol Phone Sales Rising," *The Wall Street Journal*, September 23, 2003, p. A21B.

46. Prince, Marcelo, "Dialing for Dollars," *The Wall Street Journal*, May 19, 2003, p. R9.

47. Park, Andrew, "Net Phones Start Ringing Up Customers," *Business Week*, December 29, 2003, p. 45.

48. Staff, "Clear Signal for Internet Phones," *Business Week*, June 23, 2003, p. 84.

49. Roth, Daniel, "Catch Us If You Can," *Fortune*, February 9, 2004, p. 64.

50. Grant, Peter, "Ready for Prime Time," *The Wall Street Journal*, January 12, 2004, p. R7.

51. McEnvoy, Aoife, "Time to Switch to a Net Phone," *PC World*, February 2004, p. 30.

52. Spencer, Jane, "I Ordered That?" *The Wall Street Journal*, September 4, 2003, p. D1.

53. Mangalinsan, Mylene, "Yahoo Hopes New Service Turns Searchers Into Shoppers," *The Wall Street Journal*, September 23, 2003, p. B1.

54. Higgins, Michelle, "Outsourcing Your eBay Auctions," *The Wall Street Journal*, February 24, 2004, p. D2.

55. Bank, Don, "eBay to Allow Customers to Pay by Loyalty Points," *The Wall Street Journal*, May 20, 2003, p. B7.

56. Warner, Melanie, "eBay's Worst Nightmare," *Fortune*, May 26, 2003, p. 89.

57. Gomes, Lee, "RIAA Takes Off the Gloves," *The Wall Street Journal*, September 15, 2003, p. B1.

58. Hoffman, Ellen, "Office Help From Afar," *Business Week*, September 15, 2003, p. 100.

59. Luhn, Robert, "Best Free Stuff on the Web," *PC World*, August 2003, p. 88.

60. Konrad, Rachel, "Online Voting Still Faces Hurdles," *The Denver Post*, January 25, 2004, p. 4A.

61. Rubenking, Janet, "Your Library Online," *PC Magazine*, February 17, 2004, p. 54.

62. Bialik, Carl, "AOL Offers Discount Movies as a Broadband Lure," *The Wall Street Journal*, January 21, 2004, p. D4.

63. Brandt, Andrew, "Network Cameras for All," *PC World*, February 2004, p. 60.

64. Kandra, Anne, "Can You Profit from Online Networking?" *PC World*, February 2004, p. 45.

65. Kelley, Joanne, "All in the (Displaced) Family," *Rocky Mountain News*, February 17, 2004, p. 7B.

66. Schoenberger, Chana, "Everyone Is Trying to Sell Digital Convergence to Homeowners," *Forbes*, February 2, 2004, p. 90.

67. Clark, Don, "Gadgets Grow Up," *The Wall Street Journal*, January 8, 2004, p. D1.

68. Wood, Richard, "Intranet in Staff Hands," *New Zealand Infotech Weekly*, "February 9, 2004, p. 3.

69. Knights, Miya, "Intranet Consolidates Utility's Legacy Systems," *Computing*, January 29, 2004, p. 11.

70. Staff, "Bringing the Outside In," *New Media Age*, February 26, 2004, p. 25.

71. Fielding, Rachel, "WHSmith News Extranet Helps Spot Sales Trends," *Computing*, January 8, 2008, p. 14.

72. Richmond, Riva, "How To Find Your Weak Spots," *The Wall Street Journal*, September 29, 2003, p. R3.

Australian Film Studio Benefits from Broadband Networks...and Location

Sigi Goode
The Australian National University

It's been a hectic 24 hours at Fire Is a Liquid Films, a small Australian film-production studio. The firm's 18 staff members have just finished rendering scenes from three feature films, pitched ideas for two TV commercials, and delivered two milestone [progress] updates to production clients. Thanks to the power of broadband networking, the employees have accomplished most of this work without leaving their small studio. Of his company's tight deadlines and advanced technology, Toni Brasting, Managing Director, explains, "We play to our strengths in terms of location and technology.

"Our work ranges from cleaning up footage to improve colour or contrast, up to full-blown CGI modelling and rendering. We work with TV commercials right up to feature film houses, and from one or two frames up to entire scenes. We can capitalise on our geographic location. With most feature films being shot in the U.S. or Europe, [clients] can send us their digital footage before they go to bed, and we can work on the scene during our daytime. By the time they wake up, the processed footage is ready and waiting for them. We rely heavily on our broadband network infrastructure to move digital data around. Depending on the quality and duration of the footage, the upload process can take anywhere from a few minutes for a brief scene to a few hours for an entire feature at DVD quality. Most of our clients prefer to deal with individual scenes in this way because it allows them to shuffle their work flow priorities around.

"Most films and commercials used to be shot on 35mm analogue film, which can take time to process. The trend is moving more and more towards digital video because it can be easily edited and manipulated. With 35mm, if you wanted to bid on work with a film studio, it helped if you were fairly close by so [that] you could have access to the film reels for digitising purposes. With the advent of digital video and broadband networking, we can bid on projects anywhere in the world, from the *Matrix* to *28 Days Later*. The actors can get a better feel for each scene, and the director can place each scene in greater context: their vision comes to life quicker.

"For editing, we mostly use Final Cut Pro on our Macintosh systems—the Macs are easy to use, and they look good, too (which impresses clients when they tour the studio). They run Mac OS X, which was originally based on BSD Unix. Occasionally, we need to develop a new technique for a project, so it's useful to have access to software source code. Mac OS X gives us good support for our own open-source Linux tools.

"Another technique we use is laptop imaging. Clients like the feeling that we're taking care of them and that they are our focus. When one of our teams goes to meet a client for a pitch or a milestone update, they can pull a laptop off the shelf and build a customised hard-drive image right from their desktop with Norton Ghost or one of our home-brewed imaging applications. The image software 'bakes' a hard-drive image on the fly, which includes the OS, editing software, and relevant film footage. If necessary, we can also include correspondence and storyboards—we can have that client's entire relationship history right there in the meeting. We're also about to start phasing in tablet PCs with scribe pens so that we can work on potential storyboards during the meeting itself.

"The other important part of our processing platform is our render farm. It's basically a room full of identical Linux-based PCs, which work on a problem in parallel. We use commodity PC hardware and high-speed networking to allow each machine to work on the problem at once. It means we can get supercomputer performance at a fraction of the cost. We use our desktop machines to edit in draft resolution (which is of inferior quality but much faster to render or preview). When we're happy with the cut, we 'rush it' to the render farm to process the final piece of footage. We also offer render farm leasing services, so other film companies can send us their raw model files, use the render farm over the network, and then download their completed footage. We then invoice them for processing time—they never have to leave their offices."

Discussion Questions

1. How might open-source software benefit a small business? What problems might a small business encounter when using open-source software?
2. The studio profiled in this case processes tasks during a client firm's downtime. What other tasks could be accomplished more inexpensively or more efficiently by using providers in other geographic locations?
3. What other types of firms could also benefit from a render farm? What firms do you know of that actively make use of this technology?

Critical Thinking Questions

4. Are modern firms too reliant on electronic networking? If so, what can be done about this drawback? What paper-based systems could the firm use if its electronic systems fail?
5. How might wireless networking improve the operation of systems such as this firm's render farm?

Note: All names have been changed at the request of the interviewee.

Virtual Learning Environment Provides Instruction Flexibility

Vida Bayley
Coventry University, United Kingdom

One of the major components of Coventry University's teaching and learning strategy has been the introduction and use of a Virtual Learning Environment (VLE) utilising the WebCT (Web Course Tools) product platform. The Joint Information Systems Committee (JISC) in the UK has set out a definition of a Virtual Learning Environment and states that it refers to the components that support online interactions of various kinds taking place between learners and tutors.

A central aim of the university was to enhance face-to-face teaching and to offer greater flexibility to both teaching staff and students. It created a central support unit— CHED (Centre for Higher Education Development)—to introduce and develop new initiatives in teaching and learning across the university. In addition, a task force of academic staff was established to lead and facilitate these initiatives in collaboration with CHED. Through these developments and with the support of the university's Computing Services unit, WebCT was implemented as a core instrument in achieving the university's educational vision. The WebCT application allows course material and resources to be placed on the Web and provides staff and students with a set of tools that can be customised to course and module requirements. In brief, it also contains the following functions, which JISC considers to be essential features of a VLE:

- controlled access to curriculum with mapping to elements that can be individually assessed and recorded
- tracking mechanisms for student activities and performance
- support of online learning through content development
- communication and feedback mechanisms between various participants in the learning process
- incorporation of links to other internal/external systems

In line with the aims and goals of the VLE strategy, Coventry Business School has been active for a number of years in the development of curricula and modules to create innovative learning communities and to engage students in the learning process. One of the most successful programmes the Business School developed was the B.A. in Business Enterprise (B.A.B.E.) degree. The Business Enterprise course was a learning programme designed to develop graduates who would be able to proactively manage business-related tasks, solve business problems, and influence the business environment in which they work. A central element of the programme was that it would enable students to lay the foundation for lifelong learning, in which students take responsibility for continual learning and personal development through a variety of media. The incorporation of WebCT into the B.A.B.E. programme marked a significant departure from established methods of teaching, learning, and assessment in the Business School.

Although the school had previously used a variety of computer-based activities within the learning environment, their use was limited to individual modules and in many cases even restricted. In accounting and information technology courses, software products such as EQL tutorials (EQL International Ltd is a developer of e-learning and computer-based assessment solutions) were used as supplementary teaching material and for online assessment or self-test exercises. The WinEcon package (PC-based introductory economics software) had been developed with Bristol University through a consortium of eight UK economics departments and was used by the Economics subject group in course teaching. In other subject areas, the teaching staff provided students with Web URLs related to particular reading or assessment topics, which could be easily accessed via the Internet. In contrast, the Business Enterprise programme was designed for delivery via the World Wide Web, so module delivery, programme management, and staff/student communication were all built around WebCT.

This programme was the first to be offered by the university via the World Wide Web. And it was significant because it was not designed solely for distance learning but rather to make use of the range of tools offered by the WebCT platform for teaching and learning both within a classroom setting and off campus via the Net. WebCT tools were used to structure and to provide Web-based access to course and module materials, and communications facilities were adapted to dovetail with existing classroom-based teaching and learning. Of particular importance was the consideration by the teaching team of how the Web software features could add value to the traditional teaching/learning modes and methods that they would be supporting or replacing. This evaluation in turn spurred individual lecturers to examine their teaching practice, stimulated ideas about the applicability of Web-based teaching, and encouraged staff to develop their knowledge of and skills related to the university's intranet and the Internet. Once these opportunities were identified and placed within the context of the module and course objectives, lecturers then customised the WebCT features and integrated them within the overall teaching and learning framework. The benefit to students is flexibility: Course and module content is accessible from outside the university and outside of teaching hours. Course materials either can be placed within WebCT as HTML documents (so they can be edited directly online) or can be uploaded as individual applications, such as PowerPoint slides, Word documents, and Excel spreadsheets.

Effective communication and feedback are important aspects of a successful learning environment. One way of facilitating peer group communication and communication between students and staff was to use the discussion forum/bulletin board facilities provided by the software. The B.A.B.E. programme used this feature as the primary mode of delivering information to the student community. All course- and module-related announcements, such as class rescheduling, changes to assessment deadlines, and meeting arrangements, for example, were made via the main bulletin board. The flexibility that the software provides for creating discussion areas allowed lecturers to construct arenas for students to work individually on specific projects yet be involved as group members in discussions centred on common themes. Being asynchronous, these discussion groups allowed students to become involved in an ongoing exchange of ideas, reflect on their own and others' contributions, and thereby facilitate their own learning. In addition to the discussion forum, WebCT incorporates a chat facility that can enable a small group of students to participate in online, real-time conversations. An additional advantage of this type of synchronous communication is that both students and lecturers could develop group management and e-moderation skills. In addition, the Business School linked the Web e-mail facilities available in WebCT to the e-mail provided to all students by Coventry University, allowing messages to be transmitted within modules. This meant that messages were module specific and did not get lost in the larger system—especially in modules with large numbers of students.

The student presentation area of WebCT was used to allow students to create their own Web pages, which could be displayed and shared, thus providing them with an opportunity to gain new skills and an understanding of an important part of the processes involved in managing a business enterprise.

The Business School also uses STile, a portal that was developed to provide students with a range of resources that are available on the university's intranet. The portal contains an easy-to-use link to WebCT.

In the course of the development of the B.A.B.E. programme, it became apparent that for the success of new IT innovations, such as Web-based teaching, the project staff must identify and involve key stakeholders, particularly when a radical restructuring of teaching and learning methods is needed. These stakeholders include administrative personnel, teaching professionals, and also those who are at the very centre of learning activity—the student body. The university benefited greatly by actively involving students and staff in the design, use, and evaluation of their educational environment. The experience of the B.A.B.E. venture highlighted the fact that that in order to transform the students' learning experience, all the stakeholders needed to understand how to construct a fit between the needs and expectations of the learner, the skills and pedagogical tools of the practitioners, and the tools available within the VLE. The rewards have been great. The establishment of a VLE, with WebCT as its centre, has generated a new excitement in the approach to university teaching, stimulating reflection, reinvention, and transformation and creating a "journey of learning" for both students and teaching staff.

Although the B.A.B.E. degree programme has now been modified under the current course structure, many of its innovative features have been transferred and embedded into teaching and learning processes across the whole of the undergraduate business programme.

In addition, WebCT has enabled the Business School to introduce and develop with business partners new, forward-looking projects, such as the development and delivery of the Post Graduate Certificate, Post Graduate Diploma, and MA in Communications Management via the Cable and Wireless Virtual Academy. It has also played a significant role in the establishment of learning communities and development of work-based learning projects with public-sector organisations such as the National Health Service, local authorities, and private-sector enterprises.

Discussion Questions

1. Why do you think that a discussion forum would motivate you to use the Internet to access learning resources?
2. What do you think the use of e-mail facilities would contribute to your learning?

Critical Thinking Questions

3. Discuss the advantages that a VLE with Internet access could offer students in developing their learning. Are there any disadvantages?
4. Identify and discuss the different types of contributions that various stakeholders could make to a student's learning process.

PART · 3 ·

Business Information Systems

Chapter 8   Electronic Commerce
Chapter 9   Transaction Processing and Enterprise Resource Planning
                Systems
Chapter 10 Information and Decision Support Systems
Chapter 11 Specialized Business Information Systems: Artificial
                Intelligence, Expert Systems, Virtual Reality, and Other
                Specialized Systems

CHAPTER
· 8 ·

Electronic Commerce

PRINCIPLES	LEARNING OBJECTIVES
■ E-commerce is a new way of conducting business, and as with any other new application of technology, it presents both opportunities for improvement and potential problems.	■ Identify several advantages of e-commerce. ■ Outline a multistage model that describes how e-commerce works. ■ Identify some of the major challenges that companies must overcome to succeed in e-commerce. ■ Describe some of the current uses and potential benefits of m-commerce. ■ Identify several e-commerce applications.
■ E-commerce requires the careful planning and integration of a number of technology infrastructure components.	■ Outline the key components of technology infrastructure that must be in place for e-commerce to succeed. ■ Discuss the key features of the electronic payment systems needed to support e-commerce.
■ Users of e-commerce technology must use safeguards to protect themselves.	■ Identify the major issues that represent significant threats to the continued growth of e-commerce.
■ Organizations must define and execute a strategy to be successful in e-commerce.	■ Outline the key components of a successful e-commerce strategy.

INFORMATION SYSTEMS IN THE GLOBAL ECONOMY
SPRINT, UNITED STATES; AND SONY, JAPAN

Handheld Entertainment

The birth of telecommunications has revolutionized the manner in which we conduct business. Early telecommunications networks allowed businesses to carry out long-distance transactions for the first time. Similarly, the development and growth of the Internet supported widespread electronic commerce between businesses and consumers. Today, the explosion in the mobile communications market, along with the merger of voice and data networks, has brought e-commerce to consumers over their cell phones. Buying products and services has never been easier. Whenever and wherever the impulse may strike, a purchase is only a button-push away. The products that are most likely to succeed in this market are music and entertainment.

Sony has joined with Sprint to provide music products over its cellular network. Currently, the companies are focusing on a variety of ring tunes or ringers. Sprint is offering downloadable ringers created by the Sony Music Mobile Products Group over its PCS Vision network. Ringers range from animated polyphonic tones, to actual clips from popular songs, to specialized sound and voice recordings. For example, you could download a clip of your favorite rap or rhythm and blues artist or have Bugs Bunny notify you of an incoming call. The sound is accompanied by animated graphics on the phone's color digital display. Sprint customers apparently appreciate this customization; in a recent year, Sprint sold more than 20 million ringers and screen savers to a network of 3.2 million Sprint PCS Vision customers.

The companies are not stopping with ringers. Soon you may be able to use your cell phone to download and store full songs and then listen to them with headphones, or you could transfer them to your computer or entertainment center. Nokia is designing an application it calls "visual radio." Visual radio allows handsets to receive FM radio signals and matches the audio content with related pictures, graphics, and other content. The application provides a convenient means for listeners to purchase and download music they hear on the radio over their cell phone. "What we're bringing to the table with visual radio is impulse buying," says Reidar Wasenius, a senior project manager with Nokia's multimedia group. "You happen to hear something in a certain mood, and the radio station offers you the purchase opportunity. You do it there and then." Sprint is working on a similar technology that includes music videos and sports video clips.

Partnerships between record companies and telecommunications companies are multiplying, as the industries reconfigure to take advantage of this new and booming market. Music from BMG artists such as Britney Spears, Maroon5, Three Days Grace, the Strokes, Kenny Chesney, and Pink are available as ring tunes for $2.50 each.

If music is the forerunner in mobile e-commerce, then mobile games are next in line. Major electronic gaming providers such as SEGA, Namco, and THQ have partnered with Sprint to offer dozens of games that can be downloaded and played on the tiny cell phone display.

Security is another issue with wireless communications. Technologies provided by companies such as BREW and Qpass allow for safe and secure transactions between the customer and the service provider. Using a secure wallet application, customers can save and encrypt their personal and financial information on their cell phone, making purchasing as easy as entering a password.

Electronic commerce has evolved to support every type of transaction, from multimillion-dollar deals between corporate giants to micropayment transactions between people on the street. It is anticipated that by 2010 only 17 percent of transactions will be made with cash, 22 percent with checks, and 61 percent with electronic funds transfers, debit cards, and credit cards. With so much of our economy relying on e-commerce, the technology industry has stepped up its efforts to provide the necessary security and stability to support safe and efficient transactions.

As you read this chapter, consider the following:

- What opportunities and benefits has e-commerce provided businesses and consumers?
- What dangers threaten an economy that depends on telecommunications networks?

Why Learn About Electronic Commerce?

Electronic commerce has transformed many areas of our lives and careers. One fundamental change has been the manner in which companies interact with their suppliers, customers, government agencies, and other business partners. As a result, most organizations today have or are considering setting up business on the Internet. To be successful, all members of the organization need to participate in that effort. As a sales or marketing manager, you will be expected to help define your firm's e-commerce business model. Customer service employees can expect to participate in the development and operation of their firm's Web site. As a human resource or public relations manager, you will likely be asked to provide content for a Web site directed to potential employees and investors. Analysts in finance need to know how to measure the business impact of their firm's Web operations and how to benchmark that against competitors' efforts. Clearly, as an employee in today's organization, you must understand what the potential role of e-commerce is, how to capitalize on its many opportunities, and how to avoid its pitfalls. This chapter begins by providing a brief overview of the exciting world of e-commerce and defines its various components.

AN INTRODUCTION TO ELECTRONIC COMMERCE

business-to-consumer (B2C) e-commerce
A form of e-commerce in which customers deal directly with the organization, avoiding any intermediaries.

business-to-business (B2B) e-commerce
A form of e-commerce in which the participants are organizations.

Early e-commerce news profiled start-up companies that used Internet technology to compete with the traditional players in an industry. For example, Amazon.com challenged well-established booksellers Walden Books and Barnes and Noble. Like Amazon, the companies discussed in the opening vignette provide an example of **business-to-consumer (B2C) e-commerce**, in which customers deal directly with an organization and avoid any intermediaries. Other types of e-commerce are **business-to-business (B2B) e-commerce**, in which the participants are organizations, and **consumer-to-consumer (C2C) e-commerce**, which involves consumers selling directly to other consumers. General Electric Aircraft Engine is the world's leading producer of large and small jet engines for commercial and military aircraft. It operates a B2B Web site that allows users of its engines to order spare parts and maintenance supplies. eBay is an example of a C2C e-commerce site; customers buy and sell items directly to each other through the site.

Aside from the major categories of e-commerce, companies are also using Internet technologies to enhance their current operations, such as inventory control and distribution. But whatever model is used, successful implementation of e-business requires significant changes to existing business processes and substantial investment in IS technology.

Over the past few years, we have learned a lot about the practical limitations of e-commerce. It has become painfully clear that before companies can achieve profits, they must understand their business, their consumers, and the constraints of e-commerce. Although it once seemed so, selling consumer goods online in a virtual storefront may not always be a great way to compete. And inventing a new use for cutting-edge technology isn't necessarily enough to guarantee a successful business.

Still, e-commerce is not dead; it is maturing and evolving, with the focus currently shifted from B2C to B2B. E-commerce is a useful tool for connecting business partners in a virtual supply chain to cut resupply times and reduce costs. More than 80 percent of U.S. companies have experimented with some form of online procurement, although most are channeling less than 10 percent of their total procurement online, according to data from Forrester Research and the Institute for Supply Management.[1] In contrast, according to the U.S. Department of Commerce, B2C e-commerce sales accounted for just 1.9 percent of all retail sales. Yet, for 2003, B2C e-commerce sales rose by 26.3 percent—to an estimated $54.9 billion.[2] And the trend is spreading globally. Forrester Research predicts that online shopping revenue in European countries will rise from $40 billion in 2004 to $167 billion by 2009.[3]

Businesses and individuals use e-commerce to reduce transaction costs, speed the flow of goods and information, improve the level of customer service, and enable close coordination among manufacturers, suppliers, and customers. E-commerce also enables consumers and companies to gain access to worldwide markets. Dallas-based Aviall expanded to become a worldwide provider of supply chain management services for hundreds of aviation parts manufacturers and airlines. Its clients use Aviall's Web site for ordering, inventory control, and demand forecasting. To move its services online, the firm spent $40 million to develop a Web site and install several supporting information systems, including online purchasing, sales force automation, order entry, financial management, inventory control and warehouse management, product allocation, and purchasing forecasting systems. The Web site generates $60 million of the company's $800 million in annual revenue.[4]

E-commerce is not limited to use by manufacturing firms; many service firms and government agencies also use e-commerce successfully. The Department of Homeland Security (DHS) reached an agreement with the Pentagon to use the Defense Department's electronic mall to streamline the purchase of its goods and services. Established in 1998, the Pentagon's EMall is an online, one-stop-shopping center for military services and is one of the largest online government-to-business exchanges. Through the EMall, DHS agencies have access to 383 commercial catalogs containing more than 12 million items from Defense Supply Centers and the General Services Administration. Use of the Pentagon EMall is enabling DHS to buy goods and services through existing Defense Department contracts as well as contracts established exclusively for it.[5]

Business processes that are strong candidates for conversion to e-commerce are those that are paper-based and time-consuming and those that can make business more convenient for customers. Thus, some of the first business processes that companies converted to an e-commerce model were those related to buying and selling. For example, after Cisco Systems, the maker of Internet routers and other telecommunications equipment, put its procurement operation online in 1998, the company reported that it halved cycle times and saved an additional $170 million in material and labor costs. Similarly, Charles Schwab & Co. slashed transaction costs by more than half by shifting brokerage transactions from traditional channels such as retail and phone centers to the Internet.

Some companies, such as those in the automotive and aerospace industries, have been conducting e-commerce for decades through the use of electronic data interchange (EDI), which involves application-to-application communications of business data (invoices, purchase orders, etc.) between companies in a standard data format. Many companies have now gone beyond simple EDI-based applications to launch e-commerce initiatives with suppliers, customers, and employees to address business needs in new areas.

consumer-to-consumer (C2C) e-commerce
A form of e-commerce in which the participants are individuals, with one serving as the buyer and the other as the seller.

Because of the costs involved in buying new technology, the EDI capabilities of most small businesses are nonexistent or extremely limited. A few major retailers and manufacturers have enlisted the help of third parties to bring smaller firms into their EDI supply chain. For example, SPS Commerce specializes in hooking up small businesses like American Outdoor Products (25 employees) to big supply chains like Recreational Equipment Incorporated (REI). SPS built an Internet-based application that translates EDI ordering and shipping requirements so that workers can access them through a Web browser on their PC.

Multistage Model for E-Commerce

A successful e-commerce system must address the many stages that consumers experience in the sales life cycle. At the heart of any e-commerce system is the user's ability to search for and identify items for sale; select those items and negotiate prices, terms of payment, and delivery date; send an order to the vendor to purchase the items; pay for the product or service; obtain product delivery; and receive after-sales support. Figure 8.1 shows how e-commerce can support each of these stages. Product delivery may involve tangible goods delivered in a traditional form (e.g., clothing delivered via a package service) or goods and services delivered electronically (e.g., software downloaded over the Internet). A multistage model for purchasing over the Internet includes search and identification, selection and negotiation, purchasing, product or service delivery, and after-sales support, as shown in Figure 8.1.

Figure 8.1

Multistage Model for E-Commerce (B2B and B2C)

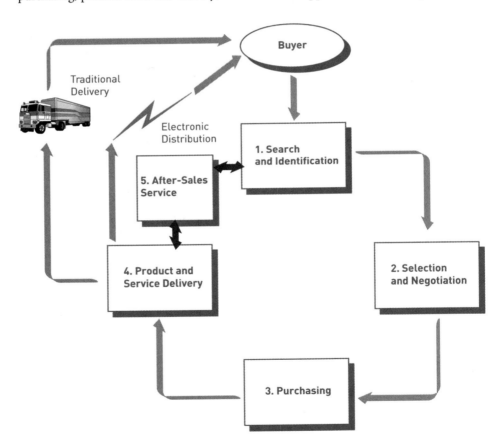

Search and Identification

An employee ordering parts for a storeroom at a manufacturing plant would follow the steps shown in Figure 8.1. Such a storeroom stocks a wide range of office supplies, spare parts, and maintenance supplies. The employee prepares a list of needed items—for example, fasteners, piping, and plastic tubing. Typically, for each item carried in the storeroom, a corporate buyer has already identified a preferred supplier based on the vendor's price competitiveness,

level of service, quality of products, and speed of delivery. The employee then logs on to the Internet and goes to the Web site of the preferred supplier.

From the supplier's home page, the employee can access a product catalog and browse until finding the items that meet the storeroom's specifications. The employee fills out a request-for-quotation form by entering the item codes and quantities needed. When the employee completes the quotation form, the supplier's Web application prices the order with the most current prices and shows the additional cost for various forms of delivery—overnight, within two working days, or the next week. The employee may elect to visit other suppliers' Web home pages and repeat this process to search for additional items or obtain competing prices for the same items. As mentioned in Chapter 7, bots are software programs that can follow a user's instructions; they can also be used for search and identification.

Selection and Negotiation

Once the price quotations have been received from each supplier, the employee examines them and indicates, by clicking on the request-for-quotation form, which of the items, if any, will be ordered from a given supplier. The employee also specifies the desired delivery date. This data is used as input into the supplier's order-processing TPS. In addition to price, an item's quality and the supplier's service and speed of delivery can be important in selection and negotiation.

Purchasing Products and Services Electronically

The employee completes the purchase order by sending a completed electronic form to the supplier. Complications may arise in paying for the products. Typically, a corporate buyer who makes several purchases from the supplier each year has established credit with the supplier in advance, and all purchases are billed to a corporate account. But when individuals make their first, and perhaps only, purchase from the supplier, additional safeguards and measures are required. Part of the purchase transaction can involve the customer providing a credit card number. However, computer criminals could capture this data and use the information to make their own purchases. To avoid this problem, some companies have developed security programs and procedures. For example, Secure Electronic Transactions (SET) is endorsed by IBM, Microsoft, MasterCard, and others. Another approach to paying for goods and services purchased over the Internet is using electronic money, which can be exchanged for hard cash; PayPal is one example.

The PayPal service of eBay is the dominant online payment system. Founded in 1998, PayPal, an eBay Company, enables any individual or business with an e-mail address to securely, easily, and quickly send and receive payments online. There are more than 40 million PayPal account members worldwide, and the service is available in 38 countries. Buyers and sellers on eBay, online retailers, online businesses, and traditional offline businesses are transacting with PayPal. To send money, you enter the recipient's e-mail address and the amount you wish to send. You can pay with a credit card or funds from a checking account. The recipient gets an e-mail that says, "You've Got Cash!" Recipients can then collect their money by clicking a link in the e-mail that takes them to *www.paypal.com*. To receive the money, the individual must have an account with PayPal. Also, because PayPal's transactions are fully electronic, the user also must have a credit card or checking account to transfer the funds to. To request money for an auction, invoice a customer, or send a personal bill, you enter the recipient's e-mail address and the amount you are requesting. The recipient gets an e-mail and instructions on how to pay you using PayPal.[6]

Product and Service Delivery

The Internet can also be used to deliver products and services, primarily software and written material. You can download software, reports on the stock market, information on individual companies, and a variety of other written reports and documents directly from the Internet. Often called *electronic distribution*, sending software, music, pictures, and written material through the Internet is faster and can be less expensive than with regular order processing. Electronic distribution can also eliminate inventory problems for manufacturers, who do not have to stock hundreds or thousands of copies of the software, reports, or documents; one copy can be downloaded to customers' computers when needed.

As more and more people use the Internet, the electronic distribution of products and services could be a major revenue source for software and publishing companies. Yet, most

products cannot be delivered over the Internet, so they are delivered in a variety of other ways: overnight carrier, regular mail service, truck, or rail. In some cases, the customer may elect to drive to the supplier and pick up the product.

Many manufacturers and retailers have outsourced the physical logistics of delivering merchandise to cybershoppers—the storing, packing, shipping, and tracking of products. To provide this service, DHL, Federal Express, United Parcel Service, and other delivery firms have developed software tools and interfaces that directly link customers' ordering, manufacturing, and inventory systems with their own system of highly automated warehouses, call centers, and worldwide shipping networks. The goal is to make the transfer of all information and inventory—from the manufacturer to the delivery firm to the consumer—fast and simple.

For example, when a customer orders a printer at the Hewlett-Packard Web site, that order actually goes to FedEx, which stocks all the products that HP sells online at a dedicated e-distribution facility in Memphis, a major FedEx shipping hub. FedEx ships the order, which triggers an e-mail notification to the customer that the printer is on its way and an inventory notice to HP that the FedEx warehouse now has one less printer in stock (see Figure 8.2). For product returns, HP enters return information into its own system, which is linked to FedEx. This signals a FedEx courier to pick up the unwanted item at the customer's house or business. Customers don't need to fill out shipping labels or package the item. Instead, the FedEx courier uses information transmitted over the Internet to a computer in his or her truck to print a label from a portable printer attached to his belt. FedEx has control of the return, and HP can monitor its progress from start to finish.

Figure 8.2

Product and Information Flow for HP Printers Ordered over the Web

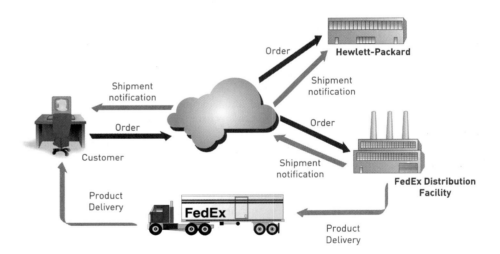

After-Sales Service

In addition to capturing the information to complete the order, comprehensive customer information is captured from the order and stored in the supplier's customer database. This information can include customer name, address, telephone numbers, contact person, credit history, and some order details. If, for example, the customer later telephones the supplier to complain that not all items were received, that some arrived damaged, or even that product use instructions are not clear, all customer service representatives can retrieve the order information from the database via a personal computer on their desks. Companies are adding the capability to answer many after-sales questions to their Web sites—how to maintain a piece of equipment, how to effectively use the product, how to receive repairs under warranty, and so on.

E-Commerce Challenges

A number of challenges must be overcome for a company to convert its business processes from the traditional form to e-commerce processes. This section summarizes a few.

The first major challenge is for the company to define an effective e-commerce model and strategy. Although a number of different approaches can be used, the most successful e-commerce models include three basic components: community, content, and commerce, as shown in Figure 8.3. Message boards and chat rooms are used to build a loyal *community* of people who are interested in and enthusiastic about the company and its products and services. Providing useful, accurate, and timely *content*—such as industry and economic news and stock quotes—is a sound approach to get people to return to your Web site time and again. *Commerce* involves consumers and businesses paying to purchase physical goods, information, or services that are posted or advertised online.

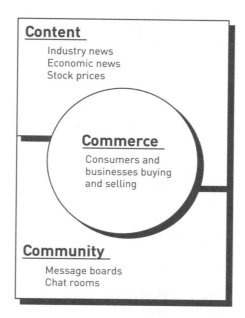

Figure 8.3

Three Basic Components of a Successful E-Commerce Model

A major challenge for companies moving to business-to-consumer e-commerce is the need to change distribution systems and work processes to be able to manage shipments of individual units directly to consumers. Traditional distribution systems send complete cases of a product to a store. The store opens the cases, takes the individual units out, and stacks them on a shelf. Then consumers walk through the aisles and pick up what they need. In business-to-consumer e-commerce, companies need a distribution system that can manage **split-case distribution**, in which cases of goods are split open on the receiving dock and the individual items are stored on shelves or in bins in the warehouse. The distribution system must also be able ship and track individual items. The demands of business-to-consumer e-commerce fulfillment are so great that many online vendors outsource the function to companies such as FedEx and UPS.

Another tough challenge for e-commerce is the integration of new Web-based order-processing systems with traditional mainframe computer–based inventory control and production planning systems. It's one thing to allow sales reps to place orders over the Web but another to let them find out what products are in (or soon to be in) inventory and available for sale. That means that front-end Web-enabled applications such as order taking need to be tightly integrated to traditional back-end applications such as inventory control and production planning, as shown in Figure 8.4.

split-case distribution
A distribution system that requires cases of goods to be opened on the receiving dock and the individual items from the cases to be stored in the manufacturer's warehouse.

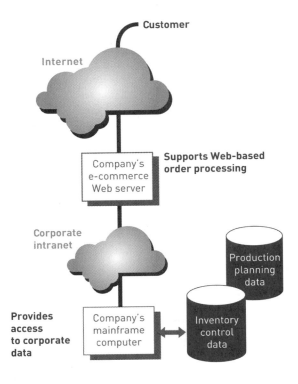

supply chain management
A key value chain composed of demand planning, supply planning, and demand fulfillment.

The E-Commerce Supply Chain

As discussed in Chapter 2, all business organizations contain a number of value-added processes. The supply chain management process is a key value chain that, for most companies, offers tremendous business opportunities if converted to e-commerce. **Supply chain management** is composed of three subprocesses: demand planning to anticipate market demand, supply planning to allocate the right amount of enterprise resources to meet demand, and demand fulfillment to fulfill demand quickly and efficiently (see Figure 8.5). The objective of demand planning is to understand customers' buying patterns and develop aggregate, collaborative long-term, intermediate-term, and short-term forecasts of customer demand. Supply planning includes strategic planning, inventory planning, distribution planning, procurement planning, transportation planning, and supply allocation. The goal of demand fulfillment is to provide fast, accurate, and reliable delivery for customer orders. Demand fulfillment includes order capturing, customer verification, order promising, backlog management, and order fulfillment.

Figure 8.5

Supply Chain Management

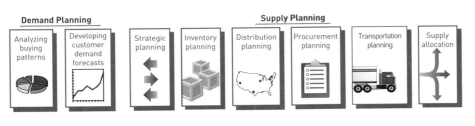

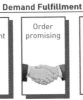

Conversion to e-commerce supply chain management provides businesses an opportunity to achieve operational excellence by increasing revenues, decreasing costs, improving customer satisfaction, and reducing inventory. But to achieve this goal requires integrating all subprocesses that exchange information and move goods between suppliers and customers, including manufacturers, distributors, retailers, and any other enterprise within the extended supply chain.

Increased Revenues and Decreased Costs

By eliminating or reducing time-consuming and labor-intensive steps throughout the order and delivery process, more sales can be completed in the same time period and with increased accuracy.

Improved Customer Satisfaction

Increased and more detailed information about delivery dates and current status can increase customer loyalty. In addition, the ability to consistently meet customers' desired delivery dates with high-quality goods and services eliminates any incentive for customers to seek other sources of supply.

Inventory Reduction Across the Supply Chain

With increased speed and accuracy of customer order information, companies can reduce the need for inventory—from raw materials, to safety stocks, to finished goods—at all the intermediate manufacturing, storage, and transportation points. For example, Nippon Steel uses supply chain management software from i2 Technologies to cut delivery lead times for products and reduce inventories for both itself and its customers. The software enables customers with Web capabilities to access ordering, production, and quality data from the company's steel materials, coil centers, and steel plant divisions. However, Nippon's supply chain management systems were put to the test when an explosion at its Nagoya steelworks in September 2003 forced the temporary shutdown of two blast furnaces that produce nearly one-fifth of its total output.[7] To ensure added protection against such unforeseen disruptions, some companies are increasing inventory levels above the minimum to make sure that they can meet changes in forecasted customer needs or unexpected production problems.

Through improved supply chain management, companies can set their sights not only on improving their profitability and service but also on transforming entire industries. For example, as mentioned in earlier chapters, radio-frequency identification (RFID) technology is improving supply chain management operations, reducing costs, lowering inventories, and providing improved customer service. The use of the technology has the potential to revolutionize supply chain operations in several industries. Major retailers and the U.S. Department of Defense have issued deadlines for their major suppliers to attach RFID tags to pallets and cases of various goods that they ship to them.[8] Airline carriers plan to use RFID tags to track air passengers' baggage.[9] Healthcare providers and pharmaceutical companies are even considering the use of RFID to track and control the use of prescription drugs.[10]

Business-to-Business (B2B) E-Commerce

Although the business-to-consumer (B2C) market grabs more of the news headlines, the business-to-business (B2B) market is considerably larger and is growing much more rapidly. Business-to-business e-commerce offers enormous opportunities. It allows manufacturers to buy at a low cost worldwide, and it offers enterprises the chance to sell to a global market right from the start. Moreover, e-commerce offers great promise for developing countries, helping them to enter the prosperous global marketplace, and hence helping reduce the gap between rich and poor countries. The entry into a global marketplace also comes with many challenges, as illustrated by the "Ethical and Societal Issues" special-interest feature.

The rapid development of e-commerce presents great challenges to society. Even though e-commerce is creating new job opportunities, it could also cause a loss of employment in traditional job sectors. Many companies may fail in the intense competitive environment of e-commerce and find themselves out of business. Therefore, it is vital that the opportunities and implications of e-commerce be understood.

Canadian Prescription Drug Web Sites

The globalization of e-commerce presents businesses with challenges above and beyond cultural adaptations. Along with convenience, it has brought fierce competition for goods and services. Online pharmacies may be the most dramatic example of this fact.

As with many industries, pharmaceutical companies price their products differently for different countries. Prices for prescription medications are determined by a country's economic well-being, its medical insurance structure, the amount of demand for medications, competition in the market, and other factors. The result is that Americans pay more for prescription drugs than consumers in most other countries. This fact became obvious to the public when Canadian pharmacies began selling prescription drugs on the Web.

Web sites such as *www.CanadaPharmacy.com*, *www.CanadaMeds.com*, and many others, showed Americans that their northern neighbors pay only 30 to 50 percent what is charged in the United States for the same brand-name drugs. Naturally, many U.S. citizens began purchasing their drugs online from Canadian pharmacies. The federal government took a stand against buying prescriptions from foreign countries, citing the lack of regulation as a potential health risk for customers. The FDA Web site states, "It is illegal for anyone, including a foreign pharmacy, to ship prescription drugs that are not approved by the FDA into the U.S. even though the drug may be legal to sell in that pharmacy's country."

Even while the FDA discouraged the practice, ever-increasing numbers of U.S. citizens began buying drugs online. New U.S.-based online businesses have also arisen, such as *www.BurlingtonDrugClub.com*, which act as intermediaries between U.S. citizens and Canadian pharmacies. Paid a percentage of sales by the Canadian pharmacies, these businesses assist Americans with finding the best price from the most reliable and trustworthy pharmacies. Several such businesses have been brought to court by the federal government but have, for the most part, survived the ordeal.

State governments, frustrated by the federal government's inability to provide their citizens with affordable prescription drugs, have begun researching ways to sidestep federal laws and obtain Canadian drugs for their state workers and senior citizens. More than 25 states are now exploring the mass purchase of Canadian drugs. The State of Wisconsin has set up a Web site that connects its citizens with Canadian pharmacies that it's found are safe.

Pharmaceutical companies are taking action to stop Canadian pharmacies from selling to anyone outside that country's borders. Such giants as Pfizer, GlaxoSmithKline, and Eli Lilly have systematically begun drug rationing in Canada. Canadians complain that the rationing barely leaves enough medicine to meet their own demand. Pfizer also cut off delivery to two Canadian prescription drug wholesalers because they were sending medicine back to the United States. Despite these efforts, U.S. prescriptions from Canada are growing by 18 percent annually and should top $1 billion in 2004.

Wisconsin Governor Jim Doyle, a Democrat, and Minnesota Governor Tim Pawlenty, a Republican, believe that the drug companies are breaking the law in drying up supply and called on U.S. Attorney General John Ashcroft to investigate. Meanwhile, some Americans are turning to countries besides Canada that also enjoy cheap drug prices. The American Drug Club Web site offers cheap prices on drugs from Britain, for instance.

Many in favor of global prescription medicine pricing are making serious accusations against the FDA. "We are here to remind the FDA they don't work for the big drug companies or their lobbyists," said Illinois Democratic Governor Rod Blagojevich.

The online prescription medicine business is indicative of the issues that we must contend with while building global e-commerce markets. How the U.S. government and the pharmaceutical companies resolve this issue will set a precedent for many other such cases that are sure to arise in the future.

Critical Thinking Questions

1. How effective can national laws be in controlling online purchases from other countries? What action could the federal government take if it wanted to put an end to reimported drugs?
2. What solution would you recommend that might satisfy the pharmaceutical companies, the FDA, and the states and their citizens?

What Would You Do?

As the promotional manager for LA Artists Inc., you are responsible for building an audience for 26 Los Angeles performers. One of the bands you represent has a huge local following, and you think it has the potential to make it really big. The band's Web site has helped build a national following, with huge sales in 20 states, which led to the band's recent national tour. LA Artists has gradually increased the price of CDs sold on the Web site to make the most of the band's recent popularity. The next step is to begin promoting the band in other countries. To build an international audience, you will need to entice those who have not yet heard the band with discounted prices on CDs. You have decided that $12 CDs sold online in the United States will be marked down to $5 for foreign sales.

3. Is it fair that U.S. citizens should be charged more than those outside the country to purchase the band's CDs?
4. How can you implement this e-commerce system so that international customers visiting the band's Web site will be charged less for CDs without offending U.S. fans? How can you be sure that U.S. citizens won't take advantage of the discounted prices?

SOURCES: Francis R. Carroll, "Rationing of Canadian Drugs Is Enough to Make You Sick," *Worcester Telegram & Gazette, Inc.*, March 29, 2004, *www.lexis-nexis.com*; "Prescription Drug Importation: Company That Helps

Purchase Canadian Drugs Opens in Oklahoma," *Health & Medicine Week*, February 16, 2004, *www.lexis-nexis.com*; Todd Richmond, "FDA Blasts Wisconsin Drug Web site," *The Associated Press*, March 18, 2004; "State of Wisconsin, Prescription Drug Resource Center" Web site, accessed April 17, 2004, *drugsavings.wi.gov*; FDA Web site, accessed April 17, 2004, *www.fda.gov*.

Business-to-Consumer (B2C) E-Commerce

E-commerce for consumers is gaining broad acceptance, although some shoppers are not yet convinced that it is worthwhile to connect to the Internet, search for shopping sites, wait for the images to download, try to figure out the ordering process, and then worry about whether their credit card numbers will be stolen by a hacker. But attitudes are changing, and an increasing number of shoppers are benefiting from the convenience of e-commerce. In time-strapped households, consumers are asking themselves, "Why waste time fighting crowds in shopping malls when from the comfort of home I can shop online anytime and have the goods delivered directly?" These shoppers have found that many goods and services are cheaper when purchased via the Web—for example, stocks, books, newspapers, airline tickets, and hotel rooms. They can also get information about automobiles, cruises, loans, insurance, and home prices to cut better deals. More than just a tool for placing orders, the Internet is emerging as a paradise for comparison shoppers. Internet shoppers can, for example, unleash shopping bots or access sites such as Google or Yahoo! to browse the Internet and obtain lists of items, prices, and merchants.

In March 2004, Yahoo! announced plans to buy European comparison-shopping site Kelkoo. France-based Kelkoo reaches about 10 percent of all European Internet users and counts more than 2,500 individual merchants among its paying customers. The site enables shoppers to compare prices on 3 million products in 25 categories, such as books, electronics, mobile phones, movies, music, and travel services. Yahoo!'s purchase of Kelkoo highlights the convergence of Web searches and online shopping, which will help Yahoo! achieve its goal to create the most comprehensive and best user experience on the Web. The purchase will also give Yahoo! another way to help marketers reach consumers by leveraging the Web's global reach. Yahoo! can now say to its biggest advertisers, companies such as Coca-Cola and General Motors, that they can reach global audiences through its portal.[11]

By using B2C e-commerce to sell directly to consumers, producers or providers of consumer services can eliminate the middlemen, or intermediaries, between them and the end consumer. In many cases, this squeezes costs and inefficiencies out of the supply chain and can lead to higher profits and lower prices for consumers. The elimination of intermediate organizations between the producer and the consumer is called **disintermediation**.

disintermediation
The elimination of intermediate organiations between the producer and the consumer.

Consumer-to-Consumer (C2C) E-Commerce

Consumer-to-consumer (C2C) e-commerce involves consumers selling directly to other consumers. Often this exchange is done through Web auction sites such as eBay, which enabled people to sell in excess of $8 billion in merchandise in one 3-month period in 2004 to other consumers by auctioning the items off to the highest bidder.[12] The growth of C2C is responsible for reducing the use of the classified pages of newspapers to advertise and sell personal items.

Global E-Commerce

The use of the Internet is growing rapidly in markets throughout Europe, Asia, and Latin America. So, e-commerce sites are broadening their focus from North American consumers.

Europe

The population of Europe is about 380 million. Although that number is expected to slowly decline, the number of Europeans who go online to purchase items is expected to increase.[13] Recently, a new value-added tax (VAT) law was enacted to level the playing field

between Europe- and U.S.-based companies selling in the European e-commerce market. Previously, U.S. e-commerce companies doing business in Europe had not charged value-added tax, putting their European counterparts at a disadvantage. Currently, Luxemburg has the lowest VAT rate, at 15 percent, and Sweden and Denmark have the highest at 25 percent. About 90 percent of online sales in Europe are B2B sales. The Nordic countries are the most wired, and 48 percent of Sweden's online population has ordered something. In stark contrast, during fall 2002, UK online sales represented a mere 4 percent of overall UK retail sales. In the United States, that figure was an even lower 1.3 percent.[14]

Asia

China represents a huge market opportunity for companies around the globe, and its population is expected to increase from 1.2 billion people today to about 1.48 billion by 2050, according to the International Institute for Applied Systems Analysis.[15] China is estimated to have 50 to 100 million Internet users. Japan has an estimated 100 million Internet users.[16] Over the next 10 to 15 years, China and India will account for nearly half of all Internet usage in the world. So far, the biggest challenge in serving the Chinese market is censorship by the Chinese government over what its people see and don't see.[17] In addition, some U.S. companies have been accused of providing the Chinese government with spyware to enable tracking of Chinese Internet users. Two of China's leading online media, communications, commerce, and mobile value-added services companies are Sohu.com and NetEase.com. Exchanging messages via cell phones and handheld devices is a common practice in China, with most people cycling through a mix of news, games, and horoscopes—all preselected on portals and then downloaded for viewing.[18]

South Korea is a leader in terms of access to broadband technology, with more than 80 percent of total households having access to high-speed Internet services at less than $30 per month. Such easy accessibility encourages almost every South Korean to go online. E-commerce is widely used in the nation's financial sector, and online share trading accounts for more than 65 percent of trades. All South Korean banks offer online- and mobile-banking services. The government declared the official opening of "e-government" in November 2002 after two years of developing an integrated Web site (*www.egov.go.kr*) that offers many public services online. The South Korean mobile-phone penetration rate is also one of the highest in the world, with more than 68 percent of the South Korean population owning cellular phones as of spring 2003.[19]

In India, a low personal computer and Internet base, an immature telecommunications infrastructure, lack of security, and high access costs slow the growth of e-commerce. Among the products Indian consumers buy online are cinema tickets, greeting cards, clothes, cassettes, books, magazines, fast food, medicines, and educational material. Credit cards are commonly used to complete transactions on the Internet in many countries, but there are only 5 million credit card users in India out of a population of 1,061 million. Internet service providers have been progressively reducing connection charges, and the anticipated spread of Internet access through cable television could expand use and reduce costs.[20]

Latin America

Brazil and Mexico, the economic giants of Latin America, are leading the way in terms of developing e-commerce businesses and the requisite infrastructure, legal, and regulatory foundations necessary to support them. Argentina, Chile, Venezuela, and Colombia make up a second tier of countries with modern urban centers, fairly large domestic economies, and institutions and markets sophisticated enough to sustain their own e-commerce industries. Costa Rica and Panama are making some progress in the e-commerce field. Bolivia, Ecuador, Guatemala, Guyana, and Honduras face substantial hurdles in developing sustainable, healthy markets. Banking and financial services organizations have been at the forefront of e-commerce development in Latin America, introducing e-banking and e-brokerage systems for their clients. Governments around the region also are actively developing and making use of e-commerce systems.[21]

Strategies for Global E-Commerce

Obviously, companies that want to succeed on the Web cannot ignore this global shift. Developing a sound global e-commerce strategy is critical for ensuring that Web sites are relevant to the consumers and businesses a company wants to reach, whether those customers are in Cleveland, Singapore, or Frankfurt.

The first step in developing a global e-commerce strategy is to determine which global markets make the most sense for selling products or services online. One approach is to target regions and countries in which a company already has online customers. Companies can track the country domains from which current users of a site are visiting, and established global companies can look to their overseas offices to help determine the languages and countries to target for their Web sites.

Once a company decides which global markets it wants to reach, it must adapt an existing U.S.-centric Web site to another language and culture—a process called *localization*. Local- ization requires companies to have a deep understanding of the country, its people, and the market, which means either building a physical presence in the country or forming partnerships so that detailed knowledge can be gathered. Companies must take painstaking steps to ensure that e-commerce customers have a local experience even though they're shopping at the Web site of a foreign company.

Some of the steps involved in localization are the following:

- Recognizing and conforming to the nuances, subtleties, and tastes of local cultures
- Supporting basic trade laws such as those covering each country's currency, payment preferences, taxes, and tariffs
- Ensuring that technological capabilities match local connection speeds

Tailoring a site to another country is not easy. When Dell launched an e-commerce site to sell PCs to consumers in Japan, it made the mistake of surrounding most of the site's content with black borders, a negative sign in Japanese culture. Japanese Web shoppers took one look at the site and fled. Also, support for Asian languages is difficult because Asian alphabets are more complex and not all Web development tools are capable of handling them. As a result, many companies choose to tackle Asian markets last. In addition, great care must be taken to choose icons that are relevant to a country. For example, mailboxes and shopping carts may not be familiar to global consumers. Users in European countries don't take their mail from large, tubular receptacles, nor do many of them shop in stores large enough for wheeled carts.

One of the most important and most difficult decisions in a company's global Web strategy is whether Web content should be generated and updated centrally or locally. Companies that expand through international partnerships may be tempted to hand control to the new international entities to take the greatest advantage of the expertise of employees in the new markets. But turning over too much control can lead to a muddle of country-specific sites with no consistency and a scattered corporate message. A mixed model of control may be best. Decisions about corporate identity, brand representation, and the technology used for the Web sites can be made centrally to minimize Web development and support effort as well as to present a consistent corporate and brand message. But a local authority can decide on content and services best tailored for given markets.

Companies must also be aware that consumers outside the United States will access sites with different devices and modify their site design accordingly. In Europe, for example, closed-system iDTVs (interactive digital televisions) are becoming popular for accessing on-line content, with some 80 million European households now using them. Such devices have better resolution and more screen space than the PC monitors that U.S. consumers use to access the Internet. So, iDTV users expect more ambitious graphics.

A new group of software and service vendors has emerged to address Web globalization issues. The group includes companies such as GlobalSight and Trados, whose multilingual software can be integrated with popular e-commerce and Web content management software from vendors such as Vignette, BroadVision, and Interwoven to build Web-based applications for enterprises with global reach. The multilingual Web-site management software can work especially well for global sites with central management.

On the Web, ultimately the only way to compete with global companies is to be a global company. Successful firms operate with a portfolio of sites designed for each target market, with shared sourcing and infrastructure to support the network of stores, and with local marketing and business development teams to take advantage of local opportunities. Service providers continue to emerge to solve the cross-border logistics, payments, and customer service needs of these global retailers.

MOBILE COMMERCE

As we discussed briefly in Chapter 1, mobile commerce (m-commerce) relies on the use of wireless devices, such as personal digital assistants, cell phones, and smart phones, to place orders and conduct business. Handset manufacturers such as Ericsson, Motorola, Nokia, and Qualcomm are working with communications carriers such as AT&T Wireless and Sprint to develop such wireless devices and their related technology.

Mobile Commerce in Perspective

Jeff Bezos, CEO of Amazon.com, has called m-commerce "the most fantastic thing that a time-starved world has ever seen."[22] Adam Zawel, an analyst at high-powered consulting firm The Yankee Group, predicts that wireless purchases will reach $4.5 billion annually by 2006.[23] Despite this optimism, m-commerce purchases have not yet taken off.

To ensure user-friendliness, the interface between the wireless device and its user must improve to the point that it is nearly as easy to purchase an item on a wireless device as it is to purchase it on a PC. In addition, network speed must improve so that users do not become frustrated. Security is also a major concern, with two major issues: the security of the transmission itself and the trust that the transaction is being made with the intended party. Encryption can be employed to provide secure transmission. Digital certificates, discussed later in this chapter, can be employed to ensure that transactions are made between the intended parties.

In geographic areas with nearly ubiquitous network coverage, such as major metropolitan areas, adoption of m-commerce is much more likely than in areas with spotty service. Similarly, regions with newer, high-speed wireless networks have a faster response time, making mobile transactions faster and more convenient. As a result, the acceptance of m-commerce is currently geographically dependent.

The market for m-commerce in North America is expected to mature much later than in Western Europe and Japan for several reasons. In North America, responsibility for network infrastructure is fragmented among many providers, consumer payments are usually done by credit card, and most Americans are unfamiliar with mobile data services. In most Western European countries, communicating via wireless devices is common, and consumers are much more willing to use m-commerce. Japanese consumers are generally enthusiastic about new technology and are much more likely to use mobile technologies for making purchases.

Technology Needed for Mobile Commerce

The handheld devices used for m-commerce do have limitations that complicate their use. Their screens are small, perhaps no more than a few square inches, and may be capable of displaying only a few lines of text. Their input capabilities are limited to a few buttons, so entering data can be tedious and error prone. They also have less processing power and less bandwidth than desktop computers, which are usually hardwired to a high-speed LAN. For these reasons it is currently impossible to directly access most Web sites with a handheld device. Web developers must rewrite Web applications so that users with handheld devices can access them.

To address the limitations of wireless devices, the industry has undertaken a standardization effort for their Internet communications. The **wireless application protocol (WAP)**, as mentioned in Chapter 7, is a standard set of specifications for Internet applications that run on handheld, wireless devices. WAP is a key underlying technology of m-commerce. WAP was conceived by four companies: Ericsson, Motorola, Nokia, and Unwired Planet (now Phone.com). It is now supported by an entire industry association of over 200 vendors of wireless devices, services, and tools. In the future, devices and service systems based on WAP will be able to interoperate. WAP has made far greater strides in Europe, where mobile devices equipped with Web-ready microbrowsers are much more common than in the United States.

wireless application protocol (WAP)
A standard set of specifications for Internet applications that run on handheld, wireless devices.

Anywhere, Anytime Applications of Mobile Commerce

Because m-commerce devices usually have a single user, they are ideal for accessing personal information and receiving targeted messages for a particular consumer. Through m-commerce, companies can reach individual consumers to establish one-to-one marketing relationships and permit communication to occur whenever it is convenient—in short, anytime and anywhere. Here are just a few examples of potential m-commerce applications:

- Banking customers can use their wireless, handheld devices to access their accounts and pay their bills.
- Clients of brokerage firms can view stock prices and company research as well as conduct trades to fit their schedules.
- Information services such as financial news, sports information, and traffic updates can be delivered to individuals whenever they want.
- On-the-move retail consumers can place and pay for orders instantaneously.
- Telecommunications service users can view service changes, pay bills, and customize their services.
- Retailers and service providers can send potential customers advertising, promotions, or coupons to entice them to try their services as they move past their place of business.

The most successful m-commerce applications suit local conditions and people's habits and preferences. For example, gamblers in Hong Kong now place bets with their mobile phones, and drivers in London can pay tolls.[24] Albertson's, the $36 billion food and drug retailer, is using m-commerce technology to improve customer service. Shoppers at some of the retailer's stores can scan the bar codes of their items with handheld devices to record purchases, tally costs, receive special offers, and check out and pay. The handheld devices run off in-store wireless networks. At checkout time, customers can elect to use an "Express Pay Station," where they scan an "end-of-trip" bar code at the station and then automatically download the contents on the scanner into the register.[25]

As with any new technology, m-commerce will only succeed if it provides users with real benefits. Companies involved in m-commerce must think through their strategies carefully and ensure that they provide services that truly meet customers' needs.

E-COMMERCE APPLICATIONS

Since B2B, B2C, C2C, and global e-commerce use is spreading, it's important to examine some of the most common current uses. E-commerce is being applied to the retail and wholesale, manufacturing, marketing, investment and finance, and auction industries.

Retail and Wholesale

E-commerce is being used extensively in retailing and wholesaling. **Electronic retailing**, sometimes called *e-tailing*, is the direct sale of goods or services by businesses to consumers through electronic storefronts, which are typically designed around the familiar electronic catalog and shopping cart model. Companies such as Office Depot, Wal-Mart, and many others have used the same model to sell wholesale to employees of corporations. There are tens of thousands of electronic retail Web sites—selling literally everything from soup to nuts. In addition, cybermalls are another means to support retail shopping. A **cybermall** is a single Web site that offers many products and services at one Internet location—similar to a regular shopping mall. An Internet cybermall pulls multiple buyers and sellers into one virtual place, easily reachable through a Web browser. For example, PC Mall is a hardware, software, and consumer electronics retailer that sells items for the home, garden, and office décor; patriotic merchandise; gifts, collectibles, toys, and games; electronics; travel accessories; business supplies; sporting goods; tools and home hardware; health and beauty products; jewelry; and more.[26]

electronic retailing (e-tailing)
The direct sale from business to consumer through electronic storefronts, typically designed around an electronic catalog and shopping cart model.

cybermall
A single Web site that offers many products and services at one Internet location.

Dell sells its products through the Dell.com Web site.

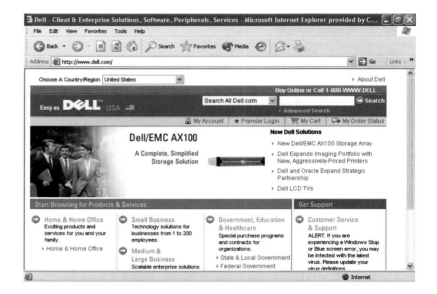

A key sector of wholesale e-commerce is spending on manufacturing, repair, and operations (MRO) goods and services—from simple office supplies to mission-critical equipment, such as the motors, pumps, compressors, and instruments that keep manufacturing facilities up and running smoothly. MRO purchases often approach 40 percent of a manufacturing company's total revenues, but the purchasing system can be haphazard, without automated controls. In addition to these external purchase costs, companies face significant internal costs resulting from outdated and cumbersome MRO management processes. For example, studies show that a high percentage of manufacturing downtime often results from not having the right part at the right time in the right place. The result is lost productivity and capacity. E-commerce software for plant operations provides powerful comparative searching capabilities to enable managers to identify functionally equivalent items, helping them spot opportunities to combine purchases for cost savings. Comparing various suppliers, coupled with consolidating more spending with fewer suppliers, leads to decreased costs. In addition, automated workflows are typically based on industry best practices, which can streamline processes.

Manufacturing

One approach taken by many manufacturers to raise profitability and improve customer service is to move their supply chain operations onto the Internet. Here they can form an **electronic exchange** to join with competitors and suppliers alike, using computers and Web sites to buy and sell goods, trade market information, and run back-office operations, such as inventory control, as shown in Figure 8.6. With such an exchange, the business center is not a physical building but a network over which business transactions occur. This approach has greatly speeded the movement of raw materials and finished products among all members of the business community, thus reducing the amount of inventory that must be maintained. It has also led to a much more competitive marketplace and lower prices. Private exchanges are owned and operated by a single company. The owner uses the exchange to trade exclusively with established business partners. Public exchanges are owned and operated by industry groups. They provide services and a common technology platform to their members and are open, usually for a fee, to any company that wants to use them.

One example of a successful exchange is the WorldWide Retail Exchange (WWRE) founded in 2000 by 17 international retailers to enable participants in the food, general merchandise, textile/home, and drugstore sectors to simplify and automate supply chain processes. Current membership consists of 64 retail industry leaders from around the world with a combined revenue of over $900 billion. Over its first four years of operation, the WWRE saved its members more than $1 billion on maintenance, repair, operating equipment, and private-label goods.[27] Nick Parnaby, global director of member development at the WorldWide Retail Exchange, says that the typical member saves 13 percent, but that

electronic exchange

An electronic forum where manufacturers, suppliers, and competitors buy and sell goods, trade market information, and run back-office operations.

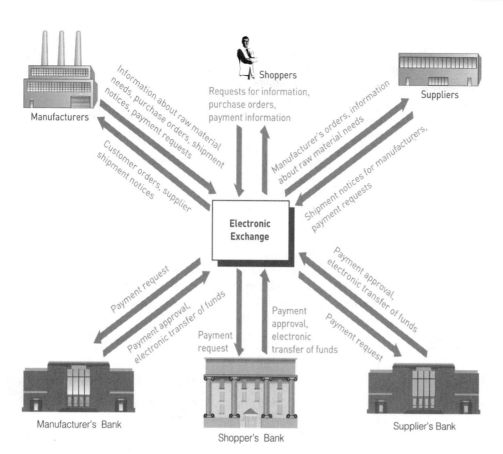

Figure 8.6

Model of an Electronic Exchange

number rises to 20 percent if member companies "pool their spend" with one another to improve economies of scale.[28]

At the turn of the twenty-first century, Internet and brick-and-mortar companies set up more than 1,000 online marketplaces in 70 industries. Many of these exchanges are no longer operating because they failed to bring their members true business benefits. Today there are about 200 exchanges in existence. Table 8.1 provides a partial list of some of the more successful electronic exchanges.

Exchange	Industry
Covisint	Automotive
Elemica	Chemical
Exostar	Defense
E2Open Inc.	Technology
Global Healthcare Exchange	Healthcare
Pantellos Group	Utility
Trade-Ranger	Energy
Transora	Consumer goods
UCCnet	Consumer goods
WorldWide Retail Exchange	Retail

Table 8.1

Some Successful Electronic Exchanges

Source: Adapted from Steve Ulfelder, "B2B Survivors," *Computerworld*, February 2, 2004, accessed at *www.computerworld.com*.

The members of successful exchanges have benefited greatly from their services. Cinergy Corp., an energy supplier in the Midwest, routes 50 percent of its purchase orders through the Pantellos exchange and has dramatically reduced the errors in its order processing. Overall, members of Pantellos have saved an aggregate $315 million since the exchange was founded. The Dow Chemical Company saved thousands of hours of effort and hopes to reduce inventories by as much as 50 percent through use of the Elemica exchange.[29]

Several strategic and competitive issues are associated with the use of exchanges. Many companies distrust their corporate rivals and fear they may lose trade secrets through participation in such exchanges. Suppliers worry that the online marketplaces and their auctions will drive down the prices of goods and favor buyers. Suppliers also can spend a great deal of money in the setup to participate in multiple exchanges. For example, more than a dozen new exchanges have appeared in the oil industry, and the printing industry is up to more than 20 online marketplaces. Until a clear winner emerges in particular industries, suppliers are more or less forced to sign on to several or all of them. Yet another issue is potential government scrutiny of exchange participants—anytime competitors get together to share information, it raises questions of collusion or antitrust behavior.

Many companies that already use the Internet for their private exchanges have no desire to share their expertise with competitors. At Wal-Mart, the world's number-one retail chain, executives turned down several invitations to join exchanges in the retail and consumer goods industries. Wal-Mart is pleased with its in-house exchange, Retail Link, which connects the company to 7,000 worldwide suppliers that sell everything from toothpaste to furniture.

Marketing

The nature of the Web allows firms to gather much more information about customer behavior and preferences than they could using other marketing approaches. Marketing organizations can measure many online activities as customers and potential customers gather information and make their purchase decisions. Analysis of this data is complicated because of the Web's interactivity and because each visitor voluntarily provides or refuses to provide personal data such as name, address, e-mail address, telephone number, and demographic data. Internet advertisers use the data they gather to identify specific portions of their markets and target them with tailored advertising messages. This practice, called **market segmentation**, divides the pool of potential customers into subgroups, which are usually defined in terms of demographic characteristics such as age, gender, marital status, income level, and geographic location.

market segmentation
The identification of specific markets to target them with advertising messages.

comScore Networks is a global information provider to large companies seeking information on consumer behavior to boost their marketing, sales, and trading strategies.

Technology-enabled relationship management is a new twist on establishing direct customer relationships made possible when firms promote and sell on the Web. **Technology-enabled relationship management** occurs when a firm obtains detailed information about a customer's behavior, preferences, needs, and buying patterns and uses that information to set prices, negotiate terms, tailor promotions, add product features, and otherwise customize its entire relationship with that customer.

DoubleClick is a leading global Internet advertising company that leverages technology and media expertise to help advertisers use the power of the Web to build relationships with customers. The DoubleClick Network is its flagship product, a collection of high-traffic and well-recognized sites on the Web including MSN, Sports Illustrated, Continental Airlines, the Washington Post, CBS, and more than 1,500 others. This network of sites is coupled with DoubleClick's proprietary DART targeting technology, which allows advertisers to target their best prospects based on the most precise profiling criteria available. DoubleClick then places a company's ad in front of those best prospects. DART powers over 60 billion ads per month and is trusted by top advertising agencies. Comprehensive online reporting lets advertisers know how their campaign is performing and what type of users are seeing and clicking on their ads. This high-level targeting and real-time reporting provide speed and efficiency not available in any other medium. The system is also designed to track advertising transactions, such as impressions and clicks, to summarize these transactions in the form of reports and to compute DoubleClick Network member compensation.

Investment and Finance

The Internet has revolutionized the world of investment and finance. Perhaps the changes have been so great because this industry had so many built-in inefficiencies and so much opportunity for improvement.

Online Stock Trading

Before the World Wide Web, if you wanted to invest in stocks, you called your broker and asked what looked promising. He'd tell you about two or three companies and then would try to sell you shares of a stock or perhaps a mutual fund. The sales commission was well over $100 for the stock (depending on the price of the stock and the number of shares purchased) or as much as an 8 percent sales charge on the mutual fund. If you wanted information about the company before you invested, you would have to wait two or three days for a one-page Standard and Poor's stock report providing summary information and a chart of the stock price for the past two years to arrive in the mail. Once you purchased or sold the stock, it would take two days to get an order confirmation in the mail, detailing what you paid or received for the stock.

The brokerage business adapted to the Internet faster than any other arm of finance. To make a trade, all you need to do is log on to the Web site of your online broker, and with a few keystrokes and a few clicks of your mouse to identify the stock and number of shares involved in the transaction, you can buy and sell securities in seconds. In addition, an overwhelming amount of free information is available to online investors—from the latest Securities & Exchange filings to the rumors spread in chat rooms. See Table 8.2 for a short list of the more valuable sites.

One indispensable tool of the online investor is a portfolio tracker. This tool allows you to enter information about the securities you own—ticker symbol, number of shares, price paid, and date purchased—at a tracker Web site. You can then access the tracker site to see how your stocks are doing. (There is typically a 15- to 20-minute delay between the price displayed at the site and the price at which the stock is actually being sold.) In addition to reporting the current value of your portfolio, most sites provide access to news, charts, company profiles, and analyst ratings on each of your stocks. You can also program many of the trackers to watch for certain events (e.g., stock price change of more than +/– 3 percent in a single day). When one of the events you specified occurs, an "alert" symbol is posted next to the affected stock. Table 8.3 lists a number of the more popular tracker Web sites.

technology-enabled relationship management
The use of detailed information about a customer's behavior, preferences, needs, and buying patterns to set prices, negotiate terms, tailor promotions, add product features, and otherwise customize the entire relationship with that customer.

Table 8.2

Web Sites Useful to Investors

Name of Site	URL	Description
The Street Dot Com	www.thestreet.com	Provides information, advice, and recommendations based on popular radio/TV personality Jim Cramer
Quote.com	http://finance.lycos.com	Enables you to get stock quotes, and price charts and access message boards for selected stocks
Kiplinger	www.kiplinger.com	Provides personal financial advice
Elite Trader	www.elitetrader.com	Virtual gathering place for day traders with bulletin boards and chat rooms
DRIP Advisor	www.dripadvisor.com	Covers the basics of dividend reinvestment programs (DRIPs), what companies offer DRIPs, and how to start a DRIP
EDGAR Online	www.edgar-online.com	Provides access to company filings with the Securities and Exchange Commission (SEC)
Federal Filings Online	www.fedfil.com	Dow Jones directory of documents filed with the federal government, including bankruptcy proceedings, initial public offering (IPO) filings, SEC reports, and court cases

Table 8.3

Popular Stock Tracker Web Sites

Name of the Web Stock Tracker Site	URL
MSN MoneyCentral	http://moneycentral.msn.com/investor
Quicken.com	www.quicken.com
The Motley Fool	www.fool.com
Yahoo!	http://quote.yahoo.com
Morningstar	www.morningstar.com

Ameritrade.com is an online brokerage site that offers information, tools, and account-management services for investors.

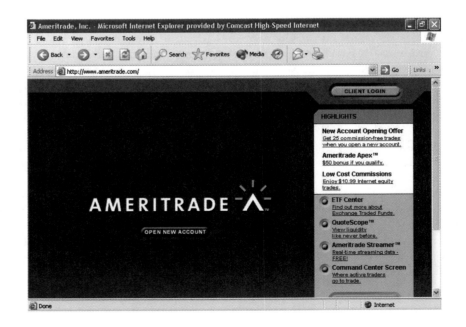

Online Banking

Online banking customers can check balances of their savings, checking, and loan accounts; transfer money among accounts; and pay their bills. These customers enjoy the convenience of not writing checks in longhand, think they have a better knowledge of their current balances, and appreciate the reduction of expenditures on envelopes and stamps. All of the nation's major banks and many of the smaller banks enable their customers to pay bills online. The number of Americans who pay at least some of their bills online is estimated to exceed 35 million.

Here's how electronic bill payment works. You first set up a list of frequent payees, along with their addresses and a code describing the type of payment, such as "home mortgage." Then, when you go online to pay your bills, you simply enter the code or name assigned to the check recipient, the amount of the check, and the date you want it paid. In many cases, the bank still prints and mails a check, so you have to time your online transactions to allow for bank processing and mail delays. But most bill-paying programs allow you to schedule recurring payments for every week, month, or quarter, which you might want to do for your auto loan or health insurance bill.

The next advance in online bill paying is **electronic bill presentment**, which eliminates all paper, right down to the bill itself. With this process, the biller posts an image of your statement on the Internet and alerts you by e-mail that your bill has arrived. You then direct your bank to pay it. Some vendors such as Edocs Inc. and Avolent Inc. offer electronic bill presentment and payment software that users must purchase and install on their computers. Other companies, such as Xign Corp. and Metavante Corp., offer electronic bill presentment and payment as a hosted service to subscribers. After deciding whether to purchase electronic bill presentment as software or as a service, customers must choose how to link the system to their existing software applications and data for automated bill payment and dispute resolution. Payless Shoe Source Inc. chose Xign's service to distribute invoices to about 45,000 vendors.[30]

electronic bill presentment
A method of billing whereby the biller posts an image of your statement on the Internet and alerts you by e-mail that your bill has arrived.

Auctions

As discussed in Chapter 7, the Internet has created many new options for C2C e-commerce, including electronic auctions, in which geographically dispersed buyers and sellers can come together. A special type of auction called *bidding* allows a prospective buyer to place only one bid for an item or a service. Priceline.com's initial business model enabled consumers to achieve significant savings by naming their own prices for goods and services. Priceline.com took these consumer offers and then presented them to sellers, who filled as much of the demand as they wished at price points determined by the buyers.

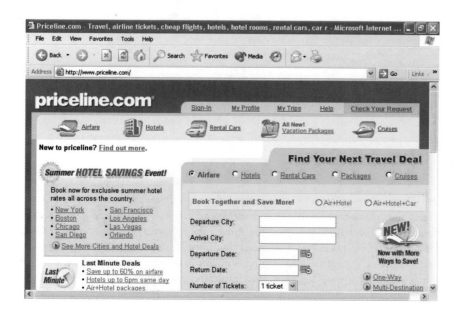

Priceline.com is a patented Internet bidding system that enables consumers to save money by naming their own price for goods and services.

eBay selected Pitney Bowes to provide its postage service, which gives its customers easy access to U.S. Postal Service (USPS) shipping services. With this new tool, eBay users can purchase postage online, pay for it using their PayPal accounts, and print their shipping label from their computer—all from the eBay Web site. Once the label is purchased, both the buyer and the seller will be able to track the delivery status of the package online from the eBay site. Simplifying the shipping part of the transaction will make trading on eBay faster and easier.[31]

Now that we've examined some of the applications of e-commerce, let's look at some of the technical issues related to information systems and technology that make it possible.

TECHNOLOGY INFRASTRUCTURE

For e-commerce to succeed, a complete technology infrastructure must be in place. These infrastructure components must be chosen carefully and integrated to support a large volume of transactions with customers, suppliers, and other business partners worldwide. Online consumers complain that poor Web site performance (e.g., slow response time, poor customer support, and lost orders) drives them to abandon some e-commerce sites in favor of those with better, more reliable performance. This section provides a brief overview of the key technology infrastructure components (see Figure 8.7).

Figure 8.7

Key Technology Infrastructure Components

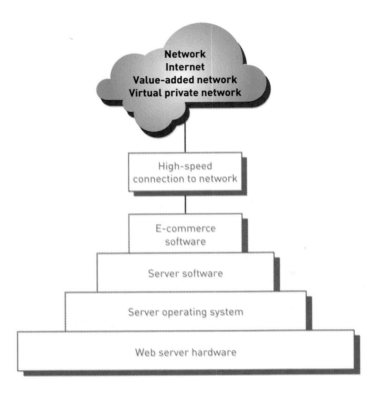

Hardware

A Web server hardware platform complete with the appropriate software is a key e-commerce infrastructure ingredient. The amount of storage capacity and computing power required of the Web server depends primarily on two things: the software that must run on the server and the volume of e-commerce transactions that must be processed. Although it is possible for IS staff to define the software to be used, sometimes they do much guesswork to estimate how much traffic the site will generate. As a result, the most successful e-commerce solutions are designed to be highly scalable so that they can be upgraded to meet unexpected user traffic.

A key decision facing new e-commerce companies is whether to host their own Web site or to let someone else do it. Many companies decide that using a third-party Web service provider is the best way to meet initial e-commerce needs. The third-party company rents space on its computer system and provides a high-speed connection to the Internet, which minimizes the initial out-of-pocket costs for e-commerce start-up. The third party can also provide personnel trained to operate, troubleshoot, and manage the Web server. Of course, a number of companies decide to take full responsibility for acquiring, operating, and supporting the Web server hardware and software themselves, but this approach requires considerable up-front capital and a set of skilled and trained individuals. Whichever approach is taken, there must be adequate hardware backup to avoid a major business disruption in case of a failure of the primary Web server.

Web Server Software

In addition to the Web server operating system, each e-commerce Web site must have Web server software to perform a number of fundamental services, including security and identification, retrieval and sending of Web pages, Web site tracking, Web site development, and Web page development. The two most popular Web server software packages are Apache HTTP Server and Microsoft Internet Information Server.

Security and Identification

Security and identification services are essential for intranet Web servers to identify and verify who is accessing the system from the Internet. Access controls provide or deny access to files based on the user name or URL. Web servers support encryption processes for transmitting private information securely over the public Internet.

In addition to managing security and identification services, Web sites must be designed to protect against malicious attacks. A denial-of-service attack is one of the most difficult Internet threats to thwart and can be costly if it knocks an e-commerce site out of commission. During such a **denial-of-service** (DOS) **attack**, the attacker takes command of many computers on the Internet and uses them to flood the target Web site with requests for data and other small tasks, preventing the target machine from serving legitimate users. Many Web sites, including Amazon, CNN, eBay, and Yahoo!, have suffered these attacks. One effective means of protecting against DOS attacks is to use Internet service providers that offer some safeguards against such attacks. After all, service providers are more able to detect and choke off traffic directed at specific IP addresses.[32] Cisco Systems, maker of the Self-Defending Network strategy, has installed an IP source tracking function in its networking equipment to make it easier to detect a DOS attack and shut down malicious traffic.[33]

denial-of-service (DOS) attack
An attack in which the attacker takes command of many computers on the Internet and uses them to flood the target Web site with requests for data and other small tasks, preventing the target machine from serving legitimate users.

Retrieving and Sending Web Pages

The fundamental purpose of a Web server is to process and respond to client requests that are sent using HTTP. In response to such a request, the Web server program locates and fetches the appropriate Web page, creates an HTTP header, and appends the HTML document to it. For dynamic pages, the server involves other programs, retrieves the results from the back-end process, formats the response, and sends the pages and other objects to the requesting client program.

Web Site Tracking

Web servers capture visitors' information, including who is visiting the Web site (the visitor's URL), what search engines and keywords they used to find the site, how long their Web browser viewed the site, what the date and time of each visit was, and which pages were displayed. This data is placed into a **Web log file** for future analysis.

Web Site Development

Web site development tools include features such as an HTML/visual Web page editor (e.g., Microsoft's FrontPage, NetStudio's NetStudio, SoftQuad's HoTMetaL Pro), software development kits that include sample code and code development instructions for languages such as Java or Visual Basic, and Web page upload support to move Web pages from a development PC to the Web site. Which tools are bundled with the Web server software depends on which Web server software you select.

Web log file
A file that contains information about visitors to a Web site.

Web site development tools
Tools used to develop a Web site, including HTML or visual Web page editor, software development kits, and Web page upload support.

Web page construction software
Software that uses Web editors and extensions to produce both static and dynamic Web pages.

static Web pages
Web pages that always contain the same information.

dynamic Web pages
Web pages containing variable information that are built in response to a specific Web visitor's request.

e-commerce software
Software that supports catalog management, product configuration, shopping cart facilities, e-commerce transaction processing, and Web traffic data analysis.

catalog management software
Software that automates the process of creating a real-time interactive catalog and delivering customized content to a user's screen.

product configuration software
Software used by buyers to build the product they need online.

Web Page Construction

Web page construction software uses Web editors and extensions to produce Web pages—either static or dynamic. **Static Web pages** always contain the same information—for example, a page that provides text about the history of the company or a photo of corporate headquarters. **Dynamic Web pages** contain variable information and are built in response to a specific Web site visitor's request. For example, if a Web site visitor inquires about the availability of a certain product by entering a product identification number, the Web server will search the product inventory database and generate a dynamic Web page based on the current product information it found, thus fulfilling the visitor's request. This same request by another visitor later in the day may yield different results due to ongoing changes in product inventory. A server that handles dynamic content must be able to access information from a variety of databases. The use of open database connectivity enables the Web server to assemble information from different database management systems, such as SQL Server, Oracle, and Informix.

E-Commerce Software

Once you have located or built a host server, including the hardware, operating system, and Web server software, you can begin to investigate and install e-commerce software. There are five core tasks that **e-commerce software** must support: catalog management, product configuration, shopping cart facilities, e-commerce transaction processing, and Web traffic data analysis.

The specific e-commerce software you choose to purchase or install depends on whether you are setting up for B2B or B2C transactions. For example, B2B transactions do not include sales tax calculations, and software to support B2B must incorporate electronic data transfers between business partners, such as purchase orders, shipping notices, and invoices. B2C software, on the other hand, must handle the complication of accounting for sales tax based on the current laws and rules in effect in the various states.

Catalog Management

Any company that provides a wide range of product offerings requires a real-time interactive catalog to deliver customized content to a user's screen. **Catalog management software** combines different product data formats into a standard format for uniform viewing, aggregating, and integrating catalog data into a central repository for easy access, retrieval, and updating of pricing and availability changes. The data required to support large catalogs is almost always stored in a database on a computer that is separate from, but accessible to, the e-commerce server machine.

Corporate Express sells furniture, paper, computer supplies, and office equipment. It maintains catalogs tailored to each customer's format, terminology, and buying practices. If certain customers don't buy office furniture from Corporate Express, office furniture is blocked out in their version of the catalog. Corporate Express also maintains a list of items that each customer orders frequently, as well as the special terms and prices that the customer has negotiated. Such attention to customization is greatly appreciated by customers because it makes their ordering process easier. Corporate buyers also appreciate the fact that their employees can only purchase prearranged items so that "maverick" buying is eliminated.[34]

Product Configuration

Customers need help when an item they are purchasing has many components and options. Product configuration software tools were originally developed in the 1980s to assist B2B salespeople to match their company's products to customer needs. Buyers use the new Web-based **product configuration software** to build the product they need online with little or no help from salespeople. For example, Dell customers use product configuration software to build the computer of their dreams. Such software is also used in the service arena as well, to help people decide what sort of consumer loan or insurance is best for them.

Shopping Cart

Today many e-commerce sites use an **electronic shopping cart** to track the items selected for purchase, allowing shoppers to view what is in their cart, add new items to it, or remove items from it, as shown in Figure 8.8. To order an item, the shopper simply clicks that item. All the details about it—including its price, product number, and other identifying information—are stored automatically. If the shopper later decides to remove one or more items from the cart, he or she can do so by viewing the cart's contents and removing any unwanted items. When the shopper is ready to pay for the items, he or she clicks a button (usually labeled "proceed to checkout") and begins a purchase transaction. Clicking the "Checkout" button displays another screen that usually asks the shopper to fill out billing, shipping, and payment method information and to confirm the order.

electronic shopping cart
A model commonly used by many e-commerce sites to track the items selected for purchase, allowing shoppers to view what is in their cart, add new items to it, and remove items from it.

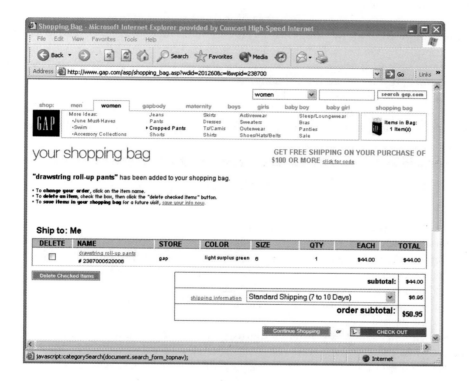

Figure 8.8

Electronic Shopping Cart

An electronic shopping cart (or bag) allows online shoppers to view their selections and add or remove items.

E-Commerce Transaction Processing

E-commerce transaction processing software connects participants in the e-commerce economy and enables communication between trading partners, regardless of their technical infrastructure. This software fully automates transaction processes from order placement to reconciliation.

Basic transaction processing software takes data from the shopping cart and calculates volume discounts, sales tax, and shipping costs to arrive at the total cost. In some cases, the software determines shipping costs by connecting directly to shipping companies such as UPS, FedEx, and DHL. In other cases, shipping costs may be a predetermined amount for each item ordered.

Anaconda Sports was founded in 1902 as Kaye Sports. Over the years it has grown to become one of the largest independent U.S. sporting goods dealers. Anaconda sells to wholesalers, mass merchandisers, and distributors, but it also sells to the general public via direct mail and the Internet. In addition, it has several retail stores in the Northeast. By 2003, Anaconda had outgrown its existing Web infrastructure, which was designed when the firm sold just 500 products, not the 4,000 products it currently stocks. The site simply could not handle the volume of user traffic and needed to upgrade fast. In May 2003 Anaconda went live with its new Web infrastructure, which is hosted by SCS, an IBM reseller and provider of information system solutions—including Web sites. The system combines IBM's DB2

e-commerce transaction processing software
Software that provides the basic connection between participants in the e-commerce economy, enabling communications between trading partners, regardless of their technical infrastructure.

database software and runs on servers using the Linux operating system. WebSphere Commerce Server and WebSphere Application Server, both running on Microsoft's Windows 2000 servers form the basis for the site's e-commerce software. (WebSphere is a set of Java-based tools from IBM that allow customers to create and manage sophisticated business Web sites.) The Web site also employs VeriSign's Payflow Pro payment-processing system and UPS's WorldShip service for automating Internet orders.[35]

More and more e-commerce companies are outsourcing their inventory management and order fulfillment process to a third party. In this situation, the e-commerce transaction processing software must actually route order information to one of the shipping companies to ship the product from inventory under their management. For example, Hewlett-Packard has an arrangement with FedEx to manage orders for HP printers.

Web Traffic Data Analysis

It is necessary to run third-party **Web site traffic data analysis software** to make sense of all the data captured in the Web log file—to turn it into useful information to improve Web site performance. For example, when someone queries a search engine for a keyword related to your site's products or services, does your page appear in the top 10 matches, or does your competitor's? If you're listed but not within the first two or three pages of results, you lose, no matter how many engines you submitted your site to.

Web search engines can also generate useful information for marketers about their potential customers, as discussed in the "Information Systems @ Work" special feature.

Web site traffic data analysis software
Software that processes and analyzes data from the Web log file to provide useful information to improve Web site performance.

ELECTRONIC PAYMENT SYSTEMS

Electronic payment systems are a key component of the e-commerce infrastructure. Current e-commerce technology relies on user identification and encryption to safeguard business transactions. Actual payments are made in a variety of ways, including electronic cash, electronic wallets, and smart, credit, charge, and debit cards.

As discussed in Chapter 7, authentication technologies are used by organizations to confirm the identity of a user requesting access to information or assets. A **digital certificate** is an attachment to an e-mail message or data embedded in a Web site that verifies the identity of a sender or Web site. A **certificate authority (CA)** is a trusted third-party organization or company that issues digital certificates. The CA is responsible for guaranteeing that the individuals or organizations granted these unique certificates are, in fact, who they claim to be. Digital certificates thus create a trust chain throughout the transaction, verifying both purchaser and supplier identities.

digital certificate
An attachment to an e-mail message or data embedded in a Web page that verifies the identity of a sender or a Web site.

certificate authority (CA)
A trusted third party that issues digital certificates.

Secure Sockets Layer

All online shoppers fear the theft of credit card numbers and banking information. To help prevent this from happening, the **Secure Sockets Layer (SSL)** communications protocol is used to secure sensitive data. The SSL communications protocol sits above the TCP layer of the OSI model discussed in Chapter 6, and other protocols such as Telnet and HTTP can be layered on top of it. SSL includes a handshake stage, which authenticates the server (and the client, if needed), determines the encryption and hashing algorithms to be used, and exchanges encryption keys. The handshake may use public key encryption. Following the handshake stage, data may be transferred. The data is always encrypted, ensuring that your transactions are not subject to interception or "sniffing" by a third party. For companies wishing to conduct serious e-commerce, such as receiving credit card numbers or other sensitive information, SSL is a must. Although SSL handles the encryption part of a secure e-commerce transaction, a digital certificate is necessary to provide server identification.

Secure Sockets Layer (SSL)
A communications protocol used to secure sensitive data.

Marketing Meets Search Engine

Web search engines draw more visitors than any other type of Web page. The keywords that people type into these search engines are a reflection of public interest. Whether users are searching for current events, academic interests, pop stars, sports heroes, or products, the sum of what is typed into the text box of search engines provides a snapshot of what is hot and what is not. This fact is influencing a number of companies with marketing interests to enter the search engine business.

One recent experiment by Market Insight Corporation attempts to provide Webizens—citizens of the Web—with a useful resource, while collecting valuable marketing information to sell to clients. Market Insight's Web tool is called MyProductAdvisor.com, and it can be used to recommend products to customers. The initial offering of MyProductAdvisor focuses on automobiles and digital cameras, with more products—including smart phones, PDAs, and home theaters—promised for the near future.

Using MyProductAdvisor.com, individuals planning to purchase a car or digital camera can find out which manufacturer and model best suits their needs. MyProductAdvisor asks some pointed questions, allowing you to respond using a sliding scale to indicate the relative importance of various features and options. For example, if you were shopping for a car and visited the page that collects information about your usage, you might move the slider to "Most Important" for "Carrying cargo such as bikes, skis, etc." and the slider for "Car pooling or transporting children" to "Least Important."

After a user replies to questions about usage, price, brands, body type, and other attributes, MyProductAdvisor provides unbiased suggestions for what vehicles best match the criteria. The word "unbiased" is most important to MyProductAdvisor. Richard Smallwood, cofounder of Market Insight states, "We want to be a sort of *Consumer Reports*, with individualized recommendations based completely on what you want, unbiased and objective." To this end, MyProductAdvisor does not accept any advertising. To further its reputation as a trusted resource, MyProductAdvisor refrains from collecting demographic information such as gender, income, or geographic information. The aggregate data that MyProductAdvisor does collect—preferences of customers who are ready to buy—is gold to companies like GM who purchase the data.

"We do have information about what you [customers] want, so we can segment the market by price range, budget, whatever. You told us how you would use the camera, so we can segment by that," Smallwood said. "Say a camera manufacturer is thinking about a camera with improved resolution. Will it get a big or a little market share? We can find out how many people think good resolution is important," said Smallwood. Market Insight is currently working with GM on various product scenarios. It will help GM estimate the size of the market and the price sensitivity of buyers.

Market Insight charges clients such as GM a fixed amount for a specified time period, usually a year. The amount includes license fees, consulting fees, training fees, and an optional exclusivity fee if a client wants exclusive access to data in a particular area.

Market Insight is just one of many B2B companies looking to capitalize on the value of the data collected by search engines. Recently, online superstore Amazon.com has come out with its own search engine, A9 (the name stands for the nine letters in the word "algorithm"), which claims to provide more relevant search results by including your personal online search habits and patterns in the search algorithm. The more you search with A9, and the more you shop at Amazon, the more accurate your search results become. Along the way, Amazon is able to build a huge database of search, Web browsing, and shopping trends that it can use in its targeted marketing and product management.

The business and marketing potential of search engines, not to mention the advertising dollars that drive search engines such as Google, have caught the attention of big players such as Microsoft, Yahoo!, and Amazon. As these Web tools are becoming recognized as rich sources of consumer information, marketing firms are turning from traditional forms of consumer research to companies such as Market Insight who are reinventing the business.

Discussion Questions

1. What advantages does Market Insight enjoy over traditional marketing firms?
2. As a marketing expert, which company would you prefer to work for, Market Insight or a firm that conducts traditional marketing research through customer focus groups and on-the-street interviews? Why?

Critical Thinking Questions

3. Market Insight provides detailed information on customer preferences collected from people who visit its Web site. Can this data be considered representative of all customers, most customers, or some customers? How can this data be use to influence product development?
4. Market Insight has nine employees. What types of skills and expertise are required of these individuals? How does the makeup of this team differ from that of a traditional marketing firm?

SOURCES: Janis Mara, "Market Insight Launches," *ClickZ News*, April 9, 2004, *www.clickz.com*; My Product Advisor Web site, accessed April 20, 2004, *www.myproductadvisor.com*; John Battelle, "Can Amazon Unplug Google?" *Business 2.0*, April 15, 2004, *www.business2.com*; A9 Search Engine Web site, accessed April 20, 2004, *www.a9.com*.

The SSL (Secure Sockets Layer encryption) communications protocol assures customers that information they provide to retailers, such as credit card numbers, cannot be viewed by anyone else on the Web.

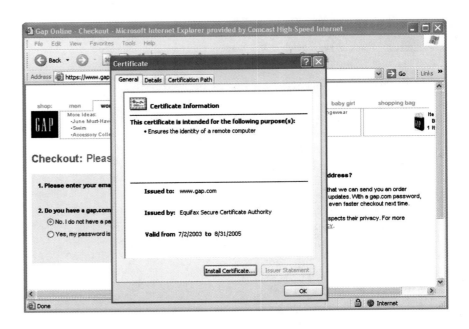

One tip to the security of a transaction is visible on-screen. Look at the bottom-left corner of your browser before sending your credit card number to an e-commerce vendor. If you use Netscape Navigator, make sure that you see a solid key in a small blue rectangle. If you use Microsoft Explorer, the words "Secure Web site" appear near a little gold lock. And, if you're worried about how secure a secure connection is, visit Netcraft at *www.netcraft.com/security*. At this site you can type in any Web site address and determine the equipment being used for secure transactions. One more tip: to ensure security, you should always use the newest release of your favorite browser; the newer the browser, the better the security.

Electronic Cash

electronic cash

An amount of money that is computerized, stored, and used as cash for e-commerce transactions.

Electronic cash is an amount of money that is computerized, stored, and used as cash for e-commerce transactions. A consumer must open an account with a bank and show some identification to establish his or her identity to obtain electronic cash. Then, whenever the consumer wants to withdraw electronic cash to make a purchase, he or she accesses the bank via the Internet and presents proof of identity—typically a digital certificate issued by a certification authority. After the bank verifies the consumer's identity, it issues the consumer the requested amount of electronic cash and deducts the same amount from the consumer's account. The electronic cash is stored in the consumer's electronic wallet on his or her computer's hard drive, or on a smart card (both are discussed later).

Consumers can spend their electronic cash when they locate e-commerce sites that accept it for payment. The consumer sends electronic cash to the merchant for the specified cost of the goods or services. The merchant validates the electronic cash to be certain that it is not forged and belongs to the customer. Once the goods or services are shipped to the consumer, the merchant presents the electronic cash to the issuing bank for deposit. The bank then credits the merchant's account for the transaction amount, minus a small service charge.

There are two distinct types of electronic cash: identified electronic cash and anonymous electronic cash (also known as *digital cash*). Identified electronic cash contains information revealing the identity of the person who originally withdrew the money from the bank. Also, in much the same manner as credit cards, identified electronic cash enables the bank to track the money as it moves through the economy. The bank can determine what people or organizations bought, where they bought it, when they bought it, and how much they paid. Anonymous electronic cash works just like real paper cash. Once anonymous electronic cash is withdrawn from an account, it can be spent or given away without leaving a transaction trail. A few of the companies that provide electronic cash mechanisms include VeriSign, Mondex (available now as part of the global *One*SMART™ MasterCard® program), and Visa Cash.

Electronic Wallets

Online shoppers quickly tire of repeatedly entering their shipment and payment information each time they make a purchase. An **electronic wallet** holds credit card information, electronic cash, owner identification, and address information. It provides this information at an e-commerce site's checkout counter. When consumers click on items to purchase, they can then click on their electronic wallet to order the item, thus making online shopping much faster and easier.

Smart, Credit, Charge, and Debit Cards

Online shoppers use credit and charge cards for the majority of their Internet purchases. A credit card, such as Visa or MasterCard, has a preset spending limit based on the user's credit limit, and each month the user can pay off a portion of the amount owed or the entire credit card balance. Interest is charged on the unpaid amount. A charge card, such as American Express, carries no preset spending limit, and the entire amount charged to the card is due at the end of the billing period. Charge cards do not involve lines of credit and do not accumulate interest charges.

Debit cards look like credit cards or automated teller machine (ATM) cards, but they operate like cash or a personal check. While a credit card is a way to "buy now, pay later," a debit card is a way to "buy now, pay now." Debit cards allow you to spend only what is in your bank account. It is a quick transaction between the merchant and your personal bank account. When you use a debit card, your money is quickly deducted from your checking or savings account. Credit, charge, and debit cards currently store limited information about you on a magnetic stripe. This information is read each time the card is swiped to make a purchase. All credit card customers are protected by law from paying any more than $50 for fraudulent transactions. At Visa, online purchases account for the highest amount of purchase fraud—24 cents for every $100 spent, compared with 6 cents for every $100 overall. Indeed, the risk of bogus credit card transactions has slowed the growth of e-commerce by exposing merchants to substantial losses and making online shoppers nervous. Based on the results of its annual survey of e-commerce crime, security company CyberSource estimates online crooks will make away with $1.6 billion of $94 billion (about 1.7 percent) of annual U.S. business-to-consumer e-commerce revenue.[36]

MasterCard estimates that 7 percent of its $922 billion in annual card purchases are now taking place on the Web. As a result, the firm now requires that high-volume merchants and payment processors conduct quarterly assessments of their Web sites, or MasterCard will stop doing business with them. Both host and network-based software offerings provide such assessments. Host-based tools reside on servers and desktops and use software agents that can log on to a host to report their findings to a management console. Network-based tools scan remotely in ways similar to a hacker probe. The agents check user access logs and the identity of those who accessed financial data on the server. However, if there is no coordinated staff approach, network administrators might think that the network is under attack because of the scans.[37]

The **smart card** is a credit card–sized device with an embedded microchip to provide electronic memory and processing capability. Smart cards can be used for a variety of purposes, including storing a user's financial facts, health insurance data, credit card numbers, and network identification codes and passwords. They can also store monetary values for spending.

Smart cards are better protected from misuse than conventional credit, charge, and debit cards because the smart card information is encrypted. Conventional credit, charge, and debit cards clearly show your account number on the face of the card. The card number, along with a forged signature, is all that a thief needs to purchase items and charge them against your card. A smart card makes credit theft practically impossible because a key to unlock the encrypted information is required, and there is no external number that a thief can identify and no physical signature a thief can forge.

electronic wallet
A computerized stored value that holds credit card information, electronic cash, owner identification, and address information.

smart card
A credit card–sized device with an embedded microchip to provide electronic memory and processing capability.

Smart cards have been around for over a decade and are widely used in Europe, Australia, and Japan, but they have not caught on in the United States. Use has been limited because there are so few smart card readers to record payments, and U.S. banking regulations have slowed smart card marketing and acceptance as well. American Express launched its Blue card smart card in 1999. You can use a smart card reader that attaches to your PC monitor to make online purchases with your American Express card. You must visit the American Express Web site to get an electronic wallet to store your credit card information and shipping address. When you want to buy something online, you go to the checkout screen of a Web merchant, swipe your Blue card through the reader, type in a password, and you're done. The digital wallet automatically tells the vendor your credit card number, its expiration date, and your shipping information.

Citing limited shopper use, retailer Target began phasing out computer chips on its Target Visa cards in March 2004, less than three years after it introduced the cards. The move dealt a setback to proponents of smart card technology. The technology allowed cardholders to download discount coupons from the Internet or in-store kiosks onto the cards and then use the coupons on shopping trips to Target stores.[38]

THREATS TO E-COMMERCE

As with any revolutionary change, a host of issues must be dealt with to ensure that e-commerce transactions are safe and consumers are protected. Many represent significant threats to its continued growth. The following sections summarize a number of these threats and present practical ideas on how to minimize their impact.

E- and M-Commerce Incidents

As mentioned previously, organizations use identification technologies to confirm the identity of a user requesting access to information or assets. Just as shoppers are concerned with the legitimacy of sites, e-businesses are concerned about the legitimacy of their customers. As a result, an increasing number of companies are investing in biometric technology to protect both parties. Biometric technology, which digitally encodes physical attributes of a person's voice, eye, face, or hand and associates them with biological attributes stored in a file, is commonly used in security organizations such as the FBI—to allow entry into a building, for instance. Currently, using the technology to secure online transactions is rare for both cost and privacy reasons. It can be expensive to outfit every customer with a biometric scanner, and it is difficult to convince consumers to supply something as personal and distinguishing as a fingerprint. In spite of this, a growing number of financial service firms (e.g., Citibank, First Financial Credit Union, Huntington Bankshares, Perdue Employees Federal Credit Union) are considering the use of biometric systems.

Heightened security concerns in the wake of the September 11 terrorist attacks have increased interest in the use of biometrics to develop faster and more effective immigrant-screening systems. The U.S. Department of Homeland Security is evaluating biometric technology to help fill gaps in U.S. border-security procedures. The DHS bought 1,000 optical-stripe read/write drives and biometric verification systems to be used in the U.S. Visitor and Immigration Status Indication Technology program. The technology is being deployed at various U.S. ports of entry. Border-crossing agents use it to read the data encoded on any of the more than 13 million permanent-resident and border-crossing cards issued by the U.S. government and then authenticate the biometric data stored on the cards. The system alerts DHS inspectors to possible counterfeit cards. The DHS is also testing 3,000 live-scan fingerprint booking stations and desktop systems to digitally capture and electronically submit fingerprint images from immigration applicants to the FBI. The fingerprints can then be used to conduct criminal background checks. [39]

U.S. Customs and Border Protection officers use the new US-VISIT biometric program to fingerprint and photograph visitors to the United States who require visas.

(Source: Erik S. Lesser/Getty Images.)

Theft of Intellectual Property

Lawsuits over **intellectual property** (music, books, inventions, paintings, and other special items protected by patents, copyrights, or trademarks) have created a virtual e-commerce war zone. From music files to patented online coupon schemes, hardly a day goes by without some new suit being filed. Every year, American businesses lose billions of dollars from the importation and sale of counterfeit goods and the infringement of copyrights, trademarks, and patents. The entertainment and travel industries have their futures riding on the outcomes of these intellectual property battles. IBM, for example, received 3,415 patents in 2003. It was the eleventh consecutive year that IBM was the world's number-one inventor.[40] Each year, IBM earns billions of dollars from its patents. Loss of this revenue would have a severe impact on the firm.

Although IBM Corp., Intel Corp., and other technology leaders have long sued or threatened to sue rivals over patent infringement, the strategy is risky. The reason is that over the last decade, tech companies have dramatically increased the number of parts and amount of services that they purchase from each other. Lawsuits could damage relationships between suppliers and vendors. In spite of this, Hewlett-Packard filed a patent infringement lawsuit against Gateway in March 2004, alleging that its rival refused to pay licensing fees on several HP-patented designs, including laptop hinges, keyboards that require passwords, and the cursor that points to icons on a computer's video display. The lawsuit could result in an award of tens of millions of dollars.[41]

CIOs must also be careful that they do not become embroiled in copyright infringements over software used to develop their company's Web sites. The SCO Group (SCO) provides software solutions for small to medium-sized businesses and their branch offices. SCO solutions include UNIX platforms for messaging, authentication, and e-business tools. SCO alleges that a major software manufacturer took confidential and copyrighted source code from SCO's Unix V operating system and used it to build their version of Linux. SCO is currently suing major users of this version of Linux, alleging copyright infringement. However, the law has not yet established that SCO's intellectual property is part of the Linux operating system. Should SCO's suit be found to have merit, CIOs everywhere will need to be concerned of pending litigation because they implemented Linux.[42]

Some argue that unscrupulous e-commerce companies threaten creativity by enabling people to steal the original works of innovators and artists. For example, the Recording Industry Association of America Inc. (RIAA) requested hundreds of subpoenas and filed 382

intellectual property
Music, books, inventions, paintings, and other special items protected by patents, copyrights, or trademarks.

civil lawsuits against people who allegedly shared music files illegally. This action was extremely unpopular with its primary consumers, youths and young adults, who believe that downloadable, shareable music files for MP3 players is the best thing to happen to recorded music since the 1983 introduction of the CD.[43]

In response to abuses and strong lobbying by owners of intellectual property, the life of patents has been increased (now lasting 20 years from date of application), copyrights now last 95 years, and more kinds of copyright infringements are now criminal offenses.

Another class of intellectual property lawsuits involves patents on business processes. For example, Overture Services filed suit against Google for allegedly infringing on a patent related to the company's bid-for-placement products. Overture's bid-for-placement search service allows companies to bid for search-result placement based on relevant keywords. Advertisers pay Overture a premium to have their firm appear higher in the list of items found in order to drive traffic to their sites. Overture also sued FindWhat.com for allegedly infringing on the same patent. Other companies have been sued because their Web site copies the "look and feel" of another company's site. Such lawsuits highlight a potentially debilitating problem. What is the proper extent of property rights in an information-based economy? The flurry of e-commerce patent lawsuits raises worries of overly aggressive protection of intellectual property rights and of stalling economic activity.

Fraud

As more people use the Internet for e-commerce, they need to know that the merchant with whom they're dealing is legitimate. The Better Business Bureau's OnLine Reliability seal helps distinguish trustworthy companies among the thousands of businesses marketing online, allowing consumers to easily identify them.

The first wave of Internet crime consisted mostly of online versions of offline hoaxes, the usual get-rich-quick schemes. For example, many people received pleas from desperate Nigerians trying to enlist their help in transferring funds out of their country. More recently, however, fraud artists have begun to exploit the Internet to execute more sophisticated ploys, using fake Web sites and spam. For example, a phony "U.S. Consumer Protection Agency" set up its own Web site, seal, and board of directors. The fake organization sent e-mails to companies saying that for $1,000 they could purchase the right to display its seal of approval.[44]

phishing
Bogus messages purportedly from a legitimate institution to pry personal information from customers by convincing them to go to a "spoof" Web site.

Phishing entails sending bogus messages purportedly from a legitimate institution to pry personal information from customers by convincing them to go to a "spoof" Web site. The spoof Web site appears to be a legitimate site but actually collects personal information from unsuspecting victims. Phishers have simulated the messages and Web sites of Citibank, Lloyds TSB Bank, NatWest, Visa, Halifax Bank, and Westpac Bank. As with all scams, phishing requires a degree of gullibility by the victims. However, a security flaw in Microsoft Internet Explorer allowed phishers to hide the true Internet address of a Web page from Internet Explorer's address bar, making it even more difficult to identify the phony site.[45] Westpac Banking of Australia was hit by a phishing scam using fake e-mail that seemed to come from Westpac to trick customers into giving up their passwords to bank accounts. The attacker carried out the attack by redirecting users to a fake version of the Westpac Web site and then opening the real Westpac Web site in a second browser window, hidden from the user.[46]

To combat Internet fraud, PC Mall, a hardware, software, and consumer electronics retailer, has implemented a system with three layers of defense for catching fraudulent orders. The first line of defense is a service from CyberSource that screens orders for suspicious entries, such as geographically mismatched customer information (an overseas IP address wth a U.S. billing address). The second line of defense is to try to match the order against a "negative database" containing fraudulent order information collected over eight years. The final level of defense compares orders being placed to historic information to see whether any information matches. With multiple stolen credit cards, a thief often will use a common ship-to address or telephone number. Each and every order must successfully pass through all three layers before it can be accepted. During its security checks, PC Mall must be careful not to alienate legitmate customers while trying to detect fraudsters.[47]

Online Auction Fraud

Online auctions brought in about $1 billion in sales in a recent year, and they represent the number-one Internet fraud, according to the National Consumers League. The majority of that fraud comes from so-called person-to-person auctions, which account for roughly half the auction sites. On these sites, it is up to the buyer and seller to resolve details of payment and delivery; the auction sites offer no guarantees. Sticking with auction sites like eBay (*www. ebay.com*) that ensure the delivery and quality of all the items up for bidding can help buyers avoid trouble.

Spam

E-mail that is sent to a wide range of people and Usenet groups indiscriminately is called **spam**. Spam allows peddlers to hawk their products instantly to thousands of people at virtually no cost. And obtaining e-mail addresses to spam is now a snap, thanks to so-called *harvester programs* that snoop Usenet chat groups and collect thousands of e-mail addresses in a single day. Do not respond to spam; responding will only confirm to the spammer that your e-mail address is accurate and active. Instead, simply delete spam as soon as you get it. If the spam is truly offensive or obviously fraudulent, forward the entire message to your Internet service provider, the Federal Trade Commission (*www.ftc.gov*), or the National Fraud Information Center (*www.fraud.org*) and ask that the sender be barred from sending additional messages.

The Federal Trade Commission (FTC) consists of 1,100 people who serve to regulate competition and act as a consumer protection agency. In addition to taking action against companies suspected of unfair business practices, the FTC has worked to shape Internet policy in a range of areas, including online disputes and resolutions, Internet-based fraud, and privacy concerns. The FTC currently has an Internet unit of investigators and lawyers who probe Internet-based complaints and surf the Web in an effort to ferret out bad behavior.[48]

For years, Internet users have complained of receiving unwanted sexually explicit e-mail messages. In April 2004, the Federal Trade Commission announced the adoption of a rule that requires sexually oriented material to include the warning "sexually explicit" to inform recipients and to facilitate filtering them out. The phrase must be included in both the subject line of any e-mail that contains sexually oriented material and in the electronic equivalent of a "brown paper wrapper" in the body of the message that the recipient initially will see when opening the message. It must include the prescribed notice, certain other specified information, and no other information or images. Images may still be present in a message but not initially viewable in a preview pane or upon first opening the e-mail.[49]

spam
E-mail sent to a wide range of people and Usenet groups indiscriminately.

Pyramid Schemes

Traditional pyramid schemes work by getting new investors to pay their recruiters to join the pyramid at the bottom. The new investors then (theoretically) get rich by recruiting additional new investors who will funnel money up the pyramid. Of course, eventually the people on the bottom of the pyramid have trouble finding new recruits, and it collapses. For that reason, pyramids are illegal. Online pyramids often include the sale of a product or service like vitamins, credit cards, or even electricity to justify downstream recruitment fees. Usually, the product that these supposed multilevel-marketing (MLM) companies hawk is so overpriced or unwanted that the company relies mainly on recruiting fees to provide its cash flow. Be wary of any company that depends on recruitment fees to pay you. Also, avoid "opportunities" that force you to buy costly inventory. Finally, check with the Better Business Bureau (*www.bbb.com*) and your state attorney general's office before joining any company you're unsure about.

Investment Fraud

The North American Securities Administrators Association estimates that up to $10 billion will be lost in a year to investment fraud, primarily through the sale of bogus investments. Not all this fraud is committed online. Some of the more colorful scams involve eel farms, imported diamonds, and super-fast-growing trees that can reach 80 feet in three years. Call your state securities board to make sure the seller and its "security" are registered. Avoid dealing with unregistered securities and brokers.

Stock Scams

Before the Internet, most individuals had a tough time spreading rumors about a stock, but now anybody can do it. Net chat rooms like Usenet's *misc.invest.stocks* make it cheap and easy for scammers to talk up (or down) a stock based on false information. The scammer buys the stock at a low price, spreads false rumors that help drive the stock price up, and then sells at an artificially high price before the bottom falls out. Always verify tips and information you get online and in chat rooms with the company directly or with reputable financial advisers. To be truly safe, avoid thinly traded stocks or penny stocks that sell for less than $1 a share, because they are the most easily manipulated.

Invasion of Consumer Privacy

clickstream data
Data gathered based on the Web sites you visit and the items you click on.

Online consumers are more at risk today than ever before. One of the primary factors causing higher risk is *online profiling*—the practice of Web advertisers' recording online behavior for the purpose of producing targeted advertising. **Clickstream data** is the data gathered based on the Web sites you visit and the items you click on. From the marketers' perspective, the use of online profiling allows businesses to market to customers electronically one to one. The benefit to customers is better, more-effective service; the benefit to providers is the increased business that comes from building relationships and encouraging customers to return for subsequent purchases. From the consumers' perspective, online profiling squeezes out anonymity, which remains crucial to privacy on the Internet. And consumers also fear that the personal information thus gathered will be shared with others without their knowledge. For example, a number of lawsuits have been filed against online marketers who intended to create massive data warehouses of consumer information by linking clickstream data with data (such as automobile registrations and product warranty registrations) from other sources.

The U.S. government has not implemented a federal privacy policy and instead relies on e-commerce self-regulation in matters of data privacy. The Better Business Bureau Online and TRUSTe are independent, nonprofit privacy initiatives dedicated to building users' trust and confidence on the Internet and supporting the accelerating growth of e-commerce. The BBB Online Privacy seal program, which helps consumers identify online businesses that honor privacy protection policies, covers more than 5,000 Web sites and has hundreds of applications in process. Figures 8.9 and 8.10 show the TRUSTe and BBB Online Privacy seals, respectively. The TRUSTe program is based on a multifaceted assurance process that establishes Web site credibility, thereby making users more comfortable when making online purchases and providing personal information.

Figure 8.9

TRUSTe Seal

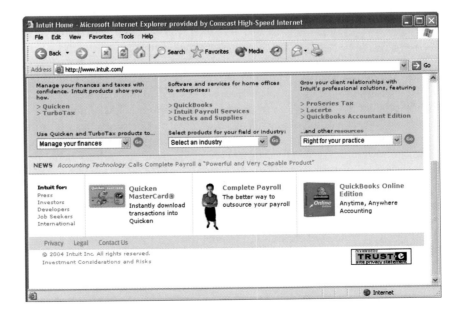

Figure 8.10

BBB Online Privacy Seal

A set of **safe harbor principles** has been established that address the issues of notice, choice, and access associated with data privacy. A company following these principles must notify consumers of the purpose of data collection, allow consumers to opt out of having their data shared with third parties, and provide users with access to their personal information to review and possibly correct it. With an opt-out policy, the Web site is free to gather and sell information about you unless you specifically tell it that it cannot, which typically involves clicking on a button. With an opt-in policy, the gathering or selling of your data is forbidden unless you specifically give the Web site permission. The safe harbor principles also require organizations using personal information to "take reasonable precautions to protect it from loss, misuse, and unauthorized access, disclosure, alteration, and destruction."

Table 8.4 lists several tips to help safeguard your privacy.

safe harbor principles
A set of principles that address the e-commerce data privacy issues of notice, choice, and access.

Table 8.4

How to Protect Your Privacy While Online

Tip	Rationale
Never give out personal information, especially your social security number.	Assume that any information you give out about yourself will appear on a database somewhere. Even something as innocuous as a warranty card may ask for income and phone numbers—don't provide such data.
Before registering with a Web site, check out its privacy policy.	Look for the BBB or TRUSTe seal. If a site doesn't explain how it plans to use your information, beware!
Be discreet in sending e-mail, posting messages to a Usenet newsgroup, and "talking" in chat rooms.	Anything you send in these forums can be viewed and used by a marketer to update your personal profile.
Update your browser software.	The newest versions of Netscape Navigator and Microsoft Internet Explorer, by far the two most popular Web browsers, have the best encryption technology available. Generally speaking, if these companies release an update, someone has found a hole in their security.
Consider the purchase of a digital ID to use in place of your real name.	This permits users to retrieve Web pages anonymously.

STRATEGIES FOR SUCCESSFUL E-COMMERCE

With all the constraints to e-commerce just covered, it is important for a company to develop an effective Web site—one that is easy to use and accomplishes the goals of the company, yet is affordable to set up and maintain. We cover several issues for a successful e-commerce site here.

Developing an Effective Web Presence

When building a Web site, the first thing to decide is which tasks the site must accomplish. Most people agree that an effective Web site is one that creates an attractive presence and that meets the needs of its visitors, including the following:

- Obtaining general information about the organization
- Obtaining financial information for making an investment decision
- Learning the organization's position on social issues
- Learning about the products or services that the organization sells
- Buying the products or services that the company offers
- Checking the status of an order
- Getting advice or help on effective use of the products
- Registering a complaint about the organization's products
- Registering a complaint concerning the organization's position on social issues
- Providing a product testimonial or idea for a product improvement or new product
- Obtaining information about warranties or service and repair policies for products
- Obtaining contact information for a person or department in the organization

Once a company determines which objectives its site should accomplish, it can proceed to the details of actually putting up a site.

As the number of e-commerce shoppers increases, and they become more comfortable—and more selective—making online purchases, it may be necessary to redefine your site's basic business model to capture new business opportunities. For example, consider the major travel sites such as Expedia, Travelocity, Cheap Tickets, Orbitz, and Priceline. These sites used to specialize in just one area of travel, inexpensive airline tickets. Now they are geared toward offering a full-range of travel products, including airline tickets, auto rentals, hotel rooms, tours, and last-minute trip packages. Expedia provides in-depth hotel details to help comparison shoppers and even offers 360-degree visual tours and expanded photo displays. It also entices flexible travelers to search for rates, compare airfares, and configure hotel and air prices at the same time. Expedia also has developed numerous hotel partnerships to reduce costs and help secure great values for consumers. Meanwhile, Orbitz has launched a special full-service program for corporate business travelers.[50]

Putting Up a Web Site

Companies large and small can establish Web sites. Previously, companies had to develop their sites in-house or find and hire contractors to develop their sites. But no longer must you learn the intricacies of HTML or Java, master Web design software, or hire someone to build your site. Web site hosting services and the use of storefront brokers are two options to designing, building, operating, and maintaining a Web site. Both options offer the advantage of getting the Web site running faster and cheaper than doing it yourself, especially for a firm with few or no experienced Web developers.

Web Site Hosting Services

Web site hosting companies such as Bigstep.com and freemerchant.com have made it possible to set up a Web page and conduct e-commerce within a matter of days and with little up-front cost. Such companies have packaged all the basic development tools and made them available for free. Using only a browser, you can choose a Web site template suited to your specific type of business, whatever it may be—clothing, collectibles, or sports equipment. Using the tools provided, you can write the descriptive text and add images of the products you want to sell. Within hours, your site is up on the Web and accessible to millions.

These companies can also provide free hosting for your store, but to allow visitors to pay for merchandise with credit cards, you need a merchant account with a bank. If your company doesn't already have one, then it must establish one. Typically, there is a monthly charge of $10 to $50 plus a transaction fee of $.05 to$.25, and it takes at least a few days to establish a merchant account. Without such an account, you must accept payment by check or money order. In many cases, your site won't have its own distinct Web address—*www.wearmyhats.com*, for instance. Instead, your store will appear as a subsection of your service provider's URL.

Web site hosting companies
Companies that provide the tools and services required to set up a Web page and conduct e-commerce within a matter of days and with little up-front cost.

Storefront Brokers

Another model for setting up a Web site is the use of a **storefront broker**, which serves as a middleman between your Web site and online merchants that have the actual products and retail expertise. At sites such as Bigstep (*www.bigstep.com*), you pick and choose what to sell to match the themes of your site and the products you think might interest your visitors. For example, you can set up to be a storefront operator with Vstore.com and build your own online store with its own unique URL. You create the product categories, choose the products, and customize the store as you see fit. You can build links to your own pages or other sites to display the products on your own Web pages, creating, in effect, a virtual storefront stocked with merchandise that is actually handled by another online merchant.

The storefront broker deals with the details of the transactions, including who gets paid for what, and is responsible for bringing together merchants and reseller sites. The storefront broker is similar to a distributor in standard retail operations, but in this case no product moves—only electronic data flows back and forth. Products are ordered by a customer at your site, orders are processed through a user interface provided by the storefront broker, and the product is shipped by the merchant.

Storefront brokers make their money by taking a commission from the merchant—anywhere from 5 percent to 25 percent—or by collecting a finder's fee per customer from the merchant. The storefront operator usually takes a commission, again in the range of 5 to 25 percent. Although these brokered sales represent a loss of margin for the original merchant, the increased volume can often make up for lost margins. By multiplying the number of outlets for the products—sometimes by the thousands—merchants stand to make more money.

Building Traffic to Your Web Site

The Internet includes hundreds of thousands of e-commerce Web sites. With all those potential competitors, a company must take strong measures to ensure that the customers it wants to attract can find its Web site. The first step is to obtain and register a domain name, and your domain name should say something about your business. For instance, stuff4u might seem to be a good catchall, but it doesn't describe the nature of the business—it could be anything. If you want to sell soccer uniforms and equipment, then you'd try to get a domain name like *www.soccerstuff4u.com*, *www.soccerequipment.com*, or *www.stuff4soccercoaches.com*. The more specific the Web address, the better.

The next step to attracting customers is to make your site search-engine-friendly by including a meta tag in your store's home page. A **meta tag** is a special HTML tag, not visible on the displayed Web page, that contains keywords representing your site's content, which search engines use to build indexes pointing to your Web site. Again, the selection of keywords is critical to attracting customers, so they should be chosen carefully.

You can also use Web site traffic data analysis software to turn the data captured in the Web log file into useful information. This data can tell you the URLs from which your site is being accessed, the search engines and keywords that find your site, and other useful information. Using this data can help you identify search engines to which you need to market your Web site, allowing you to submit your Web pages to them for inclusion in the search engine's index.

Maintaining and Improving Your Web Site

Web site operators must constantly monitor the traffic to their site and the response times experienced by visitors. Internet shoppers expect service to be better than or equal to their in-store experience, says AMR Research (a Boston-based, independent research analysis firm). In a recent year, almost 20 percent of online customers surveyed reported a negative experience with at least one site and said they would not return to that site.[51]

Nothing will drive potential customers away faster than if they experience unbearable delays while trying to view or order your products or services. To keep pace with technology and increasing traffic, it may be necessary over time to modify the software, databases, or hardware on which the Web site runs to ensure good response times. Keynote Systems offers

storefront broker
Companies that act as middlemen between your Web site and online merchants that have the products and retail expertise.

meta tag
A special HTML tag, not visible on the displayed Web page, that contains keywords representing your site's content, which search engines use to build indexes pointing to your Web site.

Web site customer-experience technology to analyze the usability of a Web site, a tool to monitor how Web site visitors use a site, and a benchmark system to provide customer scores on a site's content, ease of use, performance, brand value, and satisfaction.[52]

Web site operators must also continually be alert to new trends and developments in the area of e-commerce and be prepared to take advantage of new opportunities. For example, recent studies show that customers more frequently visit Web sites they can customize. Users who personalize such sites also more frequently subscribe to paid sites, use online bill payment services, and promote products via e-mail to their friends, all of which make them an advertiser's dream. **Personalization** is the process of tailoring Web pages to specifically target individual consumers. The goal is to meet the customer's needs more effectively and efficiently, make interactions faster and easier, and, consequently, increase customer satisfaction and the likelihood of repeat visits. Building a better understanding of customer preferences also can aid in cross-selling related products and up-selling more expensive products. The most basic form of personalization involves simply using the consumer's name in an e-mail campaign or in a greeting on the Web page. A more advanced form of personalization is that employed by Amazon.com, in which each repeat customer is greeted by name and a list of new products is recommended based on the customer's previous purchases.

Businesses use two types of personalization techniques to capture data and build customer profiles. *Implicit personalization* techniques capture data from actual customer Web sessions—primarily based on which pages were viewed and which weren't. *Explicit personalization* techniques capture user-provided information, such as information from warranties, surveys, user registrations, and contest entry forms filled out online. Data can also be gathered through access to other data sources such as the Bureau of Motor Vehicles, Bureau of Vital Statistics, and marketing affiliates (firms that share marketing data). Marketing firms aggregate this information to build databases containing a huge amount of consumer behavioral data. During each customer interaction, both types of data are analyzed in real time by use of powerful algorithms to predict the consumer's needs and interests. This analysis makes it possible to deliver new, targeted information before the customer moves on. Because personalization depends on the gathering and use of personal user information, privacy issues are a major concern.

These tips and suggestions are only a few ideas that can help a company set up and maintain an effective e-commerce site. With technology and competition changing continually, managers should read articles in print and on the Web to keep up to date on ever-evolving issues.

personalization

The process of tailoring Web pages to specifically target individual consumers.

SUMMARY

Principle

E-commerce is a new way of conducting business, and as with any other new application of technology, it presents both opportunities for improvement and potential problems.

Businesses and individuals use e-commerce to reduce transaction costs, speed the flow of goods and information, improve the level of customer service, and enable the close coordination of actions among manufacturers, suppliers, and customers. E-commerce also enables consumers and companies to gain access to worldwide markets. E-commerce offers great promise for developing countries, helping them to enter the prosperous global marketplace, and hence helping to reduce the gap between rich and poor countries.

Business-to-business (B2B) e-commerce allows manufacturers to buy at a low cost worldwide, and it offers enterprises the chance to sell to a global market right from the start. B2B e-commerce is currently the largest type of e-commerce.

Although it is gaining acceptance, business-to-consumer (B2C) e-commerce is still relatively new. By using B2C e-commerce to sell directly to consumers, a producer or provider of consumer services can eliminate the middlemen, or intermediaries, between it and its end consumer. In many cases, this squeezes costs and inefficiencies out of the supply chain and can lead to higher profits and lower prices for consumers.

Consumer-to-consumer (C2C) e-commerce involves consumers selling directly to other consumers. Online auctions are the chief method by which C2C e-commerce is currently conducted.

A successful e-commerce system must address the many stages consumers experience in the sales life cycle. At the heart of any e-commerce system is the ability of the user to search for and identify items for sale; select those items; negotiate prices, terms of payment, and delivery date; send an order to the vendor to purchase the items; pay for the product or service; obtain product delivery; and receive after-sales support.

The first major challenge to success online is for the company to define an effective e-commerce strategy. Although there are a number of different approaches to e-commerce, the most successful models include three basic components: community, content, and commerce. Another major challenge for companies moving to B2C e-commerce is the need to change their distribution systems and work processes to be able to manage shipments of individual units directly to consumers and deal with split-case distribution. A third, tough challenge for e-commerce is the integration of new Web-based order processing systems with traditional mainframe computer-based inventory control and production planning systems.

Supply chain management is composed of three subprocesses: demand planning to anticipate market demand, supply planning to allocate the right amount of enterprise resources to meet demand, and demand fulfillment to fulfill demand quickly and efficiently. Conversion to e-commerce supply chain management provides businesses an opportunity to achieve operational excellence by increasing revenues, decreasing costs, improving customer satisfaction, and reducing inventory. But to achieve this goal requires integrating all subprocesses that exchange information and move goods between suppliers and customers, including manufacturers, distributors, retailers, and any other enterprise within the extended supply chain.

The use of the Internet is growing rapidly in markets throughout Europe, Asia, and Latin America. China represents a huge market opportunity for companies around the globe because its Internet population is expected to grow rapidly. Companies that want to succeed on the Web cannot ignore the growth in non-U.S. markets.

The first step in developing a global e-commerce strategy is to determine which global markets make the most sense for selling products or services online. Then it must adapt an existing U.S.-centric site to another language and culture—a process called *localization.* A difficult decision in a company's global Web strategy is whether Web content should be generated and updated centrally or locally, and companies must remember that global consumers may access sites with different devices and modify their site design accordingly. Many manufacturers and retailers have outsourced the physical logistics of delivering merchandise to cybershoppers. To provide this service, delivery firms have developed software tools and interfaces that directly link customers' ordering, manufacturing, and inventory systems with their own system of highly automated warehouses, call centers, and worldwide shipping network. The goal is to make the transfer of all information and inventory—from the manufacturer to the delivery firm to the consumer—fast and simple.

Mobile commerce is the use of wireless devices such as PDAs, cell phones, and smart phones to facilitate the sale of goods or services—anytime, anywhere. While some industry experts predict great growth in this arena, several hurdles must be overcome, including simplifying the use of wireless devices, improving network speed, and security. The market for m-commerce in North America is expected to mature much later than in Western Europe and Japan. The wireless application protocol (WAP) is a standard set of specifications to enable development of m-commerce software for wireless devices.

Electronic retailing (e-tailing) is the direct sale from a business to consumers through electronic storefronts

designed around an electronic catalog and shopping cart model. A cybermall is a single Web site that offers many products and services at one Internet location. Manufacturers are joining electronic exchanges, where they can join with competitors and suppliers to use computers and Web sites to buy and sell goods, trade market information, and run back-office operations such as inventory control. They are also using e-commerce to improve the efficiency of the selling process by moving customer queries about product availability and prices online. The Web allows firms to gather much more information about customer behavior and preferences than they could using other marketing approaches. This new technology has greatly enhanced the practice of market segmentation and enabled companies to establish closer relationships with their customers. Detailed information about a customer's behavior, preferences, needs, and buying patterns allow companies to set prices, negotiate terms, tailor promotions, add product features, and otherwise customize a relationship with a customer. The Internet has also revolutionized the world of investment and finance, especially online stock trading and online banking. The Internet has also created many options for electronic auctions, where geographically dispersed buyers and sellers can come together.

Principle

E-commerce requires the careful planning and integration of a number of technology infrastructure components.

A number of infrastructure components must be chosen and integrated to support a large volume of transactions with customers, suppliers, and other business partners worldwide. These components include hardware, Web server software, e-commerce software, and network and packet switching.

Current e-commerce technology relies on the use of identification and encryption to safeguard business transactions. Web site operators must protect their sites against a denial-of-service (DOS) attack, in which the attacker takes command of many computers on the Internet and uses them to flood the target Web site with requests for data and other small tasks, preventing the target machine from serving legitimate users. A digital certificate is an attachment to an e-mail message or data embedded in a Web page that verifies the identity of a sender or a Web site. To help prevent the theft of credit card numbers and banking information, the Secure Sockets Layer communications protocol is used to secure all sensitive data.

Actual payments are made in a variety of ways including electronic cash, electronic wallets, and smart, credit, charge, and debit cards.

Principle

Users of e-commerce technology must use safeguards to protect themselves.

Revolutionary change always involves controversy, and e-commerce is no exception. Among the issues that must be addressed are security, intellectual property rights, fraud, and privacy. E-commerce shoppers must be on constant guard to protect their rights, security, and personal privacy.

Increasingly, organizations are using identification technologies to confirm the identity of a user requesting access to information or assets. Biometric technology, which digitally encodes physical attributes of a person's voice, eye, face, or hand and associates them with biological attributes stored in a file, is commonly used in organizations requiring a high degree of security.

Lawsuits over intellectual property have created a virtual e-commerce war zone.

Principle

Organizations must define and execute a strategy to be successful in e-commerce.

Most people agree that an effective Web site is one that creates an attractive presence and meets the needs of its visitors. E-commerce start-ups must decide whether they will build and operate the Web site themselves or outsource this function. Web site hosting services and storefront brokers provide alternatives to building your own Web site. It is also crucial to build traffic to your Web site by registering a domain name that is relevant to your business, making your site search-engine-friendly by including a meta tag in your home page, and using Web site traffic data analysis software to attract additional customers. Web site operators must constantly monitor the traffic and response times associated with their site and adjust software, databases, and hardware to ensure that visitors have a good experience when they visit the site.

Personalization is the process of tailoring Web pages to specifically target individual consumers. The goal is to meet the customer's needs more effectively and efficiently, make interactions faster and easier, and, consequently, increase customer satisfaction and the likelihood of repeat visits.

CHAPTER 8: SELF-ASSESSMENT TEST

E-commerce is a new way of conducting business, and as with any other new application of technology, it presents both opportunities for improvement and potential problems.

1. Successful implementation of e-business requires _____ and _____.

   a. changes to existing business processes; substantial investment in IS technology
   b. conversion to XML software standards; Java programming scripts
   c. implementation of tight security standards; Web site personalization
   d. market segmentation; Web site globalization

2. Amazon.com is an example of what form of e-commerce?

   a. A2B
   b. B2B
   c. B2C
   d. C2C

3. Online sales of products to consumers now exceed traditional retail sales. True or False?

4. E-commerce is pretty much limited to manufacturing firms; service firms have found it difficult to take advantage of this new business approach. True or False?

5. The two countries expected to show the fastest growth in e-commerce over the next 10 years are _____ and _____.

E-commerce requires the careful planning and integration of a number of technology infrastructure components.

6. A multistage model for purchasing over the Internet includes search and identification, _____ and _____, purchasing, product or service delivery, and after-sales support.

7. Which of the following is NOT one of three subprocesses of supply chain management?

   a. demand planning to anticipate market demand
   b. supply planning to allocate the right amount of enterprise resources to meet demand
   c. demand fulfillment to fulfill demand quickly and efficiently
   d. promotion to increase customer demand

8. The use of electronic exchanges continues to grow rapidly. True or False?

9. Which of the following is a challenge that must be overcome for a company to convert its business processes from the traditional form to business-to-consumer e-commerce processes?

   a. define an effective e-commerce model and strategy
   b. change distribution systems and work processes to be able to manage shipments of individual units directly to consumers
   c. integrate new Web-based order-processing systems with traditional mainframe computer–based inventory control and production planning systems.
   d. all of the above

10. _____ management software combines different product data formats into a standard format for uniform viewing, aggregating, and integrating catalog data into a central repository.

11. Music, books, inventions are all examples of _____.

Users of e-commerce technology must use safeguards to protect themselves.

12. Phishing entails sending bogus messages purportedly from a legitimate institution to pry personal information from consumers by convincing them to go to a "spoof" Web site. True or False?

13. Data gathered based on the Web sites you visit and what items you click on is called _____.

14. A smart card makes credit theft practically impossible because a key to unlock the encrypted information is required, and there is no external number that a thief can identify and no physical signature a thief can forge. True or False?

Organizations must define and execute a strategy to be successful in e-commerce.

15. A(an) _____ is a special HTML tag, not visible on the display Web page, that contains keywords representing your site's content, which search engines use to build indexes pointing to your Web site.

CHAPTER 8: SELF-ASSESSMENT TEST ANSWERS

(1) a (2) c (3) False (4) False (5) India and China (6) selection and negotiation (7) d (8) False (9) d (10) Catalog (11) intellectual property (12) True (13) clickstream data (14) True (15) meta tag

KEY TERMS

business-to-business (B2B)
  e-commerce 350
business-to-consumer (B2C)
  e-commerce 350
catalog management software 372
certificate authority (CA) 374
clickstream data 382
cybermall 363
consumer-to-consumer (C2C)
  e-commerce 351
denial-of-service (DOS) attack 371
digital certificates 374
disintermediation 359
dynamic Web pages 372
e-commerce software 372

e-commerce transaction processing
  software 373
electronic bill presentment 369
electronic cash 376
electronic exchange 364
electronic retailing (e-tailing) 363
electronic shopping cart 373
electronic wallet 377
intellectual property 379
market segmentation 366
meta tag 385
personalization 386
phishing 380
product configuration software 372
safe harbor principles 383

Secure Sockets Layer (SSL) 374
smart card 377
spam 381
split-case distribution 355
static Web pages 372
storefront broker 385
supply chain management 356
technology-enabled relationship
  management 367
wireless application protocol (WAP)
  362
Web log file 371
Web page construction software 372
Web site development tools 371
Web site hosting companies 384
Web site traffic data analysis software
  374

REVIEW QUESTIONS

1. Define the term *e-commerce*. Identify and briefly describe three different forms of e-commerce. Which form is the largest in terms of dollar volume?
2. What is electronic data interchange? What industries have been leaders in the use of EDI?
3. Identify the six stages consumers experience in the sales life cycle that must be supported by a successful e-commerce system.
4. What is supply chain management?
5. What is the purpose of adding message boards and chat rooms to an e-commerce Web site?
6. What benefits can a firm achieve through conversion to an e-commerce supply chain management system?
7. What is disintermediation? What is its cause?
8. Define *m-commerce*. Which forms of e-commerce can it support?
9. What are some of the special limitations that complicate the use of handheld devices for m-commerce?
10. What are some of the issues associated with the use of electronic exchanges?

11. What is the wireless application protocol?
12. Why is it necessary to continue to maintain and improve an existing Web site?
13. What role do digital certificates and certificate authorities play in e-commerce?
14. What is the Secure Sockets Layer and how does it support e-commerce?
15. Briefly explain the differences among smart, credit, charge, and debit cards.
16. What actions can you take to minimize the risk of being a victim of e-commerce fraud?
17. What actions can you take to safeguard your individual privacy as you surf the Web?
18. What is a cybermall?
19. What is technology-enabled relationship management?
20. Identify the key elements of technology infrastructure required to successfully implement e-commerce within an organization.
21. What is a denial-of-service attack? How can a Web site be defended from such an attack?

DISCUSSION QUESTIONS

1. Briefly summarize the state of global e-commerce in Europe, Asia, and Latin America.
2. Why are many manufacturers and retailers outsourcing the physical logistics of delivering merchandise to shoppers?

What advantages does such a strategy offer? Are there any potential issues or disadvantages?

3. What does it mean to localize a Web site? What are some of the steps involved in doing this?

4. Why is it said that the acceptance of m-commerce is geographically dependent?

5. Wal-Mart, the world's number-one retail chain, has turned down several invitations to join exchanges in the retail and consumer goods industries. Is this good or bad for the overall U.S. economy? Why?

6. Identify and briefly describe three potential m-commerce applications.

7. Discuss the use of e-commerce to improve spending on manufacturing, repair, and operations (MRO) goods and services.

8. Discuss the pros and cons of e-commerce companies capturing data about you as you visit their sites.

9. What is an electronic exchange? How successful have they been? What are some of the benefits and issues associated with their use?

10. Outline the key steps in developing a corporate global e-commerce strategy.

11. Describe how the DoubleClick Network works.

12. Discuss the difference between electronic bill payment and electronic bill presentment. What are the benefits and some issues associated with each?

13. Discuss how you might gather Web traffic data for analysis of your firm's Web site. What decisions might this data be useful in making?

PROBLEM-SOLVING EXERCISES

1. As a team, develop a set of criteria you would use to evaluate various business-to-consumer Web sites on the basis of ease of use, protection of consumer data, security of payment process, etc. Develop a simple spreadsheet containing these criteria. Evaluate five different popular Web sites using the criteria you developed.

2. Do research to get current data about the growth of B2B, B2C, or C2C e-commerce—either in the United States or worldwide. Use a graphics software package to create a line graph representing this growth. Extend the growth line five years beyond the available data using two different

modeling tools available with the software package. Write a paragraph discussing the issues and assumptions that affect the accuracy of your five-year projection and the likelihood that e-commerce will achieve this forecast.

3. Do research to learn more about the use of WAP and other specifications being developed to support m-commerce. Briefly describe the specifications you uncover. Who is behind the development of these standards? Which standards seem to be gaining the broadest acceptance? Prepare a one- to two-page report for your instructor.

TEAM ACTIVITIES

1. Imagine that your team has been hired as consultants to provide recommendations to boost the traffic to a Web site that sells sports apparel and equipment online. Identify as many ideas as possible for how you can increase traffic to this Web site. Next, rank your ideas from best to worst.

2. As a team, choose an idea for an e-commerce Web site—products or services you would provide. Develop an implementation plan that outlines the steps you need to take and the decisions you must make to set up the Web site and make it operational.

WEB EXERCISES

1. Find a Web site that provides investment portfolio tracking. As a team, select five stocks to make an "imaginary" purchase and allocate $100,000 among them. Check back in one week and determine the change in value in your investment. Make a list of the various features and

analyses that are available at this Web site. Compare the change in the value of your portfolio to that of other teams. Compare the various features and analyses available at your Web site to the sites used by other teams.

2. Do research and document the current status of the Recording Industry Association of America's fight to prevent the illegal sharing of music files. Write a brief report for your instructor. Include a discussion of your opinion on the legality of sharing music files.

CAREER EXERCISES

1. Do research to identify those industries where e-commerce is making the biggest impact and the least impact. Prepare a brief report about each industry.

2. For your chosen career field, describe how you might use or be involved with e-commerce. If you have not chosen a career yet, answer this question for someone in marketing, finance, or human resources.

VIDEO QUESTIONS

Watch the video clip **Find the Best Deals Online** and answer these questions:

1. What services do Web sites like Epinions.com offer consumers that are unavailable through traditional brick-and-mortar mall shopping?

2. What scam did the reporter warn about regarding e-commerce vendor reviews and ratings?

CASE STUDIES

Case One

Izumiya of Japan: E-Commerce Poster Child

Izumiya Co. of Japan is a retail chain giant that offers food, clothing, consumer electronics, furniture, and housewares. It could be called the Super Wal-Mart of Japan. Headquartered in Osaka, Izumiya employs more than 2,500 employees at 81 retail outlets. Founded in 1921 with the motto "high-quality goods at low prices," Izumiya is a popular, well-established retailer that has been quick to take advantage of both the B2C and B2B benefits of e-commerce.

Izumiya worked with IBM to develop the fastest, lowest-cost online grocery service in Japan. Izumiya customers can log on to *www.izumiya.co.jp*, place an order, and have groceries delivered to their doorstep within an hour. "By making grocery shopping almost effortless, we hoped to increase the frequency of orders and in the process boost overall sales," says Kazunobu Tanaka, E-commerce Project Team Leader.

To offer such a service, Izumiya needed to significantly change its business processes, information systems, and infrastructure. Inventory management systems, storage facilities, and distribution networks all had to be adjusted to support the new e-commerce system. Support staff were hired and equipped with handheld computers to process transactions on the customer's doorstep. An entirely new system is required to handle online orders, and it must work in harmony with the existing system that serves in-store customers.

Izumiya was able to keep the overhead low so that it could offer the service at a price that would attract the most customers. The number of customers is growing rapidly, indicating that the new Web-enabled channel will achieve a respectable return on investment for the retailer in short order.

In the meantime, Izumiya is enjoying the benefits of B2B e-commerce through its membership in the WorldWide Retail Exchange (WWRE), the premier Internet-based business-to-business exchange in the international retail marketplace. "Joining the WWRE places us in a unique position to collaborate with the world's leading retailers and manufacturers and gives us access to expertly selected technologies and standards, broad international experience, and well-tested strategies," said Norio Hayashi, President of Izumiya.

The WWRE facilitates and simplifies trading between retailers, suppliers, partners, and distributors. It consists of 62 members from Africa, Asia, Europe, North America, and

South America, with combined sales of approximately $900 billion. The exchange has saved its members more than $450 million through the use of online negotiations.

By being aggressive in its use of the Internet and experimenting with new ways to grocery shop, Izumiya has impressed its customers with its innovations and provided them with a valuable new service. In joining a global retail exchange, Izumiya is enjoying benefits of discounted supplies and expanding into the global marketplace. With such a full embrace of e-commerce, Izumiya might be considered the poster child for e-commerce.

Discussion Questions

1. How has e-commerce assisted Izumiya in gaining an edge over competitors that have not embraced e-commerce?

2. How does the WorldWide Retail Exchange (WWRE) benefit retailers and suppliers?

Critical Thinking Questions

3. What risks did Izumiya take in offering the service that allows its customers to shop online?

4. How does Izumiya's joining the WWRE bring additional benefits to the other members of the WWRE? In other words, What can Izumiya bring to the table that could be of interest to other retailers?

SOURCES: "Izumiya Joins WWRE," *Business Wire*, May 27, 2003, *www.lexis-nexis.com*; "Izumiya: Fastest, Lowest-Cost Online Grocery Service in Japan," *IBM Success Stories*, accessed April 20, 2004, *www.ibm.com*; Izumiya Web site, accessed April 20, 2004, *www.izumiya.co.jp*; The WorldWide Retail Exchange Web site, accessed April 20, 2004, *www.worldwideretailex-change.org*.

Case Two

Dylan's Candy Bar Sweetens Its Sales with E-Commerce

A new chain of upscale, trendy candy shops recently opened in Long Island, Orlando, and Houston. Owned by Dylan Lauren (daughter of fashion designer Ralph) and Jeff Rubin, Dylan's Candy Bar claims to be "the most unique and unrivaled candy and sweets store, home to thousands of candies from all over the world!" Dylan's sells much more than candy. Its inventory includes items from four categories: Candy Couture (apparel), Candy Baskets, Candy Creations (topiaries made from candy), and Candy Spa. Besides the brick-and-mortar shops that are advertised as "creating a unique and completely unmatched shopping experience in an architecturally significant and artistically awing environment," Dylan's is making strong use of e-commerce and m-commerce to sell products.

As you might assume, Dylanscandybar.com allows customers to purchase Dylan's products online and to have them delivered anywhere in the world. But Dylan's took the traditional B2C e-commerce one step further by launching what it calls an "mmm-commerce wireless campaign." Customers can purchase and download some items to their cell phones. For example, you can purchase and download candy-themed ring tones, such as "I Want Candy" by Bow Wow Wow and "Sugar, Sugar" by the Archies for $2.99. You can also send an m-commerce candygram from the Dylan's Web site. Type in a friend's cell phone number, select a message, and click the button to deliver the text message. You can use the messages to send a Dylan's gift certificate or to invite a friend to meet you at Dylan's.

Dylan's plans to expand its mmm-commerce offerings with a $3.99 package that includes a personalized text message, a related ring tone, and a lollipop gift certificate to send to a friend from Web or cell phone. Dylan's also has several candy-themed video games at its Web site, which are ideally suited for access from a smart phone.

The technology company behind Dylan's mmm-commerce campaign is the YOUIE corporation. YOUIE is in the business of creating and distributing what it calls wireless infotainment. YOUIE stands for YOU, Infotainment, and Entertainment. Shane Igoe, president of YOUIE, thinks that ring tone and text message marketing is effective with teenagers and young adults; they drive traffic not only back to the Web site but directly into the stores.

Igoe started the YOUIE corporation anticipating that markets in m-commerce will grow rapidly over the next few years. The YOUIE Web site states that "in less than four years, more people will use their phones to access wireless content than use them for voice. It is estimated that over a billion mobile phone users will be looking for new content from their mobile phones." According to industry analyst firm Yankee Group, more than 200 million people in the United States and western Europe—80 percent of all wireless phone users—will play online games using wireless devices by 2005. Probe Research predicts that there will be 450 million mobile gaming players worldwide by 2006.

The YOUIE Web site offers dozens of ring tones and cell phone graphics for $1.99 each. It also offers a number of mobile infotainment applications and is developing single and multiplayer games. YOUIE's "key to success" lies in building m-commerce applications alongside powerful partners. YOUIE's wireless games and other applications are promoted and branded in conjunction with highly visible brands, entertainment properties, and other major media events.

Discussion Questions

1. Are cell phone users likely to be annoyed when they receive text messages from Dylan's mmm-commerce service? Why or why not?

2. Why might this type of e-commerce campaign work better for a place like Dylan's Candy Bar than for a high-end retailer like Sak's Fifth Avenue?

Critical Thinking Questions

3. What other businesses might benefit from using the Dylan's m-commerce model?

4. How might customers react if they receive unsolicited text advertisements from Dylan's on their cell phone? Under what conditions might text advertising work?

SOURCES: Mark Hazlin, "Dylan Lauren's Sugar Shop: More Potent than the Super Bowl Half-Time Show," *NYU Livewire*, February 11, 2004, *journalism.nyu.edu/pubzone/livewire/000047.php*; Beth Cox, "Mmm-commerce—How Sweet Can It Be?" *e-commerce guide.com*, June 23, 2003, //www.ecommerce-guide.com/news/trends/article.php/2226311 Dylan's Candy Bar Web site, accessed April 20, 2004, *www.dylanscandybar.com*; The Youie Corporation Web site, accessed April 20, 2004, *www.youie.com*.

Case Three

RealEstate.com: Buying and Selling Homes Online

The real estate market might be considered the holy grail of B2C e-commerce. Many respectable companies have searched for it, but none have returned with the prize. Consider Cendant, one of the foremost providers of travel and real estate services in the world. In 1997, it published a Web site that automated the time-consuming process of closing a home sale, eliminating huge amounts of paperwork. That business failed. It then produced Move.com, a one-stop Web site with home listings and tools for home buyers and sellers. That idea also failed. More recently, Cendant tried to create a discount online real estate broker called Blue Edge. That, too, failed. The reason for these failures seemed to stem from the customer's unwillingness to trust important transactions to automated systems rather than competent human brokers. Cendant's frustrations with online real estate were shared by many others seeking to win the grail.

The dot-com boom spawned dozens of Internet-based brokers and other cyber real estate ideas. "Microsoft, Yahoo!, everyone and their brother poured a lot of money into real estate," said Brad Inman, a longtime publisher of real estate trade news. "They come in starry-eyed and say these Realtors are idiots. They spend a lot of money. And they realize they can't unlock the role of the Realtor. So they get out."

After ten years of failed attempts at this market, most have given up, figuring that the role of the Real Estate broker simply cannot be automated. Some people, however, find this $1.5 trillion market irresistible and continue to look for new e-commerce services that will attract home buyers. Barry Diller, CEO of InterActiveCorp, is just such a person. InterActiveCorp is the world's leading multibrand interactive commerce company. InterActiveCorp owns Expedia.com, Hotels.com, the HSN channel, Ticketmaster, Match.com, Hotwire.com, LendingTree.com, and other companies.

On the subject of online real estate, Diller commented, "If you think about real estate, with one million brokers and trillions of dollars in commerce, and information that now is able to be accessed by individuals with great ease, it is a perfect place for us to play. Right now, online adoption is tiny, and over time it will grow."

Studies show that the Web does play a role in the real estate market. Three-quarters of home buyers turn to the Internet to look at listings and learn about neighborhoods. Then they turn to realtors to visit homes and negotiate their deals. Diller wants to build a business that links Web shoppers with brokers. He bought RealEstate.com to accomplish his goal. RealEstate.com is run under the popular LendingTree.com umbrella, but unlike LendingTree, the site focuses exclusively on real estate: finding a home, finding a realtor, and finding a mortgage company.

You might wonder how a business can profit from connecting customers with realtors. The answer is that the service is provided at the expense of the realtor. While RealEstate.com isn't out to replace the realtor, it does take a substantial cut from the realtor's commission: one third to be exact. A portion of the percentage of the commission that RealEstate.com earns is given back to the customer in the form of gift cards for Home Depot. For example, if you purchase a $220,000 home through a realtor you met through RealEstate.com, you would get a $1,000 gift card for shopping at Home Depot.

Cutting into realtor's commissions, which have fallen over the last decade from 6 percent to 5.1 percent, is making RealEstate.com some enemies in the industry. No one has more at stake from real estate commissions than Cendant, which controls more than 25 percent of the brokerage business through its Century 21, Coldwell Banker, and ERA franchises. Cendant has brought charges against InterActiveCorp, owner of RealEstate.com, for false advertising. Cendant is also pushing the National Association of Realtors to adopt rules that would prevent some listings from being used on Web sites that earn most of their money from referrals. Without access to all listings in an area, RealEstate.com would lose credibility and customers.

Is InterActiveCorp worried? With the growing success of RealEstate.com, there are bound to be many legal battles to win their piece of the grail. But with its transactions exceeding $1.7 billion annually, and annual earnings of more than $16 million in fees from brokers, RealEstate.com seems to be providing a service that interests home buyers. It appears that InterActiveCorp has finally achieved what so many others have attempted but failed at—delivering a profitable e-commerce real estate service.

Discussion Questions

1. Why is it so much more difficult to sell homes over the Web than other products such as air travel and hotel rooms?

2. Besides homes, what other products are difficult to sell on the Web? Why?

Critical Thinking Questions

3. Why do people seem to distrust automated systems when it comes to real-estate transactions?

4. Do you think that the National Association of Realtors should be allowed to withhold listings from companies like RealEstate.com? Why or why not?

SOURCES: Saul Hansell, "Point. Click. En Garde!" *New York Times*, April 18, 2004, *www.nytimes.com*; "LendingTree Completes Acquisition of RealEstate.com," *PR Newswire*, December 23, 2003, *www.lexisnexis.com*; RealEstate.com Web site, accessed April 22, 2004; InterActiveCorp Web site, accessed April 22, 2004, *www.iac.com*; Cendant Web site, accessed April 22, 2004, *www.cendant.com*.

NOTES

Sources for the opening vignette: "Is That a Radio in Your Pocket?" *PR Newswire*, January 8, 2004, *www.lexis-nexis.com*; "Sprint and Sony Music Entertainment Announce Broad Strategic Partnership to Distribute Mobile Entertainment Content," *Sprint News Releases*, June 30, 2003, *www.sprint.com*; Brian Garrity, "Wireless Deals Focus on Ring Tunes," *Billboard*, April 3, 2004, *www.lexis-nexis.com*, "Sony Ericsson and Handango Announce Easy Over-the-Air Download to SmartPhones," *Hugin*, October 20, 2003, *www.lexis-nexis.com*; Rebecca Harper, "The New Currency: Kiss Your Cash Good-bye," *Wired*, April 2004, p. 59.

1. Chabow, Eric, "E-Commerce Continues to Grow Very Nicely," *InformationWeek*, May 1, 2003, accessed at *www.informationweek.com*.
2. Rosencrance, Linda, "Commerce Department: E-Commerce Sales Up Sharply in Q4," *Computerworld*, February 23, 2004, accessed at *www.computerworld.com*.
3. Regan, Keith, "Yahoo Pays $575M for E-Shopping Site," *E-Commerce Times*, March 26, 2004, accessed at *www.ecommercetimes.com*.
4. Alexander, Steve, "Web Site Adds Inventory Control and Forecasting," *Computerworld*, February 24, 2003, accessed at *www.computerworld.com*.
5. Verton, Dick, "DHS Moving Its Buying Power Online," *Computerworld*, February 6, 2004, accessed at *www.computerworld.com*.
6. Send Money and Request Money at the PayPal Web site accessed at *www.paypal.com* on April 17, 2004.
7. "Nippon Steel Back in Black," *CNN.com International*, November 6, 2003, accessed at *http://edition.cnn.com*.
8. Silva, Carol, "Retailers See RFID Ahead," *Computerworld*, January 6, 2004, accessed at *www.computerworld.com*.
9. Brewin, Bob, "Delta Air Plans RFID Bag Tag Test," *Computerworld*, June 18, 2003, accessed at *www.computerworld.com*.
10. Brewin, Bob, "FDA Backs RFID Tags for Tracking Prescription Drugs," *Computerworld*, February 23, 2004 accessed at *www.computerworld.com*.
11. Regan, Keith, "Yahoo Pays $575M for E-Shopping Site," *E-Commerce Times*, March 26, 2004, accessed at *www.ecommercetimes.com*.
12. "Update: eBay profit Nearly Doubles, Outlook Raised," *Reuters*, April 21, 2004, accessed at *http://news.moneycentral.msn.com*.
13. Diana, Alison, "Crossing Over: E-Business Transcends National Boundaries," *E-Commerce Times*, January 15, 2004, accessed at *www.ecommercetimes.com*.
14. Millard, Elizabeth, "The State of European E-Commerce," *E-Commerce Times*, July 26, 2003, accessed at *www.ecommercetimes.com*.
15. Diana, Alison, "Crossing Over: E-Business Transcends National Boundaries," *E-Commerce Times*, January 15, 2004, accessed at *www.ecommercetimes.com*.
16. Lyman, Jay, "Internet Users in China Number Nearly 80 Million," *E-Commerce Times*, January 15, 2004, accessed at *www.ecommercetimes.com*.

17. Lyman, Jay, "Internet Users in China Number Nearly 80 Million," *E-Commerce Times*, January 15, 2004, accessed at *www.ecommercetimes.com*.
18. Millard, Elizabeth, "The Mushrooming Chinese Internet Market," *E-Commerce Times*, July 19, 2004, accessed at *www.ecommercetimes.com*.
19. "Doing E-Business in South Korea," Economist Intelligence Unit accessed at *www.ebusinessforum.com* on September 2, 2004.
20. "Doing E-Business in India," Economist Intelligence Unit accessed at *www.ebusinessforum.com* on September 2, 2004.
21. Burger, Andrew, "Se Habla English, Latin America Rising," *E-Commerce Times*, March 11, 2004, accessed at *http://ecommercetimes.com*.
22. Rosencrance, Linda, "E-Commerce on the Fly," *Computerworld*, May 26, 2003, accessed at *www.computerworld.com*.
23. Rosencrance, Linda, "E-Commerce on the Fly," *Computerworld*, May 26, 2003, accessed at *www.computerworld.com*.
24. Adrian, Bradford, "Consumers Need Local Reason to Pay by Phone," Gartner Group, June 16, 2003, accessed at *www4.gartner.com*.
25. Bacheldor, Beth, "Albertson's Technology Brings Handhelds to Customers," *InformationWeek*, April 8, 2004, accessed at *www.informationweek.com*.
26. "Holiday Prep: PC Mall Fights Online Fraud with Three Tiered Defense," *Network World Fusion*, December 16, 2003, accessed at *www.nwfusion.com*.
27. WorldWide Retail Exchange Web site accessed at *www.worldwideretailexchange.org/cs/en/about_wwre/overview.htm on April 14*, 2004.
28. Ulfelder, Steve, "B2B Exchange Survivors," *Computerworld*, February 2, 2004, accessed at *www.computerworld.com*.
29. Ulfelder, Steve, "B2B Exchange Survivors," *Computerworld*, February 2, 2004, accessed at *www.computerworld.com*.
30. Scheier L., Robert, "The Price of E-Payment," *Computerworld*, May 26, 2003, accessed at *www.computerworld.com*.
31. Rosencrance, Linda "Ebay Selects Pitney Bowes," *Computerworld*, March 9, 2004 accessed at *www.computerworld.com*.
32. Vijayan, Jaikumar, "Mydoom Lesson: Take Proactive Steps to Prevent DDoS Attacks," *Computerworld*, February 9, 2004, accessed at *www.computerworld.com*.
33. Vijayan, Jaikumar, "Cisco Continues Security Push," *Computerworld*, March 15, 2004, accessed at *www.computerworld.com*.
34. Anthes, Gary H., "B2B: Corporate Express Goes Direct," *Computerworld*, September1, 2003, accessed at *www.computerworld.com*.
35. Bednarz, Ann, "Holiday Prep: Anaconda Sports Puts Web Infrastructure to the Test," *Network World Fusion*, December 18, 2003, accessed at *www.nwfusion.com*.
36. "Holiday Prep: PC Mall Fights Online Fraud with Three Tiered Defense", *Network World Fusion*, Deember 16, 2003, accessed at *www.nwfusion.com*.
37. Messmer, Ellen, "Feeling Vulnerable? Try Assessment Tools," *Network World Fusion*, April 5, 2004, accessed at *www.nwfusion.com*.

38. Maestri, Nicole, "Target to Phase Out Smart Visa Cards," *Computerworld*, March 3, 2004, accessed at *www.computerworld.com*.

39. Verton, Dan, "DHS Broadens Biometrics Use for Border Control," *Computerworld*, October 13, 2003, accessed at *www.computerworld.com*.

40. Thibodeau, Patrick, "HP Moves to Bolster Intellectual Property Rights Enforcement," *Computerworld*, January 12, 2004, accessed at *www.computerworld.com*.

41. Konrad, Richard, "HP Sues Gateway For Patent Infringement," *InformationWeek*, March 25, 2004, accessed at *www.informatineek.com*.

42. Friedenberg, Michael, "Showdown at The SCO Corral," *InformationWeek*, March 15, 2004, accessed at *www.informationweek.com*.

43. Fox, Pimm, "RIAA Stance Sounds Off Key," *Computerworld*, September 22, 2003, accessed at *www.computerworld.com*.

44. Weisman, Robyn, "FTC Commissioner on the Future of E-Commerce," *E-Commerce Times*, October 2, 2003, accessed at *www.ecommercetimes.com*.

45. Mello Jr., John P., "Big Bank Customers Targeted by Internet Scammers," *E-Commerce Times*, December 16, 2003, accessed at *www.ecommercetimes.com*.

46. Messmer, Ellen, "Craftier Web Threats Hit Finance Firms," *Network World Fusion*, March 15, 2004, accessed at *http://nwfusion.com*.

47. "Holiday Prep: PC Mall Fights Online Fraud with Three Tiered Defense," *Network World Fusion*, December 16, 2003, accessed at *www.nwfusion.com*.

48. Weisman, Robyn, "FTC Commissioner on the Future of E-Commerce," *E-Commerce Times*, October 2, 2003, accessed at *www.ecommercetimes.com*.

49. Claburn, Thomas, "New FTC Rules Requires Warning on Sexually Explicit Email," *InformationWeek*, April 13, 2004 , accessed at *www.informationweek.com*.

50. Millard, Elizabeth, "Winners and Losers in the Online Travel Market," E-Commerce Times, December 2, 2003, accessed at *www.ecommercetimes.com*.

51. Bednarz, Ann, "Holiday Prep," Urban Outfitters Revs Web Site Performance," *Network World Fusion*, December 19, 2003, accessed at *www.nwfusion.com*.

52. Travis, Paul, "Web-Experience Acquisition," *InformationWeek*, April 12, 2004, accessed at *www.informationweek.com*.

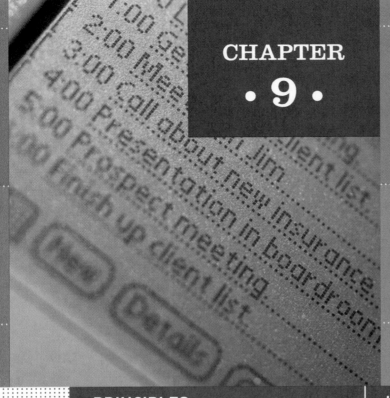

CHAPTER · 9 ·

Transaction Processing and Enterprise Resource Planning Systems

PRINCIPLES

- An organization's TPS must support the routine, day-to-day activities that occur in the normal course of business and help a company add value to its products and services.

- TPSs help multinational corporations form business links with their business partners, customers, and subsidiaries.

- Implementation of an enterprise resource planning system enables a company to achieve numerous business benefits through the creation of a highly integrated set of systems.

LEARNING OBJECTIVES

- Identify the basic activities and business objectives common to all transaction processing systems.

- Explain some key control and management issues associated with transaction processing systems.

- Describe the inputs, processing, and outputs for the transaction processing systems associated with order processing, purchasing, and accounting business processes.

- Identify the challenges that multinational corporations must face in planning, building, and operating their TPSs.

- Discuss the advantages and disadvantages associated with the implementation of an enterprise resource planning system.

INFORMATION SYSTEMS IN THE GLOBAL ECONOMY ❯❯
McDONALD'S, UNITED STATES

Self-Serve—I'm Lovin' It

Picture this scenario: An overwrought dad pulls into McDonald's and drops the kids off in the play area. Dreading the long line while straining to remember four complicated lunch requests, he turns to head for the counter to place his order. As he does so, he notices a new addition—a kiosk. Upon closer examination, he realizes that he can now place food orders from the playground. Following on-screen menus, and with a little help from the kids, the father places his order and swipes his debit card. A few minutes later, a McDonald's employee arrives on the playground with the lunches.

Sound hard to believe? Believe it! Such systems are in use at six Denver locations and one in Chicago. "It's a parent saver," says Christa Small, director of restaurant innovations for McDonald's.

The self-serve option is being provided at McDonald's counters as well. Rather than waiting for personnel to become available, you can just step up to the touchscreen monitor to browse through menus (in the traditional sense of the word), view photos of items, select what you like, total your order, and pay with cash, debit card, or credit card.

The new self-serve McDonald's restaurants have experienced increased sales. Why? Because they are able to process orders more quickly, increasing the daily sales volume. Also, customers, who appreciate the speedy service, are more likely to return. Other factors may be behind the increases as well. Some analysts believe that the tantalizing photos of the food items may contribute to impulse buying. Also, some customers may feel more comfortable "pigging out" when placing an order with a machine rather than a perhaps judgmental person.

Are McDonald's self-serve systems taking jobs away from people? Not at all. Employees who used to take orders are now assisting in other areas. They help prepare orders more quickly, assist customers in learning the new ordering system, deliver food to customers at their seats or on the playground, and help maintain a cleaner restaurant.

The self-serve systems at McDonald's provide an excellent example of the benefits of effective transaction processing systems. As you'll recall from Chapter 1, a transaction processing system, or TPS, supports and records business transactions. An effective TPS provides increased labor efficiency and faster service, which in turn builds customer loyalty and provides a company with a competitive advantage. McDonald's new systems certainly seem to be having that effect.

McDonald's is not alone in providing self-service. The Home Depot is leading the way among retailers in setting up self-checkout lines. Of the 1,763 Home Depot stores in the country, about half have self-checkout lines. John Simley, a Home Depot spokesman, said the response has been positive. "It has exceeded our expectation by about double," he said. "About 35 percent of the sales receipts generated by those stores are made by the self-checkout." Wal-Mart is also experimenting with self-checkout. By adding four self-checkout lanes to the typical 16 cashier lines, Wal-Mart has reduced the length of its checkout lines and

increased its customer service in the aisles by freeing additional sales clerks to assist customers in selecting products.

From libraries to airports, automated self-service systems are assisting customers with check-in and checkout. The systems are designed to be user-friendly, and companies provide personnel to assist customers when problems arise—or to allow them to check out the traditional way. As with all technology, companies must provide a backup when automated systems fail. For example, when the newly designed, colorful $20 bills first came out, they were rejected by many automated systems. Store personnel had to intervene so that customers were not inconvenienced by the programming glitch. The smooth running of order processing systems is essential to the survival of an organization. Still, backup systems must be prepared to take over when primary systems fail.

As you read this chapter, consider the following:

- What effect do transaction processing systems have on the efficiency of an organization and its reputation with its customers?
- What risks have McDonald's, The Home Depot, and Wal-Mart taken in providing self-checkout systems and what precautions have they taken to minimize the risks?

Why Learn About Transaction Processing and Enterprise Resource Planning Systems?

All organizations have transaction processing systems. They are key systems for businesses, because they perform routine business functions, carry out key business activities, and maintain records about them. Enterprise resource planning systems represent a collection of integrated transaction processing systems that support the major business functions of the organization. Although they were initially thought to be cost-effective only for very large companies, even midsized companies are now implementing these systems to reduce costs and improve service.

In our increasingly service-oriented economy, outstanding customer service has become a goal of virtually all companies. Employees who work directly with customers—whether in sales, customer service, or marketing—require high-quality transaction processing systems to provide good customer service. Such workers may use a transaction processing system to check the inventory status of ordered items, view the production planning schedule to tell the customer when the item will be in stock, or enter data into the shipment planning system to schedule delivery to the customer.

No matter what your role, it is very likely that you will provide input to or use the output from your organization's transaction processing systems. Thus, it is important that you understand how these systems work and what their capabilities and limitations are.

As you can see in the opening vignette, businesses rely on information systems to integrate their daily transaction activities. The many business activities associated with supply, distribution, sales, marketing, accounting, and taxation can be performed quickly, while avoiding waste and mistakes. The goal of this computerization is ultimately to satisfy a business's customers and provide a competitive advantage by reducing costs and improving service.

Transaction processing was one of the first business processes to be computerized, and without information systems, recording and processing business transactions would consume huge amounts of an organization's resources. The transaction processing system (TPS) also provides employees involved in other business processes—the management information system/decision support system (MIS/DSS) and the special-purpose information systems—with data to help them achieve their goals. A transaction processing system serves as the foundation for the other systems (see Figure 9.1). Transaction processing systems perform routine operations such as sales ordering and billing, often performing the same operations daily or weekly. The amount of support for decision making that a TPS directly provides managers and workers is low.

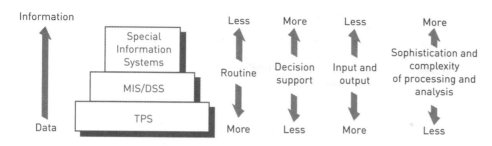

Figure 9.1

TPS, MIS/DSS, and Special Information Systems in Perspective

These systems require a large amount of input data and produce a large amount of output without requiring sophisticated or complex processing. As we move from transaction processing to management information/decision support, and special-purpose information systems, we see less routine, more decision support, less input and output, and more sophisticated and complex processing and analysis. But the increase in sophistication and complexity in moving from transaction processing does not mean that it is less important to a business. In most cases, all these systems start as a result of one or more business transactions.

AN OVERVIEW OF TRANSACTION PROCESSING SYSTEMS

Every organization has *transaction processing systems (TPSs)*, which process the detailed data necessary to update records about the fundamental business operations of the organization. These systems include order entry, inventory control, payroll, accounts payable, accounts receivable, and the general ledger, to name just a few. The input to these systems includes basic business transactions such as customer orders, purchase orders, receipts, time cards, invoices, and customer payments. The result of processing business transactions is that the organization's records are updated to reflect the status of the operation at the time of the last processed transaction. Automated TPSs consist of all the components of a computer based information system (CBIS), including databases, telecommunications, people, procedures, software, and hardware devices used to process transactions. The processing activities include data collection, data editing, data correction, data manipulation, data storage, and document production.

The U.S. Bureau of Customs and Border Protection (part of the U.S. Department of Homeland Security) has the busiest transaction processing system in the world. At peak workloads, the bureau's system processes an incredible 51,448 transactions per second, according to a study conducted by Winter Corporation, a marketing research and consulting firm. The bureau's transaction processing system uses an Advantage CA-Datacom database from Computer Associates and runs on IBM eServer zSeries hardware.[1]

For most organizations, TPSs support the routine, day-to-day activities that occur in the normal course of business that help a company add value to its products and services. Depending on the customer, value may mean lower price, better service, higher quality, or uniqueness of product. By adding a significant amount of value to their products and services, companies ensure further organizational success. Because the TPSs often perform activities related to customer contacts—such as order processing and invoicing—these information systems play a critical role in providing value to the customer. For example, by capturing and tracking the movement of each package, shippers such as Federal Express and United Parcel Service (UPS) are able to provide timely and accurate data on the exact location of a package. Shippers and receivers can access an online database and, by providing the airbill number of a package, find the package's current location. If the package has been delivered, they can see who signed for it (a service that is especially useful in large companies where packages can become "lost" in internal distribution systems and mailrooms). Such a system provides the basis for added value through improved customer service.

FedEx adds value to its service by providing timely and accurate data online on the exact location of a package.

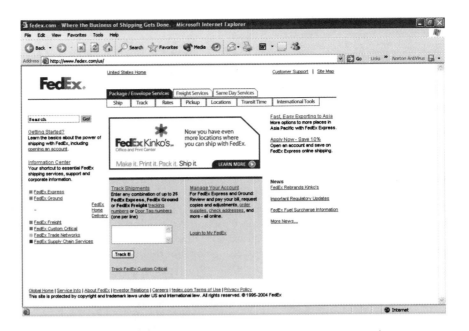

Traditional Transaction Processing Methods and Objectives

When computerized transaction processing systems first evolved, only one method of processing was available. All transactions were collected in groups, called *batches,* and processed together. With **batch processing systems**, business transactions are accumulated over a period of time and prepared for processing as a single unit or batch (see Figure 9.2a). The time period during which transactions are accumulated is whatever length of time is needed to meet the needs of the users of that system. For example, it may be important to process invoices and customer payments for the accounts receivable system daily. On the other hand, the payroll system may receive time cards and process them biweekly to create checks and update employee earnings records as well as to distribute labor costs. The essential characteristic of a batch processing system is that there is some delay between the occurrence of the event and the eventual processing of the related transaction to update the organization's records.

Today's computer technology allows another processing method called **online transaction processing** (OLTP). With this form of data processing, each transaction is processed immediately, without the delay of accumulating transactions into a batch (see Figure 9.2b). As soon as the input is available, a computer program performs the necessary processing and updates the records affected by that single transaction. Consequently, at any time, the data in an online system always reflects the current status. When you make an airline reservation, for instance, the transaction is processed, and all databases, such as seat occupancy and accounts receivable, are updated immediately. This type of processing is absolutely essential for businesses that require data quickly and update it often, such as airlines, ticket agencies, and stock investment firms. Many companies have found that OLTP helps them provide faster, more efficient service—one way to add value to their activities in the eyes of the customer. Increasingly, companies are using the Internet to perform many OLTP functions. A third type of transaction processing, called *online entry with delayed processing,* is a compromise between batch and online processing. With this type of system, transactions are entered into the computer system when they occur, but they are not processed immediately. For example, when you call a toll-free number and order a product, your order is typically entered into the computer when you make the call. However, the order may not be processed until that evening after business hours.

batch processing system
Method of computerized processing in which business transactions are accumulated over a period of time and prepared for processing as a single unit or batch.

online transaction processing (OLTP)
Computerized processing in which each transaction is processed immediately, without the delay of accumulating transactions into a batch.

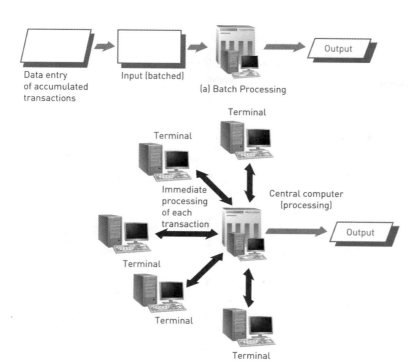

Data entry of accumulated transactions → Input (batched) → **(a) Batch Processing** → Output

Terminal

Terminal

Immediate processing of each transaction

Central computer (processing) → Output

Terminal

Terminal

Terminal

(b) Online Transaction Processing

Figure 9.2

Batch Versus Online Transaction Processing

(a) Batch processing inputs and processes data in groups. (b) In online processing, transactions are completed as they occur.

When you order a product over the phone, the vendor may use online entry with delayed processing. Your order is entered into the computer at the time of the call but is not processed immediately.

(Source: Hoby Finn/Getty Images.)

Even though the technology exists to run TPS applications using online processing, it is not done for all applications. For many applications, batch processing is more appropriate and cost-effective. Payroll transactions and billing are typically done via batch processing. Specific goals of the organization define the method of transaction processing best suited for the various applications of the company. Figure 9.3 shows the total integration of a firm's transaction processing systems.

Figure 9.3

Integration of a Firm's TPSs

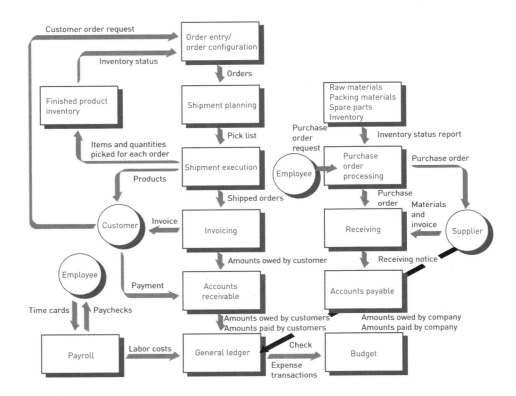

Because of the importance of transaction processing, organizations expect their TPSs to accomplish a number of specific objectives, including the following:

- *Process Data Generated By and About Transactions.* The primary objective of any TPS is to capture, process, and store transactions and to produce a variety of documents related to routine business activities. These business activities can be directly or indirectly related to selling products and services to customers. Processing orders, purchasing materials, controlling inventory, billing customers, and paying suppliers and employees are all business activities that result from customer orders. These activities result in transactions that are processed by the TPS.

  Utilities, telecommunications companies, financial-services organizations, and indeed many businesses are under enormous pressure to process ever-larger volumes of transactions in near-real time. Unfortunately, their IS infrastructures are struggling to handle the increased workload. But now improved data-management software with data store, data analysis, and transaction processing capabilities can ease the burden. Barclay's Capital Group is a leading player in the global options market and employs specialist foreign exchange teams in Hong Kong, London, Mumbai, New York, Singapore, and Tokyo. The firm realigned its Global Treasury Support through use of Aleri's Global Banking Solution (GBS), a financial software application specifically designed to deliver a cost-effective, flexible, and controlled world-class global support service for treasury, payments, and cash-management activities. Aleri's data-management software can process millions of transactions in a fraction of the time that traditional database systems take.[2]

- *Maintain a High Degree of Accuracy and Integrity.* One objective of any TPS is error-free data input and processing. Even before the introduction of computer technology, employees visually inspected all documents and reports introduced into or produced by the TPS. Because humans are fallible, the transactions were often inaccurate, resulting in wasted time and effort and requiring resources to correct them. An editing program, for example, should have the ability to determine that an entry that should read "40 hours" is not entered as "400 hours" or "4000 hours" because of a data entry error.

An important component of data integrity is to avoid fraudulent transactions. E-commerce companies face this problem when accepting credit or debit card information over the Internet. How can these companies make sure that the people making the purchases are who they say they are? One approach is to use a digital certificate. A digital certificate is a small computer file that serves as both an ID card and a signature. Some believe that digital certificates, which use complex mathematical codes, are almost fraud proof.

As the volume of data being processed and stored increases, it becomes more difficult for individuals and machines to review all input data. Doing so is critical, however, because data and information generated by the TPS are often used by other information systems in an organization. So, a company must ensure both data integrity and data accuracy.

- *Produce Timely Documents and Reports.* Manual transaction processing systems can take days to produce routine documents. Fortunately, the use of computerized TPSs significantly reduces this response time. Improvements in information technology, especially hardware and telecommunications links, allow transactions to be processed in a matter of seconds. The ability to conduct business transactions quickly can be very important for an organization's bottom line. For instance, if bills (invoices) are sent out to customers a few days earlier than usual, payment may be received earlier. A number of transaction processing systems can monitor how timely a company is when processing transactions and producing reports and documents. Some monitoring software packages can compare actual performance with corporate goals and objectives.

    Software developers at Northern Trust Corp. changed one of the bank's database systems, which created an error. If left unattended, the error could have interrupted the bank's ability to produce a handful of key reports used by the bank's customers, such as institutional investors and asset managers. Without these reports, the customers would have problems the next morning and would have made urgent calls to Northern Trust's help desk. Instead, the processing problem was noticed by Wily Technology's Introscope 5.0 application-monitoring system, installed on 16 application servers at the bank. The level of output from the application querying the database had dropped below an acceptable threshold, and night operators were notified by Introscope of that development. They in turn aroused a database administrator who was on call, found the error, and corrected it, enabling the system to process normally again and continue producing its reports on a timely basis.[3]

    Other business transaction monitoring software, Q Nami! from MQS Software, provides managers with real-time visibility and control over their critical business processes and services. It allows users to track and view the status of individual transactions. With this system, users can gain a better understanding of how all the IS components (software, hardware, network, databases) affect the delivery of key business services. Thus, the software enables managers to discover opportunities for improved transaction flow, which can lead to increased revenues, reduced costs, and enhanced operational productivity.[4]

    Timing is also crucial for related applications such as order processing, invoicing, accounts receivable, inventory control, and accounts payable. Because of electronic recording and transmission of sales information, transactions can be processed in seconds rather than overnight, thus improving companies' cash flow. Customers find credit card charges they made on the final day of the billing period on their current monthly bill.

- *Increase Labor Efficiency.* Before computers existed, manual business processes often required rooms full of clerks and equipment to process the necessary business transactions. Today, organizations have implemented TPSs to substantially reduce clerical and other labor requirements. A small minicomputer linked to a company's cash registers has replaced a room full of clerks, typewriters, and filing cabinets.

    Organizations are interested in gaining even greater labor efficiency by further streamlining business processing. For example, Aetna Inc. is a provider of healthcare, dental, pharmacy, group life, disability, and long-term care benefits in the United States. It serves approximately 13 million medical members, 11 million dental members,

7 million pharmacy members, and 12 million group insurance customers. Its annual revenues exceed $19 billion, and it employs more than 2,400 IS workers. Aetna is working with IBM to build a streamlined claims-processing system to reduce the time and effort needed to pay an insurance claim. IBM also is helping Aetna build more self-service systems for its customers. "We want to take a leadership position in creating consumer-directed health plans; IT is what's going to allow us to create and administer those plans," says Wei-Tih Cheng, CIO and senior VP at Aetna.[5]

- *Help Provide Increased Service.* Without question, we are quickly becoming a service-oriented economy. Even companies whose main activity is producing physical goods, such as household appliance makers and automobile manufacturers, realize the importance of providing superior customer service. Increased service has also hit the entertainment industry. Several Web sites, among them MovieTickets.com, Fandango, Moviefone, The Movie Secret, and Movies.com, provide a transaction processing system to enable moviegoers to buy tickets over the Internet, saving them hours of waiting in line for the hottest new shows. One objective of TPSs is to assist an organization in providing fast, efficient service. Some companies' EDI systems (see Chapter 6) allow customers to place orders electronically, thus bypassing slower and more error-prone methods of written or oral communication.

- *Help Build and Maintain Customer Loyalty.* A firm's transaction processing systems are often the means for customers to communicate. It is important that the customer interaction with these systems keeps customers satisfied and returning. IBM provides software to help automotive manufacturers meet key business challenges, including attaining more effective communications with customers and suppliers. IBM Middleware Solutions for Automotive Telematics is designed to help auto makers communicate with customers to better understand how products are used and how they can enhance the customer's experience with in-vehicle information systems as well as Internet-delivered applications, content, and services.[6]

- *Achieve Competitive Advantage.* A goal common to almost all organizations is to gain and maintain a competitive advantage. As discussed in Chapter 2, a competitive advantage provides a significant and long-term benefit for the organization. When a TPS is developed or modified, the personnel involved should carefully consider the significant and long-term benefits the new or modified system might provide.

  Wall Street firms are contending with one another to offer high-end wealth-management services to the richest clients. This strategy is highly dependent on the use of information systems to provide the services and information that affluent investors expect. Swiss firm UBS is a global wealth-management services provider, with 69,000 employees and main offices in Zurich and Basel and operations in more than 50 countries. The firm decided to use internal staff to enhance its existing TPSs' functionality and meet new goals. The new system is called ConsultWorks and is designed to eliminate boundaries between applications, according to Scott Abbey, UBS's chief technology officer. The company's 7,700 investment advisers use the system for many of their daily tasks, including checking account positions, conducting research, entering orders, and taking notes, all without having to navigate through separate applications. This enhanced system provides a competitive advantage for UBS, Abbey says. UBS justifies its use of internal staff for the system because it doesn't want to help a software vendor build a system it could later resell to UBS rivals. "We'd be asking for requirements that are in many ways unique compared to our competitors," Abbey says. "We're not looking to give away that intellectual property."[7]

  Some of the ways that companies can use transaction processing systems to achieve competitive advantage are summarized in Table 9.1.

Depending on the specific nature and goals of the organization, any of these objectives may be more important than others. By meeting these objectives, TPSs can support corporate goals such as reducing costs; increasing productivity, quality, and customer satisfaction; and running more efficient and effective operations. For example, overnight delivery companies

such as FedEx expect their TPSs to increase customer service. These systems can locate a client's package at any time—from initial pickup to final delivery. This improved customer information allows companies to produce timely information and be more responsive to customer needs and queries.

Competitive Advantage	Example
Customer loyalty increased	Use of customer interaction system to monitor and track each customer interaction with the company
Superior service provided to customers	Use of tracking systems that are accessible by customers to determine shipping status
Better relationship with suppliers	Use of an Internet marketplace to allow the company to purchase products from suppliers at discounted prices
Superior information gathering	Use of order configuration system to ensure that products ordered will meet customer's objectives
Costs dramatically reduced	Use of warehouse management system employing RFID technology to reduce labor hours and improve inventory accuracy
Inventory levels reduced	Use of collaborative planning, forecasting, and replenishment to ensure the right amount of inventory is in stores

Table 9.1

Examples of Transaction Processing Systems for Competitive Advantage

TRANSACTION PROCESSING ACTIVITIES

Along with having common characteristics, all TPSs perform a common set of basic data-processing activities. TPSs capture and process data that describes fundamental business transactions. This data is used to update databases and to produce a variety of reports people both within and outside the enterprise use (see Figure 9.4). The business data goes through a **transaction processing cycle** that includes data collection, data editing, data correction, data manipulation, data storage, and document production (see Figure 9.5).

transaction processing cycle
The process of data collection, data editing, data correction, data manipulation, data storage, and document production.

Data entry and input

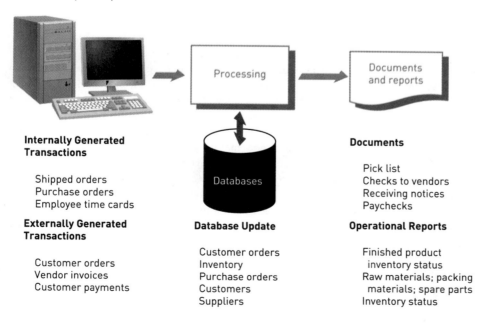

Internally Generated Transactions

Shipped orders
Purchase orders
Employee time cards

Externally Generated Transactions

Customer orders
Vendor invoices
Customer payments

Database Update

Customer orders
Inventory
Purchase orders
Customers
Suppliers

Documents

Pick list
Checks to vendors
Receiving notices
Paychecks

Operational Reports

Finished product
inventory status
Raw materials; packing
materials; spare parts
Inventory status

Figure 9.4

A Simplified Overview of a Transaction Processing System

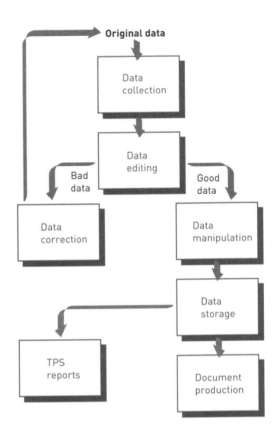

Data Collection

The process of capturing and gathering all data necessary to complete transactions is called **data collection**. In some cases it can be done manually, such as by collecting handwritten sales orders or changes to inventory. In other cases, data collection is automated via special input devices such as scanners, point-of-sale devices, and terminals.

Data collection begins with a transaction (e.g., taking a customer order) and results in the origination of data that is input to the transaction processing system. Data should be captured at its source, and it should be recorded accurately, in a timely fashion, with minimal manual effort, and in a form that can be directly entered into the computer rather than keying the data from a document. This approach is called *source data automation*. An example of source data automation is the use of automated devices at a retail store to speed the checkout process—either UPC codes read by a scanner or RFID signals picked up when the items approach the checkout stand. The use of both UPC bar codes and RFID tags is quicker and more accurate than having a clerk enter codes manually at the cash register. The product ID for each item is determined automatically, and its price is found in the item database. The point-of-sale TPS uses the price data to determine the customer's bill. The store's inventory and purchase databases record the number of units of an item purchased, the date, the time, and the price. The inventory database generates a management report notifying the store manager to reorder items that have fallen below the reorder quantity. The detailed purchases database can be used by the store or sold to marketing research firms or manufacturers for detailed sales analysis (see Figure 9.6).

Many grocery stores combine point-of-sale scanners and coupon printers. The systems are programmed so that each time a specific product—say, a box of cereal—crosses a checkout scanner, an appropriate coupon—perhaps a milk coupon—is printed. Companies can pay to be promoted through the system, which is then reprogrammed to print those companies' coupons if the customer buys a competitive brand. These TPSs help grocery stores to increase profits by improving their repeat sales and bringing in revenue from other businesses.

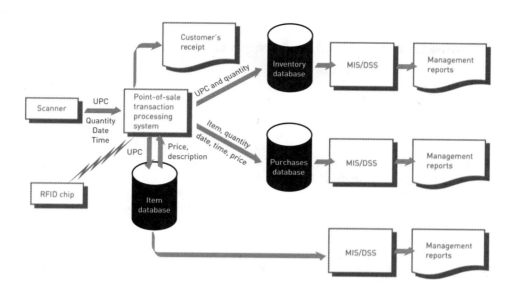

Figure 9.6

Point-of-Sale Transaction Processing System

The purchase of items at the checkout stand updates a store's inventory database and its database of purchases.

Data Editing

An important step in processing transaction data is to perform **data editing** for validity and completeness to detect any problems. For example, quantity and cost data must be numeric and names must be alphabetic; otherwise, the data is not valid. Often the codes associated with an individual transaction are edited against a database containing valid codes. If any code entered (or scanned) is not present in the database, the transaction is rejected.

data editing
The process of checking data for validity and completeness.

Data Correction

It is not enough simply to reject invalid data. The system should also provide error messages that alert those responsible for the data editing function. Error messages must specify what problem is occurring so that corrections can be made. A **data correction** involves reentering miskeyed or misscanned data that was found during data editing. For example, a UPC that is scanned must be in a master table of valid UPCs. If the code is misread or does not exist in the table, the checkout clerk is given an instruction to rescan the item or key in the information manually.

data correction
The process of reentering miskeyed or misscanned data that was found during data editing.

Data Manipulation

Another major activity of a TPS is **data manipulation**, the process of performing calculations and other data transformations related to business transactions. Data manipulation can include classifying data, sorting data into categories, performing calculations, summarizing results, and storing data in the organization's database for further processing. In a payroll TPS, for example, data manipulation includes multiplying an employee's hours worked by the hourly pay rate. Overtime calculations, federal and state tax withholdings, and deductions are also performed.

data manipulation
The process of performing calculations and other data transformations related to business transactions.

Data Storage

Data storage involves updating one or more databases with new transactions. Once the update process is complete, this data can be further processed and manipulated by other systems so that it is available for management decision making. Thus, although transaction databases can be considered a by-product of transaction processing, they have a pronounced effect on nearly all other information systems and decision-making processes in an organization.

data storage
The process of updating one or more databases with new transactions.

Document Production and Reports

document production
The process of generating output records and reports.

TPSs produce important business documents. **Document production** involves generating output records and reports. These documents may be hard-copy paper reports or displays on computer screens (sometimes referred to as *soft copy*). Paychecks, for example, are hard-copy documents produced by a payroll TPS, while an outstanding balance report for invoices might be a soft-copy report displayed by an accounts receivable TPS. Often, results from one TPS are passed downstream as input to other systems (as shown in Figure 9.6), where the results of updating the inventory database are used to create the stock exception report (a type of management report) of items whose inventory level is below the reorder point.

In addition to major documents such as checks and invoices, most TPSs provide other useful management information and decision support, such as printed or on-screen reports that help managers and employees perform various activities. A report showing current inventory is one example; another might be a document listing items ordered from a supplier to help a receiving clerk check the order for completeness when it arrives. A TPS can also produce reports required by local, state, and federal agencies, such as statements of tax withholding and quarterly income statements.

Throughout this chapter we will look at some ways companies have employed TPSs to help them meet organizational goals.

CONTROL AND MANAGEMENT ISSUES

Transaction processing systems are the backbone of any organization's information systems. They capture facts about the fundamental business operations of the organization—facts without which orders cannot be shipped, customers cannot be invoiced, and employees and suppliers cannot be paid. In addition, the data captured by the TPSs flows downstream to other systems in the organization. Like any structure, an organization's information systems are only as good as the foundation on which they are built. In fact, most organizations would grind to a screeching halt if their TPSs failed.

Business Continuity Planning

business continuity planning
Identification of the business processes that must be restored first in the event of a disaster and specification of what actions should be taken and who should take them to restore operations.

Business continuity planning identifies the business processes that must be restored first in the event of a disaster to get the business's operations restarted with minimum disruption; it also specifies the actions that must be taken and by whom to restore operations. Order processing and shipping are examples of business processes that must be resumed as quickly as possible. Disasters can be natural emergencies such as a flood, a fire, or an earthquake or interruptions in business processes such as labor unrest, terrorist activity, hacker attack, or erasure of an important file. Key actions include safe evacuation of all employees, assessment of the disaster's impact, relocation to alternate work spaces, backup and recovery of important electronic and manual business records, and use of alternate equipment.

One of the first steps of business continuity planning is to identify potential threats or problems, such as natural disasters, employee misuse of personal computers, and poor internal control procedures. Business continuity planning also involves disaster preparedness. Business managers should occasionally hold an unannounced "test disaster"—similar to a fire drill—to ensure that the disaster plan is effective.

Companies vary widely in the thoroughness and effectiveness of their business continuity planning, and some have a harder time resuming business than others. However difficult the process, companies cannot afford to go unprepared for operational outages. Recently, disasters such as the Northeast power blackout and the California wildfires wreaked havoc with U.S. businesses. Blackout losses to workers and investors were estimated to be $4.2 billion by Anderson Economic Group, while Safeco Corporation estimates business wildfire claims at $3 million.[8]

Disaster Recovery

Disaster recovery focuses on the actions that must be taken to restore computer operations and services in the event of a disaster. It includes providing for alternate computing and network facilities; the transfer of key personnel, data, and software to a backup site; and the rapid resumption of data processing and communications.

disaster recovery
Actions that must be taken to restore computer operations and services in the event of a disaster.

Companies like Iron Mountain provide a secure, off-site environment for records storage. In the event of a disaster, vital data can be recovered.

(Source: Geostock/Getty Images.)

Aeneas Internet and Telephone, an Internet service provider in Jackson, Tennessee, serves about 10,000 Internet and 2,500 telephone customers. It was hit by a ferocious tornado in May 2003 that ripped straight through Aeneas's one-story building, leaving only a pile of rubble. Aeneas lost $1 million in hardware and software that night. Fortunately, Aeneas was wise and had a disaster recovery plan. Company managers determined that the likelihood of a terrorist attack on the western Tennessee town with a population under 60,000 was slim to none. Instead, they concluded that because of the town's location in the central U.S.'s infamous Tornado Alley, the plan should address that most likely cause of disaster. The company's business continuity plan hinged on moving operations to a site shared with another firm, providing for alternative telecommunications services and making backups of critical customer and billing data. Although the tornado completely leveled Aeneas's office building, the firm was able to be up and running again in three days. This relocation was not effortless, however, because the colocation site had an insufficient number of servers. Aeneas employees had to order additional servers online, have them shipped overnight to their homes, and then install them at the new site. Furthermore, because the backups of critical information were not stored off-site, Aeneas was lucky to find a readable version of the customer records database on a destroyed computer.[9]

Transaction Processing System Audit

The accounting scandals of 2001 and 2002 spurred corporate board members, lawmakers, regulators, and stockholders to pressure corporate executives to produce accurate financial reports and do so in a timely fashion. In July 2002, the Sarbanes-Oxley Act was enacted, which set deadlines for public companies to implement procedures to ensure their audit committees could document financial data, validate earnings reports, and verify the accuracy of information. In response to the new requirements, business managers are demanding that their financial systems provide real-time data feeds and expenditure and sales updates so they can ensure their numbers are correct. Unfortunately, some organizations were unable to produce accurate reports. New World Pasta is the leading U.S. maker of dry pasta, with such brands as American Beauty, Creamette, Ideal, Ronzoni, and San Giorgio. In 2004, the firm filed for Chapter 11 bankruptcy protection, blaming two years of accounting problems caused by faulty transaction processing systems.[10]

Sarbanes-Oxley also focused attention on the security of data and systems. Regulators are interpreting the Financial Services Modernization Act (Gramm-Leach-Bliley) as requiring systems security for financial service providers, including specific standards to protect customer privacy. In the healthcare industry, the Health Insurance Portability and Accountability Act (HIPAA) defines regulations covering healthcare providers to ensure that their patient data is adequately protected.[11]

Clearly, CIOs must act to prevent the kind of accounting irregularities or loss of data privacy that can get their companies into trouble and erase investor confidence. One key step is to conduct a **transaction processing system audit** that attempts to answer four basic questions:

- Does the system meet the business need for which it was implemented?
- What procedures and controls have been established?
- Are these procedures and controls being used properly?
- Are the information systems and procedures producing accurate and honest reports?

In addition to these four basic auditing questions, other areas are typically investigated during an audit. These areas include the distribution of output documents and reports, the training and education associated with existing and new systems, and the time necessary to perform various tasks and to resolve problems and bottlenecks in the system. General areas of improvement are also investigated and reported during the audit.

Two types of audits exist. An internal audit is conducted by employees of the organization; an external audit is performed by accounting firms or companies and individuals not associated with the organization. In both internal and external audits, the auditor inspects all programs, documentation, control techniques, the disaster plan, insurance protection, fire protection, and other systems management concerns such as efficiency and effectiveness of the disk or tape library. This check is accomplished by interviewing IS personnel and performing a number of tests on the computer system. External audits are important for stockholders and others outside the company, in addition to managers and employees inside the company. A number of Internet startup companies, for example, overstated their income, which resulted in high stock evaluations in some cases. An external audit by a reputable auditing company can help uncover these reporting problems.

In establishing the integrity of the computer programs and software, an audit trail must be established. The **audit trail** allows the auditor to trace any output from the computer system back to the source documents. With many of the real-time and time-sharing systems available today, it is extremely difficult to follow an audit trail. In many cases, no record of system inputs exists; thus, the audit trail is destroyed. In such cases, the auditor must investigate the actual processing in addition to the inputs and outputs of the various programs. In an attempt to safeguard the privacy of medical records, the federal Health Insurance Portability and Accountability Act requires that healthcare organizations, healthcare providers, and insurance companies establish an audit trail for each patient record. As an individual's medical record moves, each application it touches must imprint it with an identifier that cites every person who handled it and for what purpose.

While the CIO must take an active role in ensuring the integrity of financial reporting systems, that's not the same as guaranteeing the material accuracy of financial statements. At some of the companies under scrutiny, not even the most conscientious and informed CIO could have uncovered the financial irregularities that were occurring.

TRADITIONAL TRANSACTION PROCESSING APPLICATIONS

In this section we present an overview of several common transaction processing systems that support the order processing, purchasing, and accounting business processes (see Table 9.2).

---

transaction processing system audit

An examination of the TPS to answer whether the system meets the business need for which it was implemented, what procedures and controls have been established, whether these procedures and controls are being used properly, and whether the information systems and procedures are producing accurate and honest reports.

audit trail

Documentation that allows the auditor to trace any output from the computer system back to the source documents.

Order Processing	Purchasing	Accounting
• Order entry • Sales configuration • Shipment planning • Inventory control (finished product) • Invoicing and billing • Customer interaction • Routing and scheduling	• Inventory control (raw materials, packing materials, spare parts, and supplies) • Purchase order processing • Receiving • Accounts payable	• Budget • Accounts receivable • Payroll • Asset management • General ledger

Table 9.2

The Systems That Support Order Processing, Purchasing, and Accounting Functions

Order Processing Systems

Order processing systems include order entry, sales configuration, shipment planning, shipment execution, inventory control, invoicing, customer relationship management, and routing and scheduling. The business processes supported by these systems are so critical to the operation of an enterprise that the order processing systems are sometimes referred to as the "lifeblood of the organization." Figure 9.7 is a system-level flowchart that shows the various systems and the information that flows between them. A rectangle represents a system, a line represents the flow of information from one system to another, and a circle represents any entity outside the system—in this case, the customer.

order processing systems
Systems that process order entry, sales configuration, shipment planning, shipment execution, inventory control, invoicing, customer relationship management, and routing and scheduling.

Figure 9.7

Order Processing Systems

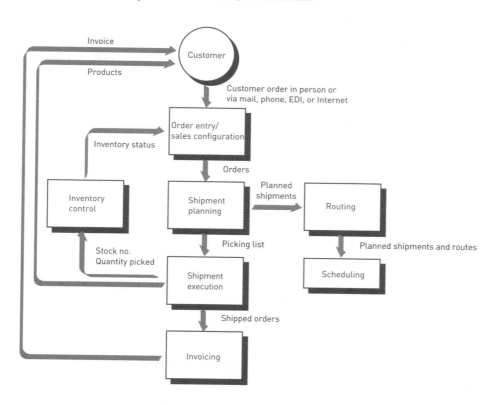

Order Entry

The **order entry system** captures the basic data needed to process a customer order. Orders may come through the mail or via a telephone ordering system, be gathered by a staff of sales representatives, arrive via EDI transactions directly from a customer's computer over a wide area network, or be entered directly over the Internet by the customer using a data entry form on the firm's Web site. Figure 9.8 is a data flow diagram of a typical order entry system. The data flow diagram shows the various business processes that are supported by a system and the flow of data between processes. A rectangle with rounded corners represents a business process.

order entry system
Process that captures the basic data needed to process a customer order.

Figure 9.8

Data Flow Diagram of an Order Entry System

Orders are received by mail, phone, EDI, or the Internet from customers or sales reps and entered into the order processing system. This application affects accounting, inventory, warehousing, finance, and invoicing applications. Note that in an integrated order processing system, order entry personnel have access to back order, inventory, and customer information from separate data files or directly through the processing mechanism.

(Source: George W. Reynolds, *Information Systems for Managers, Third Edition*, St. Paul, MN: West Publishing Co., 1995, p. 198.)

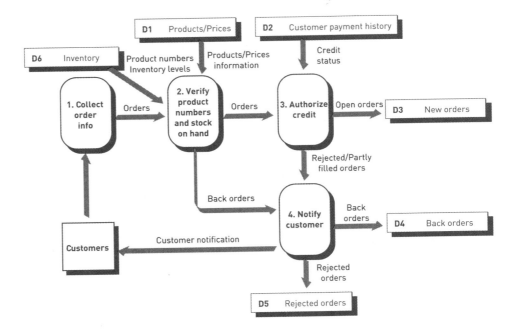

One current controversy related to online ordering systems involves access by people with physical disabilities. See the "Ethical and Societal Issues" special-interest feature for a discussion of online accessibility.

With an online order processing system, such as one used by direct retailers, the status of each inventory item (also called a *stock keeping unit*, or *SKU*) on the order is checked to determine whether sufficient finished product is available. If an order item cannot be filled, a substitute item may be suggested or a back order is created—the order will be filled later, when inventory is replenished. Order processing systems can also suggest related items for order takers to mention to promote add-on sales. Order takers also review customer payment history data from the accounts receivable system to determine whether credit can be extended.

Once an order is entered and accepted, it becomes an open order. Typically, a daily sales journal (which includes customer information, products ordered, quantity discounts, and prices) is generated.

Electronic data interchange (EDI) can be an important part of the order entry TPS. With EDI, a customer or client organization can place orders directly from its purchasing TPS into the order processing TPS of another organization. Or, the order processing TPS of the supplier companies and the purchasing TPS of the customers could be linked indirectly through a third-party clearinghouse. In any event, this computer-to-computer link allows efficient and effective processing of sales orders and enables an organization to lock in customers and lock out competitors through enhanced customer service. With EDI, orders can be placed anytime of the day or night, and immediate notification of order receipt and processing can be made. Today, more and more companies are using EDI to make paperless business transactions a reality.

TPSs for Everyone

In the rush to automate customer services, many businesses are short-changing a significant percentage of the population. One out of every seven people is estimated to suffer from some type of disability. A recent study by Britain's Disability Rights Commission (DRC) found that 81 percent of 1,000 Web sites examined fail to meet even the most basic of accessibility requirements. The most seriously disadvantaged are blind users, who are unable to navigate through poorly designed sites even with the use of automated screen readers. As a result, disabled people are unable to take advantage of Web-based services, such as making flight and hotel reservations, managing bank accounts, or buying movie tickets online. "The Web has been around for 10 years, yet within this short space of time it has managed to throw up the same hurdles to access and participation by disabled people as the physical world," stated DRC Chairman Bert Massie.

The DRC issued a stern warning to businesses that they are not complying with existing equal access laws. "Organizations that offer goods and services over the Web already have a legal duty to make their Web sites accessible to disabled people. Our investigation contains a range of recommendations to help Web site owners and developers bring down the barriers. . . . But where the response is inadequate, the industry should be prepared for disabled people to use the law to make the Web a less hostile place," said Massie.

The law to which Massie refers is Britain's Disability Discrimination Act (DDA), which states that "it is unlawful for a provider of services to discriminate against a disabled person in refusing to provide, or deliberately not providing, to the disabled person any service which he provides, or is prepared to provide, to members of the public." This includes "access to and use of information services." Most developed countries have similar laws; the United States has the Americans with Disabilities Act to prohibit discrimination on the basis of disability in employment, state and local government, public accommodations, commercial facilities, transportation, and telecommunications. The effect of this problem with Web sites is worldwide.

The World Wide Web Consortium (W3C), an organization dedicated to developing interoperable technologies for the Web, has created guidelines for creating Web sites and systems that are accessible to all called the Web Accessibility Initiative, or WAI. Some of these are:

- Provide equivalent alternatives for auditory and visual content
- Don't rely on color alone
- Provide text alternatives for all nontext content
- Use tables sparingly to make navigation simple for the blind using screen readers
- Make all functionality operable via a keyboard or a keyboard interface
- Provide clear navigation mechanisms

The guidelines also provide detailed instructions on how to follow through with the suggestions.

Besides complying with the law, other factors can motivate businesses to make their systems accessible to all. Professor Helen Petrie, who has conducted research for the DRC, points to "a usability bonus" of accessible systems. Companies that have involved blind users in usability tests of their Web sites have discovered a 35 percent rise in the speed at which certain tasks could be completed by all users. So, everyone can benefit from the modifications.

In short, accessible systems make good business sense. Inclusive design requires little to no additional investment if it is included in original system designs, and it assures businesses of reaching the largest possible customer population.

Critical Thinking Questions

1. Mr. Massie, chairman of the DRC, stated that many Web sites have hurdles for the disabled, especially the blind. What features of Web sites that you have visited might act as hurdles for a blind person?
2. Why is it that so many businesses have ignored the needs of the disabled in designing systems and Web sites?

What Would You Do?

As CEO of AceTravel.com, you have been contacted by Baird, Conley, and Finch, Attorneys at Law, with news that your company is being sued by a group of disabled users. The AceTravel.com Web site has been designed so that it is impossible for a blind person to navigate the site, even with the use of screen-reading software. An initial evaluation indicates that bringing the Web site up to the required accessibility standards would cost the company roughly $20,000. Half of this cost will go to train your Web developers; the other half to actually modifying the systems. The system upgrade will take about six months. Adding these costs to the inevitable legal costs that you are about to incur, this oversight might cost over a million dollars.

3. Is it worthwhile to give in to the demands of Baird, Conley, and Finch and their clients and modify your site and order processing systems? Or, should you fight this tooth and nail? Why?
4. If you decide to invest the $20,000 to make your Web-based services accessible to everyone, will there be other future expenses? Or, will the one-time investment suffice? What type of return can you expect on your investment?

SOURCES: "Sites Must Get Accessible or Face the Force of Law," *New Media Age*, April 22, 2004, *www.lexisnexis.com*; "DRC Web Investigation Finds Many Public Web sites 'Impossible' for Disabled People to Use," The Disability Rights

Commission Web site, accessed May 23, 2004, *www.drc-gb.org*; "Disability Discrimination Act 1995 (c. 50)," accessed May 23, 2004, at *www.legislation.hmso.gov.uk/acts/acts1995/95050—c.htm#19*; "A Guide to Disability Rights Laws," accessed May 23, 2004, at *www.usdoj.gov/crt/ada/cguide.htm#anchor62335*; "Web Accessibility Initiative," accessed May 23, 2004, at *www.w3.org/WAI*.

Proflowers uses a sophisticated order processing system that relays a consumer's online order for fresh flowers directly to the grower. Throughout the process, the Proflowers system generates a series of e-mail messages to the customer, providing up-to-date information on order status, shipping, and delivery.

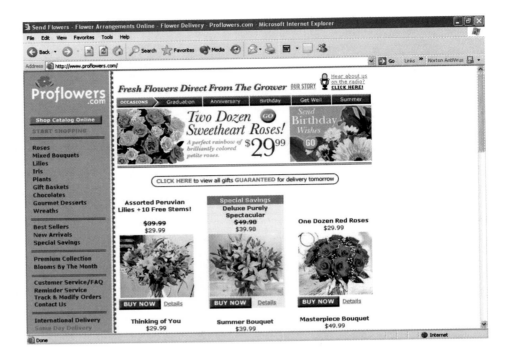

As discussed in Chapter 8, order processing is being done through e-commerce and Internet systems to a greater extent today. For example, Proflowers operates an Internet-based order processing system that delivers flowers straight from the grower to the recipient's doorstep. The direct delivery of fresh flowers from the grower bypasses the usual distribution centers and retail florists, reducing transit time and extending the life of the flowers for the recipient. The order processing cycle begins with capturing the customer order at the Proflowers Web site. The order is then relayed via the Internet to a specific grower, based on the type of flowers sought and the location of the recipient. Each participating grower has a Proflowers server set up at its site, which receives orders and prints them. FedEx shipping labels and greeting cards are also printed for each order. The grower then packs hydrated, cooled flowers into a special box that is shipped via FedEx for next-day delivery to the recipient. Throughout the process, the Proflowers system generates a series of e-mail messages to consumers—when the order is placed, when the flowers are picked up for delivery, when they arrive, and who signs for them.[12]

Sales Configuration

sales configuration system
Process that ensures that the products and services ordered are sufficient to accomplish the customer's objectives and will work well together.

Another important aspect of order processing is sales configuration. The **sales configuration system** ensures that the products and services ordered are sufficient to accomplish the customer's objectives and will work well together. For example, using a sales configuration program, a sales representative knows that a computer printer needs a certain cable and a LAN card so that it can be connected to the LAN. Without a sales configuration program, a sales representative might sell a customer the wrong cable or forget the LAN card.

Sales configuration programs also suggest optional equipment. For example, if a customer orders a handheld computer, the sales configuration program will suggest an AC adapter, backup software and communications cards to enable the user to connect wirelessly to printers, LAN, and the Internet. If a company is buying a Boeing 7E7 airplane, a sales configuration program can help the sales representative work with the company to determine

the number of seats that are needed, the most appropriate navigation systems to install, the type of landing gear that should be used, and hundreds of other available options that can be specified for the 7E7.

Sales configuration software can also solve customer problems and answer their questions. For example, a sales configuration program can determine whether a factory robot made by one manufacturer can be controlled by a computer system developed by another manufacturer. Sales configuration programs can eliminate mistakes, reduce costs, and increase revenues. You have used a sales configuration program if you have ever gone to the Web page of a major auto manufacturer and specified your dream car, complete with special options and features. For example, once you choose a basic model, the sales configuration software determines that only certain engines, exterior features, and interior colors are available.

Shipment Planning

New orders received and any other orders not yet shipped (open orders) are passed from the order entry system to the shipment planning system. The **shipment planning system** determines which open orders will be filled and from which location they will be shipped. This is a trivial task for a small company with lots of inventory, only one shipping location, and a few customers concentrated in a small geographic area. But it is an extremely complicated task for a large global corporation with limited inventory (not all orders for all items can be filled), dozens of shipping locations (plants, warehouses, contract manufacturers, etc.), and tens of thousands of customers. The trick is to minimize shipping and warehousing costs while still meeting customer delivery dates.

shipment planning system
System that determines which open orders will be filled and from which location they will be shipped.

Many companies use RFID tags to speed order processing time and improve inventory accuracy.

(Source: Spencer Grant/Photo Edit.)

The output of the shipment planning system is a plan that shows where each order is to be filled and a precise schedule for shipping with a specific carrier on a specific date and time. The system also prepares a picking list that warehouse operators use to select the ordered goods from the warehouse. These outputs may be in paper form, or they may be computer records that are transmitted electronically. The picking list document, an example of which is shown in Figure 9.9, lists the customer name, number, order number, and all items that have been ordered. A description of all items, along with the number to be shipped, is also included.

Toy maker Mattel has greatly improved its ability to meet customer demand because it has focused its attention on simplifying the software and processes in its supply chain, cutting costs, shortening cycle times, and bringing more science to the art of meeting customer demand. One major area of improvement was to install a new shipment planning system that saved it a lot of money by reducing the number of less-than-full truckload shipments and optimizing its shipping patterns.[13]

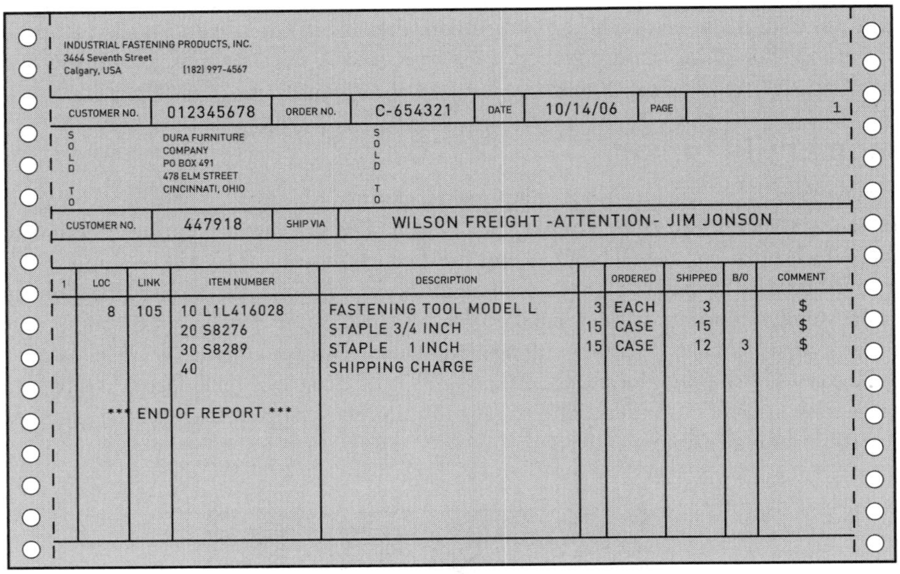

INDUSTRIAL FASTENING PRODUCTS, INC.
3464 Seventh Street
Calgary, USA (182) 997-4567

CUSTOMER NO. 012345678 ORDER NO. C-654321 DATE 10/14/06 PAGE 1

SOLD TO: DURA FURNITURE COMPANY, PO BOX 491, 478 ELM STREET, CINCINNATI, OHIO

SOLD TO:

CUSTOMER NO. 447918 SHIP VIA WILSON FREIGHT -ATTENTION- JIM JONSON

1	LOC	LINK	ITEM NUMBER	DESCRIPTION		ORDERED	SHIPPED	B/O	COMMENT
	8	105	10 L1L416028	FASTENING TOOL MODEL L	3	EACH	3		$
			20 S8276	STAPLE 3/4 INCH	15	CASE	15		$
			30 S8289	STAPLE 1 INCH	15	CASE	12	3	$
			40	SHIPPING CHARGE					
	***	END	OF REPORT ***						

Figure 9.9

A Picking List

This document guides warehouse employees in locating items to fill an order. Note the second and third columns of the slip, which instruct the warehouse workers where to locate the items. Also note that in this instance, three cases of the third item were back-ordered. Once items are picked from inventory, this data is entered into the data transaction processing system, and a packing slip and shipping notice are generated.

Shipment Execution

shipment execution system
System that coordinates the outflow of all products from the organization, with the objective of delivering quality products on time to customers.

The **shipment execution system** coordinates the outflow of all products from an organization, with the objective of delivering quality products on time to customers. The shipping department is usually responsible for physically packaging and delivering all products to customers and suppliers. This delivery system can include mail services, trucking operations, and rail service. The system receives the picking list from the shipment planning system.

Sometimes orders cannot be filled exactly as specified. One reason is "out-of-stocks," meaning the warehouse does not have sufficient quantity of an item to fill a customer's order. Shortages can be caused if a production run did not produce the expected quantity of an item because of manufacturing problems. The company policy may be not to ship any of the item, ship as many units of the item as are available and create a back-order request for the remainder, or to substitute another item. Thus, as items are picked and loaded for shipment, warehouse operators must enter data about the exact items and quantity of each that are loaded for each order. When the shipment execution system processing cycle is complete, the system passes the "shipped orders" business transactions downstream to the invoicing system. These transactions specify exactly what items were shipped, how much of each, and what person or company was sent the order. This data is used to generate a customer invoice. The shipment execution system also produces packing documents, which are enclosed with the items being shipped, to tell customers what items are being shipped, what is back-ordered, and what the exact status of all items in the order is. Soft-copy data—such as that provided by advanced shipment notices and shipment tracking systems—is also made available to other business functions.

Electronic Arts Inc. is a video-game maker that is constantly turning out new products—as many as 80 a year. The firm uses a number of software packages from various vendors that enables it to "deliver on what we call 100% accuracy by helping us to ensure we've packed the correct number of products in containers," says Marc West, senior VP and CIO. A computer-generated pick list directs warehouse workers to pick certain products, which are scanned by either handheld or fixed scanners, weighed, and then loaded. Data is collected at each point and fed into the warehouse management system. "This way, we

confirm what got packed, what got loaded onto the truck, and what's been received" by the customer, West says. "We've seen great returns on that."[14]

Inventory Control

For each item picked during the shipment execution process, a transaction providing the stock number and quantity picked is passed to the **inventory-control system**. In this way, the computerized inventory records are updated to reflect the exact quantity on hand of each stock-keeping unit. Thus, when order takers check the inventory level of a product, they receive current information.

inventory-control system
System that updates the computerized inventory records to reflect the exact quantity on hand of each stock-keeping unit.

Once products have been picked out of inventory, other documents and reports are initiated by the inventory-control application. For example, the inventory status report (see Figure 9.10) summarizes all inventory items shipped over a specified time period. It can include stock numbers, descriptions, number of units on hand, number of units ordered, back-ordered units, average costs, and related information. It is used to determine when to order more inventory and how much of each item to order, and it helps minimize "stockouts" and back orders. Data from this report is used as input to other information systems to help production and operations managers analyze the production process.

DATE 5 30 06 JANUS TOOL COMPANY INVENTORY STATUS/STOCK TAKE REPORT PAGE 2

PRD CLS	STOCK NUMBER DESCRIPTION	WH	LOCN	UNITS ON HAND	UNITS ON ORDER	UNITS RESERVED	UNIT AVERAGE COST	VALUE AVERAGE COST	PHYSICAL QUANTITY
10	1001 1 / 4* ELECTRIC DRILL . MO	1	0019	517.0000	0.0000	40.0000	29. 50000	15,251.50	
40	1001 1 / 4* ELECTRIC DRILL . MO	1	0018	80.0000	30.0000	0.0000	29.00000	2,320.00	
10	1001 1 / 4* ELECTRIC DRILL . MO	1	0017	150.0000	25.0000	10.0000	28.75000	4,312.50	
20	1001 1 / 4* ELECTRIC DRILL . MO	1	0007	410.0000	25.0000	50.0000	10.25000	4,202.50	
20	1001 1 / 4* ELECTRIC DRILL . MO	1	0008	330.0000	0.0000	0.0000	27.45000	9,058.50	
25	1001 1 / 4* ELECTRIC DRILL . MO	1	0009	14,256.0000	0.0000	1,440.0000	0.45000	6,415.20	
32	1001 1 / 4* ELECTRIC DRILL . MO	1	0003	59.0000	40.0000	5.0000	41.50000	2,445.50	
100	1001 1 / 4* ELECTRIC DRILL . MO	1	0006	2,448.0000	0.0000	0.0000	4.10000	10,036.80	
25	1001 1 / 4* ELECTRIC DRILL . MO	1	0006	0.0000	0.0000	0.0000	1.75000	0.00	
30	1001 1 / 4* ELECTRIC DRILL . MO	1	0016	314.0000	100.0000	50.0000	5.85000	1,836.90	
15	1001 1 / 4* ELECTRIC DRILL . MO	1	0015	192.0000	240.0000	0.0000	2.05000	393.60-	
20	1001 1 / 4* ELECTRIC DRILL . MO	1	0012	183.0000	50.0000	35.0000	4.75000	869.25	
XX	1001 1 / 4* ELECTRIC DRILL . MO	1	0011	105.0000	50.0000	0.0000	2.65000	278.25	
50	1001 1 / 4* ELECTRIC DRILL . MO	1	0004	109.0000	10.0000	0.0000	16.50000	1,798.50	
100	1001 1 / 4* ELECTRIC DRILL . MO	1	0001	960.0000	240.0000	0.0000	4.50000	4,320.00	
	GRAND TOTAL							62,754.80	

Figure 9.10

An Inventory Status Report

This output from the inventory application summarizes all inventory items shipped over a specified time period.

For almost all companies, inventory must be tightly controlled. One objective is to minimize the amount of cash tied up in inventory by placing just the right amount of inventory on the factory or warehouse floor.

To gain a competitive advantage, many manufacturing organizations are moving to real-time inventory-control systems based on RFID chip technology or bar coding the finished product, scanners and radio display terminals mounted on forklifts, and wireless LAN communications to track each time an item is moved in the warehouse. One significant advantage is that the inventory data is more accurate and current for people performing order entry, production planning, and shipment planning. Also, warehouse operations can be streamlined by providing directions to the forklift drivers.

In addition to being useful for physical goods such as automobiles and home appliances, inventory control is essential for industries in the service sector. Such organizations as hotels, airlines, rental-car agencies, and universities, which primarily provide services, can use inventory applications to help them monitor use of rooms, airline seats, car rentals, and classroom capacity. Airlines face an especially difficult inventory problem. Empty airline seats (inventory) have absolutely no value after a plane takes off. Yet overbooking can result in too many seats being sold and customer complaints. Sophisticated reservation systems allow airlines to quickly update and add seating assignments.

The use of RFID tags that support source data automation is revolutionizing inventory management. The RFID tag can be loaded with information such as a product's expiration date and temperature, which can be scanned from a distance of up to 30 feet. ERP software provider SAP makes software that can associate certain data gathered from reading RFID tags. For example, a certain carton of product can be associated with the shelf it sits on to give grocery stores better information about their inventories. Another type of software enables suppliers to use such data to notify a store that a shelf needs to be replenished soon or even that a perishable item is nearing its expiration date and its price should be reduced for quick sale.[15]

Casual Male is a retailer of big and tall men's apparel. It offers its merchandise to customers through diverse selling and marketing channels, including over 465 retail and outlet stores, two catalogs, and two e-commerce sites. The firm is replacing a 15-year-old mainframe inventory-control system with modern software that is integrated with its point-of-sale systems. As a result, when one of its stores sells an item, the sale will automatically flow into the inventory-control application and also be used to generate customer-demand reports that influence buying decisions. The inventory-control software employs RFID technology that employees use to scan bar-coded items as they pack them. The improved software and RFID tags are expected to help Casual Male cut costs in its distribution center by nearly 70 percent and save the retailer over $20 million—roughly 5 percent of sales revenue.[16]

Invoicing

Customer invoices are generated based on records received from the shipment execution TPS. This application encourages follow-up on existing sales activities, increases profitability, and improves customer service. Most invoicing programs automatically compute discounts, applicable taxes, and other miscellaneous charges (see Figure 9.11). Because most computerized operations contain elaborate databases on customers and inventory, many invoicing applications require only information on the items ordered and the client identification number; the invoicing application does the rest. It looks up the full name and address of the customer, determines whether the customer has an adequate credit rating, automatically computes discounts, adds taxes and other charges, and prepares invoices and envelopes for mailing.

Invoicing in a service organization can be even more complicated than invoicing in manufacturing and retail firms. The trick is to match all services rendered with a specific customer and to include all appropriate rates and charges in calculating the bill. This is especially difficult if the data needed for billing has not been accurately and completely captured in a TPS.

After floundering on the brink of bankruptcy for much of the 1980s, the Harley-Davidson Motor Company experienced rapid sales and production growth in the 1990s. However, during this period, Harley-Davidson's business practices and accounts payable systems were incompatible; the company used different payment schedules and business procedures for each of its main factories. After years of feedback that the firm was creating extra work and expenses, not just for itself but also its suppliers, Harley-Davidson embarked on a mission to integrate and standardize business practices and information systems across the company. Now suppliers such as Southwest Metal Finishing can manage payables and receivables so closely that the company gets paid within seven days of submitting an electronic invoice. Southwest Metal's own invoicing system generates an EDI-based invoice as soon as material is shipped. Because the company offers a one percent discount for the quick payment, its system monitors invoices, payments, and shipments closely.[17]

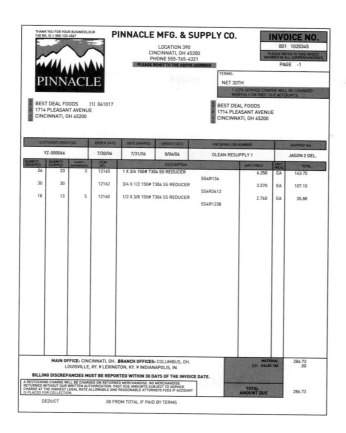

Figure 9.11

Customer Invoice

An output of the invoicing system, a customer invoice reflects the value of the current invoice, as well as which products the customer purchased.

Customer Relationship Management (CRM)

As discussed in Chapter 2, a **customer relationship management (CRM)** system helps a company manage all aspects of customer encounters, including marketing and advertising, sales, customer service after the sale, and programs to keep and retain loyal customers (see Figure 9.12). The goal of CRM is to understand and anticipate the needs of current and potential customers to increase customer retention and loyalty while optimizing the way that products and services are sold.

CRM software automates and integrates the functions of sales, marketing, and service in an organization. The objective is to capture data about every contact a company has with a customer through every channel and store it in the CRM system to enable the company to truly understand customer actions. CRM software helps an organization build a database about its customers that describes relationships in sufficient detail so that management, salespeople, customer service providers—and even customers—can access information to match customer needs with product plans and offerings, remind them of service requirements, and know what other products he or she has purchased.

customer relationship management (CRM) system
System that helps a company manage all aspects of customer encounters, including marketing and advertising, sales, customer service after the sale, and programs to retain loyal customers.

Figure 9.12

Customer Relationship Management System

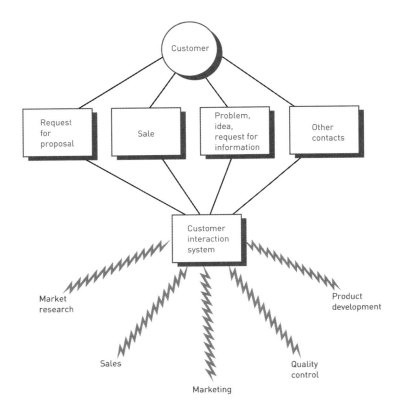

ISM, Inc. is a CRM strategic advisor that rigorously tests the available CRM packages each year. Table 9.3 lists ISM's 15 top-rated packages in alphabetical order. These packages scored the highest according to 171 selection criteria, including 105 business functions, 48 technical features, and 18 user-friendliness/support features.

Table 9.3

The 15 Top-Rated CRM Software Packages

Source: ISM Press Release, March 2, 2004, "ISM Announces Winners of Top 15 Customer Relationship Management (CRM) Software Awards for Enterprise and Medium & Small Business Solutions."

Software Package	Vendor
Amdocs ClarifyCRM version 12	Amdocs Limited
C2 CRM version 7.05	Clear Technologies, Inc.
Client Management Software 7.0	OnContact Software Corporation
E.piphany E.6 version 6.5	E.piphany, Inc.
Onyx Enterprise CRM 4.0	Onyx Software
ExSellence 4.7	Optima Technologies, Inc.
PeopleSoft CRM 8.8	PeopleSoft, Inc.
Pivotal CRM Suite 5.0	Pivotal Corporation
S1 CRM Solutions	S1 Corporation
SalesPage v 4.5	SalesPage Technologies, Inc.
mySAP CRM 4.0	SAP AG
Saratoga iAvenue version 6.2	Saratoga Systems
Siebel 7.5.3	Siebel Systems
growBusiness Solutions version 2.1	Software Innovation
Staffware Process RM version 9.0	Staffware

The focus of CRM involves much more than installing new software. Moving from a culture of simply selling products to placing the customer first is essential to a successful CRM deployment. Before any software is loaded onto a computer, a company must retrain

employees. Who handles customer issues and when must be clearly defined, and computer systems need to be integrated so that all pertinent information is available immediately, whether a customer calls a sales representative or customer service representative. In addition to using stationary computers, most CRM systems can now be accessed via wireless devices.

Hawaiian Airlines is using a CRM system to provide its managers complete views of the airline's relationships with customers such as travel agencies, corporate travel departments, and ticket wholesalers. Until now, relationships with business partners have been managed through separate accounts, depending on how they booked tickets. If a travel agency booked tickets on Hawaiian through a global reservation system such as Sabre and also through Hawaiian's Web site, managers could not easily see that. Because the CRM software aggregates that information in one view, sales and service managers can now offer their business customers volume deals based on total purchases through all the sales channels. "If we don't have a good relationship with a travel agency, they may book you on another airline," says Gordon Locke, Hawaiian's senior VP of marketing and sales. "As we enhance our view of every customer relationship and consistently apply good sales practices and great information about that customer, we should have a competitive advantage in the industry."[18]

Sometimes it can be difficult to get the data needed to support the effective use of a CRM system. This is a lesson that General Motors Corp. learned the hard way when it rolled out CRM software in Shanghai. GM sells most of its vehicles in China through its joint-venture partner Shanghai Automotive Industry Corporation and wanted to develop a more complete understanding of Chinese car buyers. However, local dealers had never heard of CRM and had no incentive to provide GM the necessary customer and sales data. "They were reluctant because it's something they never had to do before, and they just weren't used to it," says Addons Wu, CIO for GM China. So to ensure that the needed data was captured, GM built a Web portal to make it easier for dealers to order vehicles. To use it, though, the dealers must provide the essential customer information the CRM system needs. "Now, by default, when we get a vehicle order, we get the customer data at the same time," Wu says. The automaker sold nearly 390,000 vehicles in China in 2003, and year-over-year sales jumped 49 percent for the first two months of 2004.[19] Some of the increase was attributed to the CRM system.

Routing and Scheduling

Many computer manufacturers and software firms have developed specialized transaction processing systems for companies in the distribution industry. Some distribution applications are for wholesale operations; others are for retail or specialized applications. Trucking firms, beverage distributors, electrical distributors, and oil and natural gas distribution companies are only a few examples. Like airlines, distribution companies must also determine the best use of their resources. For example, a motor freight company might have 100 deliveries to make during the next week, including loads from Miami to Boston and Seattle to Salt Lake City. A **routing system** helps determine the best way to get products from one location to another.

The **scheduling system** determines the best time to pick up or deliver goods and services. For example, trucks can be scheduled to deliver automobile transmission systems from California to Michigan during the second week of September, when oil and gas prices are low. Other objectives are to carry a profitable load on the return trip and to minimize total distance traveled, which can result in lower fuel, driver, and truck maintenance costs. For these reasons, many distribution companies have designed TPSs to help determine which routes will allow for efficient service, while making cost-effective use of drivers and trucks. For firms such as these, scheduling and routing programs are connected to the organization's order and inventory TPS. FedEx uses an automated vehicle routing and scheduling system that uses the customer information to determine which packages should go on which vehicles as well as the delivery routes drivers should take and their sequence of stops. FedEx also uses a geographical information system to generate computerized maps and turn-by-turn directions for each driver. The technology ensures that drivers cover the fewest miles in the shortest time, thus saving time and reducing costs.

routing system
System that determines the best way to get products from one location to another.

scheduling system
System that determines the best time to pick up or deliver goods and services.

PURCHASING SYSTEMS

purchasing transaction processing systems
Systems that include inventory control, purchase order processing, receiving, and accounts payable.

The **purchasing transaction processing systems** include inventory control, purchase order processing, receiving, and accounts payable (see Figure 9.13).

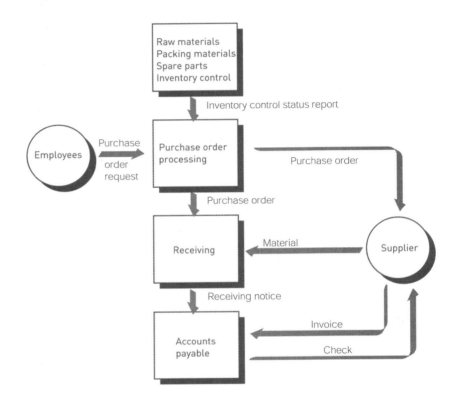

Figure 9.13

Purchasing Transaction Processing System

Inventory Control

A manufacturing firm has several kinds of inventory, such as raw materials, packing materials, finished goods, and maintenance parts. Each day manufacturers must determine how much of these items they must buy to make and ship the products customers have ordered. If they buy too much, they will tie up too much cash in idle inventory, but if they don't keep enough of these items on hand, they will be unable to meet demand and lose sales revenue. A mistake in tracking current raw material inventory can cost millions of dollars, causing the manufacturer to miss profitability targets. We have already discussed the use of an inventory-control system for finished product inventory. A similar transaction processing system can be used to manage the inventory of these items.

Carlisle Engineered Products is a supplier of rubber and plastic products primarily to automakers. The firm is systematically eliminating waste in its factory operations. In a move to eliminate extra inventory on shelves, Carlisle implemented an automated parts-replenishment system to track parts on hand and automatically reorder needed items, but only when the quantity dips too low. The system has led to simplified scheduling and reduced inventory.[20]

Purchase Order Processing

An organization's purchasing department typically has a number of employees who are responsible for all its purchasing activities. Whenever materials or high-cost items are

purchased, the purchasing department is involved. The **purchase order processing system** helps purchasing departments complete their transactions quickly and efficiently. Every organization has its own policies, practices, and procedures for purchasing supplies and equipment.

Companies are increasingly purchasing needed supplies through the Internet or an Internet exchange. The purchasing department can facilitate the buying process by keeping data on suppliers' goods and services. The increased use of telecommunications has given many purchasing departments easier access to this information. For instance, technologies like the Internet and public networks allow purchasing managers to compare products and prices listed in Internet catalogs and large-scale consumer databases. Once the supplier is selected, the suppliers' computer systems might be directly linked to the buyers' systems. Orders can be sent via EDI, reducing purchasing costs and time spent and helping companies maintain low, yet adequate, inventory levels.

Companies are also using Internet exchanges to help them purchase materials and supplies at discounted prices. As discussed previously, an Internet exchange is formed by several companies in an industry and can be open to all companies in that industry.

The Naval Sea Systems Command (NavSea) developed the SeaPort e-procurement system to buy professional services for the U.S. Navy and Marines. Professional services include program management, logistics support, engineering, and financial management, and they amount to about $500 million per year. Use of the system has substantially cut the time to complete an acquisition from 270 days to less than 42 days, mainly because paperwork has been eliminated. In addition, NavSea has reduced costs by more than 10 percent with volume discounts by working with regular bidders.[21] Here's how the SeaPort system works: The program manager logs into SeaPort and defines requirements for the purchase of professional services. Once the formal purchase requisition is approved, a contracting officer creates an electronic bidding event and invites a group of contractors to submit bids. Once bids are received and the bidding event closes, NavSea users evaluate proposals on multiple criteria, including the bidder's past performance and the price it's offering, as well as its technical capabilities. The goal is to find the best value. Once the decision is made, NavSea issues a purchase order electronically to the winning contractor.[22]

Instead of searching for the lowest prices from a list of suppliers, many companies form strategic partnerships with one or two major suppliers for important parts and materials. The partners are chosen based on prices and their ability to deliver quality products on time consistently. For example, the major automotive companies request that their suppliers have

purchase order processing system
System that helps purchasing departments complete their transactions quickly and efficiently.

Campbell's Soup Company, a global manufacturer of prepared food products, is a member of the WorldWide Retail Exchange (WWRE), an Internet-based business-to-business (B2B) exchange for retailers, manufacturers, and suppliers. The WWRE enables retailers and suppliers in the food, general merchandise, textile/home, and drugstore sectors to substantially reduce costs across product development, e-procurement, and supply chain processes.

plants or offices close to their operations in Michigan. The ability to conduct electronic commerce following EDI standards is also a key factor.

Receiving

Like centralized purchasing, many organizations have a centralized receiving department responsible for taking possession of all incoming items, inspecting them, and routing them to the people or departments that ordered them. In addition, the receiving department notifies the purchasing department when items have been received. This notification may be done using a paper form called a *receiving report* or electronically through a business transaction created by entering data into the receiving TPS.

An important function of many receiving departments is quality control by inspection. Inspection procedures and practices are set up to monitor the quality of incoming items. Any items that fail inspection are sent back to the supplier, or adjustments are made to compensate for faulty or defective products.

receiving system
System that creates a record of expected receipts.

Many suppliers now send their customers advance shipment notices. This business transaction is input to the customer's **receiving system** to create a record of expected receipts. In addition, items are shipped with a bar code on the container. At the receiving dock, the worker scans the bar code, and a transaction is sent to the receiving system, where the identification number is matched against the file of expected receipt records. This additional check improves the accuracy of the receiving process, eliminates the need to perform manual data entry, and reduces the manual effort required. As companies move to the use of RFID chips that transmit product identification information, the receiving process will speed up even more.

Accounts Payable

accounts payable system
System that increases an organization's control over purchasing, improves cash flow, increases profitability, and provides more effective management of current liabilities.

The **accounts payable system** attempts to increase an organization's control over purchasing, improve cash flow, increase profitability, and provide more effective management of current liabilities. Checks to suppliers for materials and services are the major outputs. Most accounts payable applications strive to manage cash flow and minimize manual data entry. Input from the purchase order processing system provides an electronic record to the accounts payable application that updates the accounts payable database to create a liability record showing that the firm has made a commitment to purchase a specific good or service. Once the accounts payable department receives a bill from a supplier, the bill is verified and checked for accuracy. Upon receiving notice that the goods and services have been delivered in a satisfactory manner, the data is entered into the accounts payable application. A typical check from an accounts payable application is shown in Figure 9.14. In addition to containing standard information found on any check, most accounts payable checks include the items ordered, invoice date, invoice numbers, amount of each item, any discounts, and the total amount of the check. This information allows the company to consolidate several invoices and bills into a single payment. Many companies pay their suppliers electronically using EDI, the Internet, or other electronic payment systems.

In spite of the ability to use e-commerce to buy goods and services, the reconciliation of purchase orders, supplier invoices, and shipping documents with one another so bills can be paid is still a time-intensive manual process. Global eXchange Services (GXS) operates one of the largest B2B e-commerce networks in the world, with more than 100,000 trading partners accounting for $1 trillion in goods and services. Formerly a wholly owned subsidiary of GE, GXS was acquired by Francisco Partners and now operates as an independent firm. GXS created an application that automates the reconciliation process. The software translates EDI, Web-based, and spreadsheet documents into an XML data format and then makes them available online to its trading partners. Data translation, workflow, and the storing of the actual database are all handled by GXS. The goal is to speed the purchasing process, enabling companies to achieve the quick-payment discounts and lower inventory costs that were expected from online procurement projects.[23]

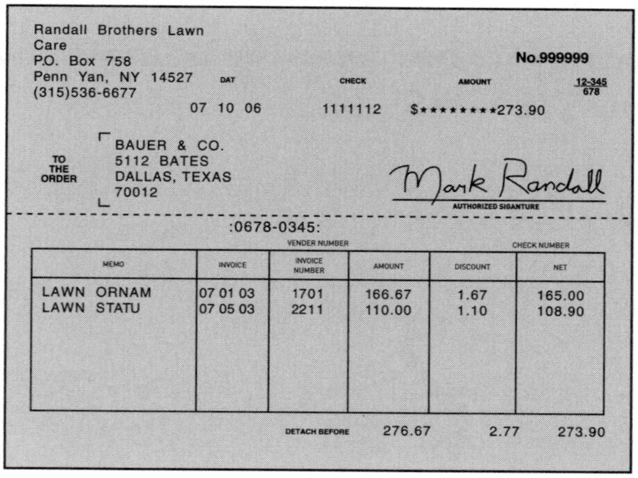

Randall Brothers Lawn Care
P.O. Box 758
Penn Yan, NY 14527
(315)536-6677

No.999999

DAT 07 10 06 CHECK 1111112 AMOUNT $*********273.90 12-345/678

TO THE ORDER BAUER & CO. 5112 BATES DALLAS, TEXAS 70012

AUTHORIZED SIGANTURE

:0678-0345:

VENDER NUMBER CHECK NUMBER

MEMO	INVOICE	INVOICE NUMBER	AMOUNT	DISCOUNT	NET
LAWN ORNAM	07 01 03	1701	166.67	1.67	165.00
LAWN STATU	07 05 03	2211	110.00	1.10	108.90
	DETACH BEFORE		276.67	2.77	273.90

Figure 9.14

A Check Generated by an Accounts Payable Application

The check stub details items ordered, invoice dates, invoice numbers, cost of each item, discounts, and the total amount of the check.

A common report produced by the accounts payable application is the purchases journal. As shown in Figure 9.15, this report summarizes all the organization's bill-paying activities for a particular period. Financial managers use this report to analyze bills that have been paid by the organization. This information is also used to help analyze current and future cash flow needs. Data is summarized for each supplier or parts manufacturer. Invoice number, description, amounts, discounts, and total checking activity are included. In addition, many purchasing reports include the total amount of checks generated by the accounts payable application on a daily, weekly, or monthly basis.

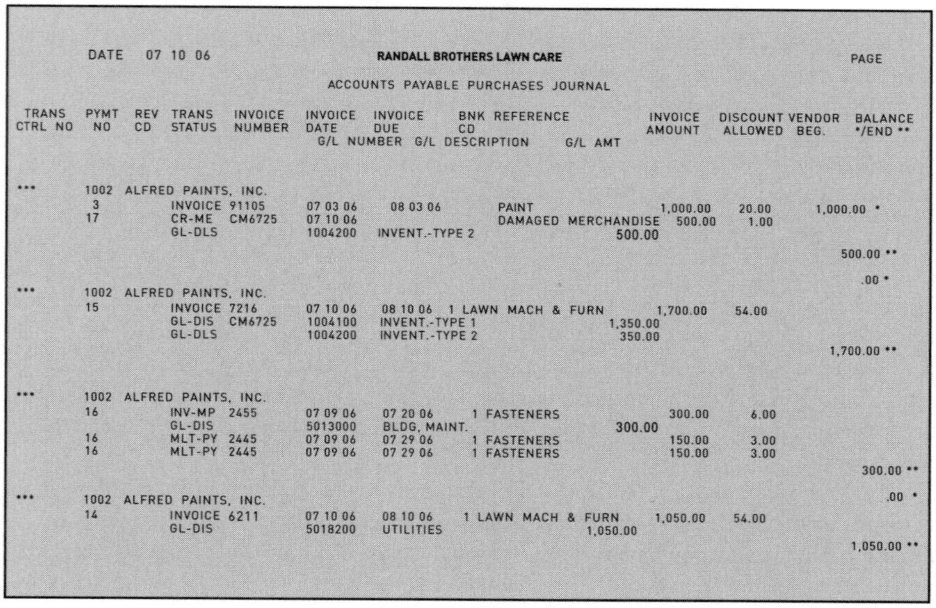

DATE 07 10 06 RANDALL BROTHERS LAWN CARE PAGE
ACCOUNTS PAYABLE PURCHASES JOURNAL
TRANS CTRL NO | PYMT NO | REV CD | TRANS STATUS | INVOICE NUMBER | INVOICE DATE | INVOICE DUE | BNK REFERENCE CD | INVOICE AMOUNT | DISCOUNT ALLOWED | VENDOR BEG. | BALANCE */END **
G/L NUMBER G/L DESCRIPTION G/L AMT

*** 1002 ALFRED PAINTS, INC.
3 INVOICE 91105 07 03 06 08 03 06 PAINT 1,000.00 20.00 1,000.00 *
17 CR-ME CM6725 07 10 06 DAMAGED MERCHANDISE 500.00 1.00
GL-DLS 1004200 INVENT.-TYPE 2 500.00
500.00 **
.00 *
*** 1002 ALFRED PAINTS, INC.
15 INVOICE 7216 07 10 06 08 10 06 1 LAWN MACH & FURN 1,700.00 54.00
GL-DIS CM6725 1004100 INVENT.-TYPE 1 1,350.00
GL-DLS 1004200 INVENT.-TYPE 2 350.00
1,700.00 **
*** 1002 ALFRED PAINTS, INC.
16 INV-MP 2455 07 09 06 07 20 06 1 FASTENERS 300.00 6.00
GL-DIS 5013000 BLDG, MAINT. 300.00
16 MLT-PY 2445 07 09 06 07 29 06 1 FASTENERS 150.00 3.00
16 MLT-PY 2445 07 09 06 07 29 06 1 FASTENERS 150.00 3.00
300.00 **
.00 *
*** 1002 ALFRED PAINTS, INC.
14 INVOICE 6211 07 10 06 08 10 06 1 LAWN MACH & FURN 1,050.00 54.00
GL-DIS 5018200 UTILITIES 1,050.00
1,050.00 **

Figure 9.15

An Accounts Payable Purchases Journal

Generated by the accounts payable application, this report summarizes an organization's bill-paying activities for a particular period.

The accounts payable application ties into other information systems, including cash flow analysis, which helps an organization ensure that sufficient funds are available for the accounts payable application and can show the best sources of funds for payments that must be made via the accounts payable application.

ACCOUNTING SYSTEMS

accounting systems
Systems that include budget, accounts receivable, payroll, asset management, and general ledger.

The primary **accounting systems** include the budget, accounts receivable, payroll, asset management, and general ledger (see Figure 9.16).

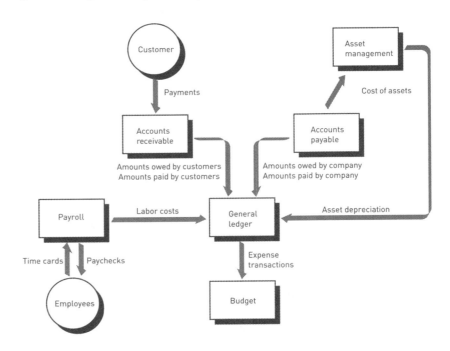

Figure 9.16

Financial Systems

Budget

In an organization, a budget is a financial plan that identifies items and dollar amounts that the organization estimates it will spend. In some organizations, budgeting can be an expensive and time-consuming process of manually distributing and consolidating information. The **budget transaction processing system** automates many of the tasks required to amass budget data, distribute it to users, and consolidate the prepared budgets. Automating the budget process allows financial analysts more time to manage it to meet organizational goals by setting enterprise-wide budgeting targets, ensuring a consistent budget model and assumptions across the organization, and monitoring the status of each department's spending.

budget transaction processing system
System that automates many of the tasks required to amass budget data, distribute it to users, and consolidate the prepared budgets.

    The highly competitive nature of today's business environment, coupled with increased demands from shareholders and federal regulators for more accurate financial guidance, is forcing companies to implement more rigorous budget-planning and financial-forecasting processes. Each month, Krispy Kreme Doughnuts develops a rolling 12-month sales and profit budget forecast using historical data and guidance from managers inside the company and financial officers at companies that operate Krispy Kreme franchises. The process was once done manually and documented in spreadsheets, but it took too much effort and did not yield timely, useful information. Now the company has implemented both new procedures and software to develop a more robust budgeting process. With the new process, Krispy Kreme is able to adjust budget forecasts to other measures such as new store openings and to compare forecasts to actual results. The process and software now yield timely, useful data for decision making.[24]

Accounts Receivable

The **accounts receivable system** manages the cash flow of the company by keeping track of the money owed the company on charges for goods sold and services performed. When goods are shipped to a customer, the customer's accounts payable system receives a business transaction from the invoicing system, and the customer's account is updated in the accounts receivable system of the supplier. A statement reflecting the balance due is sent to active customers. Upon receipt of payment, the amount due from that customer is reduced by the amount of payment.

    The major output of the accounts receivable application is monthly bills or statements sent to customers. As you can see in Figure 9.17, a bill sent to a customer can include the date items were purchased, descriptions, reference numbers, and amounts. In addition, bills can include amounts for various periods, totals, and allowances for discounts. The accounts receivable application should monitor sales activity, improve cash flow by reducing the time between a customer's receipt of items ordered and payment of bills for those items, and ensure that customers continue to contribute to profitability. Most systems can handle payment in a variety of ways, including standard bank checks, credit cards, money-wiring services, and electronic funds transfer via EDI. Increasingly, companies are using the Internet for their accounts receivable application. Using these Internet systems, customers can pay their bills while connected to the Internet on their home PC, at a retail store, or using a handheld computer with a wireless connection to the Internet.

accounts receivable system
System that manages the cash flow of the company by keeping track of the money owed the company on charges for goods sold and services performed.

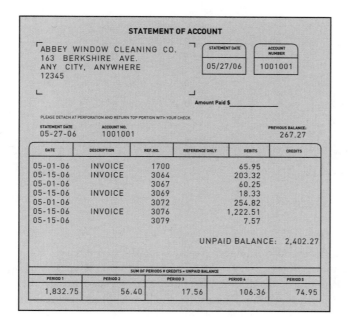

Figure 9.17

An Accounts Receivable Statement

Generated by the accounts receivable application, a bill is sent to a customer (usually monthly) and details items purchased, dates of purchase, and amounts due.

    The accounts receivable system is vital to managing the cash flow of a firm. One major way to increase cash flow is by identifying overdue accounts. Reports are generated that "age" accounts to identify customers whose accounts are overdue by more than 30, 60, or 90 days. Special action may be initiated to collect funds or reduce the customer's credit limit with the firm, depending on the amount owed and degree of lateness.

    An important function of the accounts receivable application is to identify bad credit risks. Because a sizable amount of an organization's assets can be tied up in accounts receivable, one objective of an accounts receivable application is to minimize losses due to bad debts through early identification of potential bad-debt customers. Thus, many companies routinely check a customer's payment history before accepting a new order. With advances in telecommunications, companies can search huge national databases for the names of firms and individuals who have been reported as delinquent on payments or as bad credit risks. When using external data like this in a TPS application, however, companies must be extremely cautious regarding the accuracy of the data.

The accounts receivable aging report, shown in Figure 9.18, is a valuable aspect of an accounts receivable application. In most cases, this report sorts all outstanding debts or bills by date. For unpaid bills that have remained outstanding for a predetermined amount of time, "reminder notices" can also be automatically generated. The accounts receivable aging report gives managers an immediate look into large bills that are long overdue so that they can be followed up to prevent more orders being sent to delinquent customers. This type of report can be produced on a customer-by-customer basis or in a summary format.

Figure 9.18

An Accounts Receivable Aging Report

The output from an accounts receivable application tells managers what bills are overdue, either customer by customer or in a summary format.

6588-DIAGE	TOOL DISTRIBUTORS INC.		0001 ACCOUNTS RECEIVABLE AGING ANALYSIS AS OF SEP 25 2006			SEP 25 2006	PAGE 1

TYPE OPEN-ITEM ITEM-DATE REFERENCE INVOICE/PAYMENT CURRENT 31-60 DAYS 61-90 DAYS 91-120 DAYS OVER 120 DAYS
     NUMBER   NUMBER   AMOUNT

CLASS 05   CUST 94367   NAME TOOLS OF AMERICA   3225 N PARKWAY   TEL 513-2864731   CONT B. BROWN

			CURRENT AMOUNT	31-60	61-90	91-120	OVER 120
INV 95361	03/17/06		23,058.37				23,058.37
DM 96853	03/20/06		5,589.42				1,589.42
INV 105395	07/04/06		2,923.45		2,923.45		
DM 116594	07/06/06		198.32		198.32		
INV 123984	07/15/06		23,087.28				
DSC 123984	08/17/06		1,204.36				
CA 968351	08/19/06		21,882.92				
INV 147296	07/23/06		19,709.57				
CA 83495	07/31/06		18,725.21				
CM 995473	08/19/06		984.59				
INV 149384	08/29/06		23,831.37	21,831.59			
INV 158439	09/10/06		30,086.68	30,086.68			
INV 161236	09/23/06		25,520.37	25,520.37			

TOTL RECEIVABLE FOR CUST 94367	107,208.20	79,438.64	3,121.77	0.00	0.00	24,647.79
CA	3,121.77		3,121.77			
CA	22,640.01	22,640.01				
CM	279.84	279.84				

UNAPPLIED CREDITS	26,041.62	22,919.85	3,121.77	0.00	0.00	0.00

NET RECEIVABLE FOR CUST 94367	81,166.58	56,518.79	0.00	0.00	0.00	24,647.79

TOTAL FOR CLASS   50   42   PRINTED   TOTAL RECEIVABLE   UNAPPLIED CREDITS   NET RECEIVABLE

	TOTAL RECEIVABLE	UNAPPLIED CREDITS	NET RECEIVABLE
CURRENT	161,506.12	67,832.57	93,673.55
31-60 DAYS	31,494.14	18,984.68	12,509.46
61-90 DAYS	11,569.32	4,267.91	7,301.59
91-120 DAYS	27,764.18	0.60	27,746.18
TOTAL	238,823.97	93,232.49	145,581.08

Payroll

The two primary outputs of the payroll system are the payroll check and stub, which are distributed to the employees, and the payroll register, which is a summary report of all payroll transactions. In addition, the payroll system prepares W-2 statements at the end of the year for tax purposes. Responsibility for running the TPS application can be outsourced to a service company. Other firms may rely on a purchased software application for payroll processing. Some of these packages are tailored for a specific industry; others can accommodate a wide range of uses.

The number of hours worked by each employee is collected using a variety of data entry devices, including time clocks, time cards, and industrial data-collection devices in a subsystem called *time and attendance*. In a manufacturing firm, hours worked and labor costs may be captured by job so that this information can be passed on to the manufacturing costs system. Capturing this data enables better analysis and control over labor allotment, costs, and scheduling. Banner Health, an operator of hospitals, long-term care centers, family clinics, and home-care services, realized these benefits when it deployed a new payroll application system. Banner employees log on for work at on-site terminals, via the Web, or through an interactive voice-response phone application. Their data is captured and sent to Banner's payroll system and is available to managers for decision making. Now, nursing supervisors can adjust staffing plans based on factors such as which employees have worked overtime or which skills are needed for a shift.[25]

Once collected, payroll data is used to prepare weekly, biweekly, or monthly employee paychecks (see Figure 9.19). Often payroll applications have EDI arrangements with employees' banks to make direct deposits into employees' accounts.

```
TENDER CARE DAY CARE CENTER, INC.
CINCINNATI, OH 45200                                               4207
        Vacation Taken This Check    0.000   Sick Taken This Check    0.000
        Vacation Available           0.667   Sick Available           0.667
                        Earnings                     Deductions
        Description    Hours    Amount   Description          Amount
        ------------------------------   ------------------------------
        Regular Pay   36.320   199.76    FICA Withheld         15.28
        Overtime       0.000     0.00    Fed. Tax W/H          15.00
        SICK TIME      0.000     0.00    State Tax W/H          1.67
        PERSONAL       0.000     0.00    Other W/H #1           4.19
        VACATION       0.000     0.00    Other W/H #2           0.00
        HOLIDAY PAY    0.000     0.00    Other W/H #3           0.00
        Gross Pay#4    0.000     0.00    Other W/H #4           0.00
                                -------                        -------
                Total            199.76          Total          36.14
        ------------------------------YEAR TO DATE------------------------------
                Total Earnings   397.76          FICA           30.43
```

Figure 9.19

Paycheck Stub

A typical paycheck stub details the employee's hours worked for the period, salary, vacation pay, federal and state taxes withheld, and other deductions.

As you can see in Figure 9.19, the payroll program has produced a weekly paycheck, which includes the employee's hourly rate, total hours worked, regular pay, premium pay, federal and state tax withholdings, and other deductions. In addition to paychecks, most payroll programs produce a **payroll journal**, shown in Figure 9.20. A typical payroll journal contains employees' names, the areas where employees worked during the week, hours worked, the pay rate, a premium factor for overtime pay, earnings, the earnings type, various deductions, and net pay calculations. Financial managers at the operational level use the payroll journal to monitor and control pay to individual employees. As you can see in Figure 9.20, the payroll journal also includes totals for hours worked, earnings, deductions, and net pay.

payroll journal

A report that contains employees' names, the area where employees worked during the week, hours worked, the pay rate, a premium factor for overtime pay, earnings, earnings type, various deductions, and net pay calculations.

Figure 9.20

A Payroll Journal

Generated by the payroll application, this report helps managers monitor total payroll costs for an organization and the impact of those costs on cash flow.

```
DATE  12 25 06              RANDALL BROTHERS LAWN CARE              PAGE   003

                              PAYROLL JOURNAL
                                   PR020

EMPLOYEE NAME
ERRORS        EMPLOYEE  COST                PREM           EARN          DED          PAYMENT
WARNINGS      NUMBER   CENTER HOURS  RATE FACTOR EARNINGS  TYPE DEDUCTIONS TYPE NET PAY LOCATION

  DANIEL JACKSON
                112      30   40.00  6.00        240.00    001                  44.19   CIN
                112      30   10.00  6.00  1.5    90.00    030
                112                                              5.00    220
                112                                             15.81    801
                112                                             44.19    802

  HOWARD SIMPSON                      5.25
                126      30   32.00         168.00    001                  12.82   CIN
                126                                              5.00    220
                126                                              2.00    223
                126                                             10.16    801
                126                                             12.82    802
```

Like many other transaction processing applications, the payroll application interfaces with other applications. All payroll entries are entered into the general ledger systems. Furthermore, there can be a direct link between payroll activities and production/inventory-control operations. Direct links are often used for manufacturing operations or job-shop systems because data collected on hours worked from the payroll application helps determine the total cost of completing various jobs. For example, if an employee who earns $15 per hour spends 20 hours completing a particular job, the labor costs for that job are $300. This type of information from the payroll application is useful in determining the cost to produce a product or render a service and thus in determining its profitability.

Asset Management

asset management transaction processing system
System that controls investments in capital equipment and manages depreciation for maximum tax benefits.

Capital assets represent major investments for the organization, whose value appears on the balance sheet under fixed assets. These assets have a useful life of several years or more, over which their value is depreciated, resulting in a tax reduction. The **asset management transaction processing system** controls investments in capital equipment and manages depreciation for maximum tax benefits. Key features of this application include efficient handling of a wide range of depreciation methods, country-specific tax reporting and depreciation structures for the various countries in which the firm does business, and workflow-managed processes to easily add, transfer, and retire assets.

Walt Disney World in Orlando uses asset management software to track all its capital improvements, inspections of new developments, renovations, repairs, replacements, and upgrades of just about everything on the premises. The software holds data from nearly 2 million work orders. The data is used to produce reports on historical data and trends, the causes for maintenance or repair, and analysis that helps pinpoint problems and resolve them. Without such software, companies have trouble managing their assets because they do not know how one job relates to another. With asset management software, workers have an enterprise-wide view of all the assets across the organization. That could keep a company from, say, scheduling repair of several buildings' roofs at the same time it is resurfacing a portion of its parking lot, preventing workers from reaching the buildings.[26]

General Ledger

general ledger system
System designed to automate financial reporting and data entry.

Every monetary transaction that occurs within an organization must be properly recorded. Payment of a supplier's invoice, receipt of payment from a customer, and payment to an employee are examples of monetary transactions. A computerized **general ledger system** is designed to allow automated financial reporting and data entry. The general ledger application produces a detailed list of all business transactions and activities. Reports, including profit and loss (P&L) statements, balance sheets, and general ledger statements, can be generated (see Figure 9.21). Furthermore, historical data can be kept and used to generate trend analyses and reports for various accounts and groups of accounts used in the general ledger package. Various income and expense accounts can be generated for the current period, year to date, and month to date as required. The reports generated by the general ledger application are used by accounting and financial managers to monitor the profitability of the organization and to control cash flows.

Financial reports that summarize sales by customer and inventory items can also be produced. These reports are used by marketing and financial managers to determine which customers are contributing to sales and inventory items that are selling as expected.

A key to the proper recording and reporting of financial transactions is the corporation's chart of accounts (see Table 9.4). This chart provides codes for each type of expense or revenue. By entering transactions consistent with the chart of accounts, financial data can be reported in a simple and consistent fashion across all organizations of the enterprise, even if it is a multinational corporation.

Niche Retail (*www.nicheretail.com*) manages nine boutique retail Web sites that sell such specialized items as jogging strollers, athletes' Suunto watches, inflatables that serve as spare beds, and high-end children's car seats. To keep software costs down, instead of paying more than $50,000 for sophisticated general ledger and accounting software, the company subscribes to NetLedger, Inc. This online service provider offers a set of integrated accounting applications that capture transactions directly off Niche Retail's Web sites at a cost of about $250 per month, which allows Niche Retail to avoid hiring additional data-entry personnel for multiple accounting applications.

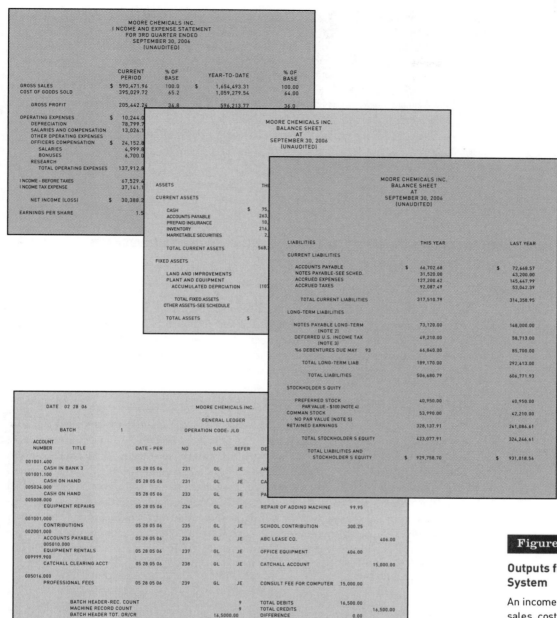

Figure 9.21

Outputs from a General Ledger System

An income statement (top) details sales, costs, and operating expenses to produce a statement of income for an organization. A balance sheet (center left and right) breaks down assets and liabilities so that managers can see at a glance whether income is covering expenses. A general ledger statement (bottom) track credits and debits.

Online service providers such as NetLedger are extremely valuable to small businesses. By offering applications as a networked service on a monthly subscription basis, they reduce the capital costs of enterprise applications.[27] Another provider of Web-based accounting services is healthcare supply chain software vendor Neoforma, Inc., with a Web-based product that integrates with healthcare providers' existing patient registration, billing, general ledger, and surgery-scheduling systems.[28] And because the Web-based system is fully hosted and offered via a monthly subscription fee, no infrastructure, maintenance, or upgrade work needs to be done by a healthcare company's internal IS organization.[29]

Table 9.4

Sample Partial Chart of Accounts

Major Account Name	Type of Expense	Subaccount Code Used to Identify Transaction
Wages and Benefits	Management salaries and benefits	MSALSB
	Nonmanagement salaries and benefits	NMALSB
	Overtime	OVT
Travel and Training	Travel-related expenses	TRAVEL
	Tuition for training classes	TUITION
Professional Services	Fees paid to consultants, contractors, trainers, and other professionals	PROFSV
Maintenance Expense	Maintenance labor	MAINTL
	Maintenance parts	MAINTP
	Maintenance supplies	MAINTS

INTERNATIONAL ISSUES

Businesses are increasingly operating across country borders or around the globe. Multinational corporations must address many issues and complexities in planning, building, and operating their TPSs. Different languages and cultures, disparities in IS infrastructure, varying laws and customs rules, and multiple currencies are among the challenges of linking all the business partners, customers, and subsidiaries of a multinational company.

Different Languages and Cultures

It is difficult to get people from several countries who speak different languages and who were raised in different cultures to agree on a single work process. In some cultures, people do not routinely work in teams in a networked environment. Despite these complications, many multinational companies are able to establish close connections with their business partners and roll out standard IS applications for all to use. However, those standard applications often don't account for all the differences among business partners and employees operating in other parts of the world. So, sometimes they require extensive and costly customization. For example, even though English has become a standard business language among executives and senior managers, many people within organizations do not speak English. As a result, software may need to be designed with local language interfaces to ensure the successful implementation of a new system. Customization may also be needed for date fields: The U.S. data format is month/day/year, the European format is day/month/year, and Japan uses year/month/day. Sometimes users may also have to implement manual processes to override established formatting to enable systems to function correctly.

Disparities in Information System Infrastructure

The lack of a robust or a common information infrastructure can also create problems. The U.S. telecommunications industry is highly competitive, with many options for high-quality service at relatively low rates. Many other countries' telecommunications services are controlled by a central government or operated as a monopoly, with no incentives to provide fast and inexpensive customer service. For example, much of Latin America lags the rest of the world in Internet usage, and online marketplaces are almost nonexistent there. This gap makes it difficult for multinational companies to get online with their Latin American business partners. Even something as mundane as the power plug on a piece of equipment built in one country may not fit into the power socket of another country.

Varying Laws and Customs Rules

Numerous laws can affect the collection and dissemination of data. For one example, labor laws in some countries prohibit the recording of worker performance data. Also, some

countries have passed laws limiting the transborder flow of data linked to individuals. Specifically, European Community Directive 95/96/EC of 1998 requires that any company doing business within the borders of the 25 European Union member nations protect the privacy of customers and employees. It bars the export of data to countries that do not have data-protection standards comparable to the EU's. Initially, the EU countries were concerned that the U.S.'s largely voluntary system of data privacy did not meet the EU directive's stringent standards. Eventually, the U.S. Department of Commerce worked out an agreement to allow American companies to import and export data. Failure to gain this compromise would have severely limited the exchange of information about employees and consumers.

Trade custom rules between nations are international laws that set practices for two or more nations' commercial transactions. They cover imports and exports and the systems and procedures dealing with quotas, visas, entry documents, commercial invoices, foreign trade zones, the payment of duty and taxes, and many other related issues. For example, the North American Free Trade Agreement (NAFTA) of 1994 created trade custom rules to address the flow of goods throughout the North American continent. The great number of these custom rules and their changes over time create nightmares for people who must keep existing TPSs consistent with the rules.

For example, new EU regulations, called the Restriction of Hazardous Substances and Waste Electrical and Electronic Equipment, require that electronics manufacturers track and manage hazardous materials in the components they use in their products. To prove a product complies with these directives, a company must have information systems and audit and reporting tools to track every part number and the quantity of chemicals each part contains. AMR Research estimates the average IS cost to support compliance with these environmental directives exceeds $2 million per manufacturer. Hewlett-Packard is working with its suppliers to ensure compliance for its 36,000 products. Having to monitor electronic waste through its supply chain also has meant changes to HP's cost structure, business model, and relationships with customers, suppliers, and retailers.[30]

Multiple Currencies

The TPSs of multinational companies must conduct transactions in multiple currencies. To do so, a set of exchange rates is defined, and the information systems apply these rates to translate from one currency to another. The systems must be current with foreign currency exchange rates, handle reporting and other transactions such as cash receipts, issue vendor payments and customer statements, record retail store payments, and generate financial reports in the currency of choice.

United Parcel Service (UPS) provides an example of a multinational company that must deal with all these issues and others as it expands its parcel delivery service into China. UPS serves both multinational companies operating in China and local businesses. The company provided its drivers with version III of its Driver Information Acquisition and Delivery (DIAD) system for handheld devices on Chinese routes to enable them to capture customer signatures and transmit delivery status information. Communications are done in real time through an internal packet-data radio that connects drivers to UPS's worldwide delivery network. UPS has no immediate plans to upgrade to version IV of the DIAD system, which supports Bluetooth and a range of other wireless technologies because those standards are not commonly used in China. UPS has developed and deployed Chinese-language versions of its WorldShip, QuantumView, and CampusShip shipping-management systems and has made them widely accessible to its customers over the Web. Introducing small businesses to automated shipping technology is key to UPS's ability to serve millions of Chinese customers. UPS will often pay for and install Web terminals at smaller customers that generate a lot of business to make sure that they continue to use UPS services. Education is another key component of UPS's expansion in China. UPS had to train the drivers to understand English so that they would be able to upload the correct shipping information into their systems. In addition, UPS runs seminars and conferences to educate business owners about the technology and its benefits. UPS also is working with Chinese businesses to help them standardize shipping data.[31]

ENTERPRISE RESOURCE PLANNING

Flexibility and quick response are hallmarks of business competitiveness. Access to information at the earliest possible time can help businesses serve customers better, raise quality standards, and assess market conditions. Enterprise resource planning (ERP) is a key factor in instant access. Although some think that ERP systems are only for extremely large companies, this is not the case. Small and midsized companies, which generally include those ranging from $50 million to $500 million in annual revenue, represent the greatest growth for ERP companies. According to Mike Dominy, senior analyst at the Yankee Group IS consulting company, "It's the largest opportunity in the market today." ERP systems are commonly used in manufacturing companies, colleges and universities, professional service organizations, retailers, and healthcare organizations.[32] A few leading vendors of ERP systems are listed in Table 9.5.

Table 9.5

Some ERP Software Vendors

Software Vendor	Name of Software
Oracle	Oracle Manufacturing
SAP America	SAP R/3
PeopleSoft	PeopleSoft
Ross Systems	iRenaissance
QAD	MFG/Pro
Lawson Software	ERP Solutions

NetERP software from NetSuite provides tightly integrated, comprehensive ERP solutions for businesses, giving them access to real-time business intelligence and thus enabling better decision making.

(Source: Courtesy of NetSuite Inc.)

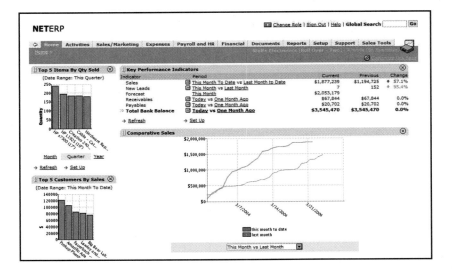

An Overview of Enterprise Resource Planning

The key to ERP is real-time monitoring of business functions, which permits timely analysis of key issues such as quality, availability, customer satisfaction, performance, and profitability. Financial and planning systems automatically receive information from manufacturing and distribution. When something happens on the manufacturing line that affects a business situation—for example, packing material inventory drops to a certain level, which affects the ability to deliver an order to a customer—a message is triggered for the appropriate person

in purchasing. The key steps in running a manufacturing organization using an ERP system are outlined here.

Develop Demand Forecast

For a manufacturing organization, the traditional planning process begins with the preparation of a long-term demand forecast. This forecast is prepared up to 18 months in advance and attempts to predict the weekly amount of each product to be purchased over this time period. The development of the forecast may require special software modules, historical data related to shipments, and discussions with members of sales, manufacturing, and finance organizations. Some organizations require years to implement an accurate, reliable demand forecasting process.

Deduct Demand Forecast from Inventory

As the forecasted demand is deducted from existing inventory to project future inventory levels, the system will display points at which additional finished products need to be produced.

Determine What Is Needed for Production

A bill of materials (BOM, a sort of "recipe" of parts and ingredients needed) for each item to be produced is used to translate finished product requirements into detailed lists of requirements for raw materials and packaging materials, which will be required to make and ship each finished product item.

Check Inventory for Needed Raw Materials

The forecasted needs for raw materials and packaging materials are subtracted from the existing inventory of these items. Again, the system eventually displays and triggers purchases of additional items required to meet production needs.

Schedule Production

The ERP production planning module uses the demand forecast and finished product inventory data to determine the week-by-week production schedules.

Assess Need for Additional Production Resources

The production schedule may reveal interesting insights, such as the need to build additional manufacturing capacity, hire additional workers, or develop new suppliers to provide sufficient raw materials. These new requirements can be input into the purchasing system and human resources modules of the ERP system and be used by managers in that area to develop future plans.

Financial Forecasting

All the generated data can be fed into the financial module of the ERP system to prepare a profit and loss forecast statement to assess the firm's future profitability. This profit forecast in turn can be used to help establish new budget limits for the upcoming year.

Lean manufacturing is a new manufacturing model some companies are considering to avoid problems with the demand-forecast approach just described. Lean manufacturing aims to improve efficiency, eliminate product backlogs, and synchronize production to *actual* customer demand rather than long-term forecasts. The major ERP vendors are now offering lean manufacturing solutions to help companies build products to actual demand, rather than forecasts. The solutions include supplier self-service portals that enable customers to log on to a Web site and enter their planned orders for weeks in advance. This enables the manufacturer to produce to actual demand rather than a long-range forecast of dubious accuracy.[33]

Implementation of a lean manufacturing ERP system from Ross Systems has enabled Michael Angelo's Gourmet Foods to dramatically reduce the time it takes to develop, test, and distribute new products. The maker of specialty frozen and refrigerated products such as lasagna and calzones used to spend six months to bring new products to market. Now the company can do so in less than 90 days. Time to market is a critical success factor for a specialty-foods maker because retailers deal with finicky consumers who change their minds frequently. With the new ERP system, Michael Angelo's is able to operate using a just-in-time production process. Its freshly made ingredients are received and products are cooked, packaged, and shipped to retail customers within 24 hours. Nearly 85 percent of

Michael Angelo's orders are scheduled via EDI. The receipt of orders initiates the whole process. When an order is received, inventory is checked in real time, so the company knows whether it has ingredients on hand to fill that order. Receipt of an order also triggers a series of data exchanges with the other applications in the ERP system, alerting buyers what to purchase and production managers what to schedule.[34]

In addition to manufacturing and finance, ERP systems can also support the human resource, sales, and distribution functions. This integration breaks through traditional corporate boundaries and can dramatically affect the entire organization. For example, the newest version of PeopleSoft's ERP software (Enterprise 8.10) is designed to help companies improve business performance by providing new software modules to support finances, asset management, recruiting, and real estate. These new modules enable companies to manage equipment maintenance, generate multiyear forecasts and budgets by units, automate job posting for real-time recruiting on the Web, comply with government guidelines and mandates (such as the financial reporting and certification requirements of the Sarbanes-Oxley Act), and provide budgeting and analysis tools for real estate transactions.[35]

Advantages and Disadvantages of ERP

Increased global competition, new needs of executives for control over the total cost and product flow through their enterprises, and ever-more-numerous customer interactions are driving the demand for enterprise-wide access to real-time information. ERP offers integrated software from a single vendor to help meet those needs. The primary benefits of implementing ERP include elimination of inefficient or outdated systems, easing adoption of improved work processes, improving access to data for operational decision making, and technology standardization. ERP vendors have also developed specialized systems for specific applications and market segments. The "Information Systems @ Work" feature describes the benefits one lending company realized by adopting an ERP system.

Elimination of Costly, Inflexible Legacy Systems

Adoption of an ERP system enables an organization to eliminate dozens or even hundreds of separate systems and replace them with a single, integrated set of applications for the entire enterprise. In many cases, these systems are decades old, the original developers are long gone, and the systems are poorly documented. As a result, the systems are extremely difficult to fix when they break, and adapting them to meet new business needs takes too long. They become an anchor around the organization that keeps it from moving ahead and remaining competitive. An ERP system helps match the capabilities of an organization's information systems to its business needs—even as these needs evolve.

Improvement of Work Processes

Competition requires companies to structure their business processes to be as effective and customer oriented as possible. ERP vendors do considerable research to define the best business processes. They gather requirements of leading companies within the same industry and combine them with research findings from research institutions and consultants. The individual application modules included in the ERP system are then designed to support these **best practices**, the most efficient and effective ways to complete a business process. Thus, implementation of an ERP system ensures good work processes based on best practices. For example, for managing customer payments, the ERP system's finance module can be configured to reflect the most efficient practices of leading companies in an industry. This increased efficiency ensures that everyday business operations follow the optimal chain of activities, with all users supplied the information and tools they need to complete each step.

best practices
The most efficient and effective ways to complete a business process.

ERP Consolidation Opens Doors for College Lender

Academic Management Services (AMS), a 500-employee academic loan provider, is a recent winner of *CIO* magazine's Enterprise Value Award. The company developed an enterprise system that so improved its work processes that it moved into new markets, increased in value, and eventually was acquired by educational lending giant Sallie Mae. To learn from this Cinderella story, let's start at the beginning.

AMS loan counselors work with clients over the phone to determine an academic loan program that best matches their needs and overall financial picture. The counselor's primary responsibility is to determine which of six categories of loans best suits the client's needs. In AMS's previous system, each loan classification had its own online system, which counselors used to determine clients' eligibility. During a phone interview, the counselor would have to navigate between several interfaces, which was not only difficult for the counselor but also inconvenient for the customers. AMS realized that if it integrated the processes and systems into a single ERP platform, it could save counselors time, and streamline operations. The ERP system the firm created was named ICE—Integrated Counseling and Enrollment.

Early ERP systems were designed primarily for manufacturing and production planning. In the mid-1990s ERP solutions expanded to include ordering systems, financial and accounting systems, asset management, and human resource management systems. By the late 1990s, ERP solutions were again broadened to include systems such as AMS's, allowing them to consolidate information across their organizations.

One of the advantages of consolidating systems into an ERP system is a reduction of data redundancy. Streamlining databases translates to significant savings in time and expense by not having to cross-reference or maintain multiple databases and by being able to respond to customer inquiries quicker. "Many companies have pockets of information relative to their customers," explains AMS CIO John Mariano. "The more that's integrated and the more business intelligence you can gain from that, the more you are able to deliver value across the company."

The consolidation at AMS greatly increased the efficiency of the loan counselors because they were able to describe, recommend, and approve the loan package that best matched the families' needs. AMS's loan agents now handle 1 million outbound and 300,000 inbound calls annually—90 percent more calls than they did prior to ICE. This increase in call volume results from each counselor's increased efficiency and the system's ability to accommodate more counselors. A family's complete financial profile is entered into ICE for processing. The software does much of the work in deciding which loan option is the best fit, presenting recommendations to the counselors for consideration. Loan counselors can authorize loans within the system, often while the potential customer is on the phone. The ability to approve a loan in this manner can reduce the waiting period for families and helps AMS loan counselors handle more calls.

Consolidating disparate systems also helped AMS's information systems budgets by consolidating and shutting down redundant systems. In implementing the ICE project, AMS saved on licensing and maintenance fees in both its systems and processes. "At one time, we had four separate [database installations] of Oracle," says CIO Mariano. "Now we've consolidated to one. It has cut our IT expenses by approximately 40 percent."

The ICE system also helped AMS enter a new line of business—consolidation loans, in which borrowers combine numerous education loans into a single, lower-interest loan. These types of loans now account for nearly 25 percent of AMS's revenue. "They've created a new market ability, created more reliable partnerships with universities and did it all for $311,500," says Paul Gaffney, executive vice president and CIO of Staples, a member of the Enterprise Value Awards judging panel. "They're the poster child for IT value."

Discussion Questions

1. How have AMS loan counselors' activities changed since the implementation of the ICE system? How has the balance shifted between time spent with computer systems and time spent with customers?

2. How has the ICE system allowed AMS loan counselors to be more effective?

Critical Thinking Questions

3. Consolidating systems into a centralized ERP system is a common activity for many businesses lately. What, do you think, has caused business systems to become so divided in years past?

4. What is the advantage of hiring large firms like SAP to design an ERP, rather than hiring smaller vendors who specialize in subsystems?

SOURCES: Lafe Low, "They Got It Together," *CIO Magazine*, February 15, 2004, *www.cio.com*; Tuition Pay Web site, accessed May 23, 2004, at *https://secure. tuitionpay.com*; Sallie Mae Web site, accessed May 23, 2004, at *www.salliemae. com/about/abt_ams.html*.

The Dial Corporation, maker of soap and other consumer goods, is installing a $35 million set of SAP AG's business applications. Dial is justifying the expense by explaining that it wants to exploit the same application functionality and best practices used by top, rival manufacturers such as Colgate-Palmolive and Procter & Gamble. Dial CIO Devon Jones said, "SAP has really got their arms around the best practices [in the consumer goods industry] and embedded it in their software." SAP has partnered with Colgate-Palmolive, P&G, and others to codevelop application features tailored for the consumer goods industry. As a result, other ERP vendors have fallen behind SAP selling to companies in the consumer products industry.[36]

Increase in Access to Data for Operational Decision Making

ERP systems operate via an integrated database, using essentially one set of data to support all business functions. So, decisions on optimal sourcing or cost accounting, for instance, can be run across the enterprise from the start, rather than looking at separate operating units and then trying to coordinate that information manually or reconciling data with another application. The result is an organization that looks seamless, not only to the outside world but also to the decision makers who are deploying resources within the organization.

The data is integrated to provide excellent support for operational decision making and allows companies to provide greater customer service and support, strengthen customer and supplier relationships, and generate new business opportunities. For example, once a salesperson makes a new sale, the business data captured during the sale is distributed to related transactions for the financial, sales, distribution, and manufacturing business functions in other departments.

Upgrade of Technology Infrastructure

An ERP system provides an organization with the opportunity to upgrade and simplify the information technology it employs. In implementing ERP, a company must determine which hardware, operating systems, and databases it wants to use. Centralizing and formalizing these decisions enables the organization to eliminate the hodgepodge of multiple hardware platforms, operating systems, and databases it is currently using—most likely from a variety of vendors. Standardization on fewer technologies and vendors reduces ongoing maintenance and support costs as well as the training load for those who must support the infrastructure.

Expense and Time in Implementation

Getting the full benefits of ERP is not simple or automatic. Although ERP offers many strategic advantages by streamlining a company's TPSs, ERP is time-consuming and is difficult and expensive to implement. Some companies have spent years and tens of millions of dollars implementing ERP systems. A survey by the Meta Group on leading ERP vendors' work for 200 user companies in 12 industries revealed that ERP investments were made over three to five years. On average, 25 percent of the cost was spent on software, 40 percent on professional services, and 25 percent on internal staff. The software cost amounted to 1 to 3 percent of the company's annual revenue, with smaller companies spending a greater percentage. The implementation time for these projects was approximately 20 months. Another seven months were needed before the benefits were realized.[37]

Difficulty Implementing Change

In some cases, a company has to radically change how it operates to conform to the ERP's work processes—its best practices. These changes can be so drastic to long-time employees that they retire or quit rather than go through the change. This exodus can leave a firm short of experienced workers.

Difficulty Integrating with Other Systems

Most companies have other systems that must be integrated with the ERP system. These systems can include financial analysis programs, Internet operations, and other applications. Many companies have experienced difficulties making these other systems operate with their ERP system. Other companies need additional software to create these links. In October 2003 Goodyear Tire & Rubber Co. disclosed that it would have to restate its financial results back to 1998 because of financial errors resulting from a faulty implementation of ERP software and a set of older applications that are used for intercompany billing. The company said that it expects to lower the net income it reported during the restatement period by as

much as $100 million. The ERP system was installed in 1999 and runs Goodyear's core accounting functions.[38]

Risks in Using One Vendor

The high cost to switch to another vendor's ERP system makes it extremely unlikely that a firm will do so. So, once a company has adopted an ERP system, the vendor knows that it has a "captive audience" and has less incentive to listen and respond to customer issues. The high cost to switch also creates a high level of risk—in the event the ERP vendor allows its product to become outdated or goes out of business. Picking an ERP system involves not just choosing the best software product but also choosing the right long-term business partner.

Risk of Implementation Failure

Implementing an ERP system is extremely challenging and requires tremendous amounts of resources, the best IS people, and plenty of management support. Failed ERP installations often result from implementation problems rather than shortcomings in the software itself. And when there are problems with an ERP implementation, it can be expensive. In 2004, Ohio's attorney general filed a lawsuit against an ERP vendor seeking $510 million in damages stemming from an allegedly faulty installation of the company's ERP and student administration applications at Cleveland State University. Cleveland State was the first school to install a full set of the ERP vendor's student administration applications. But after it began using the software in 1998, university officials blamed the technology for problems in processing financial aid, enrolling transfer students and recording grades. The problems led to more than $5 million in lost revenue because of an inability to track and collect receivables, plus additional unexpected expenditures to purchase a second mainframe and server with an Oracle database.[39]

ERP systems accommodate the different ways each company runs its business by either providing vastly more functions than one business could ever need or including customization tools that allow firms to fine-tune what should already be a close match. SAP R/3 is the undisputed king of the first approach. R/3 is easily the broadest and most feature-rich ERP system on the market. In an effort directed at capturing more customers in the fast-growing small-business market, PeopleSoft has taken a different approach. PeopleSoft completed integration of several of its applications with those of J.D. Edwards & Co. just four months after completing its 2003 $1.8 billion acquisition of Edwards.[40] PeopleSoft then developed enterprise applications that combine industry-specific business processes, including software modules for financial management, distribution, manufacturing, human resources, and project management. The software is available in four industry-specific suites: for industrial manufacturers, wholesale distributors, home builders, and construction companies. Each suite comes preconfigured with as many as 30 industry-specific business processes. For example, the construction-company suite includes a job-costing process that lets companies define a job in the software, such as building a bridge, and then identify and track all the costs associated with that job.[41]

SUMMARY

Principle

An organization's TPS must support the routine, day-to-day activities that occur in the normal course of business and help a company add value to its products and services.

Transaction processing systems (TPSs) are at the heart of most information systems in businesses today. TPSs consist of all the components of a CBIS, including databases, networks, people, procedures, software, and hardware devices to process transactions. All TPSs perform the following basic activities:data collection, which involves the capture of source data to complete a set of transactions;data editing, which checks for data validity and completeness;data correction, which involves providing feedback of a potential problem and enabling users to change the data;data manipulation, which is the performance of calculations, sorting, categorizing, summarizing, and storing data for further processing;data storage, which involves placing transaction data into one or more databases;and document production, which involves outputting records and reports.

The methods of transaction processing systems include batch, online, and online with delayed processing. Batch processing involves the collection of transactions into batches, which are entered into the system at regular intervals as a group. Online transaction processing (OLTP) allows transactions to be entered as they occur. Systems that use a compromise between batch and online processing are online with delayed entry TPSs. Transactions may be entered as they occur, but processing is not performed immediately.

Organizations expect TPSs to accomplish a number of specific objectives, including processing data generated by and about transactions, maintaining a high degree of accuracy and information integrity, compiling accurate and timely reports and documents, increasing labor efficiency, helping provide increased and enhanced service, and building and maintaining customer loyalty. In some situations, an effective TPS can help an organization gain a competitive advantage.

CIOs must take a lead role in preventing accounting irregularities, which can get their companies into trouble and erase investor confidence. One key step is to conduct a TPS audit that attempts to answer four basic questions:(1) Does the system meet the business need for which it was implemented? (2) What procedures and controls have been established? (3) Are these procedures and controls being used properly? and (4) Are the information systems and procedures producing accurate and honest reports?

Because of the importance of TPSs to ongoing operations, organizations must develop a business continuity plan that anticipates and minimizes the effects of disasters. It identifies the business processes that must be restored first to get operations restarted with minimum disruption;it also specifies the actions that must be taken and by whom to restore operations. Disaster recovery focuses on the actions that must be taken to restore computer operations and services in the event of a disaster. Although companies have known about the importance of disaster planning and recovery for decades, many do not adequately prepare.

TPS applications are seen throughout an organization. The order processing systems include order entry, sales configuration, shipment planning, shipment execution, inventory control, invoicing, customer relationship management, and routing and scheduling. Order entry captures the basic data needed to process a customer order. Once an order is entered and accepted, it becomes an open order. Sales configuration ensures that the products and services offered are sufficient for customer needs. Shipment planning determines which open orders will be filled and from which location they will be shipped. The system prepares an order confirmation notice that is sent to the customer and a picking list used by warehouse operators to fill the order. The shipment execution system is used by the warehouse operators to enter data on what was actually shipped to the customer. It passes shipped order transactions downstream to the invoicing system. For each item picked during the shipment execution process, a transaction providing stock number and quantity is passed to the finished product inventory-control system. The invoicing system generates customer invoices based on the records received from the shipment execution system. The customer relationship management system monitors and tracks each customer interaction to ensure first-quality service and maximum profits. Routing and scheduling systems are used in distribution functions to determine the best use of a company's resources.

The purchasing information systems include inventory control, purchase order processing, accounts payable, and receiving. The inventory-control system tracks the level of all packing materials and raw materials. It provides information to users on when to order additional materials. The purchase order processing system supports the policies, practices, and procedures of the purchasing department. The accounts payable system monitors and controls the outflow of funds to an organization's suppliers. The receiving system captures data about receipts of specific materials from suppliers so that approval for payment can be granted or refused.

The accounting systems include the budget, accounts receivable, payroll, asset management, and general ledger. The budget system automates many of the tasks required to amass budget data, distribute it to users, and consolidate the prepared budgets. The accounts receivable system manages the cash flow of the company by keeping track of the money owed the company. The payroll processing application processes employee paychecks and performs numerous

calculations relating to time worked, deductions, commissions, and taxes. The outputs are used to help control payroll costs and cash flows and develop reports to the federal government. The asset management system controls investments in capital equipment and manages depreciation for maximum tax benefits. The general ledger system records every monetary transaction and enables production of automated financial reporting.

Principle

TPSs help multinational corporations form business links with their business partners, customers, and subsidiaries.

Numerous complications arise that multinational corporations must address in planning, building, and operating their TPSs. These challenges include dealing with different languages and cultures, disparities in IS infrastructure, varying laws and customs rules, and multiple currencies.

Principle

Implementation of an enterprise resource planning system enables a company to achieve numerous business benefits through the creation of a highly integrated set of systems.

Enterprise resource planning (ERP) software is a set of integrated programs that manage a company's vital business operations for an entire multisite, global organization. It must be able to support multiple legal entities, multiple languages, and multiple currencies. Although the scope of an ERP system may vary from vendor to vendor, most ERP systems provide integrated software to support manufacturing and finance. In addition to these core business processes, some ERP systems are capable of supporting business functions such as human resources, sales, and distribution.

Implementation of an ERP system can provide many advantages, including elimination of costly, inflexible legacy systems; providing improved work processes; providing access to data for operational decision making; and creating the opportunity to upgrade technology infrastructure. Some of the disadvantages associated with an ERP system are that they are time-consuming, difficult, and expensive to implement.

CHAPTER 9: SELF-ASSESSMENT TEST

An organization's TPS must support the routine, day-to-day activities that occur in the normal course of business and help a company add value to its products and services.

1. Which of the following sets of characteristics are usually associated with transaction processing systems?

    a. sophisticated and complex processing and analysis
    b. process large amounts of data and produce large amounts of output
    c. batch processing only
    d. produce exception reports and support drill-down analysis

2. The primary objective of any TPS is to capture, process, and store transactions and to produce a variety of documents related to routine business activities. True or False?

3. Which of the following is NOT one of the basic components of a TPS?

    a. databases
    b. networks
    c. procedures
    d. analytical models

4. A form of TPS where business transactions are accumulated over a period of time and prepared for processing as a single unit is called _____.

5. Data should be captured at its source, and it should be recorded accurately, in a timely fashion, with minimal manual effort, and in a form that can be directly entered into the computer rather than keying the data from some type of document. True or False?

6. _____ is the process of capturing and gathering all data necessary to complete transactions.

7. Which of the following statements is true?

    a. Disaster recovery focuses on the actions that must be taken to restore computer operations and services in the event of a disaster.
    b. Business continuity planning identifies the business processes that must be restored first in the event of a disaster to get the business's operations restarted with minimum disruption; it also specifies the actions that must be taken and by whom to restore operations.
    c. Companies vary widely in the thoroughness and effectiveness of their business continuity planning.
    d. all of the above

8. The _____ _____ system captures the basic data needed to process a customer order.

9. The systems associated with order processing should be tightly integrated, with output from one becoming input to another. True or False?

10. Inventory control, purchase order processing, receiving, and accounts payable systems make up a set of systems called the _____ systems.

11. The _____ transaction processing system automates many of the tasks required to amass budget data, distribute it to users, and consolidate the prepared budgets.

12. This act was passed in July 2002 and had set deadlines for public companies to implement procedures that ensure their audit committees can document underlying financial data to validate earnings reports and meet demands for accuracy.

    a. Gramm-Leach-Bliley Act
    b. Sarbanes-Oxley Act
    c. HIPPA
    d. none of these

13. A computerized _____ system is designed to allow automated financial reporting and data entry and produces a detailed list of all business transactions and activities.

TPSs help multinational corporations form business links with their business partners, customers, and subsidiaries.

14. Many multinational companies roll out standard IS applications for all to use. However, those standard applications often don't account for all the differences among business partners and employees operating in other parts of the world. Which of the following is a frequent modification that is needed to standard software?

    a. Software may need to be designed with local language interfaces to ensure the successful implementation of a new IS.
    b. Customization may be needed to handle date fields correctly.
    c. Users may also have to implement manual processes and overrides to enable systems to function correctly.
    d. all of the above

15. The EU is still evolving due to its recent expansion to 25 members, and as a result it has had very little impact on the development and use of information systems around the globe. True or False?

Implementation of an enterprise resource planning system enables a company to achieve numerous business benefits through the creation of a highly integrated set of systems.

16. Which of the following is a primary benefit of implementing an ERP system?

    a. elimination of inefficient systems
    b. easing adoption of improved work processes
    c. improving access to data for operational decision making
    d. all of the above

17. The individual application modules included in an ERP system are designed to support the _____ _____, the most efficient and effective ways to complete a business process.

18. Because it is so critical to the operation of an organization, most companies are able to implement an ERP system without major difficulty. True or False?

19. Which ERP vendor recently integrated its applications with a former competitor and now offers ERP software in four specific suites, each preconfigured with as many as 30 industry-specific processes?

    a. Oracle
    b. PeopleSoft
    c. SAP
    d. J.D. Edwards

CHAPTER 9: SELF-ASSESSMENT TEST ANSWERS

(1) b (2) True (3) d (4) batch processing (5) True (6) Data collection (7) d (8) order entry (9) True (10) purchasing (11) budget (12) b (13) general ledger (14) d (15) False (16) d (17) best practices (18) False (19) b

KEY TERMS

accounting systems 428
accounts payable system 426
accounts receivable system 429
asset management transaction processing system 432
audit trail 412
batch processing system 402
best practices 438
budget transaction processing system 428

business continuity planning 410
customer relationship management (CRM) system 421
data collection 408
data correction 409
data editing 409
data manipulation 409
data storage 409
disaster recovery 411
document production 410

general ledger system 432
inventory-control system 419
online transaction processing (OLTP) 402
order entry system 413
order processing systems 413
payroll journal 431
purchase order processing system 425

purchasing transaction processing
    system 424
receiving system 426
routing system 423

sales configuration system 416
scheduling system 423
shipment execution system 418
shipment planning system 417

transaction processing cycle 407
transaction processing system audit
    412

REVIEW QUESTIONS

1. List several characteristics that distinguish a TPS from an MIS.
2. What basic transaction processing activities are performed by all transaction processing systems?
3. Distinguish between a batch processing system and an online processing system.
4. What specific objectives do organizations hope to accomplish through the use of transaction processing systems?
5. What is source data automation? Give an example.
6. Identify four complications that multinational corporations must address in planning, building, and operating their TPSs.
7. What is the difference between data editing and data correction?
8. A business continuity plan focuses on what two issues?
9. What is the focus of a disaster recovery plan?
10. Identify three federal acts that are causing more attention to be paid to the security and data integrity of information systems.
11. What are the two types of transaction processing system audits? How are they different?
12. What systems are included in the order processing family of systems?
13. What is the purpose of the shipment planning system?
14. What systems are included in the purchasing family of systems?
15. Why is the general ledger application key to the generation of accounting information and reports?
16. Give an example of how transaction processing systems can be used to gain competitive advantage.

DISCUSSION QUESTIONS

1. Assume that you are the owner of a small grocery store. Describe the importance of capturing complete, accurate transactions of customer purchases.
2. Your company is a medium-sized service company with revenue of $500 million per year. You've decided that the organization will implement a CRM system to capture and report information about all customer interactions. Identify key system requirements that you would like to see implemented in this system.
3. Imagine that you are the new IS manager for a *Fortune* 1000 company. An internal audit has revealed a number of problems with your firm's existing accounting systems. Prepare a brief outline of a talk you will make to senior company managers outlining the results of the audit and your next steps.
4. Compare and contrast the use of a written narrative to the use of a data flow diagram to explain how a system or procedure works.
5. What is the advantage of implementing ERP as an integrated solution to link multiple business processes? What are some of the challenges and potential problems?
6. You are the key user of the firm's inventory-control system and have been asked to perform an IS audit of this system. Outline the steps you would take to complete the audit.
7. Identify a line of business that would benefit from the use of a sales configuration system. Describe how such a system might work for this company.
8. You are in charge of a complete overhaul of your firm's purchasing systems. How would you define the requirements for this collection of systems? What features would you want to include?
9. Identify five or six key elements or pieces of information that should be included in a firm's business continuity plan. What steps would you define for your firm's disaster recovery plan?

PROBLEM-SOLVING EXERCISES

1. Assume that you are starting an online grocery store (for nonperishable items only) that will allow users to "browse" your aisles electronically via the Web and make their purchase selections. Once an order is complete, the items are pulled from a warehouse, packed, and shipped overnight. Using a graphics program, draw a diagram that shows the different ways you will interact with your customers. Use a word processing program to develop a list of key facts you would like to capture about each customer and about each contact with a customer.

2. The order processing application in your online grocery store has three main databases: grocery item, order, and customer. The grocery item database contains information about each item in stock. The order database identifies basic information about each order placed, including customer, date, and items ordered. The customer database contains each customer's ID number, address, and contact information.

3. Use a database management system to build a simple transaction processing system to support the online grocery store operation. Enter the complete database definitions into your database management software and create a data-entry screen to efficiently capture customer selections, build an order, and ship and bill for the order.

   a. Enter several items, including description and price into the item database.
   b. Enter three customers into the customer database.
   c. Enter the data necessary to represent two different orders to two different customers.

TEAM ACTIVITIES

1. Assume that your team has formed a consulting firm to perform an external audit of a firm's accounting information systems. Develop a list of at least 10 questions you would ask as part of your audit. What specific inputs and outputs from various systems would you want to see? Visit a company and perform the audit based on these questions or role play the scenario if a live visit is impossible.

2. Your team members should interview a business owner about the company's business continuity plan. What types of disasters has the company planned for? Develop a report that describes and evaluates this company's business continuity plan.

WEB EXERCISES

1. In 2004, Oracle Corporation attempted to acquire PeopleSoft to gain a larger share of the ERP system market. Do research on the Web to learn the current status of the acquisition and document some of the reasons why the proposed acquisition was so controversial. Develop a one-page report or send an e-mail message to your instructor about what you found.

2. Using the Internet, identify several companies that have implemented an ERP system in the last two years. Classify the implementations as success, partial success, or failure. What is your basis for making this classification?

CAREER EXERCISES

1. Initially thought to be cost-effective for only very large companies, CRM systems are now being implemented in mid-sized companies to reduce costs and improve service. A firm's finance and accounting personnel play a dual role in the implementation of such a system: (1) they must ensure a good payback on the investment in information systems, and (2) they must also ensure that the system meets the needs of the finance and accounting organization. Identify two or three tasks that the finance and accounting personnel need to perform to ensure that both these goals are met.

2. ERP software vendors need business systems analysts that understand both information systems and business processes. Make a list of six or more specific qualifications needed to be a strong business systems analyst.

VIDEO QUESTIONS

Watch the video clip **Army's Virtual World** and answer these questions:

1. As this report indicates, e-commerce transactions can be risky—particularly in the C2C arena of eBay auctions. What appears to be the golden rule for sellers on eBay?

2. What tools does eBay offer buyers and sellers to help assure them that they are doing business with trustworthy people?

CASE STUDIES

Case One

Blue Cross Blue Shield CRM Wins BIG Customers

When John Ounjian, senior vice president and CIO of Blue Cross and Blue Shield (BCBS) of Minnesota, approached $8 billion consumer goods giant General Mills in an attempt to win its business, he found the feature that most interested General Mills was the promise of a new state-of-the-art Web-based CRM system for insurance customers. Ounjian promised General Mills that he would soon be installing a Web-based customer service system to let subscribers manage their health benefits online. Subscribers would be able to select health plans tailored to their individual needs and wallets, calculate their own contributions to their coverage, research information on prescription drugs and other treatments, locate participating physicians, and check the status of their claims. Once the promise was made and the General Mills account won, Ounjian had to figure out how to deliver.

Other insurance companies have attempted to set up such online systems but failed. Ounjian believes the reason so many CRM projects run into problems is that they don't consider the underlying difficulties of transferring data that originates in one system and in one form to another system in a different form. Customers end up viewing data in a form that

makes sense for the company, not formatted to fit their point of view. Not wanting to make the same mistake, Ounjian decided to install an entirely new infrastructure to integrate his Web and call center operations and to provide timely, accurate information to customers. He also developed a system to move millions of bytes of data stored in back-end databases to the Web front end—massaging and reformatting the data so that consumers could understand it.

Health insurance is an industry ideally suited to online customer self-service. What person wouldn't forgo the provider's cumbersome toll-free number, with long waits on hold, in favor of a Web-based solution that allows her 24/7 access to all kinds of healthcare answers? Ounjian figures that all insurance companies will attempt to set up such a system. "We all have to have this capability; that's not what's going to differentiate us," Ounjian says. "It's how you execute these functions, how you bring the customer on board that makes all the difference."

To do it right, he and his staff devised a sound data-management strategy to overcome the problems that arose when moving raw data from back-end systems to the Web front end. If they didn't come up with a cohesive plan for moving data back and forth, they risked having customers looking at information that was out of date, was inaccurate, or varied

across the Web and call center channels. When dealing with 100 million records, attention to detail is crucial.

Once Ounjian and his staff ironed out the data issues and developed a prototype of the new Web site, they invited customers to test it in a focus group. During those initial trials, they found that the site wasn't consumer-friendly. Subjects in the test group found it difficult to navigate and confusing. So, the engineers changed the arrangement of pull-down menus to better reflect how typical customers move around a site.

The recently completed final product, named Options Blue, lets customers obtain explanations of their benefits, calculate contributions to their health coverage, and check their deductibles and out-of-pocket expenses online. Customers can now also use the Web site to order prescriptions by mail, estimate the costs of prescriptions and medical procedures, order new ID cards, and find participating pharmacies.

The rewards from the project were not insignificant. BCBS of Minnesota has not only met General Mills' expectations but also managed to beat national providers Aetna, Cigna, and Humana out of several very large accounts, such as 3M, Northwest Airlines, and Target. Ounjian says his company's membership grew by 10 percent, or 200,000 new members, largely because of its online customer self-service system. Even more remarkable, BCBS of Minnesota experienced that growth in membership at a time when several national insurers lost millions of members. Cigna, for instance, lost 10 percent of its membership due in part to a botched implementation of a new customer self-service system.

Case Two
Luxury Jet Service Develops TPS to Speed Customers on Their Way

Most of us are aware how technology has transformed commercial air travel and the process of travel planning. Fewer of us, however, have the financial means to book private jet transportation. If you had the money to hire a private jet to whisk you off to some far off location, Sentient Jet would be the company to call.

Sentient Jet is a pioneer in the private jet membership-based service industry. It provides clients with all the benefits of owning a fleet of aircraft with none of the associated costs and commitments. Sentient's program uses a nationwide fleet of safety-audited aircraft flown by private pilots. Sentient Jet members enjoy the convenience of a safe and secure executive-class aircraft with a guaranteed 10-hour response time anywhere in the United States. Sentient also offers on-demand charter flights, allowing anyone to book a roundtrip flight with a single phone call.

Unlike commercial airlines, which own and control their own fleets, Sentient Jet only arranges the flights, serving as a bridge between travelers and private jet owners to secure safe and luxurious travel accommodations. But booking such a flight is much more complex than commercial travel. When

Because Ounjian's team took the time to develop the system from the ground up, future additions or modifications won't require much effort. He uses an automotive metaphor to describe the system: "We have the chassis on which to build our investments from year to year. If my transmission needs to move from a three speed to a five speed, I don't have to redesign the whole car."

Discussion Questions

1. Why do you think General Mills found the idea of online customer self-service so appealing when deciding on a health insurance provider?
2. What did Ounjian blame for the failures of other companies who attempted to implement similar systems? Do you agree? Why, or why not?

Critical Thinking Questions

3. Why and how might data be stored differently for employee use than for customer use? How would cryptic data be translated for customers to understand?
4. Why do you think Ounjian decided to start from scratch when adding a new Web-based system? What type of problems did this approach avoid? At what cost?

SOURCES: Meredith Levinson, "Pain-free CRM," *CIO Magazine*, May 15, 2003, www.cio.com/archive/051503/crm.html; Blue Cross and Blue Shield of Minnesota Web site, accessed May 23, 2004, www.bluecrossmn.com/public/guests/index.html; Lisa Picarille, "Top Execs + CRM = Success," *CRMNews.com*, October 13, 2003, www.crmnews.com.

a customer phones and requests a flight from Munich to New York, Sentient must find a private jet near Munich, of an appropriate size for the number of passengers, that isn't needed by the owner or other clients and then calculate the price and propose it to the customer. With traditional methods of booking charter flights, this process typically takes days. Sentient can perform the task in near real time.

Sentient designed a proprietary system called Optimizer to enable customer service representatives to match would-be travelers' flight requirements with aircraft availability and then combine customer flights to maximize profit margins. "It truly has business value and offers immediate return on investment," said Tolga Erdogus, Sentient's vice president of engineering.

Optimizer is a transaction processing system that connects to multiple travel industry flight-availability databases, including the Air Charter Guide and a real-time data feed from the Federal Aviation Administration. It queries those databases and produces a color-coded Gantt project-management chart. The chart shows which flights can be combined into revenue-generating excursions. The company guarantees its clients a five-hour turnaround time for identifying an available flight.

"Commercial airlines use massive systems for overnight batch processing to conduct schedule optimization," said

Erdogus. "But we've generated a technology that does near-real-time optimization, on the order of 15 minutes." Erdogus's development team consists of himself, five developers, a database administrator, a systems administrator, a help desk administrator, and a consultant from MIT. The team effort has paid off.

For the second consecutive year, Sentient Jet has been awarded *Robb Report* magazine's "Best of the Best" award. *Robb Report* is the international authority on luxury lifestyles, and the "Best of the Best" issue is a culmination of a year's search for the most extraordinary products and services. The Optimizer system gets the credit for Sentient Jet's current award. Fluto Shinzawa, associate editor of the *Robb Report*, was particularly impressed by the Optimizer. "The tool allows Sentient to deliver the ultimate in service and safety to Sentient Jet members." Customers seem equally pleased with Sentient's innovative service. The company achieved sales growth of 82 percent the first quarter after introducing Optimizer. Sentient Jet now has more than 1,700 members.

Case Three
Point and Pay in Japan with KDDI

An important part of a TPS is its ability to handle a variety of payment mechanisms. Cash, check, credit card, or debit card are traditional methods of payment that most TPSs support. Recent trends indicate that society is moving away from paper currency in favor of electronic funds transfers. In Chapter 3 you read about new technologies that scan paper checks and translate them into electronic funds transfers. Other technologies are being developed and tested to eliminate both paper and plastic forms of currency, allowing transactions to take place anywhere, anytime.

Japanese citizens are testing a new use for their cell phone handsets. A trial service provided by Japanese cellular carrier KDDI Corp. and supported by several Japanese credit card issuers allows customers to point their cell phone handset at a cash register and push a button to pay. The technology is based on a recently released standard called Infrared for Financial Messaging (IrFM). IrFM employs the long-established Infrared Data Association (IrDA) Standard used in TV remote controls to transfer data wirelessly between devices at close range. While KDDI is providing the hardware for the trial, Harex InfoTech is providing the software.

The new point-and-pay technology includes numerous applications. A typical transaction involves having the customer work through prompts and menus on the cell phone screen to select a credit card, enter a four-digit PIN, and send the credit card information to a receiver. An Express Pay service can be used for micropayments—low-cost payments at vending machines, the subway, or bus. Express Pay allows the customer to pay by simply pressing a button on the side of the handset without having to navigate through the multistep menu routine.

Discussion Questions

1. Why would you expect that fast service is particularly important in Sentient's industry?
2. Why might a corporation decide to use Sentient rather than purchasing its own private jet, or portion of a jet?

Critical Thinking Questions

3. Why would Sentient design its own proprietary system instead of using an existing system?
4. For Optimizer to be effective, what flight information must be stored, and what types of commitments do you think pilots and jet owners make to Sentient?

SOURCES: Dan Verton, "Private Jet Service Develops IT System to Lower Costs," *Computerworld*, January 16, 2004, *www.computerworld.com*; Sentient Jet Web site, accessed May 23, 2004, at *www.sentient.com*; "Sentient Jet Lands *Robb Report*'s Prestigious 'Best of the Best' Award Two Years in a Row," *PR Newswire*, May 17, 2004.

Besides conducting business transactions, IrFM technology can be used for security clearance or for electronic attendance taking. Imagine pointing and clicking a cell phone as you enter the classroom for lecture!

Japan has more than 8 million users of infrared-enabled mobile phones. Low-cost infrared receivers connected to existing TPSs, including cash registers, point-of-sale terminals, and vending machines, communicate with IrFM handsets. Industry watchers expect the Japanese to embrace the technology quickly, and new commercial rollouts are planned in the near future. Harex is also testing point-and-pay technology in South Korea and at the University of Southern California's Marshall School of Business.

Discussion Questions

1. What conveniences are provided to the customer and vendor through IrFM?
2. Why has Japan been selected to test this new technology?

Critical Thinking Questions

3. What security concerns surround wireless transactions such as those enabled by IrFM?
4. What precautions need to be built into the IrFM system to guard against the theft of people's cell phones and consequently their identity?

SOURCES: Jørgen Sundgot, "Point-and-Pay with IrFM," *InfoSynch World*, January 9, 2003, *www.infosyncworld.com/news/n/2868.html*; "Infrared Mobile Phone Payment System, Export to Japan," *PR Newswire*, April 14, 2003, *www.lexisnexis.com*; Harex Infotech Web site, accessed May 23, 2004, at *www.mzoop.com*.

NOTES

Sources for the opening vignette: Bruce Horovitz, "It's a Do-It-Your SELF World," *USA Today*, April 27, 2004, p. 1A; Paula Ganzi Licata, "Checking Yourself Out, Quite Literally," *The New York Times*, March 7, 2004, Section 14LI, p. 3; Victor Godinez, "Cashiers Won't Lose Out to Robots Anytime Soon," *The Dallas Morning News*, May 5, 2004, *www.lexis-nexis.com*.

1. Whiting, Rick, "The World's Hardest-Working Databases," *InformationWeek*, February 13, 2004, *www.informationweek.com*.
2. Aleri Web site, Services GBS, accessed at *www.aleri.com* on April 30, 2004.
3. Babcock, Charles, "Application Monitoring Saves the Day," *Information Week*, April 29, 2004, *www.informationweek.com*.
4. MQS Software Web site accessed at *www.mqssoftware.com* on April 30, 2004.
5. McDougall, Paul, "Aetna Rejects Outsourcing, Brings IBM In House," *Information Week*, April 1, 2004, *www.informationweek.com*.
6. Rosencrance, Linda, "IBM Launches Middleware for Automotive Customers," *Computerworld*, March 22, 2004, *www.computerworld.com*.
7. Pallay, Jessica, "Build Versus Buy on Wall Street," *Information Week*, April 26, 2004, *www.informationweek.com*.
8. D'Anton, Helen, "Companies Get Ready for the Unexpected," *InformationWeek*, January 19, 2004, *www.informationweek.com*.
9. Villano, Matt, "Twisters, Hurricanes, Floods, (Oh My)," *ComputerWorld*, September 3, 2003, *www.computerworld.com*.
10. "New World Pasta Files for Ch 11," *USA Today*, May 11, 2004, page B1.
11. Hunker, Jeffrey, Ph.D., "New Security Imperative: Demonstrating Results," *Information Week*, April 12, 2004, *www.informationweek.com*.
12. Babcock, Charles, "Wait Not, Wilt Not," *Information Week*, May 10, 2004, *www.informationweek.com*.
13. Bacheldor, Beth, "Steady Supply," *Information Week*, November 4, 2003, *www.informationweek.com*.
14. Bacheldor, Beth, "Supply Chain Economics," *Information Week*, March 8, 2004, *www.informationweek.com*.
15. Kharif, Olga, "RFID: On Track for a Rapid Rise," *Business Week Online*, February 4, 2004, *www.businessweek.com*.
16. Bacheldor, Beth, "Supply-Chain Economics," *InformationWeek*, March 8, 2004, *www.informationweek.com*.
17. Sullivan, Laurie, "Ready to Roll," *Information Week*, March 8, 2004, *www.informationweek.com*.
18. Kontzer, Tony, "Airline Widens Its Customer View," *InformationWeek*, April 12, 2004, *www.informationweek.com*.
19. McDougall, Paul, "The Place to Be," *InformationWeek*, April 5, 2004, *www.informationweek.com*.
20. Sullivan, Laurie, "Delphi: Parts Maker Helps Suppliers Shape Up," *InformationWeek*, April 19, 2004, *www.informationweek.com*.
21. Hamblen, Matt, "NavSea's ROI Ship Comes In," *Computerworld*, March 24, 2003, *www.computerworld.com*.
22. Hamblen, Matt, "NavSea's ROI Ship Comes In," *Computerworld*, March 24, 2003, *www.computerworld.com*.
23. "About Us," Global eXchange Services Web site accessed at *www.gxs.com* on May 5, 2004.
24. Whiting, Rick, "Gain the Financial Advantage," *Information Week*, June 2, 2003, *www.informationweek.com*.
25. McGee, Marianne Kolbasuk, "Keep Track of Your Employees," *Information Week*, June 30, 2003, *www.informationweek.com*.
26. Bacheldor, Beth, "Show Goes On at Disney World," *Information Week*, July 14, 2003, *www.informationweek.com*.
27. Babock, Charles, "The Right Stuff," *Information Week*, August 25, 2003, *www.informationweek.com*.
28. McGee, Marianne Kolbasuk, "Helping Hand for Health Care," *InformationWeek*, June 16, 2003, *www.informationweek.com*.
29. McGee, Marianne Kolbasuk, "Neoforma Offering Tightens Medical Supply Chain," *Information Week*, June 16, 2003, *www.informationweek.com*.
30. Sullivan, Laurie, "Electronics Industry Grids for New Rules," *Information Week*, May 17, 2004, *www.informationweek.com*.
31. McDougall, Paul, "The Place to Be," *InformationWeek*, April 5, 2004, *www.informationweek.com*.
32. Bacheldor, Beth, "Midtier ERP Customers Command Attention," *Information Week*, April 26, 2004, *www.informationweek.com*.
33. Bacheldor, Beth, "PeopleSoft Aims at Lean Manufacturing," *Information Week*, March 19, 2004, *www.informationweek.com*.
34. Bacheldor, Beth, "Faster Food via ERP," *InformationWeek*, April 21, 2004, *www.informationweek.com*.
35. Bacheldor, Beth, "PeopleSoft Aims at Lean Manufacturing," *InformationWeek*, March 19, 2004, *www.informationweek.com*.
36. Songini, Marc L., "Dial Emulates Rivals, Turns to SAP Apps," *Computerworld*, July 28, 2003, *www.computerworld.com*.
37. Fox, Pimm, "The Art of ERP Done Right," *Computerworld*, May 19, 2003, *www.computerworld.com*.
38. Songini, Marc L., "Goodyear Hits $100M Bump with ERP System," *Computerworld*, November 3, 2003, *www.computerworld.com*.
39. Songini, Marc L., "University Hits PeopleSoft with $510 M Lawsuit," *Computerworld*, March 26, 2004, *www.computerworld.com*.
40. Bacheldor, Beth, "PeopleSoft Integrates Apps," *Information Week*, December 22, 2003, *www.informationweek.com*.
41. Bacheldor, Beth, "PeopleSoft Targets Small Businesses," *Information Week*, May 10, 2004, *www.informationweek.com*.

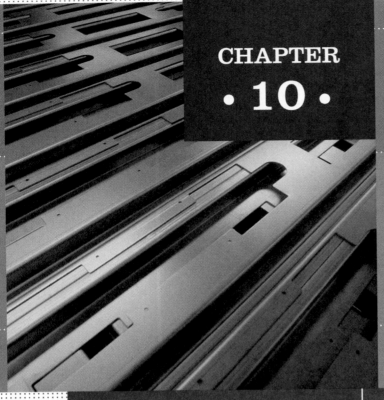

CHAPTER · 10 ·

Information and Decision Support Systems

PRINCIPLES

- **Good decision-making and problem-solving skills are the key to developing effective information and decision support systems.**

- **The management information system (MIS) must provide the right information to the right person in the right fashion at the right time.**

- **Decision support systems (DSSs) are used when the problems are unstructured.**

- **Specialized support systems, such as group support systems (GSSs) and executive support systems (ESSs), use the overall approach of a DSS in situations such as group and executive decision making.**

LEARNING OBJECTIVES

- Define the stages of decision making.
- Discuss the importance of implementation and monitoring in problem solving.

- Explain the uses of MISs and describe their inputs and outputs.
- Discuss information systems in the functional areas of business organizations.

- List and discuss important characteristics of DSSs that give them the potential to be effective management support tools.
- Identify and describe the basic components of a DSS.

- State the goals of a GSS and identify the characteristics that distinguish it from a DSS.
- Identify the fundamental uses of an ESS and list the characteristics of such a system.

INFORMATION SYSTEMS IN THE GLOBAL ECONOMY ⟫
KIKU-MASAMUNE SAKE BREWING CO., LTD, JAPAN

Ancient Company Profits from State-of-the-Art Business Intelligence System

The Kano family began brewing sake in the small town of Mikage, Kobe, Japan, in 1659. The sake wine that they brewed, named *Kiku-Masamune*—translated as chrysanthemum sake—became a staple in Japan. Over the centuries it has played an important role in Japanese culture and religion. In 1877, Kiku-Masamune became a global commodity when the company began exporting to England. Today, Kiku-Masamune sake is exported to many countries, and until recently it enjoyed a monopoly position in the sake market.

In the mid-1990s, U.S. sales for Kiku-Masamune sky-rocketed when sushi became the latest dining rage. With sushi bars opening nearly everywhere, the sake once described as "moonlight steeped in spring rain" enjoyed unprecedented sales. The company took advantage of its competitive position and began educating U.S. consumers through sake seminars in Los Angeles and New York City. The attention drawn to sake and its leading producer, Kiku-Masamune, did not go unnoticed by some ambitious entrepreneurs.

Competition in the sake industry increased both within Japan and around the world. Also, the Japanese were beginning to favor other alcoholic beverages imported from the West. With competition from an estimated 1,700 Japanese sake breweries and a dwindling client base, Kiku-Masamune was losing business both at home and abroad. The company knew that it could no longer rely solely on tradition and quality. It needed to learn more about the market—and its customers. But Kiku-Masamune had one big advantage over newcomers: its experience. The key to using its expertise lay in the millions of records it had collected over its long business history.

The company's paper-based record store was huge, with 1.9 million records for liquor vendors and 300,000 records for individual expenses, both of which were increasing by 100,000 records each year. Kiku-Masamune was in dire need of an electronic system that could manage these records and provide up-to-the-minute reports on the condition of the company, its suppliers, and customers. As Keiji Fukuda, systems manager in the company's accounting department observes, "Of course, if we'd been prepared to spend all our time trying to unearth crucial data, we could have continued indefinitely with our paper-based operations. But what if somebody needed to look at a document submitted several years ago? We simply didn't have the tools to meet those kinds of demands."

Kiku-Masamune contracted with a U.S. system developer to design an information and decision support system customized for the company's needs. "Before implementing the system, it was like we were only seeing pieces of the puzzle but never the big picture. Now we're able to identify where changes are needed in our contracts and drill down to the information we need to use our sales resources most effectively," says Yoshihiko Handa, sales and marketing director at Kiku-Masamune.

The new system produces scheduled reports annually, quarterly, monthly, weekly, and daily that provide information on key indicators such as performance by retailer and profits by product and customer. Kiku-Masamune is able

to use this information to determine which of its customers require attention and to expand its customer base. The new system also provides the company with information to better manage its manufacturing process. Kiku-Masamune can provide customers with "just in time" delivery of sake, which saves both Kiku-Masamune and its customers the expense of storing large inventories.

The system has been successful in providing Kiku-Masamune with a strong competitive advantage in the sake market. The company is working to train all 500+ employees on the new information and decision support system. The ultimate goal is to arm all employees with information to make the right decisions quickly and effectively, resulting in more profitable and efficient operations overall.

Business intelligence (BI) solutions such as the one designed for Kiku-Masamune are in increasing demand. In spite of a recent tough economic climate, the BI market experienced strong growth. Research company IDC estimates that the BI market will reach $7.5 billion in 2006. Business intelligence development projects show consistently high returns on investment and are increasingly recognized as key for business success.

BI systems are one of many types of information and decision support systems that are implemented to provide up-to-the-second information for decision making at all levels within a business. Information and decision support systems support decisions in all functional units within an organization: financial, manufacturing, marketing, human resources, and others.

Kiku-Masamune uses its new information and decision support system to price its premium sake astutely, to guide decision making, and to help optimize sales opportunities. Effective information gathering and information sharing are clearly at the center of the company's newfound calm amid the storm of competition.

As you read this chapter, consider the following:

- How can management information systems guide a company to achieve a competitive advantage within an industry?
- What characteristics of an information and decision support system make it valuable?

Why Learn About Information and Decision Support Systems?

You have seen throughout this book how information systems can make you more efficient and streamline manual systems. The true potential of information systems, however, is in helping you and your coworkers make more informed decisions. This chapter will show you how to slash costs, increase profits, and uncover new opportunities for your company. A loan committee at a bank or credit union can use a group support system to help them determine who should receive loans. Transportation coordinators can use management information reports to find the least expensive way to ship products to market and to solve bottlenecks. Store managers can use decision support systems to help them decide what and how much inventory to order to meet customer needs and increase profits. An entrepreneur who owns and operates a temporary storage company can use vacancy reports to help him or her determine what price to charge for new storage units that are being built. Everyone wants to be a better problem solver and decision maker. This chapter will show you how information systems can help. We begin with an overview of decision making and problem solving.

As seen in the opening vignette, information and decision support are the lifeblood of today's organizations. Thanks to information and decision support systems, managers and employees can obtain useful information in real time. As we saw in Chapter 9, the TPS captures a wealth of data. When this data is filtered and manipulated, it can provide powerful support for managers and employees. The ultimate goal of management information and decision

support systems is to help managers and executives at all levels make better decisions and solve important problems. The result can be increased revenues, reduced costs, and the realization of corporate goals. We begin by investigating decision making and problem solving.

DECISION MAKING AND PROBLEM SOLVING

Every organization needs effective decision making. In most cases, strategic planning and the overall goals of the organization set the course for decision making, helping employees and business units achieve their objectives and goals. Often, information systems also assist with strategic planning and problem solving. Good decision analysis, for example, can contribute millions or even billions of dollars to large chemical or photographic company's profits.

Decision Making as a Component of Problem Solving

In business, one of the highest compliments you can receive is to be recognized by your colleagues and peers as a "real problem solver." Problem solving is a critical activity for any business organization. Once a problem has been identified, the problem-solving process begins with decision making. A well-known model developed by Herbert Simon divides the **decision-making phase** of the problem-solving process into three stages: intelligence, design, and choice. This model was later incorporated by George Huber into an expanded model of the entire problem-solving process (see Figure 10.1).

decision-making phase
The first part of problem solving, including three stages: intelligence, design, and choice.

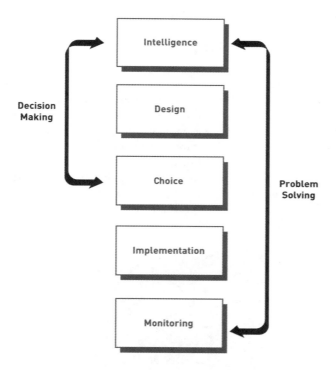

Figure 10.1

How Decision Making Relates to Problem Solving

The three stages of decision making—intelligence, design, and choice—are augmented by implementation and monitoring to result in problem solving.

The first stage in the problem-solving process is the **intelligence stage**. During this stage, potential problems or opportunities are identified and defined. Information is gathered that relates to the cause and scope of the problem. During the intelligence stage, resource and environmental constraints are investigated. For example, exploring the possibilities of shipping tropical fruit from a farm in Hawaii to stores in Michigan would be done during the intelligence stage. The perishability of the fruit and the maximum price that consumers in Michigan are willing to pay for the fruit are problem constraints. Aspects of the problem environment that must be considered in this case include federal and state regulations regarding the shipment of food products.

intelligence stage
The first stage of decision making, in which potential problems or opportunities are identified and defined.

design stage
The second stage of decision making, in which alternative solutions to the problem are developed.

choice stage
The third stage of decision making, which requires selecting a course of action.

problem solving
A process that goes beyond decision making to include the implementation stage.

implementation stage
A stage of problem solving in which a solution is put into effect.

monitoring stage
Final stage of the problem-solving process, in which decision makers evaluate the implementation.

programmed decision
Decision made using a rule, procedure, or quantitative method.

In the **design stage**, alternative solutions to the problem are developed. In addition, the feasibility of these alternatives is evaluated. In our tropical fruit example, the alternative methods of shipment, including the transportation times and costs associated with each, would be considered. During this stage, the problem solver might determine that shipment by freighter to California and then by truck to Michigan is not feasible because the fruit would spoil.

The last stage of the decision-making phase, the **choice stage**, requires selecting a course of action. In our tropical fruit example, the Hawaiian farm might select the method of shipping by air to Michigan as its solution. The choice stage would then conclude with selection of the actual air carrier. As we will see later, various factors influence choice; the apparently easy act of choosing is not as simple as it might first appear.

Problem solving includes and goes beyond decision making. It also includes the **implementation stage**, when the solution is put into effect. For example, if the Hawaiian farmer's decision is to ship the tropical fruit to Michigan as air freight using a specific air freight company, implementation involves informing the farming staff of the new activity, getting the fruit to the airport, and actually shipping the product to Michigan.

The final stage of the problem-solving process is the **monitoring stage**. In this stage, decision makers evaluate the implementation to determine whether the anticipated results were achieved and to modify the process in light of new information. Monitoring can involve feedback and adjustment. For example, after the first shipment of fruit, the Hawaiian farmer might learn that the flight of the chosen air freight firm routinely makes a stopover in Phoenix, Arizona, where the plane sits on the runway for a number of hours while loading additional cargo. If this unforeseen fluctuation in temperature and humidity adversely affects the fruit, the farmer might have to readjust his solution to include a new air freight firm that does not make such a stopover, or perhaps he would consider a change in fruit packaging.

Programmed Versus Nonprogrammed Decisions

In the choice stage, various factors influence the decision maker's selection of a solution. One such factor is whether the decision can be programmed. **Programmed decisions** are made using a rule, procedure, or quantitative method. For example, to say that inventory should be ordered when inventory levels drop to 100 units is to adhere to a rule. Programmed decisions are easy to computerize using traditional information systems. It is simple, for example, to program a computer to order more inventory when inventory levels for a certain item reach 100 units or fewer. Most of the processes automated through transaction processing systems share this characteristic: the relationships between system elements are fixed by rules, procedures, or numerical relationships. Management information systems are also used to reach programmed decisions by providing reports on problems that are routine and in which the relationships are well defined (structured problems).

Ordering more inventory when inventory levels drop to specified levels is an example of a programmed decision.

(Source: Courtesy of Symbol Technologies.)

Nonprogrammed decisions, however, deal with unusual or exceptional situations. In many cases, these decisions are difficult to quantify. Determining the appropriate training program for a new employee, deciding whether to start a new type of product line, and weighing the benefits and drawbacks of installing a new pollution control system are examples. Each of these decisions contains many unique characteristics for which the application of rules or procedures is not so obvious. Today, decision support systems are used to solve a variety of nonprogrammed decisions, in which the problem is not routine and rules and relationships are not well defined (unstructured or ill-structured problems).

nonprogrammed decision
Decision that deals with unusual or exceptional situations.

Optimization, Satisficing, and Heuristic Approaches

In general, computerized decision support systems can either optimize or satisfice. An **optimization model** will find the best solution, usually the one that will best help the organization meet its goals. For example, an optimization model can find the appropriate number of products that an organization should produce to meet a profit goal, given certain conditions and assumptions. Optimization models utilize problem constraints. A limit on the number of available work hours in a manufacturing facility is an example of a problem constraint. Some spreadsheet programs, such as Excel, have optimizing features (see Figure 10.2). An appliance manufacturer, for example, can use an optimization program to help it reduce the time and cost of manufacturing appliances and increase profits by millions of dollars. The Scheduling Appointments at Trade Events (SATE) software package is an optimization program that schedules appointments between buyers and sellers at trade shows and meetings. The optimization software also allows decision makers to explore various different alternatives. The software has been used at the Australian Tourism Exchange, the largest travel fair in the southern hemisphere.[1] Schindler, the world's largest escalator company, used an optimization technique to plan maintenance programs.[2] One important decision that Schindler and many other companies face is how much preventive maintenance to perform. Good preventive maintenance can save a company from making costly emergency repairs, which in Schindler's case involves sending maintenance teams and equipment to office buildings and retail stores to repair out-of-service escalators. However, too much preventive maintenance wastes valuable resources. Using a quantitative optimization program, Schindler was able to save about $1 million annually in total maintenance costs.

optimization model
A process to find the best solution, usually the one that will best help the organization meet its goals.

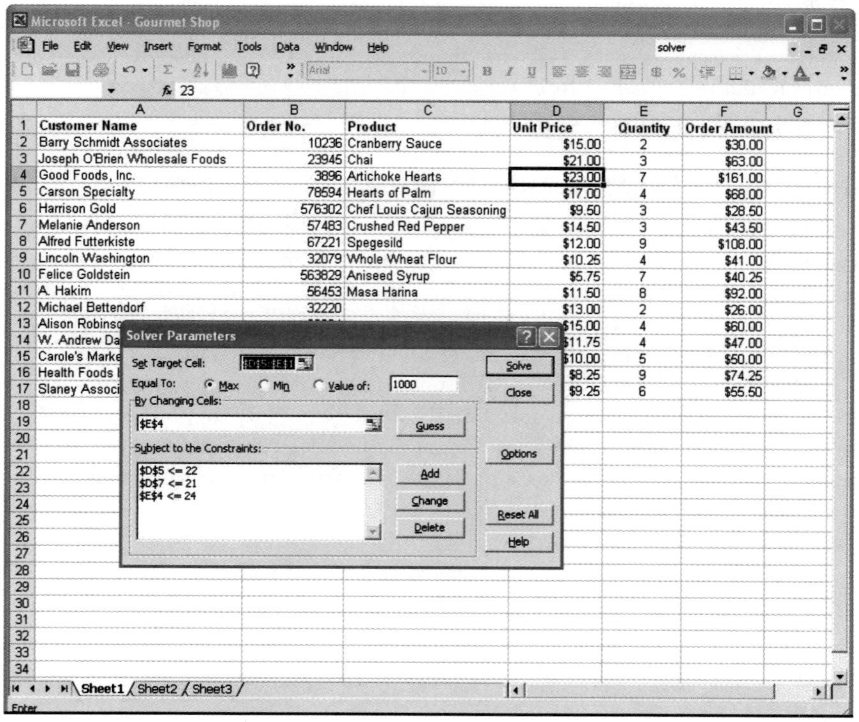

Figure 10.2

Some spreadsheet programs, such as Excel, have optimizing routines. This figure shows Solver, which can find an optimal solution given certain constraints.

satisficing model
A model that will find a good—but not necessarily the best—problem solution.

heuristics
Commonly accepted guidelines or procedures that usually find a good solution.

A **satisficing model** is one that will find a good—but not necessarily the best—problem solution. Satisficing is usually used because modeling the problem properly to get an optimal decision would be too difficult, complex, or costly. Satisficing normally does not look at all possible solutions but only at those likely to give good results. Consider a decision to select a location for a new plant. To find the optimal (best) location, you would have to consider all cities in the United States or the world. A satisficing approach would be to consider only five or ten cities that might satisfy the company's requirements. Limiting the options may not result in the best decision, but it will likely result in a good decision, without spending the time and effort to investigate all cities. Satisficing is a good alternative modeling method because it is sometimes too expensive to analyze every alternative to get the best solution.

Heuristics, often referred to as "rules of thumb"—commonly accepted guidelines or procedures that usually find a good solution—are very often used in decision making. A heuristic that baseball team managers use is to place batters most likely to get on base at the top of the lineup, followed by the power hitters who'll drive them in to score. An example of a heuristic used in business is to order four months' supply of inventory for a particular item when the inventory level drops to 20 units or fewer; even though this heuristic may not minimize total inventory costs, it may be a very good rule of thumb to avoid stockouts without too much excess inventory. Trend Micro, a provider of antivirus software, has developed an antispam product that is based on heuristics.[3] The software examines e-mails to find those most likely to be spam.

In addition to using problem-solving models, decision makers also use information systems to improve the efficiency and quality of their decisions. One such system is a management information system, discussed next.

AN OVERVIEW OF MANAGEMENT INFORMATION SYSTEMS

Management information systems (MISs) can often give companies and other organizations a competitive advantage by providing the right information to the right people in the right format and at the right time. The U.S. military coalition in Iraq, for example, used a management information system to precisely locate enemy troops and their resources.[4] The MIS used global positioning systems (GPSs), satellite mapping and surveillance, and frequently updated information obtained from troops in the field to get and display information on enemy troops and U.S. troops and equipment.

Management Information Systems in Perspective

The primary purpose of an MIS is to help an organization achieve its goals by providing managers with insight into the regular operations of the organization so that they can control, organize, and plan more effectively and efficiently. One important role of the MIS is to provide the right information to the right person in the right fashion at the right time. In short, an MIS provides managers with information, typically in reports, that supports effective decision making and provides feedback on daily operations. Figure 10.3 shows the role of MISs within the flow of an organization's information. Note that business transactions can enter the organization through traditional methods or via the Internet or an extranet connecting customers and suppliers to the firm's transaction processing systems. The use of MISs spans all levels of management. That is, they provide support to and are used by employees throughout the organization.

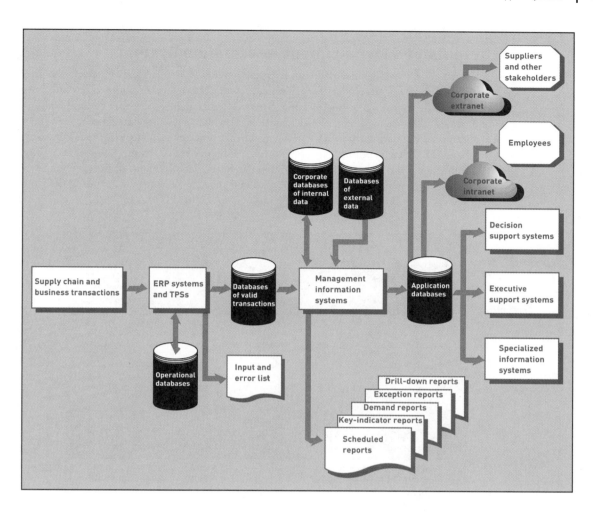

Inputs to a Management Information System

As shown in Figure 10.3, data that enters an MIS originates from both internal and external sources, including the company's supply chain, first discussed in Chapter 2. The most significant internal data sources for an MIS are the organization's various TPSs and ERP systems and related databases. As discussed in Chapter 5, companies also use data warehouses and data marts to store valuable business information. Business intelligence, also discussed in Chapter 5, can be used to turn a database into useful information throughout the organization. Other internal data comes from specific functional areas throughout the firm.

External sources of data can include customers, suppliers, competitors, and stockholders, whose data is not already captured by the TPS, as well as other sources, such as the Internet. In addition, many companies have implemented extranets to link with selected suppliers and other business partners to exchange data and information.

The MIS uses the data obtained from these sources and processes it into information more usable by managers, primarily in the form of predetermined reports. For example, rather than simply obtaining a chronological list of sales activity over the past week, a national sales manager might obtain her organization's weekly sales data in a format that allows her to see sales activity by region, by local sales representative, by product, and even in comparison with last year's sales. Providence Washington Insurance Company, for example, is using Report-Net from Cognos to reduce the number of paper reports and the associated costs.[5] According to Ed Levelle, CIO of Providence Washington, "Our intention was to replace paper reports with an Internet-based reporting process." The new reporting system creates an "executive dashboard" that shows current data, graphs, and tables to help managers make better real-time decisions.

Figure 10.3

Sources of Managerial Information

The MIS is just one of many sources of managerial information. Decision support systems, executive support systems, and expert systems also assist in decision making.

Outputs of a Management Information System

The output of most management information systems is a collection of reports that are distributed to managers. Kodak, for example, uses an MIS to send important sales information to its sales reps.[6] According to James Sanford, senior manager of sales communication and strategy, "Like a lot of companies, Kodak was good at collecting data but not very good at sharing and updating that data." Management reports can come from various company databases through data mining, first introduced in Chapter 5. Data mining allows a company to sift through a vast amount of data stored in databases, data warehouses, and data marts to produce a variety of reports, including scheduled reports, key-indicator reports, demand reports, exception reports, and drill-down reports (see Figure 10.4).

Figure 10.4

Reports Generated by an MIS

The five types of reports are (a) scheduled, (b) key-indicator, (c) demand, (d) exception, and (e–h) drill-down.

(Source: George W. Reynolds, *Information Systems for Managers*, Third Edition. St. Paul, MN: West Publishing Co., 1995.)

(a) Scheduled Report

Daily Sales Detail Report

Prepared: 08/10/06

Order #	Customer ID	Salesperson ID	Planned Ship Date	Quantity	Item #	Amount
P12453	C89321	CAR	08/12/06	144	P1234	$3,214
P12453	C89321	CAR	08/12/06	288	P3214	$5,660
P12454	C03214	GWA	08/13/06	12	P4902	$1,224
P12455	C52313	SAK	08/12/06	24	P4012	$2,448
P12456	C34123	JMW	08/13/06	144	P3214	$720
....	....	....		....	....	....

(b) Key-Indicator Report

Daily Sales Key-Indicator Report

	This Month	Last Month	Last Year
Total Orders Month to Date	$1,808	$1,694	$1,914
Forecasted Sales for the Month	$2,406	$2,224	$2,608

(c) Demand Report

Daily Sales by Salesperson Summary Report

Prepared: 08/10/06

Salesperson ID	Amount
CAR	$42,345
GWA	$38,950
SAK	$22,100
JWN	$12,350
....	....

(d) Exception Report

Daily Sales Exception Report—Orders Over $10,000

Prepared: 08/10/06

Order #	Customer ID	Salesperson ID	Planned Ship Date	Quantity	Item #	Amount
P12345	C89321	GWA	08/12/06	576	P1234	$12,856
P22153	C00453	CAR	08/12/06	288	P2314	$28,800
P23023	C32832	JMN	08/11/06	144	P2323	$14,400
....	....	....	....	....	....	....
....	....	....				

(e) First-Level Drill-Down Report			
Earnings by Quarter (Millions)			
	Actual	**Forecast**	**Variance**
2nd Qtr. 2007	$12.6	$11.8	6.8%
1st Qtr. 2007	$10.8	$10.7	0.9%
4th Qtr. 2006	$14.3	$14.5	-1.4%
3rd Qtr. 2006	$12.8	$13.3	-3.8%

(f) Second-Level Drill-Down Report			
Sales and Expenses (Millions)			
Qtr: 2nd Qtr. 2007	**Actual**	**Forecast**	**Variance**
Gross Sales	$110.9	$108.3	2.4%
Expenses	$ 98.3	$ 96.5	1.9%
Profit	12.6	$ 11.8	6.8%

(g) Third-Level Drill-Down Report			
Sales by Division (Millions)			
Qtr: 2nd Qtr. 2007	**Actual**	**Forecast**	**Variance**
Beauty Care	$ 34.5	$ 33.9	1.8%
Health Care	$ 30.0	$ 28.0	7.1%
Soap	$ 22.8	$ 23.0	-0.9%
Snacks	$ 12.1	$ 12.5	-3.2%
Electronics	$ 11.5	$ 10.9	5.5%
Total	$110.9	$108.3	2.4%

(h) Fourth-Level Drill-Down Report			
Sales by Product Category (Millions)			
Qtr: 2nd Qtr. 2007 **Division: Health Care**	**Actual**	**Forecast**	**Variance**
Toothpaste	$12.4	$10.5	18.1%
Mouthwash	$ 8.6	$ 8.8	-2.3%
Over-the-Counter Drugs	$ 5.8	$ 5.3	9.4%
Skin Care Products	$ 3.2	$ 3.4	-5.9%
Total	$30.0	$28.0	7.1%

Figure 10.4 cont.

Reports Generated by an MIS

The five types of reports are (a) scheduled, (b) key-indicator, (c) demand, (d) exception, and (e–h) drill-down.

(Source: George W. Reynolds, *Information Systems for Managers*, Third Edition. St. Paul, MN: West Publishing Co., 1995.)

Scheduled Reports

Scheduled reports are produced periodically, or on a schedule, such as daily, weekly, or monthly. For example, a production manager could use a weekly summary report that lists total payroll costs to monitor and control labor and job costs. A manufacturing report generated once a day to monitor the production of a new item is another example of a scheduled report. Other scheduled reports can help managers control customer credit, the performance of sales representatives, inventory levels, and more. Boehringer Ingelheim, a large German drug company with over $7 billion in revenues and thousands of employees in 60 countries, uses a variety of reports to allow it to respond rapidly to changing market conditions. According to the company's chief financial officer (CFO), "I want to be told where I stand and where we are heading. I like to be able to see negative trends and counter them as fast as possible." The company uses Cognos, Inc.'s Impromptu MIS to develop scheduled reports on costs for its various operations. Managers can drill down into more levels of detail to individual transactions if they want.[7] Xcel Energy has developed an online reporting system for its customers in Colorado to help them control their electricity bills.[8] The program, called InfoSmart, allows customers to analyze their energy use. "This is one instance where knowledge truly does equal power," says Debbie Mukherjee, product portfolio manager at Xcel. The scheduled reports allow customers to determine the energy efficiency of their homes and major appliances. It also shows them how their energy use compares with other homes in their area.

A **key-indicator report** summarizes the previous day's critical activities and is typically available at the beginning of each workday. These reports can summarize inventory levels, production activity, sales volume, and the like. Key-indicator reports are used by managers and executives to take quick, corrective action on significant aspects of the business.

scheduled report
Report produced periodically, or on a schedule, such as daily, weekly, or monthly.

key-indicator report
Summary of the previous day's critical activities; typically available at the beginning of each workday.

demand report
Report developed to give certain
information at someone's request.

Demand Reports

Demand reports are developed to give certain information upon request. In other words, these reports are produced on demand. For example, an executive may want to know the production status of a particular item—a demand report can be generated to provide the requested information. Suppliers and customers can also use demand reports. FedEx, for example, provides demand reports on its Web site to allow its customers to track packages from their source to their final destination. Students in Idaho and other states can go to the Rate-My-Professor Internet site to get ratings of faculty members.[9] The Laurel Pub Company, a bar and pub chain in England with over 630 outlets, uses demand reports to generate important sales data when requested.[10] The company expects to save about £500,000 over a five-year period. Other examples of demand reports include reports requested by executives to show the hours worked by a particular employee, total sales to date for a product, and so on.

Exception Reports

exception report
Report automatically produced
when a situation is unusual or
requires management action.

Exception reports are reports that are automatically produced when a situation is unusual or requires management action. For example, a manager might set a parameter that generates a report of all inventory items with fewer than the equivalent of five days of sales on hand. This unusual situation requires prompt action to avoid running out of stock on the item. The exception report generated by this parameter would contain only items with fewer than five days of sales in inventory. Detroit-based financial services provider Comerica uses exception reports to assemble a list of customer inquiries that have been open a period of time without some progress or closure.[11] The company wants to improve its customer service. According to the senior vice president of the company, "We'll be able to see who's doing real well, and reward those folks. And we'll be able to see who's not engaged yet, and help them." Exception reports are also used to help fight terrorism.[12] The Matchmaker System scans airline passenger lists and displays an exception report of passengers that could be a threat, so authorities can remove the suspected passengers before the plane takes off. The system was developed in England as part of its Defence Evaluation Research Agency.

As with key-indicator reports, exception reports are most often used to monitor aspects important to an organization's success. In general, when an exception report is produced, a manager or executive takes action. Parameters, or *trigger points*, for an exception report should be set carefully. Trigger points that are set too low may result in too many exception reports; trigger points that are too high could mean that problems requiring action are overlooked. For example, if a manager wants a report that contains all projects over budget by $100 or more, the system may retrieve almost every company project. The $100 trigger point is probably too low. A trigger point of $10,000 might be more appropriate.

Drill-Down Reports

drill-down report
Report providing increasingly
detailed data about a situation.

Drill-down reports provide increasingly detailed data about a situation. Through the use of drill-down reports, analysts are able to see data at a high level first (such as sales for the entire company), then at a more detailed level (such as the sales for one department of the company), and then a very detailed level (such as sales for one sales representative).

Developing Effective Reports

Management information system reports can help managers develop better plans, make better decisions, and obtain greater control over the operations of the firm, but in practice, the types of reports can overlap. For example, a manager can demand an exception report or set trigger points for items contained in a key-indicator report. In addition, some software packages can be used to produce, gather, and distribute reports from different computer systems. Certain guidelines should be followed in designing and developing reports to yield the best results. Table 10.1 explains these guidelines.

Guidelines	Reason
Tailor each report to user needs.	The unique needs of the manager or executive should be considered, requiring user involvement and input.
Spend time and effort producing only reports that are useful.	Once instituted, many reports continue to be generated even though no one uses them anymore.
Pay attention to report content and layout.	Prominently display the information that is most desired. Do not clutter the report with unnecessary data. Use commonly accepted words and phrases. Managers can work more efficiently if they can easily find desired information.
Use management by exception reporting.	Some reports should be produced only when there is a problem to be solved or an action that should be taken.
Set parameters carefully.	Low parameters may result in too many reports; high parameters mean valuable information could be overlooked.
Produce all reports in a timely fashion.	Outdated reports are of little or no value.
Periodically review reports.	Review reports at least once a year to make sure all reports are still needed. Review report content and layout. Determine whether additional reports are needed.

Table 10.1

Guidelines for Developing MIS Reports

Characteristics of a Management Information System

Scheduled, key-indicator, demand, exception, and drill-down reports have all helped managers and executives make better, more timely decisions. When the guidelines for developing effective reports are followed, organizations can increase revenues and lower costs. In general, MISs perform the following functions:

- *Provide reports with fixed and standard formats.* For example, scheduled reports for inventory control may contain the same types of information placed in the same locations on the reports. Different managers may use the same report for different purposes.
- *Produce hard-copy and soft-copy reports.* Some MIS reports are printed on paper and are considered hard-copy reports. Most output soft copy, using visual displays on computer screens. Soft-copy output is typically formatted in a reportlike fashion. In other words, a manager might be able to call an MIS report up directly on the computer screen, but the report would still appear in the standard hard-copy format.
- *Use internal data stored in the computer system.* MIS reports use primarily internal sources of data that are contained in computerized databases. Some MISs use external sources of data about competitors, the marketplace, and so on. The Internet is a frequently used source for external data.
- *Allow end users to develop their own custom reports.* Although analysts and programmers may be involved in developing and implementing complex MIS reports that require data from many sources, end users are increasingly developing their own simple programs to query a database and produce basic reports. This capability, however, can result in several end users developing the same or similar reports, which can increase the total time expended and require more storage, compared with having an analyst develop one report for all users.
- *Require user requests for reports developed by systems personnel.* When IS personnel develop and implement MIS reports, they typically require others to submit a formal request to the IS department. If a manager, for example, wants a production report to be used by

several people in his or her department, a formal request for the report is often required. End-user-developed reports require much less formality.

FUNCTIONAL ASPECTS OF THE MIS

Most organizations are structured along functional lines or areas. This functional structure is usually apparent from an organization chart, which typically shows vice presidents under the president. Some of the traditional functional areas are finance, manufacturing, marketing, human resources, and other specialized information systems. The MIS can be divided along those functional lines, as well, to produce reports tailored to individual functions (see Figure 10.5).

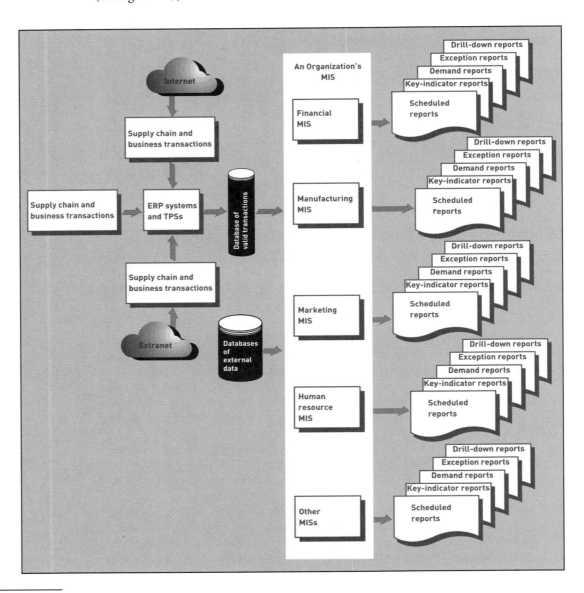

Figure 10.5

The MIS is an integrated collection of functional information systems, each supporting particular functional areas.

Financial Management Information Systems

A **financial MIS** provides financial information not only for executives but also for a broader set of people who need to make better decisions on a daily basis. Financial MISs are used to streamline reports of transactions. The 200-year-old New York Stock Exchange is investigating a new system for its electronic reports of trading.[13] The new system should cut costs and be more efficient. Other financial MISs attempt to detect stock-market fraud and abuse.[14] The Financial Services Authority in England is spending about £4 million to identify stock-market abuse and fraud with a new system called Surveillance and Automated Business Reporting Engine (SABRE). The system should be fully operational by 2006. According to one SABRE spokesperson, "SABRE is a simple database, and we use a number of predetermined and flexible reports to extract information from the system." Most financial MISs perform the following functions:

- Integrate financial and operational information from multiple sources, including the Internet, into a single system
- Provide easy access to data for both financial and nonfinancial users, often through use of a corporate intranet to access corporate Web pages of financial data and information
- Make financial data immediately available to shorten analysis turnaround time
- Enable analysis of financial data along multiple dimensions—time, geography, product, plant, customer
- Analyze historical and current financial activity
- Monitor and control the use of funds over time

Figure 10.6 shows typical inputs, function-specific subsystems, and outputs of a financial MIS, including profit and loss, auditing, and uses and management of funds.

financial MIS
An information system that provides financial information to all financial managers within an organization.

Figure 10.6

Overview of a Financial MIS

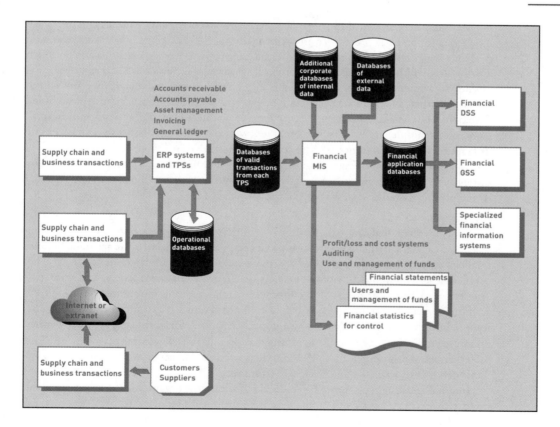

Financial institutions use information systems to shorten turnaround time for loan approvals.

(Source: Keith Brofsky/PhotoDisc/Getty Images.)

profit center
Department within an organization that tracks total expenses and net profits.

revenue center
Division within a company that tracks sales or revenues.

cost center
Division within a company that does not directly generate revenue.

auditing
Analyzing the financial condition of an organization and determining whether financial statements and reports produced by the financial MIS are accurate.

internal auditing
Auditing performed by individuals within the organization.

external auditing
Auditing performed by an outside group.

Profit/Loss and Cost Systems

Two specialized financial functional systems are profit/loss and cost systems, which organize revenue and cost data for the company. Revenue and expense data for various departments is captured by the TPS and becomes a primary internal source of financial information for the MIS.

Many departments within an organization are **profit centers**, which means that they track total expenses and net profits. An investment division of a large insurance or credit card company is an example of a profit center. Other departments may be **revenue centers**, which are divisions within the company that primarily track sales or revenues, such as a marketing or sales department. Still other departments may be **cost centers**, which are divisions within a company that do not directly generate revenue, such as manufacturing or research and development. These units incur costs with little or no direct revenues. Pharmaceutical companies, for example, are using supercomputers with over 100 processors to help accelerate drug research and development. These companies hope that such high-tech cost centers will discover new or improved drugs. Data on profit, revenue, and cost centers is gathered (mostly through the TPS but sometimes through other channels as well), summarized, and reported by the profit/loss and cost subsystems of the financial MIS.

Auditing

Auditing involves analyzing the financial condition of an organization and determining whether financial statements and reports produced by the financial MIS are accurate. Because financial statements, such as income statements and balance sheets, are used by so many people and organizations (investors, bankers, insurance companies, federal and state government agencies, competitors, and customers), sound auditing procedures are important. Auditing can reveal potential fraud, such as credit card fraud. It can also reveal false or misleading information. Fraudulent and improper accounting for healthcare provider Health-South, for example, may amount to as much as $3 billion.[15] The company operates about 1,700 clinics and hospitals, and its accounting practices and statements are now being investigated by the Securities and Exchange Commission.

Internal auditing is performed by individuals within the organization. For example, the finance department of a corporation may use a team of employees to perform an audit. Typically, an internal audit is conducted to see how well the organization is meeting established company goals and objectives—for example, ensuring that no more than five weeks of inventory is on hand, all travel reports are completed within one week of returning from a trip, and similar measures. **External auditing** is performed by an outside group, such as an accounting or consulting firm such as PricewaterhouseCoopers, Deloitte & Touche, or one of the other major, international accounting firms. The purpose of an external audit is to

provide an unbiased picture of the financial condition of an organization and uncover problems. In some cases, the financial picture from an external auditing firm may not always completely reflect the performance of the company. Critics point to the Enron bankruptcy fiasco of the early 2000s, which resulted in many employees and investors losing huge sums of money, as just one example of an external audit not showing the true picture of the company.

Uses and Management of Funds

Another important function of the financial MIS is to assist with funds usage and management. Companies that do not manage and use funds effectively often have lower profits or face bankruptcy. Some companies rely on return on investment calculations, introduced in Chapter 2, to determine when to invest in information systems and technology.[16] According to Carleton Fiorina, chairwoman and CEO of Hewlett-Packard, "CIOs will spend money on information technology only to raise ROI."

To help with funds usage and management, some banks are backing a new computerized payment system called Straight-Through Processing. The new system has the potential to clear payments in a day instead of several days or more. Outputs from the funds usage and management subsystem, when combined with those of other subsystems of the financial MIS, can locate serious cash flow problems and help organizations increase profits.

Internal uses of funds include purchasing additional inventory, updating plants and equipment, hiring new employees, acquiring other companies, buying new computer systems, increasing marketing and advertising, purchasing raw materials or land, investing in new products, and increasing research and development. External uses of funds are typically investment related. Companies often invest excess funds in such external revenue generators as bank accounts, stocks, bonds, bills, notes, futures, options, and foreign currency.

Manufacturing Management Information Systems

More than any other functional area, manufacturing has been revolutionized by advances in technology. As a result, many manufacturing operations have been dramatically improved over the last decade. Also, with the emphasis on greater quality and productivity, having an efficient and effective manufacturing process is becoming even more critical. The use of computerized systems is emphasized at all levels of manufacturing—from the shop floor to the executive suite. The use of the Internet has also streamlined all aspects of manufacturing. Figure 10.7 gives an overview of some of the manufacturing MIS inputs, subsystems, and outputs.

The manufacturing MIS subsystems and outputs monitor and control the flow of materials, products, and services through the organization. As raw materials are converted to finished goods, the manufacturing MIS monitors the process at almost every stage. European airplane manufacturer Airbus, for example, is using a manufacturing MIS to monitor and control its suppliers and parts to reduce costs.[17] New technology could make this process easier. Using specialized computer chips and tiny radio transmitters, companies can monitor materials and products through the entire manufacturing process. Procter & Gamble, Gillette, Wal-Mart, and Target have funded research into this new manufacturing MIS. Car manufacturers, which convert raw steel, plastic, and other materials into a finished automobile, also monitor their manufacturing processes. Auto manufacturers add thousands of dollars of value to the raw materials they use in assembling a car. If the manufacturing MIS also lets them provide additional service, such as customized paint colors, on any of their models, it has added further value for customers. In doing so, the MIS helps provide the company the edge that can differentiate it from competitors. The success of an organization can depend on the manufacturing function. Some common information subsystems and outputs used in manufacturing are discussed next.

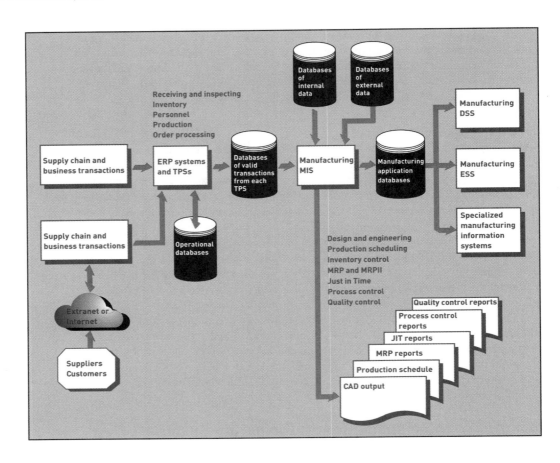

Figure 10.7

Overview of a Manufacturing MIS

Design and Engineering

During the early stages of product development, engineering departments are involved in many aspects of design. The size and shape of parts, the way electrical components are attached to equipment, the placement of controls on a product, and the order in which parts are assembled into the finished product are decisions made with the help of design and engineering departments. Increasingly, companies are involving customers in the design and engineering process.[18] In some cases, computer-assisted design (CAD) facilitates this process. CAD can model how an airplane wing or fuselage will respond to various conditions and stresses while in use. Automotive companies use CAD to help design new cars and trucks and to save millions of dollars. In one case, a company used a sophisticated CAD system equipped with powdered metals and laser welding to design and manufacture metal parts. The unique process makes metal parts from powdered metal instead of cutting the parts from solid metal blocks.

The data from design and engineering can also be used to identify problems with existing products and help develop new products. For example, Boeing uses a CAD system to develop a complete digital blueprint of an aircraft before it ever begins its manufacturing process. As mock-ups are built and tested, the digital blueprint is constantly revised to reflect the most current design. Using such technology helps Boeing reduce its manufacturing costs and the time to design a new aircraft. Computer-aided design (CAD), which enables architects and engineers to build projects digitally, has added a fourth dimension—time—in the form of scheduling software. The new 4-D CAD tools are especially useful in the design and construction of complex projects, allowing users to link 3-D building components with computer-generated work schedules for savings in time and cost.[19]

Master Production Scheduling and Inventory Control

Scheduling production and controlling inventory are critical for any manufacturing company. The overall objective of master production scheduling is to provide detailed plans for both short-term and long-range scheduling of manufacturing facilities (see Figure 10.8). Master production-scheduling software packages include forecasting techniques that determine current and future demand for products and services. After current demand has been calculated and future demand has been estimated, the master production-scheduling package determines the best way to use the manufacturing facility and all its related equipment. The result of the process is a detailed plan that reveals a schedule for every item that will be manufactured.

Figure 10.8

A master production schedule for CDs and DVDs indicates the quantity of each to be produced each week in thousands.

	week	1	2	3	4	5
CD-ROMs	amount	5	2	3	6	7

	week	1	2	3	4	5
DVDs	amount	4	1	2	2	3

An important key to the manufacturing process is inventory control. Great strides have been made in developing cost-effective inventory-control programs and software packages that allow automatic reordering, forecasting, generation of shop documents and reports, determination of manufacturing costs, analysis of budgeted costs versus actual costs, and the development of master manufacturing schedules, resource requirements, and plans. A furniture company, for example, uses an approach, called "simple, quick, and affordable (SQA)" to keep inventory levels and costs low. Once an order is received, it is broken down into the inventory parts that are needed to successfully complete the order on time. Automotive companies often use FedEx, UPS Logistics, or other shipping companies to help speed the delivery of parts to factories and finished cars to dealerships to save hundreds of millions of dollars.

economic order quantity (EOQ)
The quantity that should be reordered to minimize total inventory costs.

reorder point (ROP)
A critical inventory quantity level.

material requirements planning (MRP)
A set of inventory-control techniques that help coordinate thousands of inventory items when the demand of one item is dependent on the demand for another.

manufacturing resource planning (MRPII)
An integrated, company-wide system based on network scheduling that enables people to run their business with a high level of customer service and productivity.

just-in-time (JIT) inventory approach
A philosophy of inventory management in which inventory and materials are delivered just before they are used in manufacturing a product.

computer-assisted manufacturing (CAM)
A system that directly controls manufacturing equipment.

Many techniques are used to minimize inventory costs. Most determine how much and when to order inventory. One method of determining how much inventory to order is called the **economic order quantity (EOQ)**. This quantity is calculated to minimize the total inventory costs. The "When to order?" question is based on inventory usage over time. Typically, the question is answered in terms of a **reorder point (ROP)**, which is a critical inventory quantity level. When the inventory level for a particular item falls to the reorder point, or critical level, the system generates a report so that an order is immediately placed for the EOQ of the product. Another inventory technique used when the demand for one item is dependent on the demand for another is called **material requirements planning (MRP)**. The basic goal of MRP is to determine when finished products, such as automobiles or airplanes, are needed and then to work backward to determine deadlines and resources needed, such as engines and tires, to complete the final product on schedule. A communications company can also use MRP to plan manufacturing for the fluctuating, seasonal demand for communications products. **Manufacturing resource planning (MRPII)** refers to an integrated, company-wide system based on network scheduling that enables people to run their business with a high level of customer service and productivity, while lowering costs and inventories. MRPII is broader in scope than MRP; thus, the latter has been dubbed "little MRP." MRPII places a heavy emphasis on planning, which helps companies ensure that the right product is in the right place at the right time. MRPII often requires demand forecasting to help determine what finished products and inventory parts are needed to satisfy customer demand.

Just-in-time (JIT) inventory and manufacturing is an approach that maintains inventory at the lowest levels without sacrificing the availability of finished products. With this approach, inventory and materials are delivered just before they are used in a product. A JIT inventory system would arrange for a car windshield to be delivered to the assembly line only a few moments before it is secured to the automobile, rather than having it sit in the manufacturing facility while the car's other components are being assembled. Although JIT has many advantages, it also renders firms more vulnerable to process disruptions.

Process Control
Managers can use a number of technologies to control and streamline the manufacturing process. For example, computers can directly control manufacturing equipment, using systems called **computer-assisted manufacturing (CAM)**. CAM systems can control drilling machines, assembly lines, and more. Some operate quietly, are easy to program, and have self-diagnostic routines to test for difficulties with the computer system or the manufacturing equipment. Toyota, the Japanese automaker, is launching a manufacturing system to allow the company to build almost any type of vehicle in any of its plants.[20] The new CAM system is faster and less expensive. According to one company spokesperson, "The new line represents 50 percent less investment than the one it replaces and lets us add a different car type to the line at 70 percent lower costs than before."

Computer-assisted manufacturing systems control complex processes on the assembly line and provide users with instant access to information.

(Source: © Lester Lefkowitz/ CORBIS.)

Computer-integrated manufacturing (CIM) uses computers to link the components of the production process into an effective system. CIM's goal is to tie together all aspects of production, including order processing, product design, manufacturing, inspection and quality control, and shipping. CIM systems also increase efficiency by coordinating the actions of various production units. In some areas, CIM is used for even broader functions. For example, it can be used to integrate all organizational subsystems, not just the production systems. In automobile manufacturing, design engineers can have financial managers evaluate their ideas before new components are built to see whether they are economically viable, saving not only time but also money.

A **flexible manufacturing system (FMS)** is an approach that allows manufacturing facilities to rapidly and efficiently change from making one product to another. In the middle of a production run, for example, the production process can be changed to make a different product or to switch manufacturing materials. By using an FMS, the time and cost to change manufacturing jobs can be substantially reduced, and companies can react quickly to market needs and competition.

FMS is normally implemented using computer systems, robotics, and other automated manufacturing equipment. New product specifications are fed into the computer system, and the computer then makes the necessary changes. A large international automotive company, for example, can use an FMS to build new minivans on the same manufacturing line as its other existing minivan models.

Quality Control and Testing

With increased pressure from consumers and a general concern for productivity and high quality, today's manufacturing organizations are placing more emphasis on **quality control,** a process that ensures that the finished product meets the customers' needs. For a continuous manufacturing process, control charts are used to measure weight, volume, temperature, or similar attributes (see Figure 10.9). Then, upper and lower control chart limits are established. If these limits are exceeded, the manufacturing equipment is inspected for possible defects or potential problems.

computer-integrated manufacturing

Using computers to link the components of the production process into an effective system.

flexible manufacturing system (FMS)

An approach that allows manufacturing facilities to rapidly and efficiently change from making one product to making another.

quality control

A process that ensures that the finished product meets the customers' needs.

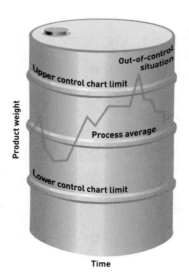

Figure 10.9

Industrial Control Chart

This chart is used to monitor product quality in an industrial chemical application. Product variances exceeding certain tolerances cause those products to be rejected.

When the manufacturing operation is not continuous, sampling can allow the producer or consumer to review and accept or reject one or more products. Acceptance sampling is used for items as simple as nuts and bolts or as complex as airplanes. The development of the control chart limits and the specific acceptance sampling plans can be fairly complex. So, quality-control software programs have been used to generate them.

Whether the manufacturing operation is continuous or discrete, the results from quality control are analyzed closely to identify opportunities for improvements. Teams using the total quality management (TQM) or continuous improvement process (see Chapter 2) often analyze this data to increase the quality of the product or eliminate problems in the manufacturing process. The result can reduce costs or increase sales.

Information generated from quality-control programs can help workers locate problems in manufacturing equipment. Quality-control reports can also be used to design better products. With the increased emphasis on quality, workers should continue to rely on the reports and outputs from this important application.

Marketing Management Information Systems

marketing MIS
Information system that supports managerial activities in product development, distribution, pricing decisions, and promotional effectiveness.

A **marketing MIS** supports managerial activities in product development, distribution, pricing decisions, promotional effectiveness, and sales forecasting. Act!, for example, is a marketing automation tool to manage sales contacts, client e-mail systems, and group sales meetings.[21] Marketing functions are increasingly being performed on the Internet. A number of companies are developing Internet marketplaces to advertise and sell products. The amount spent on online advertising is worth billions of dollars annually. Software can measure how many customers see the advertising.[22] According to a senior manager of CDW, a direct marketer of hardware and software products, "A satisfied customer is one who sees you as meeting expectations. A loyal customer, on the other hand, wants to do business with you again and will recommend you to others." CDW uses a software product called SmartLoyalty to analyze customer loyalty. Read the "Information Systems @ Work" special-interest feature to see how one golf equipment manufacturer uses a marketing MIS.

Customer relationship management (CRM) programs, available from some ERP vendors, help a company manage all aspects of customer encounters.[23] After installing a CRM system, U.S. Steel was able to increase its cash flow by millions of dollars.[24] CRM software can help a company collect customer data, contact customers, educate customers on new products, and sell products to customers through an Internet site. An airline, for example, can use a CRM system to notify customers about flight changes. Other airlines have also benefited from CRM and use Web sites to allow customers to look up possible wait times. Yet, not all CRM systems and marketing sites on the Internet are successful. Customization and ongoing maintenance of a CRM system can also be expensive. Figure 10.10 shows the inputs, subsystems, and outputs of a typical marketing MIS.

Figure 10.10

Overview of a Marketing MIS

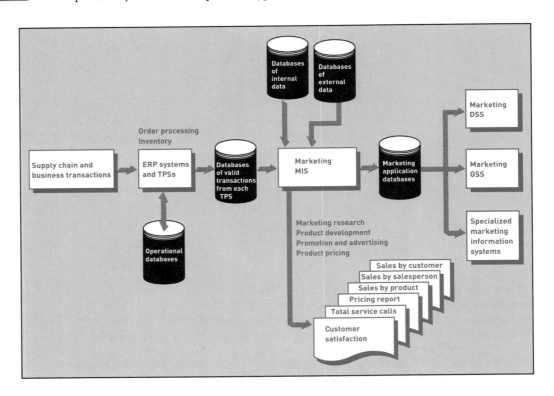

A "TaylorMade" Information and Decision Support System

For some people, being an ambassador for the best-selling golf club manufacturer would be a dream job. Traveling to all the best country clubs, playing one round of golf after another with the pros while they try out your clubs, and retiring to the clubhouse for dining and more golf chat might sound like heaven on earth. In reality, it is not as glamorous as it sounds. The sales associates for TaylorMade, makers of the number-one driver on the PGA tour, spend 60 percent of their time taking inventory in golf shops and nearly all the rest of their time traveling to visit the next customer.

Until recently, TaylorMade sales associates also spent considerable amount of time shuffling papers and entering data into an antiquated information system. They maintained records on each contact and customer so that they could review inventory records, purchasing history, and other notes prior to customer visits to determine what products to bring along on the trip. With hundreds of customers to manage, the amount of information became unmanageable.

TaylorMade's management team was struggling to keep up with the data as well. It had a group of "order management system programmers," who produced business reports for managers. The group was constantly buried under backlogged report requests. Managers were afraid to request reports because of the group's get-in-line attitude. Clearly, it was time for an IS overhaul.

TaylorMade hired a systems development team to create an information management system to provide the sales associates and managers with the information they required to make informed decisions. Scheduled key-indicator reports provide important information for guiding sales strategies. Managers in the manufacturing division rely on exception reports to let them know which types of golf clubs are running low in the warehouse. The sales force uses demand reports to find out who their best customers are and on what areas of the market they should focus. All TaylorMade employees are able to access reports from the new system through a convenient Web-based self-service interface. They no longer need to wait in line for the "order management system programmers" to fulfill report requests.

Today, TaylorMade sales associates travel equipped with much more than golf clubs. Their handheld PCs provide them with the latest sales and inventory information. They can call up charts that summarize the customer's purchases for the year and can view open orders, recent shipments, and orders in transit. They are fully equipped to answer any question that may come up. The handhelds are equipped with bar-code scanners as well. The scanners dramatically reduce the amount of time the reps spend taking inventory and entering inventory data. Once scanned, the information is automatically uploaded to the home office's database.

Several departments within TaylorMade, including operations, finance, sales, marketing, and IT, use the new system to access information on customers, inventory, and sales by product. "We use the system to build the reports that drive our business. It used to be that our sales information was somewhat invisible to us, but now we access information from four different systems," says Tom Collard, IS director. The new system is used to access, analyze, and share information from its supply chain system. It is also used to access order bookings, sales to date, and customer buying trends.

The system is available to employees over the corporate network or the Web. No matter where they are in the world, sales associates can log on instantly and access information by region, territory, or representative. A sales representative might need to know how left-handed golf clubs are selling on the West Coast versus the East or whether a harsh winter increased sales of golf jackets. "The biggest advantage will be to visually and actively manage all parameters and key metrics of our business," said Collard. "The system will help us in that we won't be stockpiling inventory three months in advance. You are actively getting signals from the sales force on what the demand truly is."

TaylorMade's new system has inspired the company to invest in other information systems. It has recently installed a new customer relationship management system that has allowed it to automate much of its customer service, reduce its staff, and improve customer satisfaction. By implementing new and effective information and decision support systems, TaylorMade has extended its reputation of high quality beyond its products into every aspect of its business.

Discussion Questions

1. How has TaylorMade increased the efficiency of its sales associates?
2. How have information systems increased TaylorMade's ability to gain a competitive advantage in the golf equipment market?

Critical Thinking Questions

3. How do you think the handheld power wielded by TaylorMade sales associates affects their level of self-esteem and job satisfaction?
4. How do you think the information system has affected relationships between the customer and sales associate now that the associate can field any question with the push of a button?

SOURCES: "KANA Honored with *CRM Magazine* 2004 Service Leader Award for Web Self-Service," *Business Wire*, March 2, 2004; Business Objects Customers in the Spotlight Web site, accessed April 26, 2004, *www.businessobjects.com/customers/spotlight/taylormade.asp*; Kana ROI Success Stories Web site, accessed April 26, 2004, *www.kana.com*; TaylorMade Web site, accessed April 26, 2004, *www.taylormadegolf.com*.

Subsystems for the marketing MIS include marketing research, product development, promotion and advertising, and product pricing. These subsystems and their outputs help marketing managers and executives increase sales, reduce marketing expenses, and develop plans for future products and services to meet the changing needs of customers.

Marketing Research

Surveys, questionnaires, pilot studies, and interviews are popular marketing research tools. Xerox, for example, uses the NewLeads software product to locate potential customers from conferences and trade shows.[25] According to a Xerox representative, "When the show's over, all you've got left are your leads." The new software helps Xerox turn curious visitors to their trade shows into real customers.

The purpose of marketing research is to conduct a formal study of the market and customer preferences. Marketing research can identify prospects (potential future customers) as well as the features that current customers really want in a good or service (such as easy-to-use plastic resealable paint cans or vanilla-flavored cola). Such attributes as style, color, size, appearance, and general fit can be investigated through marketing research. Some marketing researchers are using neuroscientists to precisely measure brain activity from customer reactions to various products and product advertising.[26] Brain activity, measured in 12 areas, is then fed into a PC for detailed analysis.

Marketing research data yields valuable information for the development and marketing of new products.

(Source: © David Young-Wolff/ Photo Edit.)

The ability to forecast demand is made possible by marketing research and sophisticated software. Some automotive companies, for example, use marketing research and software to predict the demand for car parts. Demand forecasts for products and services are also critical to make sure that raw materials and supplies are properly managed.

The Internet is changing the way that many companies think about marketing research. Conventional methods of collecting data often cost millions of dollars. For a fraction of this cost, companies can put up Internet information servers and launch discussion groups on topics that their customers care about. These information sites must be well designed or they won't be visited, but a frequently visited site can provide feedback worth a fortune. Companies that are viewed as credible, not just clever, win enormous advantages. Presence and intelligent interaction, not just advertising, are the keys that unlock commercial opportunities online. Some people, however, consider Internet marketing research to be a nuisance or even harmful.

Some companies use marketing research to advertise their products and services. For example, companies can hire people to interact with others in chat rooms, collect their preferences, and recommend the company's products and services. People visiting the chat room may not know that they are communicating with paid advertising agents for the company. Students have been hired to collect data and market credit cards to other students on college campuses. This approach is called *stealth marketing* by some.

Product Development

Product development involves the conversion of raw materials into finished goods and services and focuses primarily on the physical attributes of the product. Many factors, including plant capacity, labor skills, engineering factors, and materials are important in product development decisions. In many cases, a computer program analyzes these various factors and selects the appropriate mix of labor, materials, plant and equipment, and engineering designs. Make-or-buy decisions can also be made with the assistance of computer programs.

Promotion and Advertising

One of the most important functions of any marketing effort is promotion and advertising. Product success is a direct function of the types of advertising and sales promotion done. The size of the promotion budget and the allocation of funds among various campaigns are important factors in planning the campaigns that will be launched—everything from placing ads during the Super Bowl to offering coupons in a grocery store. Television coverage, newspaper ads, promotional brochures and literature, and training programs for salespeople are all components of these campaigns. Because of the time and scheduling savings they offer, computer programs set up the original budget and monitor expenditures and the overall effectiveness of various promotional campaigns. Computers also help consumers select the best products and services. ChoiceStream, for example, produces software that helps people find and view entertainment options online.[27] AOL, the online Internet company, uses software from ChoiceStream in its MyBestBets feature that sends people e-mail alerts about their preferred entertainment choices. Promotion and advertising can also be placed on enhanced CDs for playing on CD players or personal computers.[28] Pop singer Rachel Farris, for example, attached her mini CDs to the lids of cups at movie theaters and hopes that the enhanced disks will promote her music career when people return home and sample her music.

Richness and reach are important measures of effective advertising. *Richness* refers to the amount and level of detail that an organization can give its existing and potential customers. A company that is able to share meaningful information about its products and services with its primary customers has a high degree of richness in its advertising. With the advent of broadband Internet connections, rich media advertising is becoming more and more popular.[29] *Rich media advertising* requires broadband connections to deliver advanced audio and video effects, including cartoon characters swooping across the screen, high-quality stereo music and sound, and exclusive videos of products and people. *Reach* refers to the number of people that an organization is able to contact through advertising. A business that can successfully advertise its products and services to millions of people has a high amount of reach. In the past, some organizations could not afford to develop advertising programs with both richness and reach. With information systems and the Internet, however, companies today can achieve high levels of richness and reach in their advertising programs.

Increasingly, companies are using the Internet to deliver banner and pop-up ads. A *banner ad* is advertising that appears as sign or banner on a Web page. When you visit the Web page, you see the banner ad. A *pop-up ad*, on the other hand, automatically appears (pops up) as a separate window when you visit a Web site. Unlike a banner ad, you must close every pop-up ad window, even if you visit another Web site or exit from the Internet. Because you have to close each pop-up ad window, many people consider them a nuisance and a big waste of time. In the early 2000s, Internet advertising took a plunge.[30] In 2000, for example, companies spent about $8 billion on Internet advertising. In 2002, companies spent about $6 billion. Today, Internet advertising is again on the rise. According to the president of the Internet Advertising Bureau, "As with a lot of things related to the Internet, we all got ahead of ourselves in the beginning. Now it's showtime."

Product Pricing

Product pricing is another important and complex marketing function. Retail price, wholesale price, and price discounts must be set. A major factor in determining pricing policy is an analysis of the demand curve, which attempts to determine the relationship between price and sales. Most companies try to develop pricing policies that will maximize total sales revenues—usually a function of price elasticity. If the product is very price sensitive, a reduction in price can generate a substantial increase in sales, which can result in higher revenues. A

product that is relatively insensitive to price can substantially increase its price without a large reduction in demand. Computer programs can help determine price elasticity and various pricing policies, such as supply and demand curves for pricing analysis (see Figure 10.11). Typically, a marketing executive can alter prices on the computer system, which analyzes price changes and their impact on total revenues. The rapid feedback now possible through computer communications networks enables managers to determine the results of pricing decisions much more quickly than in the past. This ability facilitates more aggressive pricing strategies, which can be quickly adjusted to meet market needs. One critical pricing decision is when to mark down product prices. Using sophisticated software, a company can reduce the number and amount of price cuts, which has helped increase profitability.

Figure 10.11

Typical Supply and Demand Curve for Pricing Analysis

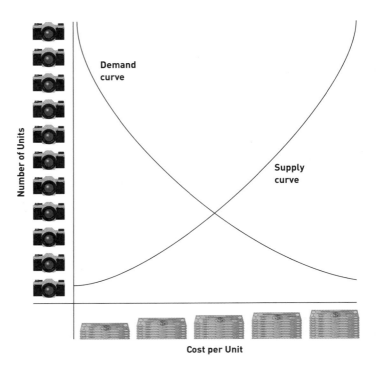

Sales analysis is also important to identify products, sales personnel, and customers that contribute to profits and those that do not. Several reports can be generated to help marketing managers make good sales decisions (see Figure 10.12). The sales-by-product report lists all major products and their sales for a period of time, such as a month. This report shows which products are doing well and which need improvement or should be discarded altogether. The sales-by-salesperson report lists total sales for each salesperson for each week or month. This report can also be subdivided by product to show which products are being sold by each salesperson. The sales-by-customer report is a tool that can be used to identify high- and low-volume customers.

Human Resource Management Information Systems

human resource MIS

An information system that is concerned with activities related to employees and potential employees of an organization, also called a *personnel MIS*.

A **human resource MIS**, also called the *personnel MIS*, is concerned with activities related to employees and potential employees of the organization. Because the personnel function relates to all other functional areas in the business, the human resource MIS plays a valuable role in ensuring organizational success. Some of the activities performed by this important MIS include workforce analysis and planning, hiring, training, job and task assignment, and many other personnel-related issues. An effective human resource MIS allows a company to keep personnel costs at a minimum, while serving the required business processes needed to achieve corporate goals. Figure 10.13 shows some of the inputs, subsystems, and outputs of the human resource MIS.

(a) Sales by Product

Product	August	September	October	November	December	Total
Product1	34	32	32	21	33	152
Product 2	156	162	177	163	122	780
Product 3	202	145	122	98	66	633
Product 4	345	365	352	341	288	1,691

(b) Sales by Salesperson

Salesperson	August	September	October	November	December	Total
Jones	24	42	42	11	43	162
Kline	166	155	156	122	133	732
Lane	166	155	104	99	106	630
Miller	245	225	305	291	301	1,367

(c) Sales by Customer

Customer	August	September	October	November	December	Total
Ang	234	334	432	411	301	1,712
Braswell	56	62	77	61	21	277
Celec	1,202	1,445	1,322	998	667	5,634
Jung	45	65	55	34	88	287

Figure 10.12

Reports Generated to Help Marketing Managers Make Good Decisions

(a) This sales-by-product report lists all major products and their sales for the period from August to December. (b) This sales-by-salesperson report lists total sales for each salesperson for the same time period. (c) This sales-by-customer report lists sales for each customer for the period. Like all MIS reports, totals are provided automatically by the system to show managers at a glance the information they need to make good decisions.

Figure 10.13

Overview of a Human Resource MIS

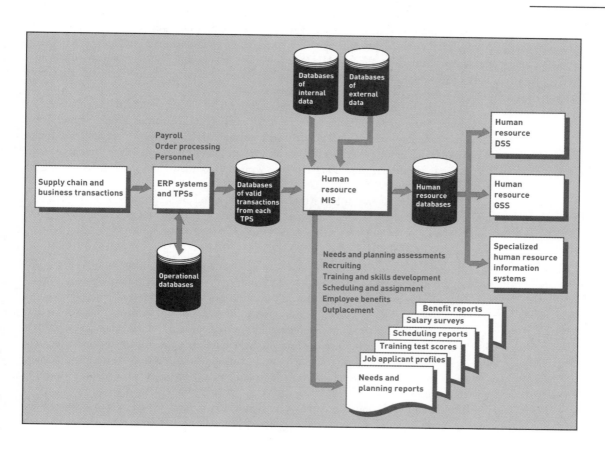

Human resource subsystems and outputs range from the determination of human resource needs and hiring through retirement and outplacement. Most medium and large organizations have computer systems to assist with human resource planning, hiring, training and skills inventorying, and wage and salary administration. Outputs of the human resource MIS include reports such as human resource planning reports, job application review profiles, skills inventory reports, and salary surveys.

Human Resource Planning

One of the first aspects of any human resource MIS is determining personnel and human needs. The overall purpose of this MIS subsystem is to put the right number and kinds of employees in the right jobs when they are needed. Effective human resource planning requires defining the future number of employees needed and anticipating the future supply of people for these jobs. For companies involved with large projects, such as military contractors and large builders, human resource plans can be generated directly from data on current and future projects.

Suppose that a construction company obtains a contract from a group of investors to build a 250-unit apartment complex. Forecasting programs and project-management software packages can be used to develop reports that describe what people are needed and when they are needed during the entire construction project. A typical output would be a human resource needs and planning report, which might specify that ten employees will be needed in August to pour concrete slabs, and eight carpenters and four painters will be needed in October. Alternatively, a factory might use this report to list total number of employees needed, broken down according to skill level, such as highly technical, technical, semitechnical, and so on.

Personnel Selection and Recruiting

If the human resource plan reveals that additional personnel are required, the next logical step is recruiting and selecting personnel. This subsystem performs one of the most important and critical functions of any organization, especially in service organizations, where employees can define the company's success. Companies seeking new employees often use computers to schedule recruiting efforts and trips and to test potential employees' skills. Some software companies, for example, use computerized testing to determine a person's programming skills and abilities. FurstPerson, for example, tests potential personnel for companies by simulating actual work situations.[31] For a call center position, for example, FurstPerson will use a simulated call center to test people applying for jobs. The testing helps companies increase employee productivity while reducing turnover.

Management information systems can also be used to help rank and select potential employees. For every applicant, the system can analyze and print the results of interviews, tests, and company visits. This report, called a *job applicant review profile*, can assist corporate recruiting teams in final selection. Some software programs can even analyze this data to help identify job applicants most likely to stay with the company and perform according to corporate standards.

Many companies now use the Internet to screen for job applicants. Applicants use a template to load their resume onto the Internet site. HR managers can then access these resumes and identify applicants they are interested in interviewing.

Training and Skills Inventory

Some jobs, such as programming, equipment repair, and tax preparation, require very specific training for new employees. Other jobs may require general training about the organizational culture, orientation, dress standards, and expectations of the organization. Today, many organizations conduct their own training, with the assistance of information systems and technology. Self-paced training involves computerized tutorials, video programs, and CD-ROM books and materials. Distance learning, where training and classes are conducted over the Internet, is also a viable alternative to more traditional training and learning approaches. This text and its supporting material, for example, can be used in a distance-learning environment.

When training is complete, employees often take computer-scored tests to evaluate their mastery of skills and new material. The results of these tests are usually given to the employee's supervisor in a training or skills inventory report. In some cases, skills inventory reports are

used for job placement. For instance, if a particular position in the company needs to be filled, managers might wish to hire internally before they recruit. The skills inventory report would help them evaluate current employees to determine their potential for the position. Such reports can also be part of employee evaluations and help determine raises or bonuses. These types of tests, however, must be valid and reliable to avoid mistakes in job placement and bonuses.

Scheduling and Job Placement

Scheduling people and jobs can be relatively straightforward or extremely complex. For some small service companies, scheduling and job placement are based on which customers walk through the door. Determining the best schedule for flights and airline pilots, placing military recruits into jobs, and determining what truck drivers and equipment should be used to transport materials across the country normally require sophisticated computer programs. In most cases, various schedules and job placement reports are generated. Employee schedules are developed for each employee, showing his or her job assignments over the next week or month. Job placements are often determined based on skills inventory reports, which show which employee might be best suited to a particular job.

Wage and Salary Administration

Another human resource MIS subsystem involves determining wages, salaries, and benefits, including medical payments, savings plans, and retirement accounts. Wage data, such as industry averages for positions, can be taken from the corporate database and manipulated by the human resource MIS to provide wage information and reports to higher levels of management. These reports, called *salary surveys*, can be used to compare salaries with budget plans, the cost of salaries versus sales, and the wages required for any one department or office. Wage and salary administration also entails designing retirement programs for employees.

Outplacement

Employees leave a company for a number of reasons. Some voluntarily leave to work for another company or retire at age 65. Others are laid off as a company reduces its workforce or are fired for poor performance or inappropriate behavior. Outplacement services are offered by many companies to help employees make the transition. *Outplacement* can include job counseling and training, job and executive search, retirement and financial planning, and a variety of severance packages and options. The outplacement MIS includes software to help employees make the transition to their lives after their association with the company ends. Computerized training programs, retirement and financial planning programs, and programs to help analyze various severance packages are often used. Some companies use computerized retirement programs to help employees gain the most from their retirement accounts and options. Scottish Life, for example, uses the Works4you software product to help employees manage their insurance needs and benefits, including retirement programs.[32] In addition, other sophisticated software packages and services can help employees find new jobs. Employment search firms, or head hunters, are used to help laid-off employees research new job opportunities, write letters with attached resumes to prospective companies, and negotiate new employment contracts.

Other Management Information Systems

In addition to finance, manufacturing, marketing, and human resource MISs, some companies have other functional management information systems. For example, most successful companies have well-developed accounting functions and a supporting accounting MIS. Also, many companies make use of geographic information systems for presenting data in a useful form.

Accounting MISs

In some cases, accounting works closely with financial management. An **accounting MIS** performs a number of important activities, providing aggregate information on accounts payable, accounts receivable, payroll, and many other applications. The organization's TPS captures accounting data, which is also used by most other functional information systems.

Some smaller companies hire outside accounting firms to assist them with their accounting functions. These outside companies produce reports for the firm using raw accounting

accounting MIS
An information system that provides aggregate information on accounts payable, accounts receivable, payroll, and many other applications.

data. In addition, many excellent integrated accounting programs are available for personal computers in small companies. Depending on the needs of the small organization and its personnel's computer experience, using these computerized accounting systems can be a very cost-effective approach to managing information.

Geographic Information Systems

Increasingly, managers want to see data presented in graphical form. A **geographic information system (GIS)** is a computer system capable of assembling, storing, manipulating, and displaying geographically referenced information, that is, data identified according to its location. A GIS enables users to pair maps or map outlines with tabular data to describe aspects of a particular geographic region. For example, sales managers may want to plot total sales for each county in the states they serve. Using a GIS, they can specify that each county be shaded to indicate the relative amount of sales—no shading or light shading represents no or little sales, and deeper shading represents more sales. While many software products have seen declining revenues, the use of GIS software is increasing by more than 10 percent per year on average.[33] Edens & Avant, a $2.3 billion real estate investment firm, uses a GIS from Environmental Systems Research Institute to increase profits and plan for potential disasters at the shopping centers it owns and manages.[34] According to David Beitz, geographic and marketing information systems manager for the company, "The cost is tremendous if you build a shopping center somewhere and a major tenant leaves. You're going to lose a lot of money."

We saw earlier in this chapter that management information systems (MISs) provide useful summary reports to help solve structured and semistructured business problems. Decision support systems (DSSs) offer the potential to assist in solving both semistructured and unstructured problems.

geographic information system (GIS)

A computer system capable of assembling, storing, manipulating, and displaying geographic information, i.e., data identified according to its location.

AN OVERVIEW OF DECISION SUPPORT SYSTEMS

A DSS is an organized collection of people, procedures, software, databases, and devices used to support problem-specific decision making and problem solving. The focus of a DSS is on decision-making effectiveness when faced with unstructured or semistructured business problems. As with a TPS and an MIS, a DSS should be designed, developed, and used to help an organization achieve its goals and objectives. Decision support systems offer the potential to generate higher profits, lower costs, and better products and services. For example, healthcare organizations use DSSs to improve patient care and reduce costs.[35] Decision support systems can also monitor and improve patient care. One diabetes Web site developed a DSS to provide customized treatment plans and reports.[36] The Web site uses powerful DSS software and grid computing, where the idle capacity of a network of computers is harnessed.

Decision support systems, although skewed somewhat toward the top levels of management, are used at all levels. To some extent, today's managers at all levels are faced with less structured, nonroutine problems, but the quantity and magnitude of these decisions increase as a manager rises higher in an organization. Many organizations contain a tangled web of complex rules, procedures, and decisions. DSSs are used to bring more structure to these problems to aid the decision-making process. In addition, because of the inherent flexibility of decision support systems, managers at all levels are able to use DSSs to assist in some relatively routine, programmable decisions in lieu of more formalized management information systems. DSSs are also used in nonprofit organizations. Cornell University, for example, used a DSS to monitor the journals and newspapers it purchases for its library.[37] The DSS showed that about 2 percent of its journal subscriptions were costing the university 20 percent of its periodicals budget. The DSS allowed the university to drop some journals that weren't being used and to pressure academic publishing companies to reduce their prices.

Characteristics of a Decision Support System

Decision support systems have many characteristics that allow them to be effective management support tools. Of course, not all DSSs work the same—some are small in scope and offer only some of these characteristics. In general, a DSS can perform the following functions:

- *Handle large amounts of data from different sources.* For instance, advanced database management systems and data warehouses have allowed decision makers to search for information with a DSS, even when some data resides in different databases on different computer systems or networks. Other sources of data may be accessed via the Internet or over a corporate intranet. Using the Internet, an oil giant can use a decision support system to save hundreds of millions of dollars annually by coordinating a large amount of drilling and exploration data from around the globe.
- *Provide report and presentation flexibility.* Managers can get the information they want, presented in a format that suits their needs. Furthermore, output can be displayed on computer screens or printed, depending on the needs and desires of the problem solvers.
- *Offer both textual and graphical orientation.* Today's DSSs can produce text, tables, line drawings, pie charts, trend lines, and more. By using their preferred orientation, managers can use a DSS to get a better understanding of a situation and to convey this understanding to others.
- *Support drill-down analysis.* A manager can get more levels of detail when needed by drilling down through data. For example, a manager can get more detailed information for a project—viewing the overall project cost or drilling down and seeing the cost for each phase, activity, and task.
- *Perform complex, sophisticated analysis and comparisons using advanced software packages.* Marketing research surveys, for example, can be analyzed in a variety of ways using programs that are part of a DSS. Many of the analytical programs associated with a DSS are actually stand-alone programs, and the DSS brings them together.
- *Support optimization, satisficing, and heuristic approaches.* By supporting all types of decision-making approaches, a DSS gives the decision maker a great deal of flexibility in computer support for decision making. For example, **what-if analysis**, the process of making hypothetical changes to problem data and observing the impact on the results, can be used to control inventory. Given the demand for products, such as automobiles, the computer can determine the necessary parts and components, including engines, transmissions, windows, and so on. With what-if analysis, a manager can make changes to problem data (the number of automobiles needed for the next month) and immediately see the impact on the parts requirements.

what-if analysis
The process of making hypothetical changes to problem data and observing the impact on the results.

Multicriteria decision making allows managers to take a number of important goals into account. A car manufacturer, for example, may want to maximize profits, while keeping all its plants open for the next few months and avoiding a labor strike. Multicriteria decision-making approaches allow managers to seek several goals at the same time.

Goal-seeking analysis is the process of determining the problem data required for a given result. For example, a financial manager may be considering an investment with a certain monthly net income, and the manager might have a goal to earn a return of 9 percent on the investment. Goal seeking allows the manager to determine what monthly net income (problem data) is needed to yield a return of 9 percent (problem result). Some spreadsheets can be used to perform goal-seeking analysis (see Figure 10.14).

goal-seeking analysis
The process of determining the problem data required for a given result.

Simulation is the ability of the DSS to duplicate the features of a real system. In most cases, probability or uncertainty are involved. For example, the number of repairs and the time to repair key components of a manufacturing line can be calculated to determine the impact on the number of products that can be produced each day. Engineers can use this data to determine which components need to be reengineered to increase the mean time between failures and which components need to have an ample supply of spare parts to reduce the mean time to repair. Drug companies are using simulated trials to reduce the need for human participants and reduce the time and costs of bringing a new drug to market. Drug companies are hoping that this use of simulation will help them identify successful drugs

simulation
The ability of the DSS to duplicate the features of a real system.

earlier in development. Corporate executives and military commanders often use computer simulations to allow them to try different strategies in different situations.[38] Corporate executives, for example, can try different marketing decisions under various market conditions. Military commanders often use computer war games to fine-tune their military strategies in different warfare conditions. The Federal Reserve Bank uses simulation to help it organize and structure work teams.[39] Toyota used 3-D simulation to design plants around the world.[40] The simulation was used to improve robotic work processes and avoid having robots collide on the factory floor.

Figure 10.14

With a spreadsheet program, a manager can enter a goal, and the spreadsheet will determine the input needed to achieve the goal.

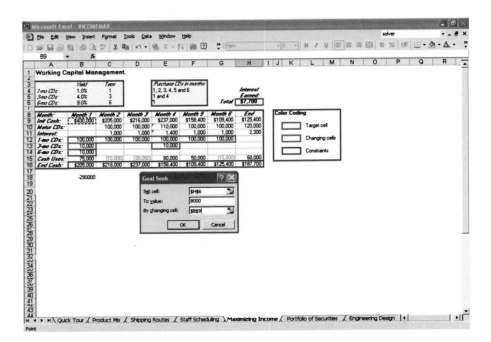

Capabilities of a Decision Support System

Developers of decision support systems strive to make them more flexible than management information systems and to give them the potential to assist decision makers in a variety of situations. Table 10.2 lists a few DSS applications. DSSs can assist with all or most problem-solving phases, decision frequencies, and different degrees of problem structure. DSS approaches can also help at all levels of the decision-making process. In this section, we investigate these DSS capabilities. A single DSS may provide only a few of these capabilities, depending on its uses and scope.

Support for Problem-Solving Phases

The objective of most decision support systems is to assist decision makers with the phases of problem solving. As previously discussed, these phases include intelligence, design, choice, implementation, and monitoring. A specific DSS might support only one or a few phases. During the implementation phase of a DSS, the state of California ordered 19 drug companies to distribute about $150 million of scarce drugs to about 150 California hospitals and clinics.[41] A DSS was developed to distribute about 90 percent of the drugs required by the court decision. This approach was able to distribute more than 125 different drugs in 20 drug categories as required by law. By supporting all types of decision-making approaches, a DSS gives the decision maker a great deal of flexibility in getting computer support for decision-making activities.

Company or Application	Description
ING Direct	The financial services company uses a DSS to summarize the bank's financial performance. The bank needed a measurement and tracking mechanism to determine how successful it was and to make modifications to plans in real time.
Cinergy Corporation	The electric utility developed a DSS to reduce lead time and effort required to make decisions in purchasing coal.
U.S. Army	It developed a DSS to help recruit, train, and educate enlisted forces. The DSS uses a simulation that incorporates what-if features.
National Audubon Society	It developed a DSS called Energy Plan (EPLAN) to analyze the impact of U.S. energy policy on the environment.
Hewlett-Packard	The computer company developed a DSS called Quality Decision Management to help improve the quality of its products and services.
Virginia	The state of Virginia developed the Transportation Evacuation Decision Support System (TEDSS) to determine the best way to evacuate people in case of a nuclear disaster at its nuclear power plants.

Support for Different Decision Frequencies

Decisions can range on a continuum from one-of-a-kind to repetitive decisions. One-of-a-kind decisions are typically handled by an **ad hoc DSS**. An ad hoc DSS is concerned with situations or decisions that come up only a few times during the life of the organization; in small businesses, they may happen only once. For example, a company might be faced with a decision on whether to build a new manufacturing facility in another area of the country. Repetitive decisions are addressed by an institutional DSS. An **institutional DSS** handles situations or decisions that occur more than once, usually several times a year or more. An institutional DSS is used repeatedly and refined over the years. Examples of institutional DSSs include systems that support portfolio and investment decisions and production scheduling. These decisions may require decision support numerous times during the year. For example, DSSs are used to solve computer-related problems that can occur multiple times throughout the day.[42] With this approach, the DSS monitors computer systems second by second for problems and takes action to prevent problems, such as slowdowns and crashes, and to recover from them when they occur. One IBM engineer believes that this approach, called *autonomic computing*, is the key to the future of computing. Between these two extremes are decisions managers make several times, but not regularly or routinely.

Support for Different Problem Structures

As discussed previously, decisions can range from highly structured and programmed to unstructured and nonprogrammed. **Highly structured problems** are straightforward, requiring known facts and relationships. **Semistructured** or **unstructured problems**, on the other hand, are more complex. The relationships among the pieces of data are not always clear, the data may be in a variety of formats, and it is often difficult to manipulate or obtain. In addition, the decision maker may not know the information requirements of the decision in advance. For example, a DSS has been used to support sophisticated and unstructured investment analysis and make substantial profits for traders and investors.[43] Some DSS trading software is programmed to place buy and sell orders automatically without a trader manually entering a trade, based on parameters set by the trader.

Support for Various Decision-Making Levels

Decision support systems can provide help for managers at different levels within the organization. Operational managers can get assistance with daily and routine decision making. Tactical decision makers can be supported with analysis tools to ensure proper planning and

Table 10.2

Selected DSS Applications

ad hoc DSS
A DSS concerned with situations or decisions that come up only a few times during the life of the organization.

institutional DSS
A DSS that handles situations or decisions that occur more than once, usually several times a year or more. An institutional DSS is used repeatedly and refined over the years.

highly structured problems
Problems that are straightforward and require known facts and relationships.

semistructured or unstructured problems
More complex problems in which the relationships among the pieces of data are not always clear, the data may be in a variety of formats, and the data is often difficult to manipulate or obtain.

control. At the strategic level, DSSs can help managers by providing analysis for long-term decisions requiring both internal and external information (see Figure 10.15).

Figure 10.15

Decision-Making Level

Strategic managers are involved with long-term decisions, which are often made infrequently. Operational managers are involved with decisions that are made more frequently.

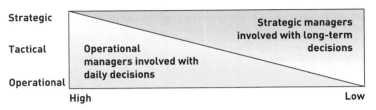

Table 10.3

Comparison of DSSs and MISs

A Comparison of DSS and MIS

A DSS differs from an MIS in numerous ways, including the type of problems solved, the support given to users, the decision emphasis and approach, and the type, speed, output, and development of the system used. Table 10.3 lists brief descriptions of these differences.

Factor	DSS	MIS
Problem Type	A DSS is good at handling unstructured problems that cannot be easily programmed.	An MIS is normally used only with more structured problems.
Users	A DSS supports individuals, small groups, and the entire organization. In the short run, users typically have more control over a DSS.	An MIS supports primarily the organization. In the short run, users have less control over an MIS.
Support	A DSS supports all aspects and phases of decision making; it does not replace the decision maker—people still make the decisions.	This is not true of all MIS systems—some make automatic decisions and replace the decision maker.
Emphasis	A DSS emphasizes actual decisions and decision-making styles.	An MIS usually emphasizes information only.
Approach	A DSS is a direct support system that provides interactive reports on computer screens.	An MIS is typically an indirect support system that uses regularly produced reports.
System	The computer equipment that provides decision support is usually online (directly connected to the computer system) and related to real time (providing immediate results). Computer terminals and display screens are examples—these devices can provide immediate information and answers to questions.	An MIS, using printed reports that may be delivered to managers once a week, may not provide immediate results.
Speed	Because a DSS is flexible and can be implemented by users, it usually takes less time to develop and is better able to respond to user requests.	An MIS's response time is usually longer.
Output	DSS reports are usually screen oriented, with the ability to generate reports on a printer.	An MIS, however, typically is oriented toward printed reports and documents.
Development	DSS users are usually more directly involved in its development. User involvement usually means better systems that provide superior support. For all systems, user involvement is the most important factor for the development of a successful system.	An MIS is frequently several years old and often was developed for people who are no longer performing the work supported by the MIS.

COMPONENTS OF A DECISION SUPPORT SYSTEM

dialogue manager
User interface that allows decision makers to easily access and manipulate the DSS and to use common business terms and phrases.

At the core of a DSS are a database and a model base. In addition, a typical DSS contains a **dialogue manager**, which allows decision makers to easily access and manipulate the DSS and to use common business terms and phrases. Finally, access to the Internet, networks, and other computer-based systems permits the DSS to tie into other powerful systems, including the TPS or function-specific subsystems. Internet software agents, for example, can

be used in creating powerful decision support systems. Figure 10.16 shows a conceptual model of a DSS.

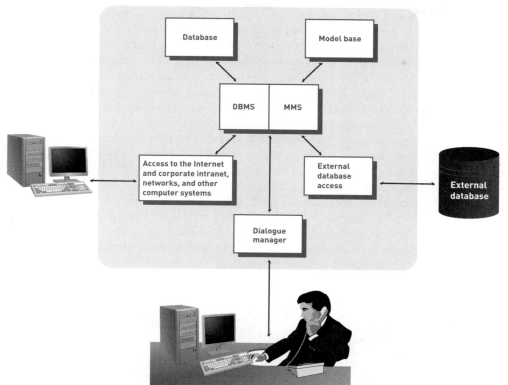

Figure 10.16

Conceptual Model of a DSS

DSS components include a model base; database; external database access; access to the Internet and corporate intranet, networks, and other computer systems; and a dialogue manager.

The Database

The database management system allows managers and decision makers to perform *qualitative analysis* on the company's vast stores of data in databases, data warehouses, and data marts, discussed in Chapter 5. A *data-driven DSS* primarily performs qualitative analysis based on the company's databases. Data-driven DSSs tap into vast stores of information contained in the corporate database, retrieving information on inventory, sales, personnel, production, finance, accounting, and other areas. Data mining and business intelligence, introduced in Chapter 5, are often used in a data-driven DSS. Airline companies, for example, use a data-driven DSS to help it identify customers for round-trip flights between major cities. The data-driven DSS can be used to search a data warehouse to contact thousands of customers who might be interested in an inexpensive flight. A casino can use a data-driven DSS to search large databases to get detailed information on patrons. It can tell how much each spends a day on gambling, and more. TUI, a travel company in Europe, uses a data-driven DSS to make better decisions to help reduce costs and increase efficiency.[44] According to the company's financial director, "We will make significant cost savings. It will allow us to become more efficient and provide us with one version of the truth."

A database management system can also connect to external databases to give managers and decision makers even more information and decision support. External databases can include the Internet, libraries, government databases, and more. The combination of internal and external database access can give key decision makers a better understanding of the company and its environment.

The Model Base

model base
Part of a DSS that provides decision makers access to a variety of models and assists them in decision making.

The **model base** allows managers and decision makers to perform *quantitative analysis* on both internal and external data. A *model-driven DSS* primarily performs mathematical or quantitative analysis. The model base gives decision makers access to a variety of models so that they can explore different scenarios and see their effects. Ultimately, it assists them in the decision-making process. For example, Bayer, the large drug company, had to choose whether to research a gene that could dispose certain people to asthma.[45] The company developed a DSS model that saved the company six months in analysis time. At the choice stage, Bayer decided to explore which genes might cause asthma. According to the head of the research and development department of a competing company, "This is revolutionary. You can change genes with the stroke of a keypad. We've never had the tools like these, to look at the complexity of disease." In the sports world, AC Milan, a football (soccer) league champion, uses a model-driven DSS to reduce injuries by 90 percent.[46] The software models help players determine their diets, training schedule, and mental attitude. According to the coach, skeptical athletes changed their minds about the program once it was in place: "When they [the players] realized that it would make them more healthy and prolong their playing careers, they became much more accepting." **Model management software** (MMS) is often used to coordinate the use of models in a DSS, including financial, statistical analysis, graphical, and project-management models. Depending on the needs of the decision maker, one or more of these models can be used (see Table 10.4).

model management software
Software that coordinates the use of models in a DSS.

Model Type	Description	Software That Can Be Used
Financial	Provides cash flow, internal rate of return, and other investment analysis	Spreadsheet, such as Excel
Statistical	Provides summary statistics, trend projections, hypothesis testing, and more	Statistical program, such SPSS or SAS
Graphical	Assists decision makers in designing, developing, and using graphic displays of data and information	Graphics programs, such as PowerPoint
Project Management	Handles and coordinates large projects; also used to identify critical activities and tasks that could delay or jeopardize an entire project if they are not completed in a timely and cost-effective fashion	Project management software, such as Project

Table 10.4

DSSs often use financial, statistical, graphical, and project-management models.

The Dialogue Manager

The dialogue manager allows users to interact with the DSS to obtain information. It assists with all aspects of communications between the user and the hardware and software that constitute the DSS. In a practical sense, to most DSS users, the dialogue manager is the DSS. Upper-level decision makers are often less interested in where the information came from or how it was gathered than that the information is both understandable and accessible.

GROUP SUPPORT SYSTEMS

The DSS approach has resulted in better decision making for all levels of individual users. However, many DSS approaches and techniques are not suitable for a group decision-making environment. Although not all workers and managers are involved in committee meetings and group decision-making sessions, some tactical- and strategic-level managers can spend more than half their decision-making time in a group setting. Such managers need assistance

with group decision making. A **group support system (GSS)**, also called a *group decision support system* and a *computerized collaborative work system*, consists of most of the elements in a DSS, plus software to provide effective support in group decision-making settings (see Figure 10.17). Lowe & Partners Worldwide, a global advertising company, uses a GSS called *swarming technology* to link experts in many diverse areas to make important advertising decisions.[47] The company electronically links account executives from Hong Kong, England, India, the United States, and other areas with experts in various industries to help craft advertising programs. According to a company executive, "We're discovering resources we didn't know existed."

group support system (GSS)

Software application that consists of most elements in a DSS, plus software to provide effective support in group decision making; also called *group support system* or *computerized collaborative work system*.

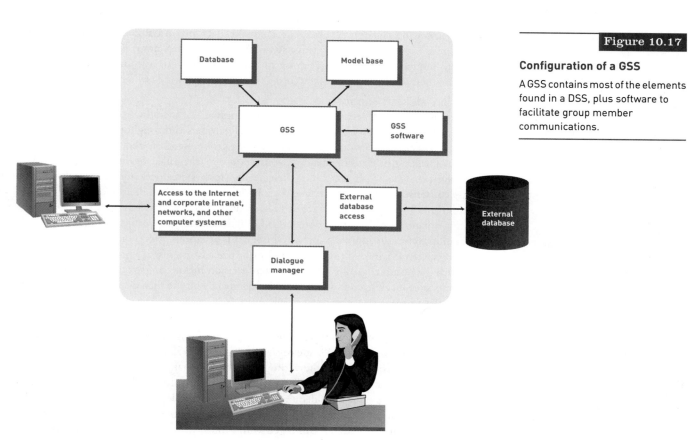

Figure 10.17

Configuration of a GSS

A GSS contains most of the elements found in a DSS, plus software to facilitate group member communications.

Group support systems are used in most industries. Architects are increasingly using GSSs to help them collaborate with other architects and builders to develop the best plans and to compete for contracts. Manufacturing companies use GSSs to link raw material suppliers to their own company systems. Engineers can use Mathcard Enterprise, another GSS.[48] The software allows engineers to create, share, and reuse calculations. Students at Carnegie Mellon University have built an experimental GSS meeting room, called the Barn, that allows interactive collaboration. The meeting room has projectors, microphones, speakers, and other devices attached to the walls and ceiling to facilitate group decision making.[49]

Characteristics of a GSS That Enhance Decision Making

It is often said that two heads are better than one. When it comes to decision making, a GSS's unique characteristics have the potential to result in better decisions. Developers of these systems try to build on the advantages of individual support systems while adding new approaches unique to group decision making. For example, some GSSs can allow the exchange of information and expertise among people without direct face-to-face interaction. The following sections describe some characteristics that can improve and enhance decision making.

Special Design

The GSS approach acknowledges that special procedures, devices, and approaches are needed in group decision-making settings. These procedures must foster creative thinking, effective communications, and good group decision-making techniques.

Ease of Use

Like an individual DSS, a GSS must be easy to learn and use. Systems that are complex and hard to operate will seldom be used. Many groups have less tolerance than do individual decision makers for poorly developed systems.

Flexibility

Two or more decision makers working on the same problem may have different decision-making styles and preferences. Each manager makes decisions in a unique way, in part because of different experiences and cognitive styles. An effective GSS not only has to support the different approaches that managers use to make decisions but also must find a means to integrate their different perspectives into a common view of the task at hand.

Decision-Making Support

A GSS can support different decision-making approaches, including the **delphi approach**, in which group decision makers are geographically dispersed throughout the country or the world. This approach encourages diversity among group members and fosters creativity and original thinking in decision making. Another approach, called **brainstorming**, in which members offer ideas "off the top of their heads," fosters creativity and free thinking. The **group consensus approach** forces members in the group to reach a unanimous decision. The Shuttle Project Engineering Office at the Kennedy Space Center has used the consensus-ranking organizational-support system (CROSS) to evaluate space projects in a group setting. The group consensus approach analyzes the benefits of various projects and their probabilities of success. CROSS is used to evaluate and prioritize advanced space projects.[50]

With the **nominal group technique**, each decision maker can participate; this technique encourages feedback from individual group members, and the final decision is made by voting, similar to a system for electing public officials.

Anonymous Input

Many GSSs allow anonymous input, where the person giving the input is not known to other group members. For example, some organizations use a GSS to help rank the performance of managers. Anonymous input allows the group decision makers to concentrate on the merits of the input without considering who gave it. In other words, input given by a top-level manager is given the same consideration as input from employees or other members of the group. Some studies have shown that groups using anonymous input can make better decisions and have superior results compared with groups that do not use anonymous input. Anonymous input, however, can result in flaming, where an unknown team member posts insults or even obscenities on the GSS.

Reduction of Negative Group Behavior

One key characteristic of any GSS is the ability to suppress or eliminate group behavior that is counterproductive or harmful to effective decision making. In some group settings, dominant individuals can take over the discussion, which can prevent other members of the group from presenting creative alternatives. In other cases, one or two group members can sidetrack or subvert the group into areas that are nonproductive and do not help solve the problem at hand. Other times, members of a group may assume they have made the right decision without examining alternatives—a phenomenon called *groupthink*. If group sessions are poorly planned and executed, the result can be a tremendous waste of time. Today, many GSS designers are developing software and hardware systems to reduce these types of problems. Procedures for effectively planning and managing group meetings can be incorporated into the GSS approach. A trained meeting facilitator is often employed to help lead the group decision-making process and to avoid groupthink.

Parallel Communication

With traditional group meetings, people must take turns addressing various issues. One person normally talks at a time. With a GSS, every group member can address issues or make comments at the same time by entering them into a PC or workstation. These comments

---

delphi approach

A decision-making approach in which group decision makers are geographically dispersed; this approach encourages diversity among group members and fosters creativity and original thinking in decision making.

brainstorming

Decision-making approach that often consists of members offering ideas "off the top of their heads."

group consensus approach

Decision-making approach that forces members in the group to reach a unanimous decision.

nominal group technique

Decision-making approach that encourages feedback from individual group members, and the final decision is made by voting, similar to the way public officials are elected.

and issues are displayed on every group member's PC or workstation immediately. Parallel communication can speed meeting times and result in better decisions.

Automated Record Keeping

Most GSSs can keep detailed records of a meeting automatically. Each comment that is entered into a group member's PC or workstation can be anonymously recorded. In some cases, literally hundreds of comments can be stored for future review and analysis. In addition, most GSS packages have automatic voting and ranking features. After group members vote, the GSS records each vote and makes the appropriate rankings.

GSS Software

GSS software, often called *groupware* or *workgroup software*, helps with joint work group scheduling, communication, and management.[51] One popular package, Lotus Notes, can capture, store, manipulate, and distribute memos and communications that are developed during group projects. It can also incorporate knowledge management, discussed in Chapter 5, into the Lotus Notes package. Some companies standardize on messaging and collaboration software, such as Lotus Notes. Microsoft's NetMeeting product supports application sharing in multiparty calls. Exchange from Microsoft is another example of groupware. This software allows users to set up electronic bulletin boards, schedule group meetings, and use e-mail in a group setting. NetDocuments Enterprise can be used for Web collaboration. The groupware is intended for legal, accounting, and real-estate businesses. A Breakout Session feature allows two people to take a copy of a document to a shared folder for joint revision and work. The software also permits digital signatures and the ability to download and work on shared documents on handheld computers. Other GSS software packages include Collabra Share, OpenMind, and TeamWare. All of these tools can aid in group decision making.

In addition to stand-alone products, GSS software is increasingly being incorporated into existing software packages. Today, some transaction processing and enterprise resource planning packages include collaboration software. Some ERP producers (see Chapter 9), for example, have developed groupware to facilitate collaboration and to allow users to integrate applications from other vendors into the ERP system of programs. Today, groupware can interact with wireless devices. Research In Motion, the maker of Blackberry software, offers mobile communications, access to group information, meeting schedules, and other services that can be directly tied to groupware software and servers.[52] In addition to groupware, GSSs use a number of tools discussed previously, including:

- E-mail and instant messaging (IM)
- Videoconferencing
- Group scheduling
- Project management
- Document sharing

GSS software allows work teams to collaborate and reach better decisions—even if they work across town, in another region, or on the other side of the globe.

(Source: © Mark Richards/Photo Edit.)

GSS Alternatives

Group support systems can take on a number of network configurations, depending on the needs of the group, the decision to be supported, and the geographic location of group members. The frequency of GSS use and the location of the decision makers are two important factors (see Figure 10.18).

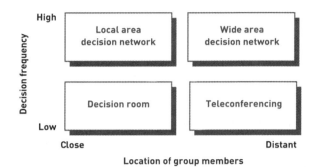

decision room

A room that supports decision making, with the decision makers in the same building, combining face-to-face verbal interaction with technology to make the meeting more effective and efficient.

The Decision Room

The **decision room** is ideal for situations in which decision makers are located in the same building or geographic area and the decision makers are occasional users of the GSS approach. In these cases, one or more decision rooms or facilities can be set up to accommodate the GSS approach. Groups, such as marketing research teams, production management groups, financial control teams, or quality-control committees, can use the decision rooms when needed. The decision room alternative combines face-to-face verbal interaction with technology-aided formalization to make the meeting more effective and efficient. Decision rooms, however, can be expensive to set up and operate. A typical decision room is shown in Figure 10.19.

Figure 10.19

The GSS Decision Room

For group members who are in the same location, the decision room is an optimal GSS alternative. This approach can use both face-to-face and computer-mediated communication. By using networked computers and computer devices, such as project screens and printers, the meeting leader can pose questions to the group, instantly collect their feedback, and, with the help of the governing software loaded on the control station, process this feedback into meaningful information to aid in the decision-making process.

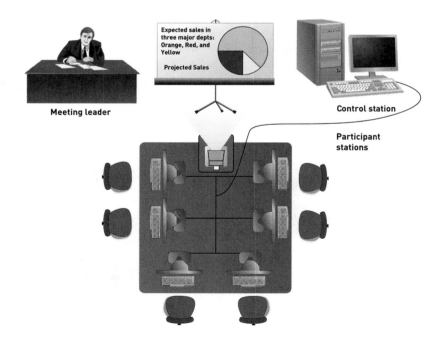

The Local Area Decision Network

The local area decision network can be used when group members are located in the same building or geographic area and under conditions in which group decision making is frequent. In these cases, the technology and equipment for the GSS approach is placed directly into the offices of the group members. Usually this is accomplished via a local area network (LAN).

The Teleconferencing Alternative

Teleconferencing is used when the decision frequency is low and the location of group members is distant. These distant and occasional group meetings can tie together multiple GSS decision-making rooms across the country or around the world. Using long-distance communications technology, these decision rooms are electronically connected in teleconferences and videoconferences. This alternative can offer a high degree of flexibility. The GSS decision rooms can be used locally in a group setting or globally when decision makers are located throughout the world. GSS decision rooms are often connected through the Internet.

The Wide Area Decision Network

The wide area decision network is used when the decision frequency is high and the location of group members is distant. In this case, the decision makers require frequent or constant use of the GSS approach. Decision makers located throughout the country or the world must be linked electronically through a wide area network (WAN). The group facilitator and all group members are geographically dispersed. In some cases, the model base and database are also geographically dispersed. This GSS alternative allows people to work in **virtual workgroups**, where teams of people located around the world can work on common problems.

virtual workgroups
Teams of people located around the world working on common problems.

The Internet is increasingly being used to support wide area decision networks. As discussed in Chapters 7 and 8, a number of technologies, including videoconferencing, instant messaging, chat rooms, and telecommuting, can be used to assist the GSS process. In addition, many specialized wide area decision networks make use of the Internet for group decision making and problem solving.

EXECUTIVE SUPPORT SYSTEMS

Because top-level executives often require specialized support when making strategic decisions, many companies have developed systems to assist executive decision making. This type of system, called an **executive support system (ESS)**, is a specialized DSS that includes all hardware, software, data, procedures, and people used to assist senior-level executives within the organization. In some cases, an ESS, also called an executive information system (EIS), supports decision making of members of the board of directors, who are responsible to stockholders. These top-level decision-making strata are shown in Figure 10.20. See the "Ethical and Societal Issues" feature to find out how these top executives are becoming involved in security for valuable information systems.

executive support system (ESS)
Specialized DSS that includes all hardware, software, data, procedures, and people used to assist senior-level executives within the organization.

An ESS can also be used by individuals at middle levels in the organizational structure. Once targeted at the top-level executive decision makers, ESSs are now marketed to—and used by—employees at other levels in the organization. In the traditional view, ESSs give top executives a means of tracking critical success factors. Today, all levels of the organization share information from the same databases. However, for our discussion, we will assume ESSs remain in the upper-management levels, where they highlight important corporate issues, indicate new directions the company may take, and help executives monitor the company's progress.

CEOs "Called to Action" Regarding Information Security

The U.S. government is pushing CEOs to take responsibility for the nation's critical information infrastructure. "In this era of increased cyber attacks and information security breaches, it is essential that all organizations give information security the focus it requires," said Amit Yoran, Director of the National Cyber Security Division, Information Analysis and Infrastructure Protection (IAIP) within the U.S. Department of Homeland Security. "[By] addressing these cyber and information security concerns, the private sector will not only strengthen its own security, but help protect the homeland as well."

As noted in a report titled *Information Security Governance: A Call to Action*, a group of CEOs observed that information security is too often treated as a technical issue and passed along to the CIO and technical department to handle. But information systems are so important that top executives and boards of directors must become involved and make security an integral part of core business operations. The report was authored by the Corporate Governance Task Force of the National Cyber Security Partnership. This CEO-led task force is responsible for identifying cybersecurity roles and responsibilities within the corporate management structure.

Typical executive support systems provide top-level managers with general corporate information and high-level indicators that assist them in strategic planning. Such systems allow executives to observe trends in corporate statistics and drill down for a more refined view.

But some suggest that CEOs also become more involved in lower-level IS operations. For example, industry observers recommend that the nation's critical infrastructure be protected by forming cybersecurity roles and responsibilities within corporate executive management structures. Also, risk-management and quality-assurance benchmarks and best practices are being developed, and forward-thinking executives are instituting industry metrics for use by companies and auditing firms. In general, executives are urged to fuse security practices into every aspect of IS development and usage.

Part of this effort involves developing best practices and metrics that bring accountability to three key elements of a cybersecurity system: people, processes, and technology. The underlying philosophy is that without cooperation from individual companies and organizations, it is impossible to secure the U.S.'s information infrastructure. Will private corporations cooperate? "There's a significant amount of compliance already," said task force cochairman Arthur Coviello, CEO and president of RSA Security, Inc. "It's hard to imagine that any CEO could not take this as a significant responsibility."

To coax companies to comply, the Department of Homeland Security is establishing an awards program for companies that meet or exceed the security guidelines. Organizations are asked to state on their Web sites that they intend to use the tools developed by the Corporate Governance Task Force to assess their performance and report the results to their boards of directors. Some believe that if there is not widespread voluntary support of the task force's recommendations, the government may need to apply added incentives. Orson Swindle, a member of the Federal Trade Commission, stated that if the task force report fails to get the attention of CEOs and boards of directors, "I have no doubt that some sort of regulation will be passed."

Critical Thinking Questions

1. How might organizations secure sensitive information from cybercriminals? Consider the security concerns and possible solutions regarding data stored in databases and also the important information that is transferred over networks and often printed in reports.

2. How could the government get 100 percent cooperation from private corporations? Is such a goal attainable? If attained, will it guarantee national security? Why or why not?

What Would You Do?

As CEO of Boston Chemical, you have been called on to audit your information systems, identify areas of security concerns, and address the concerns with remedies. The initial audit indicates that much of your information infrastructure is outdated and unable to provide security that measures up to today's standards. An overhaul will cost tens of thousands of dollars and require significant downtime for the company—neither of which the company can afford. In short, it is impossible for you to comply with the recommendations of the task force and stay in business. However, you understand the importance of complying, especially since your company deals in dangerous chemicals.

3. Considering that there are no current laws governing compliance, would it be best to simply ignore the task force recommendation? What other options might there be?

4. What might the government do for businesses such as yours that are unable to comply due to financial constraints?

SOURCES: "Corporate Governance Task Force of the National Cyber Security Partnership Releases Industry Framework," *Entrust*, April 12, 2004, *www.entrust.com*; Dan Verton, "CEOs Urged to Take Control of Cybersecurity," *Computerworld*, April 12, 2004, *www.computerworld.com*; National Cyber Security Partnership (NCSP) Web site, accessed April 25, 2004, *www.cyberpartnership.org*.

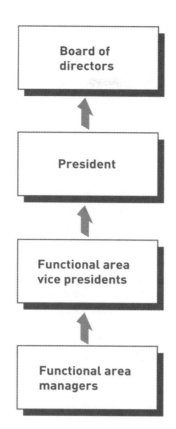

Figure 10.20

The Layers of Executive Decision Making

Executive Support Systems in Perspective

An ESS is a special type of DSS, and, like a DSS, an ESS is designed to support higher-level decision making in the organization. The two systems are, however, different in important ways. DSSs provide a variety of modeling and analysis tools to enable users to thoroughly analyze problems—that is, they allow users to *answer* questions. ESSs present structured information about aspects of the organization that executives consider important. In other words, they allow executives to *ask* the right questions.

The following are general characteristics of ESSs:

- *Tailored to individual executives.* ESSs are typically tailored to individual executives; DSSs are not tailored to particular users. An ESS is an interactive, hands-on tool that allows an executive to focus, filter, and organize data and information.
- *Easy to use.* A top-level executive's most critical resource can be his or her time. Thus, an ESS must be easy to learn and use and not overly complex.
- *Have drill-down abilities.* An ESS allows executives to drill down into the company to determine how certain data was produced. Drilling down allows an executive to get more detailed information if needed.
- *Support the need for external data.* The data needed to make effective top-level decisions is often external—information from competitors, the federal government, trade associations and journals, consultants, and so on. An effective ESS is able to extract data useful to the decision maker from a wide variety of sources, including the Internet and other electronic publishing sources such as legal and public business information from Lexis/Nexis.
- *Can help with situations that have a high degree of uncertainty.* There is a high degree of uncertainty with most executive decisions. Handling these unknown situations using modeling and other ESS procedures helps top-level managers measure the amount of risk in a decision.
- *Have a future orientation.* Executive decisions are future oriented, meaning that decisions will have a broad impact for years or decades. The information sources to support

future-oriented decision making are usually informal—from organizing golf partners to tying together members of social clubs or civic organizations.

- *Are linked with value-added business processes.* Like other information systems, executive support systems are linked with executive decision making about value-added business processes. For instance, executive support systems can be used by car-rental companies to analyze trends.

Capabilities of Executive Support Systems

The responsibility given to top-level executives and decision makers brings unique problems and pressures to their jobs. Following is a discussion of some of the characteristics of executive decision making that are supported through the ESS approach. As you will note, most of these decisions are related to an organization's overall profitability and direction. An effective ESS should have the capability to support executive decisions with components such as strategic planning and organizing, crisis management, and more.

Support for Defining an Overall Vision

One of the key roles of senior executives is to provide a broad vision for the entire organization. This vision includes the organization's major product lines and services, the types of businesses it supports today and in the future, and its overriding goals.

Support for Strategic Planning

strategic planning
Determining long-term objectives by analyzing the strengths and weaknesses of the organization, predicting future trends, and projecting the development of new product lines.

ESSs also support strategic planning. **Strategic planning** involves determining long-term objectives by analyzing the strengths and weaknesses of the organization, predicting future trends, and projecting the development of new product lines. It also involves planning the acquisition of new equipment, analyzing merger possibilities, and making difficult decisions concerning downsizing and the sale of assets if required by unfavorable economic conditions.

Support for Strategic Organizing and Staffing

Top-level executives are concerned with organizational structure. For example, decisions concerning the creation of new departments or downsizing the labor force are made by top-level managers. Overall direction for staffing decisions and effective communication with labor unions are also major decision areas for top-level executives. ESSs can be employed to help analyze the impact of staffing decisions, potential pay raises, changes in employee benefits, and new work rules.

Support for Strategic Control

Another type of executive decision relates to strategic control, which involves monitoring and managing the overall operation of the organization. Goal seeking can be done for each major area to determine what performance these areas need to achieve to reach corporate expectations. Effective ESS approaches can help top-level managers make the most of their existing resources and control all aspects of the organization.

Support for Crisis Management

Even with careful strategic planning, a crisis can occur. Major disasters, including hurricanes, tornadoes, floods, earthquakes, fires, and terrorist activities, can totally shut down major parts of the organization. Handling these emergencies is another responsibility for top-level executives. In many cases, strategic emergency plans can be put into place with the help of an ESS. These contingency plans help organizations recover quickly if an emergency or crisis occurs.

Decision making is a vital part of managing businesses strategically. IS systems such as information and decision support, group support, and executive support systems help employees by tapping existing databases and providing them with current, accurate information. The increasing integration of all business information systems—from TPSs to MISs to DSSs—can help organizations monitor their competitive environment and make better-informed decisions. Organizations can also use specialized business information systems, discussed in the next chapter, to achieve their goals.

SUMMARY

Principle

Good decision-making and problem-solving skills are the key to developing effective information and decision support systems.

Every organization needs effective decision making and problem solving to reach its objectives and goals. Problem solving begins with decision making. A well-known model developed by Herbert Simon divides the decision-making phase of the problem-solving process into three stages: intelligence, design, and choice. During the intelligence stage, potential problems or opportunities are identified and defined. Information is gathered that relates to the cause and scope of the problem. Constraints on the possible solution and the problem environment are investigated. In the design stage, alternative solutions to the problem are developed and explored. In addition, the feasibility and implications of these alternatives are evaluated. Finally, the choice stage involves selecting the best course of action. In this stage, the decision makers evaluate the implementation of the solution to determine whether the anticipated results were achieved and to modify the process in light of new information learned during the implementation stage.

Decision making is a component of problem solving. In addition to the intelligence, design, and choice steps of decision making, problem solving also includes implementation and monitoring. Implementation places the solution into effect. After a decision has been implemented, it is monitored and modified if needed.

Decisions can be programmed or nonprogrammed. Programmed decisions are made using a rule, procedure, or a quantitative method. Ordering more inventory when the level drops to 100 units or fewer is an example of a programmed decision. A nonprogrammed decision deals with unusual or exceptional situations. Determining the best training program for a new employee is an example of a nonprogrammed decision.

Decisions can use optimization, satisficing, or heuristic approaches. Optimization finds the best solution. Optimization problems often have an objective such as maximizing profits given production and material constraints. When a problem is too complex for optimization, satisficing is often used. Satisficing finds a good, but not necessarily the best, decision. Finally, a heuristic is a "rule of thumb" or commonly used guideline or procedure used to find a good decision.

Principle

The management information system (MIS) must provide the right information to the right person in the right fashion at the right time.

A management information system is an integrated collection of people, procedures, databases, and devices that provides managers and decision makers with information to help achieve organizational goals. An MIS can help an organization achieve its goals by providing managers with insight into the regular operations of the organization so that they can control, organize, and plan more effectively and efficiently. The primary difference between the reports generated by the TPS and those generated by the MIS is that MIS reports support managerial decision making at the higher levels of management.

Data that enters the MIS originates from both internal and external sources. The most significant internal sources of data for the MIS are the organization's various TPSs and ERP systems. Data warehouses and data marts also provide important input data for the MIS. External sources of data for the MIS include extranets, customers, suppliers, competitors, and stockholders.

The output of most MISs is a collection of reports that are distributed to managers. These reports include scheduled reports, key-indicator reports, demand reports, exception reports, and drill-down reports. Scheduled reports are produced periodically, or on a schedule, such as daily, weekly, or monthly. A key-indicator report is a special type of scheduled report. Demand reports are developed to provide certain information at a manager's request. Exception reports are automatically produced when a situation is unusual or requires management action. Drill-down reports provide increasingly detailed data about situations.

Management information systems have a number of common characteristics, including producing scheduled, demand, exception, and drill-down reports; producing reports with fixed and standard formats; producing hard-copy and soft-copy reports; using internal data stored in organizational computerized databases; and having reports developed and implemented by IS personnel or end users.

Most MISs are organized along the functional lines of an organization. Typical functional management information systems include financial, manufacturing, marketing, human resources, and other specialized systems. Each system is

composed of inputs, processing subsystems, and outputs. The primary sources of input to functional MISs include the corporate strategic plan, data from the ERP system and TPS, information from supply chain and business transactions, and external sources including the Internet and extranets. The primary output of these functional MISs are summary reports that assist in managerial decision making.

A financial management information system provides financial information to all financial managers within an organization, including the chief financial officer (CFO). Subsystems are profit/loss and cost systems, auditing, and use and management of funds.

A manufacturing MIS accepts inputs from the strategic plan, the ERP system and TPS, and external sources, such as supply chain and business transactions. The systems involved support the business processes associated with the receiving and inspecting of raw material and supplies; inventory tracking of raw materials, work in process, and finished goods; labor and personnel management; management of assembly lines, equipment, and machinery; inspection and maintenance; and order processing. The subsystems involved are design and engineering, master production scheduling and inventory control, process control, and quality control and testing.

A marketing MIS supports managerial activities in the areas of product development, distribution, pricing decisions, promotional effectiveness, and sales forecasting. Subsystems include marketing research, product development, promotion and advertising, and product pricing.

A human resource MIS is concerned with activities related to employees of the organization. Subsystems include human resource planning, personnel selection and recruiting, training and skills inventories, scheduling and job placement, wage and salary administration, and outplacement.

An accounting MIS performs a number of important activities, providing aggregate information on accounts payable, accounts receivable, payroll, and many other applications. The organization's ERP system or TPS captures accounting data, which is also used by most other functional information systems. Geographic information systems provide regional data in graphical form.

Principle

Decision support systems (DSSs) are used when the problems are unstructured.

A decision support system (DSS) is an organized collection of people, procedures, software, databases, and devices working to support managerial decision making. DSS characteristics include the ability to handle large amounts of data; obtain and process data from different sources; provide report and presentation flexibility; support drill-down analysis; perform complex statistical analysis; offer textual and graphical orientations; support optimization, satisficing, and heuristic approaches; and perform what-if simulations and goal-seeking analysis.

DSSs provide support assistance through all phases of the problem-solving process. Different decision frequencies also require DSS support. An ad hoc DSS addresses unique, infrequent decision situations; an institutional DSS handles routine decisions. Highly structured problems, semistructured problems, and unstructured problems can be supported by a DSS. A DSS can also support different managerial levels, including strategic, tactical, and operational managers. A common database is often the link that ties together a company's TPS, MIS, and DSS.

The components of a DSS are the database, model base, dialogue manager, and a link to external databases, the Internet, the corporate intranet, extranets, networks, and other systems. The database can use data warehouses and data marts. A data-driven DSS primarily performs qualitative analysis based on the company's databases. Data-driven DSSs tap into vast stores of information contained in the corporate database, retrieving information on inventory, sales, personnel, production, finance, accounting, and other areas. Data mining is often used in a data-driven DSS. The model base contains the models used by the decision maker, such as financial, statistical, graphical, and project-management models. A model-driven DSS primarily performs mathematical or quantitative analysis. Model management software (MMS) is often used to coordinate the use of models in a DSS. The dialogue manager provides a dialogue management facility to assist in communications between the system and the user. Access to other computer-based systems permits the DSS to tie into other powerful systems, including the TPS or function-specific subsystems.

Principle

Specialized support systems, such as group support systems (GSSs) and executive support systems (ESSs), use the overall approach of a DSS in situations such as group and executive decision making.

A group support system (GSS), also called a *computerized collaborative work system*, consists of most of the elements in a DSS, plus software to provide effective support in group decision-making settings. GSSs are typically easy to learn and use and can offer specific or general decision-making support. GSS software, also called *groupware*, is specially designed to help generate lists of decision alternatives and perform data analysis. These packages let people work on joint documents and files over a network.

The frequency of GSS use and the location of the decision makers will influence the GSS alternative chosen. The decision room alternative supports users in a single location that meet infrequently. Local area networks can be used when group members are located in the same geographic area and users meet regularly. Teleconferencing is used when decision frequency is low and the location of group members is distant. A wide area network is used when the decision frequency is high and the location of group members is distant.

Executive support systems (ESSs) are specialized decision support systems designed to meet the needs of senior management. They serve to indicate issues of importance to the organization, indicate new directions the company may take, and help executives monitor the company's progress. ESSs are typically easy to use, offer a wide range of computer resources, and handle a variety of internal and external data.

In addition, the ESS performs sophisticated data analysis, offers a high degree of specialization, and provides flexibility and comprehensive communications abilities. An ESS also supports individual decision-making styles. Some of the major decision-making areas that can be supported through an ESS are providing an overall vision, strategic planning and organizing, strategic control, and crisis management.

CHAPTER 10: SELF-ASSESSMENT TEST

Good decision-making and problem-solving skills are the key to developing effective information and decision support systems.

1. The first stage of the decision making process is:

    a. initiation stage
    b. intelligence stage
    c. design stage
    d. choice stage

2. Problem solving is one of the stages of decision making. True or False?
3. The final stage of problem solving is _____.
4. A decision that inventory should be ordered when inventory levels drop to 500 units is an example of

    a. synchronous decision
    b. asynchronous decision
    c. nonprogrammed decision
    d. programmed decision

5. A(n) _____ model will find the best solution, usually the one that will best help the organization meet its goals.
6. A satisficing model is one that will find a good problem solution, but not necessarily the best problem solution. True or False?

The management information system (MIS) must provide the right information to the right person in the right fashion at the right time.

7. What summarizes the previous day's critical activities and is typically available at the beginning of each workday?

    a. key-indicator report
    b. demand report
    c. exception report
    d. database report

8. The uses and management of funds is a subsystem of

    a. the marketing MIS
    b. the financial MIS
    c. the manufacturing MIS
    d. the auditing MIS

9. Another name for the _____ MIS is the personnel MIS because it is concerned with activities related to employees and potential employees of the organization.

Decision support systems (DSSs) are used when the problems are unstructured.

10. The focus of a decision support system is on decision-making effectiveness when faced with unstructured or semistructured business problems. True or False?
11. The process of determining the problem data required for a given result is called _____ analysis.
12. What component of a decision support system allows decision makers to easily access and manipulate the DSS and to use common business terms and phrases?

    a. the knowledge base
    b. the model base
    c. the dialogue manager
    d. the expert system

Specialized support systems, such as group support systems (GSSs) and executive support systems (ESSs), use the overall approach of a DSS in situations such as group and executive decision making.

13. In a GSS, what approach or technique uses voting to make the final decision in a group setting?

    a. group consensus
    b. groupthink
    c. nominal group technique
    d. delphi

14. A type of software that helps with joint work group scheduling, communication, and management is called _____.
15. The local area decision network is the ideal GSS alternative for situations in which decision makers are located in the same building or geographic area and the decision makers are occasional users of the GSS approach. True or False?
16. A(n) _____ supports the actions of members of the board of directors, who are responsible to stockholders.

CHAPTER 10: SELF-ASSESSMENT TEST ANSWERS

(1) b (2) False (3) monitoring (4) d (5) optimization (6) True (7) a (8) b (9) human resource (10) True (11) goal seeking (12) c (13) c (14) groupware or workgroup software (15) False (16) executive information system (EIS)

KEY TERMS

accounting MIS 479
ad hoc DSS 483
auditing 466
brainstorming 488
choice stage 456
computer-assisted manufacturing (CAM) 470
computer-integrated manufacturing (CIM) 471
cost center 466
decision room 490
decision-making phase 455
delphi approach 488
demand report 462
design stage 456
dialogue manager 484
drill-down report 462
economic order quantity (EOQ) 470
exception report 462
executive support system (ESS) 491
external auditing 466
financial MIS 465

flexible manufacturing system (FMS) 471
geographic information system (GIS) 480
goal-seeking analysis 481
group consensus approach 488
group support system (GSS) 487
heuristics 458
highly structured problems 483
human resource MIS 476
implementation stage 456
institutional DSS 483
intelligence stage 455
internal auditing 466
just-in-time (JIT) inventory approach 470
key-indicator report 461
manufacturing resource planning (MRPII) 470
marketing MIS 472
material requirements planning (MRP) 470

model base 486
model management software (MMS) 486
monitoring stage 456
nominal group technique 488
nonprogrammed decisions 457
optimization model 457
problem solving 456
profit center 466
programmed decisions 456
quality control 471
reorder point (ROP) 470
revenue center 466
satisficing model 458
scheduled report 461
semistructured or unstructured problems 483
simulation 481
strategic planning 494
virtual workgroups 491
what-if analysis 481

REVIEW QUESTIONS

1. What is a satisficing model? Describe a situation when it should be used.
2. Give an example of an optimization model.
3. What are the basic kinds of reports produced by an MIS?
4. What guidelines should be followed in developing reports for management information systems?
5. What are the functions performed by a financial MIS?
6. Describe the functions of a manufacturing MIS.
7. What is a human resource MIS? What are its outputs?
8. List and describe some other types of MISs.
9. What are the stages of problem solving?
10. What is the difference between decision making and problem solving?
11. What is a geographic information system?

12. Describe the difference between a structured and an unstructured problem and give an example of each.
13. Define *decision support system*. What are its characteristics?
14. Describe the difference between a data-driven and a model-driven DSS.
15. What is the difference between what-if analysis and goal-seeking analysis?
16. What are the components of a decision support system?
17. Describe four models used in decision support systems.
18. State the objective of a group support system (GSS) and identify three characteristics that distinguish it from a DSS.
19. Identify three group decision-making approaches often supported by a GSS.
20. What is an executive support system? Identify three fundamental uses for such a system.

DISCUSSION QUESTIONS

1. Select an important problem you had to solve during the last two years. Describe how you used the decision-making and problem-solving steps discussed in this chapter to solve the problem.

2. What is the relationship between an organization's enterprise resource planning and transaction processing systems and its management information systems? What is the primary role of management information systems?

3. How can management information systems be used to support the objectives of the business organization?

4. Describe a financial MIS for a *Fortune* 1000 manufacturer of food products. What are the primary inputs and outputs? What are the subsystems?

5. How can a strong financial MIS provide strategic benefits to a firm?

6. Why is auditing so important in a financial MIS? Give an example of an audit that failed to disclose the true nature of the financial position of a firm. What was the result?

7. What is the difference in roles played by an internal auditing group and an external auditing group?

8. You have been hired to develop a management information system and a decision support system for a manufacturing company. Describe what information you would include in printed reports and what information you would provide using a screen-based decision support system.

9. Pick a company and research its human resource management information system. Describe how its system works. What improvements could be made to its human resource MIS?

10. You have been hired to develop a DSS for a car company such as Ford or GM. Describe how you would use both data-driven and model-driven DSSs.

11. Imagine that you are the CFO for a service organization. You are concerned with the integrity of the firm's financial data. What steps might you take to ascertain the extent of problems?

12. What functions do decision support systems support in business organizations? How does a DSS differ from a TPS and an MIS?

13. How is decision making in a group environment different from individual decision making, and why are information systems that assist in the group environment different? What are the advantages and disadvantages of making decisions as a group?

14. You have been hired to develop group support software. Describe the features you would include in your new GSS software.

15. The use of ESSs should not be limited to the executives of the company. Do you agree or disagree? Why?

16. Imagine that you are the vice president of manufacturing for a *Fortune* 1000 manufacturing company. Describe the features and capabilities of your ideal ESS.

PROBLEM-SOLVING EXERCISES

1. You have been asked to select GSS software to help your company make better decisions to market a new product. Using the Internet, find and investigate three types of software that your company could use for collaborative decision making. Use your word processor to describe what you found and the advantages and disadvantages of each GSS software package.

2. Review the summarized consolidated statement of income for the manufacturing company whose data is shown here. Use graphics software to prepare a set of bar charts that shows the data for this year compared with the data for last year.

   a. This year operating revenues increased by 3.5 percent, while operating expenses increased 2.5 percent.
   b. Other income and expenses decreased to $13,000.
   c. Interest and other charges increased to $265,000.

Operating Results (in millions)

Operating Revenues	$2,924,177
Operating Expenses (including taxes)	2,483,687
Operating Income	440,490
Other Income and Expenses	13,497
Income before Interest and Other Charges	453,987
Interest and Other Charges	262,845
Net Income	191,142
Average Common Shares Outstanding	147,426
Earnings per share	1.30

If you were a financial analyst tracking this company, what detailed data might you need to perform a more complete analysis? Write a brief memo summarizing your data needs.

3. As the head buyer for a major supermarket chain, you are constantly being asked by manufacturers and distributors to stock their new products. Over 50 new items are

introduced each week. Many times these products are launched with national advertising campaigns and special promotional allowances to retailers. To add new products, the amount of shelf space allocated to existing products must be reduced or items must be eliminated altogether.

Develop a marketing MIS that you can use to estimate the change in profits from adding or deleting an item from inventory. Your analysis should include input such as estimated weekly sales in units, shelf space allocated to stock an item (measured in units), total cost per unit, and sales price per unit. Your analysis should calculate total annual profit by item and then sort the rows in descending order based on total annual profit.

TEAM ACTIVITIES

1. Have your team members select two industries. Using only the Internet, have your team work together to make a group decision to develop a list of the 10 best companies in each industry. Your team may be asked to prepare a report on your decision and how difficult it was to use only the Internet for communication.
2. Have your team make a group decision about how to solve the most frustrating aspect of college or university life. Appoint one or two members of the team to disrupt the meeting with negative group behavior. After the meeting, have your team describe how to prevent this negative group behavior. What GSS software features would you suggest to prevent the negative group behavior your team observed?
3. Imagine that you and your team have decided to develop an ESS software product to support senior executives in the music recording industry. What are some of the key decisions these executives must make? Make a list of the capabilities that such a system must provide to be useful. Identify at least six sources of external information that will be useful to its users.

WEB EXERCISES

1. Most companies typically have a number of functional MISs, such as finance. Find the site of two finance companies, such as a bank or a brokerage company. Compare these sites. Which one do you prefer? How could these sites be improved? (Hint: If you are having trouble, try Yahoo! It should have a listing for Business and Economy on its home page. From there you can go to Companies and then Finance. There will be several menu choices from there.) You may be asked to develop a report or send an e-mail message to your instructor about what you found.
2. Use the Internet to explore two or more software packages that can be used to make group decisions easier. Summarize your findings in a report.
3. Software, such as the Excel spreadsheet, is often used to find an optimal solution to maximize profits or minimize costs. Search the Internet using Yahoo!, Google, or another search engine to find other software packages that offer optimization features. Write a report describing one or two of the optimization software packages. What are some of the features of the package?

CAREER EXERCISES

1. What decisions are critical for success in a career that interests you? What specific types of reports could help you make better decisions on the job? Give three specific examples.
2. How often do you think you will be involved with group decision making for your career of choice? How can groupware be used to help your team make better decisions? What specific groupware would you like to use on the job?

VIDEO QUESTIONS

Watch the video clip **Army's Virtual World** and answer these questions:

1. The "war on terror" has ushered us into a new form of battle where individual soldiers rather than generals and platoon leaders are often responsible for making important decisions in the field. Describe the new system presented in this video and how it is being used to train soldiers to quickly make correct decisions.

2. How might this method of using virtual simulations to train decision makers be applied to more typical business-related decision support systems.

CASE STUDIES

Case One

Sensory Systems Provide Better-Tasting Products for Kraft

Have you ever wondered what makes Oreo cookies so delicious? The crunch of the cookie balances perfectly with the cool semisoft icing. The bittersweet dark chocolate flavor is offset by the sweet creamy center in perfect harmony. For those who enjoy extremes, the cookie can be twisted apart, allowing the icing to be scraped off and consumed by itself for a sugary rush followed by a mild cookie chaser. For some, Oreos may be the perfect food product.

Who is responsible for this heavenly confection? How is it that every Oreo you eat is as equally delicious as the last one or the one you ate several years ago? As you might have guessed, information and decision support systems play a large role in the production of this perfect cookie, as well as all of the other Kraft products that you enjoy.

Kraft produces hundreds of popular food products, such as Oreos, Jell-O, Post breakfast cereals, Ritz crackers, Tombstone pizza, and Grey Poupon mustard, to name just a few. Kraft products cover the full range of flavors, textures, and eating experiences. Kraft brands hold the top position in 21 of the 25 top food categories in the United States and 21 of the top 25 country categories internationally.

Kraft designs and tests its food sensations in a large lab facility in Glenview, Illinois. The product developers and sensory technologists that work there have food development and production down to a science. To ensure consistent flavor and appearance, Kraft tests its foods throughout the manufacturing process and assigns numerical measurements that quantify the flavor, color, aroma, and other attributes of each product. Concepts that most people express with words such as "chewy," "sweet," "crunchy," and "creamy" are assigned precise definitions and numerical scales to standardize product information.

The characteristics of an Oreo cookie, and all other Kraft products, are stored as a series of numbers and attributes in SENECA, the Sensory and Experimental Collection Application. SENECA assists Kraft in getting more out of its sensory data than ever before.

"SENECA has a database of information that includes everything you might be interested in, in terms of sensory tests," says Beth Knapp, lead systems developer at Kraft. Data stored in SENECA is collected through any of three diverse testing methods employed by Kraft:

- Discrimination testing compares two or more products.
- Descriptive testing employs professional sensory panelists to evaluate all aspects of a product from texture to taste.
- Consumer testing measures the personal responses and opinions of the general public.

Kraft's SENECA application takes the collected data from all of these tests and makes it available for analysis and reuse. The system builds models, histories, and trends based on consumer testing and then evaluates product changes based on discrimination and descriptive testing. Reports are generated to inform Kraft senior managers of products that are well received or perhaps falling out of favor. The system also allows them to drill down into the information to discover the qualities of the product that may be responsible. Based on this information, Kraft is able to capitalize on the qualities that people find most favorable.

SENECA is a powerful tool that provides Kraft with insight into its product line development. Kraft also employs a second information system to guide its manufacturing process. The process variation reduction (PVR) system ensures consistent flavor and appearance for every Kraft product—to make sure that every Ritz cracker tastes as good as the last. "Variation reduction also creates higher average quality," says Knapp, "which means happier customers and more repeat sales."

Kraft's PVR application boosts production, allowing the company to produce more salable cookies, crackers, and other food items per hour. The PVR application evaluates every manufacturing procedure, from recipe instructions to cookie dough shapes and sizes. Reports identify steps in the production process that create excess variation and helps executives improve processes for those areas.

What's more, the PVR system has been deployed over the Internet so that all of Kraft's manufacturing units can take advantage of it. According to Knapp, the PVR process will help Kraft increase revenues and reduce costs. "Process variation reduction has the potential to generate significant cost savings for each facility implementing these procedures," she says.

Ultimately, the SENECA and PVR applications will help ensure consistent, high-quality products for Kraft customers. "To make the ideal cookie, it has to have good shelf life and an excellent flavor," says Keith Eberhardt, a statistician at Kraft. "And if you keep it closer to the ideal, then the product that reaches customers will be even better."

Discussion Questions

1. Why is it useful for food manufacturers to represent the attributes of food products—taste, color, texture, aroma—using a numeric system?

2. Once a winning product has been designed, why is it important to mass-produce the product consistently every time?

Critical Thinking Questions

3. Can the development and manufacturing methods used at Kraft be applied to some other industrial product with equal success? Provide an example and rationale.

4. Could some products benefit from inconsistencies in production? Name some. Any food items? Why?

SOURCES: SAS Success Stories Web site, accessed April 26, 2004, *www.sas.com/success/kraft.html*; Kraft Web site, accessed April 26, 2004, *www.kraft.com*.

Case Two
The Ups and Downs of the Ladder Industry

Werner Co, the world's largest ladder manufacturer, has had its share of ups and downs. The "downs" occurred in late 2003 when Werner learned that its biggest customer, the Home Depot, was looking overseas for lower-priced ladder suppliers. David Conn, Werner's director of corporate logistics, explained that Werner initially bid against the overseas companies on a few orders that had not yet been filled but decided that it "did not support our corporate goals."

Having the rug pulled out from under it, Werner needed to react quickly to ramp down production. Thanks to its production planning and reporting system, Werner was able to turn a bad situation into new opportunities. "We thought ladders were safe from [competition coming from] over the ocean," said Bill Rippin, vice president of supply chain at Werner. "If we didn't have this stuff [information systems], it would have been a more difficult time."

Reacting to unexpected, large swings in demand is no small task for a company with revenues of $538 million per year. All of the links in the supply chain are affected. To cover all bases and calculate the best reaction requires special information and decision support systems. Werner's production and distribution planning system is based on demand-forecasting and distribution-planning applications developed by BT Smith and Associates in Butler, Pennsylvania. Once processed, demand data is extracted from the applications and imported into Microsoft Access and Excel spreadsheets for end users.

Using Excel, Werner is able to create what-if scenarios and calculate production schedules. The production plans are then fed into its manufacturing execution system, which currently includes a mix of applications developed by Mapics Inc. and J.D. Edwards & Co. This collection of systems and software provides Werner's management team with the information it needs to redirect the company through tough periods. Werner used the system to implement cost reductions, to decide which production lines to take off-line, and to shift manufacturing and distribution responsibilities among facilities. During the Home Depot phaseout, production planners and business managers at Werner were also able to use the system to prevent any service disruptions to Werner's other customers.

Werner's business savvy and cool handling of the Home Depot phaseout impressed others in the industry and earned it an important new ally. Shortly after the Home Depot pulled out, Werner struck a strategic alliance with Lowe's, the Home Depot's biggest competitor. Lowe's feels that it can benefit from advertising Werner's made-in-the-USA, high-quality reputation. "Whether getting a step up on a household project or raising the roof on a jobsite, Lowe's is committed to offering the highest-quality products for our customers," said Laurie Randolph, Lowe's vice president of merchandising. "Werner's 50 years of innovative design of top-quality climbing products and its commitment to safety are exactly the qualities that will help us best serve our customers."

Werner has now become Lowe's vendor of choice for ladders. Needless to say, Werner is delighted with its good fortune. "This alliance will also enhance the value of the Werner brand and its number-one position in the climbing-products industry, while also supporting the interests of our employees and other key constituencies," explained Dennis G. Heiner, Werner's president and chief executive officer. Meanwhile, the Werner management team is busy consulting its production and distribution planning system to see how to best support the new increase in demand. Perhaps they should invest in some additional production lines.

Discussion Questions

1. How did Werner's production and distribution planning system assist the company when production needed to be ramped down?
2. What types of what-if scenarios do you think Werner ran to determine how to react to the drop in demand?

Critical Thinking Questions

3. If demand for Werner ladders unexpectedly dropped by 50 percent, how would it affect each of the links in its value chain? (Refer to Figure 2.2.)

4. How might Werner handle its workforce under the duress of a massive drop in demand? How might an information system assist the company in finding alternatives to laying off employees?

SOURCES: Marc L. Songini, "Ladder Maker Uses Supply Chain Tools to Climb Out of Sales Hole," *Computerworld*, April 5, 2004, *www.computerworld.com*; Werner Co. Web site, accessed April 27, 2004, *www.wernerladder.com*; "Lowe's and Werner Ladder Announce Strategic Alliance," *PR Newswire*, December 24, 2003, *www.lexis-nexis.com*.

Case Three

Sanoma Magazines Follows Key Performance Indicators to Success

Sanoma Magazines ranks as the fourth largest European magazine publisher, with approximately 230 magazines published in ten countries. It is the market leader in consumer magazines in Finland, Belgium, the Czech Republic, Hungary, and the Netherlands. Its net sales totaled €1028,4 million in 2003, and it employs about 3,900 people. Sanoma Magazines is headquartered in Amsterdam.

Sanoma Magazines' Belgium division found itself facing some tough competition. It set about finding a more efficient system for reporting and analysis to lead to faster, better-informed decision making. The company's two business units, advertising and reader sales, each had its own production center and set of software applications. Since the two systems worked from separate data sources, management found it difficult, if not impossible, to track and understand information between departments and across the entire organization.

Yves Gilbert, senior systems designer at Sanoma Magazines Belgium described it this way: "We had requests for reports coming and going in all directions, and our 30-person IT staff had to handle them all. This not only created bottlenecks with IT and frustration for everyone, but it also led to reporting discrepancies and an overall lack of confidence in our reporting process." An information system that provides inaccurate information is more dangerous than no information system at all. And one that employees don't trust is useless.

The IT management team at Sanoma Magazines Belgium set out to correct the problem by asking its executives and analysts what exactly it needed an information and decision support system to provide in order to get the broader, big-picture view of the business that they desired. "Sanoma needed to define lists of key performance indicators (KPIs) that would help managers see and assess changes over time, understand the economic climate, foresee events, make and justify decisions based on accurate metrics, and analyze the impact of actions," explains Gilbert.

Sanoma Magazines Belgium chose BusinessObjects business intelligence software to build its solution. The software applications that are included in the solution are the following:

- BusinessObjects Application Foundation and BusinessObjects Analytics, an integrated suite of enterprise analytic applications, to assess the market more accurately and gather information on reader and advertiser relations.
- BusinessObjects Data Integrator to transfer corporate data sources to data marts and operate Sanoma Magazines' data flow.

Together, these applications allowed Sanoma Magazines Belgium to centralize corporate data and run high-level analytical functions that make use of key indicators to let managers know the state of the company at any given moment.

All of the units within the organization are finding the BusinessObjects reporting capabilities extremely valuable. The advertising department uses BusinessObjects Analytics to create individualized marketing programs. Employees can then drill down into a global list of KPIs for more information and more detailed reports to increasingly refine their analysis. The sales force uses the system to detect changes in the competitive environment and critical customer accounts and act accordingly. Marketing can analyze the performance of the company's magazines, pricing strategy, and marketing campaigns. They can also monitor all of these indicators over a period of time to make comparisons from one year to the next and take five-year averages.

The human resource department uses BusinessObjects HR Intelligence (Workforce Analytics module) to help manage staff, profiles, completed training programs, and costs. The finance management department uses BusinessObjects Finance Intelligence (Revenue Cycle Analytics module) to analyze expenses by cost center, as well as monitor payment periods, unpaid bills, cash generated, budgets allocated, and budgets spent.

Today, users analyzing their business activity have a coherent picture of the company and a single version of the truth. Everyone is working from the same base, with the same tool, and using the same method for analyzing business.

Discussion Questions

1. What role do key performance indicators (KPIs) play in the decision-making process? How do they assist managers in taming information overload?

2. What advantage did Sanoma Magazines Belgium gain when it centralized its data source and standardized its systems?

Critical Thinking Questions

3. How did Sanoma Magazines Belgium benefit from adopting an off-the-self solution rather than developing its own information system from scratch?

4. How have information systems, such as the one used at Sanoma Magazines Belgium, affected the magazine industry in terms of competition, workforce composition, and effectiveness?

SOURCES: Business Objects "Customers in the Spotlight" Web site, accessed April 26, 2004, *www.businessobjects.com/customers/spotlight/sanoma.asp*; Sanoma Magazines Web site, accessed April 27, 2004, *www.sanomamagazines.fi.*

NOTES

Sources for the opening vignette: "Sake Producer Provides Old-World Taste and New-World Competitive Edge," *Business Objects Success Stories*, accessed April 25, 2004, *www.businessobjects.com/customers/spotlight/kikumasamune.asp*; Kiku-Musamune Web site, accessed April 25, 2004, *www.kikumasamune.com/*; Business Objects Business Intelligence Web page, accessed April 25, 2004, *www.businessobjects.com/products/bistandardization/default.asp.*

1. Mills, P., et al., "Scheduling Appointments at Trade Events for the Australian Tourist Commission," *Interfaces*, May-June, 2003, p. 12.
2. Blakeley, Fred, et al., "Optimizing Periodic Maintenance," *Interfaces*, January, 2003, p. 67.
3. Savage, Marcia, "Trend Micro Debuts Heuristical Antispam Solution," *Asia Computer Weekly*, March 17, 2003, *www.asiacomputerweekly.com.*
4. Samuelson, Douglas, "The Netwar in Iraq," *OR/MS Today*, June 2003, p. 20.
5. MacSweeney, Greg, "ProvWash Removes Paper," *Insurance and Technology*, February 1, 2004, p. 12.
6. King, Julia, "Sealing the Deal and Collecting Their Due," *Computerworld*, March 31, 2003, p. 52.
7. Songini, Marc, "Boehringer Cures Slow Reporting," *Computerworld*, July 21, 2003, p. 30.
8. Draper, Heather, "Xcel Site Turns Up the Heat on Waste," *Rocky Mountain News*, March 3, 2003, p. 1B.
9. Giegerich, Steve, "Putting Professors to the Test," *Summit Daily News*, February 20, 2003, p. B1.
10. Thomas, Daniel, "Pub Chain Cuts a Better Deal," *Computer Weekly*, January 27, 2004, p. 10.
11. Nelson, Kristi, "Referral System Leads to Connectivity," *Bank System & Technology*, March 1, 2003, p. 16.
12. Warren, Peter, "Software Could Aid Anti-Terrorism Fight," *Computing*, January 15, 2004, p. 1.
13. Kelly, Kate, "Big Board Chief Mulls Changing Trading System," *The Wall Street Journal*, January 30, 2004, p. C1.
14. Huber, Nick, "FSA to Target Stock Market Abuse," *Computer Weekly*, January 20, 2004, p. 14.
15. Mollenkamp, Carrick, "HealthSouth Accounting Woes Grow to as Much as $4.6 Billion," *The Wall Street Journal*, January 21, 2004, p. B2.
16. Nessbaum, Bruce, "Harmony and Belly Dancing at Davos," *Business Week*, February 9, 2004, p. 38.
17. Hamblen, Matt, "Aircraft Maker Turns to Sourcing Software for New Military Plane," *Computerworld*, December 1, 2003, p. 7.
18. Prahalad, C.K., et al., "Adding Customers to the Design Team," *Business Week*, March 1, 2004, p. 22.
19. Roe, Andrew, "Building Digitally Provides Schedule, Cost Efficiencies," *Construction Management*, Vol. 24B, No. 7, p. 29.
20. Brown, Stuart, "Toyota's Global Body Shop," *Fortune*, February 9, 2004, p. 120B.
21. Ellison, Carol, "Unwire Your Sales Force," *PC Magazine*, September 2, 2003, p. 34.
22. Leon, Mark, "True-Blue Customers," *Computerworld*, August 11, 2003, p. 37.
23. Gentle, Michael, "CRM: Ready or Not," *Computerworld*, August 18, 2003, p. 40.
24. Breskin, Ira, "Rust Belt CRM," *Computerworld*, December 15, 2003, p. 42.
25. Staff, "Tricks of the Trade," *Sales & Marketing Management*, January 2004, p. 27.
26. Wells, Melanie, "In Search of the Buy Button," *Forbes*, September 1, 2003, p. 62.
27. Kirkpatrick, David, "Drowning in Media?" *Fortune*, July 21, 2003, p. 162.
28. Veiga, Alex, "Popping Open the Lid on CD Marketing," *The Wall Street Journal*, July 1, 2003, p. 20B.
29. Acohido, Byron, "Rich Media Enriching PC Ads," *USA Today*, February 25, 2004, p. 3B.
30. Mangalindan, Mylene, "After a Wave of Disappointments, The Web Lures Back Advertisers," *The Wall Street Journal*, February 25, 2004, p. A1.
31. Bennett, Julie, "Scientific Hiring Strategies," *The Wall Street Journal*, February 10, 2004, p. B7.
32. Staff, "Comprehensive Software Package Launched," *Money Management*, January 1, 2004, *www.ftadviser.com.*
33. Staff, "GIS Software Market Sees Dynamic Growth," *Electric Light & Power*, February, 2003, p. 6.
34. Mitchell, Robert, "Web Services Put GIS on the Map," *Computerworld*, December 15, 2003, p. 30.
35. Brewin, Bob, "Feds Test Network to Send Alerts," *Computerworld*, March 31, 2003, p. 19.
36. Bulkeley, William, "Diabetes Web Site to Provide Customized Treatment Plans," *The Wall Street Journal*, May 14, 2003, p. D3.
37. Goldsmith, Charles, "Reed Elsevier Feels Resistance to Web Pricing," *The Wall Street Journal*, January 10, 2004, p. B1.
38. Gomes, Lee, "How High-Tech Games Can Fail to Simulate," *The Wall Street Journal*, March 31, 2003, p. B1.

39. So, Kut, et al., "Models for Improving Team Productivity at the Federal Reserve Bank," *Interfaces*, March 2003, p. 25.

40. Brown, Stuart, "Toyota's Global Body Shop," *Fortune*, February 9, 2004, p. 120B.

41. Swamnathan, J., "Decision Support for Allocating Scarce Drugs," *Interfaces*, March 2003, p. 1.

42. Grimes, Seth, "Autonomic Computing—Major Vendors are Applying Decision Support Techniques to Service-Centric Computing," *Intelligent Enterprise*, November 15, 2002, p. 18.

43. Tan, Kopin, "Technology Transforms Options Traders," *The Wall Street Journal*, April 30, 2003, p. C14.

44. Nash, Emma, "Travel Firm Uses BI Tools to Compete," *Computing*, January 29, 2004, p. 11.

45. Barrett, Amy, "Feeding the Pipeline," *Business Week*, May 12, 2003, p. 78.

46. Thomas, Daniel, "Software Improves AC Milan's Game," *Computing*, January 29, 2004, p. 6.

47. Melymuka, Kathleen, "Swarming Technology Helps Widely Dispersed Experts," *Computerworld*, July 28, 2003, p. 35.

48. Strassbery, Dan, "Software Facilitates Sharing of Engineering Calculations," *EDN*, March 6, 2003, p. 17.

49. Anthes, Gary, "Smart Rooms," *Computerworld*, August 4, 2003, p. 29.

50. Tavana, Madjid, "CROSS for Evaluating and Prioritizing Advanced-Technology Projects at NASA," *Interfaces*, May-June 2003, p. 40.

51. Schultz, Keith, "Lotus Notes and Domino 6," *New Architect*, January 1, 2003, p. 40.

52. Dornan, Andy, "Mobile E-Mail," *Network Magazine*, February 1, 2003, p.34.

CHAPTER · 11 ·

Specialized Business Information Systems:

Artificial Intelligence, Expert Systems, Virtual Reality, and Other Specialized Systems

PRINCIPLES

- **Artificial intelligence systems form a broad and diverse set of systems that can replicate human decision making for certain types of well-defined problems.**

- **Expert systems can enable a novice to perform at the level of an expert but must be developed and maintained very carefully.**

- **Virtual reality systems have the potential to reshape the interface between people and information technology by offering new ways to communicate information, visualize processes, and express ideas creatively.**

- **Specialized systems can help organizations and individuals achieve their goals.**

LEARNING OBJECTIVES

- Define the term *artificial intelligence* and state the objective of developing artificial intelligence systems.

- List the characteristics of intelligent behavior and compare the performance of natural and artificial intelligence systems for each of these characteristics.

- Identify the major components of the artificial intelligence field and provide one example of each type of system.

- List the characteristics and basic components of expert systems.

- Identify at least three factors to consider in evaluating the development of an expert system.

- Outline and briefly explain the steps for developing an expert system.

- Identify the benefits associated with the use of expert systems.

- Define the term *virtual reality* and provide three examples of virtual reality applications.

- Discuss examples of specialized systems for organizational and individual use.

INFORMATION SYSTEMS IN THE GLOBAL ECONOMY
AMAZON.COM, UNITED STATES

Amazon Leverages Artificial Intelligence Against Fraud

The amount of information stored in and flowing through many of today's information systems has become so massive that keeping up with it is beyond human capacity. Because computers are ideally suited for quickly handling large amounts of data, many companies are designing computer programs to automate the management and interpretation of data. For these programs to be most effective, they need to be able to think and interpret trends in the data as a human being would—but much, much faster.

The ability of a computer to simulate human intelligence is referred to as *artificial intelligence,* or *AI.* AI is a vast area of computer science with many subcategories. One area in which AI is paying off is in fraud detection. Credit card fraud is a serious concern for Internet retailers. The anonymous transactions occurring on the Internet are drawing criminals of all ranks, from one-time hackers to organized crime, which tests the market's boundaries. Since most online transactions are paid for with credit cards, detecting and preventing credit card fraud has become a priority for retailers, which bear the brunt of the financial loss.

For large retailers such as Amazon.com, losses to credit card fraud are substantial. Amazon.com has 35 million customers who access its products through five Web sites adapted for different countries and languages: *www.amazon.com*, *www.amazon.fr*, *www.amazon.co.uk*, *www.amazon.de*, and *www.amazon.co.jp*. Amazon turned to SAS Institute to develop the foundation for its state-of-the-art fraud-detection system.

Fraud-detection techniques are not typically publicized—the less people know about them, the more effective they are. Even Amazon won't fully disclose the details of its AI system. The company is happy to disclose, however, the effectiveness of its new AI fraud-detection system. According to Jaya Kolhatkar, Amazon.com's director of fraud detection, the new system greatly reduced the cases of fraud at Amazon. In the first six months of the system's use, fraud rates were halved.

The system developed for Amazon uses classic AI techniques, such as decision trees and neural networks, to analyze each transaction. The neural networks work within the system to analyze patterns in the data to "learn" which patterns represent fraud—much the same way that human beings learn. The system works on multiple levels. One portion of the system analyzes transactions as they occur, while the credit card number is being approved. Another subsystem crunches data in Amazon's huge transaction database, looking for fraudulent activities in past transactions.

"Fraudsters generally follow similar patterns of behavior," says Kolhatkar. "That makes it easier to detect fraud because you can look for corresponding patterns in transaction and customer data." For example, Amazon knows that fraudsters tend to purchase goods—such as electronics—that they can dispose of easily. Also, they do not ship the goods to the same address that is used for billing, so an order not shipped to the billing address might be an indication of a fraudulent transaction. They also tend to use the fastest possible shipping

method. Of course, use of these features does not mean that fraud has definitely occurred, but combined with other indicators, they would be pointers to follow up on.

Amazon's fraud-detection system analyzes the behavioral patterns of fraudsters and builds predictive scores that indicate the likelihood of fraudulent behavior. "We run these scores against the customer database," says Kolhatkar. "We then use SAS to prioritize the results. Obviously, we have to investigate a case of potential fraud very thoroughly before beginning legal action, so we prioritize the results of running the fraud scores and begin with the highest priority cases."

Analyzing large amounts of data to turn up useful and valuable information—such as fraudulent credit card purchases, is just one of the many applications of AI. AI systems are taking over many human tasks that people find either tedious, dangerous, or too big to handle. In short, AI is becoming a valuable extension of human intelligence.

As you read this chapter, consider the following:

- How are AI systems empowering businesses to attain goals previously unattainable?
- What are the strengths and limitations of AI systems, and how do they relate to the strengths and weaknesses of human beings?

Why Learn About Specialized Information Systems?

Specialized information systems are used in almost every industry. If you are a production manager at an automotive company, you may oversee robots that attach windshields to cars or paint body panels. If you are a young stock trader, you may use a special system called a *neural network* to uncover patterns and make millions of dollars trading stocks and stock options. If you are marketing manager for a PC manufacturer, you might use virtual reality on a Web site to show customers your latest laptop and desktop computers. If you are in the military, you might use computer simulation as a training tool to prepare you for combat. If you work for a petroleum company, you might use an expert system to determine where to drill for oil and gas. You will see many additional examples of the use of these specialized information systems throughout this chapter. Learning about these systems will help you discover new ways to use information systems in your day-to-day work.

artificial intelligence (AI)
The ability of computers to mimic or duplicate the functions of the human brain.

At a Dartmouth College conference in 1956, John McCarthy proposed the use of the term **artificial intelligence (AI)** to describe computers with the ability to mimic or duplicate the functions of the human brain. Many AI pioneers attended this first conference; a few predicted that computers would be as "smart" as people by the 1960s. The prediction has not yet been realized, but the benefits of artificial intelligence in business and research can be seen today, and research continues.

Like other aspects of an information system, the overall goal of the specialized systems discussed in this chapter is to help individuals and organizations achieve their goals. In some cases, these specialized systems can help an organization achieve a long-term, strategic advantage. In this chapter, we explore artificial intelligence and many other specialized information systems, including expert systems, robotics, vision systems, natural language processing, learning systems, neural networks, genetic algorithms, intelligent agents, virtual reality, and other systems. Since many of these special-purpose systems are expensive to develop and use, organizations need to make sure that the costs are worth the effort by increasing profits, reducing costs, or achieving an important organizational goal.

AN OVERVIEW OF ARTIFICIAL INTELLIGENCE

Science fiction novels and popular movies have featured scenarios of computer systems and intelligent machines taking over the world. Steven Hawking, who is the Lucasian professor of mathematics at Cambridge University in the UK (a position once held by Isaac Newton) and author of *A Brief History of Time,* said, "In contrast with our intellect, computers double their performance every 18 months. So the danger is real that they could develop intelligence and take over the world." Computer systems such as Hal in the classic movie *2001: A Space Odyssey* and those in the movie *A.I.* are futuristic glimpses of what might be. These accounts are fictional, but in them we see the real application of many computer systems that use the notion of AI. These systems help to make medical diagnoses, explore for natural resources, determine what is wrong with mechanical devices, and assist in designing and developing other computer systems. In this chapter, we explore the exciting applications of artificial intelligence, expert systems, virtual reality, and some specialized systems to see what the future really might hold.

Science fiction movies give us a glimpse of the future, but many practical applications of artificial intelligence exist today, among them medical diagnostics and development of computer systems.

(Source: The Kobal Collection/ Amblin/Dreamworks/Stanley Kubrick/WB/David James.)

Artificial Intelligence in Perspective

Artificial intelligence systems include the people, procedures, hardware, software, data, and knowledge needed to develop computer systems and machines that demonstrate characteristics of intelligence. Researchers, scientists, and experts on how human beings think are often involved in developing these systems.

artificial intelligence systems
People, procedures, hardware, software, data, and knowledge needed to develop computer systems and machines that demonstrate the characteristics of intelligence.

The Nature of Intelligence

From the early AI pioneering stage, the research emphasis has been on developing machines with **intelligent behavior**. But machine intelligence is hard to achieve. According to Steve Grand, who developed a robot called Lucy and was given an award for his work in artificial intelligence, "True machine intelligence, let alone consciousness, is a very long way off."[1] Asimo, a robot from Honda, is making progress but has a long way to go. According to Honda's head of the European Research Institute, "Asimo is a marvelous walking machine, a masterpiece of engineering, but the next stage is to enable it to develop the ability to think for itself."[2]

intelligent behavior
The ability to learn from experiences and apply knowledge acquired from experience, handle complex situations, solve problems when important information is missing, determine what is important, react quickly and correctly to a new situation, understand visual images, process and manipulate symbols, be creative and imaginative, and use heuristics.

The *Turing Test* attempts to determine whether the responses from a computer with intelligent behavior are indistinguishable from responses from a human being. No computer has passed the Turing Test, developed by Alan Turing, a British mathematician. The Loebner Prize offers money and a gold medal for anyone developing a computer that can pass the Turing Test (see *www.loebner.net*). Some of the specific characteristics of intelligent behavior include the ability to do the following:

- *Learn from experience and apply the knowledge acquired from experience.* Being able to learn from past situations and events is a key component of intelligent behavior and is a natural ability of humans, who learn by trial and error. This ability, however, must be carefully programmed into a computer system. Today, researchers are developing systems that have this ability. For instance, computerized AI chess software can learn to improve while playing human competitors.[3] In one match, Garry Kasparov competed against a personal computer with AI software developed in Israel, called Deep Junior.[4] This match in 2003 was a 3-3 tie, but Kasparov picked up something the machine would have no interest in—$700,000. In this chapter, we explore the exciting applications of artificial intelligence and look at what the future really might hold.[5]

Computers like Deep Junior attempt to learn from past chess moves. While its predecessor, Deep Blue, was a powerful supercomputer, Deep Junior runs on standard computer hardware. Its logic system calculates around 2 to 3 million combinations per second.

(Source: REUTERS/Chip East REUTERS/Landov.)

- *Handle complex situations.* Human beings are often involved in complex situations. World leaders face difficult political decisions regarding terrorism, conflict, global economic conditions, hunger, and poverty. In a business setting, top-level managers and executives are faced with a complex market, challenging competitors, intricate government regulations, and a demanding workforce. Even human experts make mistakes in dealing with these situations. Developing computer systems that can handle perplexing situations requires careful planning and elaborate computer programming. Trying to unlock the mysteries of DNA and the human genome is a complex process.[6] Artificial intelligence can be used to make complex research decisions. According to Eric Lander, director of the Broad Institute at MIT and Harvard, "Computers looked at DNA sequences and decided the next experiment that needed to be done was to close gaps in the genome. The computer knew what experiments to order up, without human input."
- *Solve problems when important information is missing.* The essence of decision making is dealing with uncertainty. Quite often, decisions must be made even when we lack information or have inaccurate information, because obtaining complete information is too costly or impossible. Today, AI systems can make important calculations, comparisons, and decisions even when information is missing.
- *Determine what is important.* Knowing what is truly important is the mark of a good decision maker. Developing programs and approaches to allow computer systems and machines to identify important information is not a simple task.

- *React quickly and correctly to a new situation.* A small child, for example, can look over a ledge or a drop-off and know not to venture too close. The child reacts quickly and correctly to a new situation. Computers, on the other hand, do not have this ability without complex programming.

- *Understand visual images.* Interpreting visual images can be extremely difficult, even for sophisticated computers. Moving through a room of chairs, tables, and other objects can be trivial for people but extremely complex for machines, robots, and computers. Such machines require an extension of understanding visual images, called a **perceptive system**. Having a perceptive system allows a machine to approximate the way a human sees, hears, and feels objects. Military robots, for example, use cameras and perceptive systems to conduct reconnaissance missions to detect enemy weapons and soldiers.[7] Detecting and destroying them can save lives. According to Col. Bruce Jette, "I don't have any problems writing to iRobot, saying I'm sorry your robot died, can we get another? That's a lot easier letter to write than [one] to a father or mother." The military and the iRobot company are not releasing information about the PacBot robot,[8] but from 50 to 100 robots are being used in Iraq.

- *Process and manipulate symbols.* People see, manipulate, and process symbols every day. Visual images provide a constant stream of information to our brains. By contrast, computers have difficulty handling symbolic processing and reasoning. Although computers excel at numerical calculations, they aren't as good at dealing with symbols and three-dimensional objects. Recent developments in machine-vision hardware and software, however, allow some computers to process and manipulate symbols on a limited basis.

- *Be creative and imaginative.* Throughout history, some people have turned difficult situations into advantages by being creative and imaginative. For instance, when shipped a lot of defective mints with holes in the middle, an enterprising entrepreneur decided to market these new mints as Lifesavers instead of returning them to the manufacturer. Ice cream cones were invented at the St. Louis World's Fair when an imaginative store owner decided to wrap ice cream with a waffle from his grill for portability. Developing new and exciting products and services from an existing (perhaps negative) situation is a human characteristic. Few computers have the ability to be truly imaginative or creative in this way, although software has been developed to enable a computer to write short stories.

- *Use heuristics.* For some decisions, people use heuristics (rules of thumb arising from experience) or even guesses. In searching for a job, we may decide to rank the companies we are considering according to profits per employee. Today, some computer systems, given the right programs, obtain good solutions that use approximations instead of trying to search for an optimal solution, which would be technically difficult or too time-consuming.

This list of traits only partially defines intelligence. Unlike the terminology used in virtually every other field of IS research, in which the objectives can be clearly defined, the term *intelligence* is a formidable stumbling block. One of the problems in artificial intelligence is arriving at a working definition of real intelligence against which to compare the performance of an artificial intelligence system.

The Difference Between Natural and Artificial Intelligence

Since the term *artificial intelligence* was defined in the 1950s, experts have disagreed about the difference between natural and artificial intelligence. Can computers be programmed to have common sense? Profound differences exist, but they are declining in number (see Table 11.1). One of the driving forces behind AI research is an attempt to understand how human beings actually reason and think. It is believed that the ability to create machines that can reason will be possible only once we truly understand our own processes for doing so.

perceptive system
A system that approximates the way a human sees, hears, and feels objects.

Attributes	Natural Intelligence (Human)	Artificial Intelligence (Machine)
The ability to use sensors (eyes, ears, touch, smell)	High	Low
The ability to be creative and imaginative	High	Low
The ability to learn from experience	High	Low
The ability to be adaptive	High	Low
The ability to afford the cost of acquiring intelligence	High	Low
The ability to use a variety of information sources	High	High
The ability to acquire a large amount of external information	High	High
The ability to make complex calculations	Low	High
The ability to transfer information	Low	High
The ability to make a series of calculations rapidly and accurately	Low	High

Table 11.1

A Comparison of Natural and Artificial Intelligence

The Major Branches of Artificial Intelligence

AI is a broad field that includes several specialty areas, such as expert systems, robotics, vision systems, natural language processing, learning systems, and neural networks (see Figure 11.1). Many of these areas are related; advances in one can occur simultaneously with or result in advances in others.

Figure 11.1

A Conceptual Model of Artificial Intelligence

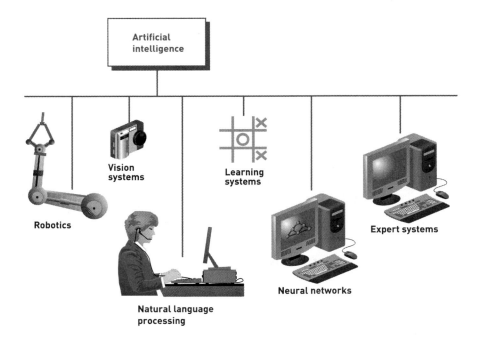

Expert Systems

An **expert system** consists of hardware and software that stores knowledge and makes inferences, similar to those of a human expert. Because of their many business applications, expert systems are discussed in more detail in the next several sections of the chapter.

expert system
Hardware and software that stores knowledge and makes inferences, similar to a human expert.

Robotics

Robotics involves developing mechanical or computer devices that can paint cars, make precision welds, and perform other tasks that require a high degree of precision or are tedious or hazardous for human beings. The NASA shuttle crash of the early 2000s, for example, has led some people to recommend using robots instead of human beings to explore space and perform scientific research.[9] Some robots, such as the ER-1, can be used for entertainment.[10] Placing a laptop computer on top, attaching a camera, and installing the necessary software allows the ER-1 to maneuver around objects. With an optional gripper arm, the robot can pick up small objects. Researchers hope that Lucy, a robot orangutan, will some day be able to help scientists determine how the brain works.[11] Paro, a harp seal robot, has been used as a therapeutic tool.[12] According to Takanore Shibata, an engineer for Japan's National Institute of Advanced Industrial Science and Technology, "We know that pet therapy helps physically, psychologically, and socially, and Paro does the same thing for people who are unable to care for a live pet." Contemporary robotics combines both high-precision machine capabilities and sophisticated controlling software. The controlling software in robots is what is most important in terms of AI. The processor in an advanced industrial robot today works at about 10 million instructions per second (MIPS)—no smarter than an insect. To achieve anything even approaching human intelligence, the robot processor must achieve 100 trillion operations per second.

robotics
Mechanical or computer devices that perform tasks requiring a high degree of precision or that are tedious or hazardous for humans.

Many applications of robotics exist, and research into these unique devices continues. Manufacturers use robots to assemble and paint products. Welding robots have enabled firms to manufacture top-quality products and reduce labor costs while shortening delivery time to their customers. Some robots, such as Honda's Asimo mentioned earlier, can shake hands, dance, and even reply to simple questions. The technology used in a robot's legs may provide improvements in mobility for people with disabilities. One robot uses technology that is based on the whiskers of a rat to allow the robot to navigate through close spaces.[13] Known as "whiskerbot," the robot uses whiskerlike sensors to determine the closeness, size, and texture of objects it touches. Other robots have been used to lay fiber-optic cable or to entertain disabled children while they receive needed therapy and exercise. A surgical robot named da Vinci performed a gall bladder operation on a 16-year-old patient in Denver, Colorado.[14] The surgical robot reduces scars, doesn't introduce bacteria that some doctors could, and can be more precise than a physician. It can also be easier and more comfortable for doctors to operate using a robot than their own hands. Some surgical robots cost more than $1 million and have multiple surgical arms and sophisticated vision systems.[15] Dr. Makimoto, chief of technology at Sony Corporation, has predicted that in 50 years, robot soccer players will be able to beat the best human soccer teams.[16] "Some think it's a crazy idea but robot engineers are very serious about this!" says Dr. Makimoto.

In military applications, robots are moving beyond movie plots to become real weapons. The U.S. Navy is developing a robot that has the appearance of a large lobster with eight legs. The underwater robot has sensors that can hear and smell. The Navy hopes that the robot will be operational by 2010 to locate mines and other underwater objects. The Air Force is developing a smart robotic jet fighter. Often called *unmanned combat air vehicles (UCAVs)*, these robotic war machines, such as the X-45A, will be able to identify and destroy targets without human pilots. UCAVs send pictures and information to a central command center and can be directed to strike military targets. These new machines extend the current Predator and Global Hawk technologies the military used in Afghanistan after the September 11 terrorist attacks.

Although robots are essential components of today's automated manufacturing and military systems, future robots will find wider applications in banks, restaurants, homes, doctors'

offices, and hazardous working environments such as nuclear stations. Microrobotics, also called *micro-electro-mechanical systems (MEMS)* that are the size of a grain of salt, are also being developed. MEMS can be used in air bags, a person's blood to monitor the body, cell phones, refrigerators, and more. A robot must not only execute tasks programmed by the user but also be able to interact with its environment.

Robots can be used in situations that are hazardous or inaccessible to humans. The Rover was a remote-controlled robot used by NASA to explore the surface of Mars.

(Source: Courtesy of NASA.)

Vision Systems

vision systems
The hardware and software that permit computers to capture, store, and manipulate visual images and pictures.

Another area of AI involves vision systems. **Vision systems** include hardware and software that permit computers to capture, store, and manipulate visual images and pictures. The U.S. Justice Department uses vision systems to perform fingerprint analysis, with almost the same level of precision as human experts. The speed with which the system can search through a huge database of fingerprints has brought quick resolution to many long-standing mysteries. Vision systems are also effective at identifying people based on facial features. In yet another application, a California wine bottle manufacturer uses a computerized vision system to inspect wine bottles for flaws.[17] The vision system saves the bottle producer both time and money. The company produces about 2 million wine bottles per day.

Vision systems can be used in conjunction with robots to give these machines "sight." A sophisticated robot and vision software are used to attach the rear windscreen on Jaguar S-Type cars.[18] The vision system guides robots to correctly and accurately install the windshields on the luxury cars. Factory robots typically perform mechanical tasks with little or no visual stimuli. Robotic vision extends the capability of these systems, allowing the robot to make decisions based on visual input. Generally, robots with vision systems can recognize black and white and some gray shades but do not have good color or three-dimensional vision. Other systems concentrate on only a few key features in an image, ignoring the rest. It may take years before a robot or other computer system can "see" in full color and draw conclusions from what it sees, the way that human beings do. Even with recent breakthroughs in vision systems, computers cannot see and understand visual images the way human beings can.[19]

Natural Language Processing

natural language processing
Processing that allows the computer to understand and react to statements and commands made in a "natural" language, such as English.

As discussed in Chapter 4, **natural language processing** allows a computer to understand and react to statements and commands made in a "natural" language, such as English. Restoration Hardware, for example, has developed a Web site that uses natural language processing to allow its customers to quickly find what they want on its site.[20] The natural language processing system corrects spelling mistakes, converts abbreviations into words and

commands, and allows people to ask questions in English. In addition to making it easier for customers, Restoration Hardware has seen an increase in revenues as a result of the use of natural language processing.

There are three levels of voice recognition: command (recognition of dozens to hundreds of words), discrete (recognition of dictated speech with pauses between words), and continuous (recognition of natural speech). For example, a natural language processing system can retrieve important information without making the user type in commands or search for key words. With natural language processing, users speak into a microphone connected to a computer and have the computer convert the electrical impulses generated from the voice into text files or program commands. With some simple natural language processors, you say a word into a microphone and type the same word on the keyboard to train the system to recognize the spoken words. The computer then matches the sound with the typed word. With more advanced natural language processors, recording and typing words is not necessary. Upstart Natural Machine is making its Java code for verbal AI available to other companies. The company hopes that everyone will benefit from making the computer code open and available.

Dragon Systems' Naturally Speaking 7 Essentials uses continuous voice recognition, or natural speech, allowing the user to speak to the computer at a normal pace without pausing between words. The spoken words are transcribed immediately onto the computer screen.

(Source: Courtesy of ScanSoft, Inc.)

Brokerage services are a perfect fit for voice-recognition technology to replace the existing "press 1 to buy or sell a stock" touchpad telephone menu system. People buying and selling stock use a vocabulary too varied for easy access through menus and touchpads but still small enough for software to process in real time. Several brokerages—including Charles Schwab & Co., Fidelity Investments, DLJdirect, and TD Waterhouse Group—offer voice-recognition services. These systems use natural language voice recognition to let customers access retirement accounts, check balances, and get stock quotes. Eventually, the technology will allow transactions to be made using voice commands over the phone and allow customers to use search engines and to have their questions answered through the brokerage firm's call center. One of the big advantages of voice recognition is that the number of calls routed to the customer service department drops considerably once new voice features are added. That is desirable to brokerages because it helps them staff their call centers correctly—even in volatile markets. While a typical person uses a vocabulary of about 20,000 words or fewer, some voice-recognition software has a built-in vocabulary of 85,000 words. Some companies claim that speech-recognition software is so good that some customers forget they are talking to a computer and start discussing the weather or sports scores. Accordant Health Services

uses natural language processing to allow insurance customers to ask questions using the company's Web site.[21] Computers convert questions, such as "Why did I get a bill from my doctor?" into answers.

Learning Systems

learning systems
A combination of software and hardware that allows the computer to change how it functions or reacts to situations based on feedback it receives.

Another part of AI deals with **learning systems**, a combination of software and hardware that allows a computer to change how it functions or reacts to situations based on feedback it receives. For example, some computerized games have learning abilities. If the computer does not win a game, it remembers not to make the same moves under the same conditions again. Learning systems software requires feedback on the results of actions or decisions. At a minimum, the feedback needs to indicate whether the results are desirable (winning a game) or undesirable (losing a game). The feedback is then used to alter what the system will do in the future.

Neural Networks

neural network
A computer system that can simulate the functioning of a human brain.

An increasingly important aspect of AI involves neural networks. A **neural network** is a computer system that can act like or simulate the functioning of a human brain. The systems use massively parallel processors in an architecture that is based on the human brain's own meshlike structure. In addition, neural network software can be used to simulate a neural network using standard computers. Neural networks can process many pieces of data at once and learn to recognize patterns. A chemical company, for example, can use neural network software to analyze a large amount of data to control chemical reactors. Fujitsu Laboratories used neural networks to improve motor coordination and movement in its robots.[22] The neural network software gives the robot a smooth walk and allows it to get up if it falls. The "Ethical and Societal Issues" feature explains yet another possible application of neural networks—predicting human behavior.

Some of the specific features of neural networks include the following:

- The ability to retrieve information even if some of the neural nodes fail
- Fast modification of stored data as a result of new information
- The ability to discover relationships and trends in large databases
- The ability to solve complex problems for which all the information is not present

Neural networks excel at pattern recognition. For example, neural network computers can be used to read bank check bar codes despite smears or poor-quality printing. Pattern recognition can be used to help prevent terrorism by analyzing and matching images from multiple cameras focusing on people or locations.[23] Many retail stores use Falcon Fraud Manager, a neural network system, to detect possible credit card fraud. Falcon Fraud Manager is used to protect more than 450 million credit card accounts. The Counter Fraud Service in England uses neural network software to detect fraud.[24] Since the late 1990s, the Counter Fraud Service has saved enough money to build three new hospitals, conduct 12,000 heart transplants, or perform 46,000 hip replacements. Fraud has dropped by 30 percent, and the money recovered has increased 700 percent—all helped by the neural network system. Some hospitals use neural networks to determine a patient's likelihood of contracting cancer or other diseases. The speed of genomic research can be increased with software that includes neural network features. Sandia Laboratories has developed a neural network system to give soldiers in a military conflict real-time advice on strategy and tactics.[25]

Biologically Inspired Algorithms Fight Terrorists and Guide Businesses

Soon people may be collaborating with intelligent machines to anticipate future human actions. While scientists have been using computers to forecast predictable natural events such as weather and earthquakes for years, the technology has recently been turned to predicting human behavior. Scientists could use such technology to explore the future of humanity, businesses can use it to anticipate the moves of their competitors and customers, and government and law enforcement can use it to catch the bad guys.

As with many truly revolutionary technologies, funding for this research is coming from the U.S. government, through the Defense Advanced Research Projects Agency (DARPA)—the same organization that produced the Internet. Research in this new intelligence technology is taking place as part of a $54 million program known as Genoa II. The goal of the project is to employ machine intelligence to anticipate future terrorist threats. Researchers hope to make it possible for humans and computers to "think together" in real time to "anticipate and preempt terrorist threats," according to official program documents.

"In Genoa II, we are interested in collaboration between people and machines," said Tom Armour, Genoa II program manager at DARPA, "We imagine software agents working with humans...and having different sorts of software agents also collaborating among themselves."

The challenge lies in getting computers to mimic the human brain's ability to reduce complexity. While computers are good at carrying out rules-based algorithms, such as those required for playing chess, they aren't so good at more complex deciphering, such as finding a word hidden in a picture. Researchers hope to change that through the use of biologically inspired algorithms. "One way to make computers more intelligent and lifelike is to look at living systems and imitate them," says Melanie Mitchell, an associate professor at Oregon Health & Science University's School of Science & Engineering in Portland.

Neural networks are one form of biologically inspired algorithms. They mimic the neurons in the human brain to identify logical patterns in data and to produce very sophisticated learning. Another form of biologically inspired algorithms, genetic algorithms, is inspired by evolution. Here, a computer program evolves a solution to a problem rather requiring a person to try to engineer one. Also, researchers are beginning to produce security applications that mimic the human immune system, which attacks other sorts of invaders.

Private-sector researchers are studying "cognitive amplifiers," which can enable software to model current situations and predict "plausible futures." They are on the verge of creating practical applications to support cognitive machine intelligence, associative memory, biologically inspired algorithms and Bayesian inference networks, which are based on a branch of mathematical probability theory that says uncertainty about the world and outcomes of interest can be modeled by combining common sense with evidence observed in the real world.

"Some of the core algorithms we are working with have been around for centuries," says Ron Kolb, director of technology at San Francisco–based Autonomy Inc., a firm that makes advanced pattern-recognition and knowledge management software. "It's just now that we're finding the practical applications for them." Kolb explains, "We're able to produce an algorithm that says here are the patterns that exist, here are the important patterns that exist, here are the patterns that contextually surround the data, and as new data enters the stream, we're able to build associative relationships to learn more as more data is digested by the system."

Grant Evans, CEO of A4Vision Inc. in Cupertino, California, and an expert in cognitive machine intelligence and biologically inspired algorithms, believes that he knows where the future of this research is leading. "Now we're integrating cognitive machine intelligence in the form of video with avatars—3-D digital renderings of real people, that can see and track you."

This is just the kind of talk that makes privacy advocates nervous. They fear that the government and businesses will go too far in collecting social information and infringe on the privacy rights of individuals. Genoa II may be shelved because of its central role in the controversial Terrorism Information Awareness program. Private-sector researchers say that many significant advances are still possible and are, in fact, already happening. Many automotive and aerospace manufacturers have used rudimentary pieces of this technology to save millions of dollars by leveraging developmental expertise across functional areas, says Kolb. "We're no longer looking for information, information is looking for us," he says.

Critical Thinking Questions

1. Why is it that computers may be able to do a better job of predicting human behavior than human beings?
2. What precautions would be wise when developing systems that study human behavior to predict future events?

What Would You Do?

Your company has implemented biologically inspired algorithms to project trends in sales for your product lines. The AI program has indicated that the popularity of one of your best-selling products will begin to decline in the first quarter of next year.

3. How might this prediction influence your treatment of the product line and your business?
4. What might you do to change the future that was predicted by the intelligent machine?

SOURCES: Dan Verton, "Using Computers to Outthink Terrorists," *Computerworld*, September 1, 2003, *www.computerworld.com*; A4Vision Web site, accessed May 31, 2004, *www.a4vision.com*.

Neural nets work particularly well when it comes to analyzing detailed trends. Large amusement parks and banks use neural networks to figure out staffing needs based on customer traffic—a task that requires precise analysis, down to the half-hour. Increasingly, businesses are firing up neural nets to help them navigate ever-thicker forests of data and make sense of myriad customer traits and buying habits. Computer Associates has developed Neugents, neural intelligence agents that "learn" patterns and behaviors and predict what will happen next. For example, Neugents can be used to track the habits of insurance customers and predict which ones will not renew, say, an automobile policy. They can then suggest to an insurance agent what changes might be made in the policy to persuade the consumer to renew it. The technology also can track individual users at e-commerce sites and their online preferences so that they don't have to input the same information each time they log on—their purchasing history and other data will be recalled each time they access a Web site.

AI Trilogy is a neural network software program that can run on a standard PC. The software can make predictions with NeuroShell Predictor and make classifications with NeuroShell Classifier. The software package also contains GeneHunter, which uses a genetic algorithm to get the best result from the neural network system.

Some pattern-recognition software uses neural networks to analyze hundreds of millions of bank, brokerage, and insurance accounts involving trillions of dollars to uncover money laundering and other suspicious money transfers. Today, partially as a result of the September 11 attacks, all transfers of $10,000 or more across U.S. borders must be reported to the federal government.

Other Artificial Intelligence Applications

A few other artificial intelligence applications exist in addition to those just discussed. A **genetic algorithm** is an approach to solving large, complex problems in which a number of repeated operations or models change and evolve until the best one emerges. The approach is based on the theory of evolution that requires (1) variation and (2) natural selection. The first step is to change or vary a number of competing solutions to the problem. This can be done by changing the parts of a program or combining different program segments into a new program, mimicking the evolution of species, in which the genetic makeup a plant or animal mutates or changes over time. The second step is to select only the best models or algorithms, which continue to evolve. Programs or program segments that are not as good as others are discarded, similar to natural selection or "survival of the fittest," in which only the best species survive and continue to evolve. This process of variation and natural selection continues until the genetic algorithm yields the best possible solution to the original problem. For example, some investment firms use genetic algorithms to help select the best stocks or bonds. Genetic algorithms are also used in computer science and mathematics.[26] Genetic algorithms can help companies determine which orders to accept for maximum profit.[27] This approach helps companies select the orders that will increase profits and take full advantage of the company's production facilities. Genetic algorithms are also being used to make better decisions in developing inputs to neural networks.[28]

An **intelligent agent** (also called an *intelligent robot* or *bot*) consists of programs and a knowledge base used to perform a specific task for a person, a process, or another program. Like a sports agent who searches for the best endorsement deals for a top athlete, an intelligent agent often searches to find the best price, the best schedule, or the best solution to a problem. The programs used by an intelligent agent can search through large amounts of data while the knowledge base refines the search or accommodates user preferences. Often used to search the vast resources of the Internet, intelligent agents can help people find information on an important topic or the best price for a new digital camera. Intelligent agents can also be used to make travel arrangements, monitor incoming e-mail for viruses or junk mail, and coordinate meetings and schedules of busy executives. In the human resource field, intelligent agents are used to help with online training. The software can look ahead in training materials and know what to start next. Staples uses intelligent agents to find job candidates.[29] The software searches a 12-million person database by title, company, gender, and other factors to find the best job candidates for companies. Intelligent agents have been used by the U.S.

---

genetic algorithm
An approach to solving large, complex problems in which a number of related operations or models change and evolve until the best one emerges.

intelligent agent
Programs and a knowledge base used to perform a specific task for a person, a process, or another program; also called *intelligent robot* or *bot*.

Army to route security clearance information for soldiers to the correct departments and individuals.[30] What used to take days when done manually now takes hours.

A new hearing aid with artificial intelligence (shown next to a fingernail) contains a tiny microprocessor that works the way the brain does in detecting and distinguishing sounds while filtering out distractions.

(Source: AP/Wide World Photos.)

AN OVERVIEW OF EXPERT SYSTEMS

As mentioned earlier, an expert system behaves similarly to a human expert in a particular field. Charles Bailey, one of the original members of the Library and Information Technology Association, developed one of the first expert systems in the mid 1980s to search the University of Houston's library to retrieve requested resources and citations.[31] Computerized expert systems have been developed to diagnose problems, predict future events, and solve energy problems. They have also been used to design new products and systems, develop innovative insurance products, determine the best use of lumber, and increase the quality of healthcare. Like human experts, computerized expert systems use heuristics, or rules of thumb, to arrive at conclusions or make suggestions. Expert systems have also been used to determine credit limits for credit cards. Soquimich, an argicultural company, uses expert systems to determine the best fertilizer mix to use on certain soils to improve crops while minimizing costs.[32] The research conducted in AI during the past two decades is resulting in expert systems that explore new business possibilities, increase overall profitability, reduce costs, and provide superior service to customers and clients.

Credit card companies often use expert systems to determine credit limits for credit cards.

(Source: © Ron Fehling/Masterfile.)

Characteristics and Limitations of an Expert System

Expert systems have a number of characteristics, including the following:

- *Can explain their reasoning or suggested decisions.* A valuable characteristic of an expert system is the ability to explain how and why a decision or solution was reached. For example, an expert system can explain the reasoning behind the conclusion to approve a particular loan application. The ability to explain its reasoning processes can be the most valuable feature of a computerized expert system. The user of the expert system thus gains access to the reasoning behind the conclusion.
- *Can display "intelligent" behavior.* Considering a collection of data, an expert system can propose new ideas or approaches to problem solving. A few of the applications of expert systems are a medical diagnosis based on a patient's condition, a suggestion to explore for natural gas at a particular location, and providing job counseling for workers.
- *Can draw conclusions from complex relationships.* Expert systems can evaluate complex relationships to reach conclusions and solve problems. For example, one proposed expert system would work with a flexible manufacturing system to determine the best use of tools. Another expert system can suggest ways to improve quality-control procedures.
- *Can provide portable knowledge.* One unique capability of expert systems is that they can be used to capture human expertise that might otherwise be lost. A classic example of this is the expert system called DELTA (Diesel Electric Locomotive Troubleshooting Aid), which was developed to preserve the expertise of the retiring David Smith, the only engineer competent to handle many highly technical repairs of such machines.
- *Can deal with uncertainty.* One of an expert system's most important features is its ability to deal with knowledge that is incomplete or not completely accurate. The system deals with this problem through the use of probability, statistics, and heuristics.

Even though these characteristics of expert systems are impressive, they have limitations, including these:

- *Not widely used or tested.* Even though successes occur, expert systems are not used in a large number of organizations. In other words, they have not been widely tested in corporate settings.
- *Difficult to use.* Some expert systems are difficult to control and use. In some cases, the assistance of computer personnel or individuals trained in the use of expert systems is required to help the user get the most from these systems. Today's challenge is to make expert systems easier to use by decision makers who have limited computer programming experience.
- *Limited to relatively narrow problems.* Whereas some expert systems can perform complex data analysis, others are limited to simple problems. Also, many problems solved by expert systems are not that beneficial in business settings. An expert system designed to provide advice on how to repair a machine, for example, is unable to assist in decisions about when or whether to repair it. In general, the narrower the scope of the problem, the easier it is to implement an expert system to solve it.
- *Cannot readily deal with "mixed" knowledge.* Expert systems cannot easily handle knowledge that has a mixed representation. Knowledge can be represented through defined rules, through comparison with similar cases, and in various other ways. An expert system in one application might not be able to deal with knowledge that combines both rules and cases.
- *Possibility of error.* Although some expert systems have limited abilities to learn from experience, the primary source of knowledge is a human expert. If this knowledge is incorrect or incomplete, it will affect the system negatively. Other development errors involve poor programming practices. Because expert systems are more complex than other information systems, the potential for such errors is greater.
- *Cannot refine its own knowledge.* Expert systems are not capable of acquiring knowledge directly. A programmer must provide instructions to the system that determine how the system is to learn from experience. Also, some expert systems cannot refine their own knowledge—such as eliminating redundant or contradictory rules.

- *Difficult to maintain.* Related to the preceding point is the fact that expert systems can be difficult to update. Some are not responsive or adaptive to changing conditions. Adding new knowledge and changing complex relationships may require sophisticated programming skills. In some cases, a spreadsheet used in conjunction with an expert system shell can be used to modify the system. In others, upgrading an expert system can be too difficult for the typical manager or executive. Future expert systems are likely to be easier to maintain and update.
- *May have high development costs.* Expert systems can be expensive to develop when using traditional programming languages and approaches. Development costs can be greatly reduced through the use of software for expert system development. **Expert system shells**, a collection of software packages and tools used to develop expert systems, can be implemented on most popular PC platforms to reduce development time and costs.
- *Raise legal and ethical concerns.* People who make decisions and take action are legally and ethically responsible for their behavior. A person, for example, can be taken to court and punished for a crime. When expert systems are used to make decisions or help in the decision-making process, who is legally and ethically responsible—the human experts used to develop the knowledge on which the system relies, the expert system developer, the user, or someone else? For example, if a doctor uses an expert system to make a diagnosis and the diagnosis is wrong, who is responsible? These legal and ethical issues have not been completely resolved.

expert system shell
A collection of software packages and tools used to develop expert systems.

When to Use Expert Systems

Sophisticated expert systems can be difficult, expensive, and time-consuming to develop. This is especially true for large expert systems implemented on mainframes. The following is a list of factors that normally make expert systems worth the expenditure of time and money. Develop an expert system if it can:

- Provide a high potential payoff or significantly reduce downside risk
- Capture and preserve irreplaceable human expertise
- Solve a problem that is not easily solved using traditional programming techniques
- Develop a system more consistent than human experts
- Provide expertise needed at a number of locations at the same time or in a hostile environment that is dangerous to human health
- Provide expertise that is expensive or rare
- Develop a solution faster than human experts can
- Provide expertise needed for training and development to share the wisdom and experience of human experts with a large number of people

Components of Expert Systems

An expert system consists of a collection of integrated and related components, including a knowledge base, an inference engine, an explanation facility, a knowledge base acquisition facility, and a user interface. A diagram of a typical expert system is shown in Figure 11.2. In this figure, the user interacts with the user interface, which interacts with the inference engine. The inference engine interacts with the other expert system components. These components must work together to provide expertise.

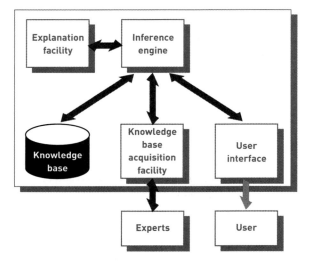

Figure 11.2

Components of an Expert System

The Knowledge Base

The **knowledge base** stores all relevant information, data, rules, cases, and relationships used by the expert system. As seen in Figure 11.3, a knowledge base is a natural extension of a database (presented in Chapter 5) and an information and decision support system (presented in Chapter 10). As discussed in Chapter 5, raw facts can be used to perform basic business transactions but are seldom used without manipulation in decision making. As we move to information and decision support, data is filtered and manipulated to produce a variety of reports to help managers make better decisions. With a knowledge base, we try to understand patterns and relationships in data as a human expert does in making intelligent decisions.

knowledge base
A component of an expert system that stores all relevant information, data, rules, cases, and relationships used by the expert system.

Figure 11.3

The Relationships among Data, Information, and Knowledge

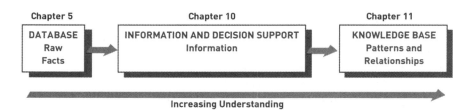

A knowledge base must be developed for each unique application. For example, a medical expert system will contain facts about diseases and symptoms. The knowledge base can include generic knowledge from general theories that have been established over time and specific knowledge that comes from more recent experiences and rules of thumb. Knowledge bases, however, go far beyond simple facts, also storing relationships, rules or frames, and cases. For example, certain telecommunications network problems may be related or linked; one problem may cause another. In other cases, rules suggest certain conclusions, based on a set of given facts. In many instances, these rules are stored as **if-then statements**, such as "If a certain set of network conditions exists, then a certain network problem diagnosis is appropriate." Cases can also be used. This technique involves finding instances, or cases, that are similar to the current problem and modifying the solutions to these cases to account for any differences between the previously solved cases stored in the computer and the current situation or problem.

Assembling Human Experts. One challenge in developing a knowledge base is to assemble the knowledge of multiple human experts. Typically, the objective in building a knowledge base is to integrate the knowledge of individuals with similar expertise (e.g., many doctors may contribute to a medical diagnostics knowledge base). A knowledge base that contains information from numerous experts can be extremely powerful and accurate in terms of its predictions and suggestions. Unfortunately, human experts can disagree on important

if-then statements
Rules that suggest certain conclusions.

relationships and interpretations of data, presenting a dilemma for designers and developers of knowledge bases and expert systems in general. Some human experts are more expert than others; their knowledge, experience, and information are better developed and more accurately represent reality. When human experts disagree on important points, it can be difficult for expert systems developers to determine which rules and relationships to place in the knowledge base.

The Use of Fuzzy Logic. Another challenge for expert system designers and developers is capturing knowledge and relationships that are not precise or exact. Computers typically work with numerical certainty; certain input values will always result in the same output. In the real world, as you know from experience, this is not always the case. To handle this dilemma, a special research area in computer science, called **fuzzy logic**, has been developed. Fuzzy logic was first applied in Japan in an automatic control system for trains. The control system allowed each train to stop within 7 centimeters (about 3 inches) of the right spot on the platform. Fuzzy logic also made train travel smoother and more efficient, saving about 10 percent in energy compared with human-controlled trains.

Instead of the usual black-and-white, yes/no, or true/false conditions of typical computer decisions, fuzzy logic allows shades of gray, or what are known as "fuzzy sets." The criteria on whether a subject or situation fits into a set are given in percentages or probabilities. For example, a weather forecaster might state, "if it is very hot with high humidity, the likelihood of rain is 75 percent." The imprecise terms of "very hot" and "high humidity" are what fuzzy logic must determine to formulate the chance of rain. Fuzzy logic rules help computers evaluate the imperfect or imprecise conditions they encounter and make "educated guesses" based on the likelihood or probability of correctness of the decision. This ability to estimate whether a condition fits a situation closely resembles the judgment a person makes when evaluating situations.

Fuzzy logic is used in embedded computer technology—for example, autofocus cameras, medical equipment that monitors patients' vital signs and makes automatic corrections, and temperature sensors attached to furnace controls. Medical facilities can use fuzzy logic to detect lung cancer earlier, offering the potential to dramatically improve survival rates from the deadly disease.

The Use of Rules. A **rule** is a conditional statement that links given conditions to actions or outcomes. As we saw earlier, a rule is constructed using if-then constructs. If certain conditions exist, then specific actions are taken or certain conclusions are reached. In an expert system for a weather forecasting operation, for example, the rules could state that if certain temperature patterns exist with a given barometric pressure and certain previous weather patterns over the last 24 hours, then a specific forecast will be made, including temperatures, cloud coverage, and the wind-chill factor. Rules are often combined with probabilities. For example, if the weather has a particular pattern of trends, then there is a 65 percent probability that it will rain tomorrow. Rules relating data to conclusions can be developed for any knowledge base. Most expert systems prevent users from entering contradictory rules. Figure 11.4 shows the use of expert system rules in helping to determine whether a person should receive a mortgage loan from a bank. In general, as the number of rules that an expert system knows increases, the precision of the expert system also increases.

The Use of Cases. As mentioned previously, an expert system can use cases in developing a solution to a current problem or situation. This process involves (1) finding cases stored in the knowledge base that are similar to the problem or situation at hand and (2) modifying the solutions to the cases to fit or accommodate the current problem or situation. Cases stored in the knowledge base can be identified and selected by comparing the parameters of the new problem with the cases stored in the computer system. For example, a company may be using an expert system to determine the best location of a new service facility in the state of New Mexico. Labor and transportation costs may be the most important factors. The expert system may identify two previous cases involving the location of a service facility where labor and transportation costs were also important—one in the state of Colorado and the other in the state of Nevada. The expert system will modify the solution to these two cases to determine the best location for a new facility in New Mexico. The result might be to locate the new service facility in the city of Santa Fe.

fuzzy logic
A special research area in computer science that allows shades of gray and does not require everything to be simple black or white, yes/no, or true/false.

rule
A conditional statement that links given conditions to actions or outcomes.

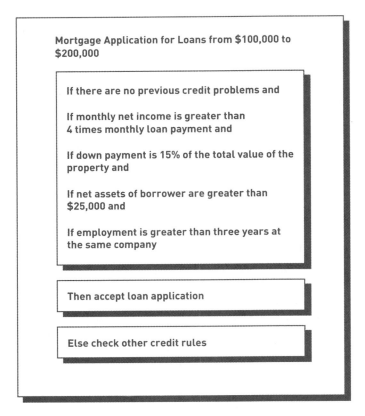

Mortgage Application for Loans from $100,000 to $200,000

If there are no previous credit problems and

If monthly net income is greater than 4 times monthly loan payment and

If down payment is 15% of the total value of the property and

If net assets of borrower are greater than $25,000 and

If employment is greater than three years at the same company

Then accept loan application

Else check other credit rules

The Inference Engine

inference engine
Part of the expert system that seeks information and relationships from the knowledge base and provides answers, predictions, and suggestions the way a human expert would.

The overall purpose of an **inference engine** is to seek information and relationships from the knowledge base and to provide answers, predictions, and suggestions the way a human expert would. In other words, the inference engine is the component that delivers the expert advice.

The process of retrieving relevant information and relationships from the knowledge base is not simple. As you have seen, the knowledge base is a collection of facts, interpretations, and rules. The inference engine must find the right facts, interpretations, and rules and assemble them correctly. In other words, the inference engine must make logical sense out of the information contained in the knowledge base, the way the human mind does when sorting out a complex situation. The inference engine has a number of ways of accomplishing its tasks, including backward and forward chaining.

backward chaining
The process of starting with conclusions and working backward to the supporting facts.

Backward Chaining. **Backward chaining** is the process of starting with conclusions and working backward to the supporting facts. If the facts do not support the conclusion, another conclusion is selected and tested. This process is continued until the correct conclusion is identified. Consider an expert system that forecasts product sales for next month. With backward chaining, we start with a conclusion, such as "Sales next month will be 25,000 units." Given this conclusion, the expert system searches for rules in the knowledge base that support the conclusion, such as "IF sales last month were 21,000 units and sales for competing products were 12,000 units, THEN sales next month should be 25,000 units or greater." The expert system verifies the rule by checking sales last month for the company and its competitors. If the facts are not true—in this case, if last month's sales were not 21,000 units or 12,000 units for competitors—the expert system would start with another conclusion and proceed until rules, facts, and conclusions match.

forward chaining
The process of starting with the facts and working forward to the conclusions.

Forward Chaining. **Forward chaining** starts with the facts and works forward to the conclusions. Consider the expert system that forecasts future sales for a product. With forward chaining, we start with a fact, such as "The demand for the product last month was 20,000 units." With the forward-chaining approach, the expert system searches for rules that contain a reference to product demand. For example, "IF product demand is over 15,000 units, THEN check the demand for competing products." As a result of this process, the expert

system might use information on the demand for competitive products. Next, after searching additional rules, the expert system might use information on personal income or national inflation rates. This process continues until the expert system can reach a conclusion using the data supplied by the user and the rules that apply in the knowledge base.

Comparison of Backward and Forward Chaining. Forward chaining can reach conclusions and yield more information with fewer queries to the user than backward chaining, but this approach requires more processing and a greater degree of sophistication. Forward chaining is often used by more expensive expert systems. Some systems also use mixed chaining, which is a combination of backward and forward chaining.

The Explanation Facility

An important part of an expert system is the **explanation facility**, which allows a user or decision maker to understand how the expert system arrived at certain conclusions or results. A medical expert system, for example, may have reached the conclusion that a patient has a defective heart valve given certain symptoms and the results of tests on the patient. The explanation facility allows a doctor to find out the logic or rationale of the diagnosis made by the expert system. The expert system, using the explanation facility, can indicate all the facts and rules that were used in reaching the conclusion. This facility allows doctors to determine whether the expert system is processing the data and information correctly and logically.

explanation facility
Component of an expert system that allows a user or decision maker to understand how the expert system arrived at certain conclusions or results.

The Knowledge Acquisition Facility

A difficult task in developing an expert system is the process of creating and updating the knowledge base.[33] In the past, when more traditional programming languages were used, developing a knowledge base was tedious and time-consuming. Each fact, relationship, and rule had to be programmed into the knowledge base. In most cases, an experienced programmer had to create and update the knowledge base.

Today, specialized software allows users and decision makers to create and modify their own knowledge bases through the knowledge acquisition facility (see Figure 11.5). The overall purpose of the **knowledge acquisition facility** is to provide a convenient and efficient means for capturing and storing all components of the knowledge base. Knowledge acquisition software can present users and decision makers with easy-to-use menus. After filling in the appropriate attributes, the knowledge acquisition facility correctly stores information and relationships in the knowledge base, making the knowledge base easier and less expensive to set up and maintain. Knowledge acquisition can be a manual process or a mixture of manual and automated procedures. Regardless of how the knowledge is acquired, it is important to validate and update the knowledge base frequently to make sure that it is still accurate.

knowledge acquisition facility
Part of the expert system that provides convenient and efficient means of capturing and storing all the components of the knowledge base.

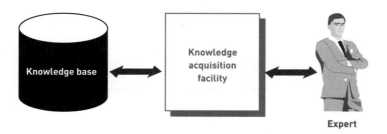

Knowledge base ⟷ **Knowledge acquisition facility** ⟷ **Expert**

Figure 11.5

The knowledge acquisition facility acts as an interface between experts and the knowledge base.

The User Interface

Specialized user interface software is employed for designing, creating, updating, and using expert systems. The main purpose of the user interface is to make the development and use of an expert system easier for users and decision makers. At one time, skilled computer personnel created and operated most expert systems; today, the user interface permits decision makers to develop and use their own expert systems. Because expert systems place more

emphasis on directing user activities than do other types of systems, text-oriented user interfaces (using menus, forms, and scripts) may be more common in expert systems than the graphical interfaces often used with DSSs. A typical user interface between a user and the expert system for the credit application shown in Figure 11.4 follows.

EXPERT SYSTEM: Are there any previous credit problems?
USER: No.
EXPERT SYSTEM: Is the applicant's net income greater than four times the monthly loan payment?
USER: Yes.
EXPERT SYSTEM: Is the down payment greater than 15% of the property value?
USER: Yes.
EXPERT SYSTEM: Are the net assets of the borrower greater than $25,000?
USER: Yes.
EXPERT SYSTEM: Has the borrower worked more than three years at the same company?
USER: Yes.
EXPERT SYSTEM: **Accept the loan application.**

Expert Systems Development

Like other computer systems, expert systems require a systematic development approach for best results (see Figure 11.6). This approach includes determining the requirements for the expert system, identifying one or more experts in the area or discipline under investigation, constructing the components of the expert system, implementing the results, and maintaining and reviewing the complete system.

Figure 11.6

Steps in the Expert System Development Process

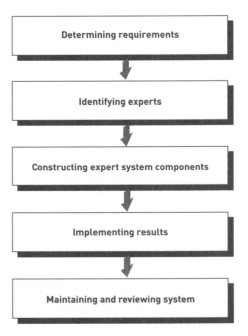

The Development Process

The Development Process

Specifying the requirements for an expert system begins with identifying the system's objectives and its potential use. Identifying experts can be difficult. In some cases, a company has human experts on hand; in other cases, experts outside the organization will be required. Developing the expert system components requires special skills. Implementing the expert system involves placing it into action and making sure that it operates as intended. Like other computer systems, expert systems should be periodically reviewed and maintained to make sure that they are delivering the best support to decision makers and users.

Many companies are only now beginning to use and develop expert systems. Expert system development is a team effort, but experienced personnel and users may be in high demand within an organization. Because development can take months or years, the cost of bringing in consultants for development can be high. It is critical, therefore, to find and assemble the right people to assist with development.

Participants in Developing and Using Expert Systems

Typically, several people are involved in developing and using an expert system (see Figure 11.7).

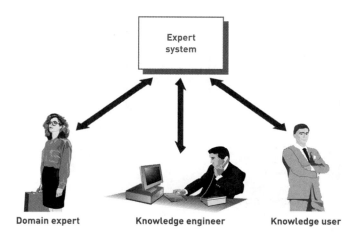

Figure 11.7

Participants in Expert Systems Development and Use

The Domain Expert. Because of the time and effort involved in the task, an expert system is developed to address only a specific area of knowledge. This area of knowledge is called the **domain**. The **domain expert** is the individual or group that has the expertise or knowledge one is trying to capture in the expert system. In most cases, the domain expert is a group of human experts. The domain expert (individual or group) usually has the ability to do the following:

- Recognize the real problem
- Develop a general framework for problem solving
- Formulate theories about the situation
- Develop and use general rules to solve a problem
- Know when to break the rules or general principles
- Solve problems quickly and efficiently
- Learn from experience
- Know what is and is not important in solving a problem
- Explain the situation and solutions of problems to others

The Knowledge Engineer and Knowledge Users. A **knowledge engineer** is an individual who has training or experience in the design, development, implementation, and maintenance of an expert system, including training or experience with expert system shells. The **knowledge user** is the individual or group that uses and benefits from the expert system. Knowledge users do not need any previous training in computers or expert systems.

Expert Systems Development Tools and Techniques

Theoretically, expert systems can be developed from any programming language. Since the introduction of computer systems, programming languages have become easier to use, more powerful, and increasingly able to handle specialized requirements. In the early days of expert systems development, traditional high-level languages, including Pascal, FORTRAN, and COBOL, were used (see Figure 11.8). LISP was one of the first special languages developed and used for artificial intelligence applications. PROLOG, a more recent language, was also developed for AI applications. Since the 1990s, however, other expert system products (such

domain
The area of knowledge addressed by the expert system.

domain expert
The individual or group who has the expertise or knowledge one is trying to capture in the expert system.

knowledge engineer
An individual who has training or experience in the design, development, implementation, and maintenance of an expert system.

knowledge user
The individual or group who uses and benefits from the expert system.

as shells) have become available that remove the burden of programming, allowing nonprogrammers to develop and benefit from the use of expert systems.

Figure 11.8

Software for expert systems development has evolved greatly since 1980, from traditional programming languages to expert system shells.

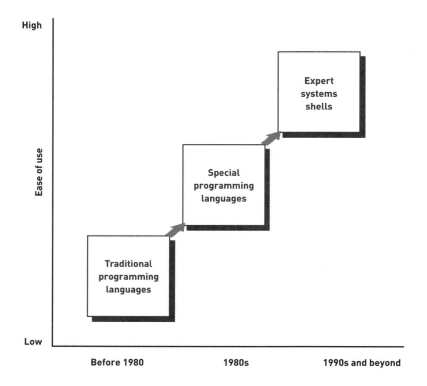

Expert System Shells and Products. As discussed, an expert system shell is a collection of software packages and tools used to design, develop, implement, and maintain expert systems. Expert system shells exist for both personal computers and mainframe systems. Some shells are inexpensive, costing less than $500. In addition, off-the-shelf expert system shells are available that are complete and ready to run. The user enters the appropriate data or parameters, and the expert system provides output to the problem or situation.

A number of expert system shells and products are available to analyze LAN networks, monitor air quality in commercial buildings, and analyze oil and drilling operations. A few expert system shells are summarized in Table 11.2.

Table 11.2

Popular Expert System Shells

Name of Shell	Application and Capabilities
Financial Advisor	Analyzes financial investments in new equipment, facilities, and the like; requests the appropriate data and performs a complete financial analysis.
G2	Assists in oil and gas operations. Transco, a British company, uses it to help in the transport of gas to more than 20 million commercial and domestic customers.
RAMPART	Analyzes risk. The U.S. General Services Administration uses it to analyze risk to the approximately 8,000 federal buildings it manages.
HazMat Loader	Analyzes hazardous materials in truck shipments.
MindWizard	Enables development of compact expert systems ranging from simple models that incorporate business decision rules to highly sophisticated models; PC based and inexpensive.
LSI Indicator	Helps determine property values; developed by one of the largest residential title and closing companies.

Expert Systems Development Alternatives

Expert systems can be developed from scratch by using an expert system shell or by purchasing an existing expert system package. A graph of the general cost and time of development alternatives is shown in Figure 11.9. It is usually faster and less expensive to develop an expert system using an existing package or an expert system shell. Note that there will be an additional cost of developing an existing package or acquiring an expert system shell if the organization does not already have this type of software.

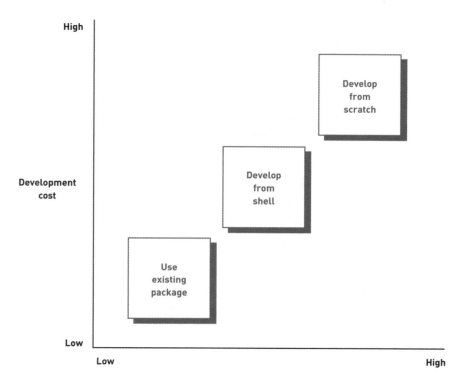

Figure 11.9

Some Expert System Development Alternatives and Their Relative Cost and Time Values

In-House Development: Develop from Scratch. Developing an expert system from scratch is usually more costly than the other alternatives, but an organization has more control over the features and components of the system. Such customization also has a downside; it can result in a more complex system, with higher maintenance and updating costs.

In-House Development: Develop from a Shell. As you have seen, an expert system shell consists of one or more software products that assist in the development of an expert system. In some instances, the same shell can be used to develop many expert systems. Developing an expert system from a shell can be less complex and easier to maintain than developing one from scratch. However, the resulting expert system may need to be modified to tailor it to specific applications. In addition, the capabilities and features of an expert system can be more difficult to control.

Off-the-Shelf Purchase: Use Existing Packages. Using an existing expert system package is the least expensive and fastest approach, in most cases. An existing expert system package is one that has been developed by a software or consulting company for a specific field or area, such as the design of a new computer chip or a weather forecasting and prediction system. The advantages of using an existing package can go beyond development time and cost. These systems can also be easy to maintain and update over time. A disadvantage of using an off-the-shelf package is that it may not be able to satisfy the unique needs or requirements of an organization.

Applications of Expert Systems and Artificial Intelligence

Expert systems and artificial intelligence are being used in a variety of ways. For example, expert systems have been used to help power plants reduce pollutants while maintaining profits.[34] They have also been used to determine the best way to distribute weight in a ferryboat to reduce the risk of capsizing or sinking.[35] Other applications of these systems are summarized next.

- *Credit granting and loan analysis.* Many banks employ expert systems to review an individual's credit application and credit history data from credit bureaus to make a decision on whether to grant a loan or approve a transaction. KPMG Peat Marwick uses an expert system called Loan Probe to review its reserves to determine whether sufficient funds have been set aside to cover the risk of some uncollectible loans.

- *Stock picking.* Some expert systems are used to help investment professionals pick stocks and other investments.

- *Catching cheats and terrorists.* Some gambling casinos use expert system software to catch gambling cheats. The CIA is testing the software to see whether it can be used to detect possible terrorists when they make hotel or airline reservations.

- *Budgeting.* Automotive companies can use expert systems to help budget, plan, and coordinate prototype testing programs to save hundreds of millions of dollars.

- *Games.* Some expert systems are used for entertainment. For example, Proverb is an expert system designed to solve standard American crossword puzzles given the grid and clues.

- *Information management and retrieval.* The explosive growth of information available to decision makers has created a demand for devices to help manage the information. Expert systems can aid this process through the use of bots. Businesses might use a bot to retrieve information from large distributed databases or a vast network like the Internet. Expert system agents help managers find the right data and information while filtering out irrelevant facts that might impede timely decision making. The Burlington Northern Santa Fe Railroad uses speech-recognition technology to help it retrieve and store shipping data automatically, as the "Information Systems @ Work" feature discusses.

- *AI and expert systems embedded in products.* The antilock braking system on today's automobiles is an example of a rudimentary expert system. A processor senses when the tires are beginning to skid and releases the brakes for a fraction of a second to prevent the skid. AI researchers are also finding ways to use neural networks and robotics in everyday devices, such as toasters, alarm clocks, and televisions.

- *Plant layout and manufacturing.* FLEXPERT is an expert system that uses fuzzy logic to perform plant layout. The software helps companies determine the best placement for equipment and manufacturing facilities. Expert systems can also be used in manufacturing.[36] According to an industry observer, "A few giants will emerge. They'll handle their customers' full logistics needs at the lowest unit cost, using a series of expert systems." Expert systems can also be used to spot defective welds during the manufacturing process.[37] The expert system analyzes radiographic images and suggests which welds could be flawed.

INFORMATION SYSTEMS @ WORK

Moving Data—and Freight—Efficiently

The Burlington Northern and Santa Fe Railway Co., more conveniently referred to as BNSF, operates one of the largest railroad networks in North America, with about 32,500 route miles covering 28 states and 2 Canadian provinces. The railway moves more freight than any other rail system in the world, is America's largest grain-hauling railroad, transports the mineral components of many of the products we depend on daily, and hauls enough coal to generate more than 10 percent of the electricity produced in the United States. Keeping track of the freight being hauled in 190,000 freight cars pulled by 5,000 locomotives is no small task and until recently was slow and error prone.

While BNSF maintained a state-of-the-art information system in its network operations center in Fort Worth, Texas, its method of retrieving information from the train crews was antiquated. As freight cars were picked up or dropped off, the crew would radio the information to the dispatcher, who would then type the information into the IBM mainframe system. Crews would start their day with written work orders and turn them in at the end of the day, interspersed with periodic calls to report cars dropped off or picked up. Jeff Campbell, BNSF's CIO, viewed this approach as outdated, cumbersome, and incapable of meeting the demands of customers and railroad management for near-real-time data. BNSF set out to design a system that could provide up-to-the-moment shipment information similar to systems used by UPS and FedEx.

BNSF developed a system, based on speech-recognition software by ScanSoft Inc., that automatically turned voice radio calls into data that could be integrated into the company's computer systems. Although speech-recognition systems are popular for automated phone systems, using speech recognition for voice radio is a new application for the technology—one that presented some challenges. "Two-way radio systems have lower fidelity than the phone lines traditionally used with IVR [interactive voice recognition]. The fidelity problem was compounded by the noisy environment of a locomotive cab," said Rob Kassel, ScanSoft's senior product manager for network speech.

ScanSoft employed special noise-filtering software to "cleanup" the signal and sampled numerous engineer radio calls to teach the software to recognize speech generated in such a noisy environment. In addition to speech-recognition technology, ScanSoft also used voice-recognition software to verify the identity of the person speaking. BNSF did not want to risk an outsider's hacking into its private microwave radio system and feeding the system false information.

Once the voice of an incoming caller is recognized and verified, the system prompts the crew for information using an interactive audio menu. The responses from the crew are digitized, interpreted, and translated into text that is automatically entered into the mainframe database. The system supports both radio calls and cell phone calls and is managed on PCs by dispatchers at BNS's network operations center.

BNSF recently rolled out its new system, which it calls the Radio Telephony Interface (RTI), and plans to take it systemwide by 2005. While the company declines to disclose the exact cost of the system, BNSF's IT budget increased from $1.5 million in 2003 to $274 million in 2004.

But the resulting system was worth it. BNSF was recognized by *Speech Technology* magazine as winner of its "Most Innovative Solutions Awards." Candidates for the award were judged on their ability to increase customer self-service, improve worker efficiency, and utilize speech in a creative format for an automated solution. Jeff Campbell says that the new system "improves customer satisfaction" by allowing BNSF to update its Transportation Support System in near-real time. The RTI allows BNSF to provide customers with more frequent information on car moves "and closer expected time of arrival."

Dan Miller, an analyst at Zelos Group Inc. in San Francisco, says radio-to-data interfaces are the next frontier for IVR systems with huge growth potential within many industries, including trucking, utilities, and field service fleet firms.

Discussion Questions

1. What effect has speech recognition had on the daily routines of BNSF crews and dispatchers?
2. What types of return on its investment could BNSF earn with its speech-recognition system?

Critical Thinking Questions

3. Many businesses in the transportation and delivery industries rely on radio communications to keep dispatchers, and even air traffic controllers in touch with drivers, engineers, and pilots. Which of these businesses might benefit from speech recognition? How?
4. What general communication scenarios are ideal for interactive voice-recognition systems? What is it about these scenarios that make them ideal?

SOURCES: Bob Brewin, "A Railroad Finds Its Voice," *Computerworld*, January 26, 2004, *www.computerworld.com*; "ScanSoft Speech System Deployments Receive Industry Innovation Awards," *Business Wire*, May 19, 2004, *www.lexis-nexis.com*; ScanSoft Inc. Web site, accessed June 2, 2004, *www.scansoft.com*; BNSF's Web site, accessed June 2, 2004, *www.bnsf.com*.

- *Hospitals and medical facilities.* Some hospitals use expert systems to determine a patient's likelihood of contracting cancer or other diseases. Hospitals, pharmacies, and other healthcare providers can use CaseAlert by MEDecision to determine possible high-risk or high-cost patients. MYCIN is an expert system developed at Stanford University to analyze blood infections. UpToDate is another expert system used to diagnose patients. A medical expert system used by the Harvard Community Health Plan allows members of the HMO to get medical diagnoses via home personal computers. For minor problems, the system gives uncomplicated treatments; for more serious conditions, the system schedules appointments. The system is highly accurate, diagnosing 97 percent of the patients correctly (compared with the doctors' 78 percent accuracy rating). To help doctors in the diagnosis of thoracic pain, MatheMEDics has developed THORASK, a straightforward, easy-to-use program, requiring only the input of carefully obtained clinical information. The program helps the less experienced to distinguish the three principal categories of chest pain from each other. It does what a true medical expert system should do without the need for complicated user input. The user answers basic questions about the patient's history and directed physical findings, and the program immediately displays a list of diagnoses. The diagnoses are presented in decreasing order of likelihood, together with their estimated probabilities. The program also provides concise descriptions of relevant clinical conditions and their presentations, as well as brief suggestions for diagnostic approaches. For purposes of record keeping, documentation, and data analysis, there are options for saving and printing cases.

Seagate Technology implemented an expert system to monitor its disk drive components-manufacturing processes and improve yields.

(Source: REUTERS/Kin Cheung/ Landov.)

- *Help desks and assistance.* Customer service help desks use expert systems to provide timely and accurate assistance. Kaiser Permanente, a large HMO, uses an expert system and voice response to automate its help desk function. The automated help desk frees up staff to handle more complex needs while still providing more timely assistance for routine calls.
- *Employee performance evaluation.* An expert system developed by Austin-Hayne, called Employee Appraiser, provides managers with expert advice for use in employee performance reviews and career development.
- *Virus detection.* IBM is using neural network technology to help create more advanced software for eradicating computer viruses, a major problem in American businesses. IBM's neural network software deals with "boot sector" viruses, the most prevalent type, using a form of artificial intelligence that mimics the human brain and generalizes by looking at examples. It requires a vast number of training samples, which in the case of antivirus software are 3-byte virus fragments.
- *Repair and maintenance.* ACE is an expert system used by AT&T to analyze the maintenance of telephone networks. IET-Intelligent Electronics uses an expert system to diagnose maintenance problems related to aerospace equipment. General Electric Aircraft Engine Group uses an expert system to enhance maintenance performance levels at all sites and improve diagnostic accuracy.

- *Shipping.* CARGEX cargo expert system is used by Lufthansa, a German airline, to help determine the best shipping routes.
- *Marketing.* CoverStory is an expert system that extracts marketing information from a database and automatically writes marketing reports.
- *Warehouse optimization.* United Distillers uses an expert system to determine the best combinations of liquor stocks to produce its blends of Scottish whiskey. This information is then supplemented with information about the location of the casks for each blend. The system optimizes the selection of required casks, keeping to a minimum the number of "doors" (warehouse sections) from which the casks must be taken and the number of casks that need to be moved to clear the way. Other constraints must be satisfied, such as the current working capacity of each warehouse and the maintenance and restocking work that may be in progress.

Integrating Expert Systems

As with the other information systems, an expert system can be integrated with other systems in an organization through a common database. An expert system that identifies late-paying customers who should not receive additional credit may draw data from the same database as an invoicing MIS that produces weekly reports on overdue bills. The same database—a by-product of the invoicing transaction processing system—might also be used by a decision support system to perform "what-if" analysis to determine the impact of late payments on cash flows, revenues, and overall profit levels.

In many organizations, these systems overlap. A TPS might be expanded to provide management information, which in turn may provide some DSS functions, and so on. In each progressive phase of this overlap, the information system assists with the decision-making process to a greater extent. Of all these information systems, expert systems provide the most support, proposing decisions based on specific problem data and a knowledge base. Understanding the capabilities and characteristics of expert systems is the first step in applying these systems to support managerial decision making and organizational goals.

VIRTUAL REALITY

The term *virtual reality* was initially coined by Jason Lanier, founder of VPL Research, in 1989. Originally, the term referred to *immersive virtual reality* in which the user becomes fully immersed in an artificial, three-dimensional world that is completely generated by a computer. Immersive virtual reality may represent any three-dimensional setting, real or abstract, such as a building, an archaeological excavation site, the human anatomy, a sculpture, or a crime scene reconstruction. Through immersion, the user can gain a deeper understanding of the virtual world's behavior and functionality.

A **virtual reality system** enables one or more users to move and react in a computer-simulated environment. Virtual reality simulations require special interface devices that transmit the sights, sounds, and sensations of the simulated world to the user. These devices can also record and send the speech and movements of the participants to the simulation program, enabling users to sense and manipulate virtual objects much as they would real objects. This natural style of interaction gives the participants the feeling that they are immersed in the simulated world. DaimlerChrysler uses virtual reality to help it simulate and design factories.[38] According to the CIO of DaimlerChrysler, "We are piloting a digital plant. As the engineers and designers are developing new products, the digital factory is simulating production." The company believes that this use of virtual reality will reduce the time it takes to move from an idea to production by about 30 percent.

virtual reality system
A system that enables one or more users to move and react in a computer-simulated environment.

Interface Devices

To see in a virtual world, often the user will wear a head-mounted display (HMD) with screens directed at each eye. The HMD also contains a position tracker to monitor the location of the user's head and the direction in which the user is looking. Using this information, a computer generates images of the virtual world—a slightly different view for each eye—to match the direction that the user is looking and displays these images on the HMD.

With current technology, virtual-world scenes must be kept relatively simple so that the computer can update the visual imagery quickly enough (at least ten times a second) to prevent the user's view from appearing jerky and from lagging behind the user's movements.

Alternative concepts—BOOM and CAVE—were developed for immersive viewing of virtual environments to overcome the often uncomfortable intrusiveness of a head-mounted display. The BOOM (Binocular Omni-Orientation Monitor) from Fakespace Labs is a head-coupled stereoscopic display device. Screens and optical systems are housed in a box that is attached to a multilink arm. The user looks into the box through two holes, sees the virtual world, and can guide the box to any position within the virtual environment. Head tracking is accomplished via sensors in the links of the arm that holds the box.

The BOOM, a head-coupled display device.

(Source: Courtesy of University of Michigan Virtual Reality Laboratory.)

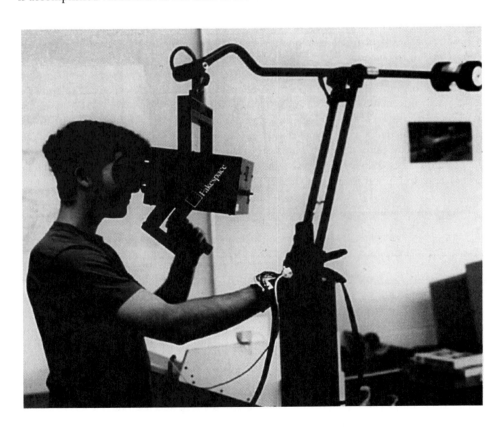

The Electronic Visualization Laboratory at the University of Illinois at Chicago introduced a room constructed of large screens on which the graphics are projected onto the three walls and the floor. The CAVE, as this room is called, provides the illusion of immersion by projecting stereo images on the walls and floor of a room-sized cube. Several persons wearing lightweight stereo glasses can enter and walk freely inside the CAVE. A head-tracking system continuously adjusts the stereo projection to the current position of the leading viewer.

Viewing the Detroit Midfield Terminal in an immersive CAVE system.

(Source: Courtesy of University of Michigan Virtual Reality Laboratory.)

Users hear sounds in the virtual world through earphones. The information reported by the position tracker is also used to update audio signals. When a sound source in virtual space is not directly in front of or behind the user, the computer transmits sounds to arrive at one ear a little earlier or later than at the other and to be a little louder or softer and slightly different in pitch.

The *haptic* interface, which relays the sense of touch and other physical sensations in the virtual world, is the least developed and perhaps the most challenging to create. Currently, with the use of a glove and position tracker, the computer locates the user's hand and measures finger movements. The user can reach into the virtual world and handle objects; however, it is difficult to generate the sensations that are felt when a person taps a hard surface, picks up an object, or runs a finger across a textured surface. Touch sensations also have to be synchronized with the sights and sounds users experience.

Forms of Virtual Reality

Aside from immersive virtual reality, which we just discussed, virtual reality can also refer to applications that are not fully immersive, such as mouse-controlled navigation through a three-dimensional environment on a graphics monitor, stereo viewing from the monitor via stereo glasses, stereo projection systems, and others.

Some virtual reality applications allow views of real environments with superimposed virtual objects. Motion trackers monitor the movements of dancers or athletes for subsequent studies in immersive virtual reality. Telepresence systems (e.g., telemedicine, telerobotics) immerse a viewer in a real world that is captured by video cameras at a distant location and allow for the remote manipulation of real objects via robot arms and manipulators. Many believe that virtual reality will reshape the interface between people and information technology by offering new ways to communicate information, visualize processes, and express ideas creatively.

Computer-generated image technology and simulation is used by companies to determine plant capacity, manage bottlenecks, and optimize production rates.

(Source: Courtesy of Flexsim Software Products, Inc.)

Virtual Reality Applications

There are literally hundreds of applications of virtual reality, with more being developed as the cost of hardware and software declines and people's imaginations are opened to the potential of virtual reality. Here is a summary of some of the more interesting ones.

Medicine

Surgeons in France performed the first successful closed-chest coronary bypass operation. Instead of cutting open the patient's chest and breaking his breastbone, as is usually done, surgeons used a virtual reality system that enabled them to operate through three tiny half-inch incisions between the patient's ribs. They inserted thin tubes to tunnel to the operating area and protect the other body tissue. Then three arms were inserted into the tubes. One was for a 3-D camera; the other two held tiny artificial wrists to which a variety of tools—scalpels, scissors, needle—were attached. The virtual reality system mimicked the movements of the surgeon's shoulders, elbows, and wrists. The surgeon sat at a computer workstation several feet from the operating table and moved instruments that controlled the ones inside the patient. The instruments inside the patient duplicated the motion of the surgeon's hands so accurately that it was possible to sew up a coronary artery as thin as a thread. The surgeon watched his progress on a screen that enlarged the artery in 3-D to the size of a garden hose.

In yet another medical application of virtual reality, researchers are trying to translate biomechanical measurements so that they can be stored in a computer database. Data from actual patients will be gathered from sensors mounted on the fingertips of virtual reality gloves to create a database that will be made available to examining physicians. When a doctor examines a new patient, a feedback system linked to the virtual reality gloves will capture sensations and compare them with data in the database. This approach could allow surgeons to go right into the operating room without having to obtain a CAT scan, thus saving time—and with many injuries, that's absolutely critical.

Virtual reality technology can also link stroke patients to their physical therapists. Patients put on special gloves and other virtual reality devices at home that are linked to the physical therapist's office. The physical therapist can then see whether the patient is performing the correct exercises without having to travel to the patient's home or hospital room. Use of virtual reality can cut travel time and costs.

Education and Training

Virtual environments are used in education to bring exciting new resources into the classroom. Students can stroll among digital bookshelves, learn anatomy on a simulated cadaver, or participate in historical events—all virtually.

Third-grade students at John Cotton Tayloe School in Washington, North Carolina, can take an exciting virtual trip down the Nile for an integrated-curriculum lesson on ancient Egypt. This interactive virtual reality computer lesson integrates social studies, geography, music, art, science, math, and language arts. The software used to design the lesson was 3-D Website Builder (created by the Virtus Corporation), which allowed the designers to create a desert landscape complete with an oasis, camels, and a pyramid. Students could view the scenes from all angles, including front and top views, and could even enter a pyramid and view the sarcophagus holding the mummy in the middle of a room, which contained various Egyptian items. On one wall was a hieroglyphic message that was part of the lesson. On another wall was artwork depicting life in ancient Egypt, and against another were pieces of Egyptian furniture and a harp.

Virtual technology has also been applied by the military. To help with aircraft maintenance, a virtual reality system has been developed to simulate an aircraft and give a user a sense of touch, while computer graphics give the senses of sight and sound. The user sees, touches, and manipulates the various parts of the virtual aircraft during training. The Virtual Aircraft Maintenance System simulates real-world maintenance tasks that are routinely performed on the AV8B vertical takeoff and landing aircraft used by the U.S. Marines. Also, the Pentagon is using a virtual reality training lab to prepare for a military crisis. The virtual reality system simulates various war scenarios.

Real Estate Marketing and Tourism

Virtual reality has been used to increase real estate sales in several powerful ways. From Web publishing to laptop display to a potential buyer, virtual reality provides excellent exposure for properties and attracts potential clients. Clients can take a virtual walk through properties and eliminate wrong choices without wasting valuable time. Virtual walkthroughs can be mailed on diskettes or posted on the Web as a convenience for nonlocal clients. A CD-ROM containing all virtual reality homes can also be sent to clients and other agents. Realatrends Real Estate Service offering homes for sale in Orange County, California (*www.realatrends.com/virtual_tours.htm*) is just one of many real estate firms offering this service. In another Web application, the U.S. government created a virtual tour of the White House while the facility was closed because of security concerns. The virtual tour allowed people to see a 360-degree view of rooms on the Internet.

Entertainment

Computer-generated image technology, or CGI, has been around since the 1970s. A number of movies used this technology to bring realism to the silver screen, including *Finding Nemo, Spider-Man II, Star Wars Episode II—Attack of the Clones*, and many others. A team of artists rendered the roiling seas and crashing waves of *Perfect Storm* almost entirely on computers using weather reports, scientific formulas, and their imagination. There was also *Dinosaur* with its realistic talking reptiles, *Titan A.E.*'s beautiful 3-D space-scapes, and the casts of computer-generated crowds and battles in *Gladiator* and *The Patriot*. CGI can also be used for sports simulation to enhance the viewers' knowledge and enjoyment of a game. A virtual reality game, for example, could allow people to experiment with decisions related to urban planning in a realistic setting.[39] Natural and man-made disasters test decisions on designing buildings and the surrounding area. Other games can display a 3-D view of the world and allow people to interact with simulated people or avatars in the game.[40] Disney's Toontown (*www.toontown.com*) is a virtual reality Internet site for children.[41] Second Life (*www.secondlife.com*) allows people to play games, interact with avatars, and build structures, such as homes. Many virtual reality entertainment sites charge a monthly fee.

OTHER SPECIALIZED SYSTEMS

In addition to artificial intelligence, expert systems, and virtual reality, a number of other interesting and exciting specialized systems have appeared. Segway, for example, is an electric scooter that uses sophisticated software, sensors, and gyro motors to transport people through warehouses, offices, downtown sidewalks, and other spaces.[42] Originally designed to transport people around a factory or around town, more recent versions are being tested by the military for gathering intelligence and transporting wounded soldiers to safety.[43] Cyberkinetics is conducting a medical trial for a small microchip that could be embedded in the brain of patients with spinal cord injuries and wired to a computer.[44] In the trial, patients will be asked to think about moving the cursor on a computer screen while the system tries to record the physiological responses to their thoughts. If successful, the chip might be able to move computer cursors or perform other tasks to help quadriplegics perform tasks they couldn't perform otherwise.[45]

Segway, a human transport device, uses sophisticated software, sensors, and gyro motors to transport people in an upright position.

(Source: © Reuters NewMedia Inc./ CORBIS.)

As mentioned previously, *radio-frequency identification (RFID)* tags that contain small chips with information about products or packages can be quickly scanned to perform inventory control or trace a package as it moves from a supplier to a company to its customers. Many companies have used RFID tags to reduce costs, improve customer service, and achieve a competitive advantage. Delta Airlines, for example, is testing RFID to track luggage through airports.[46] RFID has given the company an accuracy rate that is greater than 95 percent, although full implementation of RFID may not be possible immediately for the cash-strapped company. RFID is also being used by Metro Group, a German retailer with thousands of stores in 28 countries, to track products from warehouses to stores.[47] Tesco, an English retailer, is also using RFID to track products.[48] The Seattle Seahawks football team uses RFID in the Club Seat section to increase sales and speed food sales and purchases of other products.[49] Fans pass food and other products with RFID tags over a PowerPay reader to ring up sales. According to Mike Flood, the team's vice president, "It certainly speeds up lines, and the less wait at the concessions, the more incentive there is to buy." Sales are up by about 18 percent as a result of the RFID tags. Oracle, the large database company, is adding support for RFID technology to its database programs and applications.[50]

Another technology is being used to create "smart containers" for ships, railroads, and trucks.[51] NaviTag and other companies are developing communications systems that would allow containers to broadcast the contents, location, and condition of shipments to shipping and cargo managers. Burlington Northern and Santa Fe Railway Company uses standard radio messages to generate shipment and tracking data for customers and managers.[52] The railroad company used its radio system to link 14,000 miles of track in 27 states. The system's messages are input to a voice-response and speech-recognition system that is tied to the company's back-end processing systems, which tracks shipments.

game theory
Use of information systems to develop competitive strategies for people, organizations, or even countries.

One special application of computer technology is game theory. **Game theory** involves use of information systems to develop competitive strategies for people, organizations, or even countries. Two competing businesses in the same market can use game theory to determine the best strategy to achieve their goals. The military could also use game theory to determine the best military strategy to win a conflict against another country, and individual investors could use game theory to determine the best strategies when competing against other investors in a government auction of bonds. Groundbreaking work on game theory was pioneered by John Nash, the mathematician whose life was profiled in the book and film

A Beautiful Mind.[53] Game theory has also been used to develop approaches to deal with terrorism.

Informatics, another specialized system, combines traditional disciplines, such as science and medicine, with computer systems and technology. *Bioinformatics*, for example, combines biology and computer science. Also called *computational biology*, bioinformatics has been used to help map the human genome and conduct research on biological organisms. Using sophisticated databases and artificial intelligence, bioinformatics helps unlock the secrets of the human genome, which could eventually prevent diseases and save lives. Stanford University has a course on bioinformatics and offers a bioinformatics certification. Medical informatics combines traditional medical research with computer science. Journals, such as *Healthcare Informatics*, report current research on applying computer systems and technology to reduce medical errors and improve healthcare. The University of Edinburgh even has a School of Informatics. The school has courses in the structure, behavior, and interactions of natural and artificial computational systems. The program combines artificial intelligence, computer science, engineering, and science.

Many other specialized devices are used by companies for a variety of purposes. "Smart dust" was developed at the University of California at Berkeley with Pentagon funding.[54] Smart dust involves small networks powered by batteries and an operating system, called TinyOS. The technology can be used to monitor temperature, light, vibration, or even toxic chemicals. Small radio transceivers can be placed in other products, such as cell phones. The radio transceivers allow cell phones and other devices to connect to the Internet, cellular phone service, and other devices that use the technology. The radio transceivers could save companies hundreds of thousands of dollars annually. Microsoft's Smart Personal Objects Technology (SPOT) allows small devices to transmit data and messages over the air.[55] SPOT is being used in wristwatches to transmit data and messages over FM radio broadcast bands. The new technology, however, requires a subscription to Microsoft's MSN Direct information service. Automotive software allows cars and trucks to be connected to the Internet. The software can track a driver's speed and location, allow gas stations to remotely charge for fuel and related services, and more. Special-purpose bar codes are also being introduced in a variety of settings. For example, to manage office space efficiently, a company gives each employee and office a bar code. Instead of having permanent offices, the employees are assigned offices and supplies as needed, and the bar codes help to make sure that an employee's work, mail, and other materials are routed to the right place. Companies can save millions of dollars by reducing office space and supplies. Manufacturing experiments are also being done with inkjet printers to allow them to "print" 3-D parts.[56] The technology is being used in Iowa City to print new circuit boards using a specialized ink-jet printer. The printer sprays layers of polymers onto circuit boards to form transistors and other electronic components.

informatics
A specialized system that combines traditional disciplines, such as science and medicine, with computer systems and technology.

SUMMARY

Principle

Artificial intelligence systems form a broad and diverse set of systems that can replicate human decision making for certain types of well-defined problems.

The term *artificial intelligence* is used to describe computers with the ability to mimic or duplicate the functions of the human brain. The objective of building AI systems is not to replace human decision making completely but to replicate it for certain types of well-defined problems.

Intelligent behavior encompasses several characteristics, including the abilities to learn from experience and apply this knowledge to new experiences; handle complex situations and solve problems for which pieces of information may be missing; determine relevant information in a given situation, think in a logical and rational manner, and give a quick and correct response; and understand visual images and process symbols. Computers are better than human beings at transferring information, making a series of calculations rapidly and accurately, and making complex calculations, but human beings are better than computers at all other attributes of intelligence.

Artificial intelligence is a broad field that includes several key components, such as expert systems, robotics, vision systems, natural language processing, learning systems, and neural networks. An expert system consists of the hardware and software used to produce systems that behave as a human expert would in a specialized field or area (e.g., credit analysis). Robotics uses mechanical or computer devices to perform tasks that require a high degree of precision or are tedious or hazardous for humans (e.g., stacking cartons on a pallet). Vision systems include hardware and software that permit computers to capture, store, and manipulate images and pictures (e.g., face-recognition software). Natural language processing allows the computer to understand and react to statements and commands made in a "natural" language, such as English. Learning systems use a combination of software and hardware to allow a computer to change how it functions or reacts to situations based on feedback it receives (e.g., a computerized chess game). A neural network is a computer system that can simulate the functioning of a human brain (e.g., disease diagnostics system). There are many other artificial intelligence applications also. A genetic algorithm is an approach to solving large, complex problems in which a number of related operations or models change and evolve until the best one emerges. The approach is based on the theory of evolution, which requires variation and natural selection. Intelligent agents consist of programs and a knowledge base used to perform a specific task for a person, a process, or another program.

Principle

Expert systems can enable a novice to perform at the level of an expert but must be developed and maintained very carefully.

Expert systems can explain their reasoning or suggested decisions, display intelligent behavior, manipulate symbolic information and draw conclusions from complex relationships, provide portable knowledge, and deal with uncertainty. They are not yet widely used; some are difficult to use, are limited to relatively narrow problems, cannot readily deal with mixed knowledge, present the possibility for error, cannot refine their own knowledge base, are difficult to maintain, and may have high development costs. Their use also raises legal and ethical concerns.

An expert system consists of a collection of integrated and related components, including a knowledge base, an inference engine, an explanation facility, a knowledge acquisition facility, and a user interface. The knowledge base is an extension of a database, discussed in Chapter 5, and an information and decision support system, discussed in Chapter 10. It contains all the relevant data, rules, and relationships used in the expert system. The rules are often composed of if-then statements, which are used for drawing conclusions. Fuzzy logic allows expert systems to incorporate facts and relationships into expert system knowledge bases that may be imprecise or unknown.

The inference engine processes the rules, data, and relationships stored in the knowledge base to provide answers, predictions, and suggestions the way a human expert would. Two common methods for processing include backward and forward chaining. Backward chaining starts with a conclusion, then searches for facts to support it; forward chaining starts with a fact, then searches for a conclusion to support it. Mixed chaining is a combination of backward and forward chaining.

The explanation facility of an expert system allows the user to understand what rules were used in arriving at a decision. The knowledge acquisition facility helps the user add or update knowledge in the knowledge base. The user interface makes it easier to develop and use the expert system.

The individuals involved in the development of an expert system include the domain expert, the knowledge engineer, and the knowledge users. The domain expert is the individual or group that has the expertise or knowledge being captured for the system. The knowledge engineer is the developer whose job is to extract the expertise from the domain expert. The knowledge user is the individual who benefits from the use of the developed system.

Following is a list of factors that normally make expert systems worth the expenditure of time and money: a high potential payoff or significantly reduced downside risk, the ability to capture and preserve irreplaceable human expertise, the ability to develop a system more consistent than human experts, situations in which expertise is needed at a number of locations at the same time, and situations in which expertise is needed in a hostile environment that is dangerous to human health. The expert system solution can be developed faster than the solution from human experts. An ES also provides expertise needed for training and development to share the wisdom and experience of human experts with a large number of people.

The steps involved in the development of an expert system include determining requirements, identifying experts, constructing expert system components, implementing results, and maintaining and reviewing the system.

Expert systems can be implemented in several ways. Previously, traditional high-level languages, including Pascal, FORTRAN, and COBOL, were used. LISP and PROLOG are two languages specifically developed for creating expert systems from scratch. A faster and less-expensive way to acquire an expert system is to purchase an expert system shell or existing package. The shell program is a collection of software packages and tools used to design, develop, implement, and maintain expert systems. Advantages of expert system shells include ease of development and modification, use of satisficing, use of heuristics, and development by knowledge engineers and end users. The approach selected depends on the benefits compared with cost, control, and complexity considerations.

The benefits of using an expert system go beyond the typical reasons for using a computerized processing solution. Expert systems display "intelligent" behavior, manipulate symbolic information and draw conclusions, provide portable knowledge, and can deal with uncertainty. Expert systems can be used to solve problems in many fields or disciplines and can assist in all stages of the problem-solving process. Past successes have shown that expert systems are good at strategic goal setting, planning, design, decision making, quality control and monitoring, and diagnosis.

There are a number of applications of expert systems and artificial intelligence, including credit granting and loan analysis, catching cheats and terrorists, budgeting, games, information management and retrieval, AI and expert systems embedded in products, plant layout, hospitals and medical facilities, help desks and assistance, employee performance evaluation, virus detection, repair and maintenance, shipping, and warehouse optimization.

Principle

Virtual reality systems have the potential to reshape the interface between people and information technology by offering new ways to communicate information, visualize processes, and express ideas creatively.

A virtual reality system enables one or more users to move and react in a computer-simulated environment. Virtual reality simulations require special interface devices that transmit the sights, sounds, and sensations of the simulated world to the user. These devices can also record and send the speech and movements of the participants to the simulation program. Thus, users are able to sense and manipulate virtual objects much as they would real objects. This natural style of interaction gives the participants the feeling that they are immersed in the simulated world.

Virtual reality can also refer to applications that are not fully immersive, such as mouse-controlled navigation through a three-dimensional environment on a graphics monitor, stereo viewing from the monitor via stereo glasses, stereo projection systems, and others. Some virtual reality applications allow views of real environments with superimposed virtual objects. Virtual reality applications are found in medicine, education and training, real estate and tourism, and entertainment.

Principle

Specialized systems can help organizations and individuals achieve their goals.

A number of specialized systems have recently appeared to assist organizations and individuals in new and exciting ways. Segway, for example, is an electric scooter that uses sophisticated software, sensors, and gyro motors to transport people through warehouses, offices, downtown sidewalks, and other spaces. Originally designed to transport people around a factory or around town, more recent versions are being tested by the military for gathering intelligence and transporting wounded soldiers to safety. Radio-frequency identification (RFID) tags are used in a variety of settings. Game theory involves the use of information systems to develop competitive strategies for people, organizations, and even countries. Informatics combines traditional disciplines, such as science and medicine, with computer science. Bioinformatics and medical informatics are examples. There are also a number of special-purpose telecommunications systems that can be placed in products for varied uses.

CHAPTER 11: SELF-ASSESSMENT TEST

Artificial intelligence systems form a broad and diverse set of systems that can replicate human decision making for certain types of well-defined problems.

1. The field of artificial intelligence (AI) was developed in the 1980s after the development of the personal computer. True or False?

2. _____ are rules of thumb arising from experience or even guesses.

3. What is an important attribute for artificial intelligence?
    a. the ability to use sensors
    b. the ability to learn from experience
    c. the ability to be creative
    d. the ability to acquire a large amount of external information

4. _____ involves mechanical or computer devices that can paint cars, make precision welds, and perform other tasks that require a high degree of precision or are tedious or hazardous for human beings.

5. What branch of artificial intelligence involves a computer system that can simulate the functioning of a human brain?
    a. expert systems
    b. neural networks
    c. natural language processing
    d. vision systems

6. A(n) _____ is a combination of software and hardware that allows the computer to change how it functions or reacts to situations based on feedback it receives.

Expert systems can enable a novice to perform at the level of an expert but must be developed and maintained very carefully.

7. What is a disadvantage of an expert system?
    a. the inability to solve complex problems
    b. the inability to deal with uncertainty
    c. limitations to relatively narrow problems
    d. the inability to draw conclusions from complex relationships

8. A(n) _____ is a collection of software packages and tools used to develop expert systems that can be implemented on most popular PC platforms to reduce development time and costs.

9. An expert system heuristic consists of a collection of software and tools used to develop expert systems to reduce development time and costs. True or False?

10. What stores all relevant information, data, rules, cases, and relationships used by the expert system?
    a. the knowledge base
    b. the data interface
    c. the database
    d. the acquisition facility

11. A disadvantage of an expert system is the inability to provide expertise needed at a number of locations at the same time or in a hostile environment that is dangerous to human health. True or False?

12. What is NOT used in the development and use of expert systems?
    a. fuzzy logic
    b. the use of rules
    c. the use of cases
    d. the use of natural language processing

13. An important part of an expert system is the _____, which allows a user or decision maker to understand how the expert system arrived at certain conclusions or results.

14. In an expert system, the domain expert is the individual or group that has the expertise or knowledge one is trying to capture in the expert system. True or False?

Virtual reality systems have the potential to reshape the interface between people and information technology by offering new ways to communicate information, visualize processes, and express ideas creatively.

15. A(n) _____ enables one or more users to move and react in a computer-simulated environment.

16. What type of virtual reality is used to make human beings feel as though they are in a three-dimensional setting, such as a building, an archaeological excavation site, the human anatomy, a sculpture, or a crime scene reconstruction?
    a. chaining
    b. relative
    c. immersive
    d. visual

Specialized systems can help organizations and individuals achieve their goals.

17. _____ combines traditional disciplines, such as science and medicine, with computer science.

CHAPTER 11: SELF-ASSESSMENT TEST ANSWERS

(1) False (2) Heuristics (3) d (4) Robotics (5) b (6) learning system (7) c (8) expert system shell (9) False (10) a (11) False (12) d (13) explanation facility (14) True (15) virtual reality system (16) c (17) Informatics

KEY TERMS

artificial intelligence (AI) 508	game theory 538	knowledge user 527
artificial intelligence systems 509	genetic algorithms 518	learning systems 516
backward chaining 524	if-then statements 522	natural language processing 514
domain 527	inference engine 524	neural network 516
domain expert 527	informatics 539	perceptive system 511
expert system 513	intelligent agent 518	robotics 513
expert system shell 521	intelligent behavior 509	rule 523
explanation facility 525	knowledge acquisition facility 525	virtual reality system 533
forward chaining 524	knowledge base 522	vision systems 514
fuzzy logic 523	knowledge engineer 527	

REVIEW QUESTIONS

1. Define the term *artificial intelligence.*
2. What is a vision system? Discuss two applications of such a system.
3. What is natural language processing? What are the three levels of voice recognition?
4. Describe three examples of the use of robotics. How can a microrobot be used?
5. What is a learning system? Give a practical example of such a system.
6. What is a neural network? Describe two applications of neural networks.
7. What is an expert system shell?
8. Under what conditions is the development of an expert system likely to be worth the effort?
9. Identify the basic components of an expert system and describe the role of each.
10. What is fuzzy logic?
11. How are rules used in expert systems?
12. Expert systems can be built based on rules or cases. What is the difference between the two?
13. Describe the roles of the domain expert, the knowledge engineer, and the knowledge user in expert systems.
14. What are the primary benefits derived from the use of expert systems?
15. Identify three approaches for developing an expert system.
16. Describe three applications of expert systems or artificial intelligence.
17. Identify three special interface devices developed for use with virtual reality systems.
18. Identify and briefly describe three specific virtual reality applications.
19. Give three examples of other specialized systems.

DISCUSSION QUESTIONS

1. What are the requirements for a computer to exhibit human-level intelligence? How long will it be before we have the technology to design such computers? Do you think we should push to try to accelerate such a development? Why or why not?
2. What are some of the tasks at which robots excel? Which human tasks are difficult for them to master? What fields of AI are required to develop a truly perceptive robot?
3. Describe how natural language processing could be used in a university setting.
4. You have been hired to capture the knowledge of a brilliant attorney who has an outstanding track record for selecting jury members favorable to her clients during the pretrial jury selection process. This knowledge will be used as the basis for an expert system to enable other attorneys to have similar success. Is this system a good candidate for an expert system? Why or why not?
5. You have been hired to develop an expert system for a university career placement center. Develop five rules a student could use in selecting a career.
6. What is the purpose of a knowledge base? How is one developed?
7. What is the relationship between a database and a knowledge base?

8. Imagine that you are developing the rules for an expert system to select the strongest candidates for a medical school. What rules or heuristics would you include?

9. What skills does it take to be a good knowledge engineer? Would knowledge of the domain help or hinder the knowledge engineer in capturing knowledge from the domain expert?

10. Which interface is the least developed and most challenging to create in a virtual reality system? Why do you think this is so?

11. What application of virtual reality has the most potential to generate increased profits in the future?

PROBLEM-SOLVING EXERCISES

1. You are a senior vice president of a company that manufacturers kitchen appliances. You are considering using robots to replace up to ten of your skilled workers on the factory floor. Using a spreadsheet, analyze the costs of acquiring several robots to paint and assemble some of your products versus the cost savings in labor. How many years would it take to pay for the robots from the savings in fewer employees? Assume that the skilled workers make $20 per hour, including benefits.

2. Assume that you have just won a lottery worth $100,000. You have decided to invest half the amount in the stock market. Develop a simple expert system to pick ten stocks to consider. Using your word processing program, create seven or more rules that could be used in such an expert system. Create five cases and use the rules you developed to determine the best stocks to pick.

3. Using a graphics program, diagram the components of a virtual reality system that could be used to market real estate. Carefully draw and label each component. Use the same graphics program to make a one-page outline of a presentation to a real estate company interested in your virtual reality system.

TEAM ACTIVITIES

1. With two or three of your classmates, do research to identify three real examples of natural language processing in use. Discuss the problems solved by each of these systems. Which has the greatest potential for cost savings? What are the other advantages of each system?

2. Form a team to debate other teams from your class on the following topic: "Are expert systems superior to human beings when it comes to making objective decisions?" Develop several points supporting either side of the debate.

3. With members of your team, think of an idea for a virtual reality system for a new, exciting game. What are the main features of the game that make it unique and highly marketable?

WEB EXERCISES

1. Use the Internet to get information about the use of neural networks. Describe three examples of how this technology is used.

2. This chapter discussed several examples of expert systems. Search the Internet for two examples of the use of expert systems. Which one has the greatest potential to increase profits for the firm? Explain your choice.

3. Use the Internet to get more information about one of the specialized systems discussed at the end of chapter. Write a report about what you found. Give an example of a new special-purpose system that has great promise in the future.

CAREER EXERCISES

1. Using the Internet or a library, explore how expert systems can be used in a business career. How can expert systems be used in a nonprofit company?

2. Select and describe a special-purpose system, other than an expert system, that would be the most beneficial in your career of choice. Use the Internet to find how this special-purpose system is actually being used.

VIDEO QUESTIONS

Watch the video clip **Predicting the Future of AI—Marvin Minsky** and answer these questions:

1. What reasons does Dr. Minsky provide for the lengthy amount of time that it is taking to build an intelligent computer?

2. What does Dr. Minsky feel is necessary for a computer to be able to do in order to achieve human intelligence?

CASE STUDIES

Case One

French Burgundy Wines: The Sweet Smell of Success

Virtual reality systems strive to simulate real-world experiences by reproducing the details of an environment for our five senses. Of these senses, vision and touch tend to get the most attention. Data helmets and data gloves allow VR users to visually enter virtual environments and manipulate objects within those environments. Some virtual world designers have gone as far as including 3-D sound in their worlds. Few, however, have endeavored to re-create smells. The French trade group, the Bureau Interprofessionnel des Vins de Bourgogne (BIVB), has been working with communications giant France Télécom to produce just such a technology.

The Bureau Interprofessionnel des Vins de Bourgogne, literally translated as the Interprofessional Office of the Wines of Burgundy, wants people to be able to experience the aromas of Burgundy (a region of France), as well as the bouquet of their wines by visiting its Web page. The fruits of its IS labor have produced a computer peripheral called the Olfacom. The Olfacom consists of two foot-tall plastic columns that can be plugged into a PC's USB port to pump out aromas such as hay, flowers, and fruit, using a combination of essential oils stored in its several tanks.

BIVB developed a Web site that could take advantage of the scent capability in a virtual tour. For example, after clicking on the "winemaking" icon, animation begins on-screen, showing workers in a winery unloading pinot noir grapes. As a voice states "the first and vital task is to sort the grapes,"

the smell of black currants wafts from the Olfacom. An icon at the bottom of the screen notifies users of what they smell.

BIVB has taken its product on the road to demonstrate at a variety of trade shows including the Comdex and Apple expos. Steven Meyer-Rassow, who works for a British wine retailer and tried the Olfacom at a trade show, likened it to a remote-control air freshener.

Besides providing a more sensual environment for selling wine, the Olfacom has been used in marketing perfume. It was paired with a marketing campaign for Mania, a fragrance from Giorgio Armani. The device was said to have perfectly reproduced the fragrance from the famous Italian designer.

Some artistic endeavors have also incorporated the device. Artist Alex Sandover is touring galleries with a show entitled Synaesthia. The show re-creates life in the 1950s using video, photos, and aromas. One life-size scene shows a woman at work in a kitchen. Throughout the scene the audience can smell kitchen aromas from the 1950s, such as natural gas, a freshly lit match, fried chicken, apples, and kitchen cleaning products.

BIVB feels that the Olfacom is destined for success. With the interest of French vineyards and the tourism industry, and with the financial backing of France Télécom, the product has real potential in the French market. The device is being redesigned smaller so that it can be attached to the side of a computer. BIVB anticipates that within a few years, all French tourism sites will have Olfacom capability. "Many [wine] professionals are also showing interest," claims Nelly Blau-Picard, BIVB export marketing manager. "They want to

be a part of something that they have never seen available on the market." Just as visual virtual reality has provided a boost to e-commerce by allowing customers to manipulate 3-D products on the display, olfactory virtual reality my soon provide a similar boost to products that depend on their aroma.

Discussion Questions

1. What products do you think could benefit from the use of the Olfacom?
2. How might a device such as the Olfacom be applied to the entertainment industry?

Critical Thinking Questions

3. What are the hazards of dealing with virtual sensory information? Could you imagine an unpleasant Olfacom experience?
4. What types of devices (goggles, headphones, etc.) might be necessary for us to enter a fully simulated world as our senses experience it? What might be the most efficient method for entering a virtual world, requiring the least amount of devices?

SOURCES: Jacob Gaffney, "Smell While You Surf: Burgundy Web Site Hopes to Offer Virtual Tour with Wine Aromas," *Wine Spectator*, May 28, 2004, *www.winespectator.com*; the Olfacom Web site, accessed June 2, 2004, *www.olfacom.com*; Bureau Interprofessionnel des Vins de Bourgogne Web site, accessed June 2, 2004, *www.bivb.com*; Synaesthia, by Alex Sandover, Web site, accessed June 2, 2004, *www.alexsandover.com/peterborough.html*.

Case Two

Expert System Provides Safety in Nuclear Power Plants

TEPCO, for Tokyo Electric Power Corporation, is Japan's largest electric utility, serving more than 27 million customers in the Tokyo metropolitan region. Even with its $40 billion average annual revenue, TEPCO is being challenged to increase efficiency throughout its business. After streamlining many of its business processes and procedures, TEPCO began evaluating methods to reduce costs and improve efficiency in the reliability and safety of its power plants.

TEPCO uses a technique known as probabilistic safety analysis (PSA), which enables engineers to assess the probability and consequences of potential plant safety-related problems. PSA is a popular assessment technique in the nuclear power, chemical, and aerospace industries. PSA creates and interprets fault-tree diagrams, a graphical model illustrating the pathways within a system with logical constructs such as AND and OR. Following the pathways through the system can identify weaknesses and points of failure.

A TEPCO safety engineer might take days or even weeks to work through a fault-tree analysis. This tedious process struck TEPCO management as one that was ideal for automation. TEPCO turned to Gensym Corporation to assist it in developing an expert system to assist with PSA.

Gensym Corporation, out of Burlington, Massachusetts, provides software products and services that enable organizations to automate aspects of their operations that have historically required the direct attention of human experts. Gensym worked together with TEPCO to deploy a system they named FT-Free within the TEPCO nuclear power plant facility. The system automatically creates PSA fault-tree models that represent potential failure modes for critical plant processes. From a model, engineers apply FT-Free's built-in expertise to quickly assess the risks of various types of problems that can adversely impact the plant's reliability and safety. Through this assessment, engineers drive critical decisions about a plant's process configuration and its operational safety procedures.

The system can do in minutes what used to take days or weeks. Not only is it a huge time-saver, but it also provides consistency that was previously lacking. A consistent result is reached independently of which engineer is providing the analysis. This enables TEPCO to best utilize its safety engineers, each of whom typically requires one year of specialized training to be qualified to complete a PSA.

The new system provides "unique abilities to intuitively capture process engineering knowledge using such techniques as connectable object models and rule-based logic," said Koichi Miyata, a project manager with TEPCO Systems. "The result is that we created a solution that greatly reduces the time and effort it takes to make important plant decisions and that is broadly applicable across industries in which safety analysis is critical."

TEPCO Systems has successfully deployed FT-Free within a TEPCO nuclear power plant facility, is expanding its use within TEPCO, and is now marketing it to the electric utility industry.

Discussion Questions

1. What features are essential in an expert system responsible for the safety of something as potentially dangerous as a nuclear power plant?
2. What characteristics of the process of probabilistic safety analysis do you think led TEPCO to believe that it could be automated in an expert system?

Critical Thinking Questions

3. What other industries would benefit from a product such as FT-Free? Why?
4. Why do you think TEPCO decided to package and market its FT-Free system to other power companies? Is this a wise move in terms of competitive advantage?

SOURCES: "TEPCO Systems Expands Use of Gensym's G2 for Assessing Reliability and Safety of Electric Power Plants," *Business Wire*, February 19, 2004, *www.lexis-nexis.com*; the Gensym Web site, accessed June 2, 2004, *www.gensym.com*; TEPCO, Tokyo Electric Power Co., Web site, accessed June 2, 2004, *www.tepco.co.jp/index-e.html*.

Case Three

BankFinancial Corp. Gets a Lesson in Predictive Analytics

Predictive analytics, a relatively new form of data analysis, is becoming increasingly popular with many businesses. Predictive analytics enables companies to derive new insight or new information from existing information. It takes data mining to the next level by applying artificial intelligence techniques, such as neural networks, to customer data to predict what your customers may do in the future. This new technology has developed as a result of increasing processor power, advances in AI technology, increased amounts of customer data to draw from, and heightened demand for valuable information to guide a company's future.

By applying artificial intelligence to data mining, companies are able to predict, for example, which of their customers are most likely to leave within the next month. Imagine the value of such information! Once provided with a list of hundreds or even thousands of customers that you are likely to lose, you could take action to keep them. Predictive analytics can lead to a high return on investment if managed intelligently.

BankFinancial Corp. in Chicago began using predictive analytics to develop models so that the bank can, for example, more accurately target promotions to customers and new prospects. The software it uses employs neural networks and regression routines to create these models. You were introduced to neural nets in this chapter. Regression routines are perhaps the most widely used statistical technique to estimate relationships between independent variables and dependent variables to help understand and explain relationships and predict outcomes.

William Connerty, assistant vice president of market research for BankFinancial, anticipates that the software will reduce the time it takes the bank to develop models by 50 to 75 percent. "The biggest obstacle is getting transaction data and dealing with disparate data sources," Connerty says. Neural nets require large amounts of data for maximum accuracy. Unfortunately, BankFinancial works with data that originates in several separate bank systems in a variety of formats. Much systems integration and interface work needs to be done before the bank will see the full fruits of its predictive analytics tools. Connerty hopes the effort will be worthwhile. "We need to increase our efficiency, our ability to deliver actionable information to decision makers," he says. "I'm under a lot of pressure to deliver."

Some software tools have been developed to assist businesses in gathering and integrating data for predictive analytics. KXEN Inc., a software company in San Francisco, claims that its Analytic Framework product can greatly reduce the time it takes to define, develop, and run a model. Seymour Douglas, director of CRM and database marketing at Cox Communications in Atlanta, claims that it cuts data preparation time in half. Having clean, uniformly formatted data, however, is only the first obstacle to realizing success with predictive analytics.

Robert Berry, president and CEO of Central Michigan University Research Corp. in Mount Pleasant, thinks that lack of organization is the fundamental stumbling block that keeps businesses from capitalizing on predictive analytics. Berry says predictive modeling should involve collaboration among people who have IT, analytical, and business expertise. "You have to build a business-intelligence team," he says. "But companies are struggling with common issues like who owns it, who manages it, and so on. How do you pull the business skills, the IT skills, and the analytical skills across corporate silos and create this team? It's not easy."

Predictive analytics is a valuable technique that no doubt will earn increasing attention of businesses of all sizes. But the current analytics systems require a sizeable, consistently formatted, integrated data warehouse—plus a staff of professionals who understand its potential.

Discussion Questions

1. How can predictive analytics increase BankFinancial Corp's competitive advantage?
2. What types of data do you think must be analyzed to determine what customers are most likely to leave the bank during the coming month?

Critical Thinking Questions

3. How might predictive analytics be used by an organization such as the IRS?
4. Consider a future world where computer processing power is seemingly limitless and business analytics software is used effectively by all businesses to predict customer behavior. What will marketing look like in this future world? How will one business get a "leg up" on another?

SOURCES: Gary H. Anthes, "The Forecast Is Clear," *Computerworld*, April 14, 2003, *www.computerworld.com*; Colin Shearer, "Leveraging Predictive Analytics in Marketing Campaigns," *DMReview*, April 16, 2004, *www.dmreview.com*.

NOTES

Sources for the opening vignette: "Amazon.com Calls on SAS for Fraud Detection," *Citigate ICT PR*, May 20, 2003, *www.lexis-nexis.com*; "Growth of Internet Fraud Is Driving New Technologies to Safeguard Online Payments," *PR Newswire*, December 10, 2003; Amazon.com Web site, accessed May 31, 2004, *www.amazon.com*.

1. Sourbut, Elizabeth, "An Agreeable Android," *New Scientist*, February 28, 2004, p. 49.
2. Excell, John, "Interview – Robo Doc," *The Engineer*, March 5, 2004, p. 30.
3. Bentley, Ross, "Man Versus Machine," *Computer Weekly*, March 20, 2003, p. 24.
4. Gray, Madison, "No Winner In Chess Match Pitting Man Against Machine," *Rocky Mountain News*, February 8, 2003, p. 31A.
5. Ulanoff, Lance, "Cognitive Machines," *PC Magazine*, July 2003, p. 118.
6. Begley, Sharon, "This Robot Can Design, Perform and Interpret A Genetic," *The Wall Street Journal (Eastern Edition)*, January 16, 2004, p. A7.
7. Pope, Martin, "Robo Recon," *Rocky Mountain News*, February 3, 2003, p. 11B.
8. Staff, "Firm Cheers Loss of Robot in Iraq," *CNN Online*, April 13, 2004.
9. Lunsford, Lynn, et al., "Shuttle Crash Raises Questions about Future Manned Flights," *The Wall Street Journal*, February 3, 2003, p. A1.
10. Staff, "Get You a Beer," *PC World*, March 2003, p. 83.
11. Rosencrance, Linda, "AI Loves Lucy," *Computerworld*, November 10, 2003, p. 36.
12. Walton, Marsha, "Meet Paro, The Therapeutic Robot Seal," *CNN Online*, November 20, 2003, *www.cnn.com*.
13. Staff, "Rat's Whiskers Help to Make Better Robots," *Factory Equipment News*, May 2003.
14. Scanlon, Bill, "Robotic da Vinci Sculpts First Success on Gall Bladder," *Rocky Mountain News*, August 12, 2003, p. 8A.
15. Wysocki, Bernard, "Robots in the OR," *The Wall Street Journal*, February 26, 2004, p. B1.
16. Staff, "Robot Football Players Will Give Boot to Beckham," *Electronics Weekly*, May 14, 2003, p. 1.
17. Staff, "Machine Vision and Infrared Lighting Track Production," *Vision Systems Design*, April 2003, p. 29.
18. Staff, "A Vision of Robots Soldering On," *Automation*, April 2003, p. 14.
19. Vijayan, Jaikumar, "Captchas Eat Spam," *Computerworld*, June 16, 2003, p. 32.
20. Lunt, Penny, "Online Retailer Restores Web Sales," *Transform Magazine*, January 1, 2004, p. 32.
21. Staff, "Industry Watch," *Health Management Technology*, February 2004, p. 10.
22. Ball, Richard, "Fujitsu Walking Robot Gets Its Brains from Neural Networks," *Electronics Weekly*, April 9, 2003, p. 19.
23. Shihav, Al, et al., "Distributed Intelligence for Multiple-camera Visual Surveillance," *Pattern Recognition*, April 2004, p. 675.
24. Fielding, Rachel, "NHS Fraud Team Calls in SAS," *Accountancy Age,"* February 27, 2003, p.10.
25. Johnson, Colin, "Neural Software Could Become Soldier's Best Friend," *Electronic Engineering Times*, February 2, 2004, p. 51.
26. Ke-zhang, et al., "Recognition of Digital Curves Scanned From Paper Drawings Using Genetic Algorithms," *Pattern Recognition*, January 2003, p. 123.
27. Hyung, Rim, et al., "An Agent for Selecting Optimal Order Set in EC Marketplace," *Decision Support Systems*, March 2004, p. 371.
28. Sexton, Randall, et al., "Improving Decision Effectiveness of Artificial Neural Networks: A Modified Genetic Algorithm Approach," *Decision Sciences*, Summer 2003, p. 421.
29. Tischelle, George, "Searching For That One," *Information Week*, Febraury 17, 2003, p. 58.
30. Overby, Stephanie, "The New, New Intelligence," *CIO Magazine*, January 1, 2003, p. 35.
31. Staff, "The Imagineer," *Library Journal*, March 15, 2003, p. 18.
32. Angel, Ana Maria, et al., "Soquimich Uses a System Based on Mixed-Integer Linear Programming and Expert Systems to Improve Customer Service," *Interfaces*, July–August 2003, p. 41.
33. Lenard, Mary Jane, "Knowledge Acquisition and Memory Effects Involving an Expert System Designed as a Learning Tool for Internal Control Assessment," *Decision Sciences*, Spring 2003, p. 23.
34. Huang, Z., et al., "Development of an Intelligent Decision Support System for Air Pollution Control," *Expert Systems With Applications*, April 2004, p. 335.
35. Shaalan, K., et al., Án Expert System for the Best Weight Distribution on Ferryboats," *Expert Systems with Applications*, April 2004, p. 397.
36. Raymond, Charles, "Horizon Lines LLC," *Journal of Commerce*, January 12, p. 1.
37. Liao, T.W., "Fuzzy Reasoning Based Automatic Inspection of Radiographic Welds," *Journal of Intelligent Manufacturing*, February 2004, p. 69.
38. Saran, Cliff, "Digital Factories Use Virtual Reality to Track Car Production," *Computer Weekly*, March 2, 2004, p. 18.
39. Louderback, Jim, "Real-World Lessons," *USA Weekend*, July 25, 2003, p. 4.
40. Powers, Kemp, "In There," *Forbes*, September 29, 2003, p. 92.
41. Costa, Dan, "Virtual Worlds," *PC Magazine*, October 28, 2003, p. 158.
42. Armstrong, David, "The Segway," *The Wall Street Journal*, February 12, 2004, p. B1.
43. Staff, "Will Segways Become Battlefield Bots," *CNN Online*, December 2, 2003, *www.cnn.com*.
44. Moukheiber, Zina, "Mind Over Matter," *Forbes*, March 15, 2004, p. 186.
45. Pollack, Andrew, "With Tiny Brain Implants, Just Thinking May Make It So," *The New York Times*, April 13, 2004, *www.nytimes.com*.
46. Brewin, Bob, "Delta Says Radio Frequency ID Devices Pass First Bag-Tag Test," *Computerworld*, December 22, 2003, p. 7.
47. Sliwa, Carol, "German Retailer's RFID Effort Rivals Wal-Marts," *Computerworld*, January 13, 2004, p. 10.
48. Staff, "Riding the Wave of the Future," *MMR*, January 12, 2004, p. 38.
49. Alan Cohen, "Fast Food," *PC Magazine*, May 4, 2004, p. 76.
50. Songini, Marc, "Oracle to Add RFID Support to Its Apps," *Computerworld*, January 26, 2004, p. 54.
51. Machalaba, Daniel, et al., "Thinking Inside the Box," *The Wall Street Journal*, January 15, 2004, p. B1.
52. Brewin, Bob, "A Railroad Finds Its Voice," *Computerworld*, January 26, 2004, p. 37.
53. Begley, Sharon, "A Beautiful Science: Getting The Math Right Can Thwart Terrorism," *The Wall Street Journal*, May 10, 2003, p. B1.
54. Boyle, Matthew, "Smart Dust Kicks Up a Storm," *Fortune*, February 23, 2004, p. 76.
55. Manes, Stephen, "New Twist for the Wrist," *Forbes*, February 2, 2004, p. 94.
56. Weiss, Todd, "Printer Majic," *Computerworld*, January 26, 2004, p. 31.

Kulula.com: The Trials and Tribulations of a South African Online Airline

Anesh Maniraj Singh
University of Durban

Kulula.com was launched in August 2001 as the first online airline in South Africa. Kulula is one of two airlines that are operated by Comair Ltd. British Airways (BA), the other airline that Comair runs, is a full-service franchise operation that serves the South African domestic market. Kulula, unlike BA, is a limited-service operation aimed at providing low fares to a wider domestic market using five aircraft. Since its inception, Kulula has reinvented air travel in South Africa, making it possible for more people to fly than ever before.

Kulula is a true South African e-commerce success. The company boasts as one of its successes the fact that it has been profitable from day one. It is recognised internationally among the top low-cost airlines and participated in a conference attended by other such internationally known low-cost carriers as Virgin Blue, Ryanair, and easyJet. Kulula also received an award from the South African Department of Trade and Industry for being a Technology Top 100 company.

Kulula's success is based on its clearly defined strategy of being the lowest-cost provider in the South African domestic air travel industry. To this end, Kulula has adopted a no-frills approach. Staff and cabin crew wear simple uniforms, and the company has no airport lounges. There are no business class seats and no frequent-flyer programs. Customers pay for their food and drinks. In addition, Kulula does not issue paper tickets, and very few travel agents book its flights—90 percent of tickets are sold directly to customers. Furthermore, customers have to pay for ticket changes, and the company has a policy of "no fly, no refund." Yet, in its drive to keep costs down, Kulula does not compromise on maintenance and safety, and it employs the best pilots and meets the highest safety standards. Like all B2C companies, Kulula aims to create customer value by reducing overhead costs, including salaries, commissions, rent, and consumables such as paper and paper-based documents. Furthermore, by cutting out the middleman such as travel agents, Kulula is able to keep prices low and save customers the time and inconvenience of having to pick up tickets from travel agents. Instead, customers control the entire shopping experience.

Kulula was the sole provider of low-cost flights in South Africa until early 2004, when One Time launched a no-frills service to compete head-on with Kulula. Due to the high price elasticity of demand within the industry, any lowering of price stimulates a higher demand for flights. The increase in competition in the low-price end of the market has seen Kulula decrease fares by up to 20% whilst increasing passengers by over 40%. There, however, has been no brand switching. Kulula has grown in the market at the expense of others.

Apart from its low-cost strategy, Kulula is successful because of its strong B2C business model. As previously mentioned, 90 percent of its revenue is generated from direct sales. However, Kulula has recently ventured into the B2B market by collaborating with Computicket and a few travel agents, who can log in to the Kulula site from their company intranets. Kulula offers fares at substantial reductions to businesses that use it regularly. Furthermore, Kulula bases its success on three simple principles: Any decision taken must bring in additional revenue, save on costs, and/or enhance customer service. Technology contributes substantially to these three principles.

In its first year, Kulula used a locally developed reservation system, which soon ran out of functionality. The second-generation system was AirKiosk, which was developed in

Boston for Kulula. The system change resulted in an improvement of functionality for passengers. For example, in 2003, Kulula ran a promotion during which tickets were sold at ridiculously low prices, and the system was overwhelmed. Furthermore, Kulula experienced a system crash that lasted a day and a half, which severely hampered sales and customer service. As a result, year two saw a revamp in all technology: All the hardware was replaced, bandwidth was increased, new servers and database servers were installed, and Web hosting was changed. In short, the entire system was replaced. According to IT Director Carl Scholtz, "Our success depends on infrastructural stability; our current system has an output that is four times better than the best our systems could ever produce. " Kulula staff members are conscious of the security needs of customers and have invested in 128-bit encryption, giving customers peace of mind that their transactions and information are safe.

The success behind Kulula's systems lies in its branding—its strong identity in the marketplace, which includes its name and visual appeal. The term *kulula* means "easy," and Kulula's Web site has been designed with a simple, no-fuss, user-friendly interface. When one visits the Kulula site, one is immediately aware that an airline ticket can be purchased in three easy steps. The first step allows customers to choose destinations and dates. The second step allows customers to choose the most convenient or cheapest flight based on their need. Kulula also allows customers to book cars and accommodations in step two. Step three is the transaction stage, which allows customers to choose the most suitable payment method. The confirmation and ticket can be printed once payment has been settled. Kulula has not embraced mobile commerce yet, because the technology does not support the ability to allow customers to purchase a ticket in three easy steps. Unlike other e-commerce sites, Kulula is uncluttered and simple to understand, enhancing customer service. Kulula is a fun brand—with offbeat advertising campaigns and bright green and blue corporate and aircraft colours—but behind the fun exterior is a group of people who are serious about business.

Kulula's future is extremely promising. Technology changes continually, and Kulula strives to have the best technology in place at all times. B2B e-commerce will continue to be a major focus of the company in order to develop additional distribution channels with little or no cost. In conjunction with bank partners, Kulula is developing additional methods of payment to replace credit card payments, allowing more people the opportunity to fly. These transactions will be free. Kulula is also involving customers in its marketing efforts by obtaining their permission to promote special offers by e-mail and short message service to customers' cell phones. The Kulula Web site will soon serve as a ticketing portal, where customers can also purchase British Airways tickets, in three easy steps. The company has many other developments in the pipeline that will enhance customer service. According to Scholtz, "We are not an online airline, just an e-tailer that sells airline tickets."

Discussion Questions

1. This case does not mention any backup systems, either electronic or paper based. What would you recommend to ensure that the business runs 24/7/365?

2. It is clear from this case that Kulula is a low-cost provider. What else could Kulula do with its technology to bring in additional revenue, save on cost, and enhance customer service?

3. Does the approach taken by Kulula in terms of its strategy, its business model, and the three principles of success lend itself to other businesses wanting to engage in e-commerce?

4. Kulula flights are almost always full. Do you think that by partnering with a company such as Lastminute.com the airline could fly to capacity at all times? What are the risks related to such a collaboration?

Critical Thinking Questions

5. Kulula initially developed its systems in-house, which it later outsourced to AirKiosk in Boston. Do you think it is wise for an e-business to outsource its systems development? Is it strategically sound to outsource systems development to a company in a different country?

6. With the current trends in mobile commerce, could Kulula offer its services on mobile devices such as cellular phones? Would the company have to alter its strategic thinking to accommodate such a shift? Is it possible to develop a text-based interface that could facilitate a purchase in three easy steps?

STORAGE DATA

BYTE 0

1   2   3   4   5   6

PART
4

Systems
Development

Chapter 12   Systems Investigation and Analysis
Chapter 13   Systems Design, Implementation, Maintenance, and Review

CHAPTER · 12 ·

Systems Investigation and Analysis

PRINCIPLES	LEARNING OBJECTIVES
■ Effective systems development requires a team effort from stakeholders, users, managers, systems development specialists, and various support personnel, and it starts with careful planning.	■ Identify the key participants in the systems development process and discuss their roles. ■ Define the term *information systems planning* and list several reasons for initiating a systems project. ■ Identify important system performance requirements of transaction processing applications that run on the Internet or a corporate intranet or extranet. ■ Discuss three trends that illustrate the impact of enterprise resource planning software packages on systems development.
■ Systems development often uses tools to select, implement, and monitor projects, including net present value (NPV), prototyping, rapid application development, CASE tools, and object-oriented development.	■ Discuss the key features, advantages, and disadvantages of the traditional, prototyping, rapid application development, and end-user systems development life cycles. ■ Identify several factors that influence the success or failure of a systems development project. ■ Discuss the use of CASE tools and the object-oriented approach to systems development.
■ Systems development starts with investigation and analysis of existing systems.	■ State the purpose of systems investigation. ■ Discuss the importance of performance and cost objectives. ■ State the purpose of systems analysis and discuss some of the tools and techniques used in this phase of systems development.

INFORMATION SYSTEMS IN THE GLOBAL ECONOMY ⟫
BMW, GERMANY

Designing a New System to Ramp Up Production

The BMW Group is reputed to be the only manufacturer of automobiles and motorcycles worldwide that concentrates single-mindedly on premium manufacturing standards and outstanding quality for all its brands. Its brands include BMW, MINI, Rolls-Royce Motor Cars, and BMW Motorcycles. The BMW manufacturing plant in South Carolina is part of the BMW Group's global manufacturing network and is the exclusive manufacturing plant for the company's hotselling Z3 Roadster and X5 Sports Activity Vehicles.

The Z3 and X5 are so popular, in fact, that BMW was selling them faster than it could build them. "Customer demand for our new X5 Sport Activity Vehicle and our popular Z3 models required a solution that would successfully speed up and support streamlined production," says Hans Nowak, program manager for BMW Manufacturing Company in Spartanburg. Manufacturing each of these vehicles requires hundreds of parts from hundreds of suppliers, so coordinating purchasing and acquisition of these parts can be more complicated than constructing the vehicles. Executives at the plant recognized that increasing the efficiency of parts procurement could be key to ramping up production to meet demand.

To lay the groundwork for developing a lean and mean procurement system, BMW organized a team to investigate the problem and gather information. At the time, BMW was using a custom-built information system to manage its supplier and logistics requirements. The old system monitored parts inventory and placed orders with suppliers well before they were needed to avoid halting production. As a result, the plant had large amounts of excess parts on hand and unreliable delivery schedules. The old system was wasteful—and worrisome. The team sat down with the results of their investigation to establish objectives for a new and improved system.

An ideal system, they decided, would need to include a tight supplier network that kept parts coming to the assembly lines in just-in-time (JIT) mode to produce cars to meet customer demand. The challenge was in dealing with hundreds of different suppliers, each of which required varying amounts of time to produce and deliver the parts. It was soon clear to the team that the demands of their dream system far exceeded the capabilities of their current custom-built system. They decided to look to external software vendors for a solution.

After contacting a number of software vendors, issuing requests for proposals (RFPs), and analyzing the results, the team found one company with what appeared to be an ideal solution. Software giant SAP offered a custom-designed automotive procurement solution called mySAP Automotive. mySAP Automotive overcame the challenge of creating schedules for each of BMW's suppliers by automatically generating custom schedules for each part; the system generated schedules to match BMW's assembly-line planning and sequencing directives. Once BMW decided on the SAP solution, it was able to complete its implementation within one month.

The new system provides two types of reports to suppliers: long-horizon forecasts and short-horizon JIT delivery schedules. Larger suppliers receive the

information via electronic data interchange (EDI). Other suppliers access the mySAP Automotive Supplier Portal, where BMW posts the requirements to provide up-to-date information on its delivery needs. Using only an Internet browser, suppliers can view this information in real time, including release schedules, purchasing documents, invoices, and engineering documents.

As parts are shipped, BMW receives advance shipping notifications (ASNs), which provide exact information on parts counts and delivery dates. Parts arriving at the BMW dock are then received and transferred directly to the line.

The new system goes beyond procurement to monitor production status in real time. mySAP Automotive registers production confirmation and parts consumption information every three minutes. Parts consumed during assembly are removed from the inventory count, and costs are posted to calculate the value of work in process. Production supervisors are able to monitor the ebb and flow of parts inventory in real time, as well as look ahead and know when incoming shipments will be arriving. Hans Nowak sums it up this way: "mySAP Automotive helps us reduce order-to-delivery time, strengthens our supply chain activities in the areas of demand planning and tracking and tracing of material deliveries, and improves inventory accuracy across our Spartanburg plant—enabling us to significantly reduce time-to-customer for our popular X5 and Z3 models."

BMW also values SAP's standardized and scalable platform. The system can be expanded and modified to embrace new functionality or changes in business processes. BMW has already integrated other business processes with mySAP Automotive. Information housed in BMW systems can be easily accessed, including forecast requirements for components, sales orders, and vehicle bills of material. In addition to the industry-specific solution, BMW is using SAP applications for logistics, financials, and human resources.

As you read this chapter, consider the following:

- What is the value of continuously evaluating and improving the information systems that drive a company and connect it with its suppliers and customers?
- What are important considerations when deciding whether to implement a new information system?

Why Learn About Systems Development?

Throughout this book, you have seen many examples of the use of information systems in a variety of careers. But where do you start to acquire these systems or have them developed? How can you work with IS personnel, such as systems analysts and computer programmers, to get what you need to succeed on the job? This chapter, the first of two chapters on systems development, gives you the answer. You will see how you can initiate the systems development process and analyze your needs with the help of IS personnel. Systems investigation and systems analysis are the first two steps of the systems development process. This chapter provides specific examples of how new or modified systems are initiated and analyzed in a number of industries. In this chapter, you will learn how your project can be planned, aligned with corporate goals, rapidly developed, and much more. We start with an overview of the systems development process.

When an organization needs to accomplish a new task or change a work process, how does it do it? It develops a new system or modifies an existing one. Systems development is the activity of creating or modifying existing systems. It refers to all aspects of the process—from identifying problems to be solved or opportunities to be exploited to the implementation and refinement of the chosen solution. Even governmental agencies use systems development. The Naval Sea Systems Command used systems development to streamline its procurement process.[1] The military operation has an annual budget of $19 billion. As a result of the project,

the naval department was able to cut buying time by 85 percent—from roughly 270 days to about 30 days. In another systems development project, the U.S. Congress allocated $1.35 billion to the Internal Revenue Service (IRS) to upgrade its computer systems.[2] Congress and U.S. citizens hope that the upgrade will make it easier for people to contact and interact with the IRS.

The results of systems development can mean the success or failure of an entire organization. It can even mean life or death in some cases. Physicians, for example, were actively involved in developing a 3-D system for surgery.[3] The new system allows doctors to do practice surgeries on a patient's internal organs before actually operating, preventing mistakes and saving lives. Successful systems development has also resulted in huge increases in revenues and profits. Companies that don't innovate with new systems development initiatives or fail to successfully complete a systems development effort can lose millions. Understanding and being able to apply a systems development life cycle, tools, and techniques discussed in this chapter and the next will help ensure the success of the development projects on which you participate. It can also help your career and improve the financial success of your company. Table 12.1 summarizes some successful systems development projects.[4]

Table 12.1

A Few Successful Systems Development Projects

(Source: Data from "Best in Class," *Computerworld*, March 15, 2004, p. 3.)

Organization	Description
U.S. Air Force	The systems development effort created a new e-mail system that halved costs and allowed 2,000 U.S. Air Force personnel who were once needed for the old e-mail system to move to other jobs.
Reliant Pharmaceuticals, LLC	The large drug company developed a sales force automation system that used voice recognition. The new system reduced the time it took to respond to sales calls from eight weeks to less than a day.
Northrop Grumman, Corp.	The large defense and space company used systems development to integrate diverse applications and allow managers to collaborate on important projects. The systems development effort saved the company almost $60 million in the first year. Ten-year savings are projected to be $600 million.
Visa	The credit card company used systems development to create a system to help resolve credit card charge disputes with customers and clients. The Web-based application streamlines the processing of charge disputes, saving time and money. Visa, which manages over 300 million credit and debit cards, estimates that it will save more than $1 billion over the next five years.
State of Ohio	A new worker's compensation system was developed, and it reduced worker's compensation transaction costs by more than 40 percent.

AN OVERVIEW OF SYSTEMS DEVELOPMENT

In today's businesses, managers and employees in all functional areas work together and use business information systems. As a result, users of all types are helping with development and, in many cases, leading the way. This chapter and the next will provide you with a deeper appreciation of the systems development process and help you avoid costly failures.

Participants in Systems Development

Effective systems development requires a team effort. The team usually consists of stakeholders, users, managers, systems development specialists, and various support personnel. This team, called the *development team*, is responsible for determining the objectives of the information system and delivering a system that meets these objectives. Many development

teams use a project manager to head the systems development effort and the project management approach to help coordinate the systems development process. A *project* is a planned collection of activities that achieves a goal, such as constructing a new manufacturing plant or developing a new decision support system. All projects have a defined starting point and ending point, normally expressed as dates such as August 4th and November 11th. Most have a budget, such as $150,000. A *project manager* is the individual responsible for coordinating all people and resources needed to complete a project on time. In systems development, the project manager can be an IS person inside the organization or an external consultant hired to complete the project. Project managers need technical, business, and people skills. "It's a delicate balancing act," says Jack Probst, vice president of IT process and governance at Nationwide Insurance.[5] In addition to completing the project on time and within the specified budget, the project manager is usually responsible for controlling project quality, training personnel, facilitating communications, managing risks, and acquiring any necessary equipment, including office supplies and sophisticated computer systems. Research studies have shown that project management success factors include good leadership from executives and project managers, a high level of trust in the project and its potential benefits, and the commitment of the project team and organization to successfully complete the project and implement its results.[6]

stakeholders

Individuals who, either themselves or through the organization they represent, ultimately benefit from the systems development project.

users

Individuals who will interact with the system regularly.

In the context of systems development, **stakeholders** are individuals who, either themselves or through the area of the organization they represent, ultimately benefit from the systems development project. One systems development methodology, called *agile modeling*, calls for very active participation of customers and other stakeholders in the systems development process. **Users** are individuals who will interact with the system regularly. They can be employees, managers, or suppliers. For large-scale systems development projects, where the investment in and value of a system can be quite high, it is common to have senior-level managers, including the company president and functional vice presidents (of finance, marketing, and so on), be part of the development team.

The IS competence of managers can have a big impact on the systems development effort. Managers with more knowledge and skill in computer technology are more willing to form partnerships with IS people and to lead and participate in systems development projects. Users who do not agree with a systems development project, however, can be hostile to the project and may even try to disrupt it.

systems analyst

Professional who specializes in analyzing and designing business systems.

Depending on the nature of the systems project, the development team might include systems analysts and programmers, among others. A **systems analyst** is a professional who specializes in analyzing and designing business systems. Systems analysts play various roles while interacting with the stakeholders and users, management, vendors and suppliers, external companies, software programmers, and other IS support personnel (see Figure 12.1). Like an architect developing blueprints for a new building, a systems analyst develops detailed plans for the new or modified system. The **programmer** is responsible for modifying or developing programs to satisfy user requirements. Like a contractor constructing a new building or renovating an existing one, the programmer takes the plans from the systems analyst and builds or modifies the necessary software.

programmer

Specialist responsible for modifying or developing programs to satisfy user requirements.

The other support personnel on the development team are mostly technical specialists, including database and telecommunications experts, hardware engineers, and supplier representatives. One or more of these roles may be outsourced to outside experts or consultants. Depending on the magnitude of the systems development project and the number of IS systems development specialists on the team, the team may also include one or more IS managers. The composition of a development team may vary over time and from project to project. For small businesses, the development team may consist of a systems analyst and the business owner as the primary stakeholder. For larger organizations, formal IS staff can include hundreds of people involved in a variety of activities, including systems development. Every development team should have a team leader. This individual can be from the IS department, a manager from the company, or a consultant from outside the company. The team leader needs both technical and people skills.

Development projects place great demands on the staff who may already have day-to-day responsibilities. To keep IS personnel motivated and reduce stress, some companies have

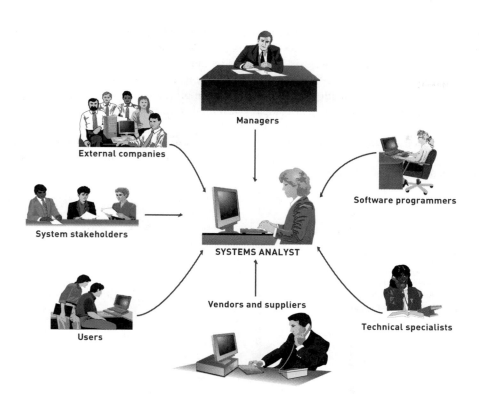

adopted a consultancy approach to development. A casino company, for example, can use this approach to help cut costs, improve productivity, and reduce burnout of its IS staff. The large casino and hotel operator can let its IS staff select upcoming projects and give systems development teams the resources they need to complete projects on time.

Regardless of the specific nature of a project, systems development creates or modifies systems, which ultimately means change. Managing this change effectively requires development team members to communicate well. It is important that you learn communication skills, because you probably will participate in systems development during your career. You may even be the individual who initiates systems development.

Initiating Systems Development

Systems development begins when an individual or group capable of initiating organizational change perceives a need for a new or modified system. Such individuals have a stake in the development of the system. Executives at an airline company, for example, can initiate a systems development project when they decide to expand the company's Web site to allow employees to log on to the site to get deeply discounted flight and travel opportunities. A financial services company can initiate a systems development effort to introduce a new customer relationship management (CRM) system to hundreds of its banks. To increase security, Pershing LLC initiated a systems development effort to build the Intrusion-Detection System (IDS). The company is a securities clearing firm that handles about 5 percent of the transactions of the New York Stock Exchange. IDS was built to make electronic securities transactions more secure because more and more transactions are conducted through computerized systems, and they must be safe from tampering.[7]

Systems development initiatives arise from all levels of an organization and are both planned and unplanned. To reduce costs and improve efficiency, the managers at the Public Broadcasting Service (PBS) launched a massive systems development effort. The project will replace expensive broadcasting equipment with less-expensive Intel computers. The new system will also increase the reliability of transmitting shows around the country. The project is scheduled to finish by 2006 and is expected to save PBS about $100 million annually.[8] In some cases, one systems development project initiates changes to other systems or companies. Aventis Pharmaceuticals, for example, embarked on a systems development project to

upgrade its e-mail system.[9] The upgrade caused a complete redevelopment of the company's global hardware and software systems, including its operating systems. The new systems development project took more than three years to complete. Peugeot Citroën initiated a systems development effort to use software to improve its car body manufacturing.[10] Like many systems development efforts, the Peugeot Citroën project had an impact on the companies that supplied its car parts. The company believes that the systems development project resulted in an additional $130 million in increased profits.

Solid planning and managerial involvement helps ensure that these initiatives support broader organizational goals. Systems development projects may be initiated for a number of reasons, as shown in Figure 12.2.

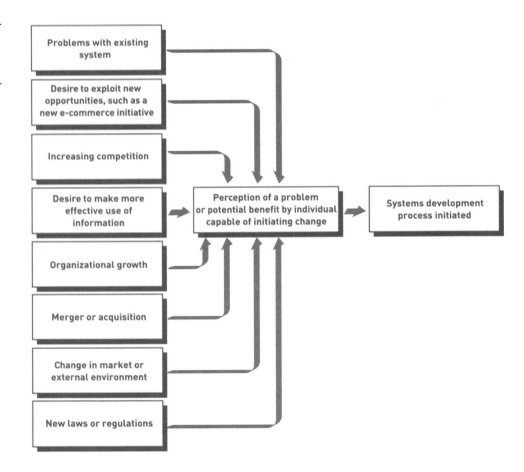

Figure 12.2

Typical Reasons to Initiate a Systems Development Project

Mergers and acquisitions can trigger many systems development projects.[11] Because the information systems for companies are usually different, a large systems development effort is typically required to unify systems. Even with similar information systems, the procedures, culture, training, and management of the information systems are typically different, requiring a realignment of the IS departments.

The federal government also fosters new systems development projects in the private sector. In the wake of recent financial scandals, the government is instituting new corporate financial reporting rules under the Sarbanes-Oxley Act. These new regulations have caused many U.S. companies to initiate systems development efforts.[12] Some companies have been overwhelmed by these new, strict reporting requirements.[13] Regis Corporation, an operator of hair salons, estimated that the cost of complying with the new act will cost it more than $100,000.[14] Some companies are building software from scratch to comply with the act, while other companies are looking for software they can purchase.[15] Other government initiatives have also initiated systems development. The CIA has funded a venture-capital firm

called In-Q-Tel, which seeks help from corporations and their systems development departments to build new spy tools for security. The Centers for Disease Control (CDC) has developed systems to help it deal with bioterrorist threats and other possible crises.

Corporate litigation, which appears to be on the rise, also has initiated systems development projects. New systems are often needed to protect companies from lawsuits and help them in court if lawsuits are brought against them.

Information Systems Planning and Aligning Corporate and IS Goals

Because an organization's strategic plan contains both organizational goals and a broad outline of steps required to reach them, the strategic plan affects the type of system an organization needs. For example, a strategic plan may identify as organizational goals a doubling of sales revenue within five years, a 20 percent reduction of administrative expenses over three years, acquisition of at least two competing companies within a year, or market leadership in a given product category. Organizational commitments to policies such as continuous improvement are also reflected in the strategic plan. Such goals and commitments set broad outlines of system performance.

Often, a section of the strategic plan lists guidelines for meeting specific goals that relate to units or departments. Examples of these guidelines might be improving customer service for luxury car buyers, expanding international distribution by purchasing existing distributors, and using a specific amount of money to buy back company stock. The strategic plan also provides general direction to the functional areas within an organization, including marketing, production, finance, accounting, and human resources. For the IS department, these directions are encompassed in the information systems plan.

Information Systems Planning

The term **information systems planning** refers to the translation of strategic and organizational goals into systems development initiatives (see Figure 12.3). The Marriott hotel chain, for example, invites its chief information officer to board meetings and other top-level management meetings. Proper IS planning ensures that specific systems development objectives support organizational goals.

information systems planning
The translation of strategic and organizational goals into systems development initiatives.

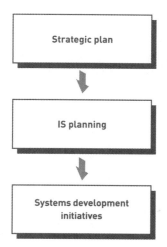

Figure 12.3

Information systems planning transforms organizational goals outlined in the strategic plan into specific system development activities.

Aligning Corporate and IS Goals

Aligning organizational goals and IS goals is critical for any successful systems development effort. Since information systems support other business activities, IS staff and people in other departments need to understand each other's responsibilities and tasks, as the "Information Systems @ Work" feature explains. Determining whether organizational and IS goals are aligned can be difficult, so researchers have increasingly tackled the problem. One measure of alignment uses five levels, ranging from ad hoc processes (Level 1 Alignment) to optimized processes (Level 5 Alignment).

Fakta Designs Financial Management System to Support Corporate Goals

The leading Danish discount retail chain, Fakta, recently became frustrated with its inability to maintain its budget. The retailer had grown beyond the capabilities of its antiquated budgeting system, and something needed to change. "We live in a world where it is crucial for us to be able to react quickly to changing circumstances because our profit margins are so small," says Financial Director C. F. Thorhauge. "If we are to earn money, we must be really tough in controlling our costs." But the old system for establishing the corporate budget was not supporting the goals of the organization.

As with all systems development projects, Fakta's IS department began by defining the goals of the system in light of the goals of the organization. They studied the details of the current system to see why it was failing. Fakta has retail outlets in 16 regions in Denmark, each of which has a manager who is financially responsible for the shops in his or her region. The system in use required each regional manager to work out his or her own budget and send it to the company headquarters in Vejle. From there, Fakta's IS department collated the budgets from all the regional managers and passed the combined information on to the finance department. Then, the finance department at the company's headquarters manually determined the overall budget by using spreadsheets or other forms of documentation, sometimes even resorting to pencil and paper.

"It was already clear to us several years ago that the situation was becoming untenable," says Jens Brinkmann, Fakta's financial controller. "It was extremely difficult to control the vast quantity of information that came from the 16 regions into [the] head office. Often, it took days or even weeks to collect all the different budget information, and there was a major risk of errors occurring during the process."

Fakta decided that the new system should provide a centralized server accessible to all regional managers to allow them to store information directly in the central system. The system would automate the calculations, create the budget, and provide useful reports to all managers within the organization. Because budgeting is common to all companies, Fakta guessed that such a solution was probably already on the market. It found its solution in a product from SAS Corporation.

SAS worked with Fakta's IS department to develop a customized system that met the organization's goals. The result? A central, user-controlled system that contains all budget-related functions from inventory turnover to production overhead costs. All regional managers input their budgets into the new system through an easy-to-use Web-based system delivered over the company's intranet. The budget is then finalized, and the numbers are reliable. As a result, everyone can quickly get an overview of the entire budget.

Fakta trained its regional managers and its finance staff on the new system to ensure that they understood the new budgeting process. Fakta's IS department's job has been simplified, too. They no longer have to worry about budgeting. The days of manually collecting and checking data are over. Today, budgeting is automated

With its new financial management system in place, the national discount chain can control business operations and handle budgeting quickly. "Where it had previously taken days or weeks to determine the budget for the whole discount chain, it can now be accomplished in just a few hours. We've put an end to all the poor excuses for mistakes in the budget," concludes Thorhauge.

Discussion Questions

1. How has Fakta's new system changed the work load of its regional managers, its IS staff, and its finance department?
2. Besides streamlining the process of collecting regional data, how has the new system improved the effectiveness of the organization?

Critical Thinking Questions

3. Now that Fakta's finance department isn't strapped with manually calculating the company's budget, how can finance employees use their additional time more productively?
4. Provide some examples of how financial management software such as that in use at Fakta might influence and change the investment and purchasing style of a company.

SOURCES: "SAS Success Stories: Fakta," accessed June 18, 2004, *www.sas.com/success/fakta.html*; SAS Financial Intelligence Web site, accessed June 18, 2004, *www.sas.com/solutions/financial/index.html*; FDB (owner of Fakta) Web site, accessed June 18, 2004, *www.fdb.dk/default.asp?id=44*.

One of the primary benefits of IS planning and alignment of goals is a long-range view of information technology's use in the organization. Specific systems development initiatives may spring from the IS plan, but the IS plan must also provide a broad framework for future success. The IS plan should guide development of the IS infrastructure over time. Another benefit of IS planning is that it ensures better use of IS resources—including funds, personnel, and time for scheduling specific projects. The steps of IS planning are shown in Figure 12.4.

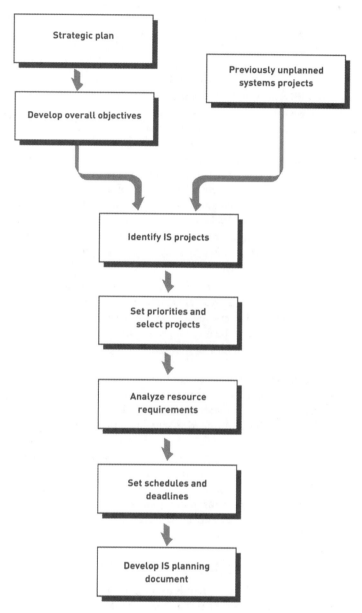

Figure 12.4

The Steps of IS Planning

Some projects are identified through overall IS objectives, whereas additional projects, called *unplanned projects*, are identified from other sources. All identified projects are then evaluated in terms of their organizational priority.

Overall IS objectives are usually distilled from the relevant aspects of the organization's strategic plan. IS projects can be identified either directly from the objectives determined in the first step or identified by others, such as managers within the various functional areas. Setting priorities and selecting projects typically requires the involvement and approval of senior management. When objectives are set, planners consider the resources necessary to complete the projects, including employees (systems analysts, programmers, and others), equipment (computers, network servers, printers, and other devices), expert advice (specialists and other consultants), and software, among others.

Developing a Competitive Advantage

In today's business environment, many companies seek systems development projects that will provide them with a competitive advantage. Thinking competitively usually requires creative and critical analysis. For example, a company may want to achieve a competitive advantage by improving its customer-supplier relationship. Linking customers and suppliers can result in superior products and services. Fuji and Kodak are hoping to achieve a competitive advantage by developing online photo-sharing services.[16] The services are aimed at camera-enabled cell phone users who don't know what to do with the photos after they snap them with their cell phone. Allstate Financial Group used the software program Visual Studio .NET to create a Web site, called AccessAllstate, to allow 350,000 sales reps to get information about Allstate insurance, retirement, and investment products.[17] Before the Web site was available, Allstate sales reps had to call the company for information. By looking at problems in new or different ways and by introducing innovative methods to solve them, many organizations have gained significant competitive advantage. In some cases, these new solutions are inspired by people and things not directly related to the problem.

Creative analysis involves the investigation of new approaches to existing problems. By looking at problems in new or different ways and by introducing innovative methods to solve them, many firms have gained a competitive advantage. Typically, these new solutions are inspired by people and events not directly related to the problem.

Critical analysis requires unbiased and careful questioning of whether system elements are related in the most effective or efficient ways. It involves considering the establishment of new or different relationships among system elements and perhaps introducing new elements into the system. Critical analysis in systems development involves the following actions:

- *Questioning statements and assumptions.* Questioning users about their needs and clarifying their initial responses can result in better systems and more accurate predictions. Too often, stakeholders and users specify certain system requirements because they assume that their needs can only be met that way. Often, an alternative approach would be better. For example, a stakeholder may be concerned because there is always too much of some items in stock and not enough of other items. So, the stakeholder might request a totally new and improved inventory control system. An alternative approach is to identify the root cause for poor inventory management. This latter approach might determine that sales forecasting is inaccurate and needs improvement or that production is not capable of meeting the set production schedule. All too often solutions are selected before a complete understanding of the nature of the problem itself is obtained.

- *Identifying and resolving objectives and orientations that conflict.* Different departments in an organization can have different objectives and orientations. The buying department may want to minimize the cost of spare parts by always buying from the lowest-cost supplier, but engineering might want to buy more expensive, higher-quality spare parts to reduce the frequency of replacement. These differences must be identified and resolved before a new purchasing system is developed or an existing one modified.

Establishing Objectives for Systems Development

The overall objective of systems development is to achieve business goals, not technical goals, by delivering the right information to the right person at the right time. The impact a particular system has on an organization's ability to meet its goals determines the true value of that system to the organization. Although all systems should support business goals, some systems are more pivotal in continued operations and goal attainment than others. These systems are called **mission-critical systems**. An order-processing system, for example, is usually considered mission critical. Without it, few organizations could continue daily activities, and they clearly would not meet set goals.

The goals defined for an organization will in turn define the objectives that are set for a system. A manufacturing plant, for example, might determine that minimizing the total cost of owning and operating its equipment is critical to meet production and profit goals.

---

creative analysis
The investigation of new approaches to existing problems.

critical analysis
The unbiased and careful questioning of whether system elements are related in the most effective or efficient ways.

mission-critical systems
Systems that play a pivotal role in an organization's continued operations and goal attainment.

Critical success factors (CSFs) are factors that are essential to the success of certain functional areas of an organization. The CSF for manufacturing—minimizing equipment maintenance and operating costs—would be converted into specific objectives for a proposed system. One specific objective might be to alert maintenance planners when a piece of equipment is due for routine preventive maintenance (e.g., cleaning and lubrication). Another objective might be to alert the maintenance planners when the necessary cleaning materials, lubrication oils, or spare parts inventory levels are below specified limits. These objectives could be accomplished either through automatic stock replenishment via electronic data interchange or through the use of exception reports. One study found that different CSFs might be important during different phases of a systems development project.

Regardless of the particular systems development effort, the development process should define a system with specific performance and cost objectives. The success or failure of the systems development effort will be measured against these objectives.

Performance Objectives

The extent to which a system performs as desired can be measured through its performance objectives.[18] Hilton Hotels, for example, invested about $13 million to improve the performance of its Web site.[19] The new Web site systems development effort allowed the company to develop specific content for its hotels' different locations and cultures. According to a senior vice president of Hilton International, "We're now much more able to take our worldwide network of hotels and market them to where our customers live." Emery, a worldwide shipping company, was able to develop a new transportation system that increased profits by $80 million in North America. System performance is usually determined by such factors as the following:

- *The quality or usefulness of the output.* Is the system generating the right information for a value-added business process or by a goal-oriented decision maker?
- *The accuracy of the output.* Is the output accurate and does it reflect the true situation? As a result of the accounting scandals of the early 2000s, when some companies overstated revenues or understated expenses, accuracy is becoming more important, and top corporate officers are being held responsible for the accuracy of all corporate reports.
- *The quality or usefulness of the format of the output.* Is the output generated in a form that is usable and easily understood? For example, objectives often concern the legibility of screen displays, the appearance of documents, and the adherence to certain naming conventions.
- *The speed at which output is generated.* Is the system generating output in time to meet organizational goals and operational objectives? Objectives such as customer response time, the time to determine product availability, and throughput time are examples.
- *The scalability of the resulting system.* As mentioned in Chapter 4, *scalability* allows an information system to handle business growth and increased business volume. For example, if a midsized business realizes an annual 10 percent growth in sales for several years, an information system that is scalable will be able to efficiently handle the increase by adding processing, storage, software, database, telecommunications, and other information systems resources to handle the growth.

In some cases, the achievement of performance objectives can be easily measured (e.g., by tracking the time it takes to determine product availability). The achievement of performance objectives is sometimes more difficult to ascertain in the short term. For example, it may be difficult to determine how many customers are lost because of slow responses to customer inquiries regarding product availability. These outcomes, however, are often closely associated with corporate goals and are vital to the long-term success of the organization. Senior management usually dictates their attainment.

critical success factors (CSFs)
Factors that are essential to the success of a functional area of an organization.

Cost Objectives

The benefits of achieving performance goals should be balanced with all costs associated with the system, including the following:

- *Development costs.* All costs required to get the system up and running should be included.
- *Costs related to the uniqueness of the system application.* A system's uniqueness has a profound effect on its cost. An expensive but reusable system may be preferable to a less costly system with limited use.
- *Fixed investments in hardware and related equipment.* Developers should consider costs of such items as computers, network-related equipment, and environmentally controlled data centers in which to operate the equipment.
- *Ongoing operating costs of the system.* Operating costs include costs for personnel, software, supplies, and such resources as the electricity required to run the system.

Balancing performance and cost objectives within the overall framework of organizational goals can be challenging. Systems development objectives are important, however, because they allow an organization to allocate resources effectively and efficiently and measure the success of a systems development effort. For PC manufacturers, for example, parts and components of a typical PC can cost under $500, which includes about $130 for the processor, $100 for a CD or DVD, $100 for memory, $45 for the Windows operating system, and the rest on other hardware parts and components.[20] Some believe these low costs will eventually lead to lower costs for PCs. The New York Stock Exchange (NYSE) has invested about $2 billion to speed trading and reduce costs.[21] Stock trades are completed with human brokers on a trading floor. In the future, electronic trading could take over, slashing trading costs to investors.

Web-Based Systems Development: The Internet, Intranets, Extranets, and E-Commerce

As seen in the many examples and cases in this and other chapters, companies are increasingly converting at least some portion of their business to run over the Internet, intranets, or extranets. Web-based systems development treats the Internet as an integral part of the development process, not just a communications tool. In some development efforts, the Internet becomes the most important aspect of the new or modified system. Applications that are being moved to the Internet include those that support selling products to customers, placing orders with suppliers, and letting customers access information about production, inventory, orders, or accounts receivable. In addition, a number of companies sell their products and services only on the Internet. Internet technology enables companies to extend their information systems beyond their boundaries to reach their customers, suppliers, and partners, allowing them to conduct business much faster, to interact with more people, and to keep one step ahead of the competition. Some companies have been willing to sustain current losses from their e-commerce initiatives to gain an advantage for the future.

Building a static Web site to display simple text and graphics is fairly straightforward. However, implementing a dynamic core business application that runs over the Web is much more complicated. Such applications must meet special business needs. They must be able to scale up to support highly variable transactions from potentially thousands of users. Ideally, they can scale up instantly when needed. They must be reliable and fault tolerant, providing continuous availability while processing all transactions accurately. They must also integrate with existing infrastructure, including customer and order databases, existing applications, and enterprise resource planning systems. Development and maintenance must be quick and easy, because business needs may require changing applications on the fly.

Tools and techniques are available for building and running Web applications. Many were discussed in Chapter 7, including HTML, XML, Java, .NET, VBScript, Web services, techniques to develop Web content, and the management and maintenance of Web content. The best tools provide components to support applications on an enterprise scale while speeding development. Several vendors, including NetDynamics, SilverStream, WebLogic, Novera Software, Netscape Communications, Microsoft, and IBM, provide what is known as an applications server to provide remote access to databases via a corporate intranet. These

tools include .NET by Microsoft and WebSphere by IBM. In addition, most of the tools and techniques discussed later in this chapter and in Chapter 13, such as CASE tools, rapid application development, joint application development, and object-oriented development, are also used to develop and maintain Web sites.

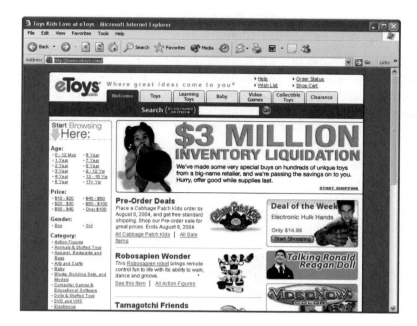

eToys.com is a company that uses the Internet as its only sales channel.

Systems Development and Enterprise Resource Planning

Enterprise resource planning (ERP) software, discussed in Chapter 9, has reached beyond business processes and is increasingly affecting systems development. Not only are planners considering different types of systems that include ERP software, but the ERP software that is already in place is also driving planners to explore and develop different types of systems. Other ERP users are moving from just using the software to run the business to trying to use it to make business decisions. ERP systems can be used to integrate an organization's transaction processing system (TPS) and provide important information and decision support to managers.

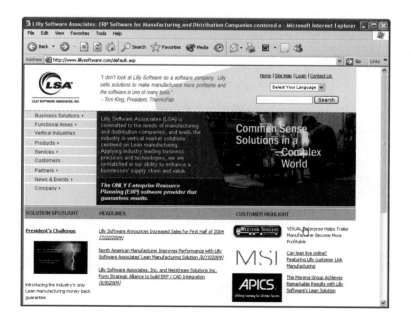

Lilly Software helps companies realize higher productivity and profitability by capturing and analyzing real-time data about all their business processes using ERP.

An important trend in systems development and the use of ERP systems is that companies wish to stay with their primary ERP vendor (SAP, Oracle, PeopleSoft, Baan, etc.) instead of looking elsewhere for answers to their data warehousing and production planning needs or developing in-house solutions. Thus, they look to their original ERP vendor to provide these solutions. In addition, many software vendors are building software that integrates with the ERP vendor's package. For example, Aspect Development, a leading supplier of electronic catalogs for high-tech components, has now entered the market for online catalogs of maintenance, repair, and operations (MRO) supplies. The company has a product called Morocco that is an MRO supplies catalog for users of SAP's ERP software. Morocco works in conjunction with the purchasing module of SAP and directs buyers to preferred suppliers specified by the purchasing department. Again, there is less in-house development and more dependence on ERP vendors and their strategic partners to provide enhancements and add-ons to the original ERP package. There are three popular approaches to acquiring an ERP system:

1. *Comprehensive ERP package.* Companies can use one ERP vendor, such as SAP, and implement one comprehensive ERP package from one vendor.
2. *Best of breed.* A second approach is to select the best parts from each ERP vendor in developing an ERP system. Although this can result in getting the best component parts, integrating the components from different vendors can be difficult or impossible.
3. *Hybrid.* With this approach, an organization will implement a comprehensive package from an ERP vendor and supplement it with other parts or components by other ERP or software vendors. This is a flexible and popular approach to implementing ERP systems.

When a systems development project involves implementing a new ERP system or modifying an existing one, a number of critical success factors come into play, including:

- Selecting the right ERP system for the organization
- Ensuring top management support
- Educating key managers and users of the new or modified ERP system
- Employing an IS staff with good technical and communications skills
- Using good project management tools and techniques to implement the ERP system

SYSTEMS DEVELOPMENT LIFE CYCLES

The systems development process is also called a *systems development life cycle* (*SDLC*) because the activities associated with it are ongoing. As each system is being built, the project has timelines and deadlines, until at last the system is installed and accepted. The life of the system continues as it is maintained and reviewed. If the system needs significant improvement beyond the scope of maintenance, if it needs to be replaced because of a new generation of technology, or if the IS needs of the organization change significantly, a new project will be initiated and the cycle will start over.

A key fact of systems development is that the later in the SDLC an error is detected, the more expensive it is to correct (see Figure 12.5). One reason for the mounting costs is that if an error is found in a later phase of the SDLC, the previous phases must be reworked to some extent. Another reason is that the errors found late in the SDLC have an impact on more people. For example, an error found after a system is installed may require retraining users once a "workaround" to the problem has been found. Thus, experienced system developers prefer an approach that will catch errors early in the project life cycle.

Several common systems development life cycles exist: traditional, prototyping, rapid application development (RAD), and end-user development. In addition, companies can outsource the systems development process. With some companies, these approaches are formalized and documented so that system developers have a well-defined process to follow; in other companies, less formalized approaches are used. Keep Figure 12.5 in mind as you are introduced to alternative SDLCs in the next section.

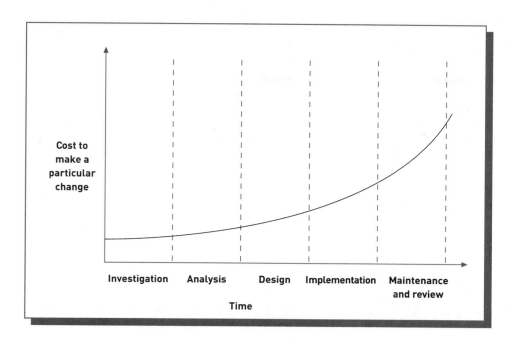

Figure 12.5

The later that system changes are made in the SDLC, the more expensive these changes become.

The Traditional Systems Development Life Cycle

Traditional systems development efforts can range from a small project, such as purchasing an inexpensive computer program, to a major undertaking. The steps of traditional systems development may vary from one company to the next, but most approaches have five common phases: investigation, analysis, design, implementation, and maintenance and review (see Figure 12.6).

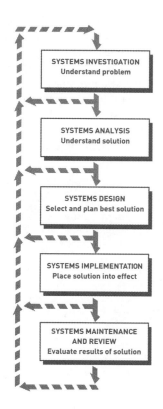

Figure 12.6

The Traditional Systems Development Life Cycle

Sometimes, information learned in a particular phase requires cycling back to a previous phase.

systems investigation
The systems development phase during which problems and opportunities are identified and considered in light of the goals of the business.

systems analysis
The systems development phase involving the study of existing systems and work processes to identify strengths, weaknesses, and opportunities for improvement.

systems design
The systems development phase that defines how the information system will do what it must do to obtain the problem solution.

systems implementation
The systems development phase involving the creation or acquiring of various system components detailed in the systems design, assembling them, and placing the new or modified system into operation.

systems maintenance and review
The systems development phase that ensures the system operates and modifies the system so that it continues to meet changing business needs.

In the **systems investigation** phase, potential problems and opportunities are identified and considered in light of the goals of the business. Systems investigation attempts to answer the questions "What is the problem?" and "Is it worth solving?" The primary result of this phase is a defined development project for which business problems or opportunity statements have been created, to which some organizational resources have been committed, and for which systems analysis is recommended. **Systems analysis** attempts to answer the question "What must the information system do to solve the problem?" This phase involves studying existing systems and work processes to identify strengths, weaknesses, and opportunities for improvement. The major outcome of systems analysis is a list of requirements and priorities. **Systems design** seeks to answer the question "How will the information system do what it must do to obtain the problem solution?" The primary result of this phase is a technical design that either describes the new system or describes how existing systems will be modified. The system design details system outputs, inputs, and user interfaces; specifies hardware, software, database, telecommunications, personnel, and procedure components; and shows how these components are related. **Systems implementation** involves creating or acquiring the various system components detailed in the systems design, assembling them, and placing the new or modified system into operation. An important task during this phase is to train the users. Systems implementation results in an installed, operational information system that meets the business needs for which it was developed. The purpose of **systems maintenance and review** is to ensure that the system operates and to modify the system so that it continues to meet changing business needs. As shown in Figure 12.6, a system under development moves from one phase of the traditional SDLC to the next.

The traditional SDLC allows for a large degree of management control. At the end of each phase, a formal review is performed and a decision is made whether to continue with the project, terminate the project, or perhaps repeat some of the tasks of the current phase. Use of the traditional SDLC also creates a great deal of documentation, such as entity-relationship diagrams. This documentation, if kept current, can be useful when it is time to modify the system. The traditional SDLC also ensures that every system requirement can be related to a business need. In addition, resulting products can be reviewed to verify that they satisfy the system requirements and conform to organizational standards.

A major problem with the traditional SDLC is that the user does not use the solution until the system is nearly complete. Quite often, users get a system that does not meet their real needs because its development was based on the development team's understanding of the needs. The traditional approach is also inflexible. Changes in user requirements cannot be accommodated during development. In spite of its limitations, however, the traditional SDLC is still used for large, complex systems that affect entire businesses, such as TPSs and MISs. It is also frequently employed on government projects because of the strengths mentioned previously. Table 12.2 lists advantages and disadvantages of the traditional SDLC.

Table 12.2

Advantages and Disadvantages of Traditional SDLC

Advantages	Disadvantages
Formal review at the end of each phase allows maximum management control.	Users get a system that meets the needs as understood by the developers; this may not be what is really needed.
This approach creates considerable system documentation.	Documentation is expensive and time-consuming to create. It is also difficult to keep current.
Formal documentation ensures that system requirements can be traced back to stated business needs.	Often, user needs go unstated or are misunderstood.
It produces many intermediate products that can be reviewed to see whether they meet the users' needs and conform to standards.	Users cannot easily review intermediate products and evaluate whether a particular product (e.g., data flow diagram) meets their business requirements.

Prototyping

Prototyping takes an iterative approach to the systems development process. During each iteration, requirements and alternative solutions to the problem are identified and analyzed, new solutions are designed, and a portion of the system is implemented. Users are then encouraged to try the prototype and provide feedback (see Figure 12.7). Prototyping begins with the creation of a preliminary model of a major subsystem or a scaled-down version of the entire system. For example, a prototype might be developed to show sample report formats and input screens. Once developed and refined, the prototypical reports and input screens are used as models for the actual system, which may be developed using an end-user programming language such as Visual Basic. The first preliminary model is refined to form the second- and third-generation models, and so on, until the complete system is developed (see Figure 12.8).

prototyping
An iterative approach to the systems development process.

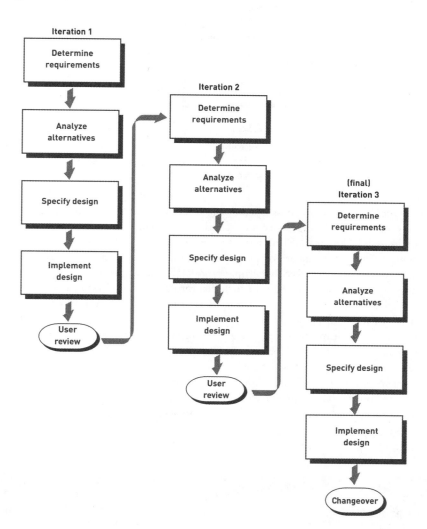

Figure 12.7

Prototyping Is an Iterative Approach to Systems Development

Prototyping is a popular technique in systems development. Each generation of prototype is a refinement of the previous generation based on user feedback.

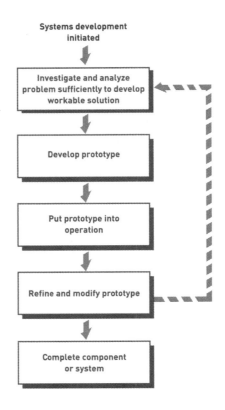

operational prototype
A functioning prototype that accesses real data files, edits input data, makes necessary computations and comparisons, and produces real output.

nonoperational prototype
A mock-up, or model, that includes output and input specifications and formats.

Table 12.3

Advantages and Disadvantages of Prototyping

Prototypes can be classified as operational or nonoperational. An **operational prototype** is a prototype that works—accesses real data files, edits input data, makes necessary computations and comparisons, and produces real output. Fully developed financial reports are examples. The operational prototype may access real files but perhaps does not edit input. A **nonoperational prototype** is a mock-up, or model. It typically includes output and input specifications and formats. The outputs include printed reports to managers and the screen layout of reports displayed on personal computers or terminals. The inputs reveal how data is captured, what commands users must enter, and how the system accesses other data files. The primary advantage of a nonoperational prototype is that it can be developed much faster than an operational prototype. Nonoperational prototypes can be discarded, and a fully operational system can be built based on what was learned from the prototypes. Research has shown that systems investigation, design, and implementation are the phases that benefit the most from the prototyping approach. According to Porus Munshi, a computer systems consultant, "Rapid prototyping is the cornerstone of innovation. In the dilemma between perfecting a design and experimenting, go for the experiment every time."[22] The advantages and disadvantages of prototyping are summarized in Table 12.3.

Advantages	Disadvantages
Users can try the system and provide contructive feedback during development.	Each iteration builds on the previous one. The final solution may be only incrementally better than the initial solution.
An operational protoype can be produced in weeks.	Formal end-of-phrase reviews may not occur. Thus, it is very difficult to contain the scope of the protoype, and the project never seems to end.
As solutions emerge, users become more positive about the process and the results.	System documentation is often absent or incomplete, since the primary focus is on development of the prototype.
Prototyping enables early detection of errors and omissions.	System backup and recovery, performance, and security issues can be overlooked in the haste to develop a prototype.

Rapid Application Development, Agile Development, Joint Application Development, and Other Systems Development Approaches

Rapid application development (RAD) employs tools, techniques, and methodologies designed to speed application development. Vendors such as Computer Associates International, IBM, and Oracle market fourth-generation languages and other products targeting the RAD market. Rational Software, a division of IBM, has a RAD tool, called Rational Rapid Developer, to make developing large Java programs and applications easier and faster.[23]

RAD reduces paper-based documentation, automatically generates program source code, and facilitates user participation in design and development activities. It makes adapting to changing system requirements easier. Other approaches to rapid development, such as *agile development* or *extreme programming (XP)*, allow the systems to change as they are being developed. Agile development requires frequent face-to-face meetings with the systems developers and users as they modify, refine, and test how the system meets users' needs and what its capabilities are. Some predict that agile programming will eventually be used by most IT departments because of the volatility of the business environment—the length of the traditional development life cycle can reduce a system's usefulness once it is finally completed. The agile development process is more fluid and flexible, but it can be complex and time-consuming. XP uses pairs of programmers who work together to design, test, and code parts of the systems they develop. The iterative nature of XP helps companies develop robust systems, with fewer errors. Sabre Airline Solutions, a $2 billion airline travel company, used XP to eliminate programming errors and shorten program development times.[24] The company manages 13 million lines of programming code in 62 software products. Sabre uses the Java programming language and XP to rapidly develop its applications.

RAD makes extensive use of the **joint application development (JAD)** process for data collection and requirements analysis. Originally developed by IBM Canada in the 1970s, JAD involves group meetings in which users, stakeholders, and IS professionals work together to analyze existing systems, propose possible solutions, and define the requirements of a new or modified system. JAD groups consist of both problem holders and solution providers. A group normally requires one or more top-level executives who initiate the JAD process, a group leader for the meetings, potential users, and one or more individuals who act as secretaries and clerks to record what is accomplished and to provide general support for the sessions. Many companies have found that groups can develop better requirements than individuals working independently and have assessed JAD as a very successful development technique. Today, JAD often uses *group support systems (GSS)* software to foster positive group interactions, while suppressing negative group behavior. Group support systems were introduced in Chapter 10.

RAD should not be used on every software development project. In general, it is best suited for DSSs and MISs and less well suited for TPSs. During a RAD project, the level of participation of stakeholders and users is much higher than in other approaches. They become working members of the team and can be expected to spend more than 50 percent of their time producing project outcomes. This time commitment can be a problem if the users are also needed to perform their normal business role. For this reason, RAD team participants are often taken off their normal assignments and put full-time on the RAD project. Because of the full-time commitment and intense schedule deadlines, RAD is a high-pressure development approach that can easily result in employee burnout. Table 12.4 lists advantages and disadvantages of RAD.

Prototyping, rapid application development, agile development, and joint application development are used to overcome some of the perceived disadvantages of the traditional systems development life cycle. These potential disadvantages include problems with the traditional project planning process, slow development times, a lower success rate, and not satisfying user needs. These approaches also take advantage of newer programming languages and techniques. In addition to these systems development approaches, there are a number other systems development approaches, including adaptive software development, lean software development, the Rational Unified Process (RUP), Feature-Driven Development

rapid application development (RAD)
A systems development approach that employs tools, techniques, and methodologies designed to speed application development.

joint application development (JAD)
Process for data collection and requirements analysis in which users, stakeholders, and IS professionals work together to analyze existing systems, propose possible solutions, and define the requirements of a new or modified system.

(FDD), and dynamic systems development method. Often created by computer vendors and authors of systems development books, these approaches all attempt to deliver better systems.

Advantages	Disadvantages
For appropriate projects, this approach puts an application into production sooner than any other approach.	This intense SDLC can burn out systems developers and other project participants.
Documentation is produced as a by-product of completing project tasks.	This approach requires systems analysts and users to be skilled in RAD system development tools and RAD techniques.
RAD forces teamwork and lots of interaction between users and stakeholders.	RAD requires a larger percentage of stakeholders' and users' time than other approaches.

Table 12.4

Advantages and Disadvantages of RAD

The End-User Systems Development Life Cycle

Systems development initiatives arise from a wide variety of organizational areas and individuals, including users. The proliferation of general-purpose information technology and the flexibility of many packaged software programs have allowed non-IS employees to independently develop information systems that meet their needs. Such employees believe that, by bypassing the formal requisitioning of resources from the IS department, they can develop systems more quickly. In addition, these individuals often believe that they have better insight into their own needs and can develop systems better suited for their purposes. In addition, many end users are increasingly developing computer systems or solving computer-related problems for others.[25] Dennis Christensen's coworkers often call him to fix computer glitches, printer problems, or e-mail stoppages. Christensen is a marketing manager for a Michigan credit bureau and not a computer systems person.

end-user systems development

Any systems development project in which the primary effort is undertaken by a combination of business managers and users.

End-user-developed systems range from the very small (e.g., a software routine to merge form letters) to those of significant organizational value (such as customer contact databases for the Web). Adnan Osmani from Dublin, Ireland, for example, developed his own Internet browser.[26] "I just wanted it faster for myself," Osmani said. Like all projects, some end-user-developed systems fail, and others are successful. Initially, IS professionals discounted the value of these projects. As the number and magnitude of these projects increased, however, IS professionals began to realize that for the good of the entire organization, their involvement with these projects needed to increase.

Today, the term **end-user systems development** describes any systems development project in which the primary effort is undertaken by a combination of business managers and users. Rather than ignoring these initiatives, astute IS professionals encourage them by offering guidance and support. Technical assistance, communication of standards, and the sharing of "best practices" throughout the organization are just some of the ways IS professionals work with motivated managers and employees undertaking their own systems development. In this way, end-user-developed systems can be structured as complementary to, rather than in conflict with, existing and emerging information systems. In addition, this open communication among IS professionals, managers of the affected business area, and users allows the IS professionals to identify specific initiatives so that additional organizational resources, beyond those available to business managers or users, are provided for its development.

There are disadvantages of end-user systems development. Some end users don't have the training to effectively develop and test a system. Multimillion-dollar mistakes, for example, can be made using faulty spreadsheets that were never tested. Some end-user systems are also poorly documented. Then, when they are updated, errors can be introduced that make the system error prone. In addition, some end users spend time and corporate resources developing systems that were already available.

Outsourcing and On Demand Computing

Many companies hire an outside consulting firm or computer company that specializes in systems development to take over some or all of its development and operations activities. As mentioned in Chapter 2, *outsourcing* and *on demand computing* are often used.[27] Sears, Roebuck, and Co., for example, is looking for an outside company to perform many of its systems development and operations activities.[28] Sears is hoping that the outsourcing company it selects will hire some of its 200 people who will be laid off. Outsourcing has also become an important economic and political issue in today's economy for companies that outsource overseas.[29] A group of U.S. companies and organizations have formed the Coalition for Economic Growth and American Jobs to try to combat the outsourcing backlash. Companies can spend millions or even billions of dollars to hire other companies to manage Web sites, network servers, data storage devices, and help desk operations. In one outsourcing deal, a major phone company agreed to pay an outsourcing company about $3 billion to provide guidance on cutting costs and improving efficiency. MCI, for example, was given an outsourcing deal worth $20 million to develop a cellular phone system in Baghdad, Iraq, by the U.S. Department of Defense.[30] Companies such as Goodyear and Saks have used on demand computing to increase services and reduce costs.[31]

A report by the U.S. Commerce Department, however, reports that the United States imports more private-sector jobs from other countries than it exports.[32] Private-sector jobs include computer programming, telecommunications, management consulting, legal work, and related jobs. The surplus of jobs coming into the United States from other countries is greater than $53 billion, according to the report. Table 12.5 describes the circumstances in which outsourcing is a good idea.

Outsourcing can involve a large number of countries and companies in bringing new products and services to market.[33] The idea for a new computer server can originate in Singapore, be approved in Houston, designed in India, engineered in Taiwan, and assembled in Australia. The chain of events can be complex.

Reducing costs, obtaining state-of-the-art technology, eliminating staffing and personnel problems, and increasing technological flexibility are reasons that companies have used the outsourcing and on demand computing approaches. Reducing costs is a primary reason for outsourcing. One American computer company, for example, estimated that a programmer with three to five years of experience in China would cost $12.50 per hour, while a programmer with similar experience in the United States would cost $56 per hour.[34]

Reason	Example
When a company believes it can cut costs.	PacifiCare outsourced its IS operations to IBM and Keane, Inc. PacifiCare hopes the outsourcing will save it about $400 million over ten years.
When there is limited opportunity for the firm to distinguish itself competitively through a particular IS operation or application.	Kodak outsourced its IS operations, including mainframe processing, telecommunications, and personal computer support, because there was limited opportunity to distinguish the company through these IS operations. Kodak kept application development and support in-house because it thought that these activities had competitive value.
When uninterrupted IS service is not crucial.	Airline reservations or catalog shopping systems are mission critical and should not be trusted to outside firms.
When outsourcing does not strip the company of technical know-how required for future IS innovation.	Firms must ensure that their IS staffs remain technically up-to-date and have the expertise to develop future applications.
When the firm's existing IS capabilities are limited, ineffective, or technically inferior.	A company might use outsourcing to help it make the transition from a centralized mainframe environment to a distributed client/server environment.
When a firm is downsizing. The decision to outsource systems development is often a response to downsizing, which reduces the number of employees or managers, equipment and systems, and even functions and departments. Outsourcing allows companies to downsize the IS department and alleviate difficult financial situations by reducing payroll and other expenses.	First Fidelity, a major bank, used outsourcing as part of a program to reduce the number of employees by 1,600 and slash expenses by $85 million.

Table 12.5

When to Use Outsourcing for Systems Development

There are disadvantages, however.[35] Companies may be asked to sign complicated and restrictive legal contracts that may be difficult to change. Internal expertise can be lost and loyalty can suffer under an outsourcing arrangement.[36] When a company outsources, key IS personnel with expertise in technical and business functions are no longer needed. Once these IS employees leave, their experience with the organization and expertise in information systems is lost. Outsourcing and on demand computing can be very costly for that reason. Because of a faulty sales ordering system that had been outsourced, aviation parts company Aviall lost about $70 million in sales.[37] So, Aviall decided to develop its own systems for $40 million and saw a $430 million increase in sales. For other companies, it can be difficult to achieve a competitive advantage when competitors are using the same computer or consulting company. When the outsourcing or on demand computing is done offshore or in a foreign country, some people have raised security concerns.[38] A Gartner, Inc., study estimates that about 80 percent of U.S. companies outsource critical activities to India, Russia, Pakistan, and China, which could jeopardize security.[39] How will important data and trade secrets be guarded? The Banking Industry Technology Secretariat (BITS) has developed a 33-page document describing how companies can develop security guidelines and procedures when outsourcing.[40] The guidelines are based on the International Organization for Standardization's ISO 17799 information security code.

A number of companies offer outsourcing and on demand computing services—from general systems development to specialized services. IBM's Global Services, for example, is one of the largest full-service outsourcing and consulting services.[41] IBM has consultants located in offices around the world. Electronic Data Systems (EDS) is another large company that specializes in consulting and outsourcing.[42] EDS has approximately 140,000 employees

in almost 60 countries and more than 9,000 clients worldwide. In one year, the company signed $31.4 billion in new contracts for consulting and outsourcing. Accenture, which was once part of Arthur Andersen, is another company that specializes in consulting and outsourcing.[43] The company has more than 75,000 employees in 47 countries.

Some companies are giving customers the option to outsource work.[44] E-Loan, an Internet loan-processing company, gives customers the choice of having their loan processed in the United States or in India. If adopted by other companies, this approach would let customers decide important issues about outsourcing jobs, costs, and quality. In some cases, companies are bringing outsourced jobs back home.[45] A software company hoped to save millions of dollars by outsourcing programming jobs to other countries. Because the foreign workers were not familiar with the company, some software features were omitted and some of the development took months instead of weeks.

FACTORS AFFECTING SYSTEMS DEVELOPMENT SUCCESS

Successful systems development means delivering a system that meets user and organizational needs—on time and within budget. For one Office Depot project, the goals were to "make it simple; make it fun; make it measurable; make it happen."[46] Systems development leaders have identified factors that can contribute to successful systems development efforts—at a reasonable cost. These factors are discussed next.

Degree of Change

A major factor that affects the quality of systems development is the degree of change associated with the project. The scope can vary from implementing minor enhancements to an existing system to major reengineering. The project team needs to recognize where they are on this spectrum of change.

Continuous Improvement Versus Reengineering

As discussed in Chapter 2, continuous improvement projects do not require significant business process or IS changes nor retraining of individuals; thus, they have a high degree of success. Typically, because continuous improvements involve minor improvements, they also have relatively modest benefits. On the other hand, reengineering involves fundamental changes in how the organization conducts business and completes tasks. The factors associated with successful reengineering are similar to those of any development effort, including top management support, clearly defined corporate goals and systems development objectives, and careful management of change. Major reengineering projects tend to have a high degree of risk but also a high potential for major business benefits (see Figure 12.9).

Managing Change

The ability to manage change is critical to the success of systems development. New systems inevitably cause change. For example, the work environment and habits of users are invariably affected by the development of a new information system. Unfortunately, not everyone adapts easily, and the increasing complexity of systems can multiply the problems, as the "Ethical and Societal Issues" special feature describes. Managing change requires the ability to recognize existing or potential problems (particularly the concerns of users) and deal with them before they become a serious threat to the success of the new or modified system. Here are several of the most common problems:

- Fear that the employee will lose his or her job, power, or influence within the organization
- Belief that the proposed system will create more work than it eliminates
- Reluctance to work with "computer people"
- Anxiety that the proposed system will negatively alter the structure of the organization

- Belief that other problems are more pressing than those solved by the proposed system or that the system is being developed by people unfamiliar with "the way things need to get done"
- Unwillingness to learn new procedures or approaches

Figure 12.9

The degree of change can greatly affect the probability of a project's success.

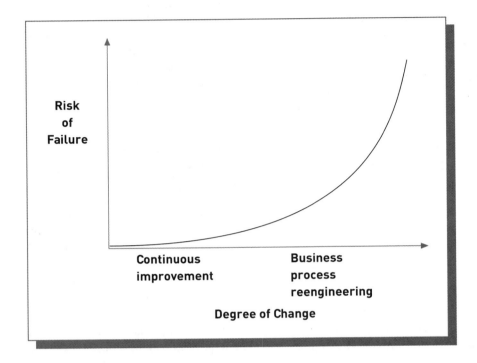

Preventing or dealing with these types of problems requires a coordinated effort from stakeholders and users, managers, and IS personnel. One remedy is simply to talk with all people concerned and learn what their biggest concerns are. Management can then deal with those concerns and try to eliminate them. Once immediate concerns are addressed, people can become part of the project team.

Complexity, Interoperability, and Control

Some IS professionals believe that technology is out of control—or at least that it has achieved a level of complexity beyond the comprehension of most human beings and, increasingly, even some professionals. One of the main sources of the complexity lies in the many systems in use, which were created by a variety of companies using differing technologies that are not easily integrated.

Technology has permeated many businesses and industries over the past ten years. In the rush to fully exploit technology, many companies have built their technology infrastructure piecemeal. They now find themselves with a hodgepodge of servers, databases, and applications from differing vendors tied together with homespun code in an incomprehensible jumble. For this reason, many of the systems development projects in process these days are centered around simplifying and consolidating overly complex systems.

Information system vendors are much to blame for the current state of affairs. In promoting the sale of their own products, many have intentionally designed systems that do not interact well with competitors' products. In 2002, Larry Ellison, CEO of Oracle Corporation, used his keynote speech at the company's annual convention to chide the information technology industry for making life difficult for corporate IS managers. He urged companies to consolidate their systems and to stop customizing their business application software. Companies, Ellison said, have collectively wasted billions of dollars over the last decade on consultants and extra staff to maintain and integrate their tangled systems. "It's a form of madness, all these separate systems," Ellison said. Implied in Ellison's speech was the message that companies would fare much better if they ran only systems designed by Oracle.

Two years later at the same gathering, Ellison preached a different message. The focus of this speech was on Oracle's new product called the Customer Data Hub, which is designed to help companies instantaneously gather and integrate information from systems designed by different—and even competing—companies. The hub uses communication standards known as Web services to talk to incompatible applications and create a "system of record" for customer data, such as orders, contracts, and service history.

Oracle has apparently changed its tune from advising companies to get on the Oracle bandwagon to a more helpful approach that provides companies with tools to help diverse systems work together. "It's a huge benefit," said Basheer Khan, senior director of IT at Vertex Systems, a Los Angeles firm specializing in Oracle consulting services. "Before, people would buy the [Oracle] suite and build their own bridges. Now Oracle is providing the bridges for them."

Oracle is one of the last of the big IS companies to jump on the interoperability bandwagon. Many of Oracle's rivals in the business applications software business, including SAP and Siebel Systems, acknowledged the problem long ago and seem to have a head start on Oracle. In Oracle's other main business area, database software, both IBM and Microsoft have already adopted Web services as a way to link incompatible systems.

Providing bridges between diverse systems will assist companies in taming some out-of-control systems, but it will not end our battles with technology's complexity. As we learn more about the potential of information systems, we continue to collect increasing amounts of data to drive increasingly complex applications for specialized services. In fact, the trend in IS development is to design systems to manage our systems.

New systems designed by HP, IBM, and Sun Microsystems work to conceal complexity by automating many system maintenance tasks. Tomorrow's information systems will further remove IS control from the human operator by providing an additional layer of software between the user and the system. With the additional automation come additional concerns over reliability and security.

Critical Thinking Questions

1. What benefits are enjoyed by using one IS vendor for all systems?
2. Why might a company decide on a multivendor IS environment?

What Would You Do?

Doug Tateson inherited a nightmare when he took the CIO position at European Tours Ltd. The entire organization was frustrated with the information systems, which were based on a hodgepodge of products adopted over the past 12 years. Doug earned the job only by assuring the president of the company that he would be able to straighten out the mess. Doug received a generous budget to revamp the information systems to a state-of-the-art design.

3. If you were Doug, how would you approach the problem? What would be your first step?
4. Would you be inclined to move to a single-vendor environment? Why or why not?

SOURCES: Alorie Gilbert, "Oracle Changes Tune on Integration," *CNET News.com*, January 27, 2004; Tom Yager, "IT Grows Out of Control," *InfoWorld*, January 30, 2004, *www.infoworld.com*; Alorie Gilbert, "Ellison Pushes IT Companies to Simplify," *CNET News.com*, November 14, 2002.

Quality and Standards

Another key success factor is the quality of project planning. The bigger the project, the more likely that poor planning will lead to significant problems. A federal jury, for example, found the maker of navigational software partly responsible for an airplane crash near Cali, Colombia. Poor systems development planning can be deadly.

Many companies find that large systems projects fall behind schedule, go over budget, and do not meet expectations. Although proper planning cannot guarantee that these types of problems will be avoided, it can minimize the likelihood of their occurrence. Good systems development is not automatic. Certain factors contribute to the failure of systems development projects. These factors and countermeasures to eliminate or alleviate the problem are summarized in Table 12.6.

Factor	Countermeasure
Solving the wrong problem	Establish a clear connection between the project and organizational goals.
Poor problem definition and analysis	Follow a standard systems development approach.
Poor communication	Communicate, communicate, communicate.
Project is too ambitious	Narrow the project focus to address only the most important business opportunities.
Lack of top management support	Identify the senior manager who has most to gain from the success of the project, and recruit this individual to champion the project.
Lack of management and user involvement	Identify and recruit key stakeholders to be active participants in the project.
Inadequate or improper system design	Follow a standard systems development approach.
Lack of standards	Implement a standards system, such as ISO 9001.
Poor testing and implementation	Plan sufficient time for this activity.
Users are unable to use the system effectively	Develop a rigorous user-training program and budget sufficient time in the schedule to execute it.
Lack of concern for maintenance	Include an estimate of employee effort and costs for maintenance in the original project justification.

Table 12.6

Project Planning Issues Frequently Contributing to Project Failure

The development of information systems requires a constant trade-off of schedule and cost versus quality. Historically, the development of application software has put an overemphasis on schedule and cost to the detriment of quality. Techniques, such as use of the ISO 9001 standards, have been developed to improve the quality of information systems. ISO 9001 is a set of international quality standards originally developed in Europe in 1987. The most recent version was published in 2000. These standards address customer satisfaction and are the only standards in the ISO 9000 family where third-party certification can be achieved. Adherence to ISO 9001 is a requirement in many international markets.[47] Many companies in the United States and around the world strive to achieve ISO 9001 certification.

Many IS organizations have incorporated ISO 9001, total quality management, six-sigma (first discussed in Chapter 2), and other quality measures into the way they produce software. IBM, the large computer and consulting company, for example, uses ISO 9001, six-sigma, the Capability Maturity Model (CMM) discussed next, and other quality approaches.[48] MasterCard International uses parts of the CMM. MasterCard's quality program has

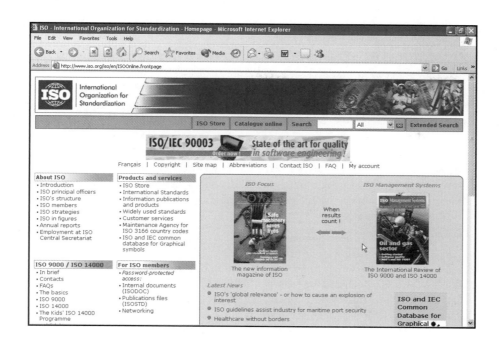

ISO 9000 is a set of international quality standards used by IS and other organizations to ensure the quality of products and services.

shortened the time it takes to develop new software, while reducing software defects. Nortel Networks uses ISO 9001. Often, to ensure the quality of the systems development process and finished product, an IS organization will form its own quality assurance groups to work with project teams and encourage them to follow established standards.

The Capability Maturity Model (CMM)

Organizational experience with the systems development process is also a key factor for systems development success.[49] The Capability Maturity Model (CMM) is one way to measure this experience. It is based on research done at Carnegie Mellon University and work by the Software Engineering Institute (SEI). CMM is a measure of the maturity of the software development process in an organization. CMM grades an organization's systems development maturity using five levels from initial to optimized (see Figure 12.10). A brief description of each level follows.

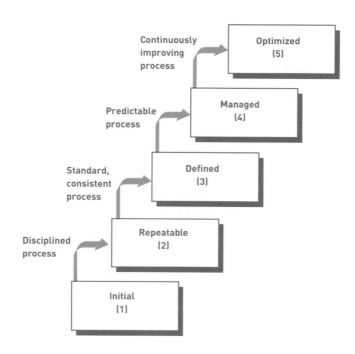

Figure 12.10

Systems Development Maturity Based on the Capability Maturity Model (CMM)

1. *Initial.* This level is typical of organizations inexperienced with software and systems development. This level often has an ad hoc or even a chaotic development process.
2. *Repeatable.* The second level tracks development costs, schedules, and functionality. The discipline to repeat previous systems development success is in place.
3. *Defined.* At the third level, organizations use documented and defined procedures. All projects done by the organization use these standardized approaches to develop software and systems. Programming standards are often used at this level.
4. *Managed.* At this level, organizations use detailed measures of the systems development process to help manage the process and improve software and systems quality.
5. *Optimized.* This is the highest level of experience and maturity. Continuous improvement is used to strengthen all aspects of the systems development process. Organizations at this level often initiate innovative projects. The goal is to optimize all aspects of the systems development effort.

The CMM model has been popular in the United States and around the world, and SEI certifies organizations as being at one of the five levels. Any organization can seek certification, and many computer-consulting companies attempt to be certified at the highest level (optimization). Wipro GE Medical, for example, received Level 5 certification. The company develops advanced medical software for computerized tomography (CT) scanners, magnetic resonance imaging (MRI) devices, and other medical equipment. Sasken Communications Company is a telecommunications and consulting company that has also achieved Level 5 certification.

Use of Project Management Tools

Project management involves planning, scheduling, directing, and controlling human, financial, and technological resources for a defined task whose result is achievement of specific goals and objectives. Even for small systems development projects, some type of project management must be undertaken. America West, for example, struggled financially after the September 11 terrorist attacks.[50] The airline industry has lost $18 billion since September 11 and may lose another $10 billion before its carriers recover. As a result, airline companies are demanding more from their systems development projects and their project management tools. According to Joe Beery, chief information officer at America West, "The rigors, constraints and milestones that you have to meet to move forward with new programs have now become much more intense."

project schedule
Detailed description of what is to be done.

project milestone
A critical date for the completion of a major part of the project.

project deadline
Date the entire project is to be completed and operational.

critical path
Activities that, if delayed, would delay the entire project.

A **project schedule** is a detailed description of what is to be done. Each project activity, the use of personnel and other resources, and expected completion dates are described. A **project milestone** is a critical date for the completion of a major part of the project. The completion of program design, coding, testing, and release are examples of milestones for a programming project. The **project deadline** is the date the entire project is to be completed and operational—when the organization can expect to begin to reap the benefits of the project. One company offers a 20 percent refund if it doesn't meet a client's project deadline. In addition, any additional work done after the project deadline is performed free of charge.

In systems development, each activity has an earliest start time, earliest finish time, and slack time, which is the amount of time an activity can be delayed without delaying the entire project. The **critical path** consists of all activities that, if delayed, would delay the entire project. These activities have zero slack time. Any problems with critical-path activities will cause problems for the entire project. To ensure that critical-path activities are completed in a timely fashion, formalized project management approaches have been developed. A number of tools, such as Project by Microsoft, are available to help compute these critical project attributes.

Program Evaluation and Review Technique (PERT)
A formalized approach for developing a project schedule.

Although the steps of systems development seem straightforward, larger projects can become complex, requiring literally hundreds or thousands of separate activities. For these systems development efforts, formal project management methods and tools become essential. A formalized approach called **Program Evaluation and Review Technique (PERT)** creates three time estimates for an activity: shortest possible time, most likely time, and longest possible time. A formula is then applied to come up with a single PERT time estimate.

A **Gantt chart** is a graphical tool used for planning, monitoring, and coordinating projects; it is essentially a grid that lists activities and deadlines. Each time a task is completed, a darkened line is placed in the proper grid cell to indicate the completion of a task (see Figure 12.11).

Gantt chart
A graphical tool used for planning, monitoring, and coordinating projects.

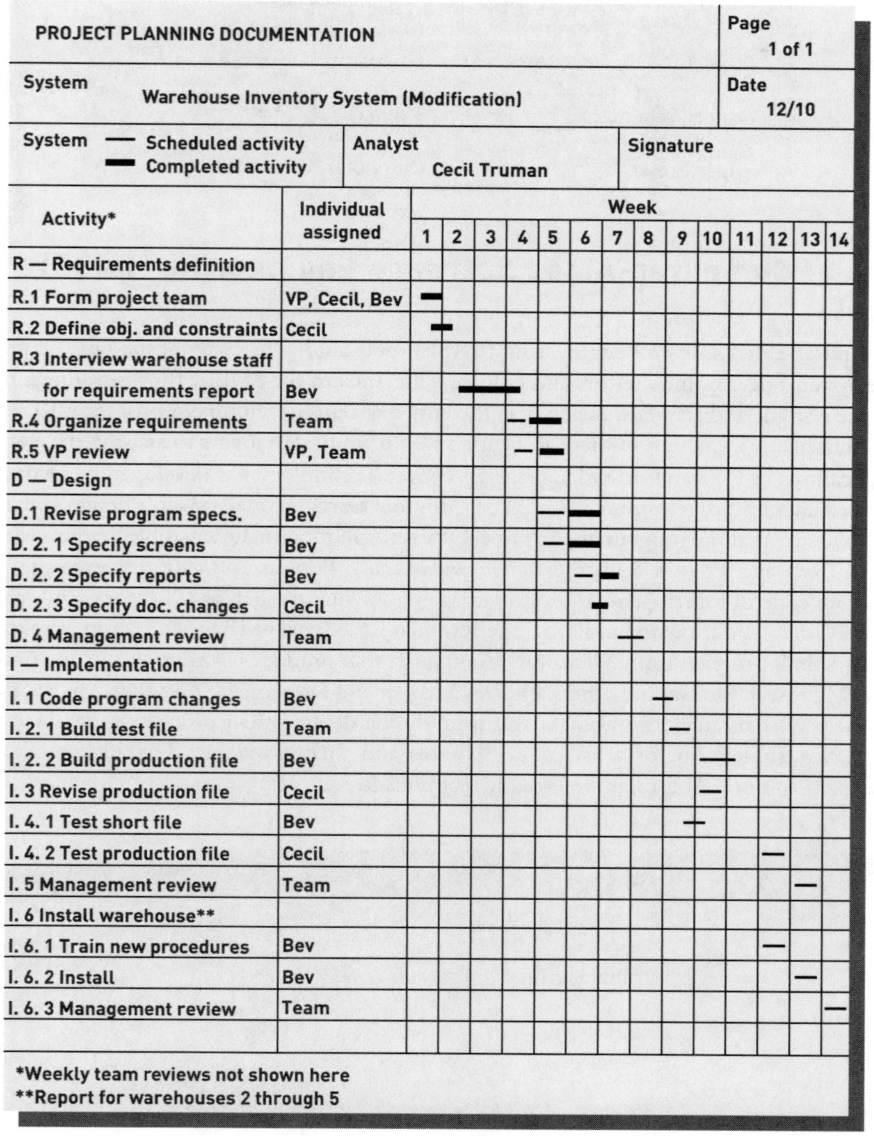

Figure 12.11

Sample Gantt Chart

A Gantt chart shows progress through systems development activities by putting a bar through appropriate cells.

PROJECT PLANNING DOCUMENTATION		Page 1 of 1
System: Warehouse Inventory System (Modification)		Date 12/10
System — Scheduled activity ▬ Completed activity	Analyst: Cecil Truman	Signature

Activity*	Individual assigned	Week
		1 2 3 4 5 6 7 8 9 10 11 12 13 14
R — Requirements definition		
R.1 Form project team	VP, Cecil, Bev	▬ (week 1)
R.2 Define obj. and constraints	Cecil	▬ (week 1)
R.3 Interview warehouse staff		
for requirements report	Bev	▬ (weeks 2–3)
R.4 Organize requirements	Team	─▬ (weeks 3–4)
R.5 VP review	VP, Team	─▬ (weeks 3–4)
D — Design		
D.1 Revise program specs.	Bev	─▬ (weeks 5–6)
D. 2. 1 Specify screens	Bev	─▬ (weeks 5–6)
D. 2. 2 Specify reports	Bev	▬ (week 6)
D. 2. 3 Specify doc. changes	Cecil	▬ (week 6)
D. 4 Management review	Team	─ (week 7)
I — Implementation		
I. 1 Code program changes	Bev	─ (week 8)
I. 2. 1 Build test file	Team	─ (week 8)
I. 2. 2 Build production file	Bev	─ (week 9)
I. 3 Revise production file	Cecil	─ (week 9)
I. 4. 1 Test short file	Bev	─ (week 8)
I. 4. 2 Test production file	Cecil	─ (week 11)
I. 5 Management review	Team	─ (week 12)
I. 6 Install warehouse**		
I. 6. 1 Train new procedures	Bev	─ (week 11)
I. 6. 2 Install	Bev	─ (week 12)
I. 6. 3 Management review	Team	─ (week 13)

*Weekly team reviews not shown here
**Report for warehouses 2 through 5

Both PERT and Gantt techniques can be automated using project management software. This software monitors all project activities and determines whether activities and the entire project are on time and within budget. Project management software also has workgroup capabilities to handle multiple projects and to allow a team to interact with the same software. Project management software helps managers determine the best way to reduce project completion time at the least cost. Many project managers, however, fear that the quality of a systems development project will suffer with shortened deadlines and think that slack time should be added back to the schedule as a result. Several project management software packages are identified in Table 12.7.

Software	Vendor
BeachBox '98	NetSQL Partners
Job Order	Management Software
OpenPlan	Welcom
Project	Microsoft
Project Scheduler	Scitor
Super Project	Computer Associates

Use of Computer-Aided Software Engineering (CASE) Tools

computer-aided software engineering (CASE)
Tools that automate many of the tasks required in a systems development effort and enforce adherence to the SDLC.

Computer-aided software engineering (CASE) tools automate many of the tasks required in a systems development effort and enforce adherence to the SDLC, thus instilling a high degree of rigor and standardization into the entire systems development process. VRCASE, for example, is a CASE tool that can be used by a team of developers to assist in developing applications in C++ and other languages.[51] Prover Technology has developed a CASE tool that searches for programming bugs. The CASE tool searches for all possible design scenarios to make sure that the program is error free. Other CASE tools include Visible Systems (*www. visible.com*) and Popkin Software (*www.popkin.com*). Popkin Software, for example, can generate code in fourth-generation programming languages, such as C++, Java, and Visual Basic. Other CASE-related tools include Rational Rose (part of IBM) and Visio, a charting and graphics program from Microsoft. Companies that produce CASE tools include Accenture, Microsoft, Oracle, and others. Oracle Designer and Developer CASE tools, for example, can help systems analysts automate and simplify the development process for database systems. See Table 12.8 for a list of CASE tools and their providers. The advantages and disadvantages of CASE tools are listed in Table 12.9.

CASE Tool	Vendor
Oracle Designer	Oracle Corporation, www.oracle.com
Visible Analyst	Visible Systems Corporation, www.visible.com
Rational Rose	Rational Software, www-306.ibm.com/software/rational
Embarcadero Describe	Embarcadero Describe, www.embarcadero.com

Advantages	Disadvantages
Produce systems with a longer effective operational life	Produce initial systems that are more expensive to build and maintain
Produce systems that more closely meet user needs and requirements	Require more extensive and accurate definition of user needs and requirements
Produce systems with excellent documentation	May be difficult to customize
Produce systems that need less systems support	Require more training of maintenance staff
Produce more flexible systems	May be difficult to use with existing systems

Object-Oriented Systems Development

The success of a systems development effort can depend on the specific programming tools and approaches used. As mentioned in Chapter 4, object-oriented (OO) programming languages allow the interaction of programming objects—that is, an object consists of both data and the actions that can be performed on the data. So, an object could be data about an employee and all the operations (such as payroll, benefits, and tax calculations) that might be performed on the data.

Developing programs and applications using OO programming languages involves constructing modules and parts that can be reused in other programming projects.[52] Chapter 4 discussed a number of programming languages that use the object-oriented approach, including Visual Basic, C++, and Java. These languages allow systems developers to take the OO approach, making program development faster and more efficient, resulting in lower costs. Modules can be developed internally or obtained from an external source. Once a company has the programming modules, programmers and systems analysts can modify them and integrate them with other modules to form new programs. GE Power developed a new system to bridge the gap between new Internet-based applications and older legacy systems. The new system allows its managers and employees to seamlessly share data and information. GE Power used object-oriented languages and approaches, such as Java classes and JavaBeans to reuse computer code and reduce systems development time and costs.[53]

Object-oriented systems development (OOSD) combines the logic of the systems development life cycle with the power of object-oriented modeling and programming. OOSD follows a defined systems development life cycle, much like the SDLC. The life cycle phases can be, and usually are, completed with many iterations. Object-oriented systems development typically involves the following:

- *Identifying potential problems and opportunities within the organization that would be appropriate for the OO approach.* This process is similar to traditional systems investigation. Ideally, these problems or opportunities should lend themselves to the development of programs that can be built by modifying existing programming modules.
- *Defining what kind of system users require.* This analysis means defining all the objects that are part of the user's work environment (object-oriented analysis). The OO team must study the business and build a model of the objects that are part of the business (such as a customer, an order, or a payment). Many of the CASE tools discussed in the previous section can be used, starting with this step of OOSD.
- *Designing the system.* This process defines all the objects in the system and the ways they interact (object-oriented design). Design involves developing logical and physical models of the new system by adding details to the object model started in analysis.
- *Programming or modifying modules.* This implementation step takes the object model begun during analysis and completed during design and turns it into a set of interacting objects in a system. Object-oriented programming languages are designed to allow the programmer to create classes of objects in the computer system that correspond to the

Table 12.9

Advantages and Disadvantages of CASE Tools

object-oriented systems development (OOSD)
Approach to systems development that combines the logic of the systems development life cycle with the power of object-oriented modeling and programming.

objects in the actual business process. Objects such as customer, order, and payment are redefined as computer system objects—a customer screen, an order entry menu, or a dollar sign icon. Programmers then write new modules or modify existing ones to produce the desired programs.

- *Evaluation by users.* The initial implementation is evaluated by users and improved. Additional scenarios and objects are added, and the cycle repeats. Finally, a complete, tested, and approved system is available for use.
- *Periodic review and modification.* The completed and operational system is reviewed at regular intervals and modified as necessary.

SYSTEMS INVESTIGATION

As discussed earlier in the chapter, systems investigation is the first phase in the traditional SDLC of a new or modified business information system. The purpose is to identify potential problems and opportunities and consider them in light of the goals of the company. In general, systems investigation attempts to uncover answers to the following questions:

- What primary problems might a new or enhanced system solve?
- What opportunities might a new or enhanced system provide?
- What new hardware, software, databases, telecommunications, personnel, or procedures will improve an existing system or are required in a new system?
- What are the potential costs (variable and fixed)?
- What are the associated risks?

Initiating Systems Investigation

systems request form
Document filled out by someone who wants the IS department to initiate systems investigation.

Because systems development requests can require considerable time and effort to implement, many organizations have adopted a formal procedure for initiating systems development, beginning with systems investigation. The **systems request form** is a document that is filled out by someone who wants the IS department to initiate systems investigation. This form typically includes the following information:

- Problems in or opportunities for the system
- Objectives of systems investigation
- Overview of the proposed system
- Expected costs and benefits of the proposed system

The information in the systems request form helps to rationalize and prioritize the activities of the IS department. Based on the overall IS plan, the organization's needs and goals, and the estimated value and priority of the proposed projects, managers make decisions regarding the initiation of each systems investigation for such projects.

Participants in Systems Investigation

Once a decision has been made to initiate systems investigation, the first step is to determine what members of the development team should participate in the investigation phase of the project. Members of the development team change from phase to phase (see Figure 12.12).

Ideally, functional managers are heavily involved during the investigation phase. Other members could include users or stakeholders outside management, such as an employee who helped initiate systems development. The technical and financial expertise of others participating in investigation would help the team determine whether the problem is worth solving. The members of the development team who participate in investigation are then responsible for gathering and analyzing data, preparing a report justifying systems development, and presenting the results to top-level managers.

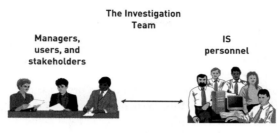

The Investigation Team

Managers, users, and stakeholders

IS personnel

• Undertakes feasibility analysis
• Establishes system development goals
• Selects system development methodology
• Prepares system investigation report

Figure 12.12

The Systems Investigation Team

The team is made up of upper- and middle-level managers, a project manager, IS personnel, users, and stakeholders.

Feasibility Analysis

A key step of the systems investigation phase is **feasibility analysis,** which assesses technical, economic, legal, operational, and schedule feasibility (see Figure 12.13). **Technical feasibility** is concerned with whether the hardware, software, and other system components can be acquired or developed to solve the problem. Technical problems, for example, were encountered in developing an automatic tool system for long-haul trucks on the German Autobahn.[54] The satellite-based systems development project may never be fully implemented.

feasibility analysis
Assessment of the technical, economic, legal, operational, and schedule feasibility of a project.

technical feasibility
Assessment of whether the hardware, software, and other system components can be acquired or developed to solve the problem.

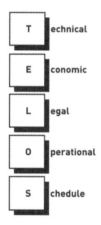

T echnical

E conomic

L egal

O perational

S chedule

Figure 12.13

Technical, Economic, Legal, Operational, and Schedule Feasibility

Economic feasibility determines whether the project makes financial sense and whether predicted benefits offset the cost and time needed to obtain them. A securities company, for example, investigated the economic feasibility of sending research reports electronically instead of through the mail. Economic analysis revealed that the new approach could save the company up to $500,000 a year. Economic feasibility can involve cash flow analysis such as that done in net present value or internal rate of return (IRR) calculations.

Net present value is an often-used approach for ranking competing projects and for determining economic feasibility. The net present value represents the net amount by which project savings exceed project expenses, after allowing for the cost of capital and the passage of time. The cost of capital is the average cost of funds used to finance the operations of the business. Net present value takes into account that a dollar returned at a later date is not worth as much as one received today, because the dollar in hand can be invested to earn profits or interest in the interim. Spreadsheet programs, such as Lotus and Excel, have built-in functions to compute the net present value and internal rate of return.

economic feasibility
Determination of whether the project makes financial sense and whether predicted benefits offset the cost and time needed to obtain them.

net present value
The preferred approach for ranking competing projects and determining economic feasibility.

legal feasibility
Determination of whether laws or regulations may prevent or limit a systems development project.

operational feasibility
Measure of whether the project can be put into action or operation.

schedule feasibility
Determination of whether the project can be completed in a reasonable amount of time.

Legal feasibility determines whether laws or regulations may prevent or limit a systems development project. For example, an Internet site that allowed users to share music without paying musicians or music producers was sued. Legal feasibility involves an analysis of existing and future laws to determine the likelihood of legal action against the systems development project and the possible consequences.

Operational feasibility is a measure of whether the project can be put into action or operation. It can include logistical and motivational (acceptance of change) considerations. Motivational considerations are very important because new systems affect people and data flows and may have unintended consequences. As a result, power and politics may come into play, and some people may resist the new system. Because of deadly hospital errors, a health-care consortium looked into the operational feasibility of developing a new computerized physician order entry system to require that all prescriptions and every order a doctor gives to staff be entered into the computer. The computer then checks for drug allergies and interactions between drugs. If operationally feasible, the new system could save lives and help avoid lawsuits.

Schedule feasibility determines whether the project can be completed in a reasonable amount of time—a process that involves balancing the time and resource requirements of the project with other projects.

Object-Oriented Systems Investigation

The object-oriented approach can be used during all phases of systems development, from investigation to maintenance and review. In addition to identifying key participants and performing basic feasibility analysis, key objects can be identified during systems investigation. Consider a kayak rental business in Maui, Hawaii, where the owner wants to computerize its operations. There are many system objects for this business, including the kayak rental clerk, renting kayaks to customers, and adding new kayaks into the rental program. These objects can be diagrammed in a use case diagram (see Figure 12.14). As you can see, the kayak rental clerk rents kayaks to customers and adds new kayaks to the current inventory of kayaks available for rent. The stick figure is an example of an *actor*, and the ovals each represent an event, called a *use case*. In our example, the actor (the kayak rental clerk) interacts with two use cases (rent kayaks to customers and add new kayaks to inventory.). The use case diagram is part of the Unified Modeling Language that is used in object-oriented systems development.

Figure 12.14

Use Case Diagram for a Kayak Rental Application

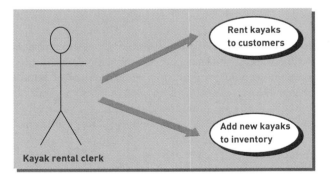

The Systems Investigation Report

systems investigation report
Summary of the results of the systems investigation and the process of feasibility analysis and recommendation of a course of action.

The primary outcome of systems investigation is a **systems investigation report**. This report summarizes the results of systems investigation and the process of feasibility analysis and recommends a course of action: continue on into systems analysis, modify the project in some manner, or drop it. A typical table of contents for the systems investigation report is shown in Figure 12.15.

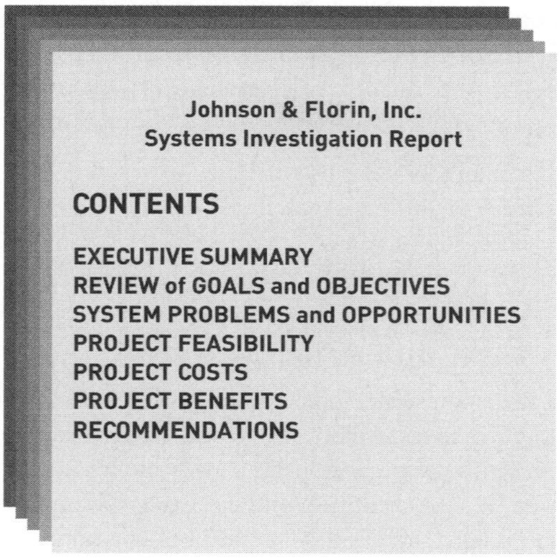

Figure 12.15

A Typical Table of Contents for a Systems Investigation Report

Johnson & Florin, Inc.
Systems Investigation Report

CONTENTS

EXECUTIVE SUMMARY
REVIEW of GOALS and OBJECTIVES
SYSTEM PROBLEMS and OPPORTUNITIES
PROJECT FEASIBILITY
PROJECT COSTS
PROJECT BENEFITS
RECOMMENDATIONS

The systems investigation report is reviewed by senior management, often organized as an advisory committee, or **steering committee,** consisting of senior management and users from the IS department and other functional areas. These individuals help IS personnel with their decisions about the use of information systems in the business and give authorization to pursue further systems development activities. After review, the steering committee might agree with the recommendation of the systems development team or suggest a change in project focus to concentrate more directly on meeting a specific company objective. Another alternative is that everyone may decide that the project is not feasible for one reason or another and cancel the project.

steering committee
An advisory group consisting of senior management and users from the IS department and other functional areas.

SYSTEMS ANALYSIS

After a project has been approved for further study, the next step is to answer the question "What must the information system do to solve the problem?" The process needs to go beyond mere computerization of existing systems. The entire system, and the business process with which it is associated, should be evaluated. Often, a firm can make great gains if it restructures both business activities and the related information system simultaneously. The overall emphasis of analysis is gathering data on the existing system, determining the requirements for the new system, considering alternatives within these constraints, and investigating the feasibility of the solutions. The primary outcome of systems analysis is a prioritized list of systems requirements.

General Considerations

Systems analysis starts by clarifying the overall goals of the organization and determining how the existing or proposed information system helps meet them. A manufacturing company, for example, might want to reduce the number of equipment breakdowns. This goal can be translated into one or more informational needs. One need might be to create and maintain an accurate list of each piece of equipment and a schedule for preventive maintenance. Another need might be a list of equipment failures and their causes.

Analysis of a small company's information system can be fairly straightforward. On the other hand, evaluating an existing information system for a large company can be a long, tedious process. As a result, large organizations evaluating a major information system normally follow a formalized analysis procedure, involving these steps:

1. Assembling the participants for systems analysis
2. Collecting appropriate data and requirements
3. Analyzing the data and requirements
4. Preparing a report on the existing system, new system requirements, and project priorities

Participants in Systems Analysis

The first step in formal analysis is to assemble a team to study the existing system. This group includes members of the original development team—from users and stakeholders to IS personnel and management. Most organizations usually allow key members of the development team not only to analyze the condition of the existing system but also to perform other aspects of systems development, such as design and implementation.

Once the participants in systems analysis are assembled, this group develops a list of specific objectives and activities. A schedule for meeting the objectives and completing the specific activities is also developed, along with deadlines for each stage and a statement of the resources required at each stage, such as clerical personnel, supplies, and so forth. Major milestones are normally established to help the team monitor progress and determine whether problems or delays occur in performing systems analysis.

Data Collection

The purpose of data collection is to seek additional information about the problems or needs identified in the systems investigation report. During this process, the strengths and weaknesses of the existing system are emphasized.

Identifying Sources of Data

Data collection begins by identifying and locating the various sources of data, including both internal and external sources (see Figure 12.16).

Figure 12.16

Internal and External Sources of Data for Systems Analysis

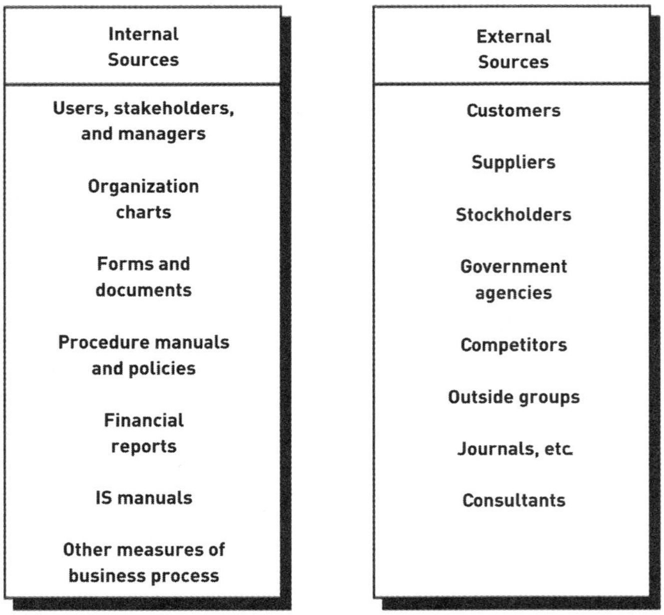

Internal Sources	External Sources
Users, stakeholders, and managers	Customers
Organization charts	Suppliers
Forms and documents	Stockholders
Procedure manuals and policies	Government agencies
Financial reports	Competitors
IS manuals	Outside groups
Other measures of business process	Journals, etc.
	Consultants

Collecting Data

Once data sources have been identified, data collection begins. Figure 12.17 shows the steps involved. Data collection may require a number of tools and techniques, such as interviews, direct observation, and questionnaires.

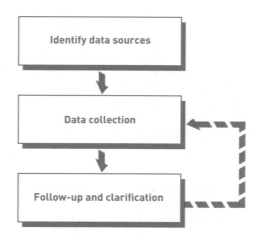

```
┌─────────────────────────────┐
│   Identify data sources     │
└─────────────────────────────┘
              │
              ▼
┌─────────────────────────────┐
│      Data collection        │◄──────┐
└─────────────────────────────┘       │
              │                        │
              ▼                        │
┌─────────────────────────────┐       │
│  Follow-up and clarification│───────┘
└─────────────────────────────┘
```

Figure 12.17

The Steps in Data Collection

Interviews may either be structured or unstructured. In a **structured interview**, the questions are written in advance. In an **unstructured interview**, the questions are not written in advance; the interviewer relies on experience in asking the best questions to uncover the inherent problems of the existing system. An advantage of the unstructured interview is that it allows the interviewer to ask follow-up or clarifying questions immediately.

With **direct observation**, one or more members of the analysis team directly observe the existing system in action. One of the best ways to understand how the existing system functions is to work with the users to discover how data flows in certain business tasks. Determining the data flow entails direct observation of users' work procedures, their reports, current screens (if automated already), and so on. From this observation, members of the analysis team determine which forms and procedures are adequate and which are inadequate and need improvement. Direct observation requires a certain amount of skill. The observer must be able to see what is really happening and not be influenced by his or her own attitudes or feelings. This approach can reveal important problems and opportunities that would be difficult to obtain using other data collection methods. An example would be observing the work procedures, reports, and computer screens associated with an accounts payable system being considered for replacement.

structured interview
An interview where the questions are written in advance.

unstructured interview
An interview where the questions are not written in advance.

direct observation
Watching the existing system in action by one or more members of the analysis team.

Direct observation is a method of data collection. One or more members of the analysis team directly observe the existing system in action.

(Source: Digital Vision/Getty Images.)

questionnaires
A method of gathering data when the data sources are spread over a wide geographic area.

When many data sources are spread over a wide geographic area, **questionnaires** may be the best approach. Like interviews, questionnaires can be either structured or unstructured. In most cases, a pilot study is conducted to fine-tune the questionnaire. A follow-up questionnaire can also capture the opinions of those who do not respond to the original questionnaire.

A number of other data collection techniques can be employed. In some cases, telephone calls are an excellent method. In other cases, activities may be simulated to see how the existing system reacts. Thus, fake sales orders, stockouts, customer complaints, and data-flow bottlenecks may be created to see how the existing system responds to these situations. **Statistical sampling**, which involves taking a random sample of data, is another technique. For example, suppose that we want to collect data that describes 10,000 sales orders received over the last few years. Because it is too time-consuming to analyze each of the 10,000 sales orders, a random sample of 100 to 200 sales orders from the entire batch can be collected. The characteristics of this sample are then assumed to apply to all 10,000 orders.

statistical sampling
Selection of a random sample of data and applying the characteristics of the sample to the whole group.

Data Analysis

The data collected in its raw form is usually not adequate to determine the effectiveness and efficiency of the existing system or the requirements for the new system. The next step is to manipulate the collected data so that it is usable by the development team members who are participating in systems analysis. This manipulation is called **data analysis**. Data and activity modeling, using data-flow diagrams and entity-relationship diagrams, are useful during data analysis to show data flows and the relationships among various objects, associations, and activities. Other common tools and techniques for data analysis include application flowcharts, grid charts, CASE tools, and the object-oriented approach.

data analysis
Manipulation of the collected data so that it is usable for the development team members who are participating in systems analysis.

Data Modeling

Data modeling, first introduced in Chapter 5, is a commonly accepted approach to modeling organizational objects and associations that employ both text and graphics. The exact way that data modeling is employed, however, is governed by the specific systems development methodology.

Data modeling is most often accomplished through the use of entity-relationship (ER) diagrams. Recall from Chapter 5 that an entity is a generalized representation of an object type—such as a class of people (employee), events (sales), things (desks), or places (Philadelphia)—and that entities possess certain attributes. Objects can be related to other objects in numerous ways. An entity-relationship diagram, such as the one shown in Figure 12.18a, describes a number of objects and the ways they are associated. An ER diagram is not capable by itself of fully describing a business problem or solution because it lacks descriptions of the related activities. It is, however, a good place to start, because it describes object types and attributes about which data may need to be collected for processing.

Activity Modeling

To fully describe a business problem or solution, it is necessary to describe the related objects, associations, and activities. Activities in this sense are events or items that are necessary to fulfill the business relationship or that can be associated with the business relationship in a meaningful way.

data-flow diagram (DFD)
A model of objects, associations, and activities that describes how data can flow between and around various objects.

Activity modeling is often accomplished through the use of data-flow diagrams. A **data-flow diagram (DFD)** models objects, associations, and activities by describing how data can flow between and around various objects. DFDs work on the premise that for every activity there is some communication, transference, or flow that can be described as a data element. DFDs describe what activities are occurring to fulfill a business relationship or accomplish a business task, not how these activities are to be performed. That is, DFDs show the logical sequence of associations and activities, not the physical processes. A system modeled with a DFD could operate manually or could be computer based; if computer based, the system could operate with a variety of technologies.

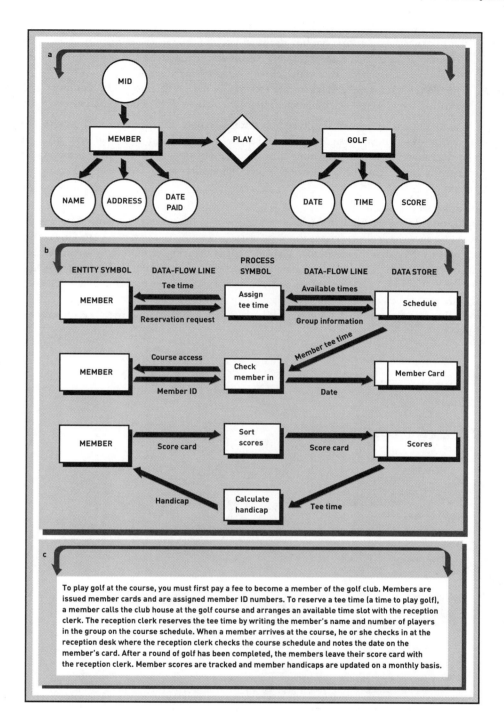

Figure 12.18

Data and Activity Modeling

(a) An entity-relationship diagram. (b) A data-flow diagram. (c) A semantic description of the business process.

(Source: G. Lawrence Sanders, *Data Modeling* (Boyd & Fraser Publishing, Danvers, MA: 1995.)

DFDs are easy to develop and easily understood by nontechnical people. Data-flow diagrams use four primary symbols, as illustrated in Figure 12.18b.

- *Data flow.* The **data-flow line** includes arrows that show the direction of data element movement.
- *Process symbol.* The **process symbol** reveals a function that is performed. Computing gross pay, entering a sales order, delivering merchandise, and printing a report are examples of functions that can be represented with a process symbol.

data-flow line
Arrows that show the direction of data element movement.

process symbol
Representation of a function that is performed.

entity symbol

Representation of either a source or destination of a data element.

data store

Representation of a storage location for data.

application flowcharts

Diagrams that show relationships among applications or systems.

Figure 12.19

A Telephone Order Process Application Flowchart

The flowchart shows the relationships among various processes.

(Source: Courtesy of SmartDraw.com.)

- *Entity symbol.* The **entity symbol** shows either the source or destination of the data element. An entity can be, for example, a customer who initiates a sales order, an employee who receives a paycheck, or a manager who gets a financial report.
- *Data store.* A **data store** reveals a storage location for data. A data store is any computerized or manual data storage location, including magnetic tape, disks, a filing cabinet, or a desk.

Comparing entity-relationship diagrams with data-flow diagrams provides insight into the concept of top-down design. Figure 12.18a and b show an entity-relationship diagram and a data-flow diagram for the same business relationship—namely, a member of a golf club playing golf. Figure 12.18c provides a brief description of the business relationship for clarification.

Application Flowcharts

Application flowcharts show the relationships among applications or systems. Assume that a small business has collected data about its order processing, inventory control, invoicing, and marketing analysis applications. Management is thinking of modifying the inventory control application. The raw facts collected, however, do not help in determining how the applications are related to each other and the databases required for each. These relationships are established through data analysis with an application flowchart (see Figure 12.19). Using this tool for data analysis makes clear the relationships among the order-processing functions.

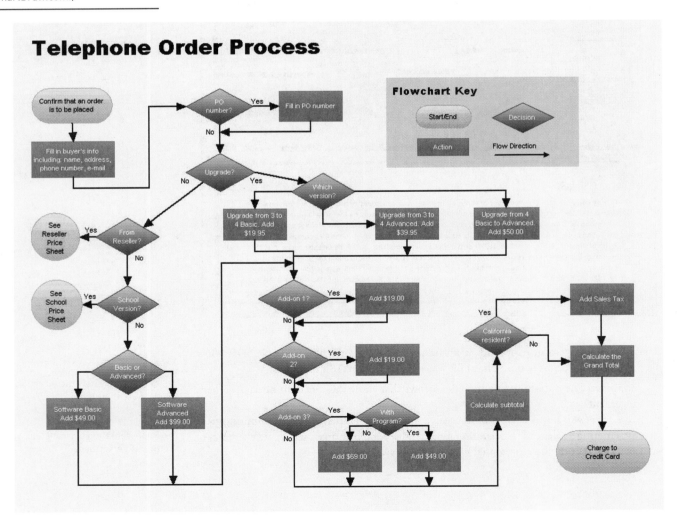

In the simplified application flowchart in Figure 12.19, you can see that the telephone order clerk provides important data to the system about items such as versions, quantities, and prices. The system calculates sales tax and order totals. Any changes made to this order-processing system could affect the company's other systems, such as inventory control and marketing.

Grid Charts

A **grid chart** is a table that shows relationships among various aspects of a systems development effort. For example, a grid chart can be used to reveal the databases used by the various applications (see Figure 12.20).

grid chart
Table that shows relationships among the various aspects of a systems development effort.

Databases → Applications	Customer database	Inventory database	Supplier database	Accounts receivable database
Order processing application	X	X		
Inventory control application		X	X	
Marketing analysis application	X	X		
Invoicing application	X			X

Figure 12.20

A Grid Chart

The chart shows the relationships among applications and databases.

The simplified grid chart in Figure 12.20 shows that the customer database is used by the order processing, marketing analysis, and invoicing applications. The inventory database is used by the order processing, inventory control, and marketing analysis applications. The supplier database is used by the inventory control application, and the accounts receivable database is used by the invoicing application. This grid chart shows which applications use common databases and reveals that, for example, any changes to the inventory control application must investigate the inventory and supplier databases.

CASE Tools

As discussed earlier, many systems development projects use CASE tools to complete analysis tasks. Most computer-aided software engineering tools have generalized graphics programs that can generate a variety of diagrams and figures. Entity-relationship diagrams, data-flow diagrams, application flowcharts, and other diagrams can be developed using CASE graphics programs to help describe the existing system. During the analysis phase, a **CASE repository**—a database of system descriptions, parameters, and objectives—will begin to be developed.

CASE repository
A database of system descriptions, parameters, and objectives.

Requirements Analysis

The overall purpose of **requirements analysis** is to determine user, stakeholder, and organizational needs. For an accounts payable application, the stakeholders could include suppliers and members of the purchasing department. Questions that should be asked during requirements analysis include the following:

requirements analysis
Determination of user, stakeholder, and organizational needs.

- Are these stakeholders satisfied with the current accounts payable application?
- What improvements could be made to satisfy suppliers and help the purchasing department?

One of the most difficult procedures in systems analysis is confirming user or systems requirements. In some cases, communications problems can interfere with the determination of these requirements. For example, an accounts payable manager may want a better procedure for tracking the amount owed by customers. Specifically, the manager would like to

have a weekly report that shows all customers who owe more than $1,000 and are more than 90 days past due on their account. A financial manager might need a report that summarizes total amount owed by customers to look at the need to loosen or tighten credit limits. A sales manager might want to review the amount owed by a key customer relative to sales to that same customer. The purpose of requirements analysis is to capture these requests in detail. Numerous tools and techniques can be used to capture systems requirements. Often, various techniques are used in the context of a JAD session.

Asking Directly

One the most basic techniques used in requirements analysis is asking directly. **Asking directly** is an approach that asks users, stakeholders, and other managers about what they want and expect from the new or modified system. This approach works best for stable systems in which stakeholders and users clearly understand the system's functions. The role of the systems analyst during the analysis phase is to critically and creatively evaluate needs and define them clearly so that the systems can best meet them.

Critical Success Factors

Another approach uses critical success factors (CSFs). As discussed earlier, managers and decision makers are asked to list only the factors that are critical to the success of their area of the organization. A CSF for a production manager might be adequate raw materials from suppliers; a CSF for a sales representative could be a list of customers currently buying a certain type of product. Starting from these CSFs, the system inputs, outputs, performance, and other specific requirements can be determined.

The IS Plan

As we have seen, the IS plan translates strategic and organizational goals into systems development initiatives. The IS planning process often generates strategic planning documents that can be used to define system requirements. Working from these documents ensures that requirements analysis will address the goals set by top-level managers and decision makers (see Figure 12.21). There are unique benefits to applying the IS plan to define systems requirements. Because the IS plan takes a long-range approach to using information technology within the organization, the requirements for a system analyzed in terms of the IS plan are more likely to be compatible with future systems development initiatives.

asking directly
An approach to gather data that asks users, stakeholders, and other managers about what they want and expect from the new or modified system.

Figure 12.21

Converting Organizational Goals into Systems Requirements

Screen and Report Layout

Developing formats for printed reports and screens to capture data and display information are some of the common tasks associated with developing systems. Screens and reports relating to systems output are specified first to verify that the desired solution is being delivered. Manual or computerized screen and report layout facilities are used to capture both output and input requirements.

Screen layout is a technique that allows a designer to quickly and efficiently design the features, layout, and format of a display screen. In general, users who interact with the screen frequently can be presented with more data and less descriptive information; infrequent users should have more descriptive information presented to explain the data that they are viewing (see Figure 12.22).

screen layout
A technique that allows a designer to quickly and efficiently design the features, layout, and format of a display screen.

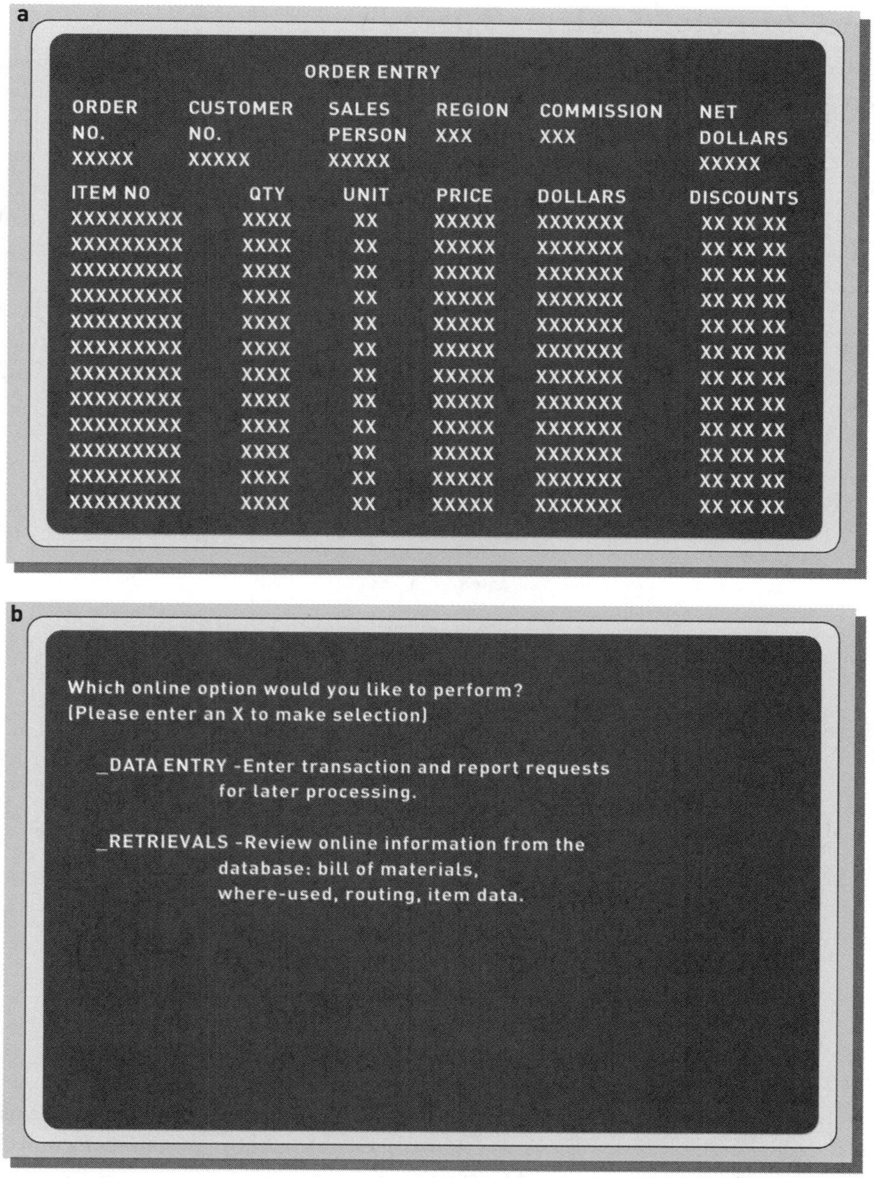

a

```
                        ORDER ENTRY
ORDER      CUSTOMER    SALES      REGION    COMMISSION    NET
NO.        NO.         PERSON     XXX       XXX           DOLLARS
XXXXX      XXXXX       XXXXX                              XXXXX

ITEM NO           QTY       UNIT     PRICE     DOLLARS       DISCOUNTS
XXXXXXXX          XXXX      XX       XXXXX     XXXXXXX       XX XX XX
XXXXXXXX          XXXX      XX       XXXXX     XXXXXXX       XX XX XX
XXXXXXXX          XXXX      XX       XXXXX     XXXXXXX       XX XX XX
XXXXXXXX          XXXX      XX       XXXXX     XXXXXXX       XX XX XX
XXXXXXXX          XXXX      XX       XXXXX     XXXXXXX       XX XX XX
XXXXXXXX          XXXX      XX       XXXXX     XXXXXXX       XX XX XX
XXXXXXXX          XXXX      XX       XXXXX     XXXXXXX       XX XX XX
XXXXXXXX          XXXX      XX       XXXXX     XXXXXXX       XX XX XX
XXXXXXXX          XXXX      XX       XXXXX     XXXXXXX       XX XX XX
XXXXXXXX          XXXX      XX       XXXXX     XXXXXXX       XX XX XX
XXXXXXXX          XXXX      XX       XXXXX     XXXXXXX       XX XX XX
XXXXXXXX          XXXX      XX       XXXXX     XXXXXXX       XX XX XX
```

b

```
Which online option would you like to perform?
(Please enter an X to make selection)

    _DATA ENTRY -Enter transaction and report requests
               for later processing.

    _RETRIEVALS -Review online information from the
                database: bill of materials,
                where-used, routing, item data.
```

Figure 12.22

Screen Layouts

(a) A screen layout chart for frequent users who require little descriptive information. (b) A screen layout chart for infrequent users who require more descriptive information.

Report layout allows designers to diagram and format printed reports. Reports can contain data, graphs, or both. Graphic presentations allow managers and executives to quickly view trends and take appropriate action, if necessary.

Screen layout diagrams can document the screens users desire for the new or modified application. Report layout charts reveal the format and content of various reports that the application will prepare. Other diagrams and charts can be developed to reveal the relationship between the application and outputs from the application.

Requirements Analysis Tools

A number of tools can be used to document requirements analysis. Again, CASE tools are often employed. As requirements are developed and agreed on, entity-relationship diagrams, data-flow diagrams, screen and report layout forms, and other types of documentation will be stored in the CASE repository. These requirements might also be used later as a reference during the rest of systems development or for a different systems development project.

report layout
A technique that allows designers to diagram and format printed reports.

Object-Oriented Systems Analysis

The object-oriented approach can also be used during systems analysis. Like traditional analysis, problems or potential opportunities are identified during object-oriented analysis. Identifying key participants and collecting data is still performed. But instead of analyzing the existing system using data-flow diagrams and flowcharts, an object-oriented approach is used.

In the section "Object-Oriented Systems Investigation," we introduced a kayak rental example. A more detailed analysis of that business reveals that there are two classes of kayaks: single kayaks for one person and tandem kayaks that can accommodate two people. With the OO approach, a class is used to describe different types of objects, such as single and tandem kayaks. The classes of kayaks can be shown in a generalization/specialization hierarchy diagram (see Figure 12.23). KayakItem is an object that will store the kayak identification number (ID) and the date the kayak was purchased (datePurchased).

Figure 12.23

Generalization/Specialization Hierarchy Diagram for Single and Tandem Kayak Classes

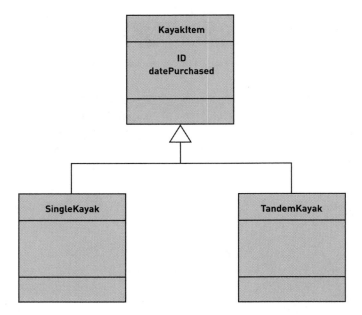

Of course, there could be subclasses of customers, life vests, paddles, and other items in the system. For example, price discounts for kayak rentals could be given to seniors (people over 65 years) and students. Thus, the Customer class could be divided into regular, senior, and student customer subclasses.

The Systems Analysis Report

Systems analysis concludes with a formal systems analysis report. It should cover the following elements:

- The strengths and weaknesses of the existing system from a stakeholder's perspective
- The user/stakeholder requirements for the new system (also called the *functional requirements*)
- The organizational requirements for the new system
- A description of what the new information system should do to solve the problem

Suppose analysis reveals that a marketing manager thinks a weakness of the existing system is its inability to provide accurate reports on product availability. These requirements and a preliminary list of the corporate objectives for the new system will be in the systems analysis report. Particular attention is placed on areas of the existing system that could be improved to meet user requirements. The table of contents for a typical report is shown in Figure 12.24.

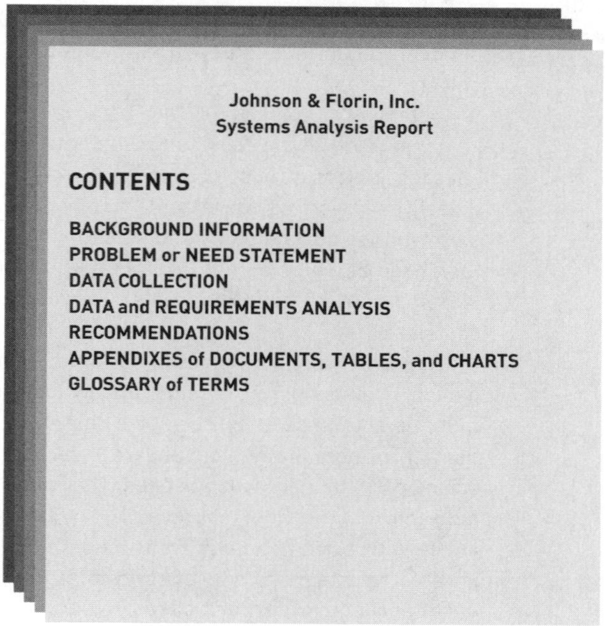

Figure 12.24

A Typical Table of Contents for a Report on an Existing System

Johnson & Florin, Inc.
Systems Analysis Report

CONTENTS

BACKGROUND INFORMATION
PROBLEM or NEED STATEMENT
DATA COLLECTION
DATA and REQUIREMENTS ANALYSIS
RECOMMENDATIONS
APPENDIXES of DOCUMENTS, TABLES, and CHARTS
GLOSSARY of TERMS

The systems analysis report gives managers a good understanding of the problems and strengths of the existing system. If the existing system is operating better than expected or the necessary changes are too expensive relative to the benefits of a new or modified system, the systems development process can be stopped at this stage. If the report shows that changes to another part of the system might be the best solution, the development process might start over, beginning again with systems investigation. Or, if the systems analysis report shows that it will be beneficial to develop one or more new systems or to make changes to existing ones, systems design, which is discussed in Chapter 13, begins.

SUMMARY

Principle

Effective systems development requires a team effort from stakeholders, users, managers, systems development specialists, and various support personnel, and it starts with careful planning.

The systems development team consists of stakeholders, users, managers, systems development specialists, and various support personnel. The development team is responsible for determining the objectives of the information system and delivering to the organization a system that meets its objectives.

Stakeholders are individuals who, either themselves or through the area of the organization they represent, ultimately benefit from the systems development project. Users are individuals who will interact with the system regularly. They can be employees, managers, customers, or suppliers. Managers on development teams are typically representative of stakeholders or may be stakeholders themselves. In addition, managers are most capable of initiating and maintaining change. For large-scale systems development projects, where the investment in and value of a system can be quite high, it is common to have senior-level managers be part of the development team.

A systems analyst is a professional who specializes in analyzing and designing business systems. The programmer is responsible for modifying or developing programs to satisfy user requirements. Other support personnel on the development team include technical specialists, either IS department employees or outside consultants. Depending on the magnitude of the systems development project and the number of IS development specialists on the team, the team may also include one or more IS managers. At some point in your career, you will likely be a participant in systems development. You could be involved in a systems development team—as a user, as a manager of a business area or project team, as a member of the IS department, or maybe even as a CIO.

Systems development projects may be initiated for a number of reasons, including the need to solve problems with an existing system, to exploit opportunities to gain competitive advantage, to increase competition, to make use of effective information, to create organizational growth, to settle a merger or corporate acquisition, and to address a change in the market or external environment. External pressures, such as potential lawsuits or terrorist attacks, can also cause an organization to initiate systems development.

Information systems planning refers to the translation of strategic and organizational goals into systems development initiatives. Benefits of IS planning include a long-range view of information technology use and better use of IS resources. Planning requires developing overall IS objectives; identifying IS projects; setting priorities and selecting projects; analyzing resource requirements; setting schedules, milestones, and deadlines; and developing the IS planning document. IS planning can result in a competitive advantage through creative and critical analysis.

Establishing objectives for systems development is a key aspect of any successful development project. Critical success factors (CSFs) can be used to identify important objectives. Systems development objectives can include performance goals (quality and usefulness of the output and the speed at which output is generated) and cost objectives (development costs, fixed costs, and ongoing investment costs).

Applications must be designed to meet special business needs. Transaction processing applications for the Internet must be able to scale up to support highly variable transactions from potentially thousands of users. Ideally, they can scale up dynamically when needed. They must be reliable and fault tolerant, providing 24 hours a day/7 days a week availability with extremely high transaction integrity. They must be able to integrate with existing infrastructure, including customer and order databases, legacy applications, and enterprise resource planning systems. Development and maintenance must be quick and easy, because business needs may require an organization to change applications on the fly.

Companies are increasingly converting at least some portion of their business to run over the Internet, intranets, or extranets. Web-based systems development treats the Internet as an integral part of the development process, not just a communications tool. In some development efforts, the Internet becomes the most important aspect of the new or modified system. The increasing use of ERP software has also affected systems development.

Principle

Systems development often uses tools to select, implement, and monitor projects, including net present value (NPV), prototyping, rapid application development, CASE tools, and object-oriented development.

The five phases of the traditional SDLC are investigation, analysis, design, implementation, and maintenance and review. Systems investigation identifies potential problems and opportunities and considers them in light of organizational goals. Systems analysis seeks a general understanding of the solution required to solve the problem; the existing system is studied in detail and weaknesses are identified. Systems design creates new or modifies existing system requirements. Systems implementation encompasses

programming, testing, training, conversion, and operation of the system. Systems maintenance and review entails monitoring the system and performing enhancements or repairs.

Advantages of the traditional SDLC include the following: it provides for maximum management control, creates considerable system documentation, ensures that system requirements can be traced back to stated business needs, and produces many intermediate products for review. Its disadvantages include the following: users may get a system that meets the needs as understood by the developers, the documentation is expensive and difficult to maintain, users' needs go unstated or may not be met, and users cannot easily review the many intermediate products produced.

Prototyping is an iterative approach that involves defining the problem, building the initial version, having users utilize and evaluate the initial version, providing feedback, and incorporating suggestions into the second version. Prototypes can be fully operational or nonoperational, depending on how critical the system under development is and how much time and money the organization has to spend on prototyping.

Advantages of the prototyping approach include the following: users get an opportunity to try the system before it is completed, useful prototypes can be produced in weeks, users become positive about the evolving system, and errors and omissions can be detected early. Disadvantages include the following: the approach makes it difficult to start over if the initial solution misses the mark widely, it is difficult to contain the scope of the project, system documentation is often absent, and key operational considerations are often overlooked.

Rapid application development (RAD) uses tools and techniques designed to speed application development. Its use reduces paper-based documentation, automates program source code generation, and facilitates user participation in development activities. RAD can use newer programming techniques, such as agile development or extreme programming. RAD makes extensive use of the joint application development (JAD) process to gather data and perform requirements analysis. JAD involves group meetings in which users, stakeholders, and IS professionals work together to analyze existing systems, propose possible solutions, and define the requirements for a new or modified system.

RAD has the following advantages: it puts an application into production quickly, documentation is produced as a by-product, and it forces good teamwork among users, stakeholders, and developers. Its disadvantages are the following: it is an intense process that can burn out the participants, it requires participants to be skilled in advanced tools and techniques, and a large percentage of time is required from stakeholders and users.

The term *end-user systems development* describes any systems development project in which the primary effort is undertaken by a combination of business managers and users.

Many companies hire an outside consulting firm that specializes in systems development to take over some or all of its systems development activities. This approach is called *outsourcing*. Reasons for outsourcing include companies' belief that they can cut costs, achieve a competitive advantage without having the necessary IS personnel in house, obtain state-of-the-art technology, increase their technological flexibility, and proceed with development despite downsizing. A number of companies offer outsourcing services, including computer vendors and specialized consulting companies.

A number of factors affect systems development success. The degree of change introduced by the project, continuous improvement and reengineering, the use of quality programs and standards, organizational experience with systems development, the use of project management tools, and the use of CASE tools and the objected-oriented approach are all factors that affect the success of a project. The greater the amount of change, the greater the degree of risk and often the amount of reward. Continuous improvement projects do not require significant business process or IS changes, while reengineering involves fundamental changes in how the organization conducts business and completes tasks. Successful systems development projects often involve such factors as support from top management, strong user involvement, use of a proven methodology, clear project goals and objectives, concentration on key problems and straightforward designs, staying on schedule and within budget, good user training, and solid review and maintenance programs. Quality standards, such as ISO 9001 and six sigma, can also be used during the systems development process. The Capability Maturity Model (CMM) can measure an organization's experience with systems development on five levels—from initial to optimized.

The use of automated project management tools enables detailed development, tracking, and control of the project schedule. Effective use of a quality assurance process enables the project manager to deliver a high-quality system and to make intelligent trade-offs among cost, schedule, and quality. CASE tools automate many of the systems development tasks, thus reducing an analyst's time and effort while ensuring good documentation. Object-oriented systems development can also be an important success factor. With the object-oriented systems development (OOSD) approach, a project can be broken down into a group of objects that interact. Instead of requiring thousands or millions lines of detailed computer instructions or code, the systems development project might require a few dozen or maybe a hundred objects.

Principle

Systems development starts with investigation and analysis of existing systems.

In most organizations, a systems request form initiates the investigation process. Participants in systems investigation can include stakeholders, users, managers, employees, analysts, and programmers. The systems investigation is designed to assess the feasibility of implementing solutions for business problems, including technical, economic, legal,

operations, and schedule feasibility. Net present value analysis is often used to help determine a project's economic feasibility. An investigation team follows up on the request and performs a feasibility analysis that addresses technical, economic, legal, operational, and schedule feasibility.

If the project under investigation is feasible, major goals are set for the system's development, including performance, cost, managerial goals, and procedural goals. Many companies choose a popular methodology so that new IS employees, outside specialists, and vendors will be familiar with the systems development tasks set forth in the approach. A systems development methodology must be selected. Objected-oriented systems investigation is being used to a greater extent today. The use case diagram is part of the Unified Modeling Language that is used to document object-oriented systems development. As a final step in the investigation process, a systems investigation report should be prepared to document relevant findings.

Systems analysis is the examination of existing systems, which begins once a team receives approval for further study from management. Additional study of a selected system allows those involved to further understand the system's weaknesses and potential areas for improvement. An analysis team is assembled to collect and analyze data on the existing system.

Data collection methods include observation, interviews, questionnaires, and statistical sampling. Data analysis manipulates the collected data to provide information. The analysis includes grid charts, application flowcharts, and CASE tools. The overall purpose of requirements analysis is to determine user and organizational needs.

Data analysis and modeling is used to model organizational objects and associations using text and graphical diagrams. It is most often accomplished through the use of entity-relationship (ER) diagrams. Activity modeling is often accomplished through the use of data-flow diagrams (DFDs), which model objects, associations, and activities by describing how data can flow between and around various objects. DFDs use symbols for data flows, processing, entities, and data stores. Application flowcharts, grid charts, and CASE tools are also used during systems analysis.

Requirements analysis determines the needs of users, stakeholders, and the organization in general. Asking directly, using critical success factors, and determining requirements from the IS plan can be used. Often screen and report layout charts are used to document requirements during systems analysis.

Like traditional analysis, problems or potential opportunities are identified during object-oriented analysis. Object-oriented systems analysis can involve using diagramming techniques, such as a generalization/specialization hierarchy diagram.

CHAPTER 12: SELF-ASSESSMENT TEST

Effective systems development requires a team effort from stakeholders, users, managers, systems development specialists, and various support personnel, and it starts with careful planning.

1. _____ is the activity of creating or modifying existing business systems. It refers to all aspects of the process—from identifying problems to be solved or opportunities to be exploited to the implementation and refinement of the chosen solution.

2. Which of the following individuals ultimately benefit from a systems development project?

    a. computer programmers
    b. systems analysts
    c. stakeholders
    d. senior-level manager

3. The quality or usefulness of the output is an example of a(n) _____.

4. Like a contractor constructing a new building or renovating an existing one, the programmer takes the plans from the systems analyst and builds or modifies the necessary software. True or False?

5. The term _____ refers to the translation of strategic and organizational goals into systems development initiatives.

6. What factors are essential to the success of certain functional areas of an organization?

    a. critical success factors
    b. systems analysis factors
    c. creative goal factors
    d. systems development factors

Systems development often uses tools to select, implement, and monitor projects, including net present value (NPV), prototyping, rapid application development, CASE tools, and object-oriented development.

7. What is the third level in the Capability Maturity Model (CMM)?

    a. Initial
    b. Optimized
    c. Managed
    d. Defined

8. System performance is usually determined by such factors as fixed investments in hardware and related equipment. True or False?

9. _____ takes an iterative approach to the systems development process. During each iteration, requirements and alternative solutions to the problem are identified and analyzed, new solutions are designed, and a portion of the system is implemented.

10. Joint application development involves group meetings in which users, stakeholders, and IS professionals work together to analyze existing systems, propose possible solutions, and define the requirements for a new or modified system. True or False?

11. What consists of all activities that, if delayed, would delay the entire project?

    a. deadline activities
    b. slack activities
    c. RAD tasks
    d. the critical path

Systems development starts with investigation and analysis of existing systems.

12. The systems request form is a document that is filled out during systems analysis. True or False?

13. Feasibility analysis is typically done during which systems development stage?

    a. Investigation
    b. Analysis
    c. Design
    d. Implementation

14. Data modeling is most often accomplished through the use of _____, while activity modeling is often accomplished through the use of _____.

15. The overall purpose of requirements analysis is to determine user, stakeholder, and organizational needs. True or False?

CHAPTER 12: SELF-ASSESSMENT TEST ANSWERS

(1) Systems development (2) c (3) performance objective (4) True (5) information systems planning (6) a (7) d (8) False (9) Prototyping (10) True (11) d (12) False (13) a (14) entity-relationship (ER) diagrams, data-flow diagrams (15) True

KEY TERMS

application flowcharts 594
asking directly 596
CASE repository 595
computer-aided software engineering (CASE) 584
creative analysis 564
critical analysis 564
critical path 582
critical success factors (CSFs) 565
data analysis 592
data store 594
data-flow diagram (DFD) 592
data-flow line 593
direct observation 591
economic feasibility 587
end-user systems development 574
entity symbol 594
feasibility analysis 587
Gantt chart 583
grid chart 595
information systems planning 561

joint application development (JAD) 573
legal feasibility 588
mission-critical systems 564
net present value (NPV) 587
nonoperational prototype 572
object-oriented systems development (OOSD) 585
operational feasibility 588
operational prototype 572
process symbol 593
Program Evaluation and Review Technique (PERT) 582
programmer 558
project deadline 582
project milestone 582
project schedule 582
prototyping 571
questionnaires 592
rapid application development (RAD) 573

report layout 597
requirements analysis 595
schedule feasibility 588
screen layout 596
stakeholders 558
statistical sampling 592
steering committee 589
structured interview 591
systems analysis 570
systems analyst 558
systems design 570
systems implementation 570
systems investigation 570
systems investigation report 588
systems maintenance and review 570
systems request form 586
technical feasibility 587
unstructured interview 591
users 558

REVIEW QUESTIONS

1. What is an IS stakeholder?
2. What is the goal of IS planning? What steps are involved in IS planning?
3. What are the typical reasons to initiate systems development?
4. What actions can be taken during critical analysis?
5. Describe the five levels of the Capability Maturity Model.
6. How is a Gantt chart developed? How is it used?
7. What is the difference between systems investigation and systems analysis? Why is it important to identify and remove errors early in the systems development life cycle?
8. Identify four reasons that a systems development project may be initiated.
9. List factors that have a strong influence on project success.
10. What tools can be used to develop Web-based applications?
11. What is the purpose of systems investigation?
12. What are the steps of object-oriented systems development?
13. Define the different types of feasibility that systems development must consider.
14. What is the purpose of systems analysis?
15. How does the JAD technique support the RAD systems development life cycle?

DISCUSSION QUESTIONS

1. Why is it important for business managers to have a basic understanding of the systems development process?
2. Briefly describe the role of a system user in the systems investigation and systems analysis stages of a project.
3. Assume that you are investigating a new sales marketing program for a clothing company. Use critical analysis to help you determine the major requirements for the new system.
4. Briefly describe when you would use the object-oriented approach to systems development instead of the traditional systems development life cycle.
5. Your company would like to develop or acquire a new sales program to help sales representatives identify new customers. Describe what factors you would consider in deciding whether to develop the application in-house or outsource the application to an outside company.
6. During the systems investigation phase, how important is it to think creatively? What are some approaches to increase creativity?
7. For what types of systems development projects might prototyping be especially useful? What are the characteristics of a system developed with a prototyping technique?

8. Assume that you work for an insurance company. Describe which applications you would place on the Internet. What tools would you use to develop Internet applications? What applications would be more appropriate to keep off the Internet?
9. How important are communications skills to IS personnel? Consider this statement: "IS personnel need a combination of skills—one-third technical skills, one-third business skills, and one-third communications skills." Do you think this is true? How would this affect the training of IS personnel?
10. Discuss three reasons why aligning overall business goals with IS goals is important.
11. Imagine that you are a highly paid consultant who has been retained to evaluate an organization's systems development processes. With whom would you meet? How would you make your assessment?
12. You are a senior manager of a functional area in which a mission-critical system is being developed. How can you safeguard this project from mushrooming out of control?

PROBLEM-SOLVING EXERCISES

1. You are developing a new information system for The Fitness Center, a company that has five fitness centers in your metropolitan area, with about 650 members and 30 employees in each location. This system will be used by both members and fitness consultants to track participation in various fitness activities, such as free weights, volleyball,

swimming, stair climbers, and aerobic and yoga classes. One of the performance objectives of the system is that it must help members plan a fitness program to meet their particular needs. The primary purpose of this system, as envisioned by the director of marketing, is to assist The Fitness Center in obtaining a competitive advantage over other fitness clubs.

Use word processing software to prepare a brief memo to the required participants in the development team for this systems development project. Be sure to specify what roles these individuals will play and what types of information you hope to obtain from them. Assume that the relational database model will be the basis for building this system. Use a database management system to define the various tables that will make up the database.

2. You are going to start a video rental business. Using the objected-oriented approach, develop a use case and a generalization/specialization hierarchy diagram using a graphics program. Use a spreadsheet to develop a budget for the video rental business, including the costs to rent a space, acquire 1,000 videos, pay for utilities, pay salaries for two full-time employees, and meet other related expenses.

TEAM ACTIVITIES

1. Systems development is more of an art and less of a science, with a wide variety of approaches in how companies perform this activity. You and the members of your team are to interview members of an IS organization's development group. List the steps that the IS group uses in developing a new system or modifying an existing one. How does the organization's approach compare with the techniques discussed in the chapter? Consider both the traditional systems development life cycle and the object-oriented approach. Prepare a short report on your findings.

2. Your team has been hired to determine the requirements and layout of the Web pages for a company that sells fishing equipment over the Internet. Using the approaches discussed in this chapter, develop a rough sketch of at least five Web pages that you would recommend. Make sure to show the important features and the hyperlinks for each page.

3. Your team has been hired to analyze the potential of developing a database of job openings and descriptions for the companies visiting your campus this year. Describe the tasks your team would perform to complete systems analysis.

WEB EXERCISES

1. A number of tools can be used to develop a new Web-based application. Describe the Web development tools you would use and the steps you would complete to implement a Web site to rent movies and games over the Internet. You may be asked to develop a report or send an e-mail message to your instructor about what you found.

2. Locate several companies on the Internet that have received CMM certification. What level was achieved? To what extent do these companies stress their CMM certification? What major outsourcing or IS consulting companies do not advertise achieving a CMM certification on the Web? In your opinion, how important is CMM certification for the success of an outsourcing or IS consulting company? Write a report describing what you found.

CAREER EXERCISES

1. Pick a career that you are considering. What type of information system would help you on the job? Perform technical, economic, legal, operational, and schedule feasibility for an information system you would like developed for you.

2. List five critical success factors (CSFs) for the system specified in the first career exercise. List the specific requirements for this system. What are the most important requirements?

VIDEO QUESTIONS

Watch the video clip **Design Your Own Video Game** and answer these questions:

1. Because of his enthusiasm for computer video games, 16-year-old Maneesh Sethi began programming in grade school. What might this story suggest as to trends in computer programming? Might computer programming some day become part of our elementary school's curriculum? Why or why not?

2. According to the video clip, what skills are required for developing computer video games besides computer programming?

CASE STUDIES

Case One

Hackensack University Medical Center Consolidates Systems for Fast Information Access

New Jersey's Hackensack University Medical Center adopted digital technologies for its diagnostic screenings shortly after they were introduced. "We're a filmless hospital," says Ed Martinez, Hackensack's director of information technology. "All radiology and ultrasound [tests are] seen on a computer screen." But digital diagnostic imaging makes up only a small percentage of the digital records that the hospital maintains.

Besides x-rays, ultrasounds, and MRIs, hospital databases store digital images of prescription medicine orders, medical test results, patients' medical histories, medical references, and patient billing and other business related data. The hospital gradually moved to the new digital technology, but as it did so, its systems became highly complex.

The hospital's information system had grown to include 300 servers from many vendors. Each patient's records were spread across several disparate and isolated systems. Radiologists typically waited from half a minute to three minutes for the system to retrieve an x-ray image. Some hospital staff were unable to get a complete set of patient data. Doctors could access medical orders and test results but not the images associated with tests. Nor could they trace a patient's medical history in hospital records. "So if you have a trauma patient in ER [the emergency room], you wouldn't know that he'd had his gall bladder taken out awhile back," explained Martinez. "Systems people were getting overwhelmed trying to manage and maintain locations of data." Martinez even questioned whether they would be able to recover data in the event of a disaster.

So, Martinez and his team set out to develop a new system that could manage and deliver the mountains of data that the hospital maintained. They found their solution in new technology from Hewlett-Packard. Storage area network (SAN) technology uses a high-speed special-purpose network to connect different data storage devices with associated data servers for a larger network of users. HP's StorageWorks enterprise virtual array (EVA) system would allow Hackensack Medical Center to treat its diverse systems as one large system, accessing all data through one interface.

Hackensack Medical Center invested $60 million to revamp its entire computing infrastructure. The hospital built two data centers from the ground up to meet data requirements that have grown tenfold in the last two years. The data centers are accessed through the HP system, which automates nearly all of the data management tasks. The systems staff no longer needs to worry about the physical location of data on the system. "If I want 500 gigs and RAID 5 for a particular application, I don't need to know the physical location," Martinez states. The software automates the process of setting up the drives and storing the data.

The system has dramatically improved daily operations at the hospital. The wait for an x-ray is down from over two minutes to a couple of seconds. When patients arrive at the ER, the attending physician can go to a computer and look up their entire medical history at the press of a button. That history includes x-rays, studies, tests, and images of ultrasounds and MRIs. Administrators are spending far less time on routine storage administration tasks, which frees them to concentrate on management of the staff. The all-digital hospital can now truly tout the benefits of digital technology and provide its patients with top-level service.

Discussion Questions

1. What lessons can be learned from Hackensack Medical Center's experience with going digital?
2. What led to the hospital's predicament in dealing with technical complexity?

Critical Thinking Questions

3. What do you think are some keys to avoiding technical complexity in large systems such as Hackensack Medical Center's?

4. How might standardization and flexibility affect a system's ability to evolve over time without becoming overly complex?

SOURCES: Elisabeth Horwitt, "Hospital Consolidates Far-flung Multi-vendor Environment," *Computerworld*, February 24, 2003, *www.computerworld. com*; Hackensack University Medical Center's Web site, accessed June 18, 2004, *www.humed.com*; HP StorageWorks Enterprise Virtual Array 5000 Web site, accessed June 18, 2004, *http://h18004.www1.hp.com/products/storage-works/enterprise/index.html*.

Case Two

PepsiCo Implements New Procurement System to Minimize Costs

PepsiCo is a world leader in manufacturing convenience foods and beverages, with annual revenues of about $25 billion and more than 142,000 employees. The company consists of the snack businesses of Frito-Lay North America and Frito-Lay International; the beverage businesses of Pepsi-Cola North America, Gatorade/Tropicana North America and PepsiCo Beverages International; and Quaker Foods North America, manufacturer and marketer of ready-to-eat cereals and other food products. PepsiCo brands are available in nearly 200 countries and territories.

The volume of supplies and ingredients purchased by PepsiCo's Frito-Lay division alone is huge. The company purchases raw materials from hundreds of vendors, which deal in everything from ingredients for potato chips to office products. To manage procurement processes across its divisions, PepsiCo implemented a system it named Purchase to Pay. Purchase to Pay tracks a variety of processes from product purchases to procurement management to vendor selection and payment.

In the company's ongoing mission to reduce costs, PepsiCo turned to the Purchase to Pay system to see whether it could assist staff in negotiating the best deal with suppliers. The investigation uncovered a considerable amount of waste in the procurement process. PepsiCo was not getting good deals on supply purchases, and in some cases it was being overcharged. It was clear that PepsiCo needed to improve the system to control its spending. The company's IS staff set out to develop a standardized system that would allow them to better track and analyze purchases. PepsiCo hoped it would be able to negotiate volume discounts with vendors and control individual, or "maverick," purchases that were above negotiated prices.

In exploring existing procurement solutions from vendors, PepsiCo decided on a system from BusinessObjects Corporation. That vendor's system would store procurement data in a database and provide business intelligence information through a Web-based user interface over the corporate intranet. "Ease of use, scalability and support are some of the reasons why we chose the Business Objects solutions to be an integral part of PepsiCo's Business Intelligence strategy," said Tien Nguyen, vice president of application services at PepsiCo Business Solutions Group. "The real value in Purchase to Pay comes in the ability to analyze our spending patterns and identify cost saving opportunities," explains Yelak Biru, Business Intelligence and Integration team member. "Business Objects is the ideal solution for this, and we can replicate the benefits as we extend it across our corporation."

One example in how BusinessObjects has improved the Purchase to Pay system can be found in the method by which PepsiCo now pays for its raw materials. PepsiCo typically pays vendors upon receipt of goods. During any month, the company may receive multiple deliveries from a vendor and will cut a check for each of those deliveries. Using the BusinessObject solution, PepsiCo can make one monthly payment that provides vendors with an itemized statement detailing each bill of lading, invoice number, the amount of each check, and the grand total, all via an extranet.

PepsiCo has minimized its expenses by streamlining the Purchase to Pay process. The company anticipates a savings of more than $10 million in the system's first year of operation and more than $43 million over the next three years. Within the next few years, the majority of business intelligence reporting will be done using solutions from Business Objects. As users learn the system and as PepsiCo develops more reports, users are expected to experiment with ad hoc queries and dig deeper into the data.

PepsiCo's new and improved Purchase to Pay system is a perfect example of the benefits of continuous improvement through the systems development process. By implementing a new system that is flexible and scalable, PepsiCo has simplified and improved its procurement processes both now and for the future as it changes to meet new challenges.

Discussion Questions

1. What stage of the systems development life cycle led PepsiCo to discover the need for improvement in its previous Purchase to Pay system?

2. How was the previous Purchase to Pay system not aligned with the goals of PepsiCo's organization?

Critical Thinking Questions

3. What are the benefits of implementing a system that allows you to custom design reports and experiment with ad hoc queries?

4. Relate the characteristics of valuable data (provided in Chapter 1, Table 1.2) to the information provided by, and the method of delivery used by, PepsiCo's new system.

SOURCES: "Customers in the Spotlight: PepsiCo," accessed June 19, 2004, *www.businessobjects.com/customers/spotlight/pepsico.asp*; WebIntelligence Web site, accessed June 19, 2004, *www.businessobjects.com/products/* *queryanalysis/webi.asp*; PepsiCo Web site, accessed June 19, 2004, *www.pepsico.com*.

Case Three

Segway Stays Light and Nimble with Outsourced Systems

Dean Kamen holds more than 150 U.S. and foreign patents related to medical devices, climate control systems, and helicopter design. In 2001, he developed a business to manufacture and market a "human transport" device that he believed would revolutionize travel. The Segway, Kamen declared, "would be to the car what the car was to the horse and buggy."

While the device has been a bit slower in taking off than Kamen had hoped, it has captured the attention of the media and transportation industry analysts. The two-wheeled Segway Human Transporter (HT) employs a unique patented "dynamic stabilization" technology.

With the assistance of optimistic investors, Segway hired a seasoned CIO to design the information systems for the new company. Patrick Zilvitis, who was CIO at The Gillette Co. before leaving to work part-time for Segway, decided that this unique company required a unique approach to information management. Zilvitis said in an interview that when he joined the company in fall 2000, he quickly decided that outsourcing would be the wisest path for a start-up that had a minimal IT infrastructure. He thought that would be the best way to hold down its technology costs.

Zilvitis thought that it would be a costly mistake for Segway to build a traditional data center and hire a big dedicated IT staff—a mistake that would slow the company's growth. Instead, he decided Segway should use software under a hosted environment that could grow with the company's needs. Outsourcing "allows us to upsize or downsize our IT infrastructure as needed," noted Scott Frock, Segway's director of finance. "For a small-to-medium-size company, there are a lot of advantages."

Outsourcing also lets companies like Segway avoid upfront investments in servers, software, and technical support staffers, said Terry Jost, a Dallas-based consultant at Cap Gemini Ernst & Young LLP.

Before Zilvitis came on board, Segway had been using Intuit Inc.'s QuickBooks accounting software. To support the company's expansion, Zilvitis believed that it should upgrade to either an accounting package designed for small and mid-sized businesses or a larger system that it could grow into. The company decided to look for a larger system to avoid a "painful and expensive" conversion down the road.

After evaluating software from Oracle and SAP AG, the company chose an uncustomized version of an Oracle suite geared to small manufacturers. The system includes manufacturing and order-management modules in addition to the finance applications. Workers at Segway's headquarters and its manufacturing plant access the software via Windows 2000 PCs.

Additionally, Zilvitis chose a finance system hosted by Fremont, California–based Appshop Incorporated, which manages the Oracle applications from a Sprint Corporation data center in Denver. So, Segway's information system includes components supplied by three different vendors in three different parts of the country.

Three years after the system's implementation, Segway is reaping the benefits of the larger Oracle system. The company recently upgraded to a newer version of its Oracle system and added software that will let workers at its customer service partner, Frazer, Pennsylvania–based DecisionOne, access product warranty data and other information via Web browsers. The Oracle system has made the integration and expansion a breeze to implement.

Maintaining data and systems off-site has provided some unanticipated telecommunications costs and savings. To speed the data from the application service provider to Segway's headquarters and manufacturing plant, Segway replaced its private branch exchange switches with Voice-over-Internet-Protocol (VoIP) equipment costing "considerably less than $20,000," Zilvitis said. He added that the transition to VoIP has provided several benefits, including the ability to set up low-cost switchboard extensions for remote employees, extra voice and data bandwidth for future growth, and cheap yet reliable connections to Sprint's data center. For instance, Segway's Internet connection to Denver costs the company $1,000 to $2,000 per month, compared with the monthly tab for a T-1 connection that would have been $10,000 to $12,000, he said.

By outsourcing its systems, Segway has been able to focus on its product rather than data centers, servers, and an extensive IT staff. Instead, the company relies on one part-time veteran CIO to manage its outsourced resources. Since the future of the company is uncertain, it makes sense for Segway to select a solution that is flexible, nimble, and can turn on a dime—just like its product.

Discussion Questions

1. What benefits is Segway enjoying by outsourcing its information infrastructure and services?
2. What benefits can a company gain by managing its own information infrastructure and services?

Critical Thinking Questions

3. What factors might influence a company to outsource its IT infrastructure and services?
4. What frustrations do you think Segway endures in dealing with outsourced vendors?

SOURCES: Thomas Hoffman, "Segway's Tech Plans Look Down the Road to Growth," *Computerworld*, January 26, 2004, *www.computerworld.com*; Segway's Web site, *www.segway.com*, accessed June 19, 2004; "A Long Road Ahead of It," *The Economist*, June 12, 2004, *www.lexis-nexis.com*.

NOTES

Sources for the opening vignette: "SAP Customer Success Story: BMW Group," accessed June 18, 2004, *www.sap.com/solutions/industry/automotive/customersuccess*; BMW Group Web site, accessed June 18, 2004, *www.bmwgroup.com*; mySAP Automotive Web site, accessed June 18, 2004, *www.sap.com/solutions/industry/automotive*.

1. Hamblen, Matt, "NavSea's ROI Ship Comes in," *Computerworld*, March 24, 2003, p. 46.
2. Weiss, Todd, "New CIO Takes Reins of IRS Tech Upgrade," *Computerworld*, June 2, 2003, p. 14.
3. Demaitre, Eugene, "The Greater Good," *Computerworld*, June 2, 2003, p. 25.
4. Staff, "Best in Class," *Computerworld*, March 15, 2004, p. 3.
5. Hoffman, Thomas, "The Resourceful Project Manager," *Computerworld*, February 6, 2004, p. 35.
6. Carter, Pamela, "The Management of Meaning: Toward a New Model of Project Management," Research Colloquium, Information and Management Science Department, College of Business, Florida State University, September 21, 2003.
7. Ulfelder, Steve, "Best in Class: Security Setup Identifies Attacks with Minimal Drag," *Computerworld*, February 24, 2003, p.48.
8. Anthes, Gary, "TV for the 21st Century," *Computerworld*, July 21, 2003, p. 32.
9. Mitchell, Robert, "E-Mail Upgrade," *Computerworld*, April 7, 2003, p. 25.
10. Patchong, Alain, et al., "Improving Car Body Production," *Interfaces*, January 2003, p. 36.
11. Thurman, Mathias, "Postmerger Audit Quashes Trust Idea," *Computerworld*, February 9, 2004, p. 34.
12. Kerstetter, Jim, "Sarbanes-Oxley Sparks a Software Boom," *Business Week*, January 12, 2004, p. 94.
13. Thurman, Mathias, "Overwhelmed by Sarbanes-Oxley," *Computerworld*, March 1, 2004, p. 33.
14. Joffman, Thomas, "Users Struggle to Pinpoint Sarbanes-Oxley IT Costs," *Computerworld*, December 1, 2003, p. 14.
15. Hoffman, Thomas, "Emcor Saves on Sarb-Ox," *Computerworld*, February 2, 2004, p. 7.
16. Bandler, James, et al., "Fuji, Kodak Creating Virtual Scrapbook," *Rocky Mountain News*, June 16, 2003, p. 2B.
17. Grimes, Brad, "Microsoft.NET," *PC Magazine*, March 25, 2003, p. 74.
18. Wallace, Linda, et al., "How Software Project Risk Affects Project Performance," *Decision Sciences*, Spring 2004, p. 289.
19. Kontzer, Tony, "Global Appeal," *InformationWeek*, January 19, 2004, *www.informationweek.com*.
20. Gomes, Lee, "Do We Get Enough In Innovation for What We Give to Microsoft," *The Wall Street Journal*, March 8, 2004, p. B1.
21. Dwyer, Paula, "Big Bang at the Big Board," *Business Week*, February 16, 2004, p. 66.
22. Munshi, Porus, "Prototyping All the Way to Perfection," *Businessline*, June 23, 2003, p. 1.
23. Sliwa, Carol, "Rational Software Set to Roll Out Rapid Development Tool," *Computerworld*, May 19, 2003, p. 7.
24. Anthes, Gary, "Sabre Takes Extreme Measures," *Computerworld*, March 29, 2004, p. 28.
25. Sandberg, Jared, "The Unofficial Techies," *The Wall Street Journal*, May 28, 2003, p. B1.
26. Marks, Debra, "Teenage Web Wiz Aspires to Create a Better Browser," *The Wall Street Journal*, May 14, 2003, p. B5B.
27. Rappa, M. A. "The Utility Business Model and the Future of Computing Services," *IBM Systems Journal*, vol. 43, No. 1, 2004, p. 32.
28. Sliwa, Carol, "Sears Plans to Outsource Part of IT Infrastructure," *Computerworld*, January 19, 2004, p. 1.
29. Schroeder, Michael, "Business Coalition Battles Outsourcing Backlash," *The Wall Street Journal*, March 1, 2004, p. A1.
30. Brewin, Bob, "WorldCom Wins $20M Bid to Build Baghdad Cell Network," *Computerworld*, May 26, 2003, p. 14.
31. Staff, "On Demand Vision," *www.IBM.com*, accessed on January 25, 2004.
32. Phillips, Michael, "More Work Is Outsourced to U.S. than Away from It, Data Shows," *The Wall Street Journal*, March 15, 2004, p. A2.
33. Buckman, Rebecca, "H-P Outsourcing: Beyond China," *The Wall Street Journal*, February 23, 2004, p. A14.
34. Bulkeley, William, "IBM Documents Give Rare Look at Sensitive Plans on Offshoring," *The Wall Street Journal*, January 19, 2004, p. A1.
35. Manes, Steven, "Electronics Jobs Outsourced—to You," *Forbes*, March 15, 2004, p. 158.
36. Hamblen, Matt, "Outsourcing IT Security Functions Can Succeed," *Computerworld*, January 19, 2004, p. 38.
37. Melymuka, Kathleen, "When The Chips Were Down," *Computerworld*, May 26, 2003, p. 41.
38. Verton, Dan, "Offshore Coding Work Raises Security Concerns," *Computerworld*, May 5, 2003, p. 1.
39. Vijayan, J., "Offshore Outsourcing Poses Privacy Perils," *Computerworld*, February 23, 2004, p. 10.
40. Mearian, Lucas, "Bank Group Offers Guidelines on Outsourcing Security Risks," *Computerworld*, January 26, 2004, p. 10.
41. IBM Web page at *www-IBM.com.services/stragegies*, accessed on July 29, 2003.
42. EDS Web page at *www.eds.com*, accessed on July 29, 2003.
43. Accenture Web page at *www.accenture.com*, accessed on July 29, 2003.
44. Drucker, Jesse, "Press 1 for Delhi, 2 for Dallas," *The Wall Street Journal*, March 9, 2004, p. B1.
45. Thurm, Scott, "Lessons in India," *The Wall Street Journal*, March 3, 2004, p. A1.
46. Anthes, Gary, "Best In Class: Data Warehouse Boosts Profits by Empowering Sales Force," *Computerworld*, February 24, 2003, p. 46.
47. Welcome to ISO Easy Web page at *www.isoeasy.com*, accessed on July 29, 2003.
48. Anthes, Gary, "Model Mania," *Computerworld*, March 8, 2004, p. 42.
49. Capability Maturity Model for Software home page at *www.sei.cmu.edu*, accessed on November 16, 2004.
50. Verton, Dan, "Struggling Airlines Scrutinize IT Projects," *Computerworld*, April 7, 2003, p. 1.
51. Liu, Xiaohua, "Multiuser Collaborative Work in a Virtual Environment Based CASE Tool," *Information and Software Technology*, April 1, 2003, p. 253.
52. Lehmann, Hans, "An Object-Oriented Architecture Model for International Information Systems," *Journal of Global Information Management*, July 2003, p. 1.
53. Vijayan, Jaikumar, "Best In Class: Application Framework Allows Easy Portal Access," *Computerworld*, February 24, 2003, p. 51.
54. Karnitsching, Matthew, "Autobahn Plan Is Stuck in Slow Lane," *The Wall Street Journal*, October 21, 2003, p. A11.

CHAPTER
· 13 ·

Systems Design,
Implementation,
Maintenance,
and Review

PRINCIPLES	LEARNING OBJECTIVES
• Designing new systems or modifying existing ones should always be aimed at helping an organization achieve its goals.	• State the purpose of systems design and discuss the differences between logical and physical systems design.
	• Describe some considerations in design modeling and the diagrams used during object-oriented design.
	• Outline key considerations in interface design and control and system security and control.
	• Define the term *RFP* and discuss how this document is used to drive the acquisition of hardware and software.
	• Describe the techniques used to make systems selection evaluations.
• The primary emphasis of systems implementation is to make sure that the right information is delivered to the right person in the right format at the right time.	• State the purpose of systems implementation and discuss the various activities associated with this phase of systems development.
	• List the advantages and disadvantages of purchasing versus developing software.
	• Discuss the software development process and some of the tools used in this process, including object-oriented program development tools.
• Maintenance and review add to the useful life of a system but can consume large amounts of resources. These activities can benefit from the same rigorous methods and project management techniques applied to systems development.	• State the importance of systems and software maintenance and discuss the activities involved.
	• Describe the systems review process.

INFORMATION SYSTEMS IN THE GLOBAL ECONOMY
THE HOME DEPOT, UNITED STATES

Digitizing The Home Depot

The Home Depot is the world's largest home improvement retailer and second largest U.S. retailer. The chain currently operates 1,640 stores in 50 U.S. states and in the District of Columbia, eight Canadian provinces, Mexico, and Puerto Rico. You would think that such a successful company would have a state-of-the-art information system, but that has not been the case. "Antiquated comes to mind," says CIO Bob DeRodes. "The wires and data switches were undersized, so we couldn't move large amounts of data." But that is being changed with the launch of The Home Depot's $400 million IS overhaul.

DeRodes likes to refer to the retailer's technological overhaul as the "digitizing of The Home Depot." And he's backing up the term with new infrastructure and applications that are drastically changing the way that The Home Depot employees—wherever they are in the sprawling enterprise—do their jobs.

DeRodes likens the net effect of the upgrade to "the difference between going online with a dial-up connection and getting broadband." Information in the form of raw data, video, and images will be zipping from PC to PC and from store to store. The new systems will provide a near 100% difference for store managers and major improvements for management at Home Depot headquarters.

Previously, each Home Depot store had only one PC, which sat on the manager's desk. Everything else in the store was a dumb terminal. "We couldn't even put browsers on them if we wanted to," said DeRodes. The new infrastructure project includes a rollout of 40 to 50 Web-ready PCs and mobile devices to each of the stores. The new PCs are situated at all the service desks in the stores: those for contractor service, customer service, customer returns, paint centers, home decorating, receiving docks, and elsewhere.

Teresa Chapman, manager of The Home Depot's Cumberland store in northwest Atlanta, has been one of the first to experience the upgrade. "I've taken back 20 hours a week that I used to spend under a pile of papers," she says. Chapman can access the new Store Manager's Workbench application from any of the 30 PCs positioned around the store—no more running back to her office to consult files. The system lets her review the performance of sales promotions, monitor security cameras positioned over the registers, and access a suite of office tools. She particularly appreciates the workload application, which tracks projects and employees who are assigned to them. As she demonstrates the application, she explains, "This entry is for the set-up of the seasonal tree-cutting area. Here's the deadline. Here's who it was assigned to. And if you look here, you'll see that it was completed and when."

The distribution of applications such as the Store Manager's Workbench to PCs throughout the stores is done centrally over the corporate intranet to ensure that everyone is using the same version of every software program. "There is no removable media on our devices, so people can't install or de-install anything," DeRodes says. They can't put in a CD. All the PCs are network managed. "We also don't want people loading data onto CDs and carrying it out," he added. "Removable media is a bad thing in a commercial environment like a store."

The upgrade also consolidates five separate point-of-sale (POS) software packages, which the staff previously used, into one. The new POS software enables cashiers to use portable scan guns for heavy items without removing them from carts. The software also includes an online cashier book, which eliminates the need for cashiers to look up items without bar codes, such as nuts and bolts, manually in a three-ring binder. "Cashiers still have to look up unbarcoded items, but they don't have to mess around with that binder," DeRodes says. "It's automated now. We guarantee that a cashier will find the code for whatever he's holding in his hand within five touches of the screen."

Consolidating the POS systems into one standardized system also makes upgrades easier to manage. "The biggest problem with having five different POS systems was that it took so long to make changes," DeRodes says. "Every time we wanted to do something new and different, we had to build it, test it on five different systems and deploy it on five different systems."

At the heart of the new system is a massive data warehouse built on IBM hardware and utilizing Big Blue's DB2 software. The warehouse holds three years of recent POS data. Populating the data warehouse with such data is becoming easier, too, as The Home Depot continues to roll out the new POS software system to all its stores. The data warehouse soon will provide valuable information and decision support to the company's human resource department, its individual store managers, and finance staff through a standard Web browser over the corporate intranet. The new data warehouse will provide a single data source for such varied functions as merchandising, price optimization, inventory management, and human resource planning. Industry insiders say ensuring that applications run off the same data makes it easier for multiple departments to execute company-wide strategies.

"When we'd go out and say we were going to digitize The Home Depot, most employees said, 'What does that mean?'" DeRodes recalls. "A lot of people thought the company had been very successful without much technology." Now that the system is being installed and realized, everyone is wondering how they ever survived without it.

As you read this chapter, consider the following:

- What type of considerations must be confronted when designing a major system renovation such as The Home Depot's?
- What disruptions to normal business practices go along with implementing new or modified systems, and how can those disruptions be minimized?

Why Learn About Systems Design, Implementation, Maintenance, and Review?

Information systems are designed and implemented for employees and managers every day. A manager at a hotel chain can use an information system to look up client preferences. An accountant at a manufacturing company can use an information system to analyze the costs of a new plant. A sales representative for a music store can use an information system to determine which CDs to order and which to discount because they are not selling. A computer engineer can use an information system to help determine why a computer system is running slowly. Information systems have been designed and implemented for almost every career and industry. This chapter shows how you can be involved in designing and implementing an information system that will directly benefit you on the job. It also shows how to avoid errors or recover from disasters. This chapter starts with how systems are designed.

The way an information system is designed, implemented, and maintained profoundly affects the daily functioning of an organization. Like investigation and analysis covered in the previous chapter, design, implementation, maintenance, and review covered in this chapter strive to achieve organizational goals, such as reducing costs, increasing profits, or improving

customer service. The goal is to develop a new or modified system to deliver the right information to the right person at the right time.

After an organization investigates and analyzes the need for a new system and approves its development, the project moves to the later stages of development: design, implementation, and maintenance and review. This chapter presents the basics of systems design, implementation, and maintenance and review. Both users and IS personnel need to be aware of these stages so that they can participate in good systems development—no matter what field their organizations are in.

SYSTEMS DESIGN

The purpose of **systems design** is to answer the question "How will the information system solve a problem?" The primary result of the systems design phase is a technical design that details system outputs, inputs, and user interfaces; specifies hardware, software, databases, telecommunications, personnel, and procedures; and shows how these components are related. The new system should overcome shortcomings of the existing system and help the organization achieve its goals. The system must also meet certain guidelines, including user and stakeholder requirements and the objectives defined during previous development phases. There, Inc. (www.there.com), a California company that lets people meet and interact on a 3-D Web site, spent more than $17 million to design an online simulation that lets people chat, play cards, and flirt with avatars, lifelike characters that appear on Internet sites.[1] The company hopes that the four-year systems development project will generate revenues from individual subscribers and companies that want to advertise their products on its Web site.

Design can range in scope from individual to multicorporate projects. Nearly all companies are continually involved in designing systems for individuals, workgroups, and the enterprise. AM General designed a new system to help to speed production of its Hummer sport-utility vehicles, as the "Information Systems @ Work" feature describes. Increasingly, companies undertake multicorporate design, where two or more companies form a partnership or an alliance to design a new system.

Systems design is typically accomplished using the tools and techniques discussed in the previous chapter. Depending on the specific application, these methods can be used to support and document all aspects of systems design. Two key aspects of systems design are logical and physical design.

Logical and Physical Design

As we discussed earlier in the database chapter, design has two dimensions: logical and physical. The **logical design** refers to what the system will do. The **physical design** refers to how the tasks are accomplished, including how the components work together and what each component does.

Logical Design

Logical design describes the functional requirements of a system. That is, it conceptualizes what the system will do to solve the problems identified through earlier analysis. Without this step, the technical details of the system (such as which hardware devices should be acquired) often obscure the best solution. Logical design involves planning the purpose of each system element, independent of hardware and software considerations. The logical design specifications that are determined and documented include output, input, process, file and database, telecommunications, procedures, controls and security, and personnel and job requirements (see Figure 13.1).

systems design
Stage of systems development that answers the question "How will the information system do what it must do to solve a problem?"

logical design
Description of the functional requirements of a system.

physical design
Specification of the characteristics of the system components necessary to put the logical design into action.

AM General Designs RFID System for Just-in-Time Deliveries

AM General Corporation, which makes the popular Hummer, had a problem as big as its vehicles. Its production lines were slow. So, it investigated methods to speed up operations. One of the primary problems that slowed production, it found, was replenishing parts on the assembly line. If, for instance, a worker ran out of #29 steering rods at station 12 on the line, production would grind to a halt while everyone waited for the forklift to arrive with another crate of rods. AM General's management team set out to find a better system that would make sure that the flow of parts to the assembly line from the warehouse was just in time—without any lags and without overstocking.

AM General, like most other auto manufacturers, used an age-old paper card–based system that originated in Japan called *Kanban* (Japanese for card signal). Under that system, as a station worker notices that parts are running low, he or she posts a paper card, which is picked up by a roaming replenishment worker, who takes the card to the warehouse. There, the supervisor dispatches a forklift driver to deliver the requested part. Although it was revolutionary when it was introduced, in today's digital world the Kanban system is cumbersome.

AM General wanted to replace the system with a computer network. However, implementing traditional network technology meant running network wires to the production line and installing terminals, which would be costly and inappropriate. Assembly-line workers didn't have time to type in orders on a terminal.

Upon examining systems from other manufacturers, AM General found the perfect solution in a system designed by WhereNet Corporation. WhereNet systems consist of radio-frequency identification (RFID) tags, fixed position antennas, and Web-enabled software to track inventory in warehouses and on production lines. The primary function of WhereNet's product is to assist companies in quickly locating inventory within the supply chain.

The WhereNet system locates and tracks inventory using extremely low-power radio-frequency tags and a communications network. Antennas positioned inside and outside the factory receive tag transmissions and deliver tracking information to a computer. The system then identifies the location of the tag within 10 feet of its exact position. While WhereNet would assist AM General with monitoring inventory through the supply chain, something more was needed to get it to the assembly line in a JIT fashion.

In working to design a production line system for Ford Motor Company, WhereNet produced a product it named WhereCall. WhereCall incorporates wireless call buttons on the production line for each inventory item in use. When supply of a specific part reaches a predetermined replenishment level, the line worker presses the WhereCall button, which sends a wireless signal to a central server. The central server locates the item being requested in the warehouse by identifying the item's RFID tag and also locates the wireless signal from the forklift nearest the item. The item is displayed on the forklift's LCD display, and the operator speeds off to gather and deliver what is needed.

The effect of the system is nearly miraculous. It cuts out two middlemen and puts production line personnel in direct communication with the forklift driver who is closest to them. Items are restocked in a matter of a few minutes rather than an hour. Not only has the system streamlined replenishment on the line, but it also feeds important information to larger systems.

AM General feeds the information from WhereNet to Cimplicity Tracker software designed by GE. Cimplicity Tracker provides supervisors with a detailed, continuous flow of information that allows them to optimize the manufacturing process, while managing inventory levels and locations, scheduling resources, and routing materials more effectively.

The flexibility and reliability of the WhereNet system has allowed AM General to operate at peak efficiency and ramp up production to meet pent-up demand for Hummers, says Deborah Cafiero, CIO and director of information systems at AM General. "Wireless on the plant floor is revolutionizing the automotive industry, and we are a perfect case in point," she said. "Through lower operating costs, AM General has already realized a return on its WhereNet investment, and the applications continue to pay big dividends every day."

Discussion Questions

1. How has wireless technology changed the daily routines of production line personnel and supervisors?
2. In what ways has the RFID technology employed in the new systems streamlined the production line at AM General?

Critical Thinking Questions

3. What jobs do you think were lost and gained by the implementation of the WhereNet system at AM General? Are such trade-offs healthy for the job market and society at large?
4. What market advantages has AM General gained by upgrading its replenishment system?

SOURCES: Linda Rosencrance, "Hummer Plant Implements Parts-Replenishment, Assembly-Tracking Systems," *Computerworld*, March 10, 2004, *www.computerworld.com*; WhereNet Web site, accessed June 20, 2004, *www.wherenet.com*; Cimplicity Tracker Web site, accessed June 20, 2004, *www.geindustrial.com/cwc/products?pnlid=2&id=hmi32*; Factory Logic Web site (for info on Kanban), accessed June 20, 2004, *www.factorylogic.com/nl_26.asp*.

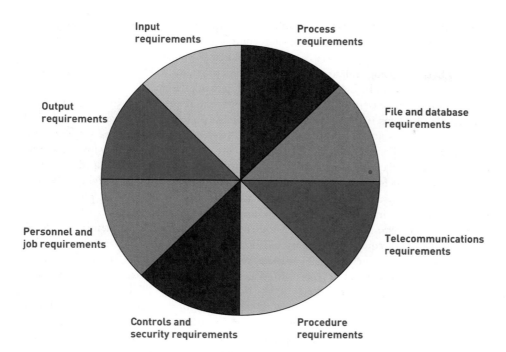

Figure 13.1

Components of Logical Design

Output Requirements. Logical design of output describes all outputs from the system and includes the types, format, content, and frequency of outputs. For example, a requirement that all company invoices must include the customer's original invoice number is a logical design specification. Screen and layout tools can be used to capture the output requirements for the system. An investment company, for example, can design its output for a new order entry system to be consistent with that of its other systems and to provide a common output format.

Input Requirements. Once output requirements have been described, input design can begin. Logical input design identifies the types, format, content, and frequency of input data. For example, the requirement that the system must capture the customer's phone number from his or her incoming call and use that to automatically look up the customer's account information is a logical design specification. A variety of diagrams and screen and report layouts can be used to reveal the type, format, and content of input data.

Process Requirements. The types of calculations, comparisons, and general data manipulations required of the system are determined during logical process design. For example, a payroll program requires gross and net pay computations, state and federal tax withholding, and various deductions and savings plans. Linux Networx, for example, has designed a cluster of more than 2,000 processors for the Pentagon.[2] The processors will form a supercomputer to be used for weapons research.

File and Database Requirements. Most information systems require files and database systems. The capabilities of these systems are specified during the logical design phase. For example, the ability to obtain instant updates of customer records is a logical design specification. In many cases, a database administrator is involved with this aspect of logical design. Data-flow and entity-relationship diagrams are typically used during file and database design.

The design of an organization's file and database systems should include the cost of file and database storage and the cost to manage the organization's storage. For every dollar spent on physical storage, a company can spend $7 on data and storage management. Because of these high costs, some companies are considering a charge-back policy, where individual departments pay for their storage and management costs. Many banks now have to capture and integrate more customer data as a result of new international banking regulations.[3] The new database regulations, which are being overseen by the Bank for International Settlements in Switzerland, could cost banks billions of dollars and must be implemented by 2006.

Telecommunications Requirements. During logical design, the general network and telecommunications requirements need to be specified. For example, a hotel might specify a

client/server system with a certain number of workstations. Sutter Health is designing a wireless telecommunications system to help comply with FDA requirements to place bar codes on drugs.[4] The design calls for wireless LAN and IP phones. From these basic requirements, a hybrid topology might be chosen. Graphics programs, CASE tools, and the object-oriented approach can be used to facilitate logical network design.

Procedure Requirements. All information systems require procedures to run applications and handle problems if they occur. These important policies are captured during logical procedures design. Once specified, procedures can be described by using text and word processing programs. For example, the steps to add a new customer account may involve a series of both manual and computerized tasks. Written procedures would be developed to provide an efficient process for all to follow. Ameritrade Holdings, for example, designed new procedures to speed systems development projects.[5] The new design calls for projects to be broken into smaller increments that can be completed in six to ten weeks.

Controls and Security Requirements. Another important part of logical design is to determine the required frequency and characteristics of backup systems. In general, everything should have a backup, including all hardware, software, data, personnel, supplies, and facilities. Planning how to avoid or recover from a computer-related disaster often starts at this stage of logical design. Security design is also important. A large PC manufacturer, for example, discovered a potential problem with the security of its Web site that would have allowed hackers to get personal information from its customers and potential customers.[6] Once discovered, the potential security problem was fixed. Designing systems that can detect and prevent viruses, invasion of privacy, and other negative aspects of information systems is also done during this phase of systems design.[7] Hidden in a worm that disrupted thousands of computers in 2003 was a message to software developers, "Stop making money and fix your software."[8] Developing safer software is an important part of controls and security requirements. Some companies and universities are also including security training in their systems design curricula.[9] One Hong Kong school teaches students how to hack into computers so that they will be better prepared for their careers. According to one instructor who teaches the course, "Of course, this is just a fake simulation."

Personnel and Job Requirements. Some new systems require additional employees; others may modify the tasks associated with one or more existing IS positions. The job titles and descriptions of these positions are identified during logical personnel and job design. Organization charts are useful during personnel design to diagram various positions and job titles. Word processing programs are also used to describe job duties and responsibilities.

Physical Design

Physical design specifies the characteristics of the system components necessary to put the logical design into action. In this phase, the characteristics of hardware, software, database, telecommunications, personnel, and procedure and control specifications must be detailed (see Figure 13.2).

Hardware Specifications. All computer equipment, including input, processing, and output devices, must be specified by performance characteristics. For example, if the logical design specified that the database must hold large amounts of historical data, then the system storage devices must have large capacity.

Software Specifications. All software must be specified by capabilities. For example, if dozens of users must be able to update the database concurrently, as required in the logical design, then the physical design must specify a database management system to allow this to occur. If the software is to be delivered or used over the Internet, this need must be considered during software design. Siebel, for example, decided to deliver its software over the Internet, so it specified an online system to perform that service.[10] Siebel makes software to help companies manage their customer accounts. Pixar, the film company started by Steve Jobs, the founder of Apple Computer, designed stunning animation software to make movies such as *A Bug's Life*, *Toy Story*, *Monsters, Inc.*, and *Finding Nemo*.[11] This software and its related computer systems have helped the company make millions of dollars. Some companies, such as FileNet and IBS America, have designed software to automate record keeping to help companies monitor their operations and to comply with new financial-reporting requirements under the Sarbanes-Oxley Act, discussed in Chapter 12.[12]

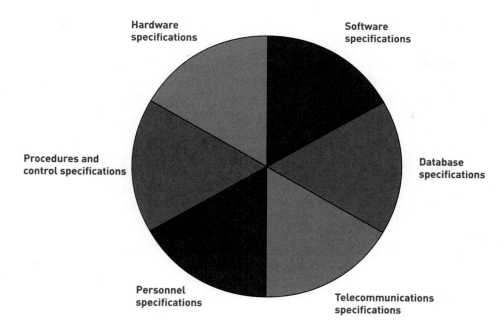

Figure 13.2

Components of Physical Design

In some cases, software can be developed internally; in others it will be purchased from an IS vendor. This is often called the "make or buy" decision. As mentioned in Chapter 4, open-source software is becoming increasingly popular. Regis Corporation, an operator of hair salons, plans to spend about $100,000 to purchase software to comply with Sarbanes-Oxley.[13] The District of Columbia decided to purchase three software modules from Hyperion to help its 68 government agencies do a better job of budgeting.[14] The million-dollar deal is part of a larger $71.5 million systems development effort that is expected to save the government more than $60 million annually. Logical design specifications for program outputs, data inputs, and processing requirements are also considered during the physical design of the software. For example, the ability to access data stored on certain disk files that the program will use is specified.

Database Specifications. The type, structure, and function of the databases must be specified. The relationships between data elements established in the logical design must be mirrored in the physical design as well. These relationships include access paths and file structure organization. Fortunately, many excellent database management systems exist to assist with this activity.

Telecommunications Specifications. The characteristics of the communications software, media, and devices must be specified. For example, if the logical design requires that all members of a department must be able to share data and run common software, then the local area network configuration and the communications software that are specified in the physical design must accommodate this capability.

Personnel Specifications. This step involves specifying the background and experience of individuals most likely to use the system, including the job descriptions specified in the logical design. In the past, companies hired a large number of people to answer questions from customers and potential customers. The costs of these employees and their training could be staggering. As a result, many companies are starting to automate this function to reduce personnel costs and provide more accurate information. A major airline company, for example, can use a knowledge management system to streamline the customer relations process. Some companies are transferring jobs and employees to other countries.[15] A London travel agency, for example, has moved many of its employees to India to save money. The travel agency promotes this program to employees as a way to see the world or as a new adventure.

Procedure and Control Specifications. How each application is to run, as well as what is to be done to minimize the potential for crime and fraud, must be specified. These specifications include auditing, backup, and output distribution methods.

Object-Oriented Design

Logical and physical design can be accomplished using either the traditional structured approach or the objected-oriented approach to systems development. Both approaches use a variety of design models to document the new system's features and the development team's understandings and agreements. Many organizations today are turning to OO development because of its increased flexibility. So, we outline a few OO design considerations and diagrams here.

Using the OO approach, we design key objects and classes of objects in the new or updated system. This process includes consideration of the problem domain, the operating environment, and the user interface. The problem domain involves the classes of objects related to solving a problem or realizing an opportunity. Refer back to our Maui, Hawaii, kayak rental shop example first introduced in Chapter 12 where we introduced the generalization/specialization hierarchy showing classes: KayakItem in Figure 12.23 is an example of a problem domain object that we will use to store information on kayaks in the rental program. The operating environment for the rental shop's system includes objects that interact with printers, system software, and other software and hardware devices. The user interface for the system includes objects that users interact with, such as buttons and scroll bars in a Windows program.

During the design phase we also need to consider the sequence of events that must happen for the system to function correctly. For example, we might want to design the sequence of events that are needed to add a new kayak to the rental program. A sequence of events is often called a *scenario*, and it can be diagrammed in a sequence diagram (see Figure 13.3)

Figure 13.3

A Sequence Diagram for the Add a New KayakItem Scenario

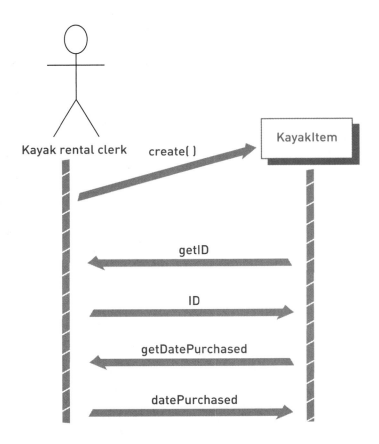

A sequence diagram is read starting from the top and moving down.

1. The Create arrow at the top is a message from the kayak rental clerk to the KayakItem object to create information on a new kayak to be placed into the rental program.
2. The KayakItem object knows that it needs the ID for the kayak and sends a message to the clerk requesting the information. See the getID arrow.
3. The clerk then types the ID into the computer. This is shown with the ID arrow. The data is stored in the KayakItem object.
4. Next, KayakItem requests the purchase date. This is shown in the getDatePurchased arrow.
5. Finally, the clerk types the purchase date into the computer. The data is also transferred to KayakItem object. This is shown in the datePurchased arrow at the bottom of Figure 13.3.

This scenario is only one example of a sequence of events. Other scenarios might include entering information about life jackets, paddles, suntan lotion, and other accessories. The same types of use case and generalization/specialization hierarchy diagrams discussed in Chapter 12 can be created for each, and additional sequence diagrams will also be needed.

Interface Design and Controls

A number of special system characteristics should be considered during both logical and physical design. These characteristics relate to how users access and interact with the system, including sign-on procedures, interactive processing, and interactive dialogue.

Procedures for Signing On

Access control methods are established during systems design. With almost any system, control problems may exist, such as criminal hackers breaking into the system or an employee mistakenly accessing confidential data. Sign-on procedures are the first line of defense against these problems. A **sign-on procedure** consists of identification numbers, passwords, and other safeguards needed for an individual to gain access to computer resources. The sign-on, also called a "logon," can identify, verify, and authorize access and usage (see Figure 13.4). Identification means that the computer identifies the user as valid. If you must enter an identification number and password when logging on to a mainframe, you have gone through the identification process. For systems and applications that are more sensitive or secure, verification is used. Verification involves entering an additional code before access is given. Finally, authorization restricts the user's access to certain parts of a system or application. Consider a credit-checking application for a major credit card company. To grant credit, a clerk may be given only basic credit information on the screen. A credit manager, however, will have an authorization code to get additional credit information about a client.

sign-on procedure
Identification numbers, passwords, and other safeguards needed for an individual to gain access to computer resources.

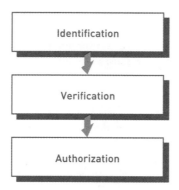

Figure 13.4

The Levels of the Sign-On Procedure

Interactive Processing

Today, most computer systems allow interactive processing. With this type of system, people directly interact with the processing component of the system through terminals or networked PCs. The system and the user respond to each other in real time, which means within a matter of seconds. Interactive real-time processing requires special design features for ease

menu-driven system
System in which users simply pick what they want to do from a list of alternatives.

Figure 13.5

A menu-driven system allows you to choose what you want from a list of alternatives.

of use, such as menu-driven systems, help commands, table lookup facilities, and restart procedures. With a **menu-driven system** (see Figure 13.5), users simply pick what they want to do from a list of alternatives. Most people can easily operate these types of systems. They select their choice or respond to questions (or prompts) from the system, and the system does the rest.

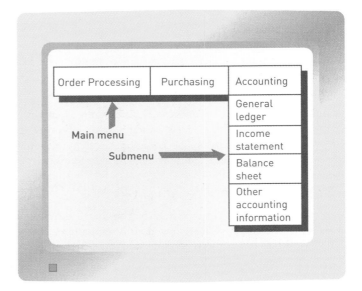

help facility
A program that provides assistance when a user is having difficulty understanding what is happening or what type of response is expected.

lookup tables
Tables containing data that are developed and used by computer programs to simplify and shorten data entry.

restart procedures
Simplified processes to access an application from where it left off.

Many designers incorporate a **help facility** into the system or applications program. When a user is having difficulty understanding what is happening or what type of response is expected, he or she can activate the help facility. The help screen relates directly to the problem the user is having with the software. The program responds with information on the program status—what possible commands or selections the user can give and what is expected in terms of data entry.

Incorporating tables within an application is another very useful design technique. **Lookup tables** can be developed and used by computer programs to simplify and shorten data entry. For example, if you are entering a sales order for a company, you simply type in its abbreviation, such as ABCO. The program will then go to the customer table, normally stored on a disk, and look up all the information pertaining to the company abbreviated ABCO that is required to complete the sales order. This information is then displayed on a screen for confirmation. The use of these tables can prevent wasting a tremendous amount of time entering the same data over and over again into the system.

If a problem occurs in the middle of an application—such as a temporary interruption of power or a printer running out of paper—the application currently being run is typically shut down. As a result, easy-to-use **restart procedures** are developed and incorporated into the design. With a restart procedure, it is very simple for an individual to restart an application where it left off.

Good Interactive Dialogue

Dialogue refers to the series of messages and prompts communicated between the system and the user. From a user's point of view, good interactive dialogue from the computer system is essential, making data entry faster, easier, and more accurate. Poor dialogue from the computer can confuse the user and result in the wrong information being entered. If the computer system prompts the user to enter ACCOUNT, does it mean account number, type of account, or something else? Table 13.1 lists some elements that help designers create good interactive dialogue. These elements should be considered during systems design.

Element	Description
Clarity	The computer system should ask for information using easily understood language. Whenever possible, the users themselves should help select the words and phrases used for dialogue with the computer system.
Response time	Ideally, responses from the computer system should approximate a normal response time from a human being carrying on the same sort of dialogue.
Consistency	The system should use the same commands, phrases, words, and function keys for all applications. After a user learns one application, all others will then be easier to use.
Format	The system should use an attractive format and layout for all screens. The use of color, highlighting, and the position of information on the screen should be considered carefully and applied consistently.
Jargon	All dialogue should be written in easy-to-understand terms. Avoid jargon known only to IS specialists.
Respect	All dialogue should be developed professionally and with respect. Dialogue should not talk down to or insult the user. Avoid statements such as "You have made a fatal error."

Design of System Security and Controls

In addition to considering the system's interface and user interactions, designers must also develop system security and controls. These key considerations involve error prevention, detection, and correction; disaster planning and recovery; and system controls.

Preventing, Detecting, and Correcting Errors

The best and least-expensive time to deal with potential errors is early in the design phase. During installation or after the system is operating, it is much more expensive and time-consuming to handle errors and related problems. Good systems design attempts to prevent errors before they occur, which involves recognizing what can happen and developing steps and procedures that can prevent, detect, and correct errors. This process includes developing a good backup system that can recover from an error.

Disaster Planning and Recovery

Disaster planning is the process of anticipating and providing for disasters. A disaster can be an act of nature (a flood, fire, or earthquake) or a human act (terrorism, error, labor unrest, or erasure of an important file). An earthquake in Seattle, for example, caused a company to lose its telecommunications capabilities for days, which forced the manufacturing company to shut down some of its operations. Disaster planning often focuses primarily on two issues: maintaining the integrity of corporate information and keeping the information system running until normal operations can be resumed. A California law that forces hospitals to be earthquake-proof has prompted huge investments in information systems.[16] Sutter Health, located in San Francisco, estimates it will spend about $1 billion at 26 hospitals to complete its disaster planning and recovery efforts to comply with the California law.

Disaster recovery is defined as the implementation of the disaster plan. Although companies have known about the importance of disaster planning and recovery for decades, many do not adequately prepare. The primary tools used in disaster planning and recovery are hardware, software and database, telecommunications, and personnel backups. The National Association of Securities Dealers (NASD), for example, was able to recover quickly from the effects of hurricane Isabel.[17] According to the chief technology officer, speaking a day after the hurricane, "We've had flooding and significant power outages, but we've not had any missed transactions. We started preparing several days ago to ensure that we had our business continuity plans in place." In another case, diesel generators at many banks, brokerages, and clearing operations in the Northeast provided automatic recovery from a massive power

---

Table 13.1

The Elements of Good Interactive Dialogue

disaster recovery
The implementation of the disaster plan.

failure that took place in 2003.[18] These disaster recovery systems meant that thousands of computers were able to process billions of dollars of transactions in the middle of the power failure.

Hardware Backup. Companies commonly form arrangements with their hardware vendor or a disaster recovery company to provide access to a compatible computer hardware system or additional hardware in the event of a disaster. A duplicate, operational hardware system that is ready to use (or immediate access to one through a specialized vendor) is an example of a **hot site**. If the primary computer has problems, the hot site can be used immediately as a backup. Another approach is to use a **cold site**, also called a *shell*, which is a computer environment that includes rooms, electrical service, telecommunications links, data storage devices, and the like. If a problem occurs with the primary computer, backup computer hardware is brought into the cold site, and the complete system is made operational. For both hot and cold sites, telecommunications media and devices are used to provide fast and efficient transfer of processing jobs to the disaster recovery facility.

A number of firms offer disaster recovery services. Sun Microsystems, for example, has developed the Americas Command Center for its customers. This large facility can monitor about 5,000 Sun clients to prevent potential problems. Business Records Management offers the Disaster Recovery Center in Pittsburgh, which provides the facilities, computer systems, and other equipment for recovery from unplanned business interruptions. To provide an effective, fully equipped workplace in time of disaster or special need, the center contains technology to ensure that a business stays connected both locally and globally, including fully equipped computer rooms for local and remote processing; wiring and public branch exchange facilities to support voice, data, facsimile, and video telecommunications; complete offices and services with conference rooms, private offices, and workstation areas to accommodate more than 250 people; and operational and technical support to assist in recovery.[19] Guardian Computer Support is an international computer support company providing a variety of computer services including on-site hardware and software maintenance, contract staffing, disaster recovery, and outsourcing services.[20]

hot site
A duplicate, operational hardware system or immediate access to one through a specialized vendor.

cold site
A computer environment that includes rooms, electrical service, telecommunications links, data storage devices, and the like; also called a *shell*.

Companies that suffer a disaster can employ a disaster recovery service, which can secure critical data backup information. These service companies can also provide a facility from which to operate and communications equipment to stay in touch with customers.

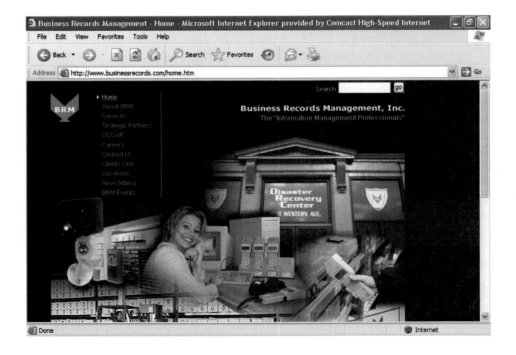

Software and Database Backup. Making duplicate copies of all programs, files, and data backs up software and databases. At least two backup copies should be made. One copy can be kept in the IS department in case of accidental destruction of the software; the other should be kept off-site in a safe, secure, fireproof, and temperature- and humidity-controlled

environment. A number of service companies provide this type of backup environment. Oracle, for example, offers its customers disaster recovery products for its database systems. A key objective is to help companies keep their e-commerce applications running. Oracle Parallel Fail Safe, one of Oracle's disaster recovery products, can switch from a failed Web site to a parallel Web site and then bring the failed Web site back online within 30 seconds of its failure.

In addition to application software, system software, such as operating systems, should also be backed up. One approach is to run multiple operating systems on the same computer system or server. Some experts estimate that this approach can save as much as 70 percent of traditional backup costs. The primary disadvantage is a slight reduction in speed and performance. Companies such as VMWare in California offer software that allows computers and servers to run up to 20 different operating systems at the same time, including Windows 3.1, Windows XP, UNIX, and others.

Backup is also essential for programs and data, including data warehouses and data marts. Some companies provide database backup services by distributing critical data to remote data storage centers. The centers are linked with fiber-optic lines to allow the rapid transfer of data to needed computer systems in case of a disaster. Companies can also uses sophisticated software to help companies manage their distributed data centers. The advent of more distributed systems, like client/server systems, means that many users now have important, and perhaps mission-critical, data and applications on their computers.

Utility packages inexpensively provide backup features for desktop computers by copying data onto CD-ROM or CD-RW disks, magnetic disk, or tape. But software and database backup can be very difficult if an organization has a large amount of data. For some companies, making a backup of the entire database could take hours. A tight budget may also prohibit backing up significant quantities of data. As a result, some companies use **selective backup**, which involves creating backup copies of only certain files. For example, only critical files might be copied every night for backup purposes.

Another backup approach is to make a copy of all files changed during the last few days or the last week, a technique called **incremental backup**. This approach to backup uses an **image log**, which is a separate file that contains only changes to applications. Whenever an application is run, an image log is created that contains all changes made to all files. If a problem occurs with a database, an old database with the last full backup of the data, along with the image log, can be used to re-create the current database.

Telecommunications Backup. Most disaster recovery plans call for backing up vital Internet and telecommunications systems. This is especially true for companies that have substantial e-commerce operations. Some plans might call for recovering whole networks. In other plans, the most critical nodes on the network are backed up by duplicate components. Using such fault-tolerant networks, which will not break down when one node or part of the network malfunctions, can be a more cost-effective approach to telecommunications backup. IBM has an information system to help monitor computer networks and the Internet. The system allows computer systems to monitor themselves and make changes that are required as they or their environment change.

Personnel Backup. Information systems personnel must also have backup, which can be accomplished in a number of ways. One of the best approaches is to provide cross-training for IS and other personnel so that each individual can perform an alternate job, if required. For example, a company might train employees in accounting, finance, or other IS departments to operate the system if a disaster strikes. The company could also make an agreement with another IS department or an outsourcing company to supply IS personnel, if necessary.

selective backup
Creating backup copies of only certain files.

incremental backup
Making a backup copy of all files changed during the last few days or the last week.

image log
A separate file that contains only changes to applications.

Systems Controls

Security lapses, fraud, and the invasion of privacy can present disastrous problems. For example, because of an inadequate security and control system, a futures and options trader for a British bank lost about $1 billion. A simple systems control might have prevented a problem that caused the 200-year-old bank to collapse. In addition, from time to time, IRS employees have been caught looking at the returns of celebrities and others. Preventing and detecting these problems is an important part of systems design. Prevention includes the following:

- Determining potential problems
- Ranking the importance of these problems
- Planning the best place and approach to prevent problems
- Deciding the best way to handle problems if they occur

Every effort should be made to prevent problems, but companies must establish procedures to handle problems if they occur.

Most IS departments establish tight **systems controls** to maintain data security. Systems controls can help prevent computer misuse, crime, and fraud by managers, employees, and others. The accounting scandals in the early 2000s have caused many IS departments to develop systems controls to make it more difficult for executives to mislead investors and employees. Some of these scandals involved billions of dollars.

Most IS departments have a set of general operating rules that help protect the system. Some IS departments are **closed shops**, in which only authorized operators can run the computers. Other IS departments are **open shops**, in which other people, such as programmers and systems analysts, are also authorized to run the computers. Other rules specify the conduct of the IS department.

These rules are examples of **deterrence controls**, which involve preventing problems before they occur. Making a computer more secure and less vulnerable to a break-in is another example. Good control techniques should help an organization contain and recover from problems. The objective of containment control is to minimize the impact of a problem while it is occurring, and recovery control involves responding to a problem that has already occurred.

Many types of systems controls may be developed, documented, implemented, and reviewed. These controls touch all aspects of the organization (see Table 13.2).

Once controls are developed, they should be documented in various standards manuals that indicate how the controls are to be implemented. They should then be implemented and frequently reviewed. It is common practice to measure the extent to which control techniques are used and to take action if the controls have not been implemented.

systems controls
Rules and procedures to maintain data security.

closed shops
IS departments in which only authorized operators can run the computers.

open shops
IS departments in which people, such as programmers and systems analysts, are allowed to run the computers, in addition to authorized operators.

deterrence controls
Rules and procedures to prevent problems before they occur.

Table 13.2

Using Systems Controls to Enhance Security

Controls	Description
Input controls	Maintain input integrity and security. Their purpose is to reduce errors while protecting the computer system against improper or fraudulent input. Input controls range from using standardized input forms to eliminating data-entry errors and using tight password and identification controls.
Processing controls	Deal with all aspects of processing and storage. The use of passwords and identification numbers, backup copies of data, and storage rooms that have tight security systems are examples of processing and storage controls.
Output controls	Ensure that output is handled correctly. In many cases, output generated from the computer system is recorded in a file that indicates the reports and documents that were generated, the time they were generated, and their final destinations.
Database controls	Deal with ensuring an efficient and effective database system. These controls include the use of identification numbers and passwords, without which a user is denied access to certain data and information. Many of these controls are provided by database management systems.
Telecommunications controls	Provide accurate and reliable data and information transfer among systems. Telecommunications controls include firewalls and encryption to ensure correct communication while eliminating the potential for fraud and crime.
Personnel controls	Make sure that only authorized personnel have access to certain systems to help prevent computer-related mistakes and crime. Personnel controls can involve the use of identification numbers and passwords that allow only certain people access to particular data and information. ID badges and other security devices (such as smart cards) can prevent unauthorized people from entering strategic areas in the information systems facility.

Many companies use ID badges to prevent unauthorized access to sensitive areas in the information systems facility.

(Source: © Noel Hendrickson/ Masterfile.)

Generating Systems Design Alternatives

When individuals or organizations require a system to perform additional functions that an existing system cannot support, they often turn to outside vendors to design and supply their new systems. Such purchases require expertise in both hardware and software. Whether an individual is purchasing a personal computer or a company is acquiring an expensive main-frame computer, the system can be obtained from a single vendor or multiple vendors.[21] Florida-based Fidelity National Financial, for example, used the single-vendor approach to acquire a centralized information system.[22] The company selected IBM for its multimillion-dollar project to speed processing for the $8 trillion of mortgages and loans it processes for large banks every day. In some cases, the vendor simply provides hardware or software. In other cases, the vendor provides additional services. Some of the factors to consider in selecting a vendor are the following:

- The vendor's reliability and financial stability
- The type of service offered after the sale
- The goods and services the vendor offers and keeps in stock
- The vendor's willingness to demonstrate its products
- The vendor's ability to repair hardware
- The vendor's ability to modify its software and its willingness to invest in continued development of its software
- The availability of vendor-offered training of IS personnel and system users
- Evaluations of the vendor by independent organizations

When additional hardware and software are not required, alternative designs are often generated without input from vendors. If the new system is complex, the original development team may want to involve other personnel in generating alternative designs. If new hardware and software are to be acquired from an outside vendor, a formal request for proposal (RFP) should be made.

Request for Proposals

request for proposal (RFP)
A document that specifies in detail required resources such as hardware and software.

The **request for proposal (RFP)** is one of the most important documents generated during systems development. It often results in a formal bid that is used to determine who gets a contract for new or modified systems. The RFP specifies in detail the required resources such as hardware and software. While it can take time and money to develop a high-quality RFP, it can save a company in the long run. Companies that frequently generate RFPs can automate the process. One company, for example, purchased a software package, called The RFP Machine from Pragmatech Software, to improve the quality of its RFPs and to reduce the time it takes to produce them. The RFP Machine stores important data needed to generate RFPs and automates the process of producing RFP documents.

In some cases, separate RFPs are developed for different needs. For example, a company might develop separate RFPs for hardware, software, and database systems. The RFP also communicates these needs to one or more vendors, and it provides a way to evaluate whether the vendor has delivered what was expected. In some cases, the RFP is made part of the vendor contract. The Table of Contents for a typical RFP is shown in Figure 13.6.

Financial Options

When it comes to acquiring computer systems, several choices are available, including purchase, lease, or rent. Cost objectives and constraints set for the system play a significant role in the alternative chosen, as do the advantages and disadvantages of each. Table 13.3 summarizes the advantages and disadvantages of these financial options.

Determining which option is best for a particular company in a given situation can be difficult. Financial considerations, tax laws, the organization's policies, its sales and transaction growth, marketplace dynamics, and the organization's financial resources are all important factors. In some cases, lease or rental fees can amount to more than the original purchase price after a few years. As a result, some companies prefer to purchase their equipment.

On the other hand, constant advances in technology can make purchasing risky. A company would not want to purchase a new multimillion-dollar computer only to have newer and more powerful computers available a few months later at a lower price, unless the

Johnson & Florin, Inc.
Systems Investigation Report

Contents

COVER PAGE (with company name and contact person)
BRIEF DESCRIPTION of the COMPANY
OVERVIEW of the EXISTING COMPUTER SYSTEM
SUMMARY of COMPUTER-RELATED NEEDS and/or PROBLEMS
OBJECTIVES of the PROJECT
DESCRIPTION of WHAT IS NEEDED
HARDWARE REQUIREMENTS
PERSONNEL REQUIREMENTS
COMMUNICATIONS REQUIREMENTS
PROCEDURES to BE DEVELOPED
TRAINING REQUIREMENTS
MAINTENANCE REQUIREMENTS
EVALUATION PROCEDURES (how vendors will be judged)
PROPOSAL FORMAT (how vendors should respond)
IMPORTANT DATES (when tasks are to be completed)
SUMMARY

Figure 13.6

A Typical Table of Contents for a Request for Proposal

Table 13.3

Advantages and Disadvantages of Acquisition Options

Renting (Short-Term Option)

Advantages	Disadvantages
No risk of obsolescence	No ownership of equipment
No long-term financial investment	High monthly costs
No initial investment of funds	Restrictive rental agreements
Maintenance usually included	

Leasing (Longer-Term Option)

Advantages	Disadvantages
No risk of obsolescence	High cost of canceling lease
No long-term financial investment	Longer time commitment than renting
No initial investment of funds	No ownership of equipment
Less expensive than renting	

Purchasing

Advantages	Disadvantages
Total control over equipment	High initial investment
Can sell equipment at any time	Additional cost of maintenance
Can depreciate equipment	Possibility of obsolescence
Low cost if owned for a number of years	Other expenses, including taxes and insurance

computer can be easily and inexpensively upgraded. Some servers, for example, are designed to be *scalable* to allow processors to be added or swapped, memory to be upgraded, and peripheral devices to be installed. Companies often employ several people to determine the best option based on all the factors. This staff can also help negotiate purchase, lease, or rental contracts.

Evaluating and Selecting a Systems Design

The final step in systems design is to evaluate the various alternatives and select the one that will offer the best solution for organizational goals. Depending on their weight, any one of these objectives may result in the selection of one design over another. For example, financial concerns might make a company choose rental over equipment purchase. Specific performance objectives—say, that the new system must perform online data processing—may result in a complex network design for which control procedures must be established. Evaluating and selecting the best design involves achieving a balance of system objectives that will best support organizational goals. Normally, the process of evaluation and selection involves both a preliminary and a final evaluation before a design is selected.

The Preliminary Evaluation

preliminary evaluation
An initial assessment whose purpose is to dismiss the unwanted proposals; begins after all proposals have been submitted.

A **preliminary evaluation** begins after all proposals have been submitted. The purpose of this evaluation is to dismiss unwanted proposals. Several vendors can usually be eliminated by investigating their proposals and comparing them with the original criteria. Those that compare favorably are asked to make a formal presentation to the analysis team. The vendors should also be asked to supply a list of companies that use their equipment for a similar purpose. The organization then contacts these references and asks them to evaluate their hardware, their software, and the vendor.

The Final Evaluation

final evaluation
A detailed investigation of the proposals offered by the vendors remaining after the preliminary evaluation.

The **final evaluation** begins with a detailed investigation of the proposals offered by the remaining vendors. The vendors should be asked to make a final presentation and to fully demonstrate the system. The demonstration should be as close to actual operating conditions as possible. Such applications as payroll, inventory control, and billing should be conducted using a large amount of test data.

After the final presentations and demonstrations have been given, the organization makes the final evaluation and selection. Cost comparisons, hardware performance, delivery dates, price, flexibility, backup facilities, availability of software training, and maintenance factors are considered. Although it is good to compare computer speeds, storage capacities, and other similar characteristics, it is also necessary to carefully analyze whether the characteristics of the proposed systems meet the company's objectives. In most cases, the RFP captures these objectives and goals. Figure 13.7 illustrates the evaluation process.

Figure 13.7

The Stages in Preliminary and Final Evaluations

Note that the number of possible alternatives decreases as the firm gets closer to making a final decision.

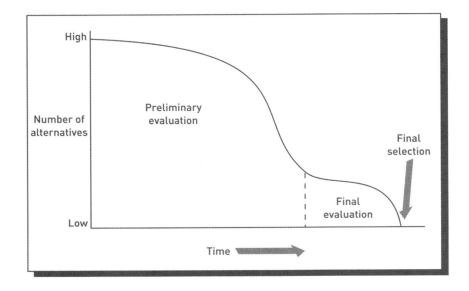

Evaluation Techniques

The exact procedure used to make the final evaluation and selection varies from one organization to the next. Some were first introduced in Chapter 2, including return on investment (ROI), earnings growth, market share, customer satisfaction, and total cost of ownership (TCO). In addition, four other approaches are commonly used: group consensus, cost/benefit analysis, benchmark tests, and point evaluation.

Group Consensus

In **group consensus**, a decision-making group is appointed and given the responsibility of making the final evaluation and selection. Usually, this group includes the members of the development team who participated in either systems analysis or systems design. This approach might be used to evaluate which of several screen layouts or reports formats is best.

Cost/Benefit Analysis

Cost/benefit analysis is an approach that lists the costs and benefits of each proposed system. Once expressed in monetary terms, all the costs are compared with all the benefits. Table 13.4 lists some of the typical costs and benefits associated with the evaluation and selection procedure. This approach is used to evaluate options whose costs can be quantified, such as which hardware or software vendor to select.

group consensus
Decision making by a group that is appointed and given the responsibility of making the final evaluation and selection.

cost/benefit analysis
An approach that lists the costs and benefits of each proposed system. Once they are expressed in monetary terms, all the costs are compared with all the benefits.

Table 13.4

Cost/Benefit Analysis Table

Costs	Benefits
Development costs	Reduced costs
Personnel	Fewer personnel
Computer resources	Reduced manufacturing costs
	Reduced inventory costs
	More efficient use of equipment
	Faster response time
	Reduced downtime or crash time
	Less spoilage
Fixed costs	**Increased revenues**
Computer equipment	New products and services
Software	New customers
One-time license fees for software and maintenance	More business from existing customers
	Higher price as a result of better products and services
Operating costs	**Intangible benefits**
Equipment lease and/or rental fees	Better public image for the organization
Computer personnel (including salaries, benefits, etc.)	Higher employee morale
	Better service for new and existing customers
Electric and other utilities	The ability to recruit better employees
Computer paper, tape, and disks	Position as a leader in the industry
Other computer supplies	System easier for programmers and users
Maintenance costs	
Insurance	

benchmark test
An examination that compares computer systems operating under the same conditions.

point evaluation system
An evaluation process in which each evaluation factor is assigned a weight, in percentage points, based on importance. Then each proposed system is evaluated in terms of this factor and given a score ranging from 0 to 100. The scores are totaled, and the system with the greatest total score is selected.

Figure 13.8

An Illustration of the Point Evaluation System

In this example, software has been given the most weight (40 percent), compared with hardware (35 percent) and vendor support (25 percent). When system A is evaluated, the total of the three factors amounts to 82.5 percent. Systems B's rating, on the other hand, totals 86.75 percent, which is closer to 100 percent. Therefore, the firm chooses system B.

Benchmark Tests

A **benchmark test** is an examination that compares computer systems operating under the same conditions. Most computer companies publish their own benchmark tests, but some forbid disclosure of benchmark tests without prior written approval. Thus, one of the best approaches is for an organization to develop its own tests, then use them to compare the equipment it is considering. This approach might be used to compare the end-user system response time on two similar systems. Several independent companies also rate computer systems. *Computerworld*, *PC Magazine*, and many other publications, for example, not only summarize various systems but also evaluate and compare computer systems and manufacturers according to a number of criteria.

Point Evaluation

One of the disadvantages of cost/benefit analysis is the difficulty of determining the monetary values for all the benefits. An approach that does not employ monetary values is a **point evaluation system**. Each evaluation factor is assigned a weight, in percentage points, based on importance. Then each proposed information system is evaluated in terms of this factor and given a score ranging from 0 to 100, where 0 means that the alternative does not address the feature at all and 100 means that the alternative addresses that feature perfectly. The scores are totaled, and the system with the greatest total score is selected. When using point evaluation, an organization can list and evaluate literally hundreds of factors. Figure 13.8 shows a simplified version of this process. This approach is used when there are many options to be evaluated, such as which software best matches a particular business's needs.

		System A			System B		
Factor's importance		Evaluation		Weighted evaluation	Evaluation		Weighted evaluation
Hardware	35%	95	35%	33.25	75	35%	26.25
Software	40%	70	40%	28.00	95	40%	38.00
Vendor support	25%	85	25%	21.25	90	25%	22.50
Totals	100%			82.5			86.75

Because many elements must be considered before an organization can make a final selection, point evaluation can include a large number of factors. Performance concerns might include speed, storage capacity, and processing capabilities. Costs might include the deposit required on contract signing, payment schedules, lease and rental arrangements, maintenance costs, and availability of leasing companies. When all these factors are added to the point evaluation system, a very large grid can result. The rows of the grid list the various factors important to the client company, and the columns of the grid represent the various vendors that responded to the request for proposal. Even if weights are not used, a chart of this type can be very helpful. Some companies just use check marks to indicate which vendors have satisfied certain factors.

Freezing Design Specifications

Near the end of the design stage, some organizations prohibit further changes in the design of the system. Freezing systems design specifications means that the user agrees in writing that the design is acceptable (see Figure 13.9). Other organizations, however, allow or even encourage design changes. These organizations often use agile or rapid systems development approaches, introduced in Chapter 12.

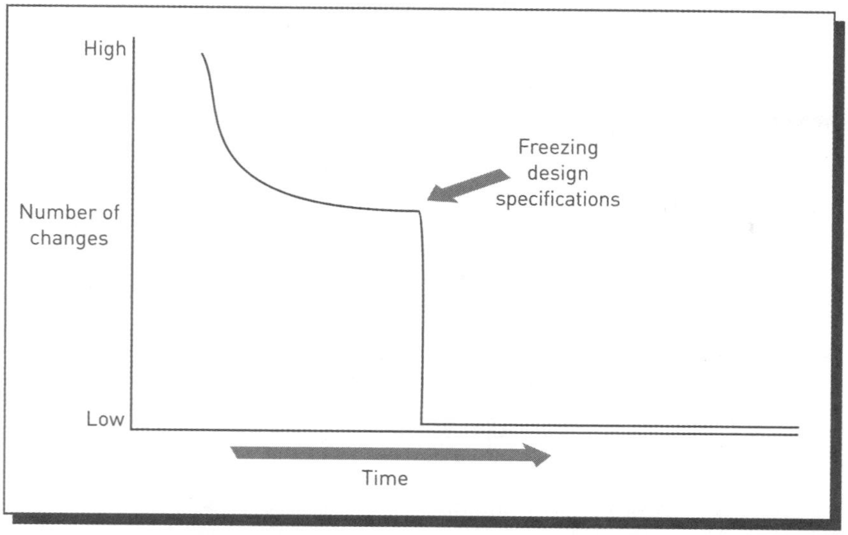

Figure 13.9

Freezing Design Specifications

The Contract

One of the most important steps in systems design is to develop a good contract if new computer facilities are being acquired. Finding the best terms where everyone makes a profit can be difficult. Most computer vendors provide standard contracts; however, such contracts are designed to protect the vendor, not necessarily the organization buying the computer equipment.

More and more organizations are using outside consultants and legal firms to help them develop their own contracts. Such contracts stipulate exactly what they expect from the system vendor and what interaction will occur between the vendor and the organization. All equipment specifications, software, training, installation, maintenance, and so on are clearly stated. Also, the contract stipulates deadlines for the various stages or milestones of installation and implementation, as well as actions that the vendor will take in case of delays or problems. Some organizations include penalty clauses in the contract, in case the vendor is unable to meet its obligation by the specified date. Typically, the request for proposal becomes part of the contract. This saves a considerable amount of time in developing the contract, because the RFP specifies in detail what is expected from the vendors.

The Design Report

System specifications are the final results of systems design. They include a technical description that details system outputs, inputs, and user interfaces, as well as all hardware, software, databases, telecommunications, personnel, and procedure components and the way these components are related. The specifications are contained in a **design report**, which is the primary result of systems design. The design report reflects the decisions made for systems design and prepares the way for systems implementation. The contents of the design report are summarized in Figure 13.10.

It is important to understand and thoroughly complete the systems development activities covered in this chapter for any new system. These phases provide the blueprints and groundwork for the rest of systems development. The activities of the next phases will be easier, faster, and more accurate and will result in a more efficient, effective system if the design is complete and well thought out.

Prior to implementation, experienced project managers place formal controls on the project scope. A key component of the process is to assess the cost and schedule impact of each requested change, no matter how small, and to decide whether to include the change. Often the users and the project team decide to hold all changes until the original effort is completed and then prioritize the entire set of requested changes.

design report
The primary result of systems design, reflecting the decisions made and preparing the way for systems implementation.

Figure 13.10

A Typical Table of Contents for a System Design Report

Johnson & Florin, Inc.
Systems Design Report

Contents

PREFACE
EXECUTIVE SUMMARY of SYSTEMS DESIGN
REVIEW of SYSTEMS ANALYSIS
MAJOR DESIGN RECOMMENDATIONS
       Hardware design
       Software design
       Personnel design
       Communications design
       Database design
       Procedures design
       Training design
       Maintenance design
SUMMARY of DESIGN DECISIONS
APPENDIXES
GLOSSARY of TERMS
INDEX

SYSTEMS IMPLEMENTATION

systems implementation
Stage of systems development that includes hardware acquisition, software acquisition or development, user preparation, hiring and training of personnel, site and data preparation, installation, testing, start-up, and user acceptance.

After the information system has been designed, a number of tasks must be completed before the system is installed and ready to operate. This process, called **systems implementation,** includes hardware acquisition, software acquisition or development, user preparation, hiring and training of personnel, site and data preparation, installation, testing, start-up, and user acceptance. The typical sequence of systems implementation activities is shown in Figure 13.11. Companies can reap great rewards after implementing new systems. United Parcel Service (UPS), for example, implemented a $30 million project to improve how packages flow and are tracked from pickup through delivery.[23] UPS's new efficiencies should cut more than 100 million delivery miles and save the company about 14 million gallons of fuel. In the banking industry, FleetBoston Financial Corporation implemented a $10 million IS project to centralize all of its network operations in one room. The project will allow Fleet-Boston to cut its IS staff and save about 3,500 square feet of office space that was used to store networking equipment.

Acquiring Hardware from an IS Vendor

To obtain the components for an information system, organizations can purchase, lease, or rent computer hardware and other resources from an IS vendor. eBay, the popular Internet auction site, for example, acquired hundreds of hardware servers to increase the speed and reliability of its computer system.[24] The old hardware couldn't prevent a crash that lasted about a day and cost eBay about $4 million in lost sales and $5 billion in the value of its stock. The newly acquired hardware is available 99.94 percent of the time.

An IS vendor is a company that offers hardware, software, telecommunications systems, databases, IS personnel, and/or other computer-related resources. Types of IS vendors include general computer manufacturers (e.g., IBM and Hewlett-Packard), small computer manufacturers (e.g., Dell and Gateway), peripheral equipment manufacturers (e.g., Epson and Cannon), computer dealers and distributors (e.g., Radio Shack and CompUSA) and leasing companies (e.g., National Computer Leasing and Paramount Computer Rentals, PLC).

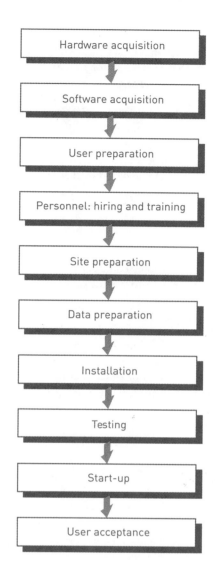

Figure 13.11

Typical Steps in Systems Implementation

Hardware acquisition

Software acquisition

User preparation

Personnel: hiring and training

Site preparation

Data preparation

Installation

Testing

Start-up

User acceptance

Computer dealers, such as CompUSA, manufacture build-to-order computer systems and sell computers and supplies from other vendors.

(Source: Courtesy of CompUSA, Inc.)

In addition to buying, leasing, or renting computer hardware, it is possible to pay only for the computing services that a company uses.[25] Called "pay-as-you-go," "on-demand," or "utility" computing, this approach requires an organization to pay only for the computer power it uses, as it would pay for a utility such as electricity. JPMorgan Chase, for example, is buying only the computer resources it needs from IBM. Hewlett-Packard offers its clients a "capacity-on-demand" approach, in which organizations pay according to the computer resources actually used, including processors, storage devices, and network facilities.[26]

It is also possible to purchase used computer equipment. This option is especially attractive to firms that are experiencing an economic slowdown. Traditional Internet auctions are often used by companies to locate used or refurbished equipment. Popular Internet auction sites sometimes sell more than $1 billion of computer-related equipment annually, and companies can purchase equipment for about 20 or 30 cents on the dollar. However, buyers need to beware: Prices are not always low, and equipment selection can be limited on Internet auction sites.

In addition, companies are increasingly turning to service providers to implement some or all of the systems they need. As discussed in Chapter 4, an application service provider (ASP) can help a company implement software and other systems. The ASP can provide both end-user support and the computers on which to run the software. ASPs often focus on high-end applications, such as database systems and enterprise resource planning packages. As mentioned in Chapter 7, an Internet service provider (ISP) assists a company in gaining access to the Internet. ISPs can also assist a company in setting up an Internet site. Some service providers specialize in specific systems or areas, such as marketing, finance, or manufacturing.

If new hardware is to be purchased or leased, old hardware may have to be discarded. Each year, millions of computers are discarded because new computers are acquired.[27] Thus, acquiring new hardware can also mean disposing of older hardware. Some hardware can be donated to schools and charitable organizations. Some companies recycle hardware. Some states are considering charging up to $10 on new PCs as a recycling fee.

Acquiring Software: Make or Buy?

As with hardware, application software can be acquired several ways. As previously mentioned, it can be purchased from external developers or developed in-house. This decision is often called the **make-or-buy decision**. Alaska Airlines, for example, decided to purchase software for its Internet search engine for fares and prices.[28] The purchased software allows its customers more choices for flight schedules and prices. To share the cost of in-house software development, Coca-Cola Enterprises and SAP, a large ERP software company, have decided to jointly develop software for beverage companies.[29] According to the CIO of Coca-Cola, "This should improve market execution, and the consumer will experience better service." In some cases, companies use a blend of external and internal software development. That is, off-the-shelf or proprietary software programs are modified or customized by in-house personnel. Software can also be rented.[30] Salesforce.com, for example, rents software online that helps organizations manage their sales force and internal staff. Increasingly, software is being viewed as a utility or service, not a product you purchase.

System software, such as operating systems or utilities, is typically purchased from a software company. Increasingly, however, companies are obtaining open-source systems software, such as the Linux operating system, which can be obtained free or for a low cost.

Externally Developed Software

Some of the reasons a company might purchase or lease externally developed software include lower costs, less risk regarding the features and performance of the package, and ease of installation. The cost of the software package is known, and there is little doubt that it will meet the company's needs. The amount of development effort is also less when software is purchased, compared with in-house development.

Open-source software, as first discussed in Chapter 4, can substantially cut the cost of software acquisition. Open-source software includes operating systems, such as Linux, and numerous application software packages, including the MySQL database package. See Chapter 4 for more information on open-source software.

make-or-buy decision
The decision regarding whether to obtain the necessary software from internal or external sources.

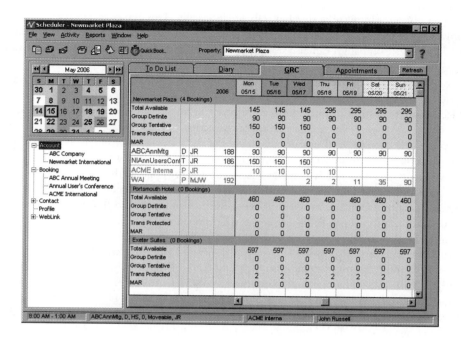

Newmarket International, which invented hospitality sales and event-management software, created Delphi, a forecasting and tracking tool that gives salespeople instant, updated on-screen room information, including transient commitments, target rates, and room type availability.

(Source: Courtesy of Newmarket International, Inc.)

If a company chooses off-the-shelf or contract software in its new systems, it must take the following steps:

- *Review needs, requirements, and costs.* It is important to analyze the program's ability to satisfy user and organizational needs. In some cases, purchased software doesn't satisfy all of an organization's requirements or might have features that are not needed. Software costs are also important. In most cases, software can be purchased for unlimited use or leased.

- *Acquire software.* Many of the approaches discussed in previous sections, including the development of requests for proposals, performing financial analysis, and negotiating the software contract, should be undertaken.

- *Modify or customize software.* Externally developed software seldom does everything the organization requires. So, externally developed software may have to be modified to satisfy user and organizational needs. Some software vendors will assist with the modification, but others may not allow their software to be modified at all.

- *Acquire software interfaces.* Usually, proprietary software requires a **software interface**, which consists of programs or program modifications that allow proprietary software to work with other software used in the organization. For example, if an organization purchases a proprietary inventory software package, software interfaces must allow the new software to work with other programs, such as sales ordering and billing programs.

- *Test and accept the software.* Externally developed software should be completely tested by users in the environment in which it is to run before it is accepted.

- *Monitor and maintain the software and make necessary modifications.* With many software applications, changes will likely have to be made over time. This aspect should be considered in advance because, as mentioned before, some software vendors do not allow their software to be modified.

software interface
Programs or program modifications that allow proprietary software to work with other software used in the organization.

When the software is not meeting organizational goals or expectations, the software can be abandoned instead of modified—often a difficult decision. Some companies have spent millions of dollars to acquire and implement an ERP package, only to encounter problems so severe that they decided to dump the software. Purchased or leased software may not meet an organization's needs, and it can be very difficult to integrate purchased software with other software the company is currently using. For example, a new accounts receivable program purchased from an outside vendor may not work well with the company's inventory control and billing programs.

In-House-Developed Software

Another option is to make or develop software internally. This approach requires the company's IS personnel to be responsible for all aspects of software development. As mentioned in Chapter 12, companies can also use outsourcing to develop part or all of their software using outside companies.[31]

Software can be developed using the object-oriented approach or more traditional approaches. The object-oriented approach is attractive if the organization can reuse existing objects from other software packages and projects. Cardinal Health, for example, spent $100 million on developing new software to reduce prescription errors. One objective was to develop reusable code for later updates or changes to the system. According to Richard Gius, senior vice president of information technology for Cardinal, "We wanted the architecture to be designed in a way that would allow us to continually upgrade the code."[32]

Some advantages inherent with in-house-developed software include meeting user and organizational requirements and having more features and increased flexibility in terms of customization and changes. Software programs developed within a company also have greater potential for providing a competitive advantage because competitors cannot easily duplicate them in the short term. The Neves Corvo mine developed sophisticated mining software that allowed the mine operators to increase the production of copper, tin, and zinc deposits in southern Portugal.[33] The software allows the geology department of Neves Corvo, a joint venture of Rio Tinto and the Portuguese government, to display 3-D models of the mines instead of the older 2-D models from the old software. In another case, NASA developed a software program called DAC for space exploration to analyze how a spacecraft enters distant environments and atmospheres.[34] The award-winning software was used for the Mars Global Surveyor and the Mars Odyssey missions.

It is possible to reuse software from other development efforts to reduce the time it takes to deliver in-house software. Bank of America, for example, reuses previously developed software to deliver new software in 90 days or less.

Chief Programmer Teams. Teams of experienced IS professionals are needed to develop software in-house. The **chief programmer team**, also called the *lead programmer team*, is a group of skilled IS professionals with the task of designing and implementing a set of programs. This team has total responsibility for building the best software possible. Although the makeup of the chief programmer team varies with the size and complexity of the computer programs to be developed, a number of functions are common for all teams. A typical team has a chief programmer, a backup programmer, one or more other programmers, a librarian, and one or more clerks or secretaries (see Figure 13.12). Chief programmers or project managers are often paid about $100,000 annually.[35] The salary for a basic programmer can range from $50,000 to $80,000 annually, which is down slightly from previous years due to outsourcing and other economic factors.

chief programmer team
A group of skilled IS professionals who design and implement a set of programs. This team has total responsibility for building the best software possible; also called *lead programmer team*.

Figure 13.12

A Hierarchy Chart Showing the Typical Structure of a Chief Programmer Team

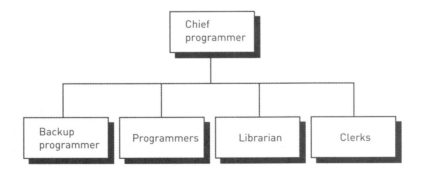

Traditionally, programmer teams consisted of employees hired by the company. Increasingly, companies are looking to other companies or even other countries to provide the important task of programming. Russian rocket scientists have turned their skills from nuclear weapons to software development. Software companies in India can make hundreds of millions of dollars by doing programming for large companies in the United States and around the world.

The Programming Life Cycle. Developing in-house software requires a substantial amount of detailed planning. A series of steps and planned activities can maximize the likelihood of developing good software. These phases make up the **programming life cycle**, as illustrated in Figure 13.13 and described next.

programming life cycle
A series of steps and planned activities developed to maximize the likelihood of developing good software.

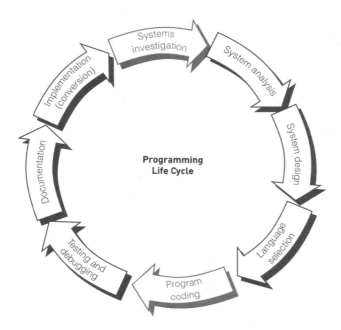

Programming Life Cycle

Figure 13.13

Steps in the Programming Life Cycle

Investigation, analysis, and design activities have already been started and completed in many cases. So the programmer usually has a detailed set of documents that describe what the system should do and how it should operate. An experienced programmer will begin with a thorough review of these documents before any code is written.

Language and software selection involves determining the best programming language for the application. Important characteristics to be considered are (1) the difficulty of the problem, (2) the type of processing to be used (batch or online), (3) the ease with which the program can be changed later, (4) the type of problem, such as business or scientific, and (5) the cost of the language. Often, a trade-off must be made between the ease of use of a language and the efficiency with which programs execute. Older programming languages are more efficient to run but more difficult and time-consuming to develop.

Program coding is the process of writing instructions in the language selected to solve the problem. Like a contractor building a house, the computer programmer follows the plans and documents developed in the previous steps. This careful attention to detail ensures that the software actually accomplishes the desired result.

Testing and debugging are vital steps in developing computer programs. In general, testing is the process of making sure that the program performs as intended; debugging is the process of locating and eliminating errors.

Documentation is the next step, and it can include technical and user documentation. **Technical documentation** is used by computer operators to execute the program and by analysts and programmers in case there are problems with the program or it needs modification. In technical documentation, the purpose of every major piece of computer code is written out and explained. Key variables are also described. **User documentation** is developed for the individuals who use the program. This type of documentation shows users, in easy-to-understand terms, how the program can and should be used. Incorporating a description of the benefits of the new application into user documentation may help stakeholders understand the reasons for the program and speed user acceptance. The software vendor often provides such documentation, or it may be obtained from a technical publishing firm. For example, Microsoft provides user documentation for its spreadsheet package Excel, but there are literally hundreds of books and manuals for this software from other sources.

technical documentation
Written details used by computer operators to execute the program and by analysts and programmers in case there are problems with the program or the program needs modification.

user documentation
Written description developed for individuals who use a program, showing users, in easy-to-understand terms, how the program can and should be used.

Implementation, or conversion, is the last step in developing new computer software. It involves installing the software and making it operational. Several approaches are discussed later in the chapter when we discuss installation.

Tools and Techniques for Software Development

If software will be developed in-house, the chief programmer team can use a number of tools, techniques, and approaches. Options include structured programming, CASE tools, object-oriented implementation, cross-platform development, integrated development environments, and structured walkthroughs.

Structured Design and Programming. Structured design and programming techniques were originally developed in the 1970s and 1980s. The basic idea is to improve the logical program flow by breaking the program into groups of statements, called *structures*. As shown in Figure 13.14, only three types of structures are allowed when using structured programming. In the **sequence structure**, there must be definite starting and ending points. After starting the sequence, programming statements are executed one after another until all the statements in the sequence have been executed. Then the program either ends or continues on to another structure. The **decision structure** allows the computer to branch, depending on certain conditions. Normally, there are only two possible branches. The final structure is the **loop structure**. Actually, there are two commonly used structures for loops. One is the do-until structure, and the other is the do-while structure. Both accomplish the same thing. In the do-until structure, the loop is repeated until a certain condition is met. In the do-while structure, the loop is repeated while a certain condition exists. Structured program code development can be the key to developing good program code. Some of the characteristics of structured programming are shown in Table 13.5.

sequence structure
A programming structure in which, after starting the sequence, programming statements are executed one after another until all the statements in the sequence have been executed. Then the program either ends or continues on to another sequence.

decision structure
A programming structure that allows the computer to branch, depending on certain conditions. Normally, there are only two possible branches.

loop structure
A programming structure with two commonly used structures for loops: do-until and do-while. In the do-until structure, the loop is repeated until a certain condition is met. In the do-while structure, the loop is repeated while a certain condition exists.

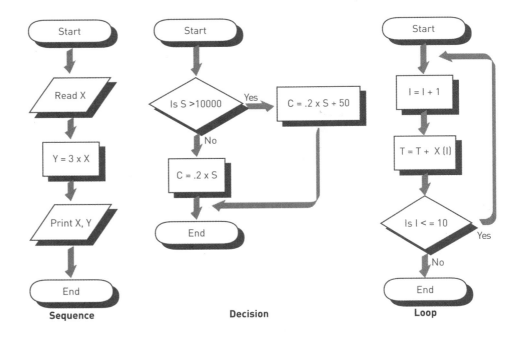

Sequence   Decision   Loop

Figure 13.14

The Three Structures Used in Structured Programming

In general, a good approach to writing a large program is to start with the main module and work down to the other modules. This is called the **top-down approach** to programming and is used in structured program design and programming. Although the concept of top-down programming is simple, its use is beneficial in untangling or avoiding coding and debugging problems. The process begins by writing the main module. Then the modules at the next level are written. This procedure continues until all the modules have been written.

top-down approach
A good general approach to writing a large program, starting with the main module and working down to the other modules.

Program code is broken into modules.
Each module has one and only one function. Such modules are said to have tight internal cohesion.
There is one and only one logic path into each module and one logic exit from each module.
The modules are loosely coupled.
GOTO statements are not allowed.

Table 13.5

Characteristics of Structured Programming

Figure 13.15 illustrates the top-down approach. In addition to program coding, the top-down approach should be used in testing and debugging. Thus, after the first, or main, module is written, it is tested and debugged. But the main module sends the computer to modules at the second level, which have not been written yet. So, simple modules at the second level are written to send the computer back to the main module so that it can be fully and completely tested. If errors are found in the main module, they are corrected immediately.

Level 1 (The main module)

a. Write the main module.
b. Write any necessary dummy modules at the second level.
c. Test the main module.
d. Debug the main module.

Level 2 (This procedure is done for each module one at a time.)

a. Write the module.
b. Write any necessary dummy modules for the next lower level.
c. Test the module (this will automatically test all the modules that are above this one in the structure chart).
d. Debug the module.

Level N (This procedure is repeated for all levels.)

Figure 13.15

The Top-Down Approach to Writing, Testing, and Debugging a Modular Program

CASE Tools. CASE tools, first discussed in Chapter 12, are often used during software development to automate some of the techniques. For example, source code can be automatically generated using CASE tools. CASE tools may also have interfaces to the code generators of other vendors' CASE tools, which allows a programmer to mix and match the program code generated. Using CASE tools can help increase programmers' accuracy and productivity, particularly in terms of time spent on maintenance.

Object-Oriented Implementation. Companies can also use the object-oriented approach to develop programs. With this approach, a collection of existing modules of code, or objects, can be used across a number of applications. In most cases, minimal coding changes are required to mesh with the developed objects or modules of code. Even though object-oriented software development does not require the use of object-oriented languages, most developers use them for the structure and ease they provide. These languages include Java, Visual Basic, and C++, and they make implementation easier and more straightforward.[36]

Being able to reuse previously developed objects speeds software development and can improve quality. For example, the owner of the kayak rental shop, discussed in an earlier example and in Chapter 12, might also have a bicycle rental shop. If the bicycle rental shop needs objects to enter new bicycles into the computer system, they can be modified from the kayak rental system and reused in the bicycle rental business. Software reuse can also reduce costs. AXA Financial Services, for example, was able to save about $55 million in developing

a system by reusing software.[37] According to the chief technology officer for AXA, "Anything we do and have to redo is a negative ROI project because we're just duplicating something we already did and spending money to do it without adding much value." JetBlue Airways used Visual Studio .NET to implement an inventory tracking system that used the object-oriented approach.[38] The application only took three months to implement and allowed employees to scan shipments using handheld computers. The scanned data was uploaded to the company's database. The application paid for itself in seven months through reduced costs.

Cross-Platform Development. In the past, most applications were developed and implemented using mainframe computers. Today, many applications are developed on personal computers by users of the systems. In response to the growth of end-user development, software vendors now offer more tools and techniques to PC users. One software development technique, called **cross-platform development**, allows programmers to develop programs that can run on computer systems that have different hardware and operating systems, or platforms. Web service tools, such as .NET by Microsoft introduced in Chapter 7, are examples. With cross-platform development, the same program might be able to run on both a personal computer and a mainframe or on two different types of PCs.

Integrated Development Environments. Software vendors also offer integrated development environments to assist with programming on personal computers. **Integrated development environments (IDEs)** combine the tools needed for programming with a programming language in one integrated package. IDEs allow programmers to use simple screens, customized pull-down menus, and graphical user interfaces. Some even use different-colored text to allow a programmer to quickly locate sections, verbs, or errors in program code. In general, IDEs can make programming software more intuitive. Combining these tools with the language itself makes it easier for programmers to develop sophisticated programs on personal computers, making them more productive.

Visual Studio .NET from Microsoft is an example of an IDE. Visual Studio .NET consists of a collection of Web tools for Microsoft's systems development strategy on the Internet. It is also considered a rapid application development tool. Oracle Designer, which is used with Oracle's database system, is another example of an IDE. The popular Eclipse Workbench supports IDEs that can be used with the C and C++ programming languages.[39] Eclipse Workbench includes a debugger and a compiler, along with other tools. Builder 6 Enterprise by Borland is another example of a product that supports the advantages of the IDE approach. Builder 6 includes tools to help programmers write, debug, compile, and run Java programs. The overall goal of Builder 6, like other IDE tools, is to increase programmer productivity and reduce software development costs.

Structured Walkthroughs. Regardless of the tools or techniques used, companies should review software throughout the development process. Companies often use a structured walkthrough technique, which is typically performed by chief programmer teams. As shown in Figure 13.16, a **structured walkthrough** is a planned and preannounced review of the progress of a program module, a structure chart, or a human procedure. The walkthrough helps team members review and evaluate the progress of components of a structured project. The structured walkthrough approach is also useful for programming projects that do not use the structured design approach.

Acquiring Database and Telecommunications Systems

Acquiring or upgrading database systems can be one of the most important steps of a systems development effort. While most companies use a relational database, some are starting to use object-oriented database systems, such as Object Store from Excelon Corporation and Objectivity from Objectivity, Inc.

Because databases are a blend of hardware and software, many of the approaches discussed earlier for acquiring hardware and software also apply to database systems. For example, an upgraded inventory control system may require database capabilities, including more hard disk storage or a new DBMS. If so, additional storage hardware will have to be acquired from an IS vendor. New or upgraded software might also be purchased or developed in-house.

cross-platform development
Development technique that allows programmers to develop programs that can run on computer systems having different hardware and operating systems, or platforms.

integrated development environments (IDEs)
Development approach that combines the tools needed for programming with a programming language into one integrated package.

structured walkthrough
A planned and preannounced review of the progress of a program module, a structure chart, or a human procedure.

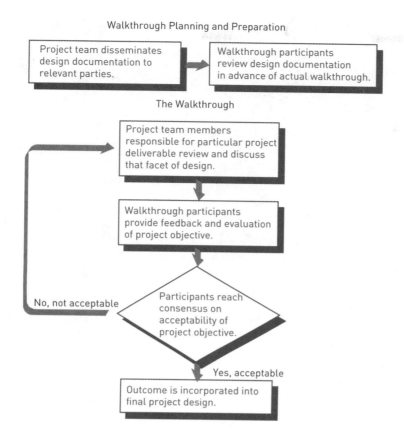

Walkthrough Planning and Preparation

Project team disseminates design documentation to relevant parties.

Walkthrough participants review design documentation in advance of actual walkthrough.

The Walkthrough

Project team members responsible for particular project deliverable review and discuss that facet of design.

Walkthrough participants provide feedback and evaluation of project objective.

Participants reach consensus on acceptability of project objective.

No, not acceptable

Yes, acceptable

Outcome is incorporated into final project design.

Figure 13.16

A structured walkthrough is a planned, preannounced review of the progress of a particular project objective.

With the increased use of e-commerce, the Internet, intranets, and extranets, telecommunications is one of the fastest-growing applications for today's businesses and individuals. The Nasdaq stock market, for example, is investing $50 million in a new network system to streamline operations and cut costs.[40] According to the chief information officer for Nasdaq, "We're going to be able to take a lot of circuits out of our network and save costs." Like database systems, telecommunications systems require a blend of hardware and software. For personal computer systems, the primary piece of hardware is a modem. For client/server and mainframe systems, the hardware can include multiplexers, concentrators, communications processors, and a variety of network equipment. Communications software will also have to be acquired from a software company or developed in-house. The Internal Revenue Service (IRS) hired a new chief information officer to oversee its large systems development effort to improve network security and other telecommunications aspects of the IRS computer system.[41] Again, the earlier discussion on acquiring hardware and software also applies to the acquisition of telecommunications hardware and software.

User Preparation

User preparation is the process of readying managers, decision makers, employees, other users, and stakeholders for the new systems. This activity is an important but often ignored area of systems implementation. A small airline might not do adequate employee training with a new software package. The result could be a grounding of most of its flights and the need to find hotel rooms to accommodate unhappy travelers who were stranded

Without question, training users is an essential part of user preparation, whether they are trained by internal personnel or by external training firms. In some cases, companies that provide software also train users at no charge or at a reasonable price. The cost of training can be negotiated during the selection of new software. Other companies conduct user training throughout the systems development process. Concerns and apprehensions about the new system must be eliminated through these training programs. Employees should be acquainted with the system's capabilities and limitations by the time they are ready to use it.

user preparation
The process of readying managers, decision makers, employees, other users, and stakeholders for new systems.

Providing users with proper training can help ensure that the information system is used correctly, efficiently, and effectively.

(Source: Photodisc/Getty Images.)

IS Personnel: Hiring and Training

Depending on the size of the new system, an organization may have to hire and, in some cases, train new IS personnel. An IS manager, systems analysts, computer programmers, data-entry operators, and similar personnel may be needed for the new system.

As with end users, the eventual success of any system depends on how it is used by the IS personnel within the organization. Training programs should be conducted for the IS personnel who will be using the computer system. These programs are similar to those for the users, although they may be more detailed in the technical aspects of the systems. Effective training will help IS personnel use the new system to perform their jobs and support other users in the organization. The CEO of PeopleSoft, for example, believes that IS personnel need training in three areas: business, technology, and an understanding of other countries and people. He says, "If you get a strong background in those three things, you'll be able to participate in the future."[42]

Site Preparation

site preparation
Preparation of the location of a new system.

The location of the new system needs to be prepared in a process called **site preparation**. For a small system, site preparation can be as simple as rearranging the furniture in an office to make room for a computer. With a larger system, this process is not so easy because it may require special wiring and air conditioning. One or two rooms may have to be completely renovated, and additional furniture may have to be purchased. A special floor may have to be built, under which the cables connecting the various computer components are placed, and a new security system may be needed to protect the equipment. For larger systems, additional power circuits may also be required.

Data Preparation

data preparation, or data conversion
Ensuring all files and databases are ready to be used with new computer software and systems.

Data preparation, or **data conversion**, involves making sure that all files and databases are ready to be used with new computer software and systems. If an organization is installing a new payroll program, the old employeepayroll data may have to be converted into a format that can be used by the new computer software or system. After the data has been prepared or converted, the computerized database system or other software will then be used to maintain and update the computer files.

Installation

Installation is the process of physically placing the computer equipment on the site and making it operational. Although normally the manufacturer is responsible for installing computer equipment, someone from the organization (usually the IS manager) should oversee the process, making sure that all equipment specified in the contract is installed at the proper location. After the system is installed, the manufacturer performs several tests to ensure that the equipment is operating as it should.

Testing

Good testing procedures are essential to make sure that the new or modified information system operates as intended. Inadequate testing can result in mistakes and problems. A popular tax preparation company, for example, implemented a Web-based tax preparation system, but people could see one another's tax returns. The president of the tax preparation company called it "our worst-case scenario." Better testing can prevent these types of problems.

Several forms of testing should be used, including testing each of the individual programs (**unit testing**), testing the entire system of programs (**system testing**), testing the application with a large amount of data (**volume testing**), and testing all related systems together (**integration testing**), as well as conducting any tests required by the user (**acceptance testing**). Figure 13.17 lists the types of testing. In addition to these forms of testing, there are different types of testing. **Alpha testing** involves testing an incomplete or early version of the system, while **beta** testing involves testing a complete and stable system by end users. Alpha-unit testing, for example, is testing an individual program before it is completely finished. Beta-unit testing, on the other hand, is performed after alpha testing, when the individual program is complete and ready for use by end users.

Unit testing is accomplished by developing test data that will force the computer to execute every statement in the program. In addition, each program is tested with abnormal data to determine how it will handle problems. Mohegan Sun Casino, for example, tested two separate programs before using them for its day-to-day operations.[43] One program allows the automatic transfer of funds to some of its slot machines. The other program provides better surveillance and data collection.

installation
The process of physically placing the computer equipment on the site and making it operational.

unit testing
Testing of individual programs.

system testing
Testing the entire system of programs.

volume testing
Testing the application with a large amount of data.

integration testing
Testing all related systems together.

acceptance testing
Conducting any tests required by the user.

alpha testing
Testing an incomplete or early version of the system.

beta testing
Testing a complete and stable system by end users.

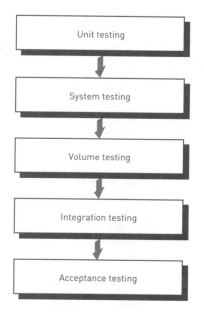

Figure 13.17

Types of Testing

System testing requires the testing of all the programs together. It is not uncommon for the output from one program to become the input for another. So, system testing ensures that the output from one program can be used as input for another program within the

system. Volume testing ensures that the entire system can handle a large amount of data under normal operating conditions. Integration testing ensures that the new programs can interact with other major applications. It also ensures that data flows efficiently and without error to other applications. For example, a new inventory control application may require data input from an older order processing application. Integration testing would be done to ensure smooth data flow between the new and existing applications. Integration testing is typically done after unit and system testing. Metaserver, a software company for the insurance industry, has developed a tool called iConnect to perform integration testing for different insurance applications and databases.[44] According to Donald Light, Celenet Communications analyst, "The difference with iConnect is its ability to deliver a return on a relatively small investment. They've taken an area that is critical with independent agents on one hand and providers of external data on the other hand and brought them together."

Finally, acceptance testing makes sure that the new or modified system is operating as intended. Run times, the amount of memory required, disk access methods, and more can be tested during this phase. Acceptance testing ensures that all performance objectives defined for the system or application are satisfied. Insurance company AON, for example, uses acceptance testing to make sure that new software is ready for use.[45] "If we deploy an application now which has 80 percent functionality but contains bugs, we can decide how to move forward," says a company representative. Involving users in acceptance testing may help them understand and effectively interact with the new system. Acceptance testing is the final check of the system before start-up. Pioneer Hi-Bred International, for example, performed acceptance testing for a new crop management system that will help farmers anticipate crop growth, disease outbreaks, and insect infestations.[46] The new system uses handheld computers and global positioning systems to collect data and feed it into a large crop database. Once collected, the database is available to farmers.

Start-Up

start-up
The process of making the final tested information system fully operational.

Start-up begins with the final tested information system. When start-up is finished, the system is fully operational. Various start-up approaches are available (see Figure 13.18). **Direct conversion** (also called *plunge* or *direct cutover*) involves stopping the old system and starting the new system on a given date. Direct conversion is usually the least desirable approach because of the potential for problems and errors when the old system is shut off and the new system is turned on at the same instant. The Tennessee Valley Authority (TVA) implemented a large IS project to improve efficiency and reduce costs. The implementation was a direct cutover that went live after a week-long implementation process. "The big-bang approach was very scary," said the senior vice president of information systems for TVA. "This meant getting everything going at the same time."[47] The **phase-in approach** is a popular technique preferred by many organizations. In this approach, sometimes called a *piecemeal approach*, components of the new system are slowly phased in while components of the old one are slowly phased out. When everyone is confident that the new system is performing as expected, the old system is completely phased out. This gradual replacement is repeated for each application until the new system is running every application. In some cases, the phase-in approach can take months or years.

direct conversion (also called *plunge* or *direct cutover*)
Stopping the old system and starting the new system on a given date.

phase-in approach
Slowly replacing components of the old system with those of the new one. This process is repeated for each application until the new system is running every application and performing as expected; also called *piecemeal approach.*

pilot start-up
Running the new system for one group of users rather than all users.

Pilot start-up involves running the new system for one group of users rather than all users. For example, a manufacturing company with a number of retail outlets throughout the country could use the pilot start-up approach and install a new inventory control system at one of the retail outlets. When this pilot retail outlet runs without problems, the new inventory control system can be implemented at other retail outlets. Carnival Cruise Lines, for example, is using a pilot start-up for a systems development project to remotely manage PCs.[48] When fully implemented, the new system will remotely manage about 4,000 PCs, including about 1,700 PCs on the company's 19 ships. The new system will allow Carnival to remotely upgrade software and perform some fixes without the need to fly a technician to a remote location or a ship.

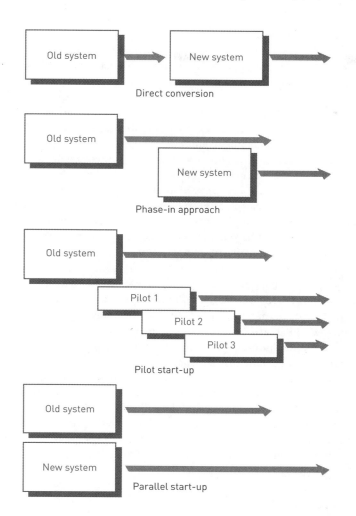

Figure 13.18

Start-Up Approaches

Parallel start-up involves running both the old and new systems for a period of time. The output of the new system is compared closely with the output of the old system, and any differences are reconciled. When users are comfortable that the new system is working correctly, the old system is eliminated.

User Acceptance

Most mainframe computer manufacturers use a formal **user acceptance document**— a formal agreement signed by the user that states that a phase of the installation or the complete system is approved. This is a legal document that usually removes or reduces the IS vendor's liability for problems that occur after the user acceptance document has been signed. Because this document is so important, many companies get legal assistance before they sign the acceptance document. Stakeholders may also be involved in acceptance to make sure that the benefits to them are indeed realized.

parallel start-up
Running both the old and new systems for a period of time and comparing the output of the new system closely with the output of the old system; any differences are reconciled. When users are comfortable that the new system is working correctly, the old system is eliminated.

user acceptance document
Formal agreement signed by the user that states that a phase of the installation or the complete system is approved.

SYSTEMS OPERATION AND MAINTENANCE

Systems operation involves all aspects of using the new or modified system. Throughout this book, we have seen many examples of information systems operating in a variety of settings and industries. Thus, we will not cover the operation of an information system in detail in this section. The operation of any information system, however, does require adequate user training before the system is used and continual support while the system is being operated.

systems operation
Use of a new or modified system.

systems maintenance
Stage of systems development that involves checking, changing, and enhancing the system to make it more useful in achieving user and organizational goals.

This training and support is required for all stakeholders, including employees, customers, and others. Companies typically provide training through seminars, manuals, and online documentation. To provide adequate support, many companies use a formal help desk. A *help desk* consists of people with technical expertise, computer systems, manuals, and other resources needed to solve problems and give accurate answers to questions. With today's advances in telecommunications, help desks can be located around the world. If you are having trouble with your PC and call a toll-free number for assistance, you might reach a help desk in India, China, or another country.

Systems maintenance involves checking, changing, and enhancing the system to make it more useful in achieving user and organizational goals. This process can be especially difficult for older software. A *legacy system* is an old system that may have been patched or modified repeatedly over time. An old payroll program in COBOL developed decades ago and frequently changed is an example of a legacy system. Legacy systems can be very expensive to maintain. At some point, it becomes less expensive to switch to new programs and applications than to repair and maintain the legacy system.

Software maintenance is a major concern for organizations.[49] In some cases, organizations encounter major problems that require recycling the entire systems development process. In other situations, minor modifications are sufficient to remedy problems. Hardware maintenance is also important. Companies such as IBM are investigating *autonomic computing*, in which computers will be able to manage and maintain themselves.[50] The goal is for computers to be self-configuring, self-protecting, self-healing, and self-optimizing. Being self-configuring allows a computer to handle new hardware, software, or other changes to its operating environment. Being self-protecting allows a computer to identify potential attacks, prevent them when possible, and recover from attacks if they occur. Attacks can include viruses, worms, identity theft, industrial espionage, and the like. Being self-healing allows a computer to automatically fix problems when they occur, and being self-optimizing allows a computer to run faster and get more done in less time. The "Ethical and Societal Issues" special feature explores how software companies are trying to enhance the trustworthiness of information systems.

Reasons for Maintenance

Once a program is written, it will need ongoing maintenance. To some extent, a program is similar to a car that needs oil changes, tune-ups, and repairs at certain times. Experience shows that frequent, minor maintenance to a program, if properly done, can prevent major system failures later. Some of the reasons for program maintenance are the following:

- Changes in business processes
- New requests from stakeholders, users, and managers
- Bugs or errors in the program
- Technical and hardware problems
- Corporate mergers and acquisitions
- Government regulations
- Change in the operating system or hardware on which the application runs
- Unexpected events, such as the terrorist attacks of September 11 or severe weather

Most companies modify their existing programs instead of developing new ones because existing software performs many important functions, and companies can have millions of dollars invested in their old legacy systems. So, as new systems needs are identified, the burden of fulfilling the needs most often falls on the existing system. Old programs are repeatedly modified to meet ever-changing needs. Yet, over time, repeated modifications tend to interfere with the system's overall structure, reducing its efficiency and making further modifications more burdensome.

ETHICAL AND SOCIETAL ISSUES

Finding Trust in Computer Systems

In June 2004, in what was heralded as "the worst mess since banks put their faith in computers," the Royal Bank of Canada was unable to tell its 10 million Canadian customers exactly how much money was in their accounts. Canada's largest bank had a problem that kept tens of millions of transactions, including every direct payroll deposit it handles, from showing up in accounts.

You can imagine the furor as millions of bank customers discovered that the paycheck they counted on hadn't yet been deposited. While many people were just inconvenienced, others were in a panic. Vacations were postponed, bill payments became delinquent, and customers were lined up at banks to try to get money that they were due for basic living expenses.

The nightmare began during what was intended to be a routine programming update. Soon afterward, the bank's entire nationwide system failed to register withdrawals and deposits against customer balances for several days. The more days that passed, the larger the backlog became, and the worse the chance of the system being able to recover.

Such computer glitches are not surprising or rare in today's computer-dependent society. Computer system bugs have become a frustrating and sometimes devastating reality that we all endure. They plague us in our jobs, when we shop, and at home on our own PCs. As the incidents have mounted, computer users have lost trust in the ability of computer systems to work without fail.

Over the past few years, computer system failures have drawn increased attention as cybercriminals have created frequent and oftentimes serious problems. Worms and viruses have ravaged the Internet's operating systems and other software. Corporate networks have been shut down, and valuable and private information has been stolen and sold on the black market. This has brought increasing pressure on systems developers and software manufacturers to invest more in product quality.

In the early days of systems development, analysts and programmers were more focused on creating the functionality of a system than on its reliability and security. As the world has become increasingly digitized, networked, and computer savvy, the reliability and security of systems have been severely tested, and consumers' patience has worn thin.

Software developers, including Microsoft's, are confronting the issue and working to regain customer trust. Microsoft recently instituted its "Trustworthy Computing Initiative." Microsoft Enterprise Technologies Director Greg Stone says, "I see Trustworthy Computing having three key elements: capability to do what you say you will, consistency of quality, and the commitment to act in the customer's best interests." Microsoft characterizes the principles of Trustworthy Computing as reliability, business integrity, privacy, and security.

Some doubt Microsoft's ability to deliver trustworthy systems in the near future. The complexity of Microsoft's programming code may make it impossible to secure. "Microsoft has done vastly more than what they've done historically, but is the state of security any better?" asks Lloyd Hession, CSO of Radianz, which operates extranets for 5,000 financial institutions. "What Microsoft has done hasn't had much of an impact. Because of the complexity of the code, you find more vulnerabilities once you start to tinker with the code."

Microsoft is investing an increasing proportion of the company's $6.8 billion research and development fund on proactive security projects, such as creating self-healing software, often called *autonomic computing*. Autonomic computing appears to be a solution that most of the big software companies are pursuing. The major systems players, HP, IBM, and Sun Microsystems, are offering products that are self-managing and self-healing. Autonomic technology is being applied to networks, databases, software, and nearly all components of information systems. The hope is that systems will be able to react on the fly to changes in demand or unexpected problems, providing a constant level of service, much like a public utility.

Without a doubt, we will see increasing efforts from all systems and software developers to deliver trustworthy systems. As our lives become increasingly dependent on information systems, the future of our society could be jeopardized. It is clear that today's systems, in all of their complexity, will require a high degree of automation and artificial intelligence capabilities to successfully manage them. Computer system stability, dependability, and security will remain a major goal for governments, industries, and technology specialists to work together to achieve.

Critical Thinking Questions

1. What social factors have arisen to make computer system stability and security an increasing concern?
2. Whose responsibility is it to ensure that computer systems are dependable and secure? Why?

What Would You Do?

As the owner of TicketBlaster, an online service that sells tickets for concerts, shows, and sporting events, you depend on technology. Over the past month, your company has experienced a total of six hours of sporadic downtime—10 minutes here, 20 minutes there. Sales figures have dropped, and you fear that customers may be turning to your competitor out of frustration.

3. What steps could you take to uncover the cause of your downtime?
4. What actions could you take to renew your customer's faith in your service?

SOURCES: John Saunders and Richard Bloom, "Bank's Clients in Limbo," *The Globe and Mail*, June 4, 2004, *www.theglobeandmail.com*; Lawrence Walsh, "Microsoft's Paradox," *Information Security*, January 2004, *www.lexis-nexis.com*; Mark Hollands, "Microsoft Struggles to Build Trust," *The Aus-* *tralian*, March 30, 2004, *www.lexis-nexis.com*; Ann Bednarz, "Autonomic Authority," *Network World*, March 22, 2004, *www.lexis-nexis.com*; "Evident Software Partners with IBM to Further Autonomic Computing Initiative," *Business Wire*, March 8, 2004, *www.lexis-nexis.com*.

Types of Maintenance

slipstream upgrade
A minor upgrade—typically a code adjustment or minor bug fix—not worth announcing. It usually requires recompiling all the code and, in so doing, it can create entirely new bugs.

patch
A minor change to correct a problem or make a small enhancement. It is usually an addition to an existing program.

release
A significant program change that often requires changes in the documentation of the software.

version
A major program change, typically encompassing many new features.

request for maintenance form
A form authorizing modification of programs.

Software companies and many other organizations use four generally accepted categories to signify the amount of change involved in maintenance. A **slipstream upgrade** is a minor upgrade—typically a code adjustment or minor bug fix. Many companies don't announce to users that a slipstream upgrade has been made. A slipstream upgrade usually requires recompiling all the code, so it can create entirely new bugs. This maintenance practice can explain why the same computers sometimes work differently with what is supposedly the same software. A **patch** is a minor change to correct a problem or make a small enhancement. It is usually an addition to an existing program. That is, the programming code representing the system enhancement is usually "patched into," or added to, the existing code. Although slipstream upgrades and patches are minor changes, they can cause users and support personnel big problems if the programs do not run as before. A new **release** is a significant program change that often requires changes in the documentation of the software. Finally, a new **version** is a major program change, typically encompassing many new features.

The Request for Maintenance Form

Because of the amount of effort that can be spent on maintenance, many organizations require a **request for maintenance form** to authorize modification of programs. This form is usually signed by a business manager, who documents the need for the change and identifies the priority of the change relative to other work that has been requested. The IS group reviews the form and identifies the programs to be changed, determines the programmer who will be assigned to the project, estimates the expected completion date, and develops a technical description of the change. A cost/benefit analysis may be required if the change requires substantial resources.

Performing Maintenance

maintenance team
A special IS team responsible for modifying, fixing, and updating existing software.

Depending on organizational policies, the people who perform systems maintenance vary. In some cases, the team who designs and builds the system also performs maintenance. This ongoing responsibility gives the designers and programmers an incentive to build systems well from the outset: If there are problems, they will have to fix them. In other cases, organizations have a separate **maintenance team**. This team is responsible for modifying, fixing, and updating existing software.

In the past, companies had to maintain each computer system or server separately. With hundreds or thousands of computers scattered throughout an organization, this task could be very costly and time-consuming. Today, the maintenance function is becoming more automated. Some companies, for example, use maintenance tools and software that will allow them to maintain and upgrade software centrally.

A number of vendors have developed tools to ease the software maintenance burden. Relativity Technologies unveiled RescueWare, a product that converts third-generation code such as COBOL to highly maintainable C++, Java, or Visual Basic object-oriented code. Using RescueWare, maintenance personnel download mainframe code to Windows NT or Windows 2000 workstations. They then use the product's graphical tools to analyze the original system's inner workings. RescueWare lets a programmer see the original system as a set of object views, which visually illustrate module functioning and program structures. IS personnel can choose one of three levels of transformation: revamping the user interface, converting the database access, and transforming procedure logic.

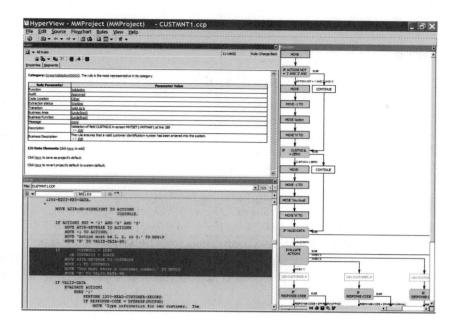

Relativity Technologies' Modernization Workbench is a PC-based software solution that enables companies to consolidate legacy or redundant systems into one, more maintainable and modern application.

(Source: Courtesy of Relativity Technologies.)

The Financial Implications of Maintenance

The cost of maintenance is staggering. For older programs, the total cost of maintenance can be up to five times greater than the total cost of development. In other words, a program that originally cost $25,000 to develop may cost $125,000 to maintain over its lifetime. The average programmer can spend more than half his or her time on maintaining existing programs instead of developing new ones. In addition, as programs get older, total maintenance expenditures in time and money increase, as illustrated in Figure 13.19. With the use of newer programming languages and approaches, including object-oriented programming, maintenance costs are expected to decline. Even so, many organizations have literally millions of dollars invested in applications written in older languages (such as COBOL), which are both expensive and time-consuming to maintain.

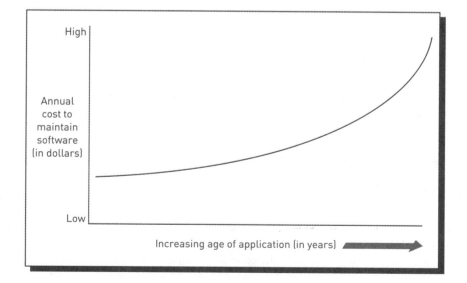

Figure 13.19

Maintenance Costs as a Function of Age

The financial implications of maintenance make it important to keep track of why systems are maintained, instead of simply keeping cost figures. This is another reason that documentation of maintenance tasks is so crucial. A determining factor in the decision to replace a system is the point at which it is costing more to fix it than to replace it.

The Relationship between Maintenance and Design

Programs are expensive to develop, but they are even more expensive to maintain. Programs that are well designed and documented to be efficient, structured, and flexible are less expensive to maintain in later years. Thus, there is a direct relationship between design and maintenance. More time spent on design up front can mean less time spent on maintenance later.

In most cases, it is worth the extra time and expense to design a good system. Consider a system that costs $250,000 to develop. Spending 10 percent more on design would cost an additional $25,000, bringing the total design cost to $275,000. Maintenance costs over the life of the program could be $1,000,000. If this additional design expense can reduce maintenance costs by 10 percent, the savings in maintenance costs would be $100,000. Over the life of the program, the net savings would be $75,000 ($100,000 − $25,000). This relationship between investment in design and long-term maintenance savings is graphically displayed in Figure 13.20.

Figure 13.20

The Value of Investment in Design

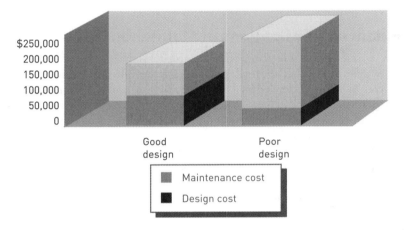

The need for good design goes beyond mere costs. There is a real risk in ignoring small system problems when they arise, but these small problems may become large in the future. As mentioned earlier, because maintenance programmers spend an estimated 50 percent or more of their time deciphering poorly written, undocumented program code, there is little time to spend on developing new, more effective systems. If put to good use, the tools and techniques discussed in this chapter will allow organizations to build longer-lasting, more reliable systems.

SYSTEMS REVIEW

systems review
The final step of systems development, involving the analysis of systems to make sure that they are operating as intended.

Systems review, the final step of systems development, is the process of analyzing systems to make sure that they are operating as intended.[51] This process often compares the performance and benefits of the system as it was designed with the actual performance and benefits of the system in operation. Hymans Robertson, an actuarial firm, reviewed an existing financial system and discovered that it no longer met the company's needs.[52] According to the finance director for the firm, "We had been using a finance system that had been implemented three years ago. We soon realised we had outgrown this system as it was not flexible enough to fit our changing needs." Problems and opportunities uncovered during systems review will

trigger systems development and begin the process anew. For example, as the number of users of an interactive system increases, it is not unusual for system response time to increase. If the increase in response time is too great, it may be necessary to redesign some of the system, modify databases, or increase the power of the computer hardware.

Internal employees, external consultants, or both can perform systems review. When the problems or opportunities are industry-wide, people from several firms may get together. In some cases, they collaborate at an IS conference or in a private meeting involving several firms.

Types of Review Procedures

There are two types of review procedures: event driven and time driven (see Table 13.6). An **event-driven review** is triggered by a problem or opportunity such as an error, a corporate merger, or a new market for products. In some cases, companies wait until a large problem or opportunity occurs before a change is made, ignoring minor problems. In contrast, some companies use a continuous improvement approach to systems development. With this approach, an organization makes changes to a system even when small problems or opportunities occur. Although continuous improvement can keep the system current and responsive, doing the repeated design and implementation can be both time-consuming and expensive.

event-driven review
Review triggered by a problem or opportunity such as an error, a corporate merger, or a new market for products.

Event Driven	Time Driven
A problem with an existing system	Monthly review
A merger	Yearly review
A new accounting system	Review every few years
An executive decision that an upgraded Internet site is needed to stay competitive	Five-year review

Table 13.6

Examples of Review Types

A **time-driven review** is performed after a specified amount of time. Many application programs are reviewed every six months to a year. With this approach, an existing system is monitored on a schedule. If problems or opportunities are uncovered, a new systems development cycle may be initiated. A payroll application, for example, may be reviewed once a year to make sure that it is still operating as expected. If it is not, changes are made.

time-driven review
Review performed after a specified amount of time.

Many companies use both approaches. A billing application, for example, might be reviewed once a year for errors, inefficiencies, and opportunities to reduce operating costs. This is a time-driven approach. In addition, the billing application might be redone if there is a corporate merger, if one or more new managers require different information or reports or if federal laws on bill collecting and privacy change. This is an event-driven approach.

Factors to Consider During Systems Review

Systems review should investigate a number of important factors, such as the following:

- *Mission.* Is the computer system helping the organization achieve its overall mission? Are stakeholder needs and desires satisfied or exceeded with the new or modified system?
- *Organizational goals.* Does the computer system support the specific goals of the various areas and departments of the organization?
- *Hardware and software.* Are hardware and software up to date and adequate to handle current and future processing needs?
- *Database.* Is the current database up to date and accurate? Is database storage space adequate to handle current and future needs?
- *Telecommunications.* Is the current telecommunications system fast enough, and does it allow managers and workers to send and receive timely messages? Does it allow for fast order processing and effective customer service?

- *Information systems personnel.* Are there sufficient IS personnel to perform current and projected processing tasks?
- *Control.* Are rules and procedures for system use and access acceptable? Are the existing control procedures adequate to protect against errors, invasion of privacy, fraud, and other potential problems?
- *Training.* Are there adequate training programs and provisions for both users and IS personnel?
- *Costs.* Are development and operating costs in line with what is expected? Is there an adequate IS budget to support the organization?
- *Complexity.* Is the system overly complex and difficult to operate and maintain?
- *Reliability.* Is the system reliable? What is the mean time between failures (MTBF)?
- *Efficiency.* Is the computer system efficient? Are system outputs generated by the right amount of inputs, including personnel, hardware, software, budget, and others?
- *Response time.* How long does it take the system to respond to users during peak processing times?
- *Documentation.* Is the documentation still valid? Are changes in documentation needed to reflect the current situation?

System Performance Measurement

system performance measurement

Monitoring the system—the number of errors encountered, the amount of memory required, the amount of processing or CPU time needed, and other problems.

system performance products

Software that measures all components of the computer-based information system, including hardware, software, database, telecommunications, and network systems.

Systems review often involves monitoring the system, called **system performance measurement**. The number of errors encountered, the amount of memory required, the amount of processing or CPU time needed, and other problems should be closely observed. If a particular system is not performing as expected, it should be modified, or a new system should be developed or acquired.

System performance products have been developed to measure all components of the computer-based information system, including hardware, software, database, telecommunications, and network systems. When properly used, system performance products can quickly and efficiently locate actual or potential problems.

A number of products have been developed to assist in assessing system performance. Candle is a leading provider of management tools that monitor mainframe performance and application availability. Its products include Candle Command Center, an advanced systems management tool for optimizing an organization's computing resources and maximizing business application availability, and OMEGAMON II performance monitors, for real-time and historical analysis of performance on a variety of systems. Precise/Pulse is a product from Precise Software Solutions that provides around-the-clock performance monitoring for Oracle database applications. It detects and reports potential problems through systems management consoles. Precise/Pulse monitors the performance of critical database applications and issues alerts about inefficiencies before they turn into application performance problems.

Measuring a system is, in effect, the final task of systems development. The results of this process may bring the development team back to the beginning of the development life cycle, where the process begins again.

SUMMARY

Principle

Designing new systems or modifying existing ones should always be aimed at helping an organization achieve its goals.

The purpose of systems design is to prepare the detailed design needs for a new system or modifications to the existing system. Logical systems design refers to the way that the various components of an information system will work together. The logical design includes data requirements for output and input, processing, files and databases, telecommunications, procedures, personnel and job design, and controls and security design. Physical systems design refers to the specification of the actual physical components. The physical design must specify characteristics for hardware and software design, database and telecommunications, and personnel and procedures design.

Logical and physical design can be accomplished using the traditional systems development life cycle or the objected-oriented approach. Using the OO approach, analysts design key objects and classes of objects in the new or updated system. The sequence of events that a new or modified system requires is often called a *scenario*, which can be diagrammed in a sequence diagram.

A number of special design considerations should be taken into account during both logical and physical system design. Interface design and control relates to how users access and interact with the system. A sign-on procedure consists of identification numbers, passwords, and other safeguards needed for individuals to gain access to computer resources. If the system under development is interactive, the design must consider menus, help facilities, table lookup facilities, and restart procedures. A good interactive dialogue will ask for information in a clear manner, respond rapidly, be consistent among applications, and use an attractive format. Also, it will avoid use of computer jargon and treat the user with respect.

System security and control involves many aspects. Error prevention, detection, and correction should be part of the system design process. Causes of errors include human activities, natural phenomena, and technical problems. Designers should be alert to prevention of fraud and invasion of privacy.

Disaster recovery is an important aspect of systems design. Disaster planning is the process of anticipating and providing for disasters. A disaster can be an act of nature (a flood, fire, or earthquake) or a human act (terrorism, error, labor unrest, or erasure of an important file). The primary tools used in disaster planning and recovery are hardware, software, database, telecommunications, and personnel backup.

Security, fraud, and the invasion of privacy are also important design considerations. Most IS departments establish tight systems controls to maintain data security. Systems controls can help prevent computer misuse, crime, and fraud by employees and others. System controls include input, output, processing, database, telecommunications, and personnel controls.

Whether an individual is purchasing a personal computer or an experienced company is acquiring an expensive mainframe computer, the system could be obtained from one or more vendors. Some of the factors to consider in selecting a vendor are the vendor's reliability and financial stability, the type of service offered after the sale, the goods and services the vendor offers and keeps in stock, the vendor's willingness to demonstrate its products, the vendor's ability to repair hardware, the vendor's ability to modify its software, the availability of vendor-offered training of IS personnel and system users, and evaluations of the vendor by independent organizations.

If new hardware or software will be purchased from a vendor, a formal request for proposal (RFP) is needed. The RFP outlines the company's needs; in response, the vendor provides a written reply. In addition to responding to the company's stated needs, the vendor provides data on its operations. This data might include the vendor's reliability and stability, the type of postsale service offered, the vendor's ability to perform repairs and fix problems, vendor training, and the vendor's reputation. Financial options to consider include purchase, lease, and rent.

RFPs from various vendors are reviewed and narrowed down to the few most likely candidates. In the final evaluation, a variety of techniques—including group consensus, cost/benefit analysis, point evaluation, and benchmark tests—can be used. In group consensus, a decision-making group is appointed and given responsibility for making the final evaluation and selection. With cost/benefit analysis, all costs and benefits of the alternatives are expressed in monetary terms. Benchmarking involves comparing computer systems operating under the same condition. Point evaluation assigns weights to evaluation factors, and each alternative is evaluated in terms of each factor and given a score from 0 to 100. After the vendor is chosen, contract negotiations can begin.

At the end of the systems design step, the final specifications are frozen and no changes are allowed, so that implementation can proceed. One of the most important steps in systems design is to develop a good contract if new computer facilities are being acquired. A final design report is developed at the end of the systems design phase.

Principle

The primary emphasis of systems implementation is to make sure that the right information is delivered to the right person in the right format at the right time.

The purpose of systems implementation is to install the system and make everything, including users, ready for its operation. Systems implementation includes hardware acquisition, software acquisition or development, user preparation, hiring and training of personnel, site and data preparation, installation, testing, start-up, and user acceptance. Hardware acquisition requires purchasing, leasing, or renting computer resources from an IS vendor. Hardware is typically obtained from a computer hardware vendor.

Software can be purchased from vendors or developed in-house—a decision termed the *make-or-buy decision*. A purchased software package usually has a lower cost, less risk regarding the features and performance, and easy installation. The amount of development effort is also less when software is purchased. Developing software can result in a system that more closely meets the business needs and has increased flexibility in terms of customization and changes. Developing software also has greater potential for providing a competitive advantage. Increasingly, companies are using service providers to acquire software, Internet access, and other IS resources.

Software development is often performed by a chief programmer team—a group of IS professionals who design, develop, and implement a software program. Programming using traditional programming languages follows a life cycle that includes investigation, analysis, design, language selection, program coding, testing and debugging, documentation, and implementation (conversion). Documentation includes technical and user documentation.

There are many tools and techniques for software development. Structured design is a philosophy of designing and developing application software. Structured programming is not a new programming language; it is a way to standardize computer programming using existing languages. The top-down approach starts with programming a main module and works down to the other modules. Other tools, such as cross-platform development and integrated development environments (IDEs), make software development easier and more thorough. CASE tools are often used to automate some of these techniques.

Fourth-generation languages (4GLs) and object-oriented languages offer another alternative to in-house development. Development using these fast and easy-to-use languages requires several steps, much like the programming life cycle. The main difference is that, with object-oriented languages, programmers must identify and select objects and integrate them into an application, instead of using step-by-step coding. Being able to reuse previously developed objects speeds software development and can improve quality when using object-oriented implementation.

Database and telecommunications software development involves acquiring the necessary databases, networks, telecommunications, and Internet facilities. Companies have a wide array of choices, including newer object-oriented database systems.

Implementation must address personnel requirements. User preparation involves readying managers, employees, and other users for the new system. New IS personnel may need to be hired, and users must be well trained in the system's functions. Preparation of the physical site of the system must be done, and any existing data to be used in the new system will require conversion to the new format. Hardware installation is done during the implementation step, as is testing. Testing includes program (unit) testing, systems testing, volume testing, integration testing, and acceptance testing.

Start-up begins with the final tested information system. When start-up is finished, the system is fully operational. There are a number of different start-up approaches. Direct conversion (also called *plunge* or *direct cutover*) involves stopping the old system and starting the new system on a given date. With the phase-in approach, sometimes called a *piecemeal approach*, components of the new system are slowly phased in while components of the old one are slowly phased out. When everyone is confident that the new system is performing as expected, the old system is completely phased out. Pilot start-up involves running the new system for one group of users rather than all users. Parallel start-up involves running both the old and new systems for a period of time. The output of the new system is compared closely with the output of the old system, and any differences are reconciled. When users are comfortable that the new system is working correctly, the old system is eliminated. Many IS vendors ask the user to sign a formal user acceptance document that releases the IS vendor from liability for problems that occur after the document is signed.

Principle

Maintenance and review add to the useful life of a system but can consume large amounts of resources. These activities can benefit from the same rigorous methods and project management techniques applied to systems development.

Systems operation is the use of a new or modified system. Systems maintenance involves checking, changing, and enhancing the system to make it more useful in obtaining user and organizational goals. Maintenance is critical for the continued smooth operation of the system. The costs of performing maintenance can well exceed the original cost of acquiring the system. Some major causes of maintenance are new requests from stakeholders and managers, enhancement requests from users, bugs or errors, technical or hardware problems, newly added equipment, changes in organizational structure, and government regulations.

Maintenance can be as simple as a program patch to correct a small problem to the more complex upgrading of software with a new release from a vendor. For older programs, the total cost of maintenance can be greater than the total cost of development. Increased emphasis on design can often reduce maintenance costs. Requests for maintenance should be documented with a request for maintenance form, a document that formally authorizes modification of programs. The development team or a specialized maintenance team may then make approved changes. Maintenance can be greatly simplified with the objected-oriented approach.

Systems review is the process of analyzing and monitoring systems to make sure that they are operating as intended. The two types of review procedures are the event-driven review and the time-driven review. An event-driven review is triggered by a problem or opportunity. A time-driven review is started after a specified amount of time.

Systems review involves measuring how well the system is supporting the mission and goals of the organization. System performance measurement monitors the system for number of errors, amount of memory and processing time required, and so on.

CHAPTER 13: SELF-ASSESSMENT TEST

Designing new systems or modifying existing ones should always be aimed at helping an organization achieve its goals.

1. _____ details system outputs, inputs, and user interfaces; specifies hardware, software, databases, telecommunications, personnel, and procedures; and shows how these components are related.
2. Determining the needed hardware and software for a new system is an example of

    a. logical design
    b. physical design
    d. interactive design
    d. object-oriented design

3. Point evaluation often results in a formal bid that is used to determine who gets a contract for a new or modified system. True or False?
4. The _____ often results in a formal bid that is used to determine who gets a contract for designing new or modifying existing systems. It specifies in detail the required resources such as hardware and software.
5. With this approach, a decision-making group is appointed and given the responsibility of making the final evaluation and selection during systems design.

    a. cost/benefit
    b. point evaluation
    c. group consensus
    d. nominal evaluation

6. Near the end of the design stage, an organization prohibits further changes in the design of the system. This is called _____.

7. In objected-oriented systems design, a sequence of events is called a *scenario* and can be diagrammed in a sequence diagram. True or False?

The primary emphasis of systems implementation is to make sure that the right information is delivered to the right person in the right format at the right time.

8. ASP is an example of an IS vendor that offers hardware and software solutions. True or False?
9. Software can be purchased from external developers or developed in house. This decision is often called the _____ decision.
10. With this type of structure, a program can branch to another part of the program, depending on certain conditions.

    a. sequence structure
    b. decision structure
    c. loop structure
    d. CASE structure

11. _____ testing involves testing the entire system of programs.
12. The phase-in approach to conversion involves running both the old system and the new system for three months or longer. True or False?

Maintenance and review add to the useful life of a system but can consume large amounts of resources. These activities can benefit from the same rigorous methods and project management techniques applied to systems development.

13. A(An) _____ is a minor change to correct a problem or make a small enhancement to a program or system.
14. Many organizations require a request for maintenance form to authorize modification of programs. True or False?
15. A systems review that is caused by a problem with an existing system is called

    a. object review
    b. structured review
    c. event-driven review
    d. critical factors review

16. Java, Visual Basic, and C++ are examples of structured programming languages. True or False?
17. Monitoring a system after it has been implemented is called _____.

KEY TERMS

acceptance testing 643
alpha testing 643
benchmark test 630
beta testing 643
chief programmer team 636
closed shops 624
cold site 622
cost/benefit analysis 629
cross-platform development 640
data preparation, or data conversion 642
decision structure 638
design report 631
deterrence controls 624
direct conversion 644
disaster recovery 621
event-driven review 651
final evaluation 628
group consensus 629
help facility 620
hot site 622
image log 623
incremental backup 623
installation 643

integrated development environments (IDEs) 640
integration testing 643
logical design 613
lookup tables 620
loop structure 638
maintenance team 648
make-or-buy decision 634
menu-driven system 620
open shops 624
parallel start-up 645
patch 648
phase-in approach 644
physical design 613
pilot start-up 644
point evaluation system 630
preliminary evaluation 628
programming life cycle 637
release 648
request for maintenance form 648
request for proposal (RFP) 626
restart procedures 620
selective backup 623
sequence structure 638

sign-on procedure 619
site preparation 642
slipstream upgrade 648
software interface 635
start-up 644
structured walkthrough 640
system performance measurement 652
system performance products 652
system testing 643
systems controls 624
systems design 613
systems implementation 632
systems maintenance 646
systems operation 645
systems review 650
technical documentation 637
time-driven review 651
top-down approach 638
unit testing 643
user acceptance document 645
user documentation 637
user preparation 641
version 648
volume testing 643

REVIEW QUESTIONS

1. What is the purpose of systems design?
2. What are procedure requirements?
3. What is interactive processing? What design factors should be taken into account for this type of processing?
4. What tools can be used to develop Web-based applications?
5. How can the object-oriented approach be used during systems design?
6. What are some of the special design considerations that should be taken into account during both the logical and physical design?

7. What are the different types of software and database backup? Describe the procedure you use to backup your homework files.
8. Identify specific controls that are used to maintain input integrity and security.
9. What is an RFP? What is typically included in one? How is it used?
10. What activities go on during the user preparation phase of systems implementation?
11. What is systems operation?
12. What are the major steps of systems implementation?

13. What are some tools and techniques for software development?
14. Give three examples of an IS vendor.
15. Explain the three types of structures allowed in structured programming.
16. What are the steps involved in testing the information system?
17. What are some of the reasons for program maintenance? Explain the types of maintenance.
18. Describe the point evaluation system for selection of the best system alternative.
19. How is systems performance measurement related to the systems review?

DISCUSSION QUESTIONS

1. Describe the participants in the systems design stage. How do these participants compare with the participants of systems investigation?
2. Assume that you are the owner of a company that is about to start marketing and selling bicycles over the Internet. Describe your top three objectives in developing a new Web site for this systems development project.
3. Assume that you want to start a new video rental business for students at your college or university. Go through logical design for a new information system to help you keep track of the videos in your inventory.
4. Assume that you are the owner of an online stock-trading company. Describe how you could design the trading system to recover from a disaster.
5. Identify some of the advantages and disadvantages of purchasing versus leasing hardware.
6. Discuss the relationship between maintenance and system design.
7. Is it equally important for all systems to have a disaster recovery plan? Why, or why not?
8. Several approaches were discussed to evaluate a number of systems acquisition alternatives. No one approach is always the best. How would you decide which approach to use for

evaluation when selecting a new personal computer and printer?
9. What are the advantages and disadvantages of the object-oriented approach to systems implementation?
10. Assume that you are starting an Internet site to sell clothing. Describe how you would design the interactive processing system for this site. Draw a diagram showing the home Web page for the site. Describe the important features of this home page.
11. Identify the various forms of testing. Why are there so many different types of tests?
12. What is the goal of conducting a systems review? What factors need to be considered during systems review?
13. What features and terms would you insist on in a software package contract?
14. What issues might you expect to arise if you initiate the use of a request for maintenance form when none had been required previously? How would you deal with these issues?
15. Assume that you have a personal computer that is several years old. Describe the steps you would use to perform systems review to determine whether you should acquire a new PC.

PROBLEM-SOLVING EXERCISES

1. You have been hired to develop a new computer system for a video rental business using the object-oriented approach. Using the information presented in this chapter and in Chapter 12, describe in a report the approach you would use. Using a graphics program, use the object-oriented approach to document the parts of the new computer system.
2. A project team has estimated the costs associated with the development and maintenance of a new system. One approach requires a more complete design and will result in a slightly higher design and implementation cost

but a lower maintenance cost over the life of the system. The second approach cuts the design effort, saving some dollars but with a likely increase in maintenance cost.

a. Enter the following data in the spreadsheet. Print the result.

The Benefits of Good Design

	Good Design	Poor Design
Design Costs	$14,000	$10,000
Implementation Cost	$42,000	$35,000
Annual Maintenance Cost	$32,000	$40,000

b. Create a stacked bar graph that shows the total cost, including the design, implementation, and maintenance costs. Be sure that the chart has a title and that the costs are labeled on the chart.

c. Use your word processing software to write a paragraph that recommends an approach to take and why.

3. You are considering purchasing a new PC. Using a database program, create a table titled "PC" that includes columns on all the important costs, including all hardware, software, Internet, printer and ink, and other costs. The primary key should be an order number. There should be a separate row for each vendor. Your database should have at least four vendors. Create a second table titled "VENDOR." This table should include the order number, sales representative, vendor name, vendor phone number, vendor address, and related information for each vendor. Using the database program, create a report that selects the PC system that minimizes total costs. The report should include all the information about the new computer in the PC table and the sales representative, phone number, and vendor name from the VENDOR table.

TEAM ACTIVITIES

1. Assume that your project team has been working for three months to complete the systems design of a new Web-based customer ordering system. Two possible options seem to meet all users' needs. The project team must make a final decision on which option to implement. The table that follows summarizes some of the key facts about each option.

   a. What process would you follow to make this important decision?
   b. Who needs to be involved?
   c. What additional questions need to be answered to make a good decision?
   d. Based on the data, which option would you recommend and why?
   e. How would you account for project risk in your decision making?

Factor	Option #1 (Millions)	Option #2 (Millions)
Annual gross savings	$1.5	$3.0
Total development cost	$1.5	$2.2
Annual operating cost	$0.5	$1.0
Time required to implement	9 months	15 months
Risk associated with project (expressed in probabilities)		
Benefits will be 50% less than expected	20%	35%
Cost will be 50% greater than expected	25%	30%
Organization will not/cannot make changes necessary for system to operate as expected	20%	25%
Does system meet all mandatory requirements?	Yes	Yes

2. Assume that your team works for a medium-sized company that trades treasury bonds in New York City. Your firm has 500 employees in a downtown location. The firm has a local area network that is tied into a global trading network with other firms. What specific recommendations would you and your team members make to the president of the company to allow your firm to recover from a potential disaster or terrorist attack? Make sure to include a detailed description of the backup procedures you would recommend. Prepare your presentation for the president and give your pitch. Document your main points using your word processing program for submission to your instructor.

3. Your team has been asked to purchase and install a network system that includes five PCs, two printers, and a wireless network for a small business. Develop an RFP that is to be sent to four PC vendors that specifies all the equipment and software that is needed.

WEB EXERCISES

1. Use the Internet to find two different companies that have recently implemented a new information system. Describe the specific steps the companies used. You may be asked to develop a report or send an e-mail message to your instructor about what you found.

2. Using the Web, search for information on structured design and programming. Also search the Web for information about the objected-oriented approach to systems design and implementation. Write a report on what you found. Under what conditions would you use these approaches to systems development and implementation?

CAREER EXERCISES

1. Describe what type of information system you would need in your chosen job. Your description should include logical and physical design. What specific steps would you include to be able to recover from a natural or man-made disaster, such as a terrorist attack?

2. What specific tasks would you perform to implement your new system described in the previous career exercise?

VIDEO QUESTIONS

Watch the video clip **Making Flowcharts Using Office XP** and answer these questions:

1. What types of organizational structures and flowcharts can be represented using Microsoft Word's Diagram and Organizational Chart tool?

2. Which of the chart styles discussed might prove useful in systems design?

CASE STUDIES

Case One

MetLife Selects the Best Technologies Around the Globe

The best ideas in the United States don't always translate well to other countries. Take, for example, Coors's marketing campaign for its beer that centered around the slogan "Turn it loose," which translated into Spanish as "Suffer from diarrhea." Equally disastrous results can occur when companies attempt to roll out a common technology infrastructure throughout a global organization.

Steve Bozzo was recently named international CIO at MetLife Incorporated, a new position that he says is a testament to MetLife's commitment to its global strategy. MetLife is a giant in the insurance industry, with offices in 11 countries. MetLife currently has a customer base of 8 million worldwide, which it hopes to build to 30 million by 2010. To provide an effective information infrastructure to all of its global offices, MetLife has taken a unique and revolutionary approach to its implementation.

Most global operations don't consider the effects that building systems at headquarters and forcing those systems onto their worldwide offices will have on their overseas operations. The reason this approach fails is because local infrastructures differ widely in countries. Countries favor different technologies, use different telecommunications companies, and implement different software and hardware options. Some countries favor non-U.S. products and are more familiar with their homegrown products.

The revolutionary part of MetLife's approach is in its method of selecting technology. MetLife has taken a truly global perspective, with equal deference to all cultures by forming an international team to analyze the needs of its worldwide offices. The team studies the unique qualities of each of MetLife's locations and designs a system that comes closest to meeting everyone's needs. Once the system is defined, the team selects the best technologies from vendors around the world and deploys those technologies as the global standard. By using this approach, MetLife has avoided the most common technology problems of global organizations: disparate systems, high complexity, and U.S.-centric thinking.

At the center of MetLife's global platform is its common knowledge database, which was developed in Brazil. The common platform is modular, so MetLife can remove outdated systems or systems that don't meet a particular country's needs and plug in new ones. MetLife has been implementing its new system one country at a time, beginning with Brazil, then moving to India, Korea, and Mexico. The system has enough flexibility to accommodate different languages, regulations, and currencies.

MetLife seems to be heading down the right path, says Prashant Palvia, president of the Global Information Technology Management Association, a research group in Greensboro, North Carolina. He praised MetLife for allowing for 10 to 20 percent customization of the global platform in different countries, as well as its decision to implement the system one country at a time. Palvia says MetLife was wise in choosing countries where the pilot's success rates would be high and in taking extra time to get the infrastructure right in those countries so that it could duplicate those successes around the world.

Throughout the global implementation, MetLife's international team holds a weekly teleconference. For team members on the U.S. East Coast, the meeting is at 8:00 P.M., which is 7:00 P.M. in Mexico City, 9:00 A.M. in Seoul, and 5:30 A.M. in Bangalore. "If we didn't do that, everything would be lost," says George Savarese, vice president of operations and technology. "When you work in international, it's literally not 9 to 5. We kind of work 12-, 13-hour days just to make sure we're communicating." Communication is the toughest part of the project, Bozzo and Savarese agree. They're learning Spanish, and foreign team members are learning English, but no one expects to master eight languages.

Travel time is another issue. Without frequent face-to-face visits, says Savarese, international projects are doomed to failure. "You really need to go there, to really be there, in their space, understanding where it's at," he advises.

Training is another important consideration that can slow implementation. MetLife has only a small team of trainers to teach the IT departments from different offices the nuances of the global platform. Since they want to get the big countries in place first, the training waiting list keeps growing. "The smaller countries are dying to get this technology, and we have to say wait," says Savarese.

The rewards of the move to a global platform are expected to be monumental. Part of the global plan is to consolidate MetLife's IT infrastructure by reducing its 300 to 400 servers to about 10 to 20 worldwide, which will yield big cost savings, Bozzo says. "And it simplifies the environment so much," he adds. "That's the key."

MetLife is also getting deep discounts on maintenance and license agreements because of its high-volume purchases. And since one team is working for 11 countries, the business can grow without simultaneously increasing the head count, according to Savarese. He says the international project's official return on investment of 26 percent is extremely conservative.

Discussion Questions

1. What challenges did MetLife face in deploying a global platform for its information systems, and how did they resolve them?
2. What are the benefits that MetLife will enjoy once all of its global offices are connected to a common global platform?

Critical Thinking Questions

3. Why might a U.S.-centric approach to technology infrastructure be problematic in other countries?
4. How can upgrading to a new global information platform assist MetLife in achieving its goal of more than tripling its customer base over the next five years?

SOURCES: Melissa Solomon, "Global Patchwork," *Computerworld*, September 22, 2003, *www.computerworld.com*; Anthony O'Donnell, "MetLife's Bozzo Takes Show on the Road," *Insurance and Technology Online*, October 29, 2003, *www.insurancetech.com*; MetLife's Web site, accessed June 20, 2004, *www.metlife.com*.

Case Two

The Hudson River Park Trust Turns to Construction ASP

The Hudson River Park Trust (HRPT) was founded in 1998 by the state of New York to create a five-mile stretch of parkland along Manhattan's West Side. One of the first challenges of the HRPT's staff was to decide how they were going to share information among the hundreds of contractors and suppliers involved in the restoration projects. It was too expensive for the organization to purchase, install, and maintain systems to store the scores of space-intensive project files, such as computer-aided design drawings and blueprints.

With an IT staff of just four people, HRPT decided that the best approach would be to find outside help. "With so many contractors involved, it would have been a nightmare to manage all of this in-house," said Michael Breen, CIO for the organization. "We've saved hundreds of thousands of dollars

in cost avoidance in terms of staffing, T1 lines, WANs, LANs, and servers."

The company that saved the HRPT so much money is an applications service provider (ASP) by the name of Constructware. An ASP is a company that "leases" software services to customers, typically over the Internet. Constructware allows the trust and the hundreds of contractors and suppliers it works with to share over the Internet 37,000 blueprints, diagrams, and other documents for various phases of construction. The ASP includes a customized portal that features a personal organizer and tools for project reporting, business development, bid management, project management, file and document management, cost management, and human resources. Constructware covers all the requirements for construction projects.

Under its $65,000 annual licensing agreement, HRPT can have an unlimited number of people access the system. All Breen does is set up a user account for each contractor and ensures that each user receives one to two hours of training using WebEx software to access the Constructware training modules. Training is done in HRPT offices or at contractors' offices. Contractors can access the system via 56K modems or DSL connections that the trust has established in construction trailers along the waterfront. Currently, 250 people are using the system.

The technology arrangement at HRPT reflects a trend among government agencies to rely more heavily on outsourcing as IT staffs become leaner, said Jim Krouse, an analyst at Reston, Virginia–based Input Inc. HRPT is now using the software to manage 165 concurrent projects, including Segment 7, the stretch of land from 46th Street to 59th Street that is known as Clinton Cove. Project work for that section, scheduled to be completed in spring 2005, "is on time and on budget, and not a lot of government organizations can say that," said Breen.

Last year, the trust used the Constructware system to manage 26 projects that were completed in Greenwich Village, said Breen. Those projects included the re-creation of a bulkhead, as well as electrical and plumbing jobs.

The trust plans to use the Constructware system to manage other phases of the park's construction through 2009, including a reconstruction of Pier 84 near the Intrepid Sea Air Space Museum. When the pier is complete, it will include a public boathouse, community garden, classroom space, and an educational/play area. A fountain, a concession facility, and a large meadow will also be included.

"I couldn't imagine managing 165 projects when I first started," said Breen. Constructware "has saved us a lot of money." He estimates that they have realized a 500 percent annualized return on investment.

Discussion Questions

1. When considering software options, what attributes of a project might lead a project leader to choose an ASP such as Constructware rather than developing something from scratch or hiring an outside company to develop a custom system?
2. What benefits has HRPT's staff experienced by going with an ASP rather than developing and maintaining the system themselves?

Critical Thinking Questions

3. Why might a large construction company prefer to build and maintain its own systems rather than using an ASP such as Constructware?
4. What do you think are the responsibilities of the four-person IT staff of HRPT since its systems are fully outsourced?

SOURCES: Thomas Hoffman, "ASP Speeds Project Management for NYC Parks Developer," *Computerworld*, June 17, 2004, *www.computerworld.com*; Constructware Web site, accessed June 21, 2004, *www.constructware.com*; The Hudson River Park Web site, accessed June 21, 2004, *www.hudsonriverpark.org*.

Case Three

Haworth Upgrades Supply Chain Systems

In the late 1990s when the U.S. economy was booming, new businesses were being established rapidly, and existing businesses were growing. Business was never better for office furnishing companies such as Haworth Inc. The demand for new office space, plus several acquisitions, propelled the 52-year-old company to $2 billion in sales in 2002, making it the world's second-largest designer and manufacturer of office furniture.

As the tide turned later the same year, the Holland, Michigan, company took a double hit. Sales stalled as companies ceased expanding, and a glut of used furniture hit the market as dot-com companies folded. Sales at Haworth plunged 40 percent. Michael Moon, vice president for global information services at Haworth, was given the mandate to cut costs in the company's supply chain systems—the warehousing and transporting of merchandise. (Supply chain management was discussed in Chapter 2.)

Haworth spent two years and $14 million overhauling its supply chain systems and achieved unprecedented success. It began by analyzing current systems to see where they were falling short. The analysis didn't take long. The existing system was mostly manual, aided by spreadsheets and driven by "tribal knowledge," Moon says. "Tribal knowledge told you that if you are going to Texas, you don't want to stop in Alabama on the way, because it's a little detour."

Moon needed to optimize the routing and warehousing of Haworth merchandise in order to store and transport merchandise economically. Because Haworth had a small IT group, Moon started looking for outside help.

After taking bids and looking at solutions from several companies, Moon found two systems that would interact to

provide the optimization he desired: a transportation management system (TMS) and a warehouse management system (WMS) that together could manage the flow of Haworth products from manufacture through delivery to the customer. Finding the perfect solution wasn't easy. "It's a tricky environment for a number of reasons," Moon says. "For example, a shipment might have to meet a strict 15-minute delivery window on a dock in New York. Moon provides another example. "You take a standard office like this," waving his arm around his modular office. "You've got the walls, the desk, the overhead files, and so forth. All these may come from different manufacturing sites, all coming together at a distribution center and then to the customer site in a sequence that allows them to install it. And maybe the customer wants to install his furniture over the weekend. You can't have missing parts off the truck, or he may not be able to move in on Monday."

The TMS consists of an optimization package called NetWorks Transport and a carrier communication module called NetWorks Carrier from Manugistics Group, Inc., in Rockville, Maryland. The system looks at customer orders, factory schedules, carrier rates and availability, and shipping costs and produces optimum, lowest-cost delivery plans. Plans are produced daily and updated every 15 minutes. The system also has an automated interface that lets Haworth negotiate deliveries with its carriers. "The goal for the TMS is to optimize deliveries from the standpoint of freight cost," says Moon. "That requires mapping out more efficient routes, minimizing 'less-than-truckload' shipments and reducing damage to goods."

The WMS tracks and controls the flow of finished goods from the receiving dock at any of Haworth's three distribution centers to the customer's site. Acting on shipping plans from the TMS, the WMS directs the movement of goods based on real-time conditions of space, equipment, inventory, and personnel. The goal of the WMS is to reduce labor costs in the warehouse by using several methods, including "cross-docking," which lets goods earmarked for a specific customer move directly from the receiving dock to the shipping dock without being checked into the system and then picked from inventory.

The TMS implementation went smoothly, partly because the system did not need customization and partly because TMS vendor had two full-time consultants on-site at Haworth for a year. The implementation of the WMS was a bit more problematic. "In the case of WMS, there were some performance concerns, and our database administrators worked closely with the vendor to help them understand where some of the flaws were," commented Brian Kovatch, manager of IS applications design. Haworth's staff helped the vendor fine-tune some SQL statements that were inappropriately coded, he says, adding that "the vendor wasn't accustomed to our volumes and need for response time." The WMS processes some 17,500 transactions per day.

Jim Rohrer, a business applications process manager and the key liaison between the IT department and the supply operations at Haworth, says the new systems haven't just optimized business processes; they have fundamentally transformed them. "The distribution centers were accustomed to getting information on labels or on screens, then deciding what to do with it and then reporting back what they did. I call that a 'signpost' system," he says. Now it's more of a "directed" system, Rohrer explains. "TMS sets up a plan and feeds it to WMS, and WMS says, 'Here's what's to be done; here's your task list.' It greatly reduces the amount of time it takes a new employee to get productive," he adds.

Moon says Haworth was counting on a 12 percent reduction in freight costs from TMS but is actually seeing 16 percent. The system paid back the initial investment in 9 months—15 months ahead of schedule. Partial-truckload shipments, which are inefficient and often lead to damaged goods, have been reduced by half. And the labor-cost savings from the WMS is "significantly beyond" the 10 to 12 percent goal.

Discussion Questions

1. What business benefits has Haworth gained by delivering products more efficiently?
2. Why might it be better for companies to purchase supply chain software from an outside company rather than develop it themselves?

Critical Thinking Questions

3. Why do you think computers are better at optimizing the details of delivering products than people are?
4. Why is it that financial hardship seems to be a high motivating factor in IS review?

SOURCES: Gary H. Anthes, "Refurnishing the Supply Chain," *Computerworld*, June 7, 2004, *www.computerworld.com*; Haworth Web site, accessed June 21, 2004, *www.haworth.com*; Manugistics Web site, accessed June 21, 2004, *www.manugistics.com*.

NOTES

Sources for the opening vignette: "Digital Depot," *Chain Store Age*, January 2004, *www.sas.com/industry/retail/rtq_0104.pdf*; The Home Depot Web site, accessed June 20, 2004, *www.homedepot.com*; "Home Depot to Upgrade Stores; Raises Predictions for 2003," *The Associated Press*, January 16, 2004, *www.lexis-nexis.com*.

1. Clark, Don, "The Affluent Avatar," *The Wall Street Journal*, January 8, 2003, p. B1.
2. Weiss, Todd, "Linux Networx to Build Cluster for Weapons Research," *Computerworld*, February 23, 2004, p. 12.
3. Merian, Lucas, "Global Standard Will Force Changes in Banks' IT," *Computerworld*, September 1, 2003, p. 11.
4. Brewin, Bob, "FDA Bar-Code Rule Provides Impetus for More WLANS," *Computerworld*, March 1, 2004, p. 14.
5. Hoffman, Thomas, "IT Execs Push New Governance Models," *Computerworld*, February 23, 2004, p. 5.
6. Gomes, Lee, "More Scary Tales Involving Big Holes in Web Site Security," *The Wall Street Journal*, February 2, 2004, p. B1.
7. Metz, Cade, "Total Security," *PC Magazine*, October 1, 2003, p. 83.
8. Guth, Robert, "Welter of Viruses Is a Wake-Up Call for Software Industry," *The Wall Street Journal*, August 26, 2003, p. B1.
9. Fong, Mei, "It Takes a Thief," *The Wall Street Journal*, January 12, 2004, p. R5.
10. Bulkeley, William, "In About-Face Siebel to Deliver Software on Net," *The Wall Street Journal*, October 2, 2003, p. B1.
11. Burrows, Peter, "Pixar's Unsung Hero," *Business Week*, June 30, 2003, p. 68.
12. Hoffman, Thomas, "Software Automates Content Tracking and Management," *Computerworld*, February 9, 2004, p. 16.
13. Hoffman, Thomas, "Big Companies Turn to Packaged Sarb-Ox Apps," *Computerworld*, March 1, 2004, p. 6.
14. Verton, Dan, "District of Columbia Melds Budgeting for 68 Agencies," *Computerworld*, January 19, 2004, p. 7.
15. Delaney, Kevin, "Outsourcing Jobs," *The Wall Street Journal*, October 13, 2003, p. B1.
16. Brewin, Bob, et al., "Earthquake Law Pushes Hospitals to Spend Big on IT," *Computerworld*, February 16, 2004, p. 1.
17. Mearian, Lucas, "IT Managers Keep Systems Running Despite Isabel's Fury," *Computerworld*, September 22, 2003, p. 10.
18. Mearian, Lucas, "IT Leads Recovery After Regional Power Failure," *Computerworld*, August 18, 2003, p. 1.
19. Disaster Recovery Services at the Web site of Business Records Management located at *www.businessrecords.com*, accessed on July 31, 2003.
20. Disaster Recovery at the Guardian Computer Support Web site at *www.guardiancomputer.com*, accessed on July 31, 2003.
21. Melymuka, Kathleen, "How Will You Manage Your Vendors," *Computerworld*, January 6, 2003, p. 32.
22. Mearian, Lucas, "Fidelity National Revamps IT with Single-Vendor Track," *Computerworld*, March 8, 2004, p. 18.
23. Brewin, Bob, "UPS Invests $30 Million in IT to Speed Package Delivery," *Computerworld*, September 29, 2003, p. 14.
24. Murphy, Victoria, "Control Freak," *Forbes*, March 29, 2004, p. 79.
25. McWilliams, Gary, "Pay As You Go," *The Wall Street Journal*, March 31, 2003, p. R8.

26. Hoffman, Thomas, "HP Takes New Pricing Path," *Computerworld*, May 26, 2003, p. 1.
27. Elgin, Ben, "The Information Age's Toxic Garbage," *Business Week*, October 6, 2003, p. 54.
28. Staff, "Alaska Airlines Selects ITA Software," *Telecomworldwire*, January 22, 2004.
29. Songini, Marc, "Coke, SAP Co-develop Bottling App," *Computerworld*, February 23, 2004, p. 8.
30. Clark, Don, "Renting Software Online," *The Wall Street Journal*, June 3, 2003, p. B1.
31. Baker, Stephen, et al., "Special Report on Software," *Business Week*, March 1, 2004, p. 84.
32. Thibodeau, Patrick, "Automated Warehouse Reduces Errors," *Computerworld*, February 24, 2003, *www.computerworld.com*.
33. Badenhorst, Colin, et al., "Operating Strategies," *Engineering and Mining Journal*, March 1, 2003, p. 3.
34. Rosenberg, Barry, "Modeling and Simulation Win NASA's Software of the Year," *Aviation Week & Space Technology*, January 13, 2003, p. 426
35. Baker, Stephen, et al., "Special Report on Software," *Business Week*, March 1, 2004, p. 84.
36. Anthes, Gary, "Code Reuse Gets Easier," *Computerworld*, July 28, 2003, p. 24.
37. King, Julia, "AXA Financial's Blueprint for Software Reuse," *Computerworld*, February 9, 2004, p. 24.
38. Dragan, Richard, "Code for the Road," *PC Magazine*, May 18, 2004, p. 124.
39. Petreley, Nicholas, "Sun Refuses to Be Eclipsed," *Computerworld*, February 9, 2004, p. 36.
40. Mearian, Lucas, "Nasdaq's CIO Looks to Streamline Systems," *Computerworld*, June 2, 2003, p. 19.
41. Weiss, Todd, "New CIO Takes Reins of IRS Tech Upgrade," *Computerworld*, June 2, 2003, p. 14.
42. Dunn, Darrell, "There's No Room for Error," *InformationWeek*, February 2, 2004, p. 34.
43. Hoffman, Thomas, "Mohegan Sun Puts Its Chips on Customer Retention Technology," *Computerworld*, August 18, 2003, p. 16.
44. Ferguson, Renee, "Server Merges Insurance Data," *eWeek*, January 19, 2004, p. 30.
45. Saran, Cliff, "Insurer Reduces Risk in Rolling Out Software," *Computer Weekly*, January 20, 2004, p. 18.
46. Staff, "Handheld Computers, GPS Help Growers See What's Coming," *The Corn and Soybean Digest*, May 8, 2003.
47. Songini, Marc, "Best In Class: Re-Engineering Drives Down Cost of Power," *Computerworld*, February 24, 2003, p. 46.
48. Brewin, Bob, "Carnival Cruise Lines Piloting Remote Management of PCs," *Computerworld*, August 18, 2003, p. 7.
49. Scheier, Robert, "Surviving Software Upgrades," *Computerworld*, May 26, 2003, p. 44.
50. Ganek, A.G., et al., "The Dawning of the Autonomic Computing Era," *IBM Systems Journal*, Volume 42, 2003, p. 5.
51. Purushothaman, D., et al., "Branch and Price Methods for Prescribing Profitable Upgrades of High Technology Products with Stochastic Demand," *Decision Sciences*, Winter 2004, p. 55.
52. Staff, "Hymans Selects CMS, Net Software," *Pensions Week*, January 26, 2004.

Brandon Trust Develops MIS Capability for Improved Operations and Services

Andy Igonor
University of the West of England

Organizations typically exist to make money and to save money. But what about those organizations registered as charities, whose goals are something other than profit making? Brandon Trust, a major player in the healthcare sector, with an annual turnover exceeding £20 million, is a registered charity in England. It does not exist to make money; rather, in the words of its chief executive officer, Steve Bennett, Brandon Trust is a "people supporting people" organization. As a visionary, Brandon Trust's CEO believes that making a difference is essential to sustaining the organization's aim and purpose.

In the quest to improve service operations, Steve Bennett met with Charles Harvey, a business strategist and expert in leadership and management practices, who also currently serves as dean of the UK's Bristol Business School. The decision to invest in information systems followed from this preliminary study of the Brandon Trust's strategy and business systems. The study revealed a lack of consistency and some redundancy in existing processes stemming from a lack of integration. Brandon Trust's situation is similar to other enterprises and was of both theoretical and practical interest to researchers at the Bristol Business School who were studying business systems and innovation. The outcome of this consultation led to a partnership agreement to help in the strategic streamlining of service operations and delivery. Sponsored by the Knowledge Transfer Partnership (KTP)—a government-funded program—the project involves the automation, rationalization, and reengineering of part of Brandon Trust's activities and procedures, with the goal of eliminating bottlenecks in information production, distribution, and use.

Brandon Trust operates community teams, employment and training units, and daycare centers spread across various areas of Bristol and the South West region of England (of which Bristol is a part). Brandon Trust also provides intensive support and supported living so that people with disabilities can live in a place of their preference. In meeting its obligations, the Trust runs a coordinated operational network, where people, both office and field staff, have to effectively communicate in meeting clients' needs. Apart from its full-time employees, the Trust also recruits part-time, hourly workers to facilitate the efficient flow of services. Brandon Trust exists in a large market for learning disability care in which purchasers demand greater efficiency and effectiveness. Brandon Trust is a market leader in providing learning disability services, and it has grown by taking over contracts. Its service area is mainly confined to Bristol, South Gloucestershire, Bath, Northeast Somerset, and North Somerset. Its competitive advantage lies in service innovation, user-centered provision, and quality management. Its main competitors are local authorities and other charitable trusts, which are typically quite small and less developed in terms of management and systems. The Trust is now trying to improve the quality of its own management systems by using innovative approaches to maximize its effectiveness.

Brandon Trust's Finance and Business Systems Director, Hilary Pearce, believes that information systems can, if efficiently applied and deployed, be used to improve facets of the organization's business. In her words, "I believe that by working with our academic partner and our KTP Associate that the Brandon Trust will derive real benefit from the project. The pooling of ideas and expertise will be vital as we analyse our information needs and data sources and identify key result areas that we need to focus on to improve the management of the Trust. I envisage an innovative and integrated business system solution that will continue to evolve and grow along with the Trust's operational needs." The

role of information systems in strengthening the Trust's service level cannot be overemphasized. Its chief executive officer, Steve Bennett, had this to say: "As an organization that has grown 500% in the ten years of its existence, it is important [for Brandon Trust] to take stock and look for continual improvements in our performance. Information and the systems that provide it are fundamental to performance management, and the importance of having systems that utilize available technology in a pragmatic and useable way cannot be overstated. I therefore am excited to address this with our academic partners and KTP associates in a two-year project, the outcome of which will be an integrated business system solution that will continue to evolve and grow with the continuing development of Brandon Trust."

The Trust's services are distributed via 50 sites, and it is essential to improve information flows for efficiency and the management of risk. This development project was therefore focused on implementing new management systems and integrating them with other core business systems, including finance, following a detailed analysis of information requirements. Lack of systems integration currently makes provision of sound information for management controls, budgeting, and costing of contract proposals difficult to achieve. Improving information and business systems across the Trust will enhance its capacity for further growth by ensuring the best possible quality and design in its services. It is anticipated that the program will significantly reduce costs through business process reengineering, resulting in savings in the region of £250,000 per annum.

One KTP associate, Abid Mohammed, was recruited to serve as Industrial Manager and to work with Andy Igonor (Academic Advisor and Project Manager) and Hilary Pearce (Director of Finance and Business Systems). Abid's initial task included a review and evaluation of all existing business systems, particularly computer systems at the Trust. While a number of systems existed in the Trust, the systems did not to talk to one another; notably, the crucial financial system needed to manage all finance-related issues from payroll to payment for contract services. The hoped-for result of the development project is a system with the following capabilities:

- An effective system capable of delivering relevant information when needed
- An integrated system with simplified information presentation and elimination of redundancy
- The development of an IS/IT audit report indicating current operating resources, which also serves as guide for future work, including identification of appropriate software and hardware for improved operations, improved access to information, improved management information systems reporting, and improved performance

Discussion Questions

1. What are the key issues to be looked into during systems investigation? How might they affect the overall success of the project?
2. Brandon Trust does not exist to make money. How relevant will this systems development project be to the organization?
3. Why is it necessary for Brandon Trust's systems to be integrated? Of what benefit would this be to the organization?

Critical Thinking Questions

4. In what ways can a management information system benefit an organization?
5. How often should reports be generated within such systems in order to fulfill managerial needs?

Note: All information provided in this case is courtesy of the Brandon Trust. Used with permission.

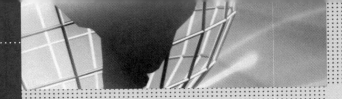

Strategic Enterprise Management at International Manufacturing Corporation (IMC)

Bernd Heesen
University of Applied Sciences Furtwangen, Germany

The International Manufacturing Corporation (IMC), with 3,000 employees worldwide, is a supplier of automotive parts and is headquartered in the United States. Given the increasing market pressure and the need to operate internationally, IMC has recently made three acquisitions. The acquisition of one company in Great Britain and two companies in Germany is expected to leverage the companies' complementary manufacturing and sales capabilities to gain a stronger market presence in North America and Europe. The subsidiaries are still managed by their former management teams, who are now reporting to the U.S. headquarters. Each of the subsidiaries still operates its own management information system because the system fulfilled the requirements in the past and at the time of the acquisition there was no visible benefit to change what was working. A couple of interfaces were created to facilitate the monthly financial reporting to the headquarters.

Recently, the CIO of IMC realized some disadvantages of the diversity of hardware and software platforms. When one IT expert from one German subsidiary terminated her employment, no other IT expert within IMC had detailed knowledge of the local system, which threatened regular operations in that plant. The company needed to seek external consulting support to maintain the system until a new IT expert could be hired. This dependency on individual experts in each location was expensive and caused problems during periods when the experts were on leave. In addition, the interfaces between the subsidiaries' systems and the headquarters' system that were created right after the acquisition had to be modified whenever a change or upgrade was made to one of the subsidiary systems. Even minor changes, such as the reorganization of product codes or sales organizations, required reprogramming, maintenance of conversion tables, and subsequent testing of the related business processes in the system. The IT departments worked overtime and still did not find the time to invest in new initiatives such as enabling mobile computing for the sales force and developing a strategic enterprise management system that would allow consolidated planning and monitoring of all key performance indicators for the organization, a project long requested by the CEO. The two German subsidiaries had recently upgraded their systems, and problems identified during the testing of the interfaces could not be corrected in time for the submission of the monthly reports to headquarters. The data was finally corrected, but the monthly reporting was several days late. This was not the first time that the monthly financial reporting had been delayed or that the data needed to be corrected because of system-related problems, which caused other problems at headquarters.

The CEO finally requested consolidation of the systems in the coming year to allow for an integrated strategic enterprise management system that would provide current information from all legal entities and support a consolidated budgeting and planning process. He compared the company's current system to the cockpit of an airplane and stated that "no one would expect a pilot to fly an airplane with malfunctioning instruments; hence nobody can expect management to run a company based on incomplete or incorrect information." The CIO was asked to develop a business case for the implementation of such a system, considering the savings from personnel, hardware, and software, as well as process improvements.

After IMC's board approved the business case to implement enterprise resource planning (ERP) software from SAP, the CIO established a project in January 1999. The

project schedule called for completion of the financial, human resource, supply chain management, and customer relationship management functions for the complete organization—called the Big Bang—in January 2000. A prototype of the application was planned to be ready by July 1999 so that the departments could test all functionality with a set of converted data extracted from the current systems. The plan also called for complete conversion of live data by the end of November 1999 so that in December both the current system and the new system could be used in parallel before making the decision to switch systems. Consultants from SAP were hired to support the implementation—for project management and configuration of the system—and an independent training organization was charged with planning the end-user training and developing computer-based training (CBT) programs for ongoing use. The project plan allowed for hiring temporary staff to help with the cutover and redundant data entry in December 1999.

The project started with a kickoff meeting in February, after which nearly all team members began their project work. The complete prototype of the application could only be made available in September, since some of the team members had to fulfill their regular jobs while providing their expertise to the project. Some of the early project decisions had to be revised to accommodate the best-practice business model from IMC, once the IMC team members better understood how the SAP system worked and the consultants had gained a better understanding of how IMC wanted to operate. The CIO still wanted to maintain the timeline and called weekly meetings of the project's team leaders for finance, human resources, supply chain management, and customer relationship management. Those meetings led to the decision to delay rollout of some of the functionality and to reduce the functionality to what was really needed (eliminating nice-to-have functionality). The prototype testing produced additional feedback from the departments that needed to be incorporated in the design of the final solution prior to the conversion. Many of the department heads, who had not been part of the core project team, only then became aware of the changes and started to talk about the new software, saying that it was not working properly yet and some of the team members were scared they would not make the deadline. In November, the data conversion revealed some additional problems (e.g., with data entry errors made by the temporary staff who were keying data into the new system in instances in which an automated conversion was not cost-effective). As a result, the parallel use of the current and new system, which depended on all data being available, was delayed. To add to the pressure, the CIO believed that delaying the project's completion could be perceived as his failure. The CIO and steering committee of the project needed to make a decision on how to move forward: (1) continue with the project's completion in January 2000 as planned or (2) delay conversion to the new system until a proper parallel test could be completed.

Discussion Questions

1. How would you develop a business case to justify an investment in an integrated strategic enterprise management system?
2. What are the essential requirements for a strategic management system?

Critical Thinking Questions

3. Which tasks would you consider when developing the original project plan for the implementation?
4. Given IMC's current situation, what would you decide if you were the CIO?

Note: The name International Manufacturing Corporation was selected to protect the identity of the real company featured in the case.

Information Systems in Business and Society

Chapter 14 Security, Privacy, and Ethical Issues in Information Systems and the Internet

CHAPTER
• 14 •

Security, Privacy, and Ethical Issues in Information Systems and the Internet

PRINCIPLES	LEARNING OBJECTIVES
▪ Policies and procedures must be established to avoid computer waste and mistakes.	▪ Describe some examples of waste and mistakes in an IS environment, their causes, and possible solutions. ▪ Identify policies and procedures useful in eliminating waste and mistakes.
▪ Computer crime is a serious and rapidly growing area of concern requiring management attention.	▪ Explain the types and effects of computer crime. ▪ Identify specific measures to prevent computer crime. ▪ Discuss the principles and limits of an individual's right to privacy.
▪ Jobs, equipment, and working conditions must be designed to avoid negative health effects.	▪ List the important effects of computers on the work environment. ▪ Identify specific actions that must be taken to ensure the health and safety of employees. ▪ Outline criteria for the ethical use of information systems.

INFORMATION SYSTEMS IN THE GLOBAL ECONOMY
COMPUTER ASSISTED PASSENGER PRESCREENING
SYSTEM, UNITED STATES

Data Privacy Concerns Ground Security System

Since the terrorist attacks of 9/11, the U.S. government and the airline industry have been urgently searching for ways to identify and thwart potential attacks before they occur. Unfortunately, these initiatives give the airlines conflicting goals—their duty and desire to help prevent terrorism and their need to maintain the privacy of customer data. The Computer Assisted Passenger Prescreening System (CAPPS) was one such initiative that highlights the dilemma.

The original CAPPS system relied on the airlines' own reservation systems to check passenger information against a government-supplied watch list. Under the proposed CAPPS II system, the passenger-screening process would be turned over to the federal government. Passengers' identities would be authenticated by matching airline passenger data such as name, address, phone number, and birth date against a Transportation Security Administration (TSA) database. Passengers would then be checked against both a federal terrorism database and lists of individuals who have outstanding warrants for violent crimes. CAPPS II would ultimately assign each airline passenger a threat level. Threat-level data would be deleted for most passengers once they reached their destinations. Data for travelers deemed high risk would be retained for an unspecified length of time. Details of exactly how the TSA would decide whether a passenger should be allowed to board a plane or questioned were never revealed.

The Air Transport Association, the trade organization for the major U.S. airlines, estimated that the new system could cost the airline industry $1 billion to change their reservation systems to provide the data required by CAPPS II. In addition to costs, the air carriers faced many other thorny issues, not the least of which was maintaining the privacy of their passengers. Precautions to protect passenger privacy would have included installing private networks between the TSA and the airlines that would pass only encrypted data, requiring the data to pass through a multitier firewall before entering the TSA system, and implementing a 24-hour audit trail that documents all access to data.

The airlines would not voluntarily turn over the data needed to implement CAPPS II, however. Although the airlines' Air Transport Association supported the concept of CAPPS II, its members wanted more privacy guarantees before they supplied data for any purpose. They wanted assurances that the TSA-collected information would pertain only to aviation security, that the information would be securely stored, that it would be destroyed as soon as travel is completed, and that passengers could access their own data and correct any errors. Airlines were especially concerned with "mission creep," in which information intended for one purpose is used for another.

The Senate Governmental Affairs Committee learned in June 2004 that at least eight airlines and airline-reservation services provided passenger data to contractors building CAPPS II for the TSA. The contractors acquired information about passengers who had reserved tickets on various airlines and who booked flights through online travel Web sites. In providing this information to

the government agency, the contractors violated the Privacy Act of 1974 by not notifying the public of what type of information their screening systems would collect and how individual passengers could find out whether their data was included in the test systems. The airline industry is now facing expensive customer class action lawsuits as a result.

In July 2004, Homeland Security Secretary Tom Ridge cited data privacy issues and system-interoperability issues as the basis for stopping further work on the CAPPS II system. Just two months later, though, the Secure Flight passenger-screening system was announced. This system would compare passenger data to watch lists held in the Terrorist Screening Database in an attempt to keep suspicious passengers from boarding domestic flights.

As you read this chapter, consider the following:

- Which issue should take precedence—maintaining the physical safety of travelers or protecting their rights?
- What actions can organizations take when the government requires them to do activities that will upset or inconvenience their customers?

| Why Learn About Security, Privacy, and Ethical Issues in Information Systems and the Internet? | A wide range of "nontechnical" issues associated with the use of information systems and the Internet provide both opportunities and threats to modern organizations. The issues span the full spectrum—from preventing computer waste and mistakes, to avoiding violations of privacy, to complying with laws on collecting data about customers, to monitoring employees. If you become a member of a human resource, information system, or legal department within an organization, you will likely be charged with leading the rest of the organization in dealing with these and other issues covered in this chapter. Also, as a user of information systems and the Internet, it is in your own self-interest to become well versed on these issues as well. You need to know about the topics in this chapter to help avoid or recover from crime, fraud, privacy invasion, and other potential problems. We begin with a discussion of preventing computer waste and mistakes. |

Earlier chapters detailed the amazing benefits of computer-based information systems in business, including increased profits, superior goods and services, and higher quality of work life. Computers have become such valuable tools that today's businesspeople would have difficulty imagining work without them. Yet the information age has also brought some potential problems for workers, companies, and society in general (see Table 14.1).

Table 14.1

Social Issues in Information Systems

• Computer waste and mistakes	• Health concerns
• Computer crime	• Ethical issues
• Privacy	• Patent and copyright violations

To a large extent, this book has focused on the solutions—not the thorny issues—presented by information systems. In this chapter we discuss some of the issues as a reminder of the social and ethical considerations underlying the use of computer-based information systems. No business organization, and hence no information system, operates in a vacuum. All IS professionals, managers, and users have a responsibility to see that the potential consequences of IS use are fully considered.

Managers and users at all levels play a major role in helping organizations achieve the positive benefits of IS. These individuals must also take the lead in helping to minimize or eliminate the negative consequences of poorly designed and improperly utilized information

systems. For managers and users to have such an influence, they must be properly educated. Many of the issues presented in this chapter, for example, should cause you to think back to some of the systems design and systems control issues we have already discussed. They should also help you look forward to how these issues and your choices might affect your future IS management considerations.

COMPUTER WASTE AND MISTAKES

Computer-related waste and mistakes are major causes of computer problems, contributing as they do to unnecessarily high costs and lost profits. Computer waste involves the inappropriate use of computer technology and resources. Computer-related mistakes refer to errors, failures, and other computer problems that make computer output incorrect or not useful, which are caused mostly by human error. In this section we explore the damage that can be done as a result of computer waste and mistakes.

Computer Waste

The U.S. government is the largest single user of information systems in the world. It should come as no surprise then that it is also perhaps the largest misuser. The government is not unique in this regard—the same type of waste and misuse found in the public sector also exists in the private sector. Some companies discard old software and even complete computer systems when they still have value. Others waste corporate resources to build and maintain complex systems never used to their fullest extent. A less dramatic, yet still relevant, example of waste is the amount of company time and money employees may waste playing computer games, sending unimportant e-mail, or accessing the Internet. Junk e-mail, also called *spam*, and junk faxes also cause waste. People receive hundreds of e-mail messages and faxes advertising products and services not wanted or requested. Not only does this waste time, but it also wastes paper and computer resources. When waste is identified, it typically points to one common cause: the improper management of information systems and resources. The growth of spam has mounted—more than half of all e-mail messages are spam. Brightmail, a developer of spam filters, scanned about 70 billion e-mail messages in October 2003, and 52 percent were identified as spam.[1] Read the "Ethical and Societal Issues" special feature to find out about more about efforts to slow the spread of spam.

Computer-Related Mistakes

Despite many people's distrust, computers themselves rarely make mistakes. Even the most sophisticated hardware cannot produce meaningful output if users do not follow proper procedures. Mistakes can be caused by unclear expectations and a lack of feedback. Or a programmer might develop a program that contains errors. In other cases, a data-entry clerk might enter the wrong data. Unless errors are caught early and prevented, the speed of computers can intensify mistakes. As information technology becomes faster, more complex, and more powerful, organizations and individuals face increased risks of experiencing the results of computer-related mistakes. Take, for example, these cases from recent news.

Since the disintegration of the space shuttle Columbia in February 2003 put NASA manned space flights on hold, the Russian Soyuz capsules have been the linchpin of the international space station's supply program. A May 2003 space mission ended in a wild ride, with the American and Russian crew going some 250 miles off course due to a computer error. A computer malfunction sent the capsule's occupants on such a steep reentry trajectory that their tongues rolled back in their mouths. Indeed, the landing was so far off target that more than two gut-wrenching hours passed before the recovery team knew the men were safe.[2]

CAN-SPAM: Deterrent or Accelerant?

Every day, millions of people worldwide receive dozens of unsolicited commercial e-mails, known popularly as spam. An individual spam e-mail may be sent to a distribution list containing millions of addresses, with the sender expecting that only a tiny number of readers will respond to the offer. For some, spam represents a minor annoyance, but many of us become so overwhelmed with spam that we are forced to switch e-mail addresses. The Yankee Group estimates that some two-thirds of e-mail qualifies as spam. The Radicati Group estimates spam costs an employer of 10,000 people nearly $500,000 just for the additional e-mail servers required to handle the load.

Spammers often take advantage of Internet technologies to conceal their identities and the location of their operations. They even resort to including the e-mail addresses of innocent third parties in the reply-to addresses of their unwanted messages or simply forging e-mail headers. Known as *spoofing*, this tactic is employed by spammers to inflict the burden of bounce-back messages (generated when spam is sent to a nonworking address) on someone other than their mail provider. The result for the recipients is effectively a denial-of-service attack, as their e-mail inboxes become overwhelmed with tens of thousands of messages.

The Controlling the Assault of Non-Solicited Pornography and Marketing Act (CAN-SPAM), which went into effect in January 2004, does not outlaw spam; it simply prohibits certain practices. Outlawed practices include using false or misleading transmission information, deceptive subject headings, and automated methods of registering multiple e-mail accounts for spamming. Senders of unsolicited commercial e-mail also must include their physical address and a way for recipients to opt out of future mailings. In addition, companies must update their marketing databases to note whether a person has opted out of solicitations. Companies that rent lists from third parties can run afoul of the law if the list lacks time-date stamps to document an individual's decision to receive messages. Although many spammers use computers outside the United States to send the spam, they can still be prosecuted under the CAN-SPAM law because their spam causes damage to U.S. Internet service providers and consumers.

America Online, EarthLink, Microsoft, and Yahoo! filed civil actions under the law in February 2004 against four individuals for sending e-mail pitches for weight-loss products that don't work and that violated the CAN-SPAM Act. The accused face up to five years in prison for illegal spamming and up to an additional 20 years for mail fraud. The alleged spammers should at least be able to afford excellent legal counsel, because it is estimated that they grossed an average of $100,000 per month from August 2003 to January 2004. Scott Richter has been called the Spam King, labeled one of the most prolific spammers in the world, and been sued for spamming by both Microsoft and the New York State attorney general. His company, OptInRealBig.com, sends more than 100 million e-mail messages every day. While it is still too soon to tell, some believe that the CAN-SPAM Act may have actually increased the volume of spam. Antispam vendor Commtouch Software Ltd. estimates that only 1 percent of spam messages in January 2004 complied with CAN-SPAM. That percentage increased to 9.5 percent in May 2004. But, coupled with a continuing rise in the volume of spam in recent months, that statistic suggests that the law has increased the amount of spam by effectively defining what "legal spam" is.

Critical Thinking Questions

1. What negative impact does spam have on corporate America? Is it simply an annoyance or a serious problem?
2. Monitor each of your e-mail accounts for a week. What percentage of your e-mail messages is spam? What percentage of your spam messages appear to conform to the CAN-SPAM law?

What Would You Do?

Your legitimate small business has relied heavily on e-mail to sell fresh organic fruit and vegetables. Recently, your company's computers were taken over by a hacker and used to send spam and viruses. As a result, your company was "blacklisted" by the antispammers and now appears on a directory of alleged offenders circulated to Internet service providers, security companies, and other businesses. As a result, most of your outgoing direct marketing messages are blocked.

3. What actions can you take to reestablish the good name of your firm and get it off the blacklist?
4. What sort of action might you take to recover your firm's losses?

SOURCES: Thomas Claburn, "Spam Law Changes Game," *InformationWeek*, December 15, 2003, *www.informationweek.com*; The Associated Press, "Anti-Spam Law Goes into Force in Europe," *InformationWeek*, October 31, 2003, *www.informationweek.com*; Thomas Claburn, "War against Spam Rages On," *InformationWeek*, May 24, 2004, *www.informationweek.com*; Thomas Claburn, "U.S. Charges Four under Can-Spam Law," *InformationWeek*, April 29, 2004, *www.informationweek.com*; Thomas Claburn, "Does Can-Spam Act Lead to More Spam?" *InformationWeek*, June 3, 2004, *www.informationweek.com*; Liane Cassavoy, "Three Minutes: The So-Called Spam King Sounds Off," *PC World*, August 2004, *www.pcworldcom*.

A Japanese company, Catena Corporation, was deluged with thousands of orders for over 100 million Apple Computer Inc. eMac computers after a glitch caused the computers to be listed on an online shopping site for a price of $25.45. The company said a code number assigned to a set of five 8X DVD-R discs, which were the products it was intending to sell, was sent to Yahoo! Japan; however, a product information database matched that code with details for the eMac computer and with the price for the DVDs. The result was a listing for the computer, part number M9461J/A, at a price of $25 rather than the usual price of more than $916.[3]

California Macy's stores agreed to a $1.2 million settlement for overcharging shoppers in the city of San Diego, Los Angeles County, and three other counties. Investigators found that while all the company's scanners were 100 percent accurate, consumers were being charged from a few cents to more than $10 over the advertised and shelf prices on some clothing and household items. The problem was tracked to difficulty in synchronizing the advertised prices, the prices on the shelf, and the prices in the (checkout register) computer. No one alleged that Macy's intentionally overcharged consumers, but the store "had a real pattern of inaccuracy all across the state," according to the Los Angeles County prosecuting attorney.[4]

Flaws in IBM Corporation's DB2 database software were responsible for a chain of glitches that turned a routine hardware repair into a weeklong operational crisis for Copenhagen-based Danske Bank. The bank, Denmark's largest, was replacing a defective electrical unit in an IBM RVA (Ramac Virtual Array) disk storage system used for DB2 data when an electrical outage in the system halted operations at one of the bank's two data-processing centers. Eventually, the system was repaired and restarted. But the next day, the bank observed problems with batch runs of collected data being processed at the affected center in Ejby. The ensuing investigation turned up four DB2 flaws, which stretched the bank's recovery time to nearly a week. During part of the week, Danske Bank's currency, securities trading, and clearing operations were inoperable.[5]

Various errors in the code of Internet Explorer (IE) have given hackers a means to compromise personal computers using the Microsoft browser. Attackers can exploit the vulnerabilities to bypass a security check in IE or to download and execute a malicious file on a user's computer. These problems with Explorer are serious, and the vulnerabilities can enable hackers to place code in a machine and run it.[6]

Thousands of patients of Kaiser Permanente Health Plan Incorporated may have received the wrong medications when its computer systems suffered a power outage that could have caused labeling errors on prescriptions dispensed to 4,700 patients. Patients could have received incorrect prescription numbers, incorrect instructions about how to take the drug, or even the wrong drugs. Kaiser Permanente quickly contacted the affected patients via automated telephone calls, courier-delivered letters, and home visits. Fortunately, there were no adverse patient reactions.[7]

PREVENTING COMPUTER-RELATED WASTE AND MISTAKES

To remain profitable in a competitive environment, organizations must use all resources wisely. Preventing computer-related waste and mistakes like those just described should therefore be a goal. Today, nearly all organizations use some type of CBIS. To employ IS resources efficiently and effectively, employees and managers alike should strive to minimize waste and mistakes. Preventing waste and mistakes involves (1) establishing, (2) implementing, (3) monitoring, and (4) reviewing effective policies and procedures.

Establishing Policies and Procedures

The first step to prevent computer-related waste is to establish policies and procedures regarding efficient acquisition, use, and disposal of systems and devices. Computers permeate organizations today, and it is critical for organizations to ensure that systems are used to their

full potential. As a result, most companies have implemented stringent policies on the acquisition of computer systems and equipment, including requiring a formal justification statement before computer equipment is purchased, definition of standard computing platforms (operating system, type of computer chip, minimum amount of RAM, etc.), and the use of preferred vendors for all acquisitions.

Prevention of computer-related mistakes begins by identifying the most common types of errors, of which there are surprisingly few (see Table 14.2). To control and prevent potential problems caused by computer-related mistakes, companies have developed policies and procedures that cover the following:

- Acquisition and use of computers, with a goal of avoiding waste and mistakes
- Training programs for individuals and workgroups
- Manuals and documents on how computer systems are to be maintained and used
- Approval of certain systems and applications before they are implemented and used to ensure compatibility and cost-effectiveness
- Requirement that documentation and descriptions of certain applications be filed or submitted to a central office, including all cell formulas for spreadsheets and a description of all data elements and relationships in a database system; such standardization can ease access and use for all personnel

Once companies have planned and developed policies and procedures, they must consider how best to implement them.

Table 14.2
Types of Computer-Related Mistakes

- Data entry or capture errors
- Errors in computer programs
- Errors in handling files, including formatting a disk by mistake, copying an old file over a newer one, and deleting a file by mistake
- Mishandling of computer output
- Inadequate planning for and control of equipment malfunctions
- Inadequate planning for and control of environmental difficulties (electrical problems, humidity problems, etc.)
- Installing computing capacity inadequate for the level of activity on corporate Web sites
- Failure to provide access to the most current information by not adding new and deleting old URL links

Sometimes computer error combines with human procedural errors to lead to the loss of human life. In March 2003, a Patriot missile battery on the Kuwait border accidentally shot down a British Royal Air Force Tornado GR-4 aircraft that was returning from a mission over Iraq. Two British pilots were killed in the incident. Defense industry experts disagreed about the possibility of a software problem being solely responsible for downing a friendly aircraft. A likely scenario combined problems with the Patriot's radar with human error to result in friendly fire.[8]

Implementing Policies and Procedures

Implementing policies and procedures to minimize waste and mistakes varies according to the business conducted. Most companies develop such policies and procedures with advice from the firm's internal auditing group or its external auditing firm. The policies often focus on the implementation of source data automation and the use of data editing to ensure data accuracy and completeness, and the assignment of clear responsibility for data accuracy within each information system. Table 14.3 lists some useful policies to minimize waste and mistakes.

- Changes to critical tables, HTML, and URLs should be tightly controlled, with all changes authorized by responsible owners and documented.

- A user manual should be available that covers operating procedures and that documents the management and control of the application.

- Each system report should indicate its general content in its title and specify the time period it covers.

- The system should have controls to prevent invalid and unreasonable data entry.

- Controls should exist to ensure that data input, HTML, and URLs are valid, applicable, and posted in the right time frame.

- Users should implement proper procedures to ensure correct input data.

Table 14.3

Useful Policies to Eliminate Waste and Mistakes

Training is another key aspect of implementation. Many users are not properly trained in developing and implementing applications, and their mistakes can be very costly. Since more and more people use computers in their daily work, it is important that they understand how to use them. Training is often the key to acceptance and implementation of policies and procedures. Because of the importance of maintaining accurate data and of people understanding their responsibilities, companies converting to ERP and e-commerce systems invest weeks of training for key users of the system's various modules.

Monitoring Policies and Procedures

To ensure that users throughout an organization are following established procedures, the next step is to monitor routine practices and take corrective action if necessary. By understanding what is happening in day-to-day activities, organizations can make adjustments or develop new procedures. Many organizations implement internal audits to measure actual results against established goals, such as percentage of end-user reports produced on time, percentage of data input errors rejected, number of input transactions entered per eight-hour shift, and so on.

The passage of the Sarbanes-Oxley Act has caused many companies to monitor their policies and procedures and to plan changes in financial information systems. These changes could profoundly affect many business activities. As mentioned in Chapter 9, the act requires public companies to implement procedures to ensure that their audit committees can document underlying financial data to validate earnings reports. Companies that fail to comply could find their top execs behind bars. In October 2004, SunTrust Banks disclosed that it is restating its earnings upward for the first two quarters of 2004 and delaying its third-quarter earnings statement because of improper accounting procedures in its auto financing division. As a result of the error, SunTrust underreported earnings for the first two quarters by $17 million and $5 million, respectively. Two executives, the chief credit officer and the controller, were put on paid leave.[9]

Reviewing Policies and Procedures

The final step is to review existing policies and procedures and determine whether they are adequate. During review, people should ask the following questions:

- Do current policies cover existing practices adequately? Were any problems or opportunities uncovered during monitoring?
- Does the organization plan any new activities in the future? If so, does it need new policies or procedures on who will handle them and what must be done?
- Are contingencies and disasters covered?

This review and planning allows companies to take a proactive approach to problem solving, which can increase productivity and improve customer service. During such a review,

companies are alerted to upcoming changes in information systems that could have a profound effect on many business activities. An example is the need for healthcare organizations to meet the requirements of the Health Insurance Portability and Accountability Act of 1996 (HIPAA). The goal of this act is to require healthcare organizations to implement cost-effective procedures for exchanging medical data. Healthcare organizations must employ standard electronic transactions, codes, and identifiers designed to enable them to fully "digitize" medical records and make it possible to use the Internet rather than expensive private networks for electronic data interchange. The regulations affect 1.5 million healthcare providers, 7,000 hospitals, and 2,000 healthcare plans. Companies had until April 2003 to comply. Now that the full details of HIPAA are becoming clear, many experts are concerned. Some fear that the HIPAA provisions are too complicated and will not meet the original objective of reducing medical industry costs and instead increase costs and paperwork for doctors without improving medical care.

Information systems professionals and users still need to be aware of the misuse of resources throughout an organization. Preventing errors and mistakes is one way to do so. Another is implementing in-house security measures and legal protections to detect and prevent a dangerous type of misuse: computer crime.

COMPUTER CRIME

Even good IS policies may not be able to predict or prevent computer crime. A computer's ability to process millions of pieces of data in less than a second can help a thief steal data worth thousands or millions of dollars. Compared with the physical dangers of robbing a bank or retail store with a gun, a computer criminal with the right equipment and know-how can steal large amounts of money from the privacy of a home. Computer crime often defies detection, the amount stolen or diverted can be substantial, and the crime is "clean" and nonviolent.

Here is a sample of recent computer crimes.

- Cyberattacks against corporations aren't under control, nor will they go away soon. The Computer Emergency Response Team (CERT) reported 137,529 security incidents in 2003, and damages attributed to Mydoom, to date the fastest and most pervasive worm attack on record, have been pegged at $40 billion worldwide.[10]
- Each year, nearly 700,000 people are victims of identity theft and other forms of computer fraud. Such a wide range of methods are used by the perpetrators of these crimes that it makes investigating them difficult. In one case, corrupt security guards at New York's Macy's department store were stopping customers as they exited the store and stealing their credit card numbers off their receipts. In another case, clerks at Bloomingdale's connected credit card scanners to handheld devices, allowing them to surreptitiously swipe the cards and collect thousands of credit card numbers.[11] A Michigan man pleaded guilty to four counts of wire fraud and unauthorized access to a computer after he and two accomplices used a vulnerable wireless network at a Lowe's store in Michigan to attempt to steal credit card numbers from the company's main computer systems in North Carolina and other U.S. Lowe's stores.[12]
- Fourteen members of an Italian hacker group known as the Reservoir Dogs were arrested by the Italian police in what became known as Operation Rootkit. The group was responsible for a series of hacking incidents that compromised more than 1,000 systems spanning the globe, affecting the likes of NASA, the U.S. Army and Navy, and various financial companies in the United States and abroad. Some members of the group were working as information security managers in big consulting firms and Internet service providers, even Italian branches of U.S. companies.[13]
- Two California defendants were indicted on fraud and money-laundering charges for setting up one of the largest Internet investment fraud cases in the United States. Their "investment club" allegedly bilked more than $60 million from 15,000 investors

worldwide. The club's Web site allegedly guaranteed investors a 120 percent annual rate of return with "no risk of losing the investor's principal investment," but the club never invested any of the victims' money. Instead, the defendants allegedly used the funds to purchase millions of dollars' worth of real estate in Mexico and Costa Rica, as well as a yacht and a helicopter.[14]

Although no one really knows how pervasive cybercrime is, the number of IT-related security incidents is increasing dramatically. The Computer Emergency Response Team Coordination Center (CERT/CC) is located at the Software Engineering Institute (SEI), a federally funded research and development center at Carnegie Mellon University in Pittsburgh, Pennsylvania. It is charged with coordinating communication among experts during computer security emergencies and helping to prevent future incidents. CERT employees study Internet security vulnerabilities, handle computer security incidents, publish security alerts, research long-term changes in networked systems, develop information and training to help organizations improve security at their sites, and conduct an ongoing public awareness campaign. The number of security problems reported to CERT increased sixfold between 2000 and 2003, as shown in Figure 14.1. As many as 60 percent of all attacks go undetected, according to security experts. What's more, of the attacks that are exposed, only an estimated 15 percent are reported to law enforcement agencies. Why? Companies don't want the bad press. Such publicity makes the job even tougher for law enforcement. Most companies that have been electronically attacked won't talk to the press. A big concern is loss of public trust and image—not to mention the fear of encouraging copycat hackers.

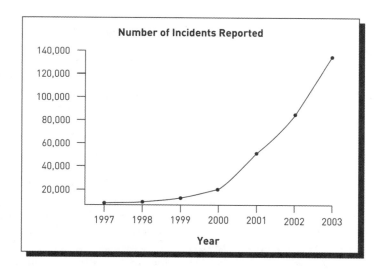

Figure 14.1

Number of Incidents Reported to CERT

(Source: Data from CERT Web site at *www.CERT.org/stats/#incients*, accessed on May 29, 2004.)

Highlights of the annual Computer Crime and Security Survey are shown in Table 14.4. The 2003 survey results are based on responses from 530 companies and government agencies. The Computer Security Institute, with the participation of the San Francisco Federal Bureau of Investigation (FBI) Computer Intrusion Squad, conducts this survey. The aim of the survey is to raise awareness of security, as well as to determine the scope of computer crime in the United States.

Today, computer criminals are a new breed—bolder and more creative than ever. With the increased use of the Internet, computer crime is now global. It's not just on U.S. shores that law enforcement has to battle cybercriminals. Regardless of its nonviolent image, computer crime is different only because a computer is used. It is still a crime. Part of what makes computer crime so unique and hard to combat is its dual nature—the computer can be both the tool used to commit a crime and the object of that crime.

Table 14.4

Summary of Key Data from 2003 Computer Crime and Security Survey

(Source: Data from Richard Power, "2003 CSI/FBI Computer Crime and Security Survey," *Computer Security Issues & Trends,* Vol. VIII, No. 1, Spring 2003.)

Incident	2003 Results
Respondents that detected computer security breaches within the last 12 months	90%
Respondents that acknowledged financial losses due to security breaches	75%
Average dollar loss of the 44% who were willing or able to quantify their financial losses	> $2.0 million
Respondents that cited their Internet connection as a frequent point of attack	73%
Respondents that cited their internal systems as a frequent point of attack	30%
Respondents that reported intrusions to law enforcement	30%
Respondents that detected computer viruses	82%

THE COMPUTER AS A TOOL TO COMMIT CRIME

A computer can be used as a tool to gain access to valuable information and as the means to steal thousands or millions of dollars. It is, perhaps, a question of motivation—many individuals who commit computer-related crime claim they do it for the challenge, not for the money. Credit card fraud—whereby a criminal illegally gains access to another's line of credit with stolen credit card numbers—is a major concern for today's banks and financial institutions. In general, criminals need two capabilities to commit most computer crimes. First, the criminal needs to know how to gain access to the computer system. Sometimes obtaining access requires knowledge of an identification number and a password. Second, the criminal must know how to manipulate the system to produce the desired result. Frequently, a critical computer password has been talked out of an individual, a practice called **social engineering**. Or, the attackers simply go through the garbage—**dumpster diving**—for important pieces of information that can help crack the computers or convince someone at the company to give them more access. In addition, there are more than 2,000 Web sites that offer the digital tools—for free—that will let people snoop, crash computers, hijack control of a machine, or retrieve a copy of every keystroke.

Also, with today's sophisticated desktop publishing programs and high-quality printers, crimes involving counterfeit money, bank checks, traveler's checks, and stock and bond certificates are on the rise. As a result, the U.S. Treasury Department redesigned and printed new currency that is much more difficult to counterfeit.

Cyberterrorism

Government officials and IS security specialists have documented a significant increase in Internet probes and server scans since early 2001. There is a growing concern among federal officials that such intrusions are part of an organized effort by cyberterrorists, foreign intelligence services, or other groups to map potential security holes in critical systems. A **cyberterrorist** is someone who intimidates or coerces a government or organization to advance his or her political or social objectives by launching computer-based attacks against computers, networks, and the information stored on them.

Even before the September 11, 2001, terrorist attacks, the U.S. government considered the potential threat of cyberterrorism serious enough that it established the National Infrastructure Protection Center in February 1998. This function was transferred to the Homeland Security Department's Information Analysis and Infrastructure Protection Directorate to serve as a focal point for threat assessment, warning, investigation, and response for threats

---

social engineering
The practice of talking a critical computer password out of an individual.

dumpster diving
Searching through the garbage for important pieces of information that can help crack an organization's computers or be used to convince someone at the company to give someone access to the computers.

cyberterrorist
Someone who intimidates or coerces a government or organization to advance his or her political or social objectives by launching computer-based attacks against computers, networks, and the information stored on them.

or attacks against our country's critical infrastructure, which provides telecommunications, energy, banking and finance, water systems, government operations, and emergency services. Successful cyberattacks against the facilities that provide these services could cause widespread and massive disruptions to the normal function of our society.

Supervisory Control and Data Acquisition systems, known as Scada systems, are used to manage and operate facilities at electric and gas utilities, opening and closing valves and switches to regulate the flow of energy. A General Accounting Office report released in April 2004 cited a number of problems, raising concern that lax security leaves these control systems vulnerable to cyber- and physical threats. Congressional auditors have recommended that the Homeland Security Department develop and implement a strategy for working with the private sector and other government agencies to improve security for Scada systems. The National Energy Reliability Council is trying to develop nationwide security standards. "Security has taken on a new meaning for us," says the council's CIO, Lynn Costantini. "Today, the ideas around security have evolved to include guards, gates, and guns, which nuclear facilities always had, as well as cybersecurity."[15]

Identity Theft

Identity theft is a crime in which an imposter obtains key pieces of personal identification information, such as Social Security or driver's license numbers, in order to impersonate someone else. The information is then used to obtain credit, merchandise, and services in the name of the victim or to provide the thief with false credentials.

In some cases, the identity thief uses personal information to open new credit accounts, establish cellular phone service, or open a new checking account to obtain blank checks. In other cases, the identity thief uses personal information to gain access to the person's existing accounts. Typically, the thief will change the mailing address on an account and run up a huge bill before the person whose identity has been stolen realizes there is a problem. The Internet has made it easier for an identity thief to use the stolen information because transactions can be made without any personal interaction.

The Federal Trade Commission received 516,740 identity-theft complaints in 2003, up from 404,000 in 2002. An FTC report issued that year estimates that more than 27 million Americans have been victims of identity theft during the past five years.[16]

Another popular method to get information is "shoulder surfing"—the identity thief simply stands next to someone at a public office, such as the Bureau of Motor Vehicles, and watches as the person fills out personal information on a form. Consumers can help protect themselves by regularly checking their credit reports with major credit bureaus, following up with creditors if their bills do not arrive on time, not revealing any personal information in response to unsolicited e-mail or phone calls (especially Social Security numbers and credit card account numbers), and shredding bills and other documents that contain sensitive information.

According to studies at Michigan State University's identity-theft research center, at least half of identity theft now results from the theft of personal information stored on business databases. Potentially as much as 70 percent of identity thefts originate in the workplace by employees or people impersonating employees, so the majority of identity thefts can be considered inside jobs. The research also showed that the majority of those identities were stolen most often from healthcare-related institutions and then from financial institutions.[17]

A high-profile case of insider identity theft broke in late 2002, when the Department of Justice charged a help desk worker at financial data company Teledata Communications Inc. with fraud and conspiracy in connection with an identity-theft scheme involving more than 30,000 victims. The worker allegedly used his insider status to access thousands of credit reports, which he sold for $60 apiece through a coconspirator.[18]

The U.S. Congress passed the Identity Theft and Assumption Deterrence Act of 1998 to fight identity theft. Under this act, the Federal Trade Commission (FTC) is assigned responsibility to help victims restore their credit and erase the impact of the imposter. It also makes identity theft a federal felony punishable by a prison term ranging from 3 to 25 years.

identity theft
A crime in which an imposter obtains key pieces of personal identification information, such as Social Security or driver's license numbers, in order to impersonate someone else. The information is then used to obtain credit, merchandise, and services in the name of the victim or to provide the thief with false credentials.

THE COMPUTER AS THE OBJECT OF CRIME

A computer can also be the object of the crime, rather than the tool for committing it. Tens of millions of dollars of computer time and resources are stolen every year. Each time system access is illegally obtained, data or computer equipment is stolen or destroyed, or software is illegally copied, the computer becomes the object of crime. These crimes fall into several categories: illegal access and use, data alteration and destruction, information and equipment theft, software and Internet piracy, computer-related scams, and international computer crime.

Illegal Access and Use

hacker
A person who enjoys computer technology and spends time learning and using computer systems.

Crimes involving illegal system access and use of computer services are a concern to both government and business. Since the outset of information technology, computers have been plagued by criminal hackers. A **hacker** is a person who enjoys computer technology and spends time learning and using computer systems. A **criminal hacker**, also called a **cracker**, is a computer-savvy person who attempts to gain unauthorized or illegal access to computer systems to steal passwords, corrupt files and programs, or even transfer money. In many cases, criminal hackers are people who are looking for fun and excitement—the challenge of beating the system. **Script bunnies** are wannabe crackers with little technical savvy—crackers who download programs called *scripts*—that automate the job of breaking into computers. **Insiders** are employees, disgruntled or otherwise, working solo or in concert with outsiders to compromise corporate systems.

criminal hacker (cracker)
A computer-savvy person who attempts to gain unauthorized or illegal access to computer systems.

script bunnies
Wannabe crackers with little technical savvy who download programs—scripts—that automate the job of breaking into computers.

insider
An employee, disgruntled or otherwise, working solo or in concert with outsiders to compromise corporate systems.

Catching and convicting criminal hackers remains a difficult task. The method behind these crimes is often hard to determine. Even if the method behind the crime is known, tracking down the criminals can take a lot of time. It took years for the FBI to arrest one criminal hacker for the alleged "theft" of almost 20,000 credit card numbers that had been sent over the Internet. Table 14.5 provides some guidelines to follow in the event of a computer security incident.

Table 14.5

How to Respond to a Security Incident

- Follow your site's policies and procedures for a computer security incident. (They are documented, aren't they?)
- Contact the incident response group responsible for your site as soon as possible.
- Inform others, following the appropriate chain of command.
- Further communications about the incident should be guarded to ensure intruders do not intercept information.
- Document all follow-up actions (phone calls made, files modified, system jobs that were stopped, etc.).
- Make backups of damaged or altered files.
- Designate one person to secure potential evidence.
- Make copies of possible intruder files (malicious code, log files, etc.) and store them offline.
- Evidence, such as tape backups and printouts, should be secured in a locked cabinet, with access limited to one person.
- Get the National Computer Emergency Response Team involved if necessary.
- If you are unsure of what actions to take, seek additional help and guidance before removing files or halting system processes.

Data Alteration and Destruction

Data and information are valuable corporate assets. The intentional use of illegal and destructive programs to alter or destroy data is as much a crime as destroying tangible goods. The most common of these programs are viruses and worms, which are software programs that, when loaded into a computer system, will destroy, interrupt, or cause errors in processing. There are more than 60,000 known computer viruses today, with over 5,000 new viruses and worms being discovered each year.

A **virus** is a computer program file capable of attaching to disks or other files and replicating itself repeatedly, typically without the user's knowledge or permission. Some viruses attach to files, so when the infected file executes, the virus also executes. Other viruses sit in a computer's memory and infect files as the computer opens, modifies, or creates the files. They are often disguised as games or images with clever or attention-grabbing titles such as "Boss, nude." Some viruses display symptoms, and some viruses damage files and computer systems. Computer viruses are written for several operating systems, including Windows, Macintosh, UNIX, and others.

Worms are parasitic computer programs that replicate but, unlike viruses, do not infect other computer program files. Worms can create copies on the same computer or can send the copies to other computers via a network. Worms often spread via IRC (Internet Relay Chat). The Mydoom worm, also known as Shimgapi and Novarg, started spreading in January 2004 and quickly became the most virulent e-mail worm ever. The worm arrives as an e-mail with an attachment that has various names and extensions, including .exe, .scr, .zip, and .pif. When the attachment is executed, the worm starts sending copies of itself to other e-mail addresses stored in the infected computer. The first version of the virus, Mydoom.A, was designed to attack The SCO Group Inc.'s Web site. A later variant, dubbed Mydoom.B, was designed to enable similar denial-of-service attacks against Microsoft Corp.'s Web site. The B variant also includes a particularly nasty feature in that it blocks infected computers from accessing sites belonging to vendors of antivirus products.[19] Infected e-mail messages carrying the Mydoom worm have been intercepted from over 142 countries and at one time accounted for 1 in every 12 e-mail messages. That surpasses the record set by the Sobig.F worm, which appeared in August 2003 and at its peak was found in 1 of every 17 messages.[20] SCO and Microsoft, in a concerted effort to discourage **malware** creators, each offered a $250,000 reward for information leading to the arrest and conviction of those responsible for releasing the Mydoom worm that targeted their Web sites. For Microsoft, the offer is part of its $5 million bounty program announced November 2003 to reward people for information relating to worm and virus authors.[21] The Web site of the Recording Industry Association of America was down for a week in March 2003, the victim of a distributed denial-of-service (DDoS) attack by the Mydoom.F virus.[22]

A **Trojan horse** program is a malicious program that disguises itself as a useful application and purposefully does something the user does not expect. Trojans are not viruses, since they do not replicate, but they can be just as destructive. Many people use the term to refer only to nonreplicating malicious programs, thus making a distinction between Trojans and viruses. In May 2004, news broke on a Russian security firm's Web site, saying that a sizable portion of Cisco Systems' Internetwork Operating System had been stolen and was circulating on the Internet. IOS is the software that controls much of Cisco's networking hardware, which many of the world's businesses and governments use to run their critical IT networks. There is concern that attackers could modify Cisco's licensing and registration mechanisms, meaning that businesses could be exposed to illegally modified copies of Cisco's software, which might contain some type of backdoor application (a computer program that bypasses security mechanisms) or Trojan-horse application that attackers could use to gain entry into systems.[23]

A **logic bomb** is a type of Trojan horse that executes when specific conditions occur. Triggers for logic bombs can include a change in a file by a particular series of keystrokes or at a specific time or date. In May 2004, authorities in Taiwan arrested 30-year-old computer engineer Wang Ping-an, who was accused of creating the Peep Trojan. Taiwan's Internet crime investigation task force, called "CIB," made the arrest. The existence of the Peep Trojan was uncovered when Taiwanese authorities discovered the theft of confidential information

virus
A program that attaches itself to other programs.

worm
An independent program that replicates its own program files until it interrupts the operation of networks and computer systems.

malware
Software that is harmful or destructive, such as viruses and worms.

Trojan horse
A program that appears to be useful but actually masks a destructive program.

logic bomb
An application or system virus designed to "explode" or execute at a specified time and date.

from government agencies, schools, and companies. Ping-an did not steal the data himself. Apparently, when he was unable to sell his data-stealing virus, he posted it on hackers' Web sites for free. Then, hackers from mainland China used the program to steal the data. Ping-an faces up to five years in prison, if convicted. More and more crackers are getting caught as the global law-enforcement community is taking these crimes more seriously. "The laws are getting better, the penalties steeper, and there is greater cooperation now that people have seen the damage these things can do," says Panda Software CTO Patrick Hinojosa.[24]

variant

A modified version of a virus that is produced by the virus's author or another person who amends the original virus code.

A **variant** is a modified version of a virus that is produced by the virus's author or another person who amends the original virus code. If changes are small, most antivirus products will also detect variants. However, if the changes are large, the variant may go undetected by antivirus software.[25] New variants of the Netsky e-mail worm are spreading on the Internet and may be the work of a different author than previous editions of that worm, according to antivirus software companies. Netsky.S appeared April 5, 2004, and Netsky.T was detected the next day. They are the 19th and 20th editions of an e-mail virus that first appeared in February 2004. Unlike earlier variants, the new Netsky strains open "back doors" on machines they infect, prompting at least one antivirus expert to declare the worm the work of a different virus author. Like its predecessors, the new Netsky variants target machines running versions of Microsoft Corp.'s Windows operating system.[26]

In some cases, a virus or a worm can completely halt the operation of a computer system or network for days or longer until the problem is found and repaired. In other cases, a virus or a worm can destroy important data and programs. If backups are inadequate, the data and programs may never be fully functional again. The costs include the effort required to identify and neutralize the virus or worm and to restore computer files and data, as well as the value of business lost because of unscheduled computer downtime.

United Parcel Service, Inc. can't allow computer viruses to disrupt the information systems that are key to the operation of its Worldport distribution hub in Louisville, Kentucky, so the company works to hunt viruses and worms down *before* an attack. The UPS technical support group, TSG, actively monitors security and hacker Internet newsgroups to check for new viruses and acts quickly when they discover a new one. When the Blaster worm hit earlier this year, a TSG technician came in to work at 4 A.M. after spotting news of the worm on newsgroups while working at home. This preemptive approach prevented Worldport systems from getting hit by the Slammer virus.[27] Table 14.6 lists the computer incidents that have had the greatest economic impact.

Table 14.6

Computer Incidents with the Greatest Worldwide Economic Impact

Year	Code Name	Worldwide Economic Impact
2000	I Love You	$8.75 billion
2001	Code Red	$2.62 billion
2001	SirCam	$1.15 billion
1999	Melissa	$1.10 billion

Macintosh computers are immune to infection from high-profile Windows worms like SoBig and Mydoom, which exploit security flaws and architectural shortcomings in Windows operating systems and software applications to cause problems. That software only runs on Windows machines. As of early 2004, experts were unaware of any Mac OS X–specific virus or worm, but it's likely only a matter of time before crackers begin developing Mac OS X viruses and worms.[28]

The F-Secure Corporation provides centrally managed security solutions, and its products include antivirus, file encryption, and network security solutions for all major platforms—from desktops to servers and from laptops to handhelds. Founded in 1988, F-Secure is headquartered in Helsinki, Finland, and listed on the Helsinki Stock Exchange. It also has offices in North America, Germany, Sweden, Japan, and the United Kingdom. F-Secure is supported by a network of resellers and distributors in more than 90 countries around the globe. F-Secure also provides real-time virus statistics on the most active viruses in the world at its Web site, *www.f-secure.com/virus-info/statistics*. During the month of November

2003, 1,418 new virus definitions were added to F-Secure Anti-Virus. In a typical month, between 500 and 1,500 entries might be added, depending on the number of new viruses discovered.[29] Table 14.7 lists the most active viruses in May 2004.

Place	Virus Name
1	W32/Sober.G@mm
2	W32/Netsky.P@mm
3	W32/Netsky.D@mm
4	W32/Bagle.Z@mm
5	W32/Lovgate.W@mm
6	W32/Netsky.B@mm
7	W32/Netsky.Z@mm
8	W32/Netsky.Q@mm
9	Html_Netsky.P
10	W32/Netsky.C@mm

Table 14.7

Most Active Viruses in the World—May 30, 2004

(Source: "F-Secure Virus Statistics," *www.f-secure.com/ virus-info/statistics*, accessed May 30, 2004. Used with permission.)

McAfee Security for Consumers is a division of Network Associates, Inc., that delivers retail and online solutions designed to secure, protect, and optimize the computers of consumers and home office users. McAfee's retail desktop products include premier antivirus, security, encryption, and desktop optimization software. McAfee delivers software through an Internet browser to provide these services to users online through its Web site *www.mcafee. com*, one of the largest paid subscription sites on the Internet with over 2 million active paid subscribers. McAfee provides a real-time map of where the latest viruses are infecting computers worldwide at *http://us.mcafee.com/virusInfo/default.asp* (see Figure 14.2). The site also provides software for scanning your computer for viruses and tips on how to remove a virus.[30] Table 14.8 provides some tips for avoiding viruses and worms.

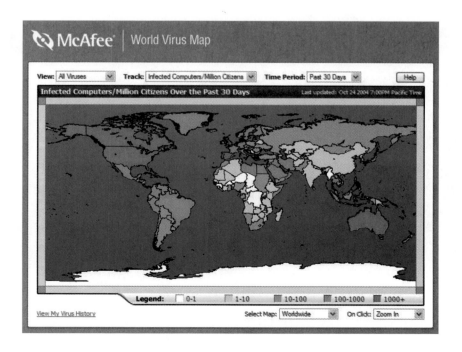

Figure 14.2

Global Virus Infection— Number of Infected Computers per Million Citizens

(Source: McAfee Security, "World Virus Map," *http://us.mcafee.com/ virusInfo/default.asp*. Courtesy of McAfee, Inc.)

Table 14.8

How to Avoid Viruses and Worms

1. Install antivirus software on your computer and configure it to scan all downloads, e-mail, and disks.

2. Update your antivirus software regularly. More than 500 viruses are discovered each month, so you need to remain current to be protected. You should download at least the product's virus signature files. You may also need to update the product's scanning engine as well.

3. Back up your files regularly. If a virus later destroys your files, you can replace them with your backup copy. You should store your backup copy in a location away from your work files, preferably not on your computer.

4. Do not open any files attached to an e-mail from an unknown, suspicious, or untrustworthy source.

5. Do not open any files attached to an e-mail unless you know what it is, even if it appears to come from a friend or someone you know. Many viruses replicate themselves and spread through e-mail.

6. Exercise caution when downloading files from the Internet. Ensure that the source is legitimate and reputable. Before you download, verify that the site has an antivirus program that checks its files. If you're uncertain, don't download the file at all or download the file to a removable disk and test it with your own antivirus software.

Using Antivirus Programs

antivirus program
Program or utility that prevents viruses and recovers from them if they infect a computer.

As a result of the increasing threat of viruses and worms, most computer users and organizations have installed **antivirus programs** on their computers. Such software runs in the background to protect your computer from dangers lurking on the Internet and other possible sources of infected files. Some antivirus software is even capable of repairing common virus infections automatically, without interrupting your work. The latest virus definitions are downloaded automatically when you connect to the Internet, ensuring that your PC's protection is current. To safeguard your PC and prevent it from spreading viruses to your friends and coworkers, some antivirus software scans and cleans both incoming and outgoing e-mail messages. Table 14.9 lists some of the most popular antivirus software.

Table 14.9

Antivirus Software

Antivirus Software	Software Manufacturer	Web Site
Symantec's Norton AntiVirus 2005	Symantec	*www.symantec.com*
McAfee Virus Scan	McAfee	*www.mcafee.com*
Panda Antivirus Platinum	Panda Software	*www.pandasoftware.com*
Vexira Antivirus	Central Command	*www.centralcommand.com*
Sophos Antivirus	Sophos	*www.sophos.com*
PC-cillin	Trend Micro	*www.trendmicro.com*

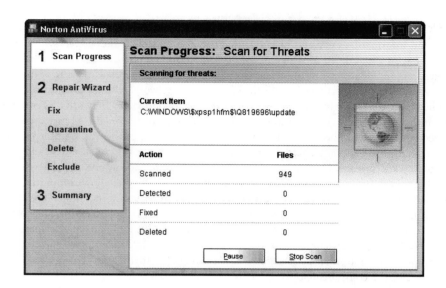

Antivirus software should be used and updated often.

Proper use of antivirus software requires the following steps:

1. *Install antivirus software and run it often.* Many of these programs automatically check for viruses each time you boot up your computer or insert a disk or CD, and some even monitor all e-mail and file transmissions and copying operations.
2. *Update antivirus software often.* New viruses are created all the time, and antivirus software suppliers are constantly updating their software to detect and take action against these new viruses.
3. *Scan all disks and CDs before copying or running programs from them.* Hiding on disks or CDs, viruses often move between systems. If you carry document or program files on disks or CDs between computers at school or work and your home system, always scan them.
4. *Install software only from a sealed package or secure Web site of a known software company.* Even software publishers can unknowingly distribute viruses on their program disks or software downloads. Most scan their own systems, but viruses may still remain.
5. *Follow careful downloading practices.* If you download software from the Internet or a bulletin board, check your computer for viruses immediately after completing the transmission.
6. *If you detect a virus, take immediate action.* Early detection often allows you to remove a virus before it does any serious damage.

Despite careful precautions, viruses can still cause problems. They can elude virus-scanning software by lurking almost anywhere in a system. Future antivirus programs may incorporate "nature-based models" that check for unusual or unfamiliar computer code. The advantage of this type of virus program is the ability to detect new viruses that are not part of an antivirus database.

Hoax, or false, viruses are another problem. Criminal hackers sometimes warn the public of a new and devastating virus that doesn't exist to create fear. Companies sometimes spend hundreds of hours warning employees and taking preventive action against a nonexistent virus. Security specialists recommend that IS personnel establish a formal paranoia policy to thwart virus panic among gullible end users. Such policies should stress that before users forward an e-mail alert to colleagues and higher-ups, they should send it to the help desk or the security team. The corporate intranet can be used to explain the difference between real viruses and fakes, and it can provide links to Web sites to set the record straight.

Information and Equipment Theft

Data and information are assets or goods that can also be stolen. Individuals who illegally access systems often do so to steal data and information. To obtain illegal access, criminal hackers require identification numbers and passwords. Some criminals try different identification numbers and passwords until they find ones that work. Using password sniffers is another approach. A **password sniffer** is a small program hidden in a network or a computer system that records identification numbers and passwords. In a few days, a password sniffer can record hundreds or thousands of identification numbers and passwords. Using a password sniffer, a criminal hacker can gain access to computers and networks to steal data and information, invade privacy, plant viruses, and disrupt computer operations. A hacker secretly installed sniffer software in 14 Kinko's stores in New York City that allowed him to capture 450 user names and passwords to access and even open bank accounts online. The hacker was caught when he used one of the stolen passwords to access a computer with GoToMyPC software, which lets individuals remotely access their own computers from elsewhere. The GoToMyPC subscriber happened to be home at the time and suddenly saw the cursor on his computer move around the screen and files open as if by themselves. To his amazement, he then saw an account being opened in his name at an online payment transfer service.[31]

In addition to theft of data and software, all types of computer systems and equipment have been stolen from offices. Computer theft is now second only to automobile theft, according to recent U.S. crime statistics. In the United Kingdom more than 30 percent of all reported thefts are computer related. Printers, desktop computers, and scanners are often targets. Portable computers such as laptops (and the data and information stored in them) are especially easy for thieves to take. In some cases, the data and information stored in these systems are more valuable than the equipment. Without adequate protection and security measures, equipment can easily be stolen. According to the Brigadoon software 2003 Computer Theft Survey, over 1.6 million computers were stolen in the United States in the past three years. Based on the responses of the 676 participants in the survey, 48 percent of the devices stolen were laptop computers, 27 percent were desktop computers, and 13 percent were some form of PDA. Unfortunately, the overwhelming majority of theft victims (88 percent) did not encrypt proprietary data on these devices.[32]

password sniffer
A small program hidden in a network or a computer system that records identification numbers and passwords.

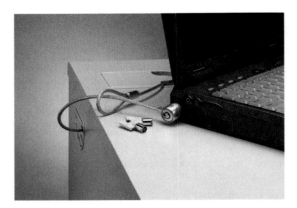

To fight computer crime, many companies use devices that disable the disk drive and lock the computer to the desk.

(Source: Courtesy of Kensington Technology Group.)

Software and Internet Software Piracy

Each time you use a word processing program or access software on a network, you are taking advantage of someone else's intellectual property. Like books and movies—other intellectual properties—software is protected by copyright laws. Often, people who would never think of plagiarizing another author's written work have no qualms about using and copying software programs they have not paid for. Such illegal duplicators are called *pirates*; the act of illegally duplicating software is called **software piracy**.

Technically, software purchasers are granted the right only to use the software under certain conditions; they don't really own the software. Licenses vary from program to program and may authorize as few as one computer or one individual to use the software or as many

software piracy
The act of illegally duplicating software.

as several hundred network users to share the application across the network. Making additional copies, or loading the software onto more than one machine, may violate copyright law and be considered piracy.

It is estimated that the software industry loses between $11 and $12 billion in revenue to software piracy annually. Half the loss comes from Asia, where China and Indonesia are the biggest offenders. In Western Europe, annual piracy losses range between $2.5 and $3 billion dollars. Although the rate of software piracy is quite high in Latin America and Central Europe, those software markets are so small that the dollar losses are considerably lower. About $2 billion in annual piracy losses come from North America. Continuing education and enforcement efforts are making a difference, however. In the United States, for example, the level of piracy was reduced from 48 percent in 1989 to 25 percent in 2002.[33]

Internet software piracy occurs when software is illegally downloaded from the Internet. It is the most rapidly expanding type of software piracy and the most difficult form to combat. The same purchasing rules apply to online software purchases as for traditional purchases. Internet piracy can take several forms, including the following:

- Pirate Web sites that make software available for free or in exchange for uploaded programs
- Internet auction sites that offer counterfeit software, which infringes copyrights
- Peer-to-peer networks, which enable unauthorized transfer of copyrighted programs

Penalties for software piracy can be severe. If the copyright owner brings a civil action against someone, the owner can seek to stop the person from using its software immediately and can also request monetary damages. The copyright owner may then choose between compensation for actual damages—which includes the amount it has lost because of the person's infringement, as well as any profits attributable to the infringement—and statutory damages, which can be as much as $150,000 for each program copied. In addition, the government can prosecute software pirates in criminal court for copyright infringement. If convicted, they could be fined up to $250,000 or sentenced to jail for up to five years, or both.[34]

Internet software piracy
Illegally downloading software from the Internet.

Computer-Related Scams

People have lost hundreds of thousands of dollars on real estate, travel, stock, and other business scams. Today, many of these scams are being perpetrated with computers. Using the Internet, scam artists offer get-rich-quick schemes involving bogus real estate deals, tout "free" vacations with huge hidden costs, commit bank fraud, offer fake telephone lotteries, sell worthless penny stocks, and promote illegal tax-avoidance schemes.

Over the past few years, credit card customers of various banks have been targeted by scam artists trying to get personal information needed to use their credit cards. The scam works by sending customers an e-mail including a link that seems to direct users to their bank's Web site. Once at the site, they are greeted with a pop-up box asking them for their full debit card numbers, their personal identification numbers, and their credit card expiration dates. The problem is that the Web site customers are directed to a fake site operated by someone trying to gain access to that information.[35] As discussed previously, this form of scam is called *phishing*. The Anti-Phishing Working Group received reports of more than 1,100 unique phishing campaigns in April 2004, a 178 percent increase from the previous month.[36] A 2004 study by the Gartner Consulting Group surveyed 5,000 adult Internet users and found that roughly 6 percent reported giving up financial data or other personal information after being drawn into phishing scams. The results suggest that as many as 30 million adults have experienced a phishing attack and that 1.78 million adults could have fallen victim to the scams. The 6 percent success rate is more than enough to ensure a continuation of such scams.[37]

The U.S. Federal Trade Commission (FTC) filed suit against a company that promotes creation of a firm's Web site over the phone and then charges the victims' phone bills without their authorization. The company calls small businesses and gets someone to agree to a 15-day trial offer to look at a Web site designed for that business. Unbeknownst to the small business, the look at the site results in a $29.95 addition to the firm's monthly telephone

bill. Often the charges on the phone bill appear as "MIS Int Serv," which people often mistake for Internet access charges. The scam has been going on for years and targets businesses all across the United States.[38]

Here is a list of tips to help you avoid becoming a scam victim.

- Don't agree to anything in a high-pressure meeting or seminar. Insist on having time to think it over and to discuss things with your spouse, your partner, or even your lawyer. If a company won't give you the time you need to check it out and think things over, you don't want to do business with it. A good deal now will be a good deal tomorrow; the only reason for rushing you is if the company has something to hide.

- Don't judge a company based on appearances. Flashy Web sites can be created and put up on the Net in a matter of days. After a few weeks of taking money, a site can vanish without a trace in just a few minutes. You may find that the perfect money-making opportunity offered on a Web site was a money maker for the crook and a money loser for you.

- Avoid any plan that pays commissions simply for recruiting additional distributors. Your primary source of income should be your own product sales. If the earnings are not made primarily by sales of goods or services to consumers or sales by distributors under you, you may be dealing with an illegal pyramid.

- Beware of shills, people paid by a company to lie about how much they've earned and how easy the plan was to operate. Check with an independent source to make sure that you aren't having the wool pulled over your eyes.

- Beware of a company's claim that it can set you up in a profitable home-based business but that you must first pay up front to attend a seminar and buy expensive materials. Frequently, seminars are high-pressure sales pitches, and the material is so general that it is worthless.

- If you are interested in starting a home-based business, get a complete description of the work involved before you send any money. You may find that what you are asked to do after you pay is far different from what was stated in the ad. You should never have to pay for a job description or for needed materials.

- Get in writing the refund, buy-back, and cancellation policies of any company you deal with. Do not depend on oral promises.

- Do your homework. Check with your state attorney general and the National Fraud Information Center before getting involved, especially when the claims about a product or potential earnings seem too good to be true.

If you need advice about an Internet or online solicitation, or if you want to report a possible scam, use the Online Reporting Form or Online Question & Suggestion Form features on the Web site for the National Fraud Information Center at *http://fraud.org*, or call the NFIC hotline at 1-800-876-7060.

International Computer Crime

Computer crime is also an international issue, and it becomes more complex when it crosses borders. As already mentioned, the software industry loses about $11 to $12 billion in revenue to software piracy annually, with about $9 billion of that occurring outside the United States.[39]

With the increase in electronic cash and funds transfer, some are concerned that terrorists, international drug dealers, and other criminals are using information systems to launder illegally obtained funds. Computer Associates International developed software called CleverPath for Global Compliance for customers in the finance, banking, and insurance industries to eliminate money laundering and fraud. Companies that are required to comply with legislation such as the USA Patriot Act and Sarbanes-Oxley Act may lack the resources and processes to do so. The software automates manual tracking and auditing processes that are required by regulatory agencies and helps companies handle frequently changing reporting regulations. The application can drill into a company's transactions and automatically detect fraud or other illegal activities based on built-in business rules and predictive analysis. Suspected fraud cases are identified and passed on to the appropriate personnel for action to thwart criminals and help companies avoid paying fines.[40]

PREVENTING COMPUTER-RELATED CRIME

Because of increased computer use today, greater emphasis is placed on the prevention and detection of computer crime. Although all states have passed computer crime legislation, some believe that these laws are not effective because companies do not always actively detect and pursue computer crime, security is inadequate, and convicted criminals are not severely punished. However, all over the United States, private users, companies, employees, and public officials are making individual and group efforts to curb computer crime, and recent efforts have met with some success.

Crime Prevention by State and Federal Agencies

State and federal agencies have begun aggressive attacks on computer criminals, including criminal hackers of all ages. In 1986, Congress enacted the Computer Fraud and Abuse Act, which mandates punishment based on the victim's dollar loss. The Department of Defense also supports the Computer Emergency Response Team (CERT), which responds to network security breaches and monitors systems for emerging threats. Law enforcement agencies are also increasing their efforts to stop criminal hackers, and many states are now passing new, comprehensive bills to help eliminate computer crimes. Recent court cases and police reports involving computer crime show that lawmakers are ready to introduce newer and tougher computer crime legislation. Several states have passed laws in an attempt to outlaw spamming. For example, an Arizona law enacted in May 2003 requires that unsolicited commercial e-mail messages include a label ("ADV:") at the beginning of the subject line, and contain an opt-out mechanism. In September 2003, legislation was approved in California that made it the second state (after Delaware) to adopt an opt-in for e-mail advertising. Under this legislation, it is illegal to send unsolicited commercial e-mail from California or to a California e-mail address.

Crime Prevention by Corporations

Companies are also taking crime-fighting efforts seriously. Many businesses have designed procedures and specialized hardware and software to protect their corporate data and systems. Specialized hardware and software, such as encryption devices, can be used to encode data and information to help prevent unauthorized use. As discussed in Chapter 7, encryption is the process of converting an original electronic message into a form that can be understood only by the intended recipients. A key is a variable value that is applied using an algorithm to a string or block of unencrypted text to produce encrypted text or to decrypt encrypted text. Encryption methods rely on the limitations of computing power for their effectiveness—if breaking a code requires too much computing power, even the most determined code crackers will not be successful. The length of the key used to encode and decode messages determines the strength of the encryption algorithm.

Public key infrastructure (PKI) enables users of an unsecured public network such as the Internet to securely and privately exchange data through the use of a public and a private cryptographic key pair that is obtained and shared through a trusted authority. PKI is the most common method on the Internet for authenticating a message sender or encrypting a message. PKI uses two keys to encode and decode messages. One key of the pair, the message receiver's public key, is readily available to the public and is used by anyone to send that individual encrypted messages. The second key, the message receiver's private key, is kept secret and is known only by the message receiver. Its owner uses the private key to *decrypt* messages—convert encoded messages back into the original message. Knowing an individual's public key does not enable you to decrypt an encoded message to that individual.

public key infrastructure (PKI)
A means to enable users of an unsecured public network such as the Internet to securely and privately exchange data through the use of a public and a private cryptographic key pair that is obtained and shared through a trusted authority.

biometrics

The measurement of one of a person's traits, whether physical or behavioral.

Using biometrics is another way to protect important data and information systems. **Biometrics** involves the measurement of one of a person's traits, whether physical or behavioral. Biometric techniques compare a person's unique characteristics against a stored set to detect differences between them. Biometric systems can scan fingerprints, faces, handprints, irises, and retinal images to prevent unauthorized access to important data and computer resources. Most of the interest among corporate users is in fingerprint technology, followed by face recognition. Fingerprint scans hit the middle ground between price and effectiveness. Iris and retinal scans are more accurate, but they are more expensive and involve more equipment.

In June 2004, the U.S. Department of Homeland Security awarded a $10 billion, five-year contract to global management consulting and technology services company Accenture to oversee the creation of a comprehensive border-control system. The system, known as the United States Visitor and Immigration Status Indicator Technology, or US-Visit, employs a combination of biometric technologies to identify visitors. Radio-frequency identification, voice- and facial-recognition, retinal- or iris-scanning, and digital-fingerprinting systems are all being tested and evaluated for inclusion in the program. A visitor to the United States might be required to undergo digital fingerprinting at a consulate in his or her home country to apply for a visa. Once in the United States, the visitor would present a smart card that is encoded with the digital fingerprint to an immigration official and undergo a second digital fingerprint scan to ensure a match. The data can be instantly cross-referenced against a database containing digital descriptors of known terrorists, criminals, and other undesirables.[41]

Fingerprint authentication devices provide security in the PC environment by using fingerprint information instead of passwords.

(Source: Courtesy of DigitalPersona.)

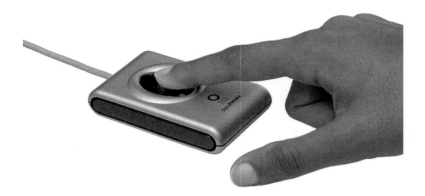

As employees move from one position to another at a company, they can build up access to multiple systems because security procedures often fail to revoke access privileges. It is clearly not appropriate for people who have changed positions and responsibilities to still have access to systems they no longer use. To avoid this problem, many organizations are creating role-based system access lists so that only people filling a particular job function can access a specific system.

Crime-fighting procedures usually require additional controls on the information system. Before designing and implementing controls, organizations must consider the types of computer-related crime that might occur, the consequences of these crimes, and the cost and complexity of needed controls. In most cases, organizations conclude that the trade-off between crime and the additional cost and complexity weighs in favor of better system controls. Having knowledge of some of the methods used to commit crime is also helpful in preventing, detecting, and developing systems resistant to computer crime (see Table 14.10). Some companies actually hire former criminals to thwart other criminals. Table 14.11 provides a set of useful guidelines to protect your computer from criminal hackers.

Methods	Examples
Add, delete, or change inputs to the computer system.	Delete records of absences from class in a student's school records.
Modify or develop computer programs that commit the crime.	Change a bank's program for calculating interest to make it deposit rounded amounts in the criminal's account.
Alter or modify the data files used by the computer system.	Change a student's grade from C to A.
Operate the computer system in such a way as to commit computer crime.	Access a restricted government computer system.
Divert or misuse valid output from the computer system.	Steal discarded printouts of customer records from a company trash bin.
Steal computer resources, including hardware, software, and time on computer equipment.	Make illegal copies of a software program without paying for its use.
Offer worthless products for sale over the Internet.	Send e-mail requesting money for worthless hair growth product.
Blackmail executives to prevent release of harmful information.	Eavesdrop on organization's wireless network to capture competitive data or scandalous information.
Blackmail company to prevent loss of computer-based information.	Plant logic bomb and send letter threatening to set it off unless paid considerable sum.

Table 14.10

Common Methods Used to Commit Computer Crimes

Even though the number of potential computer crimes appears to be limitless, the actual methods used to commit crime are limited.

- Install strong user authentication and encryption capabilities on your firewall.

- Install the latest security patches, which are often available at the vendor's Internet site.

- Disable guest accounts and null user accounts that let intruders access the network without a password.

- Do not provide overfriendly login procedures for remote users (e.g., an organization that used the word *welcome* on their initial logon screen found they had difficulty prosecuting a hacker).

- Give an application (e-mail, file transfer protocol, and domain name server) its own dedicated server.

- Restrict physical access to the server and configure it so that breaking into one server won't compromise the whole network.

- Turn audit trails on.

- Consider installing caller ID.

- Install a corporate firewall between your corporate network and the Internet.

- Install antivirus software on all computers and regularly download vendor updates.

- Conduct regular IS security audits.

- Verify and exercise frequent data backups for critical data.

Table 14.11

How to Protect Your Corporate Data from Criminal Hackers

Companies are also joining together to fight crime. The Software Publisher's Association (SPA) was the original antipiracy organization, formed and financed by many of the large software publishers. Microsoft financed the formation of a second antipiracy organization, the Business Software Alliance (BSA). The BSA, through intense publicity, has become the more prominent organization. Other software companies, including Apple, Adobe, Hewlett-Packard, and IBM, now contribute to the BSA.

The Business Software Alliance required Red Bull North America, Inc., to pay $105,000 to settle claims that it had more copies of Adobe, Microsoft, and Symantec software programs on its computers than it had licenses to support. Red Bull is based in Germany and markets a nonalcoholic energy drink in more than 70 countries. The software infringement was reported through a call to the BSA hotline, and then BSA attorneys contacted Red Bull. The company cooperated with the BSA and voluntarily conducted a self-audit. To avoid similar problems, all companies should draft policies and implement procedures to maintain an effective software management program.[42]

Using Intrusion Detection Software

intrusion detection system (IDS)

Software that monitors system and network resources and notifies network security personnel when it senses a possible intrusion.

An **intrusion detection system** (IDS) monitors system and network resources and notifies network security personnel when it senses a possible intrusion. Examples of suspicious activities include repeated failed login attempts, attempts to download a program to a server, and access to a system at unusual hours. Such activities generate alarms that are captured on log files. Intrusion detection systems send an alarm, often by e-mail or pager, to network security personnel when they detect an apparent attack. Unfortunately, many IDSs frequently provide false alarms that result in wasted effort. If the attack is real, then network security personnel must make a decision about what to do to resist the attack. Any delay in response increases the probability of damage from a hacker attack. Use of an IDS provides another layer of protection in the event that an intruder gets past the outer security layers—passwords, security procedures, and corporate firewall.

The following story is true, but the company's name has been changed to protect its identity. The ABCXYZ company employs more than 25 IDS sensors across its worldwide network, enabling it to monitor 90 percent of the company's internal network traffic. The remaining 10 percent comes from its engineering labs and remote sales offices, which are not monitored because of a lack of resources. The company's IDS worked very well in providing an early warning of an impending SQL Slammer attack. The Slammer worm had entered the network via a server in one of the engineering labs. The person monitoring the IDS noticed outbound traffic consistent with SQL Slammer at about 7:30 A.M. He contacted network operations group by e-mail and followed up with a phone call and a voice mail message. Unfortunately, the operations group gets so many e-mails that if a message is not highlighted as URGENT, the message may be missed. That is exactly what happened—the e-mail alert wasn't read, and the voice message wasn't retrieved in time to block the attack. A few hours later, the ABCXYZ company found itself dealing with a massive number of reports of network and server problems.[43]

A firm called Internet Security Systems (ISS) manages security for other organizations through its Security Incident Prevention Service. The company's IDSs are designed to recognize 30 of the most-critical threats, including worms that go after Microsoft software and those that exploit Apache Web servers and other programs. When an attack is detected, the service automatically blocks it without requiring human invention. Taking the manual intervention step out of the process enables a faster response and minimizes damage from a hacker. To encourage customers to adopt its service, ISS guaranteed up to $50,000 in cash if the prevention service failed.[44]

Using Managed Security Service Providers (MSSPs)

Keeping up with computer criminals—and with new regulations—can be daunting for organizations. Crackers are constantly poking and prodding, trying to breach the security defenses of companies. Also, such recent legislation as HIPAA, the Sarbanes-Oxley Act, and the USA Patriot Act requires businesses to prove that they are securing their data. For most

small and midsized organizations, the level of in-house network security expertise needed to protect their business operations can be quite costly to acquire and maintain. As a result, many are outsourcing their network security operations to **managed security service providers (MSSPs)** such as Counterpane, Guardent, Internet Security Services, Riptech, and Symantec. MSSPs monitor, manage, and maintain network security for both hardware and software. These companies provide a valuable service for IS departments drowning in reams of alerts and false alarms coming from virtual private networks (VPNs); antivirus, firewall, intrusion detection systems; and other security monitoring systems. In addition, some provide vulnerability scanning and Web blocking/filtering capabilities.

Merrill Lynch & Co. decided to outsource much of its network security to VeriSign Inc., including the management of its 300-some firewall and intrusion detection network security devices. "It's all about being the best that you possibly can be," says David Bauer, chief information security and privacy officer for Merrill Lynch & Co. "On the intrusion detection piece, we have a lot of network activities, as a lot of companies do, but now we're going to get analysis of all our activity in context with what else is going on in the world. It's not just about data; it's about intelligence. And with intelligence, you can make better decisions." VeriSign has such intelligence because it manages network security devices for hundreds of companies; it sees attacks occurring on the Internet and within its customers' networks. So, VeriSign can put attacks into context to decide whether an attack is random or targeted, sophisticated or unsophisticated. This determination then guides VeriSign to take the most appropriate action for its customers.[45]

managed security service provider (MSSP)
An organization that monitors, manages, and maintains network security hardware and software for its client companies.

Internet Laws for Libel and Protection of Decency

To help parents control what their children see on the Internet, some companies provide software called *filtering software* to help screen Internet content. Many of these screening programs also prevent children from sending personal information over e-mail or through chat groups. This stops children from broadcasting their name, address, phone number, or other personal information over the Internet. The two approaches used are filtering, which blocks certain Web sites, and rating, which places a rating on Web sites. According to the 2004 Internet Filter Review, the five top-rated filtering software packages are, in order, ContentProtect, Cybersitter, Net Nanny, CyberPatrol, and FilterPack.[46]

The Internet Content Rating Association (ICRA) is a nonprofit organization whose members include Internet industry leaders such as America Online, Bell South, British Telecom, IBM, Microsoft, UUNet, and Verizon. Its specific goals are to protect children from potentially harmful material, while also safeguarding free speech on the Internet. Using the **ICRA rating system,** Web authors fill out an online questionnaire describing the content of their site—what is and isn't present. The broad topics covered are the following: chat capabilities, the language used on the site, the nudity and sexual content of a site, the violence depicted on the site, and other areas such as alcohol, drugs, gambling, and suicide. Based on the authors' responses, ICRA then generates a content label (a short piece of computer code) that the authors add to their site. Internet users (and parents) can then set their browser to allow or disallow access to Web sites based on the objective rating information declared in the content label and their own subjective preferences. Reliance on Web site authors to do their own rating has its weaknesses, though. Web site authors can lie when completing the ICRA questionnaire so that their site receives a content label that doesn't accurately reflect the site's content. In addition, many hate groups and sexually explicit sites don't have an ICRA rating, so they will not be blocked unless a browser is set to block all unrated sites. Also, this option would block out so many acceptable sites that it could make Web surfing useless. For these reasons, at this time, site labeling is at best a complement to other filtering techniques.

ICRA rating system
System to protect individuals from harmful or objectionable Internet content, while safeguarding the free speech rights of others.

The Children's Internet Protection Act (CIPA) is a federal law passed in December 2000 that required federally funded libraries to use some form of prevention measure (such as Internet filters) to block access to obscene material and other material considered harmful to minors. Opponents of the law feared that it transferred power over the education process to private software companies that develop the Internet filters and define which sites are to be blocked. In June 2003, in a ruling on the consolidated cases *U.S. v. Multnomah County*

ContentProtect is a filtering software program that helps block unwanted Internet content from children and young adults.

(Source: Courtesy of ContentWatch Inc.)

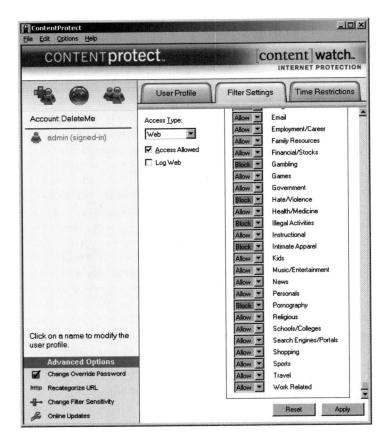

Library, et al. and *U.S. v. American Library Association*, the Supreme Court upheld the law but ruled that librarians can disable the software entirely on request and that patrons do not have to provide a reason why they want a site unblocked. The ruling also implies that patrons would not have to identify themselves to request unblocking. Some observers believe that the court's ruling left unclear the unblocking rules with respect to children.[47]

The Child Online Protection Act is a federal law passed in 1998 that made it a crime for any for-profit Web site to display material that is "harmful to minors." COPA covers written material, still images, video, audio, and even interactive dialogue about human sexuality. Violators are punished with six months in prison and/or fines of up to $50,000.[48] The law was put on hold for several years during a court challenge on behalf of artists, bookstores, an online sex therapist, a gynecological information Web site, and others. The American Civil Liberties Union (ACLU) argued the law would make criminals of anyone who offered racy or explicit material to adults. In June 2004, the Supreme Court ruled that "there is potential for extraordinary harm and a serious chill upon protected speech" if the law takes effect.[49] Material that is indecent but not obscene is protected by the First Amendment to the U.S. Constitution, which states that Congress shall make no law abridging the freedom of speech. Adults may see or buy it, but children may not. As a practical matter, this restriction is very difficult to enforce on the Internet, where most Web sites and chat rooms are available to adults and children alike. Congress has repeatedly tried to find a way to shield youngsters from "harmful material" without running afoul of the First Amendment. In June 2004, the Supreme Court in *ACLU v. Reno* upheld a lower court injunction against enforcement of the act.

With the increased popularity of networks and the Internet, libel becomes an important legal issue. A publisher, such as a newspaper, can be sued for libel, which involves publishing a written statement that is damaging to a person's reputation. Generally, a bookstore cannot be held liable for statements made in newspapers or other publications it sells. Online services, such as CompuServe and America Online, may exercise some control over who puts information on their service but may not have direct control over the content of what is published

by others on their service. So, can online services be sued for libel for content that someone else publishes on their service? Do online services more closely resemble a newspaper or a bookstore? This legal issue has not been completely resolved, but some court cases have been decided. The *Cubby, Inc. v. CompuServe* case ruled that CompuServe was similar to a bookstore and not liable for content put on its service by others. In this case, the judge stated, "While CompuServe may decline to carry a given publication altogether, in reality, once it does decide to carry a given publication, it will have little or no editorial control over that publication's content." This case set a legal precedent that has been applied in similar, subsequent cases.

Companies should be aware that publishing Internet content to the world may subject them to different countries' laws in the same way that exporting physical products does. A December 2003 ruling by the High Court of Australia found that a story published by Dow Jones & Co. on a U.S.-hosted Web site was grounds for a defamation lawsuit in Australia. The suit was brought by an Australian over the Internet version of an article in Dow Jones's *Barron's* magazine. The individual filed the suit in the Supreme Court of his home state of Victoria in Australia, saying that the article's appearance on the Internet enabled it to be accessed by people in Victoria, thereby defaming him where he is best known. "The torts of libel and slander are committed when and where comprehension of the defamatory matter occurs," agreed the Australian High Court, citing several precedents.[50]

Geolocation tools match the user's IP address with outside information to determine the actual geographic location of the online user where the customer's computer signal enters the Internet. This enables someone to identify the user's actual location within approximately 50 miles. As such tools become broadly available, Internet publishers will be able to limit the reach of their published speech to avoid potential legal risks. Use of such technology may also result in a division of the global Internet into separate content regions, with readers in Brazil, Japan, and the United States all receiving variations of the same information from the same publisher.

Individuals, too, must be careful what they post on the Internet to avoid libel charges. There have been many cases of disgruntled former employees being sued by their former employees for material posted on the Internet.

Preventing Crime on the Internet

As mentioned in Chapter 7, Internet security can include firewalls and a number of methods to secure financial transactions. A firewall can include both hardware and software that act as a barrier between an organization's information systems and the outside world. A number of systems have been developed to safeguard against crime on the Internet.

To help prevent crime on the Internet, the following steps can be taken:

1. Develop effective Internet usage and security policies for all employees.
2. Use a stand-alone firewall (hardware and software) with network monitoring capabilities.
3. Deploy intrusion detection systems, monitor them, and follow up on their alarms.
4. Monitor managers and employees to make sure that they are using the Internet for business purposes.
5. Use Internet security specialists to perform audits of all Internet and network activities.

Even with these precautions, computers and networks can never be completely protected against crime. One of the biggest threats is from employees. Although firewalls provide good perimeter control to prevent crime from the outside, procedures and protection measures are needed to protect against computer crime by employees. Passwords, identification numbers, and tighter control of employees and managers also help prevent Internet-related crime.

Mohegan Sun, a casino in Connecticut owned by the Mohegan Tribe, operates a large network with a number of critical systems that track its more than 2.5 million customers, their winnings, and their creditworthiness, as well as loyalty program points that can be used to purchase items at the casino and its shops. The casino is trying out new software from Intrusic, Inc., in an attempt to detect illegal behavior of rogue employees. Intrusic's Zephon software analyzes communications between users and the computer network, looking for violations of what it calls the "physics of networks" or fundamental laws that govern the way

legitimate network traffic looks. The casino already employs perimeter-defense products such as firewalls and intrusion detection systems, but it is increasingly concerned about insider threats or external compromise as a result of fast-moving worms and viruses.[51]

PRIVACY ISSUES

Another important social issue in information systems involves privacy. In 1890, U.S. Supreme Court Justice Louis Brandeis stated that the "right to be left alone" is one of the most "comprehensive of rights and the most valued by civilized man." Basically, the issue of privacy deals with this right to be left alone or to be withdrawn from public view. With information systems, privacy deals with the collection and use or misuse of data. Data is constantly being collected and stored on each of us. This data is often distributed over easily accessed networks and without our knowledge or consent. Concerns of privacy regarding this data must be addressed.

With today's computers, the right to privacy is an especially challenging problem. More data and information are produced and used today than ever before. When someone is born, takes certain high school exams, starts working, enrolls in a college course, applies for a driver's license, purchases a car, serves in the military, gets married, buys insurance, gets a library card, applies for a charge card or loan, buys a house, or merely purchases certain products, data is collected and stored somewhere in computer databases. Read the "Information Systems @ Work" special feature to learn more about privacy issues associated with fundamental medical and DNA data of individuals. A difficult question to answer is, "Who owns this information and knowledge?" If a public or private organization spends time and resources to obtain data on you, does the organization own the data, and can it use the data in any way it desires? Government legislation answers these questions to some extent for federal agencies, but the questions remain unanswered for private organizations.

Privacy and the Federal Government

The U.S. federal government is perhaps the largest collector of data. Close to 4 billion records exist on individuals, collected by about 100 federal agencies ranging from the Bureau of Alcohol, Tobacco, and Firearms to the Veterans Administration. Other data collectors include state and local governments and profit and nonprofit organizations of all types and sizes. In recent years, a number of actions by the federal government have caused concern for the privacy of individuals' data.

In response to the September 11 terrorist attacks, the U.S. Department of Defense and the Defense Advanced Research Projects Agency (DARPA) launched the Total Information Awareness (TIA) program to conduct research on technology's use for large-scale information gathering and analysis. The program's intent was to access a mix of government, intelligence, and commercial databases to mine electronic transactions, such as airline-ticket purchases and car rentals to look for potential terrorists and terrorist threats. In July 2003, the Senate, concerned about the lack of limits on the sources of such data gathering, voted to cancel funding for TIA. Since its inception, TIA had drawn criticism from privacy rights advocates who feared it would allow authorities to rummage though the electronic transactions of millions of law-abiding U.S. citizens in an effort to uncover the activities of suspected terrorists.[52]

UK BioBank Raises Privacy Issues

The Biobank is a £45 million (80 million U.S. dollars) project dedicated to collecting genetic, medical, and lifestyle data from 500,000 middle-aged people in the United Kingdom. The goal is to enable scientists to pinpoint specific factors involved in certain diseases, ultimately leading to new and better treatments. Volunteers chosen at random will be asked to donate DNA and answer a detailed questionnaire on their health and lifestyle. Their health will be followed for decades to help discover the roles of genes and the environment in developing cancer, heart disease, and other illnesses. Ultimately, doctors could use the databases to discover a new treatment that will work on anyone with a certain set of genetic traits. The information's true value may not be realized for some 30 years.

However, before a single piece of data has been collected, the UK Biobank has become embroiled in controversy over its goals, its costs, its underlying science, and its possibility for commercial exploitation.

Critics of the project claim that there is no evidence to suggest that investing millions of pounds in the gene bank would be as beneficial as spending it on well-targeted public health initiatives to prevent illness. They fear that the database could effectively become a permanent medical rap sheet, analogous to biological credit reports that could haunt a person's children, grandchildren, and other relatives. Ian Gibson, a member of Parliament within Britain's governing Labour party and chairman of the House of Commons Science and Technology Committee, said he has concerns that industry sectors such as insurance, which could have a

strong interest in the bank's database, might eventually obtain the medical information. Currently, in the UK there are no laws that prevent discrimination by insurers and employers.

Many practical questions remain, but the Biobank project's leaders say that the public has already expressed wide support. They have identified many people who want to participate and be part of developments in a better life.

Discussion Questions

1. What are the most controversial aspects of the Biobank project?
2. What are the potential negative impacts of these concerns on the Biobank project?

Critical Thinking Questions

3. Imagine that you are the member of a public relations firm hired to improve the public's impression of the Biobank project and overcome the arguments of its strongest critics. What specific actions would you recommend be taken by the administrators and scientists associated with the Biobank project to win over public opinion?
4. What additional actions might people outside the Biobank project take to alleviate some of the privacy concerns?

SOURCES: Kevin Davies, "First Base: Behold—BioBank," *Bio-IT World*, August 7, 2003, *www.bio-ITworld.com*; Scarlet Pruitt, "U.K. Prepares to Open World's Largest Biobank," *Bio-IT World*, November 14, 2003, *www.bio-ITWorld.com*; Pat Hagan, "Biobank Debate Heats Up," *The Scientist*, April 8, 2003, *www.biomedcentral.com*; Pat Hagan, "UK Biobank Reveals Ethics Framework," *The Scientist*, September 23, 2003, *www.biomedcentral.com*.

Also launched in response to September 11 terrorist attacks, the Matrix system lets states share criminal, prison, and vehicle information and cross-reference it with databases held by Seisint. Seisint is a provider of information products that enable organizations to extract useful knowledge from huge amounts of data. Its products are created by integrating super-computer technology, tens of billions of data records on individuals and businesses, and patent-pending data-linking methods.[53] The information contained within Matrix includes civil court records, voter registration information, and address histories going back as far as 30 years. Current participants in the project include Connecticut, Florida, Michigan, New York, Ohio, Pennsylvania, and Utah. Several other states considered the program before dropping out, citing concerns about privacy or the long-term costs.[54]

Although privacy concerns prompted Congress to kill both the CAPPS II and TIA programs, government computers are still scanning a vast array of databases for clues about criminal or terrorist activity, according to the General Accounting Office (GAO), the investigative arm of Congress. The U.S. government is managing 199 data-mining efforts, 36 of which collect personal information from the private sector. While data mining can be a useful tool for the government, "safeguards should be put in place to ensure that information isn't abused," said Nuala O'Connor Kelly, chief privacy officer at the Department of Homeland Security. Senator Daniel Akaka of Hawaii said he has asked the GAO to examine some projects more closely. "The federal government collects and uses Americans' personal information and shares it with other agencies to an astonishing degree, raising serious privacy concerns," Akaka said in a statement.[55] A growing number of privacy advocates say the Matrix database seems to be a substitute for the TIA data-mining program scrapped by the Pentagon.[56]

Most companies and computer vendors are wary of having the federal government dictate Internet privacy standards. A group called the Online Privacy Alliance is developing a voluntary code of conduct. It is backed by companies such as Apple Computer, AT&T, Boeing, Dell, DoubleClick, eBay, IBM, Microsoft, Time Warner, Verizon Communications, and Yahoo! The alliance's guidelines call on companies to notify users when they are collecting data at Web sites to gain consent for all uses of that data, to provide for the enforcement of privacy policies, and to have a clear process in place for receiving and addressing user complaints. The alliance's policy can be found at *www.privacyalliance.org/resources/ppguidelines. shtml.*

The European Union has already passed a data-protection directive that requires firms transporting data across national boundaries to have certain privacy procedures in place. This directive affects virtually any company doing business in Europe, and it is driving much of the attention being given to privacy in the United States.

Privacy at Work

The right to privacy at work is also an important issue. Currently, the rights of workers who want their privacy and the interests of companies that demand to know more about their employees are in conflict. Recently, companies that have been monitoring have raised their employee's concerns. For example, workers may find that they are being closely monitored via computer technology. These computer-monitoring systems tie directly into workstations; specialized computer programs can track every keystroke made by a user. This type of system can determine what workers are doing while at the keyboard. The system also knows when the worker is not using the keyboard or computer system. These systems can estimate what a person is doing and how many breaks he or she is taking. Needless to say, many workers consider this close supervision very dehumanizing.

E-Mail Privacy

E-mail also raises some interesting issues about work privacy. Federal law permits employers to monitor e-mail sent and received by employees. Furthermore, e-mail messages that have been erased from hard disks may be retrieved and used in lawsuits because the laws of discovery demand that companies produce all relevant business documents. On the other hand, the use of e-mail among public officials may violate "open meeting" laws. These laws, which

apply to many local, state, and federal agencies, prevent public officials from meeting in private about matters that affect the state or local area.

Privacy and the Internet

Some people assume that there is no privacy on the Internet and that you use it at your own risk. Others believe that companies with Web sites should have strict privacy procedures and be accountable for privacy invasion. Regardless of your view, the potential for privacy invasion on the Internet is huge. People wanting to invade your privacy could be anyone from criminal hackers to marketing companies to corporate bosses. Your personal and professional information can be seized on the Internet without your knowledge or consent. E-mail is a prime target, as discussed previously. Sending an e-mail message is like having an open conversation in a large room—people can listen to your messages. When you visit a Web site on the Internet, information about you and your computer can be captured. When this information is combined with other information, companies can know what you read, what products you buy, and what your interests are. According to an executive of an Internet software monitoring company, "It's a marketing person's dream."

Most people who buy products on the Web say it's very important for a site to have a policy explaining how personal information is used, and the policy statement must make people feel comfortable and be extremely clear about what information is collected and what will and will not be done with it. However, many Web sites still do not prominently display their privacy policy or implement practices completely consistent with that policy. The real issue that Internet users need to be concerned with is, What do content providers want with their personal information? If a site requests that you provide your name and address, you have every right to know why and what will be done with it. If you buy something and provide a shipping address, will it be sold to other retailers? Will your e-mail address be sold to a firm that creates a list of active Internet shoppers? And if so, you should realize that it's no different than the lists compiled from the orders you place with catalog retailers. You have the right to be taken off any mailing list.

A potential solution to some consumer privacy concerns is the screening technology called the **Platform for Privacy Preferences (P3P)** being proposed to shield users from sites that don't provide the level of privacy protection they desire. Instead of forcing users to find and read through the privacy policy for each site they visit, P3P software in a computer's browser will download the privacy policy from each site, scan it, and notify the user if the policy does not match his or her preferences. (Of course, unethical marketers can post a privacy policy that does not accurately reflect the manner in which the data is treated.) The World Wide Web Consortium, an international industry group whose members include Apple, Commerce One, Ericsson, and Microsoft, is supporting the development of P3P. Version 1.0 of the P3P was released in April 2002 and can be found at *www.w3.org/TR/P3P*.

Platform for Privacy Preferences (P3P)
A screening technology that shields users from Web sites that don't provide the level of privacy protection they desire.

The Children's Online Privacy Protection Act (COPPA) was passed by Congress in October 1998. This act was directed at Web sites catering to children, requiring them to post comprehensive privacy policies on their sites and to obtain parental consent before they collect any personal information from children under 13 years of age. Web site operators who violate the rule could be liable for civil penalties of up to $11,000 per violation.[57]

In April 2003, officials at Junkbusters Corporation held a press conference in which they discussed their decision to file a complaint with the Federal Trade Commission against eBay Inc. because of privacy concerns. The basis of the complaint was that Junkbusters thought that there was a significant inconsistency between eBay's short version of their privacy policy presented to the visitor and a more detailed policy. According to Junkbusters, the privacy policy summary states that eBay will turn over personal data to outside agencies only when absolutely necessary. However, the more detailed policy states that eBay will turn over such information at its discretion and without a warrant or subpoena. Junkbusters also claimed that eBay doesn't tell users that their e-mail addresses might be used by others to spam them and that despite efforts to protect data online at the Web site, there is a substantial risk to any private and financial information registered on eBay. eBay spokespeople rejected Junkbusters' contentions, pointing out that eBay has a privacy statement, plus a summary, a chart outlining privacy, and a FAQ link addressing privacy issues.[58]

Fairness in Information Use

Selling information to other companies can be so lucrative that many companies will continue to store and sell the data they collect on customers, employees, and others. When is this information storage and use fair and reasonable to the individuals whose data is stored and sold? Do individuals have a right to know about data stored about them and to decide what data is stored and used? As shown in Table 14.12, these questions can be broken down into four issues that should be addressed: knowledge, control, notice, and consent.

Table 14.12

The Right to Know and the Ability to Decide

Fairness Issues	Database Storage	Database Usage
The right to know	Knowledge	Notice
The ability to decide	Control	Consent

Knowledge. Should individuals have knowledge of what data is stored on them? In some cases, individuals are informed that information on them is stored in a corporate database. In others, individuals do not know that their personal information is stored in corporate databases.

Control. Should individuals have the ability to correct errors in corporate database systems? This is possible with most organizations, although it can be difficult in some cases.

Notice. Should an organization that uses personal data for a purpose other than the original purpose notify individuals in advance? Most companies don't do this.

Consent. If information on individuals is to be used for other purposes, should these individuals be asked to give their consent before data on them is used? Many companies do not give individuals the ability to decide if information on them will be sold or used for other purposes.

Federal Privacy Laws and Regulations

In the past few decades, significant laws have been passed regarding an individual's right to privacy. Others relate to business privacy rights and the fair use of data and information.

The Privacy Act of 1974

The major piece of legislation on privacy is the Privacy Act of 1974 (PA74), enacted by Congress during Gerald Ford's presidency. PA74 applies only to certain federal agencies. The act, which is about 15 pages long, is straightforward and easy to understand. The purpose of this act is to provide certain safeguards for individuals against an invasion of personal privacy by requiring federal agencies (except as otherwise provided by law) to do the following:

- Permit individuals to determine what records pertaining to them are collected, maintained, used, or disseminated by such agencies
- Permit individuals to prevent records pertaining to them from being used or made available for another purpose without their consent
- Permit individuals to gain access to information pertaining to them in federal agency records, to have a copy of all or any portion thereof, and to correct or amend such records
- Ensure that they collect, maintain, use, or disseminate any record of identifiable personal information in a manner that ensures that such action is for a necessary and lawful purpose, that the information is current and accurate for its intended use, and that adequate safeguards are provided to prevent misuse of such information
- Permit exemptions from this act only in cases where there is an important public need for such exemption, as determined by a specific law-making authority
- Be subject to civil suit for any damages that occur as a result of willful or intentional action that violates any individual's rights under this act

PA74, which applies to all federal agencies except the CIA and law enforcement agencies, also established a Privacy Study Commission to study existing databases and to recommend rules and legislation for consideration by Congress. PA74 also requires training for all federal employees who interact with a "system of records" under the act. Most of the training is conducted by the Civil Service Commission and the Department of Defense. Another interesting aspect of PA74 concerns the use of Social Security numbers—federal, state, and local governments and agencies cannot discriminate against any individual for not disclosing or reporting his or her Social Security number.

Gramm-Leach-Bliley Act

This act was passed in 1999 and required all financial institutions to protect and secure customers' nonpublic data from unauthorized access or use. Under terms of this act, it was assumed that all customers approve of the financial institutions' collecting and storing their personal information. The institutions were required to contact their customers and inform them of this fact. Customers were required to write separate letters to each of their individual financial institutions and state in writing that they wanted to opt out of the data collection and storage process. Most people were overwhelmed with the mass mailings they received from their financial institutions and simply discarded them without ever understanding their importance.

USA Patriot Act

As discussed previously, the 2001 Uniting and Strengthening America by Providing Appropriate Tools Required to Intercept and Obstruct Terrorism Act (USA Patriot Act) was passed in response to the September 11 terrorism acts. Proponents argue that it gives necessary new powers to both domestic law enforcement and international intelligence agencies. Critics argue that the law removes many of the checks and balances that previously allowed the courts to ensure law enforcement agencies did not abuse their powers. For example, under this act, Internet service providers and telephone companies must turn over customer information, including numbers called, without a court order if the FBI claims that the records are relevant to a terrorism investigation. Also, the company is forbidden to disclose that the FBI is conducting an investigation. Only time will tell how this act will be applied in the future.

Under the USA Patriot Act, Wachovia Corporation, like all financial services companies, is required to check lists provided by the Treasury Department against its own customer database to detect people who might be funneling money to terrorist organizations. To avoid unauthorized disclosure of customer data, Wachovia has a designated person within its security operations charged with the job. "You want to have a process that an individual oversees and is accountable for," says Bill Langley, the bank's chief compliance officer.[59]

Other Federal Privacy Laws

In addition to these acts, other pieces of federal legislation relate to privacy. A federal law that was passed in 1992 bans unsolicited fax advertisements. This law was upheld in a 1995 ruling by the Ninth U.S. Circuit Court of Appeals, which concluded that the law is a reasonable way to prevent the shifting of advertising costs to customers. Table 14.13 lists additional laws related to privacy.

Law	Provisions
Fair Credit Reporting Act of 1970 (FCRA)	Regulates operations of credit-reporting bureaus, including how they collect, store, and use credit information
Tax Reform Act of 1976	Restricts collection and use of certain information by the Internal Revenue Service
Electronic Funds Transfer Act of 1979	Outlines the responsibilities of companies that use electronic funds transfer systems, including consumer rights and liability for bank debit cards
Right to Financial Privacy Act of 1978	Restricts government access to certain records held by financial institutions
Freedom of Information Act of 1970	Guarantees access for individuals to personal data collected about them and about government activities in federal agency files
Education Privacy Act	Restricts collection and use of data by federally funded educational institutions, including specifications for the type of data collected, access by parents and students to the data, and limitations on disclosure
Computer Matching and Privacy Act of 1988	Regulates cross-references between federal agencies' computer files (e.g., to verify eligibility for federal programs)
Video Privacy Act of 1988	Prevents retail stores from disclosing video rental records without a court order
Telephone Consumer Protection Act of 1991	Limits telemarketers' practices
Cable Act of 1992	Regulates companies and organizations that provide wireless communications services, including cellular phones
Computer Abuse Amendments Act of 1994	Prohibits transmissions of harmful computer programs and code, including viruses
Gramm-Leach-Bliley Act of 1999	Requires all financial institutions to protect and secure customers' nonpublic data from unauthorized access or use
USA Patriot Act of 2001	Requires Internet service providers and telephone companies to turn over customer information, including numbers called, without a court order, if the FBI claims that the records are relevant to a terrorism investigation

Table 14.13

Federal Privacy Laws and Their Provisions

State Privacy Laws and Regulations

State legislatures have been considering and passing privacy legislation that is far-reaching and potentially more burdensome to business than existing federal legislation. The use of Social Security numbers, access to medical records, the disclosure of unlisted telephone numbers, the sharing of credit reports by credit bureaus, the disclosure of bank and personal financial information, and the use of criminal files are some of the issues being considered by state legislators. These state proposals could have an enormous effect on companies that do business within their borders.

- In Georgia and California, businesses are prohibited from discarding records containing personal information without shredding, erasing, or modifying the documents to make the sensitive data unreadable.

- Federal "opt out" regulations force consumers to take action if they don't want the company to sell or share information about them. In Vermont, California, New Mexico, and North Dakota, however, "opt-in" is the default, and businesses must get consumers' permission to share their data.
- In a countermeasure against identify thieves, California allows its residents to "freeze" their credit reports to prevent lenders from accessing the files and granting credit. Consumers who take this step are given a special password to unfreeze the accounts when they want to apply for new loans or credit cards.

Such state-by-state and even county-by-county exceptions to the federal law greatly complicate financial record keeping and data sharing.

Corporate Privacy Policies

Even though privacy laws for private organizations are not very restrictive, most organizations are very sensitive to privacy issues and fairness. They realize that invasions of privacy can hurt their business, turn away customers, and dramatically reduce revenues and profits. Consider a major international credit card company. If the company sold confidential financial information on millions of customers to other companies, the results could be disastrous. In a matter of days, the firm's business and revenues could be reduced dramatically. Thus, most organizations maintain privacy policies, even though they are not required by law. Some companies even have a privacy bill of rights that specifies how the privacy of employees, clients, and customers will be protected. Corporate privacy policies should address a customer's knowledge, control, notice, and consent over the storage and use of information. They may also cover who has access to private data and when it may be used.

Multinational companies face an extremely difficult challenge in implementing data-collection and dissemination processes and policies because of the multitude of differing country or regional statutes. For example, Australia requires companies to destroy customer data (including backup files) or make it anonymous once it's no longer needed. Firms that transfer customer and personnel data out of Europe must comply with European privacy laws that allow customers and employees to access data about them and let them determine how that information can be used.

A good database design practice is to assign a single unique identifier to each customer—so that each has a single record describing all relationships with the company across all its business units. That way, the organization can apply customer privacy preferences consistently throughout all databases. Failure to do so can expose the organization to legal risks—aside from upsetting customers who opted out of some collection practices. Again, the 1999 Gramm-Leach-Bliley Financial Services Modernization Act required all financial service institutions to communicate their data privacy rules and honor customer preferences.

Individual Efforts to Protect Privacy

Although numerous state and federal laws deal with privacy, privacy laws do not completely protect individual privacy. In addition, not all companies have privacy policies. As a result, many people are taking steps to increase their own privacy protection. Some of the steps that individuals can take to protect personal privacy include the following:

- *Find out what is stored about you in existing databases.* Call the major credit bureaus to get a copy of your credit report. You can obtain one free if you have been denied credit in the last 60 days. The major companies are Equifax (800-685-1111, *www.equifax.com*), TransUnion (800-916-8800, *www.transunion.com*), and Experian (888-397-3742, *www.experian.com*). You can also submit a Freedom of Information Act request to a federal agency that you suspect may have information stored on you.
- *Be careful when you share information about yourself.* Don't share information unless it is absolutely necessary. Every time you give information about yourself through an 800, 888, or 900 call, your privacy is at risk. Be vigilant in insisting that your doctor, bank, or financial institution not share information about you with others without your written consent.

- *Be proactive to protect your privacy.* You can get an unlisted phone number and ask the phone company to block caller ID systems from reading your phone number. If you change your address, don't fill out a change-of-address form with the U.S. Postal Service; you can notify the people and companies that you want to have your new address. Destroy copies of your charge card bills and shred monthly statements before disposing of them in the garbage. Be careful about sending personal e-mail messages over a corporate e-mail system. You can also get help in avoiding junk mail and telemarketing calls by visiting the Direct Marketing Association Web site at *www.the-dma.org.* Go to the Web site and look under Consumer Help-Remove Name from Lists.

- *When purchasing anything from a Web site, make sure that you safeguard your credit card numbers, passwords, and personal information.* Do not do business with a site unless you know that it handles credit card information securely (with Netscape Navigator, look for a solid blue key in a small blue rectangle; with Microsoft Explorer, look for the words "Secure Web Site"). Do not provide personal information without reviewing the site's data privacy policy.

THE WORK ENVIRONMENT

The use of computer-based information systems has changed the makeup of the workforce. Jobs that require IS literacy have increased, and many less-skilled positions have been eliminated. Corporate programs, such as reengineering and continuous improvement, bring with them the concern that, as business processes are restructured and information systems are integrated within them, the people involved in these processes will be removed.

However, the growing field of computer technology and information systems has opened up numerous avenues to professionals and nonprofessionals of all backgrounds. Enhanced telecommunications has been the impetus for new types of business and has created global markets in industries once limited to domestic markets. Even the simplest tasks have been aided by computers, making checkout lines faster, smoothing order processing, and allowing people with disabilities to participate more actively in the workforce. As computers and other IS components drop in cost and become easier to use, more workers will benefit from the increased productivity and efficiency provided by computers. Yet, despite these increases in productivity and efficiency, information systems can raise other concerns.

Health Concerns

Organizations can increase employee effectiveness by paying attention to the health concerns in today's work environment. For some people, working with computers can cause occupational stress. Anxieties about job insecurity, loss of control, incompetence, and demotion are just a few of the fears workers might experience. In some cases, the stress may become so severe that workers may sabotage computer systems and equipment. Monitoring employee stress may alert companies to potential problems. Training and counseling can often help the employee and deter problems.

Computer use may affect physical health as well. Strains, sprains, tendonitis, and other problems account for more than 60 percent of all occupational illnesses and about a third of workers' compensation claims, according to the Joyce Institute in Seattle. The cost to U.S. corporations for these types of health problems is as high as $27 billion annually. Claims relating to **repetitive motion disorder**, which can be caused by working with computer keyboards and other equipment, have increased greatly. Also called **repetitive stress injury** (**RSI**), the problems can include tendonitis, tennis elbow, the inability to hold objects, and sharp pain in the fingers. Also common is **carpal tunnel syndrome** (**CTS**), which is the aggravation of the pathway for nerves that travel through the wrist (the carpal tunnel). CTS involves wrist pain, a feeling of tingling and numbness, and difficulty grasping and holding objects. It may be caused by a number of factors, such as stress, lack of exercise, and

repetitive motion disorder (repetitive stress injury; RSI)
An injury that can be caused by working with computer keyboards and other equipment.

carpal tunnel syndrome (CTS)
The aggravation of the pathway for nerves that travel through the wrist (the carpal tunnel).

the repetitive motion of typing on a computer keyboard. Decisions on workers' compensation related to repetitive stress injuries have been made both for and against employees.

Other work-related health hazards involve emissions from improperly maintained and used equipment. Some studies show that poorly maintained laser printers may release ozone into the air; others dispute the claim. Numerous studies on the impact of emissions from display screens have also resulted in conflicting theories. Although some medical authorities believe that long-term exposure can cause cancer, studies are not conclusive at this time. In any case, many organizations are developing conservative and cautious policies.

Most computer manufacturers publish technical information on radiation emissions from their screens, and many companies pay close attention to this information. San Francisco was one of the first cities to propose a video display terminal (VDT) bill. The bill requires companies with 15 or more employees who spend at least four hours a day working with computer screens to give 15-minute breaks every 2 hours. In addition, adjustable chairs and workstations are required if employees request them.

The World Health Organization (WHO), U.S. Food and Drug Administration (FDA), and the U.S. General Accounting Office (GAO) have also analyzed health data on cell phone use, but they cannot definitively say whether cell phones pose any health risk. WHO states that gaps in knowledge need further research to better assess health risks. It expects that it will take until 2006 for the required research to be completed and evaluated, and the final results published. At this point, WHO states that radio frequency–absorbing covers or other absorbent devices on mobile phones cannot be justified on health grounds. The FDA states that while high levels of exposure to radio frequencies can produce biological damage, it is not known whether lower levels such as those associated with cell phones can cause adverse health effects.[60] A GAO report on the potential health hazards of cell phones has concluded that research conducted by the United States and international organizations shows that radio-frequency energy emitted by cell phones doesn't produce adverse health effects. But, the report noted, "There is not enough information to conclude they pose no risk." Experts recommend that if consumers are concerned, they should use a "hands-free" setup such as a headset or ear bud.

In addition to the possible health risks from radio-frequency exposure, cell phone use has raised a safety issue—an increased risk of traffic accidents as vehicle operators become distracted by talking on their cell phones (or operating their laptop computers, car navigation systems, or other computer devices) while driving. As a result, some states have made it illegal to operate a cell phone while driving.

Avoiding Health and Environmental Problems

Many computer-related health problems are minor and caused by a poorly designed work environment. The computer screen may be hard to read, with glare and poor contrast. Desks and chairs may also be uncomfortable. Keyboards and computer screens may be fixed in place or difficult to move. The hazardous activities associated with these unfavorable conditions are collectively referred to as *work stressors*. Although these problems may not be of major concern to casual users of computer systems, continued stressors such as repetitive motion, awkward posture, and eyestrain may cause more serious and long-term injuries. If nothing else, these problems can severely limit productivity and performance.

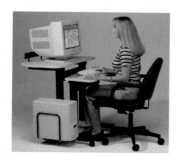

Research has shown that developing certain ergonomically correct habits can reduce the risk of RSI when using a computer.

(Source: Courtesy of Balt, Inc.)

ergonomics
The study of designing and positioning computer equipment for employee health and safety.

The study of designing and positioning computer equipment to improve worker productivity and minimize worker injuries, called **ergonomics**, has suggested a number of approaches to reducing these health problems. The objective is to have "no pain" computing. The slope of the keyboard, the positioning and design of display screens, and the placement and design of computer tables and chairs have been carefully studied. Flexibility is a major component of ergonomics and an important feature of computer devices. People come in many sizes, have differing preferences, and require different positioning of equipment for best results. Some people, for example, want to place the keyboard in their laps; others prefer it on a solid table. Because of these individual differences, computer designers are attempting to develop systems that provide a great deal of flexibility. In fact, the revolutionary design of Apple's iMac computer came about through concerns for users' comfort.

Many users feel that the ergonomics of using a notebook computer is a problem. After using one for a couple of hours on their lap, they complain of headaches and that their wrists have begun to hurt. This is probably because they sat in an awkward position bending their necks and holding their arms in an unnatural position. Although the built-in keyboard is fine for short periods, anyone who will be using a notebook for long periods should consider plugging a monitor into a docking station and using a wireless keyboard and mouse while sitting comfortably at a proper desk.

In addition to steps taken by companies, individuals can also reduce RSI and develop a better work environment. A number of excellent ideas can be found at the U.S. Department of Labor's Occupational Safety & Health Administration Web site through the Computer Workstations link at *www.osha.gov/SLTC/etools/computerworkstations/index.html* and at the Carpal Tunnel Syndrome Web page at *www.ctsplace.com*. Here is a brief set of recommendations:

- Maintain good posture and positioning. The top of the monitor should be at or just below eye level. Wrists and hands should be in line with forearms. Feet should be flat on the floor.
- In addition to good equipment, good posture and work habits can eliminate or reduce the potential of RSI.
- Don't ignore pain or discomfort. Many workers ignore early signs of RSI, and as a result, the problem becomes much worse and more difficult to treat.
- Use stretching and strengthening exercises. Often, such exercises can prevent RSI.
- Find a good physician who is familiar with RSI and ways to treat it.
- After treatment, start back slowly and pace yourself. Many people who are treated for RSI start back to work too soon and injure themselves again.

We have investigated how computers may be harmful to your health, but the computer can also be used to help prevent and treat general health problems. As discussed in Part 3 on business information systems, we have seen how computers can be used to assist doctors and other medical professionals by diagnosing medical problems and suggesting potential treatments. People can also use computers to get medical information. Special medical software for personal computers can help people get medical information and determine whether they need to see a doctor. A wealth of information is also available on the Internet on a variety of medical topics. Table 14.14 lists the top ten most useful health Web sites according to the Medical Library Association.

Table 14.14

Medical Topics on the Internet

Web Site Sponsor	URL	Description
National Cancer Institute	www.cancer.gov	Provides information about the cause, diagnosis, prevention, and treatment of cancer; rehabilitation from cancer; and the continuing care of cancer patients and the families of cancer patients.
Department for Health and Human Services Centers for Disease Control and Prevention	www.cdc.gov	Provides a wide range of health-related information.
American Academy of Family Physicians	http://familydoctor.org	Provides easy searches for information about common health problems.
U.S. Department of Health and Human Services Office of Disease Prevention and Health Promotion	www.healthfinder.gov	Resource to find government and nonprofit health and human services information on the Internet.
U.S. Center for HIV Information	http://hivinsite.ucsf.edu	Provides comprehensive, up-to-date information on HIV/AIDS prevention, treatment, and policy.
Nemours Foundation	www.kidshealth.org	Provides doctor-approved health information about children from before birth through adolescence.
Mayo Clinic	www.mayoclinic.com/index.cfm?	Provides useful and up-to-date information and tools that reflect the expertise and standard of excellence of Mayo Clinic.
Medem	http://medem.com/MedLB/medlib_entry.cfm	Online access to information and care for more than 90,000 physicians, their practices, and their patients.
U.S. National Library of Medicine	http://medlineplus.gov	Provides health information from the world's largest medical library, the National Library of Medicine.
Six New York City library organizations.	www.noah-health.org	Provides access to high quality full-text consumer health information in English and Spanish.

ETHICAL ISSUES IN INFORMATION SYSTEMS

As you've seen throughout the book in our "Ethical and Societal Issues" boxes, ethical issues deal with what is generally considered right or wrong. Some IS professionals believe that their field offers many opportunities for unethical behavior. They also believe that unethical behavior can be reduced by top-level managers developing, discussing, and enforcing codes of ethics. Information systems professionals are usually more satisfied with their jobs when top management stresses ethical behavior.

According to one view of business ethics, the "old contract" of business, the only responsibility of business is to its stockholders and owners. According to another view, the "social contract" of business, businesses are responsible to society. At one point or another in their operations, businesses may have employed one or both philosophies.

Various organizations and associations promote ethically responsible use of information systems and have developed codes of ethics. These organizations include the following:

- The Association of Information Technology Professionals (AITP), formerly the Data Processing Management Association (DPMA)
- The Association for Computing Machinery (ACM)
- The Institute of Electrical and Electronics Engineers (IEEE)
- Computer Professionals for Social Responsibility (CPSR)

The AITP Code of Ethics

The AITP has developed a code of ethics, standards of conduct, and enforcement procedures that give broad responsibilities to AITP members (see Figure 14.3). In general, the code of ethics is an obligation of every AITP member in the following areas:

- Obligation to management
- Obligation to fellow AITP members
- Obligation to society
- Obligation to college or university
- Obligation to the employer
- Obligation to country

For each area of obligation, standards of conduct describe the specific duties and responsibilities of AITP members. In addition, enforcement procedures stipulate that any complaint against an AITP member must be in writing, signed by the individual making the complaint, properly notarized, and submitted by certified or registered mail. Charges and complaints may be initiated by any AITP member in good standing.

Figure 14.3

AITP Code of Ethics

(Source: Courtesy of AITP—
www.aitp.org.)

Code of Ethics

I acknowledge:

That I have an obligation to management, therefore, I shall promote the understanding of information processing methods and procedures to management using every resource at my command.

That I have an obligation to my fellow members, therefore, I shall uphold the high ideals of AITP as outlined in the Association Bylaws. Further, I shall cooperate with my fellow members and shall treat them with honesty and respect at all times.

That I have an obligation to society and will participate to the best of my ability in the dissemination of knowledge pertaining to the general development and understanding of information processing. Further, I shall not use knowledge of a confidential nature to further my personal interest, nor shall I violate the privacy and confidentiality of information entrusted to me or to which I may gain access.

That I have an obligation to my College or University, therefore, I shall uphold its ethical and moral principles.

That I have an obligation to my employer whose trust I hold, therefore, I shall endeavor to discharge this obligation to the best of my ability, to guard my employer's interests, and to advise him or her wisely and honestly.

That I have an obligation to my country, therefore, in my personal, business, and social contacts, I shall uphold my nation and shall honor the chosen way of life of my fellow citizens.

I accept these obligations as a personal responsibility and as a member of this Association. I shall actively discharge these obligations and I dedicate myself to that end.

The ACM Code of Professional Conduct

The ACM has developed a number of specific professional responsibilities. These responsibilities include the following:

- Strive to achieve the highest quality, effectiveness, and dignity in both the process and products of professional work
- Acquire and maintain professional competence
- Know and respect existing laws pertaining to professional work
- Accept and provide appropriate professional review
- Give comprehensive and thorough evaluations of computer systems and their impacts, including analysis of possible risks
- Honor contracts, agreements, and assigned responsibilities
- Improve public understanding of computing and its consequences
- Access computing and communication resources only when authorized to do so

The mishandling of the social issues discussed in this chapter—including waste and mistakes, crime, privacy, health, and ethics—can devastate an organization. The prevention of these problems and recovery from them are important aspects of managing information and information systems as critical corporate assets. Increasingly, organizations are recognizing that people are the most important component of a computer-based information system and that long-term competitive advantage can be found in a well-trained, motivated, and knowledgeable workforce.

SUMMARY

Principle

Policies and procedures must be established to avoid computer waste and mistakes.

Computer waste is the inappropriate use of computer technology and resources in both the public and private sectors. Computer mistakes relate to errors, failures, and other problems that result in output that is incorrect and without value. Waste and mistakes occur in government agencies as well as corporations. At the corporate level, computer waste and mistakes impose unnecessarily high costs for an information system and drag down profits. Waste often results from poor integration of IS components, leading to duplication of efforts and overcapacity. Inefficient procedures also waste IS resources, as do thoughtless disposal of useful resources and misuse of computer time for games and personal processing jobs. Inappropriate processing instructions, inaccurate data entry, mishandling of IS output, and poor systems design all cause computer mistakes. More than half of all e-mail messages are spam, which represents a waste in people effort and computer and network resources.

Preventing waste and mistakes involves establishing, implementing, monitoring, and reviewing effective policies and procedures. Careful programming practices, thorough testing, flexible network interconnections, and rigorous backup procedures can help an information system prevent and recover from many kinds of mistakes. Companies should develop manuals and training programs to avoid waste and mistakes. Company policies should specify criteria for new resource purchases and user-developed processing tools to help guard against waste and mistakes.

Principle

Computer crime is a serious and rapidly growing area of concern requiring management attention.

Although no one really knows how pervasive cybercrime is, the number of IT-related security incidents is increasing dramatically—sixfold between 2000 and 2003.

Some crimes use computers as tools (e.g., to manipulate records, counterfeit money and documents, commit fraud via telecommunications links, and make unauthorized electronic transfers of money). Identity theft is a crime in which an imposter obtains key pieces of personal identification information in order to impersonate someone else. The information is then used to obtain credit, merchandise, and services in the name of the victim, or to provide the thief with false credentials.

A cyberterrorist is someone who intimidates or coerces a government or organization to advance his or her political or social objectives by launching computer-based attacks against computers, networks, and the information stored on them. A criminal hacker, also called a *cracker*, is a computer-savvy person who attempts to gain unauthorized or illegal access to computer systems to steal passwords, corrupt files and programs, and even transfer money. Script bunnies are wannabe crackers with little technical savvy. Insiders are employees, disgruntled or otherwise, working solo or in concert with outsiders to compromise corporate systems.

Computer crimes target computer systems and include illegal access to computer systems by criminal hackers, alteration and destruction of data and programs by viruses and simple theft of computer resources. A virus is a program that attaches itself to other programs. A worm functions as an independent program, replicating its own program files until it destroys other systems and programs or interrupts the operation of computer systems and networks. Malware is a general term for software that is harmful or destructive. A Trojan horse program is a malicious program that disguises itself as a useful application and purposefully does something the user does not expect. A logic bomb is designed to "explode" or execute at a specified time and date. A variant is a modified version of a virus that is produced by the virus's author or another person by amending the original virus code. A password sniffer is a small program hidden in a network or computer system that records identification numbers and passwords.

Because of increased computer use, greater emphasis is placed on the prevention and detection of computer crime. Antivirus software is used to detect the presence of viruses, worms, and logic bombs. Use of an intrusion detection system (IDS) provides another layer of protection in the event that an intruder gets past the outer security layers—passwords, security procedures, and corporate firewall. It monitors system and network resources and notifies network security personnel when it senses a possible intrusion. Many small and midsized organizations are outsourcing their network security operations to managed security service providers (MSSPs), which monitor, manage, and maintain network security hardware and software.

Software and Internet piracy may represent the most common computer crime. It is estimated that the software industry loses nearly $12 billion in revenue each year to software piracy. Computer scams have cost individuals and companies thousands of dollars. Computer crime is also an international issue.

Many organizations and people help prevent computer crime, among them state and federal agencies, corporations, and individuals. Security measures, such as using passwords, identification numbers, and data encryption, help to guard against illegal computer access, especially when supported by effective control procedures. Public key infrastructure (PKI) enables users of an unsecured public network such as

the Internet to securely and privately exchange data through the use of a public and a private cryptographic key pair that is obtained and shared through a trusted authority. The use of biometrics, involving the measurement of a person's unique characteristics such as iris, retina, or voice pattern, is another way to protect important data and information systems. Virus scanning software identifies and removes damaging computer programs. Law enforcement agencies armed with new legal tools enacted by Congress now actively pursue computer criminals.

Although most companies use data files for legitimate, justifiable purposes, opportunities for invasion of privacy abound. Privacy issues are a concern with government agencies, e-mail use, corporations, and the Internet. The Children's Internet Protection Act was enacted to protect minors using the Internet. The Privacy Act of 1974, with the support of other federal laws, establishes straightforward and easily understandable requirements for data collection, use, and distribution by federal agencies; federal law also serves as a nationwide moral guideline for privacy rights and activities by private organizations. The USA Patriot Act, passed just five weeks after the September 11 terrorist attacks, requires Internet service providers and telephone companies to turn over customer information, including numbers called, without a court order, if the FBI claims that the records are relevant to a terrorism investigation. Also, the company is forbidden to disclose that the FBI is conducting an investigation. Only time will tell how this act will be applied in the future. The Gramm-Leach-Bliley Act requires all financial institutions to protect and secure customers' nonpublic data from unauthorized access or use. Under terms of this act, it is assumed that all customers approve of the financial institutions collecting and storing their personal information.

Some states supplement federal protections and limit private organizations' activities within their jurisdictions. A business should develop a clear and thorough policy about privacy rights for customers, including database access. That policy should also address the rights of employees, including electronic monitoring systems and e-mail. Fairness in information use for privacy rights emphasizes knowledge, control, notice, and consent for people profiled in databases. Individuals should have knowledge of the data that is stored about them and have the ability to correct errors in corporate database systems. If information on individuals is to be used for other purposes, these individuals should be asked to give

their consent beforehand. Each individual has the right to know and the ability to decide. Platform for Privacy Preferences (P3P) is a screening technology that shields users from Web sites that don't provide the level of privacy protection they desire.

Principle

Jobs, equipment, and working conditions must be designed to avoid negative health effects.

Computers have changed the makeup of the workforce and even eliminated some jobs, but they have also expanded and enriched employment opportunities in many ways. Computers and related devices affect employees' emotional and physical health, especially by causing repetitive stress injury (RSI). Some critics blame computer systems for emissions of ozone and electromagnetic radiation. There is no conclusive data connecting cell phone use and cancer; however, heavy cell phone users may wish to use "hands-free" phone sets. Use of cell phones while driving has been linked to increased car accidents.

The study of designing and positioning computer equipment, called *ergonomics*, has suggested a number of approaches to reducing these health problems. Ergonomic design principles help to reduce harmful effects and increase the efficiency of an information system. The slope of the keyboard, the positioning and design of display screens, and the placement and design of computer tables and chairs are essential for good health. RSI prevention includes keeping good posture, not ignoring pain or problems, performing stretching and strengthening exercises, and seeking proper treatment. Although they can cause negative health consequences, information systems can also be used to provide a wealth of information on health topics through the Internet and other sources.

Ethics determine generally accepted and discouraged activities within a company and society at large. Ethical computer users define acceptable practices more strictly than just refraining from committing crimes; they also consider the effects of their IS activities, including Internet usage, on other people and organizations. The Association for Computing Machinery and the Association of Information Technology Professionals have developed guidelines and a code of ethics. Many IS professionals join computer-related associations and agree to abide by detailed ethical codes.

CHAPTER 14: SELF-ASSESSMENT TEST

Policies and procedures must be established to avoid computer waste and mistakes.

1. All IS professionals, managers, and users have a responsibility to see that the potential consequences of IS use are fully considered and addressed. True or False?
2. Computer-related waste and mistakes are major causes of computer problems, contributing to unnecessarily high _____ and lost _____.
3. The first step to prevent computer-related waste is to:
    a. establish policies and procedures regarding efficient acquisition, use, and disposal of systems and devices
    b. implement policies and procedures to minimize waste and mistakes according to the business conducted
    c. monitor routine practices and take corrective action if necessary
    d. review existing policies and procedures and determine whether they are adequate

Computer crime is a serious and rapidly growing area of concern requiring management attention.

4. The number of security problems reported to CERT between 2000 and 2003:
    a. increased sixfold
    b. decreased 50 percent
    c. doubled
    d. stayed about the same
5. _____ is the fastest and most pervasive worm attack on record, with damages estimated at $40 billion.
6. _____ is a crime in which an imposter obtains key pieces of personal identification information, such as Social Security or driver's license numbers, in order to impersonate someone else.
7. A person who enjoys computer technology and spends time learning and using computer systems is called a:

    a. script bunny
    b. hacker
    c. criminal hacker or cracker
    d. social engineer

8. A logic bomb is a type of Trojan horse that executes when specific conditions occur. True or False?
9. A program capable of attaching to disks or other files and replicating itself repeatedly, typically without user knowledge or permission, is called a:

    a. logic bomb
    b. Trojan horse
    c. virus
    d. worm

10. A(An) _____ is a modified version of a virus that is produced by the virus's author or another person amending the original virus code.
11. Half the loss from software piracy comes from Asia. True or False?
12. Phishing is a computer scam that seems to direct users to their bank's Web site but actually captures key personal information about its victims. True or False?

Jobs, equipment, and working conditions must be designed to avoid negative health effects.

13. CTS, or _____, is the aggravation of the pathway of nerves that travel through the wrist.
14. There is positive evidence that excessive use of cell phones increases the risk of brain cancer. True or False?
15. The study of designing and positioning computer equipment to improve worker productivity and minimize worker injuries is called _____.

CHAPTER 14: SELF-ASSESSMENT TEST ANSWERS

(1) True (2) costs, profits (3) a (4) a (5) Mydoom (6) Identity theft (7) b (8) True (9) d (10) variant (11) True (12) True (13) carpal tunnel syndrome (14) False (15) ergonomics

KEY TERMS

antivirus program 686
biometrics 692
carpal tunnel syndrome (CTS) 706
criminal hacker (cracker) 682
cyberterrorist 680
dumpster diving 680
ergonomics 708
hacker 682

ICRA rating system 695
identity theft 681
insider 682
Internet software piracy 689
intrusion detection system (IDS) 694
logic bomb 683
malware 683

managed security service provider (MSSP) 695
password sniffer 688
Platform for Privacy Preferences (P3P) 701
public key infrastructure (PKI) 691
repetitive motion disorder (repetitive stress injury, RSI) 706

script bunnies 682
social engineering 680
software piracy 688

Trojan horse 683
variant 684

virus 683
worm 683

REVIEW QUESTIONS

1. What is the CAPPS II program and what was its purpose?
2. Give two recent examples of computer mistakes causing serious repercussions.
3. Identify four broad actions that can be taken to prevent computer waste and mistakes.
4. What was the original intent of the Health Insurance Portability and Accountability Act (HIPPA)?
5. Give two examples of recent, major computer crimes.
6. What is a cyberterrorist? What evidence is there of increased cyberterrorism activity?
7. What is identity theft? What actions can you take to reduce the likelihood that you will be a victim of this crime?
8. What is a virus? What is a worm? How are they different?
9. Outline measures you should take to protect yourself against viruses and worms.

10. What are the penalties for copyright infringement? How can a software user be guilty of this crime?
11. Identify at least five tips to follow to avoid becoming a victim of a computer scam.
12. What is biometrics, and how can it be used to protect sensitive data?
13. What is the difference between antivirus software and an intrusion detection system?
14. What is the Children's Internet Protection Act?
15. What is ergonomics? How can it be applied to office workers?
16. What specific actions can you take to avoid RSI?
17. What is a code of ethics? Give an example.

DISCUSSION QUESTIONS

1. How does the CAN-SPAM Act affect those who send out spam? How might it affect those who receive spam?
2. Outline an approach, including specific techniques (e.g., dumpster diving, phishing, social engineering) that you could employ to gain personal data about the members of your class.
3. Identify and describe the different classes of computer criminals. What do you think motivates each?
4. Imagine that you are a hacker and have developed a Trojan horse program. What tactics might you use to get unsuspecting victims to load the program onto their computer?
5. Your marketing department has just opened a Web site and is requesting visitors to register at the site to enter a promotional contest where the chances of winning a prize are better than one in three. Visitors must provide the information necessary to contact them plus fill out a brief survey about the use of your company's products. What data privacy issues may arise?
6. Briefly discuss the potential for cyberterrorism to cause a major disruption in our daily life. What are some likely targets of a cyberterrorist? What sort of action could a cyberterrorist take against these targets?

7. Imagine that you are a manager in a small business organization. Identify three good reasons why your firm should hire an MSSP. Can you think of any reasons why this may not be a good idea?
8. Compare and contrast the use of the ICRA rating system and Internet filters for protecting minors from viewing inappropriate material on the Internet.
9. During 2002, a number of corporations were forced to restate earnings because they had used unethical accounting practices. Briefly discuss the extent to which you think that these problems were caused by a failure in the firm's accounting information systems.
10. Using information presented in this chapter on federal privacy legislation, identify which federal law regulates the following areas and situations: cross-checking IRS and Social Security files to verify the accuracy of information, customer liability for debit cards, individuals' right to access data contained in federal agency files, the IRS obtaining personal information, the government obtaining financial records, and employers' access to university transcripts.
11. Briefly discuss the difference between acting morally and acting legally. Give an example of acting legally and yet immorally.

PROBLEM-SOLVING EXERCISES

1. Access the Web site of one of the antivirus software providers. Get statistics on the number of viruses and their cost impact for the past four years. Use the graphics routine in your spreadsheet software to graph the increase over time.

2. Using your word processing software, write a few brief paragraphs summarizing the trends you see from reviewing the virus data for the past few years. Then cut and paste the graph from exercise 1 into your report.

3. Draft a letter to the Webmaster of the antivirus software Web site you chose in exercise 1. Share your report and request an estimate of the number of viruses for the next three years.

TEAM ACTIVITIES

1. Visit your local library and interview the librarians about their experience with the use of Internet filters. What software is used? Who is responsible for updating the list of sites off limits to minors? What level and kinds of complaints are made about the use of this technology? Do some patrons insist that the filter should be removed? Overall, what is the librarian's opinion of the use of Internet filters? Write a brief paper summarizing the interview.

2. Have each member of your team access ten different Web sites and summarize their findings in terms of the existence of data privacy policy statements: Did the site have such a policy? Was it easy to find? Was it complete and easy to understand? Did you find any sites using the P3P standard or ICRA rating method?

WEB EXERCISES

1. Search the Web for a site that provides software to detect and remove spyware. Write a short report for your instructor summarizing your findings.

2. Do research on the Web to find evidence of an increase or decrease in the amount of spam. To what is the change attributed? Write a brief memo to your instructor identifying your sources and summarizing your findings.

3. Echelon is a top-secret electronic eavesdropping system managed by the U.S. National Security Agency that is capable of intercepting and decrypting almost any electronic message sent anywhere in the world. It may have been in operation as early as the 1970s, but it wasn't until the 1990s that journalists were able to confirm its existence and gain insight into its capabilities. Do Web research to find out more about this system and its capabilities. Write a paragraph or two summarizing your findings.

CAREER EXERCISES

1. Computer forensics is a relatively new but growing field, which involves the discovery of computer-related evidence and data. It relies on formal computer evidence-processing protocols. Its findings may be presented in a court of law. Cases involving trade secrets, commercial disputes, employment discrimination, misdemeanor and felony crimes, and personal injury can be won or lost solely with the introduction of recovered e-mail messages and other electronic files and records. Computer forensics tools and methods are used extensively by law enforcement, military, intelligence agencies, and businesses. Do research to identify the experience and training necessary to become a certified computer forensics specialist. Would you consider this as a possible career field? Why or why not?

2. Your marketing organization is establishing a Web site to promote, market, and sell your firm's products. Make a list of the laws and regulations that might apply to the design and operation of this Web site. How would you confirm that your Web site is in conformance with all of them?

VIDEO QUESTIONS

Watch the video clip **Controversial Digital Bouncers**, and answer these questions:

1. Would you be comfortable with a business establishment collecting and storing identifying information about you from your driver's license, along with a snapshot of your face for security use? Why or why not?

2. Are there any laws that restrict a club from distributing the information it collects about you to other businesses?

CASE STUDIES

Case One

Working to Reduce the Number of Software Vulnerabilities

Exploiting programming flaws (called *vulnerabilities*) is the primary means by which hackers gain access to networks, computers, and applications today. Examples of commonly exploited vulnerabilities include buffer overflows, invalidated parameters, format string errors, and broken access control. The number of reported vulnerabilities increased from 171 in 1995 to 3,784 in 2003, according to Carnegie-Mellon's CERT Coordination Center. Not surprisingly, there has been a dramatic increase in the number of security breaches.

Software developers and their customers seem to be stuck in an endless cycle of costly development and deployment of faulty software, followed by the release of patches to fix newly discovered vulnerabilities, and then scrambling by end users to install the patches, ideally fast enough that crackers do not take advantage of them.

"Zero-day" attacks are hacker attacks that take advantage of programming vulnerabilities before software makers can identify them and develop patches to correct them. Fortunately, there haven't yet been any major zero-day attacks, but malicious hackers are getting better and faster at exploiting flaws. Summer 2003's Blaster worm, one of the most virulent and widespread ever, hit the Internet barely a month after Microsoft released a patch for the software flaw it exploited. In contrast, January 2003's SQL Slammer worm took eight months to appear after the vulnerability it targeted was first disclosed. The lead time for users to apply fixes is shrinking.

Fortunately, there is a small but growing group of software vendors offering tools designed to help companies identify and fix flaws in the application development stage and to eliminate software vulnerabilities before the applications are deployed. Cambridge, Massachusetts–based @stake unveiled its SmartRisk Analyzer in May 2004. This modeling and analysis tool scans computer code written in the C, C++, and Java languages for flaws that could pose security risks for customers using finished software products. The software uses a process called *static analysis*, which allows developers to identify and eliminate problems as they are writing the code. This proactive approach is opposed to so-called dynamic analysis tools, which use automated input tests to measure the response of finished applications. The Smart-Risk Analyzer product compares binary code (the zeros and ones that are the foundation of all computer languages) to an @stake database of about 400 security and code reliability rules. It can generate reports that list flaws by type or rank them by severity. A remediation module marks erroneous code and suggests ways to fix coding mistakes.

In April 2004, start-up company Fortify Software introduced Fortify Source Code Analysis, a suite of software products that lets companies compare C++ and Java code to a list of more than 500 vulnerabilities published by software quality management company Cigital. The suite also includes a capability that allows project testers and quality assurance teams to test the software just before deployment, including simulating a hacker using every trick in the book to compromise the software. The software is not inexpensive. A 25-person project team using the security platform would pay around $150,000 plus an annual $1,000 subscription to get the latest rules when they come out.

Unlike manual audits and code reviews that are time-consuming and limited in scope, an automated approach accelerates the industry's ability to deliver secure software. Software developers now can run security checks whenever they want—and against as much of the code base as they desire.

Discussion Questions

1. Why is there concern over the shrinking time between identification of a software vulnerability and the launch of an attack exploiting that vulnerability?
2. What is static analysis? What are some advantages of static analysis over dynamic analysis? Why do software developers employ both methods?

Critical Thinking Questions

3. What factors might cause a software vendor to sacrifice the quality of the underlying code of its products?

4. What measures can a software buyer take to ensure the quality of the underlying code of a software package— before purchasing and installing the software?

SOURCES: Paul Roberts, "Secure Coding Attracts Interest, Investment," *Computerworld*, May 24, 2004, *www.computerworld.com*; Jaikumar Vijavan, "Fortify Launches Tools for Security Testing during App Development," *Computerworld*, April 5, 2004, *www.computerworld.com*; Jim Wagner, "A New Approach to Fortify Your Software," *E-Security Planet*, April 5, 2004, *www.esecurityplanet.com*; Jaikumar Vijayan, "InfoSec 2003: 'Zero-Day' Attacks Seen as Growing Threat," *Computerworld*, December 11, 2003, *www.computerworld.com*.

Case Two
Beware Spyware!

Adware is any software application in which advertising banners are displayed while the program is running. The authors of these applications include additional code that delivers the ads, which can be viewed through pop-up windows or through a bar that appears on a computer screen. The developer of the application is paid an advertising fee based on how many users view the ads. This additional payment helps recover the author's programming development cost and hold down the cost for the user.

But adware can include additional code called *spyware* that captures the user's personal information and passes it on to third parties, without the user's explicit consent or knowledge. Although spyware is not illegal, computer security and privacy advocates, including the Electronic Privacy Information Center, denounce its use. Their concern is with the uses of the information that adware and spyware providers collect and the lack of control users have over what those providers "feed" them.

Spyware and adware both enter a computer system when a user opens an attachment or clicks on a Web page that allows a program to be executed on the system without the user's knowledge. Both make similar modifications to a system, such as changing Windows registry settings, adding services, and installing and executing applications. This process usually involves the tracking and sending of data and statistics via a server installed on the user's PC and the use of the user's Internet connection in the background. Even though the name may indicate it, spyware is not illegal. However, it raises certain privacy issues. The spyware may include surveillance programs, key loggers, remote-control tools, and Trojan horses that run in the background without the users' knowledge, downloading information on Web-surfing activities and uploading advertising in the background for use in pop-up ads.

Initially, organizations viewed spyware as a nuisance that was best handled by desktop support groups. But the key issue isn't system performance or productivity-sapping pop-ups—it's the uneasy feeling that these programs have opened an unauthorized communication channel that could put sensitive information at risk. CIOs worry that, in addition to downloading data on Web-surfing activity, a spyware program may capture user login and password information or that a benign adware program may provide a communications pathway that could be hijacked for uploading more malicious software or sending sensitive data to competitors.

Unfortunately, the current generation of antivirus and firewall software does not control or eliminate spyware effectively. Indeed, most users aren't even aware that they've been infected until they start noticing slow system performance or odd computer behavior. Once a system is infected, often the only way to remove the problem is to rebuild the system from scratch, a tedious and time-consuming process.

Discussion Questions

1. A number of software applications are available as freeware to help computer users search for and remove suspected spyware programs. Is it possible that a purveyor of spyware could attract unsuspecting users by promoting free spyware-removal software?
2. Do a Web search on spyware and identify specific actions recommended to identify and remove spyware.

Critical Thinking Questions

3. You are short on funds, and a friend approaches you with an opportunity to work for a company creating Web links for spyware downloads. He has been receiving checks for this work every couple of weeks. Make a list of the ethical and privacy factors you would consider in deciding whether or not to take on this opportunity.
4. Would you take a job helping companies to download spyware? Why or why not?

SOURCES: Robert L. Mitchell, "Spyware Sneaks into the Desktop," *Computerworld*, May 3, 2004, *www.computerworld.com*; John Soat, "IT Confidential: Anti-Spyware's First Step; Accenture OK," *InformationWeek*, June 21, 2004, *www.informationweek.com*; The Advisory Council, "SmartAdvice: Clean and Manage Company Data to Learn What Information You've Got," *InformationWeek*, June 21, 2004, *www.informationweek.com*; Mathias Thurman, "Spyware Gets Top Billing," *Computerworld*, July 5, 2004, *www.computerworld.com*; Emily Kumler, "Who's Seeding the Net with Spyware," *PC World*, June 15, 2004, *www.pcworld.com*.

Case Three

Cyberstalking

Unfortunately, stalkers have discovered the Internet and are using it as another means through which to exert control over their victims. Cyberstalking includes the use of e-mail, message boards, instant messaging, Web sites, chat rooms, and other electronic communication methods to stalk someone. It is analogous to physical stalking in that it incorporates persistent behaviors that instill apprehension and fear. Cyberstalking may be a precursor to more serious actions, including physical violence.

Cyberstalking is now considered a crime in many places, and as a crime, its definition varies by locale or country. One useful definition of cyberstalking from a recent study is "the repeated use of the Internet, e-mail, or related digital electronic communication devices to annoy, alarm, or threaten a specific individual or group of individuals."

Groups such as Working to Halt Online Abuse (WHOA), SafetyEd, and CyberAngels estimate that they receive as many as 400 requests for help each week from cyberstalking victims, or more than 20,000 reported cases each year. The U.S. Department of Justice estimates that there may be as many as 475,000 online victims each year. While 90 percent of the victims are women, the number of female cyberstalkers is also increasing. Sadly, growing numbers of children are cyberstalking other children. Members of certain ethnic groups, especially those from the Middle East, are increasingly being targeted. Most incidents are not related to romances gone sour; in fact, a majority of the cases involve strangers.

Cyberstalkers use a variety of techniques to locate private information about a potential victim and to conceal their own identities. Numerous Web sites offer personal information, including unlisted telephone numbers and detailed directions to homes or offices. For a fee, other Web sites promise to provide Social Security numbers, financial data, and other personal information that cyberstalkers might find useful in the pursuit of their victims. A cyberstalker's true identity can be concealed by using different Internet service providers or by adopting different screen names. More devious cyberstalkers may employ anonymous remailers, making it all but impossible to determine the true identity or source of an e-mail message.

Feeling bold in their cloak of anonymity, cyberstalkers may send unsolicited e-mail messages, which can include hate, obscene, or threatening content. They may create newsgroup or message board postings about the victim or start rumors. They may even assume the victim's persona online (such as in chat room) for the purpose of sullying the victim's reputation, posting details (whether factual or false) about the victim, or soliciting unwanted contacts from others. In addition, online harassment may include sending the victim computer viruses or electronic junk mail (spamming).

Discussion Questions

1. Do you know anyone who has been a victim of a cyberstalker? Without revealing the person's identity, what was that person's experience?
2. What do you think motivates a cyberstalker?

Critical Thinking Questions

3. What measures should you take to avoid becoming a cyberstalker's victim?
4. What measures should you take if you find yourself a victim of a cyberstalker?

SOURCES: Roy Mark, "Cyberstalking Is Increasing," *InternetNews.com*, April 18, 2003, accessed at *http://dc.internet.com/news/*; R. D'Ovidio and J. Doyle, "A Study on Cyberstalking: Understanding Investigative Hurdles," *FBI Law Enforcement Bulletin*, March 2003, vol. 72(3), pp. 10–17; Andrew Fano, "The New Internet Cops," *Computerworld*, September 8, 2003, *www.computerworld.com*; "About Cyber-stalking," Cyber-stalking.net Web site at *www.cyber-stalking.net*, accessed July 13, 2004; Douglas Schweitzer, "How to Combat Cyberstalking," *Computerworld*, July 16, 2003, *www.computerworld.com*.

NOTES

Sources for the opening vignette: Grant Gross, "U.S. Agencies Defend Data Mining Plans," *Computerworld*, May 12, 2003, *www.computerworld.com*; Larry Greenemeier, "CAPPS II Progress Raises Privacy Concerns," *InformationWeek*, February 12, 2004, *www.informationweek.com*; Dan Verton, "Airline Passenger Screening System Faces Deployment Delays," *Computerworld*, February 16, 2004, *www.computerworld.com*; Tony Kontzer, "Privacy Pressure," *InformationWeek*, March 22, 2004, *www.informationweek.com/story*; Larry Greenemeier, "Security Agency Used More Passenger Data Than First Thought," *Information-Week*, June 23, 2004, *www.informationweek.com*; Larry Greenemeier, "CAPPS II Crash-Lands," *InformationWeek*, July 19, 2004, *www.informationweek.com*.

1. Lenke, Tim, "Spam Harmed Economy More Than Hackers, Viruses," *The Washington Times*, November 10, 2003, accessed at *www.washtimes.com*.
2. Associated Press, "Soyuz Capsule Lands Safely in Kazakhstan," FoxNews.com, October 27, 2003, accessed at *www.foxnews.com*.
3. Williams, Martyn, "Japan Retailer Deluged with Orders for $25 eMacs," *Computerworld*, April 23, 2004, accessed at *www.computerworld.com*.
4. Krasnowski, Matt, "Macy's OKs Settlement in Price-Accuracy Case," *San Diego Union Tribune*, May 22, 2004, accessed at *www.signonsandiego.com*.
5. Cowley, Stacy, "Errors in IBM's DB2 Cause Outage at Denmark Bank," *Computerworld*, April 7, 2003 accessed at *www.computerworld.com*.

6. Wrolstad, Jay, "New IE Flaws Labeled Extremely Critical," *News Factor Network*, November 26, 2003, accessed at *www.newsfactor. com*.

7. Rosencrance, Linda, "Computer Glitch May Have Caused Drug Labeling Errors," *Bio-IT World*, March 19, 2003, accessed at *www.bio-ITWorld.com*.

8. Roberts, Paul, IDG News Service, "Software Bug May Cause Missile Errors," *PC World*, March 2003, *www.pcworld.com*.

9. Martin, Steve, "Lack Of IT Controls Seen as Reason for Earnings Restatement," *InformationWeek*, October 13, 2004, *www.informationweek.com*.

10. "2004 Global Information Security Survey," *InformationWeek*, June 30, 2004, accessed at *www.informationweek.com*.

11. Verton, Dan, "Criminals Using High Tech Methods for Old Style Crimes," *Computerworld*, February 13, 2003, accessed at *www.computerworld.com*.

12. Roberts, Paul, "Michigan Man Pleads Guilty to Wireless Hack into Stores," *Computerworld*, June 7, 2004, accessed at *www.computerworld.com*.

13. Verton, Dan, "Europe Battles Insiders-Turned-Hackers, EU Cybercop Says," *Computerworld*, July 31, 2003, accessed at *www.computerworld.com*.

14. Gross, Grant, "U.S. Government Cracks Down on Internet Fraud," *Computerworld*, May 16, 2003, accessed at *www.computerworld.com*.

15. Garvey, Martin J., "Utilities' Security Is Too Lax, Report Says," *InformationWeek*, April 5, 2004 accessed at *www.informationweek.com*.

16. Claburn, Thomas, "Feds Want Tougher Penalties for Insider Identity Theft," *InformationWeek*, May 24, 2004, accessed at *www.informationweek.com*.

17. Claburn, Thomas, "Feds Want Tougher Penalties for Insider Identity Theft," *InformationWeek*, May 24, 2004, accessed at *www.informationweek.com*.

18. Claburn, Thomas, "Feds Want Tougher Penalties for Insider Identity Theft," *InformationWeek*, May 24, 2004, accessed at *www.informationweek.com*.

19. Vijayan, Jaikumar, "Experts-Standard Virus Protection Best Way to Fight Mydoom," *Computerworld*, January 28, 2004, accessed at *www.computerworld.com*.

20. Roberts, Paul, "New E-Mail Worm Breaks Infection Records," *Computerworld*, January 27, 2004, accessed at *www.computerworld.com*.

21. Legard, David, "Microsoft Offers $250,000 Reward in MyDoom.B Attacks," *Computerworld*, January 30, 2004, accessed at *www.computerworld.com*.

22. Rosencrance, Linda, "RIAA Site Off-Line, Virus Blamed," *Computerworld*, March 23, 2004, accessed at *www.computerworld.com*.

23. Hulme, George V., "Cisco Still Mum on Code Theft," *InformationWeek*, May 20, 2004, accessed at *www.informationweek.com*.

24. Morphy, Erika, "Trojan Virus Author Busted for Making a Peep," *NewsFactor Network*, May 28, 2004, accessed at *www.newsfactor.com*.

25. Roberts, Paul, "New Bagel.U A Virus of Few Words," *Computerworld*, March 26, 2004, accessed at *www.computerworld.com*.

26. Roberts, Paul, "New Netsky Worms Change Their Strips," *Computerworld*, April 6, 2004, accessed at *www.computerworld.com*.

27. Brewin, Bob, "Worldport Help Desk Helps Prevent Virus Attacks," *Computerworld*, April 19, 2004, accessed at *www.computerworld.com*.

28. Cohen, Peter, "Macs and Viruses—Are Users as Safe as They Think?" *Computerworld*, February 18, 2004, accessed at *www.computerworld.com*.

29. F-Secure Web site at *www.f-secure.com/corporate/intro.shtml*, accessed on May 28, 2004.

30. Network Associates Web site at *www.networkassociates.com*, accessed on May 30, 2004.

31. Jesdandun, Anick, "Kinko's Case Highlights Internet Risks," *InformationWeek*, July 22, 2004, accessed at *www.informationweek.com*.

32. "2003 BSI Computer Theft Survey," accessed at *www.brigadoonsoftware.com/survey.php* on May 30, 2004.

33. "What Is Piracy?" SIIAA Anti-Piracy Web site, *www.spa.org/piracy/whatis.aspon*, accessed on May 30, 2004.

34. The Business Software Alliance Web site, "Piracy and the Law," *www.bsa.org/usa/antipiracy/Piracy-and-the-Law.cfm*, accessed on May 30, 2004.

35. Rosencrance, Linda, "Citibank Customers Hit with E-mail Scam," *Computerworld*, October 24, 2003, accessed at *www.computerworld.com*.

36. Roberts, Paul, "Phishing Scams Skyrocket in April," *Computerworld*, May 18, 2004, accessed at *www.computerworld.com*.

37. Roberts, Paul, "Gartner: Phishing Attacks Up Against U.S. Consumers," *Computerworld*, May 6, 2004, accessed at *www.computerworld.com*.

38. Evers, Joris, "FTC Cracks Down on Web Page Selling Scam," *Computerworld*, August 13, 2003, accessed at *www.computerworld.com*.

39. Software and Information Industry Association, "Anti-Piracy," accessed at *www.siia.net/piracy/whatis.asp*, on June 29, 2004.

40. Songini, Marc L., "CA to Push Automated Fraud Detection," *Computerworld*, March 10, 2003, accessed at *www.computerworld.com*.

41. McDougall, Paul, "Accenture's 'Virtual Border' Project," *InformationWeek*, June 7, 2004, accessed at *www.informationweek.com*.

42. Rosencrance, Linda, "Red Bull Pays $105,000 to Settle Software Piracy Claim," *Computerworld*, March 17, 2004, accessed at *www.computerworld.com*.

43. Thurman, Mathias, "Failure to Communicate," *Computerworld*, April 26, 2004, *www.computerworld.com*.

44. "ISS Backs New IPS Offering with Cash-or-Credit Guarantee," *Network World*, January 1, 2004, accessed at *www.nwfusion.com*.

45. Hulme, George V., "Merrill Lynch Hands Off Network Security to VeriSign," *InformationWeek*, May 21, 2003, accessed at *www.informationweek.com*.

46. 2004 Internet Filter Report, found at *www.internetfilterreview.com*, accessed on July 1, 2004.

47. "ACLU Disappointed in Ruling on Internet Censorship in Libraries, But Sees Limited Impact for Adults," the American Civil Liberties Union Web site, *www.aclu.org*, on June 23, 2003.

48. Text of COPA found at "Laws, Cases and Codes: U.S. Code: Title 47, Section 231," accessed at *http://caselaw.lp.findlaw.com* on July 1, 2004.

49. Gearan, Anne, "High Court Says Web Porn Law Restricts Free Speech," *Cincinnati Enquirer*, June 30, 2004, p. A-2.

50. Roberts, Paul, "I'll See Your Web Site in Court," *Computerworld*, March 7, 2003, accessed at *www.computerworld.com*.

51. Roberts, Paul, "Mohegan Sun Won't Gamble on Insider Threats," *IDG News Service*, April 30, 2004, accessed at *www.computerworld.com*.

52. Verton, Dan, "Senate Votes to Kill Funds for Antiterror Data Mining," *Computerworld*, July 21, 2003, accessed at *www.computerworld.com*.

53. From the Seisint Web site, *www2.seisint.com/aboutus/index.html*, accessed on July 3, 2004.

54. Bergstein, Brian, "Seven-State Info Store Is a Potent Repository of Personal Data," *InformationWeek*, January 23, 2004, accessed at *www.informationweek.com*.

55. Sullivan, Andy, "Government Computer Surveillance Rings Alarm Bells," *Computerworld*, May 27, 2004, accessed at *www.computerworld.com*.

56. Bergstein, Brian, "Seven-State Info Store Is a Potent Repository of Personal Data," *InformationWeek*, January 23, 2004, accessed at *www.informationweek.com*.

57. "The Children's Online Privacy Protection Act," Electronic Privacy Information Center, *www.epic.org/privacy/kids*, accessed on July 3, 2004.

58. Sullivan, Brian, "Update: EBay, Amazon Hit with Complaints from Privacy Groups," *Computerworld,* April 22, 2003, accessed at *www.computerworld.com*.

59. Kontzer, Tony, "Privacy Pressure," *InformationWeek*, March 22, 2004, accessed at *www.informationweek.com/story*.

60. "Cell Phone Facts, Consumer Information on Wireless Phones," information provided by the Federal Communications Communication at the Food and Drug Administration Web site, *www.fda.gov/cellphones/qa.html#22*, accessed On October 16, 2004.

Efforts to Build E-Government and an Information Society in Hungary

Janos Fustos
Metropolitan State College of Denver

Communication between citizens and their government has always been an important goal. State agencies and institutions must ensure open access to documents and forms and provide information about programs, services, and policies. This information must be timely, accurate, complete, and updated regularly and has to be freely and immediately available. Modern communication technology and the application of Internet services can support such governmental efforts and help establish e-government and an information society. These efforts are even more important in the former Communist countries in Eastern Europe, where the citizen-to-government communication had to be rebuilt and the governments have higher obligations to close the information gap. These countries also have to promote the use of high-tech solutions to help bridge the "information divide," keep pace with the rest of the world, and start to build the foundations of the information society.

After living under Soviet influence for 40 years, Hungary had the first free multiparty parliamentary elections in April 1990. The new National Assemblies and the coalition governments formed after the elections committed themselves to the establishment and stabilization of the political, economic, and legal foundations brought about by the systemic change. There was—and still is—a lot to do, but the gap is narrowing, as shown by the spread of telecommunications technology in the European Union (EU):

Phone lines (per 1,000 people)

	1990	1995	2002
USA	550	630	646
EU	422	492	546

Before the changes, communication was under political control—in an effort to avoid free dissemination of information and make sure it closely followed the official view. As Hungary started to move toward the Western world and a free-market economy, communication—phone lines and Internet connections—became focal points for development. By 1997, a vast majority of the previously state-owned companies and assets had been privatized and were sold to international investors that competed for the businesses.

The Hungarian government was one of the early adopters of Internet communication and solutions. In 1991, it established the Inter-Departmental Committee of Informatics and the Coordination Office at the Prime Minister's Office. These two offices coordinated official efforts for and between departments and agencies. In 1995, the government published a Strategy for Informatics, which laid out countrywide communication efforts. In 1999, the government published the *Hungarian Reply to the Challenges of the Information Society*, which discussed the technological background and requirements of an information-based society. Hungary also joined NATO in 1999, which was a major acknowledgment of the country's efforts to rebuild its political system and establish a free society. Then in 2001, the European Commission published the *eEurope + Action Plan* in conjunction with Central and East European countries. The goal of this plan was to foster the development of an information society in the countries that were joining the EU, including Hungary. That same year, the prime minister's office launched Hungary's first

e-government portal, eKormanyzat.hu, providing citizens and businesses with a user-friendly entry point to government information and services. In 2002, the Ministry of Informatics and Communications (IHM) was created and assumed responsibility for development of the information society from the Office of the Government. To establish wider involvement, the Interministerial Committee on Information Society was formed to coordinate programs with professional organizations, businesses, and economic partners. The Interministerial Committee also oversees the formation and annual renewal of Hungarian information society strategy (*http://english.itktb.hu/Engine.aspx*). As of May 1, 2004, Hungary became a full member of the EU. Recent statistics on Internet usage show the growth of Hungary's information society, in comparison with the United States and the EU overall:

Number of Internet users (per 1,000 people)

	1992	1997	2002
USA	18	150	538
EU	3	56	204
Hungary	1	19	158

Number of Internet hosts (per 1,000 people)

	1997	2002
USA	56	382
EU	12	40
Hungary	3	20

The Hungarian government also established several programs to promote the use of computers and created opportunities for citizens to use them in education and in everyday life. Higher education institutions were among the first to obtain high-speed Internet access, and it soon became fully integrated into the European academic community. At this time, all Hungarian schools have Internet connections. Students and teachers (in primary education, high school, college, and continuing education) get tax relief when buying computers, digital devices, and software. The Sulinet (school-net) program also supports research, conferences and seminars, developers, content providers, and smart classrooms. The following statistics show the growth of PC usage over a decade:

Number of PCs (per 1,000 people)

	1992	1997	2002
USA	218	409	659
EU	97	194	310
Hungary	19	58	108

The Internet is becoming more and more available as a community service all over the country. Post offices, schools, bus stations, churches, hospitals, libraries, and a network of community centers offer access for free or for a flat rate. Hungarians even invented a traffic sign containing an "@" symbol to identify Internet access points and recommended it for worldwide use.

"E-government is not 'old government' plus the Internet. E-government is the use of new technologies to transform Europe's public administrations and to improve radically the way they work with their customers, be they citizens, enterprises, or other administrations. Furthermore, e-government is now a key vehicle for the implementation and achievement of higher policy objectives," states the Commission of the European Communities.

Discussion Questions

1. Why is it important for governments to introduce and apply high-tech solutions to promote programs, services, and policies?
2. What features are implemented in your state or local government's Web portal?

Critical Thinking Questions

3. How can access to information transform a society?
4. Besides e-government efforts, what are other elements of an information society?

acceptance testing Conducting any tests required by the user.

accounting MIS An information system that provides aggregate information on accounts payable, accounts receivable, payroll, and many other applications.

accounting systems Systems that include budget, accounts receivable, payroll, asset management, and general ledger.

accounts payable system System that increases an organization's control over purchasing, improves cash flow, increases profitability, and provides more effective management of current liabilities.

accounts receivable system System that manages the cash flow of the company by keeping track of the money owed the company on charges for goods sold and services performed.

ad hoc DSS A DSS concerned with situations or decisions that come up only a few times during the life of the organization.

alpha testing Testing an incomplete or early version of the system.

analog signal A continuous, curving signal.

antivirus program Program or utility that prevents viruses and recovers from them if they infect a computer.

applet Small program embedded in Web pages.

application flowcharts Diagrams that show relationships among applications or systems.

application program interface (API) Interface that allows applications to make use of the operating system.

application service provider (ASP) A company that provides software, end-user support, and the computer hardware on which to run the software from the user's facilities.

application software Programs that help users solve particular computing problems.

arithmetic/logic unit (ALU) Portion of the CPU that performs mathematical calculations and makes logical comparisons.

ARPANET Project started by the U.S. Department of Defense (DoD) in 1969 as both an experiment in reliable networking and a means to link DoD and military research contractors, including a large number of universities doing military-funded research.

artificial intelligence (AI) A field in which the computer system takes on the characteristics of human intelligence to mimic or duplicate the functions of the human brain.

artificial intelligence systems People, procedures, hardware, software, data, and knowledge needed to develop computer systems and machines that demonstrate the characteristics of intelligence.

asking directly An approach to gathering data that asks users, stakeholders, and other managers about what they want and expect from the new or modified system.

asset management transaction processing system System that controls investments in capital equipment and manages depreciation for maximum tax benefits.

asynchronous communications Communications in which the receiver gets the message minutes, hours, or days after it is sent.

attribute A characteristic of an entity.

audit trail Documentation that allows the auditor to trace any output from the computer system back to the source documents.

auditing Analyzing the financial condition of an organization and determining whether financial statements and reports

produced by the financial MIS are accurate.

automated clearing house (ACH) A secure, private network that connects all U.S. financial institutions to one another by way of the Federal Reserve Board or other ACH operators.

backbone One of the Internet's high-speed, long-distance communications links.

backward chaining The process of starting with conclusions and working backward to the supporting facts.

bandwidth The range of frequencies that an electronic signal occupies on a given transmission medium.

batch processing system Method of computerized processing in which business transactions are accumulated over a period of time and prepared for processing as a single unit or batch.

benchmark test An examination that compares computer systems operating under the same conditions.

best practices The most efficient and effective ways to complete a business process.

beta testing Testing a complete and stable system by end users.

biometrics The measurement of a person's trait, whether physical or behavioral.

bit Binary digit—0 or 1.

bot A software tool that searches the Web for information, products, prices, and so forth.

brainstorming Decision-making approach that often consists of members offering ideas "off the top of their heads."

bridge A device used to connect two or more networks that use the same communications protocol.

broadband Telecommunications in which a wide band of frequencies is available to transmit information, allowing more information to be transmitted in a given amount of time.

budget transaction processing system System that automates many of the tasks required to amass budget data, distribute it to users, and consolidate the prepared budgets.

bus line The physical wiring that connects the computer system components.

bus network A type of topology that contains computers and computer devices on a single line; each device is connected directly to the bus and can communicate directly with all other devices on the network; one of the most popular types of personal computer networks.

business continuity planning Identification of the business processes that must be restored first in the event of a disaster and specification of what actions should be taken and who should take them to restore operations.

business intelligence The process of gathering enough of the right information in a timely manner and usable form and analyzing it to have a positive impact on business strategy, tactics, or operations.

business-to-business (B2B) e-commerce A form of e-commerce in which the participants are organizations.

business-to-consumer (B2C) e-commerce A form of e-commerce in which customers deal directly with the organization, avoiding any intermediaries.

byte (B) Eight bits that together represent a single character of data.

cache memory A type of high-speed memory that a processor can access more rapidly than main memory.

carpal tunnel syndrome (CTS) The aggravation of the pathway for nerves that travel through the wrist (the carpal tunnel).

CASE repository A database of system descriptions, parameters, and objectives.

catalog management software Software that automates the process of creating a real-time interactive catalog and delivering customized content to a user's screen.

CD-recordable (CD-R) disc An optical disc that can be written on only once.

CD-rewritable (CD-RW) disc An optical disc that allows personal computer users to replace their disks with high-capacity CDs that can be written on and edited.

central processing unit (CPU) The part of the computer that consists of three associated elements: the arithmetic/logic unit, the control unit, and the register areas.

centralized processing Processing alternative in which all processing occurs in a single location or facility.

certificate authority (CA) A trusted third party that issues digital certificates.

certification Process for testing skills and knowledge that results in a statement by the certifying authority that says an individual is capable of performing a particular kind of job.

change model Representation of change theories that identifies the phases of change and the best way to implement them.

character Basic building block of information, consisting of uppercase letters, lowercase letters, numeric digits, or special symbols.

chat room A facility that enables two or more people to engage in interactive "conversations" over the Internet.

chief programmer team A group of skilled IS professionals who design and implement a set of programs. This team has total responsibility for building the best software possible; also called *lead programmer team.*

choice stage The third stage of decision making, which requires selecting a course of action.

clickstream data Data gathered based on the Web sites you visit and the items you click on.

client/server An architecture in which multiple computer platforms are dedicated to special functions such as database management, printing, communications, and program execution.

clock speed A series of electronic pulses produced at a predetermined rate that affects machine cycle time.

closed shops IS departments in which only authorized operators can run the computers.

cold site A computer environment that includes rooms, electrical service, telecommunications links, data storage devices, and the like; also called a *shell.*

command-based user interface A user interface that requires that text commands be given to the computer to perform basic activities.

communications The transmission of a signal by way of a medium from a sender to a receiver.

communications protocol A standard set of rules that control a telecommunications connection.

communications software Software that provides a number of important functions in a network, such as error checking and data security.

compact disc read-only memory (CD-ROM) A common form of optical disc on which data, once it has been recorded, cannot be modified.

competitive advantage A significant and (ideally) long-term benefit to a company over its competition.

competitive intelligence One aspect of business intelligence limited to information about competitors and the ways that knowledge affects strategy, tactics, and operations.

competitive local exchange carrier (CLEC) A company that is allowed to compete with the LECs, such as a wireless, satellite, or cable service provider.

compiler A special software program that converts the programmer's source code into the machine-language instructions consisting of binary digits.

complex instruction set computing (CISC) A computer chip design that places as many microcode instructions into the central processor as possible.

computer literacy Knowledge of computer systems and equipment and the ways they function; it stresses equipment and devices (hardware), programs and instructions (software), databases, and telecommunications.

computer network The communications media, devices, and software needed to connect two or more computer systems and/or devices.

computer programs Sequences of instructions for the computer.

computer system architecture The structure, or configuration, of the hardware components of a computer system.

computer system platform The combination of a particular hardware configuration and systems software package.

computer-aided software engineering (CASE) Tools that automate many of the tasks required in a systems development effort and enforce adherence to the SDLC.

computer-assisted manufacturing (CAM) A system that directly controls manufacturing equipment.

computer-based information system (CBIS) A single set of hardware, software, databases, telecommunications, people, and procedures that are configured to collect, manipulate, store, and process data into information.

computer-integrated manufacturing Using computers to link the components of the production process into an effective system.

concurrency control A method of dealing with a situation in which two or more people need to access the same record in a database at the same time.

consumer-to-consumer (C2C) e-commerce A form of e-commerce in which the participants are individuals, with one serving as the buyer and the other as the seller.

content streaming A method for transferring multimedia files over the Internet so that the data stream of voice and pictures plays more or less continuously without a break, or very few of them; enables users to browse large files in real time.

continuous improvement Constantly seeking ways to improve the business processes to add value to products and services.

control unit Part of the CPU that sequentially accesses program instructions; decodes them; and coordinates the flow of data in and out of the ALU, the registers, primary storage, and even secondary storage and various output devices.

cookie A text file that an Internet company can place on the hard disk of a computer system to track user movements.

coprocessor Part of the computer that speeds processing by executing specific types of instructions while the CPU works on another processing activity.

cost center Division within a company that does not directly generate revenue.

cost/benefit analysis An approach that lists the costs and benefits of each proposed system. Once they are expressed in monetary terms, all the costs are compared with all the benefits.

counterintelligence The steps an organization takes to protect information sought by "hostile" intelligence gatherers.

creative analysis The investigation of new approaches to existing problems.

criminal hacker (cracker) A computer-savvy person who attempts to gain unauthorized or illegal access to computer systems.

critical analysis The unbiased and careful questioning of whether system elements are related in the most effective or efficient ways.

critical path Activities that, if delayed, would delay the entire project.

critical success factors (CSFs) Factors that are essential to the success of a functional area of an organization.

cross-platform development Development technique that allows programmers to develop programs that can run on computer systems having different hardware and operating systems, or platforms.

cryptography The process of converting a message into a secret code and changing the encoded message back to regular text.

culture Set of major understandings and assumptions shared by a group.

customer relationship management (CRM) system System that helps a company manage all aspects of customer encounters, including marketing and advertising, sales, customer service after the sale, and programs to retain loyal customers.

cybermall A single Web site that offers many products and services at one Internet location.

cyberterrorist Someone who intimidates or coerces a government or organization to advance his or her political or social objectives by launching computer-based attacks against computers, networks, and the information stored on them.

data Raw facts, such as an employee's name and number of hours worked in a week, inventory part numbers, or sales orders.

data administrator A nontechnical position responsible for defining and implementing consistent principles for a variety of data issues.

data analysis Manipulation of the collected data so that it is usable for the development team members who are participating in systems analysis.

data cleanup The process of looking for and fixing inconsistencies to ensure that data is accurate and complete.

data collection The process of capturing and gathering all data necessary to complete transactions.

data communications A specialized subset of telecommunications that refers to the electronic collection, processing, and distribution of data—typically between computer system hardware devices.

data correction The process of reentering miskeyed or misscanned data that was found during data editing.

data definition language (DDL) A collection of instructions and commands used to define and describe data and data relationships in a specific database.

data dictionary A detailed description of all the data used in the database.

data editing The process of checking data for validity and completeness.

data entry Process by which human-readable data is converted into a machine-readable form.

data input Process that involves transferring machine-readable data into the system.

data integrity The degree to which the data in any one file is accurate.

data item The specific value of an attribute.

data manipulation The process of performing calculations and other data transformations related to business transactions.

data manipulation language (DML) The commands that are used to manipulate the data in a database.

data mart A subset of a data warehouse.

data mining An information-analysis tool that involves the automated discovery of patterns and relationships in a data warehouse.

data model A diagram of data entities and their relationships.

data preparation, or data conversion Involves making sure all files and databases are ready to be used with new computer software and systems.

data redundancy Duplication of data in separate files.

data storage The process of updating one or more databases with new transactions.

data store Representation of a storage location for data.

data warehouse A database that collects business information from many sources in the enterprise, covering all aspects of the company's processes, products, and customers.

database An organized collection of facts and information.

database administrator (DBA) A skilled IS professional who directs all activities related to an organization's database.

database approach to data management An approach whereby a pool of related data is shared by multiple application programs.

database management system (DBMS) A group of programs that manipulate the database and provide an interface between the database and the user of the database and other application programs.

data-flow diagram (DFD) A model of objects, associations, and activities that describes how data can flow between and around various objects.

data-flow line Arrows that show the direction of data element movement.

decentralized processing Processing alternative in which processing devices are placed at various remote locations.

decision room A room that supports decision making, with the decision makers in the same building, combining face-to-face verbal interaction with technology to make the meeting more effective and efficient.

decision structure A programming structure that allows the computer to branch, depending on certain conditions. Normally, there are only two possible branches.

decision support system (DSS) An organized collection of people, procedures, software, databases, and devices used to support problem-specific decision making.

decision-making phase The first part of problem solving, including three stages: intelligence, design, and choice.

dedicated line A communications line that provides a constant connection between two points; no switching or dialing is needed, and the two devices are always connected.

delphi approach A decision-making approach in which group decision makers are geographically dispersed; this approach encourages diversity among group members and fosters creativity and original thinking in decision making.

demand report Report developed to give certain information at a person's request.

denial-of-service (DOS) attack An attack in which the attacker takes command of many computers on the

Internet and uses them to flood the target Web site with requests for data and other small tasks, preventing the target machine from serving legitimate users.

design report The primary result of systems design, reflecting the decisions made and preparing the way for systems implementation.

design stage The second stage of decision making, in which alternative solutions to the problem are developed.

desktop computer A relatively small, inexpensive, single-user computer that is highly versatile.

deterrence controls Rules and procedures to prevent problems before they occur.

dialogue manager User interface that allows decision makers to easily access and manipulate the DSS and to use common business terms and phrases.

digital certificate An attachment to an e-mail message or data embedded in a Web page that verifies the identity of a sender or a Web site.

digital computer camera Input device used with a PC to record and store images and video in digital form.

digital signal A signal represented by bits.

digital signature Encryption technique used to verify the identity of a message sender for processing online financial transactions.

digital subscriber line (DSL) A telecommunications technology that delivers high-bandwidth information to homes and small businesses over ordinary copper telephone wires.

digital versatile disk (DVD) Storage medium used to store digital video or computer data.

direct access Retrieval method in which data can be retrieved without the need to read and discard other data.

direct access storage device (DASD) Device used for direct access of secondary storage data.

direct conversion (also called *plunge* or *direct cutover*) Stopping the old system and starting the new system on a given date.

direct observation Watching the existing system in action by one or more members of the analysis team.

disaster recovery Actions that must be taken to restore computer operations and services in the event of a disaster.

disintermediation The elimination of intermediate organizations between the producer and the consumer.

disk mirroring A process of storing data that provides an exact copy that protects users fully in the event of data loss.

distance learning The use of telecommunications to extend the classroom.

distributed database A database in which the data may be spread across several smaller databases connected via telecommunications devices.

distributed processing Processing alternative in which computers are placed at remote locations but connected to each other via a network.

document production The process of generating output records and reports.

documentation Text that describes the program functions to help the user operate the computer system.

domain (1) The allowable values for data attributes. (2) The area of knowledge addressed by the expert system.

domain expert The individual or group who has the expertise or knowledge one is trying to capture in the expert system.

downsizing Reducing the number of employees to cut costs.

drill-down report Report providing increasingly detailed data about a situation.

dumpster diving Searching through the garbage for important pieces of information that can help crack an organization's computers or be used to convince someone at the company to give someone access to the computers.

dynamic Web pages Web pages containing variable information that are built in response to a specific Web visitor's request.

e-commerce Any business transaction executed electronically between parties such as companies (business-to-business), companies and consumers (business-to-consumer), consumers and other consumers (consumer-to-consumer), business and the public sector, and consumers and the public sector.

e-commerce software Software that supports catalog management, product configuration, shopping cart facilities, e-commerce transaction processing, and Web traffic data analysis.

e-commerce transaction processing software Software that provides the basic connection between participants in the e-commerce economy, enabling communications between trading partners, regardless of their technical infrastructure.

economic feasibility Determination of whether the project makes financial sense and whether predicted benefits offset the cost and time needed to obtain them.

economic order quantity (EOQ) The quantity that should be reordered to minimize total inventory costs.

effectiveness A measure of the extent to which a system achieves its goals; it can be computed by dividing the goals actually achieved by the total of the stated goals.

efficiency A measure of what is produced divided by what is consumed.

electronic bill presentment A method of billing whereby the biller posts an image of your statement on the Internet and alerts you by e-mail that your bill has arrived.

electronic cash An amount of money that is computerized, stored, and used as cash for e-commerce transactions.

electronic data interchange (EDI) An intercompany, application-to-application communication of data in standard format, permitting the recipient to perform the functions of a standard business transaction.

electronic document distribution Process that involves transporting documents—such as sales reports, policy manuals, and advertising brochures—over communications lines and networks.

electronic exchange An electronic forum where manufacturers, suppliers, and competitors buy and sell goods, trade market information, and run back-office operations.

electronic funds transfer (EFT) A system of transferring money from one bank account directly to another without any paper money changing hands.

electronic retailing (e-tailing) The direct sale from business to consumer through electronic storefronts, typically designed around an electronic catalog and shopping cart model.

electronic shopping cart A model commonly used by many e-commerce sites to track the items selected for purchase, allowing shoppers to view what is in their cart, add new items to it, and remove items from it.

electronic software distribution Process that involves installing software on a file server for users to share by signing on to the network and requesting that the software be downloaded onto their computers over a network.

electronic wallet A computerized stored value that holds credit card information, electronic cash, owner identification, and address information.

empowerment Giving employees and their managers more responsibility and authority to make decisions, take certain actions, and have more control over their jobs.

encryption. The conversion of a message into a secret code.

end-user systems development Any systems development project in which the primary effort is undertaken by a combination of business managers and users.

enterprise data modeling Data modeling done at the level of the entire enterprise.

enterprise resource planning (ERP) system A set of integrated programs capable of managing a company's vital business operations for an entire multi-site, global organization.

enterprise sphere of influence Sphere of influence that serves the needs of the firm in its interaction with its environment.

entity Generalized class of people, places, or things for which data is collected, stored, and maintained.

entity symbol Representation of either a source or destination of a data element.

entity-relationship (ER) diagrams Data models that use basic graphical symbols to show the organization of and relationships between data.

ergonomics The study of designing and positioning computer equipment for employee health and safety.

event-driven review Review triggered by a problem or opportunity such as an error, a corporate merger, or a new market for products.

exception report Report automatically produced when a situation is unusual or requires management action.

execution time (E-time) The time it takes to execute an instruction and store the results.

executive support system (ESS) Specialized DSS that includes all hardware, software, data, procedures, and people used to assist senior-level executives within the organization.

expandable storage devices Storage that uses removable disk cartridges to provide additional storage capacity.

expert system A system that gives a computer the ability to make suggestions and act like an expert in a particular field.

expert system shell A collection of software packages and tools used to develop expert systems.

explanation facility Component of an expert system that allows a user or decision maker to understand how the expert system arrived at certain conclusions or results.

Extensible Markup Language (XML) Markup language for Web documents containing structured information, including words, pictures, and other elements.

external auditing Auditing performed by an outside group.

extranet A network based on Web technologies that links selected resources of a company's intranet with its customers, suppliers, or other business partners.

feasibility analysis Assessment of the technical, economic, legal, operational, and schedule feasibility of a project.

feedback Output that is used to make changes to input or processing activities.

field Typically a name, number, or combination of characters that describes an aspect of a business object or activity.

file A collection of related records.

file server An architecture in which the application and database reside on one host computer, called the file server.

File Transfer Protocol (FTP) A protocol that describes a file transfer process between a host and a remote computer and allows users to copy files from one computer to another.

final evaluation A detailed investigation of the proposals offered by the vendors remaining after the preliminary evaluation.

financial MIS An information system that provides financial information to all financial managers within an organization.

firewall A device that sits between an internal network and the Internet, limiting access into and out of a network based on access policies.

five-forces model A widely accepted model that identifies five key factors that can lead to attainment of competitive advantage including (1) rivalry among existing competitors, (2) the threat of new entrants, (3) the threat of substitute products and services, (4) the bargaining power of buyers, and (5) the bargaining power of suppliers.

flash memory A silicon computer chip that, unlike RAM, is nonvolatile and keeps its memory when the power is shut off.

flat organizational structure Organizational structure with a reduced number of management layers.

flexible manufacturing system (FMS) An approach that allows manufacturing facilities to rapidly and efficiently change from making one product to making another.

forecasting Predicting future events to avoid problems.

forward chaining The process of starting with the facts and working forward to the conclusions.

front-end processor A special-purpose computer that manages communications to and from a computer system.

full-duplex channel A communications channel that permits data transmission in both directions at the same time thus, the full-duplex channel is like two simplex lines.

fuzzy logic A special research area in computer science that allows shades of gray and does not require everything to be simple black or white, yes/no, or true/false.

game theory Use of information systems to develop competitive strategies for people, organizations, or even countries.

Gantt chart A graphical tool used for planning, monitoring, and coordinating projects.

gateway A network point that acts as an entrance to another network.

general ledger system System designed to automate financial reporting and data entry.

general-purpose computers Computers used for a wide variety of applications.

genetic algorithm An approach to solving large, complex problems in which a number of related operations or models change and evolve until the best one emerges.

geographic information system (GIS) A computer system capable of assembling, storing, manipulating, and displaying geographic information, i.e., data identified according to their locations.

gigahertz (GHz) Billions of cycles per second.

goal-seeking analysis The process of determining the problem data required for a given result.

graphical user interface (GUI) An interface that uses icons and menus displayed on screen to send commands to the computer system.

grid chart Table that shows relationships among the various aspects of a systems development effort.

grid computing The use of a collection of computers, often owned by multiple individuals or organizations, to work in a coordinated manner to solve a common problem.

group consensus Decision making by a group that is appointed and given the responsibility of making the final evaluation and selection.

group consensus approach Decision-making approach that forces members in the group to reach a unanimous decision.

group support system (GSS) Software application that consists of most elements in a DSS, plus software to provide effective support in group decision making; also called *group decision support system* or *computerized collaborative work system*.

hacker A person who enjoys computer technology and spends time learning and using computer systems.

half-duplex channel A communications channel that can transmit data in either direction, but not simultaneously.

handheld computer A single-user computer that provides ease of portability because of its small size.

hardware Any machinery (most of which use digital circuits) that assists in the input, processing, storage, and output activities of an information system.

help facility A program that provides assistance when a user is having difficulty understanding what is happening or what type of response is expected.

hertz One cycle or pulse per second.

heuristics Commonly accepted guidelines or procedures that usually find a good solution.

hierarchical network A type of topology that uses a treelike structure with messages passed along the branches of the hierarchy until they reach their destination.

hierarchy of data Bits, characters, fields, records, files, and databases.

highly structured problems Problems that are straightforward and require known facts and relationships.

home page A cover page for a Web site that has graphics, titles, and text.

hot site A duplicate, operational hardware system or immediate access to one through a specialized vendor.

HTML tags Codes that let the Web browser know how to format text—as a heading, as a list, or as body text—and whether images, sound, and other elements should be inserted.

hub A place of convergence where data arrives from one or more directions and is forwarded out in one or more other directions.

human resource MIS An information system that is concerned with activities related to employees and potential employees of an organization, also called a *personnel MIS*.

hybrid network A network topology that is a combination of other network types.

hypermedia Tools that connect the data on Web pages, allowing users to access topics in whatever order they wish.

Hypertext Markup Language (HTML) The standard page description language for Web pages.

icon Picture.

ICRA rating system System to protect individuals from harmful or objectionable Internet content, while safeguarding the free speech rights of others.

identity theft A crime in which an imposter obtains key pieces of personal identification information, such as Social Security or driver's license numbers, in order to impersonate someone else. The information is then used to obtain credit, merchandise, and services in the name of the victim or to provide the thief with false credentials.

if-then statements Rules that suggest certain conclusions.

image log A separate file that contains only changes to applications.

implementation stage A stage of problem solving in which a solution is put into effect.

incremental backup Making a backup copy of all files changed during the last few days or the last week.

inference engine Part of the expert system that seeks information and relationships from the knowledge base and provides answers, predictions, and suggestions the way a human expert would.

informatics A specialized system that combines traditional disciplines, such as science and medicine, with computer systems and technology.

information A collection of facts organized in such a way that they have additional value beyond the value of the facts themselves.

information center A support function that provides users with assistance, training, application development, documentation, equipment selection and setup, standards, technical assistance, and troubleshooting.

information service unit A miniature IS department.

information system (IS) A set of interrelated components that collect, manipulate, store, and disseminate data and information and provide a feedback mechanism to meet an objective.

information systems literacy Knowledge of how data and information are used by individuals, groups, and organizations.

information systems planning The translation of strategic and organizational goals into systems development initiatives.

input The activity of gathering and capturing raw data.

insider An employee, disgruntled or otherwise, working solo or in concert with outsiders to compromise corporate systems.

installation The process of physically placing the computer equipment on the site and making it operational.

instant messaging A method that allows two or more individuals to communicate online using the Internet.

institutional DSS A DSS that handles situations or decisions that occur more than once, usually several times a year or more. An institutional DSS is used repeatedly and refined over the years.

instruction time (I-time) The time it takes to perform the fetch-instruction and decode-instruction steps of the instruction phase.

integrated development environments (IDEs) Development approach that combines the tools needed for programming with a programming language into one integrated package.

integrated services digital network (ISDN) A set of standards for integrating voice and data communications onto a single line via digital transmission over copper wire or other media.

integration testing Testing all related systems together.

intellectual property Music, books, inventions, paintings, and other special items protected by patents, copyrights, or trademarks.

intelligence stage The first stage of decision making, in which potential problems or opportunities are identified and defined.

intelligent agent Programs and a knowledge base used to perform a specific task for a person, a process, or another program; also called *intelligent robot* or *bot*.

intelligent behavior The ability to learn from experiences and apply knowledge acquired from experience, handle complex situations, solve problems when important information is missing, determine what is important, react quickly and correctly to a new situation, understand visual images, process and manipulate symbols, be creative and imaginative, and use heuristics.

internal auditing Auditing performed by individuals within the organization.

international network A network that links systems between countries.

Internet The world's largest computer network, actually consisting of thousands of interconnected networks, all freely exchanging information.

Internet Protocol (IP) Communication standard that enables traffic to be routed from one network to another as needed.

Internet service provider (ISP) Any company that provides individuals or organizations with access to the Internet.

Internet software piracy Illegally downloading software from the Internet.

intranet An internal network based on Web technologies that allows people within an organization to exchange information and work on projects.

intrusion detection system (IDS) Software that monitors system and network resources and notifies network security personnel when it senses a possible intrusion.

inventory-control system System that updates the computerized inventory records to reflect the exact quantity on hand of each stock-keeping unit.

Java An object-oriented programming language from Sun Microsystems based on C++ that allows small programs (applets) to be embedded within an HTML document.

joining Data manipulation that combines two or more tables.

joint application development (JAD) Process for data collection and requirements analysis, in which users, stakeholders, and IS professionals work together to analyze existing systems, propose possible solutions, and define the requirements of a new or modified system.

just-in-time (JIT) inventory approach A philosophy of inventory management in which inventory and materials are delivered just before they are used in manufacturing a product.

kernel The heart of the operating system, which controls the most critical processes.

key A field or set of fields in a record that is used to identify the record.

key-indicator report Summary of the previous day's critical activities; typically available at the beginning of each workday.

knowledge An awareness and understanding of a set of information and ways that information can be made useful to support a specific task or reach a decision.

knowledge acquisition facility Part of the expert system that provides convenient and efficient means of capturing and storing all the components of the knowledge base.

knowledge base A component of an expert system that stores all relevant information, data, rules, cases, and relationships used by the expert system.

knowledge engineer An individual who has training or experience in the design, development, implementation, and maintenance of an expert system.

knowledge management The process of capturing a company's collective expertise wherever it resides—in computers, on paper, in people's heads—and distributing it wherever it can help produce the biggest payoff.

knowledge user The individual or group who uses and benefits from the expert system.

learning systems A combination of software and hardware that allows the computer to change how it functions or reacts to situations based on feedback it receives.

legal feasibility Determination of whether laws or regulations may prevent or limit a systems development project.

linking Data manipulation that combines two or more tables using common data attributes to form a new table with only the unique data attributes.

local area network (LAN) A network that connects computer systems and devices within the same geographic area.

local exchange carrier (LEC) A public telephone company in the United States that provides service to homes and businesses within its defined geographical area, called its *local access and transport area (LATA)*.

logic bomb An application or system virus designed to "explode" or execute at a specified time and date.

logical design Description of the functional requirements of a system.

long-distance carrier A traditional long-distance phone provider, such as AT&T, Sprint, or MCI.

lookup tables Tables containing data that are developed and used by computer programs to simplify and shorten data entry.

loop structure A programming structure with two commonly used structures for loops: do-until and do-while. In the do-until structure, the loop is done until a certain condition is met. In the do-while structure, the loop is done while a certain condition exists.

machine cycle The instruction phase followed by the execution phase.

magnetic disk Common secondary storage medium, with bits represented by magnetized areas.

magnetic tape Common secondary storage medium; Mylar film coated with iron oxide with portions of the tape magnetized to represent bits.

magneto-optical (MO) disk A hybrid between a magnetic disk and an optical disc.

mainframe computer Large, powerful computer often shared by hundreds of concurrent users connected to the machine via terminals.

maintenance team A special IS team responsible for modifying, fixing, and updating existing software.

make-or-buy decision The decision regarding whether to obtain the necessary software from internal or external sources.

malware Software that is harmful or destructive, such as viruses and worms.

managed security service provider (MSSP) An organization that monitors, manages, and maintains network security hardware and software for its client companies.

management information system (MIS) An organized collection of people, procedures, software, databases, and devices used to provide routine information to managers and decision makers.

manufacturing resource planning (MRPII) An integrated, company-wide system based on network scheduling that enables people to run their

business with a high level of customer service and productivity.

market segmentation The identification of specific markets to target them with advertising messages.

marketing MIS Information system that supports managerial activities in product development, distribution, pricing decisions, and promotional effectiveness.

massively parallel processing A form of multiprocessing that speeds processing by linking hundreds or thousands of processors to operate at the same time, or in parallel, with each processor having its own bus, memory, disks, copy of the operating system, and applications.

material requirements planning (MRP) A set of inventory-control techniques that help coordinate thousands of inventory items when the demand of one item is dependent on the demand for another.

megahertz (MHz) Millions of cycles per second.

menu-driven system System in which users simply pick what they want to do from a list of alternatives.

meta tag A special HTML tag, not visible on the displayed Web page, that contains keywords representing your site's content, which search engines use to build indexes pointing to your Web site.

meta-search engine A tool that submits keywords to several individual search engines and returns the results from all search engines queried.

metropolitan area network (MAN) A telecommunications network that connects users and their computers within a geographical area larger than that covered by a LAN but smaller than the area covered by a WAN, such as a city or college campus.

microcode Predefined, elementary circuits and logical operations that the

processor performs when it executes an instruction.

middleware Software that allows different systems to communicate and transfer data back and forth.

MIPS Millions of instructions per second.

mission-critical systems Systems that play a pivotal role in an organization's continued operations and goal attainment.

mobile commerce (m-commerce) Transactions conducted anywhere, anytime.

model An abstraction or an approximation that is used to represent reality.

model base Part of a DSS that provides decision makers access to a variety of models and assists them in decision making.

model management software Software that coordinates the use of models in a DSS.

modem A device that translates data from digital to analog and analog to digital.

monitoring stage Final stage of the problem-solving process, in which decision makers evaluate the implementation.

Moore's Law A hypothesis that states that transistor densities on a single chip will double every 18 months.

MP3 A standard format for compressing a sound sequence into a small file.

multidimensional organizational structure Structure that may incorporate several structures at the same time.

multifunction device A device that can combine a printer, fax machine, scanner, and copy machine into one device.

multiplexer A device that allows several telecommunications signals to be transmitted over a single communications medium at the same time.

multiprocessing The simultaneous execution of two or more instructions at the same time.

multitasking Capability that allows a user to run more than one application at the same time.

music device A device that can be used to download music from the Internet and play the music.

natural language processing Processing that allows the computer to understand and react to statements and commands made in a "natural" language, such as English.

net present value The preferred approach for ranking competing projects and determining economic feasibility.

network Connected computers and computer equipment in a building, around the country, or around the world to enable electronic communications.

network operating system (NOS) Systems software that controls the computer systems and devices on a network and allows them to communicate with each other.

network topology Logical model that describes how networks are structured or configured.

network-attached storage (NAS) Storage devices that attach to a network instead of to a single computer.

network-management software Software that enables a manager on a networked desktop to monitor the use of individual computers and shared hardware (such as printers), scan for viruses, and ensure compliance with software licenses.

neural network A computer system that can simulate the functioning of a human brain.

newsgroups Online discussion groups that focus on specific topics.

nominal group technique Decision-making approach that encourages feedback from individual group members, and the final decision is made by voting, similar to the way public officials are elected.

nonoperational prototype A mock-up, or model, that includes output and input specifications and formats.

nonprogrammed decision Decision that deals with unusual or exceptional situations.

object-oriented database Database that stores both data and its processing instructions.

object-oriented database management system (OODBMS) A group of programs that manipulate an object-oriented database and provide a user interface and connections to other application programs.

object-oriented systems development (OOSD) Approach to systems development that combines the logic of the systems development life cycle with the power of object-oriented modeling and programming.

object-relational database management system (ORDBMS) A DBMS capable of manipulating audio, video, and graphical data.

off-the-shelf software Existing software program that is purchased.

on-demand computing (on-demand business utility computing) Contracting for computer resources to rapidly respond to an organizations's varying workflow.

online analytical processing (OLAP) Software that allows users to explore data from a number of different perspectives.

online transaction processing (OLTP) Computerized processing in which each transaction is processed immediately, without the delay of accumulating transactions into a batch.

open database connectivity (ODBC) Standards that ensure that software can be used with any ODBC-compliant database.

open shops IS departments in which people, such as programmers and systems analysts, are allowed to run the computers, in addition to authorized operators.

open-source software Software that is freely available to anyone in a form that can be easily modified.

operating system (OS) A set of computer programs that controls the computer hardware and acts as an interface with application programs.

operational feasibility Measure of whether the project can be put into action or operation.

operational prototype A functioning prototype that accesses real data files, edits input data, makes necessary computations and comparisons, and produces real output.

optical disc A rigid disc of plastic onto which data is recorded by special lasers that physically burn pits in the disc.

optical processors Computer chips that use light waves instead of electrical current to represent bits.

optimization model A process to find the best solution, usually the one that will best help the organization meet its goals.

order entry system Process that captures the basic data needed to process a customer order.

order processing systems Systems that process order entry, sales configuration, shipment planning, shipment execution, inventory control, invoicing, customer relationship management, and routing and scheduling.

organization A formal collection of people and other resources established to accomplish a set of goals.

organizational change The responses that are necessary for for-profit and nonprofit organizations to plan for, implement, and handle change.

organizational culture The major understandings and assumptions for a business, a corporation, or an organization.

organizational learning Adaptations to new conditions or alterations of organizational practices over time.

organizational structure Organizational subunits and the way they relate to the overall organization.

output Production of useful information, usually in the form of documents and reports.

outsourcing Contracting with outside professional services to meet specific business needs.

paging Process of swapping programs or parts of programs between memory and one or more disk devices.

parallel start-up Running both the old and new systems for a period of time and comparing the output of the new system closely with the output of the old system; any differences are reconciled. When users are comfortable that the new system is working correctly, the old system is eliminated.

password sniffer A small program hidden in a network or a computer system that records identification numbers and passwords.

patch A minor change to correct a problem or make a small enhancement. It is usually an addition to an existing program.

payroll journal A report that contains employees' names, the area where employees worked during the week, hours worked, the pay rate, a premium factor for overtime pay, earnings, earnings type, various deductions, and net pay calculations.

perceptive system A system that approximates the way a human sees, hears, and feels objects.

personal area network (PAN) A network that supports the interconnection of information technology within a range of 33 feet or so.

personal productivity software Software that enables users to improve their personal effectiveness, increasing the amount of work they can do and its quality.

personal sphere of influence Sphere of influence that serves the needs of an individual user.

personalization The process of tailoring Web pages to specifically target individual consumers.

phase-in approach Slowly replacing components of the old system with those of the new one. This process is repeated for each application until the new system is running every application and performing as expected; also called *piecemeal approach*.

phishing Bogus messages purportedly from a legitimate institution to pry personal information from customers by convincing them to go to a "spoof" Web site.

physical design Specification of the characteristics of the system components necessary to put the logical design into action.

pilot start-up Running the new system for one group of users rather than all users.

pipelining A form of CPU operation in which there are multiple execution phases in a single machine cycle.

pixel A dot of color on a photo image or a point of light on a display screen.

planned data redundancy A way of organizing data in which the logical database design is altered so that certain data entities are combined, summary totals are carried in the data records rather than calculated from elemental data, and some data attributes are repeated in more than one data entity to improve database performance.

Platform for Privacy Preferences (P3P) A screening technology that shields users from Web sites that don't provide the level of privacy protection they desire.

plotter A type of hard-copy output device used for general design work.

point evaluation system An evaluation process in which each evaluation factor is assigned a weight, in percentage points, based on importance. Then each proposed system is evaluated in terms of this factor and given a score ranging from 0 to 100. The scores are totaled, and the system with the greatest total score is selected.

point-of-sale (POS) device Terminal used in retail operations to enter sales information into the computer system.

Point-to-Point Protocol (PPP) A communications protocol that transmits packets over telephone lines.

policy-based storage management Automation of storage using previously defined policies.

portable computer Computer small enough to be carried easily.

predictive analysis A form of data mining that combines historical data with assumptions about future conditions to predict outcomes of events such as future product sales or the probability that a customer will default on a loan.

preliminary evaluation An initial assessment whose purpose is to dismiss the unwanted proposals; begins after all proposals have been submitted.

primary key A field or set of fields that uniquely identifies the record.

primary storage (main memory; memory) Part of the computer that holds program instructions and data.

private branch exchange (PBX) An on-premise switching system owned or leased by a private enterprise that interconnects its telephones and provides access to the public telephone system.

problem solving A process that goes beyond decision making to include the implementation stage.

procedures The strategies, policies, methods, and rules for using a CBIS.

process A set of logically related tasks performed to achieve a defined outcome.

process symbol Representation of a function that is performed.

processing Converting or transforming data into useful outputs.

product configuration software Software used by buyers to build the product they need online.

productivity A measure of the output achieved divided by the input required.

profit center Department within an organization that tracks total expenses and net profits.

Program Evaluation and Review Technique (PERT) A formalized approach for developing a project schedule.

programmed decision Decision made using a rule, procedure, or quantitative method.

programmer Specialist responsible for modifying or developing programs to satisfy user requirements.

programming languages Sets of keywords, symbols, and a system of rules for constructing statements by which humans can communicate instructions to be executed by a computer.

programming life cycle A series of steps and planned activities developed to maximize the likelihood of developing good software.

project deadline Date the entire project is to be completed and operational.

project milestone A critical date for the completion of a major part of the project.

project organizational structure Structure centered on major products or services.

project schedule Detailed description of what is to be done.

projecting Data manipulation that eliminates columns in a table.

proprietary software A one-of-a-kind program for a specific application, usually developed and owned by a single company.

prototyping An iterative approach to the systems development process.

public key infrastructure (PKI) A means to enable users of an unsecured public network such as the Internet to securely and privately exchange data through the use of a public and a private cryptographic key pair that is obtained and shared through a trusted authority.

public network services Systems that give personal computer users access to vast databases and other services, usually for an initial fee plus usage fees.

purchase order processing system System that helps purchasing departments complete their transactions quickly and efficiently.

purchasing transaction processing systems Systems that include inventory control, purchase order processing, receiving, and accounts payable.

push technology Automatic transmission of information over the Internet rather than making users search for it with their browsers.

quality The ability of a product (including services) to meet or exceed customer expectations.

quality control A process that ensures that the finished product meets the customers' needs.

questionnaires A method of gathering data when the data sources are spread over a wide geographic area.

radio-frequency identification (RFID) A technology that employs a microchip with an antenna that broadcasts its unique identifier and location to receivers.

random access memory (RAM) A form of memory in which instructions or data can be temporarily stored.

rapid application development (RAD) A systems development approach that employs tools, techniques, and methodologies designed to speed application development.

read-only memory (ROM) A nonvolatile form of memory.

receiving system System that creates a record of expected receipts.

record A collection of related data fields.

reduced instruction set computing (RISC) A computer chip design based on reducing the number of microcode instructions built into a chip to an essential set of common microcode instructions.

redundant array of independent/ inexpensive disks (RAID) Method of storing data that generates extra bits of data from existing data, allowing the system to create a "reconstruction map" so that if a hard drive fails, the system can rebuild lost data.

reengineering (process redesign) The radical redesign of business processes, organizational structures, information systems, and values of the organization to achieve a breakthrough in business results.

register High-speed storage area in the CPU used to temporarily hold small units of program instructions and data immediately before, during, and after execution by the CPU.

relational model A database model that describes data in which all data elements are placed in two-dimensional tables, called *relations*, that are the logical equivalent of files.

release A significant program change that often requires changes in the documentation of the software.

reorder point (ROP) A critical inventory quantity level.

repetitive motion disorder (repetitive stress injury; RSI) An injury that can be caused by working with computer keyboards and other equipment.

replicated database A database that holds a duplicate set of frequently used data.

report layout Technique that allows designers to diagram and format printed reports.

request for maintenance form A form authorizing modification of programs.

request for proposal (RFP) A document that specifies in detail required resources such as hardware and software.

requirements analysis Determination of user, stakeholder, and organizational needs.

restart procedures Simplified processes to access an application from where it left off.

return on investment (ROI) One measure of IS value that investigates the additional profits or benefits that are generated as a percentage of the investment in IS technology.

revenue center Division within a company that tracks sales or revenues.

ring network A type of topology that contains computers and computer devices placed in a ring or circle; there is no central coordinating computer; messages are routed around the ring from one device or computer to another.

robotics Mechanical or computer devices that perform tasks requiring a high degree of precision or that are tedious or hazardous for humans.

router A device or software in a computer that determines the next network point to which a data packet should be forwarded toward its destination.

routing system System that determines the best way to get products from one location to another.

rule A conditional statement that links given conditions to actions or outcomes.

safe harbor principles A set of principles that address the e-commerce data privacy issues of notice, choice, and access.

sales configuration system Process that ensures that the products and services ordered are sufficient to accomplish the customer's objectives and will work well together.

satisficing model A model that will find a good—but not necessarily the best—problem solution.

scalability The ability of the computer to handle an increasing number of concurrent users smoothly.

schedule feasibility Determination of whether the project can be completed in a reasonable amount of time.

scheduled report Report produced periodically, or on a schedule, such as daily, weekly, or monthly.

scheduling system System that determines the best time to piick up or deliver goods and services.

schema A description of the entire database.

screen layout A technique that allows a designer to quickly and efficiently design the features, layout, and format of a display screen.

script bunnies Wannabe crackers with little technical savvy who download programs—scripts—that automate the job of breaking into computers.

search engine A Web search tool.

secondary storage (permanent storage) Devices that store larger amounts of data, instructions, and information more permanently than allowed with main memory.

Secure Sockets Layer (SSL) A communications protocol used to secure sensitive data.

selecting Data manipulation that eliminates rows according to certain criteria.

selective backup Creating backup copies of only certain files.

semistructured or unstructured problems More complex problems in which the relationships among the pieces of data are not always clear, the data may be in a variety of formats, and the data is often difficult to manipulate or obtain.

sequence structure A programming structure in which, after starting the sequence, programming statements are executed one after another until all the statements in the sequence have been executed. Then the program either ends or continues on to another sequence.

sequential access Retrieval method in which data must be accessed in the order in which it is stored.

sequential access storage device (SASD) Device used to sequentially access secondary storage data.

Serial Line Internet Protocol (SLIP) A communications protocol that transmits packets over telephone lines.

server A computer designed for a specific task, such as network or Internet applications.

Shannon's fundamental law of information theory The law of telecommunications that states that the information-carrying capacity of a channel is directly proportional to its bandwidth—the broader the bandwidth, the more information that can be carried.

shareware and freeware Software that is very inexpensive or free, but whose source code cannot be modified.

shipment execution system System that coordinates the outflow of all products from the organization, with the objective of delivering quality products on time to customers.

shipment planning system System that determines which open orders will be filled and from which location they will be shipped.

sign-on procedure Identification numbers, passwords, and other safeguards needed for an individual to gain access to computer resources.

simplex channel A communications channel that can transmit data in only one direction.

simulation The ability of the DSS to duplicate the features of a real system.

site preparation Preparation of the location of a new system.

slipstream upgrade A minor upgrade—typically a code adjustment or minor bug fix—not worth announcing. It usually requires recompiling all the code and, in so doing, it can create entirely new bugs.

smart card A credit card–sized device with an embedded microchip to provide electronic memory and processing capability.

social engineering The practice of talking a critical computer password out of an individual.

software The computer programs that govern the operation of the computer.

software bug A defect in a computer program that keeps it from performing in the manner intended.

software interface Programs or program modifications that allow proprietary software to work with other software used in the organization.

software piracy The act of illegally duplicating software.

software suite A collection of single application programs packaged in a bundle.

source data automation Capturing and editing data where the data is initially created and in a form that can be directly input to a computer, thus ensuring accuracy and timeliness.

spam E-mail sent to a wide range of people and Usenet groups indiscriminately.

special-purpose computers Computers used for limited applications by military and scientific research groups.

sphere of influence The scope of problems and opportunities addressed by a particular organization.

split-case distribution A distribution system that requires cases of goods to be opened on the receiving dock and the individual items from the cases to be stored in the manufacturer's warehouse.

stakeholders Individuals who, either themselves or through the organization they represent, ultimately benefit from the systems development project.

star network A type of topology that has a central hub or computer system, and other computers or computer devices located at the end of communications lines that originate from the central hub or computer.

start-up The process of making the final tested information system fully operational.

static Web pages Web pages that always contain the same information.

statistical sampling Selecting a random sample of data and applying the characteristics of the sample to the whole group.

steering committee An advisory group consisting of senior management and users from the IS department and other functional areas.

storage area network (SAN) Technology that provides high-speed connections between data-storage devices and computers over a network.

storefront broker Companies that act as middlemen between your Web site and online merchants that have the products and retail expertise.

strategic alliance (strategic partnership) An agreement between two or more companies that involves the joint production and distribution of goods and services.

strategic planning Determining long-term objectives by analyzing the strengths and weaknesses of the organization, predicting future trends, and projecting the development of new product lines.

structured interview An interview where the questions are written in advance.

structured walkthrough A planned and preannounced review of the progress of a program module, a structure chart, or a human procedure.

subschema A file that contains a description of a subset of the database and identifies which users can view and modify the data items in the subset.

supercomputers The most powerful computer systems, with the fastest processing speeds.

superconductivity A property of certain metals that allows current to flow with minimal electrical resistance.

supply chain management A key value chain composed of demand planning, supply planning, and demand fulfillment.

switch A telecommunications device that routes incoming data from any one of many ports to a specific output port that will take the data toward its intended destination.

switched line A communications line that uses switching equipment to allow one transmission device to be connected to other transmission devices.

symmetrical multiprocessing (SMP) Another form of parallel processing in which multiple processors run a single copy of the operating system and share the memory and other resources of one computer.

synchronous communications Communications in which the receiver gets the message instantaneously.

syntax A set of rules associated with a programming language.

system A set of elements or components that interact to accomplish goals.

system boundary The limits of the system; it distinguishes the system from everything else (the environment).

system parameter A value or quantity that cannot be controlled, such as the cost of a raw material.

system performance measurement Monitoring the system—the number of errors encountered, the amount of memory required, the amount of processing or CPU time needed, and other problems.

system performance products Software that measures all components of the computer-based information system, including hardware, software, database, telecommunications, and network systems.

system performance standard A specific objective of the system.

system testing Testing the entire system of programs.

system variable A quantity or item that can be controlled by the decision maker.

systems analysis The systems development phase involving the study of existing systems and work processes to identify strengths, weaknesses, and opportunities for improvement.

systems analyst Professional who specializes in analyzing and designing business systems.

systems controls Rules and procedures to maintain data security.

systems design The systems development phase that defines how the information system will do what it must do to obtain the problem solution.

systems development The activity of creating or modifying existing business systems.

systems implementation The systems development phase involving the creation or acquiring of various system components detailed in the systems design, assembling them, and placing the new or modified system into operation.

systems investigation report Summary of the results of the systems investigation and the process of feasibility analysis and recommendation of a course of action.

systems investigation The systems development phase during which problems and opportunities are identified and considered in light of the goals of the business.

systems maintenance Stage of systems development that involves checking, changing, and enhancing the system to make it more useful in achieving user and organizational goals.

systems maintenance and review The systems development phase that ensures the system operates and modifies the system so that it continues to meet changing business needs.

systems operation Use of a new or modified system.

systems request form Document filled out by someone who wants the IS department to initiate systems investigation.

systems review The final step of systems development, involving the analysis of systems to make sure that they are operating as intended.

systems software The set of programs designed to coordinate the activities and functions of the hardware and various programs throughout the computer system.

team organizational structure Structure centered on work teams or groups.

technical documentation Written details used by computer operators to execute the program and by analysts and programmers in case there are problems with the program or the program needs modification.

technical feasibility Assessment of whether the hardware, software, and other system components can be acquired or developed to solve the problem.

technology acceptance model (TAM) A model that describes the factors that lead to higher levels of acceptance and usage of technology.

technology diffusion A measure of how widely technology is spread throughout the organization.

technology infrastructure All the hardware, software, databases, telecommunications, people, and procedures that are configured to collect, manipulate, store, and process data into information.

technology infusion The extent to which technology is deeply integrated into an area or department.

technology-enabled relationship management The use of detailed information about a customer's behavior, preferences, needs, and buying patterns to set prices, negotiate terms, tailor promotions, add product features, and otherwise customize the entire relationship with that customer.

telecommunications The electronic transmission of signals for communications; enables organizations to carry out their processes and tasks through effective computer networks.

telecommunications medium Anything that carries an electronic signal and interfaces between a sending device and a receiving device.

telecommuting A work arrangement whereby employees work away from the office using personal computers and networks to communicate via e-mail with other workers and to pick up and deliver results.

Telnet A terminal emulation protocol that enables users to log on to other computers on the Internet to gain access to public files.

terminal-to-host An architecture in which the application and database reside on one host computer, and the user interacts with the application and data using a "dumb" terminal.

thin client A low-cost, centrally managed computer with essential but limited capabilities that is devoid of a DVD player, floppy disk drive, and expansion slots.

time-driven review Review performed after a specified amount of time.

time-sharing Capability that allows more than one person to use a computer system at the same time.

top-down approach A good general approach to writing a large program, starting with the main module and working down to the other modules.

total cost of ownership (TCO) Measurement of the total cost of owning computer equipment, including

desktop computers, networks, and large computers.

total quality management (TQM) A collection of approaches, tools, and techniques that fosters a commitment to quality throughout the organization.

traditional approach to data management An approach whereby separate data files are created and stored for each application program.

traditional organizational structure Organizational structure in which major department heads report to a president or top-level manager.

transaction Any business-related exchange, such as payments to employees, sales to customers, and payments to suppliers.

transaction processing cycle The process of data collection, data editing, data correction, data manipulation, data storage, and document production.

transaction processing system (TPS) An organized collection of people, procedures, software, databases, and devices used to record completed business transactions.

transaction processing system audit An examination of the TPS to answer whether the system meets the business need for which it was implemented, what procedures and controls have been established, whether these procedures and controls are being used properly, and whether the information systems and procedures are producing accurate and honest reports.

Transmission Control Protocol (TCP) Widely used transport layer protocol that is used in combination with IP by most Internet applications.

Trojan horse A program that appears to be useful but actually masks a destructive program.

tunneling The process by which VPNs transfer information by encapsulating traffic in IP packets over the Internet.

Uniform Resource Locator (URL) An assigned address on the Internet for each computer.

unit testing Testing of individual programs.

unstructured interview An interview where the questions are not written in advance.

usenet A system closely allied with the Internet that uses e-mail to provide a centralized news service; a protocol that describes how groups of messages can be stored on and sent between computers.

user acceptance document Formal agreement signed by the user that states that a phase of the installation or the complete system is approved.

user documentation Written description developed for individuals who use a program, showing users, in easy-to-understand terms, how the program can and should be used.

user interface Element of the operating system that allows individuals to access and command the computer system.

user preparation The process of readying managers, decision makers, employees, other users, and stakeholders for new systems.

users Individuals who will interact with the system regularly.

utility programs Programs used to merge and sort sets of data, keep track of computer jobs being run, compress data files before they are stored or transmitted over a network, and perform other important tasks.

value chain A series (chain) of activities that includes inbound logistics, warehouse and storage, production, finished product storage, outbound logistics, marketing and sales, and customer service.

variant A modified version of a virus that is produced by the virus's author or

another person who amends the original virus code.

version A major program change, typically encompassing many new features.

videoconferencing A telecommunication system that combines video and phone call capabilities with data or document conferencing.

virtual memory Memory that allocates space on the hard disk to supplement the immediate, functional memory capacity of RAM.

virtual organizational structure Structure that employs individuals, groups, or complete business units in geographically dispersed areas.

virtual private network (VPN) A secure connection between two points across the Internet.

virtual reality The simulation of a real or imagined environment that can be experienced visually in three dimensions.

virtual reality system A system that enables one or more users to move and react in a computer-simulated environment.

virtual tape Storage device that manages less frequently needed data so that it appears to be stored entirely on tape cartridges although some parts of it may actually be located in faster, hard-disk storage.

virtual workgroups Teams of people located around the world working on common problems.

virus A program that attaches itself to other programs.

vision systems The hardware and software that permit computers to capture, store, and manipulate visual images and pictures.

voice and data convergence The integration of voice and data applications in a common environment.

voice mail Technology that enables users to leave, receive, and store verbal messages for and from other people around the world.

voice over Internet protocol (VoIP) The basic transport of voice in the form of a data packet using the Internet protocol.

voice-recognition device An input device that recognizes human speech.

volume testing Testing the application with a large amount of data.

Web auction An Internet site that matches people who want to sell products and services with people who want to purchase these products and services.

Web browser Software that creates a unique, hypermedia-based menu on a computer screen, providing a graphical interface to the Web.

Web log (blog) A Web site that people can create and use to write about their observations, experiences, and feelings on a wide range of topics.

Web log file A file that contains information about visitors to a Web site.

Web page construction software Software that uses Web editors and extensions to produce both static and dynamic Web pages.

Web services Standards and tools that streamline and simplify communication among Web sites for business and personal purposes.

Web site development tools Tools used to develop a Web site, including HTML or visual Web page editor, software development kits, and Web page upload support.

Web site hosting companies Companies that provide the tools and services required to set up a Web page and conduct e-commerce within a matter of days and with little up-front cost.

Web site traffic data analysis software Software that processes and analyzes data from the Web log file to provide useful information to improve Web site performance.

what-if analysis The process of making hypothetical changes to problem data and observing the impact on the results.

wide area network (WAN) A network that ties together large geographic regions.

wire transfer An extremely fast, reliable means to move funds from one account to another.

wireless application protocol (WAP) A standard set of specifications for Internet applications that run on handheld, wireless devices.

wordlength The number of bits the CPU can process at any one time.

workgroup Two or more people who work together to achieve a common goal.

workgroup application software Software that supports teamwork, whether in one location or around the world.

workgroup sphere of influence Sphere of influence that serves the needs of a workgroup.

workstation A more powerful personal computer that is used for technical computing, such as engineering, but still fits on a desktop.

World Wide Web (WWW or W3) A collection of tens of thousands of independently owned computers that work together as one in an Internet service.

worm An independent program that replicates its own program files until it interrupts the operation of networks and computer systems.

INDEX

Subject

A boldface page number indicates a key term and the location of its definition in the text.

A

access methods, **102**
accounting
    MIS, 479–490
    systems, **428**, 429–434
account(s)
    payable systems, 426–427
    receivable systems, 429–430
ACH (automated clearing house), **282**
activity modeling, 592–593
actors, **588**
adaptive systems, **10**
advanced Web-performance monitoring
    utilities, 157
advertising industry, **152**, 193–195, 475. *See also* banner ads; pop-up ads
Aeneas Internet and Telephone, 411
agile
    development, 573
    modeling, 558
AI (artificial intelligence), **29–31**, 44, 71, **501–502**, 503–519
airline industry, 4, 11–12, 91, 533
    changes in the structure of, 64
    CRM and, 423
    earnings growth and, 70
    Internet auctions and, 36
    inventory control and, 420
    new services in, 65
    passenger screening systems and, 671–673
    reservation systems, 65–66, 187, 423
    software and, 152
algorithms, 386, 517, **518**
alphanumeric data, **6**
ALU (arithmetic/logic unit), 92–93
analog signals, **250**
ANI (automatic number identification), 255–256
anomalies, 205–206
ANSI (American National Standards Institute), 175, 211, 265
antivirus programs, **686–687**. *See also* viruses
Apache servers, 178, **694**. *See also* servers
APIs (application program interface), **147**
applets, 174, **307**, 309–310
application flowcharts, **594–595**
application software. *See also* software
    described, **143–144**
    enterprise, **169–170**
    functions of, 158–161
    overview, 158–171
    personal, 160–168
    specialized, 171
    types of, 158–161
    workgroup, **168–169**
architecture, 59, 215
ARPANET, **297**, 298
asking directly approach, **596**

ASPs (application service providers), **160**, 634
assembly language, **172**
asset management transaction processing system, **432**
ATM (asynchronous transfer mode), 266, 285, 329
ATMs (automated teller machines), 42–43, **114**, 377
attached storage methods, **107–108**
attacks, 33–34, 328, **371**, 683. *See also* security
attributes, **197**
auctions, 36, 314, **323**, 369–370, 381
audio data, **6**
audit(s). *See also* auditing
    described, **411–412**
    trails, 412
auditing. *See also* audits
    described, 466–467
    external, **466**
    internal, **466**
authentication, 327
Auto RCA (root-cause analysis) module, 157
automobile industry, 15, 27, 89–90, 283
    AI and, 509, 513
    databases and, 219
    MIS and, 470–471, 473
    navigation systems and, 121
    software and, 179
    system design and, 613, 614
    virtual reality systems and, 533
autonomics, 72, 646

B

B2B (business-to-business) e-commerce, 21–22, 24, **350**, 351, 357, 372, 375. *See also* e-commerce
B2C (business-to-consumer) e-commerce, **350**, 351, 357, 359, 372
backbone, 299, 329. *See also* e-commerce
back-end applications, 217
background programs, 148
backups, 103, 108, **622–623**
backward chaining, **524–525**
bandwidth, **246**
banking industry, 34–36, 404, 439, 647. *See also* ATMs (automated teller machines)
    databases and, 224–225
    e-commerce and, 369, 376
    fraud and, 380–381
    hardware and, 113–115
    networks and, 281–282
banner ads, 303, 475. *See also* advertising industry
bar-code scanners, 116
batch processing systems, **402–403**
benchmark tests, **630**

best practices, **141**, **438**
BI (business intelligence), **224–225**, 454, 485
bidding, 269. *See also* auctions
biometrics, **692**
bits, **94**, **196**
Bizrate.com, 322
Bluetooth, 267, 285, 435
boot
    process, 146, 150
    sector viruses, 532
bottlenecks, 328–329
bps (bits per second), 244
brainstorming, **488**
branding, 224
bridges, **271**
broadband, **246**
brokerage companies. *See also* stock market
browsers, **307**, 124, 309, 363, 380, 675, 706
    privacy and, 383
    search engines and, 376
    spyware and, 329
buddy lists, 316
budget transaction processing systems, **428**
bugs, **176**. *See also* errors
bus
    line, 94, **95**
    networks, 263
business
    continuity planning, **510**
    information systems, **21–31**, 36
bytes, **98**, **196**

C

C (high-level language), 172, 175, 640
C++ (high-level language), 174, 175, 211, 584, 585, 639, 640, 648
C2C (consumer-to-consumer) e-commerce, 21–22, **350–351**, 359. *See also* e-commerce
cable modems, 250–251, 259
cache memory, **99–100**
CAD (computer-aided design), 36, 124, 168, 468
call centers, **276–277**
caller ID, 255–256
CAM (computer-aided manufacturing), 36, **470**
carbon nanotubes, 96–97
career opportunities, 70–77, 296, 316–317. *See also* jobs
CAs (certificate authorities), **374**
CASE (computer-aided software engineering) tools, **584–585**, 595, 597, 616, 639
CASE repository, **595**, 597
catalog management software, 372

CBIS (computer-based information system), 17–21, 90, 100, 401

CBT (computer-based training), 667. *See also* training

CDMA (Code-Division Multiple Access), 257

CD-R (CD-recordable) discs, 106

CD-ROM (compact disc read-only memory) discs, 102, **106**, 623

CD-RW (CD-rewritable) discs, 106, 107, 623

cell phones, 154–155, 316. *See also* cellular transmission; telecommunications

cellular transmission, 249–250. *See also* cell phones; telecommunications

centralized processing, 273

certifications, 76–77

CFO (chief financial officer), 75

CGA (color graphics adapter), 118

CGI (computer-generated image) technology, 537

change
managing, 577–578
models, 56
organizational, 55

channel(s)
communications, 245–246
full-duplex, 246
half-duplex, 245–246
simplex, 245

character, 196

chat rooms, 314, **320**

chief-programmer team, **636**

choice stage, 456

CICS (complex instruction set computing), 97

CIM (computer-integrated manufacturing), 36, 471

CIO (chief information officer), 75

ciphertext, 330

CISSP (Certified Information Systems Security Professional), 77

classic style, 151

CLECs (competitive local exchange carriers), **253**

clickstream data, **382**

client/server architecture, **263–265**, 285. *See also* servers

clip art, 166

clock speed, **94**

closed
shops, **624**
systems, **10**

CMM (Capability Maturity Model), 580, **581–582**

CMS (content management system), 312

coaxial cable, 246–247

COBOL, 172, 174, 211, 212, 527, 646, 648, 649

code, reusable, 174

codecs, 321

cold sites, **622**

collaborative work, **54**

COM (computer output microfilm) devices, 120

command-base user interface, **146–147**

communication(s)
asynchronous, **244**

channels, 245–246
data, 244
described, 243
effective, importance of, 284
GSS and, 488–489
software, 274–275
synchronous, 243–244
systems, 243

compact flash memory, 102

competitive
advantage, **63–66**, 564
intelligence, **225**

competitors, rivalry among, 63, 66

compilers, 173

complex systems, **10**

computational biology, 539

computer(s)
literacy, 35–36
programs, **143**
-related mistakes, 673–678
waste, 673

concurrency control, **211**

containers, 207

content streaming, 314, 322

continuous improvement, 56, 58–59, 577

contracts, 631

control
issues, 410–412
unit, **92**

cookies, 329

coprocessors, **100**

copyrights, 176–177

cost
/benefit analysis, **629**
centers, **466**
e-commerce and, 357

counterintelligence, **225**

CPUs (central processing units), 92–102, 119, 252
programming languages and, 171, 173
workstations and, 124

crackers (criminal hackers), **682**

creative analysis, **564**. *See also* creativity

creativity, 162, 193, 511. *See also* creative analysis

credit unions, 42–43

credit-card industry, 23, 125, 516, 519, 580–581
databases and, 227, 229
e-commerce and, 353, 377

credit-reporting agencies, 65

crime. *See also* terrorism; security
computer, 678–690
international, 690
prevention, 691–698

critical
analysis, **564**
path, **582**

CRM (customer relationship management), 26, 44, 49–50, **51**, **421–423**, 447–448, 472, 559. *See also* customers

cross-platform development, **640**

cruise-ship industry, 43, 218

cryptography, **330–331**

CSFs (critical success factors), 565, 596

CSMA/CD (Carrier Sense Multiple Access with Collision Detection), 285

CTO (chief technology officer), 75

CTS (carpel tunnel syndrome), **706**, 708

culture, organizational, **54–56**

currency, 55, 435

customer(s). *See also* CRM (customer relationship management)
awareness, 70
bargaining power of, 64, 66
churn, 224
privacy, invasion of, 382–383
satisfaction, 70, 357

cybermalls, **363–364**

cybersquatters, 300

cyberstalking, 719

cyberterrorism, **680–681**. *See also* terrorism

D

DASD (direct access storage), **102**

data
administrators, **214**
analysis, 592–595
cleanup, **205–206**
collection, 408–409, 590–592
conversion, **642**
correction, **409**
described, 5
dictionaries, **209**
editing, **409**
entry, **110**
-flow lines, **593**
gloves, 31
havens, 262
hierarchy, **196**
information versus, 5–7
input, **110**
integrity, **199**
items, **197**
manipulating, 203–205, 211–212, **409**
marts, **221**
mining, **221–224**, 227
modeling, **201–206**, 592
nature of, 110
preparation, **642**
redundancy, **198**, 209
-related regulations, 223
reliability, 210
security issues and, 683–686
storage, 210–211, **409**, 594
synchronization, 226
types of, 6
warehouses, **219–221**

database(s)
applications, 317–320
approach, **199–200**
backups, 622–623
creating/modifying, 208–210
data management and, 196–201
described, **19**
distributed, **225–226**
DSS and, 485
empowerment and, 52
linking, to the Internet, 217–219
normalization, 206
number of concurrent users and, 215
overview, 192–239
performance, 215–216

reasons to learn about, 194
replicated, **226**
size, 215
software, 162, 164–165, 167
special-purpose, 215
specifications, 617
systems, acquiring, 640–641
types of, 206–207
DBAs (database administrators), 74, **195**, 213–214
DBMSs (database management systems), 28, **195**, 199–201, 204, 206–217, 237–238
DDLs (data definition languages), **208–210**
debit cards, 377
decency, protection of, 695–697
decentralized processing, **273**
decision. *See also* DSS (decision support system)
making, **455**, 456–458, 487–489
nonprogrammed, **456–457**
programmed, **456–457**
rooms, **490**
structure, **638**
support, 171
decode instructions, 92–93
dedicated lines, **254**
delphi approach, **488**
demand forecasts, 437
design. *See also* systems design
reports, 631–632
stage, **456**
desktop computers, 122, **123**
deterrence controls, **624**
DFDs (data-flow diagrams), **592**, 593–594
dialing services, 255–256
dialogue
good interactive, 620–621
manager, **484**, 486
digital
cameras, 64, **112–113**
cash, 376
certificates, 374
signals, 250
signatures, **330–331**
direct
access, **102**
conversion (plunge), **644**
observation, 591
disabled individuals, 415
disaster recovery, 108, 410, 411, 494, **621–624**
disintermediation, **359**
disk mirroring, **104**
distance learning, **283**
distributed processing, **273**
DML (data manipulation language), 211
documentation. *See also* reports
described, **143**
production, 410
user, 637
user acceptance, 645
domain(s)
described, **202**, 527
experts, 527
names, 299–300
DOS (denial-of-service) attacks, 34, **371**, 683. *See also* attacks; security

dot pitch, 118
downsizing, **60–62**, 125
downstream management, 50
dpi (dots per inch), 120
DRAM (dynamic RAM), 98, 100
drill down, 219
drones, 19
DSL (Digital Subscriber Line), 253, **257**, 259, 284, 303–304
DSS (decision support system), 26–31, 480–486
ad hoc, **483**
institutional, **483**
model-driven, **486**
DTP (desktop publishing) software, 162
dual-boot feature, 146
dumpster diving, **680**
DVDs (digital versatile disks), 102, **106**, 116, 128, 675
dynamic systems, **10**

E

earnings growth, 70
eavesdropping, 34
e-commerce
applications, 363–370
challenges, 355–356
described, **21–25**, 348–396
global, 359–361
multistage model of, 352–354
payment systems, 374–378
phishing scams and, 23
software, **372**
successful, strategies for, 383–386
systems development and, 566–567
technology infrastructure, 370–374
threats to, 378–383
TPS and, 25–26
transaction processing software, **373–374**
economic feasibility, **587**
EDI (electronic data interchange), 64, 280–281, 286, 351–352, 406, 413–414, 420, 425–426, 429, 431, 438
EDRs (event data recorders), 18
education, 35–36. *See also* certifications
effectiveness, **11**, 51
efficiency
described, **11**
labor, 405–406
operations component and, 72
organizational systems and, 51, 72
EFT (electronic funds transfer), **281–282**
electoral process, 19
electronic
bill presentment, **369**
business (e-business), **24**
cash, 376
distribution, 353
document distribution, **276**
exchange, 364
management (e-management), **24–25**
procurement (e-procurement), **24**
retailing (e-tailing), **363–364**
shopping carts, 373
software distribution, 276

wallets, **377**
e-mail, 23, 44, 313–316, 700–701. *See also* viruses
embedded operating systems, 154–156
empowerment, **52**
encryption, **330–331**, 691
end-user system development, **574**
engineering, 468–469
enterprise
application software, **169–170**
data modeling, **201**
software, 145, 154, 157, **169–170**
sphere of influence, **145**
storage, **107**, 108–109
entities, **197**
entity symbol, **594**
entrants, threats of new, 63, 66
environmental issues, 707–709
EOQ (economic order quantity), **470**
EPROM (erasable programmable read-only memory), 99
equipment theft, **688**
ER (entity-relationship diagrams), 201, **202**, 204, 592, 594
ergonomics, 30, **35**, **707–709**
ERP (enterprise resource planning), 25–27, 489, **567–558**
advantages/disadvantages of, 438–441
described, **169–170**
gateways and, 272
hardware and, 89–90, 91, 97
MIS and, 459, 472
overview, 398–350, 436–441
risk and, 70
errors. *See also* bugs
preventing/correcting, 621
Windows XP and, 151
ESS (executive support system), **491–494**
Ethernet, 266, 285
ethical issues, 33–35, 709–711
E-time (execution time), 93
event-driven review, **651**
exabytes, **98**
exception reports, 27
executive information system, **28**
expandable storage devices, **107**
expert system(s)
described, 29–31, 513, 519–533
system shell, **521**, 528, 529
explanation facility, **525**
extranets
described, **20**, 326–327
management consulting firms and, 36
systems development and, 566–567

F

fairness, in information use, 702
feasibility analysis, **587–588**
feedback, 9, **16**
fetch instructions, 92–93
fiber-optic cable, **246–248**
fields, **196**, 197
file(s)
-compression utilities, 157
described, **196**
management, 149

file server(s). *See also* servers
    architecture, **263**
    described, **285**
filenames, character restrictions for, 149
filtering software, 695
final evaluation, **628**
financial
    forecasting, 437–438
    MIS, 162, **465–467**
fingerprint systems, 14, 378, 514, 692
firewalls, **35**, 274, 327, **330–331**, 697
FireWire, 267
five-forces model, **63–64**
flaming, 54
flash memory, **107**
flat-file databases, 207
flat-screen monitors, 119
floppy diskettes, 102
FMS (flexible manufacturing system), **471**
forecasting, **16**
foreground programs, 148
FORTRAN, 172, 527
forward chaining, **524–525**
frame relay, 266
frames, **266**
fraud, 224, 501, 516, 624, 678–679. *See also* security
    auditing and, 466
    e-commerce and, 377–378
    overview of, 329, 380–382
    statistics, 34
    TPS and, 405
freeware, **178**
front-end
    applications, 217
    processors, **251–252**
FTP (File Transfer Protocol), 314, **317**
functional requirements, 598
fuzzy logic, **523**

G

GaAs (gallium arsenide), 96
gambling casinos, 26–27
game theory, **538–539**
Gantt chart, **583**
gateways, **272**, 313, 320
Gbps (billions of bits per second), 244
general
    ledger system, **432–433**
    -purpose computers, 122
genetic algorithms, **518**
gigabyte, **98**
gigahertz (GHz), **94**
GIGO (garbage in, garbage out), 8
GIS (geographic information system), **480**
goal(s)
    aligning, 561–563
    examples of, 10
    -seeking analysis, **481**
    supporting, 144–145
GPS (global positioning satellite), 121, 283, 458
graphics software, 162, 165–166, 167
grid
    charts, **595**
    computing, **101–102**, 156

group consensus
    approach, **488**
    described, **629**
groupthink, 488
groupware, **28**, 168, **489**
GSS (group support systems), **28**, **486–491**, 573
GUI (graphical user interface), **147**, 211. *See also* user interfaces

H

hackers, **682**, 687, 692, 717
handheld computers, 116, **122–123**, 316, 363
handwriting recognition, 116
hard copy, 120
hard drives
    read/write heads, 104
    selecting, 127
hardware. *See also specific types*
    acquiring, from IS vendors, 632–634
    backups, 622
    CBIS and, 17–18
    components, 92–93
    described, **17–18**, **97**
    e-commerce and, 370–371
    functions, common, 146
    independence, 147
    input/output devices, 109–113
    reasons to learn about, 90
    selection, 86–136
    software and, 146
    specifications, 616
    upgrades, 126–128
    utilities, 156
harvester programs, 381
healthcare industry, 36, 117, 532
    databases and, 219, 223, 230
    DSSs and, 27, 482–483
    e-commerce and, 357
    ERP and, 25–26
    organizational structure and, 53
    portable computers and, 123
    radio-frequency chips and, 58
    regulations and, 678
    software and, 152
    TPSs and, 412, 430
    TQM and, 60
    virtual reality systems and, 536
    work environment and, 706–707
hearing devices, 121
help
    desks, 646
    facilities, **620**
hertz, **94**
heuristics, 458, 511
hierarchical networks, 263
highly structured problems, **483**
hiring, 642
HMD (head-mounted display), 30–31, 534
home pages, **304**, 312
hop, 272
hot sites, **622**
HTML (HyperText Markup Language), 305–311, 385, 677
    editors, 311–312

    tags, **305–306**
HTTP (HyperText Transfer Protocol), 371, 374
hubs, **272**, 285
human resource management, 36, 194, **476–479**
HVAC (heating, venting, and air-conditioning) systems, 125
hybrid networks, 263
hypermedia, **304–305**

I

icons, **147**
ICRA (Internet Content Rating Association) rating system, **695**
identity theft, **681**
IDEs (integrated development environments), **640**
IDS (Intrusion-Detection System), 559, **694**
if-then statements, **522**
IM (instant messaging), 19, **314**, 315–316, 489
image(s). *See also* graphics software
    data, 6
    logs, **623**
implementation stage, **456**
incremental backups, **623**
industry structure, altering, 64
inference engine, **524–525**
informatics, **539**
information. *See also* IS (information systems)
    centers, 74
    concepts, 5–8
    data versus, 5–7
    described, 5
    service units, 75
    theory, Shannon's fundamental law of, **246**, 284
    value of, 7–8
infrared transmission, 249–250
input
    described, **9**, 15
    devices, 109–113
    organizational systems and, 48–51
    productivity and, 68
    software and 146–147
insiders, **682**
installation, **643**
instructions
    decode, 92
    executing, 93
    fetch, 92
insurance industry, 69, 158
integrated application packages, 168
intellectual property
    described, **379**
    theft of, 379–380
intelligence stage, **455**
intelligent
    agents, **518–519**
    behavior, **509**, 520
international networks, **261–262**
Internet
    accessing, 300–302
    applications, 313–315

careers, 76–77
databases and, 217–219
described, **19, 296**
DSS and, 27
education and, 35
ethical issues and, 33–35
globalization and, 62
intranets and, 20
management issues and, 328
on-demand computing and, 62
overview, 294–346
phones, 314, **320–322**
privacy and, 33–35, 701–702
reasons to learn about, 296
security and, 33–35, 670–711
service/speed issues, 328–329
systems development and, 566–567
utilities, 157
virtual organizations and, 54
Internet Explorer browser. *See also* browsers
Internet2 (I2), 298
interoperability, 579
intranet(s)
described, **20, 326–327**
management consulting firms and, 36
inventory-control system, 17, 49–50, 357,
419–420, 424. *See also* JIT (just-in-time)
inventory approach
Ace Hardware and, 70
ERP and, 437
MIS and, 469–470, 478–479
investment firms, 36
invoicing, 420–421
IP (Internet Protocol), 241–242, 255, 272,
297, 371
IS (information systems). *See also*
information
components of, 15–21
computerized, 16–17
described, **4,** 33–36
functions, typical, 76–77
infrastructure, disparities in, 434
literacy, 35–36
manual, 16–17
pervasiveness of, 4–5
planning, **561**
three primary responsibilities of, 73
titles, 76–77
ISDN (Integrated Services Digital Network),
253, **256–269,** 284
ISPs (Internet Service Providers), 272,
302–304, 312, 411
I-time (instruction time), 93
IVR (interactive voice response), 112
IXCs (interexchange carriers), 252

J

JAD (joint application development),
573–574
Java, 310, 515, 585, 639, 648
applets, 174, **307,** 309–310
described, **174, 309**
software and, 154, 174–175
JIT (just-in-time) inventory approach, **470,**
555–556
job(s). *See also* career opportunities

applicant review profiles, 478
market, overview of, 70–77
joining, **203**

K

kernel, 146, 152
keyboards, types of, 111
keys, primary, **197,** 206
kilobyte, **98**
kiosks, 65, 114. *See also* ATMs (automated
teller machines)
knowledge
acquisition facility, **525**
base, **30, 522–523**
described, **6–7**
engineer, **527**
management, **225**
user, **527**

L

language differences, 434
LANs (local area networks), 75, 117, 285,
302. *See also* networks
described, **260–261**
e-commerce and, 416–417, 419
expert systems and, 528
GSS and, 490–491
mobile commerce and, 362
portable computers and, 123
standards for, 266
switching devices and, 271–272
Wi-FI and, 267
laptop computers, 34, 123
LATA (local access and transport area), **252**
latency, 272
laws. *See also* legislation
LCDs (liquid crystal displays), 12, 119. *See
also* monitors
learning
organizational, **56**
systems, **516**
LEC (local exchange carrier), **252–253**
legacy systems, 438, 646
legal feasibility, **588**
legislation, 434–435, 677–678, 690,
695–697, 702–705
libel, 695–697
licenses, 176–177
light pens, 116
line positions, 51–52
linked tables, **203–204**
Linux, 149–150, 152–155, 177, 179, 615,
634
databases and, 214, 216
e-commerce and, 374
programming languages and, 173
Star Office and, 167–168
literacy
computer, **35–36**
information systems, **35–36**
local loop, 252
localization, 361
logic bomb, **683**
logical
design, 201, **613–617**

view, 148
logon, 149, 317
long-distance carriers, **253–254**
lookup tables, **620**
loop structure, **638**
LTO-2 (Linear Tape Open), 103

M

machine
cycle, **92,** 94, 97
language, **172**
Macintosh, 91, 101, 150, 153–155
Java and, 309
PHP and, 310
programming languages and, 174
viruses and, 684
magnetic
disks, 104–105
tape, **103–104**
main memory. *See also* memory
mainframes, 157, 211, 218, 645
described, **125**
linking PCs to, 275–276
software and, 145
maintenance team, **648**
make-or-buy decision, **634**
malware, **683**
management
consulting firms, 36
issues, 410–412
skills training, 161
MANs (metropolitan area networks), **261,**
285
manufacturing MIS, 467–472
market. *See also* marketing
basket analysis, 224
segmentation, 224, **366**
share, 70
marketing. *See also* market
e-commerce and, 366–367
MIS, **472–476**
real estate, 537
research, 473
search engines and, 375
stealth, 473
mashing, 19
massively parallel processing, **100–101**
mathematical (arithmetic) models, 14
m-commerce (mobile commerce), 22, 378
medication management systems, 117
megabyte, **98**
megahertz (MHz), **94**
member numbers, 206
memory. *See also* RAM (random access
memory)
blocks, 107
cards, 106–107
characteristics, 98–100
described, **92**
devices, 92–102
functions, 98–100
management, 147–148
nonvolatile, 99
selecting, 127
storage capacity, 98
system units and, 93

menu-driven systems, **620**
message threads, 320
meta tags, 308, **385**
meta-search engines, **308**
methods, 175, 229
metropolitan services, 283
MICR (magnetic ink character recognition) devices, 113–114
microprocessors. *See also* CPUs (central processing units)
microchips, 58
microcode, **94**
microseconds, 94
microwave communications, **248–249**
middleware, 158, 406
military operations, 7, 19, 425, 556–557
 AI and, 511, 513, 518–519
 DSSs and, 482, 483
 e-commerce and, 351, 357
 MIS and, 458
 security issues and, 676, 678
 software and, 154, 176
 virtual reality systems and, 537
MIPS (millions of instructions per second), **94**, 97, 513
MIS (management information system), **26–27**, 28, 400, 458–480
 DSSs and, comparison of, 484
 financial, **465–467**
 human resource, **476–479**
 manufacturing, **467–472**
 marketing, **472–476**
mission-critical systems, **564**
MMDS (Multichannel Multipoint Distribution System), 269, 285
MMS (model management software), 28, **486**
MO (magneto-optical) disks, **106**
mobile commerce, 362–363
model(s)
 base, **486**
 described, **13–14**
modems
 cable, 250–251, 259
 described, **250**
 special-purpose, **250–252**
monitoring
 stage, **456**
 utilities, 157
monitors, 12, 118–119, 127
Moore's Law, **95–96**, 105
MP3 format, 107, 120–121, 157, 380
MPE/iX (Multiprogramming Executive with integrated POSIX), 154
MPTQM (Main Preparation Total Quality Management), 60
MRP (material requirements planning), **470**
MRPII (manufacturing resource planning), **470**
MSSPs (managed security service providers), **694–595**
multifunction devices, **121**
multiple user databases, 207
multiplexers, **251**
multiprocessing, **100**, 101
multitasking, 148
music, 314, 323–324
 devices, 120–21

file formats, 107, 120–121, 157, 380
 intellectual property issues and, 379–380
 -sharing web sites, 321
Mylar film, 103, 104

N

nanoseconds, 94
nanotechnology, 18
narrative models, **13–14**
NAS (network-attached storage), **107**, 108
natural language processing, 30, **514–516**
navigation systems, 121
net present value, 587
NetWare, 153, 180, 274
network(s). *See also specific types*
 basics, 273–275
 described, **19**, 259
 interconnecting, 265–272
 linking PCs to, 275–276
 -management software, **274–275**
 overview, 240–293
 switching devices, 271–272
 topology, **262–263**
 types of, 259–262
 utilities, 157
neural networks, **516–518**, 532
newsgroups, 314, **319–320**
NGI (Next Generation Internet), 298
noise filters, 112
nominal group technique, **488**
nonadaptive systems, **10**
NOS (network operating system), **274**
notebook computers, 123. *See also* laptop computers
nuclear energy, 30

O

object(s)
 described, **174**
 -oriented databases, **229**
 -oriented design, 618–619
 -oriented programming languages, 174, 175
 -oriented systems development (OOSD), **585–586**
 -relational database management system (ORDBMS), **229**
OCR (optical character recognition), 113
ODBC (open database connectivity), **228–229**
offshore companies, 62
off-the-shelf software, **159**
oil industry, 27–28, 126
OLAP (online analytical processing), **227**
OLED (organic light-emitting diode), 119–120
OLTP (online transaction processing), 219, 221, **402–403**
OMR (optical mark recognition), 113
on-demand computing, **60–62**, 575–577
online
 information services, 166
 profiling, 382
 services, 302

open
 shops, **624**
 systems, **10**
open-source software, **177–179**, 634
operating systems, 145–152, 154–156
 current, 149–152
 defined, **145**
operational feasibility, **588**
operations component, 72
optical
 data readers, 113
 discs, **105–106**
 processors, **96**
optimization model, **457–458**
OQL (object query language), 229
order
 entry systems, **413–416**
 processing systems, **413**
organization(s)
 classifying, by system type, 9–10
 described, **48–49**
 structure of, **51–54**
OSI (Open Systems Interconnection) model, 265, 266
outplacement, 479
output
 described, **9**, 16
 devices, 109–113, 118–121
 organizational systems and, 48–51
 productivity and, 68
 software and 146–147
outsourcing, 32, **60**, 61, 72, 575–577, 608

P

P3P (Platform for Privacy Preferences), **701**
paging, 148
Palm OS, 123, 150, 155
PANs (personal area networks), **259**, 285
parallel start-up, **645**
parameters, **11**
Pascal, 527
password sniffers, **688**
passwords, 149, 327, 688
patch, **648**
paycheck stub, **431**
payroll systems, 15, 72, 281–282, 403, 430–431, 646
 journal, **431**
 transaction processing and, 25
PBX (private branch exchange), **271**, 285, 321
PCMCIA (Personal Computer Memory Card International Association), 106–107
PDA (Personal Digital Assistants), 123, 154–156, 242
pen input devices, 116
perceptive system, **511**
performance
 -based information systems, 67–69
 databases and, 215–216
 disk drive, 56
 measurement systems, **652**
 standards, **11**
 tuning, 216
permanent systems, **10**
permanent storage. *See also* secondary storage

personalization, 386
    productivity software, 145, 160–162
    sphere of influence, 145
PERT (Program Evaluation and Review
    Technique), 582–583
petabyte, 98
pharmaceutical industry, 3, 156, 358,
    481–482, 557
    databases and, 225, 229–230
    systems development and, 557,
        559–560
phase-in approach, 644
phishing, 23, 329, 380, 690
phone(s)
    Internet, 314, 320–322
    services, 255–256
    smart, 123
PHP (Hypertext Preprocessor), 310
physical
    design, 14, 201, 613, 616–617
    models, 14
pilot start-up, 644
PIMs (personal information managers), 166
piracy, 688–689
pixels, 118
PKI (public key infrastructure), 691
plaintext, 330
plotters, 120
plug-ins, 304
point evaluation system, 630
policies, 675–678, 705
policy-based storage management, 109
pop-up ads, 157, 303, 329, 475. See also
    advertising industry
portable computers, 122, 123–124
POS (point-of-sale)
    devices, 114
    software, 612
power
    blackouts, 16
    management companies, 36
PPP (Point-to-Point Protocol), 302, 304
predictive analysis, 222
preliminary evaluation, 628
pricing, 475–476
primary storage, 92. See also memory
printers, 120, 127–128
privacy, 33–35, 327, 329, 670–711. See also
    security
    e-mail, 700–701
    invasions of, 382–383
    laws, 702–705
    RFID tags and, 270
problem solving, 455, 456, 457–458,
    482–483
procedures, 21, 675–678
process(es). See also processing
    control, 470–471
    defined, 6
    symbol, 593
processing. See also processes
    characteristics, 94–98
    described, 14, 93
    interactive, 619–620
    mechanisms, 9
    natural language, 30, 514–516
    tasks, 148–149

processors. See also CPUs (central processing
    units)
product(s)
    configuration software, 372
    creating new, 65
    delivery, 353–354
    development, 475
    improving existing, 65
    pricing, 475–476
    substitute, threat of, 64, 66
productivity
    described, 68
    paradox, 68
profit centers, 466
program
    code, 171
    trading, 16
programmers, 558
programming life cycle, 637
programming languages
    described, 171
    generations of, 171–176
    Web, 309–310
project(s)
    deadline, 582
    described, 558
    management, 162, 582–584
    managers, 558
    milestones, 582
    schedules, 582
projecting, 203
PROM (programmable read-only
    memory), 99, 268
promotion, 475
proprietary software, 159
protocols. See also specific protocols
    described, 265–267
    wireless, 267–269
prototypes
    described, 571–572
    IMC and, 667
    nonoperational, 572
    operational, 572
public
    domain software, 178
    network services, 281
publishing companies, 36
purchase order processing systems,
    424–425
purchasing transaction processing systems,
    424–427
push technology, 311
pyramid schemes, 381

Q

QBE (Query-by-Example), 211
qualitative analysis, 485
quality
    described, 59
    control, 471–472
    systems development and, 580–581
questionnaires, 592
Quicken, 141, 162, 163, 207, 325

R

RAD (rapid application development), 568,
    573–574
radio, 314, 323–324, 348. See also RFID
    (radio-frequency identification)
RAID (redundant array of independent/
    inexpensive disks), 104, 108
RAM (random access memory), 89, 118,
    127, 150. See also memory
    cache memory and, 99–100
    described, 98–99
    flash memory and, 107
receiving systems, 426
records, 196
recycling computers, 127
redundancy, planned data, 201
reengineering, 56, 57–58, 577
refreezing, 56
register, 92
regulations. See also legislation
relational database model, 201, 202,
    203–206
relations, 202
release, 648
remote logon, 317
repetitive motion disorders, 706
report(s). See also documentation
    accounts payable, 427
    accounts receivable, 429–430
    demand, 462
    drill-down, 462
    DSSs and, 480
    exception, 462
    generating, 211–212
    key-indicator, 461
    layout, 596–597
    MIS and, 460–463, 477–478
    receiving, 426
    scheduled, 27, 461–462
    system analysis, 598–599
    TPSs and, 405, 410
request for maintenance form, 648
requirements analysis, 595–597
resolution, 118, 119, 120, 127
resource-based view, 63
restart procedures, 620
restaurant industry, 64
retail companies, 14, 36, 66, 363–364
revenue centers, 466
review procedures, 651
RFID (radio-frequency identification), 58,
    65, 118, 220, 269–270, 357, 408,
    417–420, 426, 538, 614
RFP (request for proposal), 555, 626–627
richness, use of the term, 475
rightsizing. See also downsizing
ring networks, 263
RISC (reduced instruction set computing),
    97, 154
robotics, 322, 513–514, 518–519
ROI (return on investment), 67, 68–70, 629
roll up, 219
ROM (read-only memory), 99, 105
ROP (reorder point), 470
routers, 272
routing systems, 423

RSI (repetitive stress injury), **706**, 707, 708
rules
    described, **523**
    expert systems and, 523–524
RUP (Rational Unified Process), 573

S

safe harbor principles, **383**
salary surveys, 479
sales configuration systems, **416–417**
SANs (storage area networks), 107, 108
SASD (sequential access storage), **102**
satellites, 248–249. *See also* GPS (global
    positioning satellite)
satisficing model, **458**
scalability, **125**, 149, 154, 215, 565
scams, computer-related, 689–690
scanning devices, 113
scenarios, 618–618
schedule(s). *See also* scheduling system
    feasibility, **588**
    project, **582**
scheduled reports, 27, **461–462**
scheduling system, **423**, 437, 469–470, 479.
    See also schedules
schema, 207, 208
SCM (supply chain management), 49–50,
    71, **356–357**
screen layout, **596–597**
"script bunnies," **682**
SDLC (systems development life cycle),
    568–577, 584, 585
SDLT (Super Digital Linear Tape), 103
SDRAM (synchronous dynamic RAM), 98,
    102
search engines, **307–308**, 375
secondary storage, **102–103**, 104–107
security. *See also* fraud; privacy; terrorism
    attacks and, 33–34, 328, 371, 683
    authentication and, 327
    cryptography and, **330–331**
    encryption and, **330–331**, 691
    data alteration, 683–686
    fingerprint systems, 14, 378, 514, 692
    firewalls and, **35**, 274, 327, 330–331,
        697
    IDS (Intrusion-Detection System) and,
        559, **694**
    overview, 33–35, 670–711
    passwords, 149, 327, 688
    phishing and, 23, 329, **380**, 690
    system, design of, 621–625
    theft and, 683–686
    Web servers and, 371
    work environment and, 706–709
selecting, **203**
selective backups, 623
semantic Web, 219
semistructured problems, **483**
sequence
    diagrams, 618–618
    structure, **638**
sequential access, **102**
server(s). *See also* client/server architecture;
    Web servers
    databases and, 227

described, **124–125**, **263–264**
file, **285**
NAS and, 108
OLAP, 227
scalability and, 125
software and, 157
utilities, 157
service(s)
    after-sales, 254
    creating new, 65
    delivery, 353–354
    improving existing, 65
    substitute, threat of, 64, 66
Shannon's fundamental law of information
    theory, 246, 284
shareware, **178**
shipment
    execution systems, **418–419**
    planning systems, **417–418**
shopping. *See also* e-commerce
    carts, **373**
    online, 314, **322**, **373**
sign-on procedures, **619**
silicon, strained, 96
simple systems, **10**
simulation, **481–482**
single user databases, 207
site preparation, **642**
six sigma, 60
SLIP (Serial Line Internet Protocol), **302**,
    304
slipstream upgrades, **648**
smart
    cards, **377–378**
    phones, 123
    tags, 118
SMP (symmetrical multiprocessing), **101**
SNMP (Simple Network Management
    Protocol), 274
social engineering, **680**
software
    acquiring, 634–640
    backups, 622–623
    bugs, **176**
    classified by type and sphere of
        influence, 144
    described, **18**
    development, multiorganizational,
        178–179
    free, 314, 324–325
    global support for, 180
    issues/trends, 176–180
    open-source, **177–179**, 634
    overview of, 140–191
    piracy, **688–689**
    reasons to learn about, 142
    specifications, 616
    suites, 167–168
    upgrades, 180
source data automation, **110–111**
spam, 157, **381**, 673, 674
spatial data technology, 230
special-purpose
    computers, 122, 146
    devices, 154–156
sphere of influence, **144–145**
split-case distribution, 355
sponsored links, 307

spooling, 147
spreadsheet software, 162, 164, 167
spyware, 329
SQL (Structured Query Language), 173,
    211, 212
SSL (Secure Sockets Layer), **374–376**
stable systems, **10**
staff positions, 52
stakeholders, **558**
star networks, 263
start-ups, **644–645**
statistical sampling, 592
steering committee, **589**
stock market, 16, 62, 66, 142
    AI and, 515
    e-commerce and, 351, 363, 367–369,
        382
    scams and, 382
storefront brokers, **385**
Straight-Through Processing, 57
strategic
    alliances, **64**
    planning, **494**
structured
    interviews, **591**
    walkthroughs, **640**
student loans, 439
subnotebook computers, 123
subschema, **207–208**
supercomputers, 122, **126**
superconductivity, **96**
suppliers, bargaining power of, 64, 66
supply chain management (SCM), 49–50,
    71, **356–357**
support component, **74**
swarming technology, 487
switched lines, 254
switching devices, **271–272**
system(s). *See also* systems design
    analysis, 33, 558–559, 570, 589–599
    analyst, **558–559**
    architecture, **126–127**
    components, 9–10
    concepts, 8–14
    controls, **624–625**
    described, **8**
    design, 33
    hardware and, 90–93
    implementation, 33, 632, 633–645,
        570
    investigation, 33, 570, 586–589
    maintenance, 33, 645–650, 570
    modeling, 13–14
    operation, 645–650
    parameters, **11**
    performance measurement, **652**
    request form, **586**
    resources, access to, 149
    review, 33, **650–651**
    software, 143, 145–148
    types, 9–10, 122–126
    unit, **93–94**
    variables, **11**
systems design. *See also* design; systems
    alternatives, 626–268
    described, **570**, **613**
    evaluation, 628–630
    interfaces, **635**

overview, 613–632
reports, **631–632**
specifications, freezing, 630–631
systems development. *See also* system design;
systems
component, 74
described, **32–33**
life cycles (SDLC), 568–577, 584, 585
overview, 557–568
success, factors affecting, 577–586
Web-based, 566–567
syntax, 171

T

Tablet PCs, 116, **123**
TAM (technology acceptance model), **59**
tape cartridges, 102, 103
T-carrier system, 256–257, 284
TCO (total cost of ownership), 70, 629
TCP (Transmission Control Protocol), **299**, 328
TCP/IP (Transmission Control Protocol/ Internet Protocol), 265, 285, 299
technical
documentation, **637**
feasibility, **587**
technology
diffusion, **59**
-enabled relationship management, **367**
infrastructure, 17
infusion, **59**
telcos. *See also* LEC (local exchange carrier)
telecommunications
applications, 275–283
carriers, 252–256
described, **19**
medium, **244**
overview, 240–293
transmission media types, 246–260
systems, acquiring, 640–641
telecommuting, **277–278**
teleconferencing, 491
Telnet, 314, **317**, 374
temporary systems, **10**
terabyte, **98**
terminals, 113, 263
terminal-to-host architecture, **263**
terrorism, 286, 494, 646, 700. *See also* security
AI and, 510, 517, 518
biometrics and, 692
cyber-, **680–681**
databases and, 197, 221–222
disaster planning and, 621
privacy and, 671–573
testing
acceptance, **643**
alpha, **643**
beta, **643**
integration, **643**
overview, 643–644
system, **643**
unit, **643**
volume, **643**
theft, of information, 688. *See also* security

thin clients, 122, **123**
3-D Internet sites, 314, 324
3-D worlds. *See also* virtual reality
thumbnails, 151
time zones, 54
time-driven review, **651**
time-sharing, **148**
top-down approach, **638–639**
touch-sensitive screens, 116
TPS (transaction processing system), 398–450, 459, 567. *See also* transactions
audits, **411–412**
described, **25–26**
software, 373–374
TQM (total quality management), 59–60
traditional approach, to data management, **198–199**
training, 642, 667
transaction(s). *See also* TPS (transaction processing system)
described, **25**
processing applications, 412–423
processing cycle, **407–408**
transborder data flow, 262, 285
trend analysis, 224
trigger points, 462
Trojan horses, **683**
tunneling, 327
Turing Test, 510
twisted-pair cable, **246–247**

U

UDDI (Universal Discovery Description and Integration), 313
unattended systems, 276
unauthorized Internet sites, 331–332
unfreezing, **56**
UNIX, 150, 153, 173, 174, 216, 260, 309, 379, 623
unstructured
interviews, **591**
problems, **483**
UPC bar codes, 116, 408, 409
upgrades
hardware, 126–128
software, 180
URLs (Uniform Resource Locators), 299, 305, 371, 384, 385, 677
use cases, 588
Usenet, 314, **319–320**
user(s). *See also* user interfaces
acceptance document, **645**
described, **558**
documentation, **637**
preparation, **641–642**
software, 160–162
views, providing, 207–208
user interface(s). *See also* GUI (graphical user interface)
described, **146–147**, **635**
design, 619–721
expert systems and, 525–526
utility
companies, 36, 62
programs, **156–158**

V

value chain, **49**
variables, **11**
variants, **684**
version, **648**
video
-conferencing, **278–280**, 314, 320–322, 489
data, **6**
on the Internet, 314, 323–324
virtual
database systems, 230
memory, **148**
reality, **29–31**, 533–537
tape, **104**
workers, 283
workgroups, **491**
virus(es), 34, 315, 331, 616, 518
avoiding, 686–687
described, **683**
-detection utilities, 156–158
-recovery utilities, 156–157
most active, 685
variants, **684**
visa programs, 72
vision systems, **514**
visual
databases, 229–230
programming languages, 173–174
voice
and data convergence, **255**
mail, **276**
-recognition devices, **111–112**
-response devices, 121
VoIP (voice over Internet Protocol), **255**, 284, 320–321, 608
VPNs (virtual private networks), **327**, 339, 695

W

W3C (World Wide Web Consortium), 415
wage and salary administration, 479
WANs (wireless area networks), 261, 266, 285, 491. *See also* networks
WAP (wireless application protocol), 302, 362
WATS (wide-area telephone services), **256**
Web administrators, 74
Web browsers. *See also* browsers
Web logs (blogs), 314, **317–319**, 371
Web page(s)
construction software, **372**
dynamic/static, **372**
retrieving/sending, 371
Web portals, 305
Web servers, 304, 309, 371–372, 694. *See also* servers
Web services, 312–313
Web site(s)
building traffic to, 385
development tools, **371–372**
hosting services, **384–385**
maintaining/improving, 385–386
tracking, 371
traffic data analysis software, **374**

Web style, 151
Webcasting, 311
what-if analysis, **481**
Wi-Fi standard, 267–268, 316
wire transfers, **282**
wireless
    computing, 241–242, 257–259,
      267–269, 284
    mobile data services, 257–259
wiretapping, 34
word processing software, 162–164,
    167–168

wordlength, **94–95**
workgroup(s)
    described, **145**
    software and, 145, 152–154, 157,
      168–169, 489
    sphere of influence, **145**
    virtual, **491**
workstations, **124**
World Wide Web, **20**, **304**, 305–315. *See
   also* Web pages; Web sites
    applications, 313–315
    semantic Web and, 219

worms, **683**, 684, 694

X

XML (Extensible Markup Language),
    306–307, 309, 312, 313, 426
XP (extreme programming), 573

Z

zip files, 157

INDEX

Company

A

Abbey, Scott, 406
Abbott Laboratories, 118
Accenture, 77
Accordant Health Services, 515–516
Ace Hardware, 70, 219
Acer, 123
ACM (Association for Computing Machinery), 710
Advanced Learning Center, 71
Advanced Micro Devices, 95
Aetna, 405–406
Agronow, Dan, 177
Air Products and Chemicals, 91
Allstate Financial Group, 564
Altra Energy Technologies, 20
Akaka, Daniel, 700
Alaska Airlines, 187
AM General, 613, 614
Amazon.com, 76, 305, 312, 322, 362, 371, 375, 386, 501
Ameren Corporation, 253
America West, 582
American Airlines, 65–66
American Express, 378
American Outdoor Products, 352
AmeriHealth, 53
Ameritrade, 66, 616
AMR Research, 435
AMS (Academic Management Services), 439
Anaconda Sports, 373
Anderson Economic Group, 410
AOL (America Online), 162, 166, 251, 256, 283, 302, 315, 322, 475, 674, 697
AON, Inc., 644
Apple Computer, 66, 97, 106, 616, 675, 700
   P3P and, 701
   software and, 144, 151–152
Appshop Incorporated, 608
Armour, Tom, 517
ArvinMeritor, 58
ARZ (Allgemeines Rechenzentrum GmbH), 90–91
Ashcroft, John, 358
AT&T (American Telephone & Telegraph), 55, 252–254, 258, 284, 303, 362, 532, 700
Athey, Ruel, 273
@Home, 251
Atwood, Chuck, 67
Australian Tourist Exchange, 457
Autonomy, Inc., 517
Aventist Pharmaceuticals, 559–570
Aviall, 67, 70, 351
Avolent, Inc., 369
AXA Financial Services, 639–640

B

BackWeb Technologies, 3
Bailey, Charles, 519
Ballmer, Steve, 18, 179
Bank of America, 115
Bank of Montreal, 83–84
Bank of New Zealand, 326
Bank One Corporation, 282
BankFinancial Corporation, 224, 547
Barclay's Capital Group, 404
Barilla, Pietro, 241
Barilla Group, 241–242
Batstone, Joanna, 195
Bauer, David, 695
BBH (Brown Brothers Harriman & Company), 271
Bedford, Alan, 161
Beitz, David, 480
BellSouth, 253–254, 303
Ben and Jerry's, 225
Bennett, Steve, 664–665
Berners-Lee, Tim, 219
Berry, Joe, 582
Berry, Robert, 547
Best Buy, 280
Best Western, 267, 340
Bezos, Jeff, 362
Bigstep.com, 384, 385
BillPoint, 282
Biobank, 699
Bisker, Janie, 69
Black & Veatch, 75
Blagojevich, Rod, 358
Blue Cross Blue Shield, 447–448
Blue Cross of Pennsylvania, 27
BMG, 348
BMW Group, 555–557
BNSF (Burlington Northern and Santa Fe Railway), 112, 530, 531, 538
Boehringer Ingelheim GmbH, 3–5, 262
Boeing, 19, 468, 700
Bozzo, Steve, 659
Bradley, David, 146
Brandeis, Louis, 698
Brandon Trust, 664–665
Breen, Michael, 660–661
Brinkman, Jens, 562
British Broadcasting Corporation, 158
BroadVision, 312, 361
Brockwood, Nathan, 96
Broughan, John, 43
BT Group, 60
Buckman Laboratories, 20
Burlington Coat Factory, 153
Burnboch, Bill, 193
Business Application Performance Corporation, 95
Butler, David, 47

C

Cafiero, Deborah, 614
Caminer, David, 169
Campbell, Jeff, 531
Campbell's Soup Company, 425
Candle, Inc., 652
Canon, 112
Cardinal Health, 60, 118, 636
Carlile Engineered Products, 424
Carnegie Mellon University, 176, 487, 581, 717
Carnival Cruise Lines, 644
CartaSi S.p.A., 125
Catena Corporation, 675
Catholic Health, 219
CBOT (Chicago Board of Trade), 66
CDC (Centers for Disease Control), 230, 561
CDW, Inc., 472
Celenet Communications, 644
Central Ohio Technical College, 127
Centrex, 271
Cerf, Vinton, 19
Certive Corporation, 225
Cessna Aircraft, 49
Chamberlain, D. D., 211
Chapman, Teresa, 611
Charles Schwab & Co., 351, 515
Cheap Tickets, 384
Checkpoint Systems, 60
ChemConnect, 19
Cheng, Wei-Tih, 406
CHIPS (Clearing House Inter-Bank Payments System), 282
Choice Hotels International, 267
ChoiceStream, 475
Christensen, Dennis, 574
CIA (Central Intelligence Agency), 122, 560–561, 703
Cinergy Corporation, 366, 483
Cingular, 258
Cinnabon, 12
Circuit City, 63
Cisco Systems, 77, 241, 268, 272, 351, 683
Citibank, 380
Citigroup, 282
Cleveland State University, 441
Cline, Davis & Mann, Inc., 152
CNN (Cable News Network), 371
Coca-Cola, 359, 634
Cocke, John, 97
Cognos, 227, 459, 461
Cohen, Jay, 223
Colgate-Palmolive, 72
Collard, Tom, 473
Colliers, 295
Collins, Tom, 63
Columbia University, 314
Comercia, 462
Comfort Suites, 267
Commerce Department (United States), 575
Commerce One, 701
Compaq, 66
CompUSA, 633
CompuServe, 162, 283, 697

Computer Associates International, 277, 518, 573, 584, 690
Computer Shopper, 95
Con Edison Communications, 275
Conn, David, 502
Connerty, William, 547
Continental Airlines, 91, 367
Continental Technology Group, 91
Cooper Tire and Rubber Corporation, 273–274
Cornell University, 480
Corporate Express, 21, 372
Corrugated Supply, 272
Costantini, Lynn, 681
Covisint, 365
Cox Insurance Holdings, 32
Crossman, Jacob, 317
CVS, 118
CyberSource, 380

D

DaimlerChrysler, 219, 533
Darlington, Lloyd, 84
DARPA (Defense Advanced Research Projects Agency), 517, 698
Daschle, Tom, 319
DDB, Inc., 193–195
DealTime, 322
Dean, Howard, 319
Defense Department (United States), 297, 357, 575, 703
Dell, Michael, 18
Dell Computer, 18, 66, 127, 156, 700
    Business Application Performance Corporation and, 95
    call centers, 277
    disposal programs, 127
    DVDs and, 106
    e-commerce and, 361, 364, 372
    Jordan Education Initiative and, 301
    music devices and, 121
    workstations, 91
Delta Airlines, 11–12, 64, 65, 538
DeRodes, Bob, 611, 612
Deutsche Asset Management Technology, 72
DHL, 354, 373
Dial Corporation, 440
Discover Financial Services, 71
DiscoveryLink, 230
DISH Network, 12
DLJdirect, 515
Dominy, Mike, 436
DoubleClick, 367, 700
Dow Chemical, 280, 366
Dow Jones & Co., 697
Doyle, Jim, 358
Dragon Systems, 515
Dylan's Candy Bar, 393

E

E2Open, 365
Earthlink, 303, 674
Eastman Chemical Company, 338
eBay, 76, 312, 322, 323, 353, 371, 700

Eberhardt, Keith, 502
EBT, 312
Edens & Avant, 480
Edocs, Inc., 369
EDS (Electronic Data Systems), 77, 273, 476–477
Electronic Arts, 418–419
Elemica, 365
Eli Lilly, 358
Elliot, Ginger, 291
Ellis, Steve, 243
Ellison, Larry, 579
E-Loan, 325, 577
Embarcadero Describe, Inc., 584
EMC Corporation, 104, 108, 109
eMoneyMail, 282
Enron, 7, 467
Enterasys, 268
EPA (Environmental Protection Agency), 75, 127
Equifax, 705
Erdogus, Tolga, 448–449
Ericsson, 362, 701
eToys.com, 567
Evans, Grant, 517
Evans, Mary Kay, 44
Excelon Corporation, 640
Exostart, 365
Expedia, 205, 384
Experian, 705

F

Fakespace Labs, 534
Fakta Designs, 562
Fandango, 406
Farris, Rachel, 475
Fay, Linda, 326
FBI (Federal Bureau of Investigation), 197, 221, 378, 679, 682, 703
FCC (Federal Communications Commission), 253, 321
FDA (Food and Drug Administration), 117, 223, 358, 616, 707
Federal Reserve Bank, 282
FedEx, 20, 71, 113, 354–355, 373, 401–402, 407, 416, 423, 469, 531
Feldman, Terry, 298
Fidelity Investments, 515
Fidelity National Financial, 626
Field Technology, 158
First Fidelity, 576
FleetBoston Financial Corporation, 632
Flood, Mike, 538
Ford, Gerald, 702
Ford Motor Company, 66, 179
Forrester Research, 62
Forward Solutions, 107
Fost, Joshua, 295
Francisco Partners, 426
F-Secure Corporation, 684–685
FTC (Federal Trade Commission), 277, 381, 492, 681, 690–691
Fuji, 112, 564
Fujitsu, 113, 119, 123, 516
Fukuda, Keiji, 453
FundsXpress Financial Network, 42

G

Gaffney, Paul, 439
Galileo, Inc., 101, 152
Gartner Research Group, 62, 70, 576
Gary, Chares, 214
Gates, Bill, 123, 146
Gates Corporation, 97
Gateway Computers, 66, 106, 325, 379
GE Power, 585
Geek Squad, 77
Genalystics, 222
General Electric, 350, 532
General Electric Power Systems, 309–310
General Mills, 447–448
General Motors, 66, 268, 359, 423
Genex Services, 258
Gensym Corporation, 546
Georgia Institute of Technology, 121
GetThere.com, 66
Gilbert, Yves, 503
Gillette, 63, 467
Gius, Richard, 636
GlaxcoSmithKline, 152, 358
Global Healthcare Exchange, 365
GlobalSight, 361
Godiva Chocolate, 20
Goldberg, Larry, 69
Goldman Industrial Group, 20
Goodyear Tire & Rubber Company, 62, 440–441
Google, 307, 312, 316, 359, 375, 380
Grand, Steve, 509
Grigsby, Paul, 218
Guardian Computer Support, 622

H

Hackensack University Medical Center, 606
Halfords, 222
Halifax Bank, 380
Handa, Yoshihiko, 453
Handspring, 123
Harley-Davidson Motor Company, 71, 420
Harrah's Entertainment, 71
Harris Bank, 83
Harvard University, 35, 49
Hawking, Steven, 509
Hayden, Kevin, 295
Hawaiian Airlines, 423
HealthSouth, 123
Heiner, Dennis G., 502
Heliopolis Complex, 279
Hershey Foods Corporation, 71
Hession, Lloyd, 647
Hewlett-Packard, 50, 66, 89, 483, 606, 634
    Business Application Performance Corporation and, 95
    call centers, 277
    CD-RW drives, 107
    digital cameras, 112
    disposal programs, 127
    DVDs and, 106
    e-commerce and, 354
    intellectual property and, 379
    iPaq, 116
    Jordan Education Initiative and, 301
    management software, 275

MIS and, 467
music devices and, 121
NAS and, 108
servers, 91, 97
software and, 97, 150, 153–154, 156, 180
Web services and, 313
Hicks, Barry, 290–291
Hilton Hotels, 134, 194–195, 323, 565
Hinojosa, Patrick, 684
Hit Song Science, 171
Hitachi Data Systems Corporation, 108
Holland America Cruise Lines, 218
Holiday Inn, 267
Home Depot, 399, 502, 611–613
Honda, 219, 509
Hostmann, Bill, 218
Hotel Commonwealth, 19
HotJobs, 296
Hourihan, Meg, 317
Huber, George, 455
Huels, Holger, 3
Hummel, John, 291
Hymans Robertson, Inc., 650
Hyperion Solutions, 227, 617
Hyundai Motor Company, 239

I

IBM (International Business Machines), 532, 579, 623, 626, 634, 646, 700
    Business Application Performance Corporation and, 95
    call centers, 277
    CICS and, 97
    databases and, 195, 202, 207, 211, 214, 219, 223, 229–230
    disposal programs, 127
    DVDs and, 106
    e-commerce and, 353, 373–374
    encryption and, 330
    evolution of, 9–10
    handheld computers and, 123
    intellectual property and, 379
    job opportunities at, 77, 316
    Jordan Education Initiative and, 301
    LCD displays, 119
    Learning Solutions, 161
    Millipede, 104
    multiprocessing and, 101
    outsourcing and, 62, 576
    PartnerWorld, 22
    RAD and, 573
    SANs and, 108, 109
    SAP and, 47
    software and, 146, 150–154, 157, 161, 179, 180
    TPSs and, 401, 406
    Tyndall Federal and, 42
    use of strained silicon by, 96
    Web services and, 313
IDC, Inc., 68, 454
IKEA, 298
IMC (International Manufacturing Corporation), 666–667
InfoWorld Media Group, 295
ING Direct, 483

Ingersoll-Rand, 83
Inman, Brad, 394
Institute of Electrical and Electronics Engineers (IEEE), 265, 267–268
Intel, 89, 95, 101, 108, 124, 267, 301
Internal Revenue Service (United States), 557, 641
InterQual, 27
Interwoven, 295, 361
Intrusic, Inc., 698
Intuit, 141, 608
Iomega, 107
Iron Mountain, 411
ISM, Inc., 422
ISO (International Organization for Standardization), 175, 265, 580, 581
ISS (Internet Security Systems), 694
ITU (International Telecommunications Union), 269
Izumiya, 392–393

J

James Electronics, 14
Janssen Pharmaceutica, 152
Jarvis, Steve, 187
J.D. Edwards & Co., 229, 441, 502
JetBlue Airways, 640
Jette, Bruce, 511
Jevans, Dave, 23
Jobs, Steve, 616
Johnson & Johnson, 118
Jones, Devon, 440
Jones, Norah, 171
Jones Lang LaSalle, 141–142
Jost, Terry, 608
J.P. Morgan Chase & Company, 215, 282, 634
Junkbusters Corporation, 702
Jupiter Media Metrix, 329

K

Kahn, Robert, 298
Kaiser Permanente Health Plan, Inc., 675
Kamen, Dean, 608
Kasparov, Garry, 510
Kassel, Rob, 531
Kay, Alan, 147
KaySports, 373
KDDI Corporation, 449
Keane, Inc., 576
Kelkoo, 359
Kelly, Nuala O'Connor, 700
Keynote Systems, 385–386
Khan, Basheer, 579
Kiku-Masamune, 453–355
Kinko's, 688
Kodak, 64, 112–113, 119, 460, 564, 576
Kolb, Ron, 517
Kolhatkar, Jaya, 501
Kovatch, Brian, 662
KPMG Peat Markwick, 530
Kraft, 501–502
Krispy Kreme Doughnuts, 71, 428
Krouse, Jim, 661
KXEN, Inc., 547

L

Lander, Eric, 510
Lanier, Jason, 533
Laurel Pub Company, 462
Lawson Software, 436
Legato Systems, 109
Lender's, 12
LendingTree, 325
Levelle, Ed, 459
Lewin, Kurt, 56
Light, Donald, 644
Lilly Software, 567
Link, Tim, 127
LiveNote, 215
Lloyds TSB Bank, 380
Locke, Gordon, 423
Loehr, Volker, 47
Los Alamos National Laboratory, 97, 108
Lowe & Partners Worldwide, 487
Lufthansa, 533
Lundy, Frank, 214
Lyons Bakeries, 169

M

McAfee Security, 685, 686
MCandish, Ian, 215
McCarthy, John, 508
McDonald's, 267, 399–400
Macromedia, 219
Macy's, 675
Magnify (company), 222
Mallis, Laurie, 188
Marathon Oil Corporation, 276
Mariano, John, 439
Market Insight Corporation, 375
Marriott Hotels, 267
Martinez, Ed, 606
MasterCard, 353, 377, 580–581
Mattel, 418
Mazda, 283
MCI, 252–254, 284, 575
Medco Health Online, 19
Media Player, 150–151
Mendes, Andre, 312
Mental Images, 18
Mercedes Benz, 18
Merck & Company, 19
Merill Lynch & Co., 695
MetaGroup, 214
Metavante Corporation, 369
MetLife Incorporated, 659–660
MetroGroup, 270, 538
Meyer-Rassow, Steven, 545
Michael Angelo's Gourmet Foods, 437–438
Microsoft Corporation, 18, 579, 674, 700–701
    Business Application Performance Corporation and, 95
    certifications, 68, 76–77
    databases and, 162, 202, 205, 207, 212, 214, 221, 228, 502
    e-commerce and, 353
    GUIs and, 147
    Jordan Education Initiative and, 301
    .NET Framework, 312, 567
    OLAP and, 227

PDAs and, 155
Visual Basic, 571, 584, 585, 639, 648
Visual Studio .NET, 564, 640
Web services and, 313
Wi-Fi and, 267, 268
Windows operating system, 18,
149–152, 156–158, 173–174, 179,
211, 216, 276, 623, 648
Miglautsch, John, 195
Millser, Dan, 531
MineShare, 227
Minolta, 112
Mitchell, Melanie, 517
Mitsubishi Motors, 312
Mohegan Sun Casino, 643, 697–698
Molecular (company), 295
Mondex, 376
Monster.com, 76, 296, 305, 316
Moon, Michael, 661, 662
Morgan Stanley, 264
Morreale, Dan, 186–187
Motion Computing, 123
Motorola, 151, 156, 362
Moviefone, 406
Movies.com, 406
MovieTickets.com, 406
MRI Real Estate Solutions, 141, 142
Mueller, Robert, 221–222
Mukherjee, Debbie, 461
Munshi, Porus, 572
MyFamily.com, 44
MyProductAdvisor.com, 375

N

Namco, 348
Napster, 324
NASA (National Aeronautics and Space
Administration), 126, 230, 305, 313,
513–514, 673, 678
Nasdaq, 641
Natan, Mike, 69
NationalAccess, 258
Nationwide Insurance, 558
Natural Machine, 515
NatWest, 380
NCR Teradata, 222
NEC Corporation, 106, 119, 121
Neoforma, 433
Net2Phone (company), 321
Netcraft, 376
Netflix, 66
NetLedger, 433
Netscape Communications, 219, 376
Network Appliances, 71
Network Associates, 685
Network Solutions, 300
New York Stock Exchange, 465, 566
Nguyen, Tien, 607
Niche Retail, 432
Nielsen/NetRatings, 329
Nippon Steel, 357
Nokia, 348, 362
Nortel Networks, 581
North Bronx Healthcare Network, 186–187
Northern Trust, 405
Northrup Grumman, 557

Novartis, 156
Novell, 77, 179
Nowak, Hans, 555
NSF (National Science Foundation), 298

O

Objectivity, Inc., 640
Office Depot, 52, 363
Ohio State University, 127
OneBeacon, 69
OneNote, 207
Open Market, 312
Open Source Initiative, 177
Oracle, 136, 202, 207, 214, 219, 224, 441,
436, 579, 623, 640
call centers, 277
CASE tools, 584
Certified Professional, 77
E-Business Suite, 220
hardware and, 97
OLAP and, 227
ORDBMSs and, 229
RAD and, 573
Web services and, 313
Orbitz, 205, 384
O'Rourke, Pamela, 85
Osborne Computer, 9
Osmani, Adnan, 574
Ounjian, John, 447–448
Overture Services, 380
Ownes & Minor, 224–225
Oxford Bookstore, 27

P

PacifiCare, 576
Packet8, 321
Panda Software, 684
Pantellos Group, 365
Parker Adventist Hospital, 298
Parnby, Nick, 364
Partnership America, 20
Pawlenty, Tim, 358
PayPal, 282, 353
PC Mall, 363, 380
Pearce, Hilary, 664, 665
Pentax, 112
PeopleSoft, 136, 436, 438, 441, 642
PepsiCo, 607
Perry Manufacturing, 219
Pershing LLC, 559
Petrie, Helen, 415
Peugeot Citroën, 560
Pfizer, 118, 152, 358
Phillips, Jim, 71
Phone.com, 362
Pioneer Hi-Bred International, 644
Pixar, 18, 616
Pizzeria Uno, 12
Pleasant, Carroll, 338
PMC-Sierra, 97
Pocket PC, 156
Popkin Software, 584
Porsche AG, 89–90
Port City Metals Services, 196
Porter, Michael, 49, 63–64, 67

Postal Service (United States), 16, 60, 135,
370, 706
Powell, Michael, 321
PPL Corporation, 253
Pragmatech Software, 626
Priceline, 384
PricewaterhouseCoopers, Deloitte &
Touche, 466
Primerica Life Insurance, 158
Probst, Jack, 558
Procom Technology, 108
Procter & Gamble, 54, 60, 72, 118, 275,
440, 467
Prodigy, 283
Prover Technology, 584
Providence Washington Insurance
Company, 459
Purdue University, 229–230

Q

Qpass, 350
Quadstone, 222
Qualcomm, 362
Queen Mary 2 (QM2), 43
Quicken Loans, Inc., 104
Qwest Communication, 253–254s

R

Radianz, 647
Rational Software, 573, 584
Realatrends Real Estate Services, 537
RealEstate.com, 394
RealNetworks, 121
Red Bull North America, 694
Red Brick Systems, 224
Regis Corporation, 560, 617
REI (Recreational Equipment
Incorporated), 352
Relativity Technologies, 648–649
Reliant Pharmaceuticals, 557
Research In Motion, 489
Restoration Hardware, 514–515
Rhenania, 195
Rich Products, 71
Ridge, Tom, 672
Riggs, Janice, 193
Rippin, Bill, 502
Rite-Aid, 118
Rohrer, Jim, 662
Ross Systems, 436
Royal Ahold, 289–290
Royal Bank of Canada, 647
RSA Security, 492

S

SABRE reservation system, 65–66, 187, 423
SABRE (Surveillance and Automated
Business Reporting Engine), 465
Safeco Corporation, 410
Saint Luke's Health System, 71
Salesforce.com, 634
Samsung, 64, 119, 156
Sandia Laboratories, 516

Sanford, James, 460
Sanoma Magazines, 503–505
Santa Cruz Operations, 153
SAP, 3, 26, 47, 436, 440, 579, 634
Sapiens International, 69
SAS Corporation, 194, 222, 508, 562
SBC Communications, 253–254
Sbyase, 221
ScanSoft, 112
Schein, Edgar, 56
Schindler, 457
Schlotsky's Deli, 267
SchlumbergerSema, 260
SCO Group, 379, 683
Scottish Drug Enforcement Agency, 215
Scottish Life, 479
Scripps Institution of Oceanography, 18
SCS, 373
Seagate Technology, 532
Sears Roebuck, and Co., 322, 575
SEC (Securities and Exchange Commission), 223, 367, 466
SEGA, 348
Segway, 608
SEI (Software Engineering Institute), 176, 581–582, 679
Seisint, 700
Sendmail, 178
Sentient Jet, 448–449
Sentry Insurance, 339
Sesame Street Productions, 169
SETI@home, 102
7-Eleven, 220
Sezer, Esat, 47
Shanghai Automotive Industry Corporation, 423
Sharbatly, Abdulrahman, 279
Shell Oil, 84–85
Shibata, Takanore, 513
Shinzawa, Fluto, 449
Siebel Systems, 579, 616
Siemens, 20, 301
Sightward (company), 222
Simplest-Shop, 219
Skillsoft, 301
Skype, 321
Small, Christa, 399
Software AG, 221
Software Performance Systems, 71
Song Airlines, 11–12, 64
Sonofon, 269
Sony Corporation, 106, 112, 123, 348, 513
Soquimich, 519
Southwest Airlines, 64
Southwest Metal Finishing, 420
Speak With a Geek (company), 77
Specialized Bicycles, 91
SpeechWorks, 112
Spike Broadband Systems, 269
Sprint, 123, 252–254, 284, 303, 323, 348, 362
SPSS, 222
Stanford University, 532
Starbucks, 267
Starwood, 267
State Department (United States), 197
Stevenson, James, 327
Stone, Greg, 647

Stouffer, Debra, 75
Subaru, 219
Sullivan Street Bakery, 91
Sumitomo Mitsui Bank, 34–35
Sun Microsystems, 91, 97, 101, 579, 622
    benchmarks, for CPU speed, 95
    Jordan Education Initiative and, 301
    software and, 153, 174, 179
    Web services and, 313
SunTrust Banks, 677
SupplyWorks, 83
Surebridge, 136
Sutter Health, 25, 291, 621
Swindle, Orson, 492
Swith, Tom, 158
Sybase, 180, 202, 207, 214, 224
Symbol Technologies, 268
SySmark, 95

T

Tanaka, Kazunobu, 392
Tandem, 224
Target, 322, 378, 467
TaylorMade, 473
TD Waterhouse Group, 515
Telestra, 161
Tennant, Teresa, 218
TEPCO (Tokyo Electric Power Corporation), 546
Tesco, 538
There, Inc., 613
Think Outside, 65
Thomas, Craig, 62
Thorhauge, C. F., 562
THQ, 348
Time Warner, 251, 253, 700
T-Mobile, 258
TopCoder Collegiate Challenge, 74
Torvalds, Linus, 152
Toshiba, 104, 106, 112, 123, 156
Total System Services, 109
Toyota, 283
Toyota Financial Group, 32
Trade-Ranger, 365
Trados, 361
Transora, 365
Travelocity, 66, 205, 305, 384
Trend Micro, 458
Truchsess, Albrecht von, 270
Turing, Alan, 510
TVA (Tennessee Valley Authority), 58, 644
Twentieth Century Fox Home Entertainment, 116
Tyndall Federal Credit Union, 42–43

U

UBS, 406
UCAID (University Corporation for Advanced Internet Development) 298
UCC (Uniform Code Council), 116
Uioreanu, Calin, 219
United Distillers, 533
University of California, 104, 539
University of Denver, 31, 77
University of Iowa, 178

University of Memphis, 71
University of Miami, 71
Unwired Planet, 362
UPS (United Parcel Service), 188, 354–355, 373, 401, 435, 469, 531, 632, 684

V

Varsavsky, Martin, 301
Vellanki, Rao, 280
VeriSign, 374, 376, 695
Verizon Communications, 253–254, 257, 700
Vertas Software, 109
Vertex Systems, 579
ViewSonic, 123
Vignette, 312, 361
Vinton Cerf, 19
Virgin Group, 20
Virginia Tech, 101
Virtual Corporate Management System, 54
Visa International, 23, 376, 378, 380, 557
Visual Systems Corporation, 584
VMWare, 623
Voeller, John, 75
Voinovich, George, 62
Volkswagen, 219
Voutes, George, 72

W

Wachovia Corporation, 703
Walgreens, 63
Wal-Mart, 50, 66, 270, 281, 322, 363, 366, 399–400, 467
Walt Disney Company, 432, 537
Wang, Jack, 91
Weather.com, 177
Web Agency (company), 327
Webb, Richard, 327
WebPhone, 321
Wells Fargo Bank, 63, 242, 282, 312
Werley, Eric, 91
Werner, Al, 237
Werner Co, 502, 503
West, Marc, 418
Westpac Bank, 380
Whirlpool, 47
WhiteLight, 227
WHSmithNews, 327
Widgit Software, 158
Wiener, Scott, 225
Wigglesworth, Margaret, 295
Wildman, Harrold, Allen & Dixon LLP, 109
Wily Technology, 405
Wimberly, Allison, Gong & Goo, 215
Wincor Nixdorf, 42
Winslow, Raimond, 230
Winter Corporation, 401
Wipro GE Medical, 582
Wong, James, 141
Wu, Addons, 423

X

Xcel Energy, 461
Xerox, 147, 473
Xign Corporation, 369

Y

Yahoo!, 76, 296, 298–299, 305, 308,
   315–316, 322, 324, 359, 371, 375, 674,
   675, 700
Yankee Group, 436
Yoplait, 12
Yoran, Amit, 492
Young, Jocelyn, 186

Z

Zawel, Adam, 362
Zilvitis, Patrick, 608